FOUNDATIONS *of* MACROECOLOGY

FOUNDATIONS

Classic Papers with

OF MACROECOLOGY

Commentaries

EDITORS
Felisa A. Smith, John L. Gittleman, and James H. Brown

THE UNIVERSITY OF CHICAGO PRESS
Chicago & London

Felisa A. Smith is professor of biology at the University of New Mexico. John L. Gittleman is dean of the Odum School of Ecology at the University of Georgia. James H. Brown is Distinguished Professor of Biology at the University of New Mexico and past president of the International Biogeography Society.

The University of Chicago Press, Chicago 60637
The University of Chicago Press, Ltd., London

Printed in the United States of America

23 22 21 20 19 18 17 16 15 14   1 2 3 4 5
ISBN-13: 978-0-226-11533-7 (cloth)
ISBN-13: 978-0-226-11547-4 (paper)
ISBN-13: 978-0-226-11550-4 (e-book)
DOI: 10.7208/chicago/9780226115504.001.0001

Library of Congress Cataloging-in-Publication Data

Foundations of macroecology / editors Felisa A. Smith, John L. Gittleman, and James H. Brown.
pages cm
Includes bibliographical references and index.
ISBN 978-0-226-11533-7 (cloth : alkaline paper)—ISBN 978-0-226-11547-4 (paperback : alkaline paper)—ISBN 978-0-226-11550-4 (e-book) 1. Macroecology. I. Smith, Felisa A., editor. II. Gittleman, John L., editor. III. Brown, James H., 1942 September 25– editor.
QH541.15.M23F68 2014
577—dc23

2013045044

♾ This paper meets the requirements of ANSI/NISO Z39.48–1992 (Permanence of Paper).

Contents

Preface

The idea for this volume was developed during one of the first meetings of an Integrating Macroecological Pattern and Process across Scales (IMPPS) Research Coordination Network (RCN; http://biology.unm.edu/impps_rcn/) sponsored by the National Science Foundation (NSF). This ongoing group, begun in fall 2006, is devoted to examining macroecological patterns in mammalian body size over time and space. The book project was subsequently adopted by the network and most members became deeply involved. We thank all participants for their patience, assistance, and enthusiasm. Past and present RCN members include Alison Boyer, Jim Brown, Dan Costa, Tamar Dayan, Morgan Ernest, Alistair Evans, Mikael Fortelius, John Gittleman, Marcus Hamilton, Larisa Harding, Kari Lintulaakso, Kate Lyons, Christy McCain, Jordan Okie, Juha Saarinen, Richard Sibly, Felisa Smith, Patrick Stephens, Jessica Theodor, and Mark Uhen.

We are extremely grateful to macroecologists from around the world for responding to our appeal to help identify seminal papers in such a recently emerging field. We went on to tap many of them to write the introductions to the papers so identified; they are included in the list of contributors. We wish to thank Rob Colwell, Margaret Davis, Pablo Marquet, and Carsten Rahbek who provided suggestions but for one reason or another did not write a paper introduction. And we especially thank David Ackerly, Chris Carbone, Brian Enquist, Mark Lomolino, and Stuart Pimm who wrote introductions for papers that were eventually removed from the volume because of space constraints. Their understanding made our job a lot easier.

Finally, we thank our editor from the University of Chicago Press, Christie Henry, whose enthusiasm for this project was well appreciated, the many publishers and academic societies that generously gave permission to reprint these papers (often without charging!), and Fornessa Randal and Katherine Thannisch at the University of New Mexico for their assistance in the production of this volume.

Funding for this project came from the IMPPS NSF RCN (DEB-0541625); this is IMPPS publication no. 6.

Felisa A. Smith, John L. Gittleman, and James H. Brown
Santa Fe, New Mexico

Introduction: The Macro of Macroecology

Felisa A. Smith, John L. Gittleman, and James H. Brown

In many ways the idea of a "foundations" volume for a field that has existed in name for only 25 years may seem premature. Yet there are several reasons why we felt the time was ripe for a compilation of foundational macroecology papers. First, although the term *macroecology* was coined just over 20 years ago (e.g., Brown and Maurer 1989), the approach itself dates back several centuries. Arguably many of the great natural historians and biogeographers of the 1800s viewed the world through a macroecological lens (e.g., de Candolle 1855; Watson 1859; Wallace 1878; de Liocourt, 1898), as did many notable later scientists (e.g., Arrhenius 1920, 1921; J. C. Willis 1921; Kleiber 1932; Williams 1943, 1947; Hutchinson and MacArthur 1959; Preston 1962). The first part of the book, "Macroecology before Macroecology," is intended to highlight some of these early contributions.

Second, and most importantly, the field has become increasingly relevant. Over the past two decades, there has been an exponential increase in the number of publications in macroecology; they are increasing by 34% per year, much higher than the 2.5% increase in papers across all scientific disciplines (Smith et al. 2008). Presumably, this reflects a growing appreciation that many complex and pressing biological and environmental problems are best tackled by adopting a new approach and using a different tool kit than those employed in the more traditional experimental sciences. Scientists involved in global change research, for example, have been particularly proactive in employing a macroecological perspective. Recent studies relate the temporal and spatial organization of numerous morphological, physiological, behavioral, ecological, evolutionary, phonological, and phylogenetic traits to past, ongoing, and predicted future environmental conditions (e.g., Lyons 2003; Hunt and Roy 2006; Smith and Betancourt 2006; Kerr et al. 2007; K. J. Willis et al. 2007; Kuhn et al. 2008; Sekercioglu et al. 2008).

Macroecology is an approach to science that emphasizes description and explanation of patterns and processes at divergent spatial and temporal scales (Brown 1995). The prefix *macro-* (from the Greek μάκρο for "big" or "far") refers to the broad perspective such studies take, as well as the focus on the emergent statistical properties of large numbers of ecological "particles," be they genes, individuals, populations, or species. The underlying assumption is that statistical patterns that are similar across large temporal and/or spatial scales are the result of similar causal mechanisms. Although first formalized in a series of papers by James H. Brown and Brian A. Maurer (Brown and Maurer 1987, 1989; Brown 1995), this approach is implicit in earlier publications. For example, François de Liocourt (1898), studying the influence of management practices on the structure of French fir forests, characterized the distribution of tree size in three different stands. De Liocourt's findings—that in natural areas the number of trees declined exponentially with increasing trunk diameter—allowed him to draw conclusions about the influence of management practices on tree distribution patterns. Similarly, other classic macroecological patterns—including the species-area relationship (de Candolle 1855; Watson 1859; Arrhenius 1920, 1921; J. C. Willis 1921; Williams 1943), the latitudinal gradient of species richness (Wallace 1878), the relationship between body size and metabolic rate (Kleiber 1932; Hemmingsen 1960), the species-abundance distribution (Fisher et al. 1943; Williams 1943, 1947; Preston 1962), and the species body-size distribution (Hutchinson and MacArthur 1959)—were identified decades, sometimes even centuries, ago. Consequently, despite the scant 20 years that have elapsed since the term was coined, macroecology has a deep and rich history.

Because this book was an outcome of a Research Coordination Network sponsored by the National Science Foundation (NSF) on Integrating Macroecological Pattern and Process across Scales (IMPPS), which primarily dealt with mammals, we strove to involve the wider macroecological community in the selection of these foundational papers. Accordingly, we canvassed about 50 experts with specialties in different disciplines and taxonomic groups to compile our preliminary list of candidate papers. We particularly sought to include individuals who did not already have close affiliations with members of the IMPPS group, so that our choices would reflect the consensus of the macroecological community at large. Our preliminary list was vigorously vetted within the IMPPS group, using citation statistics as one, but not by any means the most important, criterion. Our goal was to identify papers that we felt had made seminal contributions by tackling new problems, employing or developing new techniques, or stimulating important new research directions or ideas.

Assembled here are 46 classic papers that have laid the foundations for modern macroecology. The papers span eight decades, from 1920 to 1998, and include divergent perspectives of space, time, and taxonomic and habitat affiliation. We have organized them into two parts: "Macroecology before Macroecology" and "Dimensions of Macroecology". The latter is further subdivided into five sections: "Allometry and Body Size," "Evolutionary Dynamics," "Abundance and Distributions," "Species Diversity," and "Methodological Advances." For each reprinted paper, we asked a macroecologist specializing in the area to write a commentary that places the paper in a broader context and explains why it is "foundational." We were delighted by the diversity of commentaries that ensued, which reflect both the personalities of the writers as well as the norms of their disciplines. To increase the usability of this volume, we added to the introductions and commentaries an asterisk before each reference list entry that corresponds to a paper printed in this volume.

Perhaps conspicuously lacking in our volume is a section on conservation, a topic we initially included. Because this is such a rapidly growing field, and one that has readily employed macroecological approaches, we realized that a much larger volume would be required to adequately represent it. Fortuitously, a parallel Foundations volume is currently being prepared. Thus, our omission of conservation is certainly not a reflection of the lack of relevance of macroecology. Rather, many of the most pressing conservation issues are closely linked to themes presented in this book. An outstanding challenge for the future of macroecology is to identify and address these urgent conservation concerns.

At present macroecology still lacks definition and unification. Several important books have appeared: among them are James H. Brown's *Macroecology* (1995), Brian A. Maurer's *Untangling Ecological Complexity: The Macroscopic Perspective* (1999), Kevin J. Gaston and Tim M. Blackburn's *Pattern and Process in Macroecology* (2000), Tim M. Blackburn and Kevin J. Gaston's *Macroecology: Concepts and Consequences* (2004), David Storch, Pablo A. Marquet, and James H. Brown's *Scaling Biodiversity* (2007), and Jon D. Witman and Kaustuv Roy's *Marine Macroecology* (2009). But these texts tend to focus on a unique perspective and/or are specialized in their coverage. So there is arguably no sufficiently broad and integrated book to serve as a textbook for a course titled "Macroecology." This volume is not intended to completely fill that role. It is intended, however, to give the field definition and historic context. We expect that it will be used mostly by students and younger scientists, in graduate seminars and discussion groups, and also by individual scholars who are studying for their qualifying exams, are interested in the history of the areas where ecology, biogeography, evolution, and paleobiology intersect, or are simply intellectually curious. To foster budding macroecologists, all royalties generated will go to a newly established fund at the International Biogeography Society to support the research and meeting travel of graduate students and postdocs.

We hope that this volume will inform and inspire graduate students, postdocs, and even established scientists, giving them an appreciation of the scope of macroecology. The rapidly increasing interest in macroecology highlights its inherent ability to synthesize across disciplines that have traditionally specialized on particular scales of space or time, or particular kinds of environments and taxa of organisms. It also emphasizes the relevance of a macroecological approach for tackling some of the big scientific problems of our time, including the effects of human-caused environmental changes on biodiversity. One thing that the foundational papers make clear, however, is that despite decades of research, there is still no consensus about the causal mechanisms underlying many large-scale patterns and processes. We hope this volume may help inspire a new generation of macroecologists to tackle some of these big, important, and still largely unanswered questions.

Literature Cited

*Arrhenius, O. 1920. Distribution of the species over the area. *Meddelanden från K. Vetenskapsakademiens Nobelinstitut* 4:1–6.

———. 1921. Species and area. *Journal of Ecology* 9:95–99.

Blackburn, T. M., and K. J. Gaston. 2004. *Macroecology: Concepts and consequences.* Blackwell, Oxford.

Brown, J. H. 1995. *Macroecology.* University of Chicago Press, Chicago.

*Brown, J. H., and B. A. Maurer. 1987. Evolution of species assemblages: Effects of energetic constraints and species dynamics on the diversification of the North American avifauna. *American Naturalist* 130:1–17.

*———. 1989. Macroecology: The division of food and space among species on continents. *Science* 243:1145–50.

de Candolle, A. 1855. *Géographie botanique raisonné.* 2 vols. V. Masson, Paris.

de Liocourt, F. 1898. De l'aménagement des sapinières. *Bulletin trimestriel* (Société forestière de Franche-Comté et Belfort) (July):396–409.

*Fisher, R. A., A. S. Corbet, and C. B. Williams. 1943. The relation between the number of species and the number of individuals in a random sample of an animal population. *Journal of Animal Ecology* 12:42–58.

Gaston, K. J., and T. M. Blackburn. 2000. *Pattern and process in macroecology.* Blackwell, Oxford.

Hemmingsen, A. M. 1960. Energy metabolism as related to body size and respiratory surfaces, and its evolution. *Reports of the Steno Memorial Hospital and the Nordisk Insulinlaboratorium* 9:7–110.

Hunt, G., and K. Roy. 2006. Climate change, body size evolution, and Cope's Rule in deep-sea ostracodes. *Proceedings of the National Academy of Sciences* 103:1347–52.

*Hutchinson, G. E., and R. H. MacArthur. 1959. A theoretical ecological model of size distributions among species of animals. *American Naturalist* 93:117–25.

Kerr, J. T., H. M. Kharouba, and D. J. Currie. 2007. The macroecological contribution to global change solutions. *Science* 316:1581–84.

Kleiber, M. 1932. Body size and metabolism. *Hilgardia* 6:315–53.

Kuhn, I., K. Bohning-Gaese, W. Cramer, and S. Klotz. 2008. Macroecology meets global change research. *Global Ecology and Biogeography* 17:3–4.

Lyons, S. K. 2003. A quantitative assessment of the range shifts of Pleistocene mammals. *Journal of Mammalogy* 84:385–402.

Maurer, B. A. 1999. *Untangling ecological complexity: The macroscopic perspective.* University of Chicago Press, Chicago.

Preston, F. W. 1962. The canonical distribution of commonness and rarity: Part I. *Ecology* 43:185–215.

Sekercioglu, C. H., S. H. Schneider, J. P. Fay, and S. R. Loarie. 2008. Climate change, elevational range shifts, and bird extinctions. *Conservation Biology* 22:140–50.

Smith, F. A., and J. L. Betancourt. 2006. Predicting woodrat (*Neotoma*) responses to anthropogenic warming from studies of the palaeomidden record. *Journal of Biogeography* 33:2061–76.

Smith, F. A., S. K. Lyons, S. K. M. Ernest, and J. H. Brown. 2008. Macroecology: More than the division of food and space among species on continents. *Progress in Physical Geography* 32:115–38.

Storch, D., P. A. Marquet, and J. H. Brown. 2007. *Scaling biodiversity.* Cambridge University Press, Cambridge.

Wallace, A. R. 1878. *Tropical nature and other essays.* Macmillan, London.

Watson, H. C. 1859. *Cybele Britannica, or British plants and their geographical relations*. Longman, London.

Williams, C. B. 1943. Area and number of species. *Nature* 152:264–67.

*———. 1947. The generic relations of species in small ecological communities. *Journal of Animal Ecology* 16:11–18.

Willis, J. C. 1921. *Age and area*. Cambridge University Press, Cambridge.

Willis, K. J., A. Kleczkowski, M. New, and R. J. Whittaker. 2007. Testing the impact of climate variability on European plant diversity: 320,000 years of water-energy dynamics and its long-term influence on plant taxonomic richness. *Ecology Letters* 10:673–79.

Witman, J. D., and K. Roy. 2009. *Marine macroecology*. University of Chicago Press, Chicago.

PAPER 1

Macroecology: The Division of Food and Space among Species on Continents (1989)

J. H. Brown and B. A. Maurer

Commentary

JAMES H. BROWN AND BRIAN A. MAURER

We did not set out to start a new subdiscipline of ecology. Like most new science, our entry into macroecology was serendipitous. In the mid-1980s Brian A. Maurer was a graduate student in a very traditional Department of Wildlife and Fisheries at the University of Arizona, where James H. Brown was a professor in the Department of Ecology and Evolutionary Biology. Maurer took Brown's biogeography course, which led to many discussions. We soon realized that there were alternatives to intensive field studies and small-scale experiments to address problems such as Maurer's thesis topic: population trends in grassland birds, many of which were rare and near the edge of their geographic range on his study site in southeastern Arizona. The standardized censuses of the North American Breeding Bird Survey, the newer, more accurate range maps in recent field guides, and J. B. Dunning's compilation of data on body weights gave us data on average abundance, geographic range size, and body size for about 380 species of land birds across the entire North American continent. We realized we could use these data to ask how the bird species divided up energy and space.

It occurred to us that what we were doing was a kind of "statistical mechanics" for ecological systems, but we were unsure what to call it since the term *statistical ecology* had been preempted for another use. We coined the term *macroecology* to reflect the idea that progress could be made by studying ensembles of many species. It is this 1989 paper that spelled out the fundamentals of our large-scale statistical approach. But nothing in science is ever really new, and the foundations of macroecology had been laid in previous work. Our paper cited some of these earlier works, including our own efforts (Brown 1981; Brown and Maurer 1987), that staked out the domain and made seminal contributions to macroecology before the term even existed. Indeed, it was really our 1987 paper that introduced the conceptual framework, empirical patterns, and analytical methods. It was the 1989 paper, however, that more clearly articulated the elements of the research program that morphed into the subdiscipline of macroecology.

It retrospect, we are gratified, surprised, and a bit awed by the subsequent development of macroecology. The explosion of macroecological research and publications is documented in F. A. Smith et al.'s (2008) article and the introduction to this volume. Back in 1989 we thought that we might be onto something potentially important, but mostly we were just two young scientists having a blast working together on some exciting new research.

From *Science* 243:1145–50. Reprinted with permission from AAAS.

Literature Cited

*Brown, J. H. 1981. Two decades of homage to Santa Rosalia: Toward a general theory of diversity. *American Zoologist* 21:877–88.

*Brown, J. H., and B. A. Maurer. 1987. Evolution of species assemblages: Effects of energetic constraints and species dynamics on the diversification of the North American avifauna. *American Naturalist* 130:1–17.

Smith, F. A., S. K. Lyons, S. K. M. Ernest, and J. H. Brown. 2008. Macroecology: More than the division of food and space among species on continents. *Progress in Physical Geography* 32:115–38.

Articles

Macroecology: The Division of Food and Space Among Species on Continents

JAMES H. BROWN AND BRIAN A. MAURER

Analyses of statistical distributions of body mass, population density, and size and shape of geographic range offer insights into the empirical patterns and causal mechanisms that characterize the allocation of food and space among the diverse species in continental biotas. These analyses also provide evidence of the processes that couple ecological phenomena that occur on disparate spatial and temporal scales—from the activities of individual organisms within local populations to the dynamics of continent-wide speciation, colonization, and extinction events.

IN THE LAST THREE DECADES, ECOLOGISTS HAVE CONFRONTED the enormous diversity and complexity of the natural world—with varying success. On the one hand, attempts to quantify biological diversity and ecological phenomena have revealed incredible variety. The total number of plant, animal, and microbial species inhabiting the earth is estimated to be between 10 million and 50 million (*1*). Each of these species has different requirements for existence and characteristic variations in abundance in space and time. Each place on earth is also distinctive and is inhabited by a particular assemblage of species. On the other hand, quantitative approaches have also revealed tantalizingly general patterns that appear to reflect the operation of natural laws that govern the organization of the ecological world. For example, there are striking regularities in the relative abundance of species within a site (*2*), in the structure of food webs (*3*), and in the way that the number of species varies with latitude (*4*). Most of these patterns have been revealed by large-scale, comparative, nonexperimental studies, and most still lack satisfactory mechanistic explanations, although they have been known for more than 20 years.

Since the early 1970s, ecology has become increasingly microscopic and experimental in its approach. As answers to the big questions remained elusive, many ecologists focused on problems that could be solved. It is possible to characterize the effects of physical conditions or of other organisms on a certain species in a particular place by means of controlled, replicated manipulations. The problem, however, is not so much in interpreting the outcome of any single experiment as in synthesizing the results of the many different studies to draw useful generalizations about the organization of the natural world. Without a complementary emphasis on large-scale phenomena, there is little basis for determining which results simply reflect the idiosyncracies of individual species and particular sites and which reflect the operation of more universal processes.

In an effort to address this deficiency, we have begun studying the ecological patterns and processes that characterize the assembly of continental biotas, specifically North American mammals and birds. The early results offer new insights into the relation between microscopic and macroscopic phenomena and into the general processes that determine the diversity, abundance, and distribution of organisms.

J. H. Brown is a professor in the Department of Biology, University of New Mexico, Albuquerque, NM 87131. B. A. Maurer is an assistant professor in the Department of Zoology, Brigham Young University, Provo, UT 84602.

A Macroecological Approach

Our goal is to understand the assembly of continental biotas in terms of how the physical space and nutritional resources of large areas are divided among diverse species. Our approach can be characterized as follows.

1) Explicitly empirical and operational. We use computer analyses of large data sets available for several kinds of organisms. These data are compiled from sources including field guides, systematic surveys, and standardized censuses (*5*). So far, we have used mainly data on all species of North American breeding land birds and nonvolant terrestrial mammals.

2) Ecologically relevant data. We have focused on variables, such as body mass, local population density, and area and shape of geographic range, that affect the allocation of space and nutritional resources (*6*). Body mass is closely correlated with the energetic requirements of individuals, local population density indicates the number of individuals that are supported within a small area, and configuration of geographic range characterizes the species distribution.

3) Multivariate analyses. We propose that the allocation of space and nutrients depends on the interaction of variables we just mentioned, as well as others that we have not yet considered. In this respect we depart from several earlier studies that analyzed univariate frequency distributions of the same variables among species (*2, 7–9*).

4) Statistical distributions of variables among species. Unlike much traditional ecology, which focuses on the attributes of just one or a few species, we draw inferences from the statistical distributions of variables among many species in a diverse biota. This enables us to characterize the pattern of variation in the entire assemblage and to assess the extent to which particular subsets of species or local areas differ from random samples of the entire biota.

5) Taxonomically defined biotas. We analyze assemblages of species of a single large taxon, such as birds or mammals. This ignores ecological relations that are not closely correlated with taxonomy, such as some trophic and competitive interactions. It has

the advantage, however, of confining the analyses to organisms within phylogenetic lineages that are subject to similar evolutionary constraints.

This approach enables us to phrase our basic question in an alternative way: can we identify ecological processes that affect the evolutionary diversification of a taxon as it exploits the geographic space and trophic resources of a large land mass?

Ecological consequences of body size. One of the well-documented and intriguing generalities about the biological world is the distribution of body sizes among species within different taxonomic groups. There are strikingly similar frequency distributions for the body masses, plotted on a logarithmic scale, for North American birds and mammals (Fig. 1, A and B). The distributions are skewed to the right, with a strong mode between 50 and 100 g. Qualitatively similar relations have been found for many other diverse taxa inhabiting large geographic areas (*8*).

Data for mammals suggest, however, that these body size distributions vary with spatial scale; that is, with the size of the area sampled. Comparable frequency distributions are shown for North American mammals for two successively smaller scales (Fig. 1, C to F). As our scope of study changes from the entire continent, to biomes (large regions with relatively uniform climate and vegetation), to small patches of very uniform habitat, the distributions change in a regular way. Each smaller scale contains a smaller subset of the species in the larger fauna and proportionately fewer small and medium-sized species. Although the distribution for the entire continent is highly modal, with a peak at approximately 100 g, the distributions for the local habitat patches are virtually flat, with an approximately equal number of species in each logarithmic size category. This pattern appears to be very general and holds for the mammal faunas of the 21 biomes and for an equal number of local habitat patches that we have analyzed.

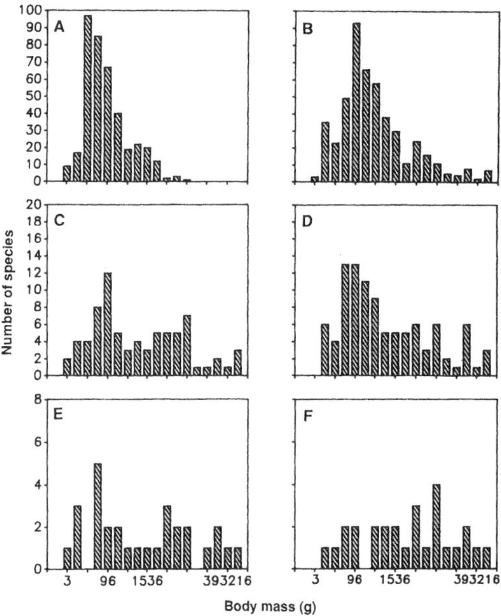

Fig. 1. Frequency distributions of body masses (on a logarithmic scale) among species of North American land birds (**A**) and land mammals (**B**) for the entire continent, for land mammals within biomes (**C**, northern deciduous forest; **D**, desert), and for land mammals within small patches of relatively uniform habitat within each of these biomes (**E**, Powdermill Reserve; **F**, Rio Grande Bosque).

We suggest that at least three kinds of mechanisms are necessary to produce this pattern. At present these should be regarded as hypotheses, supported by the following reasoning and data.

First, interspecific competition within local habitats is hypothesized to cause flat, log-uniform distributions (Fig. 1, E to F). Since the pool of species available on the continent or in the regional biome to colonize local habitats has a highly modal distribution of body sizes, some kind of strong negative interaction appears to prevent local coexistence of similar-sized species. The only process that we are aware of that can consistently have this effect is competitive exclusion. Competition strong enough to prevent coexistence should be restricted to species that are similar in their use of food or other resources. We hypothesize that local faunas are made up of several guilds, each of which uses a different food resource (for example, in the case of mammals: flesh, green vegetation, seeds, and fruit) and experiences strong competition among similar-sized species. This hypothesis is consistent with the evidence that local guilds of mammals and at least some other organisms consist of species that are more different in body size than expected from the random assembly of species from regional or continental pools (*10*).

Second, differential extinction of species of large body size is hypothesized to prevent the occurrence of numerous large species in the continental biota. Large organisms are constrained to have relatively low population densities because each individual requires large quantities of food and other resources. Since probability of extinction increases with decreasing population size (*11*), large species require large geographic ranges in order to persist for substantial time periods. This is consistent with the low frequency of large species and the much higher frequency of modal-sized species with small geographic ranges in both mammals (Fig. 2) (*12*) and birds (*6*). If extinction differentially eliminates large species with small ranges and competition tends to prevent local coexistence of similar-sized species in the same guild, these two processes together should be sufficient to account for the low frequency of large species in the continental biota.

Third, energetic constraints related to body size are hypothesized to cause the greater specialization of smaller organisms that result in the modal-sized species replacing each other with high frequency from habitat to habitat across the landscape. Such a pattern of spatial turnover is a necessary consequence of the systematic flattening of the frequency distributions of body sizes from continent to biome to local habitat patch (Fig. 1). Hutchinson and MacArthur (7), who first called attention to the highly modal distribution of mammalian body sizes for the entire North American continent, suggested that the smaller species were more specialized in their use of some essential resources than their larger relatives.

One important energetic consequence of body size seems sufficient to explain the pattern: requirements of individuals for energy and at least some nutrients scale as a fractional exponent (approximately 0.67 power), rather than linearly (1.0 power), with body mass (*13*). The physiological reasons for this are still poorly understood, but the ecological implications are profound. Because most of the variables that affect the capacity to collect and process food scale linearly with mass (*14, 15*), large animals can cover a larger area, ingest more food relative to their requirements, retain the material in the gut for a longer period, and extract a greater fraction of the energy and nutrients than small animals. This enables large species to feed on lower quality foods and to include a much wider array of items in the diet. We hypothesize that small species have smaller geographic ranges and replace each other more frequently across the landscape than their larger relatives because they confine themselves to habitats where foods of sufficiently high quality (to meet their more stringent energy and nutrient requirements) are in adequate supply.

The hypothesis that energetic constraints related to body size are important in the organization of these biotas is supported by other patterns in the organization of North American mammal and bird assemblages. One example is the low frequency in the continental bird and mammal faunas of the very smallest species, those with masses less than the modal size of approximately 100 g (Fig. 1, A and B). Insight into this pattern is offered by the relation between body mass and average population density (Fig. 3). Although there is much scatter in these data, all the points fall within well-defined bounds, which we hypothesize to reflect basic constraints. Some of these, such as the decrease in maximum population density with increasing body size, are quite straightforward. But the maximum population density also decreases with decreasing body size for birds weighing less than about 100 g. Mammals and beetles show a qualitatively similar pattern (*16*). Even the most abundant of the smallest species in the fauna are not as numerous as some of their larger relatives.

We hypothesize the following explanations for this constraint on maximum population density of the smallest species with a taxon. As mentioned earlier, daily energy requirements (E) of individuals are closely correlated with their body mass (M), such that $E = kM^{0.67}$, where k is a taxon-specific scaling constant (*13*). Over a wide range of body sizes, the area of the territory or home range used by an individual (A) is also closely correlated with size, varying as $A = cM^{1.0}$, where c is also a constant (*14*). The difference in the slopes (exponents) of these allometric equations means that the energy requirement of an individual per unit area of its territory increases with decreasing body size, $E/A = aM^{-0.33}$.

We suggest that down to some threshold body size, species are able to compensate for these increasing food requirements per unit area of territory and still maintain high local population densities. This is accomplished, at least in part, by specializing on habitats where individuals can forage efficiently. Below this threshold body size, however, individuals would no longer be able to meet their energy requirements if territory size were to continue to decrease. They can obtain adequate food only by restricting their foraging to rich, widely spaced patches of resources. These patches will also tend to be ephemeral, in part because rich patches will be depleted by the foraging of the individuals that discover them. This argument has three logical consequences. First, because the proportion of patches that are sufficiently productive to support food requirements should decrease with decreasing body size below 100 g, this will account for the apparent constraint on maximum population densities of birds and mammal species with body masses less than 100 g (Fig. 3). Second, this same reasoning will also account for the declining number of species in these smallest size classes (Fig. 1, A to B). Third, if the richest patches tend to be both widely dispersed and ephemeral, then individuals that exploit them should have to move large distances over their lifetimes.

The last of these leads to the testable prediction that lifetime territory size and movements should vary inversely with body size for birds and mammals weighing less than about 100 g. This prediction is not only counterintuitive, it also contradicts equations for the allometric scaling of territory size and movements based on birds and mammals over a wide range of body sizes (*14*). We tested our prediction with data on desert rodents (Fig. 4) (*17*). There is a great deal of scatter in the average population densities of these species as a function of their body mass, but, as in birds, species of intermediate size are most abundant (Fig. 4A). More importantly, as predicted, as body mass decreases below the threshold of approximately 100 g, individuals move longer distances over their lifetimes (Fig. 4B).

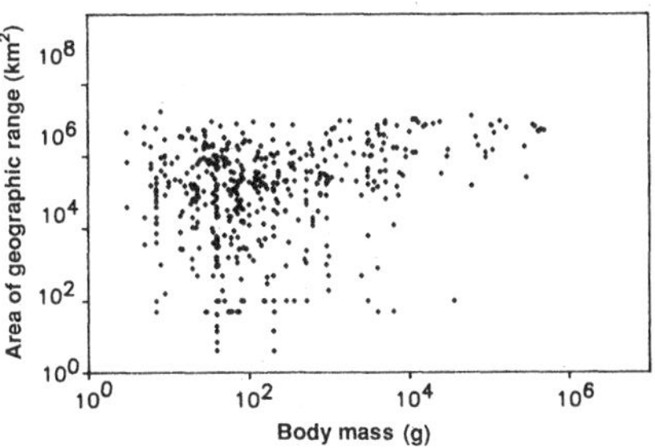

Fig. 2. Relation between area of geographic range and body mass (plotted on logarithmic axes) for the species of North American land mammals. Species of large body size tend to have large geographic ranges.

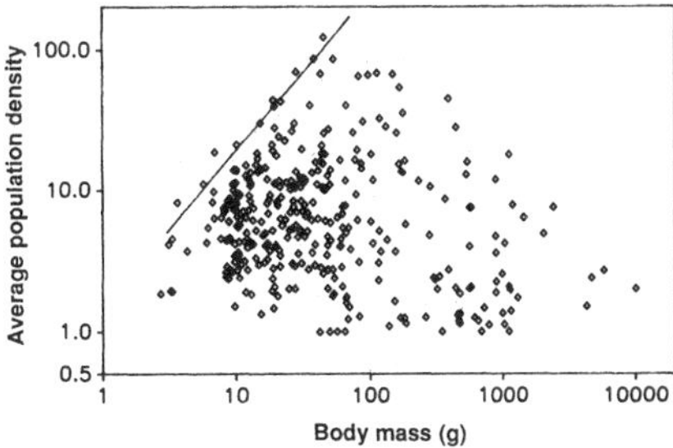

Fig. 3. Relation between average population density and body mass (plotted on logarithmic axes) for the species of North American land birds (*6*). Diagonal line, the decrease in maximum population densities of the species weighing less than approximately 100 g.

So far we have discussed how energetic constraints related to body size are important in the division of space among species. Now let us consider the allocation of energy more explicitly. Which body sizes use the most energetic and nutritional resources? Is the greater population density per species and the greater number of species of modal-sized birds and mammals compared to their larger relatives sufficient to compensate for their lower food use per individual? The prevailing widsom has been that small organisms did indeed dominate the flow of energy and nutrients through ecosystems (*18*).

We address this question on two levels. First, on a per species basis, how is food consumption related to body size? We know from the allometric equation $E = kM^{0.67}$ approximately how energy use per individual scales with body mass. Because all organisms are made of essentially the same chemical compounds, we assume that requirements for other nutrients scale similarly. Multiplying E by population density gives a good estimate of energy use per species per unit area. We made such calculations for two spatial scales, within local habitats (*19*) and for the continent as a whole (Fig. 5a) (*20*), and both give similar results. Although there is much variation among species of similar body size, large organisms consistently obtain most of the energy and nutrient resources. Second, if we sum the values for the species within a logarithmic body size class, how is total food consumption by individuals of all species related to body size? The results for birds on a continent are shown (Fig. 5B). Despite the smaller number of species and of individuals per species, large organisms consume at least as much energy and nutrients as their smaller relatives. Within local habitats the dominance of large organisms is even more pronounced, because there are more nearly equal numbers of species in all size classes (Fig. 1, E to F).

Together these data suggest a consistent view of the importance of body size–dependent energetic constraints in the ecology of mammals and birds. As body size decreases, individuals are faced with increasing energy and nutrient requirements per unit area of their territory and they become increasingly specialized to meet these demands. Initially, maximum population density increases with decreasing size, as increasing numbers of species divide up space and habitats according to their special requirements. Eventually a threshold body size is reached at which most areas are not sufficiently productive to support individuals, and then both maximum abundance per species and number of species decrease. Another consequence of these relationships is that energy and nutrient consumption, especially within local habitats, is dominated by the larger species. These energetic constraints of body size, together with intense competition between species of similar size within habitats and differential extinction of large species with small geographic ranges, seem necessary and perhaps sufficient to account for the systematic variation in body size distributions with spatial scale (Fig. 1).

Size and configuration of geographic ranges. The number and kinds of species that occur together at any scale are the result of both macroscopic and microscopic processes. On the one hand, the pool of species that are available to colonize a local area depends on the history of speciation and extinction events and on the expansion and contraction of geographic ranges. On the other hand, the origination, spread, and persistence of species in time and space depend on the effect of ecological conditions on the dynamics of local populations and the direction and rate of microevolutionary change. Valuable insights into these reciprocal relations between microscopic and macroscopic processes are offered by patterns in the sizes and configurations of geographic ranges.

The biogeographic barriers that determine the edges of a species range must be ecological limiting factors that prevent the expansion of local populations. Can we make any generalizations about the kinds of ecological variables that limit geographic distributions? We plotted the north-south dimensions of the geographic ranges against the east-west dimensions for North American terrestrial mammal species (Fig. 6A). A species with a circular or square range would fall along the diagonal line, indicating equal dimensions in each direction, and randomly distributed ranges would be dispersed equally around this line. The actual distribution of ranges exhibits a different pattern. The vast majority of the small ranges fall above the line; they are elongate in a north-south direction. In contrast, the majority of the large ranges have their long axis running east-west.

We propose the following explanation for this pattern. Species with small ranges (most of which are of small body size) are limited by habitat types and other variables that are associated with major topographic features, such as mountain ranges, river valleys, and coastlines. In North America these are oriented predominantly north-south. In contrast, species with large ranges are relatively insensitive to these variables and instead are limited by major climatic zones and biome types that are oriented predominantly east-west.

This hypothesis leads to two predictions. First, other taxonomic groups should respond similarly to mammals to the topographic and climatic features of the North American continent. This is supported by the configuration of geographic ranges for land birds in North America, which is qualitatively similar to that for mammals (Fig. 6B). Second, in Europe, where the important topographic features as well as the the major climatic belts run east-west, both the small and large geographic ranges should have their long axis oriented in this direction. This prediction is supported by the configuration of the plots of the geographic ranges of European land birds (Fig. 6C). Although there are fewer European species with very small ranges, even the smallest ones are oriented east-west, in marked contrast to ranges of comparable sizes of North American species.

These patterns show how the configuration of geographic ranges is influenced by ecological conditions that limit local populations of organisms and by continental-scale geography and geology that reflects the tectonic history and climate of the earth. This kind of coupling among disparate scales is emphasized in other studies that try to combine the macroscopic perspectives of biogeography and macroev-

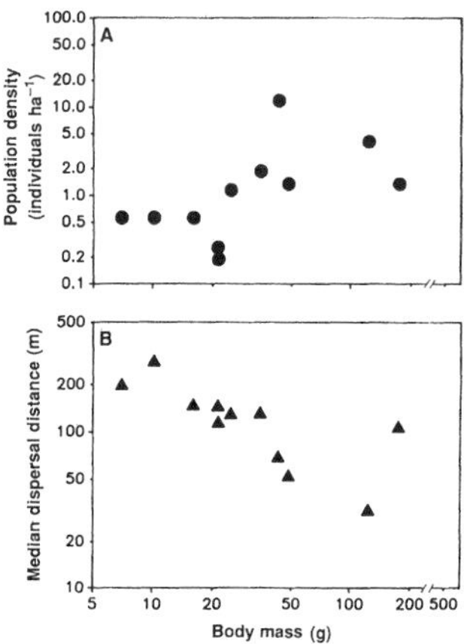

Fig. 4. Relation between (**A**) average population density and body mass and between (**B**) median lifetime dispersal distance of individuals and body mass for 11 species of desert rodents inhabiting a patch of relatively uniform Chihuahuan Desert shrub habitat. Variables are plotted on logarithmic axes from data in (*17*). The distribution of population densities is consistent with the pattern for birds shown in Fig. 3. For species weighing less than approximately 100 g, lifetime movements vary inversely with body size.

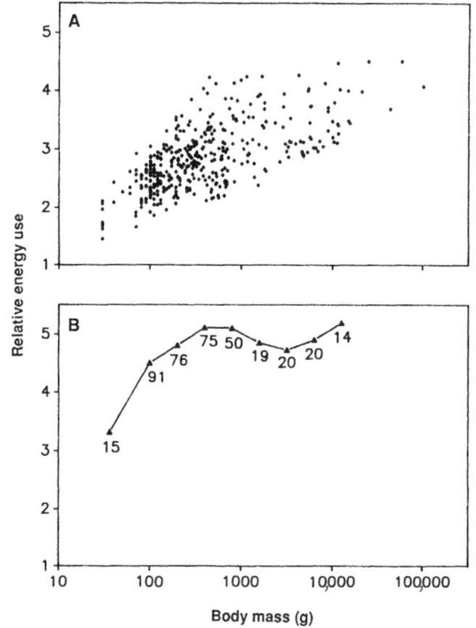

Fig. 5. Relation between energy use and body mass (on logarithmic axes) for North American land birds (*20*). (**A**) Average values for the species; (**B**) values summed for all of the species in equal-sized logarithmic body size categories. Numerals indicate the number of species in each size class. Large birds use more energy than small ones, on a per species basis and for all species within a size category.

olution with the microscopic approaches of population genetics and physiological, population, and community ecology (*21*).

Prospects for Synthesis

The patterns and explanations presented here illustrate the kinds of insights that can come from applying the questions posed by ecologists to the spatial and temporal scales normally studied by biogeographers and macroevolutionists. Our analyses suggest that the ecological and evolutionary processes that determine the assembly of continental mammal and bird faunas are reflected in regular patterns of body sizes and geographic range configurations. Comparisons of these patterns across spatial scales suggest mechanistic hypotheses that appear to be supported by available data.

Much remains to be done to assess the generality of the patterns and to test the validity of the explanatory hypotheses. To the extent that we have been able to compile and analyze appropriate data sets, the patterns and processes appear to be similar in birds and mammals, at least within North America. Although the frequency distribution of body sizes appears to be general (Fig. 1, A to B), it remains to be seen whether the other results can be generalized to other kinds of organisms and to other continents. The mechanistic hypotheses that we have proposed do not appear to depend on specific traits (such as endothermy) of birds and mammals or on the geography of North America. Therefore, although we would expect the quantitative details to vary among taxonomic groups and among continents, we predict that the same processes cause qualitatively similar patterns in other organisms and on other large land masses. The generality of patterns can be evaluated by compiling and analyzing similar data for other taxa and regions. The extent to which our mechanistic hypotheses are both necessary and sufficient to account for the patterns can be assessed by further studies designed explicitly to test their assumptions and predictions.

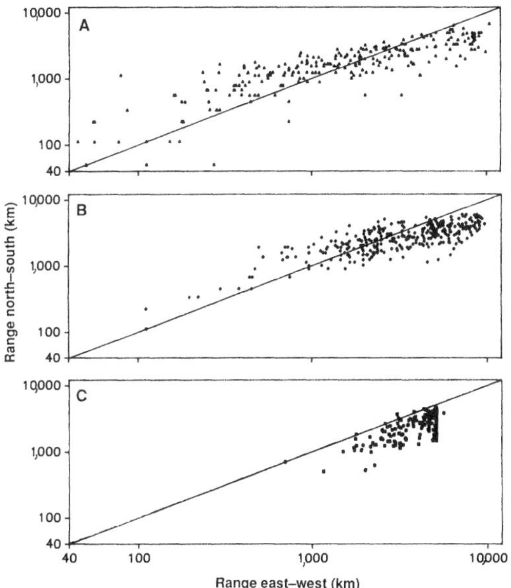

Fig. 6. Maximum north-south and east-west dimensions of the geographic ranges of North American (**A**) terrestrial mammals and (**B**) land birds and for (**C**) European land birds. Ranges of equal dimensions would fall on the diagonal line. In North America, small ranges tend to be oriented north-south whereas large ones are elongated east-west. In Europe, ranges of all sizes tend to be aligned east-west.

Our results suggest that ecological processes often provide the coupling among different levels of biological organization and among different spatial and temporal scales. Variation in the environment ultimately reflects geological, climatic, and oceanographic processes that are themselves coupled over disparate spatial and temporal scales. This environmental variation affects many different levels of biological organization, from the differential birth, death, and movement of individuals within local populations to the differential proliferation, extinction, and dispersal of species in continental and marine biotas.

Other investigators are also providing new insights into macroscopic ecological phenomena by incorporating new information from geology, climatology, and oceanography to develop syntheses between disciplines such as ecology, biogeography, systematics, macroevolution, paleontology, genetics, and microevolution (*22*). These interdisciplinary efforts promise to contribute importantly to understanding the origin and maintenance of biological diversity. For example, the fossil record studied by paleontologists and macroevolutionists documents periods of wholesale extinctions of species and higher taxa followed by periods of proliferation of new species and lineages (*23*). The fact that taxa with certain body sizes and geographic range areas differentially survive these catastrophes and speciate afterward suggests that knowledge of ecological processes that affect the assembly of contemporary biotas will contribute to understanding these historical events and vice versa. Also, new techniques are being developed to determine the phylogenetic relationships among species and to reconstruct the biogeographic histories of lineages of related species. When combined with ecological studies, these approaches offer the opportunity to better understand both the constraints of ecological processes on evolutionary events and the effects of evolutionary history on contemporary ecology (*24*).

Conclusions

The data and analyses presented here describe the division of food and space among wild species. As much as possible they represent the situation before the impact of modern humans. Within the last few centuries the exponentially growing population of *Homo sapiens* has changed the rules of resource allocation. Human beings currently use 20 to 40% of the solar energy that is captured in organic material by land plants (*25*). Never before in the history of the earth has a single species been so widely distributed and monopolized such a large fraction of the energetic resources. An ever-diminishing remainder of these limited resources is now being divided among the millions of other species. The consequences are predictable: contraction of geographic ranges, reduction of population sizes, and increased probability of extinction for most wild species; expansion of ranges and increased populations of the few species that benefit from human activity; and loss of biological diversity at all scales from local to global.

Currently, applied disciplines, such as conservation biology and natural resource management, remain focused primarily on small scales: preservation of individual endangered species, establishment of biological reserves, and management of local natural resources. But the most serious impacts of humans are global in extent and will persist for centuries or even millennia. There is great urgency to expand the spatial and temporal scale of contemporary ecology to address these problems.

REFERENCES AND NOTES

1. R. M. May, *Science* **241**, 1441 (1988).
2. F. W. Preston, *Ecology* **43**, 185 (1962); *ibid.*, p. 410; C. B. Williams, *Patterns in the Balance of Nature and Related Problems in Quantitative Ecology* (Academic Press, New York, 1964); R. M. May, in *Ecology and Evolution of Communities*, M. L. Cody and J. M. Diamond, Eds. (Harvard Univ. Press, Cambridge, MA, 1975), pp. 81–120; G. Sugihara, *Am. Nat.* **116**, 770 (1980).
3. J. E. Cohen, *Food Webs and Niche Space* (Princeton Univ. Press, Princeton, NJ, 1978); S. L. Pimm, *Food Webs* (Chapman and Hall, London, 1982); D. L. DeAngelis, W. M. Post, G. Sugihara, *Food Web Theory: Report on a Food Web Workshop* (Oak Ridge Natl. Lab., Publ. no. ORNL-5983, Oak Ridge, TN, 1983).
4. A. G. Fischer, *Evolution* **14**, 64 (1960); G. G. Simpson, *Syst. Zool.* **13**, 413 (1964); E. R. Pianka, *Am. Nat.* **100**, 33 (1967); R. E. Cook, *Syst. Zool.* **18**, 63 (1969); R. H. MacArthur, *Geographical Ecology: Patterns in the Distribution of Species* (Harper and Row, New York, 1972); J. H. Brown, in *Analytical Biogeography*, A. A. Myers and P. S. Giller, Eds. (Chapman and Hall, London, in press).
5. B. Bruun, *Birds of Europe* (McGraw-Hill, New York, 1970); E. R. Hall, *The Mammals of North America* (Wiley, New York, 1980); C. S. Robbins, B. Bruun, H. S. Zim, *Birds of North America* (Golden Press, New York, 1983); National Geographic Society, *Field Guide to the Birds of North America* (National Geographic Society, Washington, DC, 1983); J. B. Dunning, Jr., *West. Bird Banding Assoc. Monogr.* **1**, 1 (1984). We also used the North American Breeding Bird Survey, made available by D. Bystrak and C. S. Robbins of the U.S. Fish and Wildlife Service.
6. J. H. Brown and B. A. Maurer, *Am. Nat.* **130**, 1 (1987).
7. G. E. Hutchinson and R. H. MacArthur, *ibid.* **93**, 117 (1959).
8. R. M. May, in *Diversity of Insect Faunas*, L. A. Mound and N. Waloff, Eds. (Blackwell, Oxford, 1978), pp. 188–204.
9. J. C. Willis, *Age and Area* (Cambridge Univ. Press, Cambridge, 1922); E. Rapoport, *Aerography: Geographical Strategies of Species* (Pergamon, Oxford, 1982).
10. G. E. Huchinson, *Am. Nat.* **93**, 145 (1959); M. L. Rosenzweig, *J. Mammal.* **47**, 602 (1966); M. A. Bowers and J. H. Brown, *Ecology* **63**, 391 (1982); T. W. Schoener, in *Ecological Communities: Conceptual Issues and the Evidence*, D. R. Strong, Jr., D. Simberloff, L. G. Abele, A. B. Thistle, Eds. (Princeton Univ. Press, Princeton, NJ, 1984), pp. 254–281; F. A. Hopf and J. H. Brown, *Ecology* **67**, 1139 (1986).
11. R. H. MacArthur and E. O. Wilson, *The Theory of Island Biogeography* (Princeton Univ. Press, Princeton, NJ, 1967); T. W. Schoener and D. A. Spiller, *Nature* **330**, 474 (1988); D. Simberloff, *Annu. Rev. Ecol. Syst.* **19**, 473 (1988).
12. J. H. Brown, *Amer. Zool.* **21**, 877 (1981).
13. G. E. Walsberg, in *Symposium on Avian Energetics*, J. R. King, Ed. (Deutsche Ornithologen-Gesellschaft, Berlin, 1983), pp. 161–220; K. A. Nagy, *Ecol. Monogr.* **57**, 111 (1987).
14. B. K. McNab, *Am. Nat.* **97**, 113 (1963); T. W. Schoener, *Ecology* **49**, 123 (1968); A. S. Harestad and F. L. Bunnell, *ibid.* **60**, 389 (1979).
15. W. A. Calder III, *Size, Function, and Life History* (Harvard Univ. Press, Cambridge, MA, 1984); M. W. Demment and P. J. van Soest, *Am. Nat.* **125**, 641 (1985).
16. J. Damuth, *Nature* **230**, 699 (1981); D. R. Morse, J. H. Lawton, M. M. Dodson, M. H. Williamson, *ibid.* **314**, 731 (1985); D. R. Morse, N. E. Stock, J. H. Lawton, *Ecol. Entomol.* **13**, 25 (1988).
17. J. H. Brown and Z. Zeng, in preparation.
18. R. H. Peters, *The Ecological Implications of Body Size* (Cambridge Univ. Press, Cambridge 1983); P. H. Harvey and J. H. Lawton, *Nature* **324**, 212 (1987); D. Strayer, *Oecologia* **69**, 513 (1986).
19. J. H. Brown and B. A. Maurer, *Nature* **324**, 248 (1987).
20. B. A. Maurer and J. H. Brown, *Ecology* **69**, 1923 (1988).
21. M. L. Cody, in *Ecology and Evolution of Communities*, M. L. Cody and J. M. Diamond, Eds. (Harvard Univ. Press, Cambridge, MA, 1975), pp. 214–257; J. H. Brown, in *Organization of Communities: Past and Present*, J. H. R. Gee and P. S. Giller, Eds. (Blackwell, Oxford, 1987), pp. 185–203; J. Roughgarden, S. D. Gaines, S. W. Pacala, in *Organization of Communities: Past and Present*, J. H. R. Gee and P. S. Giller, Eds. (Blackwell, Oxford, 1987), pp. 491–518; T. Root, *Ecology* **69**, 330 (1988).
22. R. P. Neilson and L. H. Wullstein, *J. Biogeogr.* **10**, 275 (1983); J. Roughgarden, S. Gaines, H. Possingham, *Science* **241**, 1460 (1988).
23. S. M. Stanley, *Macroevolution: Pattern and Process* (Freeman, San Francisco, 1979); J. J. Sepkoski, Jr., *Paleobiology* **10**, 246 (1984); D. Jablonski, *Science* **231**, 129 (1986).
24. D. R. Brooks, *Ann. Mo. Bot. Gard.* **72**, 660 (1985).
25. P. M. Vitousek, P. R. Ehrlich, A. H. Ehrlich, P. A. Matson, *BioScience* **36**, 368 (1986); D. H. Wright, *J. Biometeorol.* **31**, 293 (1987).
26. We thank the many people who helped to assemble and analyze the data and those who have discussed these ideas with us. P. Nicoletto, J. F. Merrit, and J. S. Findley contributed data for Fig. 1. G. Ceballos, G. Farley, J. Findley, L. Hawkins, A. Kodric-Brown, P. Nicoletto, E. Toolson, and anonymous reviewers made helpful comments on the manuscript. The support of NSF grants BSR-8506729 and BSR-8718139 and Department of Energy grant FG02-86ER60424 is gratefully acknowledged.

PART

1 Macroecology before Macroecology

Edited by James H. Brown, S. K. Morgan Ernest, and Ethan P. White

The title of this section, "Macroecology before Macroecology," is a bit of a misnomer. The term *macroecology* was originally introduced in James H. Brown and Brian A. Maurer's (1989) paper, which provided a synthetic overview of the domain, conceptual foundations, and empirical and theoretical themes of what has emerged as the discipline of macroecology. While Brown and Maurer's (1989) paper provided a name for the discipline, it did not actually mark the beginning of macroecological research. The title of this section accurately calls attention to the fact that, for the better part of a century before the term was coined, there was a rich history of macroecological research with many seminal publications. Indeed, these earlier studies pioneered many of the themes and developed the conceptual groundwork for the explosion of macroecology in the 1990s and early 2000s.

In this section we highlight contributions of some of the first macroecologists. Before commenting on why these papers were selected as being especially noteworthy, we must stress that these are by no means the only ones that could have been chosen. Indeed, our reasons for omitting some papers are sometimes arcane and arbitrary. We elected not to include papers that are readily available in other collections, most notably the *Foundations of Ecology* and *Foundations of Biogeography* volumes published by the University of Chicago Press. Some of the earliest precursors, such as publications by A. R. Wallace (1878), J. C. Willis (1921), and R. H. MacArthur and E. O. Wilson (1967), were omitted because they were more qualitative and lengthy narrative expositions in books. Some papers were omitted and others included to deliberately showcase the broad conceptual, empirical, and methodological roots of what has become the contemporary discipline of macroecology. Indeed, it is easier to justify why these papers were selected than to explain why many others, perhaps equally worthy, were left out.

Throughout its history, much of macroecology has been inductive: starting with quantifying an empirical pattern and only later explaining it by the more deductive process of erecting and evaluating mechanistic hypotheses and developing synthetic models and theories. Olaf Arrhenius's 1920 paper on species-area relationships is a case in point. Although patterns of increasing species diversity with increasing sample area had been documented since at least the mid-1800s (Rosenzweig 1995), Arrhenius characterized them quantitatively and analytically. Furthermore, he set a precedent by fitting species-area relations with power laws. Even though Arrhenius's treatment was essentially descriptive and statistical, it laid the foundation for subsequent methodological, empirical, and theoretical studies conducted later by Arrhenius himself, C. B. Williams, Frank Preston, Robert MacArthur and E. O. Wilson, and many others. It opened up an area of macroecological research that is still very active, with much discussion about the mathematical forms and underlying causes of these pervasive relationships.

It has long been known that many characteristics of organisms vary systematically with body size. These relationships were quantified and synthesized in seminal books on allometry by D'Arcy Thompson and Julian Huxley in the early 1900s. But it was Max Kleiber's (1932) paper that attracted broad attention and stimulated subsequent research on body-size scaling relationships. Kleiber found that the metabolic rate of mammals and birds scales as the ¾ power of body mass. This challenged existing ideas that metabolic rate scaled as a ⅔ power, and other traits with characteristic third powers, reflecting the fundamentally geometric scalings of lengths, surface areas, and volumes. Kleiber's paper and subsequent book, *The Fire of Life*, ignited controversies and stimulated research efforts that have not abated in the 80 years since their publication.

C. B. Williams should perhaps be regarded as the father of macroecology. More than any of his antecedents and contemporaries, he not only documented the pervasively general patterns of abundance, distribution, and diversity of species; he also recognized that many of them had their roots in fundamental processes of resource use, demography, dispersal, speciation, extinction, and community assembly. We showcase Williams's contributions to macroecology by reprinting papers on species-abundance distributions (with R. A. Fisher and A. S. Corbett) and community assembly as reflected in species-to-genus ratios. These and other patterns and processes were presented, analyzed in detail, and discussed in Williams's book *Patterns in the Balance of Nature and Related Problems in Quantitative Ecology* (1964), which could be considered the first major synthetic publication in macroecology.

G. E. Hutchinson was arguably the most synthetic and influential ecologist of the mid-20th century, and Robert MacArthur was undoubtedly his most famous student. It is interesting, therefore, that the 1959 paper, one of only two coauthored by this teacher-student duo, has not received much attention (e.g., Google Scholar lists only 276 citations, compared to thousands for others of their papers). This is perhaps because the paper was ahead of its time. Hutchinson and MacArthur quantified the frequency distribution of body sizes among species of mammals, documented the now well-known right-skewed shape, and suggested that this was due to how the environment and its resources were allocated among niches of species. Reading the paper today shows that they were clearly calling attention to the fractal-like heterogeneity of the environment, even though the paper was published 15 years before Benoit Mandelbrot ([1975] 1977) described the phenomenon of fractality and coined the term. Studies of frequency distributions of body sizes and abundances among species have been a major theme of macroecology, and macroecologists continue to explore the ideas pioneered by Hutchinson and MacArthur.

Few scientists have a law named after them. Taylor's power law is one of the few "laws" that have endured in ecology. In the article reprinted here and several other papers, L. R. Taylor analyzed time-series data for fluctuations in abundance of many animal populations. He showed that nearly all appeared to exhibit a systematic relationship between the variance and the mean, which is well fit by a power law. Taylor's work is emblematic of macroecology. Rather than focusing on the dynamics of a single population like most contemporary studies, it compiled and analyzed multiple time series and sought to explain both the pervasive power-law scaling relationships and the considerable variation in the exponents. Following up on Taylor's power law remains an important theme of contemporary macroecological research.

Although the pioneering work of D'Arcy Thompson, Julian Huxley, and Max Kleiber had enormous influence on comparative morphology and physiology, ecologists were slow to appreciate the potential importance of allometry for ecology. One of the first to integrate allometry into ecology was BRIAN MCNAB, who realized that metabolic rate, by setting requirements for food, should influence how animals use space. In a classic macroecological study, McNab compiled and analyzed data on home ranges of mammal species, and showed that the area of space used by an individual varied as a power law with body mass and differed systematically between herbivores and carnivores. This work ushered in other studies on the ecological allometries of resource use, demography, and life history (synthesized in R. H. Peters's *Ecological Implications of Body Size* [1983] and W. A. Calder III's *Size, Function, and Life History* [1984]).

The vast majority of early research on body size and allometry was on animals, which is perhaps surprising given the enormous variation in body size of plants and its influence on many aspects of morphology, physiology, and ecology. The study by J. Yoda et al. (1963) is one of the earliest allometric studies in botany. These investigators documented a very general pattern of "self-thinning" in which plant size increases and density decreases as dense stands of seedlings give way to even-aged cohorts of larger, older plants. This phenomenon, which occurs pervasively in agricultural fields, forestry plantations, and natural successional communities, follows a power law. This pioneering study stimulated a great deal of subsequent research on size-related resource use, competition, and community organization in plants, much of which is synthesized in Karl Niklas's book *Plant Allometry* (1994).

For much of its history, the study of the fossil record was the province of paleontology, which was a branch of geology. It was only with the modern synthesis and the seminal publications of G. G. Simpson that paleontology became well integrated with biology, developed a more comprehensive, holistic approach to the history of life, and transformed into paleobiology. Now paleoecological research is a major theme of macroecology, providing an essential temporal perspective on macroecological patterns and processes. By explicitly recognizing that fossil assemblages are ecological communities, and analyzing their composition using a comparative, quantitative, and functional framework, E. C. OLSON's paper marks an important step in both paleoecology and macroecology.

C. B. Williams's book *Patterns in the Balance of Nature and Related Problems in Quantitative Ecology* (1964) called attention to the existence of very general statistical empirical patterns of abundance, distribution, and diversity. In the vast majority of these examples, the units were species. A few years later, R. W. SHELDON and his coworkers (including T. R. PARSONS) expanded the scope of macroecology by showing that some of these very general statistical patterns are independent of the Latin binomials. The abundances of pelagic organisms in the ocean exhibit a power-law distribution such that the total biomass remains essentially invariant from plankton to whales. This pioneering observation stimulated a great deal of research because of its theoretical implications for the organization of marine communities and its practical applications for fisheries and conservation.

As the disciplines of community and evolutionary ecology began to flower in the 1950s and 1960s, attention was focused on patterns and processes of species diversity. Although earlier studies had quantified and discussed the latitudinal diversity gradient (e.g., Fischer 1960; Simpson 1964), HOWARD SANDERS took this work to new levels of sophistication by collecting and analyzing samples of benthic organisms from both temperate and tropical environments. Especially noteworthy was Sanders's use of accumulation curves to quantify the increase in the number of species with increasing sample effort. This empirical work, together with the theoretical treatments of species-abundance distributions by Fisher,

Corbett, and Williams (see above) and Frank Preston (1948, 1962a, 1962b) laid the foundation for studying the interrelated patterns of diversity within and across communities that are a major theme of contemporary macroecology.

The macroecological patterns of abundance, distribution, and diversity of species play out on the abiotic template of heterogeneous geology and climate, and similar physical-chemical variation in marine and freshwater environments. An ability to quantify the relevant abiotic variables is critical for progress in understanding the mechanisms underlying the empirical patterns. These days, with the ready availability of satellite imagery, computerized databases, and geographic information systems, and the near-immediate access to data-filled papers, it is hard to appreciate the importance of the early contributions. Net primary productivity (NPP) is a critical variable for many macroecological studies, but it was not until the 1960s and 1970s that reliable estimates of rates of primary production in terrestrial and aquatic environments became available. Michael Rosenzweig's paper was pioneering in showing how NPP could be predicted from the basic climatic variables collected by the existing networks of weather stations. Follow-up work by H. Leith and R. H. Whittaker (1975), R. H. Whittaker (1975), and others laid the foundation for the current understanding of how NPP in terrestrial ecosystems varies with solar radiation, temperature, water, nutrients, successional stage, and other factors.

Although many other papers could have been included in this section, even this limited selection clearly shows that many themes of macroecological research have a long and venerable history. While the following sections of this volume will illustrate how macroecological research evolved over the decades, the papers in this section clearly demonstrate that macroecology was a vibrant and important approach for studying nature long before Brown and Maurer coined the term in 1989.

Literature Cited

*Brown, J. H., and B. A. Maurer. 1989. Macroecology: The division of food and space among species on continents. *Science* 243:1145–50.

Calder, W. A., III. 1984. *Size, function, and life history*. Harvard University Press, Cambridge, MA.

Fischer, A. G. 1960. Latitudinal variation in organic diversity. *Evolution* 14:64–81.

Kleiber, M. 1932. Body size and metabolism. *Hilgardia* 6:315–53.

Lieth, H., and R. H. Whittaker, eds. 1975. *Primary productivity of the biosphere*. Springer-Verlag, New York.

MacArthur, R. H., and E. O. Wilson. 1967. *The theory of island biogeography*. Princeton University Press, Princeton, NJ.

Mandelbrot, B. (1975) 1977. *Fractals: Form, chance, and dimension*. English edition, W. H. Freeman, San Francisco, CA.

Niklas, K. J. 1994. *Plant allometry: The scaling of plant form and function*. University of Chicago Press, Chicago.

Peters, R. H. 1983. *The ecological implications of body size*. Cambridge University Press, Cambridge.

Preston, F. W. 1948. The commonness, and rarity, of species. *Ecology* 29:254–83.

———. 1962a. The canonical distribution of commonness and rarity: Part 1. *Ecology* 43:185–215.

———. 1962b. The canonical distribution of commonness and rarity: Part 2. *Ecology* 43:410–32.

Rosenzweig, M. L. 1995. *Species diversity in space and time*. Cambridge University Press, New York.

Simpson, G. G. 1964. Species density of North American recent mammals. *Systematic Zoology* 13:57–73.

Wallace, A. R. 1878. *Tropical nature and other essays*. Macmillan, London.

Whittaker, R. H. 1975. *Communities and ecosystems*. 2nd ed. Macmillan, New York.

Williams, C. B. 1964. *Patterns in the balance of nature and related problems in quantitative ecology*. Academic Press, London.

Willis, J. C. 1921. *Age and area*. Cambridge University Press, Cambridge.

Yoda, J., T. Kira, H. Ogawa, and H. Hozumi. 1963. Self-thinning in overcrowded pure stands under cultivated and natural conditions. *Journal of Biology Osaka City University* 14:107–29.

PAPER 2

Distribution of the Species over the Area (1920)

O. Arrhenius

Commentary

ETHAN P. WHITE

The species-area relationship (SAR) is second only to the latitudinal gradient as a general ecological pattern. The study of the SAR dates back to at least 1859 (Rosenzweig 1995), but its status as a pattern of significant interest to ecologists began in earnest with a series of publications by Olaf Arrhenius in the 1920s suggesting that the SAR was well described by a simple mathematical function, the power function.

Arrhenius's (1921) paper in the *Journal of Ecology* is typically credited by authors as the origin of the power-law form of the SAR. However, the original observation that the power function represents a good description of the SAR dates to two publications in the previous year: the paper presented here and a book published in German. This body of work represents classic exploratory macroecology. Arrhenius did not have any theoretical reason for expecting that the power function would describe the SAR. He simply looked for a simple characterization of the pattern, found that the power law worked in some cases, and then continued to look at the relationship in additional locations, finding that the power law almost always provided a good description of the observed data. He even noted that the exponent of the power law maintained a remarkable consistency among sites. The reported values are equivalent to modern z values in the range of 0.2 to 0.3 and thus include F. W. Preston's widely utilized canonical value.

The power-function characterization of the species-area relationship has had far-reaching influences in the field of ecology. It motivated influential work on the SAR by Preston (1960) and the development of island biogeography theory by R. H. MacArthur and E. O. Wilson (1963, 1967). In addition, this form of the relationship has been used extensively in conservation science to estimate rates of extinction in response to habitat loss, fragmentation, and global change (e.g., McDonald and Brown 1992; Guy 1999), and to identify hotspots of species diversity (e.g., Fattorini 2006). Arrhenius's work has had a lasting influence on the field of ecology.

Literature Cited

Arrhenius, O. 1921. Species and area. *Journal of Ecology* 9:95–99.

Fattorini, S. 2006. Detecting biodiversity hotspots by species-area relationships: A case study of Mediterranean beetles. *Conservation Biology* 20:1169–80.

Guy, C. 1999. Predicting the pattern of decline of African primate diversity: An extinction debt from historical deforestation. *Conservation Biology* 13:1183–93.

MacArthur, R. H., and E. O. Wilson. 1963. An equilibrium theory of insular zoogeography. *Evolution* 17:373–87.

———. 1967. *The theory of island biogeography.* Princeton University Press, Princeton, NJ.

McDonald, K. A., and J. H. Brown. 1992. Using montane mammals to model extinctions due to global change. *Conservation Biology* 6:409–15.

Preston, F. W. 1960. Time and space and the variation of species. *Ecology* 41:611–27.

Rosenzweig, M. L. 1995. *Species diversity in space and time.* Cambridge University Press, New York.

From *Meddelanden Från K. Vetenskapsakademiens Nobelinstitut* 4:1–6.

MEDDELANDEN
FRÅN
K. VETENSKAPSAKADEMIENS NOBELINSTITUT.
BAND 4. N:o 7.

Distribution of the species over the area.

By

O. ARRHENIUS.

Communicated May 26th 1920 by S. ARRHENIUS and C. LINDMAN.

Since the knowledge of the relation between the area and the number of species i. e. the distribution of species over areas of unequal size, has very much significance in the study of plant geography, I have in a former work[1] given a short analysis of the problem. Since a more detailed discussion of the problem can be desired, I have enlarged the subject in the present paper.

A certain regularity prevails in the distribution of the species over the surface. I have shown that

$$\frac{y}{y_1} = \left(\frac{x}{x_1}\right)^n$$

gives the relation between the surface and the number of the species. In the formula y is the area upon which x species

[1] ARRHENIUS, O. Öcologische Studien in den Stockholmer Schären. Stockholm 1920, p. 12.

are found and y_1 that upon which x_1 species are coming, n is a constant. In the previous work the value 3,2 was derived for n.

Table 1—4 are summaries of this material. In the 1st column the area is given (y), in the 2nd the observed species (x), in the 3rd the calculated number of species according to the above equation and in the last column the number of observations made.

Table 1 gives the relation of species in an association, a Sesleria-meadow. The surface is here given in dm^2.

Table 2 gives the result from areas of m^2 size, taken by JACCARD[1] in meadows of the Switzerland alps.

Table 3 and 4 give values of a group of associations studied from islands in the central part of Stockholms archipelago. The area in this case is given in hectars.

Tab. 1.

Sesleria-meadow

y (dm^2)	x (obs.)	x (calc.)	Number of obs.
1	10,1	10,5	21
2	12,8	13,0	17
3	15,2	15,3	17
4	16,1	16,1	10
8	19,2	20,1	8
16	25,7	24,3	4
32	30,3	29,7	3
80	41	40,8	1

Tab. 2.

Meadows in Switzerland alps.

y (dm^2)	x (obs.)	x (calc.)	Number of obs.
1	25	24,7	52
2	30,4	30,7	7
3	33,7	35,8	7
4	34,6	37,8	7
5	36,2	39,6	5
6	39,8	43,2	6
7	44.6	44,6	7
8	49,0	47,2	2
14	55,6	56,1	3
28	75	69,8	2
52	92	84,1	1

[1] JACCARD, P. Nouvelles recherches sur la distribution florale. Bull. Soc. Vaud. sci. nat. Lausanne 1908, p. 225.

Tab. 3.

Bullerö-archipelago

y (har)	x (obs.)	x (calc.)	Number of obs.
0,25	27,6	34,5	3
1,00	53,6	53,0	3
1,43	57,8	59,2	4
2,07	60,6	64,9	3
2,85	72	73,3	2
4,8	81,4	86,2	5
7,0	84	96,9	2
22	170	135	1
62	270	190	1

Tab. 4.

Munkö-archipelago

y (har)	x (obs.)	x (calc.)	Number of obs.
0,16	25	39	2
0,6	68,3	58,6	3
1,3	105	74,4	3
3,1	115	99	2
4,25	120,5	107,5	2
10	137	140	1
21,5	173	178	2
26,3	172	188	3
95	336	351	1

It may be seen that the calculated and observed numbers agree very well with the exception in a few cases. In table 3 and 4 the variations are larger than in 1 and 2 due to the fact that in the former case one has to do with more complicated units than in the latter. Also the difficulty of calculating the exact area (0,16, 0,25 and 0,6 hectars) of the islands introduces an error since this were estimated from a map drawn to the scale of 1 : 50,000.

It is of great interest to find that the exponent n is the same over such widely separated places as Sweden's archipelago and Switzerland's mountains, and also with such widely different areas as dm^2, m^2 and hectar. Indeed this seems to be a very general rule.

I have not found the same value for n in PALMGREN's[1,2] material from Åland. The results are given in table 5.

Tab. 5.

Leaf-meadows in Åland.

y	x (obs.)	x (calc.)	Number of obs.
1	193,4	191,4	6
2	218,8	206,5	3
3	236,9	233,2	2
4	242,2	246,2	2
5	249	255,8	1
6	259	264,3	1

[1] PALMGREN, ALVAR, Studier över lövängsområdena på Åland. Acta Soc. pro Fauna et Flora Fennica. 42. 1915—17. Tables.

[2] ARRHENIUS, O., En studie över yta och arter. Sv. Bot. Tidskr. Bd. 12. 1918. Tables.

The agreement is here seen to be very good. The exponent has, however, here a value of 5,6. Still one can't draw any definite conclusions from this material since the area has varied only in the ratio of 1:6 and the species only in the ratio 1:1,38 while the variations in tables 1—4 reach to about 1:100 and 1:4. Therefore in Tab. 5 exceptional circumstances may prevail in the small interval examined, which cause an other value of *n* than the normal one.

It is of particular interest to see that the same increase of the species over the area is found in pure associations (the Sesleria-meadow), formations (the alpmeadows of JACCARD) and combinations of formations (Bullerö- and Munköarchipelagoes).

One will safely get very valuable results through continuing the work by collecting material from different lands and different associations.

The synecological transect-examination method[1] is based on the fact that the collection of the vegetation along the transect is the same as the collection over the whole surface through which the transect is laid. One measures the length of the associations crossed by the line and expresses the length of every association in percentage of the whole line. One assumes thus that this fraction of the line gives the real result corresponding to the area taken by this association. The area of each association expressed in percentage of the whole line is called the area-percent. If one lays out several such lines over the same surface one naturally obtains more accurate results. I have found that 1 line per 150 meters is sufficient for ordinary demands of accuracy for synecological purposes.

In calculating the mean deviation for the lines partly where one line is taken as an element partly when 2, 3, 4 etc. are collected to one element and compared with similar ones one obtains knowledge of how many lines are necessary to get accurate results of the associations of the area.

In calculating the results it is apparent that the mean deviation expressed in parts — as in table 6 generally percent — of the area-percent is larger the smaller the association studied.

[1] FRIES, THORE C. E., Den synekologiska linietaxeringsmetoden. Vetenskapliga och praktiska undersökningar i Lappland. Medd. fr. Abisko nat. vet. stn. Nr. 2. 1919.

Tab. 6.

Examination made with

Area-percent	1 Line obs.	1 Line Deviation in % calc$_1$	1 Line Deviation in % calc$_2$	Area-percent	2 Lines obs.	2 Lines Deviation in % calc$_1$	2 Lines Deviation in % calc$_2$	Area-percent	4 Lines obs.	4 Lines Deviation in % calc$_1$	4 Lines Deviation in % calc$_2$
30,5	22	31,9	32	30,2	20	22	23	31,8	19	16	15
11,0	56	53	54	10,1	30,5	38	40	10,1	25	28	28
8,7	67	60	61	7,1	54	46	47	7,0	40	34	33
6,1	82	71	73	4,9	57	54	57	4,9	40	40	39
3,6	92	92	94	2,2	97	81	85	2,1	75	62	59
2,5	110	111	114	1,8	70	90	94	1,8	46	67	64
1,6	157	139	142	1,2	120	111	115	1,2	65	82	78
1,2	148	161	164	0,8	125	134	140	0,9	91	101	90
0,9	172	185	190	0,2	150	268	280	0,2	240	230	191
0,7	210	210	214								
0,5	230	249	254								
0,4	150	278	284								
0,1	300	550	568								

It seems that one may express this relation through the formula $x = \frac{k}{y^n}$ where x is the mean deviation, k and n are constants and y is the area-percent.

Table 6 gives the results of examinations with 1, 2 and 4 lines used as one element laid out over the same area, Munkö. In the table besides the values of the area-percent, the values observed and calculated are entered.

For n I have found the value 0,55 from the observations. As this figure lies very near to the value 0,5, which seems more probable on theoretical grounds—see below — I have also calculated figures on the assumption that $n = 0{,}5$.

Under $calc_1$ the values are given, which correspond to $n = 0{,}55$, under $calc_2$ values corresponding to $n = 0{,}5$.

The calculated values tabulated under $calc_1$ and $calc_2$ agree nearly equally well with the observed values. Therefore I have assumed $n = 0{,}5$.

This means that the relation between the linelength examined (area-percent) and the mean deviation is that which one should expect according to the calculus of probability, namely that the mean deviation is nearly inversely proportional to the square root (since $n = 0{,}5$) of the number of observations i. e. in this case the number of observed meters.

This regularity is namely analogous to the law that the probable error of a given observed quantity is inversely proportional to the square root of the number of observations, which have served for the determination of the said quantity, if the observations possess the same weight.

This paper shows that the increase of species with the area can be expressed by a mathematical approximation and that this holds true for spaces as far apart as Switzerland, central Sweden and Åland. It also holds for areas of unequal size.

It seems also as if the mean deviation between the lines used by the transect-examination can be calculated according to the law of probability.

Tryckt den 8 september 1920.

Uppsala 1920. Almqvist & Wiksells Boktryckeri-A.-B.

PAPER 3

The Relation between the Number of Species and the Number of Individuals in a Random Sample of an Animal Population (1943)

R. A. Fisher, A. S. Corbet, and C. B. Williams

Commentary

ETHAN P. WHITE

This paper is actually three small papers rolled into one. The most well known of these contributions is R. A. Fisher's derivation of the log-series species-abundance distribution (SAD). Fisher assumed that the expected abundances of species followed a simple and general mathematical form (a gamma distribution) and that observed abundances of species represented random (Poisson) realizations of their expected abundances. He then recognized that the SAD should be defined only for one or more individuals "since by itself the collection gives no indication of the number of species which are not found in it," and he assumed that one of the parameters in the resulting distribution was equal to zero based on patterns in observed data. Like Arrhenius's use of the power function to describe the species-area relationship, Fisher's log series was largely empirical in nature, starting with a simple underlying distribution that made reasonable mathematical sense. The form of this distribution also yielded the first simple measure of species diversity, now known as Fisher's alpha. Beyond the simple characterization of species-abundance distributions, Fisher's distribution has motivated a large body of research aimed at understanding the biological processes underlying observed SADs, including the development of alternative statistical models, niche partitioning approaches, and neutral theory (e.g., Preston 1948; Tokeshi 1996; Hubbell 2001). Steven Corbet's piece provides a simple evaluation of Fisher's predictions using butterflies, and the observed distributions demonstrate remarkable consistency with the predictions. Williams's contribution also presents tests of Fisher's distribution using insect data from the Rothamsted Experimental Station. Assessments of the log-series predictions over short timescales conformed well with Fisher's predictions. However, when Williams looked at the predicted increases in the number of species with sample size by combining multiple years of samples from the Rothamsted data, he realized that in many groups the number of species increased more rapidly than would be expected from simply sampling the log-series distribution more intensively. As such, this paper represents the first plotting of a species-time relationship (though the concept was discussed earlier by J. Grinnell [1922]) and the first test comparing this macroecological pattern to the null expectation of random sampling.

Literature Cited

Grinnell, J. 1922. The role of the "accidental." *Auk* 39:373–80.

Hubbell, S. P. 2001. *The unified neutral theory of biodiversity and biogeography*. Princeton University Press, Princeton, NJ.

Preston, F. W. 1948. The commonness, and rarity, of species. *Ecology* 29:254–83.

Tokeshi, M. 1996. Power fraction: A new explanation of relative abundance patterns in species-rich assemblages. *Oikos* 75:543–50.

From *Journal of Animal Ecology* 12:42–58. Reprinted with permission from Wiley-Blackwell.

[42]

THE RELATION BETWEEN THE NUMBER OF SPECIES AND THE NUMBER OF INDIVIDUALS IN A RANDOM SAMPLE OF AN ANIMAL POPULATION

By R. A. FISHER (*Galton Laboratory*), A. STEVEN CORBET (*British Museum, Natural History*) AND C. B. WILLIAMS (*Rothamsted Experimental Station*)

(With 8 Figures in the Text)

CONTENTS

PART 1. RESULTS OBTAINED WITH MALAYAN BUTTERFLIES

By A. STEVEN CORBET (*British Museum, Natural History*)

It is well known that the distribution of a series of biological measurements usually conforms to one of three types:

(*a*) the binomial distribution, where the frequencies are represented by the successive terms of the binomial $(q+p)^n$;

(*b*) the normal distribution, in which the results are distributed symmetrically about the mean or average value, and which is the special case of (*a*) when p and q are equal;

(*c*) the Poisson series, in which the frequencies are expressed by the series

$$e^{-m}\left(1+m+\frac{m^2}{2!}+\frac{m^3}{3!}+\ldots\right),$$

where m is the mean and e is the exponential base 2·7183.

The usual practice of calculating the arithmetic mean in a set of measurements of a biological nature, such as wing length of a butterfly, assumes a distribution showing no wide departure from normality, although it does not appear that this procedure has been vindicated.

It is the usual experience of collectors of species in a biological group, such as the Rhopalocera, that the species are not equally abundant, even under conditions of considerable uniformity, a majority being comparatively rare while only a few are common. As far as we are aware, no suggestion has been made previously that any mathematical relation exists between the number of individuals and the number of species in a random sample of insects or other animals. Recently, it has been found (Corbet, 1942) that, leaving out of account the commoner species of which no attempt was made to collect all individuals seen, the number of species S of butterflies of which n individual specimens were collected by a single collector in Malaya was given closely by the expression

$$S=C/n^m,$$

where C and m are constants.* When m is unity, as is the case with the Malayan collection, and has since been found to be a condition which obtains with collections of butterflies from Tioman Island and the Mentawi Islands in which the relation between S and n follows the above equation, the number of species of which 1, 2, 3, 4, ... specimens were obtained was very close to a series in harmonic progression. Thus, the series can be written

$$C\,(1+\tfrac{1}{2}+\tfrac{1}{3}+\ldots).$$

Although this relation holds accurately with the rarer species, there is less agreement in the region of the common species; in fact, theoretical considerations preclude an exact relationship here.

Prof. Fisher (see Part 3) has evolved a logarithmic series which expresses accurately the relation between species and individuals in a random sample throughout the whole range of abundance:

$$S=n_1\left(1+\frac{x}{2}+\frac{x^2}{3}+\ldots\right),$$

where S is the total number of species in the sample, n_1 the number of species represented by single specimens, and x is a constant slightly less than unity but approaching this value as the size of the sample is increased.

The total number of individual specimens, N, in the sample at all levels of abundance is given by $N=n_1/(1-x)$ and $N(1-x)/x$ is a constant α independent of the size of the sample. As the size of the sample is increased, n_1 approaches α.

The Fisher series has been established for all the entomological collections tested in which there was

* This equation may be written

$$\log S=\log C+m\log n,$$

so that the plot of the logarithms of S and n is a straight line.

reason to believe that the collecting had been unselective (see Table 1). It is clear that when S and N are known, as is usually the case, the statistics n_1, x and α can be calculated. It is a curious fact that the number of uniques should approach a constant value with increasing size of collection. It is important to ascertain how far this type of distribution of individuals among species holds in other zoological groups, for it would appear that we have here an effective means of testing whether a collection has been made under conditions approaching random sampling or whether some degree of selection has been exercised, a consideration which is often of some importance in faunistic studies.

Table 1. *Entomological collections examined showing the relation between the numbers of species and individuals*

	Observations				Calculations			
	Total individuals	No. between 1 and 24	Total species	No. between 1 and 24	x	n_1 calc.	n_1 found	α
Malayan Rhopalocera	—	3306	620	501	0·997	135·05	118	135·47
Rhopalocera from Tioman Island (east coast of Malaya) (Malay collector, 1931)	157	—	41	—	0·887	15·96	19	18·00
Rhopalocera from Mentawi Islands (excluding Hesperiidae) (C. Boden Kloss and N. Smedley, 1924)	1,878	890	135	110	0·983	32·77	37	33·35
Karakorum Rhopalocera (Mme J. Visser-Hooft; *vide* Evans, 1927)	403	195 (1–28)	27	24 (1–28)	0·984	6·42	6	6·52
Mexican Elmidae (Col.) (H. E. Hinton; *vide* Hinton, 1940)	11,798	—	35	—	0·9998	4·72	4	4·72

How far the results obtained with any particular collection can be regarded as representative of the distribution of the same species group in the area in which the collection was made must obviously depend on the uniformity or otherwise of the conditions prevailing when the collection was made and to the extent to which these conditions are representative of the habitat. In an equatorial forest-clad island with no mountain heights above 2000 ft., conditions are very uniform as far as such orders as Leipidoptera are concerned; although even here some allowance must be made for the fact that collections of butterflies made in such regions are usually poor in the crepuscular species. In temperate climates, it is evident that results obtained during one period of the year are usually inapplicable to other seasons or to the year as a whole. It would appear that the results obtained with the moth trap at Harpenden (see Part 2) can be regarded as giving an accurate picture of the distribution frequencies of the phototropic moths in the area, and it is probable that the same is true of the collection of Mexican Elmidae.

With many collectors, and for a variety of reasons, the collecting of common species is discontinued once a certain number of specimens of these are obtained. In the case of the Malayan Rhopalocera cited, collecting of all individuals seen was not continued after 24 specimens had been taken. In such cases, we have the following information:

The total number of species under 25 individuals per species:

$$S_{(1-24)} = n_1\left(1 + \frac{x}{2} + \frac{x^2}{3} + \ldots + \frac{x^{23}}{24}\right).$$

The total number of individuals at frequencies below 25 per species:

$$N_{(1-24)} = n_1\,(1 + x + x^2 + \ldots + x^{23}).$$

Table 2, which gives the results obtained with the Malayan butterflies, is based on these considerations, and shows the very close relation between the observed and the calculated results.

Table 2. *Calculated and observed distribution frequencies of butterflies collected in Malaya*

The values in the second column are obtained from the Fisher series given on p. 42, taking $x = 0·997$.

n	S (calc.)	S (found)	Deviations
1	135·05	118	17·05
2	67·33	74	−6·77
3	44·75	44	0·75
4	33·46	24	9·46
5	26·69	29	−2·31
6	22·17	22	0·17
7	18·95	20	−1·05
8	16·53	19	−2·47
9	14·65	20	−5·35
10	13·14	15	−1·86
11	11·91	12	−0·09
12	10·89	14	−3·11
13	10·02	6	4·02
14	9·28	12	−2·72
15	8·63	6	2·63
16	8·07	9	−0·93
17	7·57	9	−1·43
18	7·13	6	1·13
19	6·74	10	−3·26
20	6·38	10	−3·62
21	6·06	11	−4·94
22	5·77	5	0·77
23	5·50	3	2·50
24	5·25	3	2·25
			0·82

According to the calculated values, the total species and total individuals at levels between $n=1$ and $n=24$, are 501·92 and 3132·24 respectively. The actual values found for the total species and total individuals between $n=1$ and 24 are 501 and 3306 respectively.

For the above series in Table 2, $\chi^2=19{\cdot}270$, degrees of freedom = 23, and P is between 0·90 and 0·80, showing the deviations of the observed results from the calculated values are not significant.

REFERENCES

Corbet, A. S. (1942). 'The distribution of butterflies in the Malay Peninsula.' Proc. R. Ent. Soc. Lond. (A), 16: 101–16.

Evans, W. H. (1927). 'Lepidoptera-Rhopalocera obtained by Mme J. Visser-Hooft of the Hague (Holland) during an exploration of previously unknown country in the Western Karakorum, N.W. India.' Tijds. Ent. 70: 158–62.

Hinton, H. E. (1940). 'A monographic revision of the Mexican water beetles of the family Elmidae.' Nov. Zool. 42: 217–396.

PART 2. RESULTS OBTAINED BY MEANS OF A LIGHT-TRAP AT ROTHAMSTED

By C. B. Williams

This gives an account of the application of Fisher's series (see Part 3 of this paper) for the frequency of occurrence of species of different levels of abundance in a random sample, to collections of nocturnal Lepidoptera made by means of a light-trap at Rothamsted Experimental Station, Harpenden, Herts, England, during the four years 1933–6. The trap and insects caught in it have already been discussed in a series of papers (Williams, 1939, 1940).

It was necessary to choose, from the material collected, groups in which all or nearly all of the specimens had been identified to species. For this certain families of Lepidoptera were most suited, and the discussion below deals with the captures in the Sphingidae, Noctuidae, Arctiidae, Geometridae and a few other related families. In the Geometridae the genus *Eupithecia* was omitted owing to difficulties of identification. Altogether 15,609 individuals belonging to 240 species were captured. The names and details of numbers for each species will be found in Williams (1939, Tables 6–8).

The frequency of species of different abundance

Table 3 shows the frequency distribution of the species for the four years added together. It will be seen that 35 species were represented by a single individual each; that 85 (including the 35 above) were represented by 5 or fewer individuals; 115 by 10 or fewer; and 205 species by 100 or fewer individuals; leaving therefore 35 species with over 100 individuals per species. The highest total of one species was 2349 individuals of *Agrotis exclamationis*. The results up to 50 individuals per species are represented diagrammatically by the vertical lines in Fig. 1A, giving a curve closely resembling a hyperbola.

If, however, the log number of species is plotted against the log number of individuals as in Fig. 2A it will be seen that, while the straight-line distribution expected for a hyperbola holds approximately true for the rarer species, the number of commoner species is distinctly below the hyperbolic expectation and falls rapidly away from it at higher numbers of individuals per species.

Fisher suggests (see Part 3) that the true series is represented not by

$$n_1, \frac{n_1}{2}, \frac{n_1}{3}, \frac{n_1}{4}, \ldots,$$

which would be the hyperbolic series and which would require an infinite number of species and an infinite number of individuals; but by the series

$$n_1 \ldots \frac{n_1x}{2} \ldots \frac{n_1x^2}{3} \ldots \frac{n_1x^3}{4} \ldots,$$

when n_1 is the number of species with 1 individual and x is a number less than 1.

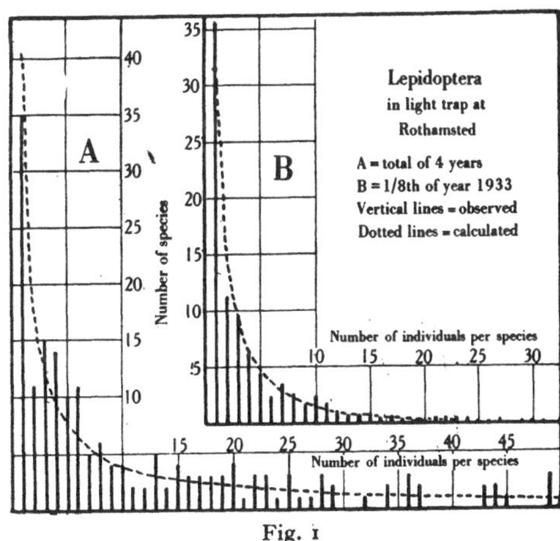

Fig. 1

If this is correct he shows that (1) the total number of individuals (N) is finite and $=\frac{n_1}{1-x}$, (2) the total number of species (S) is finite and

$$=n_1\frac{-\log_e(1-x)}{x}.$$

Table 3. *Observed and calculated captures of Macrolepidoptera in a light-trap at Rothamsted*

No. of individuals per species	(1) All 4 years		(2) Average of single year		(3) Average of ⅛ year		(4) Capsidae, all 4 years	
	Obs.	Calc.	Obs.	Calc.	Obs.	Calc.	Obs.	Calc.
1	35	40·14	39·75	37·55	35·63	31·77	18	11·84
2	11	20·03	20·25	18·60	11·38	14·74	8	5·87
3	15	13·32	16·00	12·28	9·63	9·12	3	3·88
4	14	9·96	8·75	9·12	6·24	6·35	4	2·89
5	10	7·95	8·75	7·22	4·50	4·71	2	2·29
	85	91·42	93·50	84·77	67·38	66·68	35	26·77
6	11	6·66	6·25	5·96	2·38	3·64	3	1·89
7	5	5·65	4·25	5·06	3·50	2·90	2	1·61
8	6	4·93	6·00	4·39	2·63	2·55	1	1·40
9	4	4·37	4·00	3·86	1·75	1·93	1	1·23
10	4	3·92	3·50	3·44	2·13	1·62	—	1·10
	30	25·48	24·00	22·70	12·00	12·44	7	7·22
11	2	3·56	2·00	3·10	1·75	1·37	1	0·99
12	2	3·25	2·50	2·81	1·13	1·16	—	0·90
13	5	3·00	2·75	2·57	0·75	0·99	—	0·82
14	2	2·77	2·25	2·36	0·75	0·86	—	0·76
15	4	2·58	3·25	2·19	0·75	0·74	2	0·70
	15	15·16	10·75	13·03	4·18	5·12	3	4·17
16	3	2·42	2·00	2·03	0·50	0·64	—	0·65
17	3	2·27	2·75	1·89	0·63	0·56	—	0·61
18	3	2·14	1·25	1·77	0·38	0·49	—	0·57
19	3	2·02	0·50	1·66	0·25	0·43	—	0·54
20	4	1·92	2·75	1·56	0·25	0·38	—	0·50
	16	10·78	9·25	8·91	2·00	2·53	0	2·87
21	1	1·82	2·75	1·47	0·50	0·34	—	0·48
22	3	1·73	1·75	1·39	0·50	0·30	1	0·45
23	3	1·65	0·75	1·32	0·50	0·27	1	0·43
24	1	1·58	1·00	1·25	0·50	0·24	—	0·41
25	3	1·51	0·50	1·19	0·13	0·21	—	0·38
	11	8·28	6·75	6·62	2·13	1·35	2	2·14
Total to 25	157	151·12	144·25	136·03	87·69	88·12	47	43·17
26	1	1·45	0·25	1·13	0·13	0·19	—	
27	1	1·39	0·25	1·08	0·38	0·17	2	
28	3	1·34	0·75	1·03	—	0·15	—	
29	2	1·29	1·50	0·99	0·13	0·13	—	
30	—	1·24	0·50	0·95	0·25	0·12	—	0·30
	7	6·70	3·25	5·18	0·88	0·76	2	
31	—	1·20	—	0·91	0·13	0·11	—	
32	1	1·16	1·00	0·87	—	0·10	—	
33	—	1·12	1·00	0·83	0·25	0·09	—	
34	2	1·09	0·75	0·80	—	0·08	—	
35	—	1·05	0·25	0·77	—	0·07	—	
	3	5·61	3·00	4·18	0·38	0·51	0	
36	3	1·02	0·75	0·74	0·25	0·06	—	
37	2	0·99	0·25	0·72	0·13	0·06	—	
38	—	0·96	1·00	0·69	—	0·05	—	
39	—	0·93	0·50	0·67	0·13	0·05	1	
40	—	0·91	1·50	0·64	0·25	0·04	—	0·21
	5	4·81	4·00	3·46	0·75	0·26	1	
Total to 40					89·70	89·65		
41	—	0·88	0·75	0·62			—	
42	—	0·86	0·25	0·60			—	
43	2	0·84	0·25	0·58			—	
44	2	0·81	0·25	0·56			—	
45	1	0·80	—	0·54			—	
	5	4·19	1·50	2·91			0	
46	—	0·78	0·25	0·53			—	
47	—	0·76	—	0·51			—	
48	—	0·74	1·50	0·50			—	
49	3	0·72	—	0·48			—	
50	—	0·71	—	0·47			—	0·15
	3	3·71	1·75	2·48			0	
Total to 50	180	176·14	157·75	154·24				

46 *Relation between numbers of species and individuals in samples*

Table 3 (*continued*)

Values above 50

(1) Total of 4 years: 51 (4), 52, 53, 54 (2), 57, 58 (2), 60 (3), 61, 64, 67, 73, 76 (2), 78, 84, 89, 96, 99, 109, 112, 120, 122, 129, 135, 141, 148, 149, 151, 154, 177, 181, 187, 190, 199, 211, 221, 226, 235, 239, 244, 246, 282, 305, 306, 333, 464, 560, 572, 589, 604, 743, 823, 2349 = 15,609 individuals of 240 species.

(2) Average of 1 year: 51 (0·75), 52 (0·50), 53 (0·50), 54 (0·25), 58 (0·50), 60 (0·50), 61 (0·25), 64 (0·50), 65 (0·25), 69 (0·25), 73 (0·75), 75 (0·50), 76 (0·25), 77 (0·25), 80 (0·25), 82 (0·25), 83 (0·25), 87 (0·25), 88 (0·50), 90 (0·25), 93 (0·25), 99 (0·50), 100 (0·50), 104 (0·25), 105 (0·50), and the following all (0·25) each: 107, 109, 110, 111, 115, 126, 132, 138, 139, 141, 144, 145, 153, 159, 165, 173, 179, 197, 200, 201, 219, 223, 232, 275, 294, 323, 329, 603, 1799 = 3902 individuals of 176 species.

(3) Average of ⅛ year: 79 (0·13), 109 (0·13) = 440 individuals of 89·875 species.

(4) Capsidae = 53, 80, 85, 158, 206, 237, 298 = 1414 individuals of 57 species.

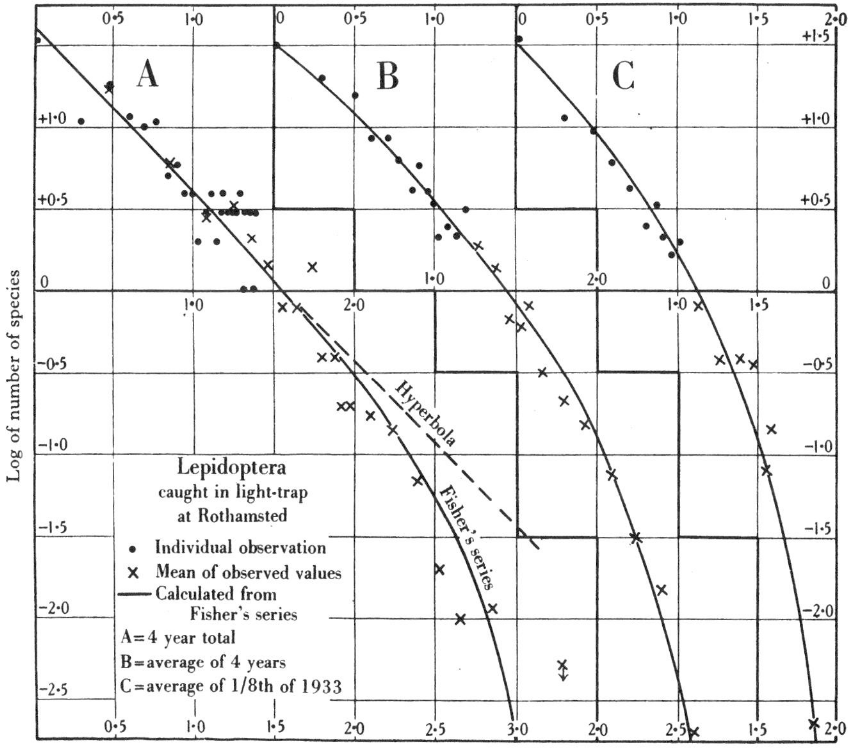

Fig. 2

Thus if N and S are known both n_1 and x can be calculated, and hence the whole series is known.

If N is large n_1 tends to become a constant value α, and Fisher shows that for all levels of sampling for the same population $x = \dfrac{N}{N+\alpha}$.

Applying these formulae to our Macrolepidoptera with 15,609 individuals of 240 species we find

$$\alpha = 40{\cdot}24,\ n_1 = 40{\cdot}14,\ x = 0{\cdot}997429.$$

The calculated values for n_2, n_3, etc., are shown in Table 3 opposite the observed figures and also by curves in Figs. 1 A and 2 A. In the latter the log of the frequency is plotted against the log of the number of individuals per species.

It will be seen that for both common and rare species the calculated values are very close indeed to the observed. The calculated number of species with one individual is slightly larger (40·14) than the observed (35). By calculation there should be 116·9 species with 10 or fewer individuals and the observed number was 115. The close resemblance at higher frequencies is best seen in Fig. 2 A.

If, instead of adding the four years together, we take the average number of individuals in each year and the average number of species with 1, 2, 3, etc., individuals we find the observed results in the trap to be as in the second column of Table 3. There are 3902 individuals of 176 species. Calculations from these figures show that $x=0.990356$ and $n_1=37.55$. The calculated series is shown up to 50 individuals per species in Table 3 and diagrammatically in Fig. 2 B. The calculated n_1 is slightly smaller than the observed.

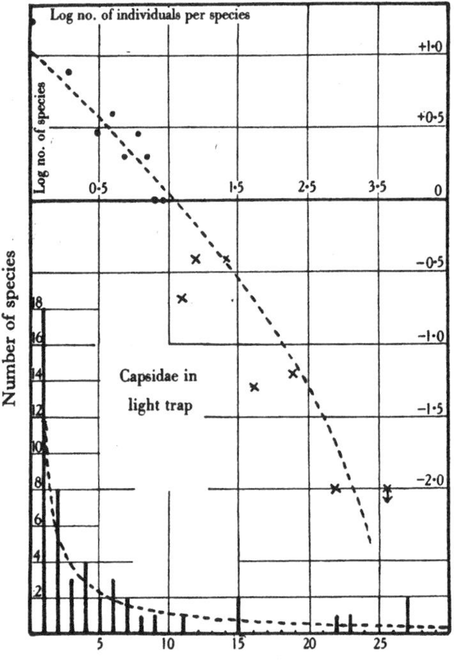

Fig. 3

A still smaller sample was obtained by dividing the catches in the year 1933 into 8 samples, the first including those insects caught in the 1st...9th...17th day, the second on the 2nd...10th...18th day, etc. The average for the 8 samples is shown in the third column of Table 3, together with the calculated figures. The number of individuals was 440, the number of species 89.9, from which $n_1=31.77$ and $x=0.9278$. The calculated n_1 is slightly smaller than the observed. Results are shown diagrammatically in Figs. 1 B and 2 C. The close resemblance in both the two series of the observed and calculated results is very striking.

Thus in the Macrolepidoptera caught in the light-trap at Harpenden samples as large as 15,609 individuals and as small as 440 individuals both agree very closely with the results calculated from Fisher's series.

The only other group of insects captured in the light-trap in which the majority of the specimens were identified was the family Capsidae of the Heteroptera. These were identified by D. C. Thomas (1938). In the four years 1414 individuals of 57 species were captured. The observed and calculated results are shown in the fourth column of Table 3 and in Fig. 3. The calculated n_1 is 11.87 and $x=0.9916$.

It will be seen that the fit is not so good as in the Lepidoptera. The observed number of species represented by only one individual is 18 and the calculated less than 12. In general there are rather more of the rarer species than the calculated series indicates, and fewer of the commoner species.

Table 4 shows a summary of all the various samples taken from the Lepidoptera and the Capsidae. It will be seen that there is a slight general tendency for the calculated value of n_1 to be below the observed. In the Macrolepidoptera for one or more years this is not so obvious, but it is distinct in the one-eighth year samples. In the Noctuidae only there are four results below and two practically equal to the observed figure; but in the Capsidae there are five below and only one equal. This requires further investigation.

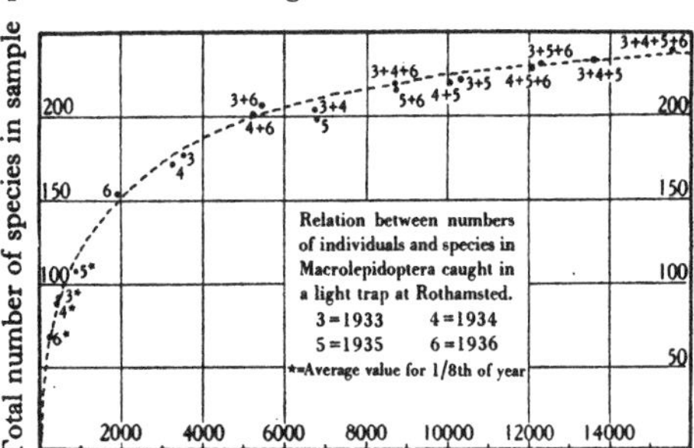

Fig. 4

The relation between the number of species and total number of individuals in the sample

In Tables 4 and 5 are shown a number of samples of different size taken from the light-trap captures with the number of individuals in the sample and the number of species represented.

Fig. 4 shows diagrammatically the number of species in relation to the number of individuals for all the different samples of Lepidoptera. It will be seen that there is a very regular relation between the observed numbers. This form of diagram is, however, not very suitable for showing large samples,

48 *Relation between numbers of species and individuals in samples*

Table 4. *Calculations of x, α and n_1 for various light-trap samples*

	Individuals	Species	x	α	Species with 1 individual: Calc.	Obs.
			Macrolepidoptera			
1933	3,540	178	0·98903	39·15 ± 1·45	38·62	32
1934	3,275	172	0·98834	38·64	38·19	34
1935	6,817	198	0·994425	38·19	37·94	37
1936	1,977	154	0·980595	39·05	38·30	50
4 years: Total	15,609	240	0·997429	40·24 ± 1·06	40·14	35
Average	3,902	176	0·990380	37·90	37·54	39·75
⅛ of 1933	561	105	0·93638	38·12	35·69	38
,,	581	104	0·94028	36·90	34·70	39
,,	458	86	0·93609	32·27	29·27	34
,,	357	84	0·91157	34·63	31·57	41
,,	344	80	0·91306	32·75	29·90	35
,,	282	75	0·8941	33·40	29·86	28
,,	503	101	0·92968	38·05	35·37	38
,	427	81	0·93725	29·26	27·42	36
Average of 8 × ⅛	440	89·5	0·9278	34·24 ± 2·67	31·77	35·63
			Noctuidae only			
1933	1,636	84	0·9887	18·70 ± 1·02	18·49	18
1934	1,894	73	0·9922	14·89	14·77	14
1935	5,413	94	0·997	16·29	16·23	21
1936	1,362	71	0·9886	15·71	15·53	20
4 years: Total	10,304	112	0·9983	17·56 ± 0·63	17·52	20
Average	2,576	80	0·994	15·55	15·46	18·25
			Capsidae			
1933	341	26	0·981	6·60 ± 0·71	6·5	10
1934	479	38	0·98	9·77	9·6	17
1935	446	24	0·988	5·41	5·45	5
1936	148	23	0·95	7·79	7·40	9
4 years: Total	1,414	57	0·9916	11·98 ± 0·76	11·88	18
Average	354	27·75	0·98047	7·04	6·90	10·25

Table 5. *Number of individuals and species in different sized samples from the light-trap*

	Macrolepidoptera: Indiv.	Sp.	Noctuidae only: Indiv.	Sp.	Capsidae: Indiv.	Sp.
1933	3,540	178	1,636	84	341	26
1934	3,275	172	1,894	73	479	38
1935	6,817	198	5,413	94	446	24
1936	1,977	154	1,362	71	148	23
Average of single year	3,902	176	2,576	80·5	353·5	27·75
1933 + 1934	6,815	204	3,530	94	820	46
1933 + 1935	10,357	222	7,049	106	787	37
1933 + 1936	5,517	206	2,998	95	489	32
1934 + 1935	10,092	220	7,307	100	925	45
1934 + 1936	5,252	200	3,256	88	627	48
1935 + 1936	8,794	216	6,775	101	594	31
Average of years in pairs	7,804	211·3	5,152	97·3	707	39·8
1933 + 1934 + 1935	13,632	234	8,943	109	1,266	52
1933 + 1934 + 1936	8,792	219	4,892	299	968	54
1933 + 1935 + 1936	12,334	232	8,411	109	935	42
1934 + 1935 + 1936	12,069	229	8,669	105	1,073	51
Average of years in threes	11,706	228·5	7,728	105·5	1,060·5	49·75
All 4 years	15,609	240	10,305	112	1,414	57

and in Fig. 5 (heavy line) the same data are shown with the number of individuals in the sample expressed as a logarithm.

According to Fisher's theory, any population should have a constant value of α for samples of any size taken from it under identical conditions. Thus it follows that for any population, if the value of α is obtained from one sample, the values of

$$x\left(=\frac{N}{N+\alpha}\right)$$

and hence of S can be calculated for other samples of different sizes from the same population. In

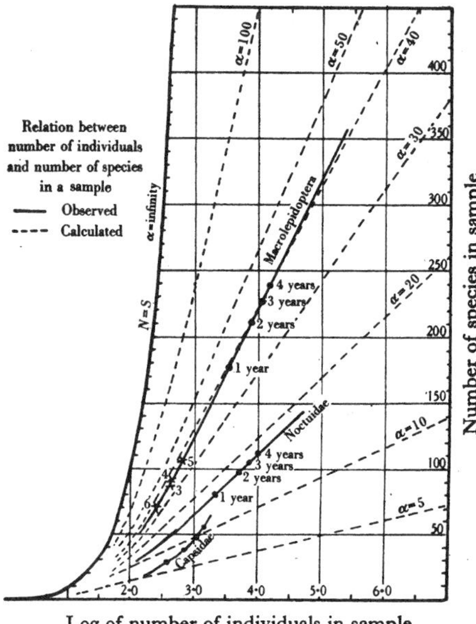

Log of number of individuals in sample

Fig. 5

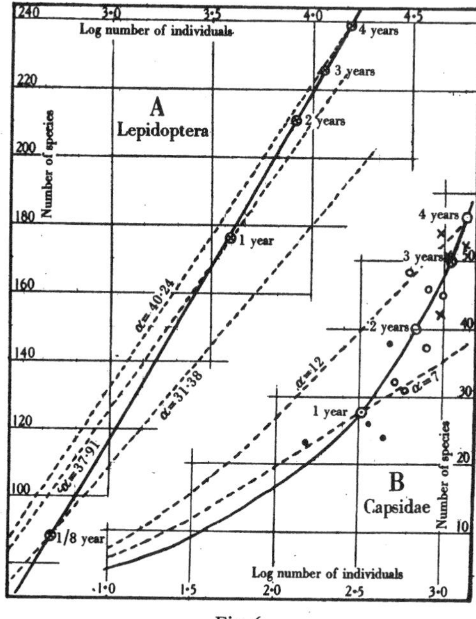

Fig. 6

Fig. 5 there are shown as dotted lines typical calculated relations between the number of individuals and number of species for different values of α. It will immediately be seen that the observed values for the Lepidoptera fit very closely to the calculated relation for $\alpha = 40$. The slight difference between the observed and calculated figures will be discussed below.

It will also be seen that the observed values for the Noctuidae alone fit very closely to a value of $\alpha = 17{\cdot}5$, but that the observed values for the Capsidae cut rather rapidly across the α lines giving too few species in small samples and too many in large samples (when they are taken, as in the light-trap, in more than one year). The various calculated values of α in the different samples together with their standard deviation as calculated from Fisher's formula are given in Table 4. It is suggested that the parameter α should be known as the 'index of diversity' of the population.

The differences between the observed and calculated values for Lepidoptera and for Capsidae are shown diagrammatically on a larger scale in Fig. 6. In the Lepidoptera (Fig. 6A) it will be seen that more species are obtained in a sample spread over several years than would be expected in a sample of the same size taken in a single year. Thus when a sample of 15,609 (log 4·19) Lepidoptera was captured in four years 240 species were represented. If the same number had been caught in one year it will be seen (by finding where the $\alpha = 37{\cdot}91$ line cuts log 4·19) that only about 226 species would have been expected. Otherwise about 14 extra species have been captured owing to the new biological conditions introduced by sampling over a longer period. The same effect on an even larger scale is obvious between samples captured throughout a whole year and those captured in one-eighth of the year. Thus the departure of the observed from the calculated results is immediately explicable by the alteration of conditions introduced by spreading sampling over a longer period of time.

The Capsidae (Fig. 6B) show the same effect in a more extreme form. With 1414 individuals caught in four years 57 species were obtained, whereas if the same-sized sample had been taken in one year we would only have expected about 36 species. There is thus a considerable increase of species by spreading

50 *Relation between numbers of species and individuals in samples*

the cat:h over a longer period. This is indicated also by a fact that I have already pointed out (Williams, 1939, p. 96) that in the Capsidae 50% of the 57 species occurred only in a single year, as compared with a figure of only 19% in the Macrolepidoptera. The Capsidae fauna therefore is much more variable from year to year than is the Lepidoptera.

Since with high number of N the relation between the number of species and log number of individuals becomes practically a straight line (both theoretically and observationally) it is possible to extrapolate the observed curve for the Macrolepidoptera with some interesting and suggestive results. In the first place it should be noted that the number of British species in the families of Macrolepidoptera dealt with is 681, and the number of species already recorded for Hertfordshire (where the trap was situated) was 461.

Thus in four years we have captured approximately 16,000 individuals representing 240 species. After ten years we would have captured approximately 40,000 individuals which would contain about 275 species. After 50 years the 200,000 individuals would include 340 species. After 100 years the 400,000 would include 365 species or only 25 new species in 50 years.

To get the Hertfordshire fauna (461 species) we would have to collect over three million insects, and this would take, at the present rate, nearly 1000 years. When this level was reached only one new species would be added in 20 years' trapping. To get the British fauna (681 species) we would have to trap about 1800 million insects which would take nearly half a million years!

Doubling the number of insects caught (and hence the time of trapping) at any level, except for very small samples, always adds about 30 species to the total.

These figures are based on the curve allowing for captures of 4000 insects per year. If the whole number were caught in one year (assuming that to be possible) the number of species would be less in each case. So that 400,000 individuals would give 350 species (instead of 368), the Hertfordshire fauna would require 8 million insects, and the British fauna 3200 million.

The seasonal changes of α in the Macrolepidoptera

If the Macrolepidoptera captured in the light-trap during the four years are tabulated month by month the numbers of individuals and species in each month (excluding the five winter months when numbers are too small) are shown in Table 6 together with the approximate value of α. The changes in α are shown in Fig. 7.

There is a regular seasonal change which is almost identical in each year; the value of α rises from a low value in April to a maximum in July and back to a low value in October. There is a very much greater difference between the α values for two different months than there is for the same month in two different years. For example, the number of insects caught in July 1935 was almost seven times as great as the number in July 1936, and yet the values of α are almost identical; but rather more insects in August 1936 than in July gives a considerably smaller value, and an almost identical number of insects in September 1936 gives a value of α only about one-third as great.

There is no doubt whatever that there is a seasonal

Table 6. *Number of individuals and of species of Macrolepidoptera in each of the summer months of the four light-trap years, with the approximate values of α*

	1933			1934			1935			1936		
	N	S	α	N	S	α	N	S	α	N	S	α
Apr.	70	12	4·0	55	8	2·5	24	9	5·0	16	5	3·0
May	318	38	11·0	131	27	10·0	119	26	10·0	38	16	10·0
June	693	73	20·5	518	53	14·5	1178	67	15·5	434	45	13·0
July	987	90	24·0	1059	97	25·5	3136	119	24·5	449	68	22·5
Aug.	701	59	15·5	591	51	13·5	1920	84	18·0	491	57	16·5
Sept.	389	30	7·5	528	31	7·0	647	36	8·0	448	33	8·0
Oct.	220	14	3·0	152	14	3·5	171	18	5·0	59	14	5·0

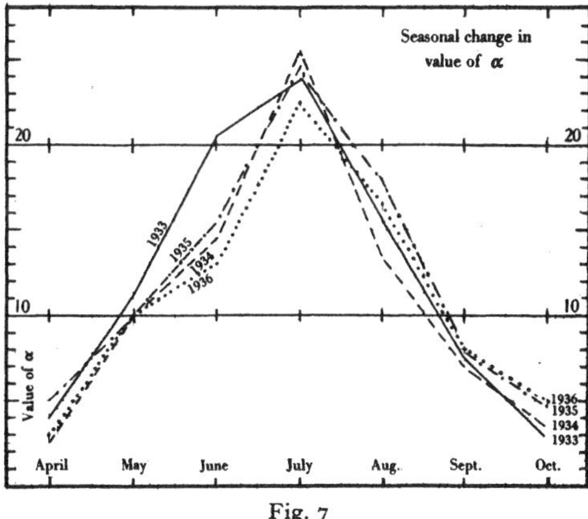

Fig. 7

change in α or in the 'richness of species' quite independent of the seasonal change in numbers of insects.

Value of α for a single night in a month

In the month of July 1935, which was a month of high catches of Lepidoptera in the light-trap, 2,586 Noctuidae were captured, belonging to 56 species. This gives α (for Noctuidae only) = 10·09.

The following were the captures on seven selected nights in the month:

Date	N	log N	S
1	429	2·63	18
6	143	2·16	10
11	394	2·60	18
16	11	1·04	7
21	47	1·67	17
26	64	1·80	11
31	53	1·72	10
	Mean	1·95	13

From these figures $\alpha = 4{\cdot}2$. Thus it is considerably smaller for a short period of sampling than for a longer period. This would be expected, as the variety of species available for catching on a single night must be less than that on a series of different nights.

Comparison of results from two light-traps a short distance apart

During the autumn of 1933 two light-traps were kept running simultaneously from 8 August to 31 October. One was in the fields at Rothamsted (the standard trap used in all the previous calculations) and the other was about 400 yards away on the roof on the Entomology building at a height of about 35 ft. from the ground. This overlooked a more varied environment with gardens and mixed vegetation.

The captures of Noctuidae in the two traps were as follows:

	Field trap		Roof trap	
1933	N	S	N	S
8–31 Aug.	419	26	706	36
Sept.	355	18	593	32
Oct.	155	9	557	19
	929	40	1856	58
	$\alpha = 8{\cdot}51 \pm 0{\cdot}7$		$11{\cdot}38 \pm 0{\cdot}7$	

The difference between the two values of α is 2·8 and the standard error of the difference approximately 1·0, so that difference is probably significant. The roof trap has caught a larger number of individuals but a still larger number of species. This is probably due, as mentioned above, to the considerably more varied vegetation that it overlooked.

The effect on number of species of increasing the size of a sample

Fisher has shown that

$$S = \alpha \log_e \left(1 + \frac{N}{\alpha}\right).$$

If the sample is large so that N is large compared with α we can neglect the 1 in comparison with N/α. Hence for large samples

$$S = \alpha \log_e \frac{N}{\alpha},$$

hence

$$S_{2N} - S_N = \alpha \left(\log_e \frac{2N}{\alpha} - \log_e \frac{N}{\alpha}\right) = \alpha \log_e 2.$$

So the number of species added to a large sample by doubling it is $\alpha \log_e 2x = 0{\cdot}693\alpha$. Similarly, if the sample were increased 10 times the number of new species added $= \alpha \log_e 10 = 2{\cdot}3\alpha$ approximately. On the other hand, if the size of the sample is multiplied by e ($= 2{\cdot}7183$) the additional number of species is $\alpha \log_e e$, which is equal to α.

For example, with the Macrolepidoptera in the light-trap α = approximately 40, so doubling a large catch will at any level add approximately 28 new species, multiplying it by 10 will add approximately 92 species.

To add one new species to a large sample $\alpha \log_e z$ must equal 1, when z is the factor by which the sample must be multiplied, i.e.

$$\log_e z = 1/\alpha;$$

for example, if $\alpha = 40$ as in the Macrolepidoptera above

$$\log_e z = 0{\cdot}025, \quad z = 1{\cdot}0255;$$

hence the number of insects must be increased by just over 1/39 of the original sample to add one additional species.

Diagram showing the interrelation of N, S, α and the standard error of α

From the formulae provided by Prof. Fisher α and the error of α can be calculated if N and S are known, and similarly S can be calculated if N and α are known. These calculations can be made to any number of decimal places. For many purposes, however, a close approximation to α and its error is all that is necessary, and if this can be obtained rapidly without calculation so much the better.

For this purpose Fig. 8 has been drawn up. In this the vertical axis is the number of species and the horizontal the log of the number of individuals in the sample. A large number of values of α have been inserted as diagonals, with the result that if any two of these three factors is known the third can be easily estimated. Thus with a sample of 1000 individuals (log = 3·0) containing 60 species $\alpha = 14$. Or with $\alpha = 25$ a sample of 10,000 individuals should

52 *Relation between numbers of species and individuals in samples*

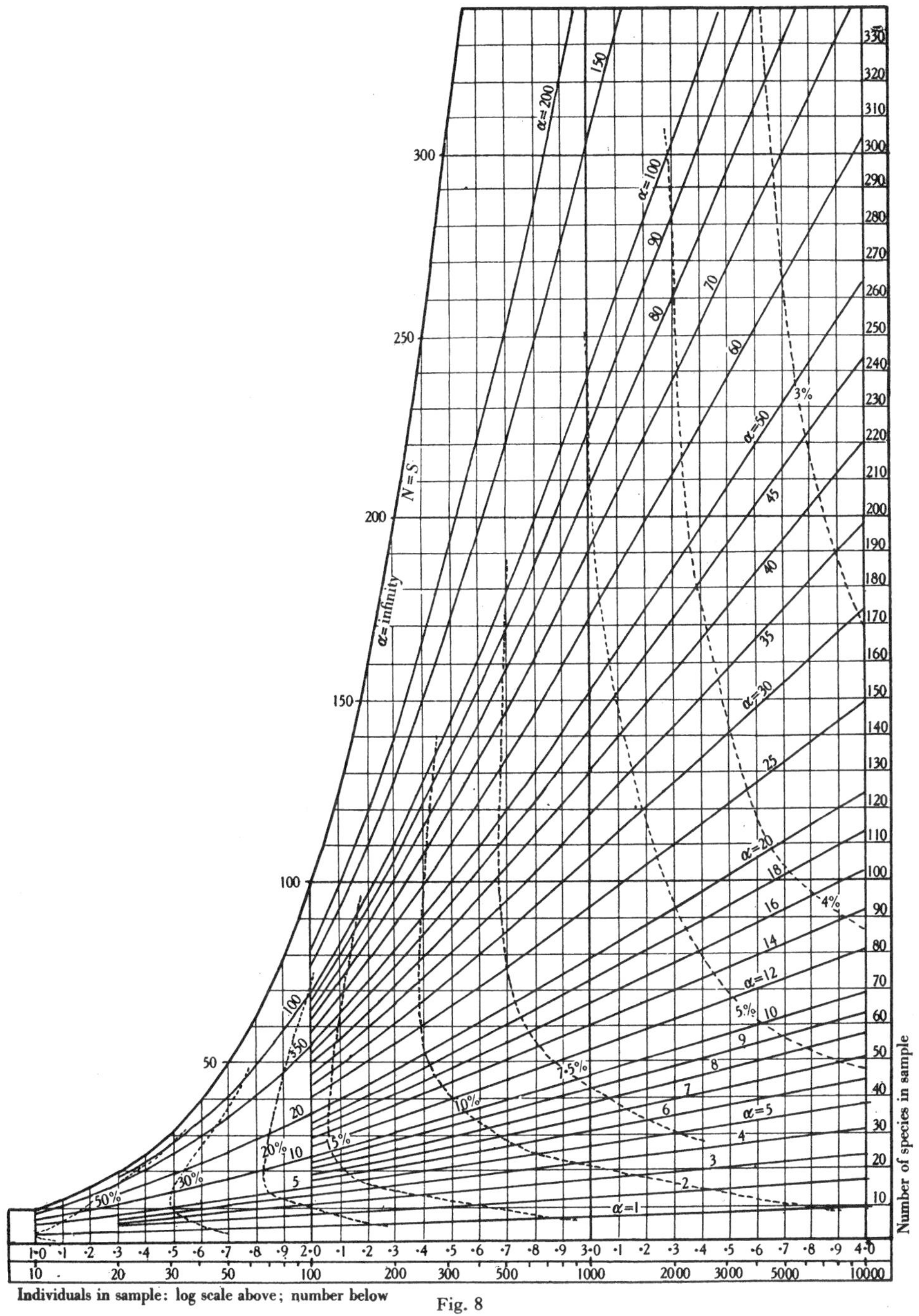

Fig. 8

Table 7. *Values of S for different combinations of N and α*

α \ N	10	20	50	100	200	500	1000	2000	5000	10,000	100,000
1	2·40	3·00	3·93	4·62	5·30	6·22	6·91	7·60	8·35	9·21	—
2	3·58	4·80	6·52	7·86	9·23	11·05	12·43	13·82	15·65	17·03	—
3	4·40	6·08	8·62	10·61	12·64	15·37	17·44	19·51	22·26	24·34	—
4	5·01	7·17	10·41	13·03	15·73	19·34	22·10	24·87	28·53	31·30	—
5	5·49	8·04	11·99	15·03	18·57	23·08	26·52	29·97	34·54	38·01	49·57
6	5·89	8·80	13·40	17·23	21·21	26·60	30·73	34·87	40·36	44·51	—
7	6·22	9·45	14·68	19·09	23·71	29·98	34·78	39·61	46·01	50·86	—
8	6·49	10·02	15·85	20·82	26·06	33·08	38·79	44·21	51·51	57·06	—
9	6·72	10·53	16·92	22·45	28·30	36·32	42·47	48·67	56·90	63·14	—
10	6·93	10·98	17·92	23·98	30·45	39·32	46·15	53·03	62·17	69·08	92·10
12	7·27	11·76	19·70	26·80	34·46	45·02	53·21	61·46	72·42	80·72	—
14	7·55	12·45	21·28	29·36	38·21	50·44	59·96	69·57	82·33	92·02	—
15	7·67	12·71	21·99	30·42	39·93	53·03	63·22	73·50	87·18	97·56	—
16	7·76	12·98	22·67	31·70	41·65	55·58	66·16	77·38	91·97	103·02	—
18	7·96	13·45	23·92	33·84	44·89	60·48	72·63	84·76	101·36	113·80	—
20	8·11	13·86	25·06	35·84	47·96	65·16	78·64	92·30	110·52	124·34	170·35
25	8·43	14·70	27·48	40·25	54·93	76·13	92·85	109·87	132·58	149·86	—
30	8·61	15·33	29·43	43·98	60·84	86·16	106·05	126·45	153·66	174·36	—
35	8·82	15·82	31·08	47·25	66·64	95·52	118·55	142·21	173·92	198·07	—
40	8·92	16·24	32·44	50·12	71·68	104·12	130·32	157·28	193·44	221·04	—
45	9·04	16·52	33·62	52·65	76·23	112·23	141·53	172·72	212·36	243·36	—
50	9·12	16·85	34·65	54·95	80·45	119·90	152·25	185·7	230·75	265·15	380·07
60	9·27	17·22	36·36	58·86	87·96	133·98	172·32	212·1	266·04	307·32	—
70	9·35	17·64	37·73	62·16	94·50	146·79	191·03	237·1	299·81	347·83	—
80	9·42	17·84	38·80	64·88	100·24	158·48	208·24	260·6	330·80	386·88	—
90	9·48	18·09	39·78	67·23	105·30	169·20	224·46	283·1	363·15	424·71	—
100	9·53	18·20	40·60	69·31	109·9	179·2	239·8	304·5	392·2	461·5	690·9
150	—	—	—	—	—	219·9	304·2	399·3	530·3	632·3	—
200	9·76	19·06	44·60	81·10	138·6	250·6	258·11	479·6	651·6	786·4	—

contain about 149 species. Or in a population with $\alpha=40$, 180 species would be produced by a sample of 1000 individuals.

Further, by following along one of the α curves it is possible to find the number of species in samples of different sizes from the same population. Thus if a sample of 1000 individuals for a population produced 73 species, then a sample of 10,000 from the same population should produce about 112 species, but a sample of 100 individuals would only contain about 34 species ($\alpha=18$). In addition, on the diagram the standard error (as a percentage of the value of α) has been indicated by dotted contours. Thus a sample containing 100 individuals and 24 species gives $\alpha=10\pm$ about 17%, say $10\pm1·7$; a sample of 1000 individuals and 46 species gives $\alpha=10\pm0·75$ (i.e. 7·5%); while a sample of 10,000 individuals containing 69 species gives $\alpha=10\pm0·45$ (4·5%). Thus the size of sample necessary to give α to any required degree of accuracy can rapidly be obtained from the figure.

Table 7 shows values of S correct to two decimal places for various combinations of α and N, and Table 8 shows the standard error of α to three significant figures for various values of N and α.

Table 8. *Standard error of α for different values of N and α*

α \ N	10	100	1000	10,000	100,000
1	0·504	0·288	0·141	0·091	—
5	2·785	0·860	0·430	0·282	0·209
10	6·46	1·60	0·719	0·445	0·321
20	15·82	3·19	1·52	0·712	0·495
50	49·87	8·79	2·67	1·359	0·891
100	153·7	20·42	5·04	2·27	1·41

REFERENCES

Corbet, A. S. (1941). 'The distribution of butterflies in the Malay Peninsula.' Proc. R. Ent. Soc. Lond. (A), 16: 101–16.

Fisher, R. A. (1941). 'The negative binomial distribution.' Ann. Eugen., Lond., 11: 182–7.

Thomas, D. C. (1938). 'Report on the Hemiptera-Heteroptera taken in the light trap at Rothamsted Experimental Station, during the four years 1933–36.' Proc. R. Ent. Soc. Lond. (A), 13: 19–24.

Williams, C. B. (1939). 'An analysis of four years captures of insects in a light trap. Part I. General survey; sex proportion; phenology; and time of flight.' Trans. R. Ent. Soc. Lond. 89: 79–131.

Williams, C. B. (1940). 'An analysis of four years captures of insects in a light trap. Part II. The effect of weather conditions on insect activity; and the estimation and forecasting of changes in the insect population.' Trans. R. Ent. Soc. Lond. 90: 227–306.

PART 3. A THEORETICAL DISTRIBUTION FOR THE APPARENT ABUNDANCE OF DIFFERENT SPECIES

By R. A. Fisher

(1) *The Poisson Series and the Negative Binomial distribution*

In biological sampling it has for some time been recognized that if successive, independent, equal samples be taken from homogeneous material, the number of individuals observed in different samples will vary in a definite manner. The distribution of the number observed depends only on one parameter, and may be conveniently expressed in terms of the number *expected*, m, in what is known as the *Poisson Series*, given by the formula

$$e^{-m}\frac{m^n}{n!}. \qquad (1)$$

Here n is the variate representing the number observed in any sample, m is the parameter, the number expected, which is the average value of n, and need not be a whole number. Obviously, m will be proportional to the size of the sample taken, and to the density of organisms in the material sampled. For example, n might stand for the number of bacterial colonies counted on a plate of culture medium, m for the average number in the volume of dilution added to each plate. The formula then gives the probability of obtaining n as the number observed.

The same frequency distribution would be obtained for the numbers of different organisms observed in one sample, if all were equally frequent in the material sampled.

If the material sampled were heterogeneous, or if unequal samples were taken, we should have a mixture of distributions corresponding to different values of m. The same is true of the numbers of different organisms observed in a single sample, if the different species are not equally abundant.

An important extension of the Poisson series is provided by the supposition that the values of m are distributed in a known and simple manner. Since m must be positive, the simplest supposition as to its distribution is that it has the Eulerian form (well known from the distribution of χ^2) such that the element of frequency or probability with which it falls in any infinitesimal range dm is

$$df = \frac{1}{(k-1)!}p^{-k}m^{k-1}e^{-m/p}\,dm. \qquad (2)$$

If we multiply this expression by the probability, set out above, of observing just n organisms, and integrate with respect to m over its whole range from 0 to ∞, we have

$$\int_0^\infty \frac{1}{(k-1)!}p^{-k}m^{k-1}e^{-m/p}e^{-m}\frac{m^n}{n!}\,dm,$$

which, on simplification, is found to have the value

$$\frac{(k+n-1)!}{(k-1)!\,n!}\,\frac{p^n}{(1+p)^{k+n}}, \qquad (3)$$

which is the probability of observing the number n when sampling from such a heterogeneous population. Since this distribution is related to the negative binomial expansion

$$\left(1-\frac{p}{1+p}\right)^{-k} = \sum_{n=0}^{\infty}\frac{(k+n-1)!}{(k-1)!\,n!}\left(\frac{p}{1+p}\right)^n,$$

it has become known as the *Negative Binomial distribution*. It is a natural extension of the Poisson series, applicable to a somewhat wider class of cases.

The parameter p of the negative binomial distribution is proportional to the size of the sample. The expectation, or mean value of n, is pk. The second parameter k measures in an inverse sense the variability of the different expectations of the component Poisson series. If k is very large these expectations are nearly equal, and the distribution tends to the Poisson form. If heterogeneity is very great k becomes small and approaches its limiting value, zero. This second parameter, k, is thus an intrinsic property of the population sampled.

(2) *The limiting form of the negative binomial, excluding zero observations*

In many of its applications the number n observed in any sample may have all integral values including zero. In its application, however, to the number of representatives of different species obtained in a collection, only frequencies of numbers greater than zero will be observable, since by itself the collection gives no indication of the number of species which are not found in it. Now, the abundance in nature of different species of the same group generally varies very greatly, so that, as I first found in studying Corbet's series of Malayan butterflies, the negative binomial, which often fits such data well, has a value of k so small as to be almost indeterminate in magnitude, or, in other words, indistinguishable from zero. That it is not really zero for collections of wild species follows from the fact that the total number of species, and therefore the total number not included in the collection, is really finite. The real situation, however, in which a large number of species are so rare that their chance of inclusion is small, is well represented by the limiting form taken by the negative binomial distribution, when k tends to zero.

The limiting value $k=0$ cannot occur in cases where the frequency at zero is observable, for the distribution would then consist wholly of such cases. If, however, we put $k=0$ in expression (3), write x for $p/(p+1)$, so that x stands for a positive number less than unity, varying with the size of the sample, and replace the constant factor $(k-1)!$ in the denominator, by a new constant factor, α, in the numerator, we have an expression for the expected number of species with n individuals, where n now cannot be zero,

$$\frac{\alpha}{n} x^n. \qquad (4)$$

The total number of species expected is consequently

$$\sum_{n=1}^{\infty} \frac{\alpha}{n} x^n = -\alpha \log_e (1-x),$$

so that our distribution is related to the algebraic expansion of the logarithm, as the negative binomial distribution is to the binomial expansion. Next, it is clear that the total number of individuals expected is

$$\sum_{n=1}^{\infty} \alpha x^n = \frac{\alpha x}{1-x}.$$

These two relationships enable the series to be fitted to any series of observational data, for if S is the number of species observed, and N the number of individuals, the two equations

$$S = -\alpha \log_e (1-x), \quad N = \alpha x/(1-x),$$

are sufficient to determine the values of α and x. The solution of the equations is, however, troublesome and indirect, so that to facilitate the solution in any particular case I have calculated a table (Table 9) from which, given the common logarithm of N/S, we may obtain that of N/α. Five-figure logarithms are advisable, such as those in *Statistical Tables*. If x be eliminated from the two equations, it appears that

$$N = \alpha (e^{S/\alpha} - 1), \quad S = \alpha \log_e \left(1 + \frac{N}{\alpha}\right),$$

and

$$\frac{N}{S} = (e^{S/\alpha} - 1) \div S/\alpha,$$

from which Table 9 has been constructed.

Table 9. *Table of $\log_{10} N/\alpha$ in terms of $\log_{10} N/S$, for solving the equation*

$$S = \alpha \log_e \left(1 + \frac{N}{\alpha}\right), \textit{ given } S \textit{ and } N$$

$\log_{10} N/S$	0	1	2	3	4	5	6	7	8	9
0·4	0·61121	63084	65023	66939	68832	70701	72551	74382	76195	77990
0·5	0·79766	81526	83271	85002	86717	88417	90105	91779	93442	95092
0·6	0·96730	98356	99973	1·01579	03174	04759	06335	07902	09460	11010
0·7	1·12550	14220	15813	17331	18772	20136	21631	23120	24602	26077
0·8	1·27546	29008	30465	31916	33361	34801	36234	37663	39087	40506
0·9	1·41920	43329	44733	46133	47528	48919	50305	51688	53066	54440
1·0	1·55810	57177	58539	59898	61254	62605	63954	65299	66640	67979
1·1	1·69314	70646	71975	73301	74623	75943	77261	78575	79886	81195
1·2	1·82501	83805	85106	86404	87700	88994	90285	91574	92860	94144
1·3	1·95426	96706	97984	99259	2·00532	01804	03073	04340	05605	06869
1·4	2·08130	09389	10647	11902	13156	14409	15659	16908	18155	19400
1·5	2·20644	21886	23126	24365	25602	26838	28072	29305	30536	31766
1·6	2·32994	34221	35446	36670	37893	39114	40334	41553	42770	43986
1·7	2·45201	46414	47627	48838	50048	51256	52464	53670	54875	56079
1·8	2·57282	58484	59684	60884	62083	63280	64476	65672	66866	68059
1·9	2·69252	70443	71633	72822	74011	75198	76385	77570	78755	79939
2·0	2·81121	82303	83484	84664	85843	87022	88199	89376	90552	91727
2·1	2·92901	94075	95247	96419	97590	98760	99930	3·01099	02267	03434
2·2	3·04600	05766	06931	08095	09259	10422	11584	12745	13906	15066
2·3	3·16225	17384	18542	19699	20856	22012	23168	24323	25477	26630
2·4	3·27783	28936	30087	31238	32389	33539	34688	35837	36985	38133
2·5	3·39280	40426	41572	42717	43862	45006	46150	47293	48436	49578
2·6	3·50719	51860	53001	54141	55280	56419	57558	58696	59833	60970
2·7	3·62106	63242	64378	65513	66648	67782	68915	70048	71181	72313
2·8	3·73445	74577	75707	76838	77968	79097	80227	81355	82484	83611
2·9	3·84739	85866	86992	88119	89244	90370	91495	92619	93743	94867
3·0	3·95991	97114	98236	99358	4·00480	01602	02723	03843	04964	06084
3·1	4·07203	08322	09441	10560	11678	12795	13913	15030	16147	17263
3·2	4·18379	19494	20610	21725	22839	23954	25068	26181	27295	28408
3·3	4·29520	30632	31744	32856	33967	35079	36189	37300	38410	39520
3·4	4·40629	41738	42847	43956	45064	46172	47280	48387	49494	50601
3·5	4·51707	52814	53920	55025	56131	57236	58340	59445	60549	61653

(3) *Fitting the series*

The use of the table is shown, using Williams's extensive data for the Macrolepidoptera at Harpenden (total catch for four years). Symbols + and − are used to indicate numbers to be added and subtracted respectively.

	Symbol	Number	Common logarithm
	S	240	−2·38021
	N	15609	+4·19338
	N/S	—	1·81317
From the table		log (N/S)	log (N/α)
		−1·81	−2·58484
		+1·82	+2·59684
	Difference	0·01	0·01200
Proportional parts		0·00317	0·00380
		1·81317	2·58864
Then		Number	Common logarithm
	N/α	—	−2·58864
	N	—	+4·19338
	α	40·248	1·60474

For constructing the distribution we should then calculate

$$x=\frac{N}{N+\alpha}=\frac{15609}{15649{\cdot}248}=0{\cdot}9974281.$$

The quantity α is independent of the size of sample, and is proportional to the number of species of the group considered, at any chosen level of abundance, relative to the means of capture employed. Values of α from different samples or obtained by different methods of capture may therefore be compared as a measure of richness in species. To this end we shall need to know the sampling errors by which an estimate of α may be affected.

(4) *Variation in parallel samples*

Whatever method of capture may be employed, it is to be expected that a given amount of activity devoted to it, e.g. a given number of hours exposure of a light-trap, or a given volume of sea water passed through a plankton filter, will yield on different occasions different numbers of individuals and of species, and, consequently, varying estimates of α. The amount of variation of these kinds attributable to chance must form the basis of all conclusions as to whether variations beyond chance have occurred in the circumstances in which two or more samples were made.

In strictly parallel samples, i.e. equivalent sampling processes applied to homogeneous material, the numbers caught of each individual species will be distributed in a Poisson series, and it easily follows that the same is true of the aggregate number, N, of all species. Since N is a large number of hundreds or thousands, this is equivalent to N being normally distributed with a variance equal to its mean, so that to any observed value N we may attach a standard error (of random sampling) equal to $\pm\sqrt{N}$.

For the variation of S we must obtain the distribution of species according to the number m expected in the sample; modifying expression (2) in the same way as (3) has been modified, this is found to be

$$\alpha e^{-\alpha m/N}dm/m. \tag{5}$$

The probability of missing any species is e^{-m}, so that the contribution to the sampling variance of S due to any one species being sometimes observed and sometimes not, is

$$e^{-m}(1-e^{-m}).$$

Multiplying this by the frequencies in (5) and integrating over all values of m, we have

$$\alpha\int_0^{\infty}e^{-m(N+\alpha)/N}\left(1-\frac{m}{2}+\frac{m^2}{6}-\dots\right)dm=\alpha\log_e\left(\frac{2N+\alpha}{N+\alpha}\right),$$

which is the sampling variance of S. For large samples this is approximately (0·6931) α.

Variations of S and N in parallel samples are not, however, independent. When present, a species must contribute on the average $m/(1-e^{-m})$ individuals, which exceeds the expectation in all samples by

$$\frac{me^{-m}}{1-e^{-m}},$$

and as the frequency of occurrence is $1-e^{-m}$, each species must contribute $m.e^{-m}$ to the covariance of S and N. The covariance is thus found to be

$$\frac{\alpha N}{N+\alpha}.$$

From these three values it is possible by standard methods to find the sampling variance of S in samples having a given number of specimens N, which is

$$V(S),\text{ given }N,=\alpha\log_e\frac{2N+\alpha}{N+\alpha}-\frac{\alpha^2N}{(N+\alpha)^2}$$

and, the variance of α,

$$V(\alpha)=\frac{\alpha^3\left\{(N+\alpha)^2\log_e\dfrac{2N+\alpha}{N+\alpha}-\alpha N\right\}}{(SN+S\alpha-N\alpha)^2}.$$

We may, therefore, complete the example of the last section by calculating the standard error of α. Using the values obtained, the variance comes to 1·1251, of which the square root is 1·0607.

The estimate obtained for α, 40·248, has, therefore, a standard error of 1·0607, available for comparison with like estimates.

(5) *Test of adequacy of the limiting distribution*

From the manner in which the distribution has been developed it appears that we never have theoretical grounds for supposing that k is actually zero;

but, on the contrary, must generally suppose that in reality it has a finite, though perhaps a very small, value. Our reasons for supposing this small value to be negligible must always be derived from the observations themselves. It is, therefore, essential to be able to test any body of data in respect to the possibility that in reality some value of k differing significantly from zero might fit the data better than the value zero actually assumed.

The most sensitive index or score by which any departure of the series of frequencies observed from those expected can be recognized, is found by the general principles of the Theory of Estimation, as, for example, in the author's *Statistical Methods for Research Workers*, to be

$$S\left\{a_n\left(1+\frac{1}{2}+\frac{1}{3}+\ldots+\frac{1}{n-1}\right)\right\},$$

when a_n is a number of species observed with n individuals in each. If the values of a_n conformed accurately with expectation, the total score would be equal to

$$\frac{S^2}{2\alpha}.$$

If, on the contrary, the series were better fitted by a negative binomial with a value of k differing from zero, we should expect the difference

$$S\left\{a_n\left(1+\frac{1}{2}+\frac{1}{3}+\ldots+\frac{1}{n-1}\right)\right\}-\frac{S^2}{2\alpha}$$

to show a positive discrepancy.

Applying this test to Williams's distribution for 240 species of Macrolepidoptera, one finds, after a somewhat tedious calculation,

$S\left\{a_n\left(1+\frac{1}{2}+\frac{1}{3}+\ldots+\frac{1}{n-1}\right)\right\}$	724·86
$\frac{S^2}{2\alpha}$	715·57
Difference	+9·29

The series, therefore, shows a deviation in the direction to be expected for the negative binomial, though apparently quite a small one. In order to test the significance of such discrepancies, I give in Table 10, for the same range of observable values of the average number of specimens in each species N/S, the values of i/S, where i is the quantity of information, in respect of the value of k, which the data supply.

Table 10. *The amount of information respecting k, supposed small, according to the numbers of individuals (N) and species (S) observed*

$\log_{10} N/S$	i/S	$\log_{10} N/S$	i/S
0·4	0·1971		
0·5	0·2882		
0·6	0·3914	2·1	3·1047
0·7	0·5054	2·2	3·3606
0·8	0·6295	2·3	3·6260
0·9	0·7639	2·4	3·9009
1·0	0·9076	2·5	4·1854
1·1	1·0608	2·6	4·4791
1·2	1·2232	2·7	4·7825
1·3	1·3950	2·8	5·0954
1·4	1·5762	2·9	5·4178
1·5	1·7665	3·0	5·7498
1·6	1·9661	3·1	6·0912
1·7	2·1751	3·2	6·4421
1·8	2·3934	3·3	6·8026
1·9	2·6211	3·4	7·1726
2·0	2·8582	3·5	7·5521

Entering the table with our value 1·81317 for $\log_{10} N/S$ we have $i/S=2·4656$, or $i=591·7$. This quantity may now be used for two purposes. In the first place it is the sampling variance of the discrepancy observed, so that, taking its square root, the standard error is found to be 24·33. This suffices to test the significance of the discrepancy, since $9·29 \pm 24·33$ is clearly insignificant.

If, on the contrary, a significant discrepancy had been found, an estimate of the value of k required to give a good fit to the data could be made by dividing the discrepancy by i. In fact

$$\frac{9·29}{591·7}=0·016$$

would have been the value of k indicated by the data, if any value other than zero had been required.

REFERENCES

Fisher, R. A. & Yates, F. (1943). 'Statistical tables for biological, agricultural and medical research' (2nd ed.). Edinburgh.

Fisher, R. A. (1941). 'Statistical methods for research workers' (8th ed.). Edinburgh.

SUMMARY

Part 1. It is shown that in a large collection of Lepidoptera captured in Malaya the frequency of the number of species represented by different numbers of individuals fitted somewhat closely to a hyperbola type of curve, so long as only the rarer species were considered. The data for the commoner species was not so strictly 'randomized', but the whole series could be closely fitted by a series of the logarithmic type as described by Fisher in Part 3. Other data for random collections of insects in the field were also shown to fit fairly well to this series.

Part 2. Extensive data on the capture of about 1500 Macrolepidoptera of about 240 species in a light-trap at Harpenden is analysed in relation to Fisher's mathematical theory and is shown to fit extremely closely to the calculations.

The calculations are applied first to the frequency of occurrence of species represented by different numbers of individuals—and secondly to the number of species in samples of different sizes from the same population.

The parameter 'α', which it is suggested should be called the 'index of diversity', is shown to have a regular seasonal change in the case of the Macrolepidoptera in the trap. In addition, samples from two traps which overlooked somewhat different vegetation are shown to have 'α' values which are significantly different.

It is shown that, provided the samples are not small, 'α' is the increase in the number of species obtained by increasing the size of a sample by e (2·718). A diagram is given (Fig. 8) from which any one of the values, total number of species, total number of individuals and index of diversity (α), can be obtained approximately if the other two are known. The standard error of α is also indicated on the same diagram.

Part 3. A theoretical distribution is developed which appears to be suitable for the frequencies with which different species occur in a random collection, in the common case in which many species are so rare that their chance of inclusion is small.

The relationships of the new distribution with the negative binomial and the Poisson series are established.

Numerical processes are exhibited for fitting the series to observations containing given numbers of species and individuals, and for estimating the parameter α representing the richness in species of the material sampled; secondly, for calculating the standard error of α, and thirdly, for testing whether the series exhibits a significant deviation from the limiting form used.

Special tables are presented for facilitating these calculations.

PAPER 4

The Generic Relations of Species in Small Ecological Communities (1947)

C. B. Williams

Commentary

NICHOLAS J. GOTELLI

C. B. Williams's paper is a deserved classic, as it represents one of the first applications of null-model analysis in ecology. In this paper, Williams took a novel approach to the analysis of species-to-genus (S/G) ratios. In the previous year, J. Elton (1946) had demonstrated that S/G ratios in small island communities were consistently lower than for regional species lists from the surrounding areas. From this result, Elton concluded competition was more severe in small local communities (where resources are more likely to be limiting).

Williams recognized that, even in the absence of species interactions, the S/G ratio might be sensitive to the number of species in the assemblage. Thus the correct comparison is not small versus large assemblages but small assemblages versus randomly sampled assemblages of the same species number. For both an analytical model of the log-series distribution and a randomization test, Williams demonstrated that the observed S/G ratios in Elton's (1946) data closely matched the values expected by random sampling. When observed S/G ratios did deviate from the null expectation, they often did so in the opposite direction: there was a slight tendency for more congeneric coexistence in small communities (Simberloff 1970), which contradicted Elton's competition hypothesis.

Perhaps because Williams did not devote any text to the novel statistical (and philosophical) implications of his analysis, his work was largely ignored, as was an earlier demonstration of the same pattern by European plant ecologists (reviewed in Järvinen 1982). Subsequent authors continued to misinterpret taxonomic diversity indices (Grant 1966; Moreau 1966; MacArthur and Wilson 1967; Cook 1969), even though Williams repeated and extended his analyses (Williams 1951, 1964). But this first paper presaged the explosive controversy over null models that would dominate ecology and biogeography in the 1980s (reviewed in Gotelli and Graves 1996). Sadly, many macroecologists have not heeded the message in Williams's paper. Recent debates surrounding the mid-domain effect (Colwell et al. 2004) and the neutral model (Gotelli and McGill 2006) demonstrate the continued importance of Williams's null-model perspective in macroecology.

Literature Cited

Colwell, R. K., C. Rahbek, and N. J. Gotelli. 2004. The mid-domain effect and species richness patterns: What have we learned so far? *American Naturalist* 163:E1–E23.

Cook, R. E. 1969. Variation in species density of North American birds. *Systematic Zoology* 18:63–84.

Elton, C. 1946. Competition and the structure of ecological communities. *Journal of Animal Ecology* 1554–68.

From *Journal of Animal Ecology* 16:11–18. Reprinted with permission from Wiley-Blackwell.

Gotelli, N. J., and G. R. Graves. 1996. *Null models in ecology*. Smithsonian Institution Press, Washington, DC.
Gotelli, N. J., and B. J. McGill. 2006. Null versus neutral models: What's the difference? *Ecography* 29:793–800.
Grant, P. R. 1966. Ecological compatibility of bird species on islands. *American Naturalist* 100:451–62.
Järvinen, O. 1982. Species-to-genus ratios in biogeography: A historical note. *Journal of Biogeography* 9:363–70.
MacArthur, R. H., and E. O. Wilson. 1967. *The theory of island biogeography*. Princeton University Press, Princeton, NJ.
Moreau, R. E. 1966. *The bird faunas of Africa and its islands*. Academic Press, New York.
Simberloff, D. 1970. Taxonomic diversity of island biotas. *Evolution* 24:23–47.
Williams, C. B. 1951. Intra-generic competition as illustrated by Moreau's records of East African bird communities. *Journal of Animal Ecology* 20:246–53.
———. 1964. *Patterns in the balance of nature and related problems in quantitative ecology*. Academic Press, New York.

[11]

THE GENERIC RELATIONS OF SPECIES IN SMALL ECOLOGICAL COMMUNITIES

By C. B. WILLIAMS, *Rothamsted Experimental Station, Harpenden*

1. INTRODUCTION

The problem of the importance of competition between two or more species in the same genus, in determining whether they can survive side by side in the same community, is of great ecological interest. The following is a statistical approach to the problem based largely on the fact that in most animal and plant communities the numbers of genera with one, two, three, etc., species appear to form a mathematical series very close to the 'logarithmic series'.

The species in any group of animals and plants that are found living side by side in any relatively small community have presumably been selected in the course of time from all the species in the surrounding areas which have been able to reach the smaller area in question. Those that survive are those which are capable of existing in the physical environment of the area and also in association with, or in competition with, the other members of the community.

Such a natural selection could conceivably be brought about under three different conditions: (1) without reference to the generic relations of the species, (2) more or less *against* species in the same genus, (3) more or less *in favour of* species in the same genus. Extremes of either (2) or (3) would result in only one species to each genus on the one hand, or in all the species in each genus on the other hand being represented. We know, however, that biologically, neither of these extremes is correct.

It is, however, most important to consider in detail what exactly happens when a selection of a relatively small number of species is made from a larger fauna or flora, without reference to their generic relations (no. (1) above), as a true interpretation can only be made by comparing the observed data with the results of a selection of the same size made at random. If the average number of species per genus in a small community is smaller than that expected by random sampling, then there is evidence of a selection against generically related species; if the number of species per genus is larger than would be expected by random sampling, then there is evidence of selection in favour of species in the same genus.

Before we can discuss the problem of sampling, it is, however, necessary to consider what is the general structure of the relative numbers of genera with 1, 2, 3, etc., species in any group of animals or plants.

2. GENERIC CLASSIFICATION AND THE LOGARITHMIC SERIES

I have shown recently (Williams, 1944) that in a number of classifications of particular families or orders of plants, insects and other animals, the number of genera with 1, with 2, with 3, etc., species forms a series which can be represented very closely by the logarithmic series.

This will be discussed more fully below, but for the moment it should be noted that if we know the number of species and the number of genera in any group, a logarithmic series of the genera with 1, 2, 3 species (which we will call n_1, n_2, n_3, etc.) can be calculated and this can be checked against the observed frequencies. If a close fit is found, it is evidence of the likelihood that the logarithmic series is a correct interpretation, and we can then use for further argument known properties of the logarithmic series.

I have shown, as stated above, that the logarithmic series gives a very close fit to published classifications of large groups, such as the flowering plants of Britain, the Mantidae of the world, the birds of Great Britain, and many other systematic and geographical groups. It is, however, important to know if the same principle is found in the classification of small communities. For this purpose some evidence brought forward recently by Elton (1946) is of great value.

Table 1 shows particulars of ten of his animal communities and three of his plant communities, together with the total (reduced to an average) of his 49 animal and of his 27 plant communities; with the number of genera with 1, 2 and 3 species in each case as given by Elton and as calculated by the logarithmic series.

The communities analysed here are all comparatively simple ones, because reliable surveys of more complex communities such as woodland do not yet exist; some of them, however, are quite large in area.

No one can deny the extremely close approximation of the observed and calculated figures in nearly all the examples and particularly in the averages. Thus in the average figure for 49 animal communities the calculated number of genera with one species is 38·1—the observed 39; in the average of 27 plant communities the calculated number of genera with one species is 30·7 and the observed 31·1. The fit for n_2 and n_3 is almost equally good.

Thus we have very strong evidence that the number of genera with 1, with 2 and with 3, etc., species is arranged in a logarithmic series in quite small or simple ecological communities as well as in large systematic groups and in large areas.

where the successive terms are the numbers of groups (in this case genera) with 1, with 2, with 3, etc., units (in this case species).

n_1 (which $= \alpha x$) is the number of genera with one species; x is a constant less than unity, α is a constant which we have called the Index of Diversity. If we know the number of species (N) and the number of genera (S) in a population which is arranged in a log series, we can calculate both n_1 and x, and

Table 1

	Elton's community	No. of genera	No. of species	No. of genera with			Index of diversity
				1 species	2 species	3 species	
	Animal	S	N	n_1	n_2	n_3	α
9	Temperate woodland (logs)	126	136	118	6	2	
				116	*8·7*	*0·9*	773·
10	Subarctic bog	19	21	17	2	0	
				19·1	*1·6*	*0·2*	94
11	Subarctic bog	29	32	26	3	0	
				26·3	*2·2*	*0·25*	155
12	Temperate fresh-water pond	26	32	21	4	1	
				21·4	*3·5*	*0·8*	65
13	Temperate fresh-water pond	72	90	61	8	1	
				59	*11*	*2·2*	171
14	Temperate lake benthos	46	59	38	6	0	
				36·8	*6·9*	*1·7*	98·1
15	Temperate lake benthos	8	13	6	1	0	
				5	*1·8*	*0·7*	8·3
16	Temperate lake benthos	58	86	44	8	1	
				40	*10·8*	*3·9*	74
20	Temperate river	99	131	82	7	6	
				79·6	*15·6*	*4·1*	203
22	Temperate river	59	87	47	6	3	
				42	*10·8*	*3·7*	81·5
	Average of 49 animal communities	45·3	54·41	39·0	4·6	1·12	
				38·1	*5·7*	*1·14*	127
	Plant						
1	Arctic rocky soil	22	31	16	4	1	
				16·4	*3·9*	*1·2*	35
3	Arctic heath	37	51	28	7	0	
				27·8	*6·3*	*1·9*	62
4	Subarctic heath	32	42	26	3	0	
				26	*4·9*	*1·35*	68
	Average of 27 plant communities	37·0	95·1	31·1	4·4	1	
				30·7	*4·9*	*1·05*	96

Data from Elton (1946). Numbers in col. 1 are those of communities given in his Tables 1 and 4.

In the figures for n_1, n_2 and n_3 the upper line is the observed number, the lower line (in italics) that calculated on the basis of the log series.

3. THE PROPERTIES OF THE LOGARITHMIC SERIES

The logarithmic series will be found fully discussed in Fisher, Corbet & Williams (1943), Williams (1944) and Williams (1947).

The series can be represented in two ways, either

$$n_1, \quad \frac{n_1 x}{2}, \quad \frac{n_1 x^2}{3}, \quad \frac{n_1 x^3}{4},$$

or

$$\alpha x, \quad \alpha\frac{x^2}{2}, \quad \alpha\frac{x^3}{3}, \quad \alpha\frac{x^4}{4},$$

hence the whole series, and also the Index of Diversity.

If a random sample is taken from a population which is arranged in a logarithmic series, a new logarithmic series is found in the sample, with a new n_1 and a new x. But for all samples of whatever size from one population the ratio of n_1 to x, which we have called 'α', is constant. In other words, this is a property of the population and not of one particular sample. It is a measure of the extent to which the species are grouped into genera, or the genera divided into species. It is high when there are a large number of genera in relation to the number of species, and low when there are a small number of genera relative to the number of species, and it is independent of the size of the sample. It is thus a measure of the generic diversification of the species population, and for this reason we have called it the 'Index of Diversity'. It is, indeed, an index of exactly the property which is at present at issue. If a small sample of a larger population has the same Index of Diversity as the larger population, then it has been selected without reference to the generic relationships of the species. If the smaller sample has a larger Index of Diversity, it is evidence that there has been selection against species in the same genus. If it has a smaller Index of Diversity, it is evidence that there has been selection in favour of species in the same genus.

The average number of species per genus (N/S), the proportion of genera with one species (n_1/S), and the proportion of species in genera with one species (n_1/N) are all dependent on the size of the sample (see Table 2); but the Index of Diversity is the same for all samples from the same population provided that they have been randomized for generic relationships.

It is, however, important to note for the present study that the error of estimation of the Index of Diversity increases rapidly as the sample gets smaller (see Fisher *et al.* 1943, pp. 53 and 56). Thus for populations with 1000 species in 100 genera the error (standard deviation) of α is about 6%; but with 100 species in 70 genera it is 20%, and with 10 species in 9 genera it is just under 100%. The error is particularly high when the average number of species per genus is very low, which is usually the case in small communities.

An example of the application of the logarithmic series is as follows: In Bentham & Hooker's *British Flora*, 1906 edition there are 1251 species of plants classified into 479 genera. These are arranged reasonably closely to a logarithmic series and give an Index of Diversity = 284 (Williams, 1944, p. 30).

If samples of different numbers of species are taken from the above flora, *without any reference to generic relationships*, they must each have the same index of diversity; and from this and the number of species we can calculate that the expected number of genera would be, as shown in Table 2.

Table 2. *Calculated properties of samples of different numbers of species taken from Bentham & Hooker's 'British Flora'* (1906 *ed.*), *without any bias with respect to generic relations. All samples have an Index of Diversity of* 284.

No. of species	Expected no. of genera	Average no. of species per genus	Expected no. of genera with 1 species	% of genera with 1 species
N	S	N/S	n_1	$100\, n_1/S$
		Original population		
1251	479	2·61	231	48
		Samples		
1000	428	2·34	222	52
500	288	1·74	181	63
200	151	1·32	117	78
100	86	1·16	74	86
50	46	1·09	42·5	92
30	28·5	1·05	—	—
20	19·3	1·04	18·7	95
10	9·8	1·02	—	—

It will be seen that the average number of species per genus steadily falls, and the percentage of genera with one species each steadily rises as the sample gets smaller. In a random sample of only 20 species of British flowering plants less than one genus with more than one species would be expected to be present.

It should be noted that if our biologically selected community contains as few as 30 species, the only possible high values for n_1 (since fractions do not exist in nature), are 30; 29; 28; 27; 26; 25. These give respectively Indexes equal to infinity; 420; 270; 195; and 150, etc. so that no community as small as this will give statistical data sufficient to distinguish between small changes in α, or small biases in favour of, or against, species in the same genus.

The data brought forward by Elton are not suitable for further inquiry along these lines as firstly the numbers of species are in general too small, and secondly we have no available lists of the number of species and genera (and that is to say of the Index of Diversity) in the larger groups, floras or faunas, from which these communities have been selected by nature. What we require is a series of natural groups of different sizes selected by nature from one much larger group; all the species and genera in each series being of course in exactly the same classification.

For example, if we could take the butterflies of the world, the butterflies of England, the butterflies of an English county, the butterflies of a small area in this county, and the butterflies of a single ecological community in the small area (all in the same classification), then we would have a series of observed facts which could be analysed to see if the Index of Diversity in each sample indicated greater or less generic diversification as the samples got smaller.

It has not been possible to get such a perfect series of data, but in what follows I give the evidence I have been able to find that appears to be suitable for study.

4. COMPARISONS OF SMALL WITH LARGE SERIES

(*a*) *Plants on Broadbalk Wilderness, Rothamsted Experimental Station, Hertfordshire*

Brenchley and Adam (1915) have published a list of the species of flowering plants that have been found on this piece of abandoned wheat field which covers an area of about half an acre. In their table on p. 198 (emended to agree with the nomenclature of Bentham & Hooker's *British Flora*, 1906 ed.), they mention 73 species which were found in four surveys carried out between 1867 and 1913. These are classified into 59 genera. Of these, 50 genera have one species; six have two species; one has three and one has four. This gives an average number of species per genus of 1·24; and an Index of Diversity of 147.

If only the plants observed in 1913 are considered, the numbers are 65 species in 52 genera (41 with one species, ten with two, one with three). The average number of species per genus is 1·25 and the Index of Diversity 134.

I have already published (Williams, 1944, p. 30) data on the whole classification of Bentham & Hooker's 1906 edition which gives an average number of species per genus of 2·61 and an Index of Diversity = 284.

The two Broadbalk Communities thus have—as would be expected from the small size of the sample—a lower average number of species per genus; but the Index of Diversity is in each case lower than the whole British flora, and the smaller sample has a lower value than the larger. Thus the evidence is that there is less generic diversification in the Broadbalk Wilderness flora, or a smaller number of genera than would be expected by a random sample including the same number of species.

In fact a random sample of 73 species from the British flora, as classified by Bentham & Hooker, would have about 65 genera instead of the 59 observed.

(*b*) *Flowering plants of Scolthead Island, Norfolk*

Chapman (1934) gives a list of the flowering plants of Scolthead Island, an island of sand-dunes which covers an area of about one and a half square miles.

His list, when altered to the nomenclature of Bentham & Hooker's *British Flora* (1906 ed.), gives the following results (Table 3).

Table 3

Species N	Genera S	Species per genus N/S		No. of genera with 1 species n_1	2 species n_2	3 species n_3	4 species n_4	5 species n_5	Index of Diversity
161	114	1·41	Obs.	82	22	7	1	2	
			Cal.	85·3	20·1	6·3	2·26	0·8	182

It will be seen that the observed numbers conform very closely to a logarithmic series as checked by the calculated values of n_1, n_2, and n_3; and that the Index of Diversity is 182, as compared with 284 for the whole British flora.

In a purely random sample of 161 species from the British flora, irrespective of generic relations, we would have expected to find representatives of approximately 126 genera. Thus the observed number is smaller than the calculated, indicating a selection in favour of generically related species rather than against them.

(*c*) *Flora of Park Grass Plots, Rothamstead Experimental Station, Hertfordshire*

At Rothamsted there are a series of plots of grass, of areas varying from ½ to ⅛ acre, which have been manured in special ways for many years, but in which no further interference has been made in the natural flora which develops under such conditions of manuring and soil.

Brenchley (1924) has given a series of tables from which Table 4 is extracted; the classification has been altered slightly to agree with Bentham & Hooker's *British Flora* (1906 ed.).

Table 4

	No. of species N	No. of genera S	Average species per genus N/S	Index of Diversity
All plots all years	59	53	1·11	263
	1919 Survey only			
Plot 3. Unmanured:				
Without lime	30	27	1·11	134
With lime	30	27	1·11	134
Plot 13. Farmyard manure:				
Without lime	20	18	1·11	89
With lime	23	21	1·10	111

The original Index of Diversity for the whole of Bentham & Hooker was 284 so that all the smaller floras have a smaller Index of Diversity. The error of estimation of α is however very high.

Plots 9 and 10, with sulphate of ammonia for many years, gave (with and without potash) from 9 to 14 species each in its own genus. It is not possible to calculate a finite value of α from such data. It is, however, interesting to note that in a random sample of ten species from Bentham & Hooker's *British Flora* ($\alpha = 284$) one would expect 9·8 genera! So the observed figure of 10 cannot be taken as evidence of any extreme departure from randomization.

(d) Lepidoptera of Wicken Fen, Cambridgeshire

Farren (1936) gives a list of the Macrolepidoptera of Wicken Fen according to the classification of Meyrick's *Handbook of the British Lepidoptera* (1895). In the whole group 368 species are listed in 135 genera. This gives an average of 2·73 species per genus and an Index of Diversity of 76·4. The number of species in the same families for the whole of Great Britain, as listed in Meyrick's handbook, is 788 in 212 genera, which gives an average of 3·72 species per genus and an Index of Diversity of 96·5.

For the family Caradrinidae (Agrotidae) alone, there are 146 species in 29 genera for Wicken Fen (average per genus 5·04; Index of Diversity 10·8); and 273 species in 39 genera for the whole of Great Britain (average per genus 7·00; Index of Diversity 12·7). In the family Plusiidae there are 24 species in 12 genera for Wicken Fen (average 2·00; $\alpha = 9{\cdot}3$) and 54 species in 22 genera for the British Isles (average 2·45, $\alpha = 14{\cdot}0$).

So we see that for the whole Macrolepidoptera and also for two separate families, one large and one small, the Index of Diversity of the local fauna is smaller than that of the larger area. In other words, there are fewer genera in the local fauna than would be expected by a random selection, thus giving evidence of a selection in favour of generically related species.

(e) Coleoptera of Windsor Forest, Berkshire

Donisthorpe (1939) gives a list of 1825 species of Coleoptera found in Windsor Forest. They are grouped into 553 genera, according to the classification of Beare & Donisthorpe (1904). The list gives a good approximation to a logarithmic series. There is an average of 3·30 species per genus and an Index of Diversity of 278. The original classification of the Coleoptera of the British Isles gave 3268 species in 804 genera, with an average of 4·60 species per genus and an Index of Diversity of 341 (see Williams, 1944, p. 24). Thus the Coleoptera of Windsor Forest have, as expected from the smaller number of species, a lower average number of species per genus; but have also a smaller Index of Diversity, indicating a selection in favour of species of the same genus rather than against.

(f) Capsidae (= Miridae) (Heteroptera) of Hertfordshire

China (1943) gives a list of the Heteroptera of the British Isles. The family Capsidae (= Miridae) in this includes 186 species in 76 genera as shown in Table 5. The Index of Diversity is 48.

Table 5

No. of species N	No. of genera S	Average sp. per gen. N/S		No. of genera with: 1 species n_1	2 species n_2	3 species n_3	4 species n_4	5 species n_5	Index of Diversity
			British						
186	76	2·45	Obs.	40	16	11	3	1	
			Cal.	38·1	15·2	8	4·8	—	48
			Hertfordshire						
127	60	2·12	Obs.	39	12	3	1	0	
			Cal.	32·5	12·1	6	—	—	43·7

16 *Generic relations of species in small communities*

Bedwell (1945) has given a list of the Hemiptera of Hertfordshire; the particulars of the Capsidae from this are given in the second half of Table 5. It will be seen that the Index of Diversity is slightly below that of the British fauna, but probably not outside the limits of error.

It shows, however, no evidence of selection against species of the same genus.

Table 6

Plants	No. of species observed	No. of genera		Av. no. of species per genus		Index of Diversity	
		obs.	calc.	obs.	calc.	orig.	sample
Broadbalk: 4 surveys	73	59	65·0	1·24	1·12	284	147
1913 survey	65	52	58·5	1·25	1·11	284	1·34
Scolthead Island	161	114	126	141	1·28	284	182
Park Grass, all plots	59	53	55·9	1·11	1·055	284	263
Plot 3 (1919)	30	27	28·5	1·11	1·05	284	134
Plot 13 (1919)	20	18	19·3	1·11	1·04	284	89
Plot 13 (1919)	23	21	22·2	1·10	1·04	284	111
Insects							
Wicken Fen:							
Macrolepidoptera	368	135	151·6	2·73	2·43	96·5	76·4
Caradrinidae	146	29	32·1	5·04	4·55	12·7	10·8
Plusiidae	24	12	14·0	2·00	1·71	14·0	9·3
Windsor Forest:							
Coleoptera	1825	553	630·0	3·30	2·90	341·0	278
Hertfordshire:							
Miridae	127	60	62·1	2·12	2·05	48	43·7
Macrolepidoptera	561	186	188·4	3·02	2·98	99·6	99
Caradrinidae	180	33	38·31	5·15	4·70	14·9	13·2

Table 7

No. of species in sample	No. of genera in sample			Av. no. of genera in sample	Observed no. in natural samples of same
	1	2	3		
20	20	19	20	19·7	18
23	23	22	22	22·3	21
30	29	28	29	28·7	27
59	56	50	54	53·3	53
65	60	54	59	59·7	52
73	67	60	66	64·3	59
161	128	117	126	123·7	114

(g) *Lepidoptera of Hertfordshire*

Foster (1937) gives a list of the Lepidoptera recorded for Hertfordshire. The classification used is that of Meyrick's, *Revised handbook of British Lepidoptera* (1927). He enumerates 561 species in 186 genera. This gives an average of 3·02 species per genus and an Index of Diversity of about 99. In Meyrick's *Revised handbook* the figures for the same families for the whole British Isles are 806 species in 218 genera. This is an average of 3·7 species per genus and an Index of Diversity of 99·6. In the Family Caradrinidae (Agrotidae) alone, there are 180 species in 33 genera in Hertfordshire (average 5·15, $\alpha = 13{\cdot}2$), and 290 species in 45 genera for the British Isles (average 6·44; $\alpha = 14{\cdot}9$). Thus the differences indicated that the smaller faunas have an Index of Diversity just equal to or smaller than the larger areas.

It is interesting to note that Meyrick wrote his first *Handbook* in 1895, and his 'Revised' edition in 1927. The first gives an Index of Diversity of 96·5, the second of 99·6; which shows how little his ideas on the scope of genera had changed in the 32 years between the two editions.

5. DISCUSSION

In Table 6 is a summary of the evidence brought forward in seven plant communities and seven animal communities. None of them is very small, as we have already explained that in the very small communities the chance of getting two species in one genus is too low to be the basis of discussion.

The table shows the number of genera observed in each natural sample and also the number of genera calculated on the assumption that it is a random sample from a larger flora or fauna which is arranged in a logarithmic series. It will be seen that in practically every case the observed number of genera is smaller than the calculated. It follows that the observed average number of species per genus is larger than that calculated on the assumption of a random selection from a logarithmic series, although much smaller than that in the larger area used for comparison.

These figures are calculated on the acceptance of the logarithmic series. If this is not accepted, it is still possible to make a practical demonstration that the same difference still exists between observed figures and a random sample. Bentham & Hooker's *British Flora* contains 1251 species. A series of numbers up to 1251 was typed on small cards and extremely well mixed. Three complete sets of random samples were then drawn from the complete set, each set being returned before the next set was selected. At intervals during the draw the number of genera represented in the species already selected was checked up. The results were as shown in Table 7. It will be seen again that in every case except one the observed natural number of genera is smaller than any of the three sets of mechanically randomized samples taken from the original flora. The amount of data at present available is insufficient for an overwhelming proof that the number of genera is really smaller, but at least it contains no evidence whatever that the number of genera is larger in a natural sample than in one selected absolutely independent of generic relationship, which would be expected if competition between species of the same genus was a major factor in determining survival.

From all this evidence it will be seen that a statistical treatment of the number of genera and species of different groups of animals or plants, in small and in large communities, indicates that the 'diversification' of species into genera is smaller in the small samples than in the larger. In other words, the smaller samples have fewer genera than would be expected in a random sample of the same number of species taken from the larger fauna or flora. This can only be interpreted as a natural selection—in the course of time—in favour of species in the same genus rather than against them.

It is possible to suggest reasons for this—for example, if one species in a genus is capable of survival in a given physical environment, it seems likely that other species in the same genus might be more likely to have a similar genetic make-up than species in another genus, and so might also have a good chance of survival.

If two or more species in the same genus are each capable of surviving in a particular ecological niche, the problem of their joint survival seems to be a question of the balance between the advantages of suitability to the physical environment and the disadvantages of the increased number of competitors with very similar habits—not only individuals of the same species, but of other species very closely related.

There are undoubtedly increasing difficulties to a species when the numbers increase beyond a certain level and in some cases the individuals of two closely related species might act almost as individuals of the same species, and so bring all the difficulties of increased numbers (e.g. competition for available food) without any compensating advantages.

The evidence at our disposal, however, seems to indicate that on an average, the advantages of increased suitability to the environment outweighs the disadvantages of any possible intensification of competition due to close relationship. Such increased competition possibly, or even probably, exists, but its effect cannot be estimated by study of the present type, as it is undoubtedly overshadowed by other factors acting in a reverse direction.

6. SUMMARY

1. Evidence of conditions being more favourable or less favourable to species of the same genus, as compared with species on different genera (intrageneric versus inter-generic competition), can be found in the relative number of species and genera in small and in large natural communities of animals or plants.

2. It is, however, insufficient to show that the average number of species per genus is smaller in the smaller communities than in the larger as this is a mathematical result of taking a smaller sample from a larger group. It is necessary to show that the proportion of genera to species in the smaller communities is *smaller* or *larger* than would have been expected in a randomized sample of the same number of species, selected without reference to generic relationships from the larger fauna or flora.

3. Evidence had previously been brought forward to show that in large groups of animals or plants the number of genera with 1, 2, and with 3 and more species are closely represented by the mathematical 'logarithmic series'. New evidence is here given to show that the same order exists in the genera and species of quite small ecological communities.

4. The logarithmic series has several mathematical properties of biological interest, including the possibility of calculating, for any population or sample, a factor known as the 'Index of Diversity' which is common to all random samples from a single population. It is a measure of the extent to which the species are grouped into genera, and it is inde-

pendent of the size of the sample. If the Index of Diversity is high, there are many genera in relation to the number of species; if the Index is low, there are fewer genera in relation to the number of species.

5. It thus becomes possible to compare the Index of Diversity in natural small or simple ecological communities, with that of the larger population from which these have been selected by nature. This has been done for a number of cases, including both animal and plant communities, and in every case from which significant results can be obtained the Index of Diversity in the small community is smaller than that of the larger fauna or flora.

6. The result therefore indicates that there are fewer genera in a small or simple community than would be expected in a sample of the same number of species selected at random—that is, independent of genera relationships—from the larger series. In other words, the evidence brought forward by this method indicates a selection by nature in favour of more than one species in the same genus rather than in favour of single species in different genera.

REFERENCES

Beare, T. H. & Donisthorpe, H. St J. K. (1904). 'Catalogue of British Coleoptera.' London.

Bedwell, E. C. (1945). 'The county distribution of the British Hemiptera-Heteroptera.' Ent. Mon. Mag. 81: 253–73.

Brenchley, W. E. (1924). 'Manuring of grass land for hay.' London.

Brenchley, W. E. & Adam, H. (1915). 'Recolonisation of cultivated land allowed to revert to natural conditions.' J. Ecol. 3: 193–210.

Chapman, V. J. (1934). 'Appendix II. Floral list.' In J. A. Steers 'Scolthead Island....' Cambridge. Pp. 229–34.

China, W. E. (1943). 'The generic names of the British Hemiptera-Heteroptera, with a check list of the British species.' In 'The generic names of British Insects,' Part 8: 217–316. London: R. Ent. Soc.

Donisthorpe, H. St J. K. (1939). 'A preliminary list of the Coleoptera of Windsor Forest.' London.

Elton, C. (1946). 'Competition and the structure of ecological communities.' J. Anim. Ecol. 15: 54–68.

Farren, W. (1936). 'A list of Lepidoptera of Wicken and the neighbouring Fens.' In J. S. Gardener, 'The natural history of Wicken Fen', Part III: 258–66. Cambridge.

Fisher, R. A., Corbet, A. S. & Williams, C. B. (1943). 'The relation between the number of species and the number of individuals in a random sample of an animal population.' J. Anim. Ecol. 12: 42–58.

Foster, A. H. (1937). 'A list of the Lepidoptera of Hertfordshire.' Trans. Herts. Nat. Hist. Soc. 20: 171–279.

Williams, C. B. (1944). 'Some applications of the logarithmic series and the Index of Diversity to ecological problems.' J. Ecol. 32: 1–44.

Williams, C. B. (1947). 'The logarithmic series and its application to biological problems.' J. Ecol. 34: 253–72.

PAPER 5

A Theoretical Ecological Model of Size Distributions among Species of Animals (1959)

G. E. Hutchinson and R. H. MacArthur

Commentary

S. K. MORGAN ERNEST

In G. E. Hutchinson and R. H. MacArthur's paper, the authors posit a niche-based theory for the generation of species-level body-size distributions. Their theory predicted that, across a broad taxonomic assemblage, there should be few species in the smallest size categories, an intermediate mode, and a slow decline in the number of species as one approached the largest members of the group, generating a very characteristic right-skewed distribution. They found support for this prediction in the size distribution of mammals in Europe. While most ecologists now would be hard pressed to recall the theory Hutchinson and MacArthur developed, the pattern they identified has endured as a macroecological mystery. In addition to mammals, many other taxonomic groups have since been shown to exhibit the same characteristic distribution (e.g., Van Valen 1973; May 1978; Dial and Marzluff 1988; Blackburn and Gaston 1994). Exceptions to this pattern appear to be generally related to either more narrowly restricting the group of species included—taxonomically or ecologically (e.g., Fenchel 1993; Gaston and Blackburn 1995; Maurer 1998)—or examining smaller land masses or spatial scales (e.g., Eadie et al. 1987; Brown and Maurer 1989; Brown and Nicoletto 1991). However, scientists are no closer to understanding the processes that could generate such a regular pattern across such evolutionarily distinct groups (Gaston and Blackburn 2000). Despite this, Hutchinson and MacArthur's pattern has become an important macroecological tool for many diverse areas of ecology: from paleoecology (Jablonski and Raup 1995) to spatial ecology (Brown and Nicoletto 1991).

Literature Cited

Blackburn, T. M., and K. J. Gaston. 1994. The distribution of body sizes of the world's bird species. *Oikos* 70:127–30.

*Brown, J. H., and B. A. Maurer. 1989. Macroecology: The division of food and space among species on continents. *Science* 243:1145–50.

*Brown, J. H., and P. F. Nicoletto. 1991. Spatial scaling of species composition: Body masses of North American land mammals. *American Naturalist* 138:1478–1512.

*Dial, K. P., and J. M. Marzluff. 1988. Are the smallest organisms the most diverse? *Ecology* 69:1620–24.

Eadie, J. M., L. Broekhoven, and P. Colgan. 1987. Size ratios and artifacts: Hutchinson's rule revisited. *American Naturalist* 129:1–17.

Fenchel, T. 1993. There are more small than large species? *Oikos* 68:375–78.

Gaston, K. J., and T. M. Blackburn. 1995. The frequency distributions of bird body weights: Aquatic and terrestrial species. *Ibis* 137:237–40.

———. 2000. *Pattern and process in macroecology*. Blackwell Science, Oxford.

Jablonski, D., and D. M. Raup. 1995. Selectivity of end-Cretaceous marine bivalve extinctions. *Science* 268:389–91.

Maurer, B. A. 1998. The evolution of body size in birds. I. Evidence for non-random diversification. *Evolutionary Ecology* 12:925–34.

*May, R. M. 1978. The dynamics and diversity of insect faunas. In *Diversity of insect faunas*, edited by L. A. Mound and N. Waloff, 188–204. Royal Entomological Society of London Symposium 9. Blackwell Scientific, Oxford.

Van Valen, L. M. 1973. Body size and numbers of plants and animals. *Evolution* 29:87–94.

Vol. XCIII, No. 869 The American Naturalist March-April, 1959

A THEORETICAL ECOLOGICAL MODEL OF SIZE DISTRIBUTIONS AMONG SPECIES OF ANIMALS

G. E. HUTCHINSON AND ROBERT H. MacARTHUR

Osborn Zoological Laboratory, Yale University, New Haven, Connecticut.
Edward Grey Institute of Field Ornithology, Oxford University,
England, and Division of Biology, University of
Pennsylvania, Philadelphia, Pennsylvania

If we examine the fauna of any area we find that the groups containing the largest numbers of species are for the most part groups of small animals, whereas large animals are represented mainly by genera containing a few species. This is in part doubtless an expression of the Eltonian pyramid of numbers and sizes; there will usually be few species of groups in which there are few individuals, so that rare, large carnivores, high in the pyramid, are unlikely to show much specific diversity. This however cannot be the sole explanation; in any fauna the carabid beetles for instance are likely to be richer in species than the carnivorous mammals and birds, while the largest terrestrial animals are herbivorous mammals in practically all non-insular undisturbed faunas. It would seem intuitively that the environment does not provide adequate room for a very large number of species of large animals while there is much more room for an abundance of smaller species. Moreover it is quite obvious in many cases that the large species roam about over a number of biotopes specific for smaller species. A large ungulate may require a water hole, a grazing area and some degree of cover; the wet marginal area of the water hole, the open grazing area and the cover might provide specific biotopes for three species of rodents.

In order to clarify somewhat these ideas, the following model has been developed.

STATEMENT OF THEORY

Consider an environment composed of an indefinite number of equal-sized mosaic elements of r different kinds arranged in a random way. Select from this environment n contiguous elements. In any selection there will be x kinds of elements where x can have the values 1, 2 . . . n (if $n < r$) or 1, 2 . . . r (if $n \geq r$). The probability of $p\{x, n\}$ of any value of $x \leq r$, for any value of n is given by*

$$p\{0, 0\} = p\{1, 1\} = 1$$
$$p\{0, 1\} = p\{1, 0\} = 0$$
$$p\{x, n\} = p\{(x-1)(n-1)\}\frac{r - x + 1}{r} + p\{x, (n-1)\}\frac{x}{r}$$

*This simple recursive form is the only practical one; the general solution involves a complicated generating function which can only be evaluated by more arithmetic than is involved in use of the recursive expression.

while the probability $p_c\{x, n\}$ of any qualitatively distinct combination will be

$$p_c\{x, n\} = 1\Big/\binom{r}{x} \cdot p\{x, n\}$$

An evaluation of the expression for $p\{x, n\}$, for $r = 5$ is given in table 1. While $p\{1, 1\}$ is unity and $p_c\{1, 1\}$ is $1/r$, for all higher values of n, $p\{1, n\}$ falls steadily. For values of x when $1 < x < r$, $p\{x, n\}$ first rises and then falls indefinitely; for $x = r$, $p\{n, x\}$ rises continuously, converging on unity as all the other probabilities fall.

TABLE 1

PROBABILITY TABLE, UNCERTAINTY AND NUMBER OF NICHES FOR $r = 5$, $n = 1 - 11$

	x = 1	2	3	4	5		
	$\binom{r}{x} = 5$	10	10	5	1	$\sum I_{x,n}$	No. niches $p_l = 0.001$
n = 1	1.000,00	...	...	...	...	1.61	5
2	0.200,00	0.800,00	...	...	...	2.66	15
3	0.040,00	0.480,00	0.480,00	...	...	3.10	25
4	0.008,00	0.224,00	0.576,00	0.192,00	...	3.13	30
5	0.001,60	0.096,00	0.480,00	0.384,00	0.038,40	3.08	28
6	0.000,32	0.036,68	0.345,60	0.499,20	0.115,20	2.84	26
7	0.000,06	0.016,13	0.231,17	0.537,60	0.215,04	2.57	26
8	0.000,01	0.006,50	0.148,38	0.522,55	0.322,56	2.22	26
9	0.000,00	0.002,60	0.092,93	0.477,39	0.427,08	1.94	21
10	0.000,00	0.001,04	0.057,32	0.419,08	0.522,56	1.68	17
11	0.000,00	0.000,42	0.035,02	0.358,19	0.606,38	1.45	16
...	...	...	...	...	...	...	
Asymptotic to	0	0	0	0	1	0	1

We now interpret the size of the selection n, as a quantity dependent on the size of the organism inhabiting the environment. We also assume that each qualitatively different combination of elements corresponds to a different niche* and that each niche can be occupied by a single species, the specific properties of the niche being defined by the existence of interfaces between different kinds of mosaic elements. Thus if $r = 5$, then for $n = 1$, $x = 1$ there are five possible elementary niches (a, b, c, d, e) each having an equal probability of occurrence ($p_c = 0.2$). For $r = 5$, $n = 2$, there will be ten possible composite niches (ab, ac, ad, ae, bc, bd, be, cd, ce, and de); the probability of the selection comprising one of these arrangements is .800, the probability that it comprises any specified one is 0.080.

*We do not intend to offer this model as a complete theory of the niche but rather consider that each kind of mosaic element contributes various ranges to the values of the parameters of the niches defined in an intensive way, as for example, in Hutchinson (1958).

INFORMATION-THEORETIC INTERPRETATION

We may now compute the uncertainty $\Sigma I_{x,n}$ associated with any array of niches for a given value of n, where

$$I_{x,n} = -\binom{r}{x} p_c\{x, n\} \ln p_c\{x, n\}$$

or

$$\Sigma I_{x,n} = -\Sigma p\{x, n\} \cdot \ln \cdot p_c\{x, n\}$$

where summation is over the rows of the table. A plot of $\Sigma I_{x,n}$ against n gives a curve rising to a maximum at some value of n slightly less than r and then falling asymptotically to zero. This relationship is a formal quantitative expression of the fact that when $n = 1$ we know we have an equal chance of getting any one of the r kinds of elements and can get nothing more; when n is very large we are almost certain to get every kind of r elements in every selection, but for intermediate values of n, near $n = r$, we shall have great qualitative diversity. If we assume for the purpose of argument that the niches can be analyzed into mosaic elements, we are more likely to get a great diversity of niches and so of species, when we are dealing with organisms to which some intermediate size, measured as n, can be associated, than when the measure n is very small or very large.

EMPIRICAL INTERPRETATION

In order to put the theory in a form that can be compared with actual observations we have to consider three further specific points in interpretation. If we assume that potentially each niche available can support but a single species, and that n is some monotonic function of the size of the individuals of that species, the quantity n is still an unsatisfactory measure of the function. This is easily seen when we consider high values of n for which $p_c\{x, n\}$ approaches unity, implying that a single species can be present. It would be quite unrealistic to suppose that for each ascending value of n this single species is replaced by another. It is reasonable to adopt a geometric measure of the size function. This may be obtained by taking the modal size n_m and then regraduating in terms of $\phi^q n_m$, obtaining ... $\phi^{-3}n_m$, $\phi^{-2}n_m$, $\phi^{-1}n_m$, n_m, ϕn_m, $\phi^2 n_m$, $\phi^3 n_m$ ... as the intervals of the new measure, which is taken to be a measure of length.

The value of ϕ may be obtained by considering the phenomenon of *character displacement.* (Brown and Wilson, 1956). As is pointed out elsewhere by one of us (Hutchinson, *forthcoming*) for birds and mammals this phenomenon suggests that a value of $\phi = 1.3$ would ordinarily prevent complete competition between species.

Now considering some fairly large value of q, for which $p_c\{x, n\}$ is almost unity, we should intuitively expect but a single species almost certainly to be present, as is indicated by the information-theoretic approach. If we accept the assumption that but a single species is present, this in fact implies that there is a limiting probability p_l such that when $p_c\{(r-1), n\} < p_l$, only the rth column is of any significance. Intuitively it seems clear that the limiting probability must be smaller for a large fauna than for a small fauna, but its value cannot be determined deductively. In any given case p_1 represents an empirical constant involved in fitting actual data to the theoretical model.

Another empirical constant is also needed, namely the value of r. Provided p_1 is sufficiently small (i.e. $p_1 \lesssim 0.01$ for $r \geq 3$) the numbers of available niches will increase to a maximum at n_m ($\simeq r$) and then after a shorter or longer plateau will decline irregularly to unity. The determination of r depends on the properties of the distribution up to the measure of modal size n_m. In table 2 the number of niches in the modal class (M), and the total number up to and including (A) or excluding (C) this class, is given for various values of r. We select from table 2 a value of r appropriate to

TABLE 2

r	$M = 2r - 1$	$A = (r + 2)2^{r-1} - r - 1$	$C = A - M$
1	1	1	0
2	3	5	2
3	7	16	9
4	15	43	28
5	31	106	75
6	63	249	186
7	127	566	339
8	255	1,271	1,016

the modal group and to the whole array of submodal groups. Since in dealing with part of a fauna, which will usually be necessary, there is always a possibility that niches can be apparently unfilled owing to competition of the smallest members of the part under consideration with the largest members of some other group (for example, small shrews and large carabid beetles), no great exactitude can be expected; all we can hope is to choose r in such a way to provide reasonably concordant values of M and B. Having obtained r, the appropriate probability table is constructed. Then by comparing the probabilities in the columns for $x < r$ with the observed distribution, an appropriate value of p_1 can be selected. Knowing p_1 we assume that all niches are filled if $p_1 < p_c\{x, n\}$, and that $\binom{r}{x}\frac{p_1}{p_c(x, n)}$ niches are filled if $p_1 > p_c\{x, n\}$. Fractional niches are rounded off to whole numbers or if $p_1 > 0.5\ p_c\{x, n\}$, discarded.

EMPIRICAL SIZE DISTRIBUTION CURVES AND THE FITTING OF AN ACTUAL CASE

For several groups that have been tried (birds of Massachusetts, dragonflies of N. E. United States, terrestrial Heteroptera of Britain, Chaetognatha of the world ocean) the size distribution by species appears to be a roughly symmetrical logarithmic distribution curve without the kind of tail predicted by the theory. The case of the dragonflies is particularly instructive, because the separate distributions for the Zygoptera and Anisoptera are skewed in opposite directions, but when both groups are considered together a close approach to symmetry is obtained (figure 1). The two suborders obviously compete in the region of the mode for the whole order. Only in the case of mammal faunas is there a distribution that is somewhat like that provided by the model. The ordinary symmetrical logarithmic curve we take to imply that a given type of organization implies a certain

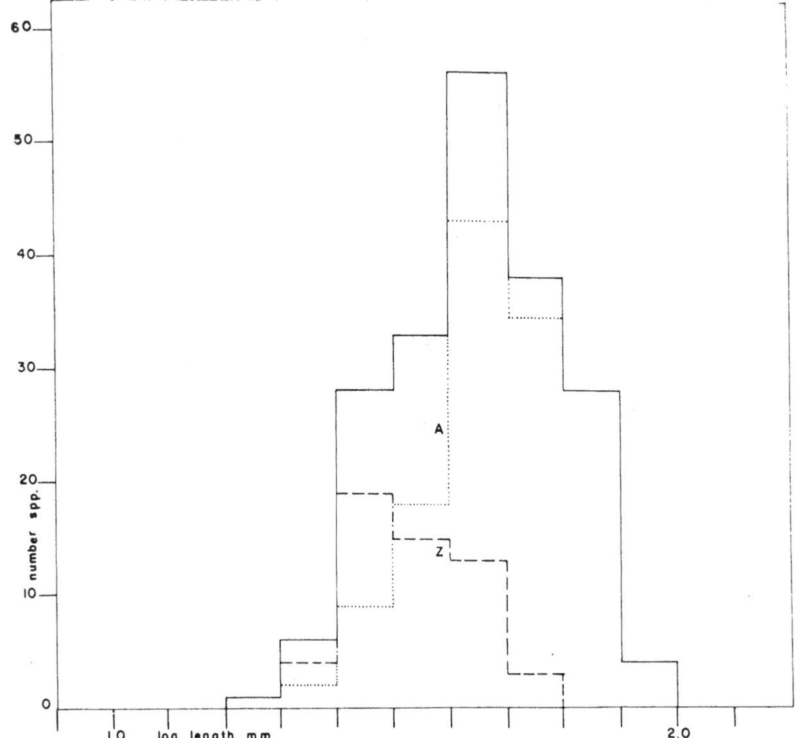

FIGURE 1. Size distribution of the species of Odonata of the northeastern United States (New England, New York, New Jersey, and Pennsylvania); A. Anisoptera separately Z. Zygoptera separately (data from Needham and Heywood (1929) for Zygoptera; Needham and Westfall (1955) for Anisoptera).

class of particularly appropriate niches and that as more extreme niches are occupied the amount of evolutionary displacement needed is about as easy to achieve, with a given organization, by a change in size of a given factor in either direction. In the case of the mammals the great diversity of structure appears to permit a far greater size range, even within the herbivorous level. We have proceeded with the analysis of two mammal faunas in the hopes that this class may provide a sort of model of the whole animal kingdom above a certain size. In selecting a fauna to analyze it is essential to choose a region small enough to exclude allopatric species, as these seem to be most developed in the families of small mammals. It is also essential that the small species be well known, and that the total fauna be sufficiently large.

The first example analyzed is that of the western part of continental Europe, roughly Belgium and northwest France, corrected for species known to have been introduced since the Middle Ages, or exterminated within historic time* in either Britain or the adjacent continent (figure 2). The data on sizes (head and body length) were obtained from Miller (1912) and from Barrett-Hamilton and Hinton (1911–1921). The value of r was taken as 3, the modal interval $n_m - \phi n_m$ as 70–91 cms, with $\phi = 1.3$, while p_l was taken as 0.005. In the second fauna (figure 3), namely that of the originally wooded part (excluding *Bison bison*) of Michigan (Burt, 1946) the same values for r and ϕ are used, but p_l is evidently lower and was taken as 0.00001. It is realized that with the rather large number of constants involved, any sufficiently asymmetrical distribution would accord with the theory at least as well as the mammal fauna in question. The interest of the method is solely in showing how a mammal fauna of the observed kind *could* be built up with a very small amount of initial diversity if a sufficient size range is permissible.

Interpretation of ϕn_m *and* n. In the case of the mammals of Michigan for which the home ranges (R) are approximately known, it appears (figure 4) that omitting the very vagrant marten, the area of the home range (Burt, 1946) varies roughly as the square of the head and body lengths (B). This suggests that the values of the linear size categories are really measures of categories of $\sqrt{R}$.

The interpretation of n is more difficult. The Michigan data suggest that n_m, corresponding to $n = 3$, also corresponds to $R = 0.38$ acres or 1538 square meters. Though nothing has been postulated about the dimensions of the unit of n, it is evident that in this case one unit of n corresponds to a length of $\sqrt{1538/3}$ or 13 meters; each mosaic element may be thought of as having the area of a square of this side length. Superficially it is by no means obvious from what we know about mammalian ecology, how biotopes would be built up of mosaics of three different kinds of elements of this size. Since the theory does not require that the elements be simple, but merely different and repeated, this objection is not fundamental.

The interpretation of randomness. The most obvious criticism that can be made of the model is that the different kinds of mosaic elements are distributed at random according to the postulates on which the model is built. If we consider a case in which the kinds of elements are arranged in a highly superdispersed manner, the distribution will have no effect on the qualitative number of choices when n is small, but will greatly increase the probability of more or less uniform arrangements when n is large. The superdispersion in fact, to a first approximation, lowers the value of p_l. It is perhaps reasonable to find that p_l is evidently lower in Michigan, which

**Rattus norvegicus* excluded; *Castor fiber*, *Ursus arctos*, *Canis lupus*, *Lynx lynx*, *Martes foina*, *M. martes*, *Felis silvestris* and *Sus scrofa* included.

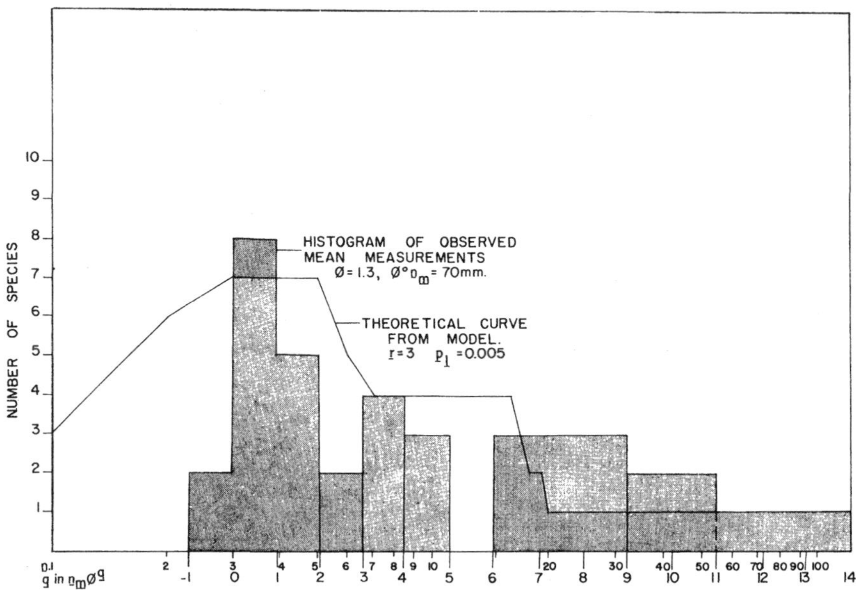

FIGURE 2. Mammals of western Continental Europe, with model curve for $r = 3$, $\phi = 1.3$ $p_1 = 0.005$.

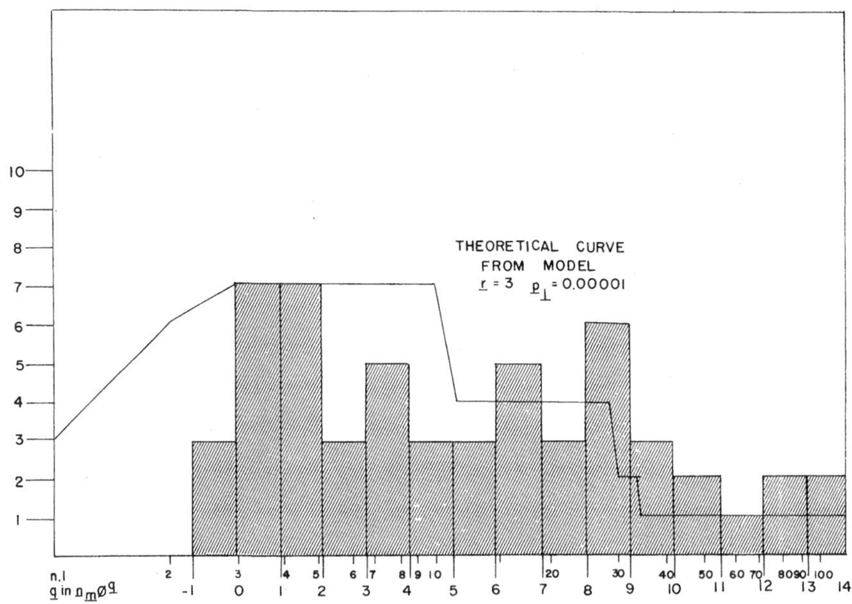

FIGURE 3. Mammals of Michigan, with model curve for $r = 3$, $\phi = 1.3$ $p_1 =' 0.00001$.

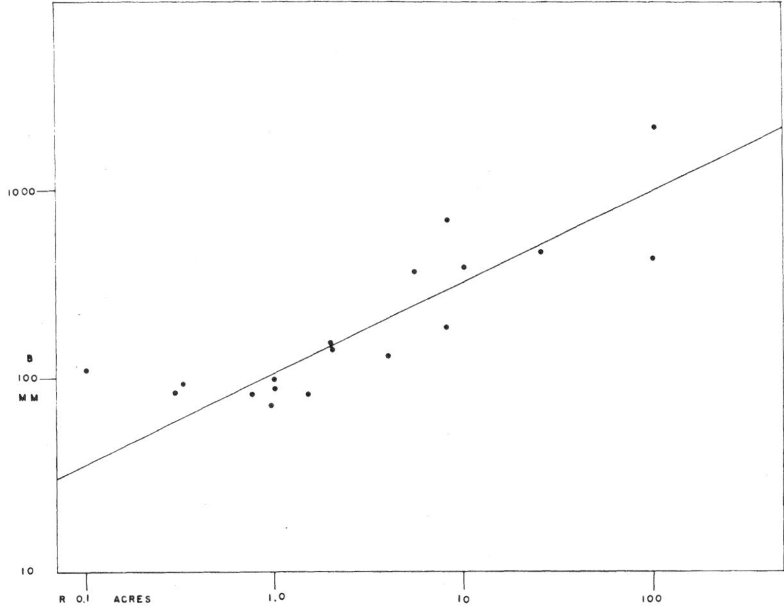

FIGURE 4. Relationship of logarithm of (B) mean head and body length (mm.) to logarithm of (R) home range (acres) for mammalian species in Michigan for which home-range is known; *Martes americana* (length 495 mm., range 6000–9000 acres) omitted. Data from Burt (1946). The straight line is fitted by eye with a slope of 1 : 2.

represented a much less disturbed habitat during the early 19th century than France and Belgium have probably provided at any time since the late Neolithic. It is quite likely that in reconstructing the natural western European fauna more large extinct species such as the aurochs, and possibly the tarpan, should have been added. Extinction of such large animals may in fact be attributed in a general way to a human randomization of the habitat.

GENERAL APPLICATION

In a certain sense the application of the theory to mammals tells us little that we did not know before, and is therefore trivial. The significance of the approach is in the fact that the mammalia, being a highly diversified group, may perhaps provide a model of the terrestrial fauna of a locality as a whole. Such an approach, however, involves a new difficulty. The postulated size of a mammalian mosaic element is so large that it must greatly exceed the mosaic elements for small insects. The size of the mosaic elements is indeed likely to be a continuous function of the size of the organisms involved. It is possible that some theoretical modification of the model involving the size measure might be developed, though at present we have not seen how to do this. In its present form the theory does at least suggest one possible if partial explanation of the way that sizes seem to be distributed throughout whole faunas, and suggests the desirability of studying such distributions more intensively than in the past. It might be possible to gain further information by an intensive examination of the incidence of related sympatric species of similar size in various groups of large and small animals in different kinds of environment.

SUMMARY

Even within a given level of the food web there appear to be fewer species of large than of small animals. A model can be constructed in which the properties of the niches of different species are defined by the numbers of kinds of interface between a limited number of sorts of randomly distributed environmental mosaic elements. This model implies few very small species, rapid increase in number of species up to a modal size and a slow decline in number, ideally asymptotic to unity, as the size increases. Structurally uniform groups of animals such as the Odonata do not show this distribution, but in mammal faunas it is approached. It is suggested that if a really complete faunistic list for a given biotope could be constructed the size distribution by species would approximate the form given by the model.

ACKNOWLEDGMENT

This work was in part done during the tenure of a John Simon Guggenheim Memorial Fellowship by one author and a Natural Science Foundation Fellowship by the other.

LITERATURE CITED

Barrett-Hamilton, G. E. H., and M. A. C. Hinton, 1911-1921, A history of British mammals. Vol. 2. Gurney and Jackson, London.

Brown, W. L., and E. O. Wilson, 1956, Character displacement. Systematic Zoology 5: 49–64.

Burt, W. H., 1946, The mammals of Michigan. University of Michigan Press.

Hutchinson, G. E., 1958, Concluding remarks. Cold Spring Harbor Symp. Quant. Biol. 22: 415–427.

1959, Homage to Santa Rosalia *or* Why are there so many different kinds of animals? Amer. Nat. (in press).

Miller, G. S., 1912, Catalogue of the mammals of western Europe. British Museum, London.

Needham, J. G., and H. B. Heywood, 1929, A handbook of the dragonflies of North America. Charles C Thomas, Springfield, Illinois, and Baltimore, Maryland.

Needham, J. G., and M. J. Westfall, 1955, A manual of the dragonflies of North America (Anisoptera). University of California Press.

PAPER 6

Aggregation, Variance and the Mean (1961)

L. R. Taylor

Commentary

FORD BALLANTYNE IV

L. R. Taylor's first paper describing the mean-variance scaling of animal abundance, were it to be published today, would certainly be classified as macroecology. Using a purely statistical approach, he characterized the spatial aggregation of individuals in populations for a taxonomically diverse set of species over a range of spatial scales. Although his initial aims were to propose a general metric for describing the aggregation exhibited by natural populations and to present a general data transformation for analysis of variance, this paper has had significant influence on how ecologists view populations and abundance.

Taylor's synoptic view of populations identified a regular pattern, the scaling of variance in abundance with mean abundance, which emerges only by comparing statistics of abundance across populations. Virtually all population biologists prior to the publication of this paper studied single populations and were thus unaware that a common set of processes and constraints appeared to be regulating abundance for all populations of a single species, as suggested by the linear scaling for each panel in figure 1. Scaling exponents vary considerably, from approximately one to over two, across species, indicating that different species exhibit differing degrees of spatial aggregation, likely the result of unique combinations of processes that tend to aggregate and disperse individuals. Thus, parameters that specify what is now known as Taylor's power law, the exponent and the normalization constant, can be used for interspecific comparisons of population-level variability. The ubiquity of Taylor's power law begged for an explanation and prompted the development of rich and diverse theory attempting to explain the general phenomenon of mean-variance scaling.

Taylor's work has had two major impacts on the field of ecology and the development of macroecology. First, he illustrated the power of a statistically based comparative approach, a hallmark of the macroecological approach, for understanding interspecific variability in abundance across space and through time. And second, he demonstrated how scaling relationships can be used to study phenomena across scales. This paper is one of the first examples in the ecological literature that highlights how inference about local dynamics can be made from a macroecological pattern.

Reprinted by permission from Macmillan Publishers Ltd: *Nature* 189:732–35, copyright 1961.

732 NATURE March 4, 1961 VOL. 189

that the thyroid hormones owe their ability to moderate the process of oxidative metabolism to their high affinity for free electrons is supported by the following circumstantial evidence: the biological activities of trichloro-, tribromo- and tri-iodo-thyronine[13] are respectively: 1, 37 and 420. The electron affinities of chloro, bromo- and iodo-benzene are respectively: 1, 6, 370. Although the structures of the thyronines are very different from those of the halogen-substituted benzenes, the comparison is justified by the fact that the electron affinity conferred by halogen substitution is frequently independent of the nature of the molecule substituted; provided, of course, no other electrophores are also present. The resemblance between the biological activity of the thyroid hormones and that of dinitrophenol is further evidence in support of this interpretation.

Another reaction in biochemistry thought for some time to involve electron transfer is that between carcinogenic hydrocarbons and protein. A stumbling block to the understanding of the nature of this reaction has been the persistently held assumption that the hydrocarbon is the donor and the protein the acceptor of electrons; this has been difficult to reconcile with theory or observation. It was suggested tentatively[14,15] that these difficulties could be resolved if the hydrocarbon were in fact the acceptor of electrons. The unpopularity of the notion of a hydrocarbon so behaving in that way seems to have prevented the general acceptance of the suggestion. However, as shown in Table 1, complex hydrocarbons are very ready to function as electron acceptors. Because of their slight volatility, it has not yet been possible to test the carcinogenic hydrocarbons themselves, although it is of interest that stilbene and azobenzene both are able to capture electrons and give rise to simple derivatives with carcinogenic activity.

In general, where a similar toxic action is shared by compounds differing only in substitution with halogen- or nitro-groups, and where a disturbance of electron transport is thought to occur, then the toxicity of the compounds could be due to their ability to trap electrons. One important difference between normal intermediates which can trap electrons and the toxic substances could be in the type of reaction which takes place. Trapping of electrons is potentially most harmful when it is irreversible; in these circumstances, the potential energy of the captured electron may be completely removed from the system. Furthermore, if dissociative capture occurs, and it is likely to occur with halogen- and nitro-compounds, then a highly reactive and potentially harmful free-radical may be formed. In addition to those classes of toxic agent already mentioned, other substances which may function in a similar way include halogenated insecticides, inhibitors of plant growth, certain fungistatic iodo- and nitro-compounds[16] and certain antibiotics such as chloramphenicol. Some of the limitations of the evidence provided from measurements of electron affinity arise from the need to conduct the observations on volatile derivatives in the gas phase. There is no reason in principle why similar measurements to those described should not be made in the liquid or even solid phases. The solution of the practical problems in this direction will greatly extend the range of measurement of this potentially important molecular property, and perhaps provide the evidence necessary for the confirmation or denial of the hypotheses outlined in this article.

I am indebted to my colleagues T. Nash and N. L. Gregory for the stimulus of their suggestions.

1 Townsend, J. S., *Electrons in Gases* (Hutchinsons, London, 1947).
2 Massey, H. S. W., *Negative Ions* (Camb. Univ. Press, London, 1950).
3 Pritchard, H. O., *Chem. Revs.*, **52**, 529 (1953).
4 Lovelock, J. E., and Lipsky, S. R., *J. Amer. Chem. Soc.*, **82**, 431 (1960).
5 Lovelock, J. E., *Nature*, **187**, 49 (1960).
6 Lovelock, J. E., *Nature*, **188**, 401 (1960).
7 Szent-Györgyi, A., *Science*, **93**, 609 (1941).
8 Bradley, D. F., and Calvin, M., *Proc. U.S. Nat. Acad. Sci.*, **42**, 710 (1956).
9 Platt, J. R., *Science*, **129**, 372 (1959).
10 Green, D. E., *Disc. Farad. Soc.*, **27**, 206 (1959).
11 George, P., and Griffith, J. S., in *The Enzymes*, ed. by Boyd, P. D., *et al.*, 397 (Academic Press, New York, 1959).
12 Rabinowitch, E., *Disc. Farad. Soc.*, **27**, 161 (1959).
13 Mussett, M. V., and Pitt-Rivers, R., *Metabolism*, **6**, 18 (1957).
14 Mason, R., *Disc. Farad. Soc.*, **27**, 129 (1959).
15 Nash, T., *Nature*, **179**, 868 (1957).
16 Grove, J. F., and McGowan, J. C., *Chem. and Indust.*, 647 (1959).

AGGREGATION, VARIANCE AND THE MEAN

By L. R. TAYLOR
Department of Entomology, Rothamsted Experimental Station, Harpenden, Herts

IN populations where the individuals are distributed at random, that is, are independent of each other, the variance (s^2) at each population density is equal to the mean (m). Individuals in natural populations are not, however, independent of each other; mutual attraction leads to aggregation which makes variance more than the mean ($s^2 > m$), and mutual repulsion leads to regularity which makes variance less than the mean. Hence, in Nature, true randomness is only one of a continuous series of possible distributions (both in a spatial and a statistical sense) from regular to highly aggregated, and is consequently rare.

This property of natural populations, to show some degree of aggregation, is commonly thought to be highly specific; for example, soil zoologists know that leather-jackets are 'more highly aggregated' than wireworms, yet the meaning of this is intangible, because no consistent measure of aggregation is at present available. In addition, many kinds of statistical analysis require that the variance should be made independent of the mean and there is no rational transformation available covering all grades of aggregation. Aggregation has been measured most widely using k in the expression:

$$s^2 = m + km^2 \qquad (1)$$

derived from the negative binomial distribution. Unfortunately k is not always independent of m (ref. 5), and it is plainly desirable that an index of population structure should be the same at different population densities, unless some actual change in behaviour is involved.

In all the sets of samples I have examined, variance appears to be related to the mean by a simple power law ; variance is proportional to a fractional power of the mean :

$$s^2 = am^b \qquad (2)$$

where a and b are characteristic of the population in question. This is most clearly demonstrated by a log × log plot of s^2 on m as in Fig. 1, which includes some classic population surveys. Table 1 gives a and b for these and other populations, namely, virus lesions, macro-zooplankton, worms and symphylids in soil, insects in soil, on plants and in the air, mites on leaves, ticks on sheep and fish in the sea.

Table 1

	Name	Site and sample	Range of m	Range of s^2	N	a	b	Transformation	Observer (refs. in brackets)
1	Shellfish on seashore, *Tellina tenuis* da Costa, Eulamellibranchiata : Mollusca	Sand, 63 units, various sizes	0·72–45·7	0·49–8·0	5	0·50	0·70	$x^{0\cdot65}$	Holme (13)
2	European chafer larvæ, *Amphimallon majalis* Raz. (= *Melolontha melolontha* L.), Coleoptera : Insecta	Pasture soil, 25 units, each 1 ft. sq.	0·20–9·72	0·26–14·93	75	1·15	1·07	$x^{0\cdot46}$	Burrage and Gyrisco (6)
3	Flying insects, various orders : Insecta	Open air, 16–104 units aerial density	1·9–238	1·7–606	24	1·0	1·17	$x^{0\cdot41}$	L. R. Taylor (unpublished)
4	Wireworms, *Agriotes* spp. mainly *obscurus*, Coleoptera : Insecta	Grassland soil 20 units, 4 in. cores	0·20–4·65	0·40–17·80	2,272	2·75	1·19	$x^{0\cdot40}$	Yates and Finney (23)
5	Wireworms, *Agriotes* as above	Arable land soil 20 units, 4 in. cores	0·20–4·65	0·39–22·50	525	2·85	1·26	$x^{0\cdot37}$	Yates and Finney (23)
6	Wireworms, *Limonius* spp., Coleoptera : Insecta	Arable land soil, 175 units, each 1 ft. sq.	0·39–10·89	0·58–60·42	24	2·0	1·33	$x^{0\cdot33}$	Jones (5, 14)
7	Gall midge larvæ, *Jaapiella medicaginis* (Rub.), Diptera : Insecta	Lucerne field soil, 10 units, 4 in. cores	0·22–5·6	0·19–13·82	1[illegible]	1·3	1·33	$x^{0\cdot33}$	Heath (12)
8	Spruce budworm larvæ, *Choristoneura fumifera* (Clem.), Lepidoptera : Insecta	Fir foliage, 25 units, larvæ/twig	0·48–15·04	0·343–51·8[illegible]	[illegible]	1	1·40	$x^{0\cdot30}$	Waters (4, 22)
9	Virus lesions, tobacco necrosis virus	Bean leaves, 4 units, lesions/half leaf	15·38–237·56	65·13–3,265·[illegible]	12[illegible]	1	1·40	$x^{0\cdot30}$	Kleczkowski (5, 15)
10	Colorado beetle adults, *Leptinotarsa decemlineata* Say., Coleoptera : Insecta	Potato foliage, 2,304 counts : insects/2 ft. row	1·0–12·1	1·4–37·7	16 16 5	1	1·48	$x^{0\cdot26}$	Beall (2, 3, 5, 19)
11	Japanese beetle larvæ, *Popillia japonica* New., Coleoptera : Insecta	Soil, 10,000 units each 1 ft. sq.*	2·76–1,914	5·90–210,000	36	1·3	1·52	$x^{0\cdot24}$	Fleming and Baker (10)
12	Macro-zooplankton	Water, net collection, slide count, 10 areas	95–1,750	1,444–194,480	4	1·0	1·57	$x^{0\cdot21}$	Littleford, Newcombe and Shepherd (16)
13	Macro-zooplankton	As above, 50 areas	93–1,822	169–18,769	4	0·14	1·57	$x^{0\cdot21}$	Littleford, Newcombe and Shepherd (16)
14	Ticks, *Ixodes ricinus* L., Acarina : Arachnida	Sheep, 20–86 units, ticks/sheep	0·84–85·6	2·5–2,292	10	1·0	1·66	$x^{0\cdot17}$	Milne (18)
15	Enchytraeid worms, mainly *Fridericia bisetosa* (Lev.), Enchytraeidae : Annelida	Pasture, 60–150 units, 3·6 cm. cores	40·8–381·5	441–29,584	4	1·0	1·66	$x^{0\cdot17}$	Nielsen (20)
16	Corn borer larvæ, *Pyrausta nubilalis* (Hubn.), Lepidoptera : Insecta	Maize stalks, 2 stage†	6·7–970	88·5–180,220	1,054	3·0	1·66	$x^{0\cdot17}$	Meyers and Patch (17)
17	Thrips, *Thrips imaginis* (Bagnall), Thysanoptera : Insecta	Rose flowers, 20 units, thrips/rose	20·6–137·5	223–7,900	16	1·0	1·80	$x^{0\cdot10}$	Davidson and Andrewartha (8)
18	Leather-jackets, *Tipula* spp., Diptera : Insecta	Soil, 2 units, Nos./sq. ft.	3·10–63·50	3·5–409·0	36	0·2	1·85	$x^{0\cdot07}$	Bartlett (1, 5)
19	Earthworms, all stages, *Allolobophora chlorotica* (Sav.), Oligochaeta : Annelida	Grassland, 4 units, 18 in. sq.	2·8–71·3	2·9–1,410	54	0·2	2·00	log x	Gerard (11)
20	Red spider mite, eggs and adults, *Metatetranychus ulmi* (Koch), Acarina : Arachnida	Apple leaves, 20 units, mites/leaf	8·5–216	90–55,100	162	0·4	2·19	$1/x^{0\cdot10}$	Daum and Dewey (7)
21	Haddock, *Melanogrammus aeglefinus*, Gadidae : Pisces	Sea, 4–47 units, Nos./trawl	4·09–288	39·1–1,239,500	15	1·0	2·35	$1/x^{0\cdot18}$	C. C. Taylor (5, 21)
22	Earthworms, all stages, *Allolobophora caliginosa* (Sav.), Oligochaeta : Annelida	Grassland, 4 units, 18 in. sq.	4·3–44·8	4·0–500	42	0·05	2·54	$1/x^{0\cdot27}$	Gerard (11)
23	Symphyla, *Symphylella* spp., Symphyla : Myriapoda	Various soils, 60–120 units, 2½ in. cores	4·5–31·8	0·49–690	5	0·06	2·75	$1/x^{0\cdot38}$	Edwards (9)
24	Symphyla, *Scutigerella* spp., Symphyla : Myriapoda	Various soils, 60–140 units, 2½ in. cores	1·3–31·4	0·64–1,250	6	0·035	3·08	$1/x^{0\cdot54}$	Edwards (9)

* Units combined to cover areas varying in size from 1 sq. ft. to 100 sq. ft. † Larvæ/stalk × infested stalks/100 : this introduces some bias in the variances.

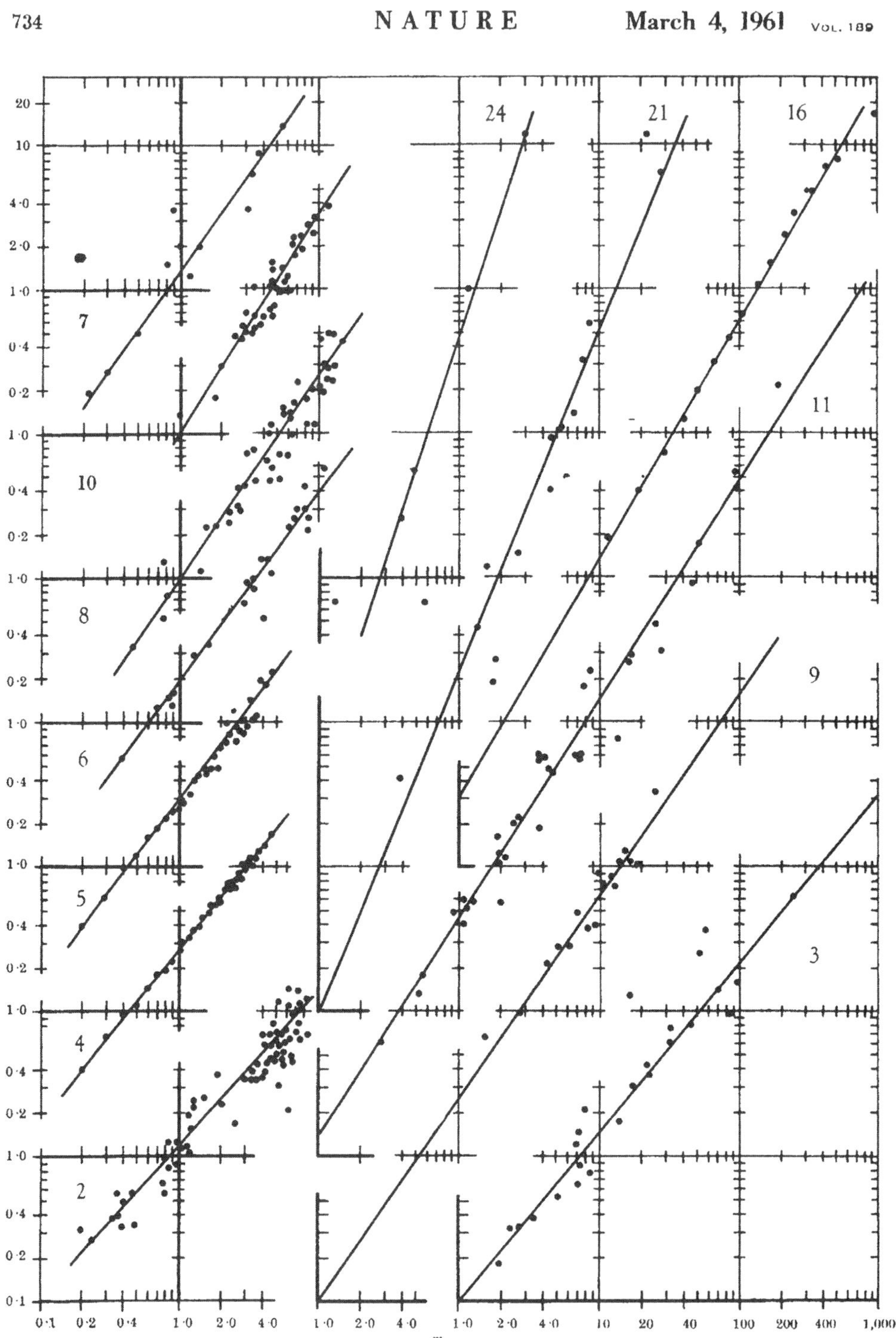

Fig. 1. Variance × mean on log × log scale for material numbered as in Table 1. The axes (heavy lines) are staggered to accommodate as much material as possible

No. 4766 March 4, 1961 NATURE 735

a is evidently largely, possibly wholly, a sampling or computing factor, depending upon the size of sampling unit and on which estimate of variance is used; this was not clearly stated in some of the analyses. b appears to be a true population statistic, an 'index of aggregation' describing an intrinsic property of the organisms concerned and with a continuous graduation from near-regular ($b \to 0$), through random ($b = 1$) to highly aggregated ($b \to \infty$).

To use the analysis of variance technique raw data must first be transformed to make s^2 independent of m. Using expression (2), the general transformation:

$$f(x) = Q\int m^{-b/2}\,dm \qquad (3)$$

where Q is a constant, is obtained. Integration gives a square root transformation when $b = 1$ and, logically, a logarithmic transformation when $b = 2$, as are commonly used; for highly aggregated populations ($b > 2$) it gives a negative fractional power, for example, a reciprocal, as is appropriate to the powerful transformation required.

The power law appears to hold good down to low densities ($m < 1$) in the material examined here. This implies that populations aggregated at high density tend to become regular when density diminishes, and vice versa; the law may break down eventually or, alternatively, perhaps the concept of aggregation ($s^2 > m$) is inappropriate at these low densities. In any event samples consisting mainly of 0's and 1's are likely to need special statistical treatment.

1 Bartlett, M. S., *J. R. Statist. Soc.*, Supp., **3**, 185 (1936).
2 Beall, G., *Biometrika*, **30**, 422 (1939).
3 Beall, G., *Biometrika*, **32**, 243 (1942).
4 Bliss, C. I., *Proc. Tenth Int. Congr. Entom.*, 1956, **2**, 1015 (1958).
5 Bliss, C. I., and Owen, A. R. G., *Biometrika*, **45**, 37 (1958).
6 Burrage, R. H., and Gyrisco, G. G., *J. Econ. Ent.*, **49**, 179 (1956).
7 Daum, R. J., and Dewey, J. E., *J. Econ. Ent.*, **53**, 892 (1960).
8 Davidson, J., and Andrewartha, H. G., *J. Anim. Ecol.*, **17**, 193 (1948).
9 Edwards, C. A., *Ent. Exp. and App.*, **1**, 308 (1958).
10 Fleming, W. E., and Baker, F. E., *J. Agric. Res.*, **53**, 319 (1936).
11 Gerard, B. M., Ph.D. thesis, Univ. London (1960).
12 Heath, G. W., Ph.D. thesis, Univ. London (1956).
13 Holme, N. A., *J. Mar. Biol. Assoc.*, **29**, 268 (1950).
14 Jones, E. W., *J. Agric. Res.*, **54**, 123 (1937).
15 Kleczkowski, A., *Ann. App. Biol.*, **36**, 139 (1949).
16 Littleford, R. A., Newcombe, C. L., and Shepherd, B. B., *Ecology*, **21**, 309 (1940).
17 Meyers, M. T., and Patch, L. H., *J. Agric. Res.*, **55**, 849 (1937).
18 Milne, A., *Ann. App. Biol.*, **30**, 240 (1943).
19 Milne, A., *Biometrics*, **15**, 270 (1959).
20 Nielsen, C. O., *Oikos*, **5**, 167 (1954).
21 Taylor, C. C., U.S. Fish Wildlife Service, *Fishery Bull.*, **54**, 145 (1953).
22 Waters, W. E., *For. Sci.*, **1**, 68 (1955).
23 Yates, F., and Finney, D. J., *Ann. App. Biol.*, **29**, 156 (1942).

SURFACE ORIENTATION AND FRICTION OF GRAPHITE, GRAPHITIC CARBON AND NON-GRAPHITIC CARBON

By Dr. J. W. MIDGLEY and D. G. TEER

Mechanical Engineering Laboratory, English Electric Co. Ltd., Whetstone, near Leicester

THE effect, if any, of surface orientation on the friction of various carbons has not yet been clearly established. Indeed, previous reports on the orientation developed by rubbing are conflicting[1-4]. To clarify the position and hence to improve our understanding of the mechanism of the friction of wear of these materials, an attempt has been made to provide fresh evidence, notably by a comparison of the behaviour of a carbon with a very low degree of graphitization[5] with graphite and graphitic carbon. These materials have been rubbed unidirectionally and the surface orientation studied by electron diffraction.

When graphite or highly graphitic carbon is rubbed unidirectionally against itself, mild steel or paper, the surface crystallites rapidly become orientated with their (00l) basal planes at a small angle (5–10°) to the surface with the basal plane normals tilted against the direction of motion of the opposing surface, as shown by the electron diffraction pattern of Fig. 1. The deviation of this orientation from the parallel orientation previously reported[1-3] is small, but nevertheless significant, since, as the coefficient of friction is in the range 0·1–0·2 the present texture is consistent with a compression texture with its axis parallel to the resultant of the normal reaction and the friction forces. Similar textures have been recently reported by Porgess and Wilman[6] to be developed by graphite when abraded with emery.

The following factors could contribute to the formation of this texture: (a) basal plane slip; (b) the plate-like shape of the crystallites; (c) preferential removal of crystallites in other orientations.

Concerning the first of these, an electron microscope study has confirmed that, as previously reported

Fig. 1. Electron diffraction pattern of rubbed graphitic carbon. A, direction of motion of opposing surface; B, basal plane normals make angle of 5–10° to the surface normal

PAPER 7

Bioenergetics and the Determination of Home Range Size (1963)

B. K. McNab

Commentary

JAMES H. BROWN AND JOHN L. GITTLEMAN

In the early 1930s independent studies by M. Kleiber (1932) and S. Brody and R. C. Procter (1932) showed that the rate of metabolism, or whole-organism energy use, of mammals and birds scales with approximately the ¾ power of body size. Subsequent studies confirmed the generality of this finding for many animal groups. This seminal finding set the stage for Brian McNab's creative and pioneering study of home range. If the rate of energy demand increases as the ¾ power with increasing body size, how does the area of space used by an animal to supply its energy needs scale? McNab assembled the data, did the allometric analysis, and obtained the answer that the home range area of mammals also scales as approximately the ¾ power of body mass.

Several aspects of McNab's study are noteworthy. First, it is explicitly macroecological. It uses data from multiple species to perform a broad quantitative comparative analysis; obtains these data from the literature, using multiple studies by different original investigators; and exploits the power of allometry, using the orders of magnitude variation in body size, and applying regression analysis to logarithmically transformed data to fit the data with a power function.

Second, McNab's paper explores multiple sources of variation around the fitted scaling relationship in insightful ways. It recognizes that in such a study there will likely be variation due to measurement error, differences in methodology among investigators, uncontrolled variables such as environmental productivity and habitat type, and apparently stochastic differences among generally similar mammals. It points out that such sources of variation are unlikely to obscure important patterns, because they are small compared to the orders of magnitude scale of the allometric analysis. The variation is sufficiently large, however, that the slope of the fitted regression, 0.63, is actually not statistically different from 0.75. Most importantly, however, the analysis of variation reveals major differences due to foraging mode and diet. Insightfully, McNab contrasts "hunters" that forage for rare, dispersed energy-rich foods, and therefore include not only insectivores and carnivores but also herbivores that feed primarily on seeds and fruit, with "croppers" that feed on abundant green vegetation. Fitting regressions separately to data for these two categories gives roughly parallel scaling lines separated by almost an order of magnitude. So the "hunters" require substantially larger areas than the "croppers" to supply sufficient energy to meet their metabolic needs.

Third, like many seminal works, McNab's study is more noteworthy for opening new areas of research than for providing the definitive study. Indeed, McNab's paper was followed by later studies that used more and better data to more precisely quantify the scaling of home range area. Especially noteworthy is T. W. Schoener's (1968) analysis of territory size in birds and mammals, which explored many

From *American Naturalist* 97:133–40.

of the same themes but found that the scaling exponent is approximately 1 rather than the ¾ suggested by McNab. Subsequent research has shown that such linear scaling holds for mammals as well as birds. The mismatch between the linear scaling of the home range area that supplies the food and the ¾-power scaling of metabolic rate and energy needs has generated considerable fruitful research, including the now well-documented increase in size of social groups with increasing body size.

Finally, this paper is noteworthy for extending an allometric and metabolic perspective to the quantitative analysis of ecologically relevant traits. The vast majority of previous studies focused on the scaling of basal metabolic rate and related physiological traits, such as heart and respiratory rates. The implication of McNab's study was that the scaling of energy use with body size could powerfully influence many aspects of ecology. Indeed, within a few years there followed influential papers applying a metabolic perspective to the allometric scaling of population growth (e.g., Fenchel 1974), life-history traits (e.g., Gould 1977), and ecosystem organization (e.g., Sheldon and Parsons 1967). The two decades of ecological allometry since McNab's pioneering study were synthesized in Peters's (1983) influential book on the ecological implications of body size.

Literature Cited

Brody, S., and R. C. Procter. 1932. XXIII. Relation between basal metabolism and mature body weight in different species of mammals and birds. In *Growth and development with special reference to domestic animals,* 89–101. University of Missouri Agricultural Experimental Station Research Bulletin, no. 166. University of Missouri, Columbia.

*Fenchel, T. 1974. Intrinsic rate of natural increase: The relationship with body size. *Oecologia* 14:317–26.

Gould, S. J. 1977. *Ontogeny and phylogeny.* Harvard University Press, Cambridge, MA.

Kleiber, M. 1932. Body size and metabolism. *Hilgardia* 6:315–53.

Peters, R. H. 1983. *The ecological implications of body size.* Cambridge University Press, Cambridge.

*Sheldon, R. W., and T. R. Parsons. 1967. A continuous size spectrum for particulate matter in the Sea. *Journal of the Fisheries Research Board of Canada* 24:900–925.

Schoener, T. W. 1968. Sizes of feeding territories among birds. *Ecology* 49:123–41.

THE
AMERICAN NATURALIST

Vol. XCVII May-June, 1963 No. 894

BIOENERGETICS AND THE DETERMINATION OF HOME RANGE SIZE

BRIAN K. McNAB

Department of Biology, University of Florida, Gainesville, Florida

All animals are limited in the extent of their daily wanderings. This is notably true of mammals, for which many estimates of home range exist. It is obvious that the size of a mammal will greatly affect the maximum area that can be covered and, therefore, will influence the size of the home range. But body size has another, independent influence: larger species must collect more energy to supply their requirements than small species. Generally, a large energy demand will necessitate a large area for food gathering, unless the food exists in superabundance. Many other considerations may also alter the size of the home range (Brown, 1962), but the object of this paper is to examine only the relation between the size of home range and body size, since body size would appear at first glance to be the most important factor in the determination of home range size.

DATA

Data from mammals are used in this paper, since this group has had the most active attention, both in terms of estimates of home range size and in measurements of body weight (which is used in this study to indicate body size). The data that form the basis for this analysis were principally taken from Burt (1948) and Fitch (1958), but were supplemented by data from Mohr (1947), Blair (1951), Calhoun and Casby (1958), and Ingles (1961). These papers do not represent a complete survey of the literature, but they contain much of the data on the smaller species of mammals.

The methods used for the estimation of home range size vary greatly, and, more importantly, give different results, as was pointed out by Mohr (1947), Hayne (1949), Calhoun and Casby (1958), Fitch (1958), and Brown (1962). I shall not enter into a discussion of the limitations of the methods; nevertheless, when comparisons are made between the results of different methods (Mohr, 1947), the individual estimates for one species are usually of the same order of magnitude so that the errors involved are probably relatively small over a wide range in body weight. The values for home range are in acres and those for body weight are in kilograms (table 1).

TABLE 1.

The size of home range and weight in some mammals

Species	Weight (kilograms)	Home Range (acres)	Authority
"Hunters"			
Didelphis virginiana	3.63	25	Burt
Didelphis virginiana	3.63	28.9	Fitch
Sorex vagrans	0.005	0.09	Ingles
Blarina brevicauda	0.018	0.5	Burt
Blarina brevicauda	0.018	0.87	Fitch
Mustela rixosa	0.05	2	Burt
Procyon lotor	10.89	386	Fitch
Vulpes fulva	5.45	25	Mohr
Tamias striatus	0.09	1.6	Burt
Tamiasciurus hudsonicus	0.185	6.3	Burt
Sciurus niger	0.953	25	Burt
Sciurus niger	0.953	13.7	Fitch
Glaucomys volans	0.061	4	Burt
Peromyscus maniculatus	0.02	1.85	Burt
Peromyscus maniculatus	0.02	0.74	Fitch
Peromyscus leucopus	0.02	0.4	Fitch
Peromyscus leucopus	0.02	0.25	Burt
Reithrodontomys megalotis	0.01	0.58	Calhoun and Casby
Reithrodontomys megalotis	0.01	0.52	Fitch
Reithrodontomys montanus	0.01	0.50	Fitch
Zapus hudsonicus	0.02	0.90	Fitch
Zapus hudsonicus	0.02	2.00	Fitch
Napaeozapus insignis	0.024	1.5	Burt
Mus musculus	0.018	0.41	Fitch
"Croppers"			
Castor canadensis	20.41	8	Burt
Synaptomys cooperi	0.02	0.3	Burt
Microtus pennsylvanicus	0.045	0.25	Burt
Microtus ochrogaster	0.03	0.18	Fitch
Pitymys pinetorum	0.028	0.30	Burt
Pitymys pinetorum	0.028	0.09	Fitch
Sigmodon hispidus	0.12	0.55	Fitch
Sylvilagus floridanus	1.58	8.34	Fitch
Sylvilagus floridanus	1.58	15	Mohr
Sylvilagus floridanus	1.36	5	Burt
Lepus americanus	1.587	10	Burt
Alces americana	358.3	100	Mohr

RESULTS AND DISCUSSION

We have seen that body weight influences the rate of energy expenditure, and that this, in turn, influences the amount of food that a species must harvest to supply its requirements. Since the food that a species harvests is collected within its home range, it is natural to ask how the size of the home range is related to body weight.

When the $\log_{10}$ of the basal rate of metabolism is plotted against the $\log_{10}$ of body weight, a linear relationship is found; the slope of the curve is the power (0.75) of the function

(1) $$M = k_1 W^{0.75} = 70\, W^{0.75}$$

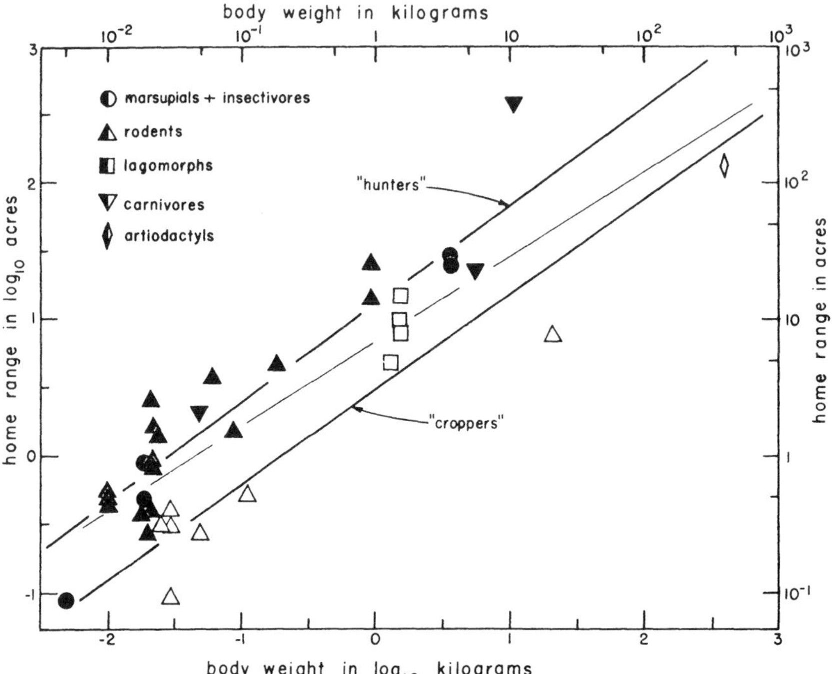

FIGURE 1. Relationship of home range to body weight in some mammals. Filled symbols are from "hunters"; open symbols are from "croppers." The two heavy lines are the fitted curves for the hunters and the croppers; the thin line is the fitted curve for the pooled data.

where M is the basal rate of metabolism (in kcal/day), W is the body weight (in kilograms), and k_1 is a constant equal to 70 (Kleiber, 1961).

When the $\log_{10}$ of the home range is plotted against the $\log_{10}$ of the body weight (figure 1), the data can also be fitted by a linear curve. The slope of the curve, fitted by the method of least squares, is 0.63.

Therefore, the size of the home range, R, can be expressed as a function of body weight by

$$R = k_2 W^{0.63} = 6.76\, W^{0.63} \tag{2}$$

where k_2 is a constant equal to 6.76.

There is no significant difference between 0.63 and 0.75 (the power of the metabolic function) with a t-test (t = 0.94, 34 degrees of freedom). Thus, the variability of the data is sufficient to allow $W^{0.63}$ to be replaced by $W^{0.75}$, and

$$R = k_3 W^{0.75} = 8.51\, W^{0.75} \tag{3}$$

where k_3 is a constant equal to 8.51. By substitution of equation (3) into equation (1)

$$R = (k_3/k_1) M = 0.12\, M \tag{4}$$

The size of the home range in mammals, accordingly, is determined by the rate of metabolism. A large mammal has a larger home range than a small mammal, because it uses more energy and, therefore, needs a greater area in which to find this energy. It is possible that the agreement between the fitted regression slope and 0.75 is simply fortuitous, but this seems unlikely, since it would require (at least) two other functions whose interaction (multiplication or division) produces a slope of 0.63. No two such functions intuitively appear.

One caution should be mentioned. We have used an expression for the basal metabolism to represent the rate of energy exchange of mammals. This is not correct, because what is important is not the *basal* rate of energy expenditure, but, rather, the *total* daily expenditure. However, there are almost no available data on the daily expenditures of any mammals (McNab, 1963), but considering the high expenditures of energy of small mammals for temperature regulation, the exponent would be less than 0.75. As a further complication, a function relating the daily energy expenditure to weight would be temperature dependent, while the function concerned with basal metabolism is temperature independent. Obviously, more data on home ranges and daily energy expenditures are needed for a complete analysis.

As important as is the rate of metabolism, it is not the only factor that influences the home range size, since there is considerable scatter about the mean curve. For example, at a weight of about 20 grams there is a variation in the home range from 0.1 to 3 acres. At this weight the mice of the genera Zapus, Napaeozapus, and Peromyscus have large home ranges (average = 1.09 acres; SE = 0.07; n = 7), while the genera Microtus, Pitymys, and Synaptomys have small home ranges (average = 0.22 acres; SE = 0.002; n = 5). The former group (cricetine) feeds upon seeds, fruits, and insects, and can be characterized as having to hunt for its food. On the other hand, the microtines feed on grass, and where these mice are usually found there is abundant food with no need to hunt. The type of food of a species appears to be the determining factor for the size of home range, for Sigmodon, which is usually classified as a cricetine (Simpson, 1945), feeds on grass and has a small home range.

A distinction between the hunters and the non-hunters is found among large mammals as well as among small mammals. Rabbits tend to have small home ranges for their body size (figure 1) and are nonhunters, while tree squirrels (Sciurus) at the same size, are hunters and have larger home ranges. For its weight the beaver, Castor, has a very small home range, which is probably based on its non-hunting economy. The moose, Alces, also has a relatively small home range for its size; it is not a hunter.

If these two groups, the "hunters" (including those species that are either granivorous, frugivorous, insectivorous, or carnivorous) and what we might call the "croppers" (including those species that are either grazers or browsers), are separated and if individual curves are calculated for them, the adjusted means for the two curves are significantly different at the 95 per cent confidence level with the covariance test (F = 17.3; 1 and 32 d.f.). The two fitted equations, $R_h = 12.6\,W^{0.71}$ (hunters) and $R_c = 3.02\,W^{0.69}$ (croppers), show powers that are much closer to that of the metabolism function than does the power of the fitted, pooled curve. There is no significant difference between the powers of the curves (F = 0.03; 1 and 32 d.f.) or between each of the powers and 0.75 (at the 95 per cent level). Therefore, we obtain the following equations by substitution:

(5) $R_h = 0.20\,M$ (for the hunters),

(6) $R_c = 0.05\,M$ (for the croppers).

The home range of hunters is about four times that of croppers at the same body weight (figure 1). The basis for the smaller home ranges of croppers is the greater concentration of their food materials within a given area.

The energy requirements of a species not only help to determine the size of the home range, but determine, in part, the kinds of food that can be utilized. For instance, as hunters increase in size they must change food sources: a small carnivore may depend almost exclusively on insects, but larger species can and must rely upon energy sources that occur in larger packets, unless they can collect small prey with great efficiency. But mobility, too, is a factor. A large predator generally feeds on large prey species, and this requires a large home range of the predator to maintain a prey population of sufficient size to support the predator. Ultimately, there may be a restriction on home range size, because of structural limitations to body size. A restriction on home range size may stimulate a predator to diversify its food sources to use its home range more efficiently.

Large carnivores apparently need to use their home ranges efficiently. The largest of the terrestrial carnivores, the bears, have a diversified diet; they are omnivorous during the spring and summer, when they eat berries, nuts, and buds. (The largest cats, such as the lion and tiger, may be near the present maximum size limit for a "pure" carnivore. This limit is not altered by the presence at an earlier time of larger terrestrial carnivores among the dinosaurs. Since these large carnivorous dinosaurs generally fed on larger "croppers," the question reduced to what permitted such large croppers.) Cetaceans show a similar trend: the smaller dolphins and porpoises are strict carnivores, while the largest (baleen) whales are grazers. There are, of course small bears and small baleen whales, but small size does not prevent omnivorous or grazing habits; the important point is that the *largest* species have these habits. Although we have dealt exclusively with mammals, we might note that sharks parallel the trend of whales. Small and medium-sized sharks are carnivorous, while "... the two largest species (Whale Shark and Basking Shark) subsist wholly on minute planktonic forms, chiefly crustacea, and on small schooling fishes" (Bigelow and Schroeder, 1948; Nicol, 1960, p. 230).

If the main determinant of home range size is the rate of metabolism, we would expect adverse environmental conditions to influence home range size. For example, one would expect that the home range is larger in cold climates than in warm climates and larger during cold seasons than during warm seasons, because of the increased energy expenditures demanded of the mammals. A seasonal trend may not be clear, however, for it may be opposed by a tendency during the breeding season for the females to have large home ranges to feed their young. Water deficiency, as well as poor soil conditions, should also increase the home range size through its action on the distribution and abundance of plants. The meager evidence available agrees with this conclusion: the home range is larger in the mammals from desert regions (Bourlière, 1954), and the beach mouse, *Peromyscus polionotus*, has a large home range (about six acres; Blair, 1951), which may be a reflection of a poor, sandy environment.

If a factor acts primarily through plants, mammal species may be differentially affected. Those species that feed on seeds or berries may be influenced more severely by environmental fluctuations than those that feed on the vegetation. A poor seed crop, for example, may greatly reduce a Peromyscus population without any effect on a Microtus population in the same area. A similar conclusion was reached by Odum, Connell, and Davenport (1962): "...food is more likely to be limiting to granivores than to foliage-consuming herbivores in the old-field ecosystem...." Therefore, I agree with these authors that the analysis of Hairston, Smith, and Slobodkin (1960) must differentiate between the population regulation mechanisms of the croppers (which would be predator limited) and the granivores (which would be food limited).

Species that require large home ranges usually cannot maintain locally dense populations, because of the limited amount of energy available within a given area. A table of "economic densities" (in grams/acre) of small mammals tends to show this (Mohr, 1947): Carnivores average about 50, granivores about 300, insectivores about 500, and croppers ("herbivores") about 3,000 g/acre. The ability to maintain a locally dense population would depend either on being a cropper, and thus having small areal requirements, or on being a very mobile hunter. Since the establishment of elaborate social behavior within a population depends on a relatively high population density, so too it depends on the population's food source. It would appear impossible for elaborate social behavior to evolve in a hunting species that lacks mobility, generally, that is, in populations whose individuals are small. The mammals that have some form of complex social behavior seem to agree with this analysis: bats (small, hunters, highly *mobile*); prairie dogs (*croppers*); ungulates (croppers and *mobile*); porpoises, hyaenas, and wolves (hunters and *mobile*); and higher primates (hunters and croppers, *mobile*). No small, immobile social hunters are found, although there are many small, immobile, non-social hunters.

SUMMARY

The size of the home range is examined in mammals. It is determined, mainly, by the amount of energy expended by the species, and, therefore, the home range area may vary according to the direct and indirect influences of weather and climate on the animal. But the kind of food that is utilized will also influence home range size. Those species that must hunt for their food need larger areas for food gathering than those species that feed on the vegetation. As a result the largest hunters appear to have their food habits regulated by considerations of the efficient use of the food materials in their home range. Finally, the home range size affects population density, which in turn influences the behavior in the population.

ACKNOWLEDGMENTS

I wish to thank the members of the ecology seminar for their comments on the ideas expressed within this paper, and most especially Rodger Mitchell, who has critically examined this manuscript several times.

LITERATURE CITED

Blair, W. F., 1951, Population structure, social behavior, and environmental relations in a natural population of the beach mouse (*Peromyscus polionotus leucocephalus*). Contrib. Lab. Vert. Biol. Univ. Mich. No. 48: 1-47.

Bigelow, H. B., and W. C. Schroeder, 1948, Sharks. *In* Fishes of the western North Atlantic. pp. 59-576. Mem. No. 1, Sears Foundation Marine Research, pt. 1.

Bourlière, F., 1954, The natural history of mammals. Alfred Knopf, New York.

Brown, L. E., 1962, Home range in small mammal communities. *In* B. Glass [ed.], Survey of biological progress 4: 131-179. Academic Press, New York.

Burt, W. H., 1948, The mammals of Michigan. Univ. Mich. Press, Ann Arbor, Mich.

Calhoun, J. B., and J. N. Casby, 1958, The calculation of home range and density of small mammals. U. S. Dept. Health, Education, and Welfare, Public Health Monograph No. 1958, pp. 1-24.

Fitch, H. S., 1958, Home ranges, territories, and seasonal movements of vertebrates on the natural history reservation. Univ. Kansas Publ. Museum Nat. Hist. 11: 63-326.

Hairston, N. G., F. E. Smith and L. B. Slobodkin, 1960, Community structure, population control, and competition. Am. Naturalist 94: 421-425.

Hayne, D. W., 1949, Calculation of size of home range. J. Mammalogy 30: 1-18.

Ingles, L. G., 1961, Home range and habits of the wandering shrew. J. Mammalogy, 42: 455-462.

Kleiber, M., 1961, The fire of life. John Wiley, New York.

McNab, B. K., 1963, A model of the energy budget of a wild mouse. Ecology (in press).

Mohr, C. O., 1947, Table of equivalent populations of North American small mammals. Am. Midland Naturalist, 37: 223-249.

Nicol, J. A. C., 1960, The biology of marine animals. Interscience, New York.

Odum, E. P., C. E. Connell and L. E. Davenport, 1962, Population energy flow of three primary consumer components of old-field ecosystems. Ecology 43: 88–96.

Simpson, G. G., 1945, The principles of classification and a classification of mammals. Bull. Am. Museum Nat. Hist., 85: 1–350.

PAPER 8

Community Evolution and the Origin of Mammals (1966)

E. C. Olson

Commentary

MARK D. UHEN AND JESSICA THEODOR

E. C. Olson's work was critical to the development of paleoecology in a number of ways: he established the notion of a chronofauna as a coherent group of fossil taxa with temporal continuity and shared ecological context (Olson 1952); he brought taphonomy, the study of fossil preservation's effects on the data in the fossil record, to the forefront in North American paleontology; and he and his students pioneered the use of synthetic, taxon-free approaches to community structure.

This paper introduced to the ecological community the notion that communities had a temporal as well as a spatial extent—that they could be traced through geologic time. It also introduced to paleontology the idea of reconstructing not simply the diet of a single species but the food web of entire assemblages, provided that one understood and accounted for the taphonomic biases of the data.

Olson made the important point that the communities he described were defined not by the taxa that they contained, but rather by the ecological interactions of the organisms and the flow of energy that these interactions facilitated. Thus, the taxa in each of these communities could change over time while the interactions remained stable. In this way, whole communities could evolve as well as their component taxa.

Olson also hypothesized that different communities that existed in the Permian gave rise to different radiations of vertebrates. He suggested that the mammalian radiation came from a community that was tied to the aquatic realm at the base of its food chain, and that the lepidosaur-archosaur radiation originated from a community in which insects formed the primary consumers. Whether these hypotheses are broadly supported remains a matter of debate, but Olson's more general approach to the study of paleocommunity analysis strongly influenced the next generation of vertebrate paleontologists to go beyond lists of taxa to more integrative modes of thinking about paleoecology.

Literature Cited

Olson, E. C. 1952. The evolution of a Permian vertebrate chronofauna. *Evolution* 6:181–96.

From *Ecology* 47:291–302. Reprinted with permission from the Ecological Society of America.

Horton, J. S., and C. J. Kraebel. 1955. Development of vegetation after fire. Ecology **36:** 244-260.

Howard, W. E., R. L. Fenner, and H. E. Childs, Jr. 1959. Wildlife survival on brush burns. J. Range Mgmt. **12:** 230-234.

Hutchinson, C. B., and E. I. Kotok. 1942. The San Joaquin experimental range. Agric. Exp. St. Bull. **663:** 3-145.

Love, R. M., and B. J. Jones. 1952. Improving California brush ranges. Calif. Agric. Exp. Sta. Circ. **371:** 4-38.

Murie, M. 1963. Homing and orientation of deermice. J. Mamm. **44:** 338-349.

Neiland, B. J. 1958. Forest and adjacent burn in the Tillamook burn area of northeastern Oregon. Ecology **39:** 660-671.

Pearson, O. P. 1959. A traffic survey of *Microtus-Reithrodontomys* runways. J. Mammal. **40:** 169-180.

Pruitt, W. O., Jr. 1953. An analysis of some physical factors affecting the local distribution of the shorttail shrew *Blarina brevicauda* in the northern part of the lower peninsula of Michigan. Univ. Mich. Misc. Publ. Zool. No. 79, 39 p.

Quast, J. C. 1954. Rodent habitat preferences on foothill pastures in California. J. Mamm. **35:** 515-521.

Sampson, A. W. 1944. Plant succession on burned chaparral lands. Univ. Calif. Agric. Exper. Sta. Bull. No. **685:** 1-144.

Scheffer, T. H. 1931. Habits and economic status of pocket gopher. U.S.D.A. Tech. Bull. No. 224: 1-9.

Sweeney, J. R. 1956. Responses of vegetation to fire; a study of the herbaceous vegetation following chaparral fires. Univ. Calif. Publ. Bot. **28:** 143-250.

Tevis, L., Jr. 1956. Effect of slash burn on forest mice. J. Wildl. Mgmt. **20:** 405-409.

Thornthwaite, C. W. 1940. Atmospheric moisture in relation to ecological problems. Ecology **21:** 17-28.

Wieslander, E. E., and Clark H. Gleason. 1954. Major brushland areas of the coast ranges and Sierra Cascade foothills in California. Misc. paper No. 15. Southeast Forest and Range Exp. Sta., Berkeley, California.

Williams, O. 1955. Distribution of mice and shrews in a Colorado montane forest. J. Mammal. **36:** 221-231.

COMMUNITY EVOLUTION AND THE ORIGIN OF MAMMALS[1]

EVERETT C. OLSON

Department of Geophysical Sciences, University of Chicago, Chicago, Illinois

Abstract. The evolutionary course from primitive pelycosaurian reptiles through therapsids to mammals can be profitably studied in relationship to modifications of the structure of the communities in which these reptiles existed. For this purpose the community is defined in very broad terms. Three types of communities are recognized upon the basis of the nature of the food chain. Each has an important tetrapod component.

Early phases of the evolution that culminated in mammals took place in communities that were strongly tied to water by the structure of the food chain. The physiological bases of the development of mammals appear to have been related to this environmental restriction. In successive pulses, however, the pelycosaur–therapsid communities developed terrestrial reptilian herbivores and thereby broke with the water-based food chain. More strictly terrestrial communities developed concurrently, with the insects, which were a food source for the reptiles, as the principal herbivores. From this sort of community came the terrestrial lepidosaurian–archosaurian reptilian radiation.

The terrestrial communities so developed came into competition. In this competition the therapsid lines were temporarily unsuccessful, leaving only small, but very mammal-like, representatives in the late Triassic. After a long period with relatively little adaptive radiation, these remnants provided the basis for the radiations of mammals that led to the great successes of the Cenozoic era.

INTRODUCTORY EXPLANATIONS

Studies of vertebrate evolution centered around the concept of faunal modifications have constituted one of the major fields of interest of the writer over the last decade and a half. A number of publications which have resulted from this interest, as cited specifically in the following text, have stressed the changes of communities with the passage of geological time. The present paper represents a continuation and extension of this kind of work. Most of the data upon which it is based have been included in the earlier studies, but the synthesis is somewhat more general than any attempted previously, and the interpretations are more directly ecological.

Studies, such as this one, which involve broad areas of paleoecology are necessarily cast at rather different levels from those of most neoecological investigations. Naturally, as well, a strong element of speculation must enter in, for assumptions of a rather sweeping nature are necessary and conclusions often must be based on complexly interwoven threads of evidence. Yet the resulting insights into the relationships of ecology and evolutionary processes are such that, even though crude, the interpretations are stimulating in them-

[1] The research leading to this paper was supported by NSF grants 19093 and 2543.

selves and frequently direct investigations into new paths.

The kinds of data that enter into these studies defy a short and concise description. They have been presented in some detail, along with consideration of the problems that they pose in two earlier papers (Olson 1958, 1962). The most reliable information on fossils, sediments and associations of the materials comes from day to day field studies made in the course of assembling collections of fossils, with ecological analyses as one of the principal goals.

Sites containing fossil vertebrates may yield materials ranging from fragments of bones and single bones, through all stages to the complete skeleton. Specimens may occur in assemblages where many individuals are associated in various states of preservation. All types of occurrences have meaning for ecological studies. They assume greater significance when their materials are related to the types of sediments in which they occur and the ways that the specimens are preserved in those sediments. Time relationships, and geographical distribution are equally significant.

A brief consideration of the nature of the record used in the present study may aid in clarifying the nature of the evidence. Many of the sites that have yielded vertebrates from the Permian and Triassic contain sediments that were deposited under conditions that can be broadly defined as deltaic. On the basis of sediment characteristics such as composition and size distributions of particles, the gross shape of a deposit and relationships of a bed to those adjacent to it, the characteristics of the environment of deposition can be reasonably well determined. In deltaic deposits sediments formed in stream channels, in lakes and ponds, and on flood plains can generally be recognized without great difficulty (see Olson 1962, for a full discussion of this matter). Other kinds of depositional sites may be determined in many cases, but generally with less ease and less certainty.

The animals themselves, of course, are good indicators of conditions of their own existence, but raise problems with respect to the relationships of the death assemblage and the living populations. In an extensive study of the related problems of the death and accumulation of animals and the nature of deposition of the containing sediments, Efremov (1950) treated in detail a process which he called taphonomy, which concerns the whole spectrum of events during the passage of animals from living populations to death assemblages. Many fossil associations are what he terms taphonomic assemblages, those in which the animals and plants which were available to the processes of erosion and transportation that produced a particular deposit were not all members of the same life assemblages. Such associations are characteristic of accumulations produced by stream deposits, which may tap a variety of life zones along the course of the stream and its tributaries. Such deposits are of great use in determining the general nature of the life of a given time, but are not as useful in paleoecological studies.

Assemblages that do reflect the associations that existed during life are especially instructive, although frequently difficult to recognize. Pond deposits may be of this type. Assemblages formed in standing water in the Clear Fork deposits of Texas, for example, show a remarkable consistency. It has been possible, on the basis of considerable field experience in the area, to specify with about 75 to 80% accuracy the generic content of particular examples of such deposits in the field by observation of the beds alone, before examination of their fossil content.

Various assemblages also tell a good deal about the life habits of the organisms. A small, somewhat worm-like amphibian of the Clear Fork Permian of Texas, *Lysorophus* occurs characteristically in high concentration, with a hundred or more individuals more or less evenly distributed over roughly circular areas with diameters generally from 10 to 20 ft. There is little question that these represent an aestivating phase of the life of the animal. Furthermore, study of size distributions shows that each of these assemblages tends to have individuals of very limited size range. Each aestivating swarm appears to represent an age group, probably an annual group. *Gnathorhiza,* an extinct lung fish, similarly was an aestivator, spending part of its life in a cylindrical burrow (Romer and Olson 1954). These, and various other examples, are taken up in reports on the details of studies which have contributed to this synthesis (Olson 1958, and a series of 13 preliminary papers cited in that reference).

To the extent that collectors have documented their materials with regard to conditions of preservation, associations, sedimentation, stratigraphy and so on, general museum collections may be used for such interpretations and may be of considerable importance. For the current study, the collections of fossils and field notes made by the writer and his associates over the last 25 years in the course of work in the Permian of Texas and Oklahoma have been the most useful. Much weight has also been given to collections from older Permian beds made by those collectors who kept records of the physical and biological associations of their materials. Very useful, as well, have been the collections from the Kazanian of the Russian Permian housed in the Museum of Paleontology in Moscow, U.S.S.R. These in many instances reflect the faunal point of view of I. A. Efremov and his colleagues and predecessors, in particular Sushkin and Bystrow.

Potentially most promising but unfortunately of less than full value are collections from the Permian and Triassic series of South Africa. Until rather recently less than sufficient heed has been given to associations of the fossil organisms, to the physical conditions of their occurrence, and to stratigraphy. The lack of published data on these critical matters, and my own lack of first hand field study and of study of some of the larger collections are reflected in the fact that the weakest part of the analysis that follows is in those sections dealing with these materials.

A first study along the lines followed in the present paper dealt with the lower Permian Clear Fork beds of Texas and their faunas (Olson 1952). This was elaborated and somewhat modified in a later summary work cited earlier (Olson 1958). More recently other publications have broadened the temporal and evolutionary scope of the investigations and have attempted to delineate some of the evolutionary implications of community structures (Olson 1961, 1962). Some matters taken up in the following pages were first developed in the last two cited references.

The nature of the communities

One of the problems that inevitably arises is the coordination of the meaning of terms that are convenient and useful in paleoecology with those used in neoecological studies. At best only an approximation of common meaning is possible. The term community as used in this paper represents such a case. As employed here the term covers a very extensive unit of biological association, and lies at one limit of the spectrum of range of usages. Furthermore, as used here the term is basically conceptual but in some instances, as is clear in context, it refers to actual systems that existed at given times over finite geographic areas.

References are made later to a community type, which is conceptual, or to an example of a type which consists of a community that existed at a given time and place, for instance the Clear Fork community of the early Permian of Texas.

It is common practice among paleontologists to give a taxonomic designation to an assemblage of animals that is encountered repeatedly in the record. Such assemblages are often called complexes and the designation is generally taken from a common element of the complex. These assemblages are usually considered to include only the animals, or only the invertebrate or the vertebrate animals, and thus are less comprehensive in scope than the community as used in this paper. They do, however, provide convenient nomenclatural units to express some temporal or taxonomic stage of a community type. Efremov, whose work has been used extensively in later parts of this paper, has employed phrases such as deinocephalian complex, based on a taxon of infraordinal rank, and pareiasaur complex, where the term of designation is generic. In the discussion of the sequence of communities in evolution, particular units are specified by these and similar terms.

There is an additional complicating factor, one not present in the same sense in the communities of neoecology. The actual communities of paleoecology had duration in time (Olson 1952). The property of temporal duration through some appreciable segment of geological time means, of course, that the existence of a community is not dependent upon persistence of its individual biological components, the animals and plants of a particular time, for these vanish to be replaced either by their own kind or by ecological equivalents. The key rather is the pattern of interaction expressed in the structure of the community. It is the persistence of this pattern that provides the continuity of existence. The biological, geographical and temporal limits are thus necessarily somewhat indefinite and are in no sense independent variables that can be treated in separate contexts. Conceptually, however, the units here designated as communities are, in spite of the variations with time, roughly determinable. They form a suitable framework for the type of work presented in this paper. The breadth of this framework is essential at the level at which such studies must be conducted.

There undoubtedly existed a large number of community types of this general sort within the major habitats available during the Permian and Triassic Periods. Presumably the community units had various and temporally varying limits of discreteness, some with little dependence upon other communities and others with a large measure of dependence. Our primary concern will be with those systems that had a terrestrial vertebrate component, and as a result the classification necessary in this study is quite limited. Among such communities three types have been distinguished, and representatives of each type seem to have existed as relatively independent entities over appreciable spans of geological time.

There are, of course, many different ways in which the community types might be designated and different arrangements emerge from application of different criteria. With interest focused upon the vertebrates, the following two aspects of the system appear to be the most important:

1. The contribution by its vertebrates, considered as ecological types, to the composition of the community. Taxonomic composition is secondary, but to the extent that the taxonomic position of a particular vertebrate relates directly to its contribution to the community pattern, taxonomy may be considered as important.
2. The paths of flow of food-derived energy through the community. The focus is upon the structure of the food chain, especially as it relates to sources of energy of the vertebrates.

Consideration of the many problems associated with determination of diets and feeding patterns among extinct organisms, pertinent to the second point is much too complex for treatment here. Some points, however, are particularly important. At best, interpretations tend to stretch the capacities of rational inference to the limit. The primary data, of course, are those of morphology. The dangers of using such data alone for determination of diets are quite evident among living animals. Yet some broad limits can be drawn.

Another source of information is in the analysis of the potentialities and limitations of feeding interactions imposed by temporary, environmental and geographic distributions. The presence of organisms may indicate that they were a potential food source for an appropriate consumer, but the absence of some types from the record does not necessarily mean that they were not living in the community from which the sample of the organisms was drawn. The inferred role of insects in the Permian Clear Fork community is a good case in point. No insects have been found in association with the plants and animals that have been collected. In adjacent areas, from sites of about the same age, well-preserved insects have been found. These sites contain no vertebrate remains. The differences between sites are related, without much question, to conditions of preservation. Thus I freely infer the existence of insects as components of the Clear Fork community, and call upon them as a food source of the vertebrates.

Coprolites, or fossil feces, are occasionally useful sources of information on diet. Very often plant and animal remains are preserved in coprolites and in some cases identifications of the contained organic materials can be made. Some kinds of rather general information can be had. It is notable, for example, that in the Clear Fork almost all coprolites contain a wealth of debris from animals but very little from plants. When more specific information is sought, the difficulty of association of the coprolite with its donor assumes importance. In the Clear Fork this has been possible, with some assurance, for only two genera, *Dimetrodon,* a large predaceous pelycosaurian reptile, and *Xenacanthus,* a fresh water shark. From Isheevo, a Permian site in the U.S.S.R., comes a series of coprolites that seem to pertain to a large, predaceous amphibian known in the same deposits. Their yield of remains of crossopterygians, lung fish and predaceous reptiles, as well as of their own kind, gives considerable insight into the feeding habits of the presumed donor.

In the analysis of diet, as in all such analyses, it is the bits of evidence from many sources that finally may be woven into a reasonably satisfactory picture.

Community evolution

The meaning of the phrase community evolution is a complex and somewhat nebulous matter. The way that it is viewed in the present context will help to set the stage for analysis of the relationships of the three community types considered later. Community evolution is thought of as crudely analogous to organic evolution. If not carried to an extreme, particularly with respect to mechanism, the analogy presents a convenient and comprehensible model for analysis. Stability of structure over a long period of time, splitting, convergence and threshold phenomena are all evident. Communities also may merge and this can, by some stretch of the imagination, be analogized to hybridization. At this point, however, the model of a population is unsuitable and such a concept clouds rather than clarifies the situation.

Community types, again like populations of organisms, are limited in their possible spectra of development by the various conditions that are functions of the time in which they exist. These are in large part products of the generally emergent nature of evolutionary processes and are thus both a product and function of the very evolution that is being studied. Any particular time provides a limited suite of settings in which both physical and biological restrictions limit the types of communities that exist. Each community is itself a part of the existing array, supplying one element of the self-limiting complex. The evolution of a complex of communities is related to changes of each of its components, as their modifications alter living conditions of communities with which they have some sort of association.

During the Permian period communities that included terrestrial vertebrates suffered rather severe restrictions, partly as a result of the relatively recent emergence of vertebrates onto land, partly as a function of the type of floral covering of the land, and partly in relationship to the level of biological organization of the vertebrates at the stage of evolution reached at that time. The relaxation of these restrictions is one of the important events that we trace as we follow the semiterrestrial and terrestrial complexes through the Permian and Triassic periods.

Community types of the Permian and Triassic

Three distinct community types which include terrestrial vertebrates can be identified in the known Permian deposits. Each is reasonably discrete, but geographic and ecological overlaps do occur. Two of these types appear to persist into the Triassic, although evidence for one of them is not too satisfactory. The three types, which are designated as I, II, and III, are as follows:

Type I (Figs. 1, 2): This type of community has its food chain based primarily in aquatic plants.

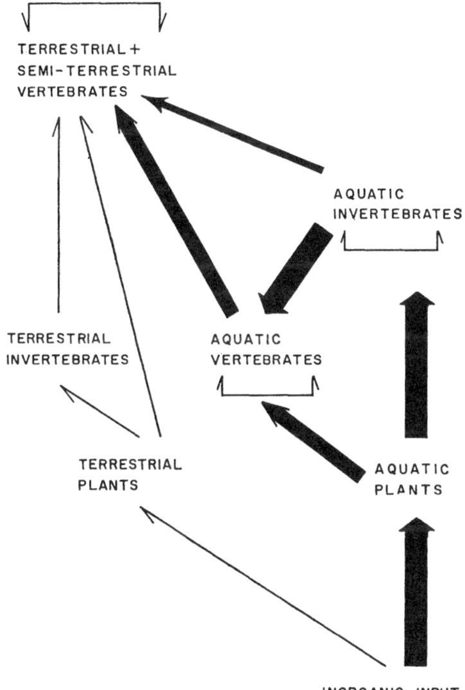

FIG. 1. Type I community with the food chain based primarily in aquatic plants, consumed by invertebrates and vertebrates, the food sources for both aquatic and terrestrial vertebrates. The direction of flow (by feeding) is indicated by the arrows. Main avenues, heavy arrows; subsidiary avenues, light arrows. Arrows on brackets leading to the same group indicate that some members of the particular group fed upon others of the same group.

These provide the primary source of food energy, supplied to the community by synthesis of organic matter from inorganic. Terrestrial and semiterrestrial invertebrates are included, but are of only minor importance. Aquatic and semiaquatic plants occupy a dominant position, with terrestrial plants of much less significance.

Following the primary conversion of inorganic substances to organic, chiefly by aquatic plants, energy follows a course that leads through small aquatic invertebrates and vertebrates to aquatic, semiaquatic and terrestrial vertebrates. Once the community has been established, organic debris, supplied from many sources, presumably became an important food source. The course of energy flow is complex, with multiple channels and alternative routes. An attempt to map out some of these is shown in Figure 2, the case for the Permian complex in the lower part of the Clear Fork (Arroyo formation). The arrows indicate the direction of flow of energy based upon presumed feeding relationships. A complicated structure, such as this, tends to make for high stability, not only in the persistence of the basic ecological types, but also in temporally extended occupancy of primary roles by particular genera and species. *Dimetrodon* is a pertinent example, as illustrated in the upper right hand corner of Figure 2. This large pelycosaurian predator persisted in its role as principal predator with no evident morphological change in a slowly modifying community for a period of several million years. Its taxonomic position and those of other genera as well as of higher categories, are summarized in the appendix.

The community type I was derived from a more ancient, essentially aquatic type of community, by expansion into the terrestrial realm without important modification of the preexisting structure. It can be recognized from the base of the Permian, with some Pennsylvanian (Carboniferous) traces that suggest its earlier existence, through to the end of the Kazanian (or possibly early Tatarian) of the upper Permian of the Soviet Union. The latest clear-cut example is from the vicinity of Isheevo, Tatar, U.S.S.R. No clearly identifiable examples of this type are discernable in the Permo-Triassic Beaufort series of South Africa, although some of the assemblages are suggestive.

Type I communities appear to have provided the framework within which the members of the lines of synapsid reptiles that eventually lead to mammals existed during the early and middle stages of this evolutionary history. Ophiacodonts, sphenacodonts, phthinosuchids, gorgonopsians and possibly therocephalians are known examples of

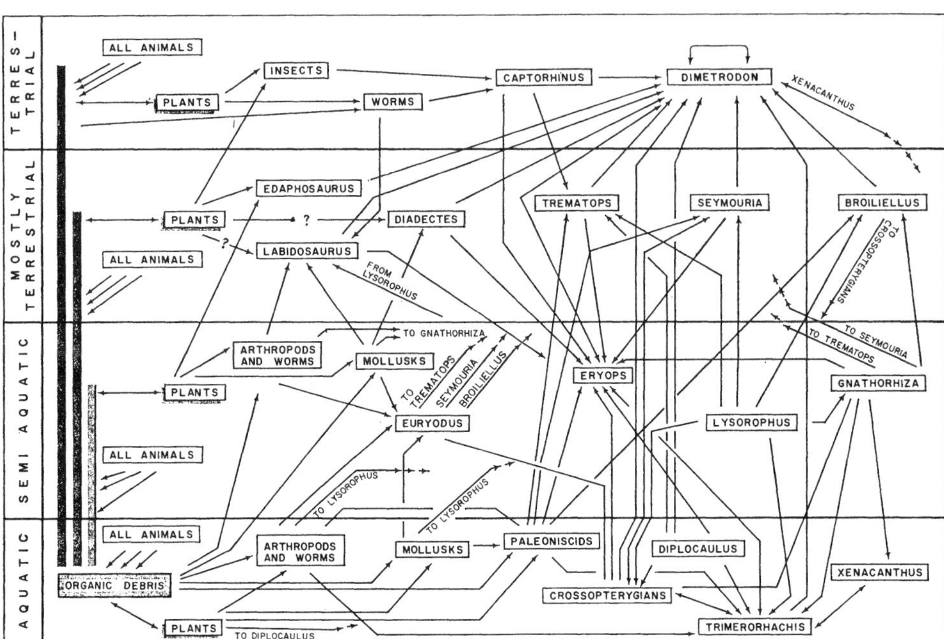

FIG. 2. An example of the complex patterns of flow of food-derived energy in a particular community of type I. The diagram is based primarily upon the well-known vertebrates of the Arroyo (Clear Fork) early Permian community in northern Texas. It is, of course, overly simple as compared to conditions that must have actually existed. Fish: *Xenacanthus* (predaceous shark), *Gnathorhiza* (lung fish), crossopterygians (lobe fins), paleoniscoids (primitive actinopterygians). Amphibians: *Lysorophus, Diplocaulus, Trimerorhachis, Euryodus, Trematops, Broiliellus, Seymouria, Diadectes* (see Romer 1964) Reptiles: *Edaphosaurus, Labidosaurus, Dimetrodon, Captorhinus.*

such lines. Many features of mammals, both morphological and physiological, seem to find reasonable explanations in the long duration of aquatic ties of this community type. Tatarinov (1959) in particular has emphasized the role of aquatic ancestors in the development of mammalian structure. Many of the morphological and physiological features that are considered reptilian because they are present in modern reptiles may never have existed in the reptilian stocks which gave rise to mammals, but themselves became extinct late in the Triassic period.

Evolving within community type I, but playing a relatively minor role, were certain elements that several times broke with their aquatic heritage. Community types II and III, considered in the following paragraphs, probably had their origin within these minor components. Final dissolution of type I appears to have taken place in the very late Permian, as the bonds to water were progressively broken and durable representatives of other community types emerged. Later, during the Triassic, new versions of type I appeared once more. With the development of birds and mammals, the chances of such communities were reduced. Type I, however, is approximated in some of the largely fish-based communities that exist in subtropical swamp regions. The assemblages of the Florida Everglades, in their unmodified form, depart only moderately from this type of community.

Type II: This community type (Fig. 3) is dominated by terrestrial vertebrates, invertebrates and plants. The primary sources of food energy of the vertebrates are terrestrial plants. These are consumed directly by terrestrial herbivorous vertebrates. The invertebrate component, which may be highly significant numerically and in terms of biomass, plays a secondary role in the vertebrate food chain, although it is, of course, important in the totality of the functions of the community type.

The structure of this type of community can be extremely complex, as witnessed by representative cases among some of the mammal-dominated communities of type II in existence today. Among modern communities, of course, extensive subdivision is possible. Such opportunities are more limited in older communities, both because of a

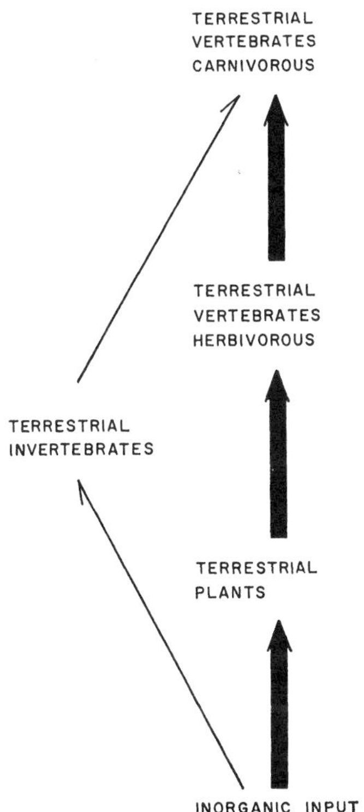

FIG. 3. Type II community. Dominantly terrestrial with food sources in terrestrial plants consumed directly by terrestrial, vertebrate herbivores. Symbols as for Figure 1.

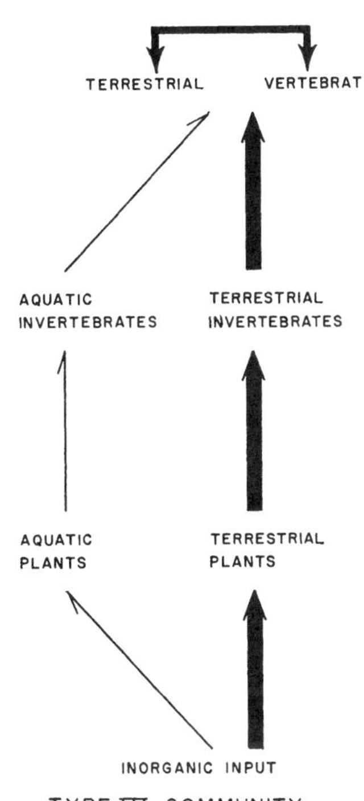

FIG. 4. Type III community. Dominantly terrestrial food sources in terrestrial plants consumed primarily by terrestrial invertebrates which in turn are principal food source of terrestrial vertebrates. The important differences from type II is the near absence of terrestrial vertebrate herbivores, and the primary role of invertebrates as plant feeders. Symbols as in Figure 1.

presumably greater simplicity, especially before the Cenozoic, and because of the nature of the evidence available for their interpretation.

The first definitely recognized occurrence of a community of type II is in the very early part of the late Permian of North America, in the San Angelo-Flowerpot formation complex. This seems to be a continuation of an earlier stage, from the Hennessey formation of Oklahoma, but the fauna of the Hennessey is not well enough known for a definite conclusion.

Communities of type II have risen many times and from many sources. The late Permian and early Triassic communities of the Beaufort Series of South Africa seem to be mostly of this type, derived in part from type I and in part from type III. Prior to the Triassic, South African assemblages show some features reminiscent of type I communities; and the degree of independence from aquatic ties, in particular from aquatic food ties, needs further investigation.

Type III: This (Fig. 4) is a type of community that consists predominantly of terrestrial organisms. The primary conversion of inorganic to organic matter is accomplished by terrestrial plants and the principal consumers of plant food are terrestrial invertebrates, primarily insects. Vertebrates are small to moderate-sized insect eaters and predators. The difference in the paths which plant food follows in providing food energy to the animal components of the community constitutes the principal distinction between types II and III. In the former, the primary herbivores are vertebrates; in the latter, they are invertebrates. Type II communities may develop either from type I or type III, but the resultant structure will initially, at least, be distinctly different.

Type III is a restricted community which could exist today only under special conditions of isolation from the dominant type II communities. Type III has been specifically identified only in the Kazanian of the Russian Permian in this study. The small number of deposits preserving this type of community results largely from the fact that uplands were the common abode. Almost all existing deposits of the Permian and early Triassic periods formed either under marine circumstances or in lowland areas marginal to the seas. There are essentially no truly upland deposits known. Occurrence of assemblages of strictly terrestrial organisms are unlikely. Members of upland communities that were carried to lower levels lost their life associations and these can be reconstructed only with serious reservations.

At the time that upland deposits first appear in the record, later in the Triassic, some of them, specifically those characterized by lepidosaurs and archosaurs, show that there had been a fairly long history of strictly terrestrial evolution before their formation. The composition of the vertebrate assemblages that they contain suggests that these communities originated from type III communities.

The role of type III communities in tetrapod evolution of the Permian and Triassic, as far as direct and unequivocal evidence is concerned, remains somewhat problematical until more adequate data are at hand. It seems very likely, however, that the lepidosaur-archosaur radiations of the Mesozoic era had their origins in stocks that gained freedom from aquatic ties by the development of such a community structure. The known examples of type III, from the Russian Kazanian, have taxonomic compositions only partially appropriate to the occupants of this ancestral position. From one site, the Mezen locality in northern European Russia, there is evidence of an eosuchian. But the others lack this element.

The fact that less than enough is known of the early stages of the lepidosaur-archosaur radiation has led to strong differences of opinion concerning the ultimate source of the stocks. Some elements that seem to have been involved, millerettids and younginids (eosuchians), occur associated with early therapsid, anomodont-theriodont communities of type II. The time of occurrence, the ecological associations, and the lack of any marked trend among these eosuchians toward archosaurian diapsids all seem to suggest that the sources of the lepidosaur-archosaurian radiations did not lie within the anomodont-theriodont complexes, but were developing elsewhere under other circumstances.

It is possible, of course, that these occurrences represent a stage of sorting out, in which the eosuchians were developing an independence from the early therapsid complexes that led to type III communities. Or more probably, as the timing suggests, such ancestral communities were already well established, and the eosuchians in the early therapsid complexes were immigrants from more stable type III communities, which also included the initial stages of the archosaurian radiation. It cannot be determined from the published information just how intimate the actual associations of the eosuchians and anomodont-theriodont therapsid complexes were.

We shall assume in the synthesis in the next section that a community of type III provided the base of the lepidosaur-archosaur radiation. We shall assume also, for want of evidence of a better solution, that this base had its source in the terrestrial, presumably insect-feeding captorhinomorphs of the type I communities of the early Permian. Whether these assumptions are precisely true or not, the critical point is that nearly complete freedom from aquatic environments must have been attained by a potential lepidosaur-archosaurian assemblage or by two assemblages independently, earlier and in a manner different from the emancipation of the therapsid communities. The reptilian characters of morphology and physiology in the radiation are considered to reflect the results of an early and complete abandonment of aquatic ties. Later successes are believed to have resulted from a consequent high level of adaptation to land.

Evolution of communities leading to mammals

The essence of the scheme of faunal evolution suggested here is shown in Figure 5. The situation is complex and the fact that the phylogeny involves communities rather than the more usual systematic units may add to the problems of following it through. The names used for communities are of three types, to conform with designations commonly used in the literature on fossil vertebrates. For the North American Permian, where the communities are known over rather wide areas, stratigraphical terms have been used, for example Wichita and Clear Fork. Some of the Russian sites are local, and place names, such as Yeshovo and Shikhovo-Chirki, are entered. However, Bashkirian and Kargalian, have a stratigraphic connotation, once again because of the extensive area over which the collections have been taken. .The names of certain animal complexes are entered in Figure 5, covering several sites and stratigraphic terms in some instances, to indicate the relationship of the community types to such designated units. In the upper Permian and Triassic and in entries for more recent times, taxonomic designations are used, since these communities cannot be readily designated either in units of space or of time. The taxonomic terms are given in systematic form in the appendix, but

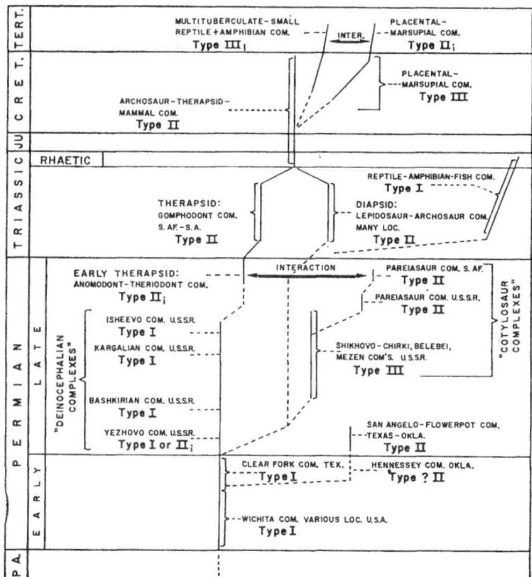

FIG. 5. A diagram of the scheme of interrelationships of community types as represented by particular communities in the series of events leading to the origin and early development of mammals. Essentially this is a phylogeny of communities. It is described in the text in more detail. The various representatives of community types are indicated by locality names, stratigraphic names and taxonomic names as appropriate to accord with the circumstances of preservation and common usage in the literature concerning each. The solid lines indicate presumed continuous spans of duration of a particular community type, as indicated. In some instances an incipient community of another type is indicated by the subscript i. The specifically noted instances of community types, for example Clear Fork or Shikhovo-Chirki, in essence represent samples taken from the geographic and temporal ranges of the community type which they represent. Complexes, such as the deinocephalian complex, may have considerable duration, as indicated by the extent of the brackets. The dashed lines show the presumed line of descent between the community types, portraying the phylogeny of the community systems. Thus the Wichita community of type I led to the Clear Fork community of type I and also, provisionally, to the community type II of the Hennessey formation. This in turn is the probable progenitor of the San Angelo-Flowerpot type II community. The type I is continued through the U.S.S.R. samples represented by sites and stratigraphically named assemblages. These are broadly grouped under a series of deinocephalian complexes, as indicated. This general pattern of interpretation carries through the rest of the phylogeny, with supplemental explanation in the text.

for convenience are also entered in the caption for Figure 5.

The early Permian communities that contained terrestrial vertebrate elements were all of type I. This was an extremely persistent type which probably began in the Pennsylvanian. There is considerable evidence of it in the later parts of this period, but none of the assemblages is complete enough for clear interpretation.

Within the early Permian communities there can be identified small segments that had capacities for breaking from the parent type. The edaphosaurids and incipient caseids, primitive herbivorous pelycosaurian reptiles, provided potential terrestrial herbivores. Neither appears to have been very abundant in the known assemblages. Captorhinomorphs, rather lizard-like anapsid reptiles, probably were the main feeders on invertebrates. Of interest in this regard is a recent paper by Eaton (1964) in which he has presented evidence, based on what appears to be an actual case of predation caught in the record, that captorhinomorphs were to some extent at least predators on small vertebrates. It seems very probable that these reptiles, like modern ones, ate in large part what was available to them, within limits of their capabilities of ingesting and utilizing the food. Thus it may be supposed that captorhinomorphs fed on insects, small vertebrates, and any other creatures that they encountered and could handle.

In the latest Clear Fork there is a suggestion of a type II community in the Hennessey formation of Oklahoma, but it is in the beds composing the San Angelo and Flowerpot formations of Texas and Oklahoma that a complex example of this type can first be authenticated. It is based on plant feeding terrestrial vertebrates, caseid in large part, and carnivores, mainly advanced sphenacodonts and phthinosuchids. Large captorhinomorphs played a numerically important role. Caseids, especially the large genus *Cotylorhynchus*, outnumber all other members of the fauna by a ratio of about 10 : 1 in the collections. They were very massive and up to 20 ft long. In mass they probably ran as high as 20 or 30 times the rest of the fauna. There is some evidence of the existence of such a community type in the upper Kazanian of Russia, but associations are poor and the record is far from unequivocal.

The main line of post Clear Fork evolution is found in the continuation of the type I community, as seen in a succession of sites in the Kazanian (late Permian) of Russia (Yeshovo, Bashkirian, Kargalian and Isheevo). These are the deinocephalian complexes of Efremov, so-named because they usually include specimens of these large, primitive therapsids. Even the youngest of these occurrences, that at Isheevo, maintains a full array of the properties manifest in the early Permian.

From several places in the Soviet Union have come assemblages that give a tangible basis for recognition of communities of type III; for example, Mezen, Shikhovo-Chirki and Belebei. These represent the cotylosaur complex of Efremov (see Efremov and Vjushkov 1954; Olson 1958, 1962) and include primitive procolophons (the cotylosaurs), carnivorous phthinosuchids and, at Mezen, an eosuchian. Amphibians occur in the same beds, but the association is to be considered taphonomic rather than ecological; that is, the death assemblage has sampled different life zones along the stream course and has brought together animals that did not form a life assemblage. It is significant that the cotylosaur complexes of type III, and the deinocephalian complexes, which are of type I, are contemporary but geographically separated. Members of the two are not mixed in the record.

From this type of community, as represented in the materials from Belebei and Shikhovo-Chirki, emerged the Russian pareiasaur complex, classified as type II. This shows strong ecological resemblances in its major constituents to the earlier San Angelo-Flowerpot community of North America, characterized by an animal complex including *Cotylorhynchus*, sphenacodonts and phthinosuchids. The taxonomic position of the predominant herbivores, however, was very different. Presumably the Russian and also the South African pareiasaur complexes had a common origin, but the latter clearly includes important increments from a type I community.

By gradual relinquishment of aquatic ties, type I communities, such as those seen in the Russian Kazanian, gave rise to type II communities, as seen in the upper Permian of South Africa. For a time these emerging communities were in close association and perhaps merged with pareiasaurian communities.

From this gradual shift emerged the early therapsid anomodont-theriodont complex of the uppermost Permian of South Africa. Such communities surely were developing elsewhere in the world, but there is no other good record of them at this time in earth history. In South Africa, during the late Permian, there appears to have existed a complex community, dependent primarily upon food energy synthesized from inorganic sources by terrestrial plants. Early in the Triassic period internal evolutionary advances, it would seem, resulted in transference of the primary herbivorous role from the anomodonts to gomphodont cynodonts, and similar creatures. This represented a change in constitution but one that did not result in fundamental alteration of the type II community. It is striking that strictly terrestrial diapsids are not a component of this community type, although they surely were in existence and undergoing a strong evolutionary development. Some diapsids, such as rhynchosaurids, specialized rhynchocephalians, contribute to the community, but these somewhat aquatic creatures are not part of the main stream of diapsid evolution.

Small, primitive diapsids, as has been noted, were constituents of the anomodont-theriodont complex of the later Permian. If they came from the millerettids, as is generally believed, and if the millerettids came from captorhinomorphs, as Parrington (1958) among others including the writer have argued, then they may be thought of as direct descendants and continuations of the insectivorous increment represented earlier by the captorhinomorphs. This stock, so interpreted, forms the base of the terrestrial radiation of diapsids. Initially its terrestrial role was acquired as a progenitor of a type III community while technically a part of a very early (Pennsylvanian) community of type I, the same type as that in which the synapsids began their early radiation. The best-known genus of the captorhinomorphs (*Captorhinus*) did not itself occupy this position, for it is too late and structurally too specialized. Rather less specialized genera from the very early Permian or Pennsylvanian approach the actual ancestry more closely.

Whatever the final solution of this problem may be, it is evident that at present we lack much-needed information about the origin and early history of the diapsids and the places of their evolution. It is significant, as noted earlier, that our records tap only relatively lowland areas, close to the sea, prior to the middle or late part of the Triassic. The important fact concerning the diapsids, even emphasized by the poor record, is that their development took place independently of that of the communities that contain the lines of therapsid evolution.

Through much of the Triassic, the highly terrestrial diapsid (lepidosaur-archosaur) communities of type II continued independently of the contemporary therapsid communities. In view of their apparently long history of independence from aquatic environments, it may be supposed that the diapsid communities had become more highly adapted to terrestrial existence than were the contemporary therapsid complexes. As the latter pushed into more strictly terrestrial environments, during the middle and late Triassic, and the lepidosaur-archosaur lines continued to deploy in these environments and increasingly to exploit less strictly terrestrial, semiaquatic zones, the two great community arrays, each of which had attained a type II organization independently of the other, came into close contact and undoubtedly into competition. For the first time the two are found intimately associated in the deposits; this occurs more or less simultaneously in several widely separated regions.

Rapid impoverishment of the therapsid stocks followed, and most of the large, prominent types became extinct. Some of its elements, however, were incorporated into the more successful lepidosaur-archosaur complexes. These were for the most part small insectivorous and herbivorous remnants, those that we consider to have been extremely advanced mammal-like reptiles, such as the trityolodonts, and primitive mammals, the triconodonts, symmetrodonts, and morganucodontids. In beds of Rhaetic age, uppermost Triassic, this type of community, type II, is found in many places in the world. Perhaps the finest known examples are in the Lu Feng deposits in Yunnan, China.

Too little is known to permit speculation upon the exact role of the Mesozoic mammals in the communities of the Jurassic and Cretaceous. At length the "insectivorous mammals," incipient carnivores, broke from the dominant communities, very probably to form a community type resembling what has been defined earlier as type III. Eventually there emerged from this a primitive mammalian community of type II, the ancestral type from which the familiar mammalian assemblages characteristic of the Cenozoic arose. In this was reconstituted much of the structure of the type II therapsid community of earlier times, but it is clear that the mammalian community structure, while similar, did not derive directly from the very mammal-like reptilian community of the Triassic.

Conclusion

The framework that has been established and in which the evolution of mammals has been cast is obviously based on less-than-sufficient evidence at many levels. On the other hand, there is good evidence for part of it and more detailed studies may add strength elsewhere. Some parts, such as the evolution of initial stages of the lepidosaur-archosaur lines may never be known. Much of what has been reconstructed is an extrapolation from the relatively extensive information about parts of the Permian. The rationale of evolution of the mammals that emerges fits the facts as I now understand them. The emergence of mammals is related to the persistence of aquatic habits which are thought to be reflected in many of the mammalian features of structure, physiology and behavior, in particular the premium on activity. The long, slow maturing of mammalian features during the interim period from the beginning Jurassic to late Cretaceous, during which they play an inconspicuous role in ecology, probably gave full measure of gestation to the incipient mammalian characters. The stage was thus set for the rapid and vast radiation that commenced in the latest Cretaceous and reached full swing early in the Cenozoic.

Literature Cited

Eaton, T. H. Jr. 1964. A captorhinomorph predator and its prey (Cotylosauria). Amer. Mus. Nat. Hist. Novitates, no. 2169, 3 p.

Efremov, I. A. 1950. Taphonomy and the geological record. Tr. Paleon. Inst. Acad. Sci. U.S.S.R., **24:** 3-177. (In Russian)

Efremov, I. A., and Vjushkov, B. P. 1955. Catalogue of Permian and Triassic terrestrial vertebrates in the territories of the U.S.S.R. Tr. Paleon. Inst. Acad. Sci. U.S.S.R., **46:** 1-185. (In Russian)

Olson, E. C. 1952. The evolution of a Permian vertebrate chronofauna. Evolution **6:** 181-196.

———. 1957. Catalogue of localities of Permian and Triassic terrestrial vertebrates of the territories of the U.S.S.R. J. Geol., **65:** 196-226.

———. 1958. Fauna of the Vale and Choza: 14. Summary, review and integration of the geology and the faunas. Fieldiana: Geology **10:** 397-448.

———. 1961. The food chain and the origin of mammals., Internat. Colloq. on Evol. Mammals. Kon. Vlaamse Acad. Wetensch. Lett. Sch. Kunsten Belgie, Brussels, 1961, pt. I, p. 97-116.

———. 1962. Late Permian terrestrial vertebrates, U.S.A., and U.S.S.R. Trans. Amer. Philos. Soc., n.s., 52 pt. 2, p 224

Parrington, F. R. 1958. The problem of the classification of reptiles. J. Linn. Soc. London, **44:** 99-115.

Romer, A. S. 1964. *Diadectes* an amphibian? Copeia **1964:** 718-719.

Romer, A. S. and E. C. Olson. 1954. Aestivation in a Permian lungfish. Mus. Comp. Zool., Harvard, Breviora, no. 30. 8 p.

Tatarinov, L. P. 1959. Origin of reptiles and some principles of their classification. Paleon. J., Acad. Sci., U.S.S.R., no. 4. p. 66-84. (In Russian)

Appendix

Class Reptilia
- Subclass Anapsida
  - Order Captorhinomorpha
    - *Captorhinus*
    - *Labidosaurus*
  - Order Procolophonia
  - Order Pareiasauria
- Subclass Diapsida
  - Infraclass Lepidosauria
    - Order Eosuchia
      - Family Millerettidae
      - Family Younginidae
    - Order Rhynocephalia
      - Family Rhynchosaurida
  - Infraclass Archosauria
    - Order Thecodontia
- Subclass Synapsida
  - Order Pelycosauria
    - Suborder Ophiacodontia
      - *Ophiacodon*

Suborder Sphenacodontia
Dimetrodon
Suborder Edaphosauria
Edaphosaurus
Suborder Caseosauria
Family Caseidae
Cotylorhynchus[1]

Order Theriodonta
Suborder Theriodonta

Infraorder Gorgonopsia
Family Phthinosuchidae
Family Gorgonopsidae
Infraorder Therocephalia
Infraorder Cynodontia
Family Gomphodontidae

Suborder Anomodontia
Suborder Tritylodontia

Class Mammalia
Subclass Allotheria
Order Multituberculata

Subclass(es) Uncertain
Order Triconodonta
Order Symmetrodonta
Order Docodonta
Family Morganucodontidae

Subclass Theria
Order Pantotheria
Order Marsupialia
Order Placentalia

PAPER 9

A Continuous Size Spectrum for Particulate Matter in the Sea (1967)

R. W. Sheldon and T. R. Parsons

Commentary

S. K. MORGAN ERNEST

In 1967, R. W. Sheldon and T. R. Parsons published a paper showing how the use of a relatively new piece of equipment, the Coulter counter, could provide more detailed data on particle size in seawater samples. This paper, "A Continuous Size Spectrum for Particulate Matter in the Sea," provided a precise look at the distribution of particle size in aquatic ecosystems. This relationship between the size of a particle and the number of particles of that size, often called the size or biomass spectrum, has gone on to be one of the foundational macroecological patterns in aquatic ecology.

Since Sheldon and Parsons, size-spectrum research has focused on how the shape of the distribution reflects the processes structuring aquatic communities. Most examples of aquatic size spectra are approximately power laws (Kerr and Dickie 2001). The aquatic size spectrum is similar to another foundational macroecological pattern: the self-thinning law in trees (see Yoda et al. 1963). Both relationships focus on the frequency distribution of size in an ecosystem. However, while self-thinning focuses on size within a single trophic level (i.e., trees), the aquatic size spectrum includes disparate taxa and multiple trophic groups (e.g., phytoplankton, zooplankton, herbivorous and predaceous fish). These differences have resulted in very different approaches to explaining the processes creating size-abundance relationships in these systems. Literature on self-thinning has focused on resource competition. Size-spectra literature has focused on the combined importance of individual-level processes such as rates of food intake and maintenance and the structuring effects of predation. Regardless of the differences between the studies of self-thinning and the studies of size spectra, the lasting influences of the foundational papers by J. Yoda et al. (1963) and Sheldon and Parsons reflect how important the frequency distribution of size has been to understanding how ecological systems are structured.

Literature Cited

Kerr, S. R., and L. M. Dickie. 2001. *The biomass spectrum*. Columbia University Press, New York.

Yoda, J., T. Kira, H. Ogawa, and H. Hozumi. 1963. Self-thinning in overcrowded pure stands under cultivated and natural conditions. *Journal of Biology Osaka City University* 14:107–29.

A Continuous Size Spectrum for Particulate Matter in the Sea[1]

By R. W. Sheldon and T. R. Parsons

Fisheries Research Board of Canada
Pacific Oceanographic Group, Nanaimo, B.C.

ABSTRACT

The size spectrum of particulate material in seawater can easily be expressed as total particle volume versus the logarithm of particle diameter. This appears to be the most informative way to present the data and it is also aptly suited to the classical divisions of nanno-, micro-, and macroplankton.

A realistic measure of the volume of irregularly shaped particles such as phytoplankton chains could be made with a Coulter Counter. Particle volume measurements were in good agreement with estimates based on microscopic determination of particle diameter. There were also highly significant correlations between total particle volume, as indicated by the counter, and particulate carbon and nitrogen.

INTRODUCTION

Methods which have been generally employed to estimate the quantity and size distribution of particulate material suspended in seawater include filtration using filters of different pore size (Mullin, 1965), counting and sizing with a microscope (Riley, 1963), and measurement of light transmission and scattering (Jerlov, 1961). The technical operation of the Coulter Counter[2] for measuring the size distribution of particulate matter in the sea has also been described (Sheldon and Parsons, MS, 1966). The latter technique appears to have two distinct advantages in that it is more rapid and gives more detailed results than any previous method. In this report results are presented from field observations which clearly demonstrate that the Coulter Counter can be of considerable value for determining the particle content of seawater.

METHODS

In the following discussion the results of counts made with the Coulter Counter are given in terms of particle diameter. However, as the instrument actually measures particle volume it should be understood that the particle diameter given refers to that of a sphere with the same volume as the particle.

The data presented in this paper were obtained as part of a 2-month time-series study carried out in Saanich Inlet, British Columbia, during June and July 1966. Seawater samples were analysed immediately after collection

[1]Received for publication September 26, 1966.

[2]A patented instrument manufactured by Coulter Electronics, 590 West 20th St., Hialeah, Florida.

J. Fish. Res. Bd. Canada, 24(5), 1967.
Printed in Canada.

for particle size distributions and for total particulate carbon and nitrogen. The samples for particulate carbon and nitrogen analyses were collected by filtration onto glass fibre filters (Whatman GFC), and the carbon and nitrogen contents were determined by total combustion using Coleman carbon and nitrogen analysers.

OBSERVATIONS AND DISCUSSION

A characteristic feature of suspended particulate matter in seawater is that particles of widely different sizes are present. In order to extract the maximum amount of information from a given set of data it is necessary to carefully select the correct grade scale for particle size distributions. The easiest application of the Coulter Counter is to count particles within selected limits of particle volume. This is convenient for certain types of closely graded particle distributions, and has been used in studies on blood and various other kinds of cells. In marine work particle volume has been used in studies of both unicellular (El-Sayed and Lee, 1963; Hastings et al., 1962) and multicellular (Parsons, 1965) planktonic organisms. However, for measurements of particulate matter in the sea the most informative arrangement of data is obtained when particle counts are expressed as volume (or biomass) on a logarithmic scale of particle diameter.

This is illustrated in Fig. 1 where three curves have been plotted using the same data from a sample of seawater in which a bloom of *Skeletonema costatum* occurred. In Fig. 1A the data are shown as they would be most easily read from the counter (i.e. numbers of particles in selected ranges of particle volume). Except for the smallest particle volume setting, the counts in each grade were too small to show clearly on the graph. Figure 1B gives numbers of particles in selected ranges of particle diameter. Although this figure was compiled from data obtained with the Coulter Counter, particle counts made with a microscope would be most easily expressed in the same way. In Fig. 1C, the concentration of particles on a logarithmic scale of particle diameter is given. It is only in this curve that the presence of a diatom bloom becomes apparent. Presentation of data in the form shown in Fig. 1C rather than that shown in Fig. 1A is not difficult if the counts are made on a previously defined logarithmic scale such as the one suggested by Sheldon and Parsons (MS, 1966).

In Fig. 2, examples are given of distributions of suspended matter at two depths. The curves giving number of particles indicate only that fewer small particles were present at depth, but the curves giving concentration show that the total biomass was less over the whole size range at the greater depth. It is further possible to identify exactly where in the size spectrum the greatest decrease in biomass occurred by noting the presence of two peaks in Fig. 2B and the marked absence of the first peak in Fig. 2D.

Figure 3 shows the continuous size distribution of particulate material, with diameters from 3 to 1000 μ, which occurred during June in Saanich Inlet. In order to obtain this size spectrum it was necessary to employ three aperture

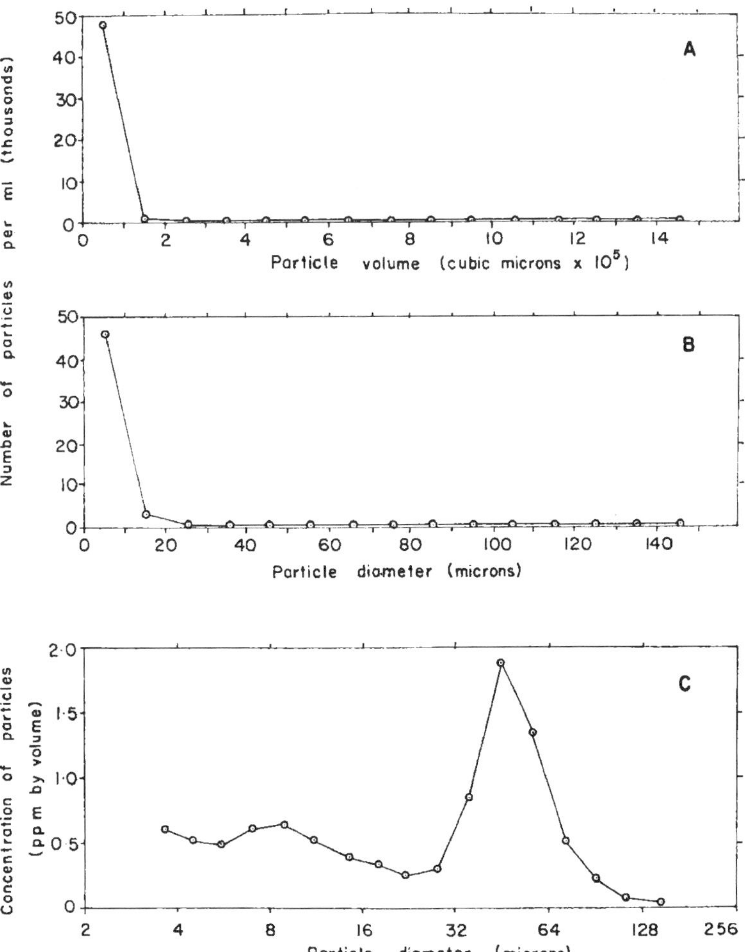

Fig. 1. Particulate matter in a single sample of seawater taken during a bloom of the diatom *Skeletonema costatum*. A, numbers of particles versus particle volume; B, numbers of particles versus particle diameter; C, concentration of particles versus particle diameter on a logarithmic scale.

sizes on the Coulter Counter. The larger particles (with diameters greater than 300 μ) were filtered onto 300-μ netting and then resuspended in membrane-filtered seawater at a suitable concentration. After measuring the size distribution of particles, the results were recalculated to give concentration of large particles in the original sample. In this particular distribution the traditional classification of nannoplankton, microplankton, and macroplankton is shown, following the definitions of Dussart (1965, corrected for a typographical error in the original; Dussart 1966, personal communication). By combining the logarithmic scale suggested here with the plankton divisions given by Dussart (1965), the latter divisions are given equal space on the logarithmic

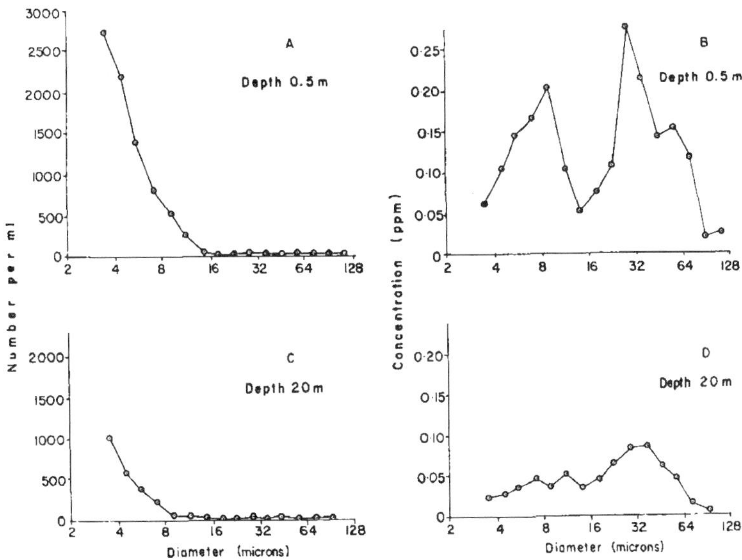

FIG. 2. Particulate matter in two samples of seawater taken at two depths. A and C, numbers of particles versus particle diameter on a logarithmic scale; B and D, concentration of particles versus particle diameter on a logarithmic scale.

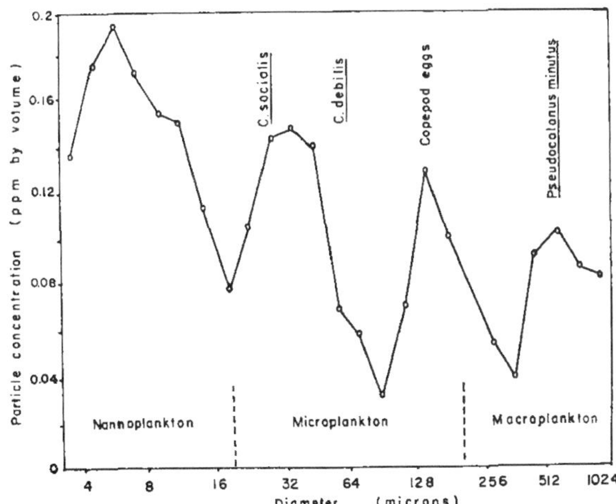

FIG. 3. Continuous size spectrum of particulate matter from Saanich Inlet, June 1966, for particles with diameters between 3 μ and 1 mm.

scale used in Fig. 3 (i.e. nannoplankton, 2–20 μ; microplankton, 20–200 μ; macroplankton, 200–2000 μ, etc.) At the same time the arithmetic progression of cell volumes on which this scale is based is ideally suited to electronic sizing.

The greatest biomass in Fig. 3 occurs among the nannoplankton (unidentified), and a mixture of the diatom species *Chaetoceros socialis* and *C.*

debilis makes up a sizeable fraction of the boimass of the microplankton. The size distribution of copepod eggs extends from micro- to macroplankton, and a portion of the macroplankton is represented by the copepod *Pseudocalanus minutus* which forms a peak at a diameter of about 600 μ. The dimensions of *P. minutus* were approximately 1.0 by 0.4 mm along its major and minor axes which gives a volume of 0.084 mm^3 if the animal's shape is assumed to approximate that of a prolate spheroid. It is of interest to note that the peak of the size distribution for this animal is only slightly more than the radius of a sphere having a volume of 0.084 mm^3.

The difficulty of deciding exactly what the counter records in terms of biomass has been one of the more controversial points to resolve when working with this type of instrument (e.g. see discussion by Hastings et al., 1962), but for most practical purposes the response of the instrument to irregularly shaped particles appears to be essentially proportional to particle volume. For example, in Fig. 3 the peak occurring between particle diameters of 20 and 80 μ represents a mixed population of the two chain-forming diatom species *Chaetoceros socialis* and *C. debilis*. Measurements made with a microscope showed that *C. socialis* cells were approximately 10 μ in diameter with about 30 cells in an average chain, and *C. deblis* cells had a diameter of approximately 20 μ with 20 cells per chain. If, as a first approximation, we assume that each individual cell was spherical, then an average *C. socialis* chain would have a total volume of about 16,000 μ^3, and a *C. debilis* chain would be about 84,000 μ^3. These would be shown on Fig. 3 as particles with diameters of about 31 and 54 μ. These results are not unreasonable. From this approximation it appears that a realistic measure of absolute biomass can be obtained with the instrument, even when dealing with irregularly shaped particles.

More direct evidence of the above point is to be found in Fig. 4 and 5. In these more than 70 measurements of total particle concentration measured by the Coulter Counter (i.e. found by adding together the concentrations for each grade) have been correlated against determinations of particulate carbon and nitrogen. The correlation coefficients for particle concentration against particulate carbon (0.88) and particle concentration against particulate nitrogen (0.52) are both significant ($P < .01$). If a density of unity is assumed for the particulate material and a factor of 0.1 is used to convert wet to dry weight, then the equations (determined as the reduced major axes; Kermack and Haldane, 1950; Imbrie, 1956) give a carbon to dry weight ratio of 0.52 and a nitrogen to dry weight ratio of 0.093. These values are within the upper limits for pure cultures of phytoplankton (Parsons et al., 1961), but are probably high for natural populations. This might reflect an underestimation of absolute volume as measured by the Coulter Counter, or it could indicate that there were significant amounts of inorganic material in suspension, the density of which would be greater than unity. It should also be considered that the amounts of carbon and nitrogen in organic matter are variable, and although a high degree of correlation between these constituents and the total volume of organic matter might be expected, the relationships shown in Fig. 4 and 5

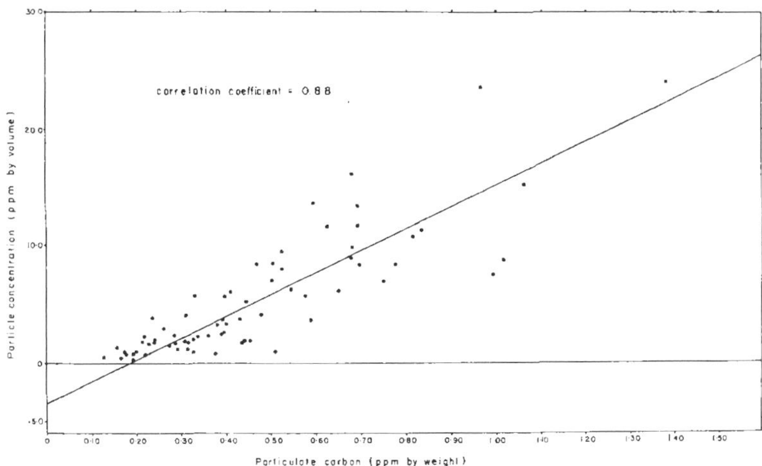

FIG. 4. Relationship between concentration of particulate matter as measured by the Coulter Counter and particulate carbon measured chemically. Particle concentration by volume = 19 (carbon concentration by weight) — 3.56.

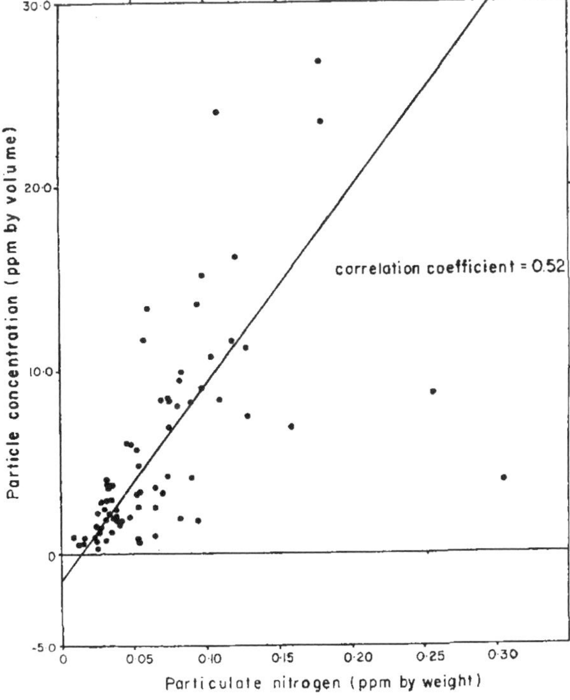

FIG. 5. Relationship between concentration of particulate matter as measured by the Coulter Counter and particulate nitrogen measured chemically. Particle concentration by volume = 107 (nitrogen concentration by weight) — 1.4.

are probably a characteristic of the particular environment studied. Furthermore, it should be emphasised that the relationships in Fig. 4 and 5 are probably linear over the range shown. But if a greater range had been considered the relationships would have been logarithmic because larger organic particles contain relatively smaller proportions of carbon and nitrogen (Mullin et al., 1966).

Other features of interest in Fig. 4 and 5 are the intercepts on the axes for particulate carbon and nitrogen. These were obtained simply by extrapolating each line beyond the range of observations to apparently negative values of particle concentration. The fact that carbon and nitrogen were present in particles which were not detected by the counter is not surprising. The filters used in the determinations of particulate carbon and nitrogen retained all particles with diameters greater than 2 μ, but the data on total volume measured by the counter were only for particles with diameters between 3 and 200 μ, as this was the range within which most of the phytoplankton occurred and over which counts were routinely made. Thus the intercepts on the graphs are a measure of the particulate material not estimated by the Coulter Counter.

ACKNOWLEDGMENTS

The work of one of us (R. W. S.) was carried out during the tenure of a National Research Council Postdoctoral Fellowship.

REFERENCES

DUSSART, B. M. 1965. Les différentes catégories de plancton. Hydrobiologia, **26**: 72–74.

EL-SAYED, S. Z., AND B. D. LEE. 1963. Evaluation of an automatic technique for counting unicellular organisms. J. Marine Res., **21**: 59–73.

HASTINGS, J. W., B. M. SWEENEY, AND M. M. MULLIN. 1962. Counting and sizing of unicellular marine organisms. Ann. N.Y. Acad. Sci., **99**: 280–289.

IMBRIE, J. 1956. Biometrical methods in the study of invertebrate fossils. Bull. Am. Museum Nat. Hist., **108**: 217–252.

JERLOV, N. G. 1961. Optical measurements in the eastern North Atlantic. Goteborgs Kgl. Vetenskaps- Vitterhets- Samhall. Handl., Ser. B., **8**: 1–39.

KERMACK, K. A., AND J. B. S. HALDANE. 1950. Organic correlation and allometry. Biometrika, **37**: 30–41.

MULLIN, M. M. 1965. Size fractionation of particulate organic carbon in the surface waters of the western Indian Ocean. Limnol. Oceanog., **10**: 459–462.

MULLIN, M. M., P. R. SLOAN, AND R. W. EPPLEY. 1966. Relationship between carbon content, cell volume and area in phytoplankton. Ibid., **11**: 307–311.

PARSONS, T. R. 1965. An automated technique for determining the growth rate of chain forming phytoplankton. Ibid., **10**: 598–602.

PARSONS, T. R., K. STEPHENS, AND J. D. H. STRICKLAND. 1961. On the chemical composition of eleven species of marine phytoplankters. J. Fish. Res. Bd. Canada, **18**: 1001–1016.

RILEY, G. A. 1963. Organic aggregates in sea-water and the dynamics of their formation and utilization. Limnol. Oceanog., **8**: 372–381.

SHELDON, R. W., AND T. R. PARSONS. MS, 1966. On some applications of the Coulter Counter to marine research. Fish. Res. Bd. Canada, MS Rept. Ser. (Oceanog. Limnol.). No. 214. 36 p.

PAPER 10

Net Primary Productivity of Terrestrial Communities: Prediction from Climatological Data (1968)

M. L. Rosenzweig

Commentary

S. K. MORGAN ERNEST

Before M. L. Rosenzweig, the idea that ecosystem productivity drives many different broadscale patterns, from the latitudinal diversity gradient to body-size distributions, was pervasive but relatively untested (e.g., Connell and Orias 1964; Hutchinson 1959). The problem was not only how to measure productivity but how to obtain productivity estimates from numerous divergent ecosystems across the globe in a practical and useful fashion. Net primary productivity, the measure of productivity most closely tied to mechanisms ecologists were proposing to explain broadscale patterns, is laborious and time intensive to collect for a single location, much less across broad spatial scales. Rosenzweig's paper is a foundational contribution to macroecological research because it demonstrates that actual evapotranspiration—a value easily calculable from latitude and monthly averages of precipitation and temperature—can serve as a reasonable proxy for net primary productivity. Since climate data is much more widely collected and available than direct measurements of ecosystem productivity, this allows for ideas to be tested with potentially hundreds of data points instead of with only a few or even none at all.

Nowadays, with the easy availability of satellite imagery, which provides ecosystem productivity proxies for locations all over the world, the foundational nature of Rosenzweig's paper is easy to take for granted. However, this paper made two critically important contributions to the future of macroecology. It set the precedent for the acceptability of using indirect measures as surrogates for harder-to-quantify values, a principle still employed when "greenness indices" from satellite imagery are used instead of directly measured plant production values. Furthermore, by showing that actual evapotranspiration and net annual aboveground productivity were highly correlated, Rosenzweig set the stage for an explosion of empirical macroecological tests of concepts long hypothesized but never rigorously assessed.

Literature Cited

Connell, J. H., and E. Orias. 1964. The ecological regulation of species diversity. *American Naturalist* 98:399–414.

Hutchinson, G. E. 1959. Homage to Santa Rosalia; or, Why there are so many kinds of animals. *American Naturalist* 93:145–59.

Vol. 102, No. 923 The American Naturalist January–February, 1968

NET PRIMARY PRODUCTIVITY OF TERRESTRIAL COMMUNITIES: PREDICTION FROM CLIMATOLOGICAL DATA

MICHAEL L. ROSENZWEIG

Division of Biology, University of Pennsylvania, Philadelphia, Pennsylvania, and
Department of Biology, Bucknell University, Lewisburg, Pennsylvania 17837

INTRODUCTION

A major research problem facing ecologists today is the quantification of energy flow in natural communities. Discussion of other ecological problems revolving around productivity would certainly benefit if some readily available quantitative estimate of productivity existed. Pianka (1966), for example, discussed the relationship of species diversity to productivity; unfortunately, he had no quantitative data to aid him. In my own work involving an analysis of the evolutionary causes of body size clines in certain terrestrial carnivores, I long ago appreciated the needs for some quantitative estimate of energy flow.

In 1947 Holdridge published a scheme which underlined the great degree to which the abiotic environment determines somehow the characteristics of the mature vegetation of terrestrial communities. This set me to wondering whether some environmental variable might exist which correlates well enough with primary terrestrial production to be usable as its predictor.

Major (1963) has noted that the actual evapotranspiration (AE) in a terrestrial environment is qualitatively related to the amount of vascular plant activity. AE may be defined as precipitation, minus runoff, minus percolation (Sellers, 1965). AE may be thought of as the reverse of rain. It is the amount of water actually entering the atmosphere from the soil and the vegetation during any period of time, that is, the evaporation plus the transpiration (Sellers, 1965). Obviously, this atmospheric entry simultaneously requires water and sufficient energy to make the phase transfer of the water possible. Thus, AE is a measure of the simultaneous availability of water and solar energy in an environment during any given period of time. Note that Holdridge's (1947) scheme depended on two variables similar to the latter, that is, precipitation and biotemperature. Further, Holdridge (1959) later showed that biotemperature and potential evapotranspiration (PE) are linearly related. PE may be defined as the amount of evapotranspiration that could occur if the soil of a large area having "vegetation typical of the surroundings" (Sellers, 1965, p. 163) were kept constantly wet, that is, at or above field capacity (Sellers, 1965). In fact, PE is the solar energy estimate used in calculating AE. Hence, it is quite relevant to wonder if AE is an environmental variable usable for the purpose of predicting production.

METHODS AND CRITERIA OF DATA COLLECTION

Thornthwaite and Mather (1957) have published a useful reference enabling an estimate of AE to be calculated from a knowledge of the latitude of a place and its mean temperature, month by month, and its precipitation, month by month. I have used such estimates of AE in this study.

Because of the fact that the special requirements of this analysis (see below) render an already scarce kind of measurement (production) scarcer, it was necessary to select as the dependent variable, net above-ground productivity—the commonest kind of production estimate. Study of other sorts of production data in the future will undoubtedly prove interesting.

Quite clearly, each production value in our collection of data (Table 1) must be accompanied by a fairly reliable estimate of the AE for the same period of time. Ideally, both productivity and AE are measured simultaneously. I know of not even one case where this happy coincidence has occurred. Fortunately, several ecologists have had the foresight to include some meteorological data along with their productivity work. Unfortunately, such data are only rarely convertible into an AE. Since this proscribes almost all work done previously, I simply made the convention of using an average annual AE garnered from some nearby weather station by the Laboratory of Climatology, Centerton, New Jersey (Thronthwaite Associates, 1962, 1964).

Due to a lack of the necessary climatological data for most tundra studies, AE could be estimated for only two constantly moist tundras where AE = PE.

In arid environments, where PE is greater than precipitation, over a broad area annual AE is very nearly equal to annual precipitation. Local runoff is common, however, and creates extremely variable plot-to-plot conditions (Hillel and Tadmor, 1962). These depend largely on local topographical relief. Creosote bushes are generally not inhabitants of desert slopes or gullies, but instead cover relatively broad, flat stretches of desert. Thus, point *A*, the creosote bush desert, was usable. Also, Pearson (1966) specifically states that there is little or no runoff from his sand dune associations. This is in agreement with the results of Hillel and Tadmor (1962). Hence, the precipitation Pearson gives for these dunes may be taken as their AE, and we obtain our point *N*.

Except for the one value used here, all tall grass prairie productions I have seen were obtained by a fall harvest method. Such a procedure does not take into account the amount of material that dies during the growing season. By completely denuding his prairie plots before beginning his harvests, Penfound (1964) showed the importance of such omissions in creating a gross underestimate of production. Penfound's was, therefore, the only tall grass prairie productivity used (point *D*). This work had the additional advantage of including simultaneous monthly temperatures and precipitations.

The data of Whittaker (1963, 1966) were obtained in the Great Smoky Mountains. Because of adiabatic cooling of the atmosphere, altitude is a

TABLE 1

SUMMARY OF PRODUCTION DATA USED IN REGRESSION

Code	Environment	Place	Log NAAP	Log AE	Reference
A	Creosote bush desert*	Nye Co., Nevada, U.S.	1.60	2.10	Odum, 1959
B	Arctic moist tundra†	Cape Thompson, Alaska, U.S.	2.16	2.30	Rickard, 1962
C	Alpine moist tundra‡	Mt. Washington, N. H., U.S.	2.16	2.37	Hadley and Bliss, 1964
D	Tall grass prairie*	Norman, Oklahoma, U.S.	2.75	2.79	Penfound, 1964
E	Heath Bald (Leiophyllum)§	Great Smoky Mts., Tenn., U.S.	2.66	2.58	Whittaker, 1963
F	Heath Bald (Rhododendron)§	Great Smoky Mts., Tenn., U.S.	2.61	2.58	Whittaker, 1963
G	Heath Bald (Rhododendron)§	Great Smoky Mts., Tenn., U.S.	2.69	2.58	Whittaker, 1963
H	Mixed Heath (Peregine Peak)§	Great Smoky Mts., Tenn., U.S.	2.58	2.72	Whittaker, 1963
J	Mixed Heath (Rocky Spur)§	Great Smoky Mts., Tenn., U.S.	2.80	2.69	Whittaker, 1963
K	Beech-maple forest*	Toronto, Ontario, Canada	2.98	2.75	Bray, 1964
L	Secondary tropical forest	Kade, Ghana	3.34	3.09	Nye, 1961
M	Tropical forest	Yangambi, Congo (Leopoldville)	3.46	3.12	Bartholomew, Meyer, and Laudelout, 1953
N	Cool desert sand dunes‡	Near Rexburg, Idaho, U.S.	2.24	2.34	Pearson, 1966
O	Oak-hickory forest§	Oak Ridge, Tenn., U.S.	3.08	2.92	Whittaker, 1966
P	Cheatgrass§	Hanford Reservation, Washington, U.S.	2.01	2.25	Rickard, 1962
Q	Fraser fir forest§	Great Smoky Mts., U.S.	2.75	2.61	Whittaker, 1966
R	Spruce fir forest (Mt. Mingus)§	Great Smoky Mts., Tenn., U.S.	2.97	2.68	Whittaker, 1966
S	Spruce fir forest (Mt. Collins)§	Great Smoky Mts., Tenn., U.S.	3.01	2.64	Whittaker, 1966
T	Gray beech forest§	Great Smoky Mts., Tenn., U.S.	2.96	2.69	Whittaker, 1669
U	Gray beech forest§	Great Smoky Mts., Tenn., U.S.	2.82	2.69	Whittaker, 1966
V	Hemlock—mixed forest§	Great Smoky Mts., Tenn., U.S.	3.07	2.82	Whittaker, 1966
W	Upper cove forest§	Great Smoky Mts., Tenn., U.S.	3.04	2.74	Whittaker, 1966
X	Deciduous cove forest§	Great Smoky Mts., Tenn., U.S.	3.09	2.85	Whittaker, 1966
Z	Hemlock—rhododendron forest§	Great Smoky Mts., Tenn., U.S.	3.01	2.75	Whittaker, 1966

* Judged climax, assumed stable from Espenshade (1957).
† Judged stable by A. W. Johnson *et al.* (1966).
‡ Assumed stable.
§ Judged stable by original author(s).

most important cause of AE change in mountains. Whittaker recorded the altitude of each of his plots; therefore, I was able to estimate AE for his plots by interpolating their altitudes on a graph of altitude versus mean AE. The graph was drawn using mean AE data obtained at nine regional meteorological stations (Thornthwaite Associates, 1964), and coupling them with three empiral observations: Shanks' (quoted in Whittaker, 1966) observation that there is an average 4.1°C decrease in temperature for every thousand-meter increase in altitude in the Smokies; Holdridge's (1962) observation that PE = 58.93 B (where B is "biotemperature") and Major's (1963) and Thornthwaite Associates' (1964) observation that in the Smokies AE is always very nearly equal to PE. From these facts, I assumed that a 1,000-m increase in altitude was accompanied by a 4.1° decrease in biotemperature, hence by a 241.6-mm decrease in AE. The meteorological data confirmed this slope and determined the intercept.

As mentioned above, Holdridge (1947) had been successful in arranging only climax vegetation into his scheme. Prairie vegetation might be temporarily successful as a sere in an area that would later be forest. Clearly, different growth forms in the same region show widely different net productivities (Odum, 1960; Whittaker, 1963; Ovington, Heitkamp, and Lawrence, 1963). Clearly, also, one environmental value cannot be made to predict more than one productivity. One might suspect, therefore, that energy flow values in communities possessing the growth form of the local climax vegetation are the only energy flow values predictable from any general climatic variable.

Further, we have the clear and strong admonition of Whittaker (1966, p. 116) that "Unstable stands should not be compared with stable ones in the study of environmental effects on production; and unstable stands may not, unless carefully chosen for comparison, indicate reliably the effects of environment on production." Restriction of data to those from mature communities seemed an ideal strategy, too, in view of the ultimate purpose of the effort, that is, to study evolutionary pressure in widely scattered vertebrate populations which had been subjected mostly to mature vegetation for centuries past. Hence I did attempt to so restrict data.

The primary and most often used criterion for rejection or acceptance of the data was a relevant statement made by the original reporter of the data. When doubtful cases occurred, the vegetation maps of A. W. Küchler as they appear in *Goode's World Atlas* (Espenshade, 1957; pp. 16, 17, 52, 53) were used to determine which data were or were not acceptable. For example, the interesting and often cited data of Ovington (1963) obtained on the Cedar Creek Natural History Reservation, Minnesota, could not be used here. One plot was prairie; a second, savanna; a third, oakwood. The authors do not identify any of their plots as being climax, and Küchler denotes this area of Minnesota as being either a maple-basswood or a maple–yellow birch–hemlock–pine climax.

In the final analysis, I must disclaim any attempt or success at collecting all usable records of production. Rather, my search was for data from as wide a variety of environments as possible. This is really the only justification for including the two rather insecure but currently available estimates of production in tropical forests (L and M). Indeed, point L does not even represent a fully mature association; Nye (1961) calls it an old secondary forest. I hope that researchers in tropical forests will soon provide ecology with more secure estimates of production in this extremely interesting type of environment.

RESULTS

All data were transformed to common logarithms. Using the method of least squares, linear regression of the productivity on the AE was performed. The data and regression line appear in Figure 1. The productivity prediction equation, including 5% confidence intervals for the slope and intercept is:

$$\log_{10}\text{NAAP} = (1.66 \pm 0.27) \log_{10}\text{AE} - (1.66 \pm 0.07), \qquad (1)$$

where AE is the annual actual evapotranspiration in millimeters, and NAAP is the net annual above-ground productivity in grams per square meter.

DISCUSSION

The fact that AE is a measure of the simultaneous availability of water and energy in an environment suggests to me an explanation as to why it should be a successful predictor of production. Gross productivity may be defined as the integral of the rate of photosynthesis throughout the year. The rate of photosynthesis depends on the concentrations of its raw materials, and water and solar radiational energy are two of these. In terrestrial environments, the third, CO_2, is a more or less constant 0.029% (Sellers, 1965). Thus, the AE is a measure of the two most variable photosynthetic resources.

If AE is a good predictor of NAAP because it is a good predictor of the rate of photosynthesis, then the ratio of shoot to root and of gross-to-net production must be approximately constant in all the communities included in this study. According to Bray (1963), it is fairly accurate to assume a constant shoot-to-root ratio (roots about 20% to 50% of shoots) for these plant communities. And, according to Muller (1962), the net-to-gross statement is also fairly accurate (net about 40% to 60% of gross).

Also necessary for production is the array of biochemicals and minerals that form the photosynthetic machinery. However, unless there is a severe shortage of one or more of the essential nutrients necessary for their synthesis, this production apparatus might be expected to be synthesized in optimal amounts at just the right times by a natural plant community subject to evolutionary pressures. Perhaps plants can even evolve to sidestep some regional nutrient deficiencies. In any case, we should not expect production to be commonly limited by anything under the control of the plant organism.

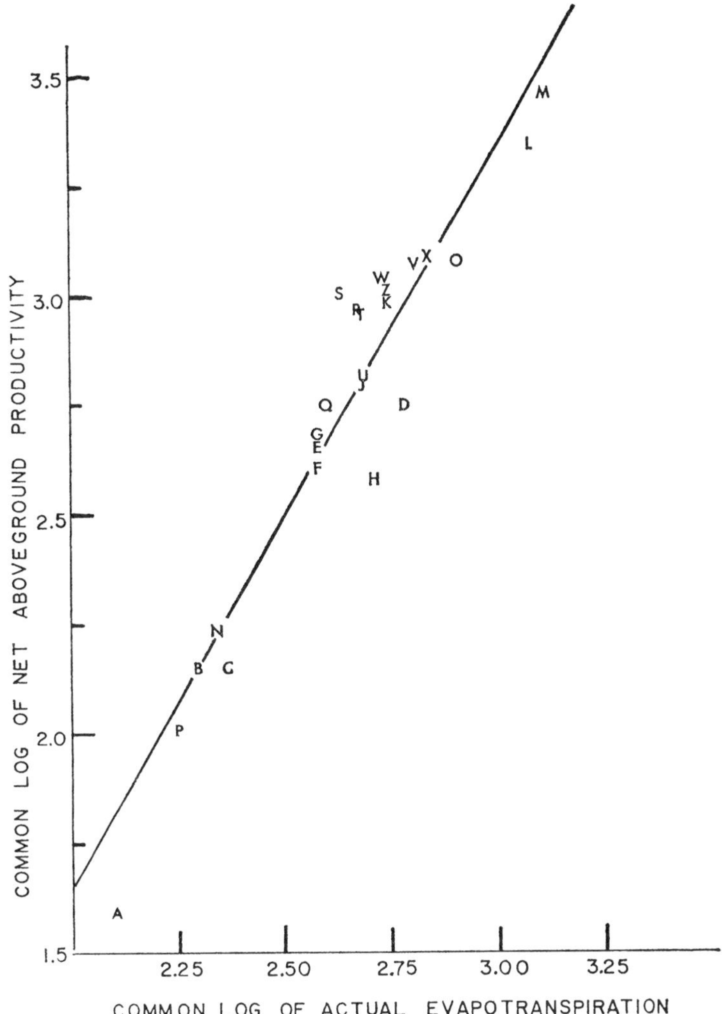

FIG. 1.—Net above-ground productivity in grams of dry matter per square meter graphed against actual evapotranspiration in millimeters. The regression line (see text) is included. See Table 1 for identities of coded points.

As I have hinted above, there are bound to be special local conditions which no general continental model can take into account. Such would be shallow soil on a mountainside, or a local deficiency of some absolutely essential nutrient like phosphorus. According to Wilson (1948), there might be local conditions of frequent fog which would raise productivity greatly by raising CO_2 concentration; AE will not predict this either.

The relationship of the productivity of successional vegetational stages to AE seems worth attention. However, this is absolutely impossible to investigate without simultaneous measurements of AE. Holdridge (1962) pointed out that the real AE (as opposed to the Thornthwaite estimated AE) is not unaffected by the character of the vegetation. Lower growths of foliage than the mature height should have a smaller AE, due presumably to decreased transpiration surface and water-holding capacity in the lower vegetation. Similarly, from his formula (Holdridge, 1962) one can conclude that higher successional stages should have greater AE than that over the mature vegetation. By using Thornthwaite estimates of AE and mature communities, I have avoided this problem. It should not be avoided forever. Interesting, indeed, would be the discovery, using real AE values, that production and real AE are closely correlated regardless of the sere. This would be strong evidence for the causality of the relationship. Further, it would be of interest to students of the evolution of efficiency in natural communities. Holdridge (1962) assumes that evaporation and transpiration are approximately equal in all types of environments. However, Hillel and Tadmor (1962) estimate that transpiration varies from 44% to about 90% of the AE in their four environments. Perhaps such variation is peculiar to deserts. It is interesting to speculate that there exists some optimal efficiency of water utilization in any environment and that it is approached as both short-term succession and long-term evolution proceed.

I cannot overemphasize the hypothetical nature of the foregoing explanation for the correlation of AE and productivity. Still, the relative usefulness of any prediction is determined by its precision. Hence, the accuracy of my explanation might be nil, but the fact that the correlation exists would remain of interest. Perhaps others can, as I have, use it in studies of the broad effects of production variation on other natural phenomena. As to my hypothesis of the cause of this correlation, that may yet prove useful, but only insofar as it can stimulate the experimental analysis that replaces it or proves it.

SUMMARY

Actual evapotranspiration (AE) is shown to be a highly significant predictor of the net annual above-ground productivity in mature terrestrial plant communities. Communities included ranged from deserts and tundra to tropical forests. It is hypothesized that the relationship of AE to productivity is due to the fact that AE measures the simultaneous availability of water and solar energy, the most important rate-limiting resources in photosynthesis.

ACKNOWLEDGMENTS

I thank Drs. R. H. MacArthur, W. J. Smith, and J. Preer, Jr., for advice and criticism. Thanks, also, are due the anonymous referees, whose criticisms were helpful in improving the manuscript. Much of the work reported on was done while I held a Cooperative-Graduate Fellowship from the National Science Foundation at the University of Pennsylvania. Material for this paper was taken, in part, from my Ph.D. dissertation.

LITERATURE CITED

Bartholomew, W. V., J. Meyer, and H. Laudelout. 1953. Mineral nutrient immobilization under forest and grass fallow in the Yangambi region (Belgian Congo). Pub. Inst. Etude Agron. Congo Belge 57.

Bray, J. R. 1963. Root production and the estimation of net productivity. Can. J. Bot. 41:65–72.

———. 1964. Primary consumption in three forest canopies. Ecology 45:165–167.

Espenshade, E. B., Jr. (ed.). 1957. Goode's world atlas. 10th ed. Rand McNally, Chicago. 272 p.

Hadley, E. B., and L. C. Bliss. 1964. Energy relationships of Alpine plants on Mt. Washington, N.H. Ecol. Monogr. 34:331–357.

Hillel, D., and N. Tadmor. 1962. Water regime and vegetation in the central Negev highlands of Israel. Ecology 43:33–41.

Holdridge, L. R. 1947. Determination of world plant formations from simple climatic data. Science 105:367–368.

———. 1959. Simple method for determining potential evapotranspiration from temperature data. Science 130:572.

———. 1962. The determination of atmospheric water movements. Ecology 43:1–9.

Johnson, A. W., L. A. Viereck, R. E. Johnson, and H. Melchior. 1966. Vegetation and flora, p. 277–354. *In* N. J. Wilimovsky and J. N. Wolfe [eds.], Environment of the Cape Thompson region, Alaska. U.S. Atomic Energy Comm., Div. Tech. Information, Oak Ridge, Tennessee.

Major, Jack. 1963. A climatic index to vascular plant activity. Ecology 44:485–498.

Muller, D. 1962. Wie gross ist der prozentuale Anteil der Nettoproduktion an der Bruttoproduktion? p. 26–28. *In* H. Leith, Die Stoffproduktion der Pflanzendecke. Gustav Fischer, Stuttgart. 156 p.

Nye, P. H. 1961. Organic matter and nutrient cycles under moist tropical forest. Plant and Soil 13:333–346.

Odum, E. 1959. Fundamentals of ecology. Saunders, Philadelphia. 546 p.

———. 1960. Organic production anh turnover in old-field succession. Ecology 41:34–49.

Ovington, J. D., D. Heitkamp, and D. B. Lawrence. 1963. Plant biomass and productivity of prairie, savanna, oakwood and maize field ecosystems in central Minnesota. Ecology 44:52–63.

Pearson, L. C. 1966. Primary productivity in a northern desert area. Oikos 15:211–218.

Penfound, W. T. 1964. Effects of denudation on the productivity of grasslands. Ecology 45:838–845.

Pianka, E. R. 1966. Latitudinal gradients in species diversity: a review of concepts. Amer. Natur. 100:33–46.

Rickard, W. H. 1962. Comparison of annual harvest yields in an arctic and a semi-desert plant community. Ecology 43:770–771.

Sellers, W. D. 1965. Physical climatology. Univ. Chicago Press, Chicago. 272 p.

Thornthwaite Associates. 1962. Average climatic water balance data of the continents, Africa. Lab. Climatol., Pub. Climatol. 15:115–287.

———. 1964. Average climatic water balance data of the continents, North America. Lab. Climatol., Pub. Climatol. 17:231–615.

Thornthwaite, C. W., and J. R. Mather. 1957. Instructions and tables for computing potential evapotranspiration and the water balance. Drexel Inst. Technol., Lab. Climatol., Pub. Climatol. 10:181–311.

Whittaker, R. H. 1963. Net production of heath balds and forest heaths in the Great Smoky Mountains. Ecology 44:176–182.

———. 1966. Forest dimensions and production in the Great Smoky Mountains. Ecology 47:103–121.

Wilson, C. C. 1948. Fog and atmospheric carbon dioxide as related to apparent photosynthetic rate of some broadleaf evergreens. Ecology 29:507–508.

PAPER 11

Marine Benthic Diversity: A Comparative Study (1968)

H. L. Sanders

Commentary

ANDREW CLARKE

The sea is vast and difficult to access. In consequence, knowledge of even quite fundamental aspects of marine ecology can lag behind that of land ecology. While the intertidal has been a fruitful source of inspiration for many ecologists, as well as the site of some classic experimental work, the open ocean and the deep sea remain among the least-known parts of the planet. By the 1960s, ecologists were aware of broadscale patterns in marine diversity, such as the richness of tropical reef habitats or the low species richness of some polar habitats, but as with so much of marine ecology, hard data were difficult to come by. A key problem had been identified by Gunnar Thorson, who suggested in the 1950s that there was evidence for a latitudinal cline in the richness of hard substratum assemblages, but not for soft sediments. It was the latter problem that Howard Sanders tackled in this classic paper.

Sanders took up the challenge of comparing the diversity of soft sediment assemblages from around the world. In doing so, he devised a novel method (rarefaction) for comparing different samples of marine benthos, drew important distinctions between assemblages where the predominant structuring forces were physical or biological, and introduced the influential stability-time hypothesis to explain why habitats such as the deep sea can be unexpectedly diverse. With its clear recognition of the importance of scale, care to ensure comparability of sampling, concentration on important but taxonomically well-known groups, statistical rigor, and exemplary setting of the results in the intellectual context of the time, this seminal paper is a model of how marine diversity work should be done.

The legacy of this work is immense, for it introduced techniques still in use and laid out ideas that continue to inform scientists today. The deep sea remains a difficult and intractable place, but the subsequent work by Fred Grassle, Nancy Maciolek, Mike Rex, and their colleagues in the United States, and John Gage and Paul Tyler in the United Kingdom, can be traced back directly to this seminal paper. It remains essential and inspirational reading for any student setting out to tackle problems in marine diversity.

From *American Naturalist* 102:243–82. *The American Naturalist* © 1968 The University of Chicago. Reprinted with permission from The University of Chicago.

Vol. 102, No. 925 The American Naturalist May–June, 1968

MARINE BENTHIC DIVERSITY: A COMPARATIVE STUDY*

HOWARD L. SANDERS

Woods Hole Oceanographic Institution, Woods Hole, Massachusetts 02543

INTRODUCTION

One of the major features of animal communities is their diversity, that is, the number of species present and their numerical composition. It has long been recognized that tropical regions, by and large, support a more diverse fauna than do regions of higher latitude. In the aquatic medium it is also evident that the marine habitats contain a greater wealth of species than do brackish regions. The reasons why certain environments harbor many kinds of organisms while others support a very limited number of species are still unclear. Various theories based on time (Fischer, 1960; Simpson, 1964), climatic stability (Klopfer, 1959; Fischer, 1960; Dunbar, 1960), spatial heterogeneity (Simpson, 1964), competition (Dobzhansky, 1950; Williams, 1964), predation (Paine, 1966), and productivity (Connell and Orias, 1964) have been proposed to explain these differences.

In the present paper, data collected from soft-bottom marine and estuarine environments of a number of differing regions will be used in a comparative study of within-habitat diversity. A new diversity measurement will be presented that is independent of sample size, and a hypothesis will be proposed to explain the observed patterns of diversity as well as to provide a framework for interpreting other diversity studies.

MATERIALS AND METHODS

The stations used in this study are as follows:

RH-14: Arabian Sea off Cochin, Kerala State, India (14 m).
RH-26: Bay of Bengal off Porto Novo, Madras State, India (20 m).
RH-28: Vellar River estuary, Porto Novo, Madras State, India (2 m).
RH-30: Bay of Bengal off Madras, India (15 m).
RH-33: Kakinada Bay, Andhra State, India (2 m).
RH-36: Bay of Bengal off Kakinada, Andhra State, India (37 m).
RH-41: Arabian Sea off Bombay, India (20 m).
RH-51: Indian Ocean off Hellville, Nossi Bé, Madagascar (18 m).
C#1: Outer continental shelf south of New England (40°27.2'N 70°47'W, 97 m).
S1.3: Upper continental slope south of New England (39°58.4'N 70°40.3'W, 300 m).
D#1: Upper continental slope south of New England (39°54.5'N 70°35'W, 487 m).
F#1: Lower continental slope south of New England (39°47'N 70°45'W, 1,500 m).

*Contribution No. 1959 from the Woods Hole Oceanographic Institution, Woods Hole, Massachusetts 02543.

G#1: Lower continental slope south of New England (39°42′N 70°39′W, 2,086 m).
GH#1: Abyssal rise south of New England (39°25.5′N 70°35′W, 2,500 m).
DR-12: Continental slope off northeast South America (07°09′S 34°25.5′W, 790 m).
DR-33: Continental slope off northeast South America (07°53.5′N 54°33.3′W, 535 m).
POC 1, 2, 3, 4: Pocasset River, Cape Cod, Massachusetts (0.5 m).
R Series: Buzzards Bay, Massachusetts (20 m).

All samples were collected with an Anchor dredge or a Higgins meiobenthic sled (the RH series of samples). The sediments were processed through a fine-meshed screen with 0.4 mm apertures, and the animals were carefully picked out and sorted in the laboratory. Sanders, Hessler, and Hampson (1965) gave details on the Anchor dredge and the methodology of processing.

THE RAREFACTION METHODOLOGY AND RESULTS

Since most diversity measurements are affected by sample size (see later discussion), it would be most useful to have a procedure which will allow one to compare directly samples of differing sizes. If this can be achieved, it may then be possible to perceive more clearly the factors influencing biological diversity. The rarefaction method, which permits each sample to generate a line, was developed to achieve this end. This methodology was applied to benthic marine samples collected from boreal estuary, boreal shallow marine, tropical estuary, tropical shallow marine, and deep-sea environments. The usual difficulty inherent in comparing samples of different sizes is that as sample size increases, individuals are added at a constant arithmetic rate but species accumulate at a decreasing logarithmic rate. The rarefaction method, instead, is dependent on the shape of the species abundance curve rather than the absolute number of specimens per sample. In all cases, the sediments were soft oozes and, therefore, comparable in regard to particle size.

The comparison was based on the polychaete-bivalve fraction of the samples rather than the entire fauna. Since these two groups comprise about 80% of the animals by number in most of the samples (Fig. 1), one can feel justified in generalizing from whatever results may be found. This study shows a systematic pattern of diversity that can be correlated with the variability of the physical environment.

With the method of diversity analysis developed for this study, samples with different numbers of specimens and from different regions of the world were compared directly. The procedure was to keep the percentage composition of the component species constant but reduce the sample size, that is, to artificially create the results that would have been obtained had smaller samples with the identical faunal composition been taken. Using this technique, the expected number of species present in populations of different sizes, that is, numbers of species per 10, 25, 50, 100, 200, . . ., 1,000, 2,000, etc., was determined.

In order to evaluate the validity of this method, one must understand how the values are obtained. The species are ranked by abundance, and

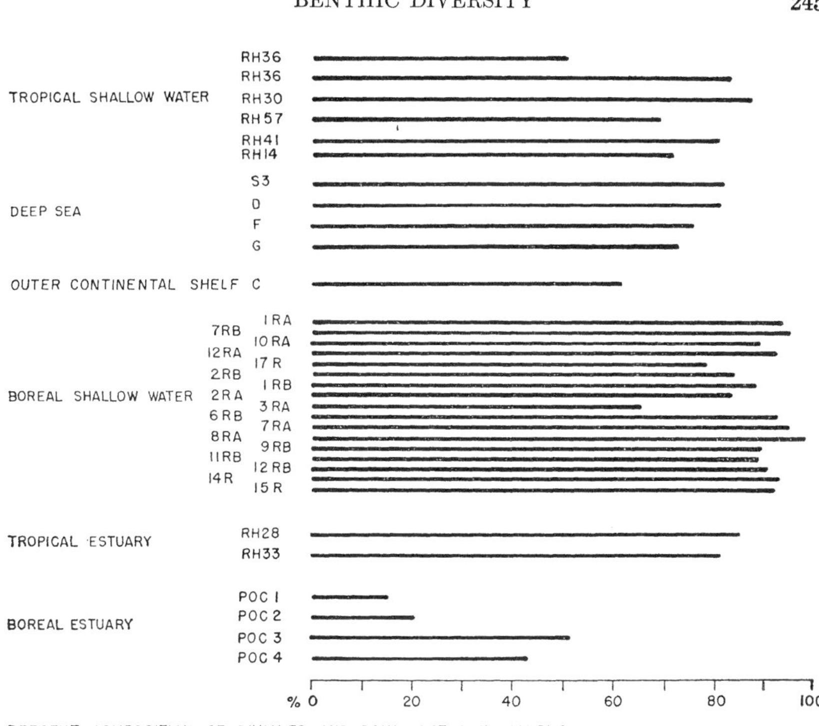

FIG. 1.—Percentage composition of the polychaete-bivalve fraction of the soft-bottom samples used in the analysis.

the percentage composition of each species and the cumulative percentage are plotted. In a hypothetical sample (Table 1) there are 1,000 individuals and 40 species. As an example, the number of species at the 25-individual level will be determined. The percentage composition is the same as in the original sample, but the number of individuals is reduced to 25. Since 25 specimens in this reduced sample represent 100% of the individuals present, then each individual specimen forms 4% of the sample. In the original sample, seven species each comprise 4% or more, and in total they compose 76% of the sample by number. Therefore, each of these seven species will be present in the reduced sample. This leaves a residue of 24% of the original sample comprising the remaining 33 species. Because none of these species forms more than 4% of the original sample, those species of this group that will appear in the reduced sample cannot be represented by more than one individual. Since one specimen comprises 4% of the reduced sample, therefore 24%/4% = 6 species; 7 + 6 = 13 species present per 25 individuals.

The determination of species per 100 individuals is as follows: (1) Since each individual represents 1% of the sample, then (2) 15 species in Table 1 each comprise $\geqq$ 1.0% of the fauna and cumulatively = 92.1% of the

TABLE 1

HYPOTHETICAL SAMPLE WITH 1,000 INDIVIDUALS AND 40 SPECIES

Rank of Species by Abundance	Number of Individuals	% of Sample	Cumulative of Sample %
1	365	36.5	36.5
2	112	11.2	47.7
3	81	8.1	55.8
4	61	6.1	61.9
5	55	5.5	67.4
6	46	4.6	72.0
7	40	4.0	76.0
8	38	3.8	79.8
9	29	2.9	82.7
10	23	2.3	85.0
11	21	2.1	87.1
12	15	1.5	88.6
13	13	1.3	89.9
14	12	1.2	91.1
15	10	1.0	92.1
16	8	0.8	92.9
17	7	0.7	93.6
18	7	0.7	94.3
19	6	0.6	94.9
20	6	0.6	95.5
21	5	0.5	96.0
22	5	0.5	96.5
23	5	0.5	97.0
24	4	0.4	97.4
25	3	0.3	97.7
26	3	0.3	98.0
27	3	0.3	98.3
28–33	2 each	0.2 each	99.3
34–40	1 each	0.1 each	100.0
Total Number	1,000		

sample. (3) The residue = 7.9% of the sample; 7.9%/1.0% = 7.9 species. (4) 15 + 7.9 = 22.9 species per 100 individuals.

For species per 200 individuals: (1) Each individual forms 0.5% of the sample, and (2) 23 species each represent ≧ 0.5% of the fauna and cumulatively = 97.0% of the sample. (3) The residue = 3.0%; 3.0%/0.5% = 6.0 species. (4) 23 + 6.0 = 29.0 species per 200 individuals.

Using this technique, we have made, in Figure 2, arithmetic plots of the number of species at different population levels up to the total number of individuals for samples from high latitude, low latitude, shallow water, deep sea, estuarine, and marine regions. The curvilinear nature of the lines is due to the fact that individuals are being added at a constant rate but the progressively rarer species are added at a continuously decreasing rate. The circles or the termination of the lines in Figure 2 give the actual number of individuals and species present in the samples. The curves themselves give the interpolated number of species at the different population levels.

What is significant is that each environment seems to have its own characteristic rate of species increment. Lowest diversity, that is, the fewest number of species per unit number of individuals, is found in the boreal

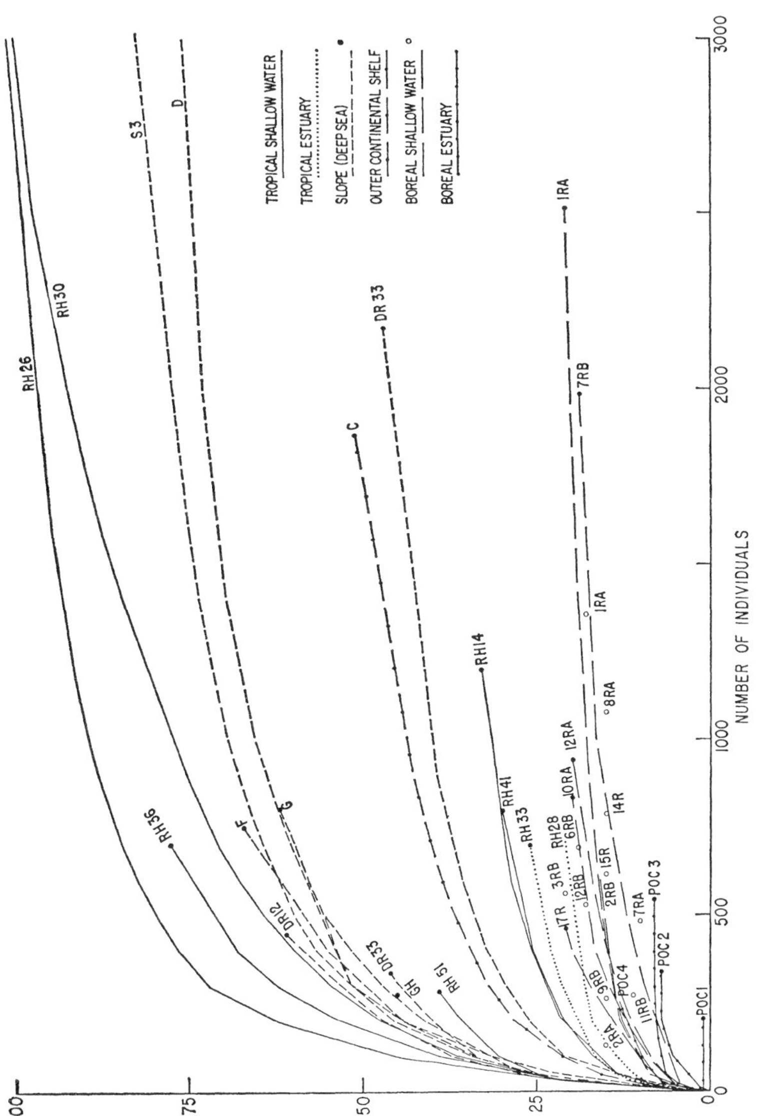

FIG. 2.—Arithmetical plot of the number of species at different population levels using the rarefaction methodology for stations from differing regions. The termination of a curve gives actual number of individuals and species found in the sample. The remainder of the curve is interpolated by the rarefaction methodology. The circles with station numbers are actual samples that have not been rarefied.

estuary as represented by the Pocasset River, Cape Cod, Massachusetts. The station highest up the estuary, POC 1 in Figure 2, with a mean sediment salinity of 7‰ and a range of 9.5‰ per tidal cycle, has the lowest diversity within the series. POC 2, next highest up the estuary, with a mean sediment salinity of 17‰ and with a 3‰ salinity range per tidal cycle, has the next lowest diversity. The diversity increases at POC 3, with a mean salinity of 20.7 and a tidal variation of 1.4‰. Still higher diversity is present at POC 4, where the mean salinity rises to 22.9‰ and the tidal variation is 1.8‰ (for details on faunal distribution in the Pocasset estuary and its relationship to salinity, see Sanders, Mangelsdorf, and Hampson, 1965). Besides the low and variable sediment salinities in the Pocasset River, there are, as well, large seasonal temperature changes.

Somewhat higher diversity values occur in tropical estuaries. These are represented by stations RH-28, the Vellar River estuary at Porto Novo, Madras State, India, and RH-33, at the mouth of the Godavari estuary at Kakinada, Andhra State, India. During the periods of heavy rainfall of October and November, the salinities in these shallow bodies of water are reduced to zero. Yet, in the dry season of May and June, the salinities are more than 34‰ (Jacob and Rangarajan, 1959). The probable reason for the greater diversity values in low- as compared with high-latitude estuaries is that it is easier to tolerate reduced salinities at high temperatures than at low temperatures (Panikkar, 1940). As a result, more marine forms are able to invade estuaries in the tropics than in higher latitudes.

Among the boreal shallow marine samples (the R-station series) diversity is modest. These samples were taken from Buzzards Bay, Massachusetts, in 20 m of water at all seasons of the year (Sanders, 1960). Here the annual temperature change is more than 23°C, with winter temperatures often less than −1.0°C and summer temperatures of more than 22°C. Such appreciable changes in annual temperature are as large as that found in any marine region of the world. This pronounced seasonal temperature variation probably accounts for the low diversity values. Note that within the R series of samples in Figure 2, the actual samples with lower densities, instead of being widely scattered throughout the graph, are clustered about the interpolated curves derived from the larger samples, thus verifying the validity of our methodology. This same clustering demonstrates that diversity values are sample-size independent when derived by the rarefaction technique. At the outer edge of the continental shelf, station C, the amplitude of temperature change has been reduced to 10°C and the faunal diversity has increased (Sanders, et al., 1965).

The most diverse values are found among the tropical shallow marine samples, although there is appreciable spread in the position of the curves. The highest values are from the three Bay of Bengal samples: RH-26, off Porto Novo, Madras State, India, in 20 m of water; RH-36, off Kakinada, Andhra State, India, in 37 m depth; and RH-30, off the city of Madras, in 15 m depth. All of these stations are too deep to be affected by the freshening of the surface water during the monsoons (LaFond, 1958;

Murty, 1958; Ramamurthy, 1953). Station RH-51, from a depth of 18 m off Nossi Bé, Madagascar, gives an intermediate value. Lowest values are from two stations in the Arabian Sea, RH-14 in 14 m off Cochin, India, and RH-41, in 20 m depth off Bombay, India. The probable cause for these modest diversity values is the low-oxygen minimum layer found throughout the northern Arabian Sea at the 100 to 200 meter depth. During the southwest monsoons, this low-oxygen water is pushed onto the continental shelf off India, creating a severe stress condition for the bottom fauna which is probably reflected in the reduced number of species present. Banse (1959) found, at almost precisely the site and depth of our Cochin station, oxygen values of only 5% saturation during the southwest monsoon, and Carruthers, Gogate, Naidu, and Laevastu (1959) obtained similar low-oxygen values at a location of equivalent depth near our Bombay station.

Our deepwater diversity curves, derived from stations Sl.3, D#1, F#1, and G#1 from the continental slope, station GH#1 on the abyssal rise (all south of New England), and stations DR-12 and DR-33 from the continental slope off northeastern South America, with but a single exception, are confined to a narrow sector of the graph. The physical factors in this environment are rigidly constant, with low temperatures, high salinity (see Sanders, et al., 1965), and high oxygen values. The single exception to our deepwater diversity pattern, DR-33, is due to the aggregation of a single polychaete species which forms more than 85% of the sample. If this species is arbitrarily excluded, the residual diversity, DR-33′, is similar to that found in other deep-sea samples. Thus the deep-sea benthic fauna appears to possess a relatively high diversity of the same general order as that present in tropical shallow seas.

It should be clearly pointed out here that this method of measuring diversity is valid when the fauna is randomly or evenly distributed but not aggregated. Even in cases of aggregation, it may be possible to uncover inherent diversity (DR-33′) by using this methodology. On the other hand, in samples with little aggregation, the diversity is only slightly increased by eliminating the most abundant species.

Applying confidence limits to the curves for certain of the environments is meaningless. In cases where the number of samples included is small, the confidence limits will be broad. The Bay of Bengal series, the pair of stations from the shallow depths of the Arabian Sea off India, and the two tropical estuarine stations suffer from this weakness.

The best that can be done is to represent the ranges in Figure 2 as bands of values (Fig. 3). The number of samples is given at each of the rarefied sample sizes from 100 individuals and larger. Environments and sample sizes with single samples, the Pocasset series, and aberrant station DR-33 are excluded. The clear separation of the environmental bands strongly indicate that these diversity differences are real.

Limitations of Methodology

The rarefaction method for measuring diversity must be used with dis-

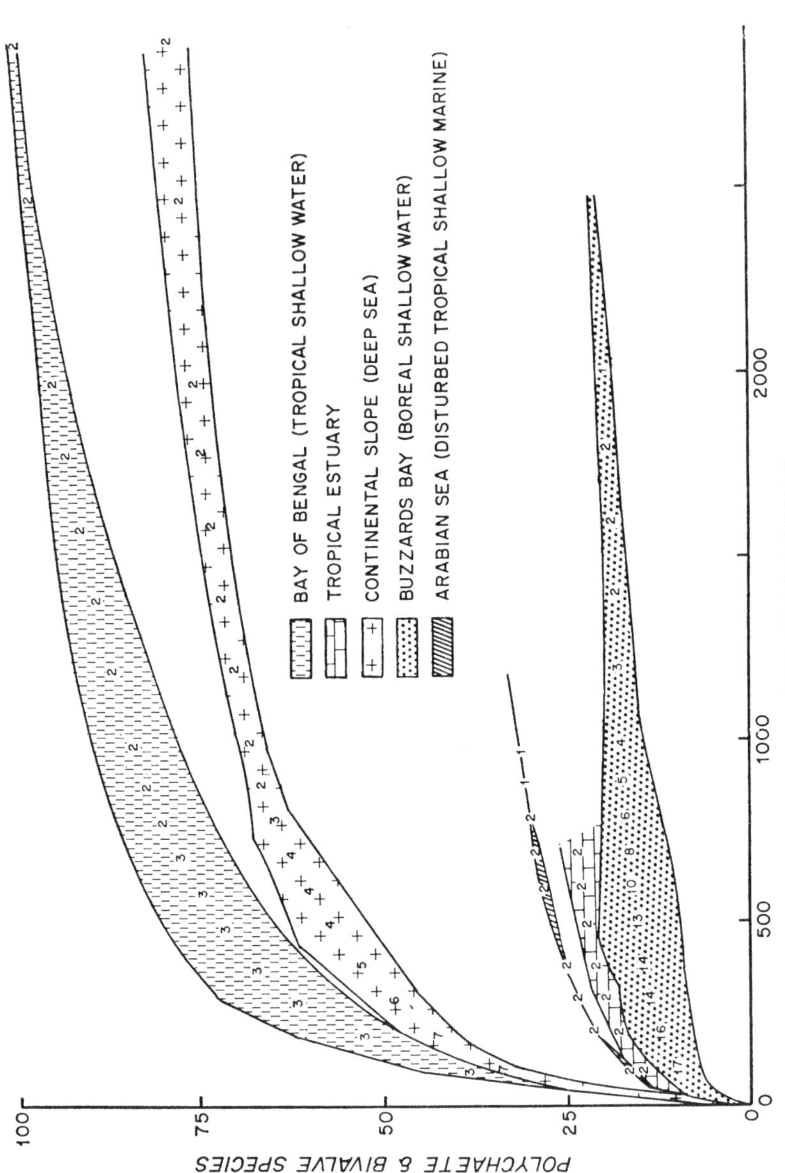

Fig. 3.—Range for diversity values found for a number of the regions included in this study.

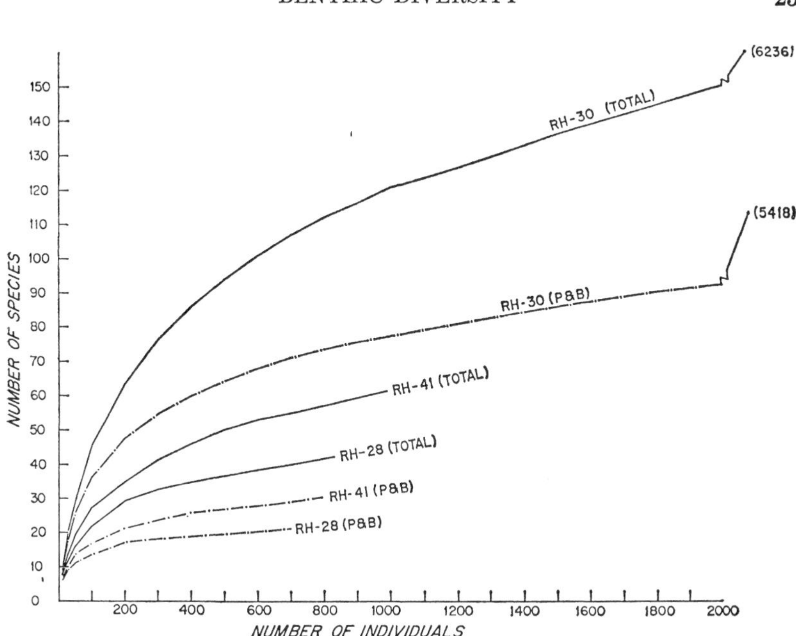

FIG. 4.—Diversity values for the total fauna compared with the polychaete-bivalve component of the same samples. Numbers in parentheses for station RH-30 are the actual numbers of specimens in the total sample and in the polychaete-bivalve fraction of that sample.

crimination to be meaningful. Such a technique is valid only when the same groups of organisms are compared and contrasted. With the inclusion of additional groups, the diversity values for a given faunal density increase. This phenomenon can be clearly observed with the few representative samples used in Figure 4. In each case, the diversity for the total fauna is decidedly higher than that of the polychaete-bivalve component of the sample.

Another requisite is that all the habitats sampled be similar; that is, the comparison must be made among a within-habitat (MacArthur, 1965) series of environments (in the present situation, the soft estuarine and marine oozes). Differing habitats from the same geographic region have differing diversity values (between-habitat comparison [MacArthur, 1965]). Thus the sand bottom fauna in Buzzards Bay is more diverse than the mud bottom fauna (Fig. 5). (Probably the fauna of stable sand bottoms will always be inherently more diverse because of the greater variety of microhabitats.)

In order to have the data comparable, it is necessary that the sampling procedures such as the type of gear used, the methodology utilized in processing the sample, and the screen size employed in washing the samples should be approximately similar.

Finally, this method does not specify which species taken from the residue will be present, and it can be used only to interpolate, not to extrapolate.

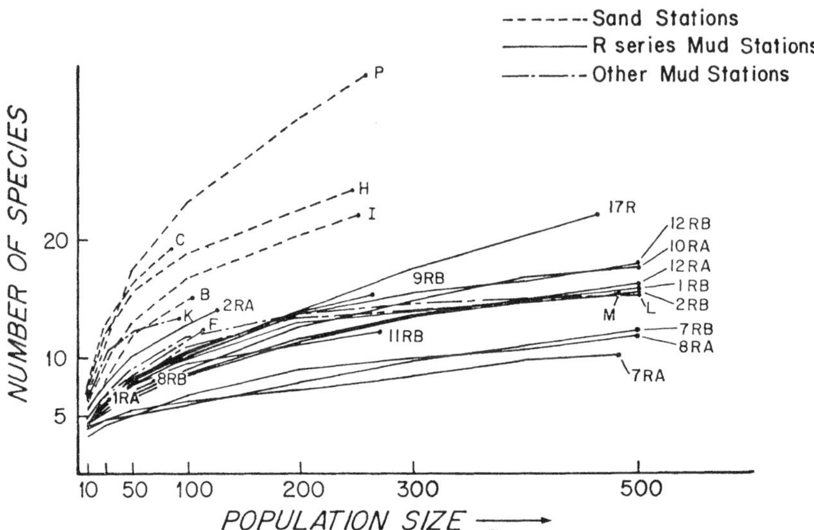

FIG. 5.—Comparison of diversity between mud- and sand-bottom samples from Buzzards Bay, Massachusetts, up to population sizes of 500 individuals.

THE PHYSICALLY CONTROLLED AND THE BIOLOGICALLY ACCOMMODATED COMMUNITIES

The interpretation of the curves in Figure 2 and the more general analyses of the total fauna might best be understood by describing two contrasting types of communities, both of which are abstractions. One can be called the "physically controlled community." In environments harboring this kind of community, the physical conditions fluctuate widely and the animals are exposed to severe physiological stress.

In the physically controlled community the adaptations are primarily to the physical environment. Examples of such communities are those found in hypersaline bays, high arctic terrestrial environments, and deserts. The physically controlled communities are always eurytopic and are characterized by a small number of species. A similar paucity of species occurs in environments of recent past history, such as most freshwater lakes.

The other extreme condition might be called the "biologically accommodated community." These communities are present where physical conditions are rather constant and uniform for long periods of time. Because of the historic constancy of the physical environment, physical conditions are not critical in controlling the success or failure of the species. With time, biological stress (intense competition, nonequilibrium conditions in prey-predator relationships, simple food web, etc.) is gradually mediated through biological interactions resulting in the evolution of biological accommodation. The resulting stable, complex, and buffered assemblages are always characterized by a large number of stenotopic species. The deep-sea regions, tropical shallow-water marine regions, and tropical rain forests best represent such conditions.

There is no such thing as a "pure" physically controlled or biologically accommodated community. All communities are the result of both their physical and biological components and are therefore somewhat intermediate between these extreme types. What determines the structure of any community is the relative proportions of these two parts.

In predominantly physically controlled communities there can be no close coupling of a species to its environment, as would be the case in the predominantly biologically accommodated communities. This is due to the variations in the amplitude of environmental factors; that is, there is no precise reproducibility from year to year. For example, one year the temperature may be slightly higher, so that one species is favored regarding breeding, which results in the "year-class" phenomenon, that is, a tremendous increase in the number of the new year class. (The year-class phenomenon probably is a characteristic feature of the predominantly physically controlled communities.) The next year, the temperature may be slightly lower, so that the same species is adversely affected, resulting in an unsuccessful breeding. At the same time, another species is favored in its breeding by the reduced temperature. The same would be true for the effects of other environmental variables, such as salinity and oxygen, on growth, breeding, metabolism, etc. Therefore, animal species of the physically controlled communities must adapt to a broad spectrum of physical fluctuations which does not allow the biological interrelationships to develop very far. (In some intertidal environments, the prey may be biologically controlled while its predator is physically controlled [Connell, 1961]).

From the concepts summarized above, it is possible to present the stability-time hypothesis in Figure 6. Where physiological stresses have been historically low, biologically accommodated communities have evolved. As the gradient of physiological stress increases, resulting from increasing physical fluctuations or by increasingly unfavorable physical conditions regardless of fluctuations, the nature of the community gradually changes

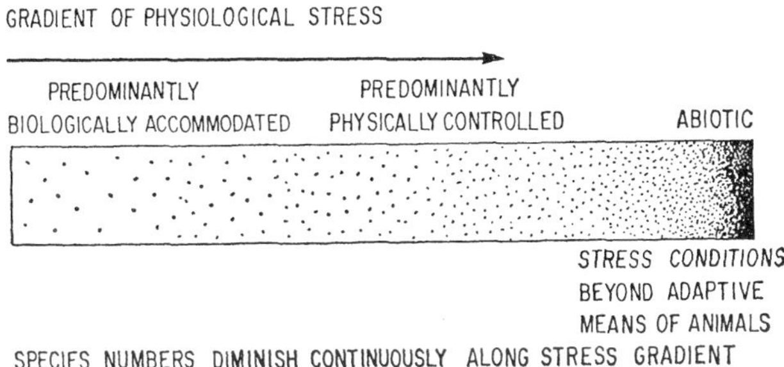

Fig. 6.—Bar graph representation of the stability-time hypothesis.

from a predominantly biologically accommodated to a predominantly physically controlled community. Finally, when the stress conditions become greater than the adaptive abilities of the organisms, an abiotic condition is reached. The number of species present diminishes continuously along the stress gradient.

When the stability-time hypothesis is applied to Figure 2, the closer the curves approach the ordinate, as shown with the shallow tropical marine and deep-sea samples, the nearer they approximate the biologically accommodated community. They describe assemblages in which there are large numbers of species per unit number of individuals (high diversity). In these environments, physical conditions are constant and have remained constant for a long period of time. The closer the curves approach the abscissa, as in the cases of the boreal estuary and boreal shallow marine samples, the greater are the physiological stress conditions imposed by the physical environment. In these assemblages there is a small number of species per unit number of individuals (low diversity). Here the physical conditions are highly variable and approach the idealized physically controlled community.

Thus each environment in Figure 2 appears to have its own unique family of curves. Such lack of randomness implies biological organization, with the nature of the organization or structure differing in different environments. Such organization is determined by the degree of stability of the physical environment and the past history of the physical environment, that is, to what degree an animal association is physically controlled and to what degree it is nonphysically regulated or biologically accommodated.

TIME

It requires appreciable time to evolve a highly diverse fauna, and the time component of our stability-time hypothesis is perhaps best illustrated with lakes. Most lakes are of a relatively transitory nature, or of recent geologic origin. It has been 10,000 years or less since the last glaciation, and the aquatic fauna from such recently glaciated regions shows limited diversification. However, there are a few ancient lakes—for example, the rift-valley lakes of Africa and Lake Baikal in Russian Siberia.

Lake Baikal was formed either about 30 million years ago in the middle Tertiary or at the end of the Tertiary and early Quaternary periods about one million years ago. (For references, see Kohzov, 1963.) This lake, in common with other ancient lakes, is characterized by a highly diverse fauna. One of the most diverse faunal elements are the gammarid amphipods, represented by 239 endemic and one nonendemic species.

To appreciate the full significance of such diversity, in all of what was glaciated North America there are no more than 28 species of gammarid amphipods (Bousfield, 1958). Certain of these crustaceans are confined to streams near the sea, can tolerate brackish water, and have recently evolved from closely related marine forms. Others are restricted to cave streams and springs. A few are limited to ponds. Still others occur in sloughs and temporary bodies of water. Only seven gammarid species are confined to lakes and rivers, and it is these few amphipods, distributed throughout vast areas of North America, which are the ecologic equivalents of the 240 species present in ancient Lake Baikal. (It should be mentioned that the diversity effects of time on a physically fluctuating environment of constant magnitude and periodicity remain unanswered.)

Lake Baikal has further implication to the concepts proposed in this paper. Two distinct and essentially separate faunas exist there. One element, broadly distributed through much of Siberia, is confined to the shal-

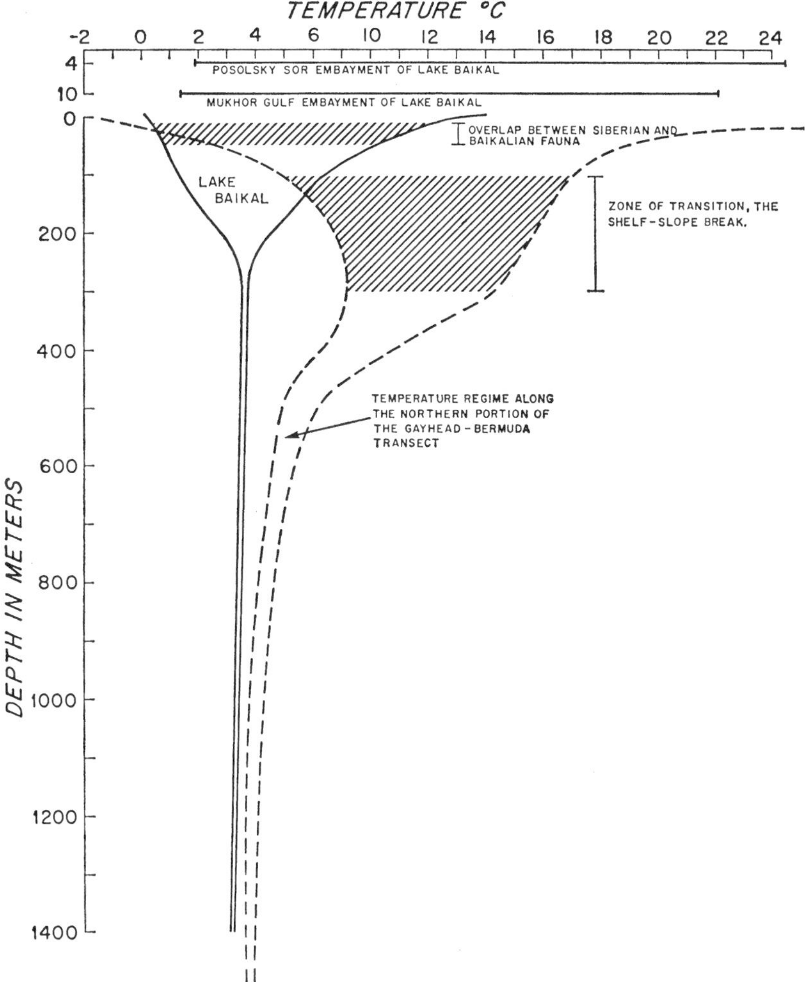

Fig. 7.—Temperature regime with depth in Lake Baikal and the northern part of the Gayhead–Bermuda transect.

low gulfs or bays and is not found deeper than 20 m in the lake. This group is represented by 137 species of nonendemic, free-living, benthic or pelagobenthic macroinvertebrates, 54% of which are insects. The other, an entirely endemic Baikalian fauna of 580 species of which are only 4% are insects, avoids the shallowed depths and embayments, and representative species are found down to 1,620 m, the maximum depth. (Lake Baikal is the deepest freshwater lake in the world.) The ecotone or region of overlap between these two faunas is very narrow (15 to 50 m depth), and but 22 species occur there (Fig. 7).

The Siberian eurytopic component exists in an environment of highly variable seasonal temperatures (see Fig. 7, Posolosky Sor and Mukhor Gulf) and low, fluctuating oxygen content (Kohzov, 1963). The deeper-dwelling Baikalian element, on the other hand, lives under physical conditions that are hardly varying. Such stable conditions, with time, have allowed the evolution of this highly diverse endemic stenotopic fauna, while the less diverse, nonendemic, eurytopic Siberian fauna remains confined to the shallow and physically more variable parts of the lake.

Similar rigidly stable physical conditions are encountered in the greater depths of Great Slave Lake in the Northwest Territories of Canada (Rawson, 1953). Yet, within the depth range from 200 to 600 m, only four species of macrofaunal benthic invertabrates were collected from this large "post-glacial" lake, again demonstrating the significance of the time component of diversity.

These findings from Lake Baikal are entirely analogous to the conditions occurring in the boreal region of the Gayhead-Bermuda transect. The continental shelf, particularly in the shoaler depths, harbors an impoverished fauna. The continental slope, however, supports a benthic fauna of high diversity. The region of very rapid and pronounced faunal change occurs somewhere between the depth range of 100 to 300 m and is the most marked zoogeographical boundary encountered along the transect. Not only are there specific and generic differences, but in some groups these changes are of familial and even ordinal significance. Conceivably, the changeover from the diverse stenotopic deep-sea fauna, which is physiologically adapted to constant temperature conditions (as well as other constant environmental factors), to that of the relatively depauperate eurytopic boreal littoral fauna, which exists under a varying temperature regime (other environmental factors tend also to fluctuate here), occurs at that depth where seasonal changes in temperature become large (see Fig. 7).

The same interpretation might be applied to the faunal changes in Lake Baikal, although the zone of rapid transition from the eurytopic assemblages of limited diversity to the highly diverse fauna of deeper water takes place at shallower depths (Fig. 7). Note that the range of seasonal temperature change in the transitional zone is approximately the same in both regions (Fig. 7).

Both limnetic Lake Baikal and the marine area of study south of New England are in boreal regions dominated by a continental climate.

In both situations, the pronounced seasonal changes in temperature are imposed on the shallow-water fauna while the benthos of greater depths are insulated from these changes (Fig. 7). With time, similar patterns of diverse stenotopic faunal assemblages have evolved in the physically stable deeper waters while the highly unstable shallow waters continue to support a rather impoverished fauna. Thus these two unrelated freshwater and marine faunas, molded by similar physical forces and time, have evolved and diversified in a parallel and analogous manner.

Such an interpretation is not at variance with the stability-time hypothesis as shown diagramatically in Figure 6. A continuous diminution in stress conditions certainly takes place from shallow to deep depths, but over a spatially restricted portion of this gradient the rate of change is very great. The region of abrupt change represents the transitional zone from the predominantly physically controlled to the predominantly biologically accommodated community.

Fischer (1960) concluded that the greater diversity in the tropics occurs not only because of the greater stability and longer history of that environment but also because the temperature is nearer the midpoint of the temperature range that protoplasm can endure. High temperature, per se, does not play a critical role in promoting diversity, for the highly diverse assemblages of the deep-sea and the endemic fauna of Lake Baikal evolved in relatively low-temperature environments. The critical factors appear to be time and environmental stability.

EXTRAPOLATIONS FROM THE STABILITY-TIME HYPOTHESIS

Our proposed hypothesis has another use. It allows us to predict. Hutchins (1947) pointed out that in the Northern Hemisphere a much greater seasonal change in water temperature takes place along the western edges of oceans at temperate latitudes than along the eastern edges. Such temperature conditions result from the prevailing west-to-east wind patterns in the middle latitudes. Thus the coastal boreal regions of eastern United States and parts of eastern Asia are dominated by a continental climate of high summer temperature and low winter temperature, while the outer European coasts and the western coast of North America are dominated by a maritime climate of appreciably less seasonal temperature change.

On this basis we can predict that two distinct types of boreal shallow-water communities exist. One can be termed the "continental climate boreal community" and would be exemplified by our Buzzards Bay series of samples. This community will be characterized by low faunal diversity. Furthermore, many, if not most, of the infaunal species will cease to grow and become inactive during the cold winter months (personal data and observations). The other boreal marine shallow-water community can be called the "maritime climate boreal community," characterized by greater faunal diversity and without the complete cessation of growth among the infauna during the winter months.

The model also suggests that the great upwelling regions of the oceans, such as the areas off southwest Africa and the Peruvian and Chilean coasts of South America, will typically show low benthic diversity. The abundant organic matter depletes the available oxygen as it sinks, so that the bottom water contains little or no oxygen. The stress condition resulting from the reduced available oxygen will be reflected in low diversities and, with further oxygen depletion, in low faunal densities or ultimately in abiotic conditions. Indeed, data by Gallardo (1963) from the upwelling areas off northern Chile provide impressive evidence for this interpretation. He found that the oxygen content of the bottom water was less than 5% saturated and that the sediments were reduced at water depths from 50 to 400 m. The benthic samples yielded few individuals, averaging only 6 to 7 per cubic meter.

SPATIAL VARIATION AND TEMPORAL VARIATION

Let us now consider a possible mechanism that would give the few though often numerically abundant species in predominantly physically unstable environments and the many species in the physically stable environments. We must consider two types of variation: *temporal variation,* which has already been discussed in some detail, and *spatial variation,* or habitat diversity.

When temporal variation is large (physiological stress conditions in physically unstable environments), it masks the effects of spatial variation (i.e., the wide range of habitats utilized by lemmings in the high latitudes of North America; see also MacArthur [1965] on within- and between-habitat avian diversities below). When temporal variation is small (minimal stress conditions in physically stable environments), the effects of spatial variations are realized, resulting in the progressive division of species with time. Thus with spatial variation occurring within the distributional range of a species, different selective forces will be acting on the species in different parts or habitats of its range. Initially, this process may result in the formation of separate subspecies and, if the gene flow is sufficiently attenuated, of separate species.

MacArthur (1965) pointed out that in a new environment (this may be comparable to environments of large temporal variation or high physiological stress), such as the initial stages in the colonization of an island by birds, few species are present but are distributed through a number of habitats. With time (this may be comparable to environments of decreasing temporal variation or reduced physiological stress) the number of species increases, but this enrichment is reflected in the *between-habitat* or β diversity of Whittaker (1965) (increase in the total number of species for all habitats) rather than the *within-habitat* or α diversity of Whittaker (1965) (number of species in a specific habitat remains constant).

A concept somewhat similar to MacArthur's constancy of within-habitat diversity is the earlier "parallel community hypothesis" postulated by

Thorson (1952) for the marine infauna. He contended that while there is a continuous gradient of species diversity from the arctic to the tropics for the epifauna, the number of infaunal species remains approximately the same. The findings in the present investigation, which is clearly a study of within-habitat diversity, give diametrically opposite results. Samples from historically stable environments of long duration and low physiological stress give high within-habitat diversity values, while samples from historically recent and/or variable environments yield low within-habitat diversity values.

Thorson, at the time he suggested his concept (1952), had only a very limited amount of data on the tropical marine infauna. At least one tropical locality upon which this interpretation is based, the Persian Gulf, with very high salinities and temperatures, represents a stress environment. Our own findings in the present study show that diversity can be quite variable in the tropics. Regions of low stress, such as the Bay of Bengal (RH-26, RH-30, and RH-36), support a very diverse infauna. Conversely, tropical areas of high stress, as exemplified by the shallow-water samples from the west coast of India (RH-14, RH-41), give much reduced values. Thorson (1966) recently found very high tropical infaunal diversities at shelf depths off the west coast of Thailand and he now feels that the applicability of the parallel community hypothesis to tropical environments should be carefully scrutinized.

There still appears to be an underlying difference between avian populations and benthic infaunal invertebrates. Among birds, with time and a physically stable environment, an increase in species occurs. This species enrichment takes place entirely as between-habitat diversity, while within-habitat diversity remains unchanged. With our infaunal benthic organisms, on the other hand, species enrichment is reflected both in the between- and the within-habitat diversities.

A lucid genetic interpretation for the relationship of environmental stability to diversity has been given by Grassle (1967). He pointed out that populations present in physically stressed and unpredictable environments show broad adaptations to these conditions by maintaining a high degree of genetic variability. Thus, even though the stress may be expressed in a variety of ways, a portion of the polymorphic population will probably survive. These genetically flexible species are opportunistic and cosmopolitan, and they have little tendency to speciate.

The price paid for this variability is "the genetic load or loss of fitness relative to the maximum in a more uniform environment." In stable environments "the expression of deleterious genes outweighs the advantages obtained from maintaining genetic flexibility." Therefore, in stable and predictable environments, such genetic variability will be selected against.

Diversity differences found in the present study between stable and unstable environments can be interpreted on the basis of genetic variability. The flexibility needed for survival in an unstable environment necessitates a larger utilization of the environment by each species. Thus diversity and genetic flexibility would be inversely related. (For a comprehensive discussion of the genetic basis of benthic diversity, see Grassle [1967].)

THE DEFINITIONS AND MEANINGS OF DIVERSITY

We might pause here and ask what we precisely mean by the word "diversity." It is apparent from looking through the literature that there are two definitions. This has resulted in some confusion.

One kind of diversity is the *numerical percentage composition* of the various species present in the sample. The more the constituent species are represented by equal numbers of individuals, the more diverse is the fauna. The less numerically equal the species are, the less diverse the sample is or, conversely, the greater is the dominance in the sample. This is a measure of how equally or unequally the species divide the sample, and the number of species involved is immaterial. Diversity measurements of this kind include the MacArthur "Broken Stick" model (1957), the Preston lognormal distribution (1948), and the Simpson index (1949). Such diversity might be designated after Whittaker (1965) as *dominance diversity.*

The other kind of diversity is determined by the *number of species.* The more species in a sample or the more species present in a species list for a given environment, the greater the diversity. Measurements of this sort are the α values of Fisher, Corbet, and Williams (1943), Margalef's d values (1957), the methodologies of Gleason (1922) and of Hessler and Sanders (1967), and the rarefaction technique used in the present paper. Such diversity can be designated after Whittaker (1965) as *species diversity.*

Since the number of species present and the relative dominance or lack of dominance in a sample are both measures of diversity, one might assume that they must be highly correlated with one another. Thus, a large number of species per unit number of individuals reflects low dominance; alternatively, a small number of species indicates high dominance. In Table 2 we will test this assumption.

Eighteen of the stations used in Figure 2 are included in the analysis. Each station is represented by a single sample, except station R, which is a composite of 15 samples taken from the same locality in Buzzards Bay. (The Pocasset series of samples are excluded because of the very few species present.)

The samples are ranked by dominance diversity from highest (lowest dominance) to lowest (highest dominance) diversity. These values are determined by plotting the percentage composition of the species along the ordinate and ranking the species by abundance along the abscissa (Fig. 8). The resultant cumulative frequency curve is used as a measure of dominance diversity.

Maximum diversification occurs when all the species in a sample are represented by exactly the same number of individuals, and the cumulative frequency curve, in this case, is a diagonal straight line that can be described by the formula $x = y$. Such a straight line forms the base line (Fig.

TABLE 2

MEASUREMENT OF THE CORRELATION BETWEEN SPECIES DIVERSITY AND DOMINANCE DIVERSITY AT DIFFERENT FAUNAL DIVERSITY LEVELS USING THE PEARSON PRODUCT MOMENT CORRELATION FOR 18 STATIONS INCLUDED IN THIS STUDY

Sample Size	N	r Value	Critical r Value at 5% Level	Critical r Value at 1% Level
Spp./10 ind.	18	.974	.456	.575
Spp./25 ind.	18	.948	.456	.575
Spp./50 ind.	18	.894	.456	.575
Spp./100 ind.	18	.766	.456	.575
Spp./200 ind.	18	.642	.456	.575
Spp./300 ind.	16	.636	.482	.606
Spp./400 ind.	15	.649	.479	.623
Spp./500 ind.	14	.622	.514	.641
Spp./600 ind.	14	.623	.514	.641
Spp./700 ind.	13	.748	.532	.661
Spp./800 ind.	9	.832	.632	.765
Spp./900 ind.	8	.769	.666	.798
Spp./1,000 ind.	8	.819	.666	.798
Spp./1,200 ind.	8	.817	.666	.798
Spp./1,400 ind.	7	.832	.707	.834
Spp./1,600 ind.	7	.833	.707	.834
Spp./1,800 ind.	7	.832	.707	.834
Spp./2,000 ind.	6	.801	.754	.874
Spp./2,500 ind.	4	.572	.878	.959
Spp./3,000 ind.	4	.519	.878	.959

8). What is measured is the deviations in percentage composition of a given sample from a hypothetical sample containing the same number of equally abundant species, that is, the degree of departure from the base line. The greater the departure, the greater the dominance and, conversely, the smaller the diversity. (See also Sanders, 1963, pp. 87 and 88.)

The stations are also ranked by species diversity from highest (most species per unit number of individuals) to lowest (least species per unit number of individuals) diversity at various population levels from 10 to 3,000 individuals. The dominance-diversity ranking is compared to the species-diversity ranking at each of the derived sample-size levels, using the rarefaction method.

This relationship is measured using the Pearson product moment correlation. The findings are given in Table 2. The square of the correlation coefficient, r, gives the approximate correlation, that is, the variance accounted for by the correlation.

This analysis reveals that species diversity and dominance diversity are indeed correlated at the 5% significance level, except with the largest samples. Using the 1% significance level, a consistent correlation exists only with the smaller-size samples (10 to 400 individuals). At intermediate sample sizes (500 to 1,800 individuals) the relationship is marginal, with the correlation values fluctuating around the critical correlation value. Among the largest sample sizes (2,000 to 3,000 individuals), species diversity and dominance diversity are not significantly correlated.

Thus a correlation, although not a particularly intimate one, exists between species diversity and dominance diversity. This conclusion is in agreement with the recent suggestion of Whittaker (1965) that the rela-

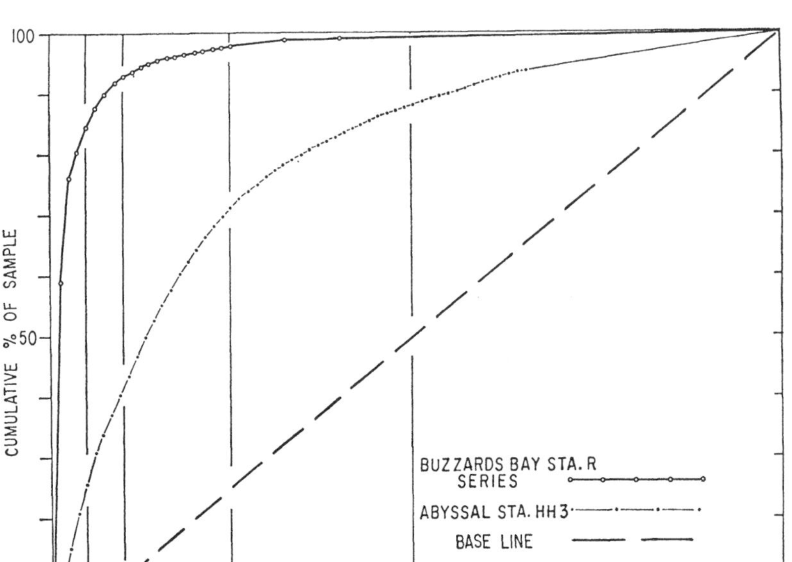

FIG. 8.—Degree of dominance of a sample related to numerical percentage composition of the included species plotted cumulatively. For explanation of figure, see text.

tionship between these two diversity measurements is weak. The degree of correlation appears to be sample-size dependent. A strong correlation is found with small sample sizes, a weaker correlation occurs among intermediate sample sizes, and either a weak or no significant relationship exists among the largest sample sizes.

Only at the smallest sample size does this correlation account for most of the variance (about 95% at the 10-individual level, about 90% per 25 individuals, and about 80% per 50 individuals). At such small population sizes only the most abundant species would normally be present, and one might expect a high level of correlation between dominance and species diversities. As sample size becomes larger, the less common species begin to appear and the percentage of variance accounted for by the correlation diminishes.

What, then, determines this relationship? As mentioned earlier, dominance diversity is independent of the number of species present. However, all species-diversity indexes are affected not only by the number of species but also by how a sample is divided among these species (percentage composition). There is always a dominance-diversity component in all species-diversity measurements because, while the height of the curve is determined by the number of species present, the shape of the curve is set

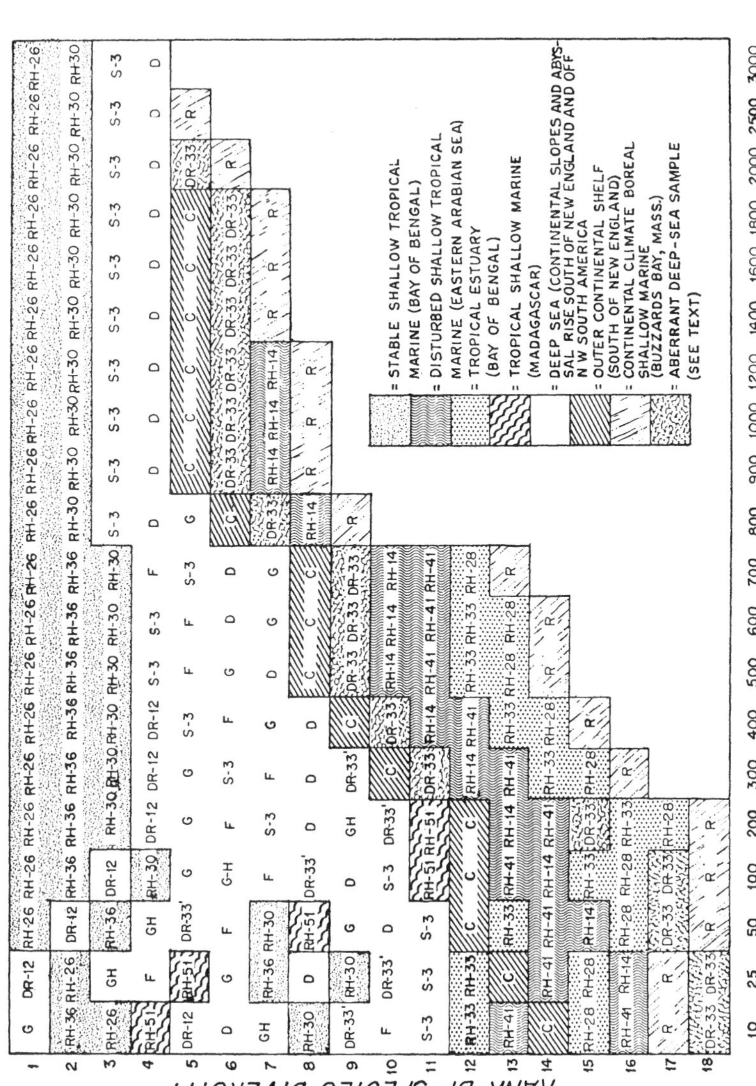

FIG. 9.—Ranking of stations by species diversity at different sample sizes.

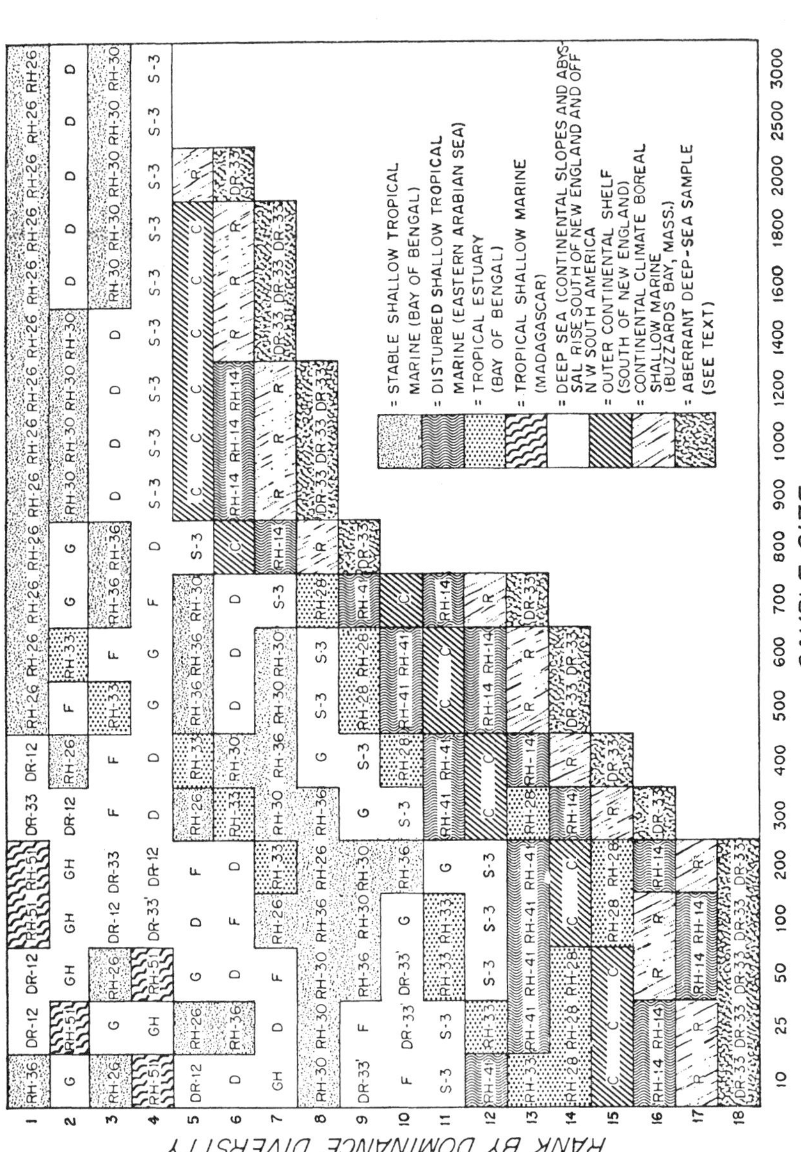

Fig. 10.—Ranking of stations by dominance diversity at different sample sizes.

by dominance. Probably it is the dominance dependency of species-diversity measurements that is primarily responsible for the percentage of variance accounted for by the correlation. Such an interpretation is consistent with the observation that the percentage of variance is very high with smallest sample sizes when only the numerically dominant species are present. Of critical importance, then, is the ecological significance of these two measures of diversity.

If these stations are ranked by both dominance diversity and species diversity, much better clustering of stations by environments is obtained within the species-diversity series (Figs. 9 and 10). Except at low densities, the species-diversity pattern remains stable by environment over the spectrum of population sizes. The stations with the highest diversity values are from the shallow marine depths of the Bay of Bengal (stations RH-26, RH-36, and RH-30). Then, except for aberrant station DR-33, there is a block of samples which includes all the deep-sea stations. They, in turn, are followed by the single shallow marine sample (RH-51) from Madagascar and the single station from the outer continental shelf (C). Next comes the previously mentioned atypical deep-sea station DR-33, the two stations from the stress shallow waters of the Arabian Sea off India (RH-14 and RH-41), the two tropical estuarine samples from India (RH-28 and RH-33), and, finally, the R series of samples from Buzzards Bay, Massachusetts.

Within the dominance-diversity series (Fig. 10), no such clear-cut groupings are present. Stations from a single environment often are widely separated. The Bay of Bengal stations (RH-26, RH-36, and RH-30) with the highest species-diversity values often show wide variability in ranking, both within a specific population size and among differing population sizes. The deep-sea stations do not form a solid block, but have stations from other environments interspersed among them. At those population sizes where the Madagascar station (RH-51) achieves first ranking in the dominance-diversity series, it does no better than eleventh rank using species-diversity criteria. The two tropical estuary stations, RH-33 and RH-28, have appreciably differing dominance-diversity values, yet their species-diversity values are almost identical. Station RH-33 gives intermediate to high dominance-diversity but low species-diversity values. The disturbed tropical stations, RH-14 and RH-41, show reasonable agreement in ranking when both diversity measurements are compared. The values are always low; yet the two stations are usually contiguous by species-diversity ranking and are always separated by dominance-diversity analyses. Outer continental shelf station C usually shows higher species- than dominance-diversity ranking. The R series of samples from Buzzards Bay give low values by both diversity methods. It occupies the lowest species-diversity rank over most of the sample-size spectrum and the next to lowest rank by dominance-diversity criteria. Aberrant deep-sea station DR-33 has last ranking in the dominance-diversity series, but its diversity values increase with sample size using species-diversity ranking. (The pronounced effect of the single numerically dominant species, forming 84.43% of this sample, is gradually mediated as the population size increases and the inherent species diversity begins to emerge. This effect is so overwhelming by dominance-diversity standards that station DR-33 is restricted to lowest ranking throughout the entire range of population sizes.) The only good agreement found between species- and dominance-diversity rankings occurs at the smallest population sizes, where only the most common species would be present.

In brief, then, the stations in the species-diversity series clearly and sharply sort themselves by environment and generally follow the gradient from the biologically accommodated to the physically controlled environmental situations. Within the dominance-diversity series, no such clear-cut groupings are present. Stations from a single environment often are widely separated. Further, there is often little agreement in the position of a station in one series as compared with the other.

These findings can only be interpreted to mean that the high level of agreement between environment and species diversity indicates that such a measure is a conservative and, therefore, ecologically powerful tool. On the other hand, the much poorer fit with dominance diversity suggests that this type of diversity is more variable in its relationship to the physical environment.

THE RELATIONSHIP OF OTHER DIVERSITY CONCEPTS TO THE STABILITY-TIME HYPOTHESIS

Pianka (1966) has summarized the various hypotheses advanced to explain the causes of latitudinal diversity gradients. He was able to separate them into six more or less distinct groupings, although most of the hypotheses contain components of more than one grouping. Presented in their most elemental form, they are:

a) *The time theory* (Simpson, 1964).—All communities tend to diversify with time. Older communities, therefore, are more diverse than younger communities.

b) *The theory of spatial heterogeneity* (Simpson, 1964).—The more heterogeneous and complex the physical (topographic) environment, the more complex and diverse its flora and fauna become.

c) *The competition theory* (Dobzhansky, 1950; Williams, 1964).—Natural selection in higher latitudes is controlled by the physical environment, while in low latitudes biological competition becomes paramount.

d) *The predation hyopthesis* (Paine, 1966).—There are more predators in the tropics who intensively crop the prey populations. As a result, competition among prey species is reduced, allowing more prey species to coexist.

e) *The theory of climatic stability* (Klopfer, 1959; Fischer, 1960; Dunbar, 1960; Connell and Orias, 1964).—Because of the greater constancy of resources, environments with stable climates have more species than environments of variable or erratic climates.

f) *The productivity theory* (Connell and Orias, 1964).—All other things being equal, the greater the productivity, the greater the diversity.

To fit better into the context of the present paper, two of these theories are rephrased as follows:

e') The *theory of climatic stability* is generalized to the *theory of environmental stability*. The more stable the environmental parameters—such as temperature, salinity, oxygen—the more species present.

c') The *competition theory* is altered to state that in environments of high physiological stress, selection is largely controlled by the physical variables, but in historically low stress environments, natural selection results in biologically accommodated communities derived from past biological interactions and competition.

How do the data from the marine benthic samples presented earlier in this paper and the derived *stability-time hypothesis* fit these theories? The *time theory* and the *theory of environmental stability* are most directly applicable to the *stability-time hypothesis*. They, in turn, determine the expression of the *competition theory* and the *theory of spatial heterogeneity*. Biologically accommodated communities resulting from past biological interactions (including competition) are realized in physically stable environments of long temporal continuity. Similiarly, the potentials of spatial heterogeneity can be achieved only under these same environmental conditions.

Neither the *predation theory* nor the *productivity theory* can readily be explained by the *stability-time hypothesis*. The *predation theory* was recently postulated by Paine (1966) for rocky intertidal marine organisms, although he feels it may have wider application. In the intertidal environment, the epibenthic animals experience alternating periods to exposure and immersion. Therefore, these organisms are subjected to desiccation, high salinity imposed by evaporation, exposure to freshwater rain, and air temperatures that are often significantly higher or lower than the seawater temperature. Thus, all rocky intertidal assemblages, independent of latitude, especially at the higher intertidal levels (see Connell, 1961), must be considered predominantly physically regulated communities, and the adaptations are primarily to the physical environment and the biological interactions are poorly developed.

Conceivably, the *productivity theory*, which says that the more food produced, the greater the diversity, may have some validity. Yet this effect is readily masked by numerous environmental variables (Hessler and Sanders, 1967). High productivity itself, from the sheer amount of organic matter produced, can create severe stress conditions and low diversity. For example, in some upwelling areas, high production is responsible for low oxygen content of the water on and just above the ocean floor. Similarly, the highly productive eutrophic lakes often have bottom water devoid of oxygen. In contrast, the high diversity values for the deep-sea benthos, shown by Hessler and Sanders (1967) and in the present paper, come from regions of low productivity.

COMPARISON USING CERTAIN OTHER FAUNAL INDEXES

Numerous indexes have been formulated to measure diversity. Odum, Cantlon, and Kornicker (1960) pointed out that in all types of presentation, logarithmic functions are involved. They recognize four categories. Three have pertinence to our paper:

1. *Cumulative species versus logarithm of abundance.*—This was exemplified by Gleason (1922), Fisher et al. (1943), and Margalef (1957). Such an index is obtained by determining the rate of species increase as additional samplings are made from the same population.

2. *Number of species of particular abundance versus logarithm of abundance.*—This method was formulated by Preston (1948) and is based on the premise that, if the presence of all species found in a given habitat can be revealed, the abundance distribution would follow a lognormal curve. Since such a complete revelation is usually impossible, the resulting curve is truncated at its rarer end. However, the shape of the curve allows one to approximate the total number of species in the habitat, including those as yet undiscovered.

3. *Abundance versus logarithm of rank.*—This type of index was proposed by MacArthur (1957). The observed abundances are compared with theoretical abundances derived from a model containing contiguous, nonoverlapping niches.

One of the fundamental drawbacks of most diversity indices is that they are sample-size or density dependent. Hairston and Byers (1954), in an analysis of cumulative samples of soil arthropods from a singel habitat by both the logarithmic series of Fisher, et al. (1943) and the lognormal distribution of Preston (1948), found that the results depended on the size of the total sample. Margalef (1957) pointed out that his diversity measurement, in common with other diversity indexes, increases with enlarged samples. A similar finding was reported by Williams (1964) for the Simpson diversity index (1949). Hairston, in a later paper (1959), demonstrated that the MacArthur model (1957) is also density dependent. From the analysis of his own data, he interpreted this phenomenon of increased diversity with increased sample size to mean that rare species are clumped. With repeated samples, there will be a greater likelihood of obtaining a new rare species than of obtaining a member of a rare species already collected. He concluded "that an increase in heterogeneity with an increase in sample size lies in the spatial distribution of the species concerned, and the inverse relationship between clumping and abundance."

In our study, single samples were collected from comparable sediment environments. Both small samples and large samples from the same environment (Fig. 2, boreal shallow marine and deep-sea) fall along the same diversity curve. If we apply the rarefaction method to a series of 15 samples of greatly differing sizes (35 to 2,514 individuals) which were carefully selected for sediment homogeneity and taken during the course of a 2-year period from a single locality in Buzzards Bay (Sanders, 1960), we find in the semilogarithmic plot in Fig. 11 no tendency for smaller samples to be

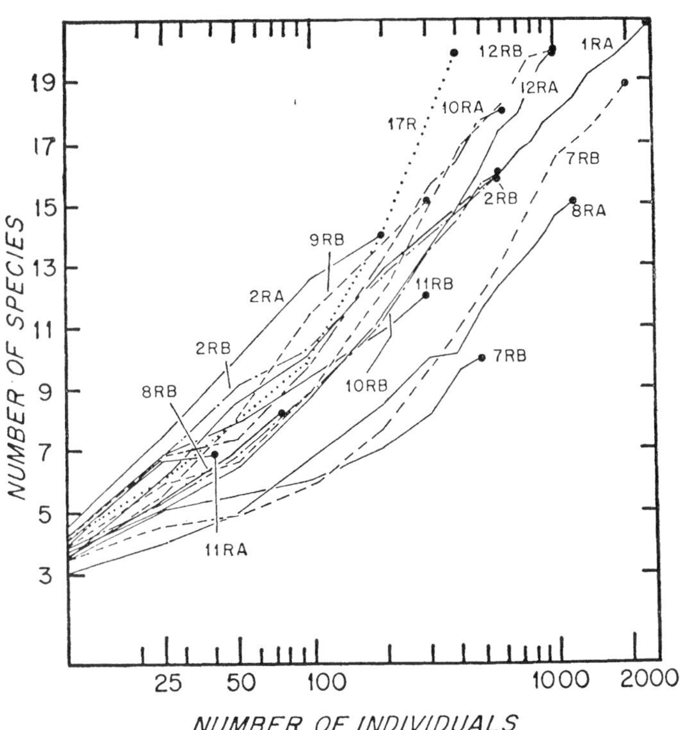

Fig. 11.—Rarefaction curves for the 15 station R samples from Buzzards Bay, Massachusetts.

less diverse than larger ones—that is, smaller-sized samples do not show a tendency to rise more steeply than larger-sized samples. Thus, both under the conditions of our sampling program and by using the rarefaction method of measuring diversity, no increase in heterogeneity takes place with increasing sample size. After all, what could be more homogeneous than a series of different-sized subsamples, each with the same percentage composition as the original samples? This, in essence, is what the rarefaction method does.

Now that we have demonstrated the effectiveness of the proposed rarefaction methodology in obtaining constancy of diversity at all population sizes in our samples, can similar stability be achieved when other diversity formulas are applied to the identical data whose internal homogeneity has been demonstrated over the entire range of sample size? We will attempt to answer this question by applying a number of diversity measurements to the data presented in Figure 2.

With the Preston truncated lognormal distribution analysis, the numbers of species with abundances of 1 to 2, 2 to 4, 4 to 8, etc., are plotted as points. The resulting curve is assumed to approximate a lognormal distribution. Such an estimate is essentially independent of sample size because a dou-

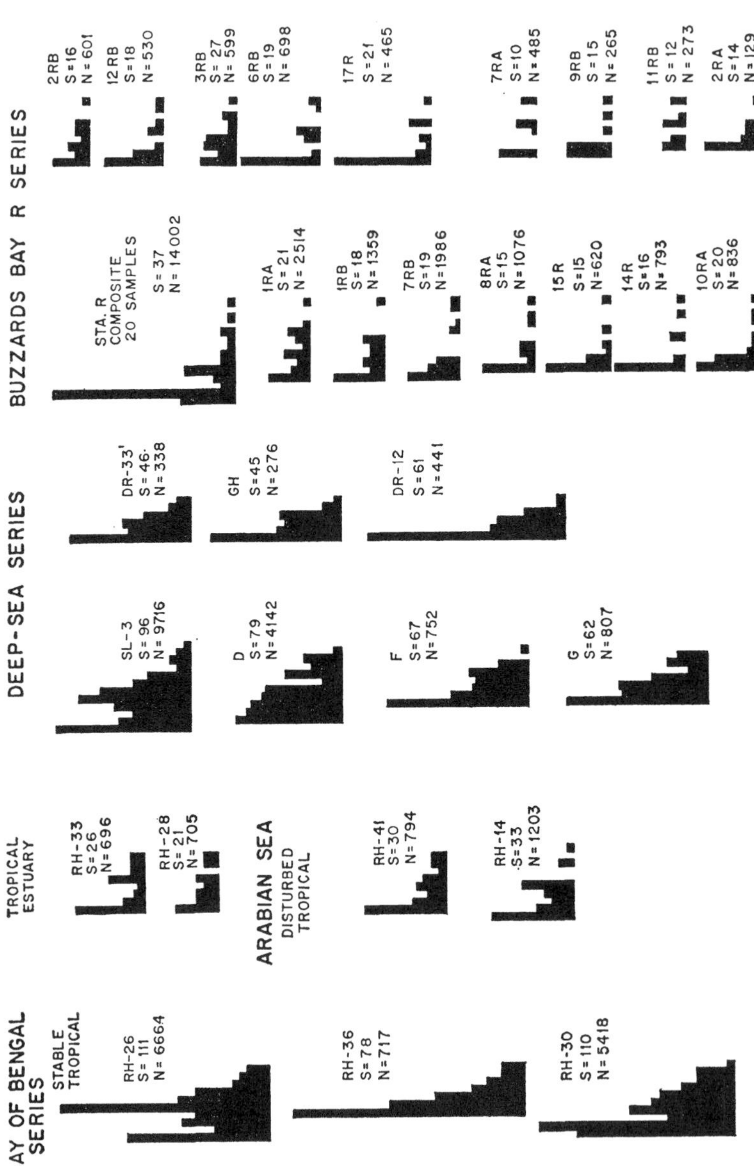

FIG. 12.—Histogram representation of samples included in this study plotted according to the Preston lognormal distribution.

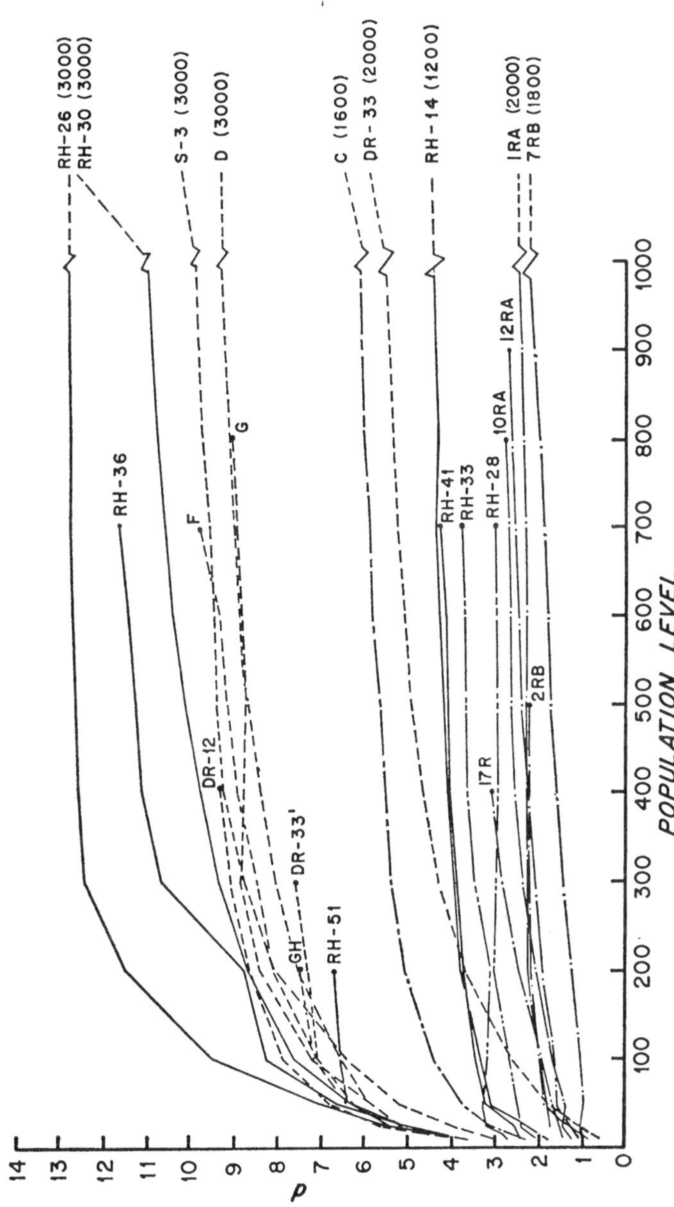

FIG. 13.—Relationship of the Margalef index to rarefaction data. Numbers in parentheses after the stations are the largest population levels to which the samples have been rarefied.

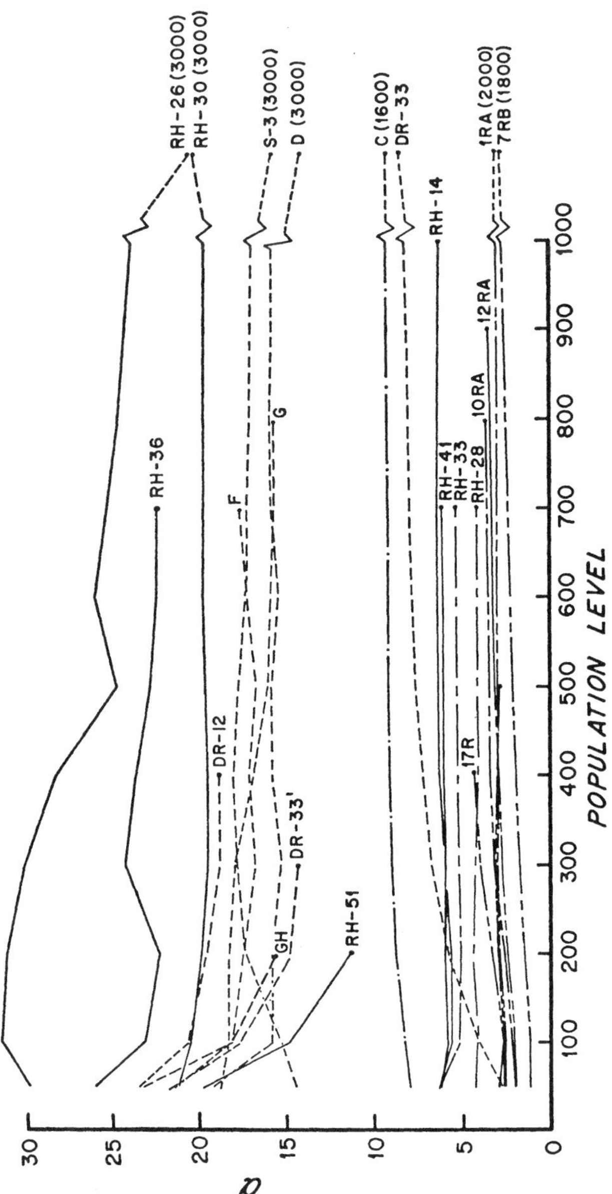

Fig. 14.—Relationship of the logarithmic series of Fisher, Corbet, and Williams to the rarefaction data. Numbers in brackets after the stations are the largest population levels to which the samples have been rarefied.

bling of individuals simply displaces the curve one unit to the right and adds a new unit to the left or rare end of the curve. When the entire suite of species is finally revealed, a lognormal rather than a truncated lognormal curve describes the situation.

Plotting our data by this method in Figure 12 gives histograms that are difficult to interpret. For example, the rarefaction curves for the two disturbed tropical shallow marine samples, RH-14 and RH-41, are essentially identical. The histograms derived for these same stations by using the Preston methodology are totally unlike (Fig. 11), and it would take a great amount of ingenuity to fit the histogram of station RH-14 to a truncated lognormal distribution. While the rarefaction curves for the two tropical shallow marine samples, RH-26 and RH-30, are somewhat alike, the Preston histograms are very different. The histogram for RH-26 also does not remotely fit a truncated lognormal pattern.

No attempt will be made to consider each of the histograms in Figure 12. It seems evident from the examples already chosen that the truncated lognormal distribution pattern cannot convincingly be made to fit these samples. Extrapolations of the total number of expected species made from such fitted truncated lognormal curves by adding to the left tail and converting them into normal distribution curves give results that are unrealistic. By such analyses, samples taken from the same environment and displaying similar rarefaction curves often have normal distribution curves containing appreciably differing numbers of species.

Using Margalef's (1957) index (which is essentially the same as Gleason's [1922] formulation), $d = (S - 1)/\ln N$, where S = number of species, N = number of individuals, and d = index of diversity, we find (Fig. 13) that d in all samples is initially low. As sample size becomes larger, the d value rapidly increases. At high densities (large sample sizes), the rate of increase gradually diminishes. This effect is most pronounced in samples with high diversities—the shallow waters of the Bay of Bengal and the deep sea. In high-stress environments—the Buzzards Bay series, tropical estuaries, and the disturbed tropical shallow marine—this effect is less pronounced. These data indicate that diversity indexes of the Margalef and Gleason types are influenced by sample size, even when such samples are internally homogeneous.

When the same samples are plotted using the α values of Fisher et al. (1943), $S = \alpha \ln (N/\alpha + 1)$, almost opposite results are obtained (Fig. 14). In the more diverse samples, the α values are highest at low densities, rapidly decrease as density increases, and then more slowly decrease until an approximate equilibrium is reached. With low-diversity samples, the tendency for higher α values at low density is either absent, poorly developed, or weakly opposed. When internally homogeneous samples are used, this diversity index also is not independent of sample size. In comparison with the d values, the α indexes tend to stabilize at lower faunal densities.

Application of the Simpson diversity index (1949), $C = (y/N)^2$, where

y = number of individuals in species N = total number of individuals, and C = measure of concentration of dominance, to these data does show an increase in diversity with sample size. The rate of change is decidedly less than when α or d indexes are used. This does not mean that the Simpson index is necessarily a more valid diversity measurement. The formula for calculating the diversity index, as shown by Williams (1964), greatly exaggerates the contributions of the few abundant species, while the influence of the many species with few individuals is insignificant. Thus there can be little increase in the diversity index as additional rare species are added. This is the explanation for the relatively low rate of diversity change with increasing sample size.

The degree of exaggeration by this methodology can be demonstrated by comparing the Simpson index per 100 individuals with the number of species for 100 individuals as determined by the rarefaction method for a number of the samples used in our study (Fig. 15). The maximum value by the rarefaction method is 4.93 times greater than the minimum value. Yet the maximum-to-minimum ratio, using the Simpson index, is 16.5:1.00.

We also compared the species numbers using the rarefaction method with the MacArthur model (1957):

$$\frac{N}{S} = \sum_{i=1}^{r} \frac{1}{(S - i + 1)},$$

where N = number of individuals, S = number of species, i = interval between successively ranked species and rarest, and r = rank in rareness. What we measured was the deviation in percentage composition of our actual and rarefied samples from the expected percentage composition derived from the MacArthur model for the same number of species. Like Hairston (1959) and others, we found the common species to be more common and the rare species to be rarer than expected from the model.

As shown in Table 3, there was a strong tendency within each environ-

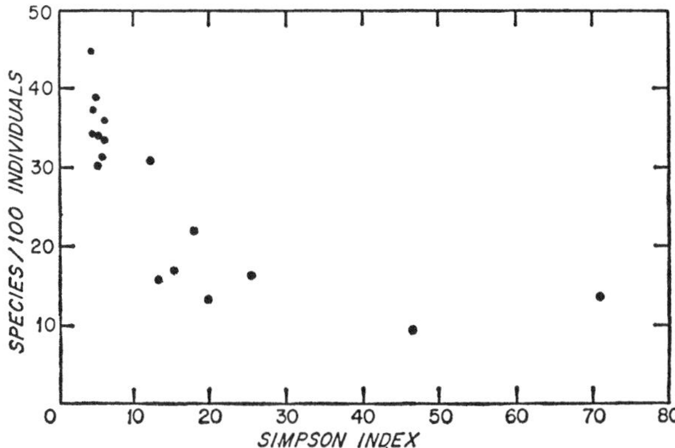

FIG. 15.—Measure of the exaggeration obtained by the Simpson index as compared with rarefaction methodology at the 100-individual level.

TABLE 3

Percentage Deviation of the Rarefaction Data from the MacArthur Model for the Entire Sample and at the 100-Individual Level for Certain of the Stations

Station	Species Number	Number of Individuals	Deviation for Total Sample	Deviation/100 Individuals	Type of Environment
RH-26...	111	6,664	64.58	35.77	Shallow tropical marine
RH-30...	110	5,418	76.28	26.62	Shallow tropical marine
RH-36...	78	717	55.42	33.82	Shallow tropical marine
RH-51...	39	285	17.90	9.04	Shallow tropical marine
RH-14...	33	1,203	98.43	64.86	Stress shallow tropical marine
RH-41...	30	794	72.27	41.21	Stress shallow tropical marine
RH-28...	21	705	66.39	46.93	Tropical estuary
RH-33...	26	696	47.79	21.43	Tropical estuary
S1-3.....	96	9,716	89.32	49.23	Deep sea
D.......	79	4,142	63.12	18.51	Deep sea
DR-33*..	47	2,171	156.76	124.28	Deep sea
G.......	62	807	50.59	19.77	Deep sea
F.......	67	752	49.75	32.78	Deep sea
DR-12...	61	449	34.63	15.84	Deep sea
DR-33′...	46	338	36.87	29.07	Deep sea
GH......	45	276	21.67	17.14	Deep sea
C.......	51	1,861	89.69	56.55	Outer continental shelf
1RA.....	21	2,514	119.61	97.27	Shallow boreal marine
7RB.....	19	1,976	109.05	44.41	Shallow boreal marine
12RA....	20	959	104.27	64.79	Shallow boreal marine
10RA....	20	836	103.35	63.70	Shallow boreal marine
2RB.....	16	465	98.17	83.20	Shallow boreal marine
17R......	21	465	93.23	70.67	Shallow boreal marine

* The gross difference between this sample and the other deep-sea samples is caused by the pronounced aggregation of a single species (for further comments, see text).

ment for progressively smaller samples to show better agreement (less deviation) with the model. The rarefied samples containing 100 individuals always gave a better fit than the actual samples from which they were derived. Finally, when a large number of rarefied samples of different sample sizes from a common sample were compared (Fig. 16), the larger the sample size (except at very low numbers), the greater the departure from the model.

Thus even in homogeneous environments, the MacArthur model is markedly density dependent. For a given faunal density in Figure 16 and Table 3, samples from high-stress environments showed greater deviations from the model than samples from low-stress environments. Such effects are readily masked by sample size. In this regard, it is more than a coincidence that one of the few studies giving a good fit to the MacArthur model (Kohn, 1959) was based on small samples from a stable, low-stress environment. In all fairness to MacArthur, it should be pointed out that he has recently (1966) disavowed the validity of this index.

Lloyd and Ghelardi (1964) proposed the equitability concept as a measure of how a sample is apportioned among its constituent species. Because

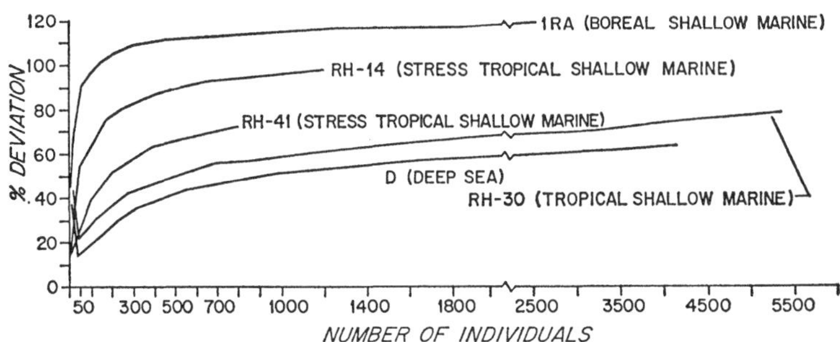

FIG. 16.—Deviation from the MacArthur model using the rarefaction method for certain stations of this study.

numerical equality is never achieved in practice, they suggested "equitability" rather than "evenness" as a more realistic standard. The Shannon-Wiener information function:

$$H(s) = -\sum_{r=1}^{s} p_r \log_2 p_r,$$

where s = total number of species and p_r = observed proportion of individuals that belong to the rth species ($r = 1, 2, \ldots, s$), provides the basis for this measure when combined with some theoretical distribution of abundances, in their case, the MacArthur model shown above.

From this model, given an s, a hypothetical diversity function can be calculated:

$$M(s) = -\sum_{r=1}^{s} \pi_r \log_2 \pi_r,$$

where π_r is the theoretical proportion of individuals in the rth species ranked in order of increasing abundance from $i = 1$ to s. By setting $M(s') = H(s)$, a calculation for the number of hypothetical "equitably distributed" species, s', is obtained. Equitability (ϵ) is the ratio of the "equitably distributed" species (s') to the actual number of observed species (s), $\epsilon = s'/s$. Higher values mean greater equitability; lower values, less equitable apportionment within the sample.

In Figure 17 we have plotted the equitability value (ϵ) for a range of population sizes for a number of our samples. Since in the rarefaction procedure the percentage composition of the original sample is unaltered, we should expect the equitability value to remain constant throughout the spectrum of population sizes. Figure 17 clearly shows that this is not the case. In every sample there is a continuous decrease in the equitability value with increasing sample size. This decrease is most pronounced at the smaller population sizes, particularly in samples from physically unstable environments. Thereafter, there is a more gradual reduction in the magnitude of decrease with increasing sample size. At larger population

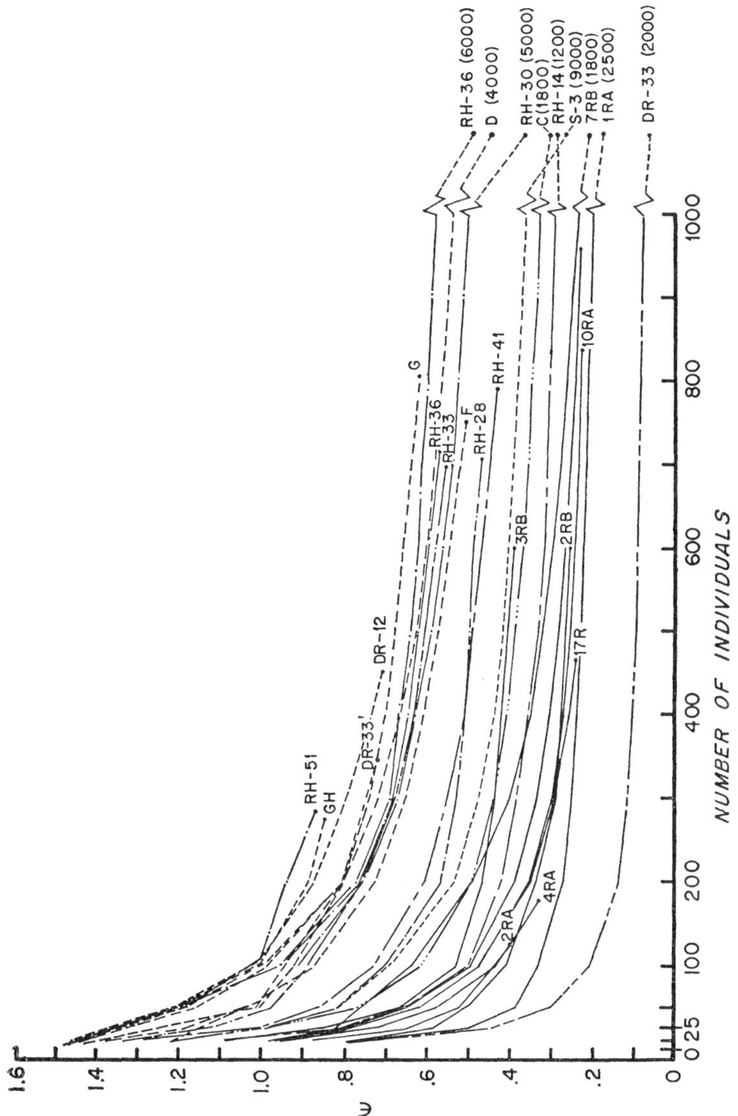

Fig. 17.—Relationship of the equitability value to the rarefaction data for a number of stations included in the study. Numbers in parentheses after the stations are the largest population levels to which the samples have been rarefied.

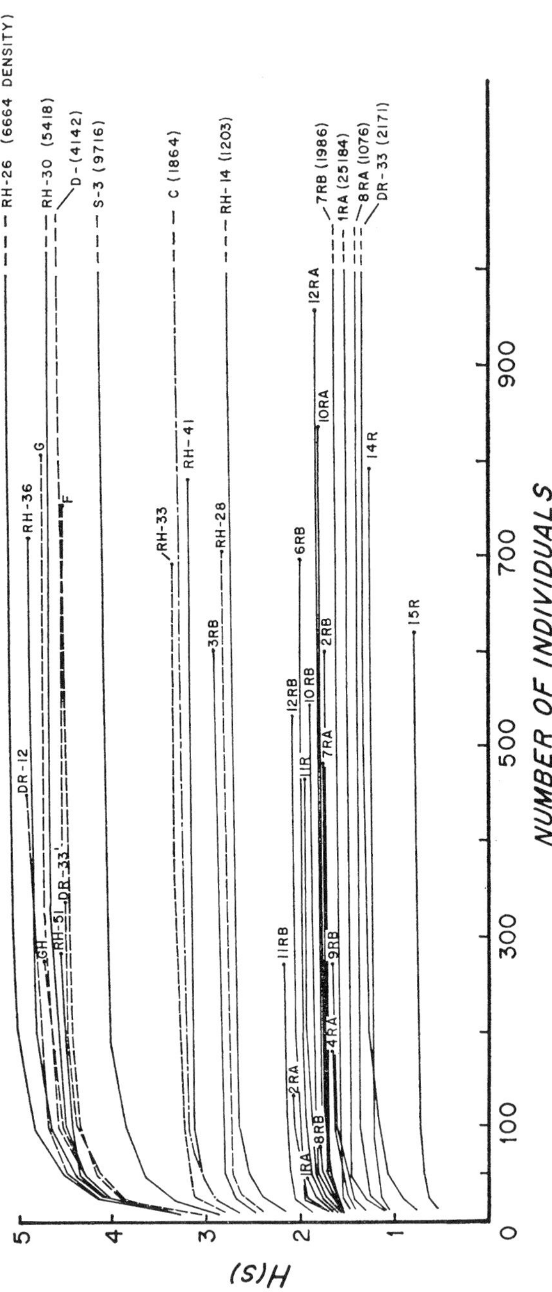

FIG. 18.—Relationship of the information function to the rarefaction data. Numbers in parentheses after the stations are the actual numbers of individuals obtained at those stations.

levels, the magnitude of decrease remains larger for samples from physically stable environments (shallow tropical marine and the deep sea).

Thus equitability is a measurement that is markedly sample-size dependent. This is not surprising when we remember that it is intimately related to the MacArthur model, which we have already shown to be highly sensitive to sample-size differences.

As a final diversity index, we will consider the Shannon-Wiener information function $H(s)$ itself. This function has the attribute of being influenced by both the number of species present and how evenly or unevenly the individuals are distributed among the constituent species. In other words, $H(s)$ is sensitive to both species and dominance diversities.

When the information function is plotted against the rarefaction data (Fig. 18), it very rapidly reaches a stable value and remains essentially constant over a broad spectrum of population sizes. Such stability is achieved at population sizes of about 200 individuals for the high-stress environments (boreal shallow marine, tropical estuary, and disturbed tropical shallow marine) and about 400 individuals for low-stress environments (tropical shallow marine and the deep sea).

Therefore, unlike the other diversity indexes tested, the information function is relatively sample-size independent, and samples of differing sizes, except at lowest faunal densities, can be directly compared.

In summary, when our series of samples were reduced to lower homogeneous population sizes by using the rarefaction method and then compared with various proposed diversity indexes, we found that most of these indexes were decidedly affected by sample size. On the other hand, the Shannon-Wiener information function, except when applied to small-sized samples, possesses the critical characteristic of a useful diversity index, that of being relatively sample-size independent.

SUMMARY

In this paper a methodology is presented for measuring diversity based on rarefaction of actual samples. By the use of this technique, a within-habitat analysis was made of the bivalve and polychaete components of soft-bottom marine faunas which differed in latitude, depth, temperature, and salinity. The resulting diversity values were highly correlated with the physical stability and past history of these environments. A stability-time hypothesis was invoked to fit these findings, and, with this hypothesis, predictions were made about the diversities present in certain other environments as yet unstudied. The two types of diversity, based on numerical percentage composition and on number of species, were compared and shown to be poorly correlated with each other. Our data indicated that species number is the more valid diversity measurement. The rarefaction methodology was compared with a number of diversity indexes using identical data. Many of these indexes were markedly influenced by sample size. Good agreement was found between the rarefaction methodology and the Shannon-Wiener information function.

ACKNOWLEDGMENTS

The ideas presented in this paper have been discussed with numerous individuals. I particularly would like to thank R. R. Hessler, J. H. Connell, and L. B. Slobodkin for their comments and criticisms. M. Rosenfeld and D. W. Spencer generously devoted many hours to both the statistical aspects of this paper and the application of appropriate computer programs. A General Electric 225 computer was used in the analyses.

Support for the acquisition of the various data used came from various sources. The tropical shallow water and estuarine samples from the Indian Ocean were collected as a result of support by the National Science Foundation as a part of the U.S. Program in Biology, International Indian Ocean Expedition. The boreal shallow-water samples from Buzzards Bay, Massachusetts, were collected during the period from 1956 to 1958 under grant NSF G-4812. Support for the collection of deep-sea samples was obtained under grants NSF G-15638, GB-3269, and GB-563. Grant NSF GB-563 also provided support for the collections of boreal estuarine samples from the Pocasset River, Massachusetts.

LITERATURE CITED

Banse, K. 1959. On upwelling and bottom-trawling off the southwest coast of India. J. Marine Biol. Ass. India 1:33–49.

Bousfield, E. L. 1958. Fresh-water amphipod crustaceans of glaciated North America. Can. Field-Natur. 72:55–113.

Carruthers, J. N., S. S. Gogate, J. R. Naidu, and T. Laevastu. 1959. Shoreward upslope of the layer of oxygen minimum off Bombay: Its influence on marine biology, especially fisheries. Nature 183:1084–1087.

Connell, J. H. 1961. Effects of competition, predation by *Thais lapillus,* and other factors on natural populations of the barnacle *Balanus balanoides.* Ecol. Monogr. 31:61–104.

Connell, J. H., and E. Orias. 1964. The ecological regulation of species diversity. Amer. Natur. 98:399–414.

Dobzhansky, T. 1950. Evolution in the tropics. Amer. Sci. 38:209–221.

Dunbar, M. J. 1960. The evolution of stability in marine environments. Natural selection at the level of the ecosystem. Amer. Natur. 94: 129–136.

Fischer, A. G. 1960. Latitudinal variations in organic diversity. Evolution 14:64–81.

Fisher, R. A., A. S. Corbet, and C. B. Williams. 1943. The relation between the number of species and the number of individuals in a random sample of an animal population. J. Anim. Ecol. 12: 42–58.

Gallardo, A. 1963. Notas sobre la densidad de la fauna bentonica en el sublitoral del norte de Chile. Gayana Zool. 10:3–15.

Gleason, H. A. 1922. On the relation between species and area. Ecology 3:158–162.

Grassle, J. F. 1967. Influence of environmental variation on species diversity in benthic communities on the continental shelf and slope. Unpublished Ph.D. dissertation. Duke Univ., Durham, N.C.

Hairston, N. G. 1959. Species abundance and community organization. Ecology 40:404–416.

Hairston, N. G., and G. W. Byers. 1954. The soil arthropods of a field in southern Michigan. A study in community ecology. Contrib. Lab. Vertebrate Biol. Univ. Michigan 64:1–37.

Hessler, R. R., and H. L. Sanders. 1967. Faunal diversity in the deep-sea. Deep-Sea Res. 14:65–78.

Hutchins, L. W. 1947. The basis for temperature zonation in geographical distribution. Ecol. Monogr. 17:325–335.

Jacob, J., and K. Rangarajan. 1959. Seasonal cycles of hydrological events in the Vellar estuary. First All-India Congr. Zool., Proc., Part 2, Scientific Papers, p. 329–350.

Klopfer, P. H. 1959. Environmental determinants of faunal diversity. Amer. Natur. 93: 337–342.

Kohn, A. J. 1959. The ecology of *Conus* in Hawaii. Ecol. Monogr. 29:47–90.

Kohzov, M. 1963. Lake Baikal and its life. Monogr. Biol. 11. 352 p.

LaFond, E. C. 1958. Seasonal cycle of the sea surface temperatures and salinities along the east coast of India. Andhra Univ. Mem. Oceanogr. 2:12–21.

Lloyd, M., and R. J. Ghelardi. 1964. A table for calculating the "Equitability" component of species diversity. J. Anim. Ecol. 33:217–225.

MacArthur, R. H. 1957. On the relative abundance of bird species. Nat. Acad. Sci. Proc. 43:293–295.

———. 1965. Patterns of species diversity. Biol. Rev. 40:510–533.

———. 1966. Note on Mrs. Pielou's comments. Ecology 47:1074.

Margalef, R. 1957. La teoria de la informacion en ecologia. Memorias de la real academia de ciencias y artes (Barcelona) 33:373–449.

Murty, C. B. 1958. On the temperature and salinity structures of the Bay of Bengal. Current Sci. 27:249.

Odum, H. T., J. E. Cantlon, and L. S. Kornicker. 1960. An organizational hierarchy postulate for the interpretation of species individual distributions, species entropy, ecosystem evolution and the meaning of a species variety index. Ecology 41:395–399.

Paine, R. T. 1966. Food web complexity and species diversity. Amer. Natur. 100: 65–75.

Panikkar, N. K. 1940. Influence of temperature on osmotic behavior of some crustacea and its bearing on problems of animal distribution. Nature 146:366–367.

Pianka, E. R. 1966. Latitudinal gradients in species diversity: A review of concepts. Amer. Natur. 100:33–46.

Preston, F. W. 1948. The commonness and rarity of species. Ecology 29:254–283.

Ramamurthy, S. 1953. Hydrobiological studies in Madras coastal waters. J. Madras Univ., Series B. 23:148–163.

Rawson, D. S. 1953. The bottom fauna of Great Slave Lake. J. Fisheries Res. Board Can. 10:486–520.

Sanders, H. L. 1960. Benthic studies in Buzzards Bay. III. The structure of the soft-bottom community. Limnol. Oceanogr. 5:138–153.

———. 1963. Components of ecosystems, p. 86–91. *In* Gordon A. Riley [ed.] Marine Biology I. First Int. Interdisciplinary Conf., Proc. Port City Press, Baltimore.

Sanders, H. L., R. R. Hessler, and G. R. Hampson. 1965. An introduction to the study of the deep-sea benthic faunal assemblages along the Gay Head–Bermuda transect. Deep-Sea Res. 12:845–867.

Sanders, H. L., P. C. Mangelsdorf, Jr., and G. R. Hampson. 1965. Salinity and faunal distribution in the Pocasset River, Massachusetts. Limnol. Oceanogr. 10 (Suppl.):R216–R228.

Simpson, E. H. 1949. Measurement of diversity. Nature 163:688.

Simpson, G. G. 1964. Species density of North American recent mammals. Syst. Zool. 13:57–73.

Thorson, G. 1952. Zur jetzigen Lage der marinen Bodentier-Ökologie: Verhandlungen Deut. Zool. Ges. Wilhelmshaven 1952, p. 276–327.

———. 1966. Some factors influencing the recruitment and establishment of marine benthic communities. Neth. J. Sea Res. 3:267–293.

Whittaker, R. H. 1965. Dominance and diversity in land plant communities. Science 147:250–260.

Williams, C. B. 1964. Patterns in the balance of nature. Academic, New York. 324 p.

PART

2 Dimensions of Macroecology

ALLOMETRY AND BODY SIZE

Edited by Alistair Evans, Daniel P. Costa, Karl J. Niklas, Richard M. Sibly, and Felisa A. Smith

Body size has long captivated the attention of scientists. Some of the earliest scientific treatments speculated on factors influencing the body mass of organisms, the consequence of larger (or smaller) size, and the role of organism size in communities and ecosystems (e.g., Galilei [1638] 1914; Haldane 1928; Thompson [1917] 1942). This fascination stems not only from the ability to clearly see and characterize body size but also from the obvious importance of size for biological structure and function. Size clearly matters in biology.

Body size is a key variable for many macroecological studies. One of the earliest scientists to rigorously investigate patterns of size across taxa was M. Kleiber (1932). Kleiber demonstrated that the energy used by an organism depends on its mass raised to the ¾ power. This "mouse-to-elephant" curve, as it came to be called, was a seminal contribution to the biological understanding of metabolism, but it also raised many intriguing questions. What is the mechanistic underpinning of the relationship? What are the consequences for an individual's ability to reproduce and function? What are the larger consequences at the levels of populations and communities? In this section, we introduce and reprint key papers that have tackled these questions. Our compilation is fairly mammal biased, in part due to the availability of data, but also includes data for invertebrates and land plants.

We begin by considering the mechanistic basis of Kleiber's law. The first plausible theory was provided, not surprisingly, by a mechanical engineer. T. A. McMahon's paper derived from a long tradition of interest in the effect of body size on form and function in animals and plants. This topic was developed earlier by D'Arcy Wentworth Thompson ([1917] 1942) who, in his classic book *On Growth and Form*, investigated the physical origin of biological shape, and by Julian Huxley (1932) who popularized allometric methods. McMahon illustrated the type of theory needed to understand Kleiber's law. Although his theory fit facts available at the time, it is not generally accepted as a universal explanation today. Nevertheless, McMahon heralded a new dawn in considering the physical origins of physiological phenomena such as metabolic scaling.

Kleiber's law has ecological consequences at the level of the individual, the population, and the ecosystem. At the level of the individual, the fact that larger organisms operate on less power per unit body mass means that they do not reproduce as fast as smaller ones. Consequently, their maximum rates of population growth are lower. The first person to demonstrate this was T. Fenchel. He showed that the "intrinsic rate of natural increase" was highly influenced by body size for an extremely wide range of organisms. He was clearly aware of the importance of his finding (sometimes called Fenchel's law) and began the discussion of its wider ecological implications at the end of his paper.

Population growth rate is fundamental to the ecological success of organisms. It results from various components of life history—for example, the timing of reproductive events, the birthrates then achieved, and the age-specific death rate. Given that population growth is dependent on size, it is perhaps not surprising that life-history traits are too. Just how life-history traits vary with body size was first studied in mammals by John Millar (1977) and David Western (1979) who showed that birth intervals increase with size but that birthrates decrease. Both of these relationships have turned out to be very general (e.g., Brown et al. 2004). In turn, life-history traits depend on physiological processes such as renal clearance and blood supply to tissues. Such scaling of important physiological processes was analyzed by S. L. Lindstedt and W. A. Calder III. They discovered, inter alia, that heart rate appears to be related to life span, such that there are a relatively constant number of heartbeats per lifetime regardless of body size. This astounding observation linked body size to metabolic rate and physiological timing.

Kleiber's law has further consequences at the population level. Kleiber demonstrated that individual metabolic rate varies as body mass$^{3/4}$, but J. Damuth showed that individuals space themselves out such that population density scales as body mass$^{-3/4}$. Consequently, energy flux per unit area scales as body mass$^{3/4-3/4}$, which equals body mass^{0}. In other words, energy flux per unit area is invariant with size. This has been called the "energy equivalence rule" and represents another benchmark contribution, if still somewhat controversial.

The spacing mechanisms in plants are more direct than those in animals. Plants cannot move out of the way of larger or better competitors. Instead they stay where they are and, if unlucky, die when outcompeted for resources. The resultant decline in plant population density as body size increases follows a power law, body mass$^{-2/3}$. How this "self-thinning" comes about mechanistically is the subject of R. Å. Norberg's paper. In part building on the McMahon paper included in this volume, Norberg linked plant size and geometry to population density to derive the $-2/3$-scaling rule of plant size to plant number.

Finally, we include two papers relating the effects of body size to the role that species play in communities. That there must be some relationship is not surprising, but the extent to which body size influences community structure perhaps is. It took some time for the empirical patterns to be clarified and theoretical explanations for these to appear. R. M. May (1978) proposed that the number and abundance of species could only be properly understood in relation to size. What was needed to test May's idea was to collect data on a number of species along with their abundance and body size for individual communities. D. R. Morse et al. were the first to gather such data for an entire community (in this instance, beetles). They showed that beetles of intermediate body size had the highest abundance, but small, rare species did exist. However, the question still remained as to whether such community composition varied with spatial scale. J. H. Brown and P. F. Nicoletto, working with mammals, showed that in fact it does. Local communities were relatively uniform in their body-mass distribution, but at increased spatial scales distributions, they became progressively more peaked.

It was obviously not possible to include books in this compilation. However, there are several that should be mentioned. While not as technical as the volumes that followed, McMahon and J. T. Bonner's *On Size and Life* (1983) provided one of the first synoptic treatments of body size. In short order this was followed by K. Schmidt-Nielsen's *Scaling* (1984), which provided a discussion of the importance of body size to functional processes in organisms from an adaptationist perspective, and Calder's *Size, Function, and Life History* (1984), which provided a complete and exhaustive review of allometric relationships. R. H. Peters's *The Ecological Implications of Body Size* (1983) provided not only a review of allometric relationships but also an insightful synthetic analysis of the importance of allometry to ecology. These volumes, all published within a short time frame, were largely responsible for the resurgence of interest in the topic of allometry over the past several decades. More recently, K. J. Niklas's

Plant Allometry (1994) has provided large data sets for land plants and algae.

There are many papers that we could or perhaps should have included in this collection. Some were omitted due to constraints of space; still others, because they are recent and their effect has yet to be realized. For example, L. Von Bertalanffy (1957) examined the role of body size and metabolism in the growth of organisms. This work was followed by an analysis carried out by S. J. Gould (1966) that provided an extensive discussion of allometry and laid the conceptual framework that was later more quantitatively presented in the Lindstedt and Calder paper reprinted here. Other papers have extended concepts, showing that processes or relationships can be applied to other species or systems. For example, M. Hemmingsen (1960) extended Kleiber's mouse-to-elephant curve to include unicells and ectotherms as well as other endotherms. W. R. Stahl (1967) provided one of the first comprehensive reviews of scaling and respiratory parameters, which was followed by a series of reviews by Calder (1981, 1983a, 1983b). P. F. Brodie (1975) demonstrated that whether whales are resident or migratory is related to metabolic scaling. While many of these early studies were based entirely on measurements of animals in the laboratory, work using isotopic tracer methods has shown that the ¾-power scaling of metabolic rate also applies to mammals and birds freely ranging in nature (Nagy 1987, 2005).

Increasingly, an awareness is developing of the importance of body size and allometry in understanding all levels of biological phenomena. New theories continue to be advanced while older concepts are refined or expanded. For example, G. B. West et al. (1997) recently provided an explanation of the ¾-power law for metabolic rates that was based on the transport of materials through organisms as a function of the geometry of their circulatory systems. This paper has had tremendous impact and has stimulated considerable dissent and discussion (West et al. 1999, 2001; Brown et al. 2005; Packard and Birchard 2008), while revitalizing the field of metabolic scaling and extending allometry to a wide variety of disciplines from molecular biology to urban landscape design.

Nevertheless, conceptual and practical difficulties in describing size-dependent relationships remain. Consider the problem of describing variation by means of a simple bivariate plot with a large number of data points. Linear or curvilinear regression often yields a highly significant, statistically descriptive relationship, even if it accounts for only a small portion of the variation and violates parametric assumptions about the homogeneity of variance (Harvey and Mace 1982). Alternative multivariate techniques can be applied to such data. However, additional variables have either not been measured or their identities remain problematic. And when such variables have been measured, their interpretation can be difficult and laden with assumptions. Although the field of allometry has seen resurgence in the development of new statistical techniques with a finer appreciation of the subtleties required to examine variance in light of mechanistic models of size-dependent relationships, considerable work remains to be done to test the efficacy of these techniques to reveal the mechanisms underlying the ecological and evolutionary processes.

Literature Cited

Brodie, P. F. 1975. Cetacean energetics, an overview of intraspecific size variation. *Ecology* 56:152–61.

Brown, J. H., J. F. Gillooly, A. P. Allen, V. M. Savage, and G. B. West. 2004. Toward a metabolic theory of ecology. *Ecology* 85:1771–89.

Brown, J. H., G. B. West, and B. J. Enquist. 2005. Yes, West, Brown and Enquist's model of allometric scaling is both mathematically correct and biologically relevant. *Functional Ecology* 19:735–38.

Calder, W. A., III. 1981. Scaling of physiological processes in homeothermic animals. *Annual Review of Physiology* 43:301–22.

———. 1983a. Ecological scaling: Mammals and birds. *Annual Review of Ecology and Systematics* 14:213–30.

———. 1983b. Scaling of osmotic regulation in mammals and birds. *American Journal of Physiology* 244:R601–R606.

———. 1984. *Size, function, and life history.* Harvard University Press, Cambridge, MA.

Galilei, G. (1638) 1914. *Dialogues concerning two new sciences*. Translated by H. Crew and A. de Salvio. MacMillian, New York.

Gould, S. J. 1966. Allometry and size in ontogeny and phylogeny. *Biological Reviews* 41:587–638.

Haldane, J. B. S. 1928. *Possible worlds and other papers*. Harper and Brothers, New York.

*Harvey, P. H., and G. M. Mace. 1982. Comparisons between taxa and adaptive trends: Problems of methodology. In *Current problems in sociobiology*, edited by King's College Sociobiology Group, pp. 343–61. Cambridge University Press, Cambridge.

Hemmingsen, A. M. 1960. Energy metabolism as related to body size and respiratory surfaces in evolution. *Reports of the Steno Memorial Hospital and the Nordisk Insulinlaboratorium* 9:7–110.

Huxley, J. S. 1932. *Problem of relative growth*. Methuen, London.

Kleiber, M. 1932. Body size and metabolism. *Hilgardia* 6:315–53.

*May, R. M. 1978. The dynamics and diversity of insect faunas. In *Diversity of insect faunas*, edited by L. A. Mound and N. Waloff, 188–204. Royal Entomological Society of London Symposium 9. Blackwell Scientific, Oxford.

McMahon, T., and J. T. Bonner. 1983. *On size and life.* W. H. Freeman, New York.

Millar, J. 1977. Adaptive features of mammal evolution. *Evolution* 31:370–86.

Nagy, K. A. 1987. Field metabolic rate and food requirement scaling in mammals and birds. *Ecological Monographs* 57:111–28.

———. 2005. Field metabolic rate and body size. *Journal of Experimental Biology* 208:1621–25.

Niklas, K. J. 1994. *Plant allometry: The scaling of form and process.* University of Chicago Press, Chicago.

Packard, G. C., and G. F. Birchard. 2008. Traditional allometric analysis fails to provide a valid predictive model for mammalian metabolic rates. *Journal of Experimental Biology* 211:3581–87.

Peters, R. H. 1983. *The ecological implications of body size*. Cambridge University Press, Cambridge.

Schmidt-Nielsen, K. 1984. *Scaling: Why is animal size so important?* Cambridge University Press, New York.

Stahl, W. R. 1967. Scaling respiratory variables in mammals. *Journal of Applied Physiology* 22:453–60.

Thompson, D. W. (1917) 1942. *On growth and form.* Cambridge University Press, Cambridge.

Von Bertalanffy, L. 1957. Quantitative laws in metabolism and growth. *Quarterly Review of Biology* 32:217–31.

West, G. B., J. H. Brown, and B. J. Enquist. 1997. A general model for the origin of allometric scaling laws in biology. *Science* 276:122–26.

———. 1999. The fourth dimension of life: Fractal geometry and allometric scaling of organisms. *Science* 284:1677–79.

———. 2001. A general model for ontogenetic growth. *Nature* 413:628–31.

Western, D. 1979. Size, life history and ecology in mammals. *African Journal of Ecology* 17:185–204.

PAPER 12

Size and Shape in Biology (1973)

T. A. McMahon

Commentary

RICHARD M. SIBLY AND KARL J. NIKLAS

This brilliant paper, written by a mechanical engineer, was the first to provide a compelling mechanistic explanation of Kleiber's law. Applying basic engineering principles, T. A. McMahon showed that for structurally idealized trees to resist bending loads caused by gusts of wind and gravity, it is necessary that the length, l, of both the trunk and the branches should be proportional to the ⅔ power of stem diameter, d. Accordingly, the optimal geometry requires an $l \propto d^{2/3}$ scaling relationship, which has become known as the principal of elastic self-similarity, as opposed to either the $l \propto d$ scaling relationship, which is referred to as geometric self-similarity, or the $l \propto d^{1/2}$ scaling relationship, which is referred to as stress self-similarity. Arguing that maximum metabolic rate, B, depends on the cross-sectional area of muscles when they are at full stretch, d^2, McMahon further concluded that $B \propto d^2$. However, because body mass, M, scales as ld^2, and $l \propto d^{2/3}$, it follows that $d \propto M^{3/8}$. So maximum metabolic rate ought to scale as $B \propto d^2 \propto M^{3/4}$, which appears to offer a plausible mechanistic explanation of Kleiber's law.

Despite the beauty and simplicity of this explanation, it has not proved to have the universality originally hoped for. K. J. Niklas (1995) showed that no single principle of optimal design, elastic or otherwise, holds along the entire range of tree trunks or throughout the lifetime of *Robinia pseudoacacia* trees. In addition, data indicate that tree geometry is as much dictated by hydraulic as by mechanical constraints. Indeed, a simple mathematical model based on the effects of hydraulics on tree growth reveals log-log nonlinear scaling relationships for tree height and mass with respect to trunk diameter (Niklas and Spatz 2004). Likewise, animal limbs tend to scale according to geometric as opposed to elastic self-similarity, and under any circumstances, aquatic plants and animals are rarely subjected to gravitational loads. The reasons for these failures of structural engineering theory are, however, still debated, and still with reference to McMahon's pioneering work.

Literature Cited

Niklas. K. J. 1995. Size-dependent allometry of tree height, diameter and trunk-taper. *Annals of Botany* 75:217–27.

Niklas. K. J., and H.–C. Spatz. 2004. Growth and hydraulic (not mechanical) constraints govern the scaling of tree height and mass. *Proceedings of the National Academy of Sciences* 101:15661–63.

From *Science* 179:1201–4. Reprinted with permission from AAAS.

and *Society* (Wiley-Interscience, New York, (1971); *Sci. Amer.* **224**, 224 (Sept. 1971).

7. A. Coale, *Science* **170**, 132 (1970).
8. I. Taeuber [in *Man's Place in the Island Ecosystems*, F. R. Fosberg, Ed. (Univ. of Hawaii Press, Honolulu, 1961), pp. 226–262] analyzes how absorption into expanding societies affects the age and sex composition of populations in formerly isolated social systems.
9. O. D. Duncan, *Handbook for Modern Sociology*, R. E. L. Faris, Ed. (Rand McNally, Chicago, 1964), pp. 36–82.
10. K. Boulding describes the human ecosystem as the "totality of human organizations" [*The Organizational Revolution* (Quadrangle, Chicago, 1952), p. xxii] and O. D. Duncan notes that the cycling of information is a unique feature of the human ecosystem (9, pp. 40–42).
11. R. Freedman, *Pop. Index* **31**, 417 (1965).
12. S. S. Kuznets, *Proc. Amer. Philos. Soc.* **111**, 170 (1967).
13. M. Abramowitz, *Amer. Econ. Rev.* **46**, 8 (1965).
14. J. Krebs and J. Spengler, in *Technology and the American Economy* (Report of the National Commission on Technology, Automation, and Economic Progress) (Government Printing Office, Washington, D.C., 1966), vol. 2, pp. 359–360.

Size and Shape in Biology

Elastic criteria impose limits on biological proportions, and consequently on metabolic rates.

Thomas McMahon

Observers of living organisms since Galileo have recognized that metabolic activities must somehow be limited by surface areas, rather than body volumes. Rubner (*1*) observed that heat production rate divided by total body surface area was nearly constant in dogs of various sizes, and proposed the explanation that metabolically produced heat was limited by an animal's ability to lose heat, and thus total body surface area. When more precise methods of measurement became available, Kleiber (*2*) noticed that when rate of heat production is plotted against body weight on logarithmic scales for animals over a size range from rats to steers, the points fall extremely close to a straight line with slope 0.75 (Fig. 1). The result has since been confirmed for animals as different in size as the mouse and the elephant (*3–5*), and has been verified for other metabolically related variables, such as rate of oxygen consumption (*6*). Excellent reviews of the problem are available (*7–10*).

While it is often true that biological laws are not derivable from physical laws in any simple sense, Kleiber's rule may be one of those fortuitous exceptions which D'Arcy Thompson (*11*) suggests lie at the basis of a fundamental "science of form." Plants as well as animals must be built strongly enough to stand under their own weight. In the following, a general rule is derived for the changing proportions of idealized trees as a function of scale, and later the results are applied to animals.

The author is assistant professor of applied mechanics in the division of engineering and applied physics, Harvard University, Cambridge, Massachusetts 02138.

Buckling

Consider a tall, slender cylindrical column of length l and diameter d loaded by the force P, representing the total weight of the column, acting at the center of mass. Such a column will fail in compression if the applied stress P/A, where $A = \pi d^2/4$, exceeds the maximum compressive stress, σ_{max}. Provided that the column is slender enough, it may also fail in what is known as elastic buckling, whereby a small lateral displacement (caused, for example, by the smallest gust of wind), allows the weight P to apply a toppling moment which the elastic forces of the bent column below are not sufficient to resist. In this case, "slender enough" means that l/d is greater than 25, a range which includes virtually all trees (*12*). The critical length for buckling is related to the diameter by:

$$l_{cr} = 0.851 \left| \frac{E}{\rho} \right|^{1/3} d^{2/3} \qquad (1)$$

where ρ is the weight per unit volume and E is the elastic modulus of the material. The mathematician Greenhill (*13*) showed that when the force due to weight is distributed over the total extent of the column instead of being taken as acting at the center of mass, the critical height becomes:

$$l_{cr} = 0.792 \left| \frac{E}{\rho} \right|^{1/3} d^{2/3} \qquad (2)$$

This result is identical to Eq. 1, with only a change in the numerical constant. It may be demonstrated that another change in the constant occurs when the solid cylinder is made hollow, provided that the thickness of the wall is proportional to the diameter. Greenhill further showed that if the shape of the column is taken as a cone, or a paraboloid of revolution, the result is again only to change the numerical constant. Recently, Keller and Niordson (*14*) have derived that the tallest self-supporting homogeneous tapering column is 2.034 times as tall as a cylindrical column made of the same volume of the same material, and that the distance to the top of such a tapering column above any cross section is proportional to the diameter of that cross section raised to the ⅔ power. The rule requiring height to go as diameter to the ⅔ power is thus independent of many details of the model proposed for the elastic stability of tree trunks.

Bending

The limbs of trees must also be proportioned to endure the bending forces produced by their own weight. If a branch is considered to be a cantilever beam built into the trunk, there exists a particular beam length l_{cr} for which the tip of the branch extends the greatest horizontal distance away from the trunk (*15*). Branches longer than l_{cr} droop so much that their tips actually come closer to the trunk. Suppose that the purpose of branches is to carry their leaves out of the shadow of higher branches, and therefore to achieve a maximum lateral displacement from the trunk. Then the limb should grow no longer than l_{cr}, where

$$l_{cr} = C\left|\frac{E}{\rho}\right|^{1/3} d^{2/3} \qquad (3)$$

and C depends only on the droop angle θ_D, which in turn depends only on the angle at which the limb leaves the trunk (*15*). The result may be made general for a tapered or hollow limb exactly as was done for the buckling problem. Comparing Eqs. 1, 2, and 3, it is apparent that elastic criteria set length proportional to the ⅔ power of diameter in both the trunk and the branches.

It should be possible to check the validity of these results by measuring the proportions of trees of different scale. Such a check would be arduous if it were necessary to know E and ρ for each species; fortunately, the ratio E/ρ is quite accurately constant in green woods (*16, 17*). In Fig. 2, the trunk diameter 1.525 meters from the ground is plotted against the total height for 576 individual trees, representing nearly every species found in the United States. The data, taken primarily from the American Forestry Association's "Social register of big trees" (*18*), include specimens both very slender and very stout, since trees are eligible for this list according to their bigness, an index depending on the sum of their circumference and height (*19*). A solid line representing Eq. 2 is also shown in Fig. 2; it was calculated for $E = 1.05 \times 10^5$ kilograms per square meter and $\rho = 6.18 \times 10^2$ kilograms per cubic meter (*16*). The broken line, which fits near the center of the data points, has the same slope as the solid line but represents a sequence of trees whose height in each case is only one-fourth of the critical buckling height. The conclusion seems to be that the proportions of trees are limited by elastic criteria, since there are no data points to the left of the solid line.

Animal Proportions

Just as trees must assume thicker proportions with increasing size, so must animals adjust their shape with scale. The argument has long been offered that animals could not remain geometrically similar from the small to the large because their limbs, whose cross-sectional area increases as the square of characteristic body dimension L, must then support a weight which increases as L^3 (*7*). The difficulty with these arguments based on strength criteria is the inevitable conclusion that animals may grow no larger than a size which makes the applied stress equal to the yield stress of their materials. Animals larger than this size would have to increase supporting areas directly with weight, so that no increases in height could be tolerated, only increases in width. If yield stress were the only criterion, an animal with slender proportions like the bobcat should be capable of attaining the same absolute height as the lion. In fact, it is widely found that some animals grow larger than others, and animals of small scale are relatively more slender than those of large scale (see cover). Perhaps this transformation occurs, as in differently sized trees, for reasons based on elastic rather than strength criteria.

In the following, we consider comparisons between animals of the same family, so that their shape is grossly similar. The only change in shape permitted is for lengths to bear a specified relationship to diameters: all lengths will be proportional to one another, as will be all diameters. Each limb, bone, or muscle will thus have a length l and diameter d, where length will be taken as a measurement parallel to the direction of tension or compression and diameter will be measured perpendicular to this direction. Thus, the length of the trunk is the distance between shoulder and hip whether the animal is bipedal or quadrapedal (Fig. 3a, bottom).

When a quadruped is standing at rest, the four limbs will be exposed primarily to buckling loads, but the vertebral column and its musculature must withstand bending loads. When the same animal runs, the situation is substantially reversed in those phases of the motion where the limbs are providing their maximum propulsive effort. At these moments, the limbs are supporting bending loads, while the vertebral column is receiving an end thrust and thus a buckling load. The fact that the loads are dynamic rather than static is not a consideration: the maximum deflection of a structure suddenly loaded under its own weight is

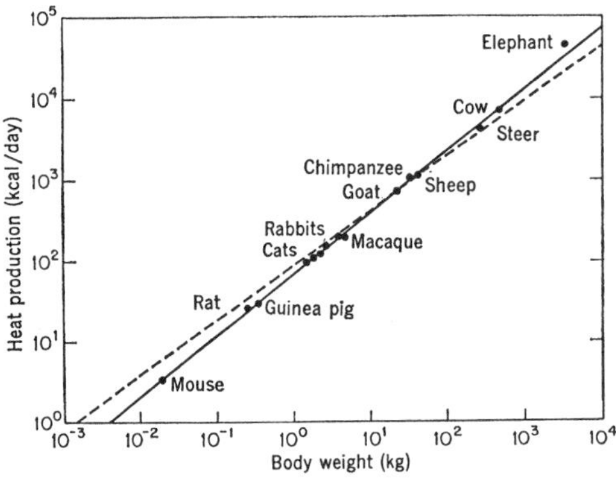

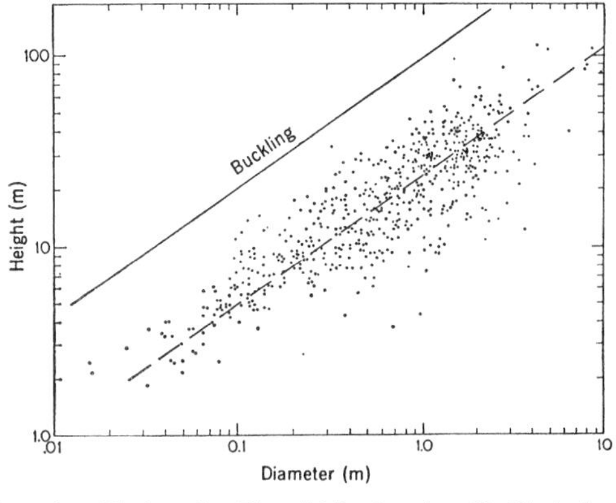

Fig. 1 (left). Metabolic heat production plotted against body weight on logarithmic scales. The solid line has slope ¾. The broken line, which does not fit the data, has slope ⅔ and represents the way surface area increases with weight for geometrically similar shapes [adapted from (*2*)]. Fig. 2 (right). Tree height plotted against trunk base diameter on logarithmic scales for record trees representing nearly every American species. The trunk proportions are limited by elastic buckling criteria, since no points lie to the left of the solid line. Data from (*18, 19*).

just twice the static deflection when the load is gradually applied (*12*). The true instantaneous loading condition for each of the quasi-cylindrical elements is thus some complicated sum of buckling, bending, and torsional loads, but fortunately the elastic criteria predict the same result independently of the type of gravitational self-loading, namely that every l should be proportional to the ⅔ power of the equivalent d.

Rashevsky (*20*) assumed that the trunk of an animal was a uniformly loaded beam, and used the linearized theory of beam bending to calculate the same result, that trunk length should go as diameter to the ⅔. Rashevsky's model additionally required the cross-sectional area of the animal's limbs to be proportional to the weight of the trunk, leading to a different set of rules for determining limb proportions from those for trunk proportions. In the present model all the proportions of an animal would change with size in the same way. If W is the total body weight, the weight of any limb is a specified fraction of W, and:

$$W \propto ld^2 \qquad (4)$$

but if l^3 is proportional to d^2, then

$$l \propto W^{1/4}; \quad d \propto W^{3/8} \qquad (5)$$

Comparative zoologists have long been aware that the gross dimensions of many species bear a power law relation to body weight. Brody (*4*) measured the chest girth G and the height at withers H of more than 3000 Holstein cattle. His data fit the present model well: he empirically found G proportional to $W^{0.36}$ ($W^{0.375}$ predicted), while H goes as $W^{0.24}$ ($W^{0.25}$ predicted).

In a study of primates whose weights ranged from 0.28 to 22 kg, Stahl and Gummerson (*21*) reported many of the important somatic and skeletal dimensions, x, as power functions of body weight, $x = aW^b$. Figure 3a, reproduced from their paper, shows that chest circumference in primates is proportional to $W^{0.37}$ with a correlation of .995. Agreement with the proposed model is excellent for most of his measurements: b is 0.28 for trunk height (0.25 predicted) and 0.38 for maximum thigh girth (0.375 predicted).

Let us return to the question of external body surface area. If the surface area of each of the quasi-cylindrical elements that make up the whole animal in the proposed model is calculated, we find

$$\text{surface area} \propto ld + d^2/2 \qquad (6)$$

where the second term is due to the ends of each cylindrical element, so that it is absent or halved in the case of many of the elements. For most limbs and many of the trunks under consideration, l/d is approximately 10, so that the second term is only 5 percent of the first and may be neglected. In this case, total body surface area is proportional to ld and thus to $W^{1/4}W^{3/8}$, or $W^{5/8}$. Hemmingsen (*8*) presented a plot of body surface area against weight for animals in a weight range of 1 to 10^6 grams, and he also included points representing defoliated beech trees. In his figure, only one solid line appears, that appropriate to the surface area of a sphere of density 1.0 g/cm^3. His figure is reproduced in Fig. 3b, with an additional line representing the proposed model of a cylinder whose surface area is three times the sphere area when both sphere and model weigh close to 8 g, but only twice the sphere area when both weigh about 70 kg. The slope of the line for this stretched cylinder is 0.63, while the slope of the line for the sphere, and thus all geometrically similar structures, is 0.67. Although Hemmingsen argues that the data points are well fitted by an imaginary line running parallel to that of the sphere, it is apparent that a good fit is obtained by the present model. In data spanning the range from rats to humans, Stahl (*22*) found that surface area increases as the 0.65 power of body weight. Thus, the present model agrees with experimental observations of body surface area as well as body proportions.

Metabolic Rate

Our ideas describing how size determines shape are now complete, and we may return to the original question concerning metabolism and Kleiber's law. Suppose a muscle, whose cross-

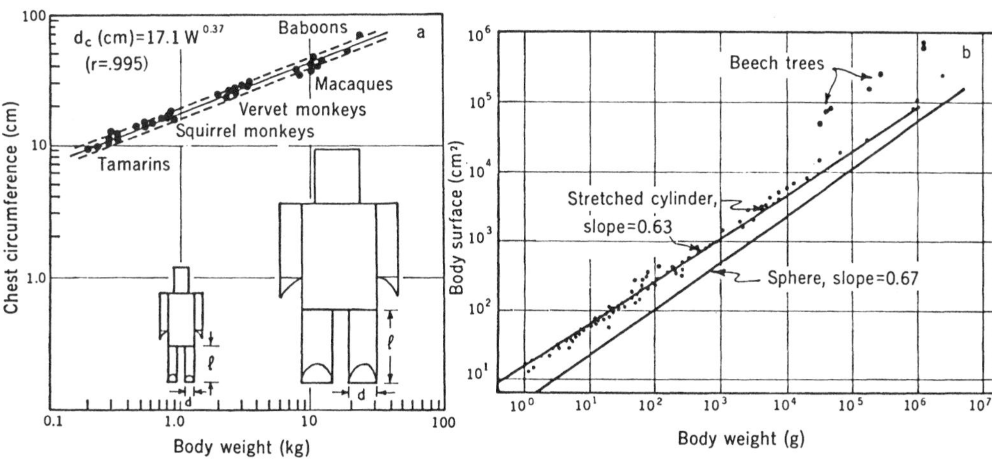

Fig. 3. (a) Chest circumference, d_c, plotted against body weight, W, for five species of primates. The broken lines represent the standard error in this least-squares fit [adapted from (*21*)]. The model proposed here, whereby each length, l, increases as the ⅔ power of diameter, d, is illustrated for two weights differing by a factor of 16. (b) Body surface area plotted against weight for vertebrates. The animal data are reasonably well fitted by the stretched cylinder model [adapted from (*8*)].

sectional area is A, shortens a length Δl against force σA in time Δt. The power this muscle expends is $\sigma A \Delta l / \Delta t$, where σ is the tensile stress developed, and is in general a function of the shortening velocity $\Delta l / \Delta t$. Hill (*23*) reported that "the inherent strength of a contracting voluntary muscle fiber is roughly constant, being of the order of a few kilograms per square centimeter of cross-section." He also presented arguments and experimental data to prove that the speed of shortening, $\Delta l / \Delta t$, is a constant in any particular muscle from species to species. If we understand from the work of Hill and others that both σ and $\Delta l / \Delta t$ may be taken as constant, then the power output of a particular muscle and hence all the metabolic variables involved in maintaining the flow of energy to that muscle depend only on its cross-sectional area. But this area is proportional to d^2, and hence

$$\text{maximal power output} \propto (W^{3/8})^2 = W^{0.75} \qquad (7)$$

This is precisely the statement of Kleiber's law we were looking for, provided we have some confidence that maximal energy metabolism exceeds basal metabolic rate by a factor, the metabolic "scope," which is invariant with respect to scale. Hemmingsen (*8*) has presented evidence to this effect.

According to the model proposed here, if lung volume goes as W (*4, 21*) but alveolar ventilation goes as $W^{0.75}$, then respiratory frequency must scale as $W^{-0.25}$. The identical argument may be made for ventricular stroke volume, cardiac output, and heart rate. In fact, Adolph (*24*) reported that b for respiratory frequency in mammals is -0.28 [Tenney (*10*) independently gave the same number]. For heart rate, b has been reported as -0.27 (*25*) and -0.25 (*22*). Stahl (*9*) observed that the ratio of many physiological periods to one another is found to be nearly constant, independent of scale. Thus, the ratio of gut pulsation time to pulse time is nearly the same in all mammals, and each animal lives for approximately the same number of heartbeats or breath cycles. Other authors have discussed the importance of this conclusion in arriving at the "physiological age" of living organisms.

Summary and Conclusions

Arguments based on elastic stability and flexure, as opposed to the more conventional ones based on yield strength, require that living organisms adopt forms whereby lengths increase as the ⅔ power of diameter. The somatic dimensions of several species of animals and of a wide variety of trees fit this rule well.

It is a simple matter to show that energy metabolism during maximal sustained work depends on body cross-sectional area, not total body surface area as proposed by Rubner (*1*) and many after him. This result and the result requiring animal proportions to change with size amount to a derivation of Kleiber's law, a statement only empirical until now, correlating the metabolically related variables with body weight raised to the ¾ power. In the present model, biological frequencies are predicted to go inversely as body weight to the ¼ power, and total body surface areas should correlate with body weight to the ⅝ power. All predictions of the proposed model are tested by comparison with existing data, and the fit is considered satisfactory.

In *The Fire of Life*, Kleiber (*5*) wrote "When the concepts concerned with the relation of body size and metabolic rate are clarified, . . . then compartive physiology of metabolism will be of great help in solving one of the most intricate and interesting problems in biology, namely the regulation of the rate of cell metabolism." Although Hill (*23*) realized that "the essential point about a large animal is that its structure should be capable of bearing its own weight and this leaves less play for other factors," he was forced to use an oversimplified "geometric similarity" hypothesis in his important work on animal locomotion and muscular dynamics. It is my hope that the model proposed here promises useful answers in comparisons of living things on both the microscopic and the gross scale, as part of the growing science of form, which asks precisely how organisms are diverse and yet again how they are alike.

References and Notes

1. M. Rubner, *Z. Biol. Munich* **19**, 535 (1883).
2. M. Kleiber, *Hilgardia* **6**, 315 (1932).
3. F. G. Benedict, *Vital Energetics: A Study in Comparative Basal Metabolism* (Carnegie Institution of Washington, Washington, D.C., 1938).
4. S. Brody, *Bioenergetics and Growth* (Reinhold, New York, 1945).
5. M. Kleiber, *The Fire of Life* (Wiley, New York, 1961).
6. S. M. Tenney and J. E. Remners, *Nature* **197**, 54 (1963); A. C. Guyton, *Amer. J. Physiol.* **150**, 70 (1947); E. Zeuthen, *Quart. Rev. Biol.* **28** (No. 1), 1 (1953).
7. K. Schmidt-Nielsen, *Proc. Amer. Physiol. Soc.* **29**, 1524 (1970); A. M. Hemmingsen, *Rep. Steno Mem. Hosp. Nord. Insulin Lab.* **4**, 1 (1950); S. J. Gould, *Amer. Natur.* **105**, 113 (1971); W. R. Stahl, *Science* **137**, 205 (1962).
8. A. M. Hemmingsen, *Rep. Steno Mem. Hosp. Nord. Insulin Lab.* **9**, 1 (1960).
9. W. R. Stahl, *Advan. Biol. Med. Phys.* **9**, 355 (1963).
10. S. M. Tenney, *Circ. Res.* **20-21** (Suppl. I), I-7 (1967).
11. D. W. Thompson, *On Growth and Form* (Cambridge Univ. Press, London, 1917).
12. S. Timoshenko, *Elements of Strength of Materials* (Van Nostrand, Princeton, N.J., 1962).
13. G. Greenhill, *Proc. Cambridge Phil. Soc.* **4**, 65 (1881).
14. J. B. Keller and F. I. Niordson, *J. Math. Mech.* **16** (No. 5), 433 (1966).
15. T. A. McMahon, in preparation.
16. T. A. McElhanney and R. S. Perry, *Forest Service Bulletin No. 78* (Forest Products Laboratories of Canada, Ottawa, 1927).
17. G. A. Garratt, *The Mechanical Properties of Wood* (Wiley, New York, 1931).
18. "Social register of big trees," *Amer. Forests* **72**, 16 (May 1966); *ibid.* **77**, 25 (January 1971).
19. Other data used in Fig. 2 are from the following books: Royal Horticultural Society, *Conifers in Cultivation* (Royal Horticultural Society, London, 1932), p. 316; W. Fry and J. R. White, *Big Trees* (Stanford Univ. Press, Stanford, Calif., 1930).
20. N. Rashevsky, *Mathematical Biophysics* (Dover, New York, 1960), vol. 2.
21. W. R. Stahl and J. Y. Gummerson, *Growth* **37**, 21 (1967).
22. W. R. Stahl, *J. Appl. Physiol.* **22**, 453 (1967).
23. A. V. Hill, *Sci. Progr. London* **38** (No. 150), 209 (1950).
24. E. F. Adolph, *Science* **109**, 579 (1949).
25. A. J. Clark, *Comparative Physiology of the Heart* (Cambridge Univ. Press, Cambridge, 1927).

26. The author is grateful to B. Budiansky, R. E. Kronauer, S. J. Gould, C. R. Taylor, and G. P. DeWolf for assistance through helpful discussions.

PAPER 13

Intrinsic Rate of Natural Increase: The Relationship with Body Size (1974)

T. Fenchel

Commentary

RICHARD M. SIBLY

How fast a species can increase in numbers is one of its defining characteristics, and this paper was the first to show how maximum population growth rate is related to body size. T. Fenchel, a Danish marine biologist, reveals in figure 1 the allometry between maximum population growth rate and body size. This shows quantitatively how much faster small-bodied populations grow than those with larger bodies. Today this pivotal relationship—sometimes called Fenchel's law—is taken for granted, and is important because maximum population growth rate constrains many aspects of a species' biology. It limits how fast populations can recover from disturbance, so only small species are found in environments dominated by large perturbations where populations have to grow fast enough to recover before further catastrophes knock the population down again. And it is no coincidence that pest species with the reproductive power to escape man's control have small bodies.

Fenchel's paper appeared in 1974 in the wake of the interest elicited by R. H. MacArthur and E. O. Wilson's bold classification of life-history strategies along a continuum from *r*-selected to *K*-selected species, which Fenchel refers to in his introduction. However, while the results are relevant to the *r*-*K* debate, the relationship to Kleiber's law is also intriguing, and this is how Fenchel opens the discussion. The relationship between metabolic rate and productivity is still of research interest today. Since the paper first appeared its central result has been repeatedly tested and confirmed (e.g., Duncan et al. 2007).

Sadly the linguistic confusion caused by the existence of synonyms that Fenchel identified in his first paragraph continues to this day. How attractive if all authors could use the simple descriptive phrase *population growth rate*, abbreviated *pgr*, to refer to the per capita rate at which a population grows. For now, however, it is necessary to remember there are a number of alternative terms. There will generally be little difference between the maximum rate at which a population can grow and its growth rate in the absence of density-dependent limitations, and Fenchel's law is generally considered to apply in both situations. Maximum population growth rate depends not only on body size but also on the environment (food, temperature, etc.), as Fenchel emphasized.

Literature Cited

Duncan R. P., D. M. Forsyth, and J. Hone. 2007. Testing the metabolic theory of ecology: Allometric scaling exponents in mammals. *Ecology* 88:324–33.

Oecologia (Berl.) 14, 317—326 (1974)

Intrinsic Rate of Natural Increase: The Relationship with Body Size

Tom Fenchel

Laboratory of Ecology, Zoological Institute, University of Åarhus, Åarhus

Received November 20, 1973

Summary. The relationship between previously published values of "the intrinsic rate of natural increase" (r_m) and body weight is studied. When organisms covering a wide range of body weights are compared, a correlation is found which can be described by the equation $r_m = aW^n$ where r_m is the intrinsic rate of natural increase per day and W is the average body weight in grams; a is a constant which takes three different values for unicellular organisms and heterotherm and homoiotherm animals respectively. The constant n has a value of about -0.275 for all three groups. This result is compared to the previously found relationship between the metabolic rate per unit weight and body size. It is shown that r_m can be interpreted as the productivity of an exponentially growing population and thus must correlate with metabolic rate. The values of the constants a and n, however, show that for each of the three groups, unicellular organisms, heterotherms and homoiotherms the ratio of energy used for maintenance to that used for production increases with increasing body size and that the evolution from protozoa to metazoa and the evolution from heterotherm to homoiotherm animals in both cases resulted in not only an increased metabolic rate, as shown previously, but also in a decreased population growth efficiency. It is shown that the increase in reproductive potential of homoiotherms relative to that of heterotherms is due to a shorter prereproductive period in the former group.

Previous estimates of r_m for different species and comparisons between these values in relation to their ecology are discussed in context with the found "r_m-body weight" relationship. Attempts to show that such comparisons will be more meaningful when body size is included in the considerations are made. It is suggested that the found relationship may represent the maximum values r_m can take rather than average values for all species, since it is likely that the species used for laboratory population-studies are biased in favor of species with high reproductive potentials.

Introduction

It has long been recognized that in a constant environment all populations tend asymptotically to grow at an exponential rate described by the equation $dN/dt = r_m N$, where N is population size. The constant r_m, which measures the growth rate in absence of density dependent limitation of a population which has attained a stable age distribution (Birch, 1948), is usually designated the "Malthusian parameter", "intrinsic rate of natural increase" or "innate capacity for increase". Some authors (e.g. Cole, 1954) have implied that natural selection will always, and in one sense exclusively, favor an increase of r_m, a problem which is not, however, trivial. More recently this problem has been discussed in terms of "r-" and "K-selection" (MacArthur, 1962). According to this idea, selection will always favor an increase in r_m of populations which are dominantly controlled by density independent factors, i.e. populations living in unstable and unpredictable environments or occurring early in ecological successions. For populations of stable and predictable environments and predominantly limited by density dependent factors, however, selection will tend to increase the number of individuals which can be sustained at equilibrium conditions, i.e. at the "carrying capacity" (K) of the environment. In such populations selection will favor that a larger part of the available resources are spent on competitive ability, defense mechanisms and a more efficient utilization of limiting resources at the expense of features which increase the reproductive potential in absence of density dependent limitation, i.e., r_m. All real populations will, of course, be found somewhere in between these two extremes. These ideas, which are related to the concept of "colonizing" or "opportunistic" versus "equilibrium" species, have recently attracted much interest and a large number of theoretical and experimental studies have been carried out (e.g. Clarke, 1972; Cody, 1966; Gadgil and Bossert, 1970; Gadgil and Solbrig, 1972; Hairston *et al.*, 1970; Heron, 1972b; Lewontin, 1965; Mertz, 1971; Pianka, 1970; Roughgarden, 1971).

The study of these concepts implies for a large part comparisons of "reproductive potential" or "reproductive effort" between different species or between genotypes within one species. However, it has previously been recognized that when organisms with different body sizes are compared there is a tendency for r_m to decrease with increasing body size (Fenchel, 1968; Pianka, 1970; Smith, 1954) and comparisons between r_m of different organisms are therefore not easily interpreted independently of body sizes of the species in question.

The purpose of the present paper is to report a study on the relationship between r_m and body size using the available data from the literature. The found relationship should make comparisons between reproductive potentials of different species more meaningful.

The value of r_m for any population is, of course, dependent on a large number of environmental factors (food, temperature, humidity, etc.) and any stated value of r_m must therefore be followed by a specification of such factors. However, at least one set of environmental factors will for any species correspond to a maximum value of r_m. Such maximum values of r_m can often be closely approximated in experimental populations. In many previous works (e.g. Evans and Smith, 1952; Heron, 1972b; Smith, 1954) this maximum value is in reality implied when comparisons of r_m

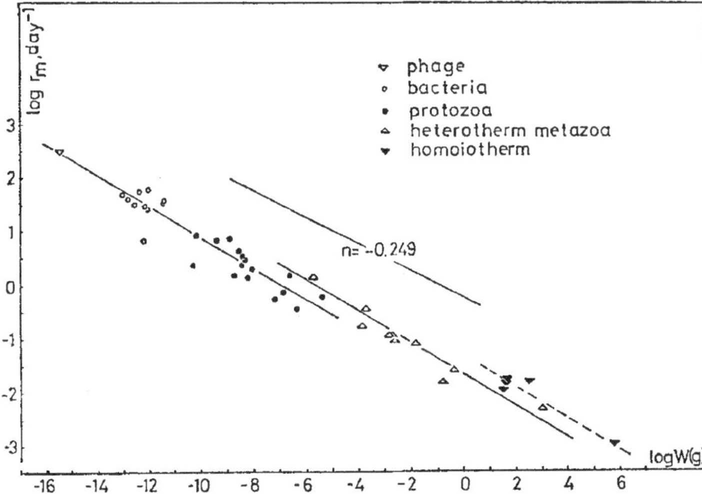

Fig. 1. The relationship between r_m and body weight for 42 species. Included is also the slope $y = K \cdot x^{-0.249}$ characteristic for the relation between body weight and metabolic rate per unit weight

between different species are made. Also in the present paper the stated values of r_m are to be considered as approximations to maximum values in order to make comparisons possible.

Most literature on the subject is centered on demographic problems, i.e. how different lifehistory phenomena affect r_m and also on attempts to explain different features of lifehistories in relation to the ecology of a species. Some ecologists (e.g. Edmondson, 1960; see also Petrusewicz and MacFadyen, 1970) have also used demographic data for estimating the productivity of field populations. In the present paper an energetic interpretation of r_m is also attempted on the basis of the found relationship with body size.

Results

It is often possible to make crude estimates of r_m for animals from data on the length of the prereproductive period, spacing between litters and litter sizes. However, it was chosen only to use data from laboratory studies directed towards estimating r_m. Such estimates for metazoa are rare in the literature since the necessary population parameters (a survivorship function and age specific fecundity) are tedious to obtain. The used data are shown in Table 1 together with references. Values of r_m for the cod and for the cow are crude estimates derived from Smith (1954); they have been included here in order to obtain a wider span of body weights. Body weights were taken from Altman and Dittmer (1962),

Table 1. r_m and body weights of 43 species

Species	r_m (day^{-1})	Reference	Body weight (g)
T-phage	300	Smith (1954)	3.3×10^{-16}
Bacillus megatherium	32.2	Altman and Dittmer (1962)	3.8×10^{-12}
Corynebact. diphteriae	28.2	Altman and Dittmer (1962)	7.5×10^{-13}
Escherichia coli	58.7	Brock (1966)	9.6×10^{-13}
Streptococcus phaseoli	31.4	Brock (1966)	3×10^{-13}
Xanthomonas	6.7	Altman and Dittmer (1962)	6.4×10^{-13}
Aerobacter aerogenes	55	Altman and Dittmer (1962)	4.0×10^{-13}
Azotobacter chroococcum	37	Altman and Dittmer (1962)	3.6×10^{-12}
Serratia marcescens	26.9	Altman and Dittmer (1962)	9.8×10^{-13}
Staphyllococcus aureus	37.0	Altman and Dittmer (1962)	1.5×10^{-13}
Diplococcus pneumoniae	48.6	Altman and Ditterm (1962)	1.0×10^{-13}
Chilomonas paramecium	2.42	Altman and Dittmer (1962)	3.9×10^{-9}
Polytomella ulvella	3.05	Altman and Dittmer (1962)	5.2×10^{-9}
Saccharomyces cerevisiae	8.32	Brock (1966)	7.1×10^{-11}
Chlorella ellipsoidea	2.48	Brock (1966)	3.4×10^{-11}
Euglena gracilis	1.53	Brock (1966)	2×10^{-9}
Stentor coerulus	0.63	Brock (1966)	0.5×10^{-6}
Paramecium caudatum	1.59	Brock (1966)	3×10^{-7}
Tetrahymena pyriformis	7.51	Brock (1966)	1.5×10^{-9}
Uronema marina	6.65	Fenchel (1968)	4.2×10^{-10}
Philasteridae sp.	4.00	Fenchel (1968)	3.0×10^{-9}
Aspidisca angulata	3.5	Fenchel (1968)	4.0×10^{-9}
Litonotus lamella	1.3	Fenchel (1968)	5.9×10^{-9}
Euplotes vannus	2.0	Fenchel (1968)	9.8×10^{-9}
Diophrys scutum	0.79	Fenchel (1968)	1.4×10^{-7}
Keronopsis rubrum	0.55	Fenchel (1968)	6.5×10^{-8}
Condylostoma patulum	0.36	Fenchel (1968)	5.5×10^{-7}
Thalia democratica	0.47–0.91	Heron (1972b)	?
Hydra	0.34	Stiven (1962)	2×10^{-4}
Dugesia tigrina	0.09	Root (1960)	1.7×10^{-2}
Euchlanis dilatata	1.40	King (1967)	2×10^{-6}
Daphnia pulex	0.30	Frank *et al.* (1957)	1.6×10^{-4}
Physa gyrina	0.027	DeWitt (1954)	5×10^{-1}
Calandra oryzae	0.109	Birch (1953)	2.4×10^{-3}
Rhizopertha dominica	0.109	Birch (1953)	?
Tribolium castaneum	0.101	Leslie and Park (1949)	2.5×10^{-3}
Pediculus humanus	0.111	Evans and Smith (1952)	1.3×10^{-3}
Melanoplus sanguinipes[a]	1.29×10^{-2}	Pfadt and Smith (1972)	1.6×10^{-1}
Gadus morrhua	6.5×10^{-3}	Smith (1954)	1.5×10^{3}
Microtus orcadensis	1.03×10^{-2}	Leslie *et al.* (1955)	14–46
M. agrestis	1.26×10^{-2}	Leslie and Ranson (1940)	19–52
Mastomys natalensis	1.67×10^{-2}	Oliff (1953)	40
Rattus norvegicus	1.48×10^{-2}	Leslie (1945)	275–520
Bos taurus	10^{-3}	Smith (1954)	6×10^{5}

[a] Pfadt and Smith (1972) give the value $r_m = 1.21 \times 10^{-2}$ per day but this is under the assumption of non-overlapping generations and $T = 1$ year. Recalculation of their data assuming a stable age structure and continuous reproduction gives $T = 0.94$ years and $r_m = 1.29 \times 10^{-2}$ per day.

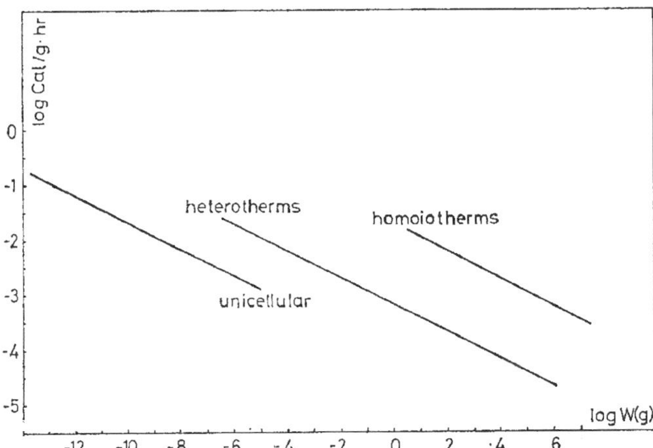

Fig. 2. The relationship between body weight and metabolic rate per unit weight for the animal kingdom. Data from Hemmingsen (1960)

Brock (1966) and Fenchel (1968) or were estimated from linear dimensions. These estimates are probably in some cases not very accurate and, as explained above, estimates of the maximum value of r_m will always be an approximation to the "true" value. Such objections may be considered of minor importance since body sizes with a span of about 21 decades and values of r_m with a span of more than 5 decades are compared.

The results are shown graphically in Fig. 1. The following interpretation is offered; the relationship between r_m and body weight can be described by the equation $r_m = aW^n$ where a and n are constants and a takes three different values characteristic for unicellular organisms and heterotherm and homoiotherm metazoa respectively. The following least squares regression equations could be calculated:

For unicellular organisms; $\log r_m = -1.9367 - 0.2796 \log W$ and for heterotherm metazoa; $\log r_m = -1.6391 - 0.2738 \log W$ where r_m has the dimension day^{-1} and W is measured in grams. With regard to the homoiotherm metazoa, the few data do not warrant the calculation of a regression line. However, the points indicate a higher value of a (about -1.4) and do not contradict the existence of a slope of the same magnitude as that characteristic for the two other groups (i.e. about -0.275).

Discussion

The results shown in Fig. 1 are strongly reminescent of the well known relationship between body weight and metabolic rate per unit weight (Fig. 2) for which, however, the exponent n takes the somewhat higher value of -0.249 when organisms of a wide span of body sizes are compared (Hemmingsen, 1960) and the relative distances between the three lines differ.

The following argument renders the correlation between r_m and body weight intelligible. Consider a population growing exponentially at the rate r_m and further that at a certain moment it consists of N' individuals. If we now continuously remove individuals at the rate of $N'r_m$ individuals per time unit in a way as not to disturb the age distribution (i.e. by a continuous dilution such as done in a chemostat) the population will be kept at a steady state size of N' individuals. The yield or production will be $N'r_mW$ per time unit, where W ist the average body weight of the population. Thus r_m can be interpreted as a measure of the potential productivity per unit weight of a population. The fraction "r_m/metabolic rate per unit weight" measures how much energy an organism spends for production relative to how much it spends for maintenance and this fraction can probably only vary within certain limits. The ratio "production/respiration" of populations takes a maximum value of the magnitude 1–2; i.e. 1/2–2/3 of the assimilated energy is spent for production, values which are found in microorganisms, whereas higher organisms are known to have lower values for this ratio. The fact that the exponent n is somewhat higher in the "metabolic rate—body weight" relationship (—0.249) than in the "r_m-body weight" relation (< -0.27) indicates that the fraction of assimilated energy used for maintenance tends to increase with increasing body size, and thus increasing structural complexity, within each of the three groups: unicellular organisms and heterotherm and homoiotherm metazoa.

When unicellular organisms and heterotherm metazoa of the same sizes are compared the latter will have a metabolic rate which on average is 8.3 times higher than that of the former (Hemmingsen, 1960; Zeuthen, 1947; see also Fig. 2). With regard to r_m it can be seen that this value is only about twice as large for a metazoan as for a protozoan of similar size. Thus the evolution of metazoans led to forms with a higher metabolic rate per unit weight but also to more complex forms which require a relatively higher proportion of the assimilated energy for maintenance. Similarly it can be seen when comparing heterotherm and homoiotherm metazoans of identical sizes that the latter have a metabolic rate which is 28 times higher than the former whereas the values of r_m of homoiotherms are only about 1.7 times higher than those of heterotherms. This is without doubt to a large part related to the energy expenditure of homoiotherms for maintaining a higher body temperature relative to the surroundings.

McNeill and Lawton (1970) compared a large number of published data on annual production and respiration for field populations of different metazoans. They found that for populations of different species

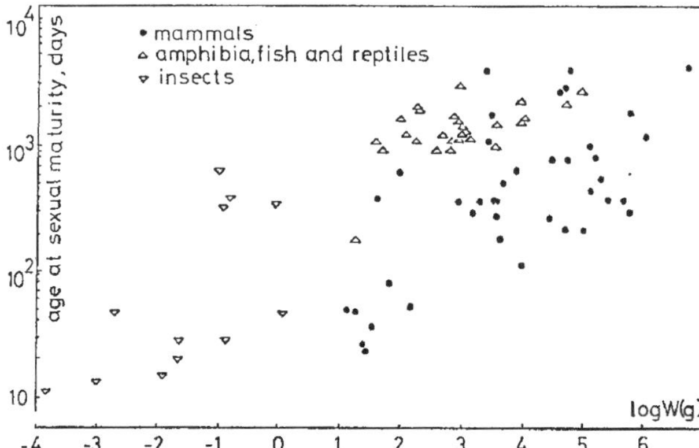

Fig. 3. The prereproductive period plotted against body weight for homoiotherm animals (mammals) and for some heterotherm groups (insects, fish, amphibians and reptiles). Data from Altman and Dittmer (1962) and Muus and Dahlstrøm (1967)

of both heterotherms and therms there was nearly linear relationship between population respiration and population production. However, whereas heterotherms showed a value of respiration which was on average about 2.3 times higher than production, the respiration of homoiotherm populations was on average about 55 times higher than their production, i.e. the ratio between productivity and respiration is about 23 times higher for heterotherms than for homoiotherms. Accepting r_m as a measure of potential population productivity the present results would indicate that the ratio "production/metabolic rate" is about 28/1.7 or 16.5 times higher for heterotherms than for homoiotherms.

It has been shown that among the lifehistory phenomena influencing the values of r_m a change in the length of the prereproductive period is the factor which has by far the strongest effect (Lewontin, 1965). In Fig. 3 the length of the prereproductive period is plotted against body weight for a number of representatives of heterotherm metazoans (insects, fish, amphibians and reptiles) and for mammals. The graph supports the assumption that homoiotherms have a higher average reproductive potential than heterotherms of similar size and that this is accomplished by a shortening of development time.

It can finally be asked whether values of r_m for all species will follow the relationship indicated in Fig. 1. This is not necessarily the case. Most of the species which have been used for experimental laboratory populations can probably be classified as "opportunistic" or "colonizing" species. In part this is due to the fact that such species are easier to grow under laboratory conditions. Also species which are of practical importance for man, notably "pests" play a large role as experimental organisms. Thus the lines in Fig. 1 are perhaps to be considered as approximations to the maximum values for r_m rather than average values for all species. So far there is only little evidence in the literature which can throw light on this question. Mertz (1971) showed that r_m of the California condor cannot exceed $2.6-3.6 \times 10^{-4}$ day $^{-1}$ and Leslie (1966) found a value of r_m for the guillemot of about 3×10^{-4} day^{-1}. The authors noted the low reproductive potentials of these species and in fact the values fall far below the predictions of Fig. 1. Mertz (*op. cit.*) showed that the low value of r_m for the condor is related to adaptations to life in a predictable environment and very low mortality rates, i.e. it is a "*K*-strategist". Similar considerations have been carried out for oceanic birds (Cody, 1966) and probably also apply to the guillemot. Another species which has a value of r_m which falls considerably below the homoiotherm line of Fig. 1 is man.

There is one example of an estimated value of r_m which probably is somewhat higher than would be predicted from the relationship found. Recently Heron (1972a, b) in a thorough study of the demography of the salp *Thalia democratica* found a value of r_m between 0.47 and 0.91 day^{-1}, i.e. a value which among hitherto studied metazoans is only exceeded by a rotifer (see Table 1). Heron describes *Thalia democratica* as a colonizing species which quickly utilizes algal blooms by a rapid increase in numbers. This finding has not been included in Fig. 1 since it was not found possible to assign a body weight to this species with its complex lifecycle involving stages of very different body sizes and an unusual body structure. However, although *Thalia* is a small metazoan, there cannot be any doubt about the fact that this species does have a very high reproductive potential.

References

Altman, P. L., Dittmer, D. S. (Eds.): Growth. Biological Handbooks. Washington, D.C.: Fed. Amer. Soc. Exper. Biol. 1962

Birch, L. C.: The intrinsic rate of natural increase of an insect population. J. Anim. Ecol. **17**, 15–26 (1948)

Birch, L. C.: Experimental background to the study of the distribution and abundance of insects. I. The influence of temperature, moisture and food on the innate capacity for increase of three grain beetles. Ecology **34**, 698–711 (1953)

Brock, T. O.: Principles of microbial ecology, 306 pp. Englewood Cliffs, N.J.: Prentice-Hall 1966

Clarke, B.: Density-dependent selection. Amer. Naturalist **106**, 1–13 (1972)

Cody, M. L.: A general theory of clutch size. Evolution **20**, 174–184 (1966)

Cole, L. C.: The population consequences of life history phenomena. Quart. Rev. Biol. **29**, 103–137 (1954)

DeWitt, R. M.: The intrinsic rate of natural increase in a pond snail (*Physa gyrina* Say). Amer. Naturalist **88**, 353–359 (1954)

Edmondson, W. T.: Reproductive rates of rotifers in natural populations. Mem. Ist. Ital. Idrobiol. **12**, 21–77 (1960)

Evans, F. C., Smith, F. E.: The intrinsic rate of natural increase for the human louse, *Pediculus humanus*. Amer. Naturalist **86**, 299–310 (1952)

Fenchel, T.: The ecology of marine microbenthos III. The reproductive potentials of ciliates. Ophelia **5**, 123–136 (1968)

Frank, P. W., Boll, C. D., Kelly, R. W.: Vital statistics of laboratory cultures of *Daphnia pulex* De Geer as related to density. Physiol. Zool. **30**, 287–305 (1957)

Gadgil, M., Bossert, W. H.: Life historical consequences of natural selection. Amer. Naturalist **104**, 1–24 (1970)

Gadgil, M., Solbrig, O. T.: The concept of *r*- and *K*-selection: evidence from wild flowers and some theoretical considerations. Amer. Naturalist **105**, 14–31 (1972)

Hairston, N. G., Tinkle, T. W., Wilbur, H. M.: Natural selection and the parameters of population growth. J. Wildlife Manage **34**, 681–690 (1970)

Hemmingsen, A. M.: Energy metabolism as related to body size and respiratory surfaces, and its evolution. Rep. Steno Hosp., Copenh. **9**, 1–110 (1960)

Heron, A. C.: Population ecology of a colonizing species: The pelagic tunicate *Thalia democratica*. I. Individual growth rate and generation time. Oecologia (Berl.) **10**, 269–293 (1972a)

Heron, A. C.: Population ecology of a colonizing species: The pelagic tunicate *Thalia democratica*. II. Population growth rate. Oecologia (Berl.) **10**, 294–312 (1972b)

King, C. E.: Food, age, and the dynamics of a laboratory population of rotifers. Ecology **48**, 111–128 (1967)

Leslie, P. H.: On the use of matrices in certain population mathematics. Biometrika **33**, 183–212 (1945)

Leslie, P. H.: The intrinsic rate of increase and the overlap of successive generations in a population of guillemots. J. Anim. Ecol. **25**, 291–301 (1966)

Leslie, P. H., Park, T.: The intrinsic rate of natural increase of *Tribolium castaneum* Herbst. Ecology **30**, 469–477 (1949)

Leslie, P. H., Ranson, R. M.: Mortality, fertility and rate of natural increase in the vole (*Microtus agrestis*). J. Anim. Ecol. **32**, 221–231 (1940)

Leslie, P. H., Tener, J. S., Vizoso, M., Chitty, H.: The longevity and fertility of the Orkney vole *Microtus orcadensis*, as observed in the laboratory. Proc. zool. Soc. Lond. **125**, 115–125 (1955)

Lewontin, R. C.: Selection for colonizing ability. In: H. G. Baker, G. L. Stebbins, Eds., The genetics of colonizing species. New York: Academic Press 1965

MacArthur, R. H.: Some generalized theorems of natural selection. Proc. nat. Acad. Sci. (Wash.) **48**, 1893–1897 (1962)

McNeill, S., Lawton, J. H.: Annual production and respiration in animal populations. Nature (Lond.) **225**, 472–474 (1970)

Mertz, D. B.: The mathematical demography of the California condor population. Amer. Naturalist **105**, 437–453 (1971)

Muus, B. J., Dahlstrøm, P.: Europas ferskvandsfisk, København: G. E. C. Gads Forlag 1967

Oliff, W. D.: The mortality, fecundity and intrinsic rate of natural increase of the multimammate mouse, *Rattus (Mastomys) natalensis* (Smith) in the laboratory. J. Anim. Ecol. **22**, 217–226 (1953)

Petrusewicz, K., MacFadyen, A.: Productivity of terrestrial animals. IBP Handbook No. 13. Oxford: Blackwell 1970

Pfadt, R. E., Smith, D. S.: Net reproductive rate and capacity for increase of the migratory grasshopper, *Melanoplus sanguinipes sanguinipes* (F.). Acrida **1**, 149–165 (1972)

Pianka, E.: On r and K selection. Amer. Naturalist **104**, 592–597 (1970)
Root, R. B.: An estimate of the intrinsic rate of natural increase in the planarian, *Dugesia tigrina*. Ecology **41**, 369–372 (1960)
Roughgarden, J.: Density-dependent natural selection. Ecology **52**, 453–468 (1971)
Smith, F. E.: Quantitative aspects of population growth. In: Dynamics of growth processes, E. Boell, Ed., p. 274–294. Princeton: Princeton University Press 1954
Smith, F. E.: Population dynamics in *Daphnia magna*. Ecology **44**, 651–663 (1963)
Stiven, A. E.: The effect of temperature and feeding on the intrinsic rate of increase of three species of hydra. Ecology **43**, 325–328 (1962)
Zeuthen, E.: Body size and metabolic rate in the animal kingdom. C. R. Lab. Carlsberg, Sér. chim. **26**, 15–161 (1947)

Dr. Tom Fenchel
Laboratory of Ecology
Zoological Institute
University of Åarhus
DK-8000 Åarhus C
Denmark

PAPER 14

Population Density and Body Size in Mammals (1981)

J. Damuth

Commentary

ALISTAIR EVANS

Is there a competitive advantage to being big? One may expect that a species will allocate its resources into large-bodied organisms if doing so permits a greater abundance in the community and the acquisition of a higher proportion of available resources. In a more general sense, do all body sizes contribute equally to community processes? From an ecological management and modeling point of view, there would also be a great advantage to being able to predict abundance from body mass if a strong association existed. In this highly cited and influential paper, John Damuth considers this key macroecological relationship between a species' body mass and its abundance.

Using literature data for 307 mammalian herbivore species, Damuth showed that there was an inverse relationship between abundance and body mass, where "ecological" density scaled with $M^{-0.75}$. The striking aspect of this finding is that, because individual metabolic rate is related to $M^{0.75}$, the total energy use by the population is independent of body size, as the scaling coefficients cancel one another: energy $\propto M^{-0.75} \times M^{0.75} = M^0$. Therefore, the amount of energy used by a local population is independent of its body size, and there is no advantage in energy gain to be a large or a small mammalian herbivore. However, a substantial amount of variation around the regression is evident, perhaps an order of magnitude on each side. This finding has become known as the "energy equivalence" rule or hypothesis (Nee et al. 1991).

The replication and explanation of this pattern remains a hotly debated aspect of macroecological research. Many studies have found results consistent with the energy equivalence rule. However, some have found that certain small species have lower densities than predicted by energy equivalence, forming a polygonal pattern rather than a simple scattered regression. The differences may depend in part on the source of the data, whether from large-scale compilations (more linear) or from all members of a taxon from a particular area (polygonal; Blackburn and Gaston 1997). In addition, the trophic level of the organisms affects the position and shape of the relationship. More general support for some influence of energy equivalence has been obtained from modeling (Carbone et al. 2007) and evolutionary simulations (Damuth 2007).

Literature Cited

Blackburn, T. M., and K. J. Gaston. 1997. A critical assessment of the form of the interspecific relationship between abundance and body size in animals. *Journal of Animal Ecology* 66:233–49.

Carbone, C., J. M. Rowcliffe, G. Cowlishaw, and N. J. B. Isaac. 2007. The scaling of abundance in consumers and their resources: Implications for the energy equivalence rule. *American Naturalist* 170:479–84.

Damuth, J. 2007. A macroevolutionary explanation for energy equivalence in the scaling of body size and population density. *American Naturalist* 169:621–31.

Nee, S., A. F. Read, J. J. D. Greenwood, and P. H. Harvey. 1991. The relationship between abundance and body size in British birds. *Nature* 351:312–13.

Nature Vol. 290 23 April 1981 699

Received 1 September 1980; accepted 26 January 1981.

1. Hoering, T. C. in *Researches in Geochemistry* (ed. Abelson, P. H.) (Wiley, New York, 1967).
2. Rumble, D., Hoering, T. C. & Grew, E. S. *Yb. Carnegie Instn Wash.* **76**, 623 (1977).
3. McKirdy, D. M. & Powell, T. G. *Geology* **2**, 591 (1974).
4. Barghoorn, E. S., Knoll, A. H., Dembicki, H. & Meinschein, W. G. *Geochim. cosmochim. Acta* **41**, 425 (1977).
5. Smith, J. W., Schopf, J. W. & Kaplan, I. R. *Geochim. cosmochim. Acta* **34**, 659 (1970).
6. Degens, E. T. in *Organic Geochemistry* (eds Eglington, G. & Murphy, M. T. J.) 304 (1969).
7. Galimov, E. in *Kerogen Insoluble Organic Matter from Sedimentary Rocks* (ed. Durand, B.) 271 (Edition Technique, Paris, 1980).
8. Goodwin, A. M., Monster, J. & Thode, H. G. *Econ. Geol.* **71**, 870 (1976).
9. Krogh, T. E. & Davis, G. L. *Yb. Carnegie Instn Wash.* **70**, 241 (1971).
10. Nunes, P. D. & Thurston, P. D. *Can. J. Earth Sci.* **17**, 710 (1980).
11. Hoering, T. C. *Yb. Carnegie Instn Wash.* **61**, 190 (1962).
12. Cloud, P. E. Jr, Gruner, J. W. & Hagen, H. *Science* **148**, 1713 (1965).
13. Oehler, D. Z., Schopf, I. W. & Kvenvolden, K. A. *Science* **175**, 1246 (1972).
14. Perry, E. C. Jr, Tan, F. C. & Morey, G. B. *Econ. Geol.* **68**, 1110 (1973).
15. Schidlowski, M., Appel, P. W. U., Eichmann, R. & Junge, C. E. *Geochim. cosmochim. Acta* **43**, 189 (1979).
16. Eichmann, R. & Schidlowski, M. *Geochim. cosmochim. Acta* **39**, 585 (1975).
17. Schoell, M. & Hartmann, M. *Mar. Geol.* **14**, 1 (1973).
18. Tietze, K., Geyh, M., Müller, H., Schröder, L. & Stahl, W. *Geol. Rdsch.* **69**, 452–472 (1980).
19. Brooks, T. M., Bright, T. J., Bernard, B. B. & Schwab, C. R. *Limnol. Oceanogr.* **24**, 735 (1979).
20. Sackett, W. M. *et al. Earth planet. Sci. Lett.* **44**, 73 (1979).
21. Coleman, D. D., Risatti, J. & Schoell, M. (in preparation).
22. Barnes, R. O. & Goldberg, E. D. *Geology* **4**, 297 (1976).
23. Martens, C. S. & Berner, R. A. *Limnol. Oceanogr.* **22**, 10 (1977).
24. Fallon, R. D., Havrits, S., Hanson, R. S. & Brock, T. D. *Limnol. Oceanogr.* **25**, 357 (1980).

Population density and body size in mammals

John Damuth

Committee on Evolutionary Biology, University of Chicago, 1103 E. 57 Street, Chicago, Illinois 60637, USA

There seems to be an inverse relationship between the size of an animal species and its local abundance. Here I describe the interspecific scaling of population density and body mass among mammalian primary consumers (herbivores, broadly defined). Density is related approximately reciprocally to individual metabolic requirements, indicating that the energy used by the local population of a species in the community is independent of its body size. I suggest that this is a more general rule of community structure.

Figure 1 shows the logarithm of mean population density plotted against the logarithm of mean adult body mass for 307 species of mammalian primary consumers. As far as possible, the density values represent 'ecological' densities, that is, those which apply to the habitat area actually used by the species. The relationship is linear, with a slope of −0.75, and species densities seem to be restricted to varying within about one order of magnitude from the value predicted by the regression line.

This analysis combines data from a wide variety of habitats throughout the world, and it is important that we know whether this overall pattern accurately reflects that found within individual communities; few, if any, mammal communities have been completely studied. To obtain a representative sample, I extracted from my data those sets yielding densities of three or more species within the same habitat type; these 'constructed' communities are shown in Table 1, with regression statistics. They include representatives of almost the whole range of habitat structures and primary productivity levels encountered by terrestrial mammals. None of the individual slopes, which vary from −0.56 to −0.95, differ significantly from −0.75, and Levene's test[1] and an analysis of covariance[2] reveal no significant differences in variance about the regression, in slope, or in intercept. Pooling the data from the communities gives an estimated slope of −0.70, which is not significantly different from the value obtained for the overall regression (t-test). Thus, the overall trend gives a reasonable representation of that which we are likely to find within individual communities.

Knowledge of population density scaling in communities allows us to consider the relative energy use of species among primary consumers of different sizes. Individual basal metabolic requirements are related to body mass by the power of 0.75 (refs 3, 4). Estimates of the metabolic requirements of free-living mammals in natural habitats roughly parallel basal requirements, but at a higher level, varying between ~1.5 and 3.0 times basal values[5–13]. Thus they will also be related to body mass by the power of ~0.75. The energy used by the local population of a species equals the population density (D) multiplied by individual metabolic requirements (R), which yields the following relationship to body mass (W): $DR \propto W^{-0.75}W^{0.75}$. The exponents of W cancel each other, which gives the important result that the amount of energy that a species population uses in the community is independent of its body size. No mammal herbivore species, on an ecological time scale, has an energetic advantage over any other solely as a result of size differences.

There are two important corollaries of this relationship. First, because secondary productivity for a given amount of assimilated energy is independent of body size in mammals (Kleiber's Law)[3], it follows from the above that the secondary productivity of a herbivore species' local population, and hence the energy that it yields to the next higher trophic level, is also independent of body size. Second, the standing-crop biomass of a species (population density multiplied by body mass) is positively related to body mass by the ~0.25 power; this is a

Table 1 Habitat types and constructed communities

Community/habitat type	a	s.e.	r	b	n
Sonoran desert, USA	−0.63	0.16	−0.76	3.37	14
Mesquite grassland, USA and Mexico	−0.56	0.086	−0.95	3.73	7
Boreal and subalpine forest, N. America	−0.79	0.080	−0.95	4.43	12
Lowland tropical rainforest, Malaysia	−0.60	0.089	−0.90	2.61	13
Transvaal lowveld (woodland–savanna), S. Africa	−0.61	0.089	−0.93	3.78	10
Mixed temperate forest, Poland	−0.79	0.080	−0.97	4.33	8
Temperate grassland, USA	−0.67	0.13	−0.92	3.28	7
Tropical grassland, Rwenzori N.P., Uganda	−0.79	0.17	−0.81	3.59	13
Tropical grassland, Sri Lanka*	−0.79	0.19	−0.92	4.59	5
Ichu grassland, Altiplano, Peru*	−0.82	0.34	−0.99	3.93	4
High arctic tundra, Canada*	−0.95	0.21	−0.96	4.71	4
Sub-arctic birch forest and meadows, Norway*	−0.83	0.72	−0.99	4.41	3
Southern pine–hardwood forest, USA*	−0.91	0.46	−0.89	4.23	3
Northern hardwoods, USA*	−0.57	0.11	−0.98	3.11	3
Oaks and chapparal, USA*	−0.72	0.021	−0.99	4.83	3
Means of statistics	−0.74	0.20	−0.93	3.93	—
Pooled data from above regressions	−0.70	0.040	−0.86	3.71	109

Statistics are for standard least-squares regression equations: $\log D = a(\log W) + b$; s.e., standard error of the slope (a); r, correlation coefficient; n, number of species. Note that in the pooled regression, as some species are found in more than one habitat type, $n = 109$ includes only 92 separate species.

* Due to small sample size and/or a single point at one end of the size range, the individual values for these regressions are not very reliable.

0028–0836/81/170699—02$01.00

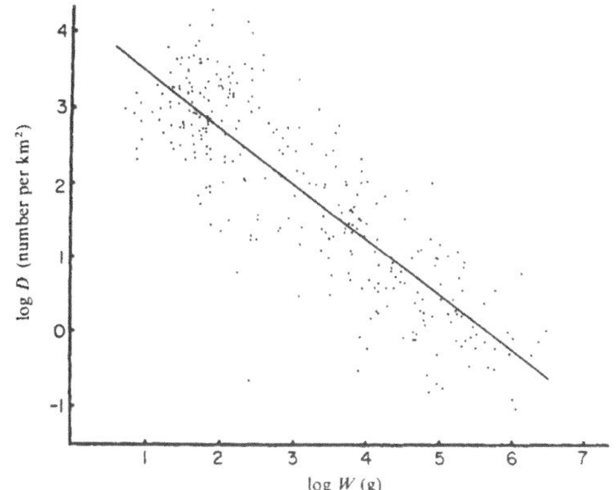

Fig. 1 Population density (D) compared with the mean adult body mass (W) for 307 mammal primary consumers; each point represents one species. Density values for each species are the mean of the means from each locality from which data were reported for the particular species. (Data are from the literature for the years 1950–79, derived from 115 journals and numerous books, ~650 references in all.) The line represents the least-squares regression line, $\log D = -0.75 (\log W) + 4.23$; $r = -0.86$, standard error of the slope = 0.026.

quantitative measure of a qualitative relationship that has been known for some time but only sporadically appreciated[14]. Species of small mammals are able to produce, on average, the same amount of biomass over time as do species of large mammals, whereas at a given moment their standing-crop biomass is considerably less, because the population turnover rates and individual growth rates per unit weight of small species are much greater[15–17].

The widespread occurrence of an approximately reciprocal relationship between population density and individual metabolic requirements among mammalian herbivores suggests that a general principle is involved. The independence of species energy control and body size revealed by this reciprocal relationship implies that random environmental fluctuations and interspecific competition act over evolutionary time to keep energy control of all species within similar bounds. It is unlikely that the occurrence of competition among species of different body sizes is restricted to herbivorous mammals. Along with the ubiquity of allometric scaling of metabolic rate with body size[4], this suggests that a reciprocal relationship, and particularly a value of ~ -0.75, characterizes a broader range of taxa and trophic levels. These points will be discussed more fully elsewhere.

I thank I. L. Heisler, V. C. Maiorana, D. M. Raup, V. L. Roth and L. Van Valen for comments, and acknowledge support from the University of Chicago and Field Museum of Natural History.

Received 6 October 1980; accepted 27 January 1981.

1. Van Valen, L. *Evol. Theory* **4,** 33–44 (1978).
2. Brownlee, K. A. *Statistical Theory and Methodology in Science and Engineering* 2nd edn (Wiley, New York, 1965).
3. Kleiber, M. *The Fire of Life* 2nd edn (Krieger, New York, 1975).
4. Hemmingsen, A. M. *Rep. Steno meml Hosp.* **4,** 7–58 (1950).
5. Brody, S. *Bioenergetics and Growth* (Reinhold, New York, 1945).
6. Chew, R. M. & Chew, A. E. *Ecol. Monogr.* **40,** 1–21 (1970).
7. Gessaman, J. A. in *Ecological Energetics of Homeotherms* (ed. Gessaman, J. A.) (Utah University Press, 1973).
8. King, J. R. in *Avian Energetics* (ed. Paynter, R. A. Jr) (Nuttall Ornithological Club, Massachusetts, 1974).
9. McKay, G. M. *Smithson. Contr. Zool.* **125,** 1–113 (1973).
10. Moen, A. N. *Wildlife Ecology: an Analytical Approach* (Freeman, San Francisco, 1973).
11. Mullen, R. K. *Comp. Biochem. Physiol.* **39,** 379–390 (1971).
12. Mullen, R. K. & Chew, R. M. *Ecology* **54,** 633–637 (1973).
13. Nagy, K. A. & Milton, K. *Ecology* **60,** 475–480 (1979).
14. Odum, E. P. *Fundamentals of Ecology* 3rd edn (Saunders, Philadelphia, 1971).
15. Fenchel, T. *Oecologia* **14,** 317–326 (1974).
16. Western, D. *Afr. J. Ecol.* **17,** 185–204 (1978).
17. Case, T. J. *Q. Rev. Biol.* **53,** 243–282 (1978).

0028–0836/81/170700—03$01.00

Competitive ability influences habitat choice in marine invertebrates

Richard K. Grosberg

Department of Biology, Yale University, 260 Whitney Avenue, PO Box 6666, New Haven, Connecticut 06511, USA

Patterns of distribution and abundance of sessile marine epibenthic invertebrates are controlled by three factors: (1) the presence and abundance of larvae which are competent to settle, (2) the choice of settling sites by recruiting larvae, and (3) the biotic and physical events occurring during and after settlement. Although there is much information on the distribution of larvae and seasons of recruitment[1–3], substratum selection[4–6] and post-settlement events[7–15], very little is known of the ecological and evolutionary relationships between these factors[8,16–18]. Natural selection acts on entire life cycles, thus information about these relationships is essential for understanding patterns of recruitment and survival. For example, sessile organisms can modify the course of post-recruitment events by selective settlement and directional growth[19]. Here, I present evidence that the larvae of several taxa of marine invertebrates avoid substrata where there is a high probability of death caused by a superior spatial competitor.

I found, as did Grave[20], that in the Eel Pond at Woods Hole, Massachusetts (USA), the two compound ascidians *Botryllus schlosseri* (Pallas) Savigny and *Botrylloides leachii* (Savigny) cover, by area, over 50%, and sometimes 100%, of hard substrata during the spring and summer. This period includes the season of recruitment for most other colonizers. Thus, settling larvae are likely to contact a botryllid ascidian at some time during their lives. A common result of contacts between colonists is overgrowth of one organism by another[21–29]. In this way, overgrowth is a consequence of competition for space (and perhaps other resources[30]), and therefore can be an important factor in the distribution and abundance of encrusting species.

B. schlosseri is the most successful overgrower (along with *B. leachii*) in the Eel Pond (Table 1) and is rarely overgrown by other species. As it is such an important member of the Eel Pond epibenthic assemblage (considering the post-settlement events of overgrowth and coverage), I investigated the effects of its density on the recruitment of other taxa.

B. schlosseri collected from the Eel Pond were placed in finger-bowls where they released larvae which were then transferred to 10-ml polycarbonate dishes. Two hours after settlement on the sides of the dishes, the juveniles were teased off and re-attached to 5×5 cm glass plates. I set up three densities of juvenile *Botryllus* with three replicates of each: (1) 15 regularly spaced *Botryllus* juveniles, with all subsequent botryllid settlement removed daily; (2) 5 regularly spaced *Botryllus* juveniles with all subsequent botryllid recruits removed daily; and (3) no *Botryllus* juveniles with daily removal of all botryllid recruits. Three additional glass plates which initially carried no juveniles, but on which all recruits were allowed to remain, served as controls.

Experimental and control plates were positioned horizontally in a Latin square arrangement[31] below a floating dock at 1 m depth. The experiment was started on 20 July, 24 h after the larvae had been transplanted. The lower side of each plate (which carried the juvenile *Botryllus*) was censused non-destructively until 28 July on a daily basis—this allowed me to distinguish between failure to settle and all but the earliest post-settlement mortality.

Figure 1 shows the numbers of each species that settled on the glass plates. To determine if there were any differences between each of the three density treatments and the controls, the data for each species were analysed using the null hypothesis that

PAPER 15

Body Size, Physiological Time, and Longevity of Homeothermic Animals (1981)

S. L. Lindstedt and W. A. Calder III

Commentary

DANIEL P. COSTA

Max Kleiber's 1932 paper clearly showed the importance of body size to metabolism. It was followed by a host of papers that examined the scaling of body constituents and other metabolic processes (Adolph 1949; Stahl 1967; McMahon 1973; Alexander et al. 1979). However, it took some time for the ecological importance of this observation to be appreciated (McNab 1963; Gould 1966; Western 1979). The importance of "Body Size, Physiological Time, and Longevity of Homeothermic Animals" by Stan Lindstedt and William Calder III was that it provided the first concise synthesis clearly detailing the linkages between physiological processes and characteristics, such as metabolic rate, renal clearance, heart rate, spleen size, heart size, and cardiac output, with important ecological characteristics, such as life span, age of reproductive maturity, growth rate, gestation, and life span. The relationships described in this paper were soon further developed in a series of papers (Calder 1981, 1983, 1987; Hofman 1983) and books (Peters 1983; Calder 1984; Schmidt-Nielsen 1984).

This paper shows that there are emergent properties of body size that determine the timing of an organism that are quite rigid. The timing of physiological processes that scale with $M^{0.25}$ will always be out of sync with the timing of environmental processes. Thus, small organisms always have limited endurance capabilities, and large organisms live longer based on body size alone. A mouse will always face a greater challenge fasting over a cold winter day than a moose. The importance of this observation is that all organisms must somehow deal with a 24-hour day and seasons that change over a 365-day yearly cycle. Given that these constraints are directly associated with body size, selection cannot operate directly on them but must act on body size itself or by developing other adaptations (torpor, hibernation, increases in body fat, etc.). Given the fundamental mismatch between physiological and environmental time, organisms needed to develop an endogenous biological clock that enables synchronizing these otherwise asynchronous processes. Further, the paper examines longevity of organisms in relation to body size and finds some intriguing patterns, such as a relationship between life span and brain size and that organisms have an approximately equivalent number of timing periods over a lifetime. That is, there is a finite number of heart beats, or breaths, that occur over an organism's life, which is independent of body size. Thus, a mouse with its more rapid heart rate would have a shorter life span because it uses its absolute number of heart beats sooner than an elephant, which has a slower heart rate. However, Lindstedt and Calder also found that there are fundamental differences between taxonomic groups such that birds have a fundamentally longer life span than mammals.

Literature Cited

Adolph, E. F. 1949. Quantitative relations in the physiological constitutions of mammals. *Science* 109:579–85.

Alexander, R. M., A. S. Jayes, G. M. O. Maloiy, and E. M. Wathuta. 1979. Allometry of the limb bones of mammals from shrews *Sorex* to elephant *Loxodonta*. *Journal of Zoology* 189:305–14.

Calder, W. A., III. 1981. Scaling of physiological processes in homeothermic animals. *Annual Review of Physiology* 43:301–22.

———. 1983. Ecological scaling: Mammals and birds. *Annual Review of Ecology and Systematics* 14:213–30.

———. 1984. *Size, function, and life history*. Harvard University Press, Cambridge, MA.

———. 1987. Scaling energetics of homeothermic vertebrates: An operational allometry. *Annual Review of Physiology* 49:107–20.

Gould, S. J. 1966. Allometry and size in ontogeny and phylogeny. *Biological Reviews* 41:587–638.

Hofman, M. A. 1983. Energy metabolism, brain size and longevity in mammals. *Quarterly Review of Biology* 58:495–512.

Kleiber, M. 1932. Body size and metabolism. *Hilgardia* 6:315–53.

*McMahon, T. 1973. Size and shape in biology. *Science* 179:1201–4.

*McNab, B. K. 1963. Bioenergetics and the determination of home range size. *American Naturalist* 97:133–40.

Peters, R. H. 1983. *The ecological implications of body size*. Cambridge University Press, Cambridge.

Schmidt-Nielsen, K. 1984. *Scaling: Why is animal size so important?* Cambridge University Press, Cambridge.

Stahl, W. R. 1967. Scaling respiratory variables in mammals. *Journal of Applied Physiology* 22:453–60.

Western, D. 1979. Size, life history and ecology in mammals. *African Journal of Ecology* 17:185–204.

VOL. 56, NO. 1 MARCH, 1981

THE QUARTERLY REVIEW
of BIOLOGY

BODY SIZE, PHYSIOLOGICAL TIME, AND LONGEVITY OF HOMEOTHERMIC ANIMALS

S. L. LINDSTEDT

Department of Zoology and Physiology,
University of Wyoming, Laramie WY 82071 USA

W. A. CALDER, III

Department of Ecology & Evolutionary Biology,
University of Arizona, Tucson, Arizona 85721 USA

ABSTRACT

The concept of physiological time is extended from contributions of Adolph, Hill, and Stahl to include a wider range of events in the life histories of mammals and birds. The durations of physiological, developmental, and ecological cycles show nearly parallel exponential relationships to body size (mass$^{1/4}$).

Maximum life span approximates a fixed multiple of shorter events, such as muscle-twitch time, circulation and filtration of blood plasma, respiratory and metabolic cycles, embryonic development, growth, sexual maturation, and even minimum population-doubling time.

Endothermy evolved independently in birds and mammals, which share the approximate $M^{1/4}$ allometry of physiological time, although with different absolute values of cycle times. Birds develop faster than similar-sized mammals, live longer, and their slower hearts actually beat more before expiring.

A time-scale proportional to $M^{1/4}$ may be difficult to synchronize with environmental cycles. Thus, the mouse experiences more events in a day's fast or a winter' cold than a moose (a long cold winter being relatively longer for mouse than moose). This asynchrony of physiological time and astronomical time has resulted in the need for biological clocks.

Life span, like the other scales of physiological time, may be regarded, not as a direct product of natural selection, but rather as an allometric consequence of other characteristics subjected to natural selection.

INTRODUCTION

THE HISTORY of animal evolution is classically viewed as a phylogenetic tree branching into different forms, with examples of adaptive radiation such as Darwin's finches at the tips of the branches. Within a single class of animals, however, the diversity

Offprints of this article are available only from the office of *The Quarterly Review of Biology*. Send $2.50 (payment in advance required) to *The Quarterly Review of Biology*, State University of New York, Stony Brook, NY 11794 USA.

of body sizes is as impressive as the diversity of forms. Terrestrial mammals span six orders of magnitude in body mass, and living birds span nearly five. The natural selection for a particular body size includes more than external dimensions; it has profound effects on the entire biology of each species. Hill pointed out that the size of an animal dictates its "rate of living." Despite its importance, however, this relationship has too often been overlooked by a number of biologists.

Body mass, more than any other single descriptive feature, is the primary determinant of ecological opportunities, as well as of the physiological and morphological requirements of an animal. To preserve function at all body sizes, there has been a differential and usually non-linear scaling of each of the details of body form and life history (Hill, 1950; Stahl, 1962, 1965, 1967; Gould, 1966; McMahon, 1973, 1974, unpub.; Calder, 1974, 1981; Gunther, 1975; Schmidt-Nielsen, 1977). This scaling can be expressed and analyzed as power functions of body mass, such as the familiar equation describing the dependence of standard metabolic rate of mammals ($\dot{H}_{smr}$) upon body mass (M) (Kleiber, 1961):

$$\dot{H}_{smr} = k\,M^{0.73} \qquad (1)$$

Although body mass is the primary factor that determines physical requirements, other factors, such as age, leanness or height-diameter proportions, and relative brain size, may also be important. Incorporation of these other factors into multiple regressions for metabolism, lifespan, etc., may improve goodness of fit, explain more fully the variability in the data, or better assess the influence of body surface area or degree of cephalization (Tanner, 1949; Kleiber, 1950; von Schelling, 1954; Sacher, 1959, 1978; Sacher and Staffeldt, 1974). However, it is body mass, and not the mass of particular organs, that determines basic support costs and control objectives. Hence, we consider the singular power function of body mass to be the most useful descriptive tool for understanding the evolution of size.

In order to examine the evolution of body size and its requirements and consequences, the biologist must find patterns. Though we acknowledge that pattern-seeking can become quite subjective, the exponents (b) which relate physiological and morphological variables (Y) to body mass (M) in allometric equations of the form:

$$Y = a\,M^{b} \qquad (2)$$

seem to fall into quantitative clusters (Adolph, 1949; Stahl, 1967; Calder, 1974; McMahon, 1975). These clusters include: physical support ($b > 1$), control ($b < 1$), capacities of transport organs ($b \simeq 1$), cyclic frequencies ($b \simeq -\frac{1}{4}$), and volume-rates such as metabolic rate and cardiac output ($b \simeq \frac{3}{4}$). In the following paragraphs we present these exponents as ranges of values derived by different authors (e.g., $M^{0.84-0.92}$).

Physical Support

To maintain skeletal strength, the cross-sectional area of the bones must increase more rapidly than their length. Hence, the skeleton becomes relatively more robust as body mass is increased (Kayser and Heusner, 1964; McMahon, 1973; Reynolds and Karlotski, 1977; Schmidt-Nielsen, 1977; Prange, Anderson, and Rahn, 1979; Anderson, Rahn, and Prange, 1979). A 3 g shrew has a skeleton that accounts for less than 4 per cent of its body mass, whereas the skeleton of man contributes 15 per cent of body mass, and the skeleton of a 3,000 kg elephant over 20 per cent. The scaling of skin mass departs nearly equally from linearity with body mass but in the opposite direction from skeletal mass. We have derived the following relationship between skin mass (kg) and body mass, based on four points (mouse, hamster, cat, cow): $M_{skin} = 0.129\,M^{0.90}$; corr. coef. = 0.999. This is similar to the relationship for the four rodents plus rabbit derived by Pace, Rahlman, and Smith, 1979: $M_{skin} = 0.139\,M^{0.94}$, corr. coef. = 0.993. Hence, the combined mass of skeleton and skin in mammals is close to a constant percentage of body mass ($\simeq 25\%$), skin contributing a much larger portion in small animals.

Control

With increase in size the number of cellular units increases, but the number of functions served does not. Therefore, a control circuit or regulating organ need not increase linearly with body size. (Thus, the headlight switch on a large bus may be only slightly heavier than that on a compact car.) For instance, mammalian brain mass is proportional to $M^{2/3}$ (Brody, 1945; Sacher, 1959; Jerison, 1961). Thus, the brain of a 3 g mammal constitutes 6 to 9 per cent of its body mass, whereas the brain of an adult elephant contributes less than 1 per cent to its body mass. The cause of this two-thirds scaling of the brain is still unclear and should be investigated by neuroanatomists.

The mass of mammalian endocrine glands also bears a less than linear proportionality to body mass: mass of pituitary $\propto M^{0.56-0.76}$; mass of thyroid and adrenal glands each $\propto M^{0.80-0.92}$; and mass of pancreas $\propto M^{0.91}$. The organs that control the composition of circulating blood have similar mass exponents: mass kidneys $\propto M^{0.85}$ and mass liver $\propto M^{0.87}$ (Adolph, 1949; Stahl, 1965).

Capacities of Organs

Qualitatively, the basic vertebrate visceral plan was preserved during evolutionary size changes. Because heart, lung, gut, blood volume, etc. were not dispensable, one of them could not usurp the space of another. Hence, those organs which function volumetrically scale essentially linearly to body size. The heart mass scales to $M^{0.98}$, blood volume to $M^{0.99-1.02}$, cardiac stroke volume to $M^{1.06}$, spleen to $M^{1.02}$, lung mass to $M^{0.99}$, and respiratory vital capacity and tidal volume to $M^{1.03-1.04}$ (listed by or derived from Brody, 1945; Adolph, 1949; Stahl, 1965, 1967).

Volume-Rates

As described almost a half-century ago by Max Kleiber (Kleiber, 1932), the metabolic rates of eutherian mammals scale to $M^{3/4}$. Parallel relationships (having similar exponents but different *Y*-intercepts, i.e., vertically displaced on log-log plots) have been described for passerine and nonpasserine birds and for marsupial mammals (Schmidt-Nielson, 1977). The physiological meaning of the *Y*-intercepts is yet to be explained. The earlier metabolic work and attempts to explain the ¾ exponent have been discussed elsewhere (Kleiber, 1961; McMahon, 1973).

Empirically, not only are the standard metabolic rates (measured as oxygen consumption rates) approximately proportional to $M^{3/4}$, but other volume-rates (volume $\cdot$ time^{-1}) of associated support functions, such as cardiac output, minute respiratory ventilation, renal glomerular filtration rate, and renal plasma flow, all scale in approximate proportion to $M^{3/4}$ (Adolph, 1949; Edwards, 1975).

Frequencies

The basal rates of cyclic events are inversely related to approximately the ¼ power of *M*. The cardiac rate of mammals $\propto M^{-0.25}$ and of birds $\propto M^{-0.23}$. This scaling must extend to the endogenous depolarization-repolarization characteristics of the sinoatrial node as well as to the mechanical efficiency of the cardiac muscle at the natural frequency for the mass and tension characteristics of the heart muscle. The relation is consistent with a model of elastic similarity of scaling, discussed below (McMahon, 1975).

Similar exponents describe respiratory frequency: $M^{-0.26}$ for mammals (Stahl, 1967), $M^{-0.28}$ for passerine birds and $M^{-0.31}$ for nonpasserine birds (Calder, 1968).

Adolph (1949) treated frequencies as reciprocals of time (time = 1/frequency): breath duration $\propto M^{0.28}$, heartbeat duration $\propto M^{0.27}$, and gut beat duration $\propto M^{0.31}$ for mammals. Many other physiological periods or times have since been found to scale near $M^{1/4}$.

INTERRELATIONSHIPS

The basic maintenance of an animal's internal composition is a logistical problem of obtaining the right amounts (in regard to body mass) at the right times (in regard to mass-dependent time scales), for all of the many simultaneous and interrelated functions of the living state. Not all species have the same requirements. In meeting these requirements the evolution of body size entails quantitative proportional adjustment of the duration of many cycles, ranging from msec to years on the absolute time scale. Regardless of their length, these cycles scale approximately as $M^{1/4}$. In view of this apparent universality, we might expect that body size and the intrinsic time scale which it requires would be dominating characteristics of the life history. Such appears to be the case (Blueweiss et al., 1978; Western, 1979).

Is there any explanation for why animals' use of time should vary in proportion to the fourth root of their body mass? Dimensional analysis can be used to derive the relationship between time and body mass with regard to muscle stress. If vertebrates are designed in such a manner that support structures retain the same relative strength regardless of body size, then an evolutionary increase in length must be accompanied by a disproportionately greater increase in diameter. McMahon (1973) found that the characteristic lengths (l) of bones, muscles, etc. are proportional to $M^{1/4}$ whereas their diameters (d) scale in proportion to $M^{3/8}$. He derived these proportionalities for scaling by elastic similarity in birds and mammals. According to this model, each cross-sectional area (d^2) is proportional to $M^{3/4}$ and the mass of each muscle, limb, or entire animal is $\propto (l \cdot d^2)$. Hence, muscle forces (mass × acceleration) can be described as $(l \cdot d^2)(l \cdot t^{-2})$. The maximum stress (i.e., force divided by cross-sectional area) generated in homologous muscles is roughly constant (Hill, 1950). Therefore:

$$\frac{\text{Mass} \times \text{acceleration}}{\text{area}} \propto \frac{(l \cdot d^2)(l \cdot t^{-2})}{d^2}$$
$$= \frac{l^2}{t^2} = \frac{M^{1/2}}{t^2} = \text{constant};\ t \propto M^{1/4} \qquad (3)$$

Hence, if l is proportional to $M^{1/4}$ and if muscle stress is constant, time (t) scales directly as length or as $M^{1/4}$. The above model, derived for static stresses, assumes skeletal failures occur by buckling. Prange (1977) has derived a model bsed on dynamic forces in which he finds $d \propto l^{1.0-1.25}$. This model would therefore predict $t \propto l \propto M^{0.285-0.33}$. Recent data from Alexander, Jayes, Maloiy, and Wathuta (1979) on the scaling of limb bones, supports this "non-elastic" scaling.

In addition, and perhaps even more fundamental, the time animals require to perceive and respond to their surroundings seems to scale to $M^{1/4}$. An animal's neurons form its communication system both within the animal and between the animal and its environment. The diameters of homologous neurons of all mammals, and therefore their conduction velocities, are nearly constant.

Nerve conduction velocities do vary, but not systematically with body size. For example, nerves of slow-moving skunks and sloths have low conduction velocities (Goffart, 1971; Van de Graaff, Frederick, Williamson, and Goslow, 1977) relative to man or baboon (Koeze, 1973; Mayer and Mawdsley, 1968), whereas those of the fast-moving cat are relatively high (Goslow, Cameron, and Stuart, 1977). However, over a broad size-range of mammals nervous messages travel along homologous nerves at nearly the same velocity. The amount of time necessary for a nervous impulse to reach its target is equal to the distance it must travel divided by its velocity. Since velocity along homologous nerves is roughly constant, the time required for an afferent message to arrive with information about the animal's environment, or the time for an efferent message to reach the muscles or organs, is directly proportional to the distance those messages must travel. That distance is proportional to the characteristic length (l), or again $M^{1/4}$. Hormonal messages within the body likewise scale close to $M^{1/4}$, as the time (minutes) to circulate the entire blood volume in mammals is $0.35\ M^{0.21}$ (Stahl, 1967). Delcomyn (1980) has reviewed the evidence that rhythmic timing of cyclic movements, as in walking, swimming, and breathing, are intrinsic properties of the central nervous system. From this we must conclude that these intrinsic properties have been scaled allometrically as well.

These properties may have set the physiological time scales of birds and mammals in proportion to $M^{1/4}$. In fact, most cyclic events do transpire on time scales approximately proportional to $M^{1/4}$. Consequently, it may not be necessary to invoke special meaning or unique explanations to describe how one rate or another scales to body mass. One need not speculate that evolution has selectively dealt with any single biological cycle. It may be meaningless to consider the evolution of isolated traits. Gould and Lewontin (1979) argue against such an adaptationist approach. Hence, we propose no specific evolutionary explanation for the scaling of physiological time. It seems more likely to us that the fundamental unit of physiological time, $M^{1/4}$, was an inevitable consequence of the geometry of changes in body size.

Brody (1945) used the term "physiological time" to denote a variable time scale among organisms of different size. A. V. Hill (1950) refined that concept, by suggesting that such things as power per unit weight, time for growth to maturity, and gestation period may be constant in all animals when compared per unit of physiological time; that is to say, 10 seconds for a large animal may be equivalent (physiologically) to 1 second for a small animal.

These ideas of physiological time have been confirmed during the past three decades. Adolph (1949) examined 34 morphological and physiological variables and related them allometrically to body size. The resulting "quantitative orderliness" suggested to him that organisms "may be pictured as systems of precise multiple interrelationships." To examine those interrelationships, Stahl (1962, 1963, 1965, 1967) combined allometric equations as ratios to form dimensional and dimensionless "criteria of similarity." For example, he calculated breath time (yrs) as $4.7 \times 10^{-5} M^{0.28}$ and pulse time (yrs) as $1.2 \times 10^{-5} M^{0.27}$. Therefore, dividing breath time by pulse times yields the dimensionless ratio of $3.9\ M^{0.01}$. As the residual mass exponent is small, Stahl reasoned that there are about 3.9 heart beats per respiratory cycle in all mammals, regardless of size. By similar analysis, he estimated that all mammals have a basal energy use of about $8 \times 10^5\ kj \cdot kg^{-1}$ per lifetime. This idea of a finite total lifetime metabolism has been discussed further as "absolute metabolic scope" (terminology that can be confused with prior usage) (Boddington, 1978). Metabolic time on the scale of embryonic lifespan in birds has been analyzed by Rahn and Ar (1980).

Many additional biological periods have been examined and described as allometric functions of body mass since Stahl's work. McMahon (1975) has cited a number of these. Table 1 and Fig. 1 show these equations and many others which we have derived from available data for both birds and mammals. These cycle times span twelve orders of magnitude, from very rapid twitch contraction times to the periodic event with the very lowest frequency, lifespan. To facilitate comparisons, all the lines have been extrapolated from 1 g to 10^3 kg to correspond roughly to the size range of terrestrial mammals (2 g to 7000 kg). This then shows the empirical basis of the "intrinsic time . . . converted to a standard clock time . . . by a metrical function depending upon constitutive parameters" which was treated only theoretically by Richardson and Rosen, 1979. The slopes are strikingly similar even though the lines themselves reflect function in: (1) individual organs (clearance of inulin, cardiac cycle); (2) systems of organs (metabolism of fat stores); (3) the entire animal (growth to maturity, lifespan); and (4) even populations of animals (minimum population doubling time).

For instance, the maximum rate of population growth is described by the variable r, the intrinsic rate of natural increase. The carrying capacity, K, represents the maximum steady-state population number the environment can support. These variables are incorporated in the simplest equation which describes density-dependent population growth, the logistic equation:

$$\frac{dN}{dt} = \frac{rN\,(K - N)}{K} \qquad (4)$$

Species with high r values undergo rapid population growth and are thought to be "colonizing" species. These "r-selected" animals differ from "K-selected" species, which have lower r values (MacArthur and Wilson, 1967). One can calculate a theoretical minimum time required for a population to increase by a constant fraction of K if equation (4) is integrated and solved for t (after Crow and Kimura, 1970):

$$t = \frac{1}{r} \ln \frac{N_t\,(K - N_o)}{N_o\,(K - N_t)} \qquad (5)$$

Fenchel (1974) demonstrated that r varies as a regular exponential function of body size ($M^{-0.25}$; corr. coef. = 0.98) in animals spanning 22 orders of magnitude of body mass. For mammals, r (days^{-1}) = $6.3 \times 10^{-3}\ M^{-0.26}$. Hence, substituting into equation (5):

TABLE 1

Equations relating cycle length (minutes) to body mass (kg)

(A) mammals; (B) birds. The mean mass exponent for all 27 equations is 0.247.

Funcational Scale	Period (Cycle Length) Minutes	C.I.*	Corr.† Coef.	Life Span Cycle Time	Reference††
	A. MAMMALIAN BIOLOGICAL PERIODS				
Life span, in captivity	$6.10 \times 10^{6} M^{0.20}$		0.77	1	Sacher, 1959
98% growth time	$6.35 \times 10^{5} M^{0.26}$			$9.61 M^{-0.06}$	Stahl, 1962
Min. time, population doubling	$3.16 \times 10^{5} M^{0.26}$	0.11–0.41	0.94	$1.93 \times 10^{1} M^{-0.06}$	Fenchel, 1974
Time to reproductive maturity	$2.93 \times 10^{5} M^{0.18}$		0.64	$2.08 \times 10^{1} M^{0.02}$	Hafez et al., 1972
50% growth time	$1.85 \times 10^{5} M^{0.25}$			$3.30 \times 10^{1} M^{-0.05}$	Stahl, 1962
Gestation period	$9.40 \times 10^{4} M^{0.25}$	0.22–0.28	0.85	$6.49 \times 10^{1} M^{-0.06}$	Sacher & Staffeldt, 1974
	$9.54 \times 10^{4} M^{0.26}$		0.72	$6.39 \times 10^{1} M^{-0.06}$	Blueweiss et al., 1978
Time to metabolize fat stores equal to 0.1% body mass	$1.70 \times 10^{2} M^{0.26}$	0.24–0.28	1.00	$3.58 \times 10^{4} M^{-0.04}$	Kleiber, 1932
Half-life of drug (methotrexate)	$5.8 \times 10^{1} M^{0.19}$		0.98	$1.05 \times 10^{5} M^{0.01}$	Dedrick, Bischoff, & Zaharko, 1977
Plasma clearance, inulin	$6.51 M^{0.27}$	0.23–0.31	0.98	$9.37 \times 10^{5} M^{-0.05}$	Stahl, 1967; Edwards, 1975
Plasma clearance, para-aminohippurate	$1.70 M^{0.22}$	0.14–0.30	0.98	$3.59 \times 10^{6} M^{-0.02}$	Stahl, 1967; Edwards, 1975
Time for circulation of blood volume	$3.5 \times 10^{-1} M^{0.21}$		0.98	$1.74 \times 10^{7} M^{-0.01}$	Stahl, 1967
Gut beat duration	$4.75 \times 10^{-2} M^{0.31}$			$1.28 \times 10^{8} M^{-0.11}$	Adolph, 1949
Respiratory cycle	$1.87 \times 10^{-2} M^{0.26}$		0.91	$3.26 \times 10^{8} M^{-0.06}$	Stahl, 1967
Cardiac cycle	$4.15 \times 10^{-3} M^{0.25}$		0.88	$1.47 \times 10^{9} M^{-0.05}$	Stahl, 1967
	$3.05 \times 10^{-3} M^{0.28}$	0.23–0.33	0.93	$2.00 \times 10^{9} M^{-0.08}$	Calder, 1968
Twitch contraction time, soleus	$1.06 \times 10^{-3} M^{0.39}$	0.29–0.49	0.99	$5.75 \times 10^{9} M^{-0.19}$	Syrovy & Gutman, 1975
Twitch contraction time, extensor digitorum longus	$3.14 \times 10^{-4} M^{0.21}$	0.12–0.31	0.99	$1.94 \times 10^{10} M^{-0.01}$	Syrovy & Gutman, 1975
	B. AVIAN BIOLOGICAL PERIODS				
Life span, in captivity	$1.49 \times 10^{7} M^{0.19}$		0.70	1	Lindstedt & Calder, 1976
Life span, wild	$9.25 \times 10^{6} M^{0.20}$	0.17–0.22	0.78	$1.61 M^{-0.01}$	Lindstedt & Calder, 1976
Incubation period	$4.16 \times 10^{4} M^{0.17}$		0.86	$3.58 \times 10^{2} M^{0.02}$	Rahn & Ar, 1974
Respiratory cycle	$5.37 \times 10^{-2} M^{0.33}$	0.30–0.36	0.93	$2.77 \times 10^{8} M^{-0.14}$	Calder, 1968
Cardiac cycle	$6.42 \times 10^{-3} M^{0.23}$	0.17–0.29	0.85	$2.32 \times 10^{9} M^{-0.04}$	Calder, 1968
	Passerine Species				
Life span, wild	$1.14 \times 10^{7} M^{0.26}$	0.20–0.32	0.76	$1.31 M^{-0.07}$	Lindstedt & Calder, 1975
Time to metabolize fat stores equal to 0.1% body mass	$1.06 \times 10^{2} M^{0.28}$	0.23–0.32	0.98	$1.41 \times 10^{5} M^{-0.09}$	Lasiewski & Dawson, 1975
Respiratory cycle	$4.38 \times 10^{-2} M^{0.28}$	0.16–0.41	0.69	$3.40 \times 10^{8} M^{-0.09}$	Calder, 1968
	Procellariformes				
Time to first breeding	$2.32 \times 10^{6} M^{0.22}$	0.11–0.32	0.91	$6.42 M^{-0.05}$	Lack, 1968

* C.I., 95% confidence interval of slope.

† A few equations were derived by combining other regression equations; in those cases, a combined correlation coefficient has been estimated as a product of the individual correlation coefficients.

†† Source of equation or data used to calculate equation.

** The slopes of the lines relating gestation period to body mass are lower if examined within individual orders of mammals (see also Kihlström, 1972).

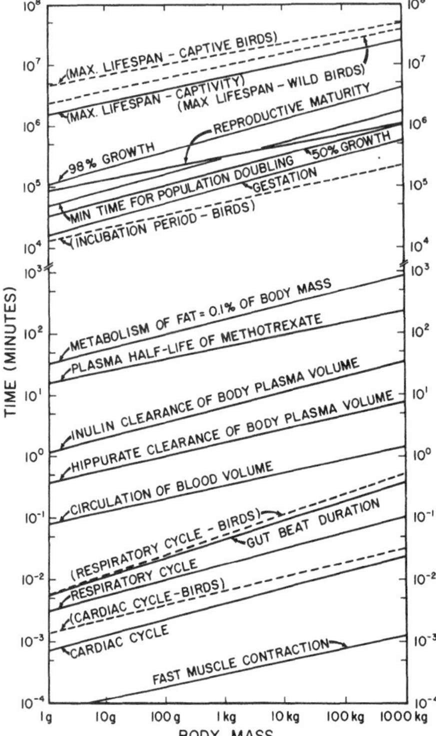

FIG. 1. THE RELATION BETWEEN BODY MASS AND LENGTHS OF BIOLOGICAL PERIODS OR CYCLES

For mammals (solid lines) and for birds (dashed lines). Individual equations (from Table 1) have been extrapolated to span the range of body sizes shown.

$$t = 159\,M^{0.26} \ln \frac{N_t\,(K - N_o)}{N_o\,(K - N_t)} \qquad (6)$$

The time required for a population to increase in number from $\frac{N_o}{K}$ to $\frac{N_t}{K}$ is thus roughly proportional to the fourth root of body mass. In Fig. 1 we have used equation (6) to calculate the minimum time required for a population to double in number. Although *r* is difficult to estimate and the values given by Fenchel (1974) may be inadequate in some cases, they should not, however, be skewed toward either large or small body size.

A much shorter biological period is the half-life of drugs. Dedrick, Bischoff, and Zaharko (1977) derived a theoretical relationship which predicts plasma half-life ($t_{1/2}$) of drugs to vary with the fourth root of body mass. They determined $t_{1/2}$ (minutes) of methotrexate for six mammalian species and found:

$$t_{1/2} = 38\,M^{0.19}$$

Weiss, Sziegoleitt, and Foster (1977) independently predicted that $t_{1/2}$ should be proportional to $M^{0.25-0.27}$ on a basis of biological similarity criteria.

The above are examples of many cycles that vary approximately as $M^{1/4}$. The large number of biological time scales that adhere to this pattern appears to confirm Stahl (1962), who wrote:

> In the analysis of growth and related matters it has become clear that two principles play a key role in biological analysis: conservation of volume (see above) and synchronism of times. . . . Corresponding processes are expected to occur in corresponding times. It appears that the time scale is uniformly that of $M^{0.25}$ to $M^{0.30}$ in mammals and probably in many other organisms (p. 209).

This observation is confirmed in Fig. 1, which suggests that identical biological cycles transpire about seven times longer in man, and twenty times longer in elephants, than in mice.

Although few of the slopes in Fig. 1 are exactly 0.25, they are all very nearly parallel. In fact, the mean of all the body mass exponents is 0.24. Unfortunately, the sources *in lit.* usually provide neither error estimates for many of these equations nor the data to calculate them. However, when available, the 95 per cent confidence intervals of most of the slopes include 0.25 (see table 1). Although variability in exponents may represent the variance in the biological periods themselves, it must also reflect the difficulty of quantifying them accurately. Certainly the longest cycles, such as growth to maturity or life span, cannot be determined with the same accuracy as heart rate; and even heart rate has been reported by different investigators as varying between $M^{0.25}$ and $M^{0.28}$ (Adolph, 1949; Calder, 1968). Little meaning should therefore be ascribed to slight differences in reported exponents. To examine the effects of these exponent deviations over a broad size range, we have selected several equations from Table 1 and combined pairs of them (after Stahl) in Fig. 2. The resultant derived equations only estimate intercepts and exponents. Thus, they may differ slightly from direct measurements. Residual mass exponents from $M^{-0.04}$ to $M^{0.04}$ have relatively little effect on the value of each ratio, over a size range from shrew to elephant. According to these equations, 100 heart beats transpire in six seconds for a shrew and in 2.5 minutes for an elephant. As a consequence of those 100 cardiac cycles, the shrew's total blood volume should circulate once, while the elephant's should circulate one and one-half times. Thus, although their body sizes differ by a factor of one million, the ratio of blood circulation time to heart contraction time varies only slightly (see Fig. 2), and the apparent difference may only be "noise" in the measurements or in the power-law equations themselves. Since the 95 per cent confidence intervals of nearly all the equations include 0.25, the observed differences in exponents may not be statistically meaningful. Whether the variance in exponents is real or artificial, small residual mass exponents have relatively little effect over a wide range of bird and mammal body sizes. Hence, biological events or cycles do appear to transpire (1) as constant multiples of one another and (2) in a time proportional to body mass raised to about the ¼ power.

While allometric relationships are not biological laws, they do describe patterns as well as identify those animals which differ from the patterns. Thus far, our focus has been on the patterns themselves rather than on the deviants. Shrews are of interest because of their small sizes and metabolic intensities. Of the subfamilies of shrews, the Crocidurinae (white-toothed shrews) are characterized by slightly elevated metabolism (relative to weight-predicted values), whereas the Soricinae (red-toothed shrews) have markedly elevated metabolic rates (Vogel, 1976; Lindstedt, 1980). Vogel (1980) reports that, among similar-sized shrews, life spans and gestation periods are shorter in the Soricinae; hence the products of metabolic rate and other periods may be similar in both groups. Physiological time may transpire even more rapidly than predicted for the Soricinae, thus *all* their biological periods may be affected. Additional examples of such interactions with metabolic rate were cited by McNab (1980).

The resting oxygen consumption of the lethargic sloths is 35 to 50 per cent lower than that predicted for their body mass. Their nerve conduction velocities and heart and respiratory rates are also lower, and the

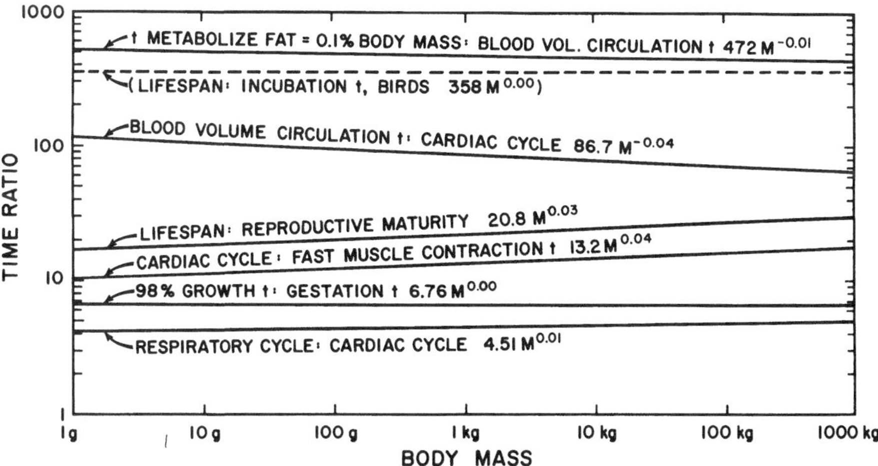

FIG. 2. PAIRS OF EQUATIONS FROM TABLE 1 COMBINED TO FORM DIMENSIONLESS RATIOS OF PERIODS
As the residual mass exponents are small, the resultant ratios are nearly the same for small and large mammals alike.

plasma half-life of thyroxin is longer than in similar-sized "typical" mammals (Goffart, 1971). Maximum longevity of two-toed sloths is 47 per cent longer than that to be predicted from Fig. 1. These data suggest that the physiological time scales of sloths have been prolonged in comparison with eutherian norms for this body-size range.

PHYSIOLOGICAL TIME AND METABOLIC RATE

Since the observation was first made that basal metabolism scales disproportionately with body mass, various explanations of that observation have been proposed. Rubner (1883) measured metabolism in dogs ranging in size from 3 to 30 kg. He found that the metabolic rates of all the dogs were essentially constant if expressed per unit body surface area. This "surface law" survived until measurements of metabolic rate were made over a broad enough size range to deviate significantly from the ⅔ surface exponent. An exponent of ¾ is, of course, now apparent from data not only for birds and mammals, but for a wide variety of vertebrates and invertebrates (Hemmingsen, 1960). Metabolic rate is only one example of a large number of physiological variables that are expressions of volume divided by time. Others, such as glomerular filtration rate, minute volume, and cardiac output also scale near $M^{3/4}$

Attempts to explain the ¾ exponent have focused on the exponent itself, perhaps as if nature had selected the volume-rate scaling. An alternative explanation may be that the scaling of biological volume-rates is the result of physiological time scales. Because of the inevitable consequences of changes in body size, physiological time scales as $M^{1/4}$. Volumes and organs of capacity have relatively little opportunity to scale except in direct proportion to body mass ($M^{1.0}$) among animals of the same design (see above). Hence, volume-rates must scale as $M^{3/4}$:

$$\frac{\text{volume}}{\text{time}} \propto \frac{M^{1.0}}{M^{1/4}} = M^{3/4} \qquad (7)$$

From the foregoing it is obvious that evolution of a change in body size will affect every aspect of physiology and morphology. Hence, one need not hypothesize that specific biological rates or volume-rates, e.g., metabolism, have been selected as isolated phenomena, but that the physiological time scale makes $M^{3/4}$ volume-rates inevitable.

PHYSIOLOGICAL TIME, ECOLOGY, AND EVOLUTION

An organism's physiology must respond to its environment, including the temporal cycles in temperature and light imposed by the earth's daily rotation and the revolutions of moon around earth and of earth around sun, all obviously independent of the animal's size (M^0). The simplest coupling of animal and environment would therefore be size-independent, but physiological time scales are size-dependent ($M^{\cong 1/4}$). It is difficult to envision the natural selection of such an incompatibility; and this difficulty suggests to us that physiological time allometry is not a direct outcome of natural selection, but rather an inevitable consequence of something else. As Hill (1950) observed, the "physiological time-scale . . . has to compromise with the constant time-scale of the external world." The compromise seems to us to be manifested in the biological clocks which synchronize animal and environment.

While circadian and circennial rhythms are useful for coupling physiological and environmental time scales, they cannot completely solve the endurance problems of small animals. As indicated previously, conservation of body plan requires that the capacity of organs be in linear proportion to body mass, so that the energy reserves stored in the volume of the gut or as body fat constitute an essentially constant fraction of body mass ($\propto M^{1.0}$), while the rate of expenditure from these reserves is a function of $M^{3/4}$. Energy amount divided by rate of utilization, $M^1 \div M^{3/4}$, gives endurance time as $\propto M^{1/4}$. Availability of food, or of conditions suitable for safe and effective foraging, is usually on a daily or seasonal clock, not a $M^{1/4}$ clock. If an animal's day is an alternation of feeding and fasting, then problems could arise in maintaining energy balance, in acquiring and storing enough during the activity or availability phase to make it through the fast. The animals most likely to have problems are the smallest homeotherms, hummingbirds and shrews.

Hummingbirds, the more easily observable of these two groups, are pressed to use all of the daylight available except at the highest latitudes of hummingbird distribution, where summer days are the longest (Calder, 1975, 1976b). The limitations of time for attaining energy balance are observed in broad-tailed hummingbirds (*Selasphorus platycercus*) of the Colorado Rockies, where cool temperatures put a significant price on thermoregulation even in the summer. When feeding time is lost during rainstorms, which keep the females on their nests, there is a high likelihood of a negative energy balance for the 24-hour period. This deficit necessitates entry into torpor to conserve energy (Calder and Booser, 1973; Calder, 1975), as if calculations had integrated the remaining energy reserves, the remaining time, and the rates of consumption at normal temperature and in hypothermic torpor.

While the environmental day is equally long, physiologically, for the fast-running shrews as it is for hummingbirds, the largely underground or undercover habitats of shrews should ameliorate the disparity between environmental and physiological time scales. The behavior pattern of stock-piling food that is observed in captive shrews (Lindstedt, unpub.) probably represents a mechanism to permit several sleep-wake cycles per day (Crowcroft, 1954), even though most of the foraging must be done at night.

The dimension of time has been neglected in more than one case of analysis of body size and ecology. For example, McNab (1963) correlated the home range of mammals with their body size, and by one means of calculation found home range proportional to $M^{0.63}$, an exponent statistically indistinguishable from that for metabolic rates ($\propto M^{0.75}$). This finding suggested that home range requirement was proportional to metabolic rate. Metabolic rate, however, has dimensions of power (energy per unit time). An area of home range can contain a certain amount of harvestable energy, but missing from the proportionality is a time function. Reanalyses (Buskirk in Calder, 1974; Harestad and Bunnell, 1979) showed that the home range of primary consumers was more closely related to $M^{1.0}$, not $M^{0.63}$. If $M^{1.0}$ is divided by metabolic rate, $M^{3/4}$ (energy divided by power), we again get a physiological time scale, $\propto M^{1/4}$.

Similarly, Fretwell (1972) proposed a model which limited minimum, maximum, and optimal body sizes of sparrows, as determined by the intersections of allometric curves for digested energy, captured energy, and metabolic rate (energy ÷ time). When these have different dimensions, their crossings are merely a consequence of plotting scales and can have no biological significance. A model could have been constructed in a dimensionally correct manner by converting all lines to the same dimensions, either energy or power (through use of appropriate physiological time scales if such information were available: turnover time for gut contents, foraging time, etc.).

PHYSIOLOGICAL TIME AND LONGEVITY

The longest period of an animal's physiology is its lifespan. Why is life-span scaling also approximately proportional to $M^{1/4}$? We examine three possible explanations which are not mutually exclusive:

(1) Natural selection may have incorporated a safety factor of time beyond reproductive maturity that insures leaving progeny. (The animal that dies too soon has no "reproductive fitness".)

(2) Death comes when vital parts or their interaction no longer function adequately. ("The old ticker has just so many beats.")

(3) Since there is a high correlation between longevity and brain mass, the larger the brain, the greater the control over homeostasis and consequently the longer life conditions can be precisely maintained.

Safety Factor for Evolution of Life Span

Natural selection acts on the reproductive fitness of an animal, and not on its life span directly. Once an individual is post-reproductive, it is out of the gene pool, and has no direct effect on evolution, except as it competes with or altruistically helps individuals still contributing to the gene pool. Thus, life span may be viewed as including a safety factor. The average animal must attain reproductive maturity with enough time for successful reproduction before biosenescence proceeds to a detrimental extent (Cutler, 1978). Thus, the expected life span would exceed the reproductive period with some margin for error or accident, just as McMahon (1973) found that trees grew in height-diameter proportions with a standard safety factor four times the critical buckling dimensions. Vertebrate bones likewise have a tenfold safety factor over static load stresses (Schmidt-Nielsen, 1977).

We can examine the hypothesis of a "safety factor" for insuring reproduction by deriving, as a first approximation, an allometric equation for age at sexual maturity of $n = 22$ for mammals other than primates (using data in Hafez, Asdell, and Blandau, 1972, and typical species weights). The equation predicts (corr. coef. = 0.639, $p < 0.001$) days to sexual maturity as

$$t_{\text{sex mat.}} = 2.93 \times 10^5 \, M^{0.18} \tag{8}$$

Sacher's (1959) equation for maximum life span in captivity ($6.10 \times 10^6 \times M^{0.20}$) would confer a safety factor of 21 $M^{-0.02}$ times sexual maturity (i.e., essentially independent of size). To be more realistic one should add gestation period plus time for a replacement number of newborn to attain independence, and then use mean life span in the wild (for which few data are available), rather than maximum longevity in captivity. Thus, the actual safety factor is considerably less than 21, but is, in all likelihood, independent of body mass.

Death due to Failure of Vital Organs

From Fig. 1 it appears that time scales are all approximately proportional to $M^{1/4}$. The parallelism of the various time-mass functions suggests, in effect, that independent of body mass, a life span consists of multiples of shorter time periods in the biology of the organism, $1.5 \times 10^9 \times M^{-0.05}$ heartbeats, $0.33 \times 10^9 \times M^{-0.06}$ breaths, $17 \times 10^6 \times M^{-0.01}$ completed circulations of total blood volume, $36 \times 10^3 \times M^{-0.04}$ g fat equivalent metabolized per kg body mass, $65 \times M^{-0.06}$ times the gestation period, etc.

Of course, the times for circulatory, respiratory, and metabolic turnovers are basal figures; hence, allowing for activity, there are actually more heartbeats, breaths, and g fat catabolized. Maximum aerobic capacity, however, is apparently a fixed (size-independent) multiple of standard metabolism (Prothero, 1979; Taylor, et al. in press). Because of the size-independent metabolic expansion, the assumption that life span is a multiple of the shorter cycles of that life span is still valid when activity is considered. Thus a lifespan consists of about the same number of cardiac contractions whether the mammal is large or small. A parallel relationship obtains for birds, but the slower bird heart gets 38 per cent more basal heartbeats per lifespan.

Consequently, in this explanation, one or more vital organs for body logistics eventually fail (see Harrison, 1978, for parallel at cellular level). Repair and resynthesis eventually reach limitations, and the heart, lungs, or kidney fail. Without proper CO_2/O_2 exchanges, provided by healthy cardiopulmonary systems, the body succumbs.

Statistical analysis of the causes of death in animals of a wide range in body size appears to be warranted. When one examines the proportional decline in physiological capacities with age in humans (Strehler, 1959), the steepest decline is in maximal breathing capacity. This decline, together with steep declines in vital capacity and cardiac index, could be regarded as a voluntary decline attributable to decreased activity associated with sedentary employment, amusement, and retirement. On the other hand, renal plasma and glomerular filtration rates show the greatest involuntary declines. Perhaps significant in this regard is the fact that kidney tissue has the highest in-vitro metabolic rate (per mg dry mass, von Bertalanffy and Pirozynski, 1951; Malzahn, 1974). If the concept of lifetime energy expenditure (Sacher, 1959, 1977; cf. metabolic curve, Fig. 1 of this study) has a bearing on the limits to maximum life span of the animal, could it not have a bearing on the functional life span of specific organs? Therefore we might expect the intensely active kidney to burn out first. The effects of aging on the kidney have been reviewed recently (Epstein, 1979), but tissue metabolic rates were not considered.

Life Span and Brain Mass

There are three correlations from which it may be concluded that life span is prolonged as a function of greater brain mass (Sacher,

1959, 1977; Fischer, 1968; Mallouk, 1975, 1976).

First, the correlation between life span and brain mass is tighter (accounting for 79% of variance in life span) than the correlation between life span and body mass, which accounts for 60 per cent of the variance in life span (Sacher, 1959). Thus, brain mass by itself is a better predictor of life span than is body mass.

Second, our species has a maximum life span about four times what the mammalian equation predicts from body mass, but our brains are about six times as heavy as predicted from body mass of a typical eutherian mammal (Sacher, 1959; Fischer, 1968). Hence our maximum life span is only twice that predicted from brain mass (Sacher, 1959). However, a larger brain, together with manual dexterity, has provided tools, strategies, and cultural practices which have improved the longevity of *Homo sapiens* compared to that of our wild ancestors. Cutler (1978) suggested that maximum life span potential may be related to ability to learn and teach, abilities enhanced by greater brain size.

The third argument is one of reason rather than quantitative empiricism. The brain is the master regulator or controller of physiological processes through neural and hormonal mediation. The more precise the regulation, the better the "milieu interieur" for the preservation of life. A larger brain has more cells, more capacity to regulate, and therefore to preserve.

Although a good correlation certainly suggests a causal relationship, it does not establish that such a relationship exists. For example, spleen mass and adrenal mass are also better predictors (higher correlation coefficients) of lifespan than is body mass (Calder, 1976a; Economos, 1980). We must, therefore, look carefully at the relative strengths of correlations.

We argue that the reason for the better prediction of life span from brain mass is that brain mass has less variability than body mass. The body mass of any individual or species can vary widely, depending upon its energy balance situation. If food intake exceeds metabolic demands, fat is stored and body mass increases; if feeding does not meet those demands, stores are depleted, and emaciation can result.

In contrast, the brain is limited in size by a bony exoskeleton, the sutures of which are fused in the adult. Thus, the brain is a less variable index of body size than is body mass. Table 2 shows that within species, or when the mammals are treated together, the coefficient of variation for brain mass is less than half that for body mass. It becomes obvious why brain mass can give a better correlation with life span than can body mass.

Finally, we must consider the relative influences of central control capacity and the requirements and characteristics of body size in limiting life span. Brain mass scales quite reliably as $M^{0.67}$ for mammals (Sacher, 1959; Jerison, 1961) and $M^{0.60}$ for birds (Alexander, 1971). Life span is proportional to $M^{0.20}$ for mammals (Sacher, 1959) and $M^{0.18-0.26}$ for birds (Lindstedt and Calder,

TABLE 2
Variation in brain (g) and body (kg) masses
In mammals, body mass is over twice as variable as brain mass.

Species	n	Brain (g) $\bar{x}$	Brain (g) SD	CV*	Body (kg) $\bar{x}$	Body (kg) SD	CV	CV Brain / CV Body
Mammals[a]	–	–	–	0.06–0.07	–	–	0.12–0.15	~ 0.48
Man[b]	45	1312	129.1	0.10	61.5	15.00	0.24	0.42
Horse[b]	15	646	75.8	0.12	452	148.3	0.33	0.36

n, sample size.
* Coefficient of variation.
[a] From Yablokov, 1966.
[b] Calculated from data in Quiring, 1950.
SD, standard deviation.

1976). A ratio of life span to brain mass expressed as common allometric powers of body mass yields:

$$\text{life span/brain mass} \propto M^{0.20}/M^{0.67} = M^{-0.47} \quad (9)$$

Therefore, the amount of gain in maximum life span per unit increase in brain mass actually decreases as the animal's size increases. If an increase in brain size directly increases life span, why does the proportionate effect decrease with size? Although the residual mass exponent would be reduced to – 0.25, the case for brain-dependence would not be improved qualitatively if brain surface, rather than mass, were to be employed.

Additional insight may be gained from the following comparison. On a basis of equal body mass, birds have smaller brains but longer life spans than mammals. The ratios of maximum life span in captivity (years) to brain mass are as follows:

$$\text{mammals: } \frac{11.6\ M^{0.20}}{11.3\ M^{0.67}} = 1.03\ M^{-0.47} \quad (10)$$

$$\text{birds: } \frac{28.3\ M^{0.19}}{8.2\ M^{0.60}} = 3.45\ M^{-0.41} \quad (11)$$

Accordingly, a larger brain mass *per se* has not endowed mammals with greater longevity.

As regards homeostatic regulation, birds can tolerate larger ranges in body temperature, plasma osmotic concentrations, and blood pH than most mammals (Braun and Dantzler, 1972; Calder and Schmidt-Nielsen, 1966, 1968; Calder and King, 1974). Consequently, we should reconsider the hypothesis that longer life necessitates finer homeostatic regulation.

Correlations have been improved when life span is expressed as a multiple regression on body and brain masses, index of cephalization, etc. (Sacher, 1959, 1977), but the fact remains that this relationship is no proof that large brain and (or) large body and (or) such indices are either cause or explanation for longevity.

We doubt that it is brain size, as such, that sets a limit to animal longevity. We interpret the allometry of longevity as a further manifestation of Stahl's "synchronism" of times (Stahl 1962, 1963). Excellent correlations of life span and brain mass probably reflect the low variance in brain masses rather than express a causal relationship. We consequently suggest that answers to gerontological problems will probably lie in other theories of aging than those based upon relative cephalization.

If there is a lession to be drawn from comparative gerontology, it is not a suggestion of how to increase the life span, which is itself a second- or third-stage consequence of metabolic logistics. Life ends when these vital functions fail. The goal of gerontology should not be to prolong life in quantity of years, but to prolong the living of a full life of unimpaired function during the alloted $M^{1/4}$ span of years.

We live in deeds, not years; in thoughts, not breaths;
In feelings, not in figures on a dial.
We should count time by heart-throbs.
He most lives
Who thinks most—feels the noblest—acts the best.
—Philip James Bailey, *A Country Town*

ACKNOWLEDGMENTS

We are grateful to Howard Haines for providing data on mouse skin mass.

This work was supported in part by NIH AM05738 (SLL) and NSF DEB79-03689 (WAC).

LIST OF LITERATURE

ADOLPH E. F. 1949. Quantitative relations in the physiological constituents of mammals. *Science*, 109: 579–585.

ALEXANDER, R. M. 1971. *Size and Shape*. Edward Arnold, London.

ALEXANDER, R. McN., A. S. JAYES, G. M. O. MALOIY, and W. M. WATHUTA. 1979. Allometry of the limb bones of mammals from shrews (*Sorex*) to elephant (*Loxodonta*). *J. Zool. Lond.*, 189: 305–314.

ANDERSON, J. F., H. RAHN, and H. D. PRANGE. 1979. Scaling of supportive tissue mass. *Q. Rev. Biol.*, 54: 139–148.

BLUEWEISS, L., H. FOX, F. KUDZMA, D. NAKA-

SHIMA, R. PETERS, and S. SAMS. 1978. Relationships between body size and some life history parameters. *Oecologia*, 37: 257-272.

BODDINGTON, M. J. 1978. An absolute metabolic scope for activity. *J. Theor. Biol.*, 75: 443-449.

BRAUN, E. J., and W. H. DANTZLER, 1972. Function of mammalian-type and reptilian-type nephrons in kidney of desert quail. *Am. J. Physiol.*, 222: 617-629.

BRODY, S. 1945. *Bioenergetics and Growth.* Hafner, New York.

CALDER, W. A., III. 1968. Respiratory and heart rates of birds at rest. *Condor*, 70: 358-365.

——. 1974. Consequences of body size for avian energetics. In R. A. Paynter, Jr. (ed.), *Avian Energetics*, pp. 86-151. Nuttall Ornithological Club, Cambridge.

——. 1975. Daylight and the hummingbirds' use of time. *Auk*, 92: 81-97.

——. 1976a. Aging in vertebrates: allometric considerations of spleen size and lifespan. *Fed. Proc.*, 35: 96-97.

——. 1976b. Energetics of small body size and high latitude: the Rufous hummingbird in costal Alaska. *Int. J. Biometeor.*, 20: 23-35.

——. 1981. Scaling of physiological processes in homeothermic animals. *Annu. Rev. Physiol.*, 43: 301-322.

CALDER, W. A., III, and J. BOOSER. 1973. Hypothermia of broad-tailed hummingbirds during incubation in nature with ecological correlations. *Science*, 180: 751-753.

CALDER, W. A., III, and J. R. KING. 1974. Thermal and caloric relations of birds. In D. S. Farner and J. R. King (eds.), *Avian Biology*, Vol. 4, p. 259-413. Academic Press, New York.

CALDER, W. A., III, and K. SCHMIDT-NIELSEN. 1966. Evaporative cooling and respiratory alkalosis in the pigeon. *Proc. Natl. Acad. Sci. U.S.A.*, 55: 750-756.

——, and ——. 1968. Panting and blood carbon dioxide in birds. *Am. J. Physiol.* 215: 477-482.

CROW, J. F., and M. KIMURA. 1970. *An Introduction to Population Genetics Theory.* Harper and Row, New York.

CROWCROFT, P. 1954. The daily cycle of activity in British shrews. *Proc. Zool. Soc. Lond.* 123: 715-729.

CUTLER, R. G. 1978. Evolutionary biology of semescence. In J. A. Behnke, C. E. Finch and G. B. Moment (eds.), *The Biology of Aging*, p. 311-360. Plenum, New York.

DEDRICK, R. L., K. B. BISCHOFF, and D. S. ZAHARKO. 1970. Interspecies correlation of plasma concentration history of methotrexate. *Cancer Chemother. Rep.*, 54: 95-101.

DELCOMYN, F. 1980. Neural basis of rhythmic behavior in animals. *Science*, 210: 492-498.

ECONOMOS, A. C. 1980. Brain-lifespan conjecture: a reevaluation of the evidence. *Gerontology*, 26: 82-89.

EDWARDS, N. A. 1975. Scaling of renal function in mammals. *Comp. Biochem. Physiol.*, 52A: 63-66.

EPSTEIN, M. 1979. Effects of aging on the kidney. *Fed. Proc.*, 38: 168-172.

FENCHEL, T. 1974. Intrinsic rate of natural increase: the relationship with body size. *Oecologia*, 14: 317-326.

FISCHER, R. 1968. On the steady state nature of evolution, learning, perception, hallucination and dreaming. In A. Locke (ed.), *Quantitative Biology of Metabolism*, p. 245-256. Springer-Verlag, New York.

FRETWELL, S. D. 1972. *Populations in a Seasonal Environment.* Monographs in Population Biology, Volume 5. Princeton University Press, Princeton, N. J.

GOFFART, M. 1971. *Function and Form in the Sloth.* Pergamon Press, Oxford.

GOSLOW, G. E., JR., W. E. CAMERON, and D. G. STUART, 1977. The fast twitch motor units of cat ankle flexors. 2. Speed-force relations and recruitment order. *Brain Res.*, 134: 47-57.

GOULD, S. J. 1966. Allometry and size in otogeny and phylogeny. *Biol. Rev.*, 41: 587-640.

GOULD, S. J., and R. C. LEWONTIN. 1979. The spandrels of San Marco and the Panglossian Paradigm: A critique of the adaptationist programme. *Proc. Roy. Soc. Lond.* (B), 205: 581-598.

GUNTHER, B. 1975. On theories of biological similarity. In W. Beir (ed.), *Fortschritte der exper. u. Theor. Biophysik*, Vol. 19, p. 7-111. G. Thieme, Leipzig.

HAFEZ, E. S. E., S. A. ASDELL, and R. J. BLANDAU. 1972. Propagation: Mammals. In P. L. Altman and D. S. Dittmer (eds.), *Biology Data Book*, Vol. I, p. 138-139. Fed. Am. Soc. Exp. Biol., Bethesda, Md.

HARESTAD, A. S., and F. L. BUNNELL, 1979. Home range and body weight—a reevaluation. *Ecology*, 60: 389-402.

HARRISON, D. E. 1978. Is limited cell proliferation the clock that times aging? In J. A. Behnke, C. E. Finch and G. B. Moment (eds.), *The Biology of Aging*, p. 33-35. Plenum, New York.

HEMMINGSEN, A. M. 1960. Energy metabolism as related to body size and respiratory surfaces in evolution. *Rep. Steno. Mem. Hosp. Nord. Insulin Lab*, 9: 7-110.

HILL, A. V. 1950. The dimensions of animals and their muscular dynamics. *Sci. Progr.*, London, 38: 209-230.

JERISON, H. J. 1961. Quantitative analysis of evolution of the brain in mammals. *Science*, 133: 1012-1014.

KAYSER, C., and A. HEUSNER. 1964. Étude comparative du métabolisme énergétique dans la série animale. *J. Physiol.*, Paris, 56: 489-524.

KIHLSTOM, J. E. 1972. Period of gestation and body weight in some placental mammals. *Comp. Biochem. Physiol.*, 43A: 673-679.

KLEIBER, M. 1932. Body size and metabolism. *Hilgardia*, 6: 315-353.

——. 1950. Physiological meaning of regression equations. *J. Appl. Physiol.*, 2: 417-423.

——. 1961. *The Fire of Life.* Wiley, New York.

KOEZE, T. H. 1973. Muscle spindle afferent studies in the baboon. *J. Physiol.*, 229: 297-317.

LACK, D. 1968. *Ecological Adaptations for Breeding in Birds.* Methuen, London.

LASIEWSKI, R. C., and W. R. DAWSON. 1967. A re-examination of the relation between standard metabolic rate and body weight in birds. *Condor* 69: 13-23.

LINDSTEDT, S. L. 1980. The smallest insectivores: Coping with scarcities of energy and water. In K. Schmidt-Nielsen, L. Bolis, and C. R. Taylor (eds.), *Comparative Physiology: Primitive Mammals*, p. 163-169. Cambridge Univ. Press, Cambridge.

LINDSTEDT, S., and W. A. CALDER. 1976. Body size and longevity in birds. *Condor* 78: 91-94.

MACARTHUR, R. H., and E. O. WILSON. 1967. *The Theory of Island Biogeography.* Princeton Univ. Press, Princeton, N. J.

MALLOUK, R. S. 1975. Longevity in vertebrates is proportional to relative brain weight. *Fed. Proc.* 34: 2102-2103.

——. 1976. "Author's reply." *Fed. Proc.* 35 97-98.

MALZAHN, E. 1974. Tissue metabolism in the common shrew and the bank vole. *Bialowieza* 19: 301-314.

MAYER, R. F., and C. MAWDSLEY. 1968. *J. Neurol. Neurosurg. Psychiat.* 28: 201-211.

——. 1973. Size and shape in biology. *Science* 179: 1201-1204.

MCMAHON, T. A. 1974. Scaling stride frequency and gait to animal size: mice to horses. *Science* 186: 1112-1113.

——. 1975. Using body size to understand the structural design of animals: quadrupedal locomotion. *J. Appl. Physiol.* 39: 619-627.

MCNAB, B. K. 1963. Bioenergetics and the determination of home range size. *Am. Nat.* 65: 133-139.

——. 1980. Food habits, energetics, and the population biology of mammals. *Am. Nat.*, 116: 106-124.

PACE, N., D. E. RAHLMAN, and A. H. SMITH. 1979. Scale effects in the musculoskeletal system, viscera and skin of small terrestrial mammals. *Physiologist*, 22: S-51-S-52.

PRANGE, H. D. 1977. The scaling and mechanics of arthropod exoskeletons. In T. J. Pedley (ed.), *Scale Effects in Animal Locomotion*, p. 169-181. Academic Press, London.

PRANGE, H. D., J. F. ANDERSON, and H. RAHN. 1979. The allometric relationship between supportive tissue mass and total body mass. *Am. Nat.*, 113: 103-122.

PROTHERO, J. W. 1979. Maximal oxygen consumption in various animals and plants. *Comp. Biochem. Physiol.*, 64A: 463-466.

QUIRING, D. P. 1950. *Functional Anatomy of the Vertebrates.* McGraw-Hill, New York.

RAHN, H., and A. AR. 1974. The avian egg: incubation time and water loss. *Condor* 76: 147-152.

——, and ——. 1980. Gas exchange of the avian egg: time, structure, and function. *Am. Zool.*, 20: 477-484.

REYNOLDS, W. W., and W. J. KARLOTSKI. 1977. The allometric relationships of skeleton weight to body weight in teleost fishes: A preliminary comparison with birds and mammals. *Copeia*, 1977: 160-163.

RICHARDSON, E. W., and R. ROSEN. 1979. Aging and the metrics of time. *J. Theor. Biol.*, 79: 415-423.

RUBNER, M. 1883. Ueber den Einfluss der Körpergrösse auf Stoff- und Kraftweschesel. *Z. Biol.*, 19: 535-562.

SACHER, G. A. 1959. Relation of lifespan to brain weight and body weight. In G. E. W. Wolstenholme and M. O'Connor (eds.), *The Lifespan of Animals*, p. 115-141. Little Brown, Boston.

——. 1977. Life table modification and life prolongation. In C. E. Finch and L. Hayflick (eds.), *Handbook of the Biology of Aging*, p. 582-638. Van Nostrand, New York.

——. 1978. Longevity and aging in vertebrate evolution. *BioScience*, 28: 497-501.

SACHER, G. A., and E. F. STAFFELDT. 1974. Relation of gestation time to brain weight for placental mammals: Implications for the theory of vertebrate growth. *Am. Nat.*, 108: 593-615.

SCHMIDT-NIELSEN, K. 1977. Problems of scaling. Locomotion and physiological correlates. In T. J. Pedley (ed.), *Scale Effects in Animal Locomotion*, p. 1-21. Academic Press, London.

STAHL, W. R. 1962. Similarity and dimensional methods in biology. *Science*, 137: 205-212.

——. 1963. The analysis of biological similarity. In J. H. Lawrence and J. W. Gofman (eds.), *Advances in Biological and Medical Physics*, Vol. 9, p. 355-464. Academic Press, New York.

——. 1965. Organ weights in primates and other mammals. *Science*, 150: 1039-1041.

——. 1967. Scaling of respiratory variables in mammals. *J. Appl. Physiol.*, 22: 453-460.

STREHLER, B. L. 1959. Origin and comparison of

the effects of time and high-energy radiations on living systems. *Q. Rev. Biol.*, 34: 117–142.

SYROVY, I., and E. GUTMANN. 1975. Myosin from fast and slow skeletal and cardiac muscles of mammals of different size. *Physiol. Bohemoslov.*, 24: 325–334.

TANNER, J. M. 1949. Fallacy of per-weight and per-surface area standards, and their relation to spurious correlation. *J. Appl. Physiol.*, 2: 1–15.

TAYLOR, C. R., G. M. O. MALOIY, E. R. WEIBEL, V. A. LANGMAN, J. M. Z. KAMAU, H. J. SEEHERMAN, and N. C. HEGLUND. In press. Structure-function correlations in the respiratory systems: scaling maximum aerobic capacity to body mass—wild and domestic animals. *Respir. Physiol.*

VAN DE GRAFF, K. M., E. C. FREDERICK, R. G. WILLIAMSON, and G. E. GOSLOW, JR. 1977. Motor unit and fiber types of primary ankle extensors of the skunk (*Mephitis mephitis*). *J. Neurophysiol.*, 40: 1424–1431.

VOGEL, P. 1976. Energy consumption of European and African shrews. *Acta Theriol.*, 21: 195–206.

——. 1980. Metabolic levels and biological strategies in shrews. In K. Schmidt-Nielsen, L. Bolis, and C. R. Taylor (eds.), *Comparative Physiology: Primitive Mammals*, p. 170–180. Cambridge Univ. Press, Cambridge.

VON BERTALANFFY, L., and W. J. PRIOZYNSKI. 1951. Tissue respiration and body size. *Science*, 113: 599–600.

VON SCHELLING, H. 1954. Mathematical deductions from empirical relations between metabolism, surface area, and weight. *Ann. N. Y. Acad. Sci.*, 56: 1143–1164.

WEISS, M. W., W. SZIEGOLEITT, and W. FORSTER. 1977. Dependence of pharmacokinetic parameters on the body weight. *Int. J. Clin. Pharmacol.*, 15: 572–575.

WESTERN, D. 1979. Size, life history, and ecology in mammals. *Afr. J. Ecol.*, 17: 185–204.

YABLOKOV, A. V. 1966. *Variability of Mammals.* Nauka, Moscow.

PAPER 16

Species Number, Species Abundance and Body Length Relationships of Arboreal Beetles in Bornean Lowland Rain Forest Trees (1988)

D. R. Morse, N. E. Stork, and J. H. Lawton

Commentary

ALISTAIR EVANS

Throughout the 1970s and 1980s, many ecologists became increasingly interested in collecting and interpreting data on various important aspects of community structure, including number of species, species abundance, and body size. Often the relationships between pairs of these variables had been considered, but R. M. May (1978) recognized that these three variables should be interrelated. However, the development of a framework to explain the interrelations between them required a data set for all three variables for the same community, which was lacking. In addition, there had been an unfortunate paucity of data regarding the majority of animals (e.g., invertebrates).

To address this, D. R. Morse, N. E. Stork, and J. H. Lawton measured the number of species, species abundance, and body length for beetle assemblages inhabiting Bornean lowland rain forest trees. The resulting three-dimensional surface in figure 7 of the paper reprinted here is relatively smooth but shows a peak in abundance in mid-range body size and the presence of small, rare species. This latter point differs from compilation-type studies (e.g., Damuth 1981) but corresponds well with some other studies (e.g., Brown and Maurer 1987).

The right-hand tail for number of species versus body length shows a very steep drop-off, around –2.6, coming close to May's (1978) predictions of –2.0. With the expected fractal nature of plant surfaces, a slope of –3.25 is expected in the number of individuals versus body length (Morse et al. 1985), but the coleopteran data of Morse, Stork, and Lawton in this paper give slopes of approximately –4.0. This may be an artifact of including only beetles, as D. H. Janzen's whole assembly data (1973) show that this slope is much greater for the Coleoptera alone as opposed to all invertebrates. Morse, Stork, and Lawton's study has invoked a number of more extensive investigations into insect diversity patterns (e.g., Loder et al. 1997; Finlay et al. 2006).

Literature Cited

*Brown, J. H., and B. A. Maurer. 1987. Evolution of species assemblages: Effects of energetic constraints and species dynamics on the diversification of the North American avifauna. *American Naturalist* 130:1–17.

*Damuth, J. 1981. Population density and body size in mammals. *Nature* 290:699–700.

Finlay, B. J., J. A. Thomas, G. C. McGavin, T. Fenshel, and R. T. Clarke. 2006. Self-similar patterns of nature: Insect diversity at local to global scales. *Proceedings of the Royal Society of London B* 273:1935–41.

Janzen, D. H. 1973. Sweep samples of tropical foliage insects: Description of study sites, with data on species abundances and size distributions. *Ecology* 54:659–86.

Loder, N., T. M. Blackburn, and K. J. Gaston. 1997. The slippery slope: Towards an understanding of the body size frequency distribution. *Oikos* 78:195–201.

*May, R. M. 1978. The dynamics and diversity of insect faunas. In *Diversity of insect faunas*, edited by L. A. Mound and N. Waloff, 188–204. Royal Entomological Society of London Symposium 9. Blackwell Scientific, Oxford.

Morse, D. R., J. H. Lawton, M. M. Dodson, and M. H. Williamson. 1985. Fractal dimension of vegetation and the distribution of arthropod body lengths. *Nature* 314:731–33.

Ecological Entomology (1988) **13**, 25–37

Species number, species abundance and body length relationships of arboreal beetles in Bornean lowland rain forest trees

D. R. MORSE, N. E. STORK* and J. H. LAWTON† The Computing Laboratory, The University, Canterbury, *Department of Entomology, British Museum (Natural History), and †Department of Biology, University of York

ABSTRACT. 1. The relationships between number of species, abundance per species, and body length are examined for 859 species of beetles in samples of arthropods collected from ten Bornean lowland forest trees by insecticide fogging. Similar relationships are examined for different feeding guilds of these beetles, and for those beetles from different species of trees.

2. The data are used to construct four interrelated graphs, namely species:abundance, species:body length, population abundance:body length and total number of individuals:body length distributions.

3. In contrast to a number of previous studies, no consistent linear relationship between population density and body length was found for the Bornean beetles and it is suggested that, as in birds, the added dispersal ability of flight reduces critical population densities necessary for persistence in small species. Previous relationships between body weight and population abundance may also be artefacts of the way in which data were gathered.

4. Despite large samples, we failed to locate the mode in plots of the number of species in each abundance category (species:abundance distribution).

5. Species:body length and total number of individuals:body length plots were similar to those found in previous studies, although using data for Coleoptera alone may have produced a steeper decline in the total number of individuals as body size increases than is apparent in samples of all arthropods.

6. We present the first three-dimensional graph relating numbers of species, body lengths and population abundances. The surface of this three-dimensional relationship is relatively simple.

Key words. Species abundance, body length, beetles, Borneo, tree canopy, rain forest, insecticide fogging.

Correspondence: Mr D. R. Morse, The Computing Laboratory, The University, Canterbury, Kent CT2 7NF.

Introduction

The structure of animal communities can be described in a number of ways; the simplest is the number of coexisting species. More subtle descriptions of samples of organisms from a community include: the number of species in each abundance category (the so-called species:abundance distribution, or species: frequency distribution, e.g. Preston, 1948; Williams, 1953; May, 1975; Southwood, 1978; Sugihara, 1980); the number of species of different body sizes (species:body-size distribution) (e.g. Hemmingsen, 1934; Hutchinson & MacArthur, 1959; May, 1978, 1986; Griffiths, 1986); and the abundance of each species versus body size (population abundance:body size relationship) (e.g. Damuth, 1981; Peters, 1983; Peters & Raelson, 1984; Peters & Wassenberg, 1983; Brown & Maurer, 1986). Summing individuals for all species of one body size yields a fourth pattern, the total number of individuals of a particular size, irrespective of species (total number of individuals:body size, or if biomass rather than number is used, total biomass:body size relationship, e.g. Janzen & Schoener, 1968; Janzen, 1973; Morse *et al.*, 1985; Griffiths, 1986; Rodriguez & Mullin, 1986; Strayer, 1986).

Each of these relationships has its own literature, and theoretical explanations for particular patterns (*loc. cit.*). But, as Fig. 1 makes plain, they are interrelated. For example, fixing any two of the basic relationships in Fig. 1 (i) (species:abundance, species:body size, or population abundance:body size) automatically defines the limits of the third (Harvey & Lawton, 1986). A growing number of papers consider relationships between two of the dis-

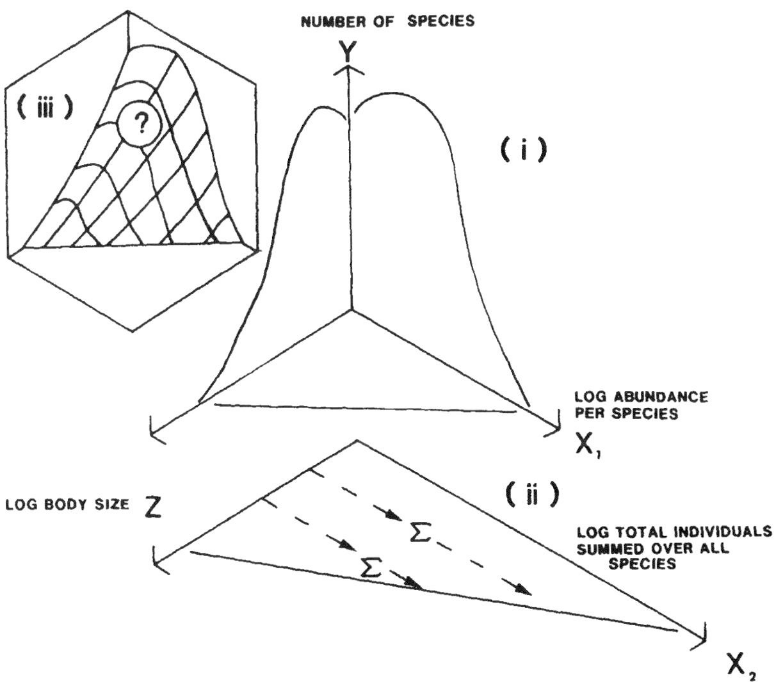

FIG. 1. Generalized relationship between community variables. The upper (3-D) figure (i) illustrates relationships between three distributions. In the YX_1 plane is the species:abundance distribution, here illustrated for convenience as a log-normal distribution (Preston, 1962). The YZ plane defines the species:body size distribution, and the ZX_1 plane the population abundance:body size relationship. Summing individuals for all species of one body size gives the projected ZX_2 plane in (ii). Focusing on (i), the question is, what does the full, three-dimensional surface (iii) look like in real communities?

tributions (e.g. Janzen, 1973; Griffiths, 1986; Strayer, 1986; Harvey & Godfrey, 1987), but because data are almost always presented in two dimensions (for example species: abundance distributions are presented by summing across individuals of all body sizes), development of a unifying theory is hampered by ignorance about the shape of the full three-dimensional surface depicted in Fig. 1. May (1978) is one of the few to have realized that all these community patterns are interrelated. The surface defined by Fig. 1 will not necessarily be simple, and may differ from community to community. For example, attempts to predict observed numbers of insect species (S) in different body size classes, knowing only the total number of individuals of each body size (N), using the empirical relationship $S \propto N^{0.25}$, derived from Preston's log-normal distribution of species abundances (May, 1978), fail (Lawton, 1986). One possible explanation is that different species abundance distributions hold for species of different body size classes.

The present work is the first to establish empirical relationships between the variables depicted in Fig. 1, for an assemblage of species collected at one place and time, namely adult, arboreal beetles sampled from the canopy of rain forest trees using knockdown insecticides (Stork, 1987a, b; and unpublished). Our purpose is to establish what the patterns look like, not to develop theories to explain the patterns; the end result is the three-dimensional surface depicted in Fig. 7. As will become apparent, the data are not ideal, but they are the best currently available, based on samples of 23,000 individuals and approximately 3000 species of arthropods. Although full information on body lengths, numbers of species and numbers of individuals are only available for the Coleoptera in these samples, the data nevertheless consist of 859 species and nearly 4000 individuals. We are, however, aware that the Coleoptera may represent a biased sample of the relationships displayed by the entire arthropod assemblage of the canopy. We have attempted to discover whether this bias might be serious by reanalysing Janzen's (1973) sweep net samples from the understorey of tropical forest, for total arthropods and the Coleoptera alone. We return to the problem of taxonomic and other sampling biases in the Discussion. Finally, as a first step in understanding the biological basis underpinning the patterns revealed by the data, we have analysed not only the total collection of Coleoptera, but also different feeding guilds, and the species associated with different species of trees.

Methods

Samples of arthropods were collected by fogging the canopy of trees with a synthetic pyrethroid insecticide in an area of lowland rain forest near Bukit Sulang, Brunei (Borneo) in September 1982. Full details of the techniques used will be given elsewhere (Stork, unpublished). Ten trees were chosen for sampling; trees 1–4, *Shorea johorensis* Foxwood; trees 5–6, *S. macrophylla* (De Vriese) [Dipterocarpaceae]; trees 7–8, *Pentaspadon motleyi* Hook [Anacardiaceae]; tree 9,

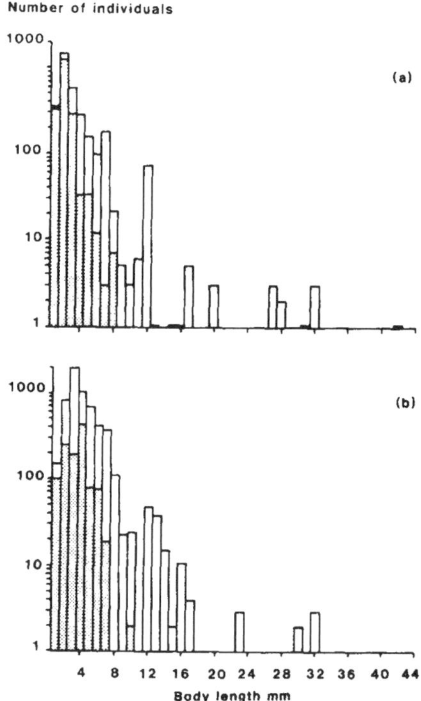

FIG. 2. Histograms of the number of individuals of all arthropods in different 1 mm size classes (i.e. graphs in the ZX_2 plane of Fig. 1(ii) for sweep samples of understorey vegetation from (a) Taboga primary riparian vegetation and (b) Osa secondary vegetation, Costa Rica (from Janzen, 1973). The shaded bars represent the number of beetles of different body lengths.

Castanopsis sp. [Fagaceae]; and tree 10, unidentified species and family.

The insect samples were sorted to orders and for most groups, to morpho-species (for simplicity, referred to as species from here on). The 859 species (3919 individuals) of adult Coleoptera were assigned to four guilds, herbivores (384 species), predators (200), scavengers (120) and fungivores (141) (guild assignments as in Stork (1987a), but with the fungivores separated from the scavengers); a further fourteen species could not be confidently assigned to guilds.

We used body length as a measure of the size of each species since length is considerably easier to measure than weight, and for arthropods the two measures are highly correlated (Rogers *et al.*, 1976, 1977; Schoener, 1980). Measurements of the body lengths of species were taken from the hind-most tip of the abdomen or the elytra to the most forward part of the head (excluding the antennae) using a dissecting microscope with an eyepiece graticule. For most species of beetle in the samples, body length varies little, the smallest individual usually not being more than 10% shorter than the longest. Mean body lengths, therefore, were estimated from a maximum sample of ten specimens per species (or all available specimens of species represented by less than ten individuals) and grouped into a series of length classes arranged on a logarithmic scale.

Results

Analysis of Janzen's Costa Rican data

In Figs. 2(a–b) total numbers of individuals are plotted against log body length classes, for the arthropods in sweep samples from two tropical lowland sites in Costa Rica. In both samples the Coleoptera comprise a major part of the smaller arthropods but are almost totally lacking in the larger size classes; hence the rate of decline from the mode in number of individuals is much steeper for the Coleoptera than for all arthropods combined.

Species:abundance distribution

The species:abundance distribution for all individuals combined is shown in Fig. 3(a). The most striking feature of the graph is the number

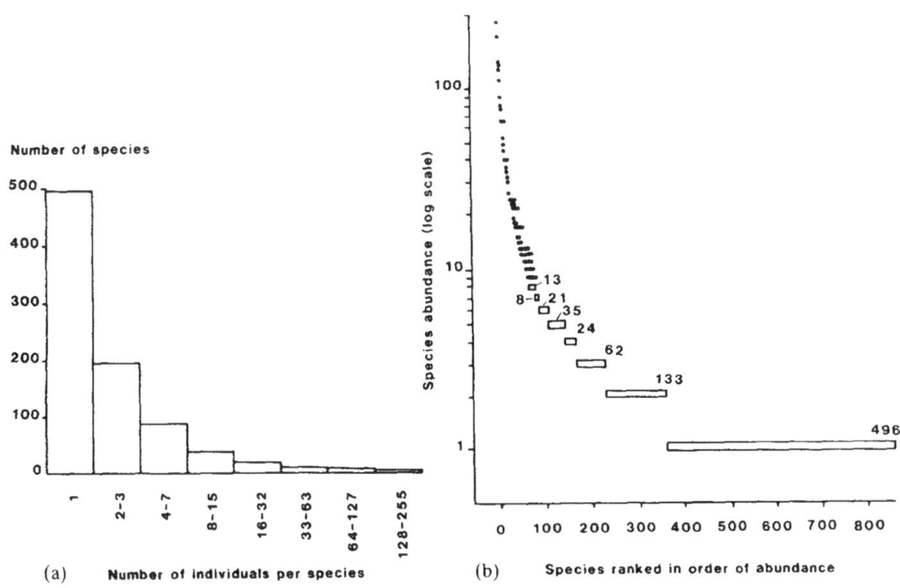

FIG. 3. (a) Species:abundance distributions for all the beetles in ten Bornean tree samples (i.e. the YX_1 plane of Fig. 1). (b) The same data as (a), plotted as abundance (on a log scale) versus species rank. Numbers are given where data points in each block are too numerous to illustrate.

TABLE 1. Equations of regression lines fitted through the species rank-abundance plots of Fig. 3(b) for the beetles collected from three different species of tree. The regression lines were fitted by ordinary least squares techniques through the data after it had been logarithmically transformed. All regression lines are significant at the 0.1% level. Also shown are the mean numbers of species and individual beetles found on each tree.

Tree species	Slope	SE	Intercept	SE
Shorea johorensis	−0.2810	0.0031	2.1546	0.0215
Shorea macrophylla	−0.2747	0.0058	1.8313	0.0352
Pentaspadon motleyi	−0.1780	0.0069	1.0176	0.0362

Tree species	Mean no. per tree	
	Species	Individuals
Shorea johorensis	175.5	466.0
Shorea macrophylla	143.0	513.0
Pentaspadon motleyi	72.0	112.0

of rare species, 58% of the species in the samples being represented by single individuals. We have not attempted to fit a log-normal distribution to the data, because it is clear that we have not yet discovered the mode of the distribution (parameters estimated for log-normal distributions without a mode are probably meaningless (Hughes, 1986)). Instead, in order to compare samples from the different tree species, we replotted the data (Fig. 3b) as a rank–abundance graph (e.g. Williamson, 1973). The equation of the line fitted to the data in Fig. 3(b) is:

$$\text{abundance} = 470 * \text{rank}^{-0.96} \ (P<0.001)$$

Similar plots for samples taken from the different species of tree show a linear relationship between species–rank and species–abundance on a double logarithmic plot, but their slopes are different (Table 1) (the data from trees 9 and 10 were not sufficient to justify separate analyses). The slopes of the lines through the rank abundance data for beetles from *Shorea johorensis* and *S. macrophylla* are very similar. However, that for *Pentaspadon motleyi* is approximately half the value of the slopes for either *Shorea* species, and the intercept for *P. motleyi* is lower.

Species : body length distributions

In Fig. 4(a) the number of species is plotted against body length class. The maximum number of species occurs in size class 3 (a body length of about 2 mm). The upper tail of the distribution from size class 5 (a body length of about 3 mm) shows a near linear decline, indicating a power-law relationship between number of species and body length. May (1978) provides a theoretical argument for the expected shape of the upper tail of this distribution (quantitative theoretical predictions about the shape of the full distribution have not been made). Following May (1978) we have therefore fitted a regression line to the upper (right-hand) part of the distribution (taking the mid-point of each size class on the abscissa); the fitted line has a slope of −2.64 (±0.38). Equivalent data for the different guilds of beetles are in Figs. 4(b–e). It is clear that each guild contributes differentially to the overall picture in Fig. 4(a). The most species-rich size classes are in different positions for each guild and the shapes of the distributions appear to be quite different.

Population abundance : body length distributions

In Fig. 5(a) the number of individuals per species is plotted against the body length of each species on double logarithmic axes. The resulting scatter-plot has two peaks the lower of which (corresponding to body lengths of 0.7–1.4 mm and population abundances of up to fifty individuals) is comprised mainly of fungivores (Fig. 5d) and the upper (corresponding to body lengths of 3–5 mm and population abundances of up to 200 individuals) mainly of herbivores (Fig. 5b). There are no consistent relationships between size and population abundances in Figs. 5(a–e) of the form proposed by Peters (1983) and Peters & Wassenberg (1983). Significant but

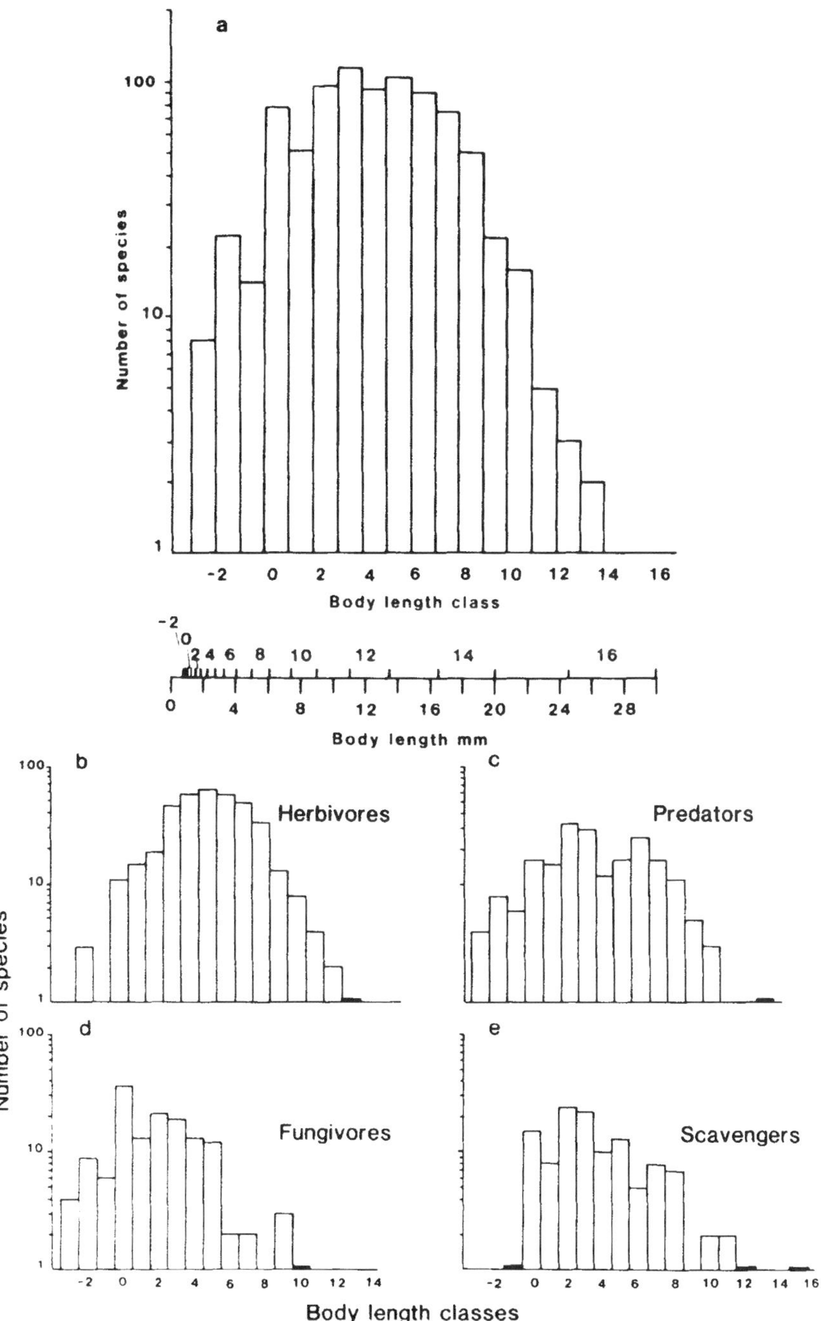

FIG. 4. Histograms of number of species in different length classes (i.e. the *YZ* plane of Fig. 1) for (a) all beetles in the Bornean samples, and for the following guilds of beetles: (b) herbivores, (c) predators, (d) fungivors, (e) scavengers. (*N.B.* In order to display the data economically, the *Y* axis of the *YZ* plane of Fig. 1 is here logarithmically transformed. The scale converting length classes to actual length is displayed below Fig. 4 (a). For example, beetles in length class 14 are between 16.4 and 20 mm long; classes are in groups of $5 * \log_e$ length.)

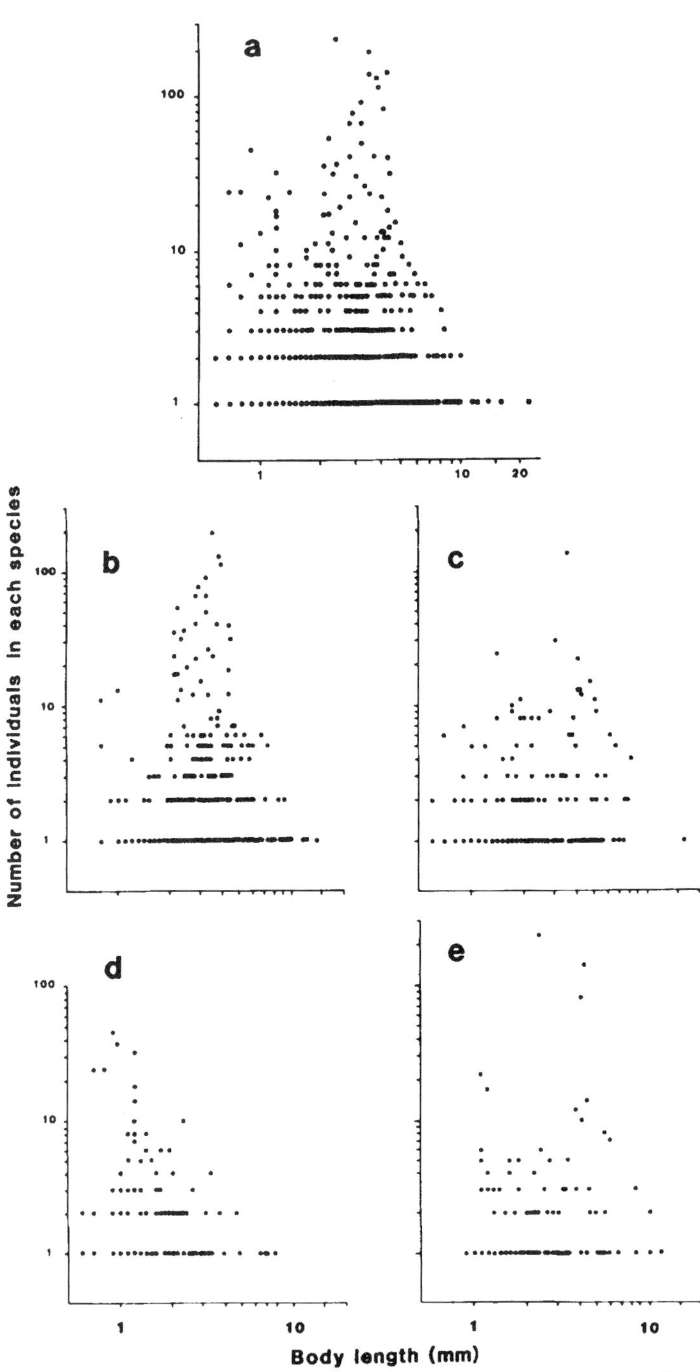

FIG. 5. Scatterplots of number of individuals in each species against body length (i.e. the ZX_1 plane in Fig. 1) for all beetles (a) and individual guilds: (b) herbivores, (c) predators, (d) fungivores, and (e) scavengers. (*N.B.* There are too many data points to illustrate all of them on the figure; each dot represents the position of one or more data.)

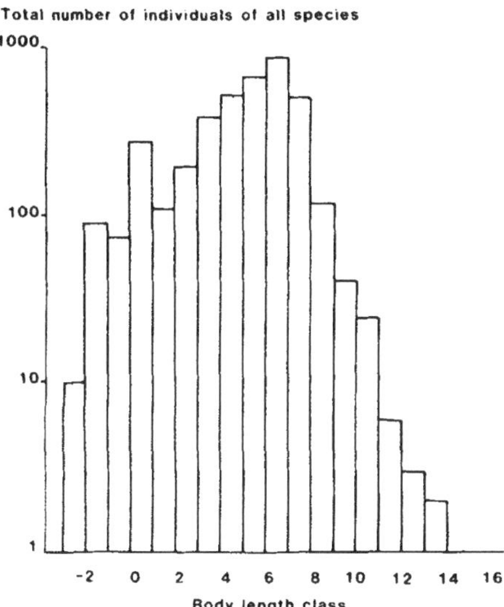

FIG. 6. Histogram of the total number of individuals in different length classes (i.e. the ZX_2 plane of Fig. 1 (ii)). The scale converting length classes to actual length is displayed below Fig. 4(a).

very weak relationships were found in the predators (which have a positive, rather than negative slope, as expected; log length=0.100 log abundance+0.325, $P<0.001$), and the fungivores (log length=−0.129 log abundance+0.229, $P<0.01$) but were not found in the other guilds, nor for all of the species when pooled. In both the predator and fungivore guilds the relationships are weak; only 4% and 0.7% of the variance is explained, respectively.

Total number of individuals: body length distributions

The total number of individuals in the different body length classes are shown in Fig. 6. The upper tail of the distribution from modal length class 6 (a body size of *c*. 3.5–4 mm) shows an almost linear decline in the number of individuals with increasing body length. Similar, but less clear-cut, patterns are evident in the individuals: body lengths graphs for each guild and in the samples taken from the different species of tree (not illustrated). The distributions for individual guilds tend to be skewed relative to the total distribution, with modes in different body length classes (Table 2). Least squares regressions fitted through the upper tails of the total individuals: length distributions (when the data have been logarithmically transformed) are in Table 2. After excluding the scavengers from the analysis, because the residual variance after regression was not homogeneous with that of other guilds (Bartlett's test for homogeneity of variances, χ^2 (with 3 df)=20.54, $P<0.001$), the slopes for the other guilds were found to differ significantly from one another (ANOVA, $F_{2,18}=21.85$, $P<0.001$). The slopes for the individual species of tree also differed significantly from one another (ANOVA, $F_{2,19}=30.91$, $P<0.001$). The slope for *Pentaspadon motleyi* is lower by a factor of more than 2.5 than the slopes of the lines through the data for either of the two *Shorea* species (again, data for trees 9 and 10 were too sparse to justify separate analyses).

TABLE 2. Equations of regression lines fitted through the upper parts of the total individuals: body length graphs for different subsets of the data (i.e. slopes to the right of the mode in Fig. 6). The regressions were fitted by ordinary least-squares techniques after the data had been logarithmically transformed. All data from and to the right of the mode of the distribution were included. The geometric mean of the limits on each size class was used for the abscissa. All regressions are significant at the 0.5% level.

Data set	Slope	SE	Intercept*	SE	Modal size-class
All species	−4.14	0.369	5.20	0.352	6
Herbivores	−4.63	0.213	5.28	0.190	6
Predators	−3.70	0.238	4.28	0.199	6
Fungivores	−2.27	0.248	2.13	0.132	0
Scavengers	−2.54	0.376	3.01	0.552	4
Shorea johorensis	−4.32	0.379	4.70	0.324	5
S. macrophylla	−4.55	0.455	4.92	0.371	6
Pentaspadon motleyi	−1.56	0.208	2.18	0.162	3

* $\log_{10}$ number of individuals at a body-size of 1 mm (based on extrapolation).

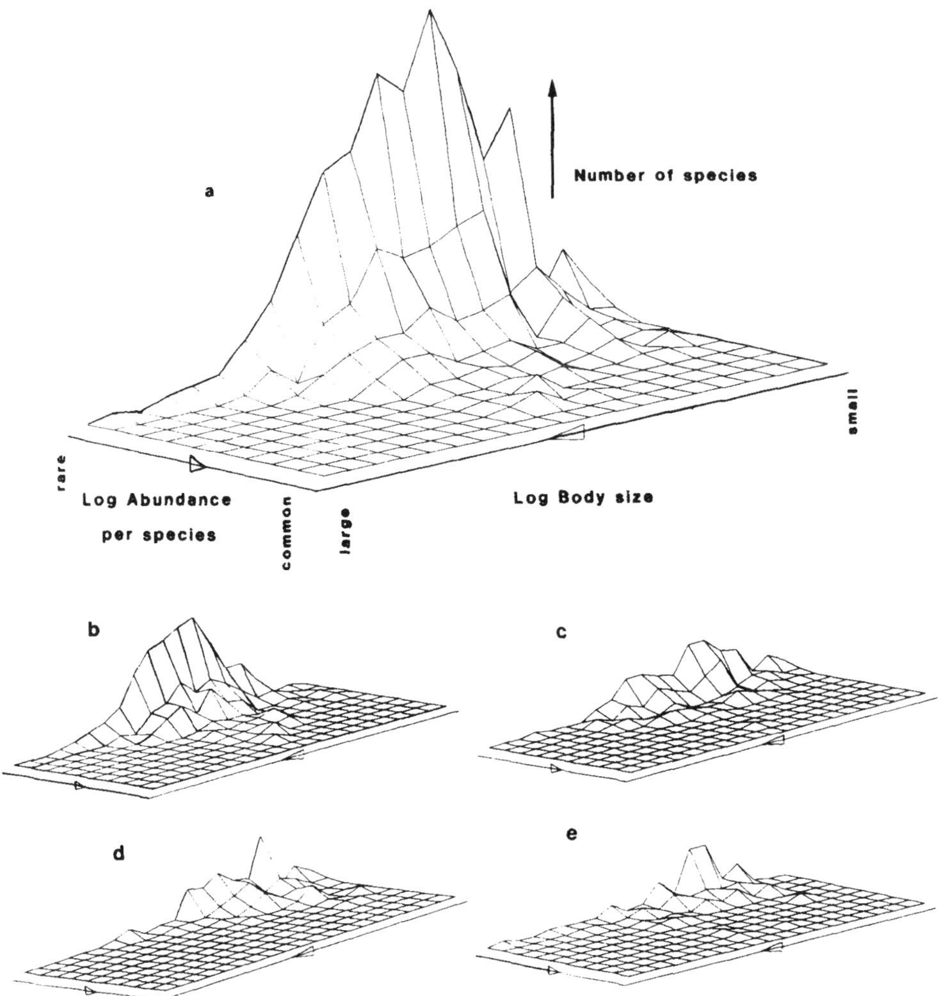

FIG. 7. (a) A three-dimensional graph plotting number of species against each log body length class and against total log abundance class for all beetles, and for each guild: (b) herbivores, (c) predators, (d) fungivores, and (e) scavengers. The height of each intersection is proportional to the number of species having a particular combination of body length in each length class (see Fig. 4(a) for the conversion scale) and abundance in each octave on the conventional $\log_2$ scale of species abundances (Preston, 1962).

Combining the variables

The full three-dimensional plots for all the beetles, and for individual guilds, are in Figs. 7(a–e). Note that, unlike Fig. 4, the *y*-axis in Fig. 7 is not logarithmic.

The full surface is reasonably smooth, although careful inspection shows three 'ridges' running roughly parallel to the X_1-axis in the YX_1 plane (Fig. 7a). These 'ridges' are due to the main and two subsidiary modes in the number of species in different body length classes, with different guilds contributing differentially to each 'ridge' (Figs. 7b–e).

Discussion

Confining the scope of this study to the Coleoptera has both advantages and disadvantages.

The chief disadvantage is lack of generality. It is difficult to extrapolate from beetles to complete arthropod communities. Examination of total number of individuals: body length distributions for arthropods in sweep net samples from two sites in Costa Rica (Fig. 2) reveals that the Coleoptera comprise the major part of the lower body length classes although sweep netting captures a biased sample of arthropods (e.g. Hespenheide, 1979). The beetles make up only 18.0% of the species and 16.5% of the individuals in the complete arthropod samples from the Bornean trees (Stork, unpublished). However, they represent major proportions of four of the guilds of arthropods most closely associated with trees. For instance, the beetles account for a mean of 79.2% of the species and 66.8% of the individual chewing herbivores. For predators the equivalent figures are 50.4% and 45.6%, and for scavengers and fungivores combined, 40.7% and 37.9% respectively (Stork, 1987a). The Coleoptera clearly represent an important cross-section, in terms of feeding habits, of the arthropod fauna in the Brunei rain forest canopy.

We are also aware that insecticide fogging as a method of collecting is not without problems but these are considerably less than for other methods of sampling from the canopy, and beetles appear to be particularly well sampled (Stork, 1987a, and unpublished). Confining attention to adult Coleoptera also has other advantages. One of the problems of examining body length relationships is that species in some groups have individuals with a wide range of body lengths. This is particularly true of Hemiptera and the orthopteroid orders where the adults and nymphs often occur together, are similar in appearance, and have similar feeding habits. For instance, 95% of the Blattodea in the samples were nymphs of a range of instars and hence sizes. Even the adults of some species can vary considerably in size. These problems are of only minor relevance to this study for several reasons. First, most of the adult beetles varied little in size within species. Second, beetle larvae are usually very different in appearance from the adults and, in some groups, are found in different habitats. Third, beetle larvae represented less than 5% of all Coleoptera in the Bornean samples (and were excluded from our analyses).

Taxonomic restrictions aside, there is one other important way in which Fig. 7, and its component parts (Figs. 3, 4 and 5) may give a distorted or indeed totally false impression of the relationships that exist between these variables in more natural and taxonomically diverse faunal assemblages. Our analyses are based on samples of selected trees (see Methods), and are not stratified according to tree abundance, and as we have indicated, different species of tree have rather different patterns (e.g. Tables 1 and 2). Only further studies, designed to overcome taxonomic and sampling biases, can reveal whether the patterns in our total data are serious distortions of patterns in real communities. At the present time we can do no more than present the data to illustrate the nature of the problem, as a spur to others to gather more and better data. The interpretations that follow must be viewed in the light of these major caveats.

It is unclear which of the many available models (see Southwood, 1978, for summary) would be the most appropriate to fit the species: abundance distribution (Fig. 3). Taylor (1978) and Hughes (1986) point out that in small samples both the logarithmic series and the upper section of the log-normal distribution fit equally well, which is also true of the data in Fig. 3(a). Since there is no mode or suggestion of a mode, the true number of beetle species in the Bornean lowland trees must be much greater than the 859 species found, if they have a log-normal distribution. Because there is no mode in Fig. 3(a), it is difficult (e.g. Hughes, 1986) to test the notion (see Introduction; Lawton, 1986) that species of different sizes may have different species: abundance distributions with modes in different abundance classes. Fig. 7(a) simply shows monotonic declines in the numbers of species in each abundance category for beetles of all sizes (i.e. 'slices' in the YX_1 plane look qualitatively similar at all points along the Z axis).

The species: body length distributions for beetles in Bornean rain forest (Figs. 4a–e) are similar to those found by other authors (e.g. Hemmingsen, 1934; Schoener & Janzen, 1968) in that they approximately conform to a log-normal distribution. The slope of -2.64 (± 0.38) 17/32.6x46.9for the upper tail of the distribution for the combined data when plotted on double-logarithmic axes shows a reasonable fit to May's (1978) predicted value of approximately -2. More recent calculations (May, 1986) based on the fractal nature of plant surfaces (Morse *et al.*, 1985) imply slopes lying between −1.5 and −3.0. Data from independent studies of tropical canopy beetles (Erwin & Scott, 1980; Erwin, 1983) and other insect communities (Terakawa & Ohsawa, 1981) also conform reasonably closely to May's (1978) prediction that the number of species scales as body length^{-2}, above some critical minimum body size (see Lawton, 1986). It is unclear whether the total number of arthropod species declines below the modal size class 3 in Fig. 4(a), or whether very small species in groups other than Coleoptera would markedly increase the total number of arthropod species, below a length of about 2 mm (see also May, 1978, 1986).

Several studies (Damuth, 1981; Peters, 1983; Peters & Raelson, 1984; Peters & Wassenberg, 1983; Brown & Maurer, 1986) have found a simple linear relationship between the population densities of individual species and body weight. The graphs presented by these authors are noticeably lacking in data for small, rare species equivalent to a complete lack of points close to the Y-axis on the ZX_1 plane of Fig. 1 (i.e. there is a 'hole' in the data on the 'floor' of Fig. 1(i), furthest from the reader). Then, by definition, a Fig. 1 style three-dimensional plot of their data would result in an incomplete surface, with two implications for the other faces of Fig. 1. First, rare species in a species:frequency distribution (the YX_1 plane) would tend mainly to be large individuals; secondly, only more abundant species would contribute to the shape of the YZ plane for small values of Z (small body size). In marked contrast to these authors, we found a considerable number of small rare species in the ZX_1 plane (Fig. 5) and the three-dimensional surface was complete (Fig. 7). There was no consistent relationship between the average abundance and body size of species (Fig. 5).

There are three possible reasons why the data in Fig. 5 differ from that of earlier studies, two biological, the other an artefact of the way in which earlier data have been assembled. Both biological reasons centre on species mobilities. For example, it is interesting to observe that both Peters & Wassenberg (1983) and Juanes (1986) found only weak and inconsistent relationships between population density and body size for birds. Perhaps the feature that sets both birds and beetles apart from the other groups analysed is that most of their species are winged, highly mobile, and therefore able to encounter one another and to mate at very low population densities (see also Juanes (1986) for alternative biological arguments). Second, high mobility will tend to yield many 'tourists' (*sensu* Moran & Southwood, 1982) in the samples, possibly boosting the number of very rare species.

It is also possible that the clear correlations that exist in earlier studies between population density and body size are sampling artefacts; many of these data were gathered from published studies on the autecology of individual species. Such species may tend to be both commoner and larger than average (few biologists choose to work with small, rare species). Hence data for smaller, rarer species may be under represented in many earlier studies (but see Brown & Maurer, 1986).

Summing individual population abundances for species of a particular size in Fig. 5(a), yields the data in Fig. 6 (see also Fig. 1, showing how the ZX_1 plane gives rise to the ZX_2 graph). Fig. 6 is again roughly log-normally distributed, with a mode at size class 6. Although there is considerable variation in the rate of decline in the total number of individuals with body length to the right of this mode for various beetle guilds (Table 2), the majority of those plots have slopes much greater, sometimes by an order of magnitude, than has been found in other studies of insect communities (Hijii, 1984; Kikuzawa & Shidei, 1967; Terakawa & Ohsawa, 1981; Lawton, 1986; Morse *et al.*, 1985). On average, the Bornean beetle data show that for an order of magnitude decrease in body length, the total number of individual beetles, summed over all species, increases by a factor of approximately 10,000. However, analysis of Janzen's (1973) data (Fig. 2) indicates that restricting our study to the Coleoptera may have had the effect of greatly increasing the slope above the mode, compared with similar slopes for entire arthropod assemblages. Data for rain forest beetles may not therefore be at variance with the model proposed by Morse *et al.* (1985) and Lawton (1986), which predicts a slope of −3.25 to the right of the mode for data of the type displayed in Fig. 6 and Table 2.

The end point of our analyses is the three-dimensional surface displayed in Fig. 7(a). Small ridges run across it, roughly parallel to the YX_1 plane, for species in the smallest body length classes; however, the bulk of the surface is relatively smooth. Each guild (Fig. 7b–e) contributes to the shape of the overall surface in a different way but we have no theoretical basis for interpreting these patterns. It is possible that they are artefacts produced by taking subsamples. Further progress is impossible without even larger samples, which might also be required to carry out separate analyses for each species of tree.

It is unclear whether widening the taxonomic base of our samples, or collecting at different seasons, will alter the shape of the surface. Future empirical and theoretical studies will have to address both points. Because there are no similar data in the literature, we cannot say whether they are typical or unusual. There are hints in the literature (e.g. Griffiths, 1986; Strayer, 1986), that surfaces for other communities may be both more complex and more irregular.

Acknowledgments

We are particularly grateful to colleagues at the BM(NH), who will be fully acknowledged in a later publication, for assisting Nigel Stork in sorting the Coleoptera to species. We thank Paul Harvey, Bob May, Stuart Pimm and Mark Wetton for valuable discussion. Nigel Stork acknowledges the permission of the Sultan of Brunei to study in Brunei, assistance from Jaya bin Sahat of the Brunei Museum, and fieldwork support from members of the Leeds University Expedition to Brunei. David Morse was supported by an NERC studentship. The insecticide was provided by Wellcome Research Laboratories through Peter Chadwick. Peter Hammond, Cliff Moran, Joe Perry and an anonymous referee made helpful comments on the manuscript.

References

Brown, J.H. & Maurer, B.A. (1986) Body size, ecological dominance and Cope's rule. *Nature*, **324**, 248–250.

Damuth, J. (1981) Population density and body size in mammals. *Nature*, **230**, 699–700.

Erwin, T.L. (1983) Beetles and other insects of tropical forest canopies at Manaus, Brazil, sampled by insecticidal fogging. *Tropical Rain Forest: Ecology and management* (ed. by S. L. Sutton, T. C. Whitmore and A. C. Chadwick), pp. 59–75. Special publication no. 2 of the British Ecological Society. Blackwell Scientific Publications, Oxford.

Erwin, T.L. & Scott, J.C. (1980) Seasonal and size patterns, trophic structure and richness of Coleoptera in the tropical arboreal ecosystem: the fauna of the tree *Luehea seemanii* Triana and Planch in the Canal Zone of Panama. *Coleopterists Bulletin*, **34**, 305–322.

Griffiths, D. (1986) Size–abundance relations in communities. *American Naturalist*, **127**, 140–166.

Harvey, P.H. & Godfray, H.C.J. (1987) How species divide resources. *American Naturalist*, **129**, 318–320.

Harvey, P.H. & Lawton, J.H. (1986) Patterns in three dimensions. *Nature*, **324**, 212.

Hemmingsen, A.M. (1934) A statistical analysis of the differences in body size of related species. *Videnskabelige Meddelelser Dansk Naturhistorisk Forening Kobenhavn*, **98**, 125–160.

Hespenheide, H.A. (1979) Are there fewer parasitoids in the tropics? *American Naturalist*, **113**, 766–769.

Hijii, N. (1984) Arboreal arthropod fauna in a forest. II. Presumed community structures based on biomass and number of arthropods in a *Chamaecyparis obtusa* Plantation. *Japanese Journal of Ecology*, **34**, 187–193.

Hughes, R.G. (1986) Theories and models of species abundance. *American Naturalist*, **128**, 879–899.

Hutchinson, G.E. & MacArthur, R.H. (1959) A theoretical ecological model of size distributions among species of animals. *American Naturalist*, **93**, 117–125.

Janzen, D.H. (1973) Sweep samples of tropical foliage insects: description of study sites, with data on species abundances and size distributions. *Ecology*, **54**, 659–686.

Janzen, D.H. & Schoener, T.W. (1968) Differences in insect abundance and diversity between wetter and drier sites during a tropical dry season. *Ecology*, **49**, 96–110.

Juanes, F. (1986) Population density and body size in birds. *American Naturalist*, **128**, 921–929.

Kikuzawa, K. & Shidei, T. (1967) On the biomass of arthropods of the Japanese red pine forest in the vicinity of Kyoto. *Bulletin of Kyoto University Forests*, **39**, 1–8.

Lawton, J.H. (1986) Surface availability and insect community structure: the effects of architecture and fractal dimension of plants. *Insects and the Plant Surface* (ed. by B. E. Juniper and T. R. E. Southwood), pp. 317–331. Edward Arnold, London.

May, R.M. (1975) Patterns of species abundance and diversity. *Ecology and Evolution of Communities* (ed. by M. L. Cody and J. M. Diamond), pp. 81–120. Harvard University Press, Cambridge, Mass.

May, R.M. (1978) The dynamics and diversity of insect faunas. *Diversity of Insect Faunas* (ed. by L. A. Mound and N. Waloff), pp. 188–204. Symposia of the Royal Entomological Society of London, No. 9. Blackwell Scientific Publications, Oxford.

May, R.M. (1986) The search for patterns in the balance of nature: advance and retreats. *Ecology*, **67**, 1115–1126.

Moran, V.C. & Southwood, T.R.E. (1982) The guild

composition of arthropod communities in trees. *Journal of Animal Ecology*, **51**, 289–306.

Morse, D.R., Lawton, J.H., Dodson, M.M. & Williamson, M.H. (1985) Fractal dimension of vegetation and the distribution of arthropod body lengths. *Nature*, **314**, 731–733.

Peters, R.H. (1983) *The Ecological Implications of Body Size*. Cambridge University Press.

Peters, R.H. & Wassenberg, K. (1983) The effect of body size on animal abundance. *Oecologia*, **60**, 89–96.

Peters, R.H. & Raelson, J.V. (1984) Relations between individual size and mammalian population density. *American Naturalist*, **124**, 498–517.

Preston, F.W. (1948) The commonness and rarity of species. *Ecology*, **29**, 254–283.

Preston, F.W. (1962) The canonical distribution of commonness and rarity. *Ecology*, **43**, 185–215 and 410–432.

Rodriguez, J. & Mullin, M.M. (1986) Relation between biomass and body weight of plankton in a steady state oceanic ecosystem. *Limnology and Oceanography*, **31**, 361–70.

Rogers, L.E., Buschbom, R.L. & Watson, C.R. (1977) Length–weight relationships of shrub-steppe invertebrates. *Annals of the Entomological Society of America*, **70**, 51–53.

Rogers, L.E., Hinds, W.T. & Buschbom, R.L. (1976) A general weight vs. length relationship for insects. *Annals of the Entomological Society of America*, **69**, 387–389.

Schoener, T.W. (1980) Length–weight regressions in tropical and temperate forest-understory insects. *Annals of the Entomological Society of America*, **73**, 106–109.

Schoener, T.W. & Janzen, D.H. (1968) Notes on environmental determinants of tropical versus temperate insect size pattern. *American Naturalist*, **102**, 207–224.

Southwood, T.R.E. (1978) *Ecological Methods with Particular Reference to the Study of Insect Populations*. Chapman & Hall, London.

Strayer, D. (1986) The size structure of a lacustrine zoobenthic community. *Oecologia*, **69**, 513–6.

Stork, N.E. (1987a) Guild structure of arthropods from Bornean rain forest trees. *Ecological Entomology*, **12**, 69–80. (Erratum, **12**, 480.)

Stork, N.E. (1987b) Arthropod faunal similarity of Bornean rain forest trees. *Ecological Entomology*, **12**, 219–226.

Sugihara, G. (1980) Minimal community structure: an explanation of species abundance patterns. *American Naturalist*, **116**, 770–87.

Taylor, L.R. (1978) Bates, Williams, Hutchinson—a variety of diversities. *Diversity of Insect Faunas* (ed. by L. A. Mound and N. Waloff), pp. 1–18. Symposium of the Royal Entomological Society of London, No. 9. Blackwell Scientific Publications, Oxford.

Terakawa, N.I. & Ohsawa, N. (1981) On the arboreal arthropods in the Inabu experimental forest of Nagoya University. II. The relationship between body length and numbers of individuals. *Japanese Forestry (Society) Discussions*, **91**, 349–350.

Williams, C.B. (1953) The relative abundance of different species in a wild population. *Journal of Animal Ecology*, **22**, 14–31.

Williamson, M. (1973) Species diversity in ecological communities. *The Mathematical Theory of the Dynamics of Biological Populations* (ed. by M. S. Bartlett and R. W. Hiorns), pp. 325–335. Academic Press, London.

Accepted 6 June 1987

PAPER 17

Theory of Growth Geometry of Plants and Self-Thinning of Plant Populations: Geometric Similarity, Elastic Similarity, and Different Growth Modes of Plant Parts (1988)

R. Å. Norberg

Commentary

KARL J. NIKLAS

Ecologists had long noted that average plant size increases as the number of plants in a population decreases as a result of natural mortality or culling/harvesting. This phenomenon, which is called self-thinning, generally but not invariably follows a $-\frac{2}{3}$-scaling rule (i.e., plant number scales as the $-\frac{2}{3}$ power of average plant size). In his paper, R. Å. Norberg clarifies the quantitative relationships among plant geometry, number of plants per unit area, and average plant size in an effort to develop predictive models for self-thinning. He provides a detailed review of the different scaling exponents reported in the literature to govern self-thinning, outlines an extended mathematical derivation for plant geometry, and shows how this geometry affects the exponent governing self-thinning. An important contribution made by this study is the recognition that plants change their geometry as they grow in size, particularly trees which have trunks best described by the rules of geometric similarity early in their ontogeny and later develop trunks described at best by elastic similarity formulas. Despite Norberg's clarification of the self-thinning rule, the debate over which (if any) scaling exponent governs the relationship between plant number and average size continues, particularly in light of claims that a single "canonical" exponent holds true.

From *American Naturalist* 131:220–56. *The American Naturalist* © 1988 The University of Chicago. Reprinted with permission from The University of Chicago.

Vol. 131, No. 2 The American Naturalist February 1988

THEORY OF GROWTH GEOMETRY OF PLANTS AND SELF-THINNING OF PLANT POPULATIONS: GEOMETRIC SIMILARITY, ELASTIC SIMILARITY, AND DIFFERENT GROWTH MODES OF PLANT PARTS

R. Åke Norberg

Department of Zoology, University of Gothenburg, Box 250 59, S-400 31 Gothenburg, Sweden

Submitted May 7, 1984; Revised January 30, 1986; Accepted November 25, 1986

In plant populations subjected to crowding, there is a strict relationship between the average plant size and the population density. The same rule, or very similar rules, seem to govern the relationships between size and density observed when comparisons are made across (1) successive growth stages of plants in a single population, (2) different growth stages of plants in different populations of the same species, and (3) populations of different species, with the plants in any growth stage, provided that all populations are crowded. This applies to terrestrial plants of profoundly different form and structure, spanning the entire size range from mosses to trees. In this paper, I treat mathematical relationships between design principles of plants for structural strength, the growth mode of individual plants, packing geometry of plants, and population thinning resulting from individual growth and ensuing competition.

In an even-aged plant population in dynamic equilibrium at some crowding density, the power function

$$V = kN^{\alpha} \tag{1}$$

relates the average volume per plant, V, to the population density, N, which declines over time as plants die off along with individual growth of survivors. A proportionality constant, k, varies with the species; the power α usually assumes values close to $-3/2$, which it does for terrestrial plants as different as mosses, ferns, grasses, herbs, and trees (Yoda et al. 1963; White and Harper 1970; Gorham 1979; White 1980). This so-called $-3/2$ thinning law describes the way the average volume of surviving plants increases as plant density in a stand decreases because of self-thinning, as a result of growth-related competition. As plants grow, the data points representing consecutive growth stages of survivors in a log-log diagram lie along this thinning line, from a high population density and small average volume at lower right, upward and leftward toward the ordinate (fig. 1). Any specific thinning line describes a law of constraints; data points for volume versus density do not occur above it.

When plant volumes are well below this thinning line of constraints, individuals grow with little or no mortality from competition. Therefore, data points for

Am. Nat. 1988. Vol. 131, pp. 220–256.

volume versus density of growing plants usually fall along trajectories much steeper than the limiting line. But when approaching the limiting thinning line, they bend off and subsequently trace it (fig. 1; White and Harper 1970; Kays and Harper 1974; Westoby 1976; White 1981). The common asymptote of these growth trajectories constitutes the limiting self-thinning line.

The dynamic equilibrium between individual growth and mortality, represented by the limiting thinning line, is identified by a lack of further flattening of growth trajectories of individual populations and a lack of further steepening of the "competition-density" lines, like $t_0 - t_3$ in my fig. 1 (term from White 1981, fig. 1), each one of which connects equal-age points for different populations. Depending on the kind of plant, the line's elevation (the y-intercept, k) in a log-log diagram may differ by a factor of eight on a linear scale (White 1981, p. 494), but the exponent usually remains close to $-3/2$. This power law applies to time trajectories for single-species populations and probably also for mixed-species populations (White and Harper 1970). Most notably, it applies to across-species regressions of different-sized plants (Yoda et al. 1963; Gorham 1979; White 1980).

The $-3/2$ thinning rule is one of the most important principles in plant ecology (White 1980), considered by J. Harper to be "the only generalization worthy of the name of a law in plant ecology" (cited in Hutchings 1983, p. 765). Derived empirically, it has subsequently been corroborated by new empirical data, even though many empirical studies have produced important deviations from the $-3/2$ exponent (Weller 1987). The theoretical explanation of the $-3/2$ thinning rule is disputed (Westoby 1977, 1984; White 1977, 1980, 1981; Miyanishi et al. 1979), and it has recently been described as an ecological law in search of a theory (Hutchings 1983). Plants of radically different structure and shape converge on the $-3/2$ thinning law, or on one or another of a few notable deviations from it, such as exponents near -1.8 for many trees, and exponents of about -1 for stands of big trees (White and Harper 1970). Therefore, the causal basis of the thinning rules does not depend on details of plant structure and shape, but rather on general, geometric principles of growth and packing.

The purpose of this paper is to clear up significant misunderstandings in previous discussions about relationships between volumetric growth and growth in linear dimensions, and to explore the causal basis for the $-3/2$ law and for deviations from it. I also show that different plant parts must often exhibit different growth modes (for functional reasons); therefore, the thinning function for whole plants must often be a combination of different functions for different plant parts (although in earlier models one growth mode has, sometimes erroneously, been applied to the whole plant). In addition, this paper attempts to derive thinning functions for stems, branches, and leaf canopies and for old trees, to derive a thinning function for plants growing according to the elastic-similarity principle (which has never been done before), and to show that when root competition drives population thinning, the geometry of aerial structures may easily be at odds with the thinning function for aboveground plant parts (which might explain perplexing discrepancies between population thinning and the allometry of aboveground plant parts). I also expose an allometry artifact from measuring practices in forestry, compare predictions from various models against

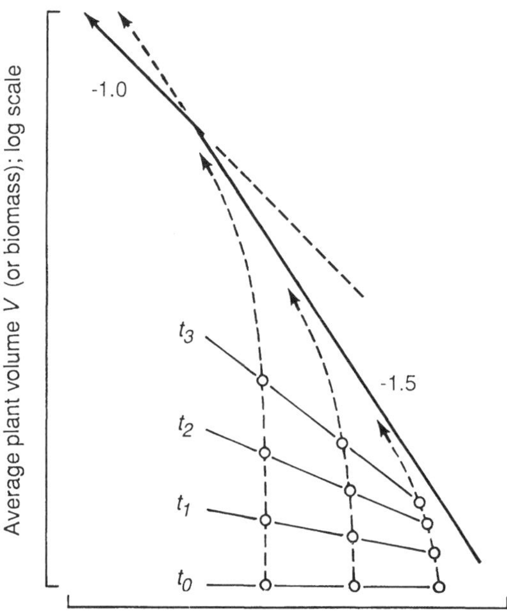

FIG. 1.—Generalized diagram showing two limiting thinning lines from the power function $V = kN^{\alpha}$, relating average plant volume, V, to plant population density, N, with α taking the values -1.5 and -1, which become gradients of the lines in this log-log plot. The -1.5 slope is consistent with geometric similarity of growth, but may be mimicked by other growth modes. ***Broken arrows***, Growth trajectories for individual populations, from various initial population densities, upward to the constraining line of slope -1.5. The "competition-density" lines labeled t_0–t_3 connect data points representing the same age of different populations with different initial densities. The -1 line represents constant yield (see legend to fig. 2). (Idealized diagram based on White and Harper 1970; Kays and Harper 1974; Westoby 1976; White 1981.)

empirical data in the literature, and finally, critically examine earlier models (and expose in some of them internal inconsistencies between empirical elements and theoretical relationships built into the same semi-empirical model). I focus on explanations of the gradient of the thinning curve in a log-log diagram (α in eq. 1), using packing theory and various growth modes, but I also consider briefly the nature of the coefficient k.

APPROACH AND BASIC ASSUMPTIONS

On the basis of the shape, structure, and growth mode of individual plants, I develop models of the self-thinning of plant populations, whereby I attempt to explain population processes from characteristics of individuals (figs. 1, 2). It is, of course, legitimate to develop a model from empirical data and then to test its predictions against the same data (as done in Mohler et al. 1978; White 1981),

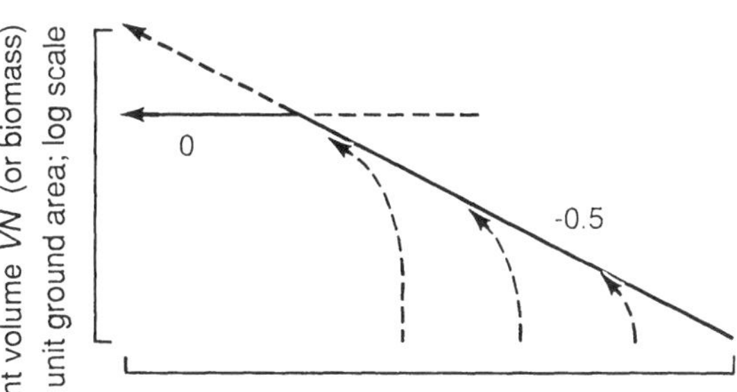

FIG. 2.—The same generalized diagram as figure 1 but with plant volume per unit of ground area plotted against population density. The slopes change from −1.5 to −0.5 and from −1 to 0. The 0 line represents constant yield; that is, the growth in biomass of individual plants is exactly compensated for by the concomitant population thinning, leaving a constant plant biomass per unit of ground area.

provided that the purpose is to determine whether the essential variables have been identified, that is, to check for internal consistency. But critical tests of the predictive power of such an empirically based model must use independent data, and the various relationships must be understood for the model to be explanatory.

With a theoretical model, it is imperative that as few empirical elements as possible are built into the equations; otherwise, it may become descriptive only, albeit with predictive power, but with little explanatory power. Empirical data should instead be reserved for tests of model-generated predictions, even though data may have exposed the problem in the first place, as with the $-3/2$ power law. Here I take a strictly theoretical approach and derive various thinning functions from mechanical and geometric principles, without recourse to empirically derived regression equations. Model-generated predictions are then critically evaluated against empirical information.

The space occupied by a plant may be thought of schematically as a hexagonal prism resulting from close packing (like a honeycomb cell; fig. 3). Neighboring plants claim similar spaces, bordering on each other. The ground area of such a space may be the projected area of the crown or root, but it might as well be any larger or smaller area from which the plant draws resources (light, water, and nutrients) to the exclusion of others. Even if branches or roots of adjacent plants intermingle in shared zones of mutual overlap, one may still assign to each plant an average space, the proportions (height-to-breadth ratio) of which are determined by the species-specific and site-dependent amount of crowding that the plant can tolerate. The ground area occupied by the average plant may be estimated simply as the plot area divided by the number of plants. The exact shapes of a plant's exclusive space and ground area need cause no concern as long as the shapes do not change systematically along with plant growth.

Henceforth, I use the terms "exclusive space" and "exclusive ground area" for the spatial volume and ground area occupied by an average plant growing in a crowded population undergoing thinning. These designations are also relevant when the projected ground areas of adjacent plants overlap. I assume that radius

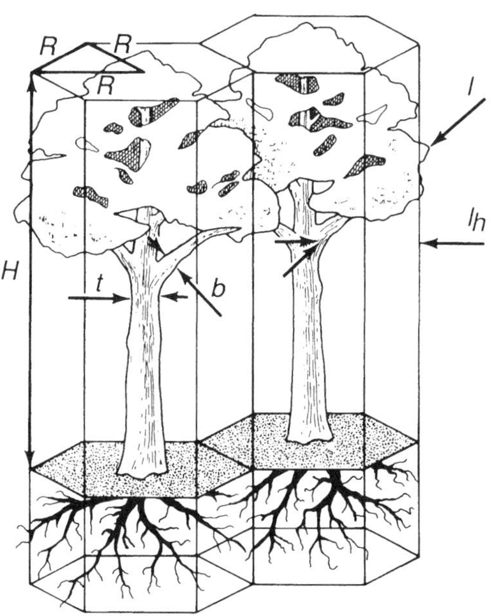

FIG. 3.—Scheme of the geometry of closely packed plants. R is the radius of the exclusive ground area occupied by an average plant in a crowded population; H is height; t, trunk diameter; b, branch diameter; l, branch length; and l_h, the horizontal projection of branch length.

R of the exclusive ground area occupied by an average plant remains directly proportional to the lateral extension of the plant's crown or root system (fig. 3). This implies a continuous canopy and/or root cover, with neighboring plants interacting competitively.

I consider the volume of an average plant in the population as a measure of plant size. Volume, rather than mass, is the natural choice because these models are based on geometry, with various linear dimensions and areas as other key variables. In general, however, plant mass M and volume V should be related linearly; $M \propto V^{1.0}$, as for trees (Ogino 1977 and Saito 1977, cited in White 1981). As long as the specific gravity of plants does not change systematically with growth, the gradients of the thinning lines will be the same, regardless of whether volume or mass is used, but the elevation changes with changes in the measuring unit. Therefore, volume and mass are interchangeable in the following treatment, albeit after adjustment of the coefficient k (the y-intercept) as required.

Whole fractions ($-3/2$, $8/3$, etc.) are commonly used for exponents in theoretical results and decimal fractions (-1.50, 2.67, etc.) for empirical relationships (McMahon and Bonner 1983, p. 55; Peters 1983, p. 40).

GEOMETRIC SIMILARITY: THE $-3/2$ SLOPE

As a convenient basis for further developments, I first give an account of close packing of geometrically similar plants based on the original idea of Yoda et al.

(1963), who derived the −1.5 thinning exponent from empirical data and also explained it geometrically (later elaborated in White and Harper 1970).

If plants retain isometry throughout growth in such a way that main structures like stems, branches, and roots remain geometrically similar, then the height-to-breadth ratio of a plant's exclusive space also remains constant; that is, height H remains a fixed multiple, h, of $2R$, the diameter of the ground area occupied per plant:

$$H = h2R \qquad \text{or} \qquad h = H/2R. \tag{2}$$

Then, in general,

$$V_1 = k_1R^3, \tag{3}$$

where V_1 is volume of plant material and k_1 is a conversion coefficient incorporating h. When plants are densely packed, their number per unit of ground area, N_1, is inversely related to the average ground area occupied per plant; thus,

$$N_1 = k_2R^{-2}, \tag{4}$$

where k_2 is a conversion coefficient that varies with the unit of area chosen for expressing density N_1. Rewriting equation (4) for our purposes,

$$R = k_2^{1/2}N_1^{-1/2}; \tag{5}$$

inserting this into equation (3) and substituting k_3 for $k_1k_2^{3/2}$, we have

$$V_1 = k_3N_1^{-3/2}, \tag{6}$$

which is the relation so often found empirically.

For equation (6) it is necessary and sufficient that plants retain geometric similarity throughout growth and that crowding prevails, leading to self-thinning. It follows that plant volume is a cubic function of any representative linear plant dimension (eq. 3). Geometric similarity may apply to most plant parts, to above-ground structures as well as to roots, but leaves and petioles of most trees do not grow with the tree and therefore violate any growth mode adopted by the trunk and branches (see the section "Leaf volume, per tree and per unit of ground area, versus tree population density").

This neat explanation, dating back to Yoda et al. (1963), has not gained general acceptance mainly because it is not compatible with the allometric relationships of trees (White 1981) and because it fails to account for certain departures from the −3/2 rule. For many tree species, the volume-versus-density line has slopes as steep as −1.8 to −1.7 during periods of high biomass productivity (White and Harper 1970, pp. 475, 477; White 1980, p. 39), whereas in stands of very old and large trees the slope tends to shift to −1 (fig. 1; White and Harper 1970, p. 478). Moreover, the −1.5 slope has been observed in prostrate plants to which it is not expected to apply (Westoby 1976).

Exploring such deviations from a general rule may be extremely useful for identifying neglected constraints. Here I suggest explanations for some of the deviant slopes (except for prostrate plants, which are treated separately elsewhere; Norberg, MS). All models are based on competition and packing theory but rely on growth modes other than geometric similarity.

The Coefficient 'k' and Exponent α

The coefficient k_1 in equation (3) is a conversion factor between volume of plant material and R, the radius of the ground area of the plant's exclusive space. The height-to-breadth ratio h (a fixed multiple of diameter $2R$; eq. 2) of a plant's exclusive space is incorporated in k_1, which also contains a packing constant. The packing constant incorporated in k reflects the proportion of the space occupied by a plant that is made up of plant biomass. Packing is determined by, for instance, the girth and taper of stem and branches as well as by the size and number of leaves. The more densely that plant parts are packed within the plant's exclusive space, and the more solid they are, the larger k_1 becomes. In equation (6), k_1 is incorporated in k_3 and $N_1^{-3/2}$ is just another expression for R^3 (eqs. 3, 5). The various k's also adjust the equations to the particular measuring units used unless they match from the beginning. In contrast to k, α in equations (1) and (6) reflects the rate by which V_1 increases with decreasing N. This is why α depends critically on the growth mode of plant parts.

In a log-log graph of average plant volume against population density, the coefficient k determines the elevation of regression lines with the same slope α. Different k's may be compared at the y-intercepts of the lines at population density 1 (since $\log 1 = 0$). As long as the slope is the same for the lines being compared, it need cause no concern that the y-intercept is often outside the observation range and therefore representing hypothetical plant volumes at population densities that may never occur; with identical slopes, the ratios between the elevations of different lines remain identical regardless of the x-coordinate chosen for a comparison. Even when slopes differ, y-intercepts are useful for descriptive purposes, but biological comparisons and interpretations must then be restricted to some specified x-coordinate, preferably within the range of observation. Differences in elevation, at a particular x-coordinate, between lines with different slopes cannot be ascribed solely to differences in k, but result from the joint action of k and α.

According to White (1981, p. 497), plant species that conform to the same limiting, thinning function $V = kN^{\alpha}$ (eq. 1), where k is common to all and α equals -1.5, may all have the same density of biomass within the spatial volume occupied per individual, be they anything from herbs to trees. But this statement is valid only under one particular condition, not realized before. This condition is that the height-to-breadth ratio (h in eq. 2) of a plant's exclusive space is identical for all the plants being compared, because, for a given population density N_1 (i.e., with a given exclusive ground area per plant), the same value of V_1 can be produced by a large number of different combinations of height-to-breadth ratios and packing constants, which are both constituent factors of k_3 in equation (6).

White (1981, p. 497), paraphrased by Lonsdale and Watkinson (1983, p. 293), suggested that the relation between population density, N, and the entire spatial volume v occupied by an average plant (as opposed to volume of plant material only) could be reduced to $v = N^{-3/2}$, "the most economical statement possible between the two variables." Again, the height-to-breadth ratio h (= $H/2R$) of the plant's exclusive space is overlooked; when considering spatial volume v, rather than volume V_1 of only plant material, the packing constant must be removed from k (as has been done), but for the remaining k_v to equal one (as implied by White's equation), h must also equal one, which is usually not true. It is strictly true that $v = N^{-3/2}$ only when plants occupy cube-shaped spaces (with side s), because $v = N^{-3/2}$ implies $v = s^3$ since $N = s^{-2}$ and, hence, $s = N^{-1/2}$. Since h usually takes values other than one (consider grasses and coniferous trees), k_v must be introduced into White's $v = N^{-3/2}$, even when measuring units match. And k_v must incorporate not only h (= $H/2R$) but also a shape factor of the exclusive ground area, even though the latter is less important, since it is near one; with a hexagonal ground area and with a height-to-breadth ratio, h, of one, the coefficient k_v is 1.24 (with R and H as defined in fig. 3 and eq. 2). Differences in the height-to-breadth ratio of the exclusive space occupied by different species thus give different elevations of the thinning line that results when the exclusive spatial volume per plant is graphed against population density. But the slope for within-population comparisons is not affected by the height-to-breadth ratio of the exclusive space or by the shape of the exclusive ground area as long as they both remain constant throughout growth (which amounts to geometric similarity of the exclusive spatial volume and a slope of $-3/2$). White's (1981, p. 497) corollary to $v = N^{-3/2}$, that the total spatial volume, V_S, occupied by plants per unit of ground area is $V_S = vN = N^{-3/2}N = N^{-1/2}$, must also be modified by the same constant k_v, such that $V_S = k_vN^{-1/2}$.

It is important to note that in an across-species regression with a given k when $\alpha = -1.5$, it is still eminently possible for the plants to be geometrically dissimilar, as indeed they often are. Data points for different plant species fall along this line despite differences in plant size, structure, and form, if only the height-to-breadth ratio is the same (throughout growth and for all species) and the same proportion of the spatial volume occupied per plant is filled with plant material (throughout growth and for all species), or if these two variables change compensatorily. In this way, the geometric-similarity effect on α (turning it into $-3/2$ in across-species regressions) may be mimicked despite geometric dissimilarities among plants. Conversely, with given k and height-to-breadth ratio, constancy of packing density (within the spatial volume occupied per plant) obtains only when the exponent is -1.5, because only then do plant parts fill the same proportion of the spatial volume occupied per plant, despite differences in plant size.

Lonsdale and Watkinson observed that grasses tend to have higher k values (y-intercepts) than dicotyledons, and coniferous trees higher k values than deciduous trees. Their explanation is that "Presumably the narrow, erect leaves of grasses . . . and the pyramidal profile of conifers or, again, their narrow leaves improve the efficiency of light interception and allow close packing of plant material into a given volume of space or the development of a greater stand height before thinning commences . . . . It appears, then, that plant geometry and in particular leaf shape may have an influence on the position of the thinning line." (1983, pp. 291, 292.) However, it is obvious that differences in height-to-breadth ratios, h, throughout the entire growth and thinning process are of overriding importance for the higher k values of grasses and conifers, as compared with dicotyledons and deciduous trees (h being contained in k). The exclusive space occupied by an average grass plant has a larger height-to-breadth ratio than most dicotyledons. Even if grass leaves may reach far laterally, leaves from neighboring plants usually intermingle in shared zones of mutual overlap, such that the height-to-breadth ratio of the exclusive space per plant remains large, a propensity probably following from foliage thinness. Moreover, conifers tend to have larger height-to-breadth ratios than deciduous trees, resulting directly in higher elevation of the thinning line (provided that the packing density of plant material is not compensatorily low).

Actual k values may differ by a factor of 8 (on a linear scale) among different plants, corresponding to a range of 0.9 on a log scale (obtained from the range 3.5–4.4 for $\log k$ with a particular set of measuring units; White 1981, pp. 475, 494). This may be considered a narrow range of k in view of the enormous diversity in plant geometry throughout the size range of plants, from mosses to trees. Constraints on the range of k probably relate to the necessity to grow in height because of competition for light, to demands on structural strength against the force of gravity (setting a lower limit on k), to the necessity for claiming a sufficiently large exclusive ground area in relation to the living biomass to be sustained with light, water, and nutrients, and to shade tolerance, which limits foliage packing and amount of overlap tolerable from neighbors (setting upper bounds on k).

Geometric Similarity and Branches

A problem, hitherto overlooked, with geometric similarity and plant growth is that the maintenance of geometric similarity throughout growth requires a constant number of branches and the subjection of individual branches, and their spatial separation, to only isometric enlargement while their proportions remain constant. But new branches obviously develop as plants grow taller, and their spacing may remain more or less constant, thereby violating the geometric-similarity criteria.

With geometric similarity maintained throughout growth, volume V_b of an individual branch varies with its length, or R, as

$$V_b \propto R^3 . \tag{7}$$

If the number of branches increases linearly with plant height, which is proportional to R, then the total branch volume ΣV_b per plant increases as

$$\Sigma V_b \propto R^4 . \tag{8}$$

With equation (5), total branch volume per plant may be related to plant population density as

$$\Sigma V_b \propto N^{-2} . \tag{9}$$

In reality, however, many plants shed low branches as new ones grow at the top, and self-thinning usually occurs among branches in a growing crown. This process is governed by light shortage near the ground and inside crowns. Therefore, after a plant has exceeded some critical minimum size and crown density, self-shading may keep the branch number down with the result that branch volume conforms roughly to the geometric-similarity model, which dictates that

$$\Sigma V_b \propto N^{-3/2} \tag{10}$$

(from eqs. 5, 7).

In conclusion, violation of the geometric-similarity constraint on branch number (increasing rather than remaining constant) might help explain why some plants that may retain geometric similarity in other respects show thinning exponents between -2 and -1.5 rather than exactly -1.5 for volume of whole plants versus population density. But this explanation does not apply to empirically observed slopes for trees, because measured tree volumes are usually for stem wood only, exclusive of branches. The elastic-similarity model seems better able to explain exponents between -2 and -1.5 in trees, as explained below.

ELASTIC SIMILARITY IN INDIVIDUAL GROWTH AND CONCOMITANT SELF-THINNING OF POPULATIONS

A Model

The elastic-similarity principle is an obvious alternative to geometric similarity (isometry) for the growth of individual plants. Elastic similarity may be expected whenever structural strength against the force of gravity is of overriding importance, as in large organisms like trees. Many tree species also seem to retain elastic, rather than geometric, similarity throughout growth (McMahon and Kronauer 1976).

When the elastic-similarity principle applies to different-sized, tapered, cantilever beams, fixed and supported at one end only, and set at the same angle, just as branches are, a characteristic feature is that the distance any one of them deflects under its own weight is always a constant fraction of its length. This is achieved by a taper ratio dictating that the diameter of a branch increases as the 3/2 power of its length (McMahon and Kronauer 1976). The same applies to the trunk: Greenhill (1881) showed that the critical diameter below which a vertical column, or a tree trunk, buckles under its own weight increases with the 3/2 power of the height, a relation later elaborated by McMahon and Kronauer (1976). The critical diameter is that for which the tendency for the trunk to buckle is exactly counterbalanced by its elastic tendency to return to an upright position (McMahon 1975, p. 99). Therefore, when elastic similarity prevails, longer branches and taller trunks are proportionately thicker than shorter ones. By maintaining this disproportionate increase in diameter as trunk height and branch length increase, the structures maintain a constant safety factor against buckling under their own weight. The elastic-similarity criterion, therefore, is a constant ratio between the critical buckling force and the gravitational force, or weight, for different-sized structures (Hokkanen 1986).

I now derive the thinning characteristics expected for populations of trees growing according to the elastic-similarity principle. The populations are assumed to be crowded and their densities subject to size- and density-dependent control.

The slope of the limiting thinning line, resulting from elastic similarity, would ideally be defined by finding a value for α in equation (1). But as I show below, it is, unfortunately, not possible to describe self-thinning under elastic similarity with the convenient power function (eq. 1) applicable to geometric similarity.

Although stem-wood biomass is the primary interest in forestry, branch size must also be considered (even if its volume is disregarded), because crown width affects tree spacing and therefore the thinning function. The longest branches are the most important since they contribute the most to canopy diameter. To render the problem accessible to a general treatment (without recourse to specific data for any particular species), some assumption must be made about the relation between trunk diameter and branch diameter, as measured at some characteristic position along their respective lengths (i.e., at some chosen proportion of their lengths away from the base). I take branch diameter b to be proportional to trunk diameter t throughout individual growth and for different-sized plants as well:

$$b = k_4 t . \tag{11}$$

A corollary of this isometry between b and t is that they have the same relative growth rate. Thus, yearly increments Δb and Δt in branch and trunk diameters (i.e., additions from an annual growth ring) are assumed to be related to branch and trunk diameters as

$$(b + \Delta b)/(t + \Delta t) = b/t; \qquad \Delta b/\Delta t = b/t . \tag{12}$$

Equations (12) state the obvious null hypothesis for the relationships between branch and trunk diameters and their yearly increments. Equation (11) is also a necessary consequence of the elastic-similarity criterion, which requires all diameters, and all lengths, of an organism to change with total size in the same way (McMahon 1984, p. 264). It must be stressed that equations (11) and (12) show assumed relations, but there is empirical support for equation (11) (Mohler et al. 1978; see the Discussion, "Branch length versus trunk diameter," below). Equation (12) is consistent with the observation that the growth rate in the diameter of tree trunks (width of an annual growth ring) is linearly related to trunk diameter in dominant trees but not in suppressed, nongrowing trees that are being outcompeted (West and Borough 1983, p. 151). It has also been shown that branches maintain elastic similarity (McMahon and Kronauer 1976), although this needs to be confirmed for trees growing in crowded populations actually undergoing self-thinning. Moreover, the assumption in equation (11) need cause no concern, because, as it turns out, the model derived is rather insensitive to branch size. Thus, for example, the predicted slope changes only from -1.97 to -1.80 when the ratio of the length of the branch to the radius of the trunk changes from 33 to 3. This property adds strongly to the generality of the following model (eq. 22).

Elastic similarity dictates that

$$H = k_5 t^{2/3} , \tag{13}$$

$$l = k_6 t^{2/3} \tag{14}$$

(from eq. 11), and

$$l_{\mathrm{h}} = k_7 t^{2/3} , \tag{15}$$

where H is plant height, l branch length, and l_h its horizontal projection (fig. 3). Therefore, branch length becomes proportional to trunk height, just as with geometric similarity (following from the assumption that branch diameter is proportional to trunk diameter; eqs. 11, 13, 14). But crown width is determined also by trunk width, and because the trunk grows disproportionately wide when elastic similarity prevails, the height-to-width ratio of the tree's outline decreases with tree growth (rather than remaining constant as with geometric similarity). The decrement in a tree's height-to-width ratio is small, however, since it is a result of the amount by which the trunk diameter associated with elastic similarity may exceed the trunk diameter associated with geometric similarity at any particular growth stage.

When elastic similarity prevails, volume V_2 of the trunk becomes $V_2 \propto t^2 H$, or, from equation (13),

$$V_2 = k_8 t^{8/3}, \tag{16}$$

and similarly for individual branches. For equation (16) to apply also to the volume of branches of a whole plant, the branch number needs to remain constant throughout growth (for an alternative, see the next section). Radius R of a plant's exclusive space is determined by trunk radius (at the level of the longest branches) and the horizontal projection of the longest branches (eq. 15),

$$R = 0.5t + k_7 t^{2/3}. \tag{17}$$

From equations (4) and (17), the crowding population density N_2 of plants maintaining elastic similarity throughout growth is

$$N_2 = k_2 R^{-2} = k_2(0.5t + k_7 t^{2/3})^{-2}. \tag{18}$$

In order to compare the respective thinning gradients associated with elastic and geometric similarity, it is necessary to know the slope $d\log V_2/d\log N_2$ of the limiting thinning curve, following upon elastic similarity, as plotted in a log-log diagram of volume versus density. By the chain rule,

$$d\ln V_2/d\ln N_2 = (d\ln V_2/dt)\,(dt/d\ln N_2). \tag{19}$$

Taking logarithms of equation (16), $\ln V_2 = \ln k_8 + (8/3)\ln t$, and differentiating with respect to t yields

$$d\ln V_2/dt = 8/(3t). \tag{20}$$

Taking logarithms of equation (18), $\ln N_2 = \ln k_2 - 2\ln(0.5t + k_7 t^{2/3})$, and differentiating with respect to t yields

$$d\ln N_2/dt = -2\,[0.5 + (2/3)k_7 t^{-1/3}]/(0.5t + k_7 t^{2/3});$$

inverting to fit equation (19), we have

$$dt/d\ln N_2 = -0.5(t + 2k_7 t^{2/3})/[1 + (4/3)k_7 t^{-1/3}]. \tag{21}$$

Combining equations (19), (20), and (21) gives

$$\frac{d\ln V_2}{d\ln N_2} = \frac{8}{3t}\left[\frac{-0.5(t + 2k_7 t^{2/3})}{1 + (4/3)k_7 t^{-1/3}}\right] = \frac{(-4/3)(t^{1/3} + 2k_7)}{t^{1/3} + (4/3)k_7}. \tag{22}$$

By the same procedure, the gradient for the volume V_2N_2 of plant material per unit of ground area versus plant population density N_2 becomes

$$\frac{d\ln(V_2N_2)}{d\ln N_2} = \frac{-(4/3)(t^{1/3} + 2k_7)}{t^{1/3} + (4/3)k_7} + 1\,, \tag{23}$$

(using eq. 18 to replace N_2 in $V_2N_2 = N_2k_8t^{8/3}$, which follows from eq. 16).

From equations (22) and (23), it follows that as t approaches 0,

$$d\ln V_2/d\ln N_2 \rightarrow -2\,, \tag{24}$$

and

$$d\ln(V_2N_2)/d\ln N_2 \rightarrow -1\,; \tag{25}$$

as t approaches infinity,

$$d\ln V_2/d\ln N_2 \rightarrow -4/3\,, \tag{26}$$

and

$$d\ln(V_2N_2)/d\ln N_2 \rightarrow -1/3\,. \tag{27}$$

For $d\ln V_2/d\ln N_2$ ever to be smaller than -2 or larger than -1.33, its derivative must equal zero in at least one point (since -2 and -1.33 are for the hypothetical, extreme t values of 0 and ∞; fig. 4). To investigate this possibility, insert $u = 2k_7t^{-1/3}$ into equation (22):

$$d\ln V_2/d\ln N_2 = (-4/3)(1 + u)/[1 + (2/3)u]\,. \tag{28}$$

Differentiation with respect to u gives

$$\frac{d}{du}\left(\frac{1 + u}{1 + (2/3)u}\right) = \frac{1 + (2/3)u - (2/3)(1 + u)}{[1 + (2/3)u]^2} = \frac{1/3}{[1 + (2/3)u]^2} \neq 0\,. \tag{29}$$

Since the derivative never equals zero, $d\ln V_2/d\ln N_2$ is confined within the range -2 to $-4/3$, where it increases monotonically from -2 (when $t \rightarrow 0$) to $-4/3$ (when $t \rightarrow \infty$; fig. 4). Likewise, the gradient $d\ln(V_2N_2)/d\ln N_2$ of the curve for total plant volume per unit of ground area versus population density increases monotonically from -1 to $-1/3$ (fig. 5).

The above derivation shows that when elastic similarity prevails throughout individual growth, self-thinning of populations proceeds along a curve, rather than along a line, in a log-log diagram of volume versus density. When trunk diameter t is small and population density N is high, the gradient of the limiting thinning curve is close to -2, with a theoretical minimum at -2 (relation 24). As trunk diameter increases and population density decreases, the gradient increases monotonically, with a theoretical maximum at -1.33 (relation 26). Gradients outside the range of -2 to -1.33 do not exist (from relation 29; fig. 4). The exact gradient for a specific population may be estimated for any t when k_7 in equation (22) is known. The k_7 depends on the branch angle with the horizontal (eq. 15) and also on k_6, containing a proportionality constant between branch length and diameter (eq. 14), as well as on k_4, a proportionality constant between branch and trunk diameters (eq. 11).

A more convenient route to predicting gradients of thinning curves for specific populations under crowding is to use the ratio ζ between average trunk radius $0.5t$ (at the height of the longest branches) and average horizontal projection $k_7t^{2/3}$ of the longest branches (eq. 15), or half the average nearest-neighbor distance (measured between the trunk surfaces at the height of the longest branches). This ratio is the only information needed to predict thinning gradients from the elastic-similarity model (eq. 22). Inserting

$$\zeta = 0.5t/k_7t^{2/3} = t^{1/3}/2k_7 \tag{30}$$

into equation (22) gives

$$\frac{d\ln V_2}{d\ln N_2} = \frac{(-4/3)(\zeta + 1)}{\zeta + 2/3}, \tag{31}$$

from which the plot in figure 4 follows. Likewise, inserting equation (30) into (23) gives

$$\frac{d\ln(V_2N_2)}{d\ln N_2} = \frac{(-4/3)(\zeta + 1)}{\zeta + 2/3} + 1, \tag{32}$$

from which the plot in figure 5 follows. As examples from figure 4, the gradient of the thinning curve is -1.97 and -1.79 when branch length (horizontal projection of longest branches) is, respectively, 33 times and 3.3 times the trunk radius (ratios 0.03 and 0.3 in fig. 4). Not until the trunk makes up two-thirds of the crown diameter (ratio 2 in fig. 4) does the gradient of the thinning curve for elastic similarity match the -1.5 slope for geometric similarity.

In a review of ratios (not allometry) between crown diameter and trunk diameter of trees, White (1981, p. 489) gave examples ranging from 13 to 36. For most trees of a general shape, my elastic-similarity model therefore predicts gradients of approximately -1.90, which is much smaller than -1.50 for geometric similarity (figs. 4, 6). This prediction is for average plant volume versus population density. When, instead, plant volume per unit of ground area is plotted against population density, the model predicts gradients of about -0.90 for trees of a general shape. This is much smaller than -0.50 for geometric similarity (figs. 5, 7). This elastic-similarity model may apply to aboveground structures like stem and branches, which must possess structural strength against the force of gravity. But it is not likely to apply to roots, because roots are not subjected to bending forces caused by gravity (except for the root bases).

Because of their convenience, power-law functions are bound to continue to predominate in fitting curves to empirical data and to be used also for testing whether the elastic-similarity model explains empirical data. When applied to elastically similar plants, power-law functions gives slopes in log-log diagrams that are simply some average of the moderately different gradients actually exhibited during various stages of the growth and thinning process of one and the same population (eqs. 22, 23, 31, 32; figs. 4, 5). The elastic-similarity model may still be eminently useful in explaining obtained slopes. But due consideration must be given to the range of ratios between trunk radius and branch length spanned by

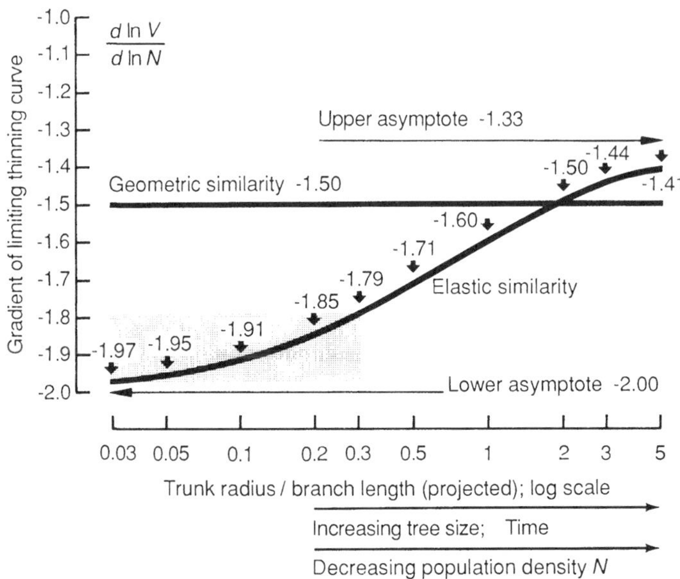

FIG. 4.—Predictions of how the gradient of limiting thinning curves in a log-log diagram, showing average plant volume (or mass) against population density (as in figs. 1, 6), varies with the ratio of trunk radius to branch length (horizontal projection) of trees growing according to either the geometric-similarity principle or the elastic-similarity principle. With geometric similarity, the ratio of trunk radius to branch length remains constant throughout tree growth, as does the -1.5 slope of the thinning line. Depending on their characteristic geometry, different plants therefore occupy different stationary positions along the horizontal -1.5 line, and any given population maintains the same position throughout the entire growth of its plants. As a plant grows taller with elastic similarity maintained, the stem contributes an increasing proportion of crown radius. Along with this individual growth (toward the right in the diagram) the population density decreases and the gradient of the limiting thinning curve increases monotonically as shown, from a lower asymptote at -2 to an upper one at -1.33. The shaded part of the curve encompasses the most realistic ratios of trunk radius to branch length for trees and the corresponding predicted gradients of the thinning curves. The curve for elastic similarity is from equation (31).

different-sized plants, because the range of these ratios defines the range of gradients (figs. 4, 5) averaged by the power-law exponent.

This elastic-similarity model was developed for two kinds of data: an intrapopulation comparison across different growth and thinning stages of one and the same population; and an interpopulation comparison across different populations at different growth and thinning stages, but with the populations otherwise similar. But it should apply also to interpopulation comparisons across populations of different species, provided that they all obey the elastic-similarity rule, have reasonably similar height-to-width ratios of the exclusive space occupied by an average plant, and have similar packing densities of plant material within this exclusive space. Here it must be stressed that differences between species, in the shape of the exclusive space that a plant occupies and in the packing density of plant material within this space, may strongly affect the vertical position of

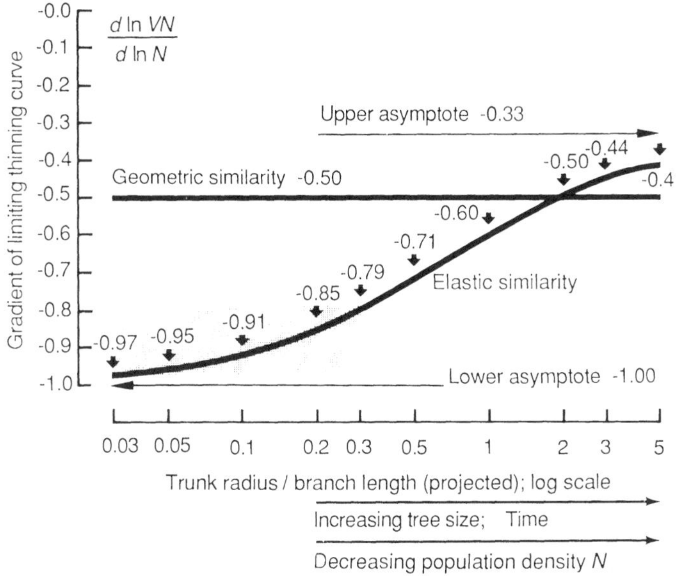

FIG. 5.—This diagram has the same horizontal axis as that in figure 4, but the vertical axis shows the gradient for the alternative form of the thinning curve, which displays total plant volume (or mass) per unit of ground area against population density (as in figs. 2, 7). With geometric similarity, different populations occupy different stationary positions along the horizontal -0.5 line throughout the entire growth of its plants. Along with individual growth, according to the elastic-similarity model (toward the right of the diagram), the population density decreases and the gradient of the limiting thinning curve increases monotonically as shown, from a lower asymptote at -1.0 to an upper one at -0.33. The shaded part of the curve encompasses the most realistic ratios of trunk radius to branch length for trees and the corresponding predicted gradients of the thinning curves. The curve for elastic similarity is from equation (32).

volume-versus-density points of individual populations, and therefore also the overall gradient of an across-species, power-law regression equation.

Elastic Similarity and Branches

The elastic-similarity model for stem-wood volume versus tree population density (eqs. 22, 31) applies also to the volume of branches of a whole tree, provided that branch number remains constant throughout tree growth. This could be accomplished by self-thinning of branches within the crown, as a result of self-shading. An alternative, idealized growth mode is a linear increase in branch number, n_b, with increasing tree height,

$$n_b \propto H \propto t^{2/3} . \tag{33}$$

The total branch volume per tree, ΣV_b, then becomes

$$\Sigma V_b \propto t^{2/3} t^2 t^{2/3} \propto t^{10/3} . \tag{34}$$

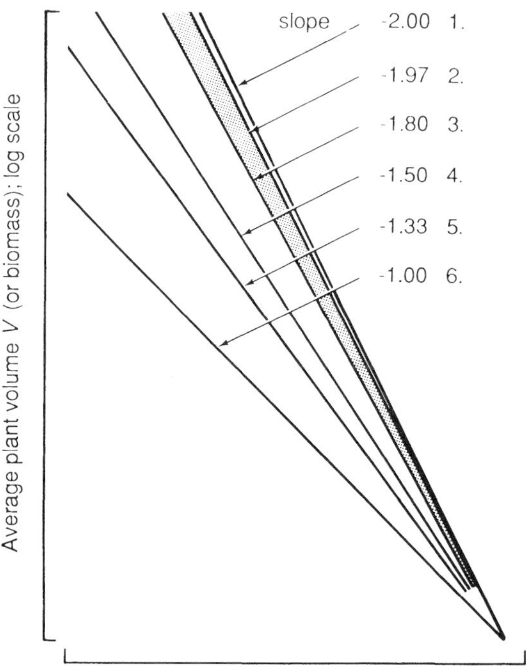

FIG. 6.—Generalized diagram of average plant volume versus population density, contrasting different thinning lines. The -1.50 line is for geometric similarity, and the -2 and -1.33 lines show the two extreme, asymptotic values for elastic similarity. With elastic similarity, population thinning proceeds along a curve, rather than a line, in a log-log diagram. For trees growing according to the elastic-similarity principle, the most realistic range of gradients is between -1.97 and -1.80 (shaded region, corresponding to shaded zone in fig. 4), which allows only a very moderate curvature of the thinning curve. The limiting thinning curve is steepest (gradient closest to -2) when trees are small (at bottom right); the steepness decreases progressively as trees grow larger and populations thin (as shown in fig. 4).

The gradient $d\ln\Sigma V_b/d\ln N_2$ for branch volume per tree versus tree population density can be obtained from the procedure outlined in equations (16)–(23). It shows that as t approaches 0,

$$d\ln\Sigma V_b/d\ln N_2 \rightarrow -5/2\,; \tag{35}$$

and as t approaches infinity,

$$d\ln\Sigma V_b/d\ln N_2 \rightarrow -5/3\,. \tag{36}$$

Even though branch number is assumed to increase linearly with tree height, the thinning gradients for total branch volume per tree versus tree population density (relations 35, 36) become only a little lower than those for a constant number of branches (relations 24, 26). When trunk and branches maintain elastic similarity but the number of branches increases linearly with plant height, then the

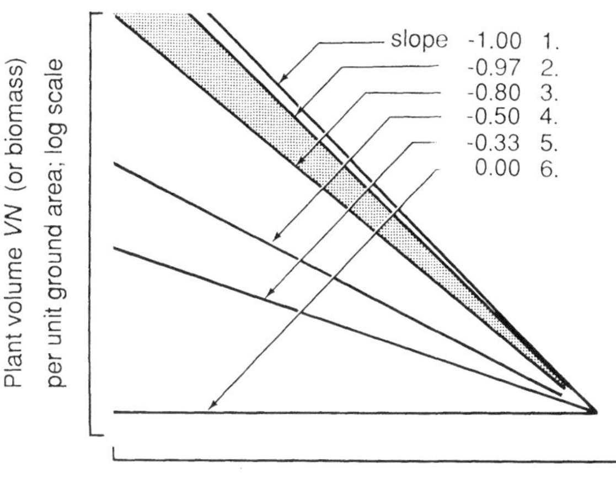

FIG. 7.—Diagram as in figure 6 but with plant volume per unit of ground area plotted against population density in analogy with figure 2. The lines are numbered to correspond to those in figure 6; the shaded region shows the most realistic range of gradients with elastic similarity of growth (corresponding to shaded zone in fig. 5).

gradient for the combined volume of trunk and branches tends to take a value only slightly lower than that of a gradient with a constant number of branches. The lower extreme value for the combined branch and trunk volume will be between -2.5 and -2 (relations 35, 24), and the upper extreme between -1.67 and -1.33 (relations 36, 26). The actual value varies with the relative volumes that trunk and branches contribute. Branch volume does not usually affect empirically observed slopes for trees, because tree volumes are usually measured for stem wood only, exclusive of branches.

Because the two alternatives considered (branch number remaining constant throughout tree growth or increasing linearly with tree height) represent the opposite extremes, the resulting thinning gradients should encompass most real cases.

ROOT COMPETITION AND SELF-THINNING OF POPULATIONS

The self-thinning gradients derived thus far could result from space competition between aerial parts of plants, requiring a complete canopy cover. But the gradients could just as well follow from joint competition between aerial parts and between roots, provided that growth geometry is the same for both, or from competition between roots only.

Consider a hypothetical case in which the growth geometry is different for aerial parts and roots. Assume, for instance, that aerial parts maintain elastic similarity, but roots maintain geometric similarity. Further assume that the ratio between the root volume and the volume of aerial parts remains constant throughout plant growth. This may be expected in view of the roots' supportive function and has also been observed in a large number of trees, in gymnosperms as well as angiosperms (Mohler et al. 1978, p. 611; O'Neill and DeAngelis 1981, pp. 431–432, fig. 7.15). Whether the thinning gradient for aerial structures becomes −1.5 (for geometric similarity) or about −1.9 (for elastic similarity) would then depend on whether competition, forcing population thinning, occurs predominantly among roots or among aerial parts. The lateral extent of structures growing according to geometric similarity would eventually outgrow the lateral spread of structures maintaining elastic similarity. Therefore, only in a transient stage would the root and crown in such a plant cover the same projected ground area. Before the transient stage, elastic similarity of aboveground parts would dictate the thinning gradient (about −1.9); after it, geometric similarity of roots would (gradient −1.5).

When competition for space takes place predominantly among roots, the thinning gradient for aerial structures would be dictated by the root growth mode, regardless of the growth geometry adopted by aerial structures (following from the assumed coupling of the respective volume of aerial parts and of roots). Where the density-limiting factors are mineral resources (in some rain forests) or water (in deserts), and when the roots maintain geometric similarity, then the roots might force a −3/2 exponent of aboveground structures, regardless of their own growth mode. Allometric relations of aboveground structures could therefore easily be at variance with the thinning gradient observed. This mechanism might explain some perplexing discrepancies between allometry of aboveground structures and the thinning gradient (see the Discussion, "Conclusions on tree allometry and thinning of forests" and "Other thinning models"). A complete root cover, with root competition alone driving population thinning, would also allow an incomplete canopy cover. This might explain the wide spacing of shrubs and trees in certain habitats (see the next section).

THE −1 SLOPE FOR OLD TREES

The thinning slope tends to shift toward −1 in populations of old and big trees (White and Harper 1970, fig. 9; the shift in slope is based on recalculated data of yield for various artificial thinning procedures in forestry). The −1 exponent for the trunk (which is what is measured in forestry) indicates that trunk volume per unit of ground area remains constant throughout tree growth and concomitant population thinning. To see this, multiply both sides of the relation $V \propto N^{-1}$ by N, yielding

$$VN \propto N^0 = 1 . \qquad (37)$$

The volume of stem wood removed with the death of a tree will therefore be only just compensated for by the combined extra growth of stem-wood biomass made by neighboring trees in response to the origin of a tree-fall gap (figs. 1, 2, 6, 7).

The shift to −1 might be caused by stagnation of height growth. This may occur after some maximum species-specific and stand-related height has been attained, whereupon trunk diameter increases in proportion to R (as with geometric similarity but in violation of elastic similarity; eq. 17). Trunk volume would then increase in proportion to R^2, and the exponent in equation (1) would take the value -1 after the breakpoint in the thinning curve (for stem wood only; from eq. 5). When the thinning exponent is -1, radius R of the exclusive space occupied by a tree must still continue to increase for competition among aboveground tree parts to drive further population thinning. But since branches also become thicker as length increases, the thinning exponent for the total branch volume per average tree versus population density assumes values lower than -1 (different values depending on which growth mode obtains; relations 9, 10, and eqs. 22, 31, 35, 36). Consequently, exponents for the combined volume of branches and trunk versus population density would take compromise values smaller than -1 (owing to the influence of the exponent for branches).

A problem with this model of only lateral growth of stems is that stem girth would increase out of proportion with respect to both geometric and elastic similarity in comparisons across different growth stages. Allometric studies across trees in different growth stages (including big, old trees) are necessary for exploring whether this occurs. In either case, it is of course possible for trunk taper of any individual tree always to conform with the elastic-similarity principle, that is, with local trunk diameters at different heights always varying as the 3/2 power of the remaining distance to the tree top. Different trees, each of which is "elastically self-similar" in this way (McMahon and Kronauer 1976, p. 446), may have different safety factors against buckling under their own weight and therefore may not comply with the elastic-similarity criterion for between-tree comparisons.

As an alternative explanation, a shift to the -1 slope would occur because the root system has ceased to grow in depth after trees have attained some large size. Many reasons could explain why it becomes unprofitable or impossible for roots to penetrate beyond a certain depth (soil structure, water, rocks, etc.). After roots begin to grow only laterally, their spread will probably soon exceed that of aerial structures, because the latter grow in three dimensions, the root system in two. From then on, root competition is likely to predominate, and the exploited soil volume will increase in proportion to R^2 (fig. 3). If root mass per unit of volume of exploited soil remains constant throughout plant growth, and if the ratio between root mass and mass of aerial parts likewise remains constant (to be expected for functional reasons), then root competition would dictate a thinning gradient of -1. This gradient would apply to roots as well as to aboveground structures. Radius R of a tree's exclusive space would then be set by the lateral spread of roots, not by crown width. Aerial structures could maintain geometric or elastic similarity, and in either case there would be progressively more open space among the trees as they trace the -1 thinning line, since crowding is restricted to the roots.

Root competition is particularly likely to predominate among trees growing on shallow soils with bedrock near the surface, on bogs where a high water table may limit depth growth, in soils with poor aeration, and in high-latitude areas where only an upper soil layer thaws in summer, subterranean permafrost limiting the depth growth of roots. Root competition may therefore explain the wide spacings usually seen between bushes and trees on such grounds; the -1 thinning gradient should then prevail.

LEAF VOLUME, PER TREE AND PER UNIT OF GROUND AREA, VERSUS TREE POPULATION DENSITY

In most trees the leaves remain about the same size throughout tree growth. This applies to the needles of conifers as well as to leaves of any size and shape in broad-leaved trees. Tree leaves thus depart drastically from any growth mode adopted by trunk and branches. This constancy of leaf size might be expected to result in radically different thinning slopes for leaves than for other plant structures, like stem and branches, that maintain geometric or elastic similarity. When total plant volume is related to the population density, different thinning slopes for different plant parts combine to form a compromise slope for the whole plant. Therefore, it is important to explore, separately, the thinning functions of total leaf volume per average tree versus tree population density.

The following derivation, leading to relations (38)–(40), refers to tree leaves that remain the same size throughout tree growth. A basic assumption is that the overall tree shape, as governed by trunk and branches, maintains geometric similarity. Nonetheless, most of the results apply also to trees whose trunk and branches grow according to the elastic-similarity principle. The total foliage volume (or mass) per tree should be roughly proportional to leaf number. As long as leaf packing and distribution within the crown remain constant and the canopy layer still grows in depth, total leaf volume (V_f), total leaf surface area (A_f), and leaf number per average tree (n) should be related to R and N as

$$V_f \propto A_f \propto n \propto R^3 \propto N^{-3/2} \tag{38}$$

(from eq. 5). A precondition for the above relationship with N is, of course, that population thinning has commenced even though the canopy layer still grows in depth. Multiplication of relation (38) by N shows that total leaf volume per unit of ground area, NV_f, varies as $N^{-1/2}$. The above proportionality between leaf surface area and population density ($A_f \propto N^{-3/2}$) has previously been derived from empirical data (Westoby 1977; but see White 1977).

If tree growth results in an increasing proportion of the leaves being subjected to diminishing light intensities in the crown's interior, then the number of thin shade leaves will increase in relation to the number of thick sun leaves. This should not affect the exponent in $A_f \propto N^{-3/2}$, but would tend to increase slightly the exponent in $V_f \propto N_1^{-3/2}$ (making leaf volume increase more slowly with population thinning than in relation 38).

Proportionality (38) applies strictly for trees whose trunk and branches maintain geometric similarity, but it is also a good approximation when trunk and branches maintain elastic similarity, so long as the trunk contributes only a negligible amount to canopy radius, R, as in young trees. The elastic-similarity principle requires all diameters of an organism to change with total size in the same way, and similarly for all lengths (McMahon 1984, p. 264), such that when R is determined primarily by branch length, then tree height, H, is approximately proportional to R (albeit strictly to branch length only). Therefore, it is approximately true that $V_f \propto R^3$ as in relation (38).

When trunk and branches maintain elastic similarity and the trunk contributes significantly to R, then the assumption that branch length approximates R leads to an overestimation of H and hence of the crown volume available for leaves to fill. With thick trunks and elastic similarity, the thinning exponent therefore takes values between $-3/2$ and -1. But with thick trunks, the foliage layer has probably ceased to increase in thickness long ago, whence relation (39) applies.

As trees grow, the thickness of the foliage layer eventually reaches some threshold, imposed by self-shading in the crown. From then on, the leafy part of the crown increases only in lateral spread, not in depth (even if the trees continue to grow taller); and therefore foliage volume (V_f), total leaf surface area (A_f), and leaf number per average tree (n) should vary as

$$V_f \propto A_f \propto n \propto R^2 \propto N^{-1} \tag{39}$$

(to be contrasted with relation [38], which applies while the canopy layer still grows in depth). Relation (39) applies when trunk and branches maintain geometric similarity, but it should be approximately true for elastic similarity also, particularly when the trunk contributes relatively little to crown diameter (eq. 17), and tree height therefore remains approximately proportional to R. When the canopy layer is closed, as is the rule with crowded forest trees, the leaves are usually restricted to the exposed top of the tree, and therefore tree height per se has little effect on crown surface area. Then relation (39) also applies strictly when trunk and branches maintain elastic similarity, regardless of how much the trunk contributes to R.

The -1 exponent in (39) indicates that the leaves removed with the death of a tree are only just compensated for by the growth of additional leaves on surviving neighboring trees as they respond by growing larger crowns. Multiplying relation (39) by N yields

$$NV_f \propto NA_f \propto Nn \propto NR^2 \propto NN^{-1} \propto 1\,. \tag{40}$$

This rule applies after the foliage layer has reached its maximum thickness. It says that total tree leaf volume, NV_f, and total tree leaf surface area, NA_f, as well, remain constant with respect to ground area throughout plant growth and concomitant population thinning (figs. 1, 2, 6, 7).

It is a common empirical observation that foliage biomass and canopy leaf area per unit of ground area increase with individual tree growth in young plants. Proportionality (38) should then apply. But after some limiting maximum value, set by light shortage deep in the canopy, has been reached, leaf biomass and leaf surface area per unit of ground area remain constant (White 1977; Mohler et al. 1978, p. 610; Waring 1983). This may happen within months for some vegetative covers, whereas it may take 40 years for others to reach such a foliage plateau (Waring 1983, p. 339). After this transition has occurred, relations (39) and (40) apply throughout all subsequent tree growth. Lines fitted to empirical data for log total leaf volume per tree versus tree population density gave the slopes -1.13

and −1.08 for *Prunus pensylvanica* and −1.01 and −0.95 for *Abies balsamea* (Mohler et al. 1978, pp. 604, 610), all close to the predicted exponent −1 in relation (39).

Perry (1984, table 1) related total leaf area (A_f) to total aboveground biomass (M) with power-law functions ($A_f \propto M^{\alpha}$). The average of four exponents for trees is 0.65 (for *Pseudotsuga menziesii*, *Pinus elliottii*, *Acer circinatum*, and *Abies* spp.). This is exactly as predicted from relation (39), provided that trunk and branches remain geometrically similar, whence $A_f \propto R^2 \propto M^{2/3}$. Since the canopy layer of these trees is likely to reach its maximum thickness early in the growth-thinning process, relation (39) rather than relation (38) should be relevant.

For plants that retain geometric similarity not only in stem and branches, but also in leaves (such that leaf length, width, and thickness increase isometrically with respect to stem and branch size), the total number of leaves should remain constant throughout plant growth, and proportionalities (38) and (39) both should break up into

$$V_f \propto R^3 \propto N^{-3/2}, \tag{41}$$

$$A_f \propto R^2 \propto N^{-1}, \tag{42}$$

with relation (41) corresponding to equation (6). As plants grow larger and their population density decreases, total leaf volume per individual plant and per unit of ground area would then increase more quickly than total leaf surface area. The maintenance of a constant leaf size throughout tree growth is thus a means of making the combined leaf surface area per tree increase at the same rate as the combined leaf volume per tree. That is, a given increment in leaf surface area requires less leaf biomass if more small leaves are grown than if existing small leaves grow larger (relations [38] and [39] for constant-sized leaves versus relations [41] and [42] for leaves growing isometrically with stem and branches).

Because current twigs are the structures that support the current leaves (Mohler et al. 1978, p. 610), and because these twigs are about the same size regardless of tree size, relations (38) and (39) should apply also to twigs. The empirically obtained exponents −0.81 and −0.95 for current twigs in *Prunus pensylvanica* and *Abies balsamea*, respectively (Mohler et al. 1978, p. 604), are near the −1 exponent predicted from proportionality (39) when the foliage layer has ceased to increase in depth.

In conclusion, with constant-sized leaves and still-increasing canopy thickness, the slope for total leaf volume and total leaf surface area per tree versus tree density should be −3/2, as for volumes of trunks and branches that maintain geometric similarity (eq. 6 and relation 38). But after the canopy has attained its maximum thickness, the slope for leaf volume and area should shift to −1 (relation 39). These slopes are predicted regardless of whether geometric or elastic similarity governs the growth of the trunk and branches carrying the leaves. (An exception is that the −3/2 exponent would tend toward −1 if the following three conditions were to apply simultaneously: the trunk contributes significantly to R, the trunk and branches maintain elastic similarity, and the foliage layer still increases in depth, which is unlikely in big trees.)

While the canopy layer still grows in depth, the thinning exponent should be −3/2 regardless of whether leaves remain a constant size (relation 38) or grow isometrically with respect to trunk and branches (relation 41). The −3/2 exponent applies despite this drastic difference in leaf growth mode because, with geometric similarity extending to the leaves, leaf number should remain constant, and empty spaces between leaves should scale up isometrically with the leaves and other plant structures. But with constant-sized leaves, the spacing between leaves may remain constant as new leaves grow, whereas crown volume increases isometrically with trunk and branches, as usual. In effect, this mimics geometric similarity; since leaf packing remains constant, total leaf volume scales isometrically with crown volume, which in turn grows isometrically with the trunk and branches. The crucial factor is that leaf number remains constant when leaves grow isometrically with the crown, whereas the number of constant-sized leaves increases with crown growth, causing leaf spacing to remain the same.

DISCUSSION

Linear Dimensions and Areas

White expressed doubts about the assumption that the ground area occupied by a plant is proportional to a linear dimension squared and asserted that the assumption is justified only if "the area is perfectly circular or perfectly square" (1981, p. 483). But so long as growth does not bring on systematic change in shape of the ground area or in cross-sectional areas of stem and branches, any such area, whatever its shape, does of course remain proportional to the square of a representative linear dimension within the plane concerned. This is the rationale behind equations like (4) and (18). But it is true that with certain growth modes (other than the one resulting in geometric similarity), there is no single plant measure whose square is proportional to the exclusive ground area of the plant, as emphasized by White (1981). With elastic similarity, for instance, the sum of trunk radius and projected branch length must be used for estimating the ground area (eqs. 17, 18).

Geometric versus Elastic Similarity of Individual Growth and Observed Self-Thinning of Populations

Gorham (1979) fitted a power function across 29 different plant species, including mosses (*Polytrichum*), ferns (*Pteridium*), *Carex* spp., grasses, herbs, and trees, and obtained a between-species thinning exponent of −1.49. Using data from various sources, White (1980, table 2.2) presented separate thinning functions for each of 36 species, growing in pure stands where density-dependent mortality occurred. About two-thirds of the species are trees, the remaining grasses, *Carex*, and herbs. The thinning exponents ranged from −1.30 to −1.80 with an average of −1.51 (SD, 0.13). Although this rough average is in striking agreement with expectations from the geometric-similarity model (eq. 6), 14 of the 23 trees were below the −1.51 average, whereas only 1 of the 13 herbs was. The average exponent for the 23 trees was −1.55 (SD, 0.13) to be sure, but trees exhibited the steeper gradients.

The tendency for populations of trees to have steeper (more negative) slopes than −1.5 is even more apparent for forests subjected to artificial thinning (for maximum yield). Six thinning lines in White and Harper (1970, fig. 8) have gradients between −1.72 and −1.82, and four others, representing different, artificial, thinning regimens for *Picea abies* take the values −1.39, −1.63, −1.77, and −2.18 (their fig. 7). But with artificial thinning, a wide range of "artificial" thinning slopes can be obtained. Since thinning practices in forestry aim at maximizing rate of stem-wood production per unit of forest area, the resulting thinning slopes need not be like those of naturally thinning populations.

West and Borough (1983) estimated thinning characteristics of several experimental populations of *Pinus radiata*, planted at different initial densities and not subjected to artificial thinning. The total mortality to age 35 yr ranged from 1% in plants with the widest initial spacing to 20% in plants with the smallest initial spacing. The slope of the relationship between the logarithm of the mean weight per tree trunk and the logarithm of the density of the various plots was computed across plots at each age, at which measurements were taken, both for all live trees and for the dominants only. These lines show what White (1981, fig. 1) called the "competition-density effect" (like lines t_0–t_3 in my fig. 1). Thinning among dominants was caused not by mortality but by a progressively increasing proportion of them becoming suppressed trees. Almost all mortality that eventually occurred was observed in this category of suppressed trees.

The slope for the dominant trees declined from −1.36 to −1.50 between population ages of 25 and 35 yr when trunk volume, V, was related to trunk diameter, t, as $V \propto t^{2.6}$ (in agreement with elastic similarity, which dictates $V \propto t^{8/3}$; the exponent 2.6 is from empirical data in White 1981, table 1). This steepening of the negative slope with increasing age and declining population density is not at variance with the flattening of the limiting thinning line predicted by my model, which is based on elastic similarity (eq. 22). West and Borough's (1983) equations are not growth trajectories from individual plots through time. Each line is instead fitted across the whole set of populations when they are all of the same age and still approaching the limiting thinning line (similar to the lines t_0–t_3 in fig. 1). These "competition-density" lines gradually approach the limiting thinning line, but do not coincide with it until self-thinning is in full operation. This dynamic equilibrium is identified by a lack of further steepening of the competition-density line with time, that is, with additional growth and thinning (fig. 1). This stage may not have been reached since the slope was still becoming steeper between population ages of 30 and 35 yr. Thus, even though the latest slope observed was −1.5, which fits geometric similarity, steeper slopes (as with elastic similarity) cannot be excluded after additional growth has taken place, forcing more intense self-thinning.

As a general rule, then, the geometric-similarity model is very successful in predicting thinning gradients, particularly those obtained by some averaging process (such as across-species regressions and means for several kinds of plants). Nevertheless, the gradients for populations of several tree species come closer to those predicted from the elastic-similarity model. I therefore now review data about tree allometry in order to explore which model of individual growth is most applicable to trees.

Allometry of Trees versus Thinning of Populations

The most direct way of exploring scaling principles of plant growth is of course to look into plant allometry itself rather than examining concomitant thinning functions. In plots of volume versus density for trees, usually only stem-wood volume is measured. But it is still essential to know the relative length of branches, since they may determine the size of the exclusive ground area occupied per tree, and thereby the thinning characteristics, provided that the roots do not do so. The geometric- and elastic-similarity models of growth will now be compared with empirical tree allometry.

Tree height versus trunk diameter.—Geometric similarity dictates that $H \propto t$ and elastic similarity that $H \propto t^{2/3}$ (eq. 13). Support for the 2/3 power was presented by McMahon (1973) and McMahon and Kronauer (1976) with data from 576 record specimen trees, representing nearly every species found in the United States. The data are from various registers of big trees; registered trees are very tall, very broad, or both (McMahon and Bonner 1983, p. 143). In a log-log diagram of tree height versus trunk diameter at breast height, the data points show enormous scatter, but nonetheless cluster around a line with slope 2/3, indicating that the tree-trunk proportions are determined by the rule of elastic similarity (McMahon 1973, fig. 2; McMahon and Kronauer 1976, fig. 5; McMahon and Bonner 1983, pp. 142, 143). Since these trees are unusually big, they probably do not belong to self-thinning populations of even-aged trees, but rather are more or less solitary. Therefore, the above data do not refer to population densities or thinning modes. They show only that, as a general rule, different-sized trees of various species have trunk proportions roughly in accord with the elastic-similarity principle (i.e., they maintain the same average safety factor against buckling under their own weight, regardless of size). Indeed, the enormous scatter in the diagram may occur largely because most of the trees were probably not subjected to crowding, at least not when unusually large, and were therefore free to develop according to the characteristics of their species. (Crowding tends to make trunk and crown shapes similar across species.) Some may also have increased in girth after stagnation of height growth, which would augment the scatter (see "The -1 slope for old trees," above). Trees growing in crowded, self-thinning populations would probably show less scatter in a height-versus-diameter diagram, because competition for light would force the trunks to grow tall and therefore to develop closer to the buckling line, leaving a narrower safety factor against failure caused by gravitational forces.

Mohler et al. (1978, table 3) listed exponents for tree height versus trunk diameter (at breast height, ca. 1.3 m) for 10 tree species. They range from 0.48 to 0.84, with an average of 0.63, which is in good agreement with 0.67 for elastic similarity. White (1981, p. 486) reviewed sources in support of the elastic-similarity model, but he also emphasized that the exponent may change along with tree growth in some species (although this might be due partly to the measuring practices in forestry as discussed below).

Trunk volume versus trunk diameter.—Geometric similarity dictates that $V \propto t^3$ and elastic similarity that $V \propto t^{8/3}$ (eq. 16). White (1981, table 1) listed 52 empirically obtained exponents, each representing one or more tree species. A rough average of these exponents is 2.46 (SD, 0.24; without weighting for the number of species each value represents). This is in better agreement with elastic similarity than with geometric similarity; 2.46 is even on the wrong side of 2.67 with respect to the geometric-similarity exponent 3. But even though the exponent 2.46 follows from empirical data for trees (and is therefore perfectly correct for practical estimates of trunk volume, using measured diameters), it is probably unrepresentatively small compared with the theoretical scaling exponents 3 and 2.67, which follow from the geometric- and elastic-similarity principles. The reason for this bias is a measuring artifact (see "An allometry artifact from measuring practices in forestry?," below).

Branch length versus trunk diameter.—Geometric similarity dictates that $l \propto t$ and elastic similarity that $l \propto t^{2/3}$ (eq. 14). Mohler et al. (1978, table 3) showed that crown spread, measured as the mean length of the longest two branches, is related to trunk diameter by exponents ranging from 0.62 to 0.71, averaging 0.66. This exactly meets expectations from the elastic-similarity model, because if branch diameter is proportional to trunk diameter and if branches remain elastically similar, then the exponent for branch length versus trunk diameter should be 0.67. The data of Mohler et al. (1978) therefore lend empirical justification to my assumptions in developing expressions (11) through (15).

Crown diameter versus trunk diameter.—Geometric similarity dictates that $2R \propto t$ and elastic similarity that $2R = t + 2k_7t^{2/3}$ (from eq. 17), which becomes $2R \propto t$ when the second (branch) term is ignored and $2R \propto t^{0.67}$ when the first (trunk) term is ignored. When the trunk is assumed to make up a realistic proportion of crown spread, the exponent should be a little larger than 0.67.

White (1981, table 2) listed 22 empirically obtained exponents, each representing one or more tree species. A rough average of these exponents is 0.85 (SD, 0.24; without weighting for the number of species each value represents). The populations from which these exponents came were not necessarily self-thinning, and crown diameter was measured as length of the longest branches in some populations and as projected crown spread in others. When crown diameter is measured as branch length, elastic similarity predicts 0.67; but when it is measured as crown spread, a little larger exponent is expected (since trunk diameter is included here, and it grows proportionately faster than branch length; see above). Regardless of which measuring practice predominated, the empirical average exponent 0.85 falls between those predicted from the geometric- and elastic-similarity models (1.0 and a little more than 0.67, respectively). However, one large study (Perez 1970, cited in White 1981), based on 4435 trees in several tropical forest types, gave the exponent 0.70 in close agreement with the elastic-similarity model.

Conclusions on tree allometry and thinning of forests.—The elastic-similarity model definitely fits the tree allometry data reviewed much better than does the geometric-similarity model. (Among other things, the discrepancy between tree allometry and the geometric-similarity model aroused doubts about that model in the first place.) For many tree species, the thinning gradient also agrees well with expectations from the elastic-similarity model, but the average thinning exponent for trees (−1.55; see above) comes closer to that for geometric similarity. It is not obvious what causes this discrepancy between allometry and population thinning in some cases. It could be that competition for space occurs predominantly among roots and that roots come closer to geometric similarity and also dictate the thinning gradient for aerial structures (see the section on root competition).

An Allometry Artifact from Measuring Practices in Forestry?

White (1981, table 1) emphasized that trunk volume (or mass), V, of trees is not the simple function $V \propto t^2H$ of trunk diameter, t, and height, H, that would be expected, but rather that $V \propto (t^2H)^a$, where, as a rule, $a < 1.0$. This comes not from theory, but from fitting curves to empirical data. He listed 22 empirically obtained values of a, averaging 0.91 (SD, 0.06).

Since t and H are measured quantities, they cannot be doubted. Also, the average exponent 2.46 reported above under "Trunk volume versus trunk diameter" is consistent with $V \propto (t^2H)^{0.91}$, provided that elastic similarity prevails. To see this, let $H \propto t^{2/3}$, as with elastic similarity and as reported from empirical observations in the subsection "Tree height versus trunk diameter." Then, $V \propto (t^2t^{2/3})^{0.91} \propto t^{2.43}$. But this relationship is confounded by the method of measuring t in forestry, as explained below.

The fact that a is less than the theoretically expected value 1.0 could indicate changes in trunk shape leading to the development of an increasingly concave taper along with growth. I suggest that this is an artifact, partly or entirely, from the measuring practices in forestry for the following reason. Trunk diameter is usually measured at breast height, about 1.3 m from the ground, regardless of tree size (White 1981, p. 483). Because measuring height is constant, the bigger the tree is, the farther down into the expanded, basal trunk zone will the trunk diameter be taken; that is, the ratio between measuring height and total tree height becomes smaller as the tree grows taller (fig. 8). This results in the diameters becoming progressively more overestimated (with respect to the actual diameter distribution along the length of the trunk) the larger the tree is, necessitating a values less than one. Therefore, the deviation of a from one may not reflect anything of interest about growth geometry of the trunk in general, even though the exponent is important practically.

There are actually two shape factors of the trunk that may contribute to overestimation of trunk diameter in big trees as compared with small ones. First, there is the ordinary downward thickening of the trunk according to the elastic-similarity rule, specifying that the local trunk diameter at any point along the trunk should vary as the 3/2 power of the remaining distance to the top. In addition, many tree trunks are much wider near the ground, exhibiting a rapid taper near the base followed by a less pronounced taper along the rest of the trunk. The trunk expansion near the transition to the roots (butt swell) is over and above that following from elastic similarity. This funnel-shaped connection provides particular structural reinforcement where trunk-bending moments are maximal. In big trees, this butt swell may reach far beyond the measuring height (fig. 8).

It must be stressed that this bias is always present provided that the trunk tapers somewhat at the measuring height, regardless of whether geometric or elastic similarity or some other criterion obtains, or whether or not there is, in addition, a

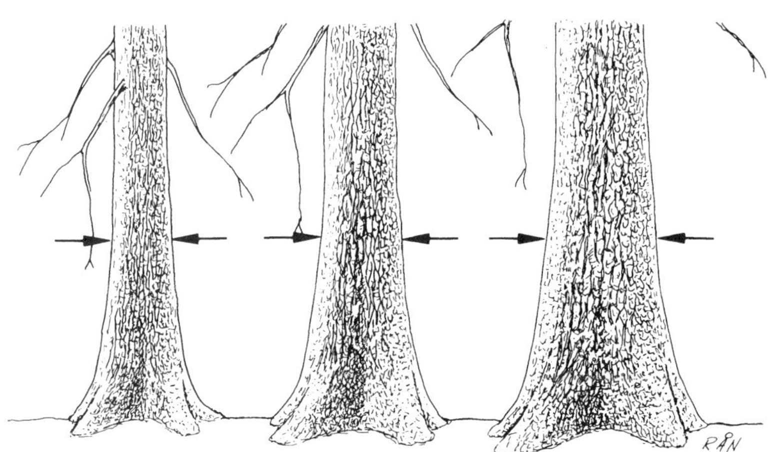

FIG. 8.—In forestry, trunk diameter is measured at a constant height (breast height, about 1.3 m above the ground) regardless of tree size. Therefore, the larger the tree, the greater the overestimation of the diameter with respect to the actual diameter distribution along the trunk. The middle figure is from a photograph of a spruce, *Picea abies*, in a closed stand (0.55-m diameter at the arrows 1.3 m above the ground). The others were obtained by scaling that trunk isometrically, such that the linear size ratios become, from left to right, 0.80:1.00:1.25 (25% increment in each step). With the taper exhibited by this particular trunk, the three diameters at 1.3 m (with respect to the middle, original trunk) become related as 0.78:1.00:1.39. Scaling according to the elastic-similarity rule, which often governs trunk growth, results in a still-greater diameter bias.

butt swell. The bias is easily accounted for in stem-wood volume calculations (see opening paragraph of this section) but it has the following unpleasant consequences for allometric studies.

If the measured trunk diameter becomes increasingly unrepresentative (too large) the larger the tree is, then the exponents obtained by regressing empirical data for trunk height, trunk volume, branch length, and crown diameter against trunk diameter will tend to be too small with respect to exponents obtained using a more representative diameter. For scaling purposes, trunk diameter should be taken at some characteristic height, such as a size-independent proportion of tree height, rather than at a given, absolute height (although this is, of course, extremely impractical in forestry). This ideal diameter is implicitly built into the various models in this paper.

Other Thinning Models

Ford (1975, p. 329) arrived at a regression model with slope -2 by using the empirically obtained relation $A \propto t^{1.5}$ between the average ground area per tree, A, and trunk diameter, t, for Douglas fir (*Pseudotsuga menziesii*; Curtis 1971) and combining it with the assumption that weight or volume, V, varies with trunk diameter as $V \propto t^3$. Adoption of $V \propto t^3$ amounts to assuming geometric similarity, whereas the use of $A \propto R^2 \propto t^{3/2}$ does not. Geometric similarity dictates $A \propto R^2 \propto t^2$, and elastic similarity $A \propto R^2 \propto (0.5t + k_7t^{2/3})^2$ (from my eq. 17), which becomes

$A \propto t^2$ when the second (branch) term is ignored and $A \propto t^{4/3}$ when the first (trunk) term is ignored. When the trunk is assumed to make up a realistic proportion of crown spread, the gradient should be a little larger than 1.33 but much smaller than 2. Therefore, the empirically obtained exponent 1.5 used for $A \propto t^{1.5}$ in Ford's model is consistent with elastic similarity, but it is appreciably smaller than the 2 for geometric similarity.

Ford (1975) thus combined a hypothetical volume relationship, resulting from geometric similarity, and an empirical area relationship, which is consistent with elastic similarity but at variance with geometric similarity. He gave no justification for combining these mutually incompatible relationships into a model, and this inconsistency makes his derivation of the -2 slope theoretically unsatisfactory. There were no underlying considerations of elastic similarity involved.

Mohler et al. (1978, p. 613) assumed that $W \propto t^2H$ and $R \propto N^{-1/2}$ (as I do in eqs. 16, 5) and combined this with empirically observed relationships between trunk height and trunk diameter and between canopy width and trunk diameter for various tree species. They took canopy radius, R, to be proportional to branch length (see "Branch length versus trunk diameter," above) and did not account for the trunk contribution to canopy spread. Although it should make only a small difference in most cases, a more correct measure is the one in my equation (17), provided that elastic similarity obtains.

Their semi-empirically derived model yields thinning exponents ranging from -2.29 to -1.75 for 10 tree species. This is in partial agreement with the gradients predicted from my theoretically derived thinning model for elastic similarity (eqs. 22, 31; fig. 4). This concordance occurs because the empirical data that they built into their model came from trees, most of which were elastically similar (see discussion above). Their model predicts thinning exponents for weight versus density (for the trunk only) of -2.17 and -1.85 for *Abies balsamea* and *Prunus pensylvanica*, whereas the empirically observed exponents are -1.30 and -1.58 (their tables 1, 2). Despite being derived from empirical allometry of trees growing in thinning populations, their model unexpectedly enough does not accurately predict the thinning exponents observed empirically for the very same populations. One possible explanation for this contradiction is that competition perhaps took place predominantly among the roots, which therefore dictated the thinning exponents (see the section on root competition).

White (1981, p. 490) also derived a model based on empirical data for trees. He used the empirically observed relation $W \propto t^{2.5}$ for trunk weight versus trunk diameter (which is similar to $W \propto t^{8/3}$ with elastic similarity) and various empirically obtained relations between canopy diameter and trunk diameter. Predicted gradients were mostly between -2 and -1.

Miyanishi et al. (1979) derived various thinning exponents for different kinds of growth. Although some of their exponents fall within the range of empirical observation, the theoretical basis of their derivation is exceedingly unrealistic, except for the geometric-similarity model (their 1d). Their plants are represented by blocks growing in different dimensions, and except for their 1d case, no account is taken of three-dimensional growth of plant parts. For growth in two horizontal dimensions (prostrate plants) their block model 1c predicts a thinning exponent of −1, predicted earlier also by Westoby (1976). But a more realistic model for prostrate plants must account for three-dimensional growth of plant parts as well as for different growth modes of different plant parts. The thinning gradient for whole plants will then be a compromise between different exponents for different plant parts (Norberg, MS). Similar considerations of three-dimensional growth of different plant parts show models 1a, 1b, and 1c of Miyanishi et al. (1979) to be as unrealistic as 1c.

White (1981, pp. 492, 493) suggested that the horizontal, two-dimensional-growth model of Westoby (1976) and Miyanishi et al. (1979, model 1c) might account for the shift to the −1 slope in big trees (stem-wood volume only) after they have ceased to grow in height. I have treated this possibility above and shown that it requires the strong assumption that trunk diameter grows in proportion to branch length and to R. Furthermore, the −1 slope can be expected to apply to the trunk only, since branches must grow in both length and width.

Pickard (1983) developed three different thinning models that predict gradients around −3/2. In deriving a thinning function for trees, "the erect woody perennial," Pickard (1983) used the relation $H \propto t^{0.66}$, which applies to elastically similar trunks. He then combined this with several other relations, which are based partly on empirical data from White (1980, 1981). The thinning function was cast in a power-law form from the outset. Its exponent comes near −1.5 and was shown to be rather insensitive to moderate variation in the constituent variables. This function thus gives a straight line with slope −1.5 in a log-log diagram. It is inconsistent with my thinning model, based on elastic similarity throughout (eq. 22), which predicts steeper gradients that level off moderately with decreasing population density.

The second of Pickard's models depends on a specific growth pattern for "the procumbent herbaceous annual." The predicted thinning slope is near −1 when plants are small and population density high and then gradually approaches −1.5 as population density declines. I have derived a different model, which considers different growth modes for different parts of prostrate plants (Norberg, MS). It predicts a decline of the thinning gradient from a theoretical maximum of −1.0 to a minimum of −2.0, with the most realistic range between −1.23 and −1.75. This agrees only partially with Pickard's prediction.

The third of Pickard's models, "the general plant," does not depend on details of growth, as do my models in this paper. Instead, an equation for individual growth rate versus population density is combined with an equation for survival rate. When empirical data are used, the model predicts gradients close to −1.5. Using an alternative approach, assuming a power-law thinning function and applying dimensional analysis, Pickard (1984) arrived at the −1.5 exponent associated with geometric similarity of growth.

Givnish (1986, p. 142) explored at what rate the minimum amount of support tissue (stem biomass or volume) per unit area of canopy must increase with increasing canopy height to prevent stem failure from buckling. He assumed a power-law thinning function from the outset; in effect, therefore, he examined factors affecting the y-intercept k in equation (1), given a constant value of the exponent α. He used the elastic-similarity constraint to identify a minimum y-

intercept k, but without deriving the associated gradient of the self-thinning curve. As I have shown above, a power-law thinning function with a constant exponent is not compatible with the principle of elastic similarity, even though the gradient changes only a little in the range of ratios of trunk radius to branch length exhibited by most plants (fig. 4). Givnish related k not only to the required structural strength against gravitational forces but also to the height-to-width ratio of the plant.

CONCLUSION

The apparently universal prevalence of self-thinning according to packing theory may be the best evidence available of the pervasive importance of size and density dependence in the regulation of population density in plants. I suggest the following conceptual framework for plant growth and population thinning.

As plants grow larger, crowding eventually takes place, whereupon competition leads to size- and density-dependent mortality. Close packing of plants is maintained throughout subsequent growth, and therefore the growth geometry of individual plants strictly governs the self-thinning process as reflected in the thinning gradient. Plant characteristics define the power-law, self-thinning function in a log-log diagram as follows. The growth mode determines the way that plant form and structure change with growth and therefore dictates the slope α. The specific form and structure of plants, and the overlap between neighbors, determine the local elevation of the thinning curve but have no direct effect on α (see below). The specific form, structure, and crowding, together with the growth mode, jointly determine the y-intercept k.

Geometric similarity and elastic similarity are two general, alternative principles of growth. The geometric-similarity model makes remarkably accurate predictions of thinning exponents near -1.5 that are obtained empirically from populations of mosses, ferns, grasses, herbs, and trees. It is particularly successful when some averaging process is involved, such as for across-species regressions or for means from several species. In many cases, it is probably also an adequate explanation, even though the $-3/2$ gradient may be mimicked by other growth modes.

With trees, the growth geometry generally conforms much better to the elastic-similarity principle. Populations of many tree species also exhibit thinning gradients near -1.9, which is in accord with the elastic-similarity model and expected from empirical tree allometry. Other tree species come closer to the geometric-similarity exponent -1.5. This incompatibility between allometry and thinning gradients for some trees might be due to root competition in combination with differences in growth mode between roots and aerial parts, and with the roots dictating the thinning gradients of aerial structures. The thinning gradient -1 for the trunk of big, old trees may be caused by cessation of growth in trunk height or root depth. Thinning exponents near -1.5 for prostrate plants may be fortuitous results of combinations of different gradients (-2 and -1) for separate plant parts; they need bear no relation to geometric similarity (Norberg, MS).

The self-thinning rule applies to two categories of data: data from single-species populations and, under certain conditions (explained below), from mixed-species ones as well; and across-species regressions of plants of different size, form, and structure. The reason for this congruence is given below.

An inherent characteristic of the power-law thinning models (eq. 1) is that the gradient is strictly dictated by the plant's growth mode. When geometric similarity obtains, growth does not cause any change in shape and structure. The slope is then absolutely insensitive to differences in shape and structure between species. With all other growth modes, however, growth and structure are interrelated since growth brings on structural change. The growth mode then dictates the range of gradients that may be taken; but for any particular combination of plant size and population density (figs. 1, 6) the precise value of the gradient is determined by the current structure of the plants, namely, the ratio of trunk radius to branch length when elastic similarity obtains (figs. 4, 5).

With elastic similarity, the gradient changes only very little when the ratio between trunk radius and projected branch length varies within the realistic range (fig. 4). The gradient is thus rather insensitive to plant structure and geometry per se. Therefore, no matter how structurally dissimilar different plant species are, they exhibit the same within-population thinning gradient, if only they obey the same growth principle ($-3/2$ with geometric similarity, and the same narrow range of potential gradients with elastic similarity). Provided that the constituent species in mixed-species populations all have the same growth mode, grow at the same relative rate, and have the same survival rate (such that each species contributes a constant proportion of the total volume of plant material per unit of ground area throughout growth), the thinning gradient is dictated simply by the growth mode, just as in single-species populations.

The local elevation of the thinning curves differs along with dissimilarities in geometry and structure among species. Nevertheless, if plants in an across-species regression span a sufficiently wide size range, then the gradient is still very close to some average gradient for single-species populations. The reason is that the height-to-breadth ratios (neglecting prostrate plants) and packing density of plant parts within the spatial volume occupied per plant (both factors included in k in eqs. 1, 6), can vary only within relatively narrow limits, whereas the possible ranges of plant size and population density are enormous. The failure to appreciate this has caused much mystification and confusion in previous discussions.

When plant population density (on the x-axis) in an across-species comparison varies by a factor of 10^6 (on a linear scale) and mean plant weight (on the y-axis) by a factor of 10^{10}, but the coefficient k in power-function (1) varies by a factor of only 8 (White 1980, p. 40; 1981, pp. 475, 494, 495), then data points are bound to fall into a narrow band. Constraints on the range of k probably come from (1) the necessity to grow in height because of competition for light, (2) the demands on structural strength against the force of gravity, (3) the necessity to claim a sufficiently large exclusive ground area in relation to the living biomass that must be sustained with light, water, and nutrients, and (4) shade tolerance, which limits foliage packing and amount of overlap tolerable from neighbors. Factor 1 tends to increase the height-to-breadth ratio, and 2 increases packing density within the volume occupied per plant; both 1 and 2 therefore force k upward and set a lower limit on k. Factor 3 tends to reduce the height-to-breadth ratio, and 4 limits packing density and also limits the height-to-breadth ratio of the average exclusive space; factors 3 and 4 both therefore place an upward limit on k. The fundamental nature of these constraints may explain why they seem to be universal among terrestrial plants (in setting strict limits on k). The factor-8 range of k dwindles in comparison with the enormous ranges of plant size and population density, which are far less constrained than k. This explains the seeming conformity among widely different plants in across-species regressions.

In a broad sense, then, gradients in thinning functions reflect size-related design principles that apply regardless of whether the plants are different species in different populations or the same individuals at different growth stages. In an across-species comparison, the gradient reflects a principle of size-related design across species, whereas in a time-sequential comparison within a population, it reflects the growth mode of individual plants.

The thinning exponents for populations of small plants, like grasses and herbs, often conform to predictions of the geometric-similarity model, whereas many trees are more compatible with the elastic-similarity principle with respect to both growth geometry and population thinning. It may be significant that long bones in small mammals seem most often to conform to geometric similarity, whereas in large mammals, long bones come closer to elastic similarity in across-species regressions (Economos 1983). Elastic similarity might therefore be a more important principle for the mechanical design of big organisms than of small ones.

After all, the interaction of gravity with large organisms calls for elastic similarity in their design. If geometric similarity applies throughout the smaller size classes of plants and animals (as seems to be the rule), then the safety factor against buckling tends to become progressively smaller with increasing weight. Selection for the elastic-similarity principle will therefore be greater among large plants (like trees) and animals than among smaller organisms.

CURRENT RESEARCH PROBLEMS

Westoby (1984) identified several problems for research about the self-thinning rule, which I now examine in light of my results. Westoby (1981, 1984) advocated the version of the rule that displays biomass per unit of ground area against population density as in my figures 2 and 7; and Weller (1987) gave several reasons why this version is better for fitting curves to empirical data. But the original version, relating volume of an average plant to population density, is the obvious choice for theoretical models based on plant shape, structure, and growth mode (as in this paper).

Westoby (1984, pp. 215–219) asked whether the self-thinning rule applies to four kinds of populations: stands in which all individuals are nearly equally large, stands showing a bimodal or multimodal size distribution among individuals, mixed-species stands, and mixed-age stands. Because the growth mode dictates the thinning gradient, the self-thinning rule may apply to such mixed populations, provided that the plants in the constituent size, age, or species classes of any one stand all adhere to the same growth mode, and provided also that competition between classes is symmetrical throughout the growth and thinning process, such that the individual growth rates and the survival rates are similar across the different classes. By contrast, when competition is asymmetrical, different classes vary continuously in their relative contribution to the mixed population and, therefore, also to the volume of the average plant in it. If the plants in the different classes are sufficiently dissimilar in size and structure, the gradient for the entire population may then fail to conform to the gradient associated with the prevailing growth mode, even if that mode is shared by all plants in all classes.

Two further problems concern the understanding of "variations between species and environments in the level, and perhaps slope, of the . . . thinning lines" and the transition from slope -1.5 to slope -1.0 (or from -0.5 to 0 when plant mass per unit of ground area is plotted against population density; figs. 1, 2) (Westoby 1984, p. 219). Variation in the local elevation of the thinning curve depends on differences in the characteristic height-to-width ratio of the average spatial volume occupied per plant, on differences in packing density of plant material in this space, and on the degree of overlap between neighboring plants. Species may differ in all three respects, and environmental differences (in, for instance, soil quality and light levels) may affect all three variables (Norberg, MS). The slope α is dictated by the growth mode of individual plants and may differ between species and environments (see "The -1 slope for old trees," above). And the y-intercept k of the thinning line is determined jointly by plant form, plant structure, overlap between neighbors, and growth mode. Transition to the -1.0 slope may result from stagnation of height growth of the trunk or stagnation of depth growth of the root system, whereupon the spatial volume occupied by the crown or root expands in two dimensions only, not in three as before.

SUMMARY

As plants grow, crowding eventually occurs and competition leads to size- and density-dependent mortality. Close packing of plants is maintained throughout, such that the growth mode of individual plants strictly dictates the exponent α in the power function $V = kN^{\alpha}$, relating average plant volume, V, to population density, N. Plant characteristics define the self-thinning line in a log-log diagram as follows. The growth mode determines the way in which plant form and structure change with growth, thereby dictating the slope α. The specific form and structure of plants, and the overlap between neighbors, determine the local elevation of the thinning curve but have no direct effect on α. The specific form and structure, crowding, and the growth mode jointly determine the y-intercept k.

Various growth modes are examined and the concomitant thinning functions derived. The geometric-similarity model makes remarkably accurate predictions of the thinning exponent -1.5 obtained empirically from populations of mosses, ferns, grasses, herbs, and some trees. But the $-3/2$ slope may be mimicked by other growth modes.

The growth geometry of most trees generally conforms much better to the elastic-similarity principle. A model based on elastic similarity predicts a thinning curve with gradients ranging from -2 to -1.33, with the most realistic range being -1.97 to -1.80. Populations of many tree species exhibit thinning gradients in accord with the elastic-similarity model (and empirical allometry), but others come closer to the geometric-similarity exponent, -1.5. The perplexing incompatibility between allometry and thinning gradients in some trees might be a result of root competition, in combination with different growth modes for roots and aerial parts, with the roots dictating the thinning gradient of aboveground structures.

Another model, based on a stagnation of the height growth of trunks or of the depth growth of roots, may explain the thinning gradient -1 for trunks of big, old trees. When root competition drives population thinning, the thinning gradient could easily be at variance with the empirical allometry of aboveground structures, and progressively more open space among plants may open up as they grow, with crowding being restricted to the roots. This may explain the wide spacings usually seen between bushes and trees growing on shallow soils where depth may be limited by bedrock or by a high water table (as in bogs); the thinning gradient -1 should then prevail.

In most trees, the leaves remain about the same size throughout the growth of the tree. This constancy of leaf size might be expected to result in radically different thinning slopes for leaves than for trunk and branches. But I show that as long as the canopy layer still grows in depth, the thinning exponent for leaf volume per tree versus density of the tree population may be -1.5, regardless of whether leaves remain of a constant size or grow isometrically with respect to trunk and branches. After the foliage layer has ceased to grow in thickness, the thinning exponent should be -1.5 for leaves that grow isometrically with stem and branches, but -1 for leaves of constant size. Gradients in thinning functions reflect size-related design principles that apply regardless of whether the plants are different species in different populations (in across-species regressions) or the same individuals at different growth stages, in which case the gradient reflects the growth mode of individual plants. The predicted thinning gradients for various growth modes are all readily testable.

The adherence to a common growth mode, or structural-design principle, explains why different plant species exhibit the same thinning gradient, despite profound dissimilarities in structure and geometry. If geometric similarity applies throughout the smaller size classes of plants and animals (as seems to be the rule), then the safety factor against buckling becomes progressively narrower with increasing weight. Selection for the elastic-similarity principle will therefore be greater among large plants (like trees) and animals than among smaller organisms.

ACKNOWLEDGMENTS

The incentive to do this study came from Hutching's (1983) stimulating article. I thank U. M. Norberg and K. Eriksson for mathematical advice and anonymous reviewers for constructive criticism. Support was obtained from the Swedish Natural Science Research Council (grant B 4450).

LITERATURE CITED

Curtis, R. O. 1971. Stand density measures for Douglas-fir. For. Sci. 17:146–159.

Economos, A. C. 1983. Elastic and/or geometric similarity in mammalian design? J. Theor. Biol. 103:167–172.

Ford, E. D. 1975. Competition and stand structure in some even-aged plant monocultures. J. Ecol. 63:311–333.
Givnish, T. J. 1986. Biomechanical constraints on self-thinning in plant populations. J. Theor. Biol. 119:139–146.
Gorham, E. 1979. Shoot height, weight and standing crop in relation to density of monospecific plant stands. Nature (Lond.) 279:148–150.
Greenhill, A. G. 1881. Determination of the greatest height consistent with stability that a vertical pole or mast can be made, and of the greatest height to which a tree of given proportions can grow. Proc. Camb. Philos. Soc. 4:65–73.
Hokkanen, J. E. I. 1986. Notes concerning elastic similarity. J. Theor. Biol. 120:499–501.
Hutchings, M. 1983. Ecology's law in search of a theory. New Sci. 98:765–767.
Kays, S., and J. L. Harper. 1974. The regulation of plant and tiller density in a grass sward. J. Ecol. 62:97–106.
Lonsdale, W. M., and A. R. Watkinson. 1983. Plant geometry and self-thinning. J. Ecol. 71:285–297.
McMahon, T. A. 1973. Size and shape in biology. Science (Wash., D.C.) 179:1201–1204.
———. 1984. Muscles, reflexes, and locomotion. Princeton University Press, Princeton, N.J.
McMahon, T. A., and J. T. Bonner. 1983. On size and life. Scientific American Library, New York.
McMahon, T. A., and R. E. Kronauer. 1976. Tree structures: deducing the principle of mechanical design. J. Theor. Biol. 59:443–466.
Miyanishi, K., A. R. Hoy, and P. B. Cavers. 1979. A generalized law of self-thinning in plant populations (self-thinning in plant populations). J. Theor. Biol. 78:439–442.
Mohler, C. L., P. L. Marks, and D. G. Sprugel. 1978. Stand structure and allometry of trees during self-thinning of pure stands. J. Ecol. 66:599–614.
Ogino, T. 1977. Pages 169–186 *in* T. Shidei and T. Kira, eds. Primary productivity of Japanese forests: productivity of terrestrial communities. University of Tokyo Press, Tokyo.
O'Neill, R. V., and D. L. DeAngelis. 1981. Comparative productivity and biomass relations of forest ecosystems. Pages 411–449 *in* D. E. Reichle, ed. Dynamic properties of forest ecosystems. Cambridge University Press, Cambridge.
Perez, J. W. 1970. Pages B105–B122 *in* H. T. Odum and R. F. Pigeon, eds. A tropical rain forest. U.S. Atomic Energy Commission. Oak Ridge, Tenn.
Perry, D. A. 1984. A model of physiological and allometric factors in the self-thinning curve. J. Theor. Biol. 106:383–401.
Peters, R. H. 1983. The ecological implications of body size. Cambridge University Press, Cambridge.
Pickard, W. F. 1983. Three interpretations of the self-thinning rule. Ann. Bot. 51:749–757.
———. 1984. The self-thinning rule. J. Theor. Biol. 110:313–314.
Saito, H. 1977. Pages 252–268 *in* T. Shidei and T. Kira, eds. Primary productivity of Japanese forests: productivity of terrestrial communities. University of Tokyo Press, Tokyo.
Waring, R. H. 1983. Estimating forest growth and efficiency in relation to canopy leaf area. Adv. Ecol. Res. 13:327–354.
Weller, D. E. 1987. A reevaluation of the $-3/2$ power rule of plant self-thinning. Ecol. Monogr. 57:23–43.
West, P. W., and C. J. Borough. 1983. Tree suppression and the self-thinning rule in a monoculture of *Pinus radiata* D. Don. Ann. Bot. (Lond.) 52:149–158.
Westoby, M. 1976. Self-thinning in *Trifolium subterraneum* not affected by cultivar shape. Aust. J. Ecol. 1:245–247.
———. 1977. Self-thinning driven by leaf area not by weight. Nature (Lond.) 265:330–331.
———. 1981. The place of the self-thinning rule in population dynamics. Am. Nat. 118:581–587.
———. 1984. The self-thinning rule. Adv. Ecol. Res. 14:167–225.
White, J. 1977. Generalization of self-thinning of plant populations. Nature (Lond.) 268:373.
———. 1980. Demographic factors in populations of plants. Pages 21–48 *in* O. T. Solbrig, ed. Demography and evolution in plant populations. Blackwell, Oxford.
———. 1981. The allometric interpretation of the self-thinning rule. J. Theor. Biol. 89:475–500.
White, J., and J. L. Harper. 1970. Correlated changes in plant size and number in plant populations. J. Ecol. 58:467–485.
Yoda, K., T. Kira, H. Ogawa, and H. Hozumi. 1963. Self-thinning in overcrowded pure stands under cultivated and natural conditions. J. Biol. Osaka City Univ. 14:107–129.

PAPER 18

Spatial Scaling of Species Composition: Body Masses of North American Land Mammals (1991)

J. H. Brown and P. F. Nicoletto

Commentary

FELISA A. SMITH

What structures ecological communities? Do patterns in the body-mass distribution of animals provide insights into the underlying processes driving community assembly? This paper was the first to use body mass—the most fundamental and arguably "integrative" or composite attribute of organisms—to examine the patterning of mammalian communities across varying spatial scale. Although distributional patterns had been previously investigated (e.g., Hutchinson and MacArthur 1959; May 1978; Bonner 1988), J. H. Brown and P. F. Nicoletto were noteworthy because of the quantitative and synthetic approach they took in analyzing multiple communities at divergent spatial scales.

Brown and Nicoletto found that at the local level the body-size structure of communities appeared to be uniform on a logarithmic scale. This suggested to them that species coexisting at a local site were a nonrandom subset of the regional species pool, differing significantly in mean, median, and skew. Moreover, in terms of community assembly, this suggested a limit to the number of species of each body size that could coexist. As sites were aggregated, higher turnover in smaller-bodied size classes (because of more specialized niches and more restricted distributions) led to increased numbers and a unimodal regional distribution, with a peak at smaller body size. Although the smallest size classes were represented at all scales, there was a reduction in the range of body size as scale decreased. This presumably resulted from the larger ranges required by large animals and the decreased likelihood that a local community might contain them. At the continental scale, distributions converged on a highly right-skewed lognormal distribution, with a mode around 100 grams. Brown and Nicoletto interpreted their results to mean that the composition of the regional biota was more influenced by historical and large-scale environmental events than that of the continental biota.

Since Brown and Nicoletto, it has become clear that patterns are somewhat more complicated than depicted and that the shape of the distribution at local scales depends not only on the definition of a "local scale" but on environmental characteristics. P. A. Marquet and H. Cofre (1999), for example, concluded in a study of mammals in South America that local and regional assemblages did not vary significantly from random draws of the species pool. Moreover, as more complete data have become available, it has also become clear that the anthropogenic-mediated extirpation of megafauna in North and South America has skewed the present distribution (Lyons et al. 2004). Nonetheless, this paper began a debate that is not yet resolved (e.g., Allen et al. 2006) and introduced a new way of looking at communities.

Literature Cited

Allen, C. R., A. S. Garmestani, T. D. Havlicek, P. A. Marquet, G. D. Peterson, C. Restrepo, C. A. Stow, and B. E. Weeks. 2006. Patterns in body mass distributions: Sifting among alternative hypotheses. *Ecology Letters* 9:630–43.

Bonner, J. T. 1988. *The evolution of complexity by means of natural selection.* Princeton University Press, Princeton, NJ.

*Hutchinson, G. E., and R. H. MacArthur. 1959. A theoretical ecological model of size distributions among species of animals. *American Naturalist* 93:117–25.

Lyons, S. K., F. A. Smith, and J. H. Brown. 2004. Of mice, mastodons and men: Human-mediated extinctions on four continents. *Evolutionary Ecology Research* 6:339–58.

Marquet, P. A., and H. Cofre. 1999. Large temporal and spatial scales in the structure of mammalian assemblages in South America: A macroecological approach. *Oikos* 85:299–309.

*May, R. M. 1978. The dynamics and diversity of insect faunas. In *Diversity of insect faunas,* edited by L. A. Mound and N. Waloff, 188–204. Royal Entomological Society of London Symposium 9. Blackwell Scientific, Oxford.

Vol. 138, No. 6 The American Naturalist December 1991

SPATIAL SCALING OF SPECIES COMPOSITION: BODY MASSES OF NORTH AMERICAN LAND MAMMALS

James H. Brown and Paul F. Nicoletto

Department of Biology, University of New Mexico, Albuquerque, New Mexico 87131

Submitted July 24, 1989; Revised October 15, 1990; Accepted November 2, 1990

Abstract.—We describe the nonrandom assembly of the North American terrestrial mammalian fauna based on body size and spatial scale. The frequency distribution of body masses among species for the entire continental fauna was highly modal and right skewed, even on a logarithmic scale; the median size of the 465 species was approximately 45 g. In contrast, comparable frequency distributions for 24 small patches of relatively homogeneous habitat were essentially uniform, with approximately equal numbers of species in each logarithmic size class; the median sizes of the 19–37 species ranged from approximately 100 to 2,500 g. Frequency distributions for 21 biomes (large regions of relatively similar vegetation) were intermediate between the continental and local assemblages. This pattern of assembly indicates that species of modal size (20–250 g) tend not to coexist in local habitat patches and they replace each other more frequently from habitat to habitat across the landscape than species of relatively large or small size. We hypothesize that three mechanisms are necessary and possibly sufficient to produce this result: competitive exclusion of species of similar size within local habitats, differential extinction of species of large size with small geographic ranges, and greater specialization of modal-sized species owing to energetic and dietary constraints.

The diversity and composition of biotas vary with spatial scale. The increase in species richness with sample area has been quantified by species-area relationships for nonisolated sites within continents, as well as for islands or insular habitat patches and continents of varying sizes (see, e.g., MacArthur and Wilson 1967; Flessa 1975; Schoener 1976; Connor and McCoy 1979; Brown and Gibson 1983; Brown 1986). Here, we address the question of whether there are other changes in the composition of the biota as species diversity increases with increasing spatial scale.

Because different processes affect biotic composition on different scales (see, e.g., Orians 1980; Ricklefs 1987), we would expect these processes to result in predictable changes in the attributes of species. On the largest scales, entire continents or large geographic regions within continents, species-level processes such as colonization, extinction, and speciation affect biotic composition. On the smallest scales, within small patches of homogeneous habitat, ecological interactions of species with each other and with the abiotic environment determine which combinations of species coexist. It is also on these small scales that microevolutionary processes of natural selection and genetic drift operate within populations. On intermediate scales, both macroscopic and microscopic processes cause changes in species composition across the landscape. However, processes

Am. Nat. 1991. Vol. 138, pp. 1478–1512.

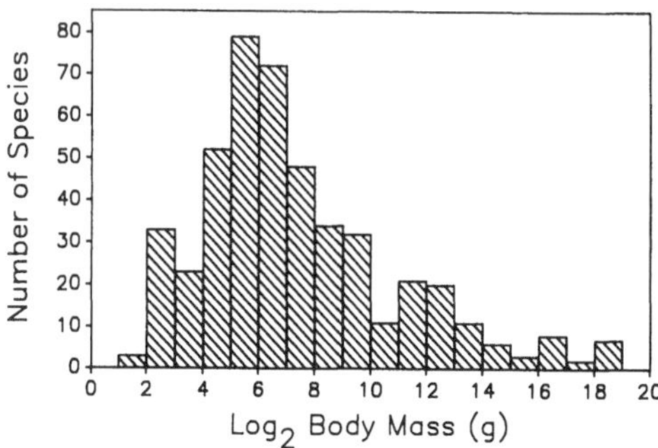

FIG. 1.—Frequency distribution, on a $\log_2$ scale, of the body masses of the 465 species of terrestrial North American mammals. Note the characteristic modal, right-skewed shape.

operating on disparate scales are partially coupled. Macroscopic biogeographic and evolutionary processes affect microscopic community structure because local ecological communities are assembled from continental and regional species pools. Conversely, microscopic ecological and evolutionary processes affect the composition of continental and regional assemblages because large-scale biotas reflect the cumulative effects of phenomena that occur in many local communities.

In this article we use the distribution of body sizes among species of North American land mammals to assess one aspect of variation in faunal composition with spatial scale. Body size is an easily measured variable that is closely correlated with many aspects of morphology, physiology, behavior, and ecology through allometric relations (Peters 1983; Calder 1984; Schmidt-Nielsen 1984; Zeveloff and Boyce 1988). In 1959, Hutchinson and MacArthur called attention to the distribution of body sizes among species of North American mammals (fig. 1). On a logarithmic scale, this distribution is highly modal and skewed toward larger body sizes. Qualitatively similar distributions have since been found for a wide range of organisms, from birds and insects (May 1978, 1988; Morse et al. 1988) to bacteria (Bonner 1988).

We characterize the variation in the distributions of body sizes among species with respect to spatial scale by comparing the frequency distributions for three scales: the entire North American continent, regional biomes, and local habitat patches. We reject the null hypothesis that the sizes of mammals co-occurring on successively smaller spatial scales are random subsamples of the larger species pools. We propose mechanistic hypotheses to account for the observed pattern of faunal assemblage.

METHODS

We compiled species lists (App. A) of terrestrial North American mammals for three different spatial scales: (1) the entire North American continent, including Mexico, (2) 21 biomes as defined and mapped by Dasman (1975), and (3) 24 small patches of relatively homogeneous habitat. The North American species list was taken from Hall (1981) and supplemented with Mexican species from Ramirez-Pulido et al. (1986). Species lists for the biomes were compiled by using the range maps in Hall (1981) to determine the occurrence of species in the biomes mapped by Dasman (1975). Species lists for the local habitats were obtained from the literature or from colleagues who contributed unpublished data from intensive field studies. We tried to ensure that the local habitats represented a small area (usually 10–1,000 ha) of uniform geology and vegetation and that the faunal list included all species that utilized the habitat at the time of colonization by European humans. The analyses included all species of native mammals except bats, pinnipeds, cetaceans, and the sea otter.

A single value of body mass was assigned to each species wherever it occurred; we ignored intraspecific geographic variation. The body masses were obtained from field guides (Burt and Grossenheider 1976; Whitaker 1980), from G. Ceballos (personal communication) for most Mexican species, and, in the few cases where masses were unavailable, from estimation based on comparisons with closely related species of similar head and body length.

The analysis focused on the frequency distributions of species in logarithmic ($\log_2$; see App. B for size categories) body-size categories. We used $\log_2$ intervals because most previous analyses have been based on log-transformed data and because base 2 divides the fauna into a convenient number of categories (see, e.g., Preston 1962). Using logarithmic size categories makes our analyses insensitive to small errors in assigning body masses to species (such as may be caused by intraspecific geographic variation).

The distributions of body masses among species for biomes and local habitats were compared with null models that assumed that species were assembled at random from appropriate larger-scale species pools. Each of the 21 biome distributions was compared with the North American distribution by drawing at random 500 times from the continental species pool the same number of species as occurred in that biome. The median was calculated for each of the 500 simulations; the number of simulated medians out of 500 was compared with the observed value to evaluate the null hypothesis that the biome was a random subsample of the continental species pool. A similar procedure was used to compare the distributions for 24 local habitats and 24 simulations, each consisting of 500 random draws from the biome distributions in which each of those habitats was located.

Inspection suggested that the North American distribution was highly modal and that the distributions for biomes and local habitat patches became progressively more uniform. We quantified this change in shape by comparing all distributions to a log-uniform distribution, with the same range as the North American distribution, by using the Kolmogorov-Smirnov *Dn* statistic and test.

If the distributions of body sizes change with spatial scale, there must be a differential replacement of certain size classes between biomes and especially between local habitats. In other words, different size classes should show different degrees of beta diversity. There are two ways species can change status from absent to present (or vice versa) between habitats. A species may be replaced because the border of its geographic range has been crossed. Alternatively, a

TABLE 1

SUMMARY STATISTICS FOR FREQUENCY DISTRIBUTIONS OF LOG_2 OF BODY MASSES (IN GRAMS) FOR MAMMALS OF NORTH AMERICA AND 21 NORTH AMERICAN BIOMES

Biome Number	Region	*N*	Median	Minimum	Maximum	Interquartile Range	Standard Skewness
All	North America	464	6.4	1.6	18.9	4.1	9.2
1	Sitkan	46	10.2	2.3	18.9	7.8	.4
2	Oregonian	77	8.1	2.8	18.9	6.5	1.9
3	Yukon taiga	46	10.1	1.6	18.8	8.9	.3
4	Canadian taiga	72	8.0	1.6	18.9	8.4	1.4
5	Eastern forest	74	8.4	1.6	18.9	7.2	1.3
6	Austroriparian	59	8.4	1.6	18.9	7.6	1.2
7	Californian	88	7.0	2.6	18.9	6.3	2.8
8	Sonoran	102	7.6	2.6	18.9	7.1	2.5
9	Chihuahuan	113	7.3	2.3	18.9	6.4	2.9
10	Tamauilipan	66	10.0	2.3	18.7	7.7	.3
11	Great Basin	95	8.1	1.6	18.9	6.4	2.3
12	Alaskan tundra	37	10.8	2.3	18.8	8.5	.0
13	Canadian tundra	38	10.6	2.3	18.8	8.6	.4
14	Grasslands	115	8.5	2.3	18.9	7.1	1.7
15	Rocky Mountains	110	7.6	1.6	18.9	6.6	2.3
16	Sierra-Cascade	108	7.1	2.3	18.9	6.4	2.8
17	Madrean-Cordilleran	182	7.1	2.3	18.9	6.2	3.4
18	Campechean	95	8.2	2.3	18.2	6.4	1.3
19	Guerreran	119	7.6	2.3	18.2	5.9	2.4
20	Sinaloan	84	7.6	2.6	17.0	7.0	2.0
21	Yucatecan	51	11.6	2.8	18.2	5.4	−.5
Biome mean		82.2	8.6	2.2	18.7	7.0	1.5

species with specialized habitat requirements may not occur in some of the habitat types within its geographic range. For each of the 24 local habitats, we compiled two frequency distributions of body sizes, one containing those species in the continental species pool whose geographic ranges did not include the habitat and the other containing those species in the continental pool whose geographic range did include the site but that did not occur in that habitat. Then we summed each kind of list with redundancy (counting each species repeatedly, as many times as it occurred) to determine the overall shapes of the distributions for species exhibiting the two kinds of replacement between habitats. For most habitats this was straightforward, because the authors of the local species lists included information on the occurrence of species in the other habitat types within that local area. In three cases, these local lists were not available and we used other published sources to determine which species had geographic ranges that included the site but were found in adjacent habitats.

RESULTS

The frequency distributions of body masses of North American terrestrial mammals varied with the spatial scale of the sample: the medians and interquartile ranges increased and the skewness decreased from continent to biome to local habitat patch (tables 1 and 2). The distribution for the entire continental fauna

TABLE 2

SUMMARY STATISTICS FOR FREQUENCY DISTRIBUTIONS OF LOG_2 OF BODY MASSES (IN GRAMS) FOR NORTH AMERICAN MAMMALS IN 24 LOCAL PATCHES OF UNIFORM HABITAT

Local Habitat Patch Number	Locality	Habitat Type	Biome	*N*	Median	Minimum	Maximum	Interquartile Range	Standard Skewness	Sources
1	White Sands, N.Mex.	Chihuahuan desert	9	34	8.4	3.2	16.8	5.8	.7	Anonymous 1987
2	Bernalillo Co., N. Mex.	Riparian forest	9	31	12.0	2.6	18.9	6.9	−.5	Hink et al. 1984
3	Deep Canyon, Calif.	Pine forest	8	18	9.7	5.1	16.8	6.9	.6	Ryan 1968
4	Deep Canyon, Calif.	Piñon/Juniper	8	23	9.8	2.6	16.8	8.7	.1	Ryan 1968
5	Deep Canyon, Calif.	Agave/Ocotillo	8	26	8.6	2.6	16.8	8.1	.8	Ryan 1968
6	Cochise Co., Ariz.	Chihuahuan desert	9	32	7.6	3.2	16.8	8.3	1.0	J. H. Brown, personal communication
7	Barge Canal, Fla.	Longleaf pine	6	19	7.6	5.1	25.7	16.7	4.5	Anonymous 1976
8	Barge Canal, Fla.	Scrub oak	6	22	10.7	2.3	17.1	7.7	.0	Anonymous 1976
9	Barge Canal, Fla.	Pine flatwoods	6	18	10.8	1.6	17.1	7.7	−.3	Anonymous 1976
10	Chamela, Jalisco, Mexico	Deciduous forest	19	27	12.0	4.8	16.8	5.3	−.5	Ceballos and Miranda 1986
11	Chamela, Jalisco, Mexico	Evergreen forest	19	28	11.8	4.8	16.8	6.8	−.1	Ceballos and Miranda 1986
12	Green Mountains, Vt.	Spruce/Fir	5	34	10.1	2.3	18.9	7.9	.1	J.F. Merritt, personal communication
13	Ligonier Valley, Pa.	Deciduous forest	5	30	8.8	1.6	18.9	8.8	.5	J.F. Merritt, personal communication
14	Cook Co., Minn.	Aspen/Birch	4	22	8.9	2.3	17.4	8.6	.3	Timm 1975
15	Cook Co., Minn.	White pine	4	25	7.6	2.3	17.4	8.4	.7	Timm 1975
16	Animas Mountains, N. Mex.	Oak forest	17	28	12.0	2.6	18.5	7.7	−.8	Cook 1986
17	Sagehen, Calif.	Jeffrey pine	16	29	9.4	2.8	18.9	8.2	.6	Morrison et al.
18	Cascade Mountains, Oreg.	Alpine tundra	16	29	8.4	2.8	18.9	9.9	.6	E.R. Brown 1985
19	Cascade Mountains, Oreg.	Sage	16	36	8.1	2.6	18.5	8.2	1.0	E.R. Brown 1985
20	Cascade Mountains, Oreg.	Ponderosa pine	16	37	10.4	4.0	18.9	7.9	.2	E.R. Brown 1985
21	Lac Qui Parle Co., Minn.	Upland prairie	14	23	11.0	2.3	18.9	10.4	.0	E. Birney and G. Nordquist, personal communication
22	Washington Co., Minn.	Shrub swamp	4	29	10.8	2.3	17.1	8.4	−.2	E. Birney and G. Nordquist, personal communication
23	Norman Co., Minn.	Willow swamp	14	32	11.2	2.3	18.8	8.6	−.1	E. Birney and G. Nordquist, personal communication
24	Konza Prairie, Kans.	Tallgrass prairie	14	31	8.5	2.3	18.9	10.7	.3	Finck et al. 1986
Habitat mean				27.6	9.8	3.8	17.3	8.4	.4	

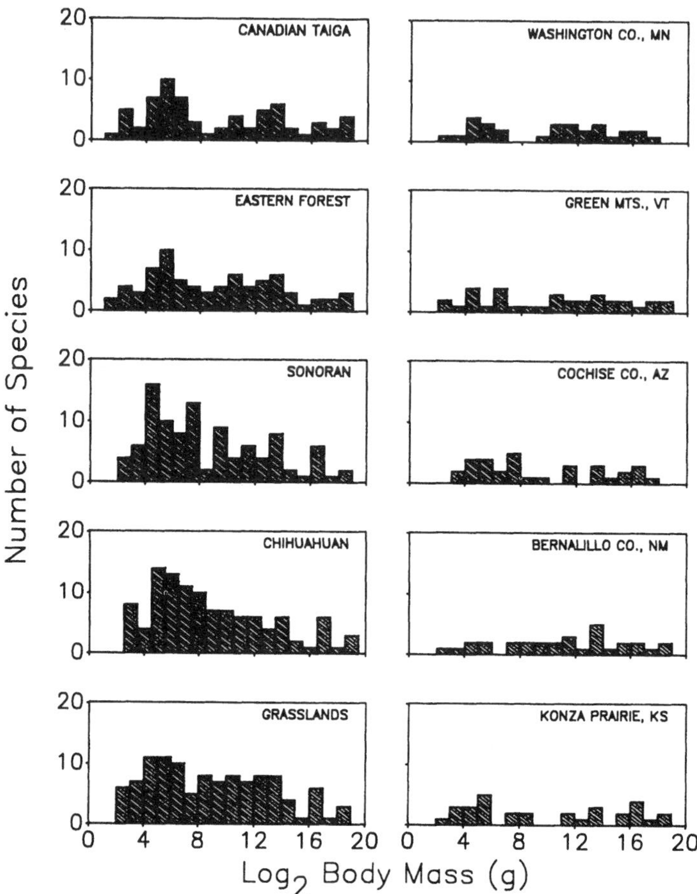

FIG. 2.—Frequency distributions, on a $\log_2$ scale, of the body masses of terrestrial mammals inhabiting five biomes and five local patches of uniform habitat within those biomes. Note that the biome distributions are all highly modal, but the habitat distributions are all nearly uniform on the logarithmic scale.

was highly modal and significantly right skewed; the mode was in size class 5, approximately 45 g (fig. 1). In contrast, the distributions for all 24 local habitats had an approximately equal number of species in each logarithmic size category (fig. 2); these distributions were all statistically indistinguishable from log-uniform distributions (table 3). The distributions for the biomes were intermediate between those for the continent and those for the local habitats (fig. 2); nine of the 21 biomes did not differ significantly from log-uniform distributions (table 3).

Neither the biomes nor the local habitats were random samples of the species pools on the next larger scale, continent and biomes, respectively. All 21 biome distributions had medians significantly ($P < .05$; <25 of 500 simulations) larger than the North American distribution. Nine of the 24 local habitat patches had medians significantly larger than the simulated medians ($P < .05$), and an addi-

TABLE 3

COMPARISONS OF THE DISTRIBUTIONS OF BODY SIZES FOR NORTH AMERICA, 21 BIOMES, AND 24 LOCAL HABITAT PATCHES WITH A LOG-UNIFORM DISTRIBUTION, USING THE KOLMOGOROV-SMIRNOV TEST

		Dn	*P* value
North America		.32	.0000
Biome:			
Sitkan		.10	.9999
Oregonian		.16	.0464
Yukon taiga		.11	.9992
Canadian taiga		.16	.0618
Eastern forest		.13	.1889
Austroriparian		.14	.2009
Californian		.23	.0001
Sonoran		.21	.0001
Chihuahuan		.22	.0001
Tamauilipan		.20	.4693
Great Basin		.16	.0116
Alaskan tundra		.08	1.0000
Canadian tundra		.11	.9999
Grasslands		.15	.0128
Rocky Mountains		.17	.0034
Sierra-Cascade		.20	.0003
Madrean-Cordilleran		.20	.0001
Campechean		.21	.0001
Guerreran		.22	.0001
Sinaloan		.18	.0079
Yucatecan		.16	.1626
	Habitat	*Dn*	*P* value
Locality:			
White Sands, N.Mex.	Chihuahuan desert	.18	.2078
Bernalillo Co., N.Mex.	Riparian forest	.12	.9999
Deep Canyon, Calif.	Pine forest	.13	.9999
Deep Canyon, Calif.	Piñon/Juniper	.16	.9999
Deep Canyon, Calif.	Agave/Ocotillo	.17	.4057
Cochise Co., Ariz.	Chihuahuan desert	.21	.1157
Barge Canal, Fla.	Longleaf pine	.19	.5303
Barge Canal, Fla.	Scrub oak	.13	.9996
Barge Canal, Fla.	Pine flatwoods	.13	.9999
Chamela, Jalisco, Mexico	Deciduous forest	.24	.0876
Chamela, Jalisco, Mexico	Evergreen forest	.19	.2611
Green Mountains, Vt.	Spruce/Fir	.07	1.0000
Ligonier Valley, Pa.	Deciduous forest	.14	.9991
Cook Co., Minn.	Aspen/Birch	.15	.9998
Cook Co., Minn.	White pine	.18	.4121
Animas Mountains, N.Mex.	Oak forest	.16	.4658
Sagehen, Calif.	Jeffrey pine	.12	.9999
Cascade Mountains, Oreg.	Alpine tundra	.17	.4053
Cascade Mountains, Oreg.	Sage	.13	.5224
Cascade Mountains, Oreg.	Ponderosa pine	.08	1.0000
Lac Qui Parle Co., Minn.	Upland prairie	.09	1.0000
Washington Co., Minn.	Shrub swamp	.10	.9999
Norman Co., Minn.	Willow swamp	.11	.9999
Konza Prairie, Kans.	Tallgrass prairie	.19	.2009

NOTE.—The North American and biome distributions were compared with a log-uniform distribution with the same range as the observed North American distribution, and the local habitats were compared to a log-uniform distribution with the same range as the biome in which they occurred. For each comparison the Kolmogorov-Smirnov *Dn* statistic and the probability value are given.

TABLE 4

RESULTS OF SIMULATIONS TO EVALUATE THE NULL HYPOTHESIS THAT THE BODY-SIZE DISTRIBUTIONS FOR EACH OF THE 24 LOCAL HABITATS ARE A RANDOM SUBSET OF THE BIOME POOL IN WHICH THAT HABITAT IS LOCATED

	MEDIAN		
HABITAT	Observed	Simulated	Proportion of Simulations That Are Less than Observed Value
White Sands, N.Mex.	8.4	7.29	.088
Bernalillo Co., N.Mex.	12.0	7.54	.001
Deep Canyon, Calif.	9.7	7.38	.096
Deep Canyon, Calif.	9.8	7.46	.048
Deep Canyon, Calif.	8.6	7.29	.156
Cochise Co., Ariz.	7.6	7.09	.386
Barge Canal, Fla.	7.6	7.12	.736
Barge Canal, Fla.	10.7	8.61	.060
Barge Canal, Fla.	10.8	8.59	.090
Chamela, Jalisco, Mexico	12.0	7.78	.001
Chamela, Jalisco, Mexico	11.8	7.77	.001
Green Mountains, Vt.	10.1	7.72	.068
Ligonier Valley, Pa.	8.8	8.80	.384
Cook Co., Minn.	8.9	7.23	.382
Cook Co., Minn.	7.6	6.69	.432
Animas Mountains, N.Mex.	12.0	8.38	.004
Sagehen, Calif.	9.4	7.16	.002
Cascade Mountains, Oreg.	8.4	6.99	.168
Cascade Mountains, Oreg.	8.1	6.99	.168
Cascade Mountains, Oreg.	10.4	7.35	.001
Lac Qui Parle Co., Minn.	11.0	8.42	.058
Washington Co., Minn.	10.8	7.36	.004
Norman Co., Minn.	11.2	8.53	.001
Konza Prairie, Kans.	8.5	7.70	.238

NOTE.—The probability of failing to reject each null hypothesis is given as the proportion of the 500 simulations that are less than the observed value for medians.

tional six patches were marginally significantly larger ($P < .10$) than the random draws from the biome in which they were located (table 4). The medians for the local habitat patches were all significantly larger than the medians of random draws from the North American distribution.

The only way to account for the progressive flattening of the frequency distributions from continental to biome to local scales is by high spatial replacement of species in the modal size classes. This is reflected in the limited habitat distributions and small geographic ranges of these species compared to their larger relatives. The vast majority of species in the smaller size categories (less than class 12, approximately 1 kg) occurred in four or fewer biomes, whereas the majority of species in the larger size categories occurred in eight or more biomes (fig. 3). An even clearer pattern is apparent for the areas of geographic ranges: a substantial proportion of the species in the smaller size categories (less than class 13) had ranges of less than 100,000 km^2, whereas all species in the larger size classes had ranges that exceeded this area (fig. 4).

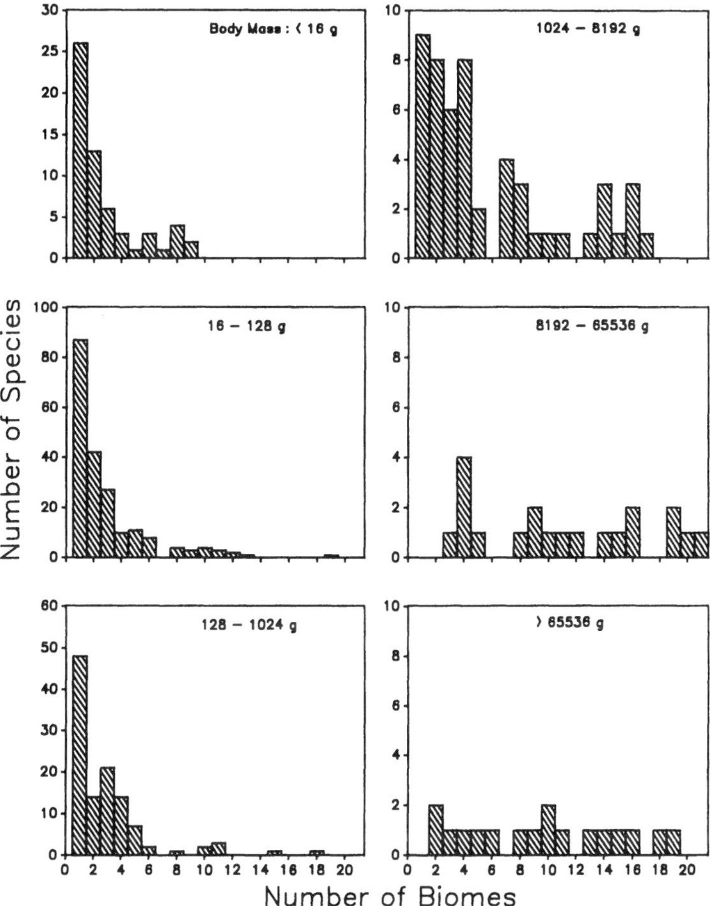

FIG. 3.—Frequency distributions of species in different body-size classes according to the numbers of biomes where they occur. Note that species of small size tend to occur in few biome types, whereas the majority of species in the larger size classes inhabit many biomes.

The composition of species may vary among habitats across the landscape for two reasons: because these species have limited geographic ranges or because they do not occur in all the habitats within their geographic ranges. It is of interest to assess the contributions of these two phenomena to the replacement of different-sized species among local habitats. For each local habitat, we determined the identity and body mass of those species in the continental species pool that were absent because they either had a geographic range that did not encompass the site or had a geographic range that included the site but did not occur in that habitat. These are illustrated in figure 5, which shows the frequency distributions of body masses of the species in these two categories summed with redundancy for the 24 local habitats. Note the general similarity in these distributions, although there were proportionately more medium-sized species (classes

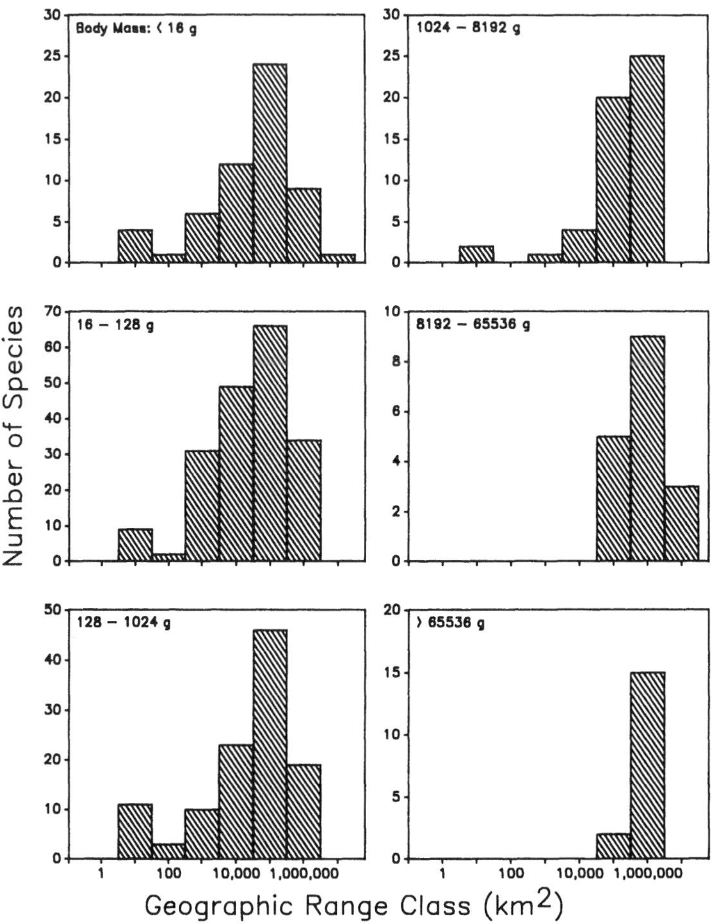

FIG. 4.—Frequency distributions of species in different body-size classes according to the areas of their geographic ranges (measured by planimetry from maps in Hall 1981). Note that many species of small size have restricted geographic ranges, whereas most species of large size are distributed over large areas.

9–13) in category 2. Species in the modal size classes replaced each other more rapidly across the landscape both because they had more restricted geographic ranges and because they occurred in a smaller proportion of habitat types within their geographic ranges.

DISCUSSION

Patterns

We have shown what appears to be a general pattern of spatial scaling of North American terrestrial mammalian assemblages with respect to body size. The distribution of body masses for the continent as a whole is highly modal and skewed

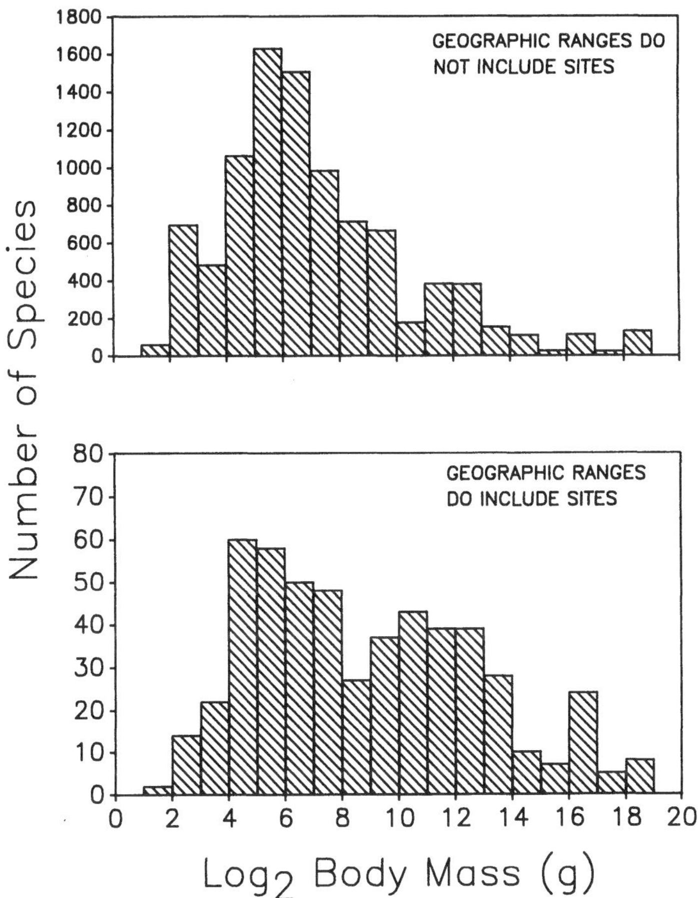

FIG. 5.—Frequency distributions of body masses summed with redundancy (see text) for the two classes of species that did not occur in the 24 local habitat patches: those whose geographic ranges did not include the local sample sites (*top*) and those whose geographic ranges did encompass the sites even though they did not occur in the habitats sampled (*bottom*). Note that both distributions are modal and right skewed.

toward larger size classes (fig. 1). The frequency distributions for small patches of homogeneous habitat have ranges of body sizes similar to those of the North American distribution, but they contain an approximately equal number of species in each logarithmic size class (fig. 2). The distributions for biomes, which are intermediate in area and habitat diversity between the continent and the local habitats, are intermediate in shape between the North American and the habitat distributions. These patterns characterize the composition of the mammalian faunas on these scales before the impact of European humans. Hunting, trapping, habitat changes, and other human impacts within the last three centuries have caused the local extirpation of a few species; human introductions of alien species have resulted in the colonization of some additional species. These recent changes in species composition were not sufficient, however, to change the shapes of the above distributions to any substantial extent.

The North American mammalian fauna experienced additional changes in the period 500–20,000 yr ago when catastrophic extinctions occurred (Martin and Klein 1984). Although the causes of these losses are hotly debated, humans may have played a substantial role. It is clear that large species suffered more extinctions (Martin 1984; Webb 1984). These extinctions were not sufficient, however, to affect the qualitative patterns reported here. Despite the number of species and genera that were lost, the frequency distribution of body masses for North American fauna prior to the Quaternary extinctions still has a mode and right skew similar to the distribution for the contemporary fauna (R. Rusler, personal communication). Similarly, the addition of species known or presumed to have become extinct within the last 20,000 yr to the local habitats would have extended the range to larger size categories in some cases but would not have substantially altered the shape of the frequency distributions.

The shape of the continental distribution, first emphasized by Hutchinson and MacArthur (1959), appears to be typical of the frequency distributions of body sizes of species in diverse taxa from large geographic regions (May 1978, 1988; Bonner 1988). An explanation for this pattern in mammals, and perhaps in other organisms, must take into account the changes in the frequency distributions with spatial scale. The continental mammalian fauna is not simply the sum of species assemblages on smaller spatial scales, and the faunas of local habitats and biomes are not just random subsamples of the species pools on large spatial scales. Species near the modal size of approximately 45 g replace each other frequently between habitats and biomes, whereas species of large size tend to have large geographic ranges and to occur in a large proportion of the habitats (figs. 3 and 4).

Hypotheses

We offer the following hypotheses to explain this pattern: (1) competitive exclusion tends to prevent local coexistence of similar-sized species with similar resource requirements, (2) differential extinction of species of large size with small geographic ranges tends to limit the number of large species in the continental fauna, and (3) allometric energetic constraints cause modal-sized species to be more specialized in their use of resources than larger species. We develop each of these hypotheses in more detail below.

Competitive exclusion among similar-sized species.—Since the faunas of local habitats contain fewer modal-sized species than random samples of assemblages from larger spatial scales, it appears that some local process prevents coexistence of species of similar size. Competitive exclusion is the most likely process to have this effect. If such competition occurs, however, its impact should be limited to those species that share requirements for the same limited resources. Competitive exclusion should occur within but not between feeding guilds. There is substantial evidence for interspecific competition among species of mammals and other organisms with similar diets (see, e.g., Hutchinson 1959; McNab 1963; Rosenzweig 1966; MacArthur 1972; Brown 1975, 1987; Pacala and Roughgarden

1982, 1985; Brown and Munger 1985; Grant 1986). In at least some of these cases, competition appears to account for the tendency of local communities to be composed of species that are more different in body size than expected from random assemblages drawn from appropriate species pools (see, e.g., Schoener 1970, 1984; Brown 1973; Simberloff and Boecklen 1981; Bowers and Brown 1982; Brown and Bowers 1985; Hopf and Brown 1986). It is hypothesized that the log-uniform distributions on the scale of habitats reflect the fact that local assemblages are composed of multiple guilds, each of which tends to have its own log-uniform distribution. For example, if mammals are divided into two very general trophic groups, herbivores and carnivores, most habitats contain representatives from each group that vary in size from less than 20 g to more than 100 kg.

Large species with small geographic ranges have high extinction probabilities.—The fact that large species exhibit low beta diversity (low replacement between habitats and geographic regions) suggests that some large-scale process prevents the accumulation of high diversity of large mammals in the continental fauna. We hypothesize that this process is the selective extinction of species with large body sizes and small geographic ranges. Individuals of large size have large resource requirements, and as a result they require large home ranges and occur at low population densities (McNab 1963; Schoener 1968; Parra 1978; Harestad and Bunnell 1979; Damuth 1981; Peters 1983; Peters and Raelson 1984). As a consequence, species of large size with small geographic ranges have small total population sizes and should be differentially susceptible to extinction (MacArthur and Wilson 1967; Brown 1981; Pimm et al. 1988; Schoener and Spiller 1988). If extinction is caused exclusively by demographic variation (in the absence of major environmental change), critical population sizes for persistence are very small, and even small geographic ranges would often contain sufficient individuals to avoid extinction. If, on the other hand, extinction is caused by environmental variation, even relatively large, dispersed populations may be at risk (Goodman 1987).

We hypothesize that environmental changes have been a major cause of mammalian extinctions and that these changes have differentially affected the species in each size class with the smallest geographic ranges. The observation that the minimum size of geographic ranges of large species is larger than that of modal-sized species (Brown 1981; Rapoport 1982; Brown and Maurer 1987, 1989; see also fig. 4) is consistent with this hypothesis, as is the observation that mammals and other organisms of large size have been differentially susceptible to historical perturbations that caused mass extinctions (Martin and Klein 1984; Webb 1984).

Large mammals may also have lower speciation rates than their smaller relatives. Large species tend to have greater vagility and broader environmental tolerances, and these should result in less isolation and genetic differentiation of populations and in lower speciation rates than in small species. While lower speciation rates might contribute to the low diversity of large mammals in the continental fauna, they are not sufficient to account for the failure to observe large species with small geographic ranges.

Specialization of modal-sized species.—A correlate of body size that is related to both of the above hypotheses is the apparently greater specialization of modal-sized species. Not only does the greatest number of mammalian species occur in the size range of 20–250 g, but species within this size range exhibit greater turnover among habitats and on the average have smaller geographic ranges (figs. 3, 4). This suggests that modal-sized species are more frequently limited by variation in the physical environment, the presence of other organisms, or both processes than their larger relatives. We hypothesize that this specialization is not simply a consequence of geographic and habitat distribution (in which case the above arguments would be circular) but is caused by allometric constraints on physiology and energetics.

It is well-known that energetic and basic nutrient requirements (D) of individuals scale as a fractional exponent of body mass (M) as $M^{0.67}$ to $M^{0.75}$ (Peters 1983; Calder 1984). Although the following qualitative argument does not depend on the precise value of the exponent, we will assume that the daily energy demand of free-living mammals scales as

$$D = c_1M^{0.75}, \tag{1}$$

where c_1 is a constant (Nagy 1987). The rate of food intake in a variety of mammals also scales as $M^{0.75}$ (Calder 1984). However, larger animals have larger and longer alimentary tracts; in both herbivorous and carnivorous mammals gut capacity (A) scales as approximately

$$A = c_2M^{1.0}, \tag{2}$$

where c_2 is another constant (Calder 1984). As a consequence, larger animals retain food in the gut for a longer period; turnover time (T) for gut contents scales as approximately

$$T = c_3M^{0.25}, \tag{3}$$

where c_3 is another constant (Calder 1984). This enables larger animals to ingest poorer-quality food and, by subjecting it to digestion for a longer period, still to extract sufficient energy and nutrients to meet their requirements. Thus, diet quality scales inversely and digestive efficiency scales directly with body size.

The food quality (Q) necessary to meet requirements can be predicted from the ratio of metabolic demand to gut capacity; it should scale as

$$Q = D/A = c_4M^{-0.25}, \tag{4}$$

where c_4 is another constant. The best data to evaluate this prediction are for ruminant mammals (Hoppe 1977; Sibly 1981; du Toit and Owen-Smith 1989). The predicted value of -0.25 is very close to those (-0.20 to -0.27) estimated by McNaughton and Georgiadis (1986). We also used data on 14 species of African ruminants from Hoppe (1977; cited in Sibly 1981) to obtain another estimate of the scaling exponent for food quality. We performed a regression of the reciprocal of reticulo-rumen contents divided by metabolic demand against body mass and obtained an exponent of -0.351 ± 0.088.

We hypothesize that these physiological constraints on food quality force smaller animals to specialize in higher-quality foods and to restrict their foraging to habitats where suitable foods are available in sufficient supply. Even small omnivorous species should be subject to these allometric relations and should restrict their diets to foods of high energetic and nutritional value. Other constraints of body size, such as those on reproductive and life-history traits (see, e.g., Eisenberg 1981; Peters 1983; Calder 1984), increase the nutritional demands on smaller organisms and tend to reinforce the selection to specialize on foods and habitats. Another mechanism must be hypothesized to account for the small number of species (and the low population densities of these species) in the smaller-than-modal classes (see Brown and Maurer 1987, 1989; Dial and Marzluff 1988).

The specialization hypothesis developed here is not independent of the competitive-exclusion and differential-extinction hypotheses presented above. Rather, the nutritional constraints and the resulting specialization provide another level of explanation for the pervasive effects of body size on resource use, habitat selection, population regulation, and distributional limits that ultimately determine small-scale species interactions and large-scale species dynamics.

Allocation of Food and Space among Species

Regardless of whether the above hypotheses are necessary and sufficient to explain the empirical patterns, the spatial scaling of the body-size distributions has important implications for the allocation of food and space among species and hence for the assembly of continental biotas. First, the nature of ecological communities depends on the spatial scale of study. The larger and more heterogeneous the area sampled (and these will tend to be correlated), the more species composition will reflect beta diversity (replacement of species between habitats) relative to alpha diversity (number of coexisting species within a habitat).

Furthermore, not only the number of species but also the kinds of species that actually and potentially interact vary with spatial scale. The limited number of species that coexist in small patches of relatively homogeneous habitat tend to be of different sizes. Because of the energetic constraints outlined above, this will tend to result in coexisting species' using different food resources or using the same resources in different ways. Many of the coexisting species of similar size are in different trophic guilds. In terrestrial mammals, carnivores and herbivores span virtually the entire range of body sizes. Although species in the same guilds may coexist locally and compete for food (especially if they are of different body sizes), often the most severe competition will be among species (of similar sizes) that rarely encounter each other because they occur in adjacent but largely nonoverlapping habitats and geographic regions.

The spatial scaling of body size emphasizes the importance of beta diversity (Cody 1975; Wilson and Shmida 1984). The high frequency of species in the modal size classes on the biome or continental scale reflects the frequent replacement of these species among local habitats. This turnover can be of two types. On the one hand, it may reflect crossing the borders of geographic ranges. On the other hand, it may reflect the fact that species do not occur in all habitats in their geographic ranges. The distribution of body sizes of species in the former category (fig. 5, *top*) is virtually identical in shape to the North American distribution. The species in the latter category (fig. 5, *bottom*) also exhibit a highly modal body-size distribution but are composed of relatively more representatives in size classes 9–13 (500 g–16 kg), many of which appear to be omnivores with relatively large geographic ranges.

Generality

Since the present analyses are exclusively of data for North American terrestrial mammals, it is reasonable to ask how general the results are. Of course it would be desirable to obtain and analyze comparable data for other groups of organisms in other geographic regions. Until this is done, we can make two comments.

First, the right-skewed distribution for the North American mammal fauna is typical of the shapes of frequency distributions of body-size measurements for many other taxa from large geographic areas. For example, May (1978, 1988), Bonner (1988), Morse et al. (1988), and Brown and Maurer (1989) present examples of qualitatively similar size distributions for a wide variety of organisms from bacteria to insects to birds. In addition, Rusler (1987) examined frequency distributions of body masses among species and genera of mammals on different continents. She found similar right-skewed distributions for all the larger continents with diverse faunas. We cannot claim that these biotas exhibit the same pattern of spatial scaling within continents without analyses of data for smaller scales; but it seems unlikely that there would be so much similarity in the large-scale patterns unless the underlying mechanisms and small-scale patterns were also similar.

Second, none of the mechanisms that we have proposed above to account for the patterns in North American mammals is inherently specific either to mammals or to the North American continent. Allometric scaling of morphological and physiological variables with body size exhibits similar exponents in a wide variety of animals, including vertebrates and invertebrates, endotherms and ectotherms (Peters 1983; Calder 1984). The relationship of body size to population density, area of geographic range, dietary specialization, and other attributes of species that affect the assembly of both local communities and continental biotas appears to be similar in the organisms that have been studied to date (see data on beetles, birds, and mammals in Brown and Maurer [1987, 1989] and Morse et al. [1988]). The generality of these relationships makes us optimistic that both the patterns of spatial scaling and the mechanistic processes that produce them are also general.

The constraints on the body sizes of species that occur together on different spatial scales can be thought of as one important component of the rules for assembling continental biotas. Although complete assembly rules would be based on other variables in addition to body size, it is clear that size, and traits that are correlated with size, profoundly affects the composition of species assemblages (see also Cody 1975; Diamond 1975; Brown 1981; Brown and Maurer 1987, 1989;

Dial and Marzluff 1988). Because of its pervasive influence on physiology, behavior, and ecology, body size influences the coexistence of species in local habitats, the turnover of species across the landscape, and the colonization, speciation, and extinction of species on continental scales.

Additional assembly rules will be required to account for differences in biotas among continents and between continents and islands. The importance of body size is also seen in the divergence of insular populations from their mainland relatives (see, e.g., for mammals, Foster 1964; Lomolino 1985) and in the systematic variation in the frequency distribution of size among species or genera in continental faunas as a function of the area of the land mass (see, e.g., Brown 1986; Rusler 1987; see also Van Valen 1973).

Implications for Conservation

The rules for assembling species on different spatial scales also have important implications for the maintenance of diversity and for the disassembly of biotas during environmental perturbations. Such disassembly has occurred naturally in the past, most dramatically during episodic mass extinctions (Martin and Klein 1984; Jablonski 1986; Raup 1986). At present, such disassembly is occurring in response to the multiple effects of the growing human population (Wilson 1988). Losses of species during past mass extinctions and in response to the impacts of modern humans are nonrandom with respect to body size and other attributes of species (see, e.g., Diamond 1984; Martin 1984; Webb 1984). For example, of the species known to have disappeared during the last three centuries from the 24 local habitats that we studied, 11 of 13 are larger than 2 kg and seven are larger than 35 kg. From the patterns of faunal assembly and these kinds of data on known extinctions, it should be possible to predict which species will be most vulnerable to perturbations (Arita et al. 1990).

Most current conservation strategies focus on efforts to preserve individual endangered species and threatened habitats. These procedures, by themselves, will be inadequate to prevent wholesale extinctions caused by human activities. It is estimated that between 5 and 30 million species of organisms currently inhabit the earth (Erwin 1983; May 1988; Wilson 1988). It will be impossible to identify which ones are endangered, much less to develop recovery plans for them, on a species-by-species basis. It will be almost equally difficult to inventory all kinds of habitats.

Rules for the assembly and disassembly of biotas suggest alternatives to present approaches to conservation biology. The statistical patterns of biotic composition on different spatial scales provide a basis for assessing vulnerability to extinction on the basis of body size and other variables, such as area of geographic range (e.g., see Brown and Maurer 1987, 1989; Pimm et al. 1988; Arita et al. 1990). Not only can these assessments identify classes of species that may be vulnerable, they can point to the spatial scales and habitat types that warrant attention. For example, reserves of different sizes, dispersions, and habitats will be differentially effective in enhancing the survival of different species, depending on their body sizes and other ecologically relevant attributes. In addition, body size and other easily measured variables provide a way of initially assessing the ecological roles of species. The patterns of species replacement within and between large areas, such as biomes and continents, suggest that there is considerable complementarity in these roles. As human impacts inevitably increase and native species inevitably become extinct, the only way to restore lost diversity and ecological functions will be through introduction of alien species. Assembly rules can be used to identify candidate species for introduction to fill missing ecological roles. For example, Janzen (1982) has advocated the use of feral livestock to replace seed dispersal, browsing, and grazing roles of large native mammals that have been extirpated from tropical reserves.

CONCLUSIONS

We conclude that processes operating over a wide range of spatial scales—from interspecific interactions that affect coexistence within local habitats to colonization, speciation, and extinction events that affect the distribution of species over the continent—interact to determine the composition of the biota at all scales, from local to continental (Cody 1975; Orians 1980; Brown 1981; Rapoport 1982; Ricklefs 1987; Brown and Maurer 1987, 1989). All of these processes are reflected in the body sizes of co-occurring species, because of the pervasive influence of size on many aspects of physiology, behavior, and ecology. Both the microscopic perspectives of physiological, population, and community ecology and of microevolution and the macroscopic perspectives of biogeography and macroevolution are required if we are to understand completely the composition of biotas on any spatial scale.

ACKNOWLEDGMENTS

We are especially grateful to the colleagues who gave us species lists of mammals for local habitat patches from their unpublished field notes or from the "gray literature" of environmental impact studies: E. Birney, G. Ceballos, J. Cook, J. Findley, E. Fink, S. Humphery, D. Kaufman, T. Lacher, J. Merritt, and R. Timm. We thank M. Taper for statistical advice and numerous other colleagues for discussions that helped us to formulate the ideas presented here. M. Boyce, D. Glazier, L. Hawkins, A. Kodric-Brown, B. Maurer, K. Schoenly, M. Taper, and an anonymous reviewer made valuable comments on an earlier draft of the manuscript. The research was supported by National Science Foundation grants BSR-8718139 and BSR-8807792 to J.H.B.

APPENDIX A

TABLE A1

List of Species, Their Body Masses, and the Biomes and Local Habitats in Which They Occur

Genus and Species	Mass (g)	Biomes	Local Habitats
Agouti paca	8,200	17,18,19,21	
Alces alces	457,000	1,3,4,5,11,12,13,14,15,16	12,23
Allouatta aplliata	7,280	17,18,19	
A. pigra	7,000	17,18,21	
Alpoex lagopus	3,000	3,4,12,13	
Ammospermophilus harrisii	150	8,17	6
A. interpres	156	9	
A. leucurus	156	7,8,9,11,15,16,20	4,5
A. nelsoni	120	7	
Antilocapra americana	68,000	7,8,9,10,11,14,15,16,17,20	1,6,19,21,24
Aplodontia rufa	950	2,16	
Ateles geofroyi	7,500	17,18,19,21	
Baiomys musculus	9	17,18,19,20	
B. taylori	9	6,10,14,17,18,20	16
Bassariscus astutus	1,400	2,6,7,8,9,10,11,14,15,16,17,18,19,20	1,3,4,5,10,16
B. sumichrasti	1,400	17,18,19,21	
Bison bison	422,000	3,4,5,6,9,10,11,14,15,16	24
Blarina brevicauda	22	4,5,6,14	7,8,9,12,13,14,15,22,23,24
B. telmalestes	22	6	
Cabassus centrali	800	17,19	
Caluromys derbianus	500	17,18	
Canis lupus	43,000	1,2,3,4,5,7,8,9,10,11,12,13,14,15,16,17,19,20	2,6,12,13,14,15,16,17,19,20,21,22,23,24
C. latrans	16,000	1,2,3,4,5,6,7,8,9,10,11,12,14,15,16,17,18,19,20	1,2,3,4,5,6,10,11,16,17,18,19,20,21,22,23,24
C. rufus	29,100	5,6,14	
Castor canadensis	24,000	1,2,3,4,5,6,7,8,9,10,11,13,14,15,16,17	2,12,14,19,20,22,23
Cervus elaphus	500,000	1,2,4,5,6,7,8,9,11,14,15,16,17	2,12,13,17,18,20,21,24
Chironectens minimus	750	17,18,19	
Clethrionomys californicus	29	2,16	
C. gapperi	29	1,2,4,5,9,11,13,14,15,16,17	12,13,14,15,22,23
C. rutilus	29	1,3,4,12,13,15	
Coendu mexicanus	5,000	17,18,19,21	
Condylura cristata	53	4,5,6	14,22,23

Conepatus leuconotus	3,500	10,14,18	1
C. mesoleucus	4,500	8,9,10,14,17,19,20	10,11,16
C. semistriutus	3,500	18,21	
Cryptotis goldmani	8	17,19	
C. goodwini	7	19	
C. magna	7	17	
C. mexicana	7	17,18,19	
C. nigrescens	7	18,19,21	
C. parva	5	5,6,9,10,14,17,18,19	7,8,9,24
Cyclopes dydactililus	300	17,18,19	
Cynomys gunnisoni	900	8,9,15	
C. leucurus	900	11	
C. ludovicianus	1,200	9,14,17	1
C. mexicanus	900	9	
C. parvidens	900	11,15	
Dama hemionus	118,100	1,2,3,4,7,8,9,10,11,14,15,16,17,20	1,2,6,16,17,18,19,20,21,24
D. virginia	106,850	2,4,5,6,8,9,10,11,14,15,16,17,18,19,20,21	3,4,5,7,8,9,10,11,12,13,16,20,22,23,24
Dasyprocta mexicana	5,000	18	
D. punctata	4,000	17,18,19,21	
Dasypus novemcinctus	7,000	6,10,14,17,18,19,20,21	7,8,10,11
Dicrostonyx groendlandicus	57	3,4,12,13	
D. hudsonius	57	13	
Didelphis marsupialis	3,000	10,17,18,19,21	10,11
D. virginiana	3,000	2,5,6,7,8,9,10,14,15,17,18,19,20,21	7,8,9,13,22,23
Dipodomys agilis	77	7,8	4
D. deserti	138	8,11	
D. elator	72	14	
D. elephantinus	72	7	
D. gravipes	92	7	
D. heermanni	72	2,7,16	
D. ingens	156	7	
D. merriami	47	8,9,11,16,17,20	1,4,6
D. microps	65	8,11,16	
D. nelsoni	100	9	
D. nitratoides	42	7	
D. ordii	72	8,9,10,11,14,15,16,17	1,6,19
D. panamintinus	57	7,11	
D. peninsularis	47	8	
D. phillipsii	41	17	
D. spectabilis	132	8,9,15,17	1,6

TABLE A1 (*Continued*)

Genus and Species	Mass (g)	Biomes	Local Habitats
D. stephensi	56	7	
D. venustus	72	7	
Eira barbara	5,000	17,18,19,21	
Erethizon dorsatum	13,000	1,2,3,4,5,7,8,9,10,11,12,13,14,15,16,17	1,2,6,12,13,14,15,16,17,19,20
Felis concolor	110,000	1,2,3,4,5,6,7,8,9,10,11,14,15,16,17,18,19,20,21	1,2,3,4,5,6,7,8,10,11,13,16,17,18,19,20,21,22,23,24
F. onca	113,000	6,8,9,10,14,17,18,19,20,21	10,11
F. pardalis	12,000	6,8,10,14,17,18,19,20,21	10,11
F. wiedii	5,000	10,14,17,18,19,20,21	10,11
F. yagouaroundi	9,000	8,10,14,17,18,19,20,21	10,11
Galictis vittata	3,000	17,18,19,21	
Geomys arenarius	354	9	1
G. bursarius	354	5,6,9,14	20,24
G. colonus	165	6	
G. fontanelus	165	6	
G. personatus	397	14	
G. pinetis	165	6	7,8
G. tropicalis	350	10	
Glaucomys sabrinus	85	1,2,4,5,7,11,14,15,16	7,12,14,15,20
G. volans	85	5,6,14,17,18	12,13
Gulo gulo	20,000	1,2,3,4,5,7,11,12,13,14,15,16	12,14,15,18,20
Habromys chinanteco	40	18	
H. lepturus	85	17,18	
H. lophurus	40	19	
H. simulatus	40	18	
Heteromys desmarestianus	85	17,18,19,21	
H. gaumeri	70	18,21	
H. goldmani	85	19	
H. lepturus	40	17,18	
H. longicaudatus	85	19	
H. nelsoni	70	18	
H. temporalis	85	18	
Hodomys alleni	220	19,20	
Homo sapiens	55,000	1,2,3,4,5,6,7,8,9,10,11,12,13,14,15,16,17,18,19,20,21	3,4,5,6,7,8,9,12,13,14,15,16,17,19,21,22,23,24
Lagurus curtatus	28	7,11,14,15,16	19

Lemmus trimucronatus	90	1,3,4,12,13,15	
Lepus alleni	3,600	8,20	
L. americanus	1,148	3,4,5,12,14,15,16,17	12,14,15,17,22,23
L. arcticus	4,750	13	
L. californicus	3,100	2,7,8,9,10,11,14,15,16,17,20	1,4,5,6,16,19,20
L. callotis	3,000	9,17	
L. flavigularis	3,000	19	
L. othus	4,250	12	
L. townsendi	3,450	11,14,15,16	19,21,24
Liomys irroratus	50	9,10,14,17,18,19	
L. pictus	65	8,17,18,19,20	10,11
L. salvini	57	19	
L. spectabilis	65	19	
Lutra canadensis	11,000	1,2,3,4,5,6,7,8,9,11,12,13,14,15,16,17	2,19,20,23
L. longicaudus	11,000	17,18,19,20,21	
Lynx canadensis	12,000	1,2,3,4,5,11,12,13,15,16	12,14,15,19,22,23
L. rufus	15,000	2,4,5,6,7,8,9,10,11,14,15,16,17,19,20	1,2,3,4,5,6,8,9,12,13,16,17,18,19,20,22,23,24
Marmosa canescens	60	17,18,19,20	10,11
M. mexicana	130	17,18,19,21	
Marmota caligata	6,300	1,2,3,12,16	
M. flaviventris	3,350	7,11,15,16	17,19
M. monax	4,200	1,3,4,5,6,14,15	13
M. olympus	6,300	2	
M. vancouverensis	6,300	1,2	
Martes americana	1,060	1,2,3,4,5,11,12,13,14,15,16	12,15,18
M. pennanti	4,800	1,2,4,5,7,11,13,14,15,16	12,15
Mazama americana	25,000	17,18,19,21	
Megadontomys thomasi	80	17,18,19	
Megasorex gigas	20	20	
Mephitis macroura	2,700	8,9,10,17,18,19,20	10,11
M. mephitis	4,000	2,4,5,6,7,8,9,10,11,13,14,15,16,17	1,2,4,6,8,9,12,13,16,21,22,23,24
Microdipodops megacephalus	14	11,16	
M. pallidus	14	16	
Microsorex hoyi	3	3,4,5,14,15	13
Microtus californicus	80	2,7,8,16	
M. canicaudus	72	2	
M. chrotorrhinus	35	4,5	15
M. guatemalensis	42	17	
M. longicaudus	55	1,2,3,7,8,9,11,14,15,16,17	17,18,19,20
M. ludovicianus	35	6	

TABLE A1 (*Continued*)

Genus and Species	Mass (g)	Biomes	Local Habitats
M. mexicanus	42	8,9,10,15,17,19	
M. miurus	41	3,12	19
M. montanus	72	11,14,15,16,17	18
M. oaxacensis	42	17	
M. ochrogaster	40	4,5,6,14	21,24
M. oeconomus	53	3,4,12,13	
M. oregoni	23	2,16	
M. pennsylvanicus	71	1,3,4,5,6,9,11,12,13,14,15,16,17	14,15,21,22,23
M. pinetorum	32	5,6,14	13
M. quasiater	40	18	
M. richardsoni	85	11,15,16	18
M. townsendi	72	1,2	
M. umbrosus	42	17	
M. xanthognathus	142	3,4,15	
Mustela erminea	114	1,2,3,4,5,9,11,12,13,14,15,16	12,14,15,22,23
M. frenata	340	1,2,4,5,6,7,8,9,10,11,14,15,16,17,18,19,20,21	1,2,3,6,7,8,11,12,13,17,18,19,20,21
M. nigripes	584	8,9,14,15,17	18
M. nivalis	45	1,3,4,5,12,13,14,15	13,20,22,23,24
M. vison	1,350	1,2,3,4,5,6,7,9,11,12,13,14,15,16	2,14,15,19,20,22,23
Myrmecophaga tridactyla	32,000	17,18,19,21	
Napeozapus insignis	22	4,5	12,13,15
Nasua nasua	9,000	8,9,10,14,17,18,19,20,21	10,11,16
Nelsonia neotomodon	80	19	
Neofiber alleni	213	6	
Neotoma albigula	198	8,9,14,15,17,20	1,4,5,16
N. alleni	200	19,20	10
N. augustopalata	198	10	
N. cinerea	300	1,2,3,8,9,11,15,16	17,18,19,20
N. devia	200	8	
N. floridana	328	5,6,14	24
N. fuscipes	267	2,7,8,16	
N. goldmani	198	9	
N. lepida	200	7,8,11,15	3,5
N. mexicana	185	8,9,15,17,19,20	1,16

N. micropus	310	9,10,14,17	1
N. nelsoni	198	18	
N. palatina	198	19	
N. phenax	198	20	
N. stephensi	200	9,15,17	
Neotomodon alstoni	40	17	
Neurotrichus gibbsii	10	2	
Notiosorex crawfordi	6	7,8,9,10,14,15,17,20	2,4,5,16
N. gigas	6	19	
Nyctomys sumichrasti	60	17,18,19,20	11
Ochotona collaris	118	3,4	
O. princeps	118	1,11,15,16	18
Ondatra zibethicus	1,800	1,2,4,5,6,8,9,11,12,13,14,15,16,17	2,19,22,23
Onychomys arenicola	30	9,19	
O. leucogaster	30	8,9,10,11,14,15,16,17	1,6,19,21
O. torridus	30	7,8,9,10,11,16,17,20	1,6
Oreamnos americanus	91,000	1,3,11,15,16	18
Orthogeomys cuniculus	500	19	
O. grandis	500	19	
O. hispidus	500	17,18,21	
O. lanius	500	18	
Oryzomys alfaroi	80	17,18,19	
O. caudatus	50	17	
O. couesi	80	18,21	
O. fulgens	80	9	
O. fulvescens	80	17,18,19,20,21	
O. melanotis	50	17,18,19,20,21	
O. palustris	54	5,6,10,14,17,19,20	11
Osgoodomys banderanus	50	19	
Otonyctomys hatti	40	18,21	
Ototylomys phyllotis	64	17,18,19,21	
Ovibus moschatus	286,000	12,13	
Ovis canadensis	91,000	7,8,9,11,14,15,16,17	4,5,18,20
O. dalli	73,100	3,12,15	
Pappogeomys alcorni	150	19	
P. bulleri	150	19,20	
P. castanops	330	9,10,14	1,2,11
P. fumosus	150	19	
P. gymnurus	600	19	
P. merriami	600	17,18,19	

TABLE A1 (*Continued*)

Genus and Species	Mass (g)	Biomes	Local Habitats
P. neglectus	150	17	
P. tylorhynus	600	17,19	
P. zinzeri	150	17	
Parascalops breweri	52	4,5	
Perognathus alticolus	24	8	
P. amplus	9	8	
P. apache	10	9,15	
P. arenarius	23	8,20	
P. artus	23	20	
P. baileyi	21	8,17,20	
P. californicus	23	7	
P. dalquesti	14	20	
P. fallax	21	7,8	4,5
P. fasciatus	9	14,15	
P. flavescens	11	9,14,15	
P. flavus	9	8,9,10,14,15,17	1,6,21,24
P. formosus	24	8,11	
P. goldmani	23	8,20	
P. hispidus	47	8,9,10,14,17	2,6,24
P. hooperi	36	10	
P. inornatus	13	7	
P. intermedius	18	8,9	
P. lineatus	23	9	
P. longimembris	9	7,8,11,16	4,5
P. merriami	8	14	
P. nelsoni	17	9,17	
P. parvus	24	11,15,16	17,19,20
P. penicillatus	23	8,9,17	1,6
P. pernix	17	20	
P. spinatus	24	8,20	5
P. xanthonotus	24	7	
Peromyscus attwateri	29	14	
P. aztecus	40	18	
P. banderanus	60	19,20	

P. boylii	36	2,7,8,9,10,11,14,15,16,17,19	4,5,16
P. bullatus	40	18	
P. californicus	38	2,7,8	3
P. crinitus	20	7,8,9,11,15	5,19
P. difficilis	32	9,14,15,17	
P. eremicus	40	7,8,9,10,17,20	1,5,6
P. eva	22	20	
P. evidens	40	19	
P. floridanus	29	6	7,8,9
P. furvus	33	17	
P. gossypinus	33	5,6	7,8,9
P. grandis	71	17	
P. guatemalensis	40	17,19	
P. gymnotis	40	19	
P. leucopus	20	5,6,8,9,10,14,17,18,19,21	2,12,13,22,23,24
P. maniculatus	35	1,2,3,4,5,6,7,8,9,10,11,13,14,15,16,17,18,19,20	1,2,3,4,5,6,12,13,14,15,17,18,19,20,21,24
P. megalops	71	19	
P. mekistrurus	60	18	
P. melanocarpus	60	17	
P. melanophrys	40	9,17,19	
P. melanotis	30	9,17	
P. melanurus	40	19	
P. merriami	40	8,20	
P. mexicanus	75	17,18,19	
P. nuttalli	25	5,6	8,9
P. ochraventer	40	10	
P. pectoralis	39	9,10,17	
P. perfulvus	40	19,20	10,11
P. polionotus	12	6	7,8
P. polius	40	9,17	
P. simulus	40	20	
P. spicilegus	36	17	
P. thomasi	75	17,18,19	
P. truei	37	2,7,8,9,11,14,15,16,17,20	3,4,17,19
P. winkelmani	40	20	
P. yucatanicus	40	18,21	
P. zarhinchus	40	17	
Phenacomys albipes	23	2	
P. intermedius	28	1,2,4,7,11,15,16	17,18
P. longicaudus	36	2	

TABLE A1 (*Continued*)

Genus and Species	Mass (g)	Biomes	Local Habitats
P. silvicola	36	2	
Philander opossum	500	10,17,18,19,21	
Potos flavus	5,000	17,18,19,21	
Procyon lotor	7,000	1,4,5,6,7,8,9,10,11,14,15,16,17,18,19,20,21	2,7,8,9,10,11,12,13,22,23
P. pygmeus	2,500	21	
Rangifer tarandus	169,000	3,4,5,12,13,15	12,14,15
Reithrodontomys burti	20	8,20	
R. chrysopsis	19	17,19	
R. fulvescens	28	6,8,9,10,14,17,18,19,20	6,10,11
R. gracilis	20	17,18,20,21	
R. hirsutus	20	19	
R. humulis	13	5,6	7
R. megalotis	15	2,7,8,9,11,14,15,16,17	1,2,6,19,21,24
R. mexicanus	19	17,18,19	
R. microdon	20	17,19	
R. montanus	9	8,9,14,17	24
R. raviventris	15	7	
R. sumichrasti	19	17,19	
R. tenuirostris	20	19	
Rheomys mexicanus	40	19	
R. thomasi	40	19	
Romerolagus diazi	477	17	
Scalopus anthony	100	8	
S. aquaticus	140	5,6,14	8,9
Scapanus latimanus	140	2,7,8,16	3,17
S. orarius	56	2,11,16	18,20
S. townsendi	140	2	
Sciurus aberti	900	8,9,15,17	
S. alleni	750	10	
S. arizonensis	700	8,17	
S. aureogaster	690	17,18,19	
S. carolinensis	555	2,4,5,6,14	9,12,13
S. colliaei	498	20	10,11
S. deppei	225	17,18,19,21	
S. griseus	681	7,8,16	17,20

S. kaibabensis	900	8	
S. nayaritensis	498	17	
S. niger	1,000	4,5,6,10,14	7,9
S. oculatus	750	17	
S. variegatoides	498	17,19	
S. yucatanesis	225	18,21	
Scotinomys teguina	15	17	
Sigmodon alleni	120	19	
S. arizonae	198	8,20	
S. fulviventer	120	8,9,17	
S. hispidus	198	5,6,8,9,10,14,17,18,19,21	1,2,6,7,8,9,24
S. leucotis	120	17	
S. mascotensis	120	19,20	10
S. minimus	120	9,17	
S. ochrognatus	112	9,17	16
Sorex alaskanus	14	1	
S. arcticus	9	2,3,4,5,12,13,14	14,15,23
S. arizonae	7	17	
S. bendii	16	2	18,20
S. cinereus	5	1,3,4,5,12,13,14,15	12,13,14,15,21,22,23
S. dispar	6	4,5	12,13
S. emarginatus	7	17	
S. fumeus	8	4,5	12,13
S. gaspensis	5	4	
S. juncensis	7	7	
S. longirostris	3	5,6	7,9
S. lyelli	5	16	
S. macrodon	7	18	
S. merriami	6	8,9,11,15,16,17	19
S. milleri	7	9,10,17	
S. monticolus	7	8,9,17	
S. nanus	7	8,9,11,14,15	
S. oreopolus	7	17	
S. ornatus	7	7	
S. palustris	14	3,4,5,8,11,13,14,15,16	22,23
S. preblei	3	11,15	
S. saussurei	6	9,17	
S. sclateri	7	18	
S. stizodon	7	17	
S. tenellus	7	11,16	

TABLE A1 (*Continued*)

Genus and Species	Mass (g)	Biomes	Local Habitats
S. trigonirostris	7	16	
S. trowbridgii	8	2,7,16	17
S. vagrans	7	1,2,3,4,9,12,14,15,17	17,18,20
S. ventralis	7	17	
S. veraepacis	7	17	
Spermophilus adocetus	125	19	
S. annulatus	500	20	
S. armatus	355	11,15	
S. atricapillus	275	20	
S. beecheyi	738	2,7,8,16	3,4,5,17
S. beldingi	284	11,16	17
S. brunneus	300	11	
S. columbianus	576	11,15	18,20
S. franklini	600	4,5,14	22,23
S. lateralis	223	2,7,11,15,16	17,18,19,20
S. madrensis	275	17	
S. mexicanus	340	9,10,14,17	
S. parryi	700	3,4,12,13,15	
S. perotensis	140	18	
S. richardsonii	419	11,14,15	21
S. saturatus	223	16	
S. spilosoma	125	8,9,10,14,15,17	1,6
S. tereticaudus	145	7,8,20	
S. townsendii	226	11,16	19
S. tridecemlineatus	160	4,5,9,14,15	21,24
S. variegatus	817	8,9,10,11,14,15,17,18,19,20,21	2,16
S. washingtoni	218	11	
Spilogale gracilis	900	7,11	4,5,16
S. putorius	900	2,5,6,8,9,10,11,14,15,16,17,18,19,20,21	20
S. pygmaea	320	19,20	10,11
Sylvilagus aquaticus	2,150	5,6,14	
S. audubonii	1,013	1,2,7,8,9,10,11,14,15,16,17,20	1,2,6
S. bachmani	843	2,7,8,20	5,19,20
S. brasiliensis	950	17,18,19	

S. cunicularius	3,000	17,19,20	10
S. floridanus	1,800	4,5,6,8,9,10,14,17,21	8,16
S. griseus	964	2,7	
S. idahoensis	35	11,16	19
S. insonus	3,000	19	
S. nuttalli	855	11,15,16	
S. palustris	1,600	6	8
S. transitionalis	1,048	5	12,13
Synaptomys borealis	29	3,4,12,13,15,16	
S. cooperi	36	4,5,14	15,22,23,24
Tamandua mexicana	6,000	17,18,19,21	
Tamias alpinus	39	7	
T. amoenus	51	1,7,11,15,16	17,20
T. bulleri	100	20	
T. cinereicollis	71	17	
T. dorsalis	85	8,11,15,17,20	16
T. durangae	85	17	
T. merriami	113	7,8	3
T. minimus	50	3,4,5,7,8,9,11,13,14,15,16	19
T. obscurus	100	7	
T. ochrogenys	75	2	
T. palmeri	75	8	
T. panamintinus	54	7,16	
T. quadrimaculatus	85	7,16	
T. quadrivataus	71	9,15	
T. ruficaudus	60	11,15	
T. senex	75	2,7,16	
T. siskiyou	75	2	
T. sonomae	75	2	
T. speciosus	67	7,16	
T. striatus	103	4,5,6,14	12,13,14,15
T. townsendii	75	2,16	
T. umbrinus	71	11,15	
Tamiasciurus douglasii	225	1,2,7,16	
T. hudsonicus	196	1,4,5,8,9,11,12,13,14,15	12,13,14,15,20
Tapirus bairdii	300,000	17,18,19,21	
Taxidea taxus	10,000	2,4,5,7,8,9,10,11,14,15,16,17,20	1,2,4,5,6,17,18,19,20,21,24
Tayassu pecari	60,000	17,18,19	
T. tajacu	30,000	8,9,10,14,17,18,19,20,21	6,10,11,16
Thalarctos maritimi	382,500	12,13	

TABLE A1 (*Continued*)

Genus and Species	Mass (g)	Biomes	Local Habitats
Thomomys baileyi	100	9	
T. bottae	130	8,9,16,17	2,3,4,5,6
T. mazama	75	2,16	
T. merriami	113	7	
T. monticola	80	7,16	17
T. talpoides	100	4,11,14,15,16	18,19,20
T. umbrinus	130	2,7,8,9,11,15,16,17,19,20	16,19
Tylomys bullaris	280	17	
T. nudicaudus	280	17,18,19	
T. tumbalensis	280	18	
Urocyon cinereoargenteus	4,000	2,4,5,6,7,8,9,10,11,14,15,16,17,18,19,20,21	1,2,3,4,5,7,8,9,10,11,12,13,16,22,23
Ursus americanus	140,000	1,2,3,4,5,6,7,8,9,10,11,12,13,14,15,16,17,20	2,8,9,12,13,14,15,16,17,18,20,22,23,24
U. arctos	363,000	1,2,3,4,7,8,9,10,11,12,13,14,15,16,17	2,16,17,18,19,20
Vulpes macrotis	3,200	7,8,9,17,20	1,6
V. velox	2,000	7,8,9,11,14,15,16,17,20	20,21
V. vulpes	5,000	1,2,3,4,5,6,7,8,9,11,12,13,14,15,16,17	14,15,21,22,23,24
Xenomys nelsoni	130	19,20	10,11
Zapus hudsonius	18	1,3,4,5,6,12,14,15	2,22,23,24
Z. princeps	28	2,4,7,8,9,11,14,15,16	18,20
Z. trinotatus	28	2,16	
Zygogeomys trichopus	500	20	

NOTE.—See tables 1 and 2 for the keys to the numbers used to designate biomes and local habitats, respectively.

APPENDIX B

TABLE B1

LOG$_2$ BODY-SIZE CLASSES FOR NORTH AMERICAN MAMMALS, AND THE MINIMUM, MIDPOINT, AND MAXIMUM BODY MASS IN EACH CLASS

	Mass (g)		
Size Class	Minimum	Midpoint	Maximum
1	1	1.4	1.9
2	2	2.8	3.9
3	4	5.7	7.9
4	8	11.3	15.9
5	16	22.6	31.9
6	32	45.3	63.9
7	64	90.5	127.9
8	128	181	255.9
9	256	362	511.9
10	512	724	1,023.9
11	1,024	1,448	2,047.9
12	2,048	2,896	4,095.9
13	4,096	5,793	8,191.9
14	8,192	11,585	16,383.9
15	16,384	23,170	32,767.9
16	32,768	46,341	65,535.9
17	65,536	92,682	131,071.9
18	131,072	185,364	262,143.9
19	262,144	370,728	524,287.9
20	524,288	741,455	10,485,503

LITERATURE CITED

Anonymous. 1976. Cross Florida barge canal restudy report: wildlife study. Florida Game and Fresh Water Fish Commission 4:D88–D95.

———. 1987. Plants and animals of White Sands. White Sands National Monument, N.Mex.

Arita, H. T., J. G. Robinson, and K. H. Redford. 1990. Rarity in neotropical forest mammals and its ecological correlates. Conservation Biology 4:181–192.

Bonner, J. T. 1988. The evolution of complexity by means of natural selection. Princeton University Press, Princeton, N.J.

Bowers, M. A., and J. H. Brown. 1982. Body size and coexistence in desert rodents: chance or community structure? Ecology 63:391–400.

Brown, E. R. 1985. Management of wildlife and fish habitats in forests of western Oregon and Washington. U.S. Department of Agriculture, Forest Service, Pacific Northwest Division.

Brown, J. H. 1973. Species diversity of seed-eating desert rodents in sand dune habitats. Ecology 54:775–787.

———. 1975. Geographical ecology of desert rodents. Pages 315–341 *in* M. L. Cody and J. M. Diamond, eds. Ecology and evolution of communities. Harvard University Press, Cambridge, Mass.

———. 1981. Two decades of homage to Santa Rosalia: toward a general theory of diversity. American Zoologist 21:877–888.

———. 1986. Two decades of interaction between the MacArthur-Wilson model and the complexities of mammalian distributions. Biological Journal of the Linnean Society 28:231–251.

———. 1987. Variation in desert rodent guilds: patterns, processes, and scales. Pages 185–203 *in* J. H. R. Gee and P. S. Giller, eds. Organization of communities: past and present. Blackwell Scientific, Oxford.

Brown, J. H., and M. A. Bowers. 1985. On the relationship between morphology and ecology: community organization in hummingbirds. Auk 102:251–269.

Brown, J. H., and A. C. Gibson. 1983. Biogeography. Mosby, St. Louis.

Brown, J. H., and B. A. Maurer. 1987. Evolution of species assemblages: effects of energetic constraints and species dynamics on the diversification of the North American avifauna. American Naturalist 130:1–17.

———. 1989. Macroecology: the division of food and space among species on continents. Science (Washington, D.C.) 243:1145–1150.

Brown, J. H., and J. C. Munger. 1985. Experimental manipulation of a desert rodent community: food addition and species removal. Ecology 66:1545–1563.

Burt, W. H., and R. P. Grossenheider. 1976. A field guide to the mammals. Houghton Mifflin, Boston.

Calder, W. A., III. 1984. Size, function, and life history. Harvard University Press, Cambridge, Mass.

Ceballos, G., and A. Miranda. 1986. Los mamíferos de Chamela, Jalisco. Universidad Nacional Autónoma de México, Ciudad Universitaria, Mexico.

Cody, M. L. 1975. Towards a theory of continental species diversities: bird distributions over Mediterranean habitat gradients. Pages 214–257 *in* M. L. Cody and J. M. Diamond, eds. Ecology and evolution of communities. Harvard University Press, Cambridge, Mass.

Cook, J. A. 1986. The mammals of the Animas Mountains and adjacent areas, Hidalgo County, New Mexico. Occasional Papers of the Museum of Southwestern Biology, University of New Mexico 4:1–45.

Connor, E. F., and E. D. McCoy. 1979. The statistics and biology of the species-area relationship. American Naturalist 113:791–833.

Damuth, J. 1981. Population density and body size in mammals. Nature (London) 290:699–700.

Dasmann, R. 1975. Biogeographical provinces. International Union for Conservation of Nature and Natural Resources, Occasional Papers 18:1–5.

Dial, K. P., and J. M. Marzluff. 1988. Are the smallest organisms the most diverse? Ecology 69:1620–1624.

Diamond, J. M. 1975. Assembly of species communities. Pages 342–444 *in* M. L. Cody and J. M. Diamond, eds. Ecology and evolution of communities. Harvard University Press, Cambridge, Mass.

———. 1984. Historic extinction: a Rosetta Stone for understanding prehistoric extinctions. Pages 824–862 *in* P. S. Martin and R. G. Klein, eds. Quaternary extinctions. University of Arizona Press, Tucson.

du Toit, J. T., and N. Owen-Smith. 1989. Body size, population metabolism, and habitat specialization among large African herbivores. American Naturalist 133:736–740.

Eisenberg, J. F. 1981. The mammalian radiations. University of Chicago Press, Chicago.

Erwin, T. L. 1983. Beetles and other insects of tropical rain forest canopies at Manaus, Brazil, sampled by insecticidal fogging. Pages 59–75 *in* S. L. Sutton, T. C. Whitmore, and A. C. Chadwick, eds. Tropical rain forest: ecology and management. Blackwell, Edinburgh.

Finck, E. J., D. W. Kaufman, G. A. Kaufman, S. K. Gurtz, B. K. Clark, L. J. McLellan, and B. S. Clark. 1986. Mammals of the Konza Prairie research natural area, Kansas. Prairie Naturalist 18:153–166.

Flessa, K. W. 1975. Area, continental drift and mammalian diversity. Paleobiology 1:189–194.

Foster, J. B. 1964. The evolution of mammals on islands. Nature (London) 202:234–235.

Goodman, D. 1987. The demography of chance extinctions. Pages 11–34 *in* M. E. Soule, ed. Viable populations. Cambridge University Press, Cambridge, Mass.

Grant, P. R. 1986. Ecology and evolution of Darwin's finches. Princeton University Press, Princeton, N.J.

Hall, E. R. 1981. The mammals of North America. Vols. 1 and 2. 2d ed. Wiley, New York.

Harestad, A. S., and F. L. Bunnell. 1979. Home range and body weight: a re-evaluation. Ecology 60:389–402.

Hink, V. C., and R. D. Ohmart. 1984. Middle Rio Grande biological survey. Final report, Army Corp of Engineers contract no. DACW47-81-C-0015.

Hopf, F. A., and J. H. Brown. 1986. The bullseye method for testing for randomness in ecological communities. Ecology 67:1139–1155.

Hoppe, P. P. 1977. Rumen fermentation and body weight in African ruminants. Pages 141–150 *in* Proceedings of 12th Congress of Game Biologists, Atlanta, Ga.

Hutchinson, G. E. 1959. Homage to Santa Rosalia, or why are there so many kinds of animals? American Naturalist 93:145–159.

Hutchinson, G. E., and R. H. MacArthur. 1959. A theoretical ecological model of size distributions among species of animals. American Naturalist 93:117–125.

Jablonski, D. 1986. Causes and consequences of mass extinctions: a comparative approach. Pages 183–229 *in* D. K. Elliot, ed. Dynamics of extinction. Wiley, New York.

Janzen, D. H. 1982. Differential seed passage rates in cows and horses, surrogate Pleistocene dispersal agents. Oikos 38:150–156.

Lomolino, M. V. 1985. Body sizes of mammals on islands: the island rule reexamined. American Naturalist 125:310–316.

MacArthur, R. H. 1972. Geographical ecology. Harper & Row, New York.

MacArthur, R. H., and E. O. Wilson. 1967. Theory of island biogeography. Princeton University Press, Princeton, N.J.

Martin, P. S. 1984. Prehistoric overkill: the global model. Pages 354–403 *in* P. S. Martin and R. G. Klein, eds. Quaternary extinctions. University of Arizona Press, Tucson.

Martin, P. S., and R. G. Klein. 1984. Quaternary extinctions. University of Arizona Press, Tucson.

May, R. M. 1978. The dynamics and diversity of insect faunas. Pages 188–204 *in* L. A. Mound and N. Waloff, eds. Diversity of insect faunas. Blackwell, Oxford.

———. 1988. How many species are there on earth? Science (Washington, D.C.) 241:1441–1449.

McNab, B. K. 1963. Bioenergetics and the determination of home ecological significance of Bergmann's rule. Ecology 52:845–854.

McNaughton, S. J., and N. J. Georgiadis. 1986. Ecology of African grazing and browsing mammals. Annual Review of Ecology and Systematics 17:39–65.

Morrison, M. L., M. P. Yoder-Williams, D. C. Erman, R. H. Barrett, M. White, A. S. Leopold, and D. A. Airola. Natural history of vertebrates of Sagehen Creek Basin, Nevada County, California. Agricultural Experimental Station, University of California, Division of Agriculture and Natural Resources.

Morse, D. R., N. E. Stork, and J. H. Lawton. 1988. Species number, species abundance and body length relationships of arboreal beetles in Bornean lowland rain forest trees. Ecological Entomology 13:25–37.

Nagy, K. A. 1987. Field metabolic rate and food requirement scaling in mammals and birds. Ecological Monographs 57:111–128.

Orians, G. H. 1980. Micro and macro in ecological theory. BioScience 30:79.

Pacala, S., and J. Roughgarden. 1982. Resource partitioning and interspecific competition in two-species insular *Anolis* lizard communities. Science (Washington, D.C.) 217:444–446.

———. 1985. Population experiments with the *Anolis* lizards of St. Maarten and St. Eustatius (Neth. Antilles). Ecology 66:129–141.

Parra, P. 1978. Comparison of foregut and hindgut fermentation in herbivores. Pages 205–229 *in* G. Montgomery, ed. The ecology of arboreal folivores. Smithsonian Institution Press, Washington, D.C.

Peters, R. H. 1983. The ecological implications of body size. Cambridge University Press, Cambridge.

Peters, R. H., and J. V. Raelson. 1984. Relations between individual size and mammalian population density. American Naturalist 124:498–517.

Pimm, S. L., H. L. Jones, and J. Diamond. 1988. On the risk of extinction. American Naturalist 132:757–785.

Preston, F. W. 1962. The canonical distribution of commonness and rarity. Ecology 43:185–215, 410–432.

Ramirez-Pulido, J., M. C. Britton, A. Perdomo, and A. Castro. 1983. Guía de los mamíferos de México. Universidad Autónoma Metropolitana, Mexico City, Mexico.

Rapoport, E. H. 1982. Areography: geographical strategies of species. Pergamon, Oxford.

Raup, D. M. 1986. Biological extinction in earth history. Science (Washington, D.C.) 231:1528–1533.

Ricklefs, R. E. 1987. Community diversity: relative roles of local and regional processes. Science (Washington, D.C.) 235:167–171.

Rosenzweig, M. L. 1966. Community structure in sympatric Carnivora. Journal of Mammalogy 47:602–612.

Rusler, R. 1987. Frequency distributions of mammalian body size analyzed by continent. M.S. thesis. University of Arizona, Tucson.

Ryan, R. M. 1968. Mammals of Deep Canyon Colorado desert, California. Desert Museum, Palm Springs, Calif.

Schmidt-Nielsen, K. 1984. Scaling: why is animal size so important? Cambridge University Press, Cambridge.

Schoener, T. W. 1968. Sizes of feeding territories among birds. Ecology 49:123–131.

———. 1970. Size patterns in West Indian *Anolis* lizards. II. Correlations with the sizes of sympatric species—displacement and convergence. American Naturalist 104:155–174.

———. 1976. The species-area relation within archipelagos: models and evidence from island birds. Pages 629–642 *in* Proceedings of the 16th International Ornithological Congress, Canberra.

———. 1984. Size differences among sympatric, bird-eating hawks: a worldwide survey. Pages 254–281 *in* D. R. Strong, D. Simberloff, L. G. Abele, and A. B. Thistle, eds. Ecological communities: conceptual issues and the evidence. Princeton University Press, Princeton, N.J.

Schoener, T. W., and D. A. Spiller. 1988. High population persistence in a system with high turnover. Nature (London) 330:474–477.

Sibly, R. M. 1981. Strategies of digestion and defecation. Pages 109–139 *in* C. R. Townsend and P. Calow, eds. Physiological ecology. Sinauer, Sunderland, Mass.

Simberloff, D., and W. Boecklen. 1981. Santa Rosalia reconsidered: size ratios and competition. Evolution 35:1206–1228.

Timm, R. M. 1975. Distribution, natural history, and parasites of mammals of Cook County, Minnesota. Bell Museum of Natural History Occasional Papers 14:1–56.

Van Valen, L. 1973. Pattern and the balance of nature. Evolutionary Theory 1:31–40.

Webb, S. D. 1984. Ten million years of mammalian extinctions in North America. Pages 189–210 *in* P. S. Martin and R. G. Klein, eds. Quaternary extinctions. University of Arizona Press, Tucson.

Whitaker, J. O., Jr. 1980. The Audubon Society field guide to North American mammals. Chanticleer, New York.

Wilson, E. O. 1988. Biodiversity. National Academy, Washington, D.C.

Wilson, M. V., and A. Shmida. 1984. Measuring beta diversity with presence-absence data. Journal of Ecology 72:1055–1064.

Zeveloff, S. I., and M. S. Boyce. 1988. Body size patterns in North American mammal faunas. Pages 123–146 *in* M. S. Boyce, ed. Evolution of life histories of mammals. Yale University Press, New Haven, Conn.

EVOLUTIONARY DYNAMICS

Edited by Mark D. Uhen

One of the basic tenets of macroecology is that one can study ecological processes on a large scale, either geographic, interrelational, or temporal. However, most ecological processes are studied in the present, ignoring the large temporal scale over which these processes have evolved. Certainly, one of the goals of macroecology is to determine how ecological processes, organismal distributions, and modes of life came to be. The fundamental answer to this question is simple: they are as they are today, because they were that way yesterday. The more interesting answer(s) to the question of how these macroecological observations came to be is the realm of evolutionary dynamics.

The papers in this section come from three different perspectives. First is the perspective of the modern—ecologists and biologists looking at generational timescales to see how observable energetic, genetic, and microevolutionary processes have helped to shape the ecology, morphology, and distributions of modern organisms (J. H. Brown and B. A. Maurer; K. J. Gaston). This perspective has, at least potentially, the advantage of complete sampling of the floras and faunas in question and direct measurements of parameters such as gene flow, body size, and so on. The disadvantage is that the time depth is very shallow and for the most part limited to historical records.

The second perspective is that of the shallow paleontological record, usually taken from the Pleistocene to early Holocene (R. W. Graham). This perspective has some of the advantages of the modern perspective in that most organisms from the Pleistocene have the same or similar morphological and ecological limits today. In addition, even extinct organisms tend to be similar to those alive today. Thus their morphological and ecological characteristics can be estimated with more confidence than possible for organisms that went extinct in deeper time. However, some organisms are found in the Pleistocene in nonanalog associations, challenging assumptions of ecological limits based only on modern distributions (Graham).

The third perspective is that of the deep past (L. Van Valen; R. K. Bambach; D. Jablonski). This perspective takes the longest temporal view possible, from millions to even billions of years (Van Valen), and includes processes that take place over the entire Phanerozoic (Bambach). It also can take into account processes that are virtually instantaneous but episodic, occurring only occasionally in the history of life, such as mass extinctions (Jablonski). This perspective has the advantage of being able to perceive processes that take place over extremely long timescales and also those that take place only very rarely. It has the distinct disadvantage that it is often necessary to make educated guesses as to the modes of life of long-extinct organisms, and estimations of morphological and ecological parameters such as body size and diet are imperfect. It also suffers from the problem of incomplete and sometimes very inconsistent sampling of the record (Alroy et al. 2001).

The modern synthesis sought to integrate biology and paleontology under a single rubric where all change in organisms observed over time is attributable to microevolutionary processes. Clearly, some of the phenomena described by paleontologists are not conventionally thought of in microevolutionary terms either because they affect the global biota (mass extinctions), can only be observed over timescales greater than a generation (punctuated equilibrium), or are best observed at hierarchical levels above that of the population (species selection). This has led to calls for an expansion, or a postmodern interpretation of the modern synthesis (Gould 1994).

S. J. Gould (1994) criticized paleontologists of the modern synthesis such as G. G. Simpson for failing to use the strengths of paleontological data to propose novel evolutionary hypotheses that could be broadly applied and tested throughout the realm of evolutionary study. Part of this criticism seems unfounded for the simple reason that paleontologists and neontologists deal with different sets of data, and thus the "synthesis" part of the modern synthesis can only go so far. Paleontologists cannot measure gene flow among populations, so their work must remain somewhat separated from that of population geneticists (but see Yang et al. 2006). Likewise, population geneticists cannot observe their organisms for more than a few generations, so they cannot study processes that take place over millions of years, much less even hundreds of years. The basic issue can best be seen in an example. If a modern species were undergoing a punctuation event, would a population geneticist notice? Perhaps, but then again, perhaps not. The phenomenon of punctuated equilibrium could never be studied in the modern world because by definition it has a long time element of stasis associated with the rapid punctuation event. Thus it must be studied with a different tool kit than that used by population geneticists. Gould (1994) wisely called for a complimentary development of macroevolutionary theory to bond with microevolutionary theory. Such a development would leverage the strengths of paleontological data to expand the horizons of theories that have been necessarily restricted to the recent time plane.

As the papers in this section demonstrate, Gould's call has been heeded, at least in part, by the development of macroecology. Paleontologists have, for the most part, studied patterns of evolution over long periods of geologic time using the fossil record, and neontologists have studied patterns of evolution over short periods of time using data on modern organisms. While this set of papers is certainly not comprehensive, it is representative of the field. While some types of studies must be performed in either the realm of paleontology or modern biology, more integration of hypotheses across time planes would strengthen many hypotheses of evolutionary dynamics.

Van Valen's paper is something of an exception to this generalization. His observations, based mainly on the fossil record, spurred great interest among other paleontologists and neontologists alike. In fact, most (94%) of the studies on the Red Queen hypothesis to date have been performed by neontologists, rather than paleontologists, based on the distribution of journal titles publishing papers with Red Queen as a keyword. This suggests that paleontology can contribute to modern evolutionary dynamics. Contrary to Gould's view, paleoecology does not need to "hold unique turf" next to modern biology. Rather there is a continuum of processes that are not divorced from one another but that interact across broad scales of time, space, and causation.

Literature Cited

Alroy, J., C. R. Marshall, R. K. Bambach, K. Bezosko, M. Foote, F. T. Fürsich, T. A. Hansen, et al. 2001. Effects of sampling standardization on estimates of Phanerozoic marine diversification. *Proceedings of the National Academy of Sciences* 98:6261–66.

Gould, S. J. 1994. Tempo and mode in the macroevolutionary reconstruction of Darwinism. *Proceedings of the National Academy of Sciences* 91:6764–71.

Yang, S., X. Lai, S. Shi, S. Cheng, X. Zhou, X. Yang, and H. Yang. 2006. New ancient DNA sequences suggest high genetic diversity for the woolly mammoth (*Mammuthus primigenius*). *Progress in Natural Science* 16:379–86.

PAPER 19

A New Evolutionary Law (1973)

L. Van Valen

Commentary

MATTHEW A. KOSNIK

There is much folklore surrounding this paper and the publication process that caused L. Van Valen to launch his own journal (*Evolutionary Theory*) rather than make the changes suggested by an editorial machine. Yet there is no doubt that this paper, published in an unindexed journal without an Institute for Scientific Information (ISI) impact factor, is a tremendously influential paper. Even now, some 30 years after its original publication, it averages more than 40 citations per year (see Google Scholar).

The paper starts with a relatively simple observation—that applying the analysis of survivorship curves from population biology to geologic ranges leads to a disturbingly consistent relation between taxon age and the probability of extinction. Van Valen includes a mind-boggling 50 survivorship curves that fill six full pages to demonstrate this point. Van Valen then proposes a new evolutionary law, which, while providing the basis for the Red Queen hypothesis, has received comparatively little attention relative to its derivative hypothesis. One cannot help but wonder if others had followed down the path of "evolutionary laws" versus evolutionary theories, how that might have shaped the current debate about teaching biblical creation stories in public schools. I would certainly suggest that it is as unfashionable to speak of evolutionary laws now as it was then.

While the Red Queen hypothesis has a powerful grip on the imagination as an intuitive construct, a fact that is no doubt aided by the inevitable links to Lewis Carroll's popular children's stories, it is easy to gloss over the mathematical equations and quantitative predictions interwoven throughout the paper. While Van Valen claimed that quantitative predictions were unnecessary for a law, these are an important element demonstrating the seriousness and testability of his proposal.

Finally, Van Valen's paper is exceptional for its breadth of scale. Starting with analyses of extinction rates, he ends with a discussion of molecular evolution. Even today as more is known about each scale, there are precious few papers that attempt to unite evolutionary theory across such biological and temporal scales. All macroecologists can learn from this example. While many might not all be able to provide specific quantitative predictions across such vast scales, macroecologists are all well served by remembering the importance of the various scales at which "endless forms most beautiful and most wonderful have been, and are being, evolved" (Darwin 1859, 490).

Literature Cited

Darwin, C. 1859. *On the origin of species by means of natural selection, or The preservation of favoured races in the struggle for life.* John Murray, London.

From *Evolutionary Theory* 1:1–30.

A NEW EVOLUTIONARY LAW

Leigh Van Valen
Department of Biology
The University of Chicago
Chicago, Illinois 60637

ABSTRACT:

All groups for which data exist go extinct at a rate that is constant for a given group. When this is recast in ecological form (the effective environment of any homogeneous group of organisms deteriorates at a stochastically constant rate), no definite exceptions exist although a few are possible. Extinction rates are similar within some very broad categories and vary regularly with size of area inhabited. A new unit of rates for discrete phenomena, the macarthur, is introduced. Laws are appropriate in evolutionary biology. Truth needs more than correct predictions. The Law of Extinction is evidence for ecological significance and comparability of taxa. A non-Markovian hypothesis to explain the law invokes mutually incompatible optima within an adaptive zone. A self-perpetuating fluctuation results which can be stated in terms of an unstudied aspect of zero-sum game theory. The hypothesis can be derived from a view that momentary fitness is the amount of control of resources, which remain constant in total amount. The hypothesis implies that long-term fitness has only two components and that events of mutualism are rare. The hypothesis largely explains the observed pattern of molecular evolution.

* * *

Introduction

During a study (Van Valen, submitted) on the effects of extinction I wanted to show that a model I was using was oversimplified. It assumed no correlation of probability of extinction with age of the group, and I thought that generally more vulnerable groups should die out first. A test using data from Simpson (1953) showed to my astonishment that the assumption was reasonably correct in these cases. I did not believe it could be generally true and so tested these and other cases in more detail. The assumption proved to be consistent with all available data. Others (unpublished results) have now confirmed this finding for individual taxa. I will present a more extended treatment elsewhere; the present paper is condensed.

The Evidence

The method is an application of the survivorship curve of population ecology (including demography). It is a simple plot of the proportion of the original sample that survive for various intervals. In this case the sample is the set of all known subgroups of some larger group, no matter when in absolute time each subgroup originated. A logarithmic ordinate, standard in ecology, gives the property that the slope of the curve at any age is proportional to the probability of extinction at that age. Simpson (1944, 1953) compiled two well-known taxonomic survivorship curves but used an arithmetic ordinate (1-4).

The results (Figs. 1-5) for over 25,000 subtaxa show almost uniform linearity for extinct taxa except for effects attributable to sampling error (5,6). Sampling error is most noticeable at the bottom of the graphs, where

Evol. Theory 1:1-30 (July 1973)

2

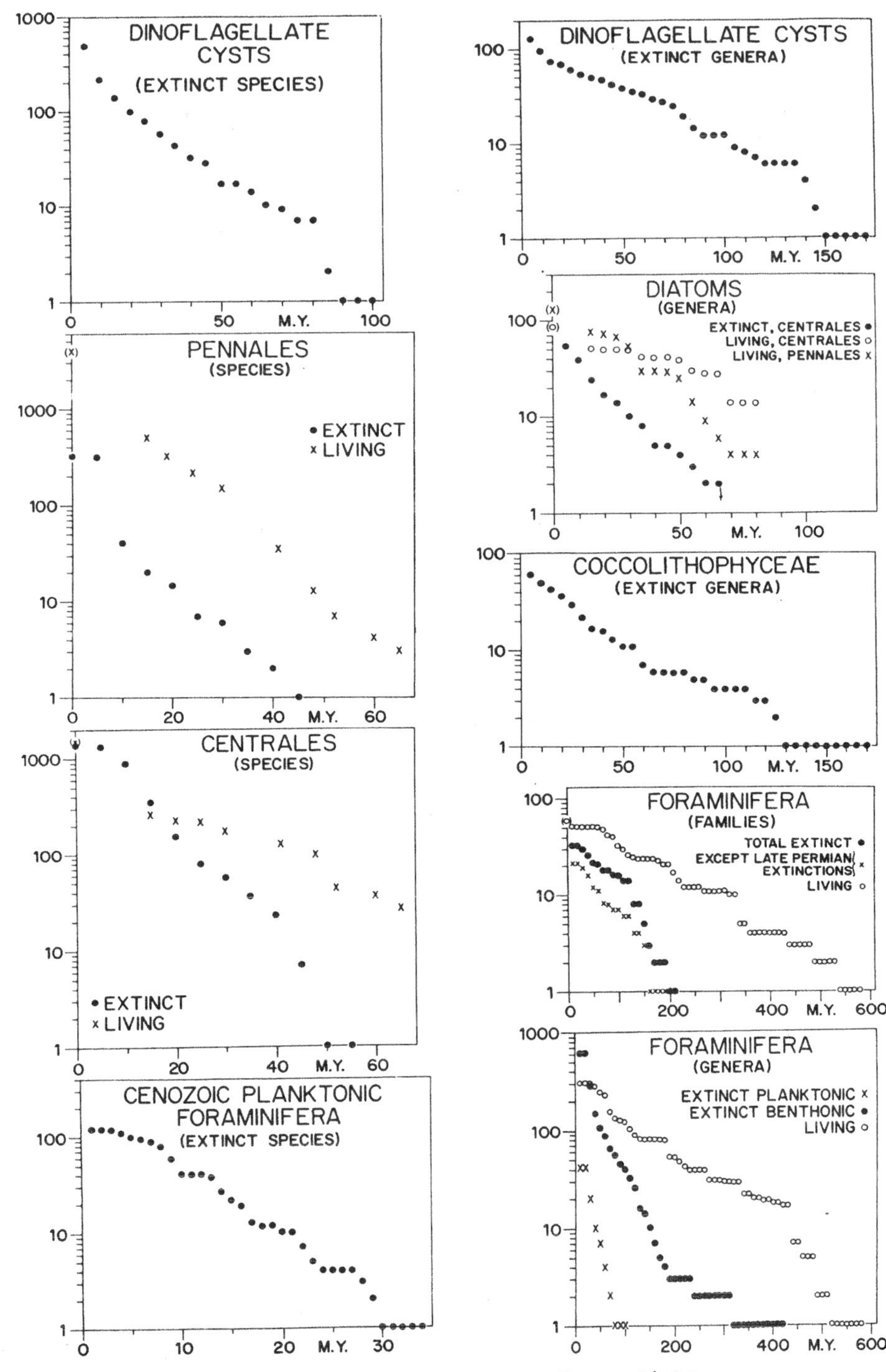

Fig. 1. Taxonomic survivorship curves for protists.

3

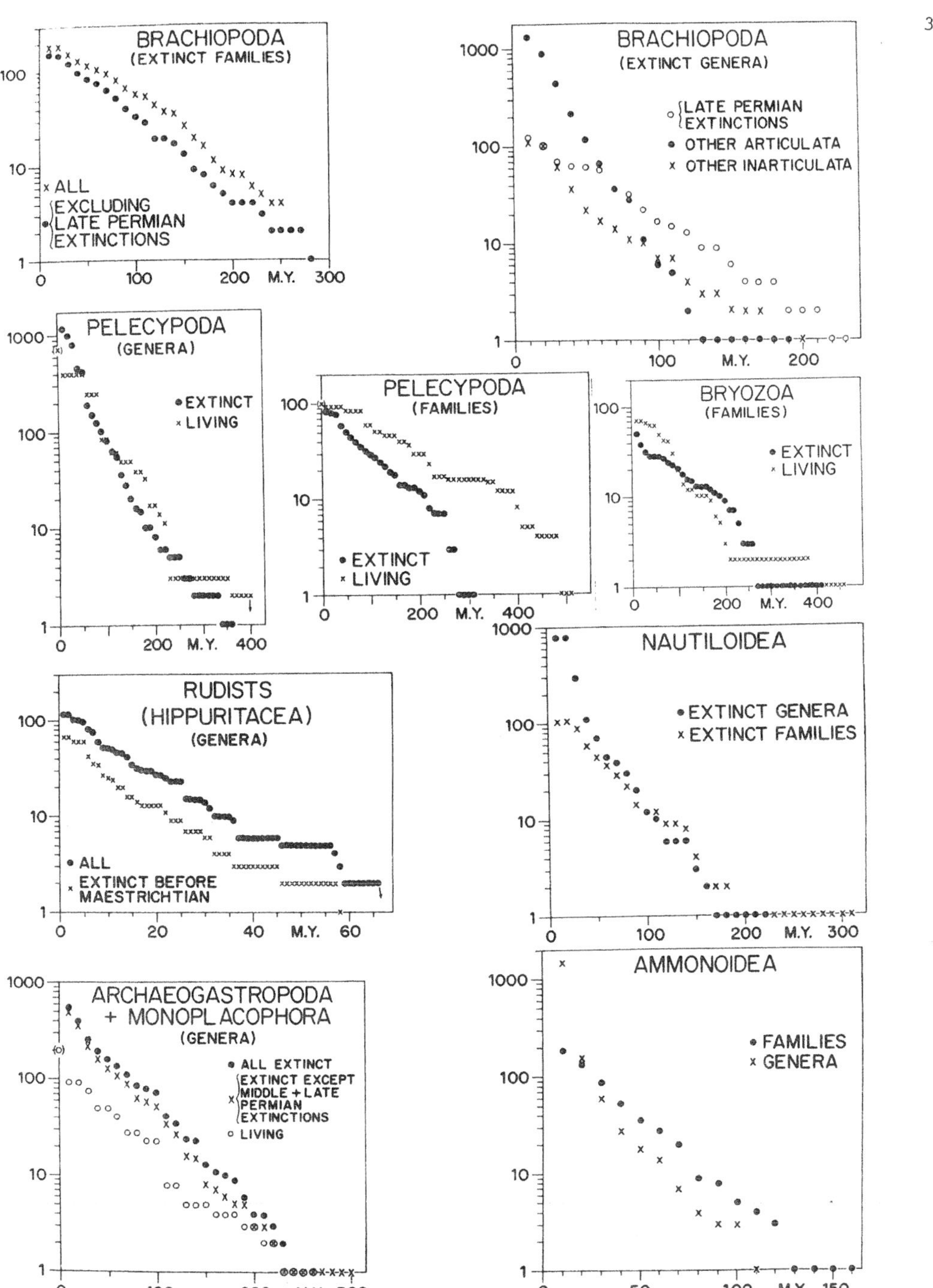

Fig. 2. Taxonomic survivorship curves for Mollusca and Brachiopoda.

4

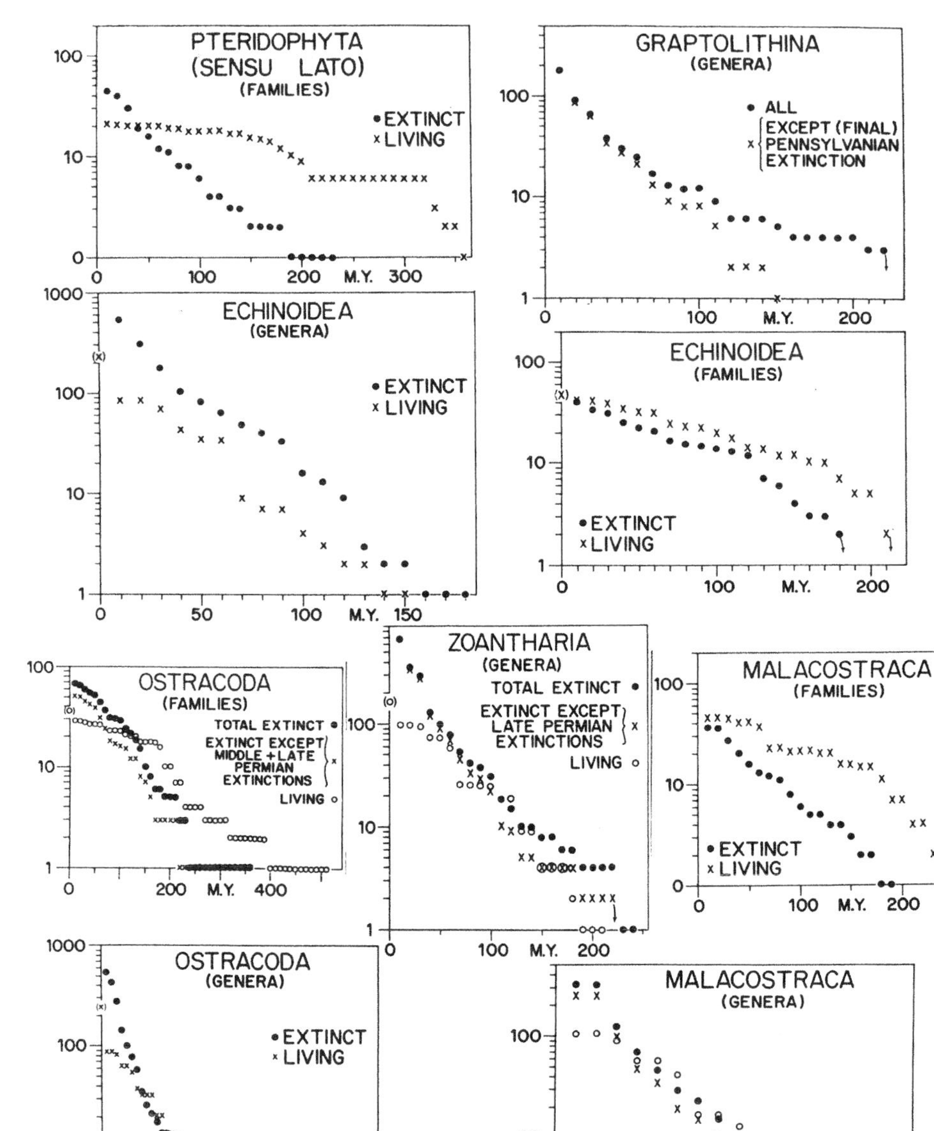

Fig. 3. Taxonomic survivorship curves for plants and invertebrates.

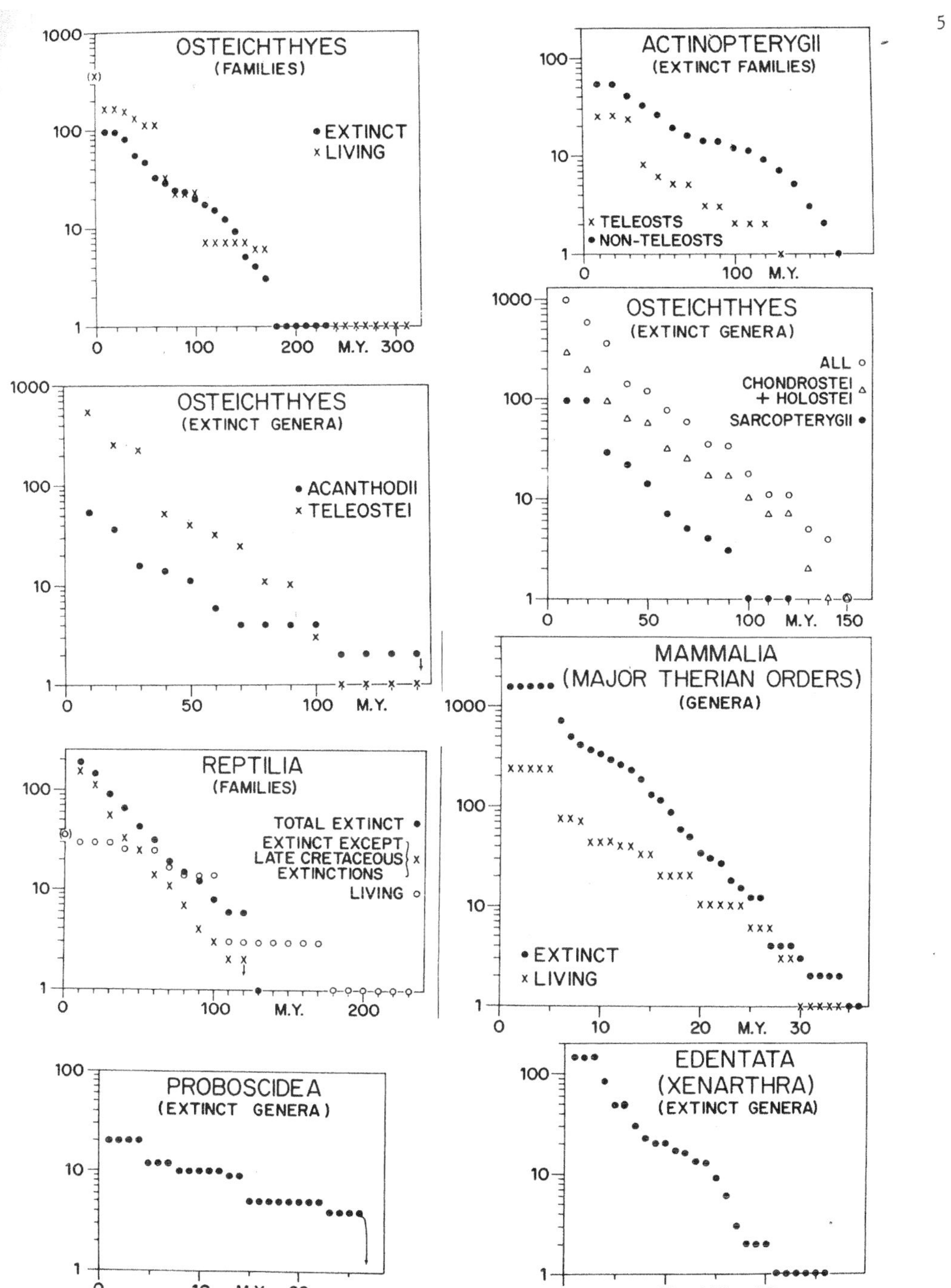

Fig. 4. Taxonomic survivorship curves for vertebrates. "Major therian orders" are those with individual plots.

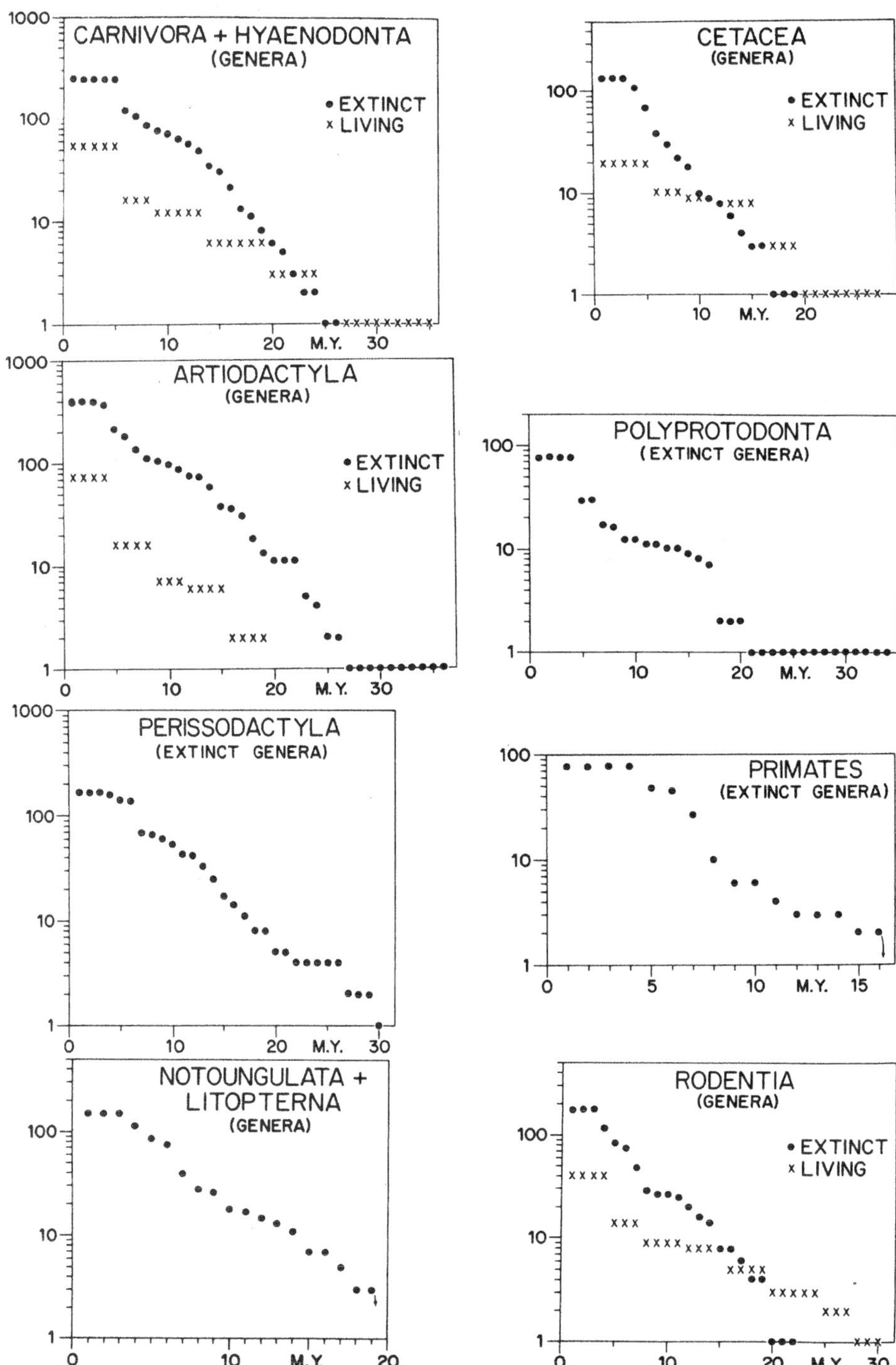

Fig. 5. Taxonomic survivorship curves for mammals. For Primates, Madagascar genera are omitted because the island lacks pre-Pleistocene fossils. Polyprotodonta includes Caenolestoidea.

a few taxa have disproportionate weight, and at the left side, where inaccuracies in dating are most important (7). Usually, the more taxa included the closer is the approach to linearity.

For living taxa linearity of the distribution requires both constant extinction and constant origination. It further requires that both be more nearly constant over absolute time than does a distribution for extinct taxa, which normally spans many overlapping half-lives of taxa, and that probability of discovery not decrease appreciably with age of strata. It is therefore surprising that many even of these distributions are linear (8).

Linearity is not an artifact of the method. Most survivorship curves for individuals are either markedly concave or markedly convex (9). Most small passerine birds are exceptional in having linear survivorship curves. The very wide diversity (biologically and stratigraphically) of groups plotted here argues against any kind of special artifact.

The sources of data usually do not distinguish between real extinction of a lineage and pseudo-extinction by evolution of one taxon into a successor taxon (Simpson, 1953). The latter proves to be negligible. Most taxa which give rise to successor taxa continue in their original form for an appreciable period after the branching. For families of mammals, for which I am familiar with the phylogeny, a maximum (and surely inflated) estimate is 20 per cent pseudo-extinction; a more likely estimate is 5 per cent. Ammonites (1) give 6 per cent (12 of 188 taxonomic extinctions for families). All other available phylogenies give similar results. The pattern of constant extinction remains unchanged whether pseudo-extinctions are included or excluded. Additionally, it is plausible that even pseudo-extinction usually implies the end of an adaptive mode and so would fit into the hypothesis given below. Pseudo-extinctions are probably more common at lower taxonomic levels, despite the counter-claim made or implied by Ruzhentsov (1963), MacGillavry (1968), and Eldredge (1971; Eldredge and Gould, 1972) that they do not occur for species. Any example of an ancestral taxon co-existing with a descendant taxon violates a basic premise of cladistic systematics (Hennig, 1966). As noted, such examples not merely exist but greatly predominate.

Apparent Exceptions

There are some real and some spurious exceptions to constant extinction rates. As noted above, the ages of living taxa are not directly comparable to those of extinct taxa. Sampling error also causes deviations from linearity.

Some short intervals of geologic time have had massive extinctions of some kinds of organisms (Newell, 1963, 1967, 1971). All graptolites became extinct in the Pennsylvanian and almost all stony corals (Zoantharia) did so in the late Permian. This is clearly a different sort of event from the usual process of extinction and does not fit the general explanation given below. In the late Permian, almost everything in the adaptive zone of stony corals was eliminated (10). The adaptive zone itself was demolished for a while. If we eliminate such extinctions as being different from events within an adaptive zone (done supplementarily in parts of Figs. 1-4), linearity is never reduced and sometimes, as with corals, increased (11). When, as with brachiopod genera, extinctions at such a time are sufficiently numerous to be plotted separately, the slope is less than for others, as for the ages of genera living today and for the same trivial reason.

There is no accepted classification of corals at the family level. Two more or less orthogonal classifications are plotted in Figure 6. One is linear, the other convex. This gives some evidence that the Treatise

8

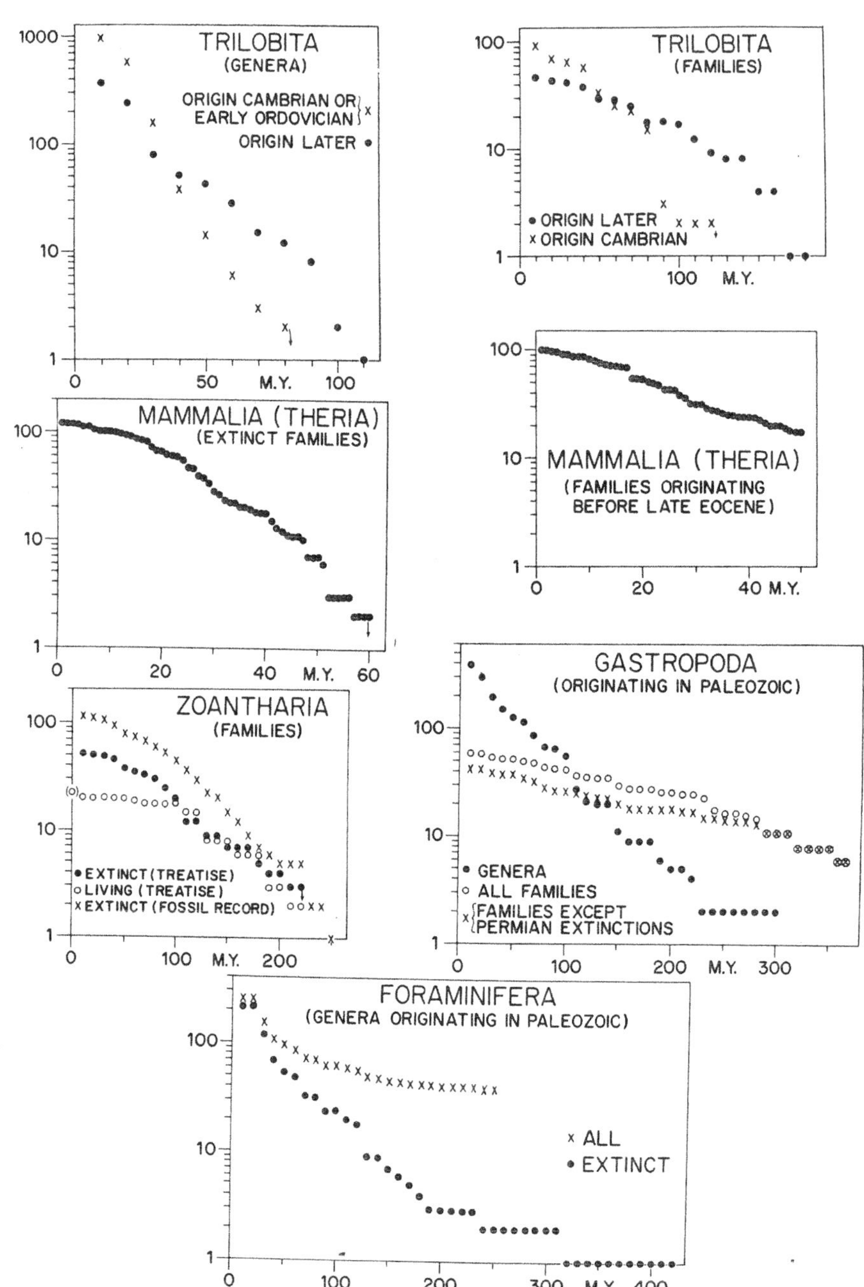

Fig. 6. Taxonomic survivorship curves for apparent exceptions to linearity. See text. Zoantharia families are from very different classifications.

classification is ecologically more realistic, but a decision must ultimately come from phylogeny.

Mammalian families also give a convex curve (Fig. 6). As shown by plotting only families (extinct and living) originating before the late Eocene, however, this is an artifact of the short duration of the Cenozoic in relation to the durations of many families. Living families seem similar to extinct ones, because they jointly determine a single linear curve. This is also true for gastropods but not for foraminiferans.

This difference in foraminiferans of some living taxa from most extinct ones also occurs elsewhere, notably in the Pteridophyta. It is indeterminable from these data whether real exceptions are involved or whether the deviant taxa occupy sufficiently different adaptive zones as not to interact appreciably with the majority of related taxa. Simpson (1944) noted this phenomenon for pelecypods and made it the basis of his bradytelic evolution.

Deviant adaptive zones are obvious for genera of most of the different mammalian orders (Van Valen, 1971) and also for rudists, coral-like pelecypods which formed reefs in the late Cretaceous and seem to have caused the extinction of many corals while doing so (12). Their extinction rate is greater than that of normal pelecypods, although also greater than that of genera of corals, which may not be taxonomically similar (13).

Ammonite and nautiloid genera, but not families, give ostensibly concave curves. When grouped into more homogeneous classes (Fig. 7), approximate linearity results. This also applies to the lengths of the terminal branches (after each final branch-point) of the phylogeny of ammonite families (14). Removal of the late Devonian extinctions has a similar effect for genera as for terminal branches but is not plotted. I did not try to subdivide the nautiloids. I further suspect that pseudo-extinction is more common for ammonite and perhaps nautiloid genera than is true elsewhere, which would increase concavity by increasing short-lived taxa at the expense of longer-lived lineages.

The need to separate Paleozoic and Mesozoic ammonites shows that there is at least a descriptively real exception here. The extinction rate is definitely not constant throughout the existence of the group. The same is true for trilobites (Fig. 6), the separation here being in the Ordovician. Linearity again holds for each segment. I do not know what caused either change, but it may be relevant that effectively all Mesozoic ammonites descended from one Paleozoic lineage and that trilobites declined greatly in the Ordovician. In each case the division time found for extinction rates corresponds to the greatest separation in the group on other criteria (15).

Moreover, because of preservational bias and incomplete collecting and study, short-lived taxa will be found less often than long-lived taxa. However, for equally frequent groups the effect of this bias will be a reduction in observed longevity by a constant absolute amount. This will leave linearity and even the slope of the curve unchanged. Rarer groups will have a greater expected reduction in observed longevity than commoner groups. Any effect of this property depends on whether rarity is correlated with longevity, and I know of no relevant data.

Any combination of subgroups with unequal constant extinction rates produces a concave resultant curve. The amount of inequality can be estimated by this concavity, but unfortunately any concavity manifests itself most in the regions of greatest sampling error. Linearity is nevertheless sufficient that exceptions must be rare or slight. Abrupt ends to distributions, as with echinoid families, occur when (as we have seen with mammal families) the stratigraphic range of the group impinges on the possibility of long-lived taxa having already become extinct. Linearity is unaffected by multiplication or addition with a logarithmic (ordinate) or arithmetic (abscissa) constant.

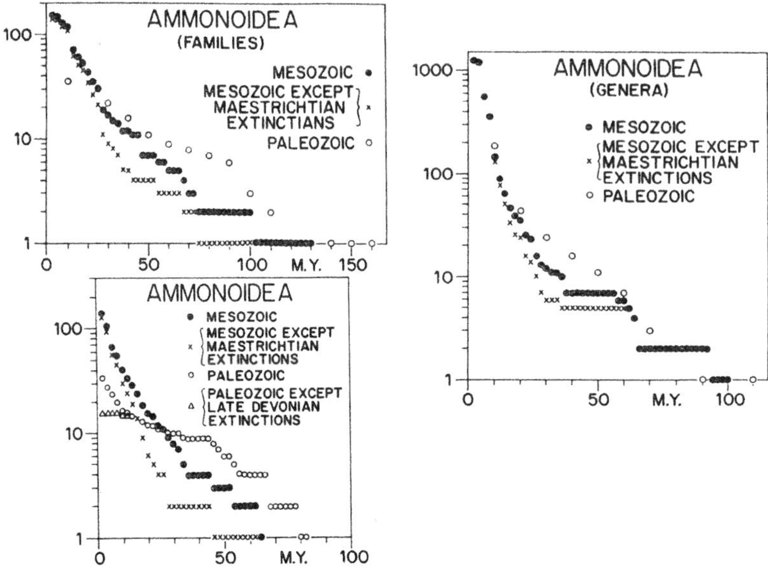

Fig. 7. Taxonomic survivorship curves for ammonites. See text. The graph on the lower left is for terminal twigs of a phylogeny of families.

* * *

We see that the exceptions are either spurious, rare, slight, doubtful, or exceptions only in part. The pattern is therefore sufficiently general that the minor exceptions are best explained by unusual circumstances peculiar to each case. There is a strong first-order effect of linearity, and it is this, rather than its perturbations by special and diverse circumstances, that deserves primary attention.

Related Phenomena

The constancy of extinction in terms of survivorship does not imply constancy over geological time, and vice versa. Any form of survivorship curve has a mean, as does any pattern of extinction rates over absolute time, and this is the only formal connection between the two phenomena. There may nevertheless be a deeper causal connection.

I give some extinction and origination curves over geological time to illustrate their variability in the measurement framework proposed here (Fig. 8) There is an extraordinary exponential decrease in origination rate of mammalian families, by two orders of magnitude, from the beginning of the Cenozoic. An inverse phenomenon occurs in diatoms, where the species of Pennales (a largely benthonic groups, unlike the Centrales) have increased exponentially, as in the log phase of bacterial culture, in the same interval (16; Fig. 1). We can look at Figure 8F inversely: there is a nearly linear increase in the proportion of families that had originated by a given time and that are now extinct.

Large foraminiferans (17) have originated about 41 times independently from smaller foraminiferans, almost always from much smaller ones. About 33 of these clades have existed since the end of the early Cretaceous, when an approximate equilibrium was established. Figure 9 shows that the numbers of clades and genera present simultaneously have followed more or less log-normal

11

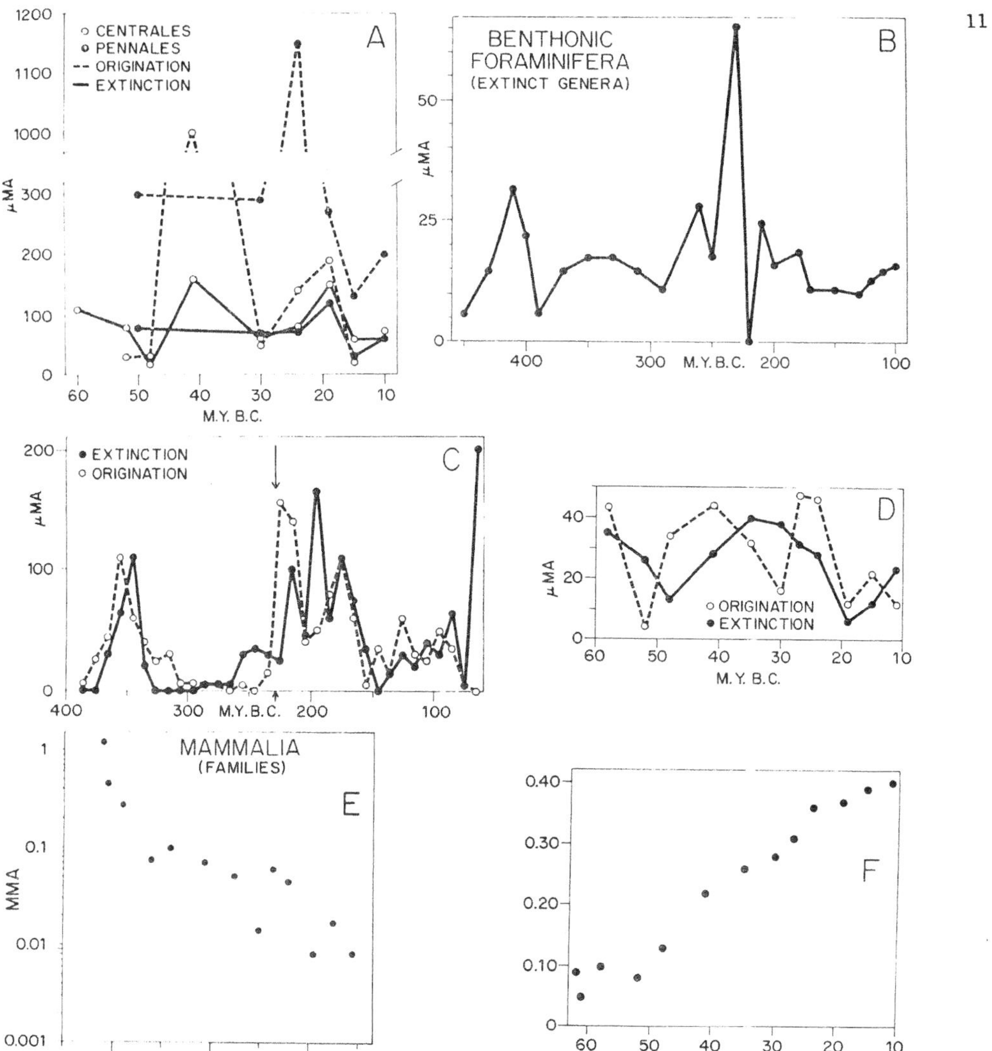

Fig. 8. A: Origination and extinction rates for species of diatoms. B: Extinction rates for genera of benthonic foraminiferans. C: Origination and extinction rates for families of ammonites. High origination rates precede high extinction rates except for the Permian and latest Cretaceous. D: Extinction rates of mammalian family lineages, and origination rates for mammalian families now surviving. E: Origination rates for all mammalian families (new families per family in previous age per unit time). F: For mammalian families, (cumulative recent families originated)/(cumulative total families originated).

12

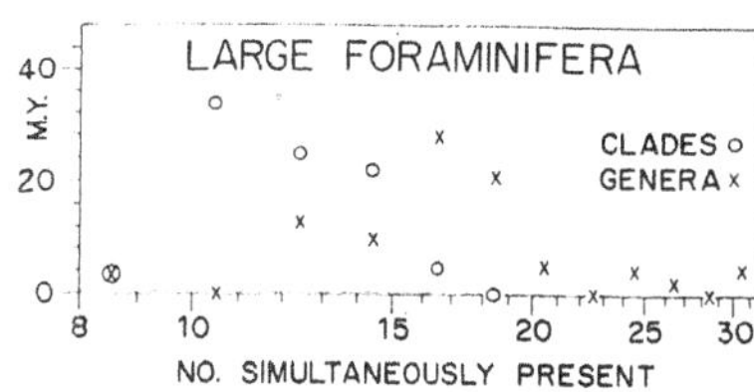

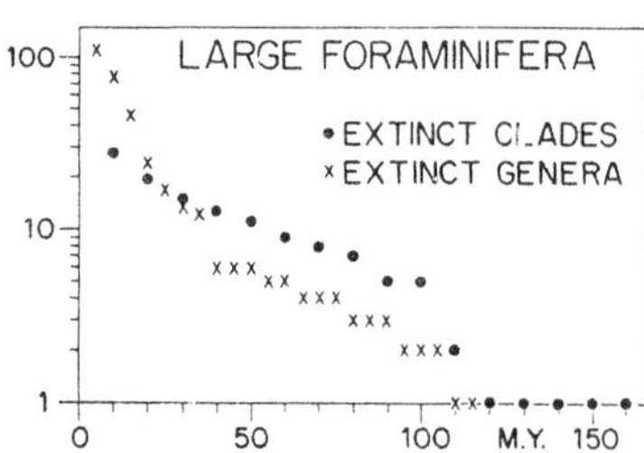

Fig. 9. Survivorship curves for genera and clades of large foraminiferans, and distribution of number of genera and clades simultaneously present after the early Cretaceous (Albian).

* * *

distributions since that time. Survivorship curves for clades and genera of this taxonomically very heterogeneous, but ecologically rather homogeneous, group are linear within sampling error and differ from the curve for genera of all benthonic foraminiferans.

It would be useful to know whether the extinction of entire lineages also occurs at a constant rate, or if some taxa can avoid it by evolving successor taxa. This appears to represent an unsolved problem in graph theory (18).

Measurement of Rates

I propose a general unit of rates for phenomena which can be treated as discrete.

A macarthur (ma) is the rate at which the probability of an event per 500 years is 0.5. Robert H. MacArthur showed the importance of extinction in ecology (19).

Let P be the probability per t thousand years.

$$ma = -\log_2(1-P^{2t}) \quad (1)$$

With respect to extinction, one macarthur is the rate of extinction (Ω) giving a half-life of 500 years. With respect to origination, one macarthur is the occurrence of one origin per thousand years per potential ancestor. With respect to molecular evolution, one millimacarthur (mma) is the rate giving one substitution per million years (20).

Apparent equilibrium extinction rates of bird species on islands that have been studied (21) are 0.5 to 10 ma. Mammal species in the late Pleistocene of Florida (Martin and Webb, in press) had a rate of regional turnover (time from immigration to local extinction) of about 7 mma (Fig. 10).

Table 1 gives estimated extinction rates for the taxa studied. The sedentary marine benthos is remarkably homogeneous, and motile marine organisms have somewhat higher rates less similar among themselves. Mammalian genera (and Mesozoic ammonites) have the highest rates, but rates for reptilian families are as high as those for mammalian families. For everything except mammals the extinction rate for families is about half that for genera. Groups in relatively new adaptive zones (at least to them) usually have higher rates than long-established groups. Not surprisingly, ecology is a better predictor of evolutionary rate than is amount of information-bearing DNA (22).

If genera of a family go extinct independently, the extinction rate of families will depend directly on the number of contemporaneous genera per family

TABLE 1: Extinction Rates (in μma).
Abbreviations: fam., family; gen., genus; sp., species

GROUP	RATE fam.	gen.	sp.
Pteridophyta	20		
diatoms		50	90
Coccolithophyceae		25	
Dinoflagellata		20	55
Foraminifera	10		
benthonic		20	
planktonic		30	100
large		50	
Ostracoda	10	25	
Graptolithina		30	
Bryozoa	7		
Brachiopoda	15		
Articulata		45	
Inarticulata		25	
Malacostraca	12	30	
Trilobita			
early	20	80	
late	10	35	
Echinoidea	10	25	
Zoantharia	10	25	
Ammonoidea			
Paleozoic	20	35	
Mesozoic	75	150	
Nautiloidea	15	30	
Archaeogastropoda + Monoplacophora		20	
Gastropoda			
Paleozoic	4	20	
Pelecypoda	8	20	
rudists		50	
Osteichthyes	15	30	
Teleostei	20	35	
Holostei + Chondrostei	12	25	
Sarcopterygii		30	
Acanthodii		30	
Reptilia	30		
Mammalia (Theria)	30	150	
Polyprotodonta		150	
Primates		220	
Rodentia		160	
Carnivora + Hyaenodonta		120	
Edentata		180	
Proboscidea		60	
Notoungulata + Litopterna		170	
Perissodactyla		120	
Artiodactyla		120	
Cetacea		200	

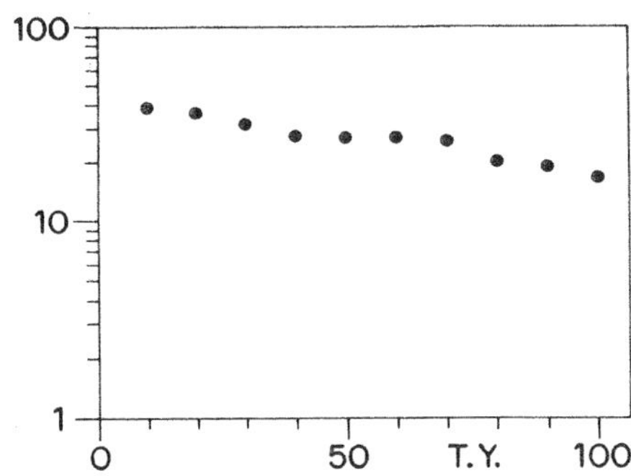

Fig. 10. Regional survivorship curve for mammalian species in Florida during the late Pleistocene.

* * *

if there is no origination of new genera in the family. Rough estimates, weighted by the number of taxa in each interval, give 2.9 contemporaneous genera per family of Ostracoda, 2.8 for Echinoidea, and 5.1 and 5.8 species per genus respectively for pennate and centric diatoms. With 2.9 genera per family, independent extinction will give an extinction rate per genus 2.6 times that per family. This ratio would be 2.5 for echinoids and 3.5 for average diatom species. I take the expected time of family extinction to be when the expected number of genera per family reaches 0.5. With branching (origination of new genera), the families will of course have longer expected longevity and so the expected ratio will increase. If the rate of branching equals that of extinction, simulation in the above range indicates roughly a doubling of the expected ratio, although a high variance among families in genera per family probably increases it more.

Table 1 shows that genera of echinoids and ostracodes have about 2.5 times the extinction rate of families, a result indistinguishable from that on the assumptions of independence and no branching. Because branching exists, some degree of correlated extinction of genera within families seems probable. The observed rate for species of diatoms is only twice that for genera, much less than the expected value of 3.5, so there is a strong correlation in extinction here even without considering branching.

Contemporaneous Subgroups

Extinction rate obviously depends on the area considered as well as on the inhabitants. Fig. 11 gives a relation between area and extinction rate, using all available data (21). Approximate linearity holds over 11 orders of magnitude on a log scale. With the extinction rate in $\log_{10}$ macarthurs and A the area in $\log_{10}$ sq. km, the regression is $\Omega = -0.66A + 1.53$ when calculate from the birds and mammals and excluding the world fauna. A regression from complete data would presumably differ somewhat. As expected, the arthropods on Simberloffia seem to require less area for the same extinction rate than do vertebrates (23). A continent seems to be the largest area in which organis can interact more or less as they do in any smaller area that is large enough for a population.

Therefore species and higher taxa occupying smaller areas, as will often be the case when they originate, have an expected rate of extinction greater than that of more widespread taxa. Similarly, as Small (1946, 1948b) noted in

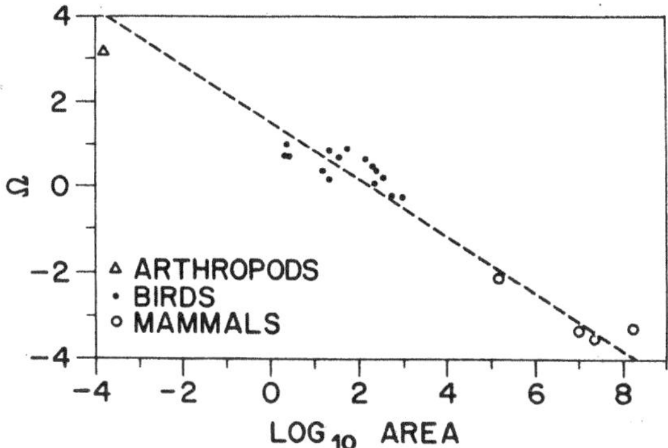

Fig. 11. Extinction rate (Ω, in $\log_{10}$ ma) of species as a function of area (sq. km.).

* * *

a different context for diatoms, genera with fewer species have a higher probability of extinction than do genera with more species. This is of course because branching at the specific level increases the probability of survival of the entire sub-tree, since the species will usually have different ecologies.

For these and other, less well-documented, reasons, there are definable and large subgroups that have a higher extinction rate than others, and moreover these subgroups should be relatively most frequent among younger taxa. Therefore a concave survivorship curve should result. How can it be that the curves are nevertheless linear?

There are two possibilities: incorrect assumptions and compensatory feedback. It may be that, by the time taxa have an appreciable probability of appearing in the fossil record, they have the same mean area occupied as older taxa. This sort of rapid increase may also be true, although it doesn't seem as likely, for number of species in a genus. Enough data are probably available to resolve the latter point. Compilations by Sloss (1950) and Simpson (1953), Small's studies (e.g. 1946) on diatoms, and data by Berggren (1969) and others on planktonic foraminiferans, suggest that there is indeed usually an appreciable lag before the number of subordinate taxa approaches its maximum value. The kinds of evidence in these papers are by no means conclusive on the present point, however. Furthermore, there may well be a threshold (like those found by MacArthur [1972] and others for population size in simplified models of colonization) above which survival is virtually assured and below which rapid extinction is likely. Other possible inadequacies of the assumptions, that the effects are too small to be detected or that they affect only a small proportion of taxa, are clearly unrealistic.

Even if an initial concavity in survivorship curves is avoided by the rapid changes invoked above, there still remains a major heterogeneity among the surviving taxa with respect to area, number of species or genera, and other factors relevant to survival. There must then be some kind of compensatory feedback to give the observed degree of linearity. Two kinds of feedback seem possible. There might be an interaction between age and area (or number of species) such that older and younger taxa have different effects of area on probability of extinction. No simple model fits this biologically implausible alternative. However, if for new taxa small area and low number of species were not harmful, or as harmful as later, the expected initial concavity would be reduced or eliminated.

The other kind of feedback is by the continual appearance at all ages of taxa of low area and other characters giving greater susceptibility to extinction. In fact this could be regarded as an aspect of the extinction process: taxa would become more susceptible at a constant average rate per taxon per time. This transformation need not be by feedback per se, but it does require that the production of greater susceptibility be part of the same process as the extinction. There is no requirement that each subordinate taxon be equally likely at any given time to become more susceptible, but the overall control must compensate for survival by greater susceptibility at some other time or for some other species. This second alternative is plausible (given linear survivorship) but largely untested. It predicts that, at least after a short initial period, criteria of susceptibility will have a steady-state distribution in survivorship time independent of the age of the taxon. Small (1946, 1947, 1948a,c) found such a steady-state distribution in real time for generic sizes of diatoms through the Cenozoic.

An Evolutionary Law

The effective environment (24) of the members of any homogeneous group of organisms deteriorates at a stochastically constant rate.

This law goes a bit beyond the observations in postulating the cause to be extrinsic rather than intrinsic, like nuclear decay. There is of course much other evidence (Simpson, 1953) for such a step. The law is also stated in terms of real time rather than survivorship time, although it applies to both and ties them together. A more neutral statement is that extinction in any adaptive zone occurs at a stochastically constant rate.

"Homogeneous" is a necessarily ambiguous word, and its meaning in a particular case must depend on the particular circumstances and on the degree of precision desired. Paleozoic and Mesozoic ammonites may differ sufficiently to be nonhomogeneous, but the entire marine benthos could be treated together almost as well as subdividing it. This does not lead to circularity; no subdivision of a group with racial senescence (25) would give a set of linear survivorship curves. The homogeneity required (and so its verification) is entirely ecological and is in terms of ultimate regulatory factors (Van Valen, 1973) of population density. Mice and fruit-eating flies might be homogeneous with each other but not with blood-sucking flies, although a scaling problem would remain. We could say that there always exists a degree of homogeneity at which the law is true, and try to measure it independently. Homogeneity does not imply equality of ability to respond to the deterioration; but counterexamples to a constant distribution of such abilities would disprove the law in its first form as a universal phenomenon (26).

Like any law (27), the effects of the law of extinction are observable only under appropriate conditions (here, persistence of the width of the adaptive zone). Unlike many laws (28), it may not be universal. If it is not universal, it should be possible to limit its domain in an objective manner. Such limitation should derive from an understanding of the causal basis for the law.

It is not fashionable to speak of laws in evolutionary biology or for historical processes generally. I think this is based on both a misunderstanding of the regularity of actual processes (29) and on an over-reaction to poorly formulated laws of earlier workers (30). Laws are propositions that specify sufficient conditions for a result; given the conditions, the result will occur, although some of the conditions (the bounds of the domain) may be implicit. The degree of confirmation of a law is of course a different matter from whether a proposition is (or represents) a law in this sense. Any general statement of the nature of a causal process states a law (31).

The Red Queen's Hypothesis (32)

The probability of extinction of a taxon is then effectively independent of its age. This suggests a randomly acting process. But the probability is strongly related to adaptive zones. This shows that a randomly acting process cannot be operating uniformly. How can it be that extinction occurs randomly with respect to age but nonrandomly with respect to ecology?

We can consider the situation in terms of an ensemble of mutually incompatible optima. It is selectively advantageous for a prey or host (including plants) to decrease its probability of being eaten or parasitized. It is often selectively advantageous for a predator or parasite species, and much more often for a predator or parasite individual, to increase its expected rate of capture of food. It is selectively advantageous for a competitor for resources in short supply (food and space in the broadest senses, and sometimes also externally supplied adjuncts to reproduction or dispersal) both to increase its own effect on its competitors and to decrease the effect of its competitors on itself (33).

Every species does the best it can in the face of these pressures. Probably all species are affected importantly by them at least over intervals of a few generations. Response to one kind of pressure may well decrease resistance to some other, at that time weaker, kind.

The various species in an adaptive zone (Van Valen, 1971), whether or not this zone is sharply delimited from others, can be considered together. We can assume as an approximation that a proportional amount w of successful response by one species produces a total negative effect of $v = w$ on other species jointly, usually less than v for any one of these species. For n species, the mean decrement of fitness per species is v/n. For m successful responses simultaneously (in an interval τ), the mean total decrement per species is mv/n. Over some interval t (in units of τ), the decrement is then mvt/n. To maintain itself as before, the species must increase its fitness by an amount mwt/n. Since each decrement generates a response, $m = n$. Most species will be able to recoup their loss, more or less while it occurs, but in doing so they jointly produce another disadvantage of v for the average species. This process of successive overlapping decrements of v may continue indefinitely. Species at least locally new may replace those for which the rate of environmental deterioration has been too great.

For the momentary fitness F of a random species,

$$\phi \equiv \frac{dF}{dt} = \frac{m(w-v)}{n} . \qquad (2)$$

For a given species, ϕ will of course vary from time to time as its specific decrements and possible responses covary. However, since $w = v$ on the average, $E(\phi) = 0$, i.e. the mean fitness in the adaptive zone is constant in the long run. If we take a standard average fitness F_1, the real average fitness will be $F_0 = F_1 - v$, because any response of one species above F_0 will bring a corresponding decrement from the counter-response of other species. v is the environmental load. Thus, even for a single species, the total selection pressure is constant over intervals that are long enough to average out irregularities.

ϕ is of course itself a variable among the species at any time, and its variation (or that of F) will determine the extinction rate in the adaptive zone. Species may (and do, but to an unknown extent) differ in the spectral threshold T (Reyment and Van Valen, 1969) of ϕ or its components to which they can no longer respond successfully and at which they therefore become extinct. Which species have which values of ϕ and T will change markedly with the nature of the stresses at a given time. The observed pattern of extinction seems to imply that the stresses are sufficiently diverse that they affect most species similarly over a very long interval. Assume, as justified in part by the Central Limit Theorem, that both ϕ and T are normally distributed, with means μ_1 and μ_2 respectively. Then, since both distributions range over the same set of species and so have the same area, the rate of extinction is the overlap of the distributions. In the case of equal variances for ϕ and T,

$$\Omega \equiv \frac{dE}{dt} = \frac{1}{\sqrt{2\pi}} e^{-\frac{(\mu_1-\mu_2)^2}{8}} . \qquad (3)$$

Uncompensated departures from normality and from equal variances in the distribution will of course affect the accuracy of this expression but not the dependence of Ω on the real area of overlap.

On this self-contained system we must now impose the physical environment and irregular biotic perturbations. Almost all changes in the physical environment of an adaptive zone will be deleterious to its inhabitants, either directly or by permitting the establishment of other species which can then survive there. Therefore regular changes in the physical environment can be treated as constant factors of ϕ. The observed constancy of Ω is evidence that such a treatment is permissible.

Important perturbations do, however, occur, from both biotic and physical causes. They can be treated together. The probability of extinction is not constant over geological time, as Fig. 8 and the extinctions at the end of the Cretaceous suffice to remind us. When major perturbations occur well within the geological range of a group, as for brachiopods in the late Permian, Paleozoic ammonites at the end of the Devonian, or coccoliths at the end of the Cretaceous, it happens that if we ignore the subgroups that became extinct then, the remainder show a constant rate of extinction (34). There is no important lasting effect of the unique event, unless total extinction occurred, nor do its effects extend detectably beyond the subgroups prematurely eliminated. Newly appearing subgroups after the event continue in the same way as those that lived through it. Smaller perturbations occur much more frequently, however (Fig. 8), and it is the very resultant of these perturbations that determines the long-term constancy.

Therefore the effects of these minor perturbations are not independent of each other over time. If they were, they should not be distributed about a constant mean. The rate of deterioration of the effective environment has what we can descriptively call the property of homeostasis. A large change in one interval is compensated for by a small one later, on the average, and vice versa; the mean itself does not undergo a random walk. It is the organisms in the adaptive zone that can link the perturbations across time in that the effects of one perturbation may depend on the effects of those before it. Species easily removed by one kind of perturbation are not there if it comes again soon, and may accumulate if it does not come for a long time. Meanwhile other kinds of perturbations are taking their own toll, more or less independently of the first kind (35).

We see here a major difference from the usual theory of genic selection. The latter depends only on the current distribution of frequencies of alleles and their interactions with each other and the environment. It does not depend at all on the process by which the current distribution was obtained. In formal language, it is a Markov process. But any process can be made Markovian by choice of a suitable level of analysis. With extinction we can see that a non-Markovian analysis of selection is appropriate. This is probably true for the general case also (36).

The Red Queen does not need changes in the physical environment, although she can accommodate them. Biotic forces provide the basis for a self-driving (at this level) perpetual motion of the effective environment and so of the evolution of the species affected by it (37).

The Red Queen's Hypothesis is a sufficient explanation of the law of extinction but not one that is yet derivable with confidence from lower-level knowledge of the causes of individual extinctions and the nature of species interactions. There may be other sufficient explanations that I have not been imaginative enough to see, and predictions by the Red Queen may be duplicated by such undiscovered explanations. Disproof of the Red Queen's Hypothesis is possible in several ways, but fully adequate confirmation must await derivation of it from what we can reasonably regard as facts.

We can go a step further by thinking of an adaptive landscape in a resource space (Van Valen, 1971). The amount of resources is fixed and can be thought of as an incompressible gel neutrally stable in configuration, supporting the peaks and ridges. If one peak is diminished there must be an equal total increase elsewhere, in one related peak or more uniformly. Similarly, increase in a peak results in an equal decrease elsewhere.

Species occupy this landscape and can be thought of as trying to maximize their share of whatever resource is scarcest relative to its use and availability. This resource will take the role of the gel, and the momentary fitness of a species will be proportional to the amount of gel under its area (the amount of the limiting resource it controls). To a sufficiently close approximation this momentary fitness seems to be what natural selection maximizes.

The landscape is changing continuously, at three levels. Species displace each other from areas of the adaptive surface. This can be by resistance to predation as well as other means, and some parts of the landscape may in this way or others (e.g. severe weather) be under-occupied. Incompatible optima maintain diversity. Secondly, the distribution of gel within the landscape changes, as with climatic change or (for herbivores) when the flora changes. Finally, the total amount of gel (or the amount in fact used) can change. This can occur by another necessary resource becoming limiting, by constraints on occupation of the adaptive zone, or by a change in the amount of the same limiting resource. The entire landscape can in this way disappear.

The first two kinds of change in the landscape lead to the the Red Queen's Hypothesis, while the third is, without compensatory changes, inconsistent with it. Empirically extinction is less independent of age when extinctions caused by major constrictions of adaptive zones are included. The observed degree of independence suggests that the third kind of change is relatively unimportant on the evolutionary time scale. We can hardly regard such a conclusion as established, but it does show that the question of variation in the size of major adaptive zones is fundamental.

Molecular Evolution

The Red Queen's Hypothesis provides reason to expect a long-term approximate constancy in the rate of evolution of individual proteins within any single adaptive zone (38). It does so without ad hoc assumptions, although such assumptions could be invoked to explain away exceptions just as they have been invoked (Clarke, 1970) to explain away constancy by the usual selective framework. Constancy of the rates of protein evolution is often regarded as the most important evidence for what King and Jukes (1969) called non-Darwinian evolution, despite serious misinterpretations (39).

A further prediction is that rates of protein evolution will be related monotonically to rates of taxonomic evolution, to the extent that the subtaxa in different groups are comparable (13,40). Some evidence (41) is available that rates are not always constant, and preliminary results I obtained long before discovering constant extinction suggest that the mean rate of protein evolution during the relatively short time while the orders of placental mammals diverged from each other was greater than that later (42). *Lingula*, a brachiopod genus present since the Ordovician or Silurian, and *Triops cancriformis*, a branchiopod (not brachiopod) species not detectably changed since at least the early Triassic (1), would be expected to have proteins more similar to those of the common ancestor than would their more divergent relatives. The evidence for approximately constant evolutionary rates of proteins comes from organisms (mainly vertebrates) which have all evolved appreciably since their separation from each other.

The Red Queen in her simplest gown also predicts a perhaps real phenomenon, which Ohta and Kimura (1971) noted as mysterious: that irregularities in the rate of molecular evolution seem to be more or less cancelled out over long intervals by a seemingly negative autocorrelation.

Implications of the Red Queen

Thoday (1953) proposed a concept of fitness as the probability of survival of a lineage to some very distant time in the future. More generally, we can take

$$F_t = \int_0^\infty w(t)P(t)dt \qquad (4)$$

where $P(t)$ is the probability (43) at time 0 of survival to time t, and $w(t)$ is a weighting function (the same for all lineages and perhaps integrating to 1 for scale) for which I would choose exponential decay at a low rate (44). Time 0 is the variable present.

The Red Queen proposes that this fitness has only two components for almost any real lineage: which adaptive zone is occupied and what the probability distribution is of new sublineages occurring by branching. Artiodactyls have largely replaced perissodactyls in the same adaptive zone, yet their extinction rates are identical (Tables 1,2). It is the rates of branching that are decisive. Monotremes still linger on in two isolated subzones, one of which has already been invaded by marsupials. The gradual extinction of multituberculates (Van Valen and Sloan, 1966) occurred by herbivorous placentals diversifying into parts of the joint adaptive zone formerly held by multituberculates. One multituberculate survived 15 or 20 million years after the rest of the extinction was completed (45); the latter had taken only about 10 million years. The Red Queen's proposal implies that extinction of lineages (subtrees) will prove not to be constant when it can be measured.

For the Red Queen, curiously, does not deny progress in evolution. By group selection such as this, as well as by individual selection, properties of communities can change in a directional manner. It may well be that the average Cenozoic species could outcompete the average Cambrian one; some information on this can probably be derived from functional analysis. It is

TABLE 2: Extinction and origination rates (in ma) for genera of two competing orders of mammals.
E: early; M: middle; L: late

	EXTINCTION		ORIGINATION	
Epoch	Perissodactyla	Artiodactyla	Perissodactyla	Artiodactyla
E. Eocene	120	160	1300	500
M. Eocene	80	90	340	920
L. Eocene	120	90	290	260
E. Oligocene	120	100	80	150
M. Oligocene	100	160	40	130
L. Oligocene	170	130	130	80
E. Miocene	90	130	130	260
M. Miocene	30	90	90	90
L. Miocene	40	60	60	100
E. Pliocene	90	130	90	320
L. Pliocene	90	80	20	70

* * *

well known (cf. Romer [1966]) that such functional progress has occurred in vertebrate evolution. The Red Queen measures environmental deterioration on a scale that is determined by the resistance of contempory species to it, so the scale and real deterioration themselves may well change without a change in the measured deterioration. Darwin's example of wolf and deer exemplifies this.

The Red Queen proposes that events of mutualism, at least on the same trophic level, are of little importance in evolution in comparison to negative interactions, although she does not consider other cases where mutualism is so great that the mutualists function as an evolutionary unit, as with lichens and perhaps chloroplasts. She considers the usual contrary view to be a result of wishful thinking, the imposition of human values on nonhuman processes.

The existence of the law of extinction is evidence for ecological significance and ecological comparability of taxa from species to family, within any adaptive zone.

We can think of the Red Queen's Hypothesis in terms of an unorthodox game theory (46). To a good approximation, each species is part of a zero-sum game against other species. Which adversary is most important for a species may vary from time to time, and for some or even most species no one adversary may ever be paramount. Furthermore, no species can ever win, and new adversaries grinningly replace the losers. This is a direction of generalization of game theory which I think has not been explored.

From this overlook we see dynamic equilibria on an immense scale, determining much of the course of evolution by their self-perpetuating fluctuations. This is a novel way of looking at the world, one with which I am not yet comfortable. But I have not yet found evidence against it, and it does make visible new paths and it may even approach reality.

Acknowledgments

I thank the National Science Foundation for regularly rejecting my (honest) grant applications for work on real organisms (cf. Szent-Györgyi, 1972), thus forcing me into theoretical work. This paper has been circulating in samizdat since December, 1972, and I have given talks based on it before and after then. I thank Dr. R. A. Martin for unpublished data and Drs. P. Billingsley, J. Cracraft,

J. F. Crow, D. H. Janzen, T. H. Jukes, S. A. Kauffman, H. W. Kerster, M. Kimura, E. G. Leigh, J. S. Levinton, R. C. Lewontin, J. F. Lynch, V. C. Maiorana, J. Maynard Smith, P. Meier, D. M. Raup, G. A. Sacher, T. J. M. Schopf, and E. O. Wilson for discussion. The Louis Block Fund of the University of Chicago paid for the preparation of the figures.

Notes

(1). The taxonomic data I used are from the following sources: For vertebrates, Romer (1966), with a few modifications from various later work. For invertebrates and genera and families of Foraminifera, R.C. Moore (ed.), Treatise on Invertebrate Paleontology (Boulder, Colorado: Geological Society of America and Univ. Kansas Press), 1953-present, including revisions (C. Teichert, ed.), with supplementation or (for Bryozoa) substitution from The Fossil Record (W. B. Harland et al., eds.; London: Geological Society of London [1967]). I deliberately ignored later pertinent work (e.g. Yochelson [1969]) on invertebrates. Pteridophytes and coccoliths are also from The Fossil Record. For species of Foraminifera, Berggren (1969). For Dinoflagellata, Sarjeant (1967). For diatoms, Small (1945,1946).

(2). The time scale I used is from the following sources: Berggren (1972); Anonymous (1964); Everndon, Savage, Curtis, and James (1964); Everndon and Curtis (1965); Kauffman (1970); and Van Valen and Sloan (1966). Experiments show that the results are robust to reasonable changes in the time scale.

(3). Various conventions are necessary in any such compilation. Some general ones I used are the following: I took the duration of a taxon as from the middle of the epoch (or other shortest interval) before the first record, to the middle of the epoch of the last record. I ignored questioned records and unrecognizable taxa. For data accurate only to period I plotted the range as ending in the middle unit of the period. I used all data regardless of degree of precision (unless imprecise beyond a period) but used them to the precision allowable. Also, summaries of ranges (even of individual taxa) are too often inaccurate, as shown by unquestioned records of subordinate taxa beyond the stated limits, so whenever possible I compiled data at the level of genus.

(4). I knew of this deficiency for many years but saw no reason to pursue it, as I expected the shape of the curves to remain concave. In hindsight one could also expect that the group as such might have a progressively higher probability of extinction as the biota around it evolves while it does not. This would give convex survivorship curves.

(5). In addition to the data plotted, I made several dozen plots of subsets of the same data, using such criteria as exclusion of a major extinction or subtaxon, or restriction of the time interval used. All important deviations from the total distribution are included in the figures given here. There is a bias in using all extinct subtaxa of a living group in that the longer-lived of subtaxa originating recently are still alive. Tests show that this effect is negligible for long-ranging groups (not, e.g., for families of mammals or echinoids) and obviously all living and extinct subtaxa cannot be combined into one useful curve (Simpson, 1953).

(6). I used all groups for which adequate data were available. Omissions are due to poverty of the fossil record or its study (e.g. Insectivora), small number of extinct taxa, error of dating being a substantial part of estimated durations (e.g. Archaeocyatha), or lack of adequate compilation (e.g. Gymnospermae).

(7). This is the reason for the apparently flat tops of some distributions.

(8). The initial figure in parentheses in distributions of most living taxa represents the total now alive, including those not known as fossils. Many living taxa in some groups are not easily fossilizable, and our knowledge of the present fauna is better than that at any other time even aside from this.

(9). Examples can be found in the following: Pearl (1928); Allee, Emerson, Park, Park, and Schmidt (1949); Kurten (1953); Odum (1971).

(10). Coral-like brachiopods of the Permian also became extinct, as did many other organisms superficially less similar to corals.

(11). Tests are necessarily weak, but data from several groups give no indication that a major extinction affects taxa of different ages to a different extent.

(12). A third (17 of 51) of the families of stony corals present in the late Cretaceous became extinct then, but only one of these did so at the major crisis at the end of the Cretaceous. *The Fossil Record* families are the only ones with adequate data on this point. Rudists diversified throughout the late Cretaceous and abruptly disappeared at its close.

(13). I give several ways of evaluating comparability of taxa elsewhere (Van Valen, in press).

(14). The phylogeny is from the *Treatise* (1) and excludes 5 families of unknown ancestry. Schindewolf (1961-1968) has given a rather different phylogeny, but parts of it are not detailed enough for this application.

(15). A brief discussion of possible factors in the crisis for trilobites can be found in *The Fossil Record* (1), p. 54. Some other curves also probably exhibit real deviations from linearity, but the ones discussed are the largest.

(16). Small (1946,1952) noticed this exponential increase for some individual genera, as Small (1950) and Tappan and Loeblich (1971) did for the entire group.

(17). I arbitrarily define these as having a maximum diameter of at least 5 mm. Most are highly complex internally, and most are discoidal, fusiform, or low conical.

(18). What is necessary is the expected distribution of the durations of all sub-trees, including parts of larger sub-trees, given a constant extinction rate and a branching rate that varies from time to time. Harris (1963, p. 32) considered the problem for a constant branching rate and found it intractable. It would further be useful if some sub-trees could be ignored as effectively infinite (extending an unknown amount beyond some absolute date such as the present). The point is directly resolvable by simulation, but this requires more money for computer use than I have.

(19). MacArthur died (at the age of 42) two weeks after the discovery of constant extinction.

(20). 1 ma is the reciprocal of a half-life of 500 years; in general, ma = (half-life in units of 500 years)$^{-1}$. Macarthurs apply also to phenomena (such as origination rates) for which half-lives are inappropriate. Applications in the text are to phenomena variously with and without replacement of the item sampled; the derivations therefore differ in detail. Some computational aids: ma = $(\log_{10}2)^{-1}(-\log_{10}(1 - p^{2\underline{t}}))$; $P_{0.5}$(1 day) = 183 kilomacarthurs (kma); $P_{0.5}$(1 minute) = 263 megamacarthurs (Mma). Given $P(\underline{t})$ of an event per interval $\underline{t}$, $P(\underline{kt}) = [P(\underline{t})]^{\underline{k}}$. The rate for interval $\underline{kt}$ is ma($\underline{kt}$) = $\underline{k}^{-1}$ma($\underline{t}$), given the same probability for both intervals. Haldane (1949) proposed an analogous measure for rates of continuous phenomena: a *darwin* is the rate giving a change by a factor of <u>e</u> per million years. Kimura (1969) defined a unit of rates for molecular evolution: a <u>pauling</u> is, in this context, the same as a millimacarthur.

(<u>21</u>). For birds, the data are for islands and of variable quality: MacArthur and Wilson (1967); Lack (1942); Diamond, (1969,1971). The islands are, in order of size, Los Coronados, Santa Barbara, Anacapa, St. Kilda, the Scillies, Krakatoa, San Miguel, San Nicolas, San Clemente, Santa Catalina, Santa Rosa, Santa Cruz, Karkar, Man, and the Orkneys. For arthropods, Simberloff and Wilson (1969) have approximate data from mangrove islets. The inaccuracy of the estimated extinction rates for birds and arthropods, in the present context, is about an order of magnitude. Estimates for both groups are probably too high, but I use available values. Mammal data are from Kurten (1968) for Europe, Martin and Webb (in press) for Florida, Webb (1969) for North America, and the data of Table 1 for the world. There are now about 4 species per genus of mammals in both North America and the world; if they go extinct independently, the extinction rate of species is three times that of genera. This inaccurate assumption, which ignores the correlation between number of species present and number of new species produced, is reasonable at the scale used. The grouping of birds and mammals into one equation is less defensible but is supported by the Florida data. The same is true for grouping data from islands, a peninsula, and two partly isolated continents.

(<u>22</u>). Treatments of evolutionary rates in paleontology not in other notes can be found in, e.g., the books by Schindewolf (1950), Zeuner (1958), and Kurten (1968); a symposium (<u>J. Paleont.</u> 26: 297-394 [1952]); and papers by Williams (1957), Kurten (1960), Bone (1963), Lerman (1965), House (1967), Newell (1967), Valentine (1969), Lipps (1970), Kurten (1971), Olsson (1972), and Cooke and Maglio (1972). DNA estimates are readily accessible in the <u>Atlas</u> (Dayhoff, 1972); various guesses exist on the proportions that are informational.

(<u>23</u>). A more extreme divergence from the regression would be expected for smaller organisms. Cairns, Dahlberg, Dickson, Smith, and Waller (1969) in fact give data for protozoans on blocks of an artificial substrate in a lake. An extinction rate of about 12 kma [P(extinction) = 0.043 per species per day] can be derived from their data. However, the substrate was a foam and so the effective area is unknown; the area of the top was 5×10^{-9} sq. km. Furthermore, the extinction rate may be higher than that in naturally occurring isolated substrates of the same effective area, and the experiment lasted only about 40 days. The glass slides that Patrick (1967) used as islands for diatoms would seem an excellent model system for such estimations, especially because the effect of area itself can be isolated from that of spatial heterogeneity and both studied together. The estimated extinction rate on Simberloffia may also be higher than a rate comparable to that for vertebrates; Simberloff and Wilson (1969) say that most "extinctions" seem to have been of species that couldn't colonize the islets under any circumstances, so no real populations of them existed to become extinct. This is a serious bias even if one difficult to overcome, and illustrates the danger of letting what we can easily measure determine what we think we want to measure, the tyranny of epistemology on ontology.

(<u>24</u>). The <u>effective environment</u> of any organism is its adaptive zone (Van Valen, 1971) plus the effects of any other organisms within that adaptive zone.

(<u>25</u>). A hypothesis recently revived in terms of DNA by Bachmann, Goin and Goin (1972).

(<u>26</u>). If we look at too narrow a part of the zone, with only a few taxa, discrete single events will be individually noticeable, as with any random process in the real world. In a causal universe a claim of randomness is a badge of ignorance. With evolutionary diversification the causes of seemingly random patterns may well be important and discoverable (Van Valen and Sloan [1972] give an example). The law of extinction is on the next level of abstraction from such causes.

(27). Gravitation does not cause an object resting on the floor to fall. Lakatos' critique (1963-1964) of mathematical proof is based on the difficulty of delimiting domains objectively.

(28). But like, e.g., Mendel's Laws or the gas laws.

(29). I have treated this subject elsewhere (Van Valen, 1972).

(30). For instance, E.D. Cope proposed a famous law in the nineteenth century that primitive taxa have a greater expected longevity than their descendants. This has never been adequately tested and should be re-formulated in terms of degree of primitiveness (assuming a threshold is absent) and a definition of primitiveness by entrance to an adaptive zone.

(31). A law need not be quantitative (although the law of extinction is). The contrary tradition is a myth derived, as Egbert Leigh has said, from physics envy.

(32). "Now here, you see, it takes all the running you can do, to keep in the same place." (L. Carroll, Through the Looking Glass.)

(33). Fisher (1930) and others, including Darwin and especially Lyell (1832), foreshadowed the Red Queen's Hypothesis but had no reason to impose the crucial constraint of constancy, and did not do so. I regard interference competition as causally a mechanism of resource competition, a proximal rather than ultimate regulator.

(34). Whether the total group does also depends on the distribution of longevities of the subgroups omitted.

(35). On one level, the probability of extinction of a group is related to its own properties because different groups go extinct for different reasons and so at different times. But on the next level, the Red Queen says that having one set of properties is not appreciably better than having another because the expected time to extinction is the same.

(36). Levine and Van Valen (1964) showed experimentally for *Drosophila* that natural selection has rather non-Markovian aspects. Lewontin (1966) later elaborated the point theoretically but without specific results.

(37). Origination-extinction equilibria are implicit in Simpson's work (1944,1953), and in Lyell's (1832). I realize that the Red Queen's Hypothesis is at least a simplification of reality. It is directly analogous to Newton's third law of motion.

(38). That the Red Queen in her simplest gown implies long-term constancy in total evolutionary rate is obvious. For any single protein we must invoke an analogue of the Central Limit Theorem: pervasive pleiotropy makes the rate for one protein roughly proportional to that for all, or linkage effects have a similar result. For instance, many proteins are to some extent attached, and the other components of the attachment may change for extraneous reasons, making the previous structure nonoptimal. Dickerson (1971) and others have made a similar point from the other end of the microscope.

(39). Stebbins and Lewontin (1972) actually think that "the entire argument is based upon a confusion between an average and a constant." What is remarkable is, however, precisely that the average rate (over shorter segments of a phylogeny) is so nearly constant (among these segments) for a given protein, rather than reflecting a branching random walk or some other process.

(40). More precisely, the prediction is of a monotone relationship of the average rate of protein evolution with the average rate of change among phenotypic characters (including the origin of new characters).

(41). Horne (1967), Kohne (1970), Ohta and Kimura (1971), Uzzell and Pilbeam, (1971), Jukes and Holmquist (1972). Also, constancy predicts the same expected number of changes in each lineage after the latest common ancestor. The data for hemoglobins in the Atlas (Dayhoff, 1972) seem inconsistent with this expectation. The assumption of total constancy leads to the expectation, presented seriously by D. Boulter (1972) and Ramshaw et al. (1972) that angiosperms originated in the early or middle Paleozoic.

(42). The approach used the protein sequence data of the Atlas, 1969 edition, and probability estimates of various alternative placental phylogenies as determined by myself.

(43). Probability in the sense of a propensity, a property of any single lineage.

(44). The rank-order of different groups with respect to F_t can depend on the choice of $w(t)$. It is an almost universal mistake to think that evolution locally maximizes fitness. Evolutionary fitness is F_t except to some population geneticists, but evolution doesn't maximize it. Selection at any level locally maximizes momentary fitness for that level, but the optima of different levels need not coincide. This is obvious between prezygotic and individual selection but is equally true for higher levels. Individual selection, the most important evolutionary force, can decrease F_t until extinction by, e.g., forcing the occupation of only a temporary niche.

(45). The latest record has now been found by J.F. Sutton, University of Kansas (talks at 1971 and 1972 meetings of the Society of Vertebrate Paleontology).

(46). Lewontin (1961), Warburton (1967), and Maynard Smith (1972) have made applications of game theory to evolution within the usual evolutionary framework.

Literature Cited

Allee, W.C., A.E. Emerson, O. Park, T. Park, and K.P. Schmidt. 1949. Principles of Animal Ecology. Philadelphia: Saunders. 837 pp.

Anonymous. 1964. Geological Society Phanerozoic time-scale 1964. Quart. Jour. Geol. Soc. London 120 S: 260-262.

Bachmann, K., O.B. Goin, and C.J. Goin. 1972. Nuclear DNA amounts in vertebrates, In evolution of Genetic Systems (H.H. Smith, ed.), pp. 419-450. New York: Gordon and Breach.

Berggren, W.A. 1969. Rates of evolution in some Cenozoic planktonic foraminifera. Micropaleontology 15: 351-365.

_____. 1972. A Cenozoic time-scale--some implications for regional geology and paleobiogeography. Lethaia 5: 195-215.

Bone, E.L. 1963. Paleontological species and human speciation. South African Jour. Sci. 59: 273-277.

Boulter, D. 1972. Protein structure in relationship to the evolution of higher plants. Sci. Prog. 60: 217-229.

Cairns, J., M.L. Dahlberg, K.L. Dickson, N. Smith, and W.T. Waller. 1969. The relationship of fresh-water protozoan communities to the MacArthur-Wilson equilibrium model. Amer. Nat. 103: 439-454.

Clarke, B. 1970. Darwinian evolution of proteins. Science 168: 1009-1011.

Cooke, H.B.S., and V.J. Maglio. 1972. Plio-Pleistocene stratigraphy in East Africa in relation to proboscidean and suid evolution. In Calibration of Hominoid Evolution (W.W. Bishop and J.A. Miller, eds.), pp. 303-329. Scottish Academic Press.

Dayhoff, M.O. 1972. Atlas of Protein Sequence and Structure 1972. Washington: National Biomedical Research Foundation. 124+382 pp.

Diamond, E.M. 1969. Avifaunal equilibria and species turnover rates on the Channel Islands of California. Proc. Natl. Acad. Sci. U.S.A. 64: 57-63.

__________. 1971. Comparison of faunal equilibrium turnover rates on a tropical island and a temperate island. Proc. Natl. Acad. Sci. U.S.A. 68: 2742-2745.

Dickerson, R.E. 1971. The structure of cytochrome c and the rates of molecular evolution. Jour. Molec. Evol. 1:26-45.

Eldredge, N. 1971. The allopatric model and phylogeny in Paleozoic invertebrates. Evolution 25: 156-167.

__________ and S.J. Gould. 1972. Punctuated equilibria: an alternative to phyletic gradualism. In Models in Paleobiology (T.J.M. Schopf, ed.), pp. 82-115. San Francisco: Freeman, Cooper.

Everndon, J.F., and G.H. Curtis. 1965. The potassium-argon dating of late Cenozoic rocks in East Africa and Italy. Cur. Anth. 6: 343-385.

Everndon, J.F., D.E. Savage, G.H. Curtis, and G.T. James. 1964. Potassium-argon dates and the Cenozoic mammalian chronology of North America. Amer. Jour. Sci. 262: 145-198.

Fisher, R.A. 1930. The Genetical Theory of Natural Selection. Oxford: Clarendon Press. 272 pp.

Haldane, J.B.S. 1949. Suggestions as to the quantitative measurement of rates of evolution. Evolution 3: 51-56.

Harris, T.E. 1963. The Theory of Branching Processes. Berlin: Springer-Verlag. 230 pp.

Hennig, W. 1966. Phylogenetic Systematics. Urbana: Univ. Illinois Press. 263 pp.

Horne, S.L. 1967. Comparisons of primate catalase tryptic peptides and implications for the study of molecular evolution. Evolution 21: 771-786.

House, M.R. 1967. Fluctuations in the evolution of Paleozoic invertebrates. In The Fossil Record (W.B. Harland et al., eds.), pp. 41-54. London: Geological Society of London.

Jukes, T.H., and R. Holmquist. 1972. Evolutionary clock: nonconstancy of rate in different species. Science 177: 530-532.

Kauffman, E.G. 1970. Population systematics, radiometrics and zonation--a new biostratigraphy. Proc. North American Paleont. Conv. (F): 612-666. Lawrence: Allen Press.

Kimura, M. 1969. The rate of molecular evolution considered from the standpoint of population genetics. Proc. Natl. Acad. Sci. U.S.A. 63: 1181-1188.

King, J.L., and T. H. Jukes. 1969. Non-Darwinian evolution. Science 164: 788-798.

Kohne, D.E. 1970. Evolution of higher-organism DNA. Quart. Rev. Biophys. 3: 327-375.

Kurtén, B. 1953. On the variation and population dynamics of fossil and recent mammal populations. Acta Zool. Fennica 76: 1-122.

_________. 1960. Rates of evolution in fossil mammals. Cold Spring Harbor Symp. Quant. Biol. 24 (for 1959): 205-215.

_________. 1968. Pleistocene Mammals of Europe. London: Weidenfeld and Nicholson. 317 pp.

_________. 1971. Time and hominid brain size. Comment Biol. Soc. Sci. Fennica 36: 1-8.

Lack, D. 1942. Ecological features of the bird faunas of British small islands. Jour. Anim. Ecol. 11: 9-36.

Lakatos, I. 1963-1964. Proofs and refutations. Brit. Jour. Philos. Sci. 14: 1-25, 120-139, 221-245, 296-342.

Lerman, A. 1965. On rates of evolution of unit characters and character complexes. Evolution 19: 16-25.
Levine, L., and L. Van Valen. 1964. Genetic response to the sequence of two environments. Heredity 19: 734-736.
Lewontin, R.C. 1961. Evolution and the theory of games. Jour. Theor. Biol. 1: 382-403.
____________. 1966. Is nature probable or capricious? BioScience 16: 25-27.
Lipps, J.H. 1970. Plankton evolution. Evolution 24: 1-21.
Lyell, C. 1832. Principles of Geology. 1st ed., vol. 2. London: J. Murray. 330 pp.
MacArthur, R.H. 1972. Geographical Ecology. New York: Harper and Row. 269 pp.
____________ and E.O. Wilson. 1967. The Theory of Island Biogeography. Princeton: Princeton Univ. Press. 203 pp.
MacGillavry, H.J. 1968. Modes of evolution mainly among marine invertebrates. Bijdr. Dierk. 38: 69-74.
Martin, R.A., and S.D. Webb. In press. Late Pleistocene mammals from Devil's Den, Levy County. In Pleistocene Mammals of Florida (S.D. Webb, ed.) Gainesville: Univ. Florida Press.
Maynard Smith, J. 1972. On Evolution. Edinburgh: Edinburgh Univ. Press. 125 pp.
Newell, N.D. 1963. Crises in the history of life. Sci. Amer. 28(2): 76-92.
__________. 1967. Revolutions in the history of life. Geol. Soc. Amer. Spec. Pap. 89: 63-91.
__________. 1971. An outline history of tropical organic reefs. Amer. Mus. Novit. 2465: 1-37.
Odum, E.P. 1971. Fundamentals of Ecology. 3rd. Ed. Philadelphia: Saunders. 574 pp.
Ohta, T., and Kimura, M. 1971. On the constancy of the evolutionary rate of cistrons. Jour. Molec. Evol. 1: 18-25.
Olsson, R.K. 1972. Growth changes in the Globorotalia fohsi lineage. Eclogae Geol. Helvetiae 65: 165-184.
Patrick, R. 1967. The effect of invasion rate, species pool, and size of area on the structure of the diatom community. Proc. Natl. Acad. Sci. U.S.A. 58: 1335-1342.
Pearl, R. 1928. The Rate of Living. New York: Knopf. 185 pp.
Ramshaw, J.A.M., D.L. Richardson, B.T. Meatyard, R.H. Brown, M. Richardson, E.W. Thompson, and D. Boulter. 1972. The time of origin of the flowering plants determined by using amino acid sequence of cytochrome c. New Phytol. 71: 773-779.
Reyment, R., and L. Van Valen. 1969. Buntonia olokundudui sp. nov. (Ostracoda, Crustacea): a study of meristic variation in Paleocene and Recent ostracods. Bull. Geol. Inst. Univ. Uppsala (N.S.) 1: 83-94.
Romer, A.S. 1966. Vertebrate Paleontology. 3rd ed. Chicago: Univ. Chicago Press. 468 pp.
Ruzhentsov, V. Ye. 1963. The problem of transition in paleontology. Paleont. Zhur. 1963(2): 3-16. (Translated 1964, Int. Geol. Rev. 6: 2204-2213).
Sarjeant, W.A.S. 1967. The stratigraphical distribution of fossil dinoflagellates. Rev. Palaeobot. Palynol. 1: 323-343.
Schindewolf, O.H. 1950. Der Zeitfaktor in Geologie und Paläontologie. Stuttgart: Schweizerbart. 114 pp.
______________. 1961-1968. Studien zur Stammesgeschichte der Ammoniten. Akad. Wiss. Lit. Mainz, Abhandl. Math.-Naturw. Kl. 1960: 635-744; 1962: 425-572; 1963: 285-432; 1965: 137-238; 1966: 323-454, 719-808; 1968: 39-209.
Simberloff, D., and E.O. Wilson. 1969. Experimental zoogeography of islands: the colonization of empty islands. Ecology 50: 278-296.
Simpson, G.G. 1944. Tempo and Mode in Evolution. New York: Columbia Univ. Press. 237 pp.

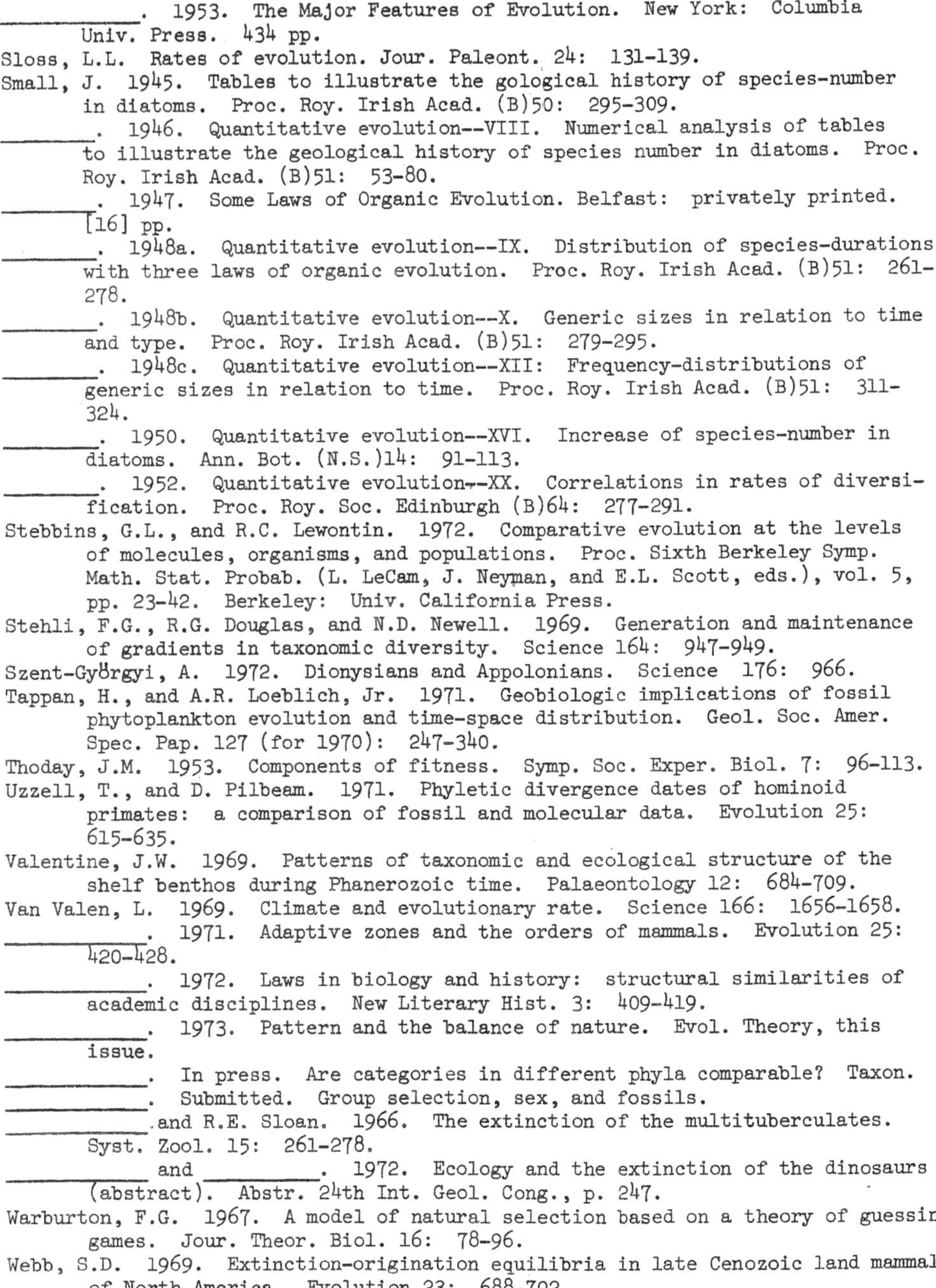

________. 1953. The Major Features of Evolution. New York: Columbia Univ. Press. 434 pp.

Sloss, L.L. Rates of evolution. Jour. Paleont. 24: 131-139.

Small, J. 1945. Tables to illustrate the gological history of species-number in diatoms. Proc. Roy. Irish Acad. (B)50: 295-309.

________. 1946. Quantitative evolution--VIII. Numerical analysis of tables to illustrate the geological history of species number in diatoms. Proc. Roy. Irish Acad. (B)51: 53-80.

________. 1947. Some Laws of Organic Evolution. Belfast: privately printed. [16] pp.

________. 1948a. Quantitative evolution--IX. Distribution of species-durations, with three laws of organic evolution. Proc. Roy. Irish Acad. (B)51: 261-278.

________. 1948b. Quantitative evolution--X. Generic sizes in relation to time and type. Proc. Roy. Irish Acad. (B)51: 279-295.

________. 1948c. Quantitative evolution--XII: Frequency-distributions of generic sizes in relation to time. Proc. Roy. Irish Acad. (B)51: 311-324.

________. 1950. Quantitative evolution--XVI. Increase of species-number in diatoms. Ann. Bot. (N.S.)14: 91-113.

________. 1952. Quantitative evolution--XX. Correlations in rates of diversification. Proc. Roy. Soc. Edinburgh (B)64: 277-291.

Stebbins, G.L., and R.C. Lewontin. 1972. Comparative evolution at the levels of molecules, organisms, and populations. Proc. Sixth Berkeley Symp. Math. Stat. Probab. (L. LeCam, J. Neyman, and E.L. Scott, eds.), vol. 5, pp. 23-42. Berkeley: Univ. California Press.

Stehli, F.G., R.G. Douglas, and N.D. Newell. 1969. Generation and maintenance of gradients in taxonomic diversity. Science 164: 947-949.

Szent-Györgyi, A. 1972. Dionysians and Appolonians. Science 176: 966.

Tappan, H., and A.R. Loeblich, Jr. 1971. Geobiologic implications of fossil phytoplankton evolution and time-space distribution. Geol. Soc. Amer. Spec. Pap. 127 (for 1970): 247-340.

Thoday, J.M. 1953. Components of fitness. Symp. Soc. Exper. Biol. 7: 96-113.

Uzzell, T., and D. Pilbeam. 1971. Phyletic divergence dates of hominoid primates: a comparison of fossil and molecular data. Evolution 25: 615-635.

Valentine, J.W. 1969. Patterns of taxonomic and ecological structure of the shelf benthos during Phanerozoic time. Palaeontology 12: 684-709.

Van Valen, L. 1969. Climate and evolutionary rate. Science 166: 1656-1658.

________. 1971. Adaptive zones and the orders of mammals. Evolution 25: 420-428.

________. 1972. Laws in biology and history: structural similarities of academic disciplines. New Literary Hist. 3: 409-419.

________. 1973. Pattern and the balance of nature. Evol. Theory, this issue.

________. In press. Are categories in different phyla comparable? Taxon.

________. Submitted. Group selection, sex, and fossils.

________ and R.E. Sloan. 1966. The extinction of the multituberculates. Syst. Zool. 15: 261-278.

________ and ________. 1972. Ecology and the extinction of the dinosaurs (abstract). Abstr. 24th Int. Geol. Cong., p. 247.

Warburton, F.G. 1967. A model of natural selection based on a theory of guessing games. Jour. Theor. Biol. 16: 78-96.

Webb, S.D. 1969. Extinction-origination equilibria in late Cenozoic land mammals of North America. Evolution 23: 688-702.

Williams, A. 1957. Evolutionary rates of brachiopods. Geol. Mag. 94: 201-211.

Yochelson, E.L. 1969. Stenothecoida, a proposed new class of Cambrian Mollusca. Lethaia 2: 49-62.

Zeuner, F.E. 1958. Dating the Past. 4th ed. London: Methuen. 516 pp.

PAPER 20

Ecospace Utilization and Guilds in Marine Communities through the Phanerozoic (1983)

R. K. Bambach

Commentary

ANDREW M. BUSH

For forty years, Richard Bambach has been a member of the "paleobiology" movement that brought ecology, evolution, and diversity change to the forefront of paleontological research. In the following paper, he develops a comprehensive, ecologically driven model for the long-term biodiversity changes documented in the marine fossil record. Beyond its immediate and influential findings, this is a central paper in paleobiology, because it shows that rigorous, broadscale analyses need not abandon rich knowledge of the ecology and function of fossil organisms. Many paleontological studies of diversity history are surprisingly abiological—in an effort to be quantitative and rigorous, paleontologists have simplified the richness of the fossil record into counts of taxa through time, with rates of origination and extinction calculated therefrom. Processes are inferred statistically from these numbers. In contrast, Bambach maintains that scientists must appreciate how organisms lived in order to understand the history of biodiversity and evolution.

The foundation for Bambach's analysis is an ecological classification of fossil animals based on attributes such as diet, mobility, attachment, and tiering (life position relative to the seafloor). He first shows how the range of ecological lifestyles utilized by marine animals expanded through their history; this analysis, though a simple description of the record, has been widely cited. The second part of the paper is theoretically more substantial. In it, Bambach classifies fossil species into narrower ecological categories (guilds) and compares the ecological composition of individual fossil assemblages through Phanerozoic time (the past 542 million years). His finding is that the number of species in a fossil assemblage is strongly related to the number of guilds, with both increasing (on average) through geological time. He concludes that long-term biodiversification had an ecological component, with habitat-level diversity increasing through geological time as new ecological lifestyles became abundant and were layered into ecosystems. The expansion of animals into new "ecospace" was thus a significant contributor to marine biodiversification.

Bambach's method of ecologically classifying fossil organisms for quantitative analysis has inspired several recent updates ("theoretical ecospaces"; Bush et al. 2007; Bambach et al. 2007; Novack-Gotshall 2007). These should dovetail with efforts to comprehensively database the fossil record (Alroy et al. 2008); as the spatiotemporal distributions and ecological properties of taxa are integrated on massive scales, Bambach's paper will serve as a template for ambitious new studies of the broad sweep of life's history.

In *Topics in Geobiology,* edited by M. J. S. Tevesz and P. L. McCall, pp. 719–46. Plenum Press, New York. Reprinted with permission from Springer Science + Business Media.

Literature Cited

Alroy, J., M. Aberhan, D. J. Bottjer, M. Foote, F. T. Fürsich, P. J. Harries, A. J. W. Hendy, et al. 2008. Phanerozoic trends in the global diversity of marine invertebrates. *Science* 321:97–100.

Bambach, R. K., A. M. Bush, and D. H. Erwin. 2007. Autecology and the filling of ecospace: Key metazoan radiations. *Palaeontology* 50:1–22.

Bush, A. M., R. K. Bambach, and G. M. Daley. 2007. Changes in theoretical ecospace utilization in marine fossil assemblages between the mid-Paleozoic and late Cenozoic. *Paleobiology* 33:76–97.

Novack-Gottshall, P. M. 2007. Using a theoretical ecospace to quantify the ecological diversity of Paleozoic and modern marine biotas. *Paleobiology* 33:273–94.

Chapter 15

Ecospace Utilization and Guilds in Marine Communities through the Phanerozoic

R. K. BAMBACH

1. Introduction

Does the ecological theater influence the staging of the evolutionary play? What roles do biotic interactions have in influencing evolutionary patterns? This chapter presents a system of ecologic analysis for paleocommunities that reveals some relationships between the utilization of ecospace and change in diversity through time. The analysis is restricted to

R. K. BAMBACH • Department of Geological Sciences, Virginia Polytechnic Institute and State University, Blacksburg, Virginia 24061.

marine shelf faunas. The data base for marine faunas is the most voluminous available. These faunas have been emphasized in community paleoecology because of the relatively complete fossil record of shallow marine habitats. Recent studies summarizing diversity patterns for the Phanerozoic have also concentrated on the marine realm.

1.1. Diversity Change through the Phanerozoic

The diversity of the marine fauna has increased in a distinctive pattern since the Late Precambrian (Sepkoski *et al.*, 1981). In the Early Paleozoic (Cambrian and Ordovician), diversity increased rapidly. This was followed by 200 m.y. (Silurian to Permian) of fluctuating diversity without any persistent trend of either increasing or decreasing diversity. The great mass extinction of the Late Permian–Early Triassic reduced diversity dramatically. Diversity has increased steadily since the Early Triassic, and the Late Cenozoic (Neogene) diversity is higher than any achieved in the Paleozoic. This pattern is observed for taxa below the ordinal level (Valentine, 1969; Sepkoski, 1981; Raup, 1976) and is also reflected both in

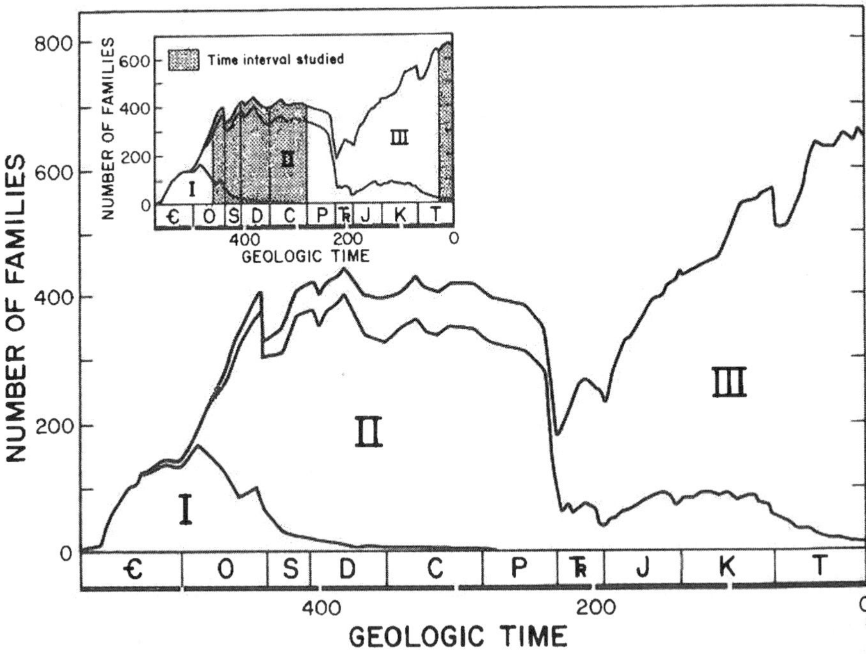

Figure 1. Diversity of well-preserved marine family taxa through the Phanerozoic (after Sepkoski, 1981). I, II, and III indicate the diversity of the three great evolutionary faunas. The shaded bands in the inset mark the time intervals used for guild analysis of communities in this study.

behavioral patterns as preserved by ichnofossils (Crimes, 1974; Seilacher, 1977) and in species richness within habitats (Bambach, 1977).

A notable feature of this pattern is the long interval in the Paleozoic without much change in diversity. Detailed study of familial diversity (Sepkoski, 1979, 1981) also reveals a significant pause or step in diversity increase during the Middle and Late Cambrian and a possible leveling off in diversity in the Neogene (Fig. 1). This latter is supported by within-habitat data for the Neogene (Bambach, 1977). There is no significant difference in median species richness between Miocene and Pleistocene level-bottom marine communities. On the other hand, the length of time involved in the Neogene is relatively short (24 m.y.) and the Pleistocene climatic perturbations may have superimposed increased extinction rates on a still diversifying fauna (Stanley and Campbell, 1981). This may create a false sense of leveling off, in the same way that the "pull of the Recent" (Raup, 1978) obscures consideration of the Late Cenozoic record because we are influenced by the extra detail we impose from our knowledge of the subtleties of the living fauna. Nonetheless, the Neogene has higher diversity than any previous time. There have been two times of different relatively constant levels of diversity (Mid-Cambrian–Early Ordovician, Silurian–Permian) and the Neogene has achieved a third higher level of diversity.

Sepkoski has modeled the pattern of diversity change in a stimulating series of papers (Sepkoski, 1978, 1979, 1981) and suggests that diversity at any time is related to the interaction between origination rates and diversity-dependent extinction rates. In this "kinetic model," as diversity increases the extinction rate rises until it balances the origination rate. This establishes an equilibrium diversity level. The overall pattern of diversity through the Phanerozoic is modeled by Sepkoski (1981) as the composite evolution of three groups of taxa, each with a different origination rate and each with a different diversity-influenced extinction rate. Sepkoski (1981) also analyzed the marine fossil record and identified the taxa that most influence the changes in and maintenance of constant diversity at different times. He concludes that three great evolutionary faunas match the three hypothetical systems in the kinetic model. The timing of their evolutionary expansion produced the three different prominent diversity levels that characterize the Phanerozoic (Sepkoski, 1980, 1981).

1.2. The Question of an Ecologic Role in Controlling Diversity Patterns

Do these three groupings of taxa differ in some way that reflects their biological or ecological response to the world? The major adaptive themes of the dominant organisms in the three evolutionary faunas should indicate the degree of ecospace exploitation by each fauna. If more ecospace is exploited by more recent faunas, then increased complexity in the utilization of the ecosystem has accompanied the change from dominance by one fauna to another. An ecologic component would be identified in the course of evolution.

Study of major adaptive strategies common in faunas of various times reveals a general sense of the ecologic complexity of these faunas. The concomitant increase in species richness within habitats that has paralleled increase in general taxonomic diversity (Sepkoski *et al.*, 1981) also resulted in a change in community structure. This report demonstrates that the increase in community species richness has not simply been by addition of species to communities with roughly the same pattern of organization but has resulted from changes in the pattern of community organization that have permitted increase in species richness.

A method of evaluating ecospace utilization within communities, guild analysis, can be used to compare community structure from different time intervals. The influence of biotic interactions in evolution may be examined by comparing the variety of modes of life (guilds) present in communities to the species richness in communities at different times. If increased species richness has resulted only in greater species packing within the same variety of modes of life (even if the modes of life were different), then little ecologic control on evolutionary diversification would be implied. If, however, an increase in the variety of modes of life has been the method by which the increase in species richness through time has been achieved, then a role for biotic interactions in controlling community structure and diversity would be implied.

The next section examines the general adaptive properties of the three evolutionary faunas and the following sections develop an analysis of the variety of modes of life (guilds) represented in communities of Paleozoic in contrast to Neogene age. In this way the influence of increased Neogene species richness on community structure will be examined.

2. General Pattern of Ecospace Utilization

2.1. Turnover of Class-Level Taxa through Time

Figure 2 represents cladograms of families within classes (after Sepkoski, 1981) arranged in order of decreasing age of maximum familial diversity. These class-level clades have a surprisingly limited variety of general shapes. A large number have never diversified much and are represented in the left column of Fig. 2 as dot-shaped clades (those with a

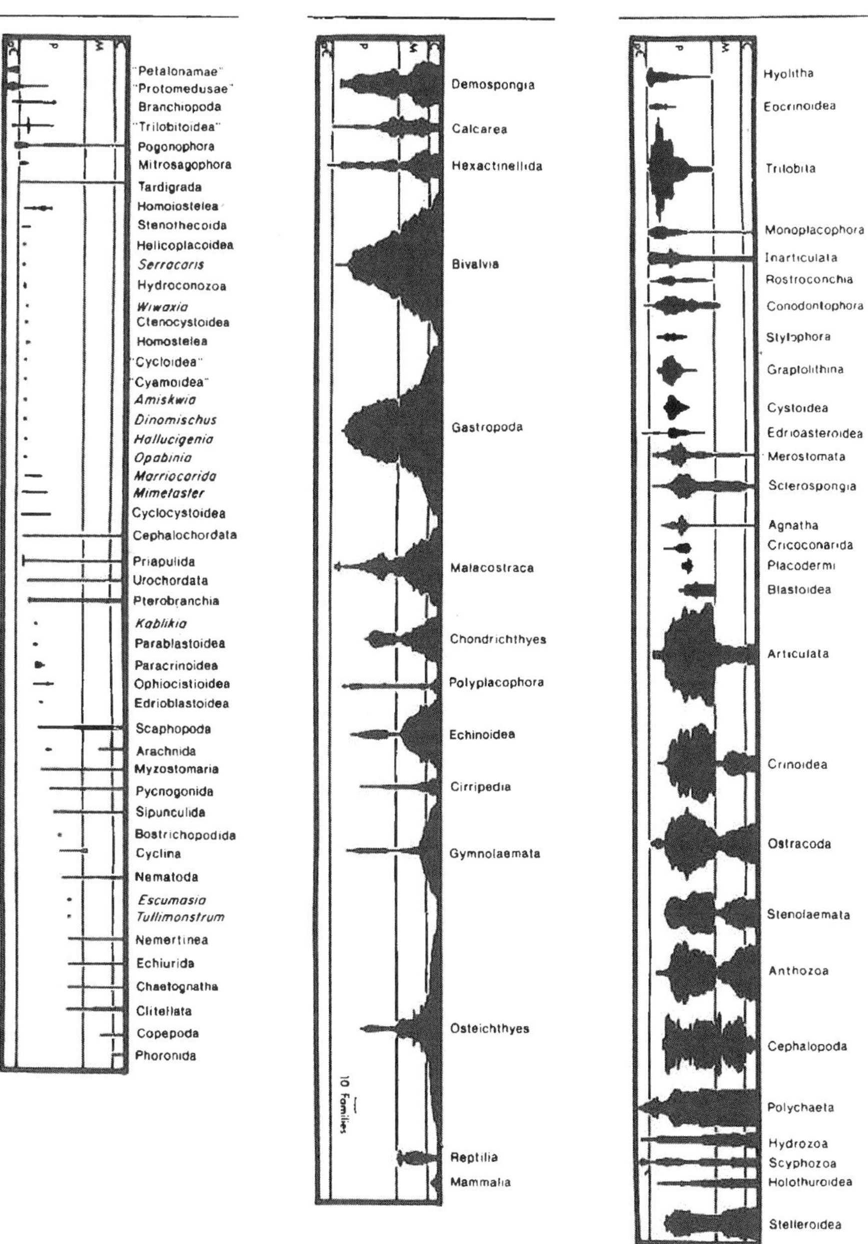

Figure 2. Cladograms of marine family diversity of class-level taxa (after Sepkoski, 1981). Classes achieving diversity of five families or more are arranged in order of decreasing age of maximum diversity. Classes not achieving significant family diversification are separately grouped in the order of time of first appearance. Time scale by eras: pC, Late Precambrian; P, Paleozoic; M, Mesozoic; C, Cenozoic.

very short geologic time range) and stick-shaped clades (those with relatively long time ranges). Those clades that have undergone significant expansion *do not* show a pattern of variation around an average symmetrical shape of maximum diversity at the midpoint of total geologic range. Rather they represent four general forms: bottom-heavy (with maximum familial diversity in the early half of the total time range), dumbbell-shaped (with a diversity constriction associated with the Permo-Triassic boundary), straight-sided (with little change in diversity after an initial diversification), and top-heavy (those with maximum diversity in the later half of their time range). By far the majority of bottom-heavy clades are most diverse in the Paleozoic. The dumbbell-shaped clades appear as if they would have been straight-sided but were severely constricted by the Permo-Triassic extinction. The top-heavy clades almost all reach maximum diversity in the Late Cenozoic.

Table I shows the number of class-level taxa of different ranges of familial diversity present at various times through the Phanerozoic. Although the total extant number of classes was highest in the Mid-Paleozoic (Silurian–Devonian), it has otherwise remained strikingly constant (at about 50) since the Late Precambrian. The number of new classes entering the fauna declined from a high in the Early Paleozoic (Cambrian–Ordovician) to low levels from the Late Paleozoic (Carboniferous–Permian) through the Cenozoic. The number of diverse classes (those with more than five families) was low in the Early Paleozoic but, as with total classes since the Precambrian, has been rather constant since the Silurian (start of Middle Paleozoic). Very diverse classes (over 30 families) remained quite constant (6 or 7) over the same time span.

Of the total of 90 class-level taxa present at some time in the Phanerozoic, only 42 diversified to any great extent (Fig. 2). The relatively

Table I. Number of Class-Level Taxa of Different Diversities at Various Times in the Phanerozoic[a]

	No. of families in class				Total diverse classes	Small classes (1–4 families)	Total classes	New classes
	5–9	10–19	20–29	30+				
Cenozoic	5	7	5	7	24	24	48	2
Mesozoic	2	2	8	7	19	29	48	2
L. Paleozoic	6	3	4	6	19	33	52	6
M. Paleozoic	8	10	4	7	29	37	66	28
E. Paleozoic	4	2		1	7	44	51	41
Pre-Cambrian						11	11	11

[a] Cambrian and Early Ordovician are the Early Paleozoic; Middle Ordovician, Silurian, and Devonian are the Middle Paleozoic; Carboniferous and Permian are the Late Paleozoic.

constant number of class-level taxa coupled with the regular appearance of new classes (Table I) results from the faunal replacement or turnover characteristic of the fossil record. Figure 2 records not only this turnover but the sequential replacement of dominant taxa that Sepkoski also discussed (Sepkoski, 1981).

This sequential replacement is also recorded in the three faunal curves (I, II, III) of Fig. 1, the three great evolutionary faunas recognized by Sepkoski (1981). The characterizing taxa of the Cambrian fauna are the first five more diverse classes (Hyolitha, Eocrinoidea, Trilobita, Monoplacophora, and Inarticulata) at the top of the right column of Fig. 2. The Middle and Late Paleozoic fauna is typified by all the other classes in the right column of Fig. 2, and the third (Mesozoic and Cenozoic) fauna is typified by all the classes in the middle column of Fig. 2 plus the nine Paleozoic dominants that retained or recovered higher diversity after the Triassic.

2.2. Change in General Ecospace Utilization

Are there differences in basic ecological preferences or adaptive strategies in the dominant taxa of the three faunas? The common adaptive strategies of the more diverse classes of each of the three faunas are recorded in Fig. 3. Ecospace parameters considered are mode of life and feeding type. Modes of life are categorized as pelagic (living in the water column as either nekton or plankton), epifaunal (living on the substratum), and infaunal (living buried in the substratum). The epifaunal mode of life is subdivided into mobile (creeping or walking), attached low (cemented, pedunculate, or byssate forms), attached erect (colonial forms or stalked forms that elevate a large portion of the colony or organism above the substratum), and reclining (passive organisms that simply lie on the substratum). The infaunal mode of life is subdivided into shallow passive and active forms (in which a part of the body of the organism is maintained at the surface) and deep passive and active forms (in which the body is located at some depth below the surface of the substratum so that contact with the overlying water mass must be by pumping water either through the sediment or through burrows or siphons). Four feeding types are considered; suspension- or filter-feeders, deposit-feeders, herbivores, and carnivores. These subdivisions of mode of life and feeding type are generalized categories that incorporate the full spectrum of realized life activities. They are intended to illustrate the common styles of life utilized by the different evolutionary faunas.

The Cambrian fauna contained predominantly epifaunal organisms with no common deep infaunal or shallow passive forms (see Fig. 3). There

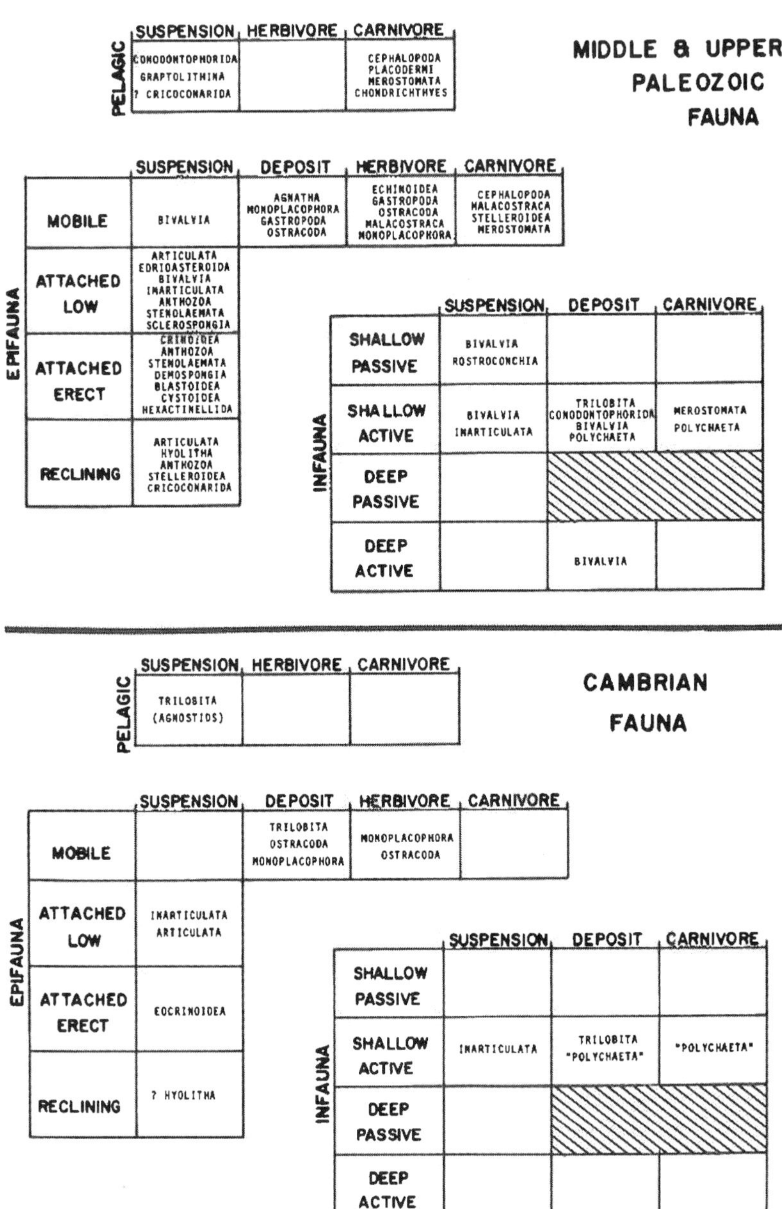

Figure 3. General adaptive strategies that typify the more diverse class of each of the three great evolutionary marine faunas of the Phanerozoic. All class-level taxa except the dot- and stick-shaped clades in Fig. 2 are recorded. The shaded boxes are not biologically practical adaptive strategies.

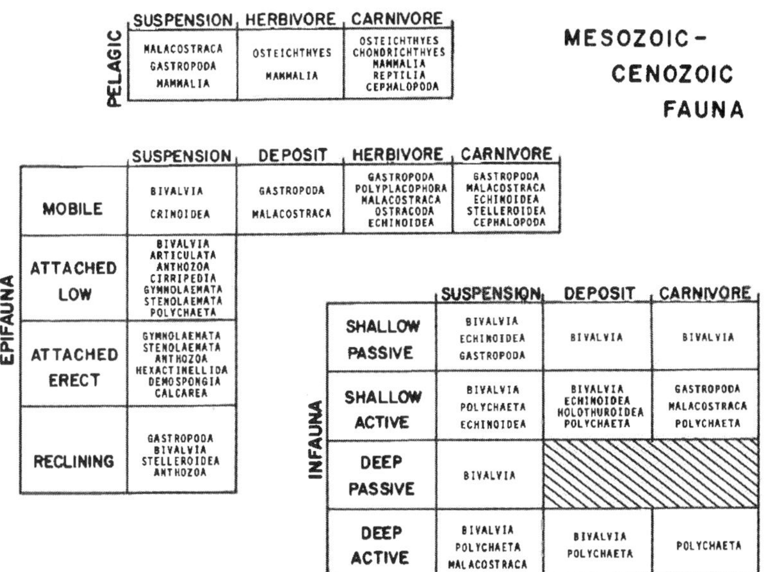

Figure 3. (*Continued*)

also were few pelagic forms in the diverse classes. Compared to the later faunas, the Cambrian fauna exploited rather little of the potential ecospace. The Middle and Upper Paleozoic fauna diversified extensively into epifaunal modes of life and also began the expansion into pelagic and infaunal niches. These latter were less intensely exploited than would be the case for the Mesozoic–Cenozoic fauna whereas the epifaunal mode of life was as fully developed as it was later. Diversification of infaunal adaptations typifies the Mesozoic–Cenozoic fauna, as does increase in carnivores.

Exploitation of more ecospace has typified the replacement of faunas through the Phanerozoic. The Cambrian fauna was not only relatively low in taxonomic diversity but it had a low diversity of adaptive types, primarily epifaunal with few infaunal or pelagic forms and few carnivores. The Middle–Upper Paleozoic fauna diversified extensively in epifaunal modes of life and began to expand into pelagic and shallow infaunal ecospace. The great Mesozoic–Cenozoic diversification is concentrated in infaunal and pelagic habits with replacement but not further diversification in epifaunal life styles. Carnivorous habits have also increased in the Mesozoic–Cenozoic fauna.

With an increase in ecospace utilization clearly associated with the replacement or turnover of the class-level taxa dominating the three great evolutionary faunas, the question now turns to species within communities. The expansion into new ecospace at the class level was reflected in changes in community structure.

3. The Guild Concept and Its Application to Paleocommunities

3.1. Extension of the Guild Concept

Root (1967, p. 335) defined a guild as "a group of species that exploit the same class of environmental resources in a similar way. This term groups together species, without regard to taxonomic position, that overlap significantly in their niche requirements." In this chapter this concept is used to recognize the combination of three factors: the major methods by which organisms utilize space and maintain themselves in the environment, the significantly different basic body plans and physiological systems appearing in a community, and the categories of resource limitation (primarily food sources) present in the community. The number of these broadly defined guilds can be used to characterize the ecologic complexity of a community. The recognition of guilds in paleocommunities should be a method of extracting similar information from the fossil record.

I take a more wide-ranging set of factors into my definition of guilds (Fig. 4) than used in the original definition by Root (1967). Although this extends the concept, I believe the term *guild* is still appropriate. The goal of this extension of the guild concept is to erect groupings that reflect not only the potentially limiting resources for the community over which competition may develop but also to reveal the scope of adaptive strategies present in the community. For this reason three general factors are used in assigning species to guilds in this study. They are food source, space utilization, and bauplan.

Food source is a standard item of interest in ecologic analysis. Food is related to the channeling of energy flow through the ecosystem and it is a limiting resource in some instances. Space utilization is important in viewing the structure of a whole community. The variety of modes of life possible in an environment are controlled by the physical conditions of the habitat. Competition for space or interference between organisms in their life activities also influences the abundance and distribution of species. The bauplan of an organism includes all aspects of its physiology, reproductive strategy, growth, and other such intrinsic features that influence the presence or success of the organism in the habitat of the community.

For purposes of guild assignment for this study, class-level taxonomy is used to designate different baupläne. Features of the bauplan are usually so different at the class level that different classes have sufficiently distinctive intrinsic properties to qualify as separate baupläne in the sense meant in this discussion. Using classes to designate baupläne is therefore

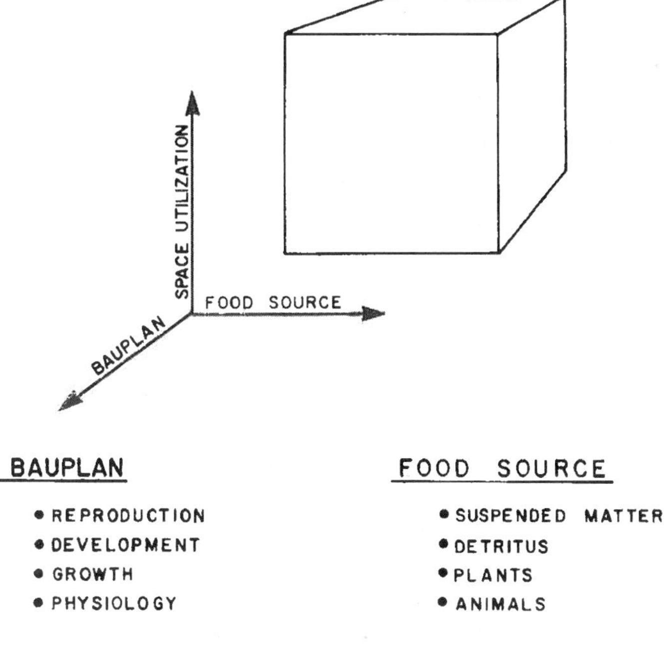

Figure 4. Diagrammatic representation of ecospace using the three general criteria used in this chapter for defining guilds as axes. A guild would be represented by the species of a community that all fall in the same place on each axis and therefore occupy one small segment of the ecospace "cube," such as a corner.

a relatively conservative method of designating baupläne and is not likely to overemphasize differences between organisms. The class level is also easily recognized, even in the fossil record. Although the guild concept should be independent of taxonomy if it is used as a method of studying total exploitation of a single resource, as Root (1967) intended, it is a common practice in many ecologic studies to examine guilds within a class [guilds of birds (Holmes *et al.*, 1979; Noon, 1981; Wagner, 1981), amphibians (Hairston, 1981), and insects (Miller, 1980)]. Therefore, recognition of class-level segregation of species does not violate most actual practice of guild study and it provides a reasonably reliable way of expressing the presence of the important differences in intrinsic features among the species of a community.

The guild approach places species in various regions of ecospace (various parts of the volume of the ecospace "cube" of Fig. 4). Figure 4 represents ecospace as defined by the three criteria used in this study. No absolute scale can be applied to the axes of the model. They simply indicate general parameters for ecospace in the same way Hutchinson defined the niche as an abstractly inhabited hypervolume (Hutchinson, 1965). A simple community structure with few guilds would have species in only a small region of ecospace, a complex community with a larger number of guilds would have species scattered more widely in ecospace.

3.2. Defining Guilds in Paleocommunities

Guilds in paleocommunities should be determined only for large single collections or at most a few closely spaced (either stratigraphically or geographically) pooled collections. The purpose of guild analysis is to examine the habitat structure of a community as it functioned in some place at some time, not to discover all the possible modes of life that were practiced everywhere in a particular environmental setting. Species lists compiled from large suites of "repetitive assemblages" or environmentally similar regions inflate the number of guilds that a community actually contained in any one setting at one time. Of course, all fossil assemblages are time-averaged (Walker and Bambach, 1971; Peterson, 1976) and can contain fossils that did not coexist in the strict sense, but that is simply an unavoidable property of the fossil record. We must accept this situation and hope that the degree of mixing from time-averaging has not changed appreciably between the Early Paleozoic and now. Collecting from single sedimentation units (discrete, separate beds) produces samples that were at least preserved together and represent the shortest available time span. Unless the fundamental processes producing stratification have changed with time, samples from single beds should be approximately comparable throughout the Phanerozoic.

The species in a paleocommunity collection are assigned to guilds by (1) recognizing the class to which each species belongs, (2) determining the feeding type and food source of each species, and (3) interpreting the life habit or life position of each species. The classification of species into classes is generally unambiguous. The categories of food source and feeding type used in this chapter are sufficiently generalized (suspension-feeder, carnivore, deposit-feeder, herbivore) that they can be determined with reasonable confidence through a combination of interpretation of functional morphology and analogy with closely related living taxa. Note that it is impossible to do such things as specify the prey organisms that a predator might specialize on, and this is *not* attempted. Such conservatism strengthens the reliability of the assignments made, but it does influence the nature of some guilds as they are recognized in fossil communities. This will be discussed in a later section (4.4). Likewise, interpretation of functional morphology plus observations on living taxa make categorization of life positions and generalized life habits (space utilization) for fossil taxa a relatively straightforward exercise.

Table II lists examples of guilds recognized for four important classes—the Trilobita, articulate Brachiopoda, Bivalvia, and Gastropoda. This listing is intended to convey an idea of the level of subdivision recognized in this chapter. Table II is not claimed to be a complete listing of all possible guilds for these groups. At this time no ideal or strictly paradigmatic method of defining guilds has been developed, nor is a fully satisfactory one likely to appear soon. The goal of this study is to strive for a reasonable approximation of ecologic grouping of species from the

Table II. Examples of Guilds

Trilobita	Gastropoda
Asymmetry of cephalon to pygidium	Cap-shaped grazers
Symmetry between cephalon and pygidium	Spired grazers
Elaborately spinose	Epifaunal predators
Blind, short thorax, often with pitted fringe on cephalon	Tiny, interstitial
Agnostid form	Infaunal suspension-feeders
	High-spired infaunal predators
	Low-predators
Articulate Brachiopoda	Bivalvia
Small (under 1 cm), strongly biconvex	Nonsiphonate shallow infaunal suspension-feeder
Large, strongly biconvex, pedunculate	Siphonate shallow infaunal suspension-feeder
Large, strongly biconvex, alate pedunculate	Sluggish siphonate deep infaunal suspension-feeder
Slightly biconvex or plano-convex, pedunculate	Rapid siphonate deep infaunal suspension-feeder
Relatively flat, free-lying	Deep infaunal mucus-tube suspension-feeder
Inflated, free-lying	Nonsiphonate deposit-feeder
Spinose	Siphonate surface deposit-/suspension-feeder
Cemented	Siphonate palp-probiscide deposit-feeder
	Endobyssate (semi-infaunal) suspension-feeder
	Erect epibyssate suspension-feeder
	Reclining epibyssate suspension-feeder
	Free-lying epifaunal suspension-feeder
	Swimming epifaunal suspension-feeder
	Cemented epifaunal suspension-feeder

fossil record using Root's concept of groups of species "that exploit the same class of environmental resources in the same way" to see if any patterns emerge. The system described is sufficiently robust that it produces consistent results (see Fig. 5, Tables III and 12 and text below) despite its imperfections.

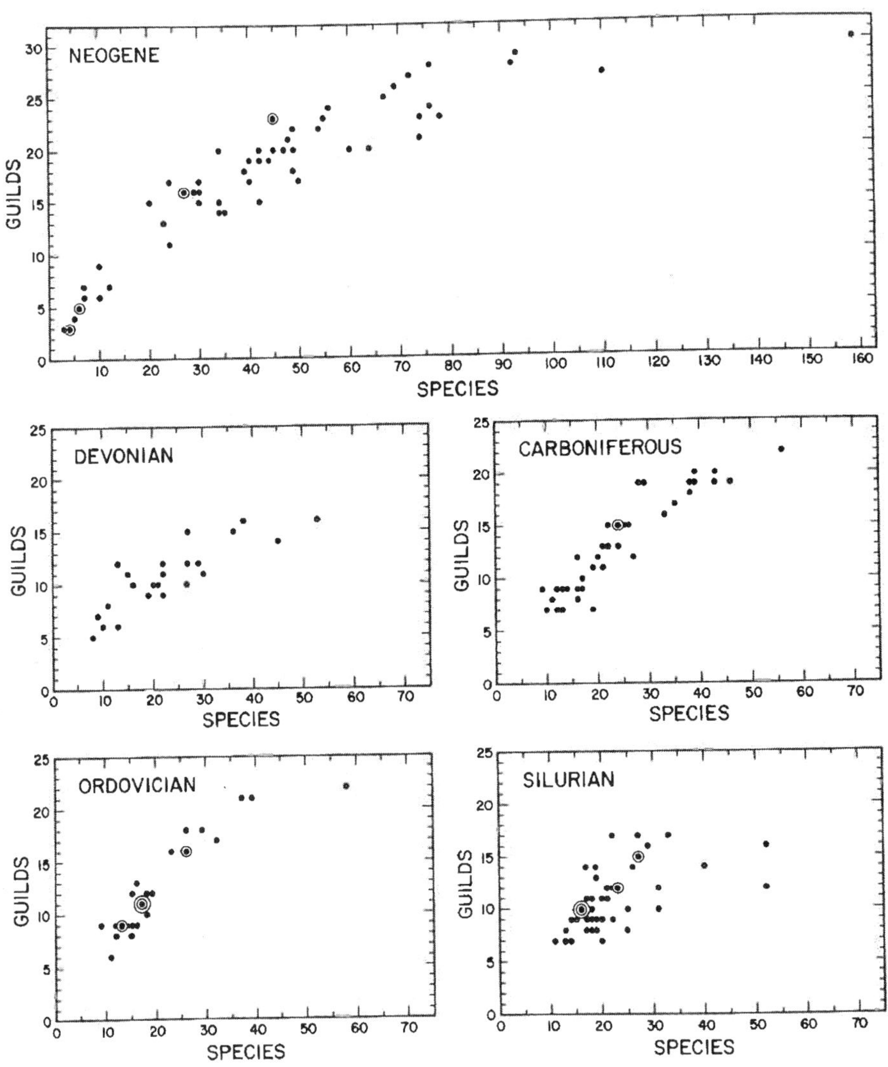

Figure 5. Scatter diagrams of guild and species-richness data. The number of species and the number of guilds are recorded for each of the 193 collections studied. Data are grouped by appropriate time interval.

4. Guilds in Paleocommunities

The number of guilds and number of species in guilds from fossil assemblages from Upper Ordovician, Silurian, Devonian, and Carboniferous rocks are compared with those in Neogene assemblages. This contrasts the guild structure of communities dominated by the second great evolutionary fauna (Middle and Late Paleozoic) of Sepkoski (see Fig. 1) with those dominated by the third (Mesozoic–Cenozoic) fauna with its markedly higher diversity both for the total fauna (Sepkoski *et al.*, 1981, and Fig. 1) and within habitats (Bambach, 1977). Data from four periods in the Paleozoic trace the nature of guild patterns through the long interval of diversity equilibrium in the Middle and Late Paleozoic.

4.1. The Data

The number of guilds and the number of species in each guild were determined for 193 collections or restricted groups of collections. Data were obtained from 29 sources (see Reference section); 29 Late Ordovician, 43 Silurian, 23 Devonian, 39 Carboniferous (12 Lower and 27 Upper Carboniferous), and 59 Neogene assemblages were analyzed. In Fig. 5 the data are plotted as number of guilds against total number of species in each assemblage. The data are for level-bottom, essentially nontiered communities (with one exception—the 22-guild 73-species Carboniferous data point).

Two potential biases against an increase in the number of guilds between Paleozoic and Neogene are incorporated in the data. Many of the Neogene collections report data only on molluscan species and some sources do not report complete faunal lists, whereas sources in which the complete fauna was listed were the only ones used for Paleozoic data. Also, only single collection lists were used for the Neogene data whereas some (about 30%) of the Paleozoic data were taken from pooled collection lists for closely spaced, environmentally related collections.

4.2. Differences in Guild Structures of Paleozoic and Neogene Communities

The histogram in Fig. 6 shows the data distribution for the Paleozoic in contrast to the Neogene. The paucity of data in the 8- to 12-guild community range in the Neogene data set is simply an artifact; few Neogene collections with 12 to 20 species were examined (see Fig. 5). Nonetheless it is obvious that the shapes of the two distributions are different. The

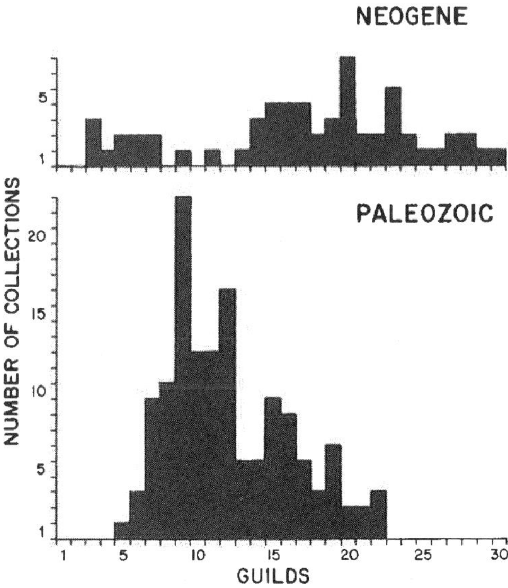

Figure 6. Histograms of the number of collections with different numbers of guilds.

median number of guilds in Paleozoic level-bottom marine communities is 11 whereas it is 18 in Neogene communities. This holds for each period studied for the Paleozoic (Table III). The median number of species in the Paleozoic communities studied is 21 whereas it is twice that (42) for the Neogene communities studied. This ratio is exactly that reported previously for the increase of within-habitat species richness (Bambach, 1977), but two-thirds of the Paleozoic communities used here were not available for use in the earlier study. The median numbers are lower in this study than those reported for open marine environments in Bambach (1977) because of the use of a large fraction of lower diversity communities from variable nearshore environments in both the Paleozoic and the Neogene data. This study is not restricted to a narrower range of environments

Table III. Median Number of Guilds and Species for Each Time Interval[a]

Guilds		Species
18	Neogene	42
13	Carboniferous	22
11	Devonian	22
11	Silurian	20
11	Late Ordovician	17

[a] Compiled from data in Fig. 5.

Table IV. Median Number of Class-Level Taxa in Collections Studied

	Range in No. of species in collections					
	1–15	16–30	31–45	46–60	61–75	76+
Neogene	2	3	3	3	3.5	5
Carboniferous	4.5	7	7.5	8		
Devonian	5	5	7	8		
Silurian	4.5	6	8.5	5.5		
Ordovician	5	8	10	11		

because full lists of species names are required so that the ecologic properties of each individual species can be determined from published descriptions. In the 1977 work all that was required was a full species count along with data for environmental placement for the community as a whole. As it turned out, complete lists of species are more frequently published for less diverse nearshore collections.

Two conclusions are (1) an increase in the number of guilds in communities as well as an increase in species richness characterizes the Neogene when compared to the Paleozoic and (2) the long Paleozoic equilibrium in diversity was matched by a relatively constant number of guilds in communities. The first conclusion suggests that the replacement of the Paleozoic by the Mesozoic–Cenozoic fauna has brought about a change in community structure that reflects the increased ecospace utilization that characterizes the Mesozoic–Cenozoic fauna. The second conclusion suggests that there was no significant alteration of general community structure during the Middle–Late Paleozoic diversity equilibrium despite the internal taxonomic turnover that took place within the Paleozoic fauna from the Late Ordovician to the Late Carboniferous.

The increase in number of guilds from the Paleozoic to the Neogene occurs even though the number of class-level taxa (one of the criteria used to erect guilds) is lower in the Neogene collections studied than it is in the Paleozoic communities (Table IV). This is true for all size communities and for all periods of the Paleozoic in comparison to the Cenozoic. One reason is that the Paleozoic data are from fully studied faunas whereas molluscan faunas alone were accepted for many of the Neogene data. This reinforces the conclusion that the increase in number of guilds in the Neogene represents the increased variety of modes of life produced by the newly dominant taxa exploiting more ecospace. It is not simply the result of an increased number of classes. It is not an artifact of high-level taxonomy.

There is a relationship between species richness and the number of guilds in a community, however (Fig. 5). As the number of species rises, the number of guilds increases. This means that there are no species-rich

Table V. Contrast between Paleozoic and Neogene in Median Number of Guilds in Collections with Different Numbers of Species

	Range in No. of species in collections					
	1–15	16–30	31–45	46–60	61–75	76+
Neogene	5	16	19	21	24	28
Paleozoic	8	11	17	18		

communities with few guilds. At no time in the Phanerozoic has it been possible to pack more species into the same number of guilds. There seems to be a regular ecologic component of additional ecospace utilization that has always accompanied increased diversity.

Table V, a comparison of the median number of guilds present in communities of particular numbers of species, shows the trend to increased number of guilds in more diverse communities. Although the Paleozoic communities of the same species richness as Neogene communities seem to have fewer guilds, the difference is not great, and is not significant.

Another difference between Paleozoic and Neogene communities is an attenuation in increase in number of guilds for larger Paleozoic communities. This is shown in Table V by the median number of guilds (18) in 46- to 60-species Paleozoic communities compared to the median number of guilds (17) for 31- to 45-species communities. The attenuation of increase in guilds in the largest Paleozoic communities suggested in Table V is borne out by examining the data distribution as diagrammed in Fig. 7. A truncation in guild increase at about 20 guilds appears in each Paleozoic field as the rise in number of guilds with increasing number of species cuts off at 15 to 20 guilds. The Neogene field also has a decreasing rate of guild addition at large species numbers, but the Neogene field clearly rises beyond the Paleozoic distribution before this has a strong effect.

On average the guild structure of Paleozoic communities was less diverse than Neogene communities, just as Paleozoic communities were less species rich. The attenuation of increase in the number of guilds in larger Paleozoic communities is not imposed by any arbitrary limitation on the possible number of guilds resulting from the system of guild analysis. It is a direct function of the limited range of adaptive strategies actually represented by the species present in the communities.

4.3. Similarities in Species Distribution within Guilds

The trend in number of guilds in communities of the same species richness is similar for both Paleozoic and Neogene until the Paleozoic attenuation starts (Fig. 7). For each grouping of communities of similar numbers of species, the proportions of guilds with the same number of species is remarkably similar in both Paleozoic and Neogene (Table VI). The mean number of species per guild is similar in Paleozoic and Neogene communities of comparable size (Table VII). Although the percentage of the species in a community that are members of guilds of the same size changes as the species richness of communities increases (more species-rich communities have a larger proportion of their species in larger guilds), both the percentage of species in guilds of any particular size and the pattern of change in distribution of species among guilds in different size communities are remarkably similar for Paleozoic and Neogene communities (Table VIII and Fig. 8). All this suggests that the guilds recognized in this study represent ecologic groupings with similar properties

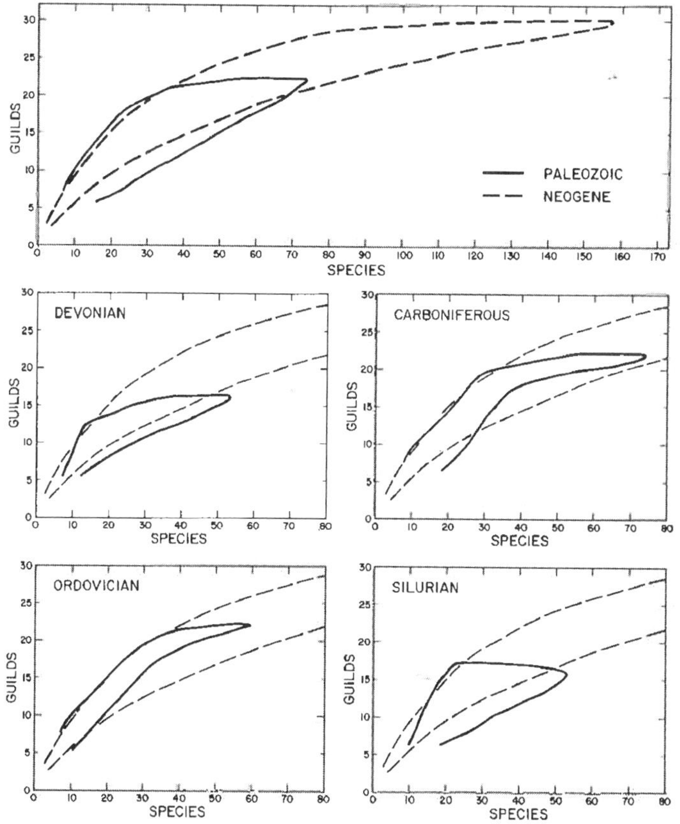

Figure 7. Outlines of the fields of data scatter in Fig. 6. One data point (12 guilds, 52 species) omitted in outline of the Silurian field. The composite Paleozoic field is superimposed on the total Neogene field. The low-species-richness end of the Neogene field is shown with the field from each Paleozoic time interval.

Table VI. Percentage of Guilds of Different Sizes

Range in No. of species in collection		No. of species in guild									
		1	2	3	4	5	6	7	8	9	10+
1–15	Neogene	79	17		3						
	Paleozoic	66	25	5	3	1		0.4			
16–30	Neogene	59	25	8	4	3.6		0.6			
	Paleozoic	59	21	10	6	3	0.7	0.1	0.5	0.1	
31–45	Neogene	50	21	12	7	3.5	2	3	0.4	0.4	0.8
	Paleozoic	53	20	11	5	4	2	1	1	0.3	1.2
46–60	Neogene	42	22	14	12	3	3	2		2	1
	Paleozoic	43	23	9	5	8	2		0.8	2	7
61–75	Neogene	41	25	9	6	6	3	1.4	2		5.7
76+	Neogene	40	20	12	7	4	3	2	2	2	8

Table VII. Contrast in Mean Number of Species in Guilds between Paleozoic and Neogene for Collections of Different Numbers of Species

	Range in No. of species in collections					
	1–15	16–30	31–45	46–60	61–75	76+
Neogene	1.3	1.7	2.2	2.5	3.0	3.7
Paleozoic	1.5	1.8	2.2	3.0		

Table VIII. Percentage of Species of Community Collections in Guilds of Different Sizes

Range in No. of species in collection		No. of species in guild									
		1	2	3	4	5	6	7	8	9	10+
1–15	Neogene	62	27		11						
	Paleozoic	44	34	11	8	2.5		2			
16–30	Neogene	34	29	14	10	10		2			
	Paleozoic	33	23	18	13	8	2	0.4	2.5	0.6	
31–45	Neogene	23	19	16	13	8	5	9	1.5	1.5	4
	Paleozoic	24	18	15	9	9	5	4	5	1.3	8
46–60	Neogene	17	17	16	19	6	8	7		3	6
	Paleozoic	14	15	9	6	13	5		2	7	29
61–75	Neogene	14	17	9	9	10	6	3	6		27
76+	Neogene	11	11	10	8	6	5	4	5	5	35

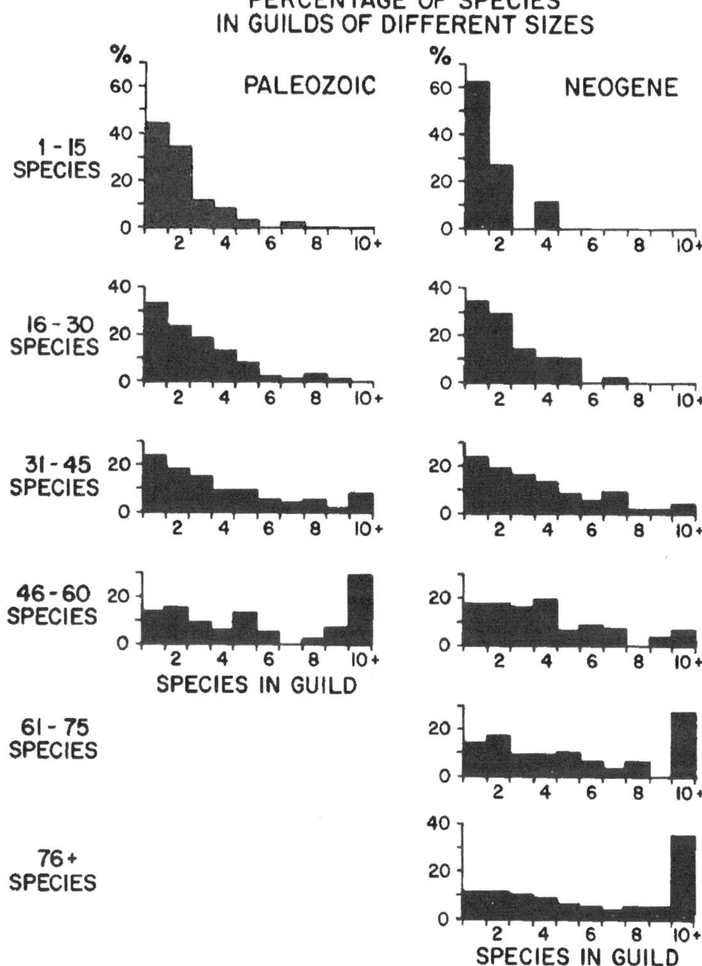

Figure 8. Comparison of proportions of species in guilds of different sizes between Paleozoic and Neogene communities of different sizes.

whether using the guild assignments interpreted for Paleozoic organisms or those made for the Neogene where more biological information is available.

As the number of species in a community increases, there is both an increase in the number of guilds (Fig. 5) and an increase in the number of guilds with several (three to nine) species each (Table VI). In all cases, however, over 90% of the guilds in a community contain fewer than 10 species (Table VI). The proportion of species in small guilds (five species or less) decreases from 100% to about 50% as community species richness increases (Table VIII and Fig. 8), and the distribution of species among guilds of different sizes becomes rather uniform in larger size communities. Increased species diversity in communities is accommodated both by adding new guilds and by packing species into established guilds. The species packing within guilds is distributed among a variety of guilds, however, as half the species are in guilds of five or fewer species in even the largest communities. This has been true, even to maintaining similar proportions in the distribution of species in different size guilds, for both Paleozoic and Cenozoic communities.

Only in the largest communities (over 45 species in the Paleozoic, over 60 species in the Neogene) is there a significant proportion of species concentrated in large guilds (guilds of 10 or more species) (Fig. 8). These large guilds raise questions about the thesis that the number of guilds increases as diversity increases. Do they indicate that at some threshold level of species richness further species are accommodated in a community by simply packing species into a few guilds and that at this point increase in number of guilds stops? No. These large (10 species or more) guilds ("superguilds") are actually guilds that could be subdivided into legitimate smaller guilds *if data* for detailed biological interpretation were available in the fossil record.

4.4. "Superguilds"

Large guilds (those containing 10 or more species) occur only in species-rich communities (Fig. 8). Table IX lists the guilds that occasionally

Table IX. Large Guilds ("Superguilds")

	No. of community collections with 10 or more species in guild
Neogene	
Epifaunal predaceous gastropods	14
Spired grazing gastropods	10
Shallow infaunal nonsiphonate bivalves	1
Paleozoic	
Strongly biconvex pedunculate articulate brachiopods	3
Flat, free-lying articulate brachiopods	2
Slightly biconvex pedunculate articulate brachiopods	1
Epifaunal low-spired ("grazing") gastropods	1
Long-stalked crinoids	1
Short-stalked crinoids	1
Cystoids	1

contained a large number of species along with the number of community collections in which those guilds contained 10 or more species. In each case it is clear that these guilds are very broadly defined. If adequate information were accessible in the fossil record, they could be subdivided into several legitimately separate guilds. For example, all epifaunal predaceous gastropods are included in a single guild in this study. It is not possible to determine the prey specificity of a gastropod from its shell and, while it is reasonable to assume a mode of attack by taxonomic analogy for Pleistocene forms, it is not certain that this would hold if one were dealing with Mesozoic ancestors. Therefore, although it is well known that some predatory gastropods specialize on various worms, others on gastropods, and others on bivalves, etc., each generalized food source being worthy of separation as a different guild (see Kohn, 1959, on *Conus*), this food source specialization cannot be specified with any degree of certainty in the fossil record. Likewise, some gastropods bore through shells, others inject poison into the prey, others simply rasp away with the radula at exposed soft tissues. These and other feeding strategies would differentiate guilds, too, if they could reliably be recognized, but they cannot.

A similar line of argument can be constructed for all the other large Neogene and Paleozoic guilds listed in Table IX. In the case of most of the large Paleozoic guilds, the guild definition is based on morphology which relates to the way the organism interacted with the substratum. Enough well-understood analyses of functional morphology are not yet available to differentiate all realistic guild types. For instance, should all strophomenids from small *Sowerbyella* to large *Rafinesquina* be grouped as "flat" free-lying articulates or should other criteria, such as size (but what size?), be used, too? Surely globose, pedunculate, smooth terebratulids and globose, pedunculate, strongly ribbed rhynchonellids differ in detailed habitat interaction that is not immediately apparent or understood. Despite these uncertainties, only 10 of the guilds recognized in the 134 Paleozoic community collections studied had 10 or more species.

In conclusion, the large guilds in this study are really composite guilds. In most cases they simply represent life forms for which adequate paleobiological data are not available to permit precise guild assignment. These "superguilds" could be subdivided into smaller guilds comparable to the others recognized if data were accessible. The fact that these large guilds are all of this type suggests that biotic interactions associated with resource subdivision and adaptive invasion of ecospace always produce relatively small guilds. Because of limitation of the fossil record, large guilds are not homogeneous ecologic entities but are just less completely defined.

5. Conclusions

Although class-level diversity has stayed relatively constant through the Phanerozoic, diversity at lower taxonomic levels has undergone stepwise increase. This has been accompanied by a matching increase in within-habitat species diversity. Different diverse class-level taxa typify the three major levels of diversity during the Phanerozoic. Each of these three evolutionary faunas had a different pattern of ecospace utilization (Fig. 3). The sequential invasion and exploitation of more ecospace characterized the evolutionary replacement of dominance by one fauna after another. The number of diverse class-level taxa has remained relatively constant, but there has been a complete replacement or turnover of dominant classes. The increase in ecospace utilization through time even though the number of diverse classes has remained constant implies that Cenozoic dominant classes are adaptively more plastic or flexible than Paleozoic dominants.

Guilds in paleocommunities can be recognized using class-level taxonomy as a method of categorizing major groups with similar intrinsic biological traits and using a combination of interpretation of functional morphology and analogy with related living forms to categorize modes of life and food resources. The number and size (species richness) of guilds are recognized by compiling data on all the individual species in communities. The actual suites of modes of life, food resources used, and class-level taxa present in a community dictate the guilds recognized, not some preconceived model of the guilds that should be present. As practiced here, the goal of guild analysis of paleocommunities is to construct a view of the suite of adaptive strategies represented in a paleocommunity. This can be a bridge that links autecologic study to the fields of synecology and evolution.

The parallelism of species distributions in guilds in Paleozoic and in Neogene assemblages suggests that the system of guild identification used here is robust. It identifies ecologically significant groups of species and provides a method for categorizing the ecologic complexity of communities. Guild analysis reveals both the nature and the style of ecospace utilization by the community and the amount of ecospace utilized by the community.

Increased species richness in Cenozoic communities seems to be a product of adding more guilds rather than simply packing more species into the same number of guilds. Therefore, Cenozoic communities exploit more ecospace than Paleozoic communities did. Although more species are added to some guilds as community species richness increases, the additional species are distributed across many guilds, not just selectively added to a few. The seeming exceptions (guilds of over 10 species) are clearly a result of imprecise guild definition. They are composite guilds representing a spectrum of modes of life or feeding types that would be separated into several guilds if adequate data were available.

Resource partitioning within any particular set of guilds has remained almost unchanged between the Paleozoic and the Cenozoic. The increase of species richness within habitats in the Cenozoic has been produced by an increase in the number of guilds (number of ways of doing things or making a living) within communities and not simply by more subtle species packing within guilds. Cenozoic communities are more diverse because more ecospace is utilized, not because of an increase in resource exploitation within established methods of ecospace utilization. Judging by the greater number of guilds in Neogene communities, total potential resources are probably more completely exploited by the Cenozoic fauna. Individual modes of life (guilds) are no more species rich in the Cenozoic than in the Paleozoic. There are just more ways of doing things available to the Cenozoic fauna.

If, as faunal turnover has proceeded, diversity has increased because of the replacement of adaptively narrow by adaptively broader groups, then we may see a link between ecologic factors and evolutionary patterns. The adaptive expansion of faunas into more ecospace may be considered as progress in evolution. It is interesting that during the long interval of diversity equilibrium dominated by one suite of diverse classes in the Middle and Late Paleozoic (Figs. 1, 2), the pattern of guilds in communities remained unaltered (Figs. 5, 7, Table III). With the replacement of the Paleozoic faunal dominants by the Mesozoic–Cenozoic dominants (and their subsequent diversification), the guild patterns in Neogene communities also changed, apparently because of the greater ecospace utilized by the newly dominant taxa of the Mesozoic–Cenozoic fauna. No increase in range of ecospace exploited nor increase in number of guilds in communities accompanied faunal evolution within the dominant classes of the Paleozoic fauna. Increased diversity and increased utilization of ecospace have occurred only as one fauna replaced another. Progress in evolution may then be seen as a consequence of adaptive breakthroughs. Trilobites remain trilobites, brachiopods remain brachiopods, and bivalves remain bivalves. But bivalves had a greater potential adaptive range than brachiopods, which had a greater potential than trilobites. It may even be that the rates of origination and the diversity-dependent rates of extinction called for by Sepkoski's kinetic model have a relationship to the complexity of adaptive flexibility and the degree of resistance to extinction it may convey. Investigation of such evolutionary and ecologic relationships is a useful way in which to view, in Hutchinson's (1965) words, "the ecological theater and the evolutionary play."

ACKNOWLEDGMENTS. Ms. Llyn Sharp assisted in data compilation. J. J. Sepkoski, Jr. kindly provided preprints of his work and extra copies of class-level cladograms as well as stimulating discussions. Enthusiastic and constructive critical discussions were held with Phil Signor, George McGhee, and Geerat Vermeij. George McGhee and Bruce Wallace kindly consented to review the manuscript. Both editors also made numerous helpful suggestions that substantially improved the presentation.

References

Bambach, R. K., 1977, Species richness in marine benthic habitats through the Phanerozoic, *Paleobiology* **3**:152–167.

Crimes, T. P., 1974, Colonization of the early ocean floor, *Nature (London)* **248**:328–330.

Hairston, N. G., 1981, An experimental test of a guild: Salamander competition, *Ecology* **62**:65–72.

Holmes, R. T., Bonney, R. E., Jr., and Pacala, S. W., 1979, Guild structure of the Hubbard Brook bird community: Multivariate approach, *Ecology* **60**:512–520.

Hutchinson, G. E., 1965, *The Ecological Theater and the Evolutionary Play*, Yale University Press, New Haven, Conn.

Kohn, A. J., 1959, The ecology of *Conus* in Hawaii, *Ecol. Monogr.* **29**:47–90.

Miller, J. C., 1980, Niche relationships among parasitic insects occurring in a temporary habitat, *Ecology* **61**:270–275.

Noon, B. R., 1981, The distribution of an avian guild along a temperate elevational gradient: The importance and expression of competition, *Ecol. Monogr.* **51**:105–124.

Peterson, C. H., 1976, Relative abundance of living and dead molluscs in two California coastal lagoons, *Lethaia* **9**:137–148.

Raup, D. M., 1976, Species diversity in the Phanerozoic: A tabulation, *Paleobiology* **2**:279–288.

Raup, D. M., 1978, Cohort analysis of generic survivorship, *Paleobiology* **4**:1–15.

Root, R. B., 1967, The niche exploitation pattern of the blue-gray gnatcatcher, *Ecol. Monogr.* **37**:317–350.

Seilacher, A., 1977, Evolution of trace fossil communities, in: *Patterns of Evolution* (A. Hallam, ed.), pp. 359–376, Elsevier, Amsterdam.

Sepkoski, J. J., Jr., 1978, A kinetic model of Phanerozoic taxonomic diversity. I. Analysis of marine orders, *Paleobiology* **4**:223–251.

Sepkoski, J. J., Jr., 1979, A kinetic model of Phanerozoic taxonomic diversity. II. Early Phanerozoic families and multiple equilibria, *Paleobiology* **5**:222–251.

Sepkoski, J. J., Jr., 1980, The three great evolutionary faunas of the Phanerozoic marine fossil record, *Geol. Soc. Am. Abstr. Progr.* **12**:520.

Sepkoski, J. J., Jr., 1981, A factor analytic description of the Phanerozoic marine fossil record, *Paleobiology* **7**:36–53.

Sepkoski, J. J., Jr., Bambach, R. K., Raup, D. M., and Valentine, J. W., 1981, Phanerozoic marine diversity and the fossil record, *Nature (London)* **293**:435–437.

Stanley, S. M., and Campbell, L. D., 1981, Neogene mass extinction of western Atlantic molluscs, *Nature (London)* **293**:457–459.

Valentine, J. W., 1969, Patterns of taxonomic and ecological structure of the shelf benthos during Phanerozoic time, *Palaeontology* **12**:684–709.

Wagner, J. L., 1981, Seasonal change in guild structure: Oak woodland insectivorous birds, *Ecology* **62**:973–981.

Walker, K. R., and Bambach, R. K., 1971, The significance of fossil assemblages from fine-grained sediments: Time-averaged communities, *Geol. Soc. Am. Abstr. Progr.* **3**:783–784.

References: Data Sources for Species Lists Used in Guild Analysis

Ager, D. V., 1963, *Principles of Paleoecology*, McGraw–Hill, New York.

Bambach, R. K., 1969, Bivalvia of the Siluro-Devonian Arisaig Group, Nova Scotia, unpublished Ph.D. dissertation, Yale University.

Bambach, R. K., Deemer, E. J., and Lewis, S. J., 1974, Community paleoecology of a turbulent environment: The basal Cloyd Conglomerate (Lower Miss.) in Montgomery and Pulaski counties, Virginia, *Geol. Soc. Am. Abstr. Progr.* **6**:330–331 (and unpublished data).

Bayer, T. N., 1967, Repetitive benthonic community in the Maquoketa Formation (Ordovician) of Minnesota, *J. Paleontol.* **41**:417–422.

Bretsky, P. W., 1970, Upper Ordovician ecology of the central Appalachians, *Peabody Mus. Nat. Hist. Yale Univ. Bull.* **34**.

Copper, P., and Grawbarger, D. J., 1978, Paleoecological succession leading to a Late Ordovician biostrome on Manitoulin Island, Ontario, *Can. J. Earth Sci.* **15**:1987–2005.

DuBar, J. R., and Beardsley, D. W., 1961, Paleoecology of the Choctawhatchee deposits (Late Miocene) at Alum Bluff, Florida, *Southeast. Geol.* **2**:155–189.

DuBar, J. R., and Chaplin, J. R., 1963, Paleoecology of the Pamlico Formation (Late Pleistocene); Nixonville Quadrangle, Horry County, South Carolina, *Southeast. Geol.* **4**:127–165.

DuBar, J. R., and Taylor, D. S., 1962, Paleoecology of the Choctawhatchee deposits, Jackson Bluff, Florida, *Trans. Gulf Coast Assoc. Geol. Soc.* **12**:349–376.

Ferguson, L., 1962, The paleoecology of a Lower Carboniferous marine transgression, *J. Paleontol.* **36**:1090–1107.

Fisher, J. N., 1970, The paleoecology of the Hamilton Group (Middle Devonian) in southeastern New York State, unpublished honors thesis (summa cum laude), Smith College.

Gernant, R. E., 1970, Paleoecology of the Choptank Formation (Miocene) of Maryland and Virginia, *Md. Geol. Surv. Rep. Invest.* **12**.

Humphreville, R., 1981, Stratigraphy and paleoecology of the Upper Mississippian Bluefield Formation, unpublished M.S. thesis, Virginia Polytechnic Institute and State University.

Hurst, J. M., 1975, Wenlock carbonate, level bottom, brachiopod dominated communities from Wales and the Welsh Borderland, *Palaeogeogr. Palaeoclimatol. Palaeoecol.* **17**:227–255.

Hurst, J. M., 1979, Evolution, succession and replacement in the type Upper Caradoc (Ordovician) benthic faunas of England, *Palaeogeogr. Palaeoclimatol. Palaeoecol.* **27**:189–246.

Lane, N. G., 1973, Paleontology and paleoecology of the Crawfordsville fossil site (Upper Osagean: Indiana), *Univ. Calif. Publ. Geol. Sci.* **99**:1–141.

Lesperance, P. J., and Sheehan, P. M., 1975, Middle Gaspe limestones communities on the Forillon Peninsula, Quebec (Siegenian, Lower Devonian), *Palaeogeogr. Palaeoclimatol. Palaeoecol.* **17**:309–326.

MacDonald, K. B., 1975, Quantitative community analysis: Recurrent group and cluster techniques applied to the fauna of the Upper Devonian Sonyea Group, New York, *J. Geol.* **82**:473–499.

Makurath, J. H., 1977, Marine faunal assemblages in the Silurian–Devonian Keyser Limestone of the central Appalachians, *Lethaia* **10**:235–256.

Mazzullo, S. J., 1973, Deltaic depositional environments in the Hamilton Group (Middle Devonian), southeastern New York State, *J. Sediment. Petrol.* **43**:1061–1071.

Shaack, G. D., 1975, Diversity and community structure of the Brush Creek marine interval (Conemaugh Group, Upper Pennsylvanian), in the Appalachian Basin of western Pennsylvania, *Bull. Fla. State Mus. Biol. Ser.* **19**:69–133.

Smith, A. B., 1959, Paleoecology of a molluscan fauna from the Trent Formation, *J. Paleontol.* **33**:855–871.

Stump, T. E., 1975, Pleistocene molluscan paleoecology and community structure of the Puerto Libertad region, Sonora, Mexico, *Palaeogeogr. Palaeoclimatol. Palaeoecol.* **17**:177–226.

Valentine, J. W., and Lipps, J. H., 1963, Late Cenozoic rocky-shore assemblages from Anacapa Island, California, *J. Paleontol.* **37**:1292–1302.

Vedder, J. G., and Norris, R. M., 1963, Geology of San Nicolas Island, California, *U.S. Geol. Surv. Prof. Pap.* **369**.

Watkins, R., 1973, Carboniferous faunal associations and stratigraphy, Shasta County, northern California, *Am. Assoc. Petrol. Geol. Bull.* **57**:1743–1764.

Watkins, R., 1978, Silurian marine communities west of Dingle, Ireland, *Palaeogeogr. Palaeoclimatol. Palaeoecol.* **23**:79–118.

Watkins, R., 1979, Benthic community organization in the Ludlow Series of the Welsh Borderland, *Bull. Br. Mus. (Nat. Hist.) Geol.* **31**(3):175–280.

Ziegler, A. M., Cocks, L. R. M., and Bambach, R. K., 1968, Composition and structure of Lower Silurian marine communities, *Lethaia* **1**:1–27.

PAPER 21

Response of Mammalian Communities to Environmental Changes during the Late Quaternary (1986)

R. W. Graham

Commentary

S. KATHLEEN LYONS

Understanding and predicting community assembly is something of a holy grail in ecology. Indeed many early papers on the topic proclaimed the existence of assembly *rules* and often argued that competition was a prime driver behind community structure. In his classic paper, R. W. Graham chastised modern ecologists for assuming that communities were relatively static sets of species that had been interacting for long periods of time, and he laid out in detail the evidence against this assumption using late Pleistocene and Holocene mammal assemblages. He first showed that even small mammals (e.g., lemmings) had extremely long distance range shifts during the last glaciation cycle. He demonstrated that species responses to climate change were individualistic and that species within communities did not respond in similar ways. Indeed, he provided several examples of communities with novel combinations of species whose ranges do not overlap today.

Although other Pleistocene workers had made many of the same points, particularly, plant specialists (Webb 1981), he expanded the arguments to mammals and showed that responses were not a simple reaction to climate. His analyses found that past range shifts could not be easily predicted from modern distributions. Species whose ranges are limited to the far north today had more southerly range limits in the late Pleistocene than species whose ranges are currently more temperate. In essence, he demonstrated that simple bioclimate envelope modeling using species' modern ranges would not be sufficient to predict their late Pleistocene distributions. However, he did it decades before anyone had thought of bioclimate modeling. He concluded that other factors such as vegetation, disease, and species interactions must have played a role in how species shifted their ranges during the last glaciation cycle.

The other important thing that Graham did in this paper was show that modern ecological theory can be tested using the Pleistocene fossil record. He argued that if temperature was the primary driver of the latitudinal gradient in species richness, temperate communities in the late Pleistocene should have fewer species than Holocene or modern communities. He reasoned that because climate zones shifted toward the equator during the expansion of the glaciers, what are now temperate communities would effectively be closer to the poles and would support fewer species. What he found was the complete opposite. Late Pleistocene communities were much more diverse than communities in the same spot in the Holocene or in the present. He argued that the effect of glaciation on temperate climates was to make the seasons less extreme; winters were warmer, and summers were cooler. The result of this more equable climate was to increase habitat heterogeneity and allow for the intermingling of species that are currently disjunct. His analyses provided evidence against the idea that temperature alone caused the increase of species in the tropics and evidence for the idea that it was the greater stability of climate in the tropics that allowed for more species. More importantly though, he set the stage for the importance of the fossil record in answering macroecological questions and did so in a forum that was guaranteed to reach large numbers of modern ecologists.

Literature Cited

Webb, T. 1981. The past 11,000 years of vegetational changes in eastern North America. *Bioscience* 31:501–6.

chapter 18

Response of Mammalian Communities to Environmental Changes During the Late Quaternary

Russell W. Graham

INTRODUCTION

During the last glacial maximum (~18,000 years ago) in North America, glacial ice covered most of Canada and much of the upper midwestern United States (Fig. 18.1). The biotas that previously inhabited these areas were physically displaced by the glacial ice sheets, and the attendant glacial climates and environments affected the distributions of organisms in unglaciated areas as well. Shortly after the glacial maximum continental ice sheets began to melt and rapidly withdraw to the north. By ~8,000 B.P. the continental glaciers occupied only a small portion of arctic Canada. The ablation of the continental glaciers opened vast areas to biotic recolonization.

This pattern has been repeated many times during the Pleistocene. For example, analysis of oxygen isotopes ($^{18}O/^{16}O$) in the shells of fossil foraminifera from deep-sea cores provides proxy measures of paleotemperatures and ice volumes for the Pleistocene (Fig. 16.1E). These data suggest that there have been at least 22 alternating stages of high and low ice volumes in the Northern Hemisphere during the last million years (Shackleton and Opdyke 1973).

Oscillations in the physical environment would have caused repeated fluctuations in the biota. How ancient, then, are the ecological communities that we see today? Also, given the environmental fluctuations of the Quaternary, is it likely, as ecologists often implicitly assume, that species that coexist today have had a long continued history of opportunity for coevolutionary adjustments to each other?

The answers to these questions depend on how species ranges shifted during the Pleistocene. If we focus on middle and high latitudes in the Northern Hemisphere, one possibility is that communities moved as a whole from north to south during glacial periods and from south to north during interglacials (Blair 1958, 1965; Dansereau 1957; Martin 1958). If this were the case, then modern communities and guilds may have had their present species composition for a long time, and there would have been substantial intervals of time for coevolution.

If, however, various species shifted their ranges for different distances, in different direc-

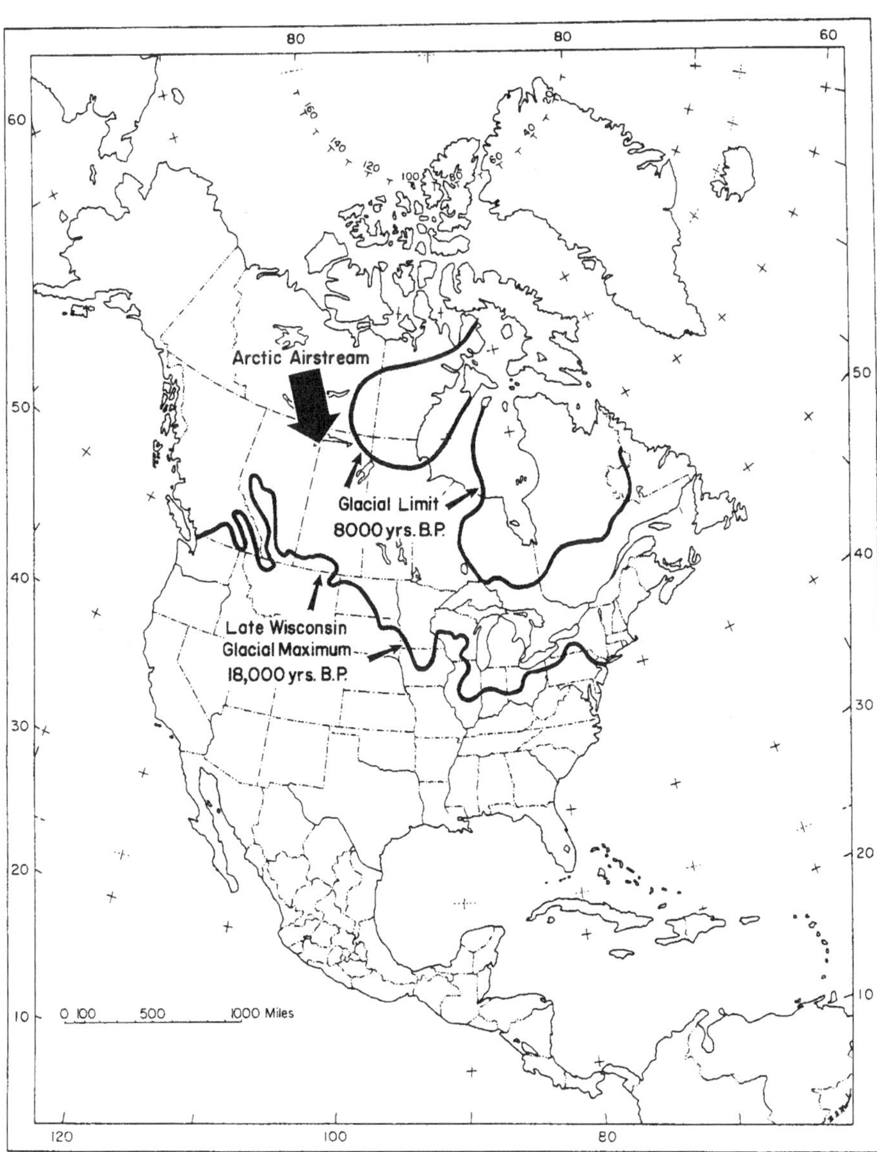

Fig. 18.1 Southern margin of continental ice masses in North America at 18,000 B.P. and 8,000 B.P. Retreat of the glacial ice northward during the late Pleistocene and early Holocene not only exposed new terrains for biotic recolonization, but it also altered climatic patterns by allowing arctic air masses to migrate southward in the early Holocene.

tions, at different times, or with different response rates, then species sets in the temperate zones would have been subjected to numerous reshufflings during the Pleistocene. Any given species could have belonged to a whole series of clusters of sympatric species through time (Cushing 1965, 1967; Davis 1976; Graham 1976, 1979; Livingstone 1967, 1975; Wright 1981). If this were the case, then coevolutionary adjustments would have had to have been rapid or would have been nonexistent (Hoffman 1979).

Species migrations during the Quaternary would have also affected species diversity for any given area. The simplest expectation would be that any site in the temperate zone supported fewer species during glacials than during interglacials. The basis for this simple assumption is that today species diversity generally increases from the poles toward the equator, and that climate zones shifted poleward during the interglacial and toward the equator during the glacial periods. However, if species responded individually to environmental fluctuations, then one might expect different diversity patterns in the Pleistocene.

All of these alternative hypotheses about biotic responses to fluctuating environments during the Quaternary can be tested by the fossil record. I shall discuss these alternatives for the Quaternary small mammal fauna of eastern North America. Small mammals, especially insectivores and rodents, are well suited for this study because of their inability to migrate large distances in a short period of time, their relatively easy and consistent identification as fossils, their occurrence in substantial numbers in fossil deposits, and their adaptations to local environments. Also, most of the late Pleistocene species are extant today and their environmental adaptations are fairly well known.

This chapter begins with a brief summary of the evidence available for reconstructing vanished communities from fossil evidence. I then compare the late Pleistocene and Holocene with respect to geographical ranges of individual small mammal species, integrity of faunas, habitat heterogeneity, and species diversity. I consider the reasons for the differences between the late Pleistocene and Holocene that emerge from these comparisons. Finally, I use these results to reexamine the extinction of the Pleistocene megafauna, the phenomenon that will first come to the minds of many readers when thinking about faunal changes at the end of the Pleistocene, but one that was actually accompanied by the many other changes in small mammals discussed in this chapter.

METHODS FOR RECONSTRUCTING PALEOCOMMUNITIES

Fossil floras and faunas are not complete, unbiased "snapshots" of biological communities that lived in the past. Instead, as discussed for fossil floras in Chapter 7, paleocommunities are derived from the accumulation and burial of fossil organisms, and these collections contain many inherent biases. One method to control these sampling biases is to compare fossil communities at different times (e.g., Pleistocene versus Holocene), since all fossil communities would have been subject to similar accumulation processes and similar sampling biases.

In the analysis of paleocommunities it is also important to demonstrate that they are not artificial samples that result from the spatial and temporal mixing of past communities. Therefore, it is essential to consider in detail how collections of fossil organisms have been formed. It is beyond the scope of this paper to provide a detailed discussion of taphonomy, the study of the processes by which living organisms are converted into fossil assemblages (see Behrensmeyer and Hill 1980). I shall instead confine myself to a brief discussion of four primary taphonomic factors (depositional environments, agents of bone accumulation, rates of sedimentation, and postdepositional disturbances).

The depositional environments of different sedimentary bodies can affect the types and size fractions of the fossils preserved. For example, in high-energy environments (such as beaches and channels of fast-running streams) only durable materials, like teeth and dense bone, will be preserved. These environments also tend to wash away the remains of smaller animals and to concentrate the fossils of larger ones. Also, high-energy depositional environments may transport fossils for some distance. Low-energy depositional environments (such as ponds, lakes, overbank stream deposits, and wind-blown deposits) generally establish a more complete sample of all components of the fossil communities, and they are also more representative of local environments. In this study I have used faunas only from low-energy environments or caves.

Caves frequently contain rich fossil deposits because they serve as focal points for bone accumulation. Bone deposits in caves are usually not the result of long-distance transport, but reflect local environments. The catchment, or area sampled for cave faunas, is partially dependent upon the size of the internal drainage system, which is usually quite small compared to open fluvial (river) systems. Also, the catchment for cave faunas is dependent upon the home ranges of predators or scavengers who serve as vectors for bone accumulation. Owls are frequently the primary vectors for fossil microvertebrate faunas, especially small mammals. Most owl species retain prey items in their digestive systems for only one day (Reed and Reed 1928), and they usually forage within a few kilometers of their roost. Thus, studying cave faunas that result from owl pellet accumulations can help minimize the likelihood of mixing communities from broad geographical areas.

Cave bone deposits may contain an overrepresentation of the remains of the predators and their prey. With regard to the prey species, these samples may be skewed by all the variables involved in prey selection by different species of predators. Bones can also be contributed to cave deposits by resident species (such as bats and snakes). In the case of pit caves (vertical shafts), the random victims of these natural traps may be still another significant component of the bone accumulation. Therefore, paleocommunities do not represent a complete spectrum of the past biota, but instead a mixture produced by several types of sampling.

For these reasons I have attempted to compensate for many of the biases by analyzing fossil faunas with similar taphonomic histories. Even though the samples may be biased, the biasing processes should be similar and the comparisons should be valid. Furthermore, because most speleological microvertebrate samples are the result of owl predation and because most owl species prefer rodents and insectivores as prey, I have restricted my comprehensive analyses of species densities to these mammalian taxa.

Depositional environments with rapid rates of sedimentation provide the greatest temporal separation of events; sites with slow rates of deposition yield the lowest resolution. For most late Quaternary sites radiocarbon dating can be used to establish rates of sedimentation. Often the only datable materials in cave deposits are bones, which generally do not give reliable dates (Land et al. 1980). However, bone dates can be accepted as minimum ages (Lundelius 1967).

Rapid deposition is frequently manifest at intact kill sites, that is, archeological sites where human hunters have killed various species of prey. If these sites are not buried rapidly, then natural processes such as decay and erosion will destroy the integrity of the association of artifacts and prey. Thus, when one finds intact kill sites, one can be confident that they represent material accumulated over short intervals of time, perhaps less than 10 years, and that paleocommunities from these sites are not the result of extensive time averaging processes.

Bone accumulations can also be altered by postdepositional processes such as chemical breakdown in soils, bioturbation or burrowing by organisms, and redeposition. Processes that cause mixing of deposits of different ages will produce artificial fossil assemblages that do not represent biological communities. These processes can frequently be identified in the field during excavation by differences in preservation of bones or by chemical analyses of bones. Obviously, mixed fossil assemblages must be excluded from paleocommunity analyses. In short, although the fossil record is not a "snapshot" of the past, it can be used with care to contribute an invaluable temporal perspective to community ecology and evolution.

LATE QUATERNARY SPECIES MIGRATIONS

One of the most common responses for mammalian species to Pleistocene environmental fluctuations was the southward displacement of northern species and the displacement of montane species to lower elevations. For example, the collared lemming (*Dicrostonyx*) today resides in the arctic tundra of extreme northern Canada (Fig. 18.2a). During the late Pleistocene it was widely distributed in the northern half of the United States, where it has been found as a fossil at elevations below 1,500 m and was not restricted to mountain tops (Kurten and Anderson 1980).

Many other boreal and montane species, such as the yellow-cheeked vole (*Microtus xanthognathus*), northern bog lemming (*Synaptomys borealis*), and arctic shrew (*Sorex arcticus*), similarly expanded their distribution to lower latitudes and altitudes. These range shifts were partly the result of an obvious effect: General boreomontane climate zones shifted southward and to lower elevations during the Pleistocene. However, they also resulted from a subtler effect: Seasonal patterns of late Pleistocene climates were quite different from modern ones, especially during the summer. Thus, boreal species whose southern or low-altitude limit is set today by warm summer temperatures were especially likely to expand their ranges during the Pleistocene, when climates were cooler and moister.

Northward migrations of southern species during the late Pleistocene are not as apparent as the southward migrations of northern species. However, species of wood rats (*Neotoma*), ground squirrels (*Spermophilus*), jaguars (*Felis onca*), and jaguarundi (*Felis yagouaroundi*) did extend their distributions further north during the late Pleistocene. These range extensions suggest that late Pleistocene winters did not suffer the intense cold extremes characteristic of modern winters. It is also interesting to note that many extinct species with southern counterparts today—tapirs (*Tapirus*), capybaras (*Hydrochoerus*), and armadillos (*Dasypus*)—ranged into the southern and central parts of the United States during the late Pleistocene.

Species that are today restricted to the deciduous and coniferous forests of the eastern United States expanded their distributions westward during the late Pleistocene (Fig. 18.2c). Likewise, species that currently inhabit the dry grassland environments of the Great Plains ranged much further eastward (Fig. 18.2d). More taxa migrated westward than eastward (Graham in press a). These westward and eastward shifts are related to the east-west moisture gradient in North America east of the Rocky Mountains. As with the temperature gradient, relaxation of the moisture gradient would allow eastern and western species to inhabit different microenvironments within the same area. In some cases the shifts in eastern and western species occurred at different times (Foley 1984, Rhodes 1984).

INDIVIDUALISTIC RESPONSE

Distributional shifts of species during the Quaternary were not mass movements of groups of species (communities). Instead, each species responded to environmental changes individually. Species migrated in different directions, at different times, and for different distances, as emphasized by Davis (Chapter 16) for plant species. As a striking example of individualistic responses, consider the collared lemming (Fig. 18.2a) and brown lemming (Fig. 18.2b). Today the brown lemming extends further south than the collared lemming, but in the late Pleistocene the collared extended at least 1,000 km further south than the brown. Even though only one Pleistocene site with brown lemming is known south of its present range, our knowledge of its southern limit during the Pleistocene is unlikely to be an artifact of sampling, since it should have been recovered from the numerous fossil sites in the northern United States if it had had a wider distribution in the late Pleistocene.

The differential migrations of these two lemmings illustrate that factors other than climatic ones were involved in some late Pleistocene range shifts. If climate was the sole limiting factor, then one would expect a direct relationship between the range limits today and those in the Pleistocene. In other words, because the brown lemming has a more southern distribution than the collared lemming today, it should have spread further south than the collared lemming during the late Pleistocene. Instead, the opposite pattern is apparent in the fossil record. Therefore, other factors, such as vegetation, disease, and species interactions, may have overridden any direct effects of climatic change.

INTERMINGLED PLEISTOCENE BIOTAS

The southern, eastern, and western migrations of species did not cause the wholesale exclusion of other species groups to isolated refugia. Instead,

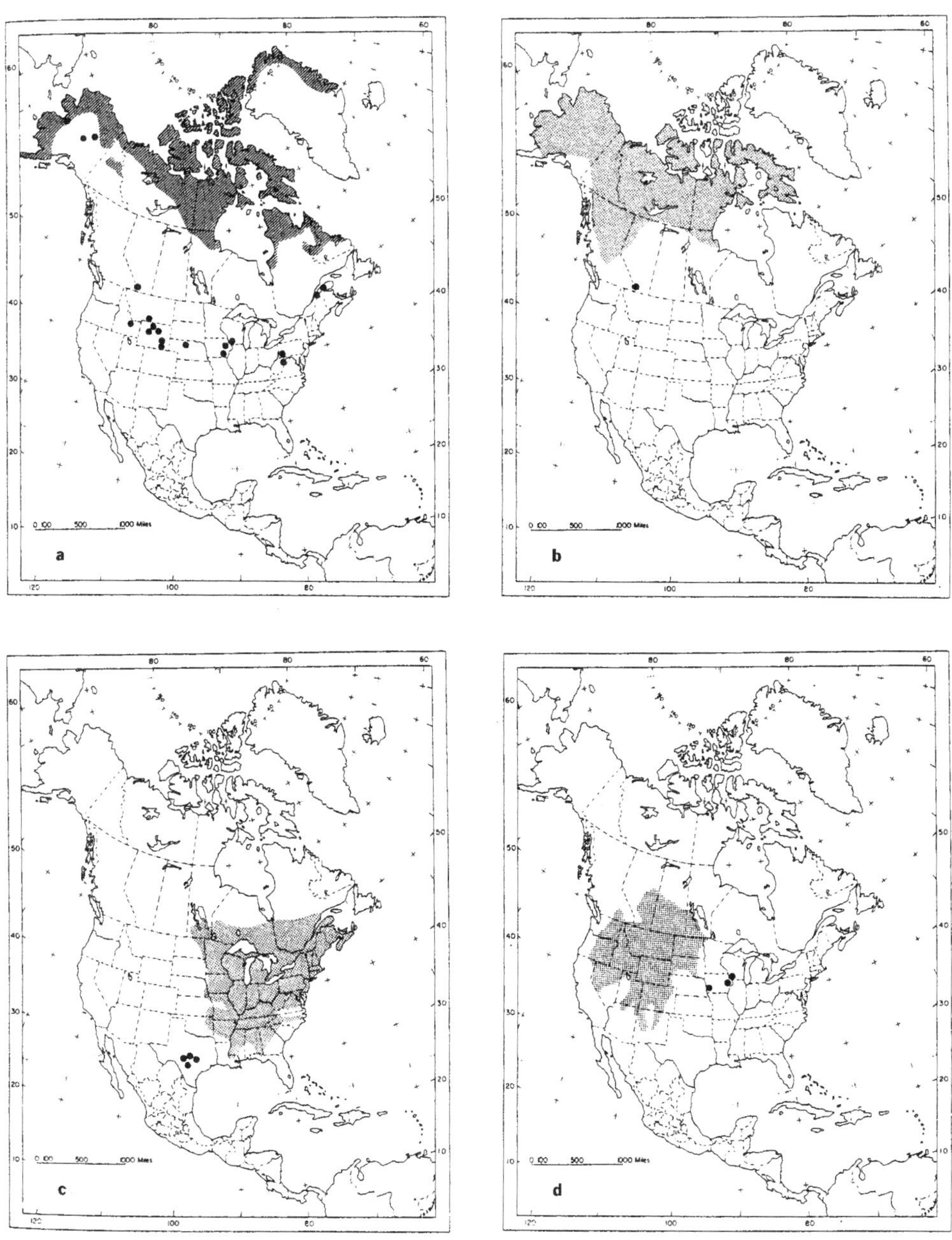

Fig. 18.2 Modern (shaded area) and late Pleistocene (black circles) distributions for four small rodent species. Compared to the modern distribution, the late Pleistocene distribution lies (a) to the south for the collared lemming (*Dicrostonyx*), (b) to the south for the brown lemming (*Lemmus*), (c) to the west for the eastern chipmunk (*Tamias striatus*), and (d) to the east for the northern pocket gopher (*Thomomys talpoides*). Note that *Dicrostonyx* is a more northerly species than *Lemmus* today but was a more southerly species in the Pleistocene.

arctic and boreal species were amalgamated with biotas from more temperate latitudes. Eastern and western species were also intermingled. The result of these individualistic responses is that Pleistocene biotas were composed of species that today live in separate habitats and remote geographical areas (Fig. 18.3). This intermingling is a crucial, worldwide characteristic of Pleistocene biotas (Fig. 18.4) that is little appreciated by ecologists studying modern communities.

The late Pleistocene biotas have been referred to as "disharmonious" because their species associations are unlike any modern assemblages (Lundelius et al. 1983, Semken 1974). This choice of term is somewhat misleading, because the late Pleistocene biotas were undoubtedly in harmony with prevailing environments. Therefore, I use the term "intermingled" to refer to such Pleistocene biotas without modern analogues.

Mixed fossil assemblages similar to the intermingled biotas from the late Pleistocene could result artifactually from numerous processes that do not reflect actual biological communities. For instance, redeposition of older fossils in younger sediments, mixing by burrowing organisms, and slow depositional rates with rapid environmental fluctuations (temporal compression) can produce intermingled patterns. These mixtures can be differentiated by careful fieldwork and laboratory procedures. However, mixed assemblages may also result from geographical and habitat mingling during deposition of the fossil deposits.

Do the intermingled Pleistocene biotas represent real communities, or did they arise as artifacts of taphonomic processes? Many types of evidence support the former conclusion.

The most compelling evidence for intermingled biotas as reflecting actual community patterns is the lack of intermingled associations in most Holocene fossil assemblages. Presumably, the taphonomic processes of specific depositional systems (such as caves, streams, and lakes) were the same in the Pleistocene and Holocene. Therefore, if the intermingled nature of the late Pleistocene biotas was a result of artificially mixed assemblages, then intermingled biotas should also appear in Holocene fossil assemblages. Semken's (1983) review of 68 Holocene fossil sites from the eastern United States ranging in age from 10,000 years ago to historic times clearly demonstrates the lack of intermingled associations in Holocene faunas. In contrast, a review of 178 late Pleistocene faunas revealed a predominance of intermingled associations (Lundelius et al. 1983).

Intermingled faunas have also been found with Pleistocene paleoindian kill sites (Johnson and Holliday 1980, Slaughter 1975). As already discussed, these archeological sites require rapid burial to maintain the integrity of the association of the human artifacts with the bones of prey. Sedimentary deposits at these sites may accumulate in tens of years. These intermingled faunas are therefore not the consequence of the temporal compression of numerous events.

Floral studies exhibit the same intermingled pattern. Late Pleistocene floras contained intermingled associations unlike any modern assemblages (Cushing 1965, 1967; Davis 1976; West 1964; Wright 1981), and these species associations had disappeared by the early Holocene in the eastern United States (Webb et al. 1983). These intermingled patterns are apparent in both the pollen and the plant megafossil (needles, cones, nuts, etc.) records. A contribution of long-distance transport to the megafossil record can generally be eliminated. Hence, these intermingled floras were not produced by mixing of materials from distant sites. Analyses of later Quaternary floras also allow a high-resolution dating and calculation of sedimentation and pollen influx rates, thereby eliminating the mixing of materials from different time levels as a cause for the intermingled biotas.

A further type of evidence is that intermingled biotas are not restricted to a certain geographical area, but appear to have a global distribution during the Pleistocene (Graham and Lundelius 1984) (see Fig. 18.4). Intermingled biotas are not limited to particular depositional environments, and they are not confined to specific taxonomic groups. Assemblages of late Pleistocene terrestrial mollusks (Miller 1976), beetles (Morgan and Morgan 1980), amphibians and reptiles (Holman 1976), and birds (Guilday et al. 1977, Lundelius et al. 1983) as well as mammals exhibit intermingled associations.

All these types of evidence leave no doubt that the late Pleistocene intermingled biotas represent communities and environments without modern analogues.

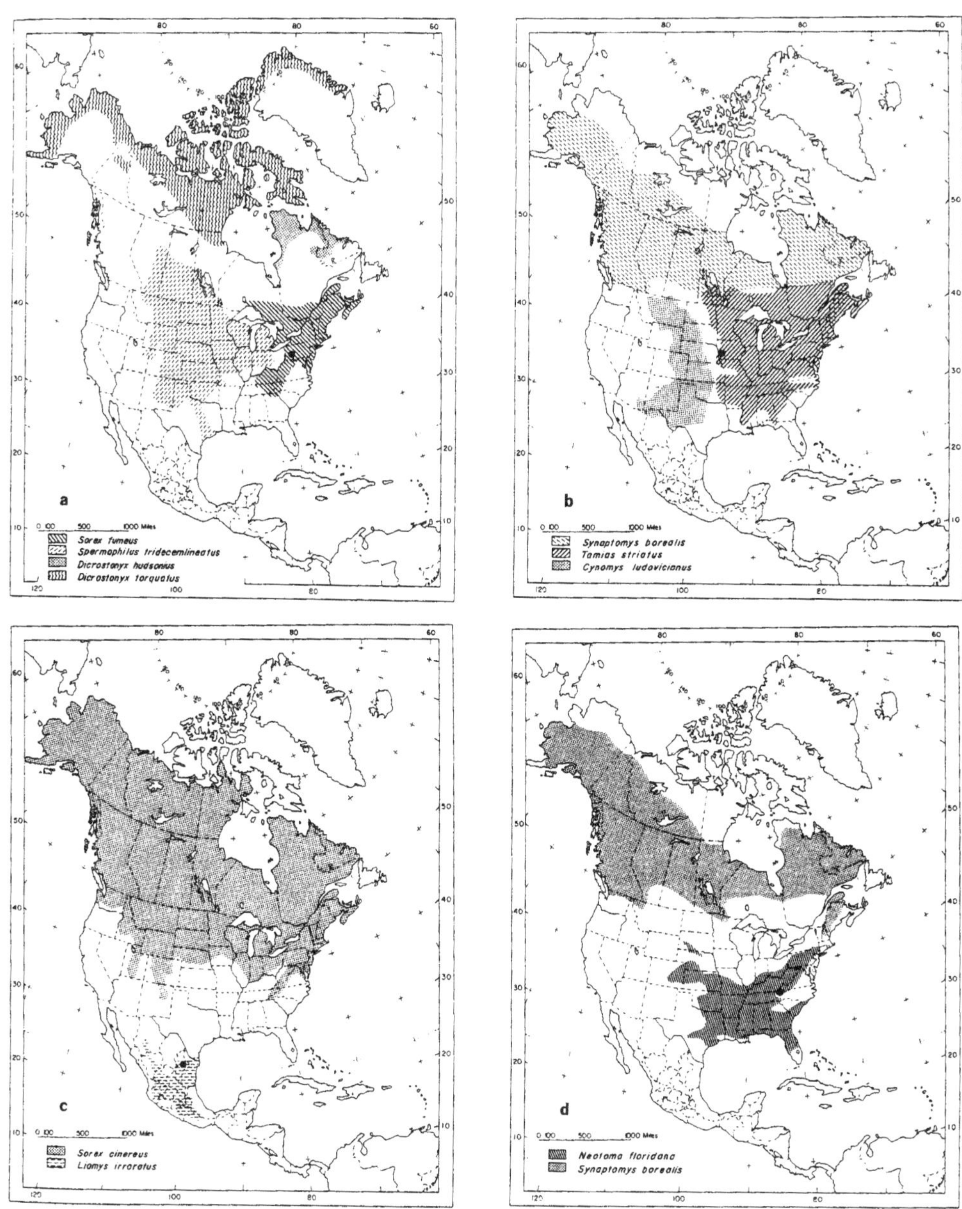

Fig. 18.3 Four examples of intermingled late Pleistocene biotas in North America. In each map the black circle shows the location of a representative site where the named species co-occur as fossils in a primary context (not mixed artificially), and the shaded areas are the modern distributions. Note that in each case the modern ranges are largely or completely disjunct, so that co-occurrence at the same site would be unlikely or completely impossible today. (a) New Paris No. 4 (11,000 B.P.), Bedford County, Pennsylvania, United States. Only one species of *Dicrostonyx* occurs in this fauna. (Guilday et al. 1964.) (b) Craigmile (23,000 B.P.), Mills County, Iowa, United States. (Rhodes 1984.) (c) San Josecito Cave (no date, late Pleistocene), Nuevo Leon, Mexico. (Jakway 1958.) (d) Baker Bluff Cave (~10,500–19,000 B.P.), Sullivan County, Tennessee, United States (Guilday et al. 1978.)

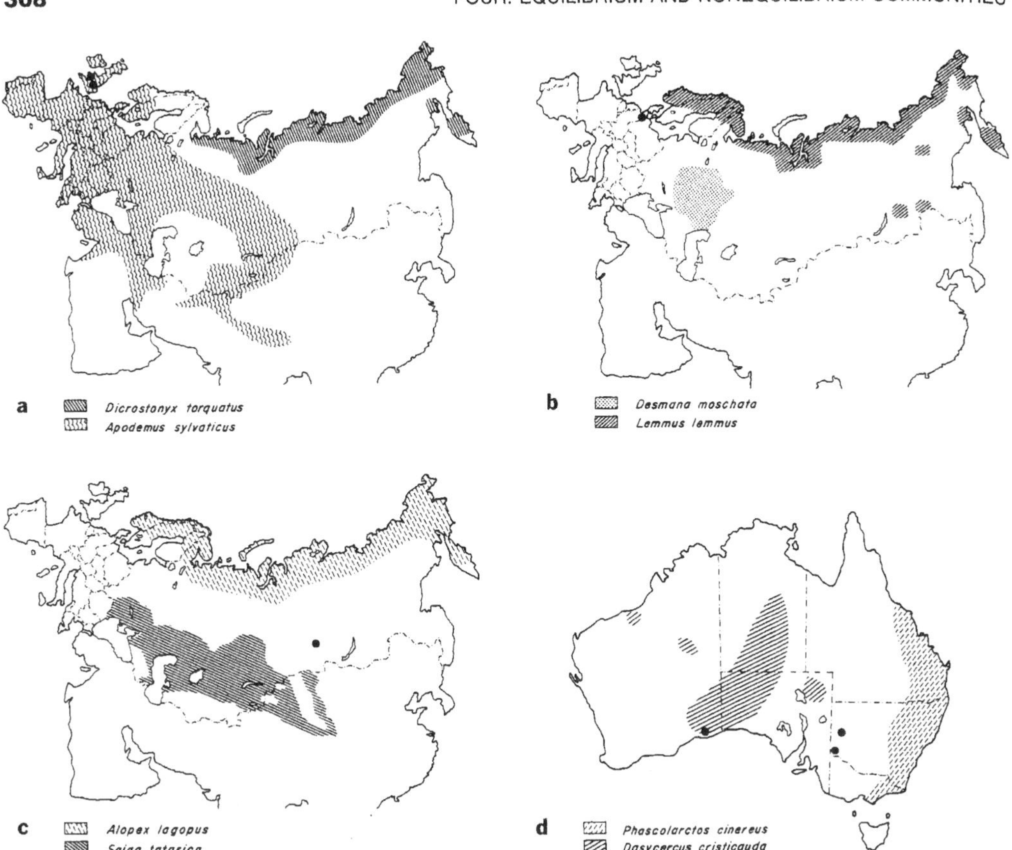

Fig. 18.4 Four examples of intermingled late Pleistocene biotas in Great Britain, Eurasia, and Australia. As in Fig. 18.3, circles mark representative fossil sites where species with largely or completely disjunct distributions today (shaded areas) co-occurred during the late Pleistocene. (a) Cat Hole, South Wales (upper Paleolithic, 15,000–10,000 B.P.) and Robin Hood's Cave, Derbyshire (upper Paleolithic, 28,500–10,600 B.P.), Great Britain. (Campbell 1977.) (b) Meiendorf (older Dryas, 12,000–11,800 B.P.) and Stellmoor (older Dryas, 12,000–11,800 B.P. and younger Dryas, 11,000–10,000 B.P.), Holstein, Germany. (Degerbøl 1964.) (c) Middle Yenisei River sites (early and late Sartan, 30,000–10,000 B.P.), Siberia, USSR. (Klein 1971.) (d) Lake Menindee, New South Wales (26,000–18,000 B.P.), Lake Victoria, New South Wales (>18,000 B.P.), and Madura Cave, Western Australia (37,880–15,600 B.P.), Australia. (Lundelius 1983.)

THE FINE-GRAINED MOSAIC OF PLEISTOCENE HABITATS

The intermingled biotas of the late Pleistocene were probably, in part, the result of greater heterogeneity in habitat types than exist today. The heterogeneous environments of the late Pleistocene were composed of a fine-grained mosaic of habitats that today occur far apart. In fact, some of the late Pleistocene habitats have no modern equivalents at all. McNaughton (1983a) has shown that spatial heterogeneity of species distribution is one of the most important features contributing to the maintenance of the Serengeti grassland ecosystem.

The biota from the mastodon bone bed at Christensen Bog (Graham et al. 1983) in central Indiana provides an excellent example. The vertebrate fauna and associated flora are from deposits that date from 13,220 B.P. to 12,060 B.P.; the age of the fauna is closer to the 13,220-year-old date at the base of the deposit. The fossils were encapsulated by a fairly rapid sedimentation rate of 10 cm/100 years.

Floral remains (pollen and macrofossils) from these deposits indicate a much more diversified forest than exists in the area today. The flora contained mixtures of spruce (*Picea*), fir (*Abies*), birch (*Betula*), larch (*Larix*), and a variety of other woody taxa such as oak (*Quercus*), elm (*Ulmus*), hazel (*Corylus*), ash (*Fraxinus*), and hickory (*Carya*) (Whitehead et al. 1982). There was also a diverse aquatic flora in the kettle lake. The fossil vertebrate fauna from the same stratigraphic horizon and temporal period also reflects the diversity of habitats. The fauna was composed, in part, of mastodon (*Mammut*), giant beaver (*Castoroides*), caribou (*Rangifer*), turtles (*Chrysemys, Trionyx, Chelydra*), ducks (cf. *Anas*), and turkey (*Meleagris*). Today, caribou inhabit boreal forest and tundra environments, whereas wild turkey and many of the turtles reside primarily in deciduous forests.

The coexistence of these diverse floral and faunal elements at Christensen Bog during the late Pleistocene was facilitated by the variety of habitats present in a fine-grained vegetational mosaic. Diverse habitat associations extend well back into the late Pleistocene and occur in many different geographical areas and physiographical settings (Graham in press a, Guthrie 1982, Lundelius et al. 1983, Rhodes 1984).

SPECIES NUMBERS

The Holocene mammalian fauna of North America has been regarded as "depauperate" or "impoverished" in comparison to the Pleistocene (Semken 1983). While the most glaring example involves the megafauna that disappeared at the end of the Pleistocene, the higher species diversity of late Pleistocene mammalian communities is also reflected in the small mammal guilds of which most of the species are extant today (Fig. 18.5). Application of conventional diversity measures, such as the Shannon-Weaver index, to fossil samples is often inappropriate because of the many inherent biases in the fossil sample. Hence my analysis instead uses species number (Graham 1976). I focused on soricid (shrew) and arvicolid (microtine rodent) species because their fossil remains are readily identifiable and because they are fairly abundant in fossil deposits. These taxa are also the common prey of many owl species that sample relatively local environments and retain prey items for a short time. Fig. 18.5 shows that local numbers of shrew and microtine species in the late Pleistocene in the midcontinent were approximately double those in the same area today.

Could these apparently higher species numbers in Pleistocene faunas be an artifact from mixing of spatially or temporally distinct sites? No, because such artifacts would presumably apply at any time in the past, yet species numbers demonstrate a decline from Pleistocene to Holocene fossil sites (Fig. 18.5). This pattern seems to be independent of depositional environments, since caves and alluvial sites exhibit the same pattern, although caves tend to have a higher species numbers (points ● vs. ▲ in Fig. 18.5). In addition, modern owl pellet samples (◆ in Fig. 18.5), the presumed source of micromammal fossil deposits, exhibit the same species number patterns as the Holocene fossil sites.

In mountainous areas, such as the western United States and the Appalachians, topographical relief probably contributed importantly to species diversity in the Pleistocene as it does today (Simpson 1969). However, these complications can be eliminated by examining flat areas. The reduction in species number from the Pleistocene to the Holocene illustrated in Fig. 18.5 is also exhibited by a chronological sequence of fossil faunas from Iowa, a state in which the highest and lowest point differ by only 363 m, and from other low-relief areas of the Great Plains (Graham in press a).

The higher species numbers for shrews and microtines are, in part, a result of the intermingling of arctic and boreal biotas with temperate biotas. In essence, the displacement of arctic and boreal biotas by glacial climates did not cause the relocation of more southern communities; instead, these biotas were amalgamated to form new communities in the temperate zone.

Higher species numbers in the late Pleistocene are not restricted to shrews and microtines, but have also been documented for other rodent fami-

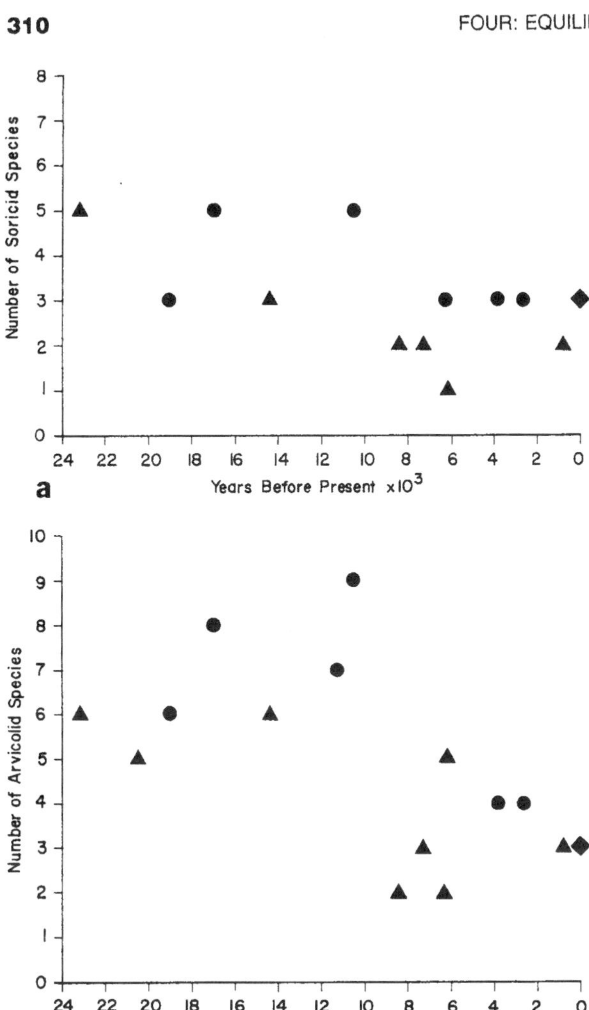

Fig. 18.5 Ordinate: number of soricid (shrew) and arvicolid (microtine rodent) species at individual fossil sites in the eastern United States. Abscissa: absolute age of the site (the Pleistocene-Holocene transition is around 10,000 B.P. Triangle = open sites; circle = cave sites; diamond = modern owl pellet sample from central Illinois (see text for explanation of site types). Sites at approximately the same absolute latitude were selected, since shrew and microtine species numbers in this area increase with latitude during the late Pleistocene as they do today (Graham 1976). In calculating arvicolid species number I lumped *Microtus ochrogaster* and *M. pinetorum* because of difficulties in identification, and I excluded *Ondatra zibethica* from all counts. Note that species numbers of both shrews and microtines decrease from late Pleistocene to Holocene fossil sites.

lies (Lundelius et al. 1983). Because of sampling and identification problems, it is uncertain whether these increased species numbers occurred universally in all terrestrial vertebrate communities. Late Pleistocene amphibian and reptilian communities also exhibit the intermingling of northern and southern biotas (Holman 1976). However, in contrast to the mammalian communities, the intermingling of herpetological communities occurs at more southerly latitudes. The reason for this difference may be that there are no herpetological equivalents for the northern mammalian communities of the tundra. Therefore, the southward displacement of northern herpetological taxa can only be detected at more southern latitudes than that of the mammalian fauna.

MAINTENANCE OF PLEISTOCENE DIVERSITY

At least three primary factors may have been involved in maintaining the intermingled biotas and higher species numbers during the late Pleistocene: equable climate, habitat diversity, and plant quality.

The nature of the late Pleistocene climate may have been one critical factor. Although the climates of glacial periods are generally characterized as being colder than at present, the actual configuration of the glacial climates was more complicated. Glacial summers were cooler and did not experience the extremes of heat characteristic of modern climates in North America. Also, winters during the glacial period exhibited less extreme cold than modern winter seasons, probably as a direct result of the glacial ice mass in North America that would have formed an orographic barrier to the southward movement of the arctic air mass during the winter (Fig. 18.1). In addition, air descending from an ice mass more than 2 km thick would have been adiabatically warmed.

Late Pleistocene climates south of the ice margin would therefore have been more equable than modern climates for these areas (Hibbard 1960, 1970). The equable climatic model is supported by independent evidence from the study of isotopic compositions of fossil wood (Yapp and Epstein 1977) and bone (Land et al. 1980). Limits of distribution for many species are controlled by seasonal climatic extremes, the southern limits of boreal species being set by hot summer temperatures, the northern limit of temperate species being set by cold winter temperatures. Hence, the relaxation of these seasonal temperature extremes during the late Pleistocene would have led naturally to higher species numbers. These patterns are consistent with MacArthur's (1975) predictions for higher species diversity in areas with reduced seasonal climatic extremes.

Higher habitat diversity in the Pleistocene would have been a second factor contributing to higher mammalian species diversity. As previously discussed, Pleistocene environments were more heterogeneous than modern ones. Greater habitat diversity would have led to more mammal species in Pleistocene sites compared to modern ones, just as it does in comparisons of different sites in the Great Plains and Rocky Mountains today (Graham in press a).

Finally, the diversity of herbivores may be a function of the effective nutritional value of the vegetation, just as Kareiva (Chapter 11) discusses for insect herbivores. Some plant species, such as black spruce (*Picea mariana*) and alder (*Alnus*), grow slowly in "poor sites" and defend their photosynthetic capital from herbivores throughout their life. Therefore, they are poor forage for herbivores. Other plant species, such as quaking aspen (*Populus tremuloides*), balsam poplar (*Populus balsamifera*), and paper birch (*Betula papyrifera*), live in better sites, grow more rapidly, tolerate more herbivory, and are relatively undefended in their mature growth form (Bryant and Kuropat 1980). Thus, they provide a more utilizable resource for herbivores.

The equable late Pleistocene environments may have selected for a better mix of palatable plant species than exists today and may thereby have afforded a greater availability of nutritious forage to mammalian herbivores. For example, the late Pleistocene spruce forests of the eastern United States contained a wider variety of palatable plant species than does the modern spruce forest (Wright 1981). Also, the late Quaternary vegetational succession in central Alaska exhibits a higher effective nutritional value in the Pleistocene than in the Holocene (Graham in press b). Indirect consequences of effective nutritional value may have also been important. Guthrie (1982) believes that reduced snowfall in unglaciated parts of Alaska during the Pleistocene may have increased effective nutritional value by allowing the herbivores to forage on frost-hardy herbs for a larger period of time in the winter.

HOLOCENE CHANGES

A rapid warming of the global climate at the beginning of the Holocene (~10,000 B.P.) significantly altered the physical and biotic environments of North America. Melting glaciers exposed new landscapes in previously glaciated terrains and eliminated old landscapes by a eustatic rise in sea level which inundated pre-Holocene terrestrial environments on the continental shelves. Retreat of the continental glaciers in response to global warming also changed climatic patterns by allowing arctic air masses to extend further southward, especially during the winter. These changes in North America created a more continental Holocene climate, marked by colder winters and hotter summers.

These climatic changes from the Pleistocene to the Holocene caused individualistic species migrations along environmental gradients. The colder winters and hotter summers of the Holocene limited the ranges of species distributions along these gradients. The ranges of many formerly coexisting species were drawn far apart from each other. Hence the fine-grained habitat mosaic of the Pleistocene was pulled apart into simpler communities that became widely separated geographically. Northward shifts of many species' ranges into deglaciated terrains lowered the diversity of small mammal communities at middle latitudes. Thus, modern mammalian communities are geographically young, less than 10,000 years old.

Throughout the Holocene, climatic fluctuations have continued to cause shifts in the ranges of individual mammal species (Hoffman and Jones 1970, Semken 1983). Also, Holocene mammalian communities have been altered by local extinctions in isolated environments like those on mountain tops (Brown 1971b, 1978). However, the magnitude of these changes appears to be small compared to the Pleistocene-Holocene transition.

EXTINCTION OF THE PLEISTOCENE MEGAFAUNA

At the end of the Pleistocene there was a major extinction of terrestrial vertebrate species, especially among birds and mammals. Although this was a global extinction event, the heaviest losses were among medium- to large-sized mammalian herbivores and carnivores in North and South America and in Australia. This extinction of the Pleistocene megafauna is the most dramatic of the changes in mammalian communities during the late Quaternary.

North America, for instance, supported during the Pleistocene a suite of large mammalian herbivores and carnivores that are now extinct. The structure of these communities in North America was probably similar, although not identical, to the modern mammalian communities in Africa. For example, it is not uncommon to find the remains of camels (*Camelops*), horses (*Equus*), mammoths (*Mammuthus*), peccaries (*Platygonus*), antelopes (*Antilocapra*), and bison (*Bison*) in Pleistocene deposits of the Great Plains. Pleistocene mammalian communities of the eastern United States frequently contain extinct large browsers such as American mastodon (*Mammut*), giant beaver (*Castoroides*), woodland musk ox (*Symbos*), and stag-moose (*Cervalces*). Large carnivores that are now either extinct or have more restricted distributions, such as short-faced bear (*Arctodus*), dire wolf (*Canis dirus*), cheetah (*Acinonyx trumani*), and lion (*Felis atrox*), were probably the primary predators on the Pleistocene large herbivores of North America.

The cause of the megafaunal extinction has been the subject of a long debate, of which both sides are presented in a recent book by Martin and Klein (1984). According to one well-known view, the first appearance of humans in North and South America and Australia was the major cause of the megafaunal extinction. According to this view humans' ability as efficient predators permitted them to overexploit the "big game" and cause its extinction.

An alternative view, however, notes that the megafaunal extinction at the end of the Pleistocene coincides with major environmental changes throughout the world (Graham and Lundelius 1984). The diverse large herbivores of the Pleistocene may have been supported by the habitat mosaic of the Pleistocene. Herbivore species could have partitioned plant resources by habitat or else by selective feeding on different plant species, plant parts, or plant growth stages (Guthrie 1982, Martin 1982). The development of continental climates at the end of the Pleistocene produced several environmental changes that would have profoundly affected the large herbivores and may have caused their extinction: destruction of Pleistocene habitats, disruption of herbivore feeding strategies (Graham and Lundelius 1984), reduction in effective nutritional value of the vegetation (Guthrie 1982, Graham in press b), and shortening of the growing season (Guthrie 1982). In this view, as in the opposite view that stresses humans as the agent of extinction, extinction of the large carnivores was a direct result of extinction of their diet, the large herbivores.

The environmental-change model of megafaunal extinction begs the question why extinction events were not associated with previous glacial-interglacial cycles. Supporters of this model must assume that the general patterns of change in the physical and biological environments were repeated many times in the past, but that the specific details of each fluctuation varied significantly. This assumption now appears to be supported. For instance, the vegetational succession of the preceding glacial-interglacial cycle (Illinoian-Sangamonian) in the Midwest consisted of a series of stages similar to those of the Wisconsinan-Holocene. However, the species compositions of the Illinoian-Sangamonian biotas were quite different from those of the Wisconsinan-Holocene, and the Illinoian biotas suggest less severe climatic conditions than in the late Wisconsinan (King and Saunders 1984). Furthermore, there is substantial evidence from the terrestrial biota throughout the world to suggest that the last interglacial had less cold winters and more equable climates than the Holocene (Graham in press a, Stuart 1982, Webb 1974).

Thus, the Holocene cannot be used as a direct analogue for the last interglacial. It is instead a somewhat individual interval in earth history, as probably was each glacial-interglacial episode. Hence the diversity patterns and species compositions for Wisconsinan and Holocene mammalian communities may not be directly applicable to previous glacial-interglacial periods. Understanding the particular qualities of the late Pleistocene-Holocene environmental change is fundamental to comprehending the late Pleistocene extinction and the evolution of our modern communities.

CONCLUSION

A myopic view of community evolution from the late Pleistocene to the present might reason as follows. With the advance of ice sheets, species retreated from north to south as whole communities and then spread north again in synchrony when ice sheets melted. Pleistocene habitats and species communities were for the most part similar to those existing today at some higher latitude. By analogy to the latitudinal gradient in species diversity that can be observed today, species number at a given site rose from the late Pleistocene to the Holocene. The only major qualitative difference between Pleistocene and Holocene communities is that the Pleistocene megafauna is now extinct, and this may have been an idiosyncratic consequence of human hunters rather than of general ecological considerations.

Reality proves to be quite different from this simple picture. Different species shifted north, south, east, or west over different distances and at different times. There are even species pairs in which the more northerly species in the late Pleistocene became the more southerly species in the Holocene. As a result, communities have been massively and repeatedly reshuffled. These facts need to be integrated into models of coevolution, which often make the implicit assumption that communities or species associations have remained intact throughout a long history. Except for land under or at the edge of the ice sheets, late Pleistocene communities had higher rather than lower species diversity of small mammals, and higher local habitat diversity, than do modern communities. Some Pleistocene habitats were entirely without modern equivalents.

No one questions that these dramatic changes in small mammal communities and habitats at the end of the Pleistocene were due to environmental changes rather than to human intervention. I therefore suspect that these same environmental changes may have also caused the extinction of the Pleistocene megafauna. Whether or not this suspicion is correct, the evidence from North American small mammals makes clear that present configurations of ecosystems are quite unlike those of much of the Pleistocene.

ACKNOWLEDGMENTS

I thank Jared Diamond for his efforts and suggestions in helping improve this paper. James Brown, Thomas Van Devender, and Andrew Knoll also provided valuable and constructive comments on an earlier draft. Neal Woodman assisted with editorial comments. I am also indebted to Mary Ann Graham for her assistance in preparation of the manuscript. Julianne Snider is thanked for drawing all of the figures. This is contribution No. 78 of the Quaternary Studies Program of the Illinois State Museum.

PAPER 22

Background and Mass Extinctions: The Alternation of Macroevolutionary Regimes (1986)

D. Jablonski

Commentary

MICHAEL FOOTE

Many aspects of the study of biological extinction were framed or brought into focus in the 1980s by statistical analyses of the times of first and last occurrence of fossil taxa, mainly marine invertebrates. For example, paleontologists documented a secular decline in the average rate of extinction, punctuated by a series of mass extinctions (Raup and Sepkoski 1982; Van Valen 1984). Some of these extinction events were so profound that they raised the possibility of nearly nonselective extinction—if the odds of success are quite small, then, as in NSF funding, there is likely to be an important chance element to who succeeds. This idea echoed earlier theoretical suggestions that differential extinction of higher taxa could result from stochastic effects even if species do not differ in their chances of survival (Raup et al. 1973; Raup 1978). But the study of mass extinctions added an interesting new twist: even if extinction at the species level was highly selective, with some species surviving preferentially because of their organism-level traits (such as body size) or species-level traits (such as geographic range), the environmental perturbations associated with these major extinction events could be very different in nature and severity from what species ordinarily faced. If so, the rules of survival might be different in times of mass extinction compared with other times when background extinction operates. The discovery of an apparent periodicity in mass extinction (Raup and Sepkoski 1984) made this possibility even more salient (Gould 1985).

The idea that the traits favoring survival might be different during mass extinctions versus other times was interesting and alluring but essentially untested until D. Jablonski's landmark comparison of differential survival during the Late Cretaceous versus the end-Cretaceous mass extinction. The following paper was one of the first of a series in which he used the model system of Cretaceous mollusks to study many important aspects of extinction and macroevolution. (Perhaps my personal favorite is his 1987 paper showing that geographic range—a species-level trait—is heritable from ancestral to descendant species, thus demonstrating the feasibility of the evolution of average clade properties by species selection, an important aspect of the hierarchical analysis of evolution by selection [see also Jablonski and Hunt 2006].) I won't give away the interesting details, but the following paper argued that the organismal and higher-level traits that make taxa longer lived during normal times do not necessarily confer preferential survival during mass extinction. If this result is quite general, it has profound implications. For example, evolutionary trends may be offset or redirected by mass extinctions, and the unfolding of evolutionary history may be less predictable and more contingent than if there were no "alternation of macroevolutionary regimes."

From *Science* 231:129–33. Reprinted with permission from AAAS.

Literature Cited

Gould, S. J. 1985. The paradox of the first tier: An agenda for paleobiology. *Paleobiology* 11:2–12.

Jablonski, D. 1987. Heritability at the species level: Analysis of geographic ranges of Cretaceous mollusks. *Science* 238:360–63.

Jablonski, D., and G. Hunt. 2006. Larval ecology, geographic range, and species survivorship in Cretaceous mollusks: Organismic versus species-level explanations. *American Naturalist* 168:556–64.

Raup, D. M. 1978. Approaches to the extinction problem. *Journal of Paleontology* 58:517–23.

Raup, D. M., S. J. Gould, T. J. M. Schopf, and D. S. Simberloff. 1973. Stochastic models of phylogeny and the evolution of diversity. *Journal of Geology* 81:525–42.

*Raup, D. M., and J. J. Sepkoski Jr. 1982. Mass extinctions in the marine fossil record. *Science* 215:1501–3.

———. 1984. Periodicity of extinctions in the geologic past. *Proceedings of the National Academy of Sciences* 81:801–5.

Van Valen, L. M. 1984. A resetting of Phanerozoic community evolution. *Nature* 307:50–52.

Other Honors and Achievements

In view of their remarkable achievements, the granting of the Nobel Prize came as the natural culmination of a series of scientific honors and prizes. They had been awarded many different scientific prizes including the Heinrich Weiland Prize, the Pfizer Award, the Passano Award, the Lounsbery Award, the Gairdner Award, the New York Academy of Sciences Award, the Lita Annenberg Hazen Award, the V. D. Mattia award, the distinguished research award of the Association of American Medical Colleges, the research achievement award of the American Heart Association, the Louisa Gross Horvitz Award, the 3-M Life Sciences Award (FASEB), the William Allan Award of the American Society of Human Genetics, and most recently the Lasker award. Both were elected to the National Academy of Sciences in 1980. They have been asked to give many prestigious lectures, were given honorary doctorates (University of Chicago and Rensselaer Polytechnic Institute), and belong to many review committees and editorial boards. Brown is a member of the Board of Scientific Advisers of the James Coffin Clinical Fund and is a senior consultant to the Lucille P. Markey Trust. Goldstein is a member of the Scientific Advisory Board of the Howard Hughes Medical Institute and a nonresident fellow of the Salk Institute. The Nobel Prize therefore came as no surprise to observers of the never-ending stream of Brown and Goldstein's important discoveries. The only question was its timing. Considering their past record, the scientific community is eagerly awaiting their future work.

Background and Mass Extinctions: The Alternation of Macroevolutionary Regimes

DAVID JABLONSKI

Comparison of evolutionary patterns among Late Cretaceous marine bivalves and gastropods during times of normal, background levels of extinction and during the end-Cretaceous mass extinction indicates that mass extinctions are neither an intensification of background patterns nor an entirely random culling of the biota. During background times, traits such as planktotrophic larval development, broad geographic range of constituent species, and high species richness enhanced survivorship of species and genera. In contrast, during the end-Cretaceous and other mass extinctions these factors were ineffectual, but broad geographic deployment of an entire lineage, regardless of the ranges of its constituent species, enhanced survivorship. Large-scale evolutionary patterns are evidently shaped by the alternation of these two macroevolutionary regimes, with rare but important mass extinctions driving shifts in the composition of the biota that have little relation to success during the background regime. Lineages or adaptations can be lost during mass extinctions for reasons unrelated to their survival values for organisms or species during background times, and long-term success would require the chance occurrence within a single lineage of sets of traits conducive to survivorship under both regimes.

THE PAST FEW YEARS HAVE SEEN A BURGEONING OF DATA and hypotheses on mass extinctions (*1, 2*). Most of this work has focused on evidence for or against extraterrestrial impacts as forcing mechanisms, particularly at the Cretaceous-Tertiary boundary, and the evolutionary role of this and other mass extinction events has been relatively neglected. A comparison of extinction patterns among bivalves and gastropods of the Gulf and Atlantic Coastal Plain region over the last 16 million years (m.y.) of the Cretaceous Period with those across the mass extinction boundary, corroborated from the literature on other taxa and extinction events, indicates that mass extinctions are not simply intensifications of processes operating during background times. Current evolutionary theory is formulated almost exclusively in terms of pattern and process during background times (*3, 4*), but if mass and background extinctions are qualitatively as well as quantitatively different in their effects, the alternation of background and mass extinction regimes shapes large-scale evolutionary patterns in the history of life.

Background extinction. For the shallow-water, bottom-dwelling marine organisms that constitute much of the fossil record, three factors affecting survivorship of species and higher taxa during times of normal background extinction are mode of larval development, geographic range (which for some groups is closely tied to larval mode), and the number of species within a taxonomic group (species richness). Each factor will be tested in turn for its effects on background and mass extinction in Late Cretaceous mollusks of the Gulf and Atlantic Coastal Plain of North America (*5*).

Speciation and extinction rates should be relatively high in species having nonplanktotrophic development because characteristically low rates of larval dispersal will be unable to maintain genetic continuity among disjunct populations; isolated populations will tend to become extinct or diverge into new species. Both rates should be lower in planktotrophs, with greater larval dispersal suppressing divergence of populations and imparting colonizing ability and broad geographic ranges that enhance species' ability to survive local extinctions (*6–10*). These predictions were verified in the fossil record for marine gastropods, in which the earliest parts of the shell preserve a record of larval development. In late Cretaceous

David Jablonski is associate professor in the Department of Geophysical Sciences, University of Chicago, Chicago, Illinois 60637.

Table 1. The synergistic effect of geographic range and species richness is lost during the end-Cretaceous extinctions. Pairwise comparisons yield binomial probabilities of 0.24 to 0.55 (see Fig. 2 for definitions).

Species richness	Species geographic range	
	Widespread (%)	Restricted (%)
Rich	59	65
Poor	60	56

(Fig. 1, A and B) (*11*), early Cenozoic (*12*), and late Cenozoic (*9, 10*) faunas, nonplanktotrophic species exhibit significantly higher speciation and extinction rates than planktotrophic species during background times.

Even for taxa in which development types are not known, or are homogeneous throughout, geographic range is closely related to geologic longevity (*12, 13*). This relation is statistically significant in the Coastal Plain Cretaceous bivalves and gastropods (*14*). Taxonomic survivorship analysis (*15*) reveals significant differences in species duration when restricted, intermediate, and broad geographic ranges are compared (Fig. 2A).

One characteristic that reportedly imparts extinction-resistance to higher taxa or clades (monophyletic evolutionary lineages) is species richness: species-poor clades are more likely to be terminated as a result of random extinction of their constituent species than are species-rich clades. Extremely volatile clades may be exceptions to

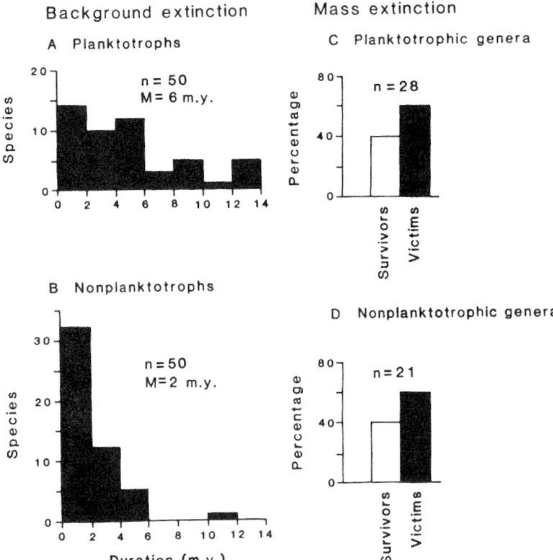

Fig. 1. Effect of larval ecology on taxonomic survivorship of Late Cretaceous gastropods during background and mass extinctions. (A and B) Background extinctions. During background times species with (A) planktotrophic larval development exhibit significantly greater duration than species with (B) nonplanktotrophic development (Kolmogorov-Smirnov test, $P < 0.01$). (B and C) Mass extinction. Mode of larval development had no statistically significant influence on survivorship of gastropod genera during the end-Cretaceous extinction event; 39 percent each of the genera inferred to have (C) planktotrophic and (D) nonplanktotrophic larvae survived. Abbreviations: M, median; m.y., millions of years (binomial probability, 0.50). For sources of data and methods for inferring modes of larval development, see (*11*).

this generalization, but the species-richness effect is commonly observed and is in accord with probabilistic models of background extinction at high taxonomic levels (*4, 16*).

In Late Cretaceous bivalves and gastropods, geologic durations of species-rich genera are significantly greater than those of species-poor genera (Fig. 3, A and B). Because a number of genera range across the Cretaceous-Tertiary boundary, it might be suspected that this result represents an integration of background and mass extinctions rather than the operation of background processes alone. However, the observed pattern is due to differences in survivorship among genera endemic to the Coastal Plain region (median durations for species-rich and species-poor genera are 23 and 10 m.y., respectively). As discussed below, endemic genera suffer such great losses during mass extinctions that their evolutionary patterns are a good index of pure background processes. In contrast, the widespread genera, whose durations reflect the integration of background and mass extinctions, fail to exhibit a significant difference between species-rich and species-poor genera (median durations, 79 and 73 m.y., respectively).

During background times, there is also a synergistic interaction between species richness and the geographic range of the individual species constituting a clade. As might be predicted from the preceding results, species-rich genera composed mainly of widespread species exhibited significantly greater geologic longevities than species-poor genera composed mainly of species with restricted geographic ranges; the other two alternatives are of intermediate duration (Fig. 2B).

Mass extinctions. In strong contrast to patterns during background times, extinction and survival among Coastal Plain mollusks during the end-Cretaceous mass extinction was influenced by neither larval development nor species richness. Instead, geographic range at the clade level, but not at the species level, appears to have been among the most favorable traits.

Gastropod genera were classified as planktotrophic or nonplanktotrophic according to the larval shell morphology of their constituent species; genera in which both modes of development were present were excluded from the analysis. Planktotrophs and nonplanktotrophs showed identical levels of extinction during the end-Cretaceous event (61 percent of 28 and 23 genera, respectively) (Fig. 1, C and D). Hansen's analyses of the early Tertiary gastropods of the Coastal Plain region (*11*) demonstrates, however, that the larval mode soon regained its prominent role in determining evolutionary patterns within the postextinction biota.

Geographic range at the species level had little influence on the survival of genera during the end-Cretaceous extinction. The frequency distributions of geographic ranges of the component species within both victims and survivors of the mass extinction are not significantly different (Fig. 2, E and F). Moving up another level in the taxonomic hierarchy reveals an unexpected difference in survivorship, however. Although species-level patterns are not influential, clades with species outside the Coastal Plain Province exhibit significantly higher survivorship than clades restricted to this biogeographic province. For bivalves (Fig. 2C), 33 percent of the extinct genera were restricted to that province, but only 3 percent of the genera that survived were endemic; viewed another way, 55 percent of the widespread genera survived, but only 9 percent of the endemics. For gastropods (Fig. 2D), 48 percent of the extinct genera were endemic, but only 11 percent of the survivors were restricted to the province; thus, 50 percent of the widespread genera survived, but only 14 percent of the endemics. This is a selectivity manifested at a different hierarchical level from those characteristic of background times.

Again in contrast to background times, species richness did not improve a genus's chances of survival, nor did species-poor genera suffer disproportionate extinction (Fig. 3, C and D). For bivalves, about 40 percent of the victims of the mass extinction were species-rich, as were 40 percent of the survivors; thus, 47 percent of the species-rich genera and 45 percent of the species-poor genera survived. Among the gastropods, about 50 percent of the victims, but only about 33 percent of the survivors were species-rich; this yields a rather surprising reversal of the background pattern: 29 percent of the species-rich genera survived, but a significantly greater 47 percent of the species-poor genera survived.

The synergistic effect of species richness and species' geographic ranges is also lost during the mass extinction. Species-rich taxa containing mainly widespread species show no higher survivorship (59 percent) than do species-poor taxa with geographically restricted species (56 percent).

Comparison to other extinctions. Patterns of extinction and survival during the end-Cretaceous extinction are discordant with the patterns of clade expansion and contraction during the times of background extinction preceding and succeeding the event. Not only did extinction rates soar during this time interval, but different kinds of taxa were vulnerable to extinction. After the mass extinction, there was a return to normal background levels and to a different constellation of traits favoring taxonomic survivorship. Comparative data for other extinctions are sparse, but the available studies are consistent with the end-Cretaceous patterns.

Among Ordovician-Jurassic bivalves (*17*), endemic genera diversified during background times but underwent preferential extinction relative to widespread and cosmopolitan genera during all four mass extinctions within the interval: end-Ordovician, late Devonian, late Permian, and end-Triassic events. Recovery of the bilvalve fauna after these extinctions derived mainly from cosmopolitan survivors. During the Pliocene molluscan extinction in the North Atlantic, species survivorship was also unrelated to traits that enhanced background survivorship (*18*). Again in common with the end-Cretaceous extinction, endemic taxa became disproportionately extinct relative to widespread taxa, regardless of species richness (*19*).

Anstey (*20*) also detected qualitative differences between background and mass extinctions in Paleozoic Bryozoa. Morphologically simple genera (interpreted as ecological generalists) had fairly steady extinction rates throughout the Paleozoic, but losses among morphologically complex genera (inferred specialists) are concentrated at mass extinctions. When the effects of mass extinctions are removed, complex taxa actually had lower background extinction rates than simple taxa did. Furthermore, because the complex genera tend to be richer in species and more geographically restricted than the simple genera (*21*), this shift in survivorship patterns during mass extinctions coincided with the one observed for end-Cretaceous mollusks.

The ammonites are such a volatile group that it is difficult to separate the effects of background and mass extinction over the course of their boom-and-bust history. In keeping with the bivalve and gastropod pattern, however, ammonite clades consisting of few, long-lived taxa tended to survive or be the last to vanish at mass extinction boundaries (*22*); these same clades tended to be geographically widespread rather than endemic (*23*).

My analysis of Alekseyev's (*24*) global compilation of 586 families and 1816 genera from marine, terrestrial, and freshwater habitats at the end of the Cretaceous shows no significant difference between the frequency distribution of genera within families for victims and survivors (contrary to Alekseyev's proposition) (Kolmogorov-Smirnov test, $0.20 < P < 0.40$). Again, taxonomic structure, albeit at a higher level, evidently had little bearing on clade survivorship during the end-Cretaceous extinction.

Macroevolutionary regimes. The fossil record near the end-

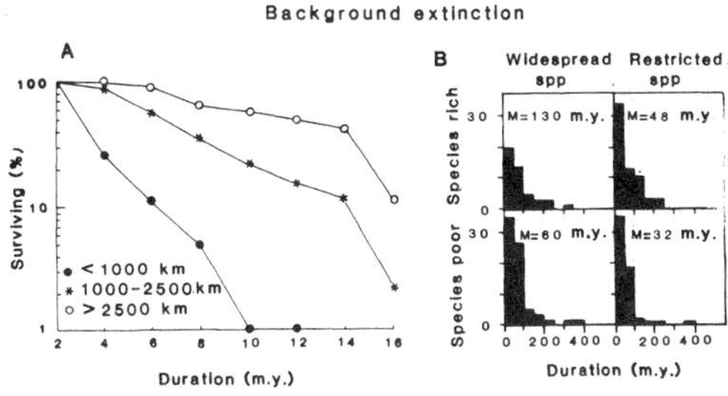

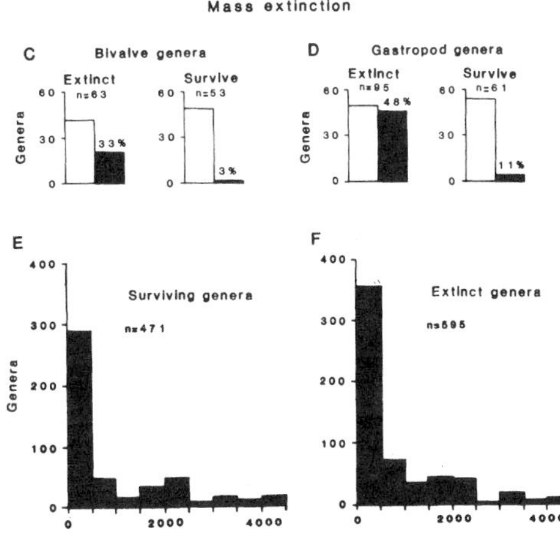

Fig. 2. Effect of geographic range on taxonomic survivorship of Late Cretaceous bivalves and gastropods during background and mass extinctions. (A) During background times, taxonomic survivorship curves are significantly different for species having narrow ranges (<1000 km), intermediate ranges (1000 to 2500 km), and broad ranges (>2500 km) (Kolmogorov-Smirnov test on original frequency distributions of durations among categories, $P < 0.01$ with Bonferroni-type correction). Taxonomic survivorship curves reflect the probability of extinction with respect to age for a given assemblage of taxa: the steeper the slope, the higher the extinction rate (*15*). (B) Also during background times, geographic ranges of constituent species influence duration of genera, in a synergistic interaction with species richness. Species-rich clades with predominantly widespread species are significantly longer lived than species-poor clades with predominantly restricted species ($P < 0.05$, Kolmogorov-Smirnov test). Species-rich, those having three or more species in the last, best known, 6 m.y. of the Late Cretaceous strata in the study region; widespread species, clades having 50 percent species with geographic ranges > 500 km along the Late Cretaceous outcrop belt of the Coastal Plain. Other cutoff limits for either trait had little effect on the calculations (see Table 1 for mass extinction pattern). During mass extinctions (C to F), geographic ranges of constituent species had no significant effect on clade survival, but geographic range of clades did have a significant effect. Frequency distribution of geographic ranges of species within genera are not significantly different for survivors (E) and victims (F) on the mass extinction ($0.50 > P > 0.40$, Kolmogorov-Smirnov test). However, endemic genera (genera restricted to the Coastal Plain province) were overrepresented among victims of the end-Cretaceous mass extinction among both bivalves (C) and gastropods (D). For bivalves, 55 percent of the widespread genera and only 9 percent of the endemic genera survived (a statistically significant difference; binomial probability = 0.00004); for gastropods, 51 percent of the widespread genera and 14 percent of the endemic genera survived (binomial probability = 0.0000005).

Table 2. Faunal replacements previously regarded as competitive that now seem to have been mediated by mass extinctions (*28*).

Taxa	Reference
End-Ordovician brachiopods	(*29*)
Late Paleozoic mammal-like reptiles	(*30*)
Post-Paleozoic benthos	(*31*)
End-Triassic tetrapods	(*32*)
End-Cretaceous tetrapods	(*33*)
End-Cretaceous reef-builders	(*34*)
Mid-Tertiary carnivorous mammals	(*35*)

Cretaceous and other extinction events suggest that large-scale revolutionary patterns are shaped by two macroevolutionary regimes. Microevolutionary and macroevolutionary processes of the background regime are occasionally disrupted, presumably by external abiotic forcing factors (*25*), and replaced by the mass extinction regime. Many traits of individuals and species that had enhanced the survival and proliferation of species and clades during background times become ineffective during mass extinctions, and other traits that were not closely correlated with survivorship differences become influential. The mass extinction regime is apparently relatively short-lived—1 or 2 m.y. at most for the end-Cretaceous event, and perhaps considerably less.

During these mass extinction events, evolution is channeled in directions that could not have been predicted on the basis of patterns that prevailed during background times. As indicated by the observed selectivity with respect to clade-level deployment, mass

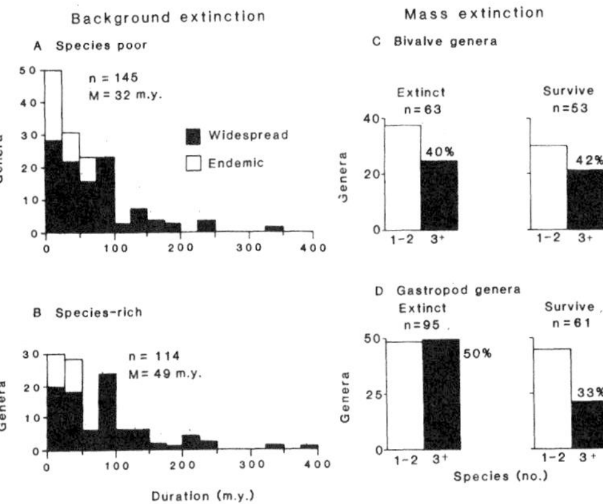

Fig. 3. Effect of species richness on taxonomic survivorship of Late Cretaceous bivalves and gastropods during background and mass extinctions. During background times, median geologic durations of species-rich bivalve and gastropod genera are greater than those of species-poor genera. The basis for this difference lies in the endemic genera, which suffer so severely during mass extinctions that their patterns of survivorship are a good index of purely background processes (Kolmogorov-Smirnov test, $P < 0.05$); there is no significant difference between species-rich and species-poor widespread genera ($0.40 > P > 0.20$). During the end-Cretaceous mass extinction, species-rich bivalve genera are equally represented among the victims and survivors (C); 47 percent of species-rich genera and 45 percent of the species-poor genera survive (binomial probability, 0.60). Among the gastropods (D), species-rich genera are disproportionately represented among the victims of the mass extinction; 29 percent of species-rich genera and 47 percent of species-poor genera survive (binomial probability, 0.0018).

extinction is not a regime of complete randomness, but this selectivity will be indifferent to many adaptations that had been valuable during the much more prolonged background regime—a process resembling what Raup (*26*) has called "nonconstructive selectivity." Particular traits captured during background times can be lost not because they are in themselves maladaptive but because they occur in clades that lack the environmental tolerances or geographic distribution necessary to survive the mass extinction regime. Thus, a broader range of hypotheses must be considered in analyzing the histories of higher taxa and major adaptations, hypotheses that recognize the potential role of extinction processes that will not be congruent during mass and background regimes, or across levels in the biological hierarchy.

At the same time, these results imply that mass extinctions play a larger role than is generally appreciated in creating opportunities for faunal change, removing dominant taxa and thereby enabling other groups—previously unimportant but with traits enhancing survivorship during mass extinctions—to undergo adaptive radiations. In recent years a number of major faunal replacements have been interpreted in this way (Table 2). The most persistent and diverse clades may therefore be those in which major new adaptations and other traits favored during background times happen to be associated with traits that favor survival under the mass extinction regime. Such an association may underlie the steady success of post-Paleozoic extratropical bivalves and gastropods relative to other marine clades.

Comparative analyses among mass extinctions are needed to test this view of macroevolution. Pairwise comparisons of survivorship patterns across a spectrum of extinction magnitudes can determine whether there is a true threshold or a gradual changeover between the background and mass extinction regimes. Some similarities in biotic patterns certainly exist among the mass extinctions (*2*), but some differences remain, such as the apparent selectivity in favor of nonplanktotrophic clades during the end-Permian event (*27*) and the lack of any larval pattern at the end of the Cretaceous. Such differences may be due to contrasts in biogeographic and other factors immediately before the various extinction events, but each particular mass extinction may prove to be an isolated excursion away from the background regime.

The patterns reported here support a hierarchical view of evolution, in which selection, drift, and other evolutionary processes operate at a variety of focal levels, with consequences both upward and downward within a genealogical hierarchy from gene to clade (*4*). Taxonomic survivorship during the end-Cretaceous extinction is a function of clade geographic range per se, rather than a direct consequence of the environmental tolerance or other features of individual organisms (as indicated by the lack of effect of individual species' ranges on clade survivorship); clade geographic range behaved in this instance as an emergent property that influenced evolutionary fates at other levels. Evolutionary processes at these other levels did not cease, but selectivity at a higher focal level became more important and had cascading effects through the biological hierarchy. However, this selectivity at the clade level cannot yet be inferred to indicate an ongoing process of evolution by clade selection. An important component of such a process—heritability of geographic range at the clade level—has not been demonstrated.

Rather than simply accelerating or emphasizing trends already manifest during background times, the end-Cretaceous mass extinction was characterized by qualitative as well as quantitative changes in patterns of extinction and survival. The few comparable data available from other mass extinctions suggest that these events, too, imposed survivorship patterns having little correspondence with those of background times. Because the traits that enhance survival during mass extinctions (for example, broad geographic range at the clade level) tend to be poorly correlated with traits that enhance survival and diversification during background times, mass extinctions will not promote the long-term adaptation of the biota. In fact, mass extinctions can break the hegemony of species-rich clades honed by millions of years of selection and thereby permit radiation of taxa little favored during the interval preceding the extinction event. The alternation of these macroevolutionary regimes disrupts any smooth extrapolation of microevolutionary or macroevolutionary processes across the sweep of geological time; a complete theory of evolution must incorporate the different sets of selective and random processes that characterize the background and mass extinction regimes.

REFERENCES AND NOTES

1. D. M. Raup, *Science*, in press; K. W. Flessa *et al.*, in *Patterns and Processes in the History of Life*, D. M. Raup and D. Jablonski, Eds. (Springer-Verlag, Berlin, in press).
2. D. Jablonski, in *Patterns and Processes in the History of Life*, D. M. Raup and D. Jablonski, Eds. (Springer-Verlag, Berlin, in press); in *Dynamics of Extinction*, D. K. Elliott, Ed. (Wiley, New York, in press).
3. G. G. Simpson, *Tempo and Mode in Evolution* (Columbia Univ. Press, New York, 1944); *Major Features of Evolution* (Columbia Univ. Press, New York, 1953).
4. S. M. Stanley, *Macroevolution* (Freeman, San Francisco, 1979); S. J. Gould, *Science* **216**, 380 (1982); in *Perspectives on Evolution*, R. Milkman, Ed. (Sinauer, Sunderland, MA, 1982), pp. 83–104; N. Eldredge, *Syst. Zool.* **31**, 338 (1982); E. S. Vrba and N. Eldredge, *Paleobiology* **10**, 146 (1984).
5. Species durations were estimated with eight stratigraphic intervals of approximately 2 m.y. each. Geographic ranges were mapped on the discontinuous, 5000-km outcrop belt to an accuracy of ±20 km. This degree of temporal and geographic resolution permitted a standardized comparison among sedimentary basins and time intervals throughout the study area [see Jablonski (*11*) for details]. During the Late Cretaceous, the Gulf and Atlantic Coastal Plain constituted a single cohesive biogeographic province [N. F. Sohl, *Proc. First N. Am. Paleontol. Conv.* **2**, 1610 (1971)]; E. G. Kauffman, in *Atlas of Palaeobiogeography*, A. Hallam, Ed. (Elsevier, Amsterdam, 1973), pp. 353–383], so that appearance and disappearance of species within the study area are likely to represent global rather than local invasion or extinction. Although the last 1 m.y. of the Cretaceous Period is unrecorded in the Coastal Plain region, prohibiting detailed analysis of the timing of the end-Cretaceous event, the excellent preservation of the diverse molluscan faunas permits comparison of survivorship across the boundary with patterns of background extinction in the immediately preceding 16 m.y. to a degree unparalleled by other stratigraphic sections in shallow-marine sediments.
6. D. Jablonski and R. A. Lutz, *Biol. Rev.* **58**, 21 (1983).
7. J. A. Pechenik, R. S. Scheltema, L. E. Eyster, *Science* **224**, 1097 (1984); C. S. Gallardo and F. E. Perron, *Malacologia* **22**, 109 (1982).
8. G. J. Vermeij, *Biogeography and Adaptation* (Harvard Univ. Press, Cambridge, 1978).
9. R. S. Scheltema, in *Concepts and Methods of Biostratigraphy*, E. G. Kauffman and J. E. Hazel, Eds. (Dowden, Hutchinson & Ross, Stroudsburg, PA, 1977), pp. 73–108; in *Marine Organisms: Genetics, Ecology and Evolution*, B. Battaglia and J. A. Beardmore, Eds. (Plenum, New York, 1978), pp. 303–332; in *Historical Biogeography, Plate Tectonics and the Changing Environment* (Oregon State Univ. Press, Corvallis, 1979), pp. 391–397.
10. T. Shuto, *Lethaia* **7**, 239 (1974); *Bull. Mar. Sci.* **33**, 536 (1983).
11. D. Jablonski, *Proc. Third N. Am. Paleontol. Conv.* **1**, 257 (1982); *Bull. Mar. Sci.*, in press.
12. T. A. Hansen, *Science* **199**, 885 (1978); *Paleobiology* **6**, 193 (1980).
13. J. B. C. Jackson, *Am. Nat.* **108**, 541 (1974); A. J. Boucot, *Evolution and Extinction Rate Controls* (Elsevier, Amsterdam, 1975); A. Hoffman and B. Szubdzda-Studencka, *Neues Jahrb. Palaeontol. Abh.* **163**, 122 (1982).
14. Simple linear regression reveals a significant linear relation between geographic range and species duration [n = 1066 species, slope = 2.00 ± 0.07 (million years per 10^3 km); $P < 0.0005$; t-test]. When species that originated in the last 2 m.y. preceding the end-Cretaceous mass extinction (and thus are most likely to suffer truncation of durations by mass extinction rather than background processes) are omitted from the analysis, n drops to 848 and 95 percent confidence limits fail to overlap [slope = 2.19 ± 0.08 (million years per 10^3 km)]. The fact that even these truncated species frequently attained broad geographic ranges before becoming extinct suggests that during background times geologic duration is a function of geographic range and not vice versa.
15. L. Van Valen, *Evol. Theory* **1**, 1 (1973); D. M. Raup, *Paleobiology* **1**, 82 (1975).
16. D. M. Raup, *ibid.* **4**, 1 (1978); *Acta Geol. Hispanica* **16**, 25 (1981); S. J. Gould *et al.*, *Paleobiology* **3**, 23 (1977); R. R. Strathmann and M. Slatkin, *ibid.* **9**, 97 (1983).
17. P. W. Bretsky, *Geol. Soc. Am. Bull.* **84**, 2079 (1973).
18. S. M. Stanley, *Proc. Third N. Am. Paleontol. Conv.* **2**, 505 (1982). It is surprising that this much smaller extinction also disrupted survival patterns relative to background times. Shifts in selective pressures owing to climatic changes, rather than a switch to the extinction regime characterizing the major mass extinctions, may have been involved.
19. G. J. Vermeij, in *Dynamics of Extinction*, D. K. Elliott, Ed. (Wiley, New York, in press).
20. R. L. Anstey, *Paleobiology* **4**, 407 (1978).
21. Of the genera for which biogeographic data were available, 23 percent of the simple genera (n = 44) are present on four or more paleocontinental blocks, but only 9 percent of the complex genera (n = 36) are so widespread (binomial probability, 0.044) [G. G. Astrova, *Akad. Nauk SSSR Paleontol. Inst. Trudy* **169**, 1 (1978); J. R. P. Ross, *Palaeontology* **21**, 341 (1978); *ibid.* **24**, 313 (1981); R. A. Robison, Ed., *Treatise on Invertebrate Paleontology*, vol. 1 part G, *Bryozoa* (Geological Society of America, Boulder, CO, revised edition, 1983).
22. P. D. Ward and P. W. Signor, III, *Paleobiology* **9**, 183 (1983).
23. J. Wiedmann, quoted by E. G. Kaufmann, in *Catastrophes and Earth History*, W. A. Berggren and J. A. Van Couvering, Eds. (Princeton Univ. Press, Princeton, NJ, 1984), pp. 151–246.
24. A. S. Alekseyev, *Int. Geol. Rev.* **26**, 1006 (1984).
25. Of course, this change in macroevolutionary regime does not necessarily confirm the hypothesis that an extraterrestrial impact triggered the end-Cretaceous mass extinction. It does indicate that a major perturbation took place, crossing a threshold beyond which normal background processes were of secondary importance. Nor does the crossing of this threshold during other extinction events necessarily require an identical forcing factor.
26. D. M. Raup, *Science* **206**, 217 (1979); in *Patterns of Change in Earth Evolution*, H. D. Holland and A. F. Trendall, Eds. (Springer-Verlag, Berlin, 1984), pp. 5–14.
27. R. R. Strathmann, *Evolution* **32**, 894 (1978); J. W. Valentine, *Bull. Mar. Sci.*, in press.
28. M. J. Benton, *Nature (London)* **302**, 16 (1983).
29. Orthacians and strophomenids replaced by spiriferids and other brachipod groups; P. M. Sheehan, *Proc. Third N. Am. Paleont. Conv.* **2**, 477 (1982).
30. Synapsids replaced by therapsids; T. S. Kemp,, *Mammal-Like Reptile and the Origin of Mammals* (Academic Press, London, 1982).
31. Articulate brachiopods replaced by bivalves; S. J. Gould and C. B. Calloway, *Paleobiology* **6**, 383 (1980); see (*27*).
32. Mammal-like reptiles replaced by dinosaurs; M. J. Benton, *Q. Rev. Biol.* **58**, 29 (1983).
33. Dinosaurs replaced by placental mammals; D. A. Russell, *Annu. Rev. Earth Planet Sci.* **7**, 163 (1979).
34. Rudist bivalves replaced by scleractinian corals; N. D. Newell, *Am. Mus. Novit. No.* 2465 (1971), p. 1; P. M. Sheehan, *Geology* **13**, 46 (1985).
35. Archaic carnivores, Orders Creodonta and Condylartha, replaced by Order Carnivora; L. B. Radinsky, *Paleobiology* **8**, 177 (1982).

36. I thank A. J. Boucot, J. H. Brown, K. W. Flessa, E. G. Kauffman, S. M. Kidwell, D. M. Raup, M. L. Rosenzweig, J. J. Sepkoski, Jr., S. Suter, J. W. Valentine, and G. J. Vermeij for valuable discussions and reviews. Supported by NSF grants EAR 81-21212 and EAR 84-17011.

14 June 1985; accepted 27 September 1985

PAPER 23

Evolution of Species Assemblages: Effects of Energetic Constraints and Species Dynamics on the Diversification of the North American Avifauna (1987)

J. H. Brown and B. A. Maurer

Commentary

DOUGLAS A. KELT

Macroecology was established as a field of endeavor in the late 1980s, via a series of articles published by James Brown and Brian Maurer. Although predating the *Science* paper (Brown and Maurer 1989) that would provide a name for this field, their 1987 paper was the first to apply "constraint envelopes" to describe energetic or environmental limits to the distribution of species in bivariate space. Although the implications were many and complex, the paper itself was notable for its simplicity. After developing straightforward predictions for associations between body mass, geographic range, and population density at continental spatial scales, Brown and Maurer documented how the North American avifauna fit these expectations, and proposed novel hypotheses to explain where data failed to fit their predictions. Apart from putting *constraint envelope* into the ecological lexicon, they distinguished between absolute and probabilistic constraints. The underlying mechanisms Brown and Maurer proposed were in some respects the same old parameters—differential speciation and extinction. What distinguished their work was the integration of ecological and evolutionary temporal scales and the explicit interaction with energetics. And while some of these constraints were well known (e.g., large species with small geographic ranges have relatively high probabilities of extinction), others led to fundamental revisions in thinking about both evolutionary patterns and ecological mechanisms. Perhaps most notable was the "energetic constraint" that appeared to preclude the smallest species from having high population densities. This has percolated into several discussions on body-size allometry. For example, R. M. May (1978, 1988) documented lower known species richness at the smallest size categories for most taxonomic groups, and K. P. Dial and J. M. Marzluff (1988) argued that this was a real pattern and not a reflection of poor data for the smallest size categories; Brown and P. F. Nicoletto (1991) suggested that this might relate to energetic constraints at the smallest size classes. Subsequently, Brown et al. (1993) developed an energetics-based model suggesting an optimal body size for mammals, explaining reduced species diversity both above and below this optimum. This was extended to birds (Maurer 1998) and bivalves (Roy et al. 2000), but failed to fully explain other groups (Boback and Guyer 2008; Jones and Purvis 1997; Symonds 1999), and has been warmly debated (Bokma 2001; Kozlowski 1996, 2002; Perrin 1998). The fact remains that numerous ecological traits exhibit both qualitative and quantitative shifts above and below a body size that is normally near but not at the lower range of body sizes (reviewed in Brown 1995; Gaston and Blackburn 2000), and refocusing investigators to emphasize ecological and evolutionary constraints over linear relations has led to a resurgence of interest in explaining trenchant patterns in nature. To a nontrivial extent this emphasis is in response to the following paper.

From *American Naturalist* 130:1–17.

Literature Cited

Boback, S. M., and C. Guyer. 2008. A test of reproductive power in snakes. *Ecology* 89:1428–35.

Bokma, F. 2001. Evolution of body size: Limitations of an energetic definition of fitness. *Functional Ecology* 15:696–99.

Brown, J. H. 1995. *Macroecology*. University of Chicago Press, Chicago.

Brown, J. H., P. A. Marquet, and M. L. Taper. 1993. Evolution of body size: Consequences of an energetic definition of fitness. *American Naturalist* 142:573–84.

*Brown, J. H., and B. A. Maurer. 1989. Macroecology: The division of food and space among species on continents. *Science* 243:1145–50.

*Brown, J. H., and P. F. Nicoletto. 1991. Spatial scaling of species composition: Body masses of North American land mammals. *American Naturalist* 138:1478–1512.

*Dial, K. P., and J. M. Marzluff. 1988. Are the smallest organisms the most diverse? *Ecology* 69:1620–24.

Gaston, K. J., and T. M. Blackburn. 2000. *Pattern and process in macroecology*. Blackwell Science, Oxford.

Jones, K. E., and A. Purvis. 1997. An optimum body size for mammals? Comparative evidence from bats. *Functional Ecology* 11:751–56.

Kozlowski, J. 1996. Energetic definition of fitness? Yes, but not that one. *American Naturalist* 147:1087–91.

———. 2002. Theoretical and empirical status of Brown, Marquet and Taper's model of species-size distribution. *Functional Ecology* 16:540–42.

Maurer, B. A. 1998. The evolution of body size in birds. II. The role of reproductive power. *Evolutionary Ecology* 12:935–44.

*May, R. M. 1978. The dynamics and diversity of insect faunas. In *Diversity of insect faunas*, edited by L. A. Mound and N. Waloff, 188–204. Royal Entomological Society of London Symposium 9. Blackwell Scientific, London.

———. 1988. How many species are there on Earth? *Science* 241:1441–49.

Perrin, N. 1998. On body size, energy and fitness. *Functional Ecology* 12:500–502.

Roy, K., D. Jablonski, and K. K. Martien. 2000. Invariant size-frequency distributions along a latitudinal gradient in marine bivalves. *Proceedings of the National Academy of Sciences* 97:13150–55.

Symonds, M. R. E. 1999. Insectivore life histories: Further evidence against an optimum body size for mammals. *Functional Ecology* 13:508–13.

Vol. 130, No. 1 The American Naturalist July 1987

EVOLUTION OF SPECIES ASSEMBLAGES: EFFECTS OF ENERGETIC CONSTRAINTS AND SPECIES DYNAMICS ON THE DIVERSIFICATION OF THE NORTH AMERICAN AVIFAUNA

JAMES H. BROWN AND BRIAN A. MAURER*

Department of Ecology and Evolutionary Biology, University of Arizona, Tucson, Arizona 85721

Submitted February 19, 1986; Revised June 18, 1986; Accepted November 21, 1986

One way to account for the diversity of living things is to derive the rules that govern the adaptive diversification of a taxonomically constrained biota within a geographical region. These rules, if they exist, should be reflected in patterns of attributes that represent the outcomes of a combination of ecological, biogeographical, and evolutionary processes. Species, the basic units that make up biotas, interact with their environment on a wide range of spatial and temporal scales. At one extreme, variation on large spatial and long temporal scales determines the biogeographical and evolutionary processes that shape the composition of the pool of species that historically have had access to a geographical region. At the other extreme, variation on small spatial scales and short time scales influences the dynamics of local populations of interacting species and determines the combinations of species that constitute local communities. The dynamics of processes at these two extremes are not independent: species cannot occur in local communities unless historical events have given them access to the region, and species do not remain in the pool unless they are able to maintain populations within local communities.

The characteristics of species that are found at any scale represent the outcome of ecological interactions and evolutionary changes within populations as well as the origination and extinction of species. At the large scale, relationships between species and their environment are mediated largely by the dynamics of evolutionary change both within and among species. Opportunities for colonization continually modify the composition of the species pool, and speciation, extinction, and evolutionary differentiation adjust the number and attributes of species to environmental conditions. On the small scale, the interactions of species with their environment are mediated by the dynamics of local populations.

To the extent that these evolutionary and ecological processes are general, we can expect them to be reflected in the attributes of species that constitute taxonomically and geographically defined biotas. In order to study these processes empirically, we must first select an assemblage of species that share a common

* Present address: Department of Zoology, Brigham Young University, Provo, Utah 84602.

Am. Nat. 1987. Vol. 130, pp. 1–17.

geographical region and close taxonomic relationships. Then, we must select for analysis those traits that influence the species' ecological and evolutionary dynamics.

The present study focuses on the almost 400 species that make up the North American terrestrial avifauna. Although this fauna results from the separate invasion and proliferation of several different lineages on the North American continent, the attributes of its species should reflect the common evolutionary history of the monophyletic class Aves and the shared ecological conditions within temperate North America. We analyze four attributes of these species that should influence the structure and dynamics of this biota: individual body size, average local population density, area of geographical range, and trophic guild. Each variable reflects different aspects of the interactions between a bird species and its environment. Because of the ubiquity and importance of allometric scaling relationships, body size, better than any other single variable, describes the attributes of the lower-level units, the individual organisms that species comprise (Peters 1983; Calder 1984). Population density, averaged over many local sites, characterizes the intensity with which the total area inhabited by the species is populated by individuals. The area of the geographical range provides a measure of the breadth of tolerances and requirements of the individual units, and it characterizes the extent to which each species is able to use the total space available to the biota. Trophic guild provides a discrete, qualitative categorization of the types of resources used by the species.

There is precedence for analyzing population density, body size, and geographical-range size as important characteristics of species assemblages. Initial efforts focused on documenting empirically the frequency distributions of each of these variables among species and explaining them theoretically: body size (e.g., Hutchinson and MacArthur 1959; Van Valen 1974; May 1978); population density (e.g., Fisher et al. 1943; Williams 1944, 1953, 1964; Kendall 1948; Preston 1948, 1962; Brian 1953; Kerner 1957, 1959; MacArthur 1957); and area of geographical range (e.g., Willis 1922; Rapoport 1982). Additional insights into the factors affecting biotic diversity and composition have come from analyzing bivariate distributions of these variables among species: population density and body mass (e.g., Mohr 1940; Damuth 1981; Peters 1983; Peters and Wassenberg 1983; Calder 1984; Peters and Raelson 1984; Brown and Maurer 1986); area of geographical range and body mass (e.g., Brown 1981); and population density and geographical range (e.g., Hanski 1982*a*,*b*; Bock and Ricklefs 1983; Bock 1984; Brown 1984). Most efforts to explain the distributions of these attributes among species have been couched in terms of traditional models of population dynamics and natural selection, where the units of primary interest have been individuals within populations. Beginning with MacArthur and Wilson's (1967) attempt to explain the composition of insular biotas, however, numerous neontologists and paleontologists have begun to recognize the potential importance of processes that operate at the level of species within biotas in determining the body sizes, population densities, geographical ranges, trophic specializations, and other attributes of living and fossil organisms (see, e.g., Stanley 1975, 1979; Vrba 1980, 1983; Arnold and Fistrup 1982; Fowler and MacMahon 1982; Gould 1982; Vrba and Eldredge 1984; Damuth 1985; Eldredge 1985).

THEORETICAL FRAMEWORK AND PREDICTIONS

The main objective of this paper is to propose a provisional set of rules that characterizes the adaptive response of a large biota of many species (the terrestrial avifauna of North America) to the limits imposed by biological constraints (in this case, the morphological, physiological, and behavioral attributes shared by the terrestrial birds), the large-scale environment (the North American continent with its limited space, resources, and other biotic and abiotic factors), and the dynamics of species origination and extinction. We develop these rules by considering species as the unit of interest and by examining the empirical patterns and underlying mechanisms in the relationships among the three variables that have been measured for each species: average population density where the species occurs, average adult body mass, and area of geographical range.

Consider the following general framework. Imagine a three-dimensional space formed by the log-transformed values of body mass, population density, and area of geographical range (fig. 1*a*). Each species of a large biota can be plotted as a point on this graph, such that the distribution of points reflects the dispersion of species with respect to the three variables. Some regions of the three-dimensional space defined by the graph contain no points because they represent impossible combinations of traits that are precluded by absolute constraints. Other regions contain only sparsely distributed points because they represent regions of species-phenotype space that have few resources, have only recently been colonized, or have a high probability of extinction. Still other regions contain densely distributed points because their attributes enable many species to exploit abundant resources. To clarify the relationships among these variables, we focus on two-dimensional representations in which the boundaries separating those regions containing points from those representing combinations of variables not possessed by species can be characterized more precisely.

We now make predictions about the distribution of species with respect to these pairwise combinations of variables, beginning with the relationship between population density and body size. First, we assume some minimum size of bird that is determined primarily by physiological constraints; however, different bird species that approach this minimum size can exhibit a wide range of population densities. Second, we hypothesize that the minimum population density reflects an increasing probability of extinction with decreasing population size. If I represents the total number of individuals in a species, N the average population density, and A the area of the geographical range, then $NA = I$, and $\log N + \log A = \log I$. We further assume that some minimum total population size, I_{min}, is necessary for long-term persistence. Since $\log A$ tends to increase with increasing body mass, M (Brown 1981; see below), then the log of minimum population density, log N_{min}, should decrease with $\log M$. Third, we predict that maximum population density, N_{max}, should also decrease with increasing body size. Because the space

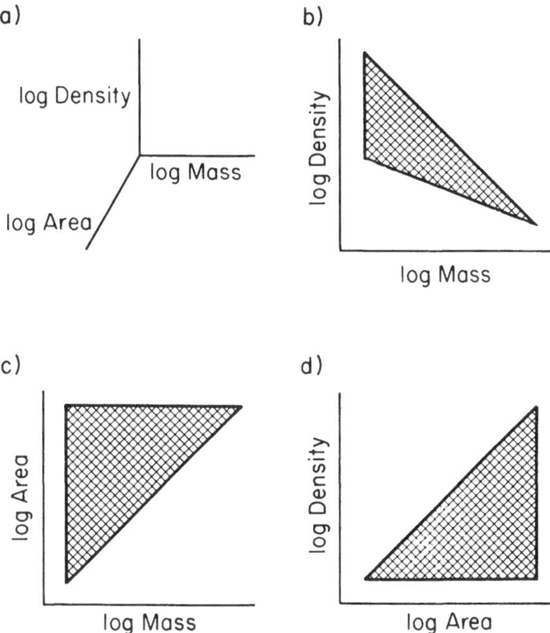

FIG. 1.—Predicted distributions of species with respect to three variables, individual body mass (M), average local population density (N), and area of geographical range (A). Conceptually, each species can be thought of as a point in three-dimensional space (*a*). Practically, the predicted relationships can be represented in two-dimensional plots (*b–d*), in which the combinations of variables expected to be possessed by species are delineated by the cross-hatched triangles. For the basis of these predictions, see the text.

and energy resources used by an individual organism are positive allometric functions of body mass, a fixed unit of space with its resources should support fewer large individuals than small individuals. Because the trade-off between population density and body mass should be steep (Damuth 1981 and Peters 1983 suggested that N scales as $M^{-1.0}$ to $M^{-0.75}$), we suggest that N_{max} decreases more rapidly with increasing body size than N_{min}. Thus, if on a log scale these boundaries are approximately linear, the relationship between $\log M$ and $\log N$ should be triangular, with the range and minimum and maximum densities all decreasing with increasing body size (fig. 1*b*).

Now consider the relationship between geographical range and body size. Two limits are straightforward: physiological constraints again set the minimum body mass; and the land area of the North American continent determines the maximal size of a geographical range. However, the minimal size of the geographical range should be related to body size (Brown 1981). Since species of large body size are constrained to have low population densities, such species with small geographical ranges should have a high probability of extinction because the total species population is small. Consequently, the minimal size of the geographic range should increase with body size (fig. 1*c*). These considerations imply that the relationship between the log of the geographical-range size ($\log A$) and $\log M$ should be roughly triangular with few or no species of large body size having small geographical-range sizes (fig. 1*c*).

The relationship between population size and area of geographical range is predicted to be as follows. Species of large body size are limited to a small region of low density and large geographical-range size. Species of small body size vary much more in both geographical-range size and density; however, many factors should limit the population growth of the species restricted to small geographical ranges, and hence they should have lower average population densities than more widely distributed species (Brown 1984). Thus, the upper density maximum should increase with geographical-range size. These factors suggest a triangular relationship between $\log A$ and $\log N$, such that few species have small geographical ranges, and these species tend not to have high average population densities (fig. 1*d*).

We consider the predicted distributions of trophic guilds with respect to these variables only briefly because they seem straightforward consequences of dietary and energetic constraints. Nectarivores should be limited to small body sizes and perhaps also to low population densities (e.g., Brown et al. 1978), but they might exhibit a wide range of geographical-range sizes. Body sizes of obligate insectivores should be relatively small, those of omnivore-insectivores should be of an intermediate size, and those of herbivores and carnivores should be relatively large. Insectivores, omnivore-insectivores, and herbivores should exhibit a wide range of population densities and geographical-range sizes for their body sizes, whereas carnivores should be limited to relatively low densities and large ranges (Brown 1981).

METHODS

We obtained estimates of average population density from the Breeding Bird Survey (BBS) conducted by the U.S. Fish and Wildlife Service and the Canadian Wildlife Service since 1968 (Bystrak 1981). Each year, birds are censused using standardized procedures along approximately 2000 BBS routes distributed across the North American continent north of the Mexican border. Each route consists of 50 census locations 0.8 km apart. At each location, observers count all birds detected during a 3-min period. Routes begin at 0.5 h before sunrise and are conducted during June of each year on a day with good weather conditions. Although the BBS data set has a number of problems (Bystrak 1981), no other long-term standardized censuses cover such a large geographical area for any other group of organisms. Thus, any limitations of the BBS must be weighed against the great value of these standardized estimates of population density. Any species-specific, observer-specific, or site-specific biases should average out over time and/or space, so that major trends in the entire data set (approximately 400 species) are not greatly biased.

Our estimate of population density was obtained by averaging the number of detections per route for each species over all routes on which it was recorded between 1968 and 1982. We assumed that this average is proportional to the actual

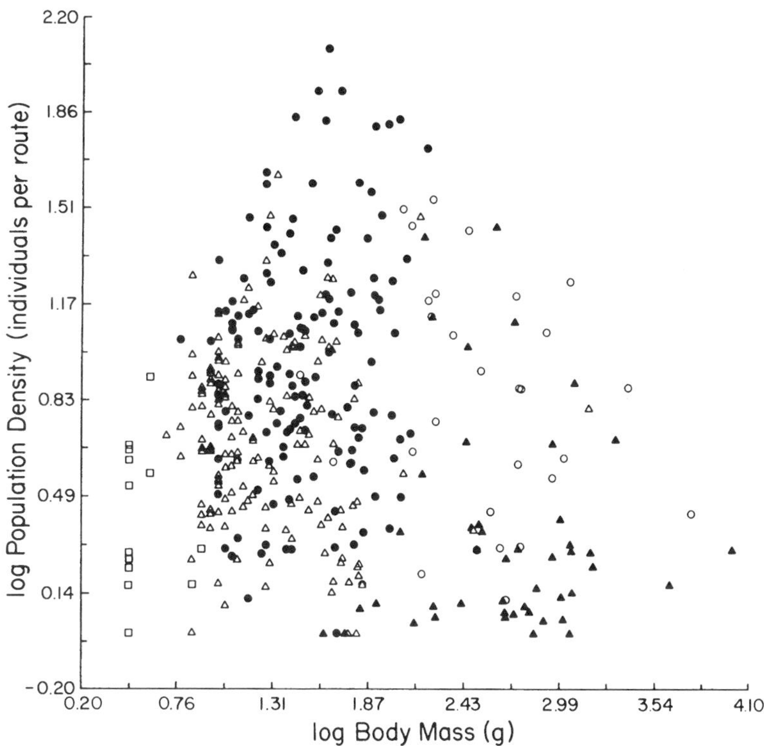

FIG. 2.—Relationship between log N (average local population density) and log M (individual body mass) for North American terrestrial bird species. Trophic guilds are depicted as follows: *open squares*, nectarivores; *open triangles*, insectivores; *open circles*, herbivores; *solid circles*, omnivore-insectivores; *solid triangles*, carnivores. Note that, contrary to our prediction, the points fall within a well-defined quadrilateral with a constant minimum density and a maximum density that is intermediate for the smallest birds, rises to a peak at an intermediate size, and then declines to the minimum with increasing body size.

average density (pairs per km^2) of each species. We obtained average adult body weights of all species from Dunning (1984), and sizes of breeding ranges by planimetry of unpublished breeding-range maps prepared by C. S. Robbins of the U.S. Fish and Wildlife Service. These range maps have since been published (Robbins et al. 1983). We measured any portion of the breeding range that extended across the Tropic of Cancer into northern Mexico, but not parts of the breeding ranges of the few species that extended farther south into the Neotropics or across the water gaps to Eurasia. We omitted from our analysis those species with breeding ranges largely in tropical America or Eurasia that occurred in only a small part of the continental United States, northern Mexico, or Canada.

We assigned each species to a trophic guild based on its most common food resources: carnivore, herbivore, omnivore-insectivore (those species that feed their young insects but consume substantial plant material outside the breeding season), insectivore, and nectarivore. These categories are broad enough that entire families could usually be assigned to a single category on the basis of

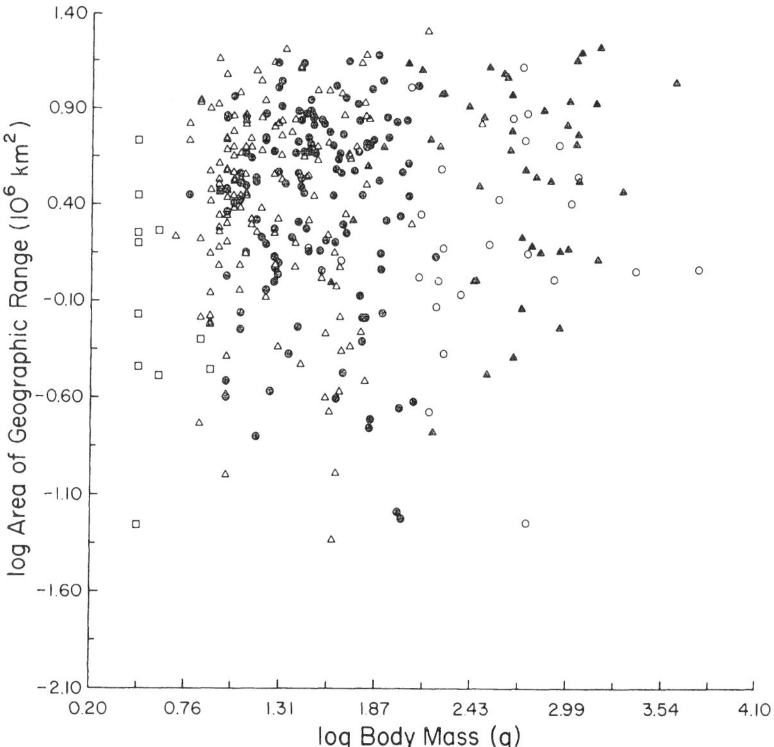

FIG. 3.—Relationship between $\log A$ (area of geographical range) and $\log M$ (individual body mass) for the land-bird species of North America. For an explanation of the symbols, see figure 2. Note that the points fall approximately within a triangle as predicted, but the lower boundary is not clearly defined.

general life history information. Bivariate plots of $\log N$, $\log M$, and $\log A$ were constructed. We were unable to obtain data for all variables for every species; we used all the values available for each pair of variables, hence the small differences in the number of points in figures 2–4. We used discriminant-function analysis to assess patterns of variation in $\log N$, $\log M$, and $\log A$ associated with differences among the trophic guilds. For the three log-transformed variables, we performed a four-group discriminant-function analysis, using all trophic guilds except nectarivores, which had many fewer species and a significantly different covariance matrix from that of the other groups.

RESULTS

Pairwise plots of log-transformed values of N, M, and A generally conformed to the hypothesized relationships among these variables (fig. 1), but with some conspicuous exceptions. Log$N$ did not attain its maximum for the smallest species as we anticipated (fig. 1*b*), but it was highest at a body weight in the range of 50–100 g, decreasing on either side (fig. 2). Moreover, minimum population

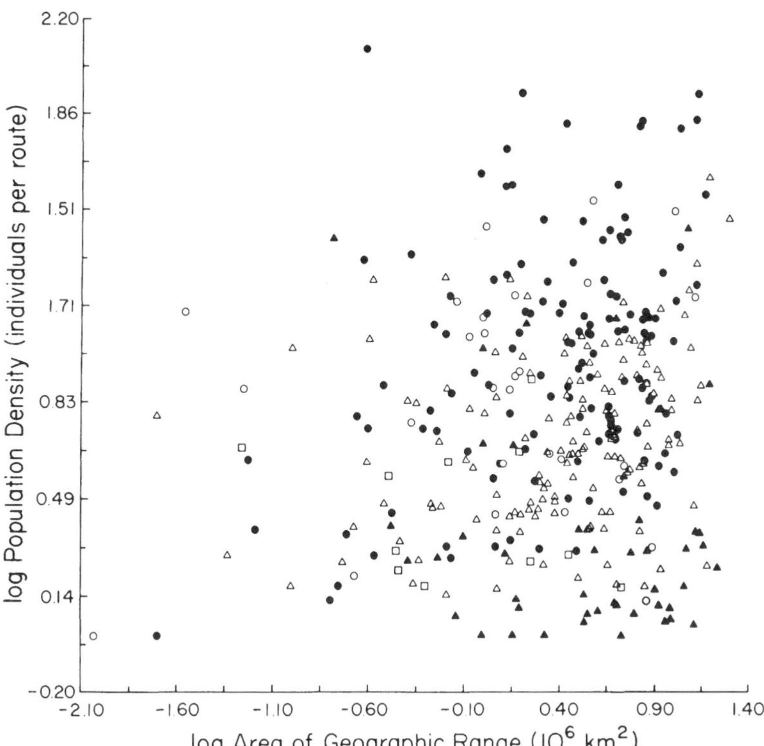

FIG. 4.—Relationship between $\log N$ (average local population density) and $\log A$ (area of geographical range) for the species of North American land birds. For an explanation of the symbols, see figure 2. Note that the points fall approximately within a triangle as predicted, but the upper boundary and the region near the origin are not clearly defined.

density did not decrease with increasing size, as we predicted, but was approximately constant, independent of size. Plots of $\log A$ against $\log M$ and of $\log N$ against $\log A$ (figs. 3 and 4, respectively) conformed quite closely to our predictions, but several of the boundaries defining the polygons enclosing the data points were not as clear-cut as those for the plot of $\log N$ against $\log M$. We also produced a three-dimensional plot of these log-transformed variables, but it is difficult to represent in a two-dimensional figure and is not presented here. It did, however, provide some additional information. The species of intermediate to large body sizes with the lowest population densities did not have small geographical ranges, whereas the species of small body size with the lowest population densities did not have large geographical ranges. The smallest geographical ranges occurred among species of small to intermediate body size and low to medium population densities, whereas the largest ranges (encompassing virtually the entire continent) were represented by species with almost all combinations of density and size except for low values of each.

The different trophic guilds tended to cluster in different parts of the two- and three-dimensional plots, but there was extensive overlap (table 1; figs. 2–4). Herbivores generally had higher densities than carnivores of the same size, whereas omnivore-insectivores had higher densities than insectivores. The maximum span of geographical-range sizes occurred in small insectivorous species (fig. 3). All trophic groups overlapped extensively in geographical range, although for their body sizes carnivores appeared to have the largest ranges and herbivores the smallest (table 1). When $\log N$ was plotted against $\log A$, we observed only a few species that had both low densities and small geographical ranges (fig. 4). Carnivores conspicuously occupied a region of low densities coupled with large ranges, whereas the other four trophic groups were more widely dispersed. Differences in the positions of the four largest trophic groups (carnivores, herbivores, omnivore-insectivores, and insectivores) in the three-dimensional space were quantified by a discriminant analysis (table 1). The results imply that despite the substantial overlap, population density, body size, and geographical-range size form three relatively independent axes that can be used to characterize differences among the four trophic groups. Body size effectively distinguishes herbivores and carnivores from insect-eating species, and density effectively separates species that eat plant materials (herbivores and omnivore-insectivores) from species that eat exclusively animals. Geographical-range size appears to distinguish herbivores from the other three groups.

INTERPRETATION AND HYPOTHESES

Our analyses establish well-defined limits to the variation among species with respect to population density, body size, area of geographical range, and trophic status. We hypothesize that these limits reflect either absolute constraints on the attributes of species or differential extinction rates and/or differential origination rates of species possessing certain characteristics. The effects of these processes should be reflected in the distributions of points in the bivariate plots. On the one hand, absolute constraints should lead to sharp, well-defined boundaries delineating attributes possessed by species from impossible combinations of variables. On the other hand, characteristics of species that affect origination and extinction rates should result in boundaries that are neither abrupt nor absolute. These attributes interact with a spatially and temporally varying environment to cause a degree of uncertainty in the outcome of colonization, speciation, and extinction processes.

Some of the absolute constraints, such as the effect of the limited area of the North American continent on the maximum size of the geographical range (figs. 3, 4), are straightforward. Others, such as the apparent upper and lower limits on population density as a function of body size (fig. 2), which appears to be just as clearly delineated, have less obvious explanations.

A similar linear decline in the log of local population density with the log of increasing body mass in mammals has been explored by Damuth (1981). He hypothesized an exact trade-off between density and body size, such that the rate of decline in density as body size increased allowed populations of different species to use the same amount of energy. Damuth claimed that this relationship

TABLE 1

Discriminant Analysis and Means for Four Trophic Guilds of North American Terrestrial Birds

Trophic Guild	No. of Species	Discriminant Variable 1	2	3	log *N*	log *M*	log *A*
Group means							
Carnivore	53	3.00	−0.61	0.09	0.35	2.66	0.57
Herbivore	29	2.36	0.94	−0.26	0.86	2.54	0.24
Insectivore	154	−0.98	−0.44	−0.06	0.70	1.27	0.43
Omnivore-insectivore	144	−0.49	0.52	0.08	1.00	1.51	0.41
Canonical correlations							
log *N*		−0.21	0.90	0.39			
log *M*		0.98	0.20	0.03			
log *A*		0.03	−0.19	0.98			
Eigenvalues		2.16	0.30	0.01			
% variation explained		87.26	12.32	0.42			

Note.—Separation of the four groups on the first two discriminant variables implies that species eating at least some insects are confined to small body sizes, whereas species that eat substantial amounts of plant material attain higher population densities than carnivores or strict insectivores.

holds for all species within assemblages of mammals, but our analysis indicates that, at least in birds, this is a boundary condition, characteristic only of the highest-density species and only those above some threshold body size. We calculated the regression of log density on log body mass for those species that achieve maximum densities for a given weight and obtained the equation $N = cM^{-0.66}$. This is consistent with Damuth's hypothesis of equal energy use because the daily individual energy requirements of free-living birds and mammals scale as $M^{0.67}$ (Kendeigh et al. 1977; Walsberg 1983).

It is important to emphasize that our interpretation differs from Damuth's in one critical respect. Damuth's view implies that a constant amount of energy is available to each species and utilized by it within a taxonomically and geographically constrained assemblage, but the species allocate their energy to different numbers and sizes of individuals according to the strict trade-off. Our interpretation is that the maximum rate of energy intake that can be attained by species within a taxon is a constant, independent of body size above some threshold body mass. However, the trade-off between population density and body size that is responsible for this constant represents only an upper limit on the rate of energy use. Consequently, between the threshold and maximum body masses, a wide range of population densities can occur.

Figure 2 suggests that below the threshold body size (approximately 50–100 g in North American birds) lies another absolute constraint that results in decreasing maximum population densities of species as body size declines. The equation characterizing this boundary is approximately $N = cM^1$. Although we did not predict this boundary (fig. 1*b*), we suggest that it also implies some fundamental energetic constraint. We hypothesize that below the threshold body size, the increasingly high energy requirements per unit of mass (which scale as approximately $M^{-0.33}$) as body size decreases requires that smaller species use more concentrated energy sources (see, e.g., Brown et al. 1978). The consequence of this limitation is that the density of usable resources declines with decreasing body size; thus, so does population density. Individuals of these small species probably are not dispersed uniformly over space; rather, we expect small birds to be restricted to local patches of habitat that are rich in the resources they require.

The apparent constancy of minimum population densities over the entire range of body sizes (the lower boundary in fig. 2) was not predicted a priori and calls for an explanation. It does not appear that this limit can be explained simply in terms of the effect resulting from the association of low total population size with a high probability of extinction, as we had conjectured. If this were the case, we would expect minimum population density to decrease with increasing body mass because species of large body size tend to have large geographical ranges. The tradeoff between local density and geographical distribution should tend to maintain total population size and allow large birds to become locally rare without becoming extinct. But the fact that species of large birds apparently attain minimum densities that are no lower than those of some small birds suggests that an energetic constraint may again be involved. Since the minimum population density scales as M^0 and the rate of daily energy use per individual scales as $M^{0.67}$, the rate of total energy use by all individuals of these species per unit of area also scales as $M^{0.67}$. Since minimum energy use per species apparently scales with body mass with exactly the same exponent as energy use per individual, this suggests that the minimum population density is a consequence of a limit on the abilities of these species, independent of their body size, to extract usable energy from the environment.

It is important to recognize that the Breeding Bird Survey (BBS) censuses sample the environment at a standardized spatial scale, and therefore minimum population densities of different species may reflect different dispersions of individuals. We suspect that large, rare species are distributed in a relatively fine-grained manner, whereas small, rare species are clumped in patches of favorable habitat. Thus, the energy attenuation hypothesized above should have an effect on dispersion that scales with body size.

These energetic constraints, which appear to limit local population densities so rigidly, are of considerable interest in their own right. They must constitute some of the "allocation rules" that designate how the resources of a region are distributed among the resident species (Brown 1981). The implications of these energetic relationships will be addressed in a later paper.

The only diffuse, poorly defined boundaries that we observed appeared to be related to the minimum sizes of geographical ranges (figs. 3, 4). The minimum area of geographical range as a function of body size (the lower boundary in fig. 3) should reflect high extinction rates. Noting a similar diffuse boundary when the same variables were plotted for North American mammalian species, Brown (1981) attributed this effect of geographical-range size to increasing extinction rates as body size increases and total population size decreases. This interpretation is supported by the fact that many of the bird species (e.g., *Tympanuchus pallidicinctus, Dendroica chrysoparia*, and *Vermivora bachmanii*) along this boundary are currently considered either endangered or threatened with extinction.

At least two diffuse boundaries appear to characterize the relationship between population density and area of geographical range (fig. 4). The first of these characterizes the region of few species with low population densities and small geographical ranges. These species should be subject to high rates of extinction because they have low total populations. Many of these same species (e.g., *D. chrysoparia* and *Crotophaga ani*) appear along the lower boundary in figure 3. Colonizing species should tend to show similar combinations of these variables. In fact, the two points nearest the origin in figure 3 represent *Acridotheres cristatellus* and *Streptopelia risoria*, two introduced species that appear to be becoming established residents. The second diffuse boundary defines a region of increasing maximum density with increasing geographical range. We cannot imagine absolute constraints that would preclude species with small ranges from attaining high densities, but unusual environmental conditions would be required for species to exhibit this combination of attributes. Some species may be specialized to exploit the resources of highly productive but spatially restricted environments. These species are an exception to the generalization that high densities within a local site usually depend on the ability to exploit a wide range of resources and tolerate a wide range of conditions, so that local abundance is positively correlated with geographical distribution (Brown 1984). Perhaps the best example of such an exceptional species is *Agelaius tricolor*, an abundant species largely restricted to the highly productive marshes of the Central Valley of California.

The distribution of trophic guilds with respect to the pairwise combinations of variables is fairly straightforward. Two points warrant brief consideration. First, within guilds, the distribution of species with respect to certain variables is strictly limited. For example, both nectarivores and insectivores exhibit only a narrow range of body sizes, whereas carnivores are restricted to relatively large geographical ranges. These patterns were predicted a priori on the basis of trophic and energetic relationships, and additional explanations seem unnecessary. Second, and more interesting, when there is sufficient variation within a guild to permit resolution, it appears that the same qualitative patterns that characterize the entire avifauna in the bivariate plots (figs. 2–4) also apply to subsets of species that are even more similar in resource use. Thus, the same absolute biological constraints and origination and extinction processes that were hypothesized to determine the limits of variation for the avifauna as a whole can also be invoked to account for the qualitatively similar, but more restricted patterns of variation among species within the individual guilds.

We propose a model, based on the empirical patterns, for adaptive evolution of the North American terrestrial avifauna. The essentials of this model are illustrated in figure 5. For the present, the limits of variation in the species attributes can be characterized by polygons. The solid lines on some sides of these polygons reflect either intrinsic properties of the species that place absolute constraints on their structure and function or extrinsic environmental factors that have general

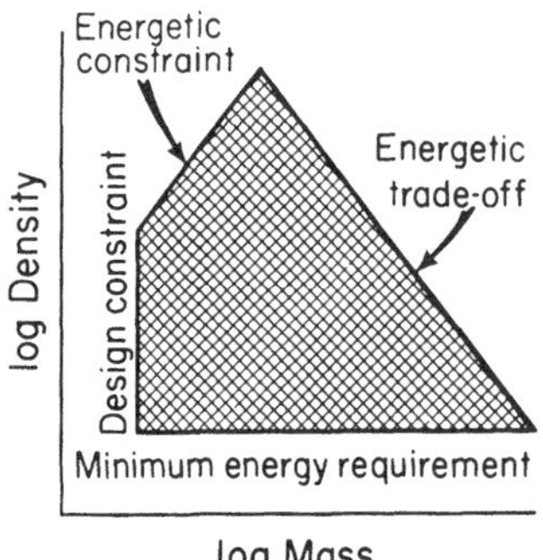

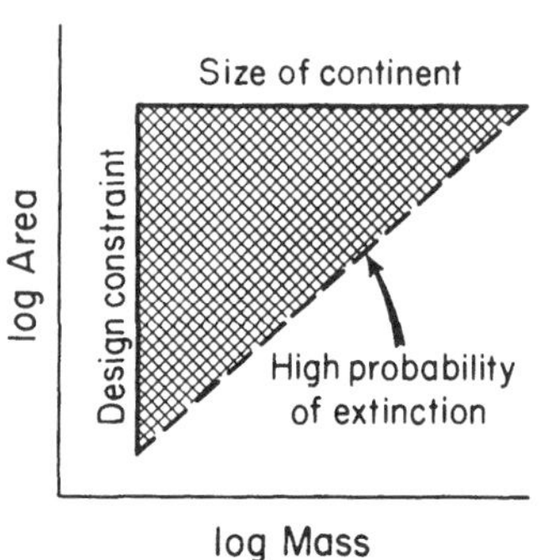

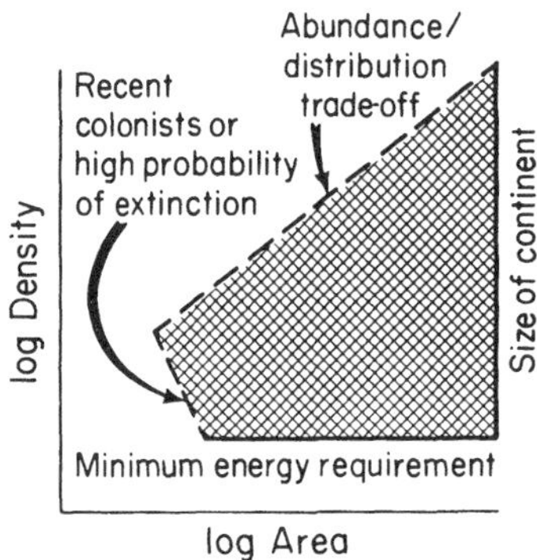

FIG. 5.—Hypothesized constraints and species dynamics to account for the relationships on log-transformed axes among local population density (N), individual body mass (M), and area of geographic range (A) for the land-bird species of North America. The cross-hatched polygons enclose the combinations of variables exhibited by the majority of species in the fauna. *Solid lines*, absolute constraints separating combinations of attributes possessed by species from impossible combinations; *dashed lines*, probabilistic effects of species origination and extinction processes limiting most species to less discretely defined boundaries. Taken together, these relationships provide an initial model for the role of intrinsic biological limits and extrinsic environmental limits in the adaptive evolution of the avifauna.

effects on all species. The dashed lines forming other sides of the polygons represent characteristics that interact with the environment in a more probabilistic way to affect the origination and extinction of species. Ultimately, it should be possible to develop a more sophisticated model that goes beyond this polygonal representation to account for the density with which the species units are distributed within the three-dimensional space defined by population density, body size, and area of geographical range.

DISCUSSION

The model presented above was developed to provide an ad hoc explanation for the static structure of the North American terrestrial avifauna. The model should be regarded as a hypothesis or, better, as a series of hypotheses about the processes that limit the adaptive differentiation of the bird species. The model can be evaluated by gathering additional and independent evidence about the energetic constraints and species dynamics that we have hypothesized. It may be possible to generalize the model from North American land birds to other biotas by making some simple quantitative and perhaps qualitative modifications. Our model can be thought of as characterizing the adaptive exploitation of resources available to a taxon within a geographical region. Since the partitioning of resources among the species of any natural assemblage occurs within limits imposed largely by a combination of absolute energetic constraints and the dynamics of origination and extinction, we suggest that the patterns and processes may be general.

Our analysis supports the suggestions of recent authors (e.g., Allen and Starr 1982; Arnold and Fristrup 1982; Vrba and Eldredge 1984; Eldredge 1985; Maurer 1985; Salthe 1985) that it is productive to view the complex organization of biological systems as a nested hierarchy, with ecological communities and geographically and taxonomically limited biotas as important upper levels. Over geographical spatial scales and evolutionary temporal scales, species assemblages exhibit dynamic behavior that is not readily predictable from the properties of the individual organisms constituting the species populations. The patterns that we have documented appear to be in large part attributable to the dynamics of species origination and extinction, and these processes depend greatly on properties, such as population density and area of geographical range, that cannot even be defined at the individual-organism level of the hierarchy.

Our analysis also supports the recent claims of "macroevolutionists" (e.g., Eldredge and Gould 1972; Stanley 1975, 1979; Gould and Eldredge 1977; Eldredge and Cracraft 1980; Vrba 1980, 1983, 1984; Arnold and Fristrup 1982; Gould 1982; Vrba and Eldredge 1984) that a dynamic selective process, analogous in many ways to the natural selection of individual organisms within populations, operates at the level of species within biotas. The combinations of attributes possessed by species appear to be determined in part by how the species-level units interact with the environment to result in selective speciation, colonization, and extinction. We agree with Damuth (1985) that this concept of species selection is most useful when patterns are analyzed and hypotheses are tested within a limited geographical and environmental context so that the dynamic processes can be identified. We believe that it is also essential to evaluate the importance of taxon-specific evolutionary constraints; this will require accurate reconstruction and appropriate analyses of phylogenetic lineages, a point seemingly ignored by Damuth.

Although most of the current interest in macroevolutionary patterns and processes has been generated by paleontologists, the present study demonstrates the value of neontological data for developing and testing macroevolutionary hypotheses. Just as studies of the patterns of phenotypic variation in relation to environmental gradients continue to provide valuable insights into the adaptive evolution of individuals, so analogous studies of species characteristics in a large-scale environmental context can be expected to shed new light on the adaptive evolution of biotic diversity at the species level and above. Data from the fossil record will be required to provide essential information on kinds and rates of changes that have actually occurred in species and biotas over the long sweep of evolutionary time. However, since species dynamics are a consequence of interactions between the evolving units and their environment, it is equally important to incorporate the ecological perspective that can come only from analyses of contemporary biotas.

SUMMARY

Characteristics of the terrestrial avifauna of North America can be viewed as adaptations by a taxonomically, geographically, and ecologically defined assemblage of many species to the constraints imposed by its own biology and by the environment. We have identified distinctive patterns in the variation among species in population density, body size, area of geographical range, and trophic status.

The patterns observed in bivariate plots of log-transformed variables can be characterized provisionally in terms of polygons that enclose combinations of the variables exhibited by species. The sides of these polygons may be either abrupt or indistinct. We suggest that sharp, clear-cut boundaries separating combinations of characteristics that species possess from those combinations that are not observed in any species are the result of absolute constraints. As a trivial example, the maximum size of the geographical range is determined by the size of the continent. A more interesting example of an apparently absolute constraint is an energetic trade-off between maximum population density and body size. Boundaries separating combinations of characteristics that species possess from those not possessed can also be diffuse and relatively poorly defined. We suggest that such boundaries result from the probabilistic processes of origination and extinction, such that the number of species declines gradually across the boundary. An example is the increase in minimum area of geographical range with increasing body size, which is hypothesized to reflect the probability of extinction. We summarize our hypotheses to account for the observed patterns in a model for the adaptive evolution of the North American terrestrial avifauna. With appropriate modifications, a similar model could be developed for any biota.

Our analyses provide neontological evidence for the kinds of patterns observed in the fossil record and used by paleontologists to argue that a process analogous to natural selection at the level of individuals within populations operates at the level of species within biotas. It is useful to view certain attributes of contemporary species as the result of some combination of absolute constraints and the dynamics of speciation, colonization, and extinction processes.

ACKNOWLEDGMENTS

We are grateful to many people whose assistance was invaluable in completing this project: D. Bystrak made available the North American Breeding Bird Survey; C. S. Robbins provided the range maps; J. B. Dunning, Jr., supplied data on body weights; M. Kurzius assisted with data recording and analysis and read the manuscript; and C. E. Bock, W. A. Calder III, M. L. Rosenzweig, and many others contributed helpful ideas and discussion. Three anonymous reviewers made many helpful comments on a previous version of the manuscript. We thank the University of Arizona for computer time, and the U.S. Department of Energy and the National Science Foundation (grants DEB-8021535 and BSR-8506729) for financial support.

LITERATURE CITED

Allen, T. F. H., and T. B. Starr. 1982. Hierarchy: perspectives for ecological complexity. University of Chicago Press, Chicago.

Arnold, J. H., and K. Fristrup. 1982. The theory of evolution by natural selection: a hierarchical expansion. Paleobiology 8:113–129.

Bock, C. E. 1984. Geographical correlates of abundance vs. rarity in some North American winter landbirds. Auk 101:266–273.

Bock, C. E., and R. E. Ricklefs. 1983. Range size and local abundance of some North American songbirds: a positive correlation. Am. Nat. 122:295–299.

Brian, M. V. 1953. Species frequencies in random samples from animal populations. J. Anim. Ecol. 22:457–464.

Brown, J. H. 1981. Two decades of homage to Santa Rosalia: toward a general theory of diversity. Am. Zool. 21:877–888.

———. 1984. On the relationship between abundance and distribution of species. Am. Nat. 124: 255–279.

Brown, J. H., and B. A. Maurer. 1986. Body size, ecological dominance and Cope's rule. Nature (Lond.) 324:248–350.

Brown, J. H., W. A. Calder III, and A. Kodric-Brown. 1978. Correlates and consequences of body size in nectar-feeding birds. Am. Zool. 18:687–700.

Bystrak, D. 1981. The North American Breeding Bird Survey. Stud. Avian Biol. 6:34–41.

Calder, W. A., III. 1984. Size, function, and life history. Harvard University Press, Cambridge, Mass.

Damuth, J. 1981. Population density and body size in mammals. Nature (Lond.) 290:699–700.

———. 1985. Selection among "species": a formulation in terms of natural functional units. Evolution 39:1132–1146.

Dunning, J. B., Jr. 1984. Body weights of 686 species of North American birds. Western Bird Banding Association Monograph 1. Eldon, Cave Creek, Ariz.

Eldredge, N. 1985. Unfinished synthesis: biological hierarchies and modern evolutionary thought. Oxford University Press, New York.

Eldredge, N., and J. Cracraft. 1980. Phylogenetic patterns and evolutionary processes. Columbia University Press, New York.

Eldredge, N., and S. J. Gould. 1972. Punctuated equilibria: an alternative to phyletic gradualism. Pages 82–115 *in* T. J. M. Schopt, ed. Models in paleobiology. Freeman, San Francisco.

Fisher, R. A., A. S. Corbet, and C. B. Williams. 1943. The relation between the number of species and the number of individuals in a random sample of an animal population. J. Anim. Ecol. 12: 42–58.

Fowler, C. W., and J. A. MacMahon. 1982. Selective extinction and speciation: their influence on the structure and functioning of communities and ecosystems. Am. Nat. 119:480–498.

Gould, S. J. 1982. Darwinism and the expansion of evolutionary theory. Science (Wash., D.C.) 216:380–387.

Gould, S. J., and N. Eldredge. 1977. Punctuated equilibria: the tempo and mode of evolution reconsidered. Paleobiology 3:115–151.

Hanski, I. 1982*a*. Distributional ecology of anthropochorous plants in villages surrounded by forest. Ann. Bot. Fenn. 19:1–15.

———. 1982*b*. Dynamics of regional distribution: the core and satellite species hypothesis. Oikos 38:210–221.

Hutchinson, G. E., and R. H. MacArthur. 1959. A theoretical ecological model of size distributions among species of animals. Am. Nat. 93:117–125.

Kendall, D. G. 1948. On some modes of population growth leading to R. A. Fisher's logarithmic series distribution. Biometrika 35:6–15.

Kendeigh, S. C., V. R. Dolnik, and V. M. Gavrilov. 1977. Avian energetics. Pages 127–204 *in* J. Pinowski and S. C. Kendeigh, eds. Granivorous birds in ecosystems. Cambridge University Press, Cambridge.

Kerner, E. H. 1957. A statistical mechanics of interacting biological species. Bull. Math. Biophys. 19:121–146.

———. 1959. Further considerations on the statistical mechanics of biological associations. Bull. Math. Biophys. 21:217–255.
MacArthur, R. H. 1957. On the relative abundance of bird species. Proc. Natl. Acad. Sci. USA 43:293–295.
MacArthur, R. H., and E. O. Wilson. 1967. The theory of island biogeography. Princeton University Press, Princeton, N.J.
Maurer, B. A. 1985. Avian community dynamics in desert grasslands: observational scale and hierarchical structure. Ecol. Monogr. 55:295–312.
May, R. M. 1978. The dynamics and diversity of insect faunas. Pages 188–204 *in* L. A. Mound and N. Waloff, eds. Diversity of insect faunas. Blackwell, Oxford.
Mohr, C. O. 1940. Comparative populations of game, fur, and other mammals. Am. Midl. Nat. 24: 581–584.
Peters, R. H. 1983. Ecological implications of body size. Cambridge University Press, Cambridge.
Peters, R. H., and J. V. Raelson. 1984. Relations between individual size and mammalian population density. Am. Nat. 124:498–517.
Peters, R. H., and K. Wassenberg. 1983. The effect of body size on animal abundance. Oecologia (Berl.) 60:89–96.
Preston, F. W. 1948. The commonness and rarity of species. Ecology 29:254–283.
———. 1962. The canonical distribution of commonness and rarity. Ecology 43:185–215, 410–432.
Rapoport, E. H. 1982. Areography: geographical strategies of species. Pergamon, Oxford.
Robbins, C. S., B. Bruun, and H. S. Zim. 1983. Birds of North America. Golden Press, New York.
Salthe, S. N. 1985. Evolving hierarchical systems, their structure and representation. Columbia University Press, New York.
Stanley, S. M. 1975. A theory of evolution above the species level. Proc. Natl. Acad. Sci. USA 72:646–650.
———. 1979. Macroevolution, pattern and process. Freeman, San Francisco.
Van Valen, L. 1974. Body size and numbers of plants and animals. Evolution 27:27–35.
Vrba, E. S. 1980. Evolution, species and fossils: how does life evolve? S. Afr. J. Sci. 76:61–64.
———. 1983. Macroevolutionary trends: new perspectives on the roles of adaptation and incidental effect. Science (Wash., D.C.) 221:387–389.
———. 1984. What is species selection? Syst. Zool. 33:318–328.
Vrba, E. S., and N. Eldredge. 1984. Individuals, processes and hierarchy: towards a more complete evolutionary theory. Paleobiology 10:146–171.
Walsberg, G. E. 1983. Avian ecological energetics. Avian Biol. 6:161–220.
Williams, C. B. 1944. Some applications of the logarithmic series and the index of diversity to ecological problems. J. Ecol. 32:1–44.
———. 1953. The relative abundance of different species in a wild animal population. J. Anim. Ecol. 22:14–31.
———. 1964. Patterns in the balance of nature. Academic Press, New York.
Willis, J. C. 1922. Age and area. Cambridge University Press, Cambridge.

PAPER 24

Species-Range Size Distributions: Products of Speciation, Extinction and Transformation (1998)

K. J. Gaston

Commentary

DAVID STORCH

Kevin Gaston's paper on evolutionary changes of geographic range sizes is surprisingly simple. Although it concerns an important quantitative macroecological pattern—that is, the skewed (almost lognormal) distribution of range sizes with prevalence of small-ranged species—it does not provide any quantitative theory of the pattern. Instead, it reviews the possibilities of changes in range size in the course of evolution, from speciation to extinction. Its strength consists in explicit consideration of the fact that a macroecological pattern may not represent the final state of a particular process, but a snapshot of various stages of a process. The species–range size distribution could be skewed toward smaller ranges simply due to the fact that ranges are mostly small at their initial stage (i.e., just after speciation) as well as in their final stage (i.e., before extinction), and only some ranges expand between these two stages. Although this does not provide any clues to the quantitative parameters of the distribution, it implies that any attempt to explain this distribution must concern evolutionary dynamics of species ranges.

The importance of the paper thus lies in the explicit consideration of a species' range as something that has its own evolution. This view raises important questions—namely, those concerning range-size heritability, an issue that is so far unresolved. Do phylogenetically closely related species have ranges of similar size? On one hand, they should—closely related species share many characteristics certainly affecting range size (i.e., body size, dispersal abilities, niche width, etc.). On the other hand, if speciation happens mostly in small peripheral populations of an ancestral species, and range sizes do not change too much after speciation, then strong asymmetries in range sizes between sister species should be expected. Evolutionary conservatism or, conversely, volatility of species range size is thus tightly connected to fundamental evolutionary processes concerning different modes of speciation, species proliferation, and extinction. A related topic comprises the existence of taxon cycles—that is, the question whether there is a predictable course of the evolution of range size associated with a repeatable pattern of subsequent habitat specialization and diversification of species.

This paper, together with Gaston's other papers on species ranges, promoted comprehensive study of range-size evolution, the determinants of species range sizes, and the effects of range sizes for species-richness patterns. Notably, it has been demonstrated that species-richness patterns are determined mainly by widely distributed species. Given that the number of species at a site is determined by the number of ranges that overlay the site, diversity patterns are necessarily intrinsically linked to all the processes affecting range-size evolution.

From *Philosophical Transactions of the Royal Society of London B* 353:219–30. Reprinted with permission from the Royal Society.

Species-range size distributions: products of speciation, extinction and transformation

Kevin J. Gaston

Department of Animal and Plant Sciences, University of Sheffield, Sheffield S10 2TN, UK (k.j.gaston@sheffield.ac.uk)

One basic summary of the spatial pattern of biodiversity across the surface of the Earth is provided by a species-range size distribution, the frequency distribution of the numbers of species exhibiting geographic ranges of different sizes. Although widely considered to be approximately log-normal, increasingly it appears that across a variety of groups of organisms this distribution systematically departs from such a form. Whatever its detailed shape, however, the distribution must arise as a product of three processes, speciation, extinction and transformation (the temporal dynamics of the range sizes of species during their life times). Considering the role potentially played by each of these processes necessitates drawing on information from a diverse array of research fields, and highlights the possible role of geographic range size as a common currency uniting them.

Keywords: speciation; transformation; geographic ranges; extinction; macroecology

1. INTRODUCTION

No species is distributed ubiquitously across the Earth. Nonetheless, global geographic range sizes still vary by at least some 12 orders of magnitude (Brown *et al.* 1996). At one extreme lie those species constrained by ecology or by history to occupy small isolated islands of habitat or very scarce sets of environmental conditions. Human activity increasingly plays a role in this constraint, but undoubtedly very many species would never have been widely distributed even in its absence. At the other extreme lie those species which are distributed across multiple biogeographic regions. Such widespread occurrence appears more frequent among marine than terrestrial species, although particularly in the case of the latter it has doubtless become more common also as a consequence of human activity (facilitating the movement of species across otherwise largely impassable barriers).

The tendency for marine species on average to be more widely distributed than terrestrial (Rapoport 1994) is associated with a lower species richness in marine than terrestrial systems (May 1994; Gaston & Williams 1996). This reflects a more general observation that variation in geographic distribution is related to many other large-scale spatial patterns of biodiversity. Latitudinal, altitudinal and depth gradients, as well as patterns of hotspots, turnover and complementarity, all follow from the fact that species exhibit ranges of differing geographic extent, which are then distributed non-randomly across the landscape and through the media which envelop it.

The interspecific variance in geographic range sizes has stimulated a host of investigations of the constraints on the occurrences of more narrowly distributed species and of the mechanisms which enable the more widely distributed to become so (e.g. for references and reviews, see Woodward 1987; Hengeveld 1990). This contrasts with the relative paucity of interest that has been directed toward broad interspecific patterns in the determinants and consequences of this variation (for reviews, see Kunin & Gaston 1993, 1997; Gaston 1994*a*; Brown 1995). Perhaps the most basic summary of, and pattern in, the variation of geographic range sizes exhibited by a taxonomic assemblage is the species-range size distribution (the frequency distribution of species with different range sizes) and its associated statistics. In stark contrast to the closely related species-abundance distribution there have been only a comparative handful of studies concerned with species-range size distributions (Gaston 1996*a*).

In this paper I consider the form of, and some of the mechanisms underlying, species-range size distributions. In so doing, I will essentially be exploring one viewpoint on the generation of spatial and temporal patterns in biodiversity. Throughout, discussion will be centred on the entire geographic range sizes of species. The frequency distributions of the occurrences of species over smaller areas have been the subject of substantial interest (e.g. Hanski 1982; Gotelli & Simberloff 1987; Williams 1988; Gaston & Lawton 1989; Collins & Glenn 1990; Maurer 1990; Gotelli 1991; Tokeshi 1992; Gaston 1994*a*). They may, however, take a rather different form from distributions based on entire geographic range sizes, with doubtless an interesting interplay between the two, and will not be addressed herein.

2. THE FORM OF THE DISTRIBUTION

The frequency distribution of the geographic range sizes of species in a taxonomic assemblage tends to be unimodal with a strong right-skew. That is, most species have relatively small range sizes, and a few have relatively large ones (figure 1; e.g. Willis 1922; Freitag 1969; Anderson 1977, 1984*a*,*b*, 1985; Rapoport 1982; McAllister *et al.* 1986; Schoener 1987; Pomeroy & Ssekabiira 1990; Pagel *et al.* 1991; Gaston 1994*a*, 1996*a*; Brown 1995; Roy *et al.* 1995;

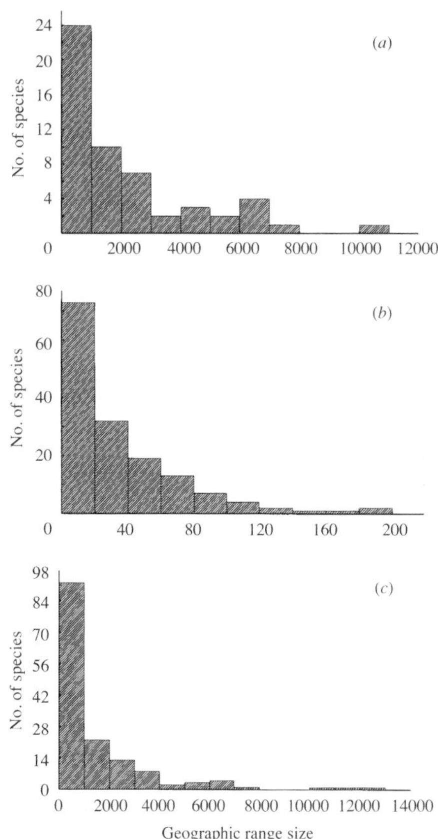

Figure 1. Species-range size distributions for: (*a*) *Harpalus* beetles in North America (maximum extent, km); (*b*) wildfowl worldwide (number of 611 000 km^2 cells occupied); and (*c*) primates worldwide (1000s km^2). From unpublished analyses and data in Noonan (1990) and Wolfheim (1983).

Blackburn & Gaston 1996; Brown *et al.* 1996; Hughes *et al.* 1996). Indeed, in virtually all published species-range size distributions in which range size is untransformed, the most left-hand range size class is also the modal class. These distributions tend toward an approximately normal distribution when geographic range sizes are subject to a logarithmic transformation (e.g. Anderson 1984*a*,*b*; McAllister *et al.* 1986; Pagel *et al.* 1991; Gaston 1994*a*, 1996*a*; Blackburn & Gaston 1996; Brown *et al.* 1996; Hughes *et al.* 1996). This implies that there are not simply disproportionately fewer species with large range sizes, but also with very small ones.

How adequate a description of species-range size distributions a log-normal distribution actually provides remains at present an open question. First, few studies have actually formally tested for significant departure from this model. Second, there have been growing hints that under a logarithmic transformation, species-range size distributions for at least some assemblages are, in some cases quite markedly, left-skewed (hereafter referred to as a left log-skew; Ruggiero 1994; Blackburn & Gaston 1996; Brown *et al.* 1996; Gaston & Blackburn 1997*b*). A perusal of published distributions suggests that this phenomenon may be widespread, a view which is supported by formal analyses of several data sets, although the coefficient of skewness is not always statistically significant (table 1).

If, as seems likely, a left log-skew is a general feature of species-range size distributions, this pattern could potentially simply be an artefact of the ways in which such distributions are assembled, with underlying distributions being log-normal. First, it could perhaps result from the rather crude fashion in which geographic range sizes are typically measured. The range sizes of species tend to be quantified either in terms of the approximate area contained within the geographic limits to their occurrence (an extent of occurrence measure) or in terms of an estimate of the numbers of cells in which a species has been recorded of a grid layed over a map of the region in which it occurs (an area of occupancy measure; both terms, *sensu* Gaston 1991, 1994*b*). For any given species, the former measure will tend to be larger than the latter, as it incorporates more areas in which individuals do not actually occur. However, in practice most assessments of either quantity are sufficiently crude that the true areal occurrence of a species is markedly overestimated. The proportional overestimation is unlikely to be similar for species of all range sizes, possibly distorting species-range size distributions markedly (Gaston *et al.* 1996). The variety of range size measures used by the studies in table 1 suggest that this is not a cause of the left log-skew, unless such distortions are shared by all such measures.

Second, many published species-range size distributions concern continental faunas. They may thus be truncated in one of two ways. Either those species in a taxonomic group whose ranges extend beyond the bounds of the continent are excluded from consideration, or their range sizes are measured only on the continent. Again such constraints may markedly distort species-range size distributions. However, such effects plainly cannot explain all left log-skewed distributions because these are also exhibited by global assemblages (table 1), and by assemblages of species endemic to continents (figure 2).

Third, the left log-skew may perhaps be a consequence of the impact of human activities on the occurrences of species, either in recent or prehistoric times, particularly in terms of increased levels of local extinction. This is difficult to discount for any assemblage. However, if this alone is the cause of the skew it would require that humans acted on the ranges of species in a rather different fashion from other agents of limitation (perhaps through the strong tendency to fragment ranges).

Assuming, as seems likely, that it is real and not an artefact, then the left log-skew pattern of species-range size distributions is reminiscent of species-abundance distributions at geographic scales, which have also been found to be left-skewed under logarithmic transformation, at least where the very rarest species have been censused adequately (Nee *et al.* 1991; Gaston 1994*a*; Gregory 1994;

Table 1. *The coefficients of skewness (g_1) associated with the species-range size distributions of various assemblages*

(In all cases, geographic range sizes are logarithmically (base 10) transformed. or, overall range; br, breeding range. Measure is that used to quantify range size: m, maximum linear extent; b, area within boundary; c, number of grid cells occupied; n.s., not significantly different from zero; $^{*}p<0.05$, $^{**}p<0.01$, and $^{***}p<0.001$.)

		measure	*n*	skewness	source
Harpalus carabids	N. America	b	54	-1.202^{***}	data in Noonan (1990)
Nebria carabids	N. America	m	55	$-0.350^{n.s.}$	data in Kavanaugh (1985)
bumblebees	global	c	241	$-0.034^{n.s.}$	P. H. Williams (unpublished data)
procellariiforms	global	c	108	-0.857^{***}	Gaston & Chown (unpublished data)
wildfowl	global (or)	c	154	$-0.36^{n.s.}$	Gaston & Blackburn (1996)
wildfowl	global (br)	c	154	$-0.31^{n.s.}$	Gaston & Blackburn (1996)
rails	global	c	132	$-0.750^{n.s.}$	data in del Hoyo *et al.* (1996)
woodpeckers	global	c	214	$-0.07^{n.s.}$	Blackburn *et al.* (1998)
birds	New World	c	3901	-0.585^{***}	Gaston & Blackburn (1997*b*)
ciconiiformes	New World	c	387	-0.923^{***}	Gaston & Blackburn (1997*b*)
sub-oscine passerines	New World	c	1107	-0.590^{***}	Gaston & Blackburn (1997*b*)
oscine passerines	New World	c	1158	-0.544^{***}	Gaston & Blackburn (1997*b*)
birds	Australia	c	573	-0.575^{***}	data in Blakers *et al.* (1984)
endemic birds	Australia	c	320	-0.531^{***}	data in Blakers *et al.* (1984)
primates†	global	b	148	-0.493^{*}	data in Wolfheim (1983)
bats	S. America	b	187	-0.825^{***}	Ruggiero (1994)
marsupials	S. America	b	77	$0.235^{n.s.}$	Ruggiero (1994)
edentates	S. America	b	28	$-0.472^{n.s.}$	Ruggiero (1994)
primates	S. America	b	57	$-0.385^{n.s.}$	Ruggiero (1994)
carnivores	S. America	b	43	-1.255^{**}	Ruggiero (1994)
artiodactyls	S. America	b	18	$0.177^{n.s.}$	Ruggiero (1994)
hystriocognath rodents	S. America	b	122	0.499^{*}	Ruggiero (1994)

†Data missing for a few very restricted species.

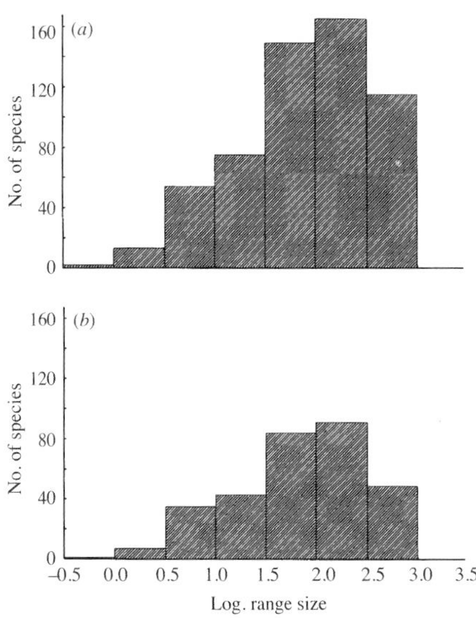

Figure 2. Species-range size distributions for birds in Australia: (*a*) all species; and (*b*) endemic species only. Seabirds and introductions are excluded in both cases. Data from Blakers *et al.* (1984).

Gaston *et al.* 1998). Given that species-body size distributions for geographic scale assemblages tend to be right-skewed under logarithmic transformation (Brown & Nicoletto 1991; Maurer *et al.* 1992; Brown *et al.* 1993; Blackburn & Gaston 1994), this would imply that none of the distributions of the three primary macroecological variables (abundance, range size, body size) for assemblages at large spatial scales are distributed strictly according to a log-normal.

In the case of species-abundance distributions, the left log-skew has been explained in terms of niche apportionment models, in which the abundances of species are considered to be associated with different processes of niche division (Nee *et al.* 1991; Tokeshi 1996). Models in which successive niche division tends to be higher for species with larger niches or higher abundances appear to fit particularly well, and Tokeshi (1996) has associated this with the greater likelihood that both higher abundances and larger geographic range sizes are more likely to generate new species (a point to which I shall return later). In so doing he raises a potential connection between the processes determining species-abundance and species-range size distributions, particularly because there is generally a positive (albeit not especially strong) interspecific relationship between abundance and range size (Hanski 1982; Brown 1984; Gaston & Lawton 1990; Hanski *et al.* 1993; Gaston 1994*a*, 1996*b*; Gotelli & Graves 1996; Gaston *et al.* 1997). However, if applicable at large spatial scales, niche apportionment models can only be acting as caricatures of the processes moulding species-abundance distributions. At such scales ultimately these distributions must be products of speciation, extinction and the temporal dynamics of the abundances of individual species. Similar processes must also ultimately determine the form of the frequency distributions of the geographic range sizes of species.

3. MECHANISMS

(a) *Speciation*

Speciation generates new species and hence additional geographic ranges, adding to a species-range size distribution. Its influence on such a distribution is, however, more complex, and depends fundamentally on the geographic mode of speciation. If we assume that speciation is predominantly allopatric, whilst acknowledging that other forms are probably more frequent or important than has often been claimed in the past (Tauber & Tauber 1989; Ripley & Beehler 1990; Schliewen *et al.* 1994; Rosenzweig 1995), then there are two essential issues to consider. First, we need to know whether species with geographic ranges of particular sizes are more likely to speciate than are those with ranges of other sizes. Second, we need to know the frequency of different asymmetries in the division of ancestral ranges as a product of speciation.

(i) *Probability of speciation*

The idea that species with larger geographic range sizes have a greater probability of speciation dates at least to Darwin (1859), and continues to attract support (e.g. Marzluff & Dial 1991; Wagner & Erwin 1995; Tokeshi 1996). Rosenzweig (1975, 1978, 1995) argues that on a purely probabilistic basis, species with larger geographic range sizes are more likely to undergo speciation, because the likelihood of their ranges being bisected by a barrier is greater than for a small range size. Differentiating, as Rosenzweig does, between two kinds of barriers, 'knives' (which have beginnings and ends) and 'moats' (which surround their isolates), then strictly this assertion is only true of moats. Very large ranges will tend to engulf knives, such that they do not engender speciation, and the probability of division will have a peak at intermediate range sizes. Rosenzweig (1995) argues that this is unlikely to occur because there are no, or virtually no, species with ranges so large that reducing them would make them an easier target for barriers. However, this view depends critically on the frequency distribution of barrier size. If most barriers are small-to-intermediate in size, relative to the range sizes of widespread species, then intermediate-sized ranges may indeed have a higher probability of speciation. Such an effect would be enhanced because barriers seem far more likely to take the form of knives than of moats.

In support of the view that species with small-to-intermediate range sizes are more likely to speciate than are those with larger range sizes, it has been argued that widespread and abundant taxa may often possess well-developed dispersal abilities (perhaps associated with them becoming widespread) and should as a consequence have a strong proclivity to maintain gene flow among populations, which will tend to inhibit speciation (e.g. Mayr 1963, 1988; Stanley 1979). Narrowly distributed and locally rare taxa with poor dispersal abilities (and patchy populations which may tend to form isolates) will tend to have higher speciation rates. The extent to which widely distributed species do indeed tend to have greater dispersal abilities remains debatable (e.g. Levinton 1988; Palumbi 1994; Gaston & Kunin 1997*a*).

In a similar vein, Chown (1997) proposes that rare, but not the rarest, species have the highest probability of speciation. Based on Stanley's (1986) 'fission effect' model, Chown envisages that speciation is a peaked function of geographic range size, which rises rapidly at small range sizes and then progressively subsides towards large range sizes. This is a substantial modification of the original fission effect model, which was intended to capture how the relative rates of speciation and extinction varied with the mean population size across all the species in an assemblage. However, equating geographic range size with population size, the underlying relationship between speciation rate and range size is not dissimilar to the full relationship modelled by Rosenzweig (1978, 1995) when the truncation effects he postulates are ignored.

A complication to assertions that small-to-intermediate range sizes are more likely to undergo speciation arises from the reduced probability of extinction associated with larger range sizes (see §3(*c*)). If species with larger range sizes are likely to persist for longer, this may enhance their probability of speciation. Even if species with smaller range sizes have a greater likelihood of speciation per unit time, species with larger ranges could potentially still be more likely to leave descendants.

I am aware of only one explicit attempt empirically to test the relationship between geographic range size and likelihood of speciation. Wagner & Erwin (1995) find that in analyses of two Neogene clades of Foraminifera and an Ordovician family of gastropods, species with larger geographic ranges are likely to leave more descendants in two cases but not in the third, and in all three cases species that have persisted for longer are likely to leave more descendants (the patterns are not consistent between cladogenetic and anagenetic modes of speciation). In the gastropod case (for which data are provided) the partial correlation between number of descendants (dependent variable) and range size, controlling for differences in longevity, is not significant ($r = 0.005$, $n = 45$, n.s.).

Regardless of whether speciation is more probable for species with small, intermediate or large range sizes, speciation will result (initially, but not necessarily eventually, see below) in the disproportionate addition to a species-range size distribution of range sizes towards smaller sizes than that of the ancestral species.

(ii) *Asymmetry of range division*

If it is ancestral species with relatively small geographic range sizes which are most likely to speciate, then the importance of the asymmetry of range division for an understanding of the form of species-range size distributions is markedly lessened. The products of any range division can only be two small ranges (be they two daughters, or a daughter and its ancestor if the latter persists). If species with large geographic range sizes are more likely to speciate, depending on the asymmetry of division, the outcome may span one large and one small range size through to two ranges each half the size of that of the ancestor at speciation. If we continue to regard patterns of speciation in terms of simple random events then, even with species with small range sizes being the more likely to speciate, the most likely immediate products of speciation are two species one of which is more widely distributed than is the other. A perfect 50:50 split is highly improbable.

The question of the degree of asymmetry in the range sizes of sister groups, immediately post-speciation, is closely related to the issue of whether allopatric speciation is best typified by a peripheral isolation model or by a vicariance model. In the former, peripheral isolates form by waif dispersal (establishment of a new population through long-distance movement across a barrier), micro-vicariance (physical division of a previously continuous distribution) or range retraction (causing peripheral populations to become isolated; Frey 1993). Here, the relatively widespread ancestral species is likely to change little while the peripheral isolate diverges (Glazier 1987). In the vicariance model, a subdivision of the range of an ancestral species occurs, such that there is cessation of contact between the two subpopulations, giving rise to new species. Here, both daughters of the ancestral range are likely to diverge, and the ancestral species will cease to exist. The two models are plainly very closely related, and may in some sense be seen as constituting points on a continuum of speciation processes. However, vicariant speciation may potentially result in any degree of asymmetry in the initial range sizes of daughter species, and is often portrayed as generating very similar-sized ranges. In contrast, peripheral isolation results, immediately post-speciation, in a highly asymmetrical split.

Both peripheral isolation (e.g. Kavanaugh 1979; Ripley & Beehler 1990; Levin 1993; Chesser & Zink 1994) and vicariance models (e.g. Cracraft 1982, 1986; Cracraft & Prum 1988; Lynch 1989) have significant support. The relative frequency of the two modes of speciation remains a point of some contention (e.g. see Bush 1975; Barton & Charlesworth 1984; Mayr 1988; Lynch 1989; Brooks & McLennan 1993; Frey 1993; Ripley & Beehler 1993; Chesser & Zink 1994; Taylor & Gotelli 1994; Wagner & Erwin 1995). Resolution of the issue rests in major part on the extent of post-speciational change in geographic range sizes (see §3(*b*)). This will to a marked degree determine the extent to which the present-day distributions of species can be used to reconstruct past patterns of speciation (Lynch 1989; Brooks & McLennan 1991). If dispersal is important, then it may entirely obscure the relative positions of, say, sister taxa at the time of their divergence. If it is not important then this will not be a problem.

Simplifying matters greatly, the probable patterns of gain and loss of small (S), intermediate (I) and large (L) range sizes can be determined under peripheral isolation and vicariance modes of speciation, if large ranges or small/intermediate ranges have the greater probability of speciation (table 2). The patterns vary perhaps most markedly in the loss of large ranges when speciation is by vicariance and large ranges have a greater likelihood of speciation.

(b) *Transformation*

The influence of speciation on the shape of species-range size distributions will rest in large part on the form the temporal dynamics of the sizes of the ranges of species subsequently take over their lifetimes (here termed 'transformations', to avoid possible confusion with other elements of the temporal dynamics of ranges). Plainly if ranges tended to remain very similar in size to those they initially attained, then speciation would be of far more importance than if they dramatically increased (or decreased) in range size shortly after (or at least on a far faster time-scale than that on which speciation operates). Speciation and transformation are not entirely independent; if an ancestral species persists after a speciation event (e.g. by peripheral isolation), the range size of that species will be reduced by that event. However, for the purposes of this discussion this complication will largely be ignored. Plainly its significance will rest on the predominant mode of speciation.

Table 2. *Probable patterns of gain and loss of small (S), intermediate (I) and large (L) range sizes under peripheral isolation and vicariance modes of speciation, if large ranges or small/intermediate ranges have the greater probability of speciation (and ignoring subsequent range size transformation)*

greater probability of speciation	peripheral isolation gain	peripheral isolation loss	vicariance gain	vicariance loss
large ranges	S	—	S/I	L
small/intermediate ranges	S	I?	S	S/I

Several models of transformation have, explicitly or implicitly, been discussed (see, also, Gaston & Kunin 1997*b*; Gaston & Blackburn 1997*a*). I will consider five, although these are not, as we shall see, necessarily of equivalent status.

(i) *Models*

Stasis I: the simplest model of range size change is one of stasis, in which the range size of a species changes little post-speciation, either until the species disappears as a result of cladogenesis or until a rapid decline to extinction (figure 3*a*). This model is plainly unrealistic. If it operated then geographic range sizes would be expected to have declined steadily, and rather rapidly, through evolutionary time, as speciation progressively subdivided them. This is not to say that mean range size may not indeed have declined through time, it seems inevitable that for terrestrial organisms declines will have accompanied continental breakup, and that, more generally, declines may have accompanied increases in species richness.

As previously noted, our ability to reconstruct past speciation events will often depend in large part on the extent of deviation from the stasis I model. Departure may, however, under some circumstances be substantial without necessarily greatly influencing reconstructions. For example, the geographic ranges of daughter species either side of a major barrier (e.g. continental separation) may change dramatically after a vicariant speciation event without necessarily obscuring that event, provided that in spreading neither daughter species crosses the barrier.

Stasis II: an alternative model of stasis is one in which post-speciation the range size of a daughter species increases rapidly, remains approximately at this level for the bulk of its existence, and then either declines rapidly to extinction or simply ceases to exist through vicariant speciation (figure 3*b*). Such a model has been argued to

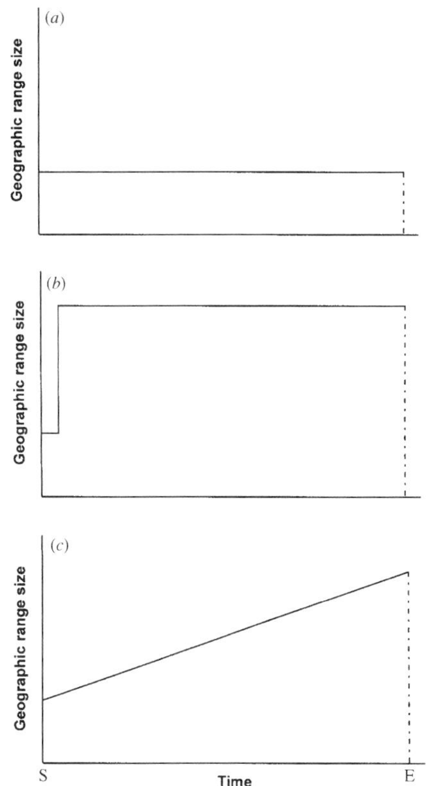

Figure 3. Three models of the temporal dynamics of the geographic range sizes of species (see text for details). (*a*) Stasis I, (*b*) stasis II, and (*c*) age and area. Solid lines represent the trajectory of range size from speciation (S) to extinction (E), with the latter either occurring through vicariant speciation or a decline in range size to zero (dashed line).

reflect well the dynamics of the geographic range sizes of species of Late Cretaceous molluscs, because the distribution of range sizes of those species that originated in the two million years preceding the end-Cretaceous extinction (whose geologic durations were thus truncated) is statistically indistinguishable from that of the species originating in the previous 14 million years (Jablonski 1987). Additional tests of this model are lacking and would be highly desirable.

This stasis II model would require that species be able to spread very rapidly postspeciation. An explicit test is difficult, but some indication of potential rates of spread may be given by the results of introducing species to areas in which historically they have not occurred (e.g. continents which lie outside their native distributions) or by human alteration of conditions enabling them to do so. In such cases rates of spread can indeed be high, with large areas being invaded in matters of decades (e.g. Wing 1943; Lynch 1989; Hengeveld 1989; Grosholz 1996; Veit & Lewis 1996; Williamson 1996).

Perceptions of the extent to which the geographic range sizes of species may remain approximately constant for long periods may have been unduly influenced by studies of the ranges of north temperate species post glaciation, which tend to exhibit great dynamism (although even then often punctuated by perhaps long periods of stasis; e.g. Woods & Davis 1989; Dennis 1993; Burton 1995; Coope 1995). These may be far from typical of all species, and it is frequently not clear to what extent such dynamism is reflected in changes in range size rather than simply range shifts.

Age and area: Willis (1922) argued that the geographic ranges of species followed a trajectory of steadily increasing size the longer they persisted, presumably culminating with a rapid decline to extinction or disappearance through cladogenesis (figure 3*c*). That is, there is a positive intraspecific relationship between evolutionary age and range size. The age and area hypothesis underwent vigorous debate, and was widely rejected (e.g. Gleason 1924; Stebbins & Major 1965; Stebbins 1978). In its simplest form it cannot be correct because there are ample examples of young species that are widely distributed and old species that are very restricted in their distribution. Nonetheless, it continues to be maintained in some circles that there is a broad positive interspecific correlation between age and area (although such a pattern need not follow from an intraspecific age and area relationship), albeit not necessarily of strong predictive value (McLaughlin 1992), and some recent analyses have reported such a pattern, at least among small numbers of closely related species (Taylor & Gotelli 1994).

Cyclic: the dynamics of the geographic range sizes of species have been regarded by some as essentially cyclic (e.g. Dillon 1966). This is most explicitly expressed in the concept of the taxon cycle (initially proposed for insular faunas but later generalized to mainland ones; Wilson 1961; Ricklefs & Cox 1972, 1978). Here, a newly evolved species expands its range (stage I), to become widespread (stage II), this range then fragments due to local extinctions (stage III), and the species becomes restricted to a small area (stage IV) (Ricklefs & Cox 1972, 1978).

Depending on the relative duration of these stages, this model could be closely caricatured by some that have already been mentioned. For example, if stage II were to be dominant, the taxon cycle would equate to the stasis II model, whereas if stage I were to be dominant the cycle would equate to an age and area model.

Evidence has been produced both in support of (Ricklefs & Cox 1972; Glazier 1980; Rummel & Roughgarden 1985; Roughgarden & Pacala 1989) and against (Pregill & Olson 1981; Liebherr & Hajek 1990; Losos 1992) the existence of taxon cycles for particular assemblages; the observed pattern is likely to be determined by the geographic pattern of isolation. Likewise, other essentially cyclic models have variously been postulated, supported and rejected (e.g. Erwin 1985; Liebherr & Hajek 1990). It certainly seems unlikely that any single cyclic model is of very general applicability.

Idiosyncratic: there need, of course, be no general pattern of change in the geographic range sizes of species

Table 3. *Spearman rank correlations between the range sizes of sister species (sharing a terminal bifurcation), assuming that they are of equal age, and based on arbitrary sequencing of species in each pair*

	No. pairs	r_s
Harpalus carabids	10	0.515^{NS}
leopard frogs	7	0.643^{NS}
albatrosses	6	0.170^{NS}
passerine birds	18	0.220^{NS}
birds	18	0.647^{**}
dabbling ducks	16	0.275^{NS}

Data sources: Madge & Burn (1988), Lynch (1989), Noonan (1990), Livezey (1991), del Hoyo *et al.* (1992), Chesser & Zink (1994), Nunn *et al.* (1996).

between speciation and extinction. Different species may exhibit entirely idiosyncratic trajectories. This would seem in keeping with the continual adaptation and change that may result from responding to the demands of the Red Queen (Van Valen 1973; Ricklefs & Latham 1992), and with the changes in the distributions of some species over the past few decades (e.g. Frey 1992; Burton 1995; Parmesan 1996). If such a pattern prevails then one might expect to see little similarity in the geographic range sizes of closely related species, unless the traits which influence range size are strongly phylogenetically conserved, in which case in climatically and ecologically similar regions the distributions of close relatives might be expected to fluctuate in parallel (Ricklefs & Latham 1992).

There are two lines of evidence regarding the similarity of geographic range sizes of closely related species. First, a significant positive correlation has been documented between the geographic range sizes of closely related species of Late Cretaceous molluscs, from which Jablonski (1987) inferred that range size is heritable at the species level (see, also, Ricklefs 1989; Ricklefs & Latham 1992). However, examination of a few contemporary data sets based on sister species does not uphold this as a generality, albeit all have comparatively small sample sizes (table 3); although note that all the correlations are positive. These results could potentially be reconciled if the geographic range sizes of species are highly labile over their lifetimes such that at any one time closely related species do not have very similar range sizes, but that when this variation is effectively summed over periods of evolutionary or geological time (as inevitably occurs in the fossil record) then strong phylogenetic patterns of interspecific variation in range sizes become apparent. Equally, the results for Late Cretaceous molluscs simply may not generalize.

Second, it also appears that most of the variation in the geographic range sizes of species is explained at low taxonomic levels (Arita 1993; Brown 1995; Gaston & Blackburn 1997*b*; for similar results for range sizes at mesoscales, see Hodgson 1993; Peat & Fitter 1994; Kelly & Woodward 1996). Indeed, for several data sets the majority of variation is explained at the level of species within genera (figure 4). This contrasts with many life history variables, where little variation is explained at this level (e.g. Read & Harvey 1989; Harvey & Pagel 1991). However, this is not to say that higher taxonomic levels are not important, as they may still account for reasonable amounts of variation (figure 4; Cotgreave & Pagel 1997).

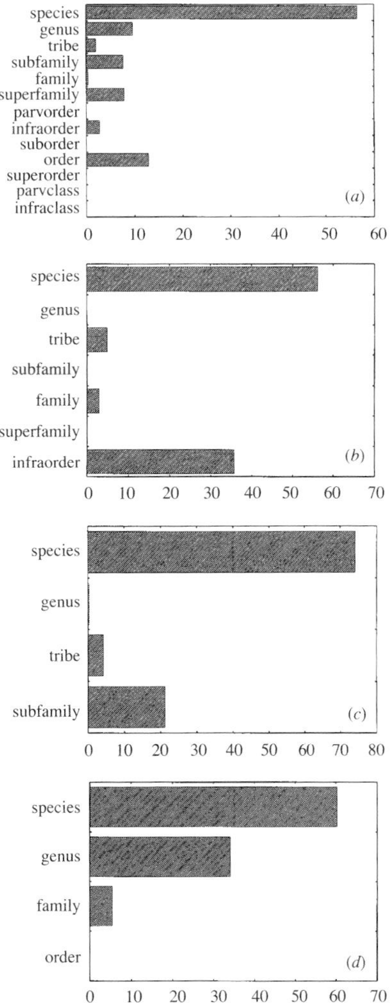

Figure 4. The proportion of variation in the geographic range sizes (logarithmically transformed) of species that is explained at different taxonomic levels for: (*a*) New World birds; (*b*) wildfowl; (*c*) woodpeckers; and (*d*) regular sea urchins. From: Gaston & Blackburn (1997*b*), Blackburn *et al.* (1998), and data in Emlet (1995) and Kier & Lawson (1978).

Overall, the weight of evidence suggests that range sizes are not strongly phylogenetically conserved, possibly favouring an idiosyncratic model of range size dynamics. However, such a pattern could equally fit several other models of such dynamics. For example, if closely related species each followed a stasis II-type model, but the maximum range sizes achieved were rather different.

In considering the potential validity of an idiosyncratic model it should be borne in mind that, if for an individual species this tends to result in the maintenance of an average range size of similar size to that immediately after its speciation, then as with the stasis I model the average range size across species may be expected to decline markedly with evolutionary time.

(ii) *Synthesis*

In summation, there is limited evidence in support of several possible models of the long-term dynamics of range sizes, but we do not appear at the present time to be in a position to conclude which are the most important. Such a conclusion seems at odds with, for example, the claims of vicariance biogeography, in which dispersal is rejected as a first-order explanation of the distribution of a group, in favour of one based on vicariance (Wiley 1988).

(c) ***Extinction***

The form of a species-range size distribution is not only potentially influenced by the addition of new ranges through speciation and the transformation of range sizes through temporal dynamics, but also by the loss of ranges through extinction, and hence by the duration for which species with different geographic range sizes persist. As with transformation, the role of extinction in shaping species-range size distributions is intimately associated with the predominant pattern of speciation. If speciation occurs primarily by vicariance, then speciation events may be accompanied by extinction (pseudo-extinction, *sensu* Wagner & Erwin 1995) of the ancestral species (Brooks & McLennan 1991). This would mean that, ignoring other extinction events, the relationship between probability of extinction and range size would be the same as that between the probability of speciation and range size.

There is empirical evidence for a positive relationship between time to extinction and range size for various paleontological species assemblages (Jackson 1974; Hansen 1978, 1980; Stanley 1979; Koch 1980; Jablonski 1986*a*,*b*; Buzas & Culver 1991; Jablonski & Raup 1995). Species with larger range sizes tend to persist for longer (an observation which dates at least to Lamarck; see McKinney 1997). The correlation appears very general, although it is not always especially strong, and may break down during periods of mass extinction (Jablonski 1986*b*; Norris 1991), which should perhaps raise concerns among conservationists given the present highly elevated levels of extinction (May *et al.* 1995). The relationship may also have an artefactual component, because species with larger range sizes have a greater probability of preservation in the fossil record and may thus appear to persist for longer (Russell & Lindberg 1988*a*,*b*; but, see Jablonski 1988).

Assuming, not unreasonably, that it is not simply an artefact, such a pattern may exist for three possible reasons, although often only the first is explicitly stated: (i) species with larger geographic range sizes may be less likely to walk randomly to extinction, and thus they persist for longer; (ii) species with traits which make them less prone to local extinction, and hence able to persist for longer, may also be enabled to maintain larger range sizes because of this extinction resistance; and (iii) species with larger range sizes may have them because they have persisted for longer (there is an age and area relationship; see above).

The first argument, that species with large range sizes are *per se* less likely to become extinct, seems inescapable, if for no other reason than that in a changing environment a widespread species is more likely to be able to continue to persist somewhere than is a narrowly distributed species. There have, however, been few empirical attempts to ascertain whether range size exerts an effect on extinction risk which is independent of local density, perhaps the most important correlate of range size. A positive correlation between geographic range size and persistence may potentially result because widespread species tend to be locally more abundant (i.e. at higher density; see earlier references), and locally more abundant species tend to be less likely to become extinct (e.g. Terborgh & Winter 1980; Pimm *et al.* 1988; Tracy & George 1992). What limited evidence there is suggests, however, that range size may indeed have an independent effect (see Rosenzweig 1995; Gaston & Blackburn 1996; Mace & Kershaw 1997).

As mentioned earlier, although there is some limited evidence for an age and area relationship it is not a general one, and seems unlikely to contribute strongly to a general relationship between geographic range size and persistence.

Whatever determines the relationship between range size and likelihood of extinction, the influence of extinction on species-range size distributions will predominantly be to remove species from the left-hand side of the distribution. If speciation tends to generate large numbers of species with very small range sizes, and extinction probability is greater for species with small ranges, this begs the question of how these incipient species manage to persist. The obvious answer is that many of them probably do not (Gorodkov (1992) suggests most). Indeed, it would seem likely that there is strong selection among newly evolved species (Glazier 1987). The existence of some very general biological differences between narrowly distributed and widely distributed species (Kunin & Gaston 1997) may, in part, be generated by this process, with only those species possessing traits which reduce their vulnerability to extinction persisting at small range sizes.

4. CONCLUSIONS

Charles Darwin (1975) described the study of geographic ranges as 'a grand game of chess with the world for a board'. The game is not a simple one. The pieces—the species—occupy different numbers of squares on the board at different times, they appear and then disappear, and many pieces may occupy the same square at the same time. Moreover, we can only gain glimpses of the past moves. Nonetheless, this is also no idle game. The pattern of moves has resulted in the patterns of biodiversity that we observe today.

The species-range size distribution provides a useful framework for considering how Darwin's 'game of chess' has generated the patterns of biodiversity and for organizing and cataloguing our thoughts about them. Determining how this distribution results requires the answers to questions that have been posed in a variety of fields of study, and whose connections have in the past not often been readily apparent. Indeed, to determine

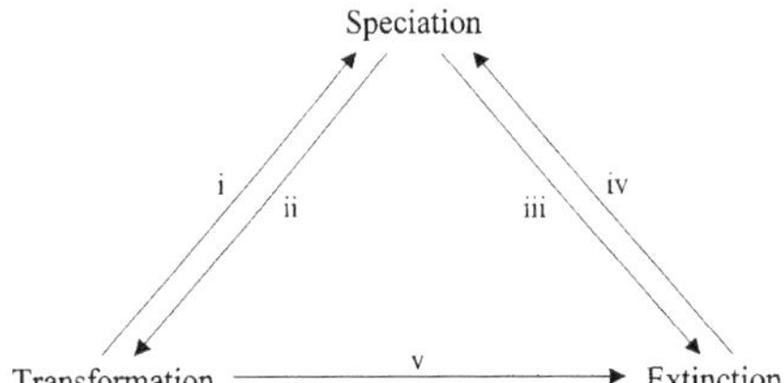

Figure 5. Some interactions between the effects of speciation, transformation and extinction with possible implications for species-range size distributions: (i) temporal changes in range size change the likelihood of speciation; (ii) speciation changes range size; (iii) under some speciation scenarios the ancestral species does not persist after speciation; (iv) lower rates of extinction may increase the likelihood of speciation; and (v) temporal changes in range size change the likelihood of extinction.

the relative importance of the processes of speciation, transformation and extinction in shaping the distribution is to forge a link between questions that are central to the study of ecology and of biogeography, with geographic range size as the common currency. On the one hand, ecology is concerned with the interactions between organisms and their environment (Begon *et al.* 1996), typically at local scales. In particular, in this context, it is concerned with the role of present conditions (abiotic and biotic) in determining distribution (transformation). On the other hand, the origins of biogeography lie in systematics (Myers & Giller 1988), and a more historical, and regional or global, view of the determinants of distributions has been a dominant theme (speciation and extinction).

If the likelihood of speciation and extinction were unbiased with respect to geographic range size and were the dominant processes, and if speciation had relatively little impact on the size of ancestral geographic ranges, then an approximately log-normal species-range size distribution would be predicted to result. However:

(i) Likelihood of speciation is probably not an unbiased function of range size. The form of the bias remains contentious, but the weight of opinion would seem to favour an approximately unimodal relationship, with small to intermediate range sizes having the greatest likelihood of speciation and large and very small range sizes the least likelihood.
(ii) Speciation may have a marked impact on the range sizes of ancestral species, unless it occurs predominantly by peripheral isolation.
(iii) Likelihood of extinction is not an unbiased function of range size. Rather, risk of extinction declines with increasing range size.
(iv) Transformation must influence species-range size distributions to some degree, because the only model which involves no change of range sizes at speciation (a stasis I model) is problematic. The limited evidence for several different models of transformation suggests that the importance of the process in shaping species-range size distributions may be very variable. For example, it would seem likely to be very significant, and to potentially swamp the effects of speciation, in regions experiencing strong environmental change. Here, areas which a species initially occupied may become untenable and new areas may become accessible.

The interactions of speciation, extinction and transformation are potentially complex (figure 5). Their outcome is a species-range size distribution which although sometimes not dissimilar from a log-normal, seems consistently to depart from such a description. This departure may ultimately prove a key to understanding the relative importance of the processes that generate it.

K.J.G. is a Royal Society University Research Fellow. I am grateful to Tim Blackburn, Steven Chown, Bob May, John Spicer, Phil Warren, Mark Williamson and an anonymous referee for helpful discussion and comments, to Bob May for his encouragement to pursue the ideas outlined herein, and to Anne Magurran and Bob May for the invitation to participate in the Royal Society discussion meeting.

REFERENCES

Anderson, S. 1977 Geographic ranges of North American terrestrial mammals. *Am. Mus. Novitates* **2629**, 1–15.
Anderson, S. 1984*a* Geographic ranges of North American terrestrial birds. *Am. Mus. Novitates* **2785**, 1–17.
Anderson, S. 1984*b* Areography of North American fishes, amphibians and reptiles. *Am. Mus. Novitates* **2802**, 1–16.
Anderson, S. 1985 The theory of range-size (RS) distributions. *Am. Mus. Novitates* **2833**, 1–20.
Arita, H. T. 1993 Rarity in Neotropical bats: correlations with phylogeny, diet, and body mass. *Ecol. Appl.* **3**, 506–517.
Barton, N. H. & Charlesworth, B. 1984 Genetic revolutions, founder effects, and speciation. *A. Rev. Ecol. Syst.* **15**, 133–164.
Begon, M., Harper, J. L. & Townsend, C. R. 1996 *Ecology*, 3rd edn. Oxford: Blackwell Science.
Blackburn, T. M. & Gaston, K. J. 1994 Body size distributions: patterns, mechanisms and implications. *Trends Ecol. Evol.* **9**, 471–474.
Blackburn, T. M. & Gaston, K. J. 1996 Spatial patterns in the geographic range sizes of bird species in the New World. *Phil. Trans. R. Soc. Lond.* B **351**, 897–912.
Blackburn, T. M., Gaston, K. J. & Lawton, J. H. 1998 Patterns in the geographic ranges of woodpeckers (Aves: Picidae). *Ibis* (In the press.)
Blakers, N., Davies, S. J. J. F. & Reilly, P. N. 1984 *Atlas of Australian birds*. Carlton: Melbourne University Press.
Brooks, D. R. & McLennan, D. A. 1991 *Phylogeny, ecology, and behavior: a research program in comparative biology*. University of Chicago Press.
Brooks, D. R. & McLennan, D. A. 1993 Comparative study of adaptive radiations with an example using parasitic flatworms (Platyhelminthes: Cercomeria). *Am. Nat.* **142**, 755–778.
Brown, J. H. 1984 On the relationship between abundance and distribution of species. *Am. Nat.* **124**, 255–279.
Brown, J. H. 1995 *Macroecology*. University of Chicago Press.
Brown, J. H., Marquet, P. A. & Taper, M. L. 1993 Evolution of body size: consequences of an energetic definition of fitness. *Am. Nat.* **142**, 573–584.
Brown, J. H. & Nicoletto, P. F. 1991 Spatial scaling of species composition: body masses of North American land mammals. *Am. Nat.* **138**, 1478–1512.
Brown, J. H., Stevens, G. C. & Kaufman, D. M. 1996 The geographic range: size, shape, boundaries, and internal structure. *A. Rev. Ecol. Syst.* **27**, 597–623.
Burton, J. F. 1995 *Birds and climate change*. London: Christopher Helm.

Bush, G. L. 1975 Modes of animal speciation. *A. Rev. Ecol. Syst.* **6**, 334–364.

Buzas, M. A. & Culver, S. J. 1991 Species diversity and dispersal of benthic foraminifera. *BioScience* **41**, 483–489.

Chesser, R. T. & Zink, R. M. 1994 Modes of speciation in birds: a test of Lynch's method. *Evolution* **48**, 490–497.

Chown, S. L. 1997 Speciation and rarity: separating cause from consequence. In *The biology of rarity: causes and consequences of rare–common differences* (ed. W. E. Kunin & K. J. Gaston), pp. 91–109. London: Chapman & Hall.

Collins, S. L. & Glenn, S. M. 1990 A hierarchical analysis of species' abundance patterns in grassland vegetation. *Am. Nat.* **135**, 633–648.

Coope, G. R. 1995 Insect faunas in ice age environments: why so little extinction? In *Extinction rates* (ed. J. H. Lawton & R. M. May), pp. 55–74. Oxford University Press.

Cotgreave, P. & Pagel, M. 1997 Predicting and understanding rarity: the comparative approach. In *The biology of rarity: causes and consequences of rare–common differences* (ed. W. E. Kunin & K. J. Gaston), pp. 237–261. London: Chapman & Hall.

Cracraft, J. 1982 Geographic differentiation, cladistics, and vicariance biogeography: reconstructing the tempo and mode of evolution. *Am. Zool.* **22**, 411–424.

Cracraft, J. 1986 Origin and evolution of continental biotas: speciation and historical congruence within the Australian avifauna. *Evolution* **40**, 977–996.

Cracraft, J. & Prum, R. O. 1988 Patterns and processes of diversification: speciation and historical congruence in some neotropical birds. *Evolution* **42**, 603–620.

Darwin, C. 1975 *Charles Darwin's natural selection: being the second part of his big species book written from 1856–1858* (ed. R. C. Stauffer). Cambridge University Press.

Darwin, C. 1859 *On the origin of species by means of natural selection, or the preservation of favoured races in the struggle for life.* London: John Murray.

del Hoyo, J., Elliott, A. & Sargatal, J. (ed.) 1992 *Handbook of the birds of the world. 1. Ostrich to ducks.* Barcelona: Lynx Edicions.

del Hoyo, J., Elliott, A. & Sargatal, J. (ed.) 1996 *Handbook of the birds of the world. 3. Hoatzin to auks.* Barcelona: Lynx Edicions.

Dennis, R. L. H. 1993 *Butterflies and climate change.* Manchester University Press.

Dillon, L. S. 1966 The life cycle of the species: an extension of current concepts. *Syst. Zool.* **15**, 112–126.

Emlet, R. B. 1995 Developmental mode and species geographic range in regular sea urchins (Echinodermata: Echinoidea). *Evolution* **49**, 476–489.

Erwin, T. L. 1985 The taxon pulse: a general pattern of lineage radiation and extinction among carabid beetles. In *Taxonomy, phylogeny and zoogeography of beetles and ants* (ed. G. E. Ball), pp. 3437–3472. Dordrecht: Junk.

Freitag, R. 1969 A revision of the species of the genus *Evarthrus* LeConte (Coleoptera: Carabidae). *Quaest. Entomol.* **5**, 89–212.

Frey, J. K. 1992 Response of a mammalian faunal element to climatic changes. *J. Mamm.* **73**, 43–50.

Frey, J. K. 1993 Modes of peripheral isolate formation and speciation. *Syst. Biol.* **42**, 373–381.

Gaston, K. J. 1991 How large is a species' geographic range? *Oikos* **61**, 434–438.

Gaston, K. J. 1994*a* *Rarity.* London: Chapman & Hall.

Gaston, K. J. 1994*b* Measuring geographic range sizes. *Ecography* **17**, 198–205.

Gaston, K. J. 1996*a* Species-range size distributions: patterns, mechanisms and implications. *Trends Ecol. Evol.* **11**, 197–201.

Gaston, K. J. 1996*b* The multiple forms of the interspecific abundance–distribution relationship. *Oikos* **75**, 211–220.

Gaston, K. J. & Blackburn, T. M. 1996 Global scale macroecology: interactions between population size, geographic range size and body size in the Anseriformes. *J. Anim. Ecol.* **65**, 701–714.

Gaston, K. J. & Blackburn, T. M. 1997*a* Evolutionary age and risk of extinction: the global avifauna. *Evol. Ecol.* **11**, 557–565.

Gaston, K. J. & Blackburn, T. M. 1997*b* Age, area and avian diversification. *Biol. J. Linn. Soc.* **62**, 239–253.

Gaston, K. J. & Kunin, W. E. 1997*a* Rare-common differences: an overview. In *The biology of rarity: causes and consequences of rare–common differences* (ed. W. E. Kunin & K. J. Gaston), pp. 12–29. London: Chapman & Hall.

Gaston, K. J. & Kunin, W. E. 1997*b* Concluding comments. In *The biology of rarity: causes and consequences of rare–common differences* (ed. W. E. Kunin & K. J. Gaston), pp. 262–272. London: Chapman & Hall.

Gaston, K. J. & Lawton, J. H. 1989 Insect herbivores on bracken do not support the core-satellite hypothesis. *Am. Nat.* **134**, 761–777.

Gaston, K. J. & Lawton, J. H. 1990 Effects of scale and habitat on the relationship between species local abundance and large scale distribution. *Oikos* **58**, 329–335.

Gaston, K. J. & Williams, P. H. 1996 Spatial patterns in taxonomic diversity. In *Biodiversity: a biology of numbers and difference* (ed. K. J. Gaston), pp. 202–229. Oxford: Blackwell Science.

Gaston, K. J., Quinn, R. M., Wood, S. & Arnold, H. R. 1996 Measures of geographic range size: the effects of sample size. *Ecography* **19**, 259–268.

Gaston, K. J., Blackburn, T. M. & Lawton, J. H. 1997 Interspecific abundance-range size relationships: an appraisal of mechanisms. *J. Anim. Ecol.* **66**, 579–601.

Gaston, K. J. & Blackburn, T. M. & Gregory, R. D. 1998 Abundance-range size relationships of breeding and wintering birds in Britain: a comparative analysis. *Ecography.* (In the press.)

Glazier, D. S. 1980 Ecological shifts and the evolution of geographically restricted species of North American *Peromyscus* (mice). *J. Biogeogr.* **7**, 63–83.

Glazier, D. S. 1987 Toward a predictive theory of speciation the ecology of isolate selection. *J. Theor. Biol.* **126**, 323–333.

Gleason, H. A. 1924 Age and area from the viewpoint of phytogeography. *Am. J. Bot.* **11**, 541–546.

Gorodkov, K. B. 1992 Dynamics of range: general approach. II. Dynamics of range and evolution of taxa (qualitative or phyletic changes of range). *Entomol. Rev.* **70**, 81–99.

Gotelli, N. J. 1991 Metapopulation models: the rescue effect, the propagule rain, and the core-satellite hypothesis. *Am. Nat.* **138**, 768–776.

Gotelli, N. J. & Graves, G. R. 1996 *Null models in ecology.* Washington: Smithsonian Institution.

Gotelli, N. J. & Simberloff, D. 1987 The distribution and abundance of tallgrass prairie plants: a test of the core-satellite hypothesis. *Am. Nat.* **130**, 18–35.

Gregory, R. D. 1994 Species abundance patterns of British birds. *Proc. R. Soc. Lond.* B **257**, 299–301.

Grosholz, E. D. 1996 Contrasting rates of spread for introduced species in terrestrial and marine systems. *Ecology* **77**, 1680–1686.

Hansen, T. A. 1978 Larval dispersal and species longevity in Lower Tertiary gastropods. *Science* **199**, 886–887.

Hansen, T. A. 1980 Influence of larval dispersal and geographic distribution on species longevities in neogastropods. *Paleobiology* **6**, 193–207.

Hanski, I. 1982 Dynamics of regional distribution: the core and satellite species hypothesis. *Oikos* **38**, 210–221.

Hanski, I., Kouki, J. & Halkka, A. 1993 Three explanations of the positive relationship between distribution and abundance of species. In *Historical and geographical determinants of community diversity* (ed. R. Ricklefs & D. Schluter), pp. 108–116. University of Chicago Press.

Harvey, P. H. & Pagel, M. D. 1991 *The comparative method in evolutionary biology.* Oxford University Press.
Hengeveld, R. 1989 *Dynamics of biological invasions.* London: Chapman & Hall.
Hengeveld, R. 1990 *Dynamic biogeography.* Cambridge University Press.
Hodgson, J. G. 1993 Commonness and rarity in British butterflies. *J. Appl. Ecol.* **30**, 407–427.
Hughes, L., Cawsey, E. M. & Westoby, M. 1996 Geographic and climatic range sizes of Australian eucalypts and a test of Rapoport's rule. *Global Ecol. Biogeog. Lett.* **5**, 128–142.
Jablonski, D. 1986*a* Causes and consequences of mass extinctions: a comparative approach. In *Dynamics of extinction* (ed. D. K. Elliot), pp. 183–229. New York: Wiley.
Jablonski, D. 1986*b* Background and mass extinctions: the alternation of macroevolutionary regimes. *Science* **231**, 129–133.
Jablonski, D. 1987 Heritability at the species level: analysis of geographic ranges of Cretaceous mollusks. *Science* **238**, 360–363.
Jablonski, D. 1988 Response [to Russell & Lindberg]. *Science* **240**, 969.
Jablonski, D. & Raup, D. M. 1995 Selectivity of end-Cretaceous bivalve extinctions. *Science* **268**, 389–391.
Jackson, J. B. C. 1974 Biogeographic consequences of eurytopy and stenotopy among marine bivalves and their evolutionary consequences. *Am. Nat.* **108**, 541–560.
Kavanaugh, D. H. 1979 Rates of taxonomically significant differentiation in relation to geographical isolation and habitat: examples from a study of the Nearctic *Nebria* fauna. In *Carabid beetles: their evolution, natural history and classification* (ed. T. L. Erwin, G. E. Ball & A. L. Halpern), pp. 35–57. The Hague: Junk.
Kavanaugh, D. H. 1985 On wing atrophy in carabid beetles (Coleoptera: Carabidae), with special reference to Nearctic *Nebria.* In *Taxonomy, phylogeny and zoogeography of beetles and ants* (ed. G. E. Ball), pp. 408–431. Dordrecht: Junk.
Kelly, C. K. & Woodward, F. I. 1996 Ecological correlates of plant range size: taxonomies and phylogenies in the study of plant commonness and rarity in Great Britain. *Phil. Trans. R. Soc. Lond.* B **351**, 1261–1269.
Kier, P. M. & Lawson, M. H. 1978 Index of living and fossil echinoids 1924–1970. *Smithsonian Contrib. Paleobiol.* **34**, 1–182.
Koch, C. F. 1980 Bivalve species duration, areal extent and population size in a Cretaceous sea. *Paleobiology* **6**, 184–192.
Kunin, W. E. & Gaston, K. J. 1993 The biology of rarity: patterns, causes, and consequences. *Trends Ecol. Evol.* **8**, 298–301.
Kunin, W. E. & Gaston, K. J. (ed.) 1997 *The biology of rarity: causes and consequences of rare-common differences.* London: Chapman & Hall.
Levin, D. A. 1993 Local speciation in plants: the rule not the exception. *Syst. Botany* **18**, 197–208.
Levinton, J. 1988 *Genetics, paleontology, and macroevolution.* Cambridge University Press.
Liebherr, J. K. & Hajek, A. E. 1990 A cladistic test of the taxon cycle and taxon pulse hypothesis. *Cladistics* **6**, 39–59.
Livezey, B. C. 1991 A phylogenetic analysis and classification of recent dabbling ducks (Tribe Anatini) based on comparative morphology. *The Auk* **108**, 471–507.
Losos, J. B. 1992 A critical comparison of the taxon-cycle and character-displacement models for size evolution of *Anolis* lizards in the Lesser Antilles. *Copeia* **1992**, 279–288.
Lynch, J. D. 1989 The gauge of speciation: on the frequencies of modes of speciation. In *Speciation and its consequences* (ed. D. Otte & J. A. Endler), pp. 527–553. Sunderland, MA: Sinauer.
Mace, G. M. & Kershaw, M. 1997 Extinction risk and rarity in an ecological timescale. In *The biology of rarity: causes and consequences of rare–common differences* (ed. W. E. Kunin & K. J. Gaston), pp. 130–149. London: Chapman & Hall.
Madge, S. & Burn, H. 1988 *Wildfowl: an identification guide to the ducks, geese and swans of the world.* London: Christopher Helm.
Marzluff, J. M. & Dial, K. P. 1991 Life history correlates of taxonomic diversity. *Ecology* **72**, 428–439.
Maurer, B. A. 1990 The relationship between distribution and abundance in a patchy environment. *Oikos* **58**, 181–189.
Maurer, B. A., Brown, J. H. & Rusler, R. D. 1992 The micro and macro in body size evolution. *Evolution* **46**, 939–953.
May, R. M. 1994 Biological diversity: differences between land and sea. *Phil. Trans. R. Soc. Lond.* B **343**, 105–111.
May, R. M., Lawton, J. H. & Stork, N. E. 1995 Assessing extinction rates. In *Extinction rates* (ed. J. H. Lawton & R. M. May), pp. 1–24. Oxford University Press.
Mayr, E. 1963 *Animal species and evolution.* Cambridge, MA: Harvard University Press.
Mayr, E. 1988 *Toward a new philosophy of biology: observations of an evolutionist.* Cambridge, MA: Harvard University Press.
McAllister, D. E., Platania, S. P., Schueler, F. W., Baldwin, M. E. & Lee, D. S. 1986 Ichthyofaunal patterns on a geographical grid. In *Zoogeography of freshwater fishes of North America* (ed. C. H. Hocutt & E. D. Wiley), pp. 17–51. New York: Wiley.
McKinney, M. L. 1997 How do rare species avoid extinction? A paleontological view. In *The biology of rarity: causes and consequences of rare–common differences* (ed. W. E. Kunin & K. J. Gaston), pp. 110–129. London: Chapman & Hall.
McLaughlin, S. P. 1992 Are floristic areas hierarchically arranged? *J. Biogeog.* **19**, 21–32.
Morell, V. 1996 Amazonian diversity: a river doesn't run through it. *Science* **273**, 1496–1497.
Myers, A. A. & Giller, P. S. 1988 Process, pattern and scale in biogeography. In *Analytical biogeography: an integrated approach to the study of animal and plant distributions* (ed. A. A. Myers & P. S. Giller), pp. 3–21. London: Chapman & Hall.
Nee, S., Harvey, P. H. & May, R. M. 1991 Lifting the veil on abundance patterns. *Proc. R. Soc. Lond.* B **243**, 161–163.
Noonan, G. R. 1990 Biogeographical patterns of North American *Harpalus* Latreille (Insecta: Coleoptera: Carabidae). *J. Biogeog.* **17**, 583–614.
Norris, R. D. 1991 Biased extinction and evolutionary trends. *Paleobiology* **17**, 388–399.
Nunn, G. B., Cooper, J., Jouventin, P., Robertson, C. J. R. & Robertson, G. C. 1996 Evolutionary relationships among extant albatrosses (Procellariiformes: Diomedeidae) established from complete cytochrome-B gene sequences. *The Auk* **113**, 784–801.
Pagel, M. P., May, R. M. & Collie, A. R. 1991 Ecological aspects of the geographic distribution and diversity of mammalian species. *Am. Nat.* **137**, 791–815.
Palumbi, S. R. 1994 Genetic-divergence, reproductive isolation, and marine speciation. *A. Rev. Ecol. Syst.* **25**, 547–572.
Parmesan, C. 1996 Climate and species' range. *Nature* **382**, 765–766.
Peat, H. J. & Fitter, A. H. 1994 Comparative analyses of ecological characteristics of British angiosperms. *Biol. Rev.* **69**, 95–115.
Pimm, S. L., Jones, H. L. & Diamond, J. 1988 On the risk of extinction. *Am. Nat.* **132**, 757–785.
Pomeroy, D. & Ssekabiira, D. 1990 An analysis of the distributions of terrestrial birds in Africa. *Afr. J. Ecol.* **28**, 1–13.
Pregill, G. K. & Olson, S. L. 1981 Zoogeography of West Indian vertebrates in relation to Pleistocene climatic cycles. *A. Rev. Ecol. Syst.* **12**, 75–98.
Rapoport, E. H. 1992 *Areography: geographical strategies of species.* Oxford: Pergamon.
Rapoport, E. H. 1994 Remarks on marine and continental biogeography: an areographical viewpoint. *Phil. Trans. R. Soc. Lond.* B **343**, 71–78.
Read, A. F. & Harvey, P. H. 1989 Life history differences among the eutherian radiations. *J. Zool.* **219**, 329–353.

Ricklefs, R. E. 1989 Speciation and diversity: integration of local and regional processes. In *Speciation and its consequences* (ed. D. Otte & J. Endler), pp. 599–622. Sunderland, MA: Sinauer.

Ricklefs, R. E. & Cox, G. W. 1972 Taxon cycles in the West Indies avifauna. *Am. Nat.* **106**, 195–219.

Ricklefs, R. E. & Cox, G. W. 1978 Stage of taxon cycle, habitat distribution and population density in the avifauna of the West Indies. *Am. Nat.* **122**, 875–895.

Ricklefs, R. E. & Latham, R. E. 1992 Intercontinental correlation of geographical ranges suggests stasis in ecological traits of relict genera of temperate perennial herbs. *Am. Nat.* **139**, 1305–1321.

Ripley, S. D. & Beehler, B. M. 1990 Patterns of speciation in Indian birds. *J. Biogeog.* **17**, 639–648.

Rosenzweig, M. L. 1975 On continental steady states of species diversity. In *Ecology and evolution of communities* (ed. M. L. Cody & J. M. Diamond), pp. 124–140. Cambridge, MA: Harvard University Press.

Rosenzweig, M. L. 1978 Geographical speciation: on range size and the probability of isolate formation. In *Proceedings of the Washington State University Conference on Biomathematics and Biostatistics* (ed. D. Wollkind), pp. 172–194. Washington State University, WA.

Rosenzweig, M. L. 1995 *Species diversity in space and time.* Cambridge University Press.

Roughgarden, J. & Pacala, S. 1989 Taxon cycle among *Anolis* lizard populations: review of evidence. In *Speciation and its consequences* (ed. D. Otte & J. A. Endler), pp. 403–432. Sunderland, MA: Sinauer.

Roy, K., Jablonski, D. & Valentine, J. W. 1995 Thermally anomalous assemblages revisited: patterns in the extraprovincial latitudinal range shifts of Pleistocene marine mollusks. *Geology* **23**, 1071–1074.

Ruggiero, A. 1994 Latitudinal correlates of the sizes of mammalian geographical ranges in South America. *J. Biogeog.* **21**, 545–559.

Rummel, J. D. & Roughgarden, J. 1985 A theory of faunal buildup for competition communities. *Evolution* **39**, 1009–1033.

Russell, M. P. & Lindberg, D. R. 1988*a* Real and random patterns associated with molluscan spatial and temporal distributions. *Paleobiology* **14**, 322–330.

Russell, M. P. & Lindberg, D. R. 1988*b* Estimates of species duration. *Science* **240**, 969.

Schliewen, U. K., Tautz, D. & Pääbo, S. 1994 Sympatric speciation suggested by monophyly of crater lake cichlids. *Nature* **368**, 629–632.

Schoener, T. W. 1987 The geographical distribution of rarity. *Oecologia* **74**, 161–173.

Stanley, S. M. 1979 *Macroevolution: patterns and process.* San Francisco, CA: W. H. Freeman.

Stanley, S. M. 1986 Population size, extinction, and speciation: the fission effect in Neogene Bivalvia. *Paleobiology* **12**, 89–110.

Stebbins, G. L. 1978 Why are there so many rare plants in California? II. Youth and age of species. *Fremontia* **6**, 17–20.

Stebbins, G. L. & Major, J. 1965 Endemism and speciation in the California flora. *Ecol. Monogr.* **35**, 1–35.

Tauber, C. & Tauber, M. J. 1989 Sympatric speciation in insects: perception and perspective. In *Speciation and its consequences* (ed. D. Otte & J. A. Endler), pp. 307–344. Sunderland, MA: Sinauer.

Taylor, C. M. & Gotelli, N. J. 1994 The macroecology of *Cyprinella*: correlates of phylogeny, body size, and geographical range. *Am. Nat.* **144**, 549–569.

Terborgh, J. & Winter, B. 1980 Some causes of extinction. In *Conservation biology: an evolutionary-ecological perspective* (ed. M. E. Soulé & B. A. Wilcox), pp. 119–133. Sunderland, MA: Sinauer.

Tokeshi, M. 1992 Dynamics of distribution in animal communities: theory and analysis. *Res. Popul. Ecol.* **34**, 249–273.

Tokeshi, M. 1996 Power fraction: a new explanation of relative abundance patterns in species-rich assemblages. *Oikos* **75**, 543–550.

Tracy, C. R. & George, T. L. 1992 On the determinants of extinction. *Am. Nat.* **139**, 102–122.

Van Valen, L. 1973 A new evolutionary law. *Evol. Theory* **1**, 1–30.

Veit, R. R. & Lewis, M. A. 1996 Dispersal, population growth, and the Allee effect: dynamics of the house finch invasion of eastern North America. *Am. Nat.* **148** 255–274.

Wagner, P. J. & Erwin, D. H. 1995 Phylogenetic patterns as tests of speciation models. In *New approaches to speciation in the fossil record* (ed. D. H. Erwin & R. L. Anstey), pp. 87–122. New York: Columbia University Press.

Wiley, E. O. 1988 Vicariance biogeography. *A. Rev. Ecol. Syst.* **19**, 513–542.

Williams, P. H. 1988 Habitat use by bumble bees (*Bombus* spp.). *Ecol. Entomol.* **13**, 223–237.

Williamson, M. 1996 *Biological invasions.* London: Chapman & Hall.

Willis, J. C. 1922 *Age and area: a study in geographical distribution and origin of species.* Cambridge University Press

Wilson, E. O. 1961 The nature of the taxon cycle in the Melanesian ant fauna. *Am. Nat.* **95**, 169–193.

Wing, L. 1943 Spread of the starling and English sparrow. *The Auk* **60**, 74–87.

Wolfheim, J. H. 1983 *Primates of the world: distribution, abundance, and conservation.* Seattle, WA: University of Washington Press.

Woods, K. D. & Davis, M. B. 1989 Paleoecology of range limits: beech in the upper peninsula of Michigan. *Ecology* **70**, 681–696.

Woodward, F. I. 1987 *Climate and plant distribution.* Cambridge University Press.

ABUNDANCE AND DISTRIBUTIONS

Edited by Kevin J. Gaston, Christy M. McCain, and S. Kathleen Lyons

Ecology is commonly characterized as being largely about the abundance and distribution of species. It is thus logical that these topics should and do also lie at the heart of current macroecology. That centrality of thinking has, however, a much longer history. Over an extensive period, numerous large-scale spatial and temporal patterns in the abundance and distribution of species and of assemblages have been proposed (for broad overviews, see Brown 1995; Gaston and Blackburn 2000; Gaston 2003). These include (1) intraspecific patterns in the frequency distribution of abundances, in the population size and density structure of geographic ranges, and in the temporal variability in local and regional abundances; (2) interspecific patterns in density, occupancy, and geographic range extent, and in the covariation between those variables; and (3) assemblage patterns in how total abundances change with latitude, elevation, depth, environment, and time. Some of these patterns have gained key roles in macroecological thinking; some have not but perhaps should; and others have been shown to be artifacts, to be secondary effects of more primary patterns, or not to have great generality.

The papers in this section variously concern important attempts to establish the nature of particularly significant patterns in abundance and distribution, to assess the generality of those patterns, and to understand the mechanisms that have given rise to them. In several cases they have achieved more than one of these ends. Some papers in other sections of this book or other volumes in the Foundations series might equally have found a home here if they had not done so elsewhere (e.g., Preston 1948; Taylor 1961; Brown and Maurer 1987; Stevens 1989).

Tracing the roots of the recognition of individual macroecological patterns in the ecological literature, and, preceding that, in the natural history and general botanical and zoological literatures, can be a fascinating journey. However, it can often also be frustrating, in that one never really knows what constitutes the destination (i.e., the earliest clear reference to the pattern), or indeed whether one has arrived (i.e., possible earlier sources always remain unread). Moreover, while those who first documented given patterns should be duly acknowledged, the initial steps were typically very small. Much more important are the more recent papers whose explicit statements about, or empirical descriptions of, particular patterns have been disproportionately significant, and we have chosen to reprint some of those here (e.g., S. J. McNaughton and L. L. Wolf; S. Anderson; J. C. Bernabo and T. Webb III; E. C. Pielou; I. Hanski). Interestingly, many such papers bear testimony to how robust macroecological patterns in abundance and distribution can often be, with the essentials they report remaining unshaken by often substantial subsequent improvements in the quality and the quantity of the data and in the bioinformatic and statistical tools available with which to examine them (e.g., as a consequence of coordinated monitoring programs and remote imagery).

There is no simple answer as to the point at

which a macroecological distribution or relationship can be regarded as sufficiently general as to constitute a genuine "pattern." At one extreme, a single exception can be regarded as enough to undermine such generalization (albeit one needs to be cautious that this is a genuine exception). At the other extreme, one might regard a generalization as established if it is displayed in at least 50% of cases (Mayr 1956). Of course, in either circumstance one would need a sufficiently large and representative set of empirical studies. Some of the papers in this section have proven pivotal in establishing that a pattern in abundance or distribution (or both) was sufficiently general as to be of interest (D. Rabinowitz; I. Hanski; J. H. Brown). Interestingly, in such cases the quantity of evidence available at the time was but a small fraction of that which now graces the literature, perhaps suggesting that having a good sense for what was likely to be ecologically important might have been as significant as the body of evidence demonstrating that importance.

Although they may be useful, even very general macroecological patterns remain rather dissatisfying to many macroecologists in the absence of a good understanding of why they exist. If those answers were available for all of the key macroecological patterns in abundance and distribution alone, then there would be little left to study in substantial portions of the field. However, this is far from being the case. The mechanistic papers that we include here (Rabinowitz; Hanski; Brown) were chosen not so much because the ideas they present were always correct, or even because it is necessarily known whether or not this is so. Rather, they were picked foremost because of the profound influence they have had on thinking about mechanisms. Part of the reason they attained this influence was that they almost invariably took a very synthetic approach to mechanism, not attempting simply to explain individual macroecological patterns in isolation but rather drawing connections between different patterns.

Particularly when considering abundance and distribution, the interdependencies of intraspecific, interspecific, and assemblage patterns become very apparent (Gaston et al. 2008). Indeed, practically they can be thought of in most cases as different expressions of the row or column means, totals and variances derived from simple species × sites matrices. Arguably, understanding of large-scale spatial and temporal patterns in abundance and distribution might most rapidly be advanced in the coming years in three ways. First, by continuing to develop understanding of the structure and behavior of such matrices. We suspect that in so doing previously overlooked major spatial and temporal patterns in abundance and distribution will yet be discovered. Second, by focusing on the interconnectedness of the patterns and the extent to which many are shared outcomes of a few basic underlying principles. Third, by improving knowledge of the abundance structure of the geographic ranges of species, which lie at the root of most large-scale patterns in abundance and distribution, albeit variously being influenced by the spatial distribution and overlaps of those ranges. In this final vein, Charles Darwin ([1856–1858] 1975, 528) observed, "I have lately been especially attending to Geograph. Distrib., & most splendid sport it is, —a grand game of chess with the world for a Board." The papers in this section concern the way in which that game has been played and the search for the rules.

Literature Cited

Brown, J. H. 1995. *Macroecology*. University of Chicago Press, Chicago.

*Brown, J. H., and B. A. Maurer. 1987. Evolution of species assemblages: Effects of energetic constraints and species dynamics on the diversification of the North American avifauna. *American Naturalist* 130:1–17.

Darwin, C. (1856–1858) 1975. *Charles Darwin's natural selection: Being the second part of his big species book written from 1856 to 1858*. Edited by R. C. Stauffer. Cambridge University Press, Cambridge.

Gaston, K. J. 2003. *The structure and dynamics of geographic ranges*. Oxford University Press, Oxford.

Gaston, K. J., and T. M. Blackburn. 2000. *Pattern and process in macroecology*. Blackwell Science, Oxford.

Gaston, K. J., S. L. Chown, and K. L. Evans. 2008. Ecogeographical rules: Elements of a synthesis. *Journal of Biogeography* 35:483–500.

Mayr, E. 1956. Geographical character gradients and climatic adaptation. *Evolution* 10:105–8.

Preston, F. W. 1948. The commonness, and rarity, of species. *Ecology* 29:254–83.

*Stevens, G. C. 1989. The latitudinal gradient in geographical range: How so many species coexist in the tropics. *American Naturalist* 133:240–56.

*Taylor, L. 1961. Aggregation, variance and the mean. *Nature* 189:732–73.

PAPER 25

Dominance and the Niche in Ecological Systems (1970)

S. J. McNaughton and L. L. Wolf

Commentary

BRIAN J. MCGILL

It has sometimes been suggested that there is not a modern, viable theory of the niche or, in R. H. Peters's (1991) words, that the niche is not "operationalized" (i.e., measurable). Ironically, 30 years ago S. J. McNaughton and L. L. Wolf laid the foundations for just such a theory. In the following paper, they outline a macroecology of dominance and niche width, going beyond the traditional macroecological variables of abundance, species richness, range size, and body size (Brown 1995). This paper is clearly macroecological in nature; echoing modern-day angst in macroecology, McNaughton and Wolf note, "We rely heavily, however, upon regression analysis and correlation. . . . We approach but do not reach descriptions of causality" (138).

McNaughton and Wolf's operationalization of niche width and dominance depends on a return to R. H. Whittaker's (1960) study of communities along environmental gradients. They visualize niche as a curve of abundance along the gradient with niche width defined by an abundance-weighted average of spatial extent. This also allows inclusion of environmental position (benign versus harsh) in their macroecological analysis. They clearly consider G. E. Hutchinson's (1957) conceptualization of the niche as a region in n-dimensional space to be elegant and desirable, but confess to not having sufficient data to handle high dimensionality. So they fall back to using a spatial measure instead, thereby achieving results that 30 more years of ecology have not achieved with the Hutchinsonian niche.

The paper is packed full of ideas, making it in some ways a difficult but rewarding read. There are multiple key findings. Distribution of abundance between species is roughly lognormal, but there is a systematic overrepresentation of rare species (which has received further study [e.g., McGill 2003]) and an underrepresentation of abundant species (which has largely gone unstudied) showing dominance. More-dominant species have wider niches. McNaughton and Wolf suggest this can be due to dominant species being generalists that outcompete other species or due to every species being a specialist with some specializing in more available environments. They prefer the second explanation, presaging the idea of the inclusive niche (Colwell and Fuentes 1975), now strongly supported (Wisheu 1998). Dominance in communities (the inverse of evenness) is greater with environmental harshness and negatively correlated with species richness. Finally, richness increases toward the benign part of a gradient, partly due to an increased carrying capacity but also due to decreased niche widths. Even evolutionary and genetic mechanisms are discussed. Ecology is long overdue for a synthetic, operational theory of the niche, and perhaps 30 years later, this paper and macroecology can provide the road map.

From *Science* 167:131–39. Reprinted with permission from AAAS.

Literature Cited

Brown, J. H. 1995. *Macroecology*. University of Chicago Press, Chicago.

Colwell, R. K., and E. R. Fuentes. 1975. Experimental studies of the niche. *Annual Review of Ecology and Systematics* 6:281–310.

Hutchinson, G. E. 1957. Concluding remarks. *Cold Spring Harbor Symposia on Quantitative Biology* 22:415–27.

McGill, B. J. 2003. Does Mother Nature really prefer rare species or are log-left-skewed SADs a sampling artefact? *Ecology Letter* 6:766–73.

Peters, R. H. 1991. *A critique for ecology*. Cambridge University Press, Melbourne, Australia.

Whittaker, R. H. 1960. Vegetation of the Siskiyou mountains, Oregon and California. *Ecological Monographs* 30:279–338.

Wisheu, I. C. 1998. How organisms partition habitats: Different types of community organization can produce identical patterns. *Oikos* 83:246–58.

9 January 1970, Volume 167, Number 3915 SCIENCE

Dominance and the Niche in Ecological Systems

Dominance is an expression of ecological inequalities arising out of different exploitation strategies.

S. J. McNaughton and L. L. Wolf

The concept of dominance, that is, the idea that certain species so pervade the ecosystem that they exert a powerful control on the occurrence of other species, is one of the oldest concepts in ecology. But, while the concept is universally employed in early texts and continues to be widespread (*1*), many recent texts have omitted it (*2*). The concept, in fact, seems to have fallen somewhat into disrepute because of its ambiguity (*3*), and some population biologists have suggested that the term is not biologically meaningful (*4*).

Current disenchantment with the idea of dominance undoubtedly arises out of an absence of rigorous proof of its occurrence together with general omission of the idea from recent theories of community diversity (*5*). The idea of dominance, however, is closely tied to species diversity and, as Whittaker has pointed out (*6*), many widely employed indices of "diversity" are actually measures of the concentration of dominance in the community. Dominance, relative abundance of species in communities, and species diversity of communities are intricately interrelated in the conceptual framework of ecology, and, while diversity and relative abundance problems have been concisely explored, their relationship to the earlier idea of dominance has not been carefully developed. It is obvious that the abundances of species in a local area vary, and that the diversities of communities are often distinct. What is not obvious is how these differences relate to the organization of communities.

The conceptual difficulty is confounded by a frequent failure to distinguish between absolute and relative abundances of a species. Since ecological interactions must be most frequent among proximal organisms, an examination of absolute abundances in a localized area, or of relative abundances if several localized areas are combined in analysis, will provide us with insight into the nature of those interactions. However, an examination of absolute abundances over a large area, although it may provide substantial insight into the overall organization of the biosphere, is unlikely to provide information about the most rigorous ecological interactions, those among species that occur together.

We propose an examination of the distribution of relative abundances in several communities to test the validity of the concept of dominance and to test the relationship of the relative abundance of species to techniques of exploiting environmental gradients, both in space and time. We are primarily interested in the relationship between relative abundance or degree of dominance in a particular community and the degree of specialization or generalization of the particular species. We propose to measure the degree of specialization as the ability of a species to exploit an environmental range, either in space or time, and hence, as the ability of a species to maintain populations in differing types of environments. Generalist and specialist species are defined relative to one another, with the former being able to maintain themselves over a broader environmental range than the latter. We are dealing with composite environments as encountered by species in nature and have made no attempt to subdivide the niche parameters of the species. Since we are using data that were used to document species composition of communities along either time or space gradients, it was impossible to separate the parameters. It is hoped that we will stimulate tests of the relevance of the model proposed to such carefully defined niches.

We assume that dominance is the appropriation of potential niche space of certain subordinate species by other dominant species and so can be manifested most clearly only within a trophic level; that is, a producer cannot dominate a decomposer or predator because the immediate sources of their energy and inorganic nutrients are not overlapping. This does not mean that these types may not be mutually limited by resources since it is obvious that what is incorporated in support structure for predators is not available for producers. We also assume that communities tend toward saturation of the environment with biotypes so that many species have overlapping niche spaces. One distinction we make is between dominant and essential species. Certain species are essential because of their important function in mineral cycling or energy flow (*7*) and, at the same time, may not be dominant in the sense of occupying niche space potentially occupied by other species in the system.

In the initial tests of dominance and the formulation of the dominance

Dr. McNaughton is associate professor of botany and Dr. Wolf is assistant professor of zoology at Syracuse University, Syracuse, New York 13210.

model, we have utilized data on plant communities from (i) Chadwick and Dalke's (*8*) data on a 100-year successional sequence on sand dunes in Nevada, (ii) McNaughton's (*9*) study of the effects of soil type and exposure on California grasslands, and (iii) Whittaker's (*10*) data on the tree species at low elevations on gabbro-derived soils in the Siskiyou Mountains of Oregon. We used these data because they provided roughly equal numbers of species occurrences including a variety of vegetation types, methods of community analysis, and environmental situations. Chadwick and Dalke analyzed shrub communities of different ages using coverage to estimate importance. McNaughton examined grasslands along a moisture gradient, including an abrupt discontinuity between sandstone and serpentine soils, and used biomass to estimate species importance. Whittaker analyzed forests along an intuitive exposure gradient with no abrupt discontinuities and used density to assess species importance.

Relative abundance of the *t*th species in the community is measured as

$$I_t = 100\ (y_t/Y) \qquad (1)$$

where y_t is the abundance of species *t*, *Y* is the sum of all abundances in the stand, and I_t is the relative abundance of *t* as a percentage of total abundance.

Distribution of Abundances in Communities

The most satisfactory general theory of the distribution of absolute species abundance in nature is Preston's (*11*) canonical system. As Whittaker (*6*) has pointed out, however, the distribution of sets of species in communities may depart from the lognormal distribution of Preston in which

$$n_R = n_0 e^{-(aR)^2} \qquad (2)$$

where n_R is the number of species in the *R*th octave from the modal octave, n_0 is the number of species in the modal octave, and *a* is a constant such that

$$a = n_0(\pi)^{\frac{1}{2}}/\Sigma n \qquad (3)$$

Our first test is an examination of the absolute abundances of species in the three plant systems to test for fit to the Preston model. The data were combined as six octaves of abundance from least- to most-abundant species

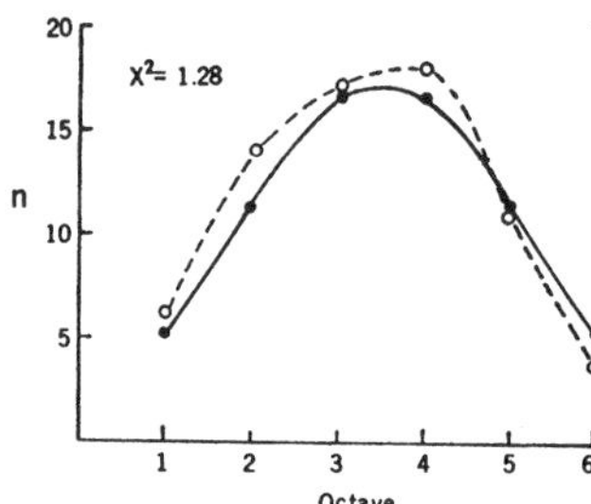

Fig. 1. Observed (O) frequencies of species absolute abundances in 26 plant communities compared with the frequencies predicted (●) by Preston's lognormal equation; *n*, number of species; *Octave*, octave of absolute abundance.

(*12*). A chi-square test of fit to Preston's model indicates that the distribution of absolute abundances of the species in these systems agrees with the predictions of the model (Fig. 1). What we ask now, from Whittaker's observation that species in the same community may not fit this model, is how the distribution of relative abundances departs from the lognormal distribution.

An octave plot of relative abundances (Fig. 2) is truncated on the right by the upper limit of 100 upon the relative abundance of species, compared to absolute abundances which have no independently definable upper limit. What we are most interested in, however, is the distribution of relative abundances around the mode. If they are symmetrical about the mode, in fit to Preston's lognormal distribution, we may assume that there are no species interactions in a community which do not occur in a universe. That is, fit of

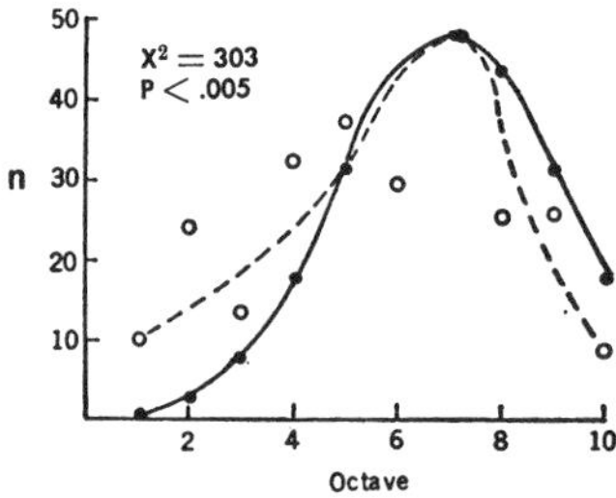

Fig. 2. Observed (O) frequencies of species relative abundances in 26 plant communities compared with the frequencies predicted (●) by Preston's lognormal equation; *n*, number of species occurrences; *Octave*, octave of relative abundance.

relative abundances to the lognormal would disprove the overlapping niche assumption stated earlier. However, if there is niche overlap among species in a common area, and species are not equally efficient in the exploitation of overlap zones, then dominance will occur, and relative abundances will be impoverished in classes above the mode and enriched in classes below the mode. We test for dominance by testing for fit to Eq. 2. Data from the three studies of vegetation confirm the occurrence of dominance (Fig. 2) with frequencies greater than predicted in the abundance classes below the mode and less than expected in the abundance classes above the mode.

We can test more thoroughly for the nature of the departure from the lognormal by the regression of n/n_R on the upper limit of the relative abundance octave where *n* is observed frequency. We find (Fig. 3) that

$$\log n/n_R = 0.481 - 0.471 \log I_u$$

where I_u is the upper limit of the octave. From this equation, $I_u = 10.5$ percent when $n = n_R$, which suggests that species with abundances greater than this will be less frequent than we would predict if communities were universes, while species with relative abundances below 5.25 percent will be more frequent than we would predict from the universe. It seems likely that species for which $I > 10.5$ percent are probably occupying niche space of species for which $I < 5.25$ percent.

Dominance and the Niche in Species

If we consider the environment as a single dimension, we can plot *I* against the environment and expect, from Whittaker (*6*), to get a series of bell-shaped curves. One of the more interesting questions in ecology is the relationship between the form of these curves and I_{max} for a species. The form of these curves relates to the problem of niche width. We propose to measure niche width as

$$W = \left[\frac{\Sigma(y_p \cdot p)^2 - (\Sigma y_p \cdot p)^2/\Sigma y}{\Sigma y}\right]^{\frac{1}{2}} \qquad (4)$$

where *p* is the position of the community in the environmental ordering (from 1 to 10), y_p is the importance of the species in that community, Σy is the total importance of the species for all of its occurrences, and *W* is niche width of the species. Another assessment of niche width, pointed out by

Levins (*13*), is the derivation from information theory with

$$H = -\Sigma r_p \log r_p \qquad (5)$$

where r_p is the proportion of the species' abundance occurring in position p, and e^H gives the number of sites on which the species would occur if it were equally distributed among all positions. This measure is closely related to W (Fig. 4) with

$$e^H = 0.873 + 2.179\ W$$

The principal differences between the two definitions is that e^H assumes that there is no information in the environmental ordering while W weights the importances for their positions in the environmental gradient. More formally, we may say that W measures constancy of relative abundance over a range of environments.

An examination of the relationship between niche width of a species and the dominance of that species as defined by its maximum contribution to community structure (Fig. 5) shows that

$$W = 0.659 + 0.025\ I_{max}$$

This indicates that those species which are most dominant, in the community where they make the maximum contribution, have the broadest niche. Species which have a high relative abundance in the community of their maximum development, also have the broadest niches. This is a somewhat different observation than Levins' (*13*) that the most abundant Puerto Rican *Drosophila* species have the broadest niches since he refers to absolute abundance rather than ecological efficiency under optimum conditions, which I_{max} estimates.

The idea of the niche was first proposed as a description of the dissimilar ecological requirements of different bird species (*14*) but was first generalized precisely by Hutchinson (*15*). The fundamental proposition of niche theory, arising out of competition experiments by Gause (*16*), is the competitive exclusion principle which states that species with identical niches cannot coexist. Most of modern niche theory (*17*) derives from efforts to relate the competitive exclusion principle to Hutchinson's n-dimensional niche. Niche theory provides two alternative explanations of the greater niche width of more dominant species. Either dominant species are generalists with adaptations to many more dimensions in their niches and, as a result, less frequently encounter a limiting dimension, or they are specialists that have evolved adaptations to a single dimension which is most likely to be limiting in the current environmental array. If the dominants are generalists, all of the species in a trophic level have a certain similarity of ecological requirements, with the relative efficiences in exploiting these requirements reflected in the species' relative abundances. The subordinate species, then, can coexist only by occupying portions of niche dimensions where the dominants are inefficient. In this explanation, the dominants are generalists while subordinate species are specialists less through genetic requirements than through being excluded, by the greater efficiency of the dominants, from some environmental dimensions they would occupy in the absence of the dominants. In the alternative explanation, all species are specialists, and the relative abundances reflect the abundances of the speciality. That is, there are a variety of environmental factors on each site likely to be limiting and the species can exploit only one of them efficiently. For instance, if a plant develops adaptations to exploit low soil nitrogen levels, it may not simultaneously exploit low potassium levels. In a site where nitrogen was generally low and potassium was infrequently low, this species would be dominant and a species specialized for the exploitation of low soil potassium would be a subordinate species restricted to spots where nitrogen was high and potassium was low.

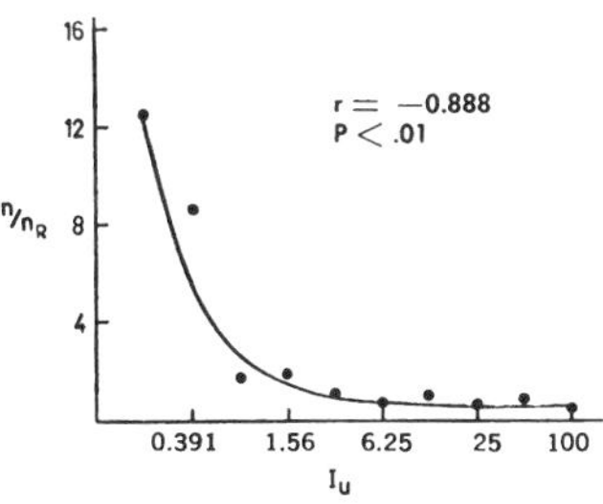

Fig. 3. Relation between observed (n) and predicted (n_R) frequencies in a relative abundance octave and the upper limit of that abundance octave (I_u).

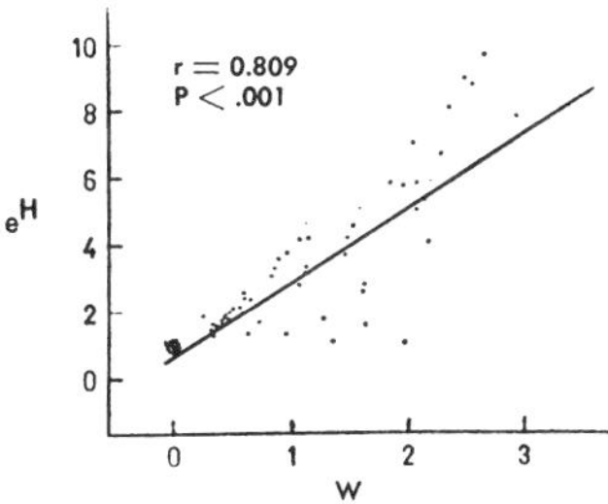

Fig. 4. Relation between the niche-width estimator proposed by Levins (e^H) and the measure proposed in this paper (W).

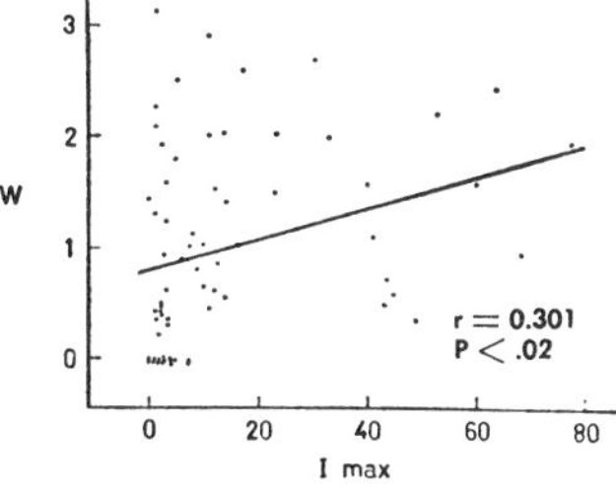

Fig. 5. Relation between the niche width of a species (W) and the contribution of the species in the community where it attains its greatest relative abundance (I_{max}).

There is some evidence to support the hypothesis that all species are specialists and that the most abundant species have specialized on a widely distributed environmental parameter which forces other species into peripheral specializations. Connell (*18*) has shown that the most abundant barnacle in marine communities of the Scottish coast is specialized for occupation of space below the high tide line. Above the high tide line it is replaced by another species capable of living under periodic desiccation. The interesting point is that the desiccation-adapted, and less abundant, barnacle, which would have a narrower realized niche, is actually capable of occupying inundated sites if the other species is removed. But the most abundant and widely distributed species cannot occupy sites above the tide line, even if competition is eliminated. Similar evidence is available from the marsh plants, cattails. The broad-leaved cattail is a widely distributed species which, except in areas with a high frequency of saline sites, is much more abundant than another species, the narrow-leaved cattail, which is generally restricted to saline habitats (*19*). From this, we would conclude that the narrow-leaved cattail is a saline habitat specialist. Experiments on salt tolerance controvert this (*20*) by indicating that whereas narrow-leaved cattail, the presumed specialist, can occupy both high and low salt conditions, the broad-leaved cattail can occupy only freshwater sites. Ecological specialization, then, arises out of physiological generalization. The narrow-leaved cattail is an ecological specialist on high salt sites because it has not physiologically specialized, as the broad-leaved cattail has, for effective competition in freshwater sites. The narrow-leaved cattail becomes an ecological specialist out of inability to compete under freshwater conditions, rather than out of inability to grow at these conditions. The broad-leaved cattail, by physiological specialization on the most widely distributed type of habitat, becomes an ecological generalist.

Dominance and the Niche in Communities

A related problem in community ecology is the relationship between dominance in the community and position of the community on a habitat gradient. To gain insight into this problem, we have examined McNaughton's (*21*) community dominance index

$$DI = 100\ (y_{1,2}/Y) \qquad (6)$$

where $y_{1,\,2}$ is the abundance of the two top species on the dominance-diversity curve (*6*), in relation to the environmental ordering of mesic, mesic-sandstone, and old stands to xeric, xeric-sandstone, and young stands in the three vegetation studies. We discover that

$$DI = 58.34 + 1.91\,B$$

where *B* is environmental position from "equitable" (*1*) to "harsh" (*10*) sites, but there is substantial scatter around the line ($r = 0.345$ for $0.1 > P > .05$ with d.f. = 24). It seems likely, however, that the environmental orderings, although internally consistent, allowing us to use them in defining niche width earlier, are not consistent with one another and thereby create scatter in this analysis. In addition, none of these orderings contain moist sites, which should be included for a conclusive analysis of community dominance properties in relation to environmental gradients.

To obviate these problems, we utilized Dix and Smeins' (*22*) analysis of marsh, meadow, and prairie vegetation in North Dakota. The moisture gradient sampled ranged from "permanently incomplete" to "excessive" drainage. In accordance with the above analysis, dominance within the community increases from mesic to dry sites. However, this analysis also indicates that dominance increases from mesic to wet sites so that the curve is antimodal and the best fit line is a second degree polynominal (Fig. 6) with

$$DI = 28.1 - 7.75\,B + 0.842\,B^2$$

where *B* is the position of the community in the ordering from 1 (dry) to 10 (wet). There is very little scatter around this line, indicating that the scatter above probably does arise out of environmental orderings which are not cross-consistent.

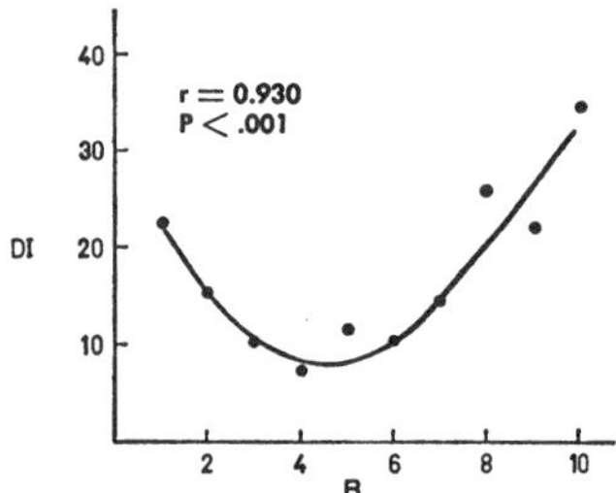

Fig. 6. Relation between the degree of dominance in a community (*DI*) and the position of that community on a habitat gradient from dry (1) to wet (10).

The analysis of North Dakota vegetation allows us to test the relationship between dominance and diversity. An inverse relationship between these two community properties has been proposed (*23*). Some authors (*24*) have argued that the documentation of this relationship elsewhere (*21*) is a mathematical artifact arising out of interdependence of definitions. Although the argument has been answered (*25*), an additional examination is allowed by the data of Dix and Smeins, in which we define diversity as its simplest component, *R*, of the diversity equation

$$d = (R - 1)/\ln N \qquad (7)$$

where *d* is diversity in bits per individual, *R* is number of species in the community, and *N* is the number of individuals in the community (*26*). That is, we compare diversity as its simplest case, floristic richness of the community, with DI. The relationship between these two properties in the North Dakota communities is linear and negative, with

$$R = 71.380 - 2.024\,DI$$

and $r = -0.890$ for $P < .001$ with $N = 10$, in accord with previous studies of terrestrial systems (*9*) and with the model (*23*).

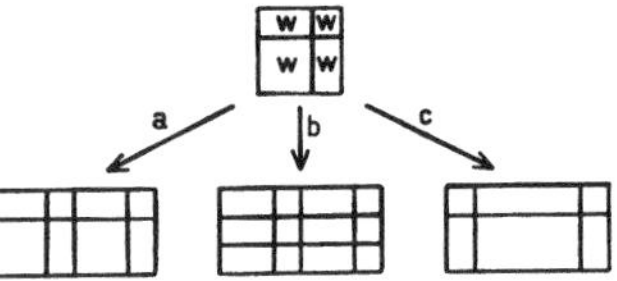

Fig. 7. Diagrammatic representation of methods by which species may be added to the community, assuming that species addition is accompanied by an increase in carrying capacity of the system (*K*) where *W* is niche width of a species (represented here as an area) and *K* is envisioned as an approximation to the sum of the niche structures of the species. That is, the model assumes a close approach to saturation of the available resources with exploitive biotypes. In alternative *a*, species are added to the system in direct proportion to *K* increase. In *b*, species are added more rapidly than *K* increases, so niches are "squeezed." And, in *c*, species are added less rapidly than *K* increases so that niches expand with increasing richness.

The intriguing observation here, however, is that at this point modern ecosystem ecology of Margalef and others converges with the classical community ecology of Clements (*27*). For the stands with minimum dominance and maximum richness, that is, climax by modern theory, are midprairies dominated by *Andropogon scoparius, Stipa spartea,* and *Sporobolus heterolepis* which occupy "by far the largest acreage of ungrazed prairie" in the area (*22*), that is, climax by classical theory. The increase in dominance and decrease in richness on both sides of this type of community on the habitat gradient suggest that the community best organized to cope with site contingencies may be objectively defined as the community on the landscape with minimum dominance. Although the argument at this point becomes circular, the conclusion generated from the North Dakota communities is in accord with early intuitive ecological theory (*27*), more recent empirically derived theory (*23*), and mathematically generated theory (*28*).

The richness property of communities brings us to one of the fundamental questions of ecology for which, through niche-width analysis, we may now propose an answer. That is, how are species added to communities? We, with Hutchinson (*15*), conceive of the niche as an *n*-dimensional hypervolume, with *n* defined by the variety of physiological requirements of the population and with the integrated volume a function of the range and efficiency of dimensional exploitation. The community also has a niche hypervolume, defined by the hypervolumes of its constituent species. For visualization, however, we simplify the model to a planar system (Fig. 7), and ask what happens during community development as species are added to the system. We assume, as the model indicates, that communities are near the innate carrying capacity of the environment, K, so that $\Sigma W = K$. Our question is how species fill the K-area as more species are added to the system. Species may be added to the system at the same rate at which K expands (alternative *a*), in which case niche width and richness are unrelated, or (alternative *b*) species may be added to the system more rapidly than K expands, generating a negative relationship between richness and mean niche width of the community constituents. To be logically consistent, although the argument is not biologically compelling, we may assume (alternative *c*) that species are added to the community less rapidly than K expands, with an increase in mean niche width as richness increases. These arguments are essentially an expansion of a previous dominance-diversity model in which K was treated as a constant (*25*). We find that alternative *b* is supported by the data, with a decrease in niche width as richness increases (Fig. 8). But the nature of the relationship is different for the forest communities of the Siskiyou Mountains and the grasslands and shrubs in California and Nevada. For the trees

$$\overline{W} = 2.789 - 0.063\,R$$

and for the shrub and grass systems

$$\overline{W} = 1.759 - 0.063\,R$$

where $\overline{W}$ is mean niche width of the community and R is the number of species in the community. The average niche width is somewhat greater in the forests, but the rate of decrease in niche width as species are added is similar in all of the systems. This indicates that niches are being "squeezed" as species are added to the system, and we interpret this to mean that diversification generates increased competition. Although we draw the niches discretely in Fig. 7, we believe that the discreteness arises out of competitive exclusion rather than out of totally distinct physiological requirements. We can also analyze mean niche width in

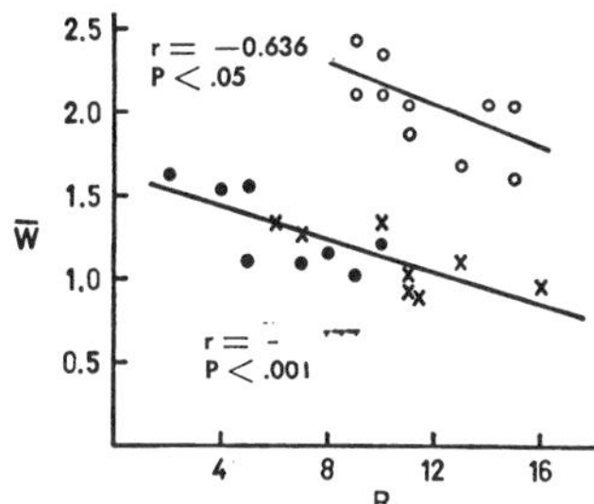

Fig. 8. Relation between mean niche width ($\overline{W}$) of the system and number of species (R) contributing to the system. Communities are shrubs (●), grasslands (X), or forests (O).

relation to community dominance and find that for the trees

$$\overline{W} = 1.18 + 0.014\ \mathrm{DI}$$

with $r = 0.787$ for $P < .01$ with $N = 10$, and for the shrub and grass communities

$$\overline{W} = 0.579 + 0.009\ \mathrm{DI}$$

with $r = 0.734$ for $P < .01$ with $N = 16$.

Although the niche characteristics of the species are similar in relation to the maximum relative abundance of the species whether the plant is a shrub, an herb, or a tree, the forest communities are organized somewhat differently than the shrub and grass systems. The niche widths are larger in the forest for a given diversity or dominance, but the change in niche width with changes in dominance-diversity relations are similar in all of the systems. The similarities of the slopes indicate generally comparable responses to the addition of species although the "starting points" are somewhat different in tree than in the shrub and grass communities.

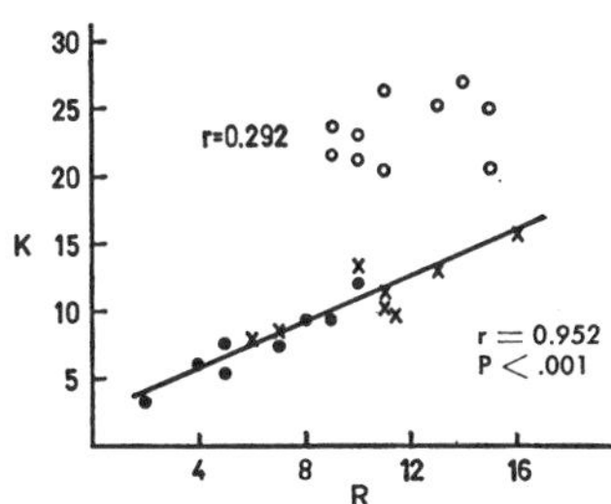

Fig. 9. Relation between carrying capacity of the system (K) and number of species (R) contributing to the system. Communities are shrubs (●), grasslands (X), or forests (O).

Finally, we ask whether K does increase, as we suppose, with increasing species diversity. We make no judgment on the relationship between the two, whether an increase in richness is a response to, or the generator of, K. That is, increasing K may allow more species to occupy the site by allowing more overlap of niches, or increasing richness may generate a larger K through more efficient utilization of total site resources. In fact, there may be no cause and effect relation definable between K and richness. We define K as ΣW, from Fig. 7, and ask whether K is related to R. For the shrub and grass systems, there is a significant relation (Fig. 9) such that

$$K = 3.11 + 0.778\,R$$

but for the forests the association is not significant even though the slope of the line,

$$K = 19.79 + 0.319\,R$$

is in the same direction. This suggests that K changes much less consistently, and less rapidly, in the forests than it does in the shrub and grassland systems (*29*). Overall, the plant systems fit alternative *b* of the model although the rate of K expansion is somewhat less in the trees than in the grass or shrub systems.

Tests of the Dominance Model

With theoretically derived models, errors are most likely to arise from the premises from which the model obtains, and the validity of these premises can be tested only on real systems. With empirically derived models, such as the present one, errors are most likely to arise from the peculiarities of the data used to generate the model. The test for validity here also is reliability of the conclusions in application to other real systems. To make this test, we have compiled data from two additional sources. One is a series of bird communities in the southeastern United States which vary along a successional sequence (*30*), and the other is a series of shrub-grass communities in the northwestern United States which vary along a broad climatic gradient with sampling stabilized for substrate and exposure (*31*).

In all systems, the trends are in the same direction (Table 1). There tend to be fewer species than expected with high relative abundance and more spe-

Table 1. Components of the dominance model tested on shrub-grass and bird communities.

	Fit of relative abundances to lognormal distribution	
Model	$X^2 = 303.2$ for $P < .005$ with $N = 10$	$a = 0.338$
Shrub-grass	$X^2 = 33.94$ for $P < .005$ with $N = 9$	$a = 0.335$
Birds	$X^2 = 12.98$ for $P < .1$ with $N = 9$	$a = 0.341$
	Actual and predicted frequency (n/n_r) *and abundance class* (I_u)	
Model	$\log n/n_r = 0.481 - 0.471 \log I_u$	$r = -0.888$ for $P < .01$ with $N = 10$
Shrub-grass	$\log n/n_r = 0.266 - 0.258 \log I_u$	$r = -0.799$ for $P < .05$ with $N = 9$
Birds	$\log n/n_r = 0.065 - 0.098 \log I_u$	$r = -0.530$ for $P > .1$ with $N = 9$
	Niche width (W) *and dominance* (I_{max})	
Model	$W = 0.659 + 0.024\, I_{max}$	$r = 0.301$ for $P < .02$ with $N = 67$
Shrub-grass	$W = 1.13 + 0.048\, I_{max}$	$r = 0.571$ for $P < .001$ with $N = 34$
Birds	$W = 0.531 + 0.011\, I_{max}$	$r = 0.333$ for $P < .05$ with $N = 35$
	Community dominance (*DI*) and *harshness* (B)	
Model	$DI = 58.34 + 1.91B$	$r = 0.345$ for $P < .1$ with $N = 26$
Shrub-grass	$DI = 55.17 + 1.19B$	$r = 0.547$ for $P < .02$ with $N = 20$
Birds	$DI = 9.30 + 9.08B$	$r = 0.811$ for $P < .01$ with $N = 9$
	Niche width (W) *and community richness* (R)	
Trees	$\overline{W} = 2.79 - 0.063R$	$r = -0.636$ for $P < .05$ with $N = 10$
Grass and shrubs	$\overline{W} = 1.76 - 0.063R$	$r = -0.777$ for $P < .001$ with $N = 16$
Shrub-grass	$\overline{W} = 4.15 - 0.127R$	$r = -0.838$ for $P < .001$ with $N = 20$
Birds	$\overline{W} = 0.924 - 0.005R$	$r = -0.304$ for $P > .1$ with $N = 9$
	Carrying capacity $(\Sigma W = K)$ *and richness* (R)	
Trees	$K = 19.79 + 0.319R$	$r = 0.291$ for $P > .1$ with $N = 10$
Grass and shrubs	$K = 3.11 + 0.778R$	$r = 0.952$ for $P < .001$ with $N = 16$
Shrub-grass	$K = 18.21 + 0.562R$	$r = 0.460$ for $P < .05$ with $N = 20$
Birds	$K = 1.066 + 0.745R$	$r = 0.972$ for $P < .001$ with $N = 9$

cies than expected with low relative abundance, as compared to the distribution of abundances predicted by the lognormal distribution. Dominance, then, seems to be a real characteristic of the organization of communities inasmuch as high measures of importance in communities are concentrated in fewer species than in universes. However, dominance seems to be less important in the bird communities than in the plant communities. This suggests that bird species are either specialists on resources that are less abundant or they are more able to subdivide the resources.

The decrease in average niche width in a community as the number of species increases is present in all cases. However, for the plants the average niche width is greater as the number of species approaches zero than it is for the birds. This suggests that the birds are more specialized when there are few species and that species which are added are only slightly greater specialists than the ones already present. Throughout the range of communities, in this case along a successional gradient, the bird species are about equally specialized at each stage. For the plants there is a significant decrease in average niche width, suggesting that the species which are added are in fact more narrow in niche hypervolume than species occurring in less rich communities. On the model, then, the birds fit alternative *a* and the plants fit alternative *b*.

Dominant species tend to have broader niches than less dominant or subordinate species. The similarity in the *Y* intercepts for the plant data are striking. Compared to birds, species of plants which contribute very little to community production are more broadly distributed on an environmental gradient. The major difficulty with all these comparisons of niche width is that *W* is in part a function of the number of stages or sites on which a species occurs and will be larger for environmental series that are more finely divided. However, it must be remembered that in one collection of data from which the model was derived, the ordering was eight stages over a period of 1000 years, far more gross than the bird divisions. The birds generally have narrower niches for the same maximum contribution than the plants. This may arise from the types of resources which the two groups utilize and the distribution of these resources.

Overall, the model of niche structure in relation to species dominance and community structure is a powerful description. It encompasses diverse systems and types of data. The signs are the same for every equation generated. The marginal confidence intervals in certain cases with the birds probably arise more out of the small size of samples available rather than out of striking functional differences. We hope that more data on other animal systems will become available for tests of the dominance and niche models at different trophic levels.

Evolution of the Niche: The Meaning of Dominance

In the systems which we have examined, we find that: (i) dominance is a characteristic of the most abundant species, (ii) dominant species have broader niches than subordinate species, (iii) species are added to the system by compression of niches or expansion of *K*, or both, and (iv) community dominance is minimum on the most equitable sites.

The observation that there tend to be few abundant species and many rare species in communities has been made frequently in the past (*32*). What we offer that is new is evidence that this phenomenon is a manifestation of species interactions that do not occur in a universe. The argument, rephrased, is: the distribution of absolute abundances of species in nature is best described by Eq. 2; if relative abundances of species in the same community (localized area) are distributed according to this equation, we may assume that there are no species interactions in a localized area that do not occur in a universe. In testing this null hypothesis, we find that all systems analyzed fail to fit the lognormal distribution. We test for the direction of departure by the regression of n/n_R on I_u and find that the slope is negative in all cases, which supports the theory of dominance and, in contradiction of Ehrlich and Holm (*4*), demonstrates that dominance is a biologically meaningful concept.

Preston (*11*) points out that the constant *a* approaches 0.2 in canonical arrangements of absolute species abundances in universes. For the arrangements of relative abundances here, the value of *a* approaches 0.34 with the 95 percent limit, by Student's *t*-test, being 0.331 to 0.345, clearly far from 0.2. The higher value of this constant indicates that the distribution of relative abundances falls off more rapidly from the modal class than the distribution of absolute abundances. Or, more formally, the larger *a* indicates increased leptokurtosis and a general depauperization of communities relative to floras and faunas. The number of species occurrences based on an *a* of 0.34 is 59 percent of the number that would be predicted from an *a* of 0.2. The reason progressively smaller samples show progressively poorer fit to Preston's model is that such samples increasingly reflect the interactions among populations. MacArthur (*33*) has recently disavowed his broken-stick model (*34*) of community organization, and it seems unlikely that models such as this which are based on randomness will provide a satisfactory description of communities because communities are grounded in order. Ecologically, our argument is quite straightforward. Individuals react primarily with their neighbors. Populations react primarily, through their individuals, with intermixed or adjacent populations. Unequal efficiency at these interactions generates dominance, thereby distorting the lognormal in a regular and predictable way.

The most important question raised by our analysis is: What determines niche width? The alternatives, restated, are (i) relative efficiencies at exploiting critical limiting factors, or (ii) frequency and carrying capacity of the exploitation specialty. We believe that the available data provide much more support for the latter proposition. However, as MacArthur points out (*35*), the statement "two species with identical niches cannot coexist" is true, but trivial, when applied to Hutchinson's *n*-dimensional niche since it is probable that no two individuals, much less species, have identical niches. There are numerous documentations of niche divergence among coexisting species, including birds (*36*), grazing ungulates (*37*), and plants (*38*). The important point, however, is that although niches diverge, they do not become distinct. These studies indicate that substantial overlap is preserved even though the centers of exploitation are nonidentical. Root's (*39*) analysis of the foliage-gleaning guild in California oak woodlands provides particularly compelling evidence of the tight packing of the exploitation volume through divergent, but overlapping, food size preferences and foraging behavior.

What, then, determines niche width? We believe it is genetic diversity. And it seems likely that the greater niche width of more abundant species would, in fact, be driven by their greater abundance (*40*). There are two major competitive interfaces in communities, the interface between individuals of the same species, and the interface between individuals of different species. Competition must arise, in part, out of contact frequency. Dispersal mechanisms will tend to maintain a somewhat higher contact frequency among individuals of the same species than among individuals of other species. This will tend to generate genetic divergence among individuals of the same species. If we assume that two species are occupying, initially, the same niche dimension, the species whose abundance is larger will, through the increased frequency of competition with members of its own species, be driven toward genetic differentiation of individuals, while the species with a slightly smaller initial abundance will be driven, by competition with members of the other species, toward increasing genetic uniformity. The latter species eventually becomes subordinate, perpetuated in the system only through specialization on another niche dimension.

To pose a reasonable test of this idea, we must have three pieces of information about a single niche dimension. We must know the rates of exploitation of the dimension by different species, the distribution of resources along the dimension, and the distribution of species exploitations along the dimension. Root's (*39*) studies of the foliage-gleaning guild provide us with these measures. He presents measurements of frequency of observation in this niche during the breeding season, as well as estimates of the distribution of the size classes of prey present in the habitat and prey sizes taken by each of the species. The dominant species should be driven, by intraspecific competition, toward diversification to the limits of the niche resources. That is, for the major exploiter of the dimension, there should be a close correlation between the distribution of prey size classes in the dimension and size classes taken. As species become less abundant along the dimension, the correlation between their preference and availability should decrease or become negative. Of the five species observed by Root, only the most abundant gleaner, the blue-grey gnatcatcher, shows a significant positive correlation ($r = 0.722$ for $P < .01$ with $N = 16$) between percentage of food taken and percentage available in the canopy with

$$P_t = 1.360 + 0.877\, P_a$$

where P_t is the percentage of the prey taken in a certain size class and P_a is the percentage of the prey available in a certain size class. It is interesting, also, that one of the least frequent utilizers of this dimension, the warbling vireo, showed a significant negative correlation ($r = -0.554$ for $P < .05$ with $N = 16$) between food taken and food available. In fact, the correlation between the slope of diet on niche resources and abundance in the niche shows a close, though not significant, association ($r = 0.849$ with $N = 6$) in support of the argument that, within the niche dimension, the most abundant species will diversify to the limits of the dimension while the less abundant species will be crowded into increasingly peripheral portions of the dimension, resulting, finally, in certain dimensionally rare species, like the warbling vireo, showing what appears to be inability to exploit the dimension.

The species which show poor association with this dimension should each show a strong correlation between food class taken and available within another dimension. The niche dimensions represent exploitation zones and each species in a community will specialize on a different zone so that community richness becomes a measure of the number of exploitation zones or, stated in other terms, the number of niche dimensions available in the system. We do not argue that all potential dimensions are occupied. Explosive evolution, of course, arises from the opening of a potential dimension through a new adaptive event. Neither do we argue that there may not be more efficient exploiters of the dimension than those currently present. Successful invasions that displace native species, like the rabbit invasion of Australia, indicate that more efficient occupants may occur in remote ecosystems.

In the species, there is a change in genetic structure from one end to another of the niche dimension exploited. The blue-grey gnatcatchers with longer bills exploit generally larger food classes, we assume, just as the warbling vireo population exploits larger food than the gnatcatcher population. The evolution of ecotypes (*41*) is a particularly easily examined case of genetic differentiation along functionally definable niche dimensions. This, of course, raises the question of temporal gradients, and we ask whether there are changes in the genetic structure of populations along successional gradients. Dobzhansky's (*42*) documentation of seasonally associated changes in *Drosophila* chromosomal inversions suggests that the answer is affirmative. This suggests that life is so wasteful of itself because of the inability of individuals to store sufficient information to deal with the information content of the environment.

There are two mechanisms for the storage of biological information. In the individual, information may be stored as heterozygosity. In the population, alleles may be stored both as heterozygosity and as heterogeneity. Small neighborhood size (*43*), self-fertilization, and strong selection (*41*) will tend to generate heterogeneous populations with a large degree of homozygosity. Large neighborhood size, outcrossing, and weak selection will tend to generate population homogeneity and individual heterozygosity. Heterozygosity generates phenotypic plasticity, physiological generalization, and ecological specialization. The species with a high degree of heterozygosity would be, like the barnacle that lives above the tide line (*18*), capable of living in a broad zone but forced, by competition, to live in a more narrow zone. Homozygosity generates phenotypic rigidity, physiological specialization, and ecological generalization.

Our analysis of niche structure of communities in relation to richness indicates that there are likely to be substantial changes in gene structures of populations through succession and in the types of species that occur in different stages. Successional species, in species-poor systems, should be heterogeneous, but their individuals should be more homozygous. Species occurring in climax communities should be less heterogeneous with the individuals more heterozygous.

What we are essentially arguing is adaptation versus acclimation as generators of niche structure. In climax species, we believe, there is more niche broadening through acclimation by the individual. In successional species, there is more niche broadening through diversification within the population. Kruckeberg (*44*) has observed that woody species occurring on both serpentine and more conventional edaphic substrates are rarely differentiated into ecotypes whereas earlier successional species commonly show striking differentiation into edaphic races. He points out the contradiction between this evidence and Baker's (*45*) observation that general-purpose genotypes are more likely to occur in early than in late successional species. The conflict arises out of Baker's failure to distinguish between the ability of the species and the ability of the individual to occupy diverse habitats. The successional species is likely to be capable of occupying a greater habitat range, while its individuals are likely to be organized for narrower portions of that range.

It is well known from studies of ecotypes that phenotypic homeostasis increases from north to south (*46*). It is well known from community ecology that diversity of communities increases from north to south (*47*). As a general conclusion, we state that niche structure of the species arises out of the conflicting demands for efficiency and adaptability, with diversification of the environment and interspecific competition driving evolution toward individual heterozygosity and population homogeneity, while uniformity of the environment and intraspecific competition drive evolution toward individual homozygosity and population heterogeneity.

A Cautionary Conclusion

We believe that our analyses of dominance structure and niche properties provide substantial insights into the organization of communities and species. We rely heavily, however, upon regression analysis and correlation, and we present the best-fit equations for what they are—initial approximations, rather than as final descriptions. In fact, the ranges of the intercepts and regression coefficients tend to be rather small, suggesting that we are close to general statements. There is no question from the close agreements of correlation analyses that we describe strongly associated variables, but the lack of identity among the regression terms indicates that we approach but do not reach descriptions of causality.

The extent to which our description departs from causality probably depends upon its distance from community analysis in energetic and entropic terms. Much has been made in ecology recently of the application of information theory to formulation of definitions, and our emphasis upon richness flies in the face of much of this application. If, however, we define information (*48*) as

$$T = k \log C \qquad (8)$$

where k is a positive constant and C is the number of possible cases from which one may be selected, then R becomes an approximation to C since it presumably describes the number of exploitation zones in a system. Wilson (*48*) has pointed out that T, rather than being negentropy as Brillouin supposed, is the same as entropy as defined in statistical mechanics. Our inability to distinguish niche broadening through heterozygosity and heterogeneity probably is the principal barrier between our analysis and regressions describing causality. A decrease in the scatter in our analysis probably depends upon true estimates of allelic frequencies in ecological systems. It would be interesting to compare allelic frequencies (*49*) in populations of a species which occur at different points along a successional gradient.

References and Notes

1. J. E. Weaver and F. E. Clements, *Plant Ecology* (McGraw-Hill, New York, 1938); G. L. Clarke, *Elements of Ecology* (Wiley, New York, 1954); A. M. Woodbury, *Principles of General Ecology* (McGraw-Hill, New York, 1954); H. J. Oosting, *The Study of Plant Communities* (Freeman, San Francisco, 1958); E. P. Odum and H. T. Odum, *Fundamentals of Ecology* (Saunders, Philadelphia, 1959); H. C. Hanson and E. D. Churchill, *The Plant Community* (Reinhold, New York, 1961); A. H. Benton and W. E. Werner, *Field Biology and Ecology* (McGraw-Hill, New York, 1966); R. L. Smith, *Ecology and Field Biology* (Harper & Row, New York, 1966); R. Daubenmire, *Plant Communities* (Harper & Row, New York, 1968).
2. A. S. Boughey, *Ecology of Populations* (Macmillan, New York, 1968); K. E. F. Watt, *Ecology and Resource Management* (McGraw-Hill, New York, 1968); E. J. Kormondy, *Concepts of Ecology* (Prentice-Hall, Englewood Cliffs, N.J., 1969).
3. P. Greig-Smith, *Quantitative Plant Ecology* (Butterworths, London, 1964); K. A. Kershaw, *Quantitative and Dynamic Ecology* (Arnold, London, 1966).
4. P. R. Ehrlich and R. W. Holm, *Science* **137**, 652 (1962).
5. R. MacArthur, *Biol. Rev.* **40**, 510 (1965); E. C. Pielou, *J. Theor. Biol.* **13**, 131 (1966).
6. R. H. Whittaker, *Science* **147**, 250 (1965).
7. E. P. Odum, *Jap. J. Ecol.* **12**, 108 (1962).
8. H. W. Chadwick and P. D. Dalke, *Ecology* **46**, 765 (1965).
9. S. J. McNaughton, *ibid.* **49**, 962 (1968).
10. R. H. Whittaker, *Ecol. Monogr.* **30**, 279 (1960).
11. F. W. Preston, *Ecology* **43**, 185, 410 (1962).
12. In this, and succeeding, statistical tests we have used standard analytical techniques. Curves were tested for goodness of fit by chi-square according to G. G. Simpson, A. Roe, and R. C. Lewontin [*Quantitative Zoology* (Harcourt, Brace, New York, 1960)]. Correlation coefficients were calculated as the geometric mean of the regression coefficients, and terms in best fit line equations were calculated by least squares [J. B. Williams, *Statistical Analysis* (Olivetti-Underwood, New York, 1968)] or Bartlett's method [R. R. Sokal and F. J. Rohlf, *Biometry* (Freeman, San Francisco, 1969)], according to appropriateness.
13. R. Levins, *Evolution in Changing Environments* (Princeton Univ. Press, Princeton, N.J., 1968).
14. J. Grinnell, *Auk* **21**, 364 (1904).
15. G. E. Hutchinson, *Cold Spring Harbor Symp. Quant. Biol.* **22**, 415 (1958).
16. G. F. Gause, *The Struggle for Existence* (Williams & Wilkins, Baltimore, 1934).
17. R. MacArthur, *Ecology* **39**, 599 (1958); R. Levins, *Amer. Natur.* **96**, 361 (1962); G. E. Hutchinson, *The Ecological Theatre and the Evolutionary Play* (Yale Univ. Press, New Haven, Conn., 1965); T. W. Schoener, *Evolution* **19**, 189 (1965); R. MacArthur and E. Pianka, *Amer. Natur.* **100**, 603 (1966); R. Levins, *Evolution in Changing Environments* Princeton Univ. Press, Princeton, N.J., 1968).
18. J. H. Connell, *Ecology* **42**, 710 (1961).
19. S. G. Smith, *Amer. Midland Natur.* **78**, 257 (1967).
20. C. McMillan, *Amer. J. Bot.* **46**, 521 (1959).
21. S. J. McNaughton, *Nature* **216**, 168 (1967).
22. R. L. Dix and F. E. Smeins, *Can. J. Bot.* **45**, 21 (1967).
23. R. Margalef, *Amer. Natur.* **97**, 357 (1963).
24. M. P. Austin, *Nature* **217**, 1163 (1968).
25. S. J. McNaughton, *ibid.* **219**, 180 (1968).
26. R. Margalef, *Mem. Real Acad. Cienc. Artes Barcelona* **32**, 373 (1957); N. Hairston, *Ecology* **40**, 404 (1959). Our use of richness here

anticipates our arguments later regarding the meaning of richness, but by this stage of the analysis we had come to believe that more elaborate measures of diversity [see, for instance, R. P. McIntosh, *Ecology* **48**, 392 (1967) for a superb development of a diversity index], although formally gratifying, may have no particularly compelling biological meaning. The meaning of such indices may be more metaphoric than functional.
27. F. E. Clements, *J. Ecol.* **24**, 252 (1936).
28. E. Leigh, *Proc. Nat. Acad. Sci. U.S.* **53**, 777 (1965).
29. Dominance and richness are inversely related in all of these systems with $R = 17.17 - 0.084$ DI ($r = -0.626$ for $P < .1$) for the trees, and $R = 23.48 - 0.209$ DI ($r = -0.888$ for $P < .001$ for the shrubs and grasslands. For the trees, there is a much closer association between K and DI ($r = -0.606$ for $P < .05$) with $K = 32.82 - 0.144$ DI than between K and R, as developed in the text. For the shrubs and grasses, the relationship is similar with $K = 23.29 - 0.189$ DI and $r = -0.879$ for $P < .001$.
30. The bird density data of D. W. Johnston and E. P. Odum [*Ecology* **37**, 50 (1956)] were converted to biomass values using weights from A. Norris and D. W. Johnston [*Wilson Bull.* **70**, 114 (1958)] and unpublished data of L. L. Wolf.
31. R. Daubenmire, *Science* **151**, 291 (1966).
32. C. Raunkiaer, *Biol. Medd. Kbh.* **7**, 1 (1928); P. Greig-Smith, *Quantitative Plant Ecology* (Butterworths, London, 1957); N. G. Hairston, *Ecology* **40**, 404 (1959).
33. R. MacArthur, *Ecology* **47**, 1074 (1966).
34. ———, *Proc. Nat. Acad. Sci. U.S.* **43**, 293 (1957).
35. ———, in *Population Biology and Evolution*, R. C. Lewontin, Ed. (Syracuse Univ. Press, Syracuse, N.Y., 1968), p. 159.
36. ———, *Ecology* **39**, 599 (1958).
37. M. D. Gwynne and R. H. V. Bell, *Nature* **22**, 390 (1968).
38. J. L. Harper, in *Population Biology and Evolution*, R. C. Lewontin, Ed. (Syracuse University Press, Syracuse, N.Y., 1968), p. 139.
39. R. B. Root, *Ecol. Monogr.* **37**, 317 (1967).
40. G. Svardson, *Oikos* **1**, 157 (1949).
41. S. K. Jain and A. D. Bradshaw, *Heredity* **21**, 407 (1966); J. L. Aston and A. D. Bradshaw, *ibid.*, p. 649; T. McNeilly, *ibid.* **23**, 99 (1968); J. Antonovics, *ibid.*, pp. 219 and 507.
42. Th. Dobzhansky, *Genetics* **28**, 162 (1943).
43. S. Wright, *Ann. Eugen.* **15**, 323 (1951); H. W. Kerster and D. A. Levin, *Genetics* **60**, 577 (1968).
44. A. R. Kruckeberg, *Brittonia* **19**, 133 (1967).
45. H. G. Baker, in *The Genetics of Colonizing Species*, H. G. Baker and G. L. Stebbins, Eds. (Academic Press, New York, 1965), p. 147.
46. C. McMillan, *Ecol. Monogr.* **29**, 285 (1959); S. J. McNaughton, *ibid.* **36**, 297 (1966); *Amer. J. Bot.* **56**, 37 (1969).
47. G. L. Clarke. *Elements of Ecology* (Wiley, New York, 1954).
48. J. A. Wilson, *Nature* **219**, 534 (1968).
49. R. C. Lewontin, *Annu. Rev. Genet.* **1**, 37 (1967).
50. Supported by NSF grants GB-8099 and GB-7611.

Mission to an Asteroid

Hannes Alfvén and Gustaf Arrhenius

Importance of Studying Asteroids

As long as the asteroids were regarded as fragments of a broken-up planet, interest in them was limited. There are now good reasons to believe that the asteroidal belt represents an intermediate stage in the formation of planets. This links the present conditions in the asteroidal region with the epoch in which the earth and the other planets were accreting from interplanetary grains. Hence, in order to understand how the solar system originated it may be essential to explore the asteroids.

We have already tangible samples of the earth and of the moon. Furthermore, meteorites have been carefully investigated. It is important to study also bodies intermediate in size between the moon and meteorites. The asteroids are such bodies. In this respect a study of an asteroid is more important than the study of Mars or Venus.

The Apollo 11 results suggest that the chemical composition of the moon may be significantly different from that of the terrestrial planets, the meteorites, and the sun. It is also possible that these differ from each other. It is therefore important to obtain samples of other bodies in order to establish the range of variation in elemental abundance in the solar system. There are indications that the chemical abundance in different bodies depends on their distance from the sun. An examination of samples from Mars and from one or several asteroids would clarify this. The data from asteroids would be easier to interpret than those from Mars since the asteroids are less likely to be differentiated.

Since a manned landing on Mars will not take place until after 1980, it is of interest to discuss whether a sample of an asteroid may be obtained in an easier way, at an earlier time, and as a technologically intermediate step.

A few asteroids have diameters of the order of 100 kilometers, but most of them have diameters as small as a few kilometers; probably there are also large numbers of microasteroids covering the entire range below the observed sizes.

A sample from an asteroid could in principle be obtained in two different ways:

1) A spacecraft could land on a large asteroid. This would be easier than a lunar landing because of the fact that the escape velocity of the asteroid is negligible. On the other hand, an asteroid mission is more difficult because of the distance to the asteroids and their large relative velocity with respect to the earth when some of them come into our neighborhood. An asteroid landing would be much easier than a landing on Mars.

2) A small asteroid could perhaps be captured and brought back to the earth, and either landed on the earth's surface or stored in orbit around the earth for later investigations. This would require that the asteroid be very small (mass less than 100 kilograms). The spacecraft need not necessarily be brought up to the full speed of the asteroid if some device could be constructed which catches and slows down the asteroid. A major problem is to detect objects that small and to compute their orbits.

Asteroids Close to the Earth

We shall confine ourselves to discussing missions—manned or unmanned—to asteroids in our close environment. There are a number of asteroids which at regular intervals come close to the earth. A landing on such an asteroid would be of special significance to the investigation of the early history of the solar system. Since such asteroids have acted as probes registering events in the neighborhood of the earth's orbit, an analysis of them could make possible a reconstruction of the essential features of the earth, the moon, and the earth-moon system as they were in the past. One could also derive clues to the history of the sun.

Dr. Alfvén is professor at the Royal Institute of Technology, Stockholm, Sweden, and visiting professor at the Department of Applied Physics and Information Science, University of California, San Diego. Dr. Arrhenius is professor at Scripps Institution of Oceanography, University of California, San Diego.

PAPER 26

Geographic Ranges of North American Terrestrial Mammals (1977)

S. Anderson

Commentary

CHRISTY M. MCCAIN

In the study of the abundance and distribution of species, the patterns and theory of geographic range sizes among species has received less attention than abundance patterns and theory. In this foundational paper, Sydney Anderson, a curator at the American Museum of Natural History, was the first to describe the frequency distribution of range sizes in mammalian faunas and to posit explanations. He subsequently extended these analyses to faunas of birds, fishes, amphibians, and reptiles (Anderson 1984a, 1984b). Anderson described the now-familiar pattern of range size–frequency distributions: most species have small ranges and large ranges are rare. He acknowledged that this pattern had been described earlier in plants by J. C. Willis (1949) who termed the decreasing frequency distribution of range sizes the "hollow curve." Anderson also noted the huge span of range sizes, as mammal ranges vary over more than six orders of magnitude.

One insight of Anderson's work is his explicit use of scale. He clearly documented that as faunal extent decreases, the linearity of the range-size distribution increases (see fig. 1). On a logarithmic scale, Anderson compared range-size distributions of all mammals, insular mammals, and several mammalian clades to show pattern variation. He contrasted these figures with the latitudinal diversity of each clade. Notably, Anderson found little support among mammalian clades, except bats, for what later became known as "Rapoport's rule"—areas of high diversity contain species with smaller ranges (Rapoport 1982; Stevens 1989). Although, in subsequent papers, Anderson found various levels of support for Rapoport's rule among the vertebrate clades. In his exploration of explanations for the range size–frequency distribution, Anderson covered statistical explanations based on the central limit theorem and explored the best-fit distributional model by testing the lognormal, broken-stick, geometric, and log-series models. He found little support for any of these distributions and posited that a new model and a contrast of other groups was necessary, which he then solidified in future papers (Anderson 1984a, 1984b). His ideas have been expanded upon and explored in great detail by recent researchers (e.g., Brown 1995; Brown et al. 1996; Gaston 1996, 2003; Gaston et al. 1998; Orme et al. 2006).

From *American Museum Novitates* 2629:1–15. Reproduced with permission from the American Museum of Natural History.

Literature Cited

Anderson, S. 1984a. Areography of North American fishes, amphibians and reptiles. *American Museum Novitates* 2802:1–16.

———.1984b. Geographic ranges of North American birds. *American Museum Novitates* 2785:1–17.

Brown, J. H. 1995. *Macroecology*. University of Chicago Press, Chicago.

Brown, J. H., G. C. Stevens, and D. M. Kaufman. 1996. The geographic range: Size, shape, boundaries, and internal structure. *Annual Review of Ecology and Systematics* 27:597–623.

Gaston, K. J. 1996. Species-range-size distributions: Patterns, mechanisms and implications. *Trends in Ecology and Evolution* 11:197–201.

———. 2003. *The structure and dynamics of geographic ranges*. Oxford University Press, Oxford.

Gaston, K. J., T. M. Blackburn, and J. I. Spicer. 1998. Rapoport's rule: Time for an epitaph? *Trends in Ecology and Evolution* 13:70–74.

Orme, C. D. L., R. G. Davies, V. A. Olson, et al. 2006. Global patterns of geographic range size in birds. *PLoS Biology* 4:1276–83.

Rapoport, E. H. 1982. *Areography: Geographical strategies of species*. Pergamon Press, Oxford.

*Stevens, G. C. 1989. The latitudinal gradient in geographical range: How so many species coexist in the tropics. *American Naturalist* 133:240–56.

Willis, J. C. 1949. The birth and spread of plants. *Boissiera* 8:1–561.

AMERICAN MUSEUM
Novitates

PUBLISHED BY THE AMERICAN MUSEUM OF NATURAL HISTORY
CENTRAL PARK WEST AT 79TH STREET, NEW YORK, N.Y. 10024
Number 2629, pp. 1-15, figs. 1-5 July 20, 1977

Geographic Ranges of North American Terrestrial Mammals

SYDNEY ANDERSON[1]

ABSTRACT

Existing theory on the geographic ranges occupied by species focuses on individual species, the density of species at different places, and not on the question as to what the size-distribution of the ranges of species in a larger fauna is or how this distribution may be explained. The ranges of North American terrestrial mammals are examined and a regular decline in the number of species having ranges in successively larger size-classes of ranges is found. The frequency distribution does not fit the lognormal or any of several other familiar distributions and further work is needed to develop a model that does fit. Other taxa and faunas should be examined also.

INTRODUCTION

One major parameter in the data or phenomena of interest to biogeographers is the size of the geographic range occupied by a species. The pattern or frequency distribution based on sizes of such ranges of the different species in a fauna is also of interest. The former is commonly considered; the latter rarely. The present paper considers the latter.

First the scientific history (not faunal history) and existing theory that bear on the question are outlined. Then, because concepts of space used by organisms are varied and sometimes confusing, the question of such concepts at different organizational levels is examined in order to relate these concepts. Upper and lower boundaries for them in the Mammalia are presented as examples. Then, using North American terrestrial mammals as an example of a fauna, I consider the methods and basic data available for examining the frequency distribution of sizes of geographic ranges and go on to examine that distribution. To put the continental pattern in perspective, the faunas of several progressively smaller areas within North America are also examined. History ends with the present. To illustrate the state of knowledge now, the opinions of several mammalogists (as to the frequency distribution being studied here) are related to the actual distribution. Logically, this might have been included at the end of the historical discussion, but the actual presentation will be more comprehensible where it is included (on page 8).

Some of the details of the actual frequency data for sizes of species ranges are outlined. Latitudes and major systematic groups are compared to range sizes. Finally, some hypotheses are dis-

[1]Curator, Department of Mammalogy, the American Museum of Natural History.

cussed, predictions made, and future work suggested. Before I began the study I had the idea that some simple and regular pattern might emerge. That idea could be regarded as a hypothesis. I do not think it is important whether hypotheses are formulated before, during, or after data are gathered and analyzed, or even whether the questions and ideas of interest are formulated linguistically as hypotheses. Often, however, such formulations clarify thought, expedite the gathering of the most relevant data, and thus focus the study.

ACKNOWLEDGMENTS

Dr. Karl F. Koopman contributed by allowing me to use his notes on the current status of North American mammals, by arguing various points with me, and by reviewing the manuscript. Other colleagues in the department and elsewhere also contributed ideas. Drs. Richard G. Van Gelder and Guy G. Musser, and my volunteer research assistant Ms. Mary Evensen, reviewed the manuscript and made helpful suggestions. I am grateful to all of these persons and to others who have helped; especially my secretary, Ms. Margaret Canning, whose usual diligence expedited this paper.

HISTORY AND EXISTING THEORY

There are few comprehensive and quantitative summaries on the sizes of geographic areas or ranges occupied by the species of any major taxa in areas of any size, let alone continental regions. Two principal reasons for this are (1) that in few cases are there adequate data on ranges for entire groups over sizable areas, and (2) that few persons have been interested in examining the data available. The latter may relate to a dearth of theory that would lead one to examine the data. As background for the present study of ranges of terrestrial mammals in North America, prior work and relevant theory are here reviewed.

Many data on ranges of taxa of plants were compiled by Willis (1949, and earlier works cited therein). He advanced the hypothesis that the area occupied by a taxon is related to its age more than to anything else, and he established that the frequency distribution of areas occupied by different species form what he called a "hollow curve." I explored the prevalence of such curves in a variety of taxonomic distributions but did not consider the sizes of areas occupied by species (Anderson, 1974). Some of the ideas developed there led to the present study.

The influence of size of an island area on the probability of survival of a newly arrived species and on the probability of extinction was treated by MacArthur and Wilson (1967) along with distances from the mainland and between islands, and other factors, but they did not deal with the question of why species occupy areas of different sizes in the absence of conspicuous discontinuities of habitat. Their theory of island biogeography has been applied in contexts other than islands of land surrounded by water. For example, Dritschilo et al. (1975) considered the ranges of host species of mice as islands occupied by species of parasitic mites.

Some ecological relationships of the theory of island biogeography have been discussed by Simberloff (1974). The mathematical relationships of species abundance distributions, principally in an ecological context, were lucidly summarized by May (1975). He compared the lognormal, broken stick, simple geometric series, and logseries distributions. When a pattern of relative abundance arises from the interplay of many independent factors, a lognormal distribution is predicted by theory and usually is found in nature. This distribution reflects the statistical Central Limit Theorem.

The lognormal distribution may be viewed as uninteresting since it does not suggest any special biological properties of the population under consideration except that many independent factors are involved. It seems to me that the presence of many such distributions in our science is interesting if for no other reason than to demonstrate how ignorant we are, how poor our predictions are liable to be, and how much work remains to be done.

In addition to the paper by May noted above, other papers in a symposium volume edited by Cody and Diamond, 1975, deal with many aspects of the relationships of such ecological and community concepts as stability or steadiness, resilience, population size, probabilities of extinction, niche formation, diversity in biotas, patchy vs. continuous distributions, effects of disturbance, species-packing level or α-diversity, productivity in ecosystems, and competitive exclusion.

In only one of the papers did an author tie any of these related phenomena and theories to the question of the sizes of ranges of species. Rosenzweig (1975) presented data on the average size of the total ranges of the species of bats found at each of some 39 different places in the United States and Canada. He demonstrated a negative correlation between average size of range and faunal diversity (number of species of bats) and suggested that this may be the result of more intense habitat selection in more diverse faunas. I offer a different explanation below.

The thesis that distribution and abundance are different aspects of the same problem was exhaustively and convincingly explored by Andrewartha and Birch (1954). Their recommended approach (p. 10) to explaining distribution and abundance included the following steps: (1) study the physiology and behavior of the species, (2) study the physiography, climate, soil, and vegetation of the area occupied, (3) experiment or observe further in the field or in the laboratory, and (4) measure the numbers of the animal as accurately as possible over a long period of time. Their approach was to consider factors affecting each species separately. The fact that other species are among these factors is acknowledged, but the emphasis is on the species separately. The question of how these separate species ranges might be distributed among all possible sizes of areas was not really addressed by these authors.

The question of influences on the extent of geographic ranges of species was addressed by Hesse, Allee, and Schmidt (1937, chapt. 8). Their definition of range was the area inhabited by a species as delimited by lines connecting the outermost known localities, even though within this range only certain habitats are occupied. This is acceptable as a working concept although problems arise in special cases, as is true with most concepts. They wrote that range depends on (1) geologic age of the taxon, for a younger taxon may have had access to fewer routes of dispersal, (2) the vagility or capacity for active dispersal or passive transport of the species, (3) the ecological valence or amplitude of the range of the conditions of life, within which an animal is able to exist, and (4) existing barriers. The dynamic nature of a range was clearly stated, ranges may expand or contract, or move from one area to another. The extent of range for a species has some minimum value below which the probability of accidental extinction, reduction in variability, and inbreeding may jeopardize its survival. The size of this range may differ with the species, but would be roughly equal to the area needed to survive by one pair multiplied by the number of pairs needed to maintain variability at some [undefined, but presumably important] level. The approach of these authors is like that of Andrewartha and Birch in focusing on the factors affecting the range of a single species. Again the question of whether this can lead to any generalization about the various ranges within a fauna is not addressed.

One of the few authors who have considered species areas in larger faunas and how these relate to other large-scale processes such as evolution and phylogeny is Boucot (1975, and earlier papers). He observed that taxa with larger geographic ranges have longer stratigraphic records in some groups of fossil invertebrates and discussed related theory on how this affects our interpretation of rates of morphological change and of taxonomic diversity.

Mammalogists have scarcely considered ranges in terms of size-frequency distributions in a continental fauna, or in any other smaller area. They have considered the numbers of species at different places (species density) and the degree of coincidence of boundaries (the delimitation of faunal areas). The question of limits to a given species has been much considered also. The relationships of the numbers of species in areas (or samples) of different sizes has been considered to some degree in the literature.

Suppose that a given number of species is postulated in the fauna of a certain space, e.g., the mammalian fauna of North America. Has any generalization been formulated that describes the frequency distribution of areas occupied by the species? Is there any theory that would explain or predict the distribution?

No one has summarized existing knowledge of areas for North American mammals nor explained in theory any pattern that may exist. Nor am I aware of any such treatment of any other group of animals. The question has been considered by botanists to some degree.

The compilation by Hall and Kelson (1959) of distributional data for mammals of North America, in the form of range maps for virtually all species, provides a valuable source of information. This source has been used in an analysis of species density, that is the numbers of species present in different parts of the continent, by Simpson (1964). The density of borders of species ranges has also been analyzed in order to better define faunal areas within the continent by Hagmeier (1966). I have commented in greater detail on these analyses elsewhere (Anderson, 1972).

The expansions of the ranges of seven species of terrestrial mammals in Europe were examined by Nowak (1975). Actual ranges are expressed in square kilometers and changes in the historically documented record are described. These seven species are only about 4 percent of the 185 in the fauna. Nowak noted (p. 112) that according to "a rough analysis the decrease of ranges of mammals and birds, in the last few hundreds of years, was not greater (both as regards the number of species and the surface area lost) than the expansion." This implies an equilibrium of sorts; however, Nowak also indicated that in his judgment the fauna has not overcome its reduction during the last glaciation, and that there are still unoccupied niches available. These views seem to imply that an equilibrium has not been reached. The main points to me are that the dynamic nature of an approximate equilibrium is suggested and that data on seven specific mammals and 21 species of other taxa are used to examine some of the processes and patterns that are involved.

THE QUESTION OF SPACE OCCUPIED AT DIFFERENT LEVELS

We have been considering the geographic areas occupied by species or their "ranges" (sometimes termed distributions). The vertical component of space occupied is negligible on the geographic scale, hence, it is meaningful to express range in square miles or square kilometers. Area or space occupied may also be examined at other levels.

The smallest biologically meaningful area occupied by a species is the area physically occupied by one individual at one time. I will call this an α-area. An aggregate of α-areas is clearly shown on an aerial photograph of a herd of wildebeest or nesting colony of flamingos, for example. Among mammals, α-areas range from about 4 cm^2 for a small shrew to 200 m^2 for a blue whale.

An individual moves in time and the area traversed defines a home range or β-area. A β-area may be measured or calculated in various ways depending on the species involved, and it may be defined in detail in somewhat different ways. The known marginal sites of occurrence may be connected by lines and all of the area enclosed measured, for example; or if a large enough number of data points in space and time for an individual are known, some density or probability function may be derived and used. Among North American mammals, β-areas range from less than .1 ha for some small rodents (Stickel, 1968) to 100 km^2 for a cougar, and many more species have ranges near the lower end of the range than near the upper end. This excludes from consideration a few migratory bats, caribou, and some widely ranging marine mammals.

The next larger meaningful area would be a composite of the β-areas of all the individuals in a contiguous population, whatever the size of this γ-area might be. This concept might also be expressed in terms of a unit of suitable habitat, which in most cases would be more or less continuously inhabited by individuals. There are degrees of contiguity and continuity, as is usual in biological phenomena, but the general concept of a γ-area is useful. The concept of a deme in population genetics is equivalent. Among mammals, γ-areas range from about 1 km^2 for the entire range of an insular species like *Microtus breweri* (confined to Muskeget Island off the Massachusetts mainland) to 10^7 km^2 for the original range of the lynx, subspecies *Felis lynx canadensis.*

The largest biologically meaningful unit of area for a species is the total species range. This area I term a δ-area, and these are the areas with which this paper chiefly deals. Knowing that the δ-area expands, contracts, and moves from one place to another during the species lifetime, we might define a larger area to encompass the total lifetime range of a taxon, but in most cases virtually nothing is known about this. Among mammals, δ-areas range from 1 km^2 for *Microtus breweri* again to 5.1×10^8 km^2 for *Homo sapiens*, assuming that this species occupies the entire surface of Earth.

Biogeographic areas larger than those of individual species are commonly used. These are composites of different species and are usually recognized, if not always defined, on the basis of congruence of species boundaries, faunal communality, and related terms. These are biomes, life-zones, and other faunal areas, and may be termed ϵ-areas. In order to depict areas at the different levels from α to ϵ it is necessary to use maps of different scales, and in order to obtain useful data it is necessary to use sampling and measurement techniques of different orders of magnitude. In figure 1, for example, as sample areas become progressively smaller relative to species ranges or δ-areas, the frequency distribution of occupied parts of the sample area changes. Finally it becomes impossible to say anything meaningful about these areas, except that they are all larger than the sample area. Conversely, in order to say anything meaningful about γ-areas one must use smaller sample areas and more refined data (i.e., measurements at a greater level of precision). Data on the γ-area or habitat level are used along with data on the

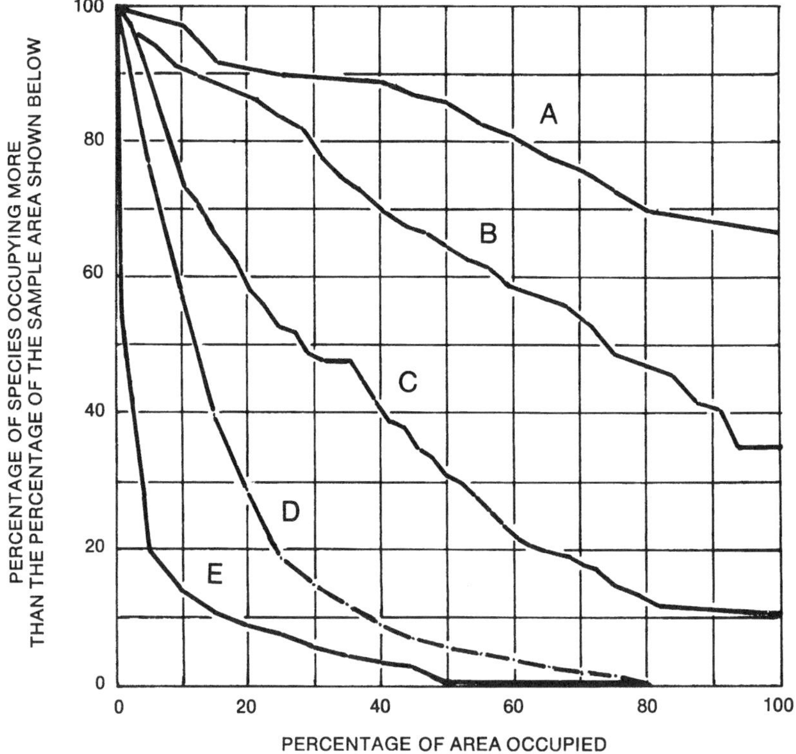

FIG. 1. The relationship of percentages of species occupying different percentages of the total area in areas of four different sizes and a comparative curve to show the average mammalogist's impression of the relationship for one of the four areas. The curves are: A. 63 species in about 12 counties in central Kansas, B. 79 species in Kansas, C. 120 species in Colorado, D. the average of the estimates of six mammalogists for all species of terrestrial mammals in North America, and E. the actual values for 714 species in North America.

α-level (localities of individual specimens) in order to decide what ranges are on the δ-level. Where there are many α-points (dots on the range map) and where there are good data on habitat requirements together with clear boundaries of habitats (γ-areas) in the area being studied, as is true for some species in many parts of Colorado, then δ-area boundaries can be drawn with a precision on the order of a few kilometers.

A point that is sometimes overlooked and is usually, of necessity, smoothed over, even by knowledgeable zoogeographers, is that the boundaries (of δ-areas) mapped for different species, even when drawn by one author and in one region, are not equivalent in accuracy. In fact, the boundary of a single species represented by a single continuous line on the map varies in its reliability from place to place. For example, on a range map for *Dipodomys merriami* published in 1959 (Hall and Kelson, p. 531) the line in Nevada distinguishes with delicacy certain tongues of unoccupied area of less than 100 km. width (even less than 10 km. in one case) and in Chihuahua the line sweeps boldly and encompasses an area of more than 200 km. in width which was later (Anderson, 1972, p. 311) shown probably to be unoccupied by the species. The scales used in mapping vary with the ranges of species also, so an error of 9×10^4 km^2 for *D. merriami* in Chihuahua is not so great relative to the total range of the species, about 2×10^6 km^2; however, some 30 percent of North American species have ranges smaller than 9×10^4 km^2 (as shown in figure 2).

METHODS AND BASIC DATA

The geographic ranges of North American mammals were measured on maps published by Hall and Kelson (1959) except that newer information on the ranges and taxonomy was included. The maps of Hall and Kelson are roughly equal-area projections, departures therefrom seemed less than 5 percent. Errors in the scales of miles drawn were probably greater than 5 percent in some cases. An error of linear scale of 10 percent would cause an error in the measurement of area of about 20 percent.

A square ruled grid was prepared on a transparent acetate film. The grid unit was read from the scale of miles on each map. The number of squares of the grid occupied by the mapped range was counted. Any square more than half occupied was counted. The unit distance was then squared and multiplied by the number of squares occupied. This gave the number of square miles, which was converted to square kilometers. The measurements finally were rounded to the nearest value of one significant figure, for example 1650 became 2000 and 64,500 became 60,000.

The ranges vary over more than six orders of magnitude, from those species known from a single locality, in which case the range was assumed to be 10 km^2, up to 2×10^7 km^2 in the case of *Lutra canadensis*, *Canis lupus*, and *Castor fiber*. The North American part of the range only was measured for species that occur also in one or more other continents. North America includes Middle America south through Panama and the Caribbean Islands south through Grenada in the Lesser Antilles. It also includes Greenland.

Historical ranges were used rather than present ranges. The ranges of some species have been much reduced by man since Europeans arrived. These species are chiefly large carnivores and artiodactyls: *Canis lupus*, *C. rufus*, *Ursus americanus*, *U. arctos*, *Gulo gulo*, *Lutra canadensis*, *Felis onca*, *F. concolor*, *F. pardalis*, *F. lynx*, *Cervus elaphus*, *Alces alces*, *Rangifer tarandus*, *Antilocapra americana*, *Bison bison*, *Oreamnos americanus*, and *Ovis canadensis*. Not only have ranges been reduced to various degrees, but populations have been reduced in areas still occupied.

The population levels of many other species have increased as a result of human activity and some species have expanded their ranges, probably in response to climatic changes as well as human activity. Among these are *Didelphis virginiana*, *Dasypus novemcinctus*, *Spilogale putorius*, *Baiomys taylori*, and *Sigmodon hispidus*.

For most species, information is either not adequate to reveal significant changes in ranges or changes have not occurred in the last hundred years or whatever other time is well enough known to judge.

The species with most drastic reductions of

ranges all had ranges of at least 1×10^6 km^2 and hence appear at the extreme right end of the curves in figures 1 and 2. The effect of human activity on geographic ranges has been to reduce those of a few (less than 5%) of the species. This would tend to move the right-hand tail of the curve slightly to the left. If the changes were made on the graph, the difference in the curve would be scarcely detectable.

Homo sapiens, the species that comes closest to occupying all of North America and is the only non-introduced species occurring also in Eurasia, North America, and South America, was omitted from my analysis. I omitted also nine introduced species with established ranges in North America, namely *Rattus norvegicus*, *Rattus rattus*, *Mus musculus*, *Dasyprocta aguti* (on St. Thomas prior to 1852), *Myocastor coypus*, *Herpestes auropunctatus*, *Cercopithecus aethiops*, *Cercopithecus mona*, and *Lepus europaeus*. I omitted also those species (69) occurring in South America and not ranging into North America north of Nicaragua. I omitted the marine species, namely those of the Cetacea, Sirenia, the pinniped Carnivora, and the genus *Enhydra*. Two species of highly dubious status (*Oryzomys fulgens*, taxonomy dubious; *Coendou prehensilis pallidus*, West Indian records dubious) were omitted.

The ranges of the remaining species of native North American terrestrial mammals were measured. The latitude of the "center of gravity" (or point at which a piece of cardboard the shape of the range would balance) for each species was estimated.

Taxonomic work from 1924 (Miller) to 1955 (Miller and Kellogg) had reduced the number of recognized species from 1399 to 1065. By 1959 Hall and Kelson recognized 1003. The actual dates would have been about two years earlier in each case because of the production time for volumes of this type. Most of these reductions in numbers of species are the result of the discovery of intergrades between previously recognized species that are thereafter recognized as one species or as subspecies thereof. A few new species, or species new to North America are still being discovered from time to time. Both of these events increase the number of recognized species. Hall and Kelson (1959, p. vi) made the above comparison and estimated that of their 995 species (eight introductions omitted), perhaps 125 will eventually be found to intergrade and hence be regarded as subspecies only. This would reduce the total to 870. Of these, they noted that approximately 170 are confined to an island or some isolated mountain mass. In each such case, the most closely related species is on the mainland or with a range separated by some barrier. If these isolated species are all to be synonymized, the total would become 700. Revisions have proceeded since 1959 and my figures for 1975 compare as follows.

Taking the 1959 figure of 1003 and subtracting nine introductions, 49 cetacea, 14 pinnipeds, one sirenian, and the sea otter, yields 929 species. My count for 1975, omitting these same groups (and three other special cases mentioned above) is 911, a net reduction of 18 recognized species. Excluding the 69 species not reaching north of Nicaragua (and occurring in South America as well as North America) leaves the 842 used in my analysis of ranges. Of these, 100 are insular. This does not include species with isolated ranges on islands or on mountaintops on the mainland, such as Hall and Kelson included in their figure of 170.

Of the 100 insular species, 26 are now extinct. These are *Geomys cumberlandius*, *Oryzomys victus*, *Elasmodontomys obliquus*, *Quemisia gravis*, *Hexolobodon phenax*, *Plagiodontia spelaeum*, *Isolobodon portoricensis*, *Aphaetreus montanus*, *Heteropsomys insularis*, *Homopsomys antillensis*, *Brotomys voratus*, *B. contractus*, *Boromys offella*, *B. torrei*, six species of *Nesophontes*, three species of ground sloths, and three species of *Megalomys*. All of these except *Geomys cumberlandius* were confined to one or more Caribbean islands, and most are known only from sub-Recent remnants found in caves. No species on the mainland has become extinct in historic times.

Most of the 15 species that occur both in North America and in Eurasia were, in 1959, regarded as specifically distinct on these two continents. The 15 species (and the names used for North American representatives in 1959) are: *Lemmus sibiricus* (*nigripes* and *trimucronatus*), *Dicrostonyx torquatus* (*groenlandicus*), *Canis lupus*, *Vulpes vulpes* (*fulva*), *Ursus arctos* (a plethora of names), *Ursus maritimus*, *Mustela nivalis* (*rixosa*), *Felis lynx* (*canadensis*), *Cervus elaphus* (*canadensis*, *merriami*, and *nannodes*), *Alces alces*, *Rangifer tarandus*, *Gulo gulo* (*luscus*), *Castor fiber* (*canadensis*), *Microtus gregalis*, and *Microtus oeconomus*.

Increasing taxonomic knowledge also can increase the number of recognized species by showing that a formerly recognized species actually consists of two or more distinct species. Recent examples are seen in Gardner (1973) who re-separated *Didelphis virginiana* from *D. marsupialis*; Thaeler (1972, and other works) who has divided several species of *Thomomys*; Schmidly (1973) who has divided *Peromyscus boylii*; and Zimmerman (1970) who has divided *Sigmodon hispidus*.

COMPARISON OF SIZES OF SPECIES RANGES IN SMALLER AREAS WITH THE CONTINENTAL PATTERN

The ranges of species in Colorado (Armstrong, 1972) and Kansas (Hall, 1955) were ascertained. The cumulative percentages of species occupying different percentages of the total area of each state are compared (curves B and C in figure 1) with two curves (D and E) for all terrestrial mammals of North America and one curve (A) for a smaller area in central Kansas. The sample area there was roughly square and included about 12 counties or 2.6×10^4 km^2 which compares with 1.8×10^5 km^2 for all of Kansas, 2.3×10^5 km^2 for all of Colorado, and 2×10^7 km^2 for all of North America. The numbers of species in these four sample areas are 63, 79, 120, and 714, respectively. Colorado is not only larger than Kansas but is much more diverse in topography and habitat than is Kansas.

I asked six mammalogists who have considerable familiarity with North American terrestrial mammals and who are interested in biogeography to draw simple frequency diagrams of their estimates of the sizes of ranges of these mammals. An average (and rounded) cumulative frequency plot of their quick and unstudied estimates is shown as curve D in figure 1 for the comparison with the actual distribution (curve E) based on my measurements. Two of the six persons estimated the frequency in the first class to be greater than in any other class, but neither of them went so far as the actual distribution. The extent of the "hollow curve" distribution is not widely known.

If the reduction in sample area is taken an order of magnitude smaller, we find that in Douglas County, in eastern Kansas, an area of about 1.2×10^3 km^2, there are 52 species of mammals and all but *Geomys bursarius* occupy the entire area, so that the curve if drawn on figure 1 would nearly coincide with the upper border of the graph.

The estimates of the six mammalogists for all North American mammals erred in the direction of the frequency distribution found in smaller areas within the continent. The bias is quite understandable. Each of these persons has worked more closely with some smaller area within a continent than with the entire continent, and might be expected to extrapolate from the local familiarity.

WHAT IS THE SIZE DISTRIBUTION OF RANGES?

In figure 2 are graphed the ranges for insular and non-insular species separately as cumulative percentages of species. The abscissa is on a log scale. The names of some familiar islands and areas on the mainland are given to aid the reader in visualizing the scale used.

For species with small ranges (up to 10^3 km^2 and including many known from only one locality), the estimates are in most cases in error (on the low side) by a greater percentage than for species with larger ranges. It is unlikely that a species known from only one location lives only at that location. It is more likely that some significant area is occupied and that this area is somewhere among the smaller ranges of species, so that inadequate sampling accounts for the relatively poor estimate. In figure 2, a broken line is drawn to show my hypothesis as to the distribution of the species with small ranges that will be approached as better data become available.

I plotted the centers of ranges of species against latitudes with different symbols for different orders, but the resulting graph is too complex for convenient reproduction here. Some dif-

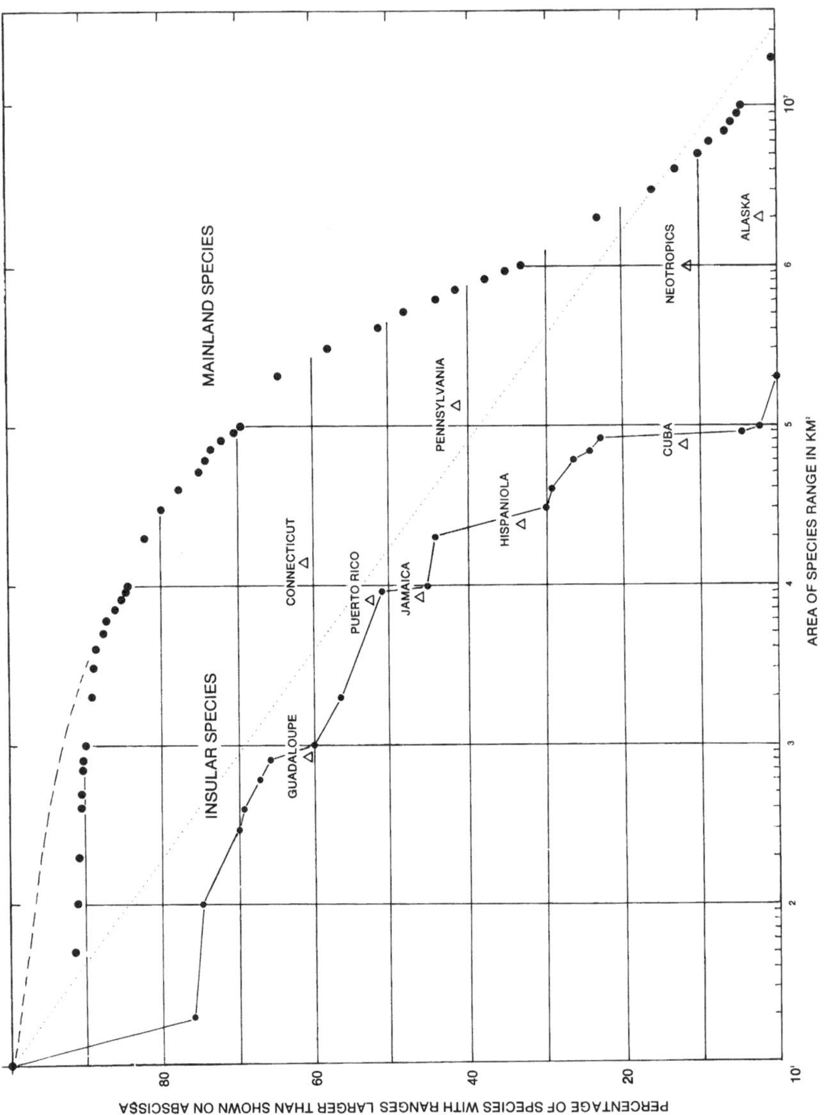

FIG. 2. Cumulative plots of ranges of North American terrestrial mammals, insular and non-insular species separately. Same data are plotted in a different way in figure 1.

ferences in both sizes of geographic ranges and latitudes of the centers of ranges for the species of the major orders are shown in figure 3. The abscissa is logarithmic for ranges. Insular species occupy smaller ranges, as would be expected, and artiodactyles and carnivores occupy larger ranges than average. Chiroptera are noticeably more southern, as are insular species. This tendency of insular species results from the presence of a more diverse fauna *and* more is-

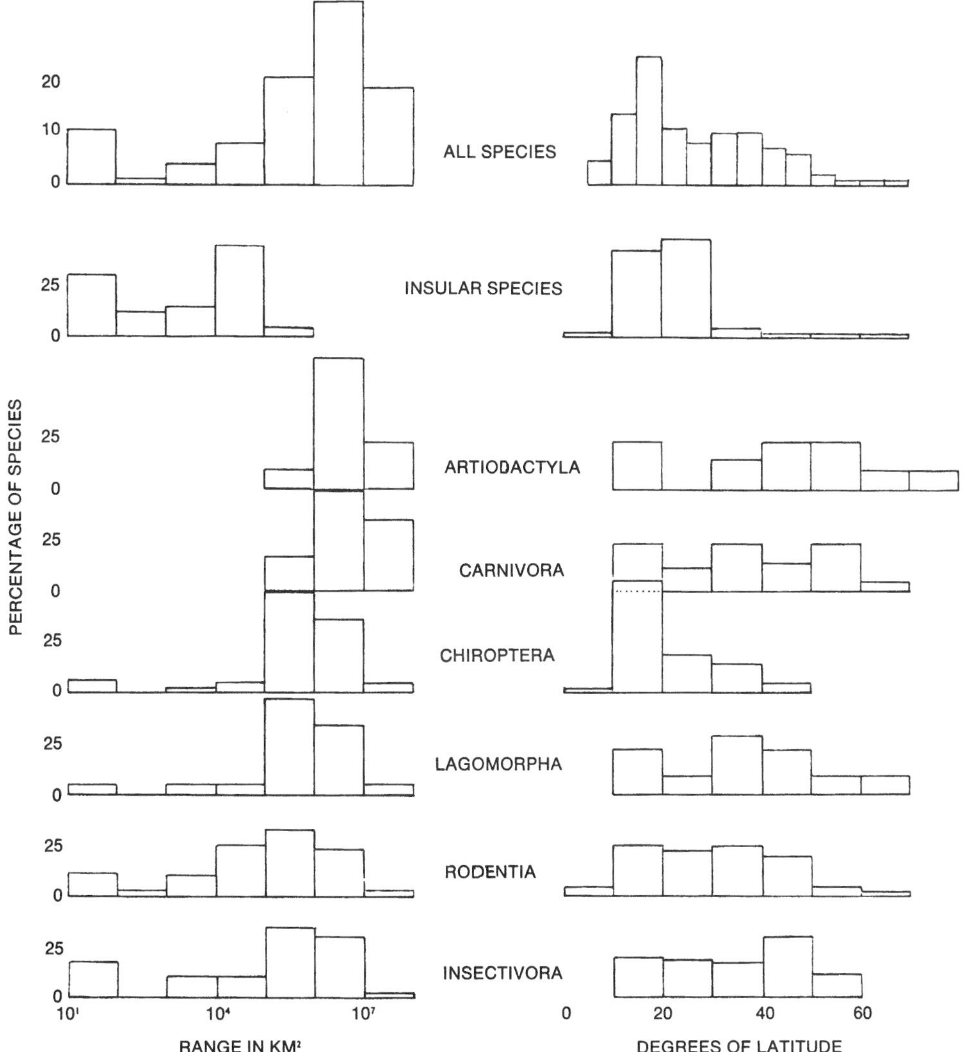

FIG. 3. Graph showing the percentage of the species of each of eight groupings whose geographic ranges fall in each of seven orders of magnitude (at left) and the centers of whose ranges fall in each of seven 10-degree ranges of latitude (at right). The top right plot is divided into 5-degree ranges.

lands in the tropical part of the continent. There are many islands in northern Canada but not much diversity.

Several observations based on the complex graph from which data for figure 3 were extracted are as follows. From 10^3 to 10^6 km^2 in range size, bimodality as to latitude is evident. There tend to be tropical and temperate species with the centers of their ranges averaging (for ranges of different sizes) from 15 to 20 degrees of latitude for the tropical species and near 35 degrees for temperate species. At ranges larger than 10^6 the tropical-temperate distinction does not exist and the average range is centered at progressively more northern latitudes as ranges become larger. The species with the largest ranges (about 2×10^7 km^2) are centered at about 49 degrees. The bimodality reflects a major faunal distinction. The other observations largely result because North America is shaped the way it is, large at the north and narrow at the south. If South America were analyzed in the same way, I suppose that the bimodality would be evident but that the species with larger ranges would predominate near the equator rather than farther away from the equator. I postulate that ranges of species in a more diverse fauna do not necessarily have smaller average ranges (contrary to the hypothesis of Rosenzweig, 1975), even though this happens to be true for North American bats. An examination of the facts for other groups, such as rodents, and for other faunas, such as South America, would be interesting.

DISCUSSION

Are the species spread evenly throughout the possible ranges? The use of a logarithmic scale for ranges makes it difficult to answer this question from graphs such as figures 1, 2, and 3. In figure 4 are plotted the numbers of species present in each 100 km^2 size class, averaged over each order of magnitude. It is clear that the species are not spread evenly, but that they are about an order of magnitude (10 times) less "concentrated" in each successively larger order of magnitude of range.

I predicted in reference to figure 2 that when better data are available there will be fewer species in the area of 10^1 km^2 range and more in ranges from 10^2 to about 3×10^3 km^2. Taking this into account, I suggest that the best estimate of the actual distribution in figure 4 would be curvilinear, more or less as shown by the broken line.

The hypothesis of equal probability of occurrence of species in all possible sizes of ranges having been examined and rejected, let us consider several other hypotheses or "distributions" developed and explicated chiefly by ecological theorists. Some of these distributions have been posited to imply or suggest possible community relationships or interactions among the components.

The Central Limit Theorem of statistics states that the means of samples drawn from a population of any distribution will approach the normal distribution as sample size increases (Sokol and Rohlf, 1969, p. 130). May (1975, p. 89) stated it in more general terms, "essentially all additive statistical distributions are asymptotically gaussian, or 'normal.' " He suggested that the lognormal reflects the Central Limit Theorem and that broken-stick, geometric, and logseries distributions may reflect features of community biology. It is my view that these three distributions may or may not reflect features of community biology, depending on a variety of circumstances in each model examined. The Central Limit Theorem may also be significantly involved.

The way in which the lognormal may reflect the Central Limit Theorem in the case of ranges of species (assuming that they had a lognormal distribution, even though they do not exactly have that distribution) is to interpret the range of a species as the result of the interaction of a variety of relatively independent environmental and internal factors or as a representation or "mean" of a sample of all these factors. Independence is relative. A deterministic philosophical view would hold that nothing is really independent and that things that seem independent are merely so poorly known or so complex in interaction that we do not perceive of or have any way of dealing with the interaction. Pragmatically we have to act as though they were independent until we figure out connections.

Do the areas for North American mammals conform to any of the four principal distribu-

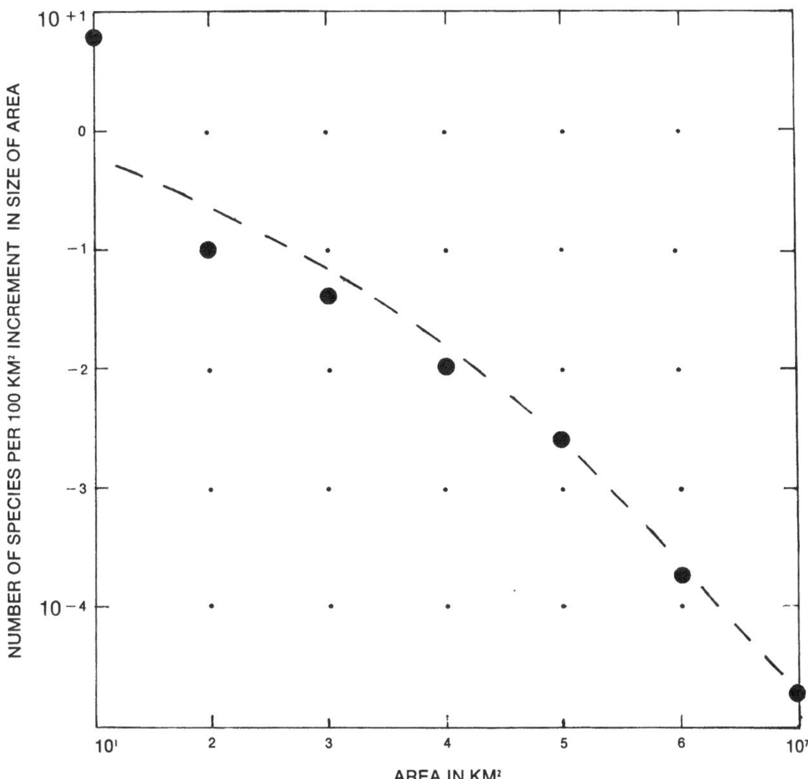

FIG. 4. Graph for North American terrestrial mammals showing the number of species (averaged for each succeeding order of magnitude) having ranges of any given size. Counts are grouped in 100 km^2 increments. The negative values on the ordinate are powers of 10, thus 10^{-4} or .0001 species per 100 km^2 increment for a range of 10^6 (1,000,000) km^2 means that there are so few species with ranges of this size that most increments or size-classes of 100 km^2 are unoccupied and, on the average, there is about one species for each 10,000 size-classes.

tions summarized by May (1975), namely the lognormal, broken stick, simple geometric, or logseries? Each of these can be examined by graphic means more easily than by computation, although some simple computation is needed to work out the graphs.

I computed a broken stick distribution for comparison. The largest ranges for the North American mammals are larger than in the "broken stick" model and the smaller ranges are smaller. This model (MacArthur, 1957, discussed by Anderson, 1975) assumes that some finite resource is divided randomly into discrete segments. Species ranges are not discrete; they overlap. Species "niches" in a more abstract sense may be discrete, however. There seems to be no theoretical reason to expect the distribution to fit this model.

In figure 5, percentages of species on a probability scale are plotted against cumulative log (X10) classes to test for conformity with a lognormal distribution. Conformity, which would be shown by a linear relationship, does not exist over all, although the upper three or four points approach linearity.

Plotting the areas occupied by the species on

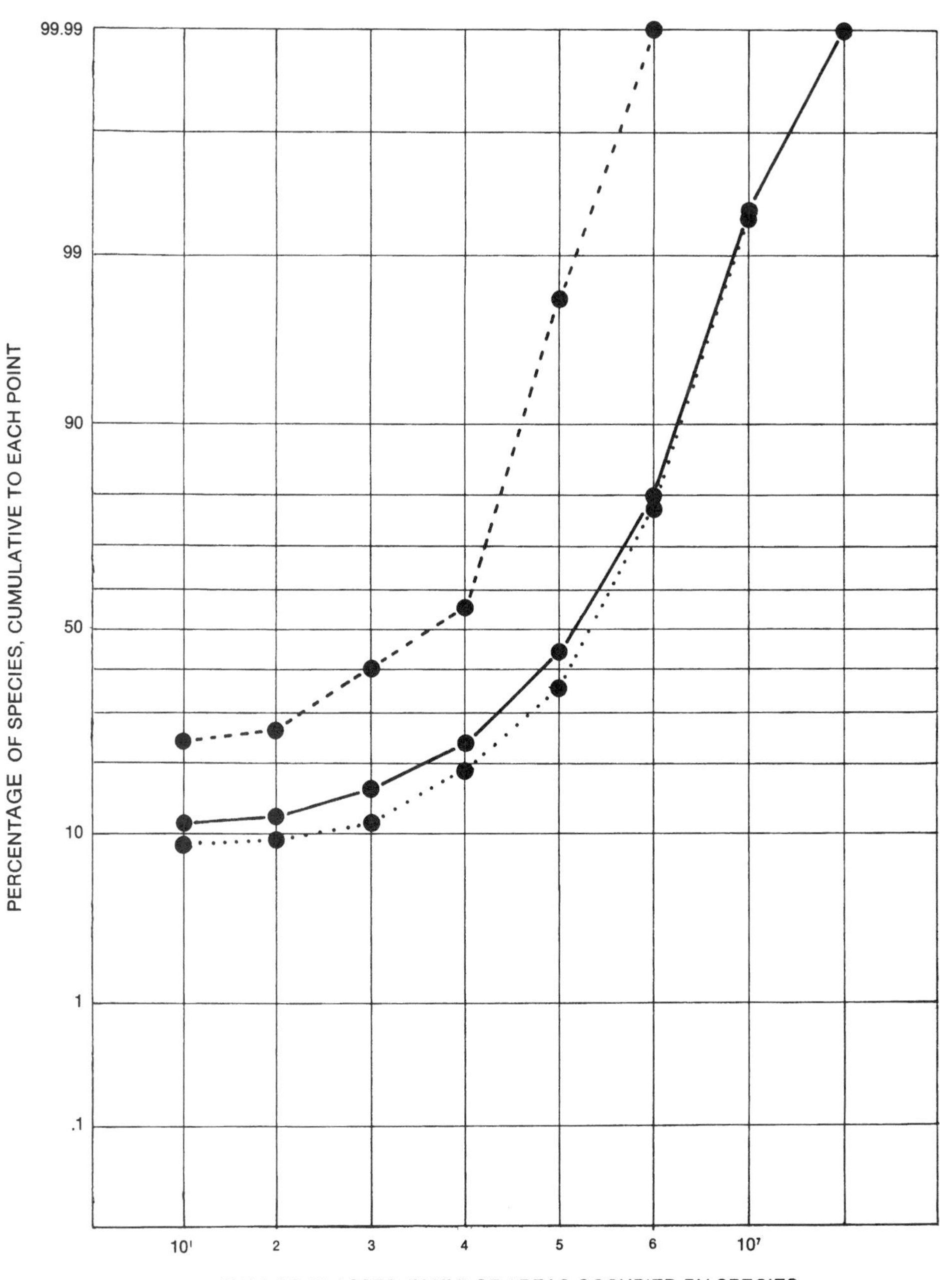

FIG. 5. Cumulative percentages of species in X10 (1-10, 11-100, 101-1000, etc.) log classes (abscissa) plotted on probability scale (ordinate) to see whether distributions are lognormal. Since the points are not on straight lines, the distributions are not lognormal. The three curves are (top to bottom), 100 insular species, 714 insular and mainland species, and 614 mainland species. Some "mainland" species also occur on islands, but the island range is usually a negligible part of the total range.

a log scale against their rank order on an arithmetic scale should give a straight line if a geometric or logseries exists. This was done and the fit was not close.

The actual distribution is not as I had suspected it might be (lognormal), nor does it fit well with any one of several distributions familiar to ecologists. It is, however, a very regular distribution and this suggests the need for further testable hypotheses or models to help explain this regularity. The study of such models will be the subject of a later paper. The examination of other groups of organisms to see whether the distributions of sizes of geographic ranges resemble that found for North American terrestrial mammals would also be interesting.

LITERATURE CITED

Anderson, Sydney
1972. Mammals of Chihuahua, taxonomy and distribution. Bull. Amer. Mus. Nat. Hist., vol. 148, art. 2, pp. 149-410, figs. 1-366, tables 1-15.
1974. Patterns of faunal evolution. Quart. Rev. Biol., vol. 49, pp. 311-332, figs. 1, 2.
1975. On the number of categories in biological classifications. Amer. Mus. Novitates, no. 2584, pp. 1-9, figs. 1-6.

Andrewartha, H. G., and L. C. Birch
1954. The distribution and abundance of animals. Chicago, Univ. Chicago Press, xvi + 782 pp.

Armstrong, David M.
1972. Distribution of mammals in Colorado. Univ. Kansas Mus. Nat. Hist. Monogr. 3, x + 415 pp.

Boucot, A. J.
1975. Standing diversity of fossil groups in successive intervals of geologic time viewed in the light of changing levels of provincialism. Jour. Paleont., vol. 49, no. 6, pp. 1105-1111, 11 text figs.

Cody, Martin L., and Jared M. Diamond (eds.)
1975. Ecology and evolution of communities. Cambridge, The Belknap Press of Harvard University, xii + 545 pp.

Dritschilo, William, Howard Cornell, Donald Nafus, and Barry O'Connor
1975. Insular biogeography: of mice and mites. Science, vol. 190, no. 4213, pp. 467-469, figs. 1, 2.

Gardner, Alfred L.
1973. The systematics of the genus *Didelphis* (Marsupialia, Didelphidae) in North and Middle America. Mus. Texas Tech Univ., special publ. no. 4, 81 pp., 14 figs.

Hagmeier, Edwin M.
1966. A numerical analysis of the distributional patterns of North American mammals. II. Re-evaluation of the provinces. Syst. Zool., vol. 15, pp. 279-299, figs. 1-5.

Hall, E. Raymond
1955. Handbook of mammals of Kansas. Univ. Kansas Publ., Mus. Nat. Hist., misc. publ. no. 7, pp. 1-303, 99 figs.

Hall, E. R., and Keith R. Kelson
1959. The mammals of North America. New York, Ronald Press, vol. 1, xxx + 546 + 79 pp.

Hesse, Richard, W. C. Allee, and Karl P. Schmidt
1937. Ecological animal geography. Univ. Chicago Publ., by John Wiley and Sons, Inc., xiv + 597 pp.

MacArthur, R. H.
1957. On the relative abundance of bird species. Proc. Natl. Acad. Sci., vol. 43, pp. 293-295.

MacArthur, R., and E. O. Wilson
1967. The theory of island biogeography. Princeton, Princeton Univ. Press, 203 pp.

May, Robert M.
1975. Patterns of species abundance and diversity. Pp. 81-120, *in* Cody, Martin L., and Jared M. Diamond (eds.), Ecology and evolution of communities. Cambridge. The Belknap Press of Harvard Univ., 545 pp.

Miller, Gerrit S., Jr.
1924. List of North American Recent mammals, 1923. U.S. Natl. Mus. Bull., vol. 128, 673 pp.

Miller, Gerrit S., Jr., and Remington Kellogg
1955. List of North American Recent mammals. U.S. Natl. Mus. Bull., vol. 205, xii + 954 pp.

Nowak, Eugeniusz
1975. The range expansion of animals and its causes (as demonstrated by 28 presently spreading species from Europe). English translation available from Natl. Tech Info. Service, Springfield, Virginia 22151. pp. 1-164, from Zeszyty Naukowe, no. 3, pp. 1-255, 1971.

Rosenzweig, Michael L.
1975. On continental steady states of species diversity. Pp. 121-140, *in* Cody, Martin L., and Jared M. Diamond (eds.), Ecology and evolution of communities. Cambridge, The Belknap Press of Harvard Univ., 545 pp.

Schmidly, David J.
1973. Geographic variation and taxonomy of *Peromyscus boylii* from Mexico and the southern United States. Jour. Mammal., vol. 54, pp. 111-130, figs. 1-6, tables 4.

Simberloff, Daniel S.
1974. Equilibrium theory of island biogeography and ecology. Ann. Rev. Ecol. Syst., vol. 5, pp. 161-182, 1 fig., 1 table.

Simpson, George Gaylord
1964. Species density of North American Recent mammals. Syst. Zool., vol. 13, pp. 57-73.

Sokal, Robert R., and F. James Rohlf
1969. Biometry. San Francisco, W. H. Freeman Co., xxi + 776 pp.

Stickel, Lucille F.
1968. Home range and travels, *in* King, John A. (ed.), Biology of *Peromyscus* (Rodentia), pp. 373-411. Amer. Soc. Mammals, Special Publ. no. 2, 593 pp.

Thaeler, C. S., Jr.
1972. Taxonomic status of the pocket gophers, *Thomomys idahoensis* and *Thomomys pygmaeus* (Rodentia, Geomyidae). Jour. Mammal, vol. 53, no. 3, pp. 417-428, 3 figs., table.

Willis, J. C.
1949. The birth and spread of plants. Boissiera, 8:x + 1-561 pp.

Zimmerman, E. G.
1970. Karyology, systematics and chromosomal evolution in the rodent genus, *Sigmodon*. Publs. Michigan State Univ. Mus., vol. 4, pp. 385-454, 9 figs., 16 pls., 4 tables.

PAPER 27

Changing Patterns in the Holocene Pollen Record of Northeastern North America: A Mapped Summary (1977)

J. C. Bernabo and T. Webb III

Commentary

JOHN W. (JACK) WILLIAMS

By mapping the shifting distributions of plant taxa from 11,000 years ago to the present, J. C. Bernabo and T. Webb III showed that mapped syntheses of networks of fossil pollen records could illuminate the complex responses of vegetation to changing climates. Prior to this paper, Quaternary paleoecology was primarily a time-oriented discipline, in which palynologists displayed their data for individual sites in pollen diagrams, and attempts at spatial syntheses often emphasized synchronous behavior among sites. Bernabo and Webb's maps revealed that the broadscale vegetation dynamics involved the time-transgressive movements of major ecotones across northeastern North America as the composition and location of ecosystems changed in response to the changing patterns of temperature, moisture, and seasonality during the Holocene.

Bernabo and Webb's article was a groundbreaking paper, but it was not alone: Margaret Davis, Brian Huntley, and John Birks, among others, were beginning to use fossil pollen data sets to explore spatiotemporal patterns in vegetation history. What distinguished Bernabo and Webb's paper was its rich variety of mapped visualizations, its thoughtful analysis of how various ecological phenomena were highlighted by these maps, and its careful linking of the patterns of pollen abundances to the distributions of vegetation formations, past and present. Moreover, this paper was a first step in a path that ultimately led to the establishment of the North America Pollen Database, still widely used today, and to the Cooperative Holocene Mapping Project (COHMAP), an interdisciplinary team of terrestrial paleoecologists, oceanographers, and climatologists, which provided pathbreaking insights into the drivers of late Quaternary environmental change. The most recent descendent of these maps is Pollen Viewer, a series of online animated maps of the changing distributions of pollen taxa.

Webb, Bernabo's major advisor, received a PhD in meteorology and spent his career in a geology department. Thus, Webb's greatest contribution to paleoecology and macroecology was arguably his meteorologist's perspective, and in particular his training in the meteorological tradition of using spatial networks of synoptic data to map and study emergent phenomena. It is also worth noting that Bernabo and Webb's paper was possible only because the palynological community was willing to share its data—unusual for the time. So the success of this paper and subsequent syntheses rests on both the innovative approach of the authors and the ongoing generosity of the broader community.

QUATERNARY RESEARCH 8, 64–96 (1977)

Changing Patterns in the Holocene Pollen Record of Northeastern North America: A Mapped Summary

J. CHRISTOPHER BERNABO AND THOMPSON WEBB III

Department of Geological Sciences, Brown University, Providence, Rhode Island 02912

Received February 5, 1976

By mapping the data from 62 radiocarbon-dated pollen diagrams, this paper illustrates the Holocene history of four major vegetational regions in northeastern North America. Isopoll maps, difference maps, and isochrone maps are used in order to examine the changing patterns within the data set and to study broad-scale and long-term vegetational dynamics. Isopoll maps show the distributions of spruce (*Picea*), pine (*Pinus*), oak (*Quercus*), herb (nonarboreal pollen groups excluding Cyperaceae), and birch + maple + beech + hemlock (*Betula, Acer, Fagus, Tsuga*) pollen at specified times from 11,000 BP to present. Difference maps were constructed by subtracting successive isopoll maps and illustrate the changing patterns of pollen abundances from one time to the next. The isochrone maps portray the movement of ecotones and range limits by showing their positions at a sequence of times during the Holocene. After 11,000 BP, the broad region over which spruce pollen had dominated progressively shrank as the boreal forest zone was compressed between the retreating ice margin and the rapidly westward and northward expanding region where pine was the predominant pollen type. Simultaneously, the oak-pollen-dominated deciduous forest moved up from the south and the prairie expanded eastward. By 7000 BP, the prairie had attained its maximum eastward extent with the period of its most rapid expansion evident between 10,000 and 9000 BP. Many of the trends of the early Holocene were reversed after 7000 BP with the prairie retreating westward and the boreal and other zones edging southward. In the last 500 years, man's impact on the vegetation is clearly visible, especially in the greatly expanded region dominated by herb pollen. The large scale changes before 7000 BP probably reflect shifts in the macroclimatic patterns that were themselves being modified by the retreat and disintegration of the Laurentide ice sheet. Subsequent changes in the pollen and vegetation were less dramatic than those of the early Holocene.

INTRODUCTION

The Holocene pollen record is a rich source of information about temporal and spatial variations in the vegetation on the local and regional scales. Standard presentations of pollen data in pollen diagrams, however, have emphasized the temporal changes at individual sites and combine both local and regional information about the vegetation. Displaying the data on maps shifts the focus to the temporally changing spatial patterns in the pollen record and aids in distinguishing local variations from regional changes. Cartographic techniques also enable identification of vegetational features such as range boundaries and ecotones. Mapping thus facilitates interpreting the data in terms of past biological and environmental changes and allows summarization of vast amounts of pollen data into a useful form for anthropologists, biologists, geologists, meteorologists, and geographers.

McIntyre *et al.* (1976) demonstrated the usefulness of mapping marine micropaleontological data in the study of Pleistocene paleo-oceanography. These researchers then used the statistical techniques described in Imbrie and Kipp (1971) and Kipp (1976) in order to construct paleotemperature and paleosalinity maps for the world ocean (CLIMAP, 1976). LaMarche and Fritts (1971) have similarly produced synoptic maps of tree ring data from the southwestern United States, and Fritts *et al.* (1971) have transformed these data into a series of estimated surface pressure maps for the past

ISSN 0033-5894

200 years. We take the first step in producing a similar type of paleoclimatic reconstruction by mapping pollen data in northeastern North America. The methods described in Webb and Bryson (1972), Bryson and Kutzbach (1974), and Webb and Clark (1976) can eventually be applied to produce the paleoclimatic reconstructions. Ritchie's summary (1976) from the western interior of Canada and Nichol's summary (1975) from northern Canada also open the possibility for paleoclimatic mapping in these areas.

Our paper uses isopoll, difference, and isochrone maps in order to study some major aspects of the Holocene vegetational history of northeastern North America. The isopoll maps illustrate the distributions of five major pollen groups including spruce, pine, oak, herbs, and the BAFT group (birch, maple, beech, and hemlock). Difference maps are employed to show the patterns of change between each consecutive pair of isopoll maps and the isochrone maps trace the movement of pollen-defined ecotones and range limits.

The remainder of this paper is divided into five sections. The first describes the data base, and the second explains the methods used in constructing the maps. The second section also presents maps comparing the modern distributions of pollen and tree inventory data for spruce, pine, and oak. Section three describes the sequence of isopoll and difference maps in chronologic order from 11,000 BP to present. The fourth section discusses four major vegetational trends by tracing the decline of spruce pollen, the changing distribution of pine and oak pollen, the increasing dominance of birch, maple, beech, and hemlock trees within the conifer–hardwood forest, and the movement of the prairie/forest border. The concluding section provides a general synopsis of the results and examines their paleoclimatic implications as well as discussing the perspective gained by mapping the pollen record.

THE DATA BASE

Pollen Cores

The maps and other figures presented in this study are all based on the data from pollen analyses of 62 cores with radiocarbon dates in northeastern North America (Fig 1, Table 1). In most instances, the data were obtained from published sources, but some investigators have also kindly made unpub-

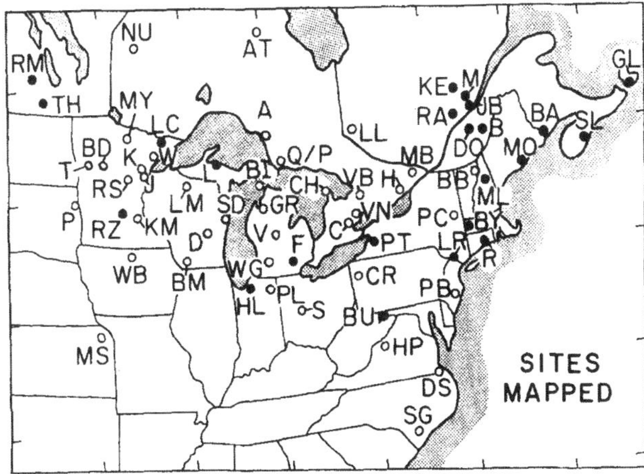

FIG. 1. Site map showing the location of the 62 pollen cores used in mapping. Letter codes refer to Table 1 which lists the sites and references. Solid dots indicate sites where information on pollen concentration or influx data was available.

TABLE 1

Code	Site	Reference
A	Alfies Lake, Ont.	Saarnisto, 1975
AT	Attawapiskat Lake, Ont.	Terasmae, 1968
B	St. Benjamin Lake, Que.	Richard, 1973c
BA	Basswood Rd. Lake, N.B.	Mott, 1975
BB	Bugbee Bog, Vt.	McDowell *et al.*, 1971
BD	Bog Pond D, Minn.	McAndrews, 1966
BI	Beaver Island, Mich. (Barney Lake)	Kapp *et al.*, 1969
BM	Blue Mounds Creek, Wis.	Davis, A., 1975
BU	Buckle's Bog, Md.	Maxwell and Davis, 1972
BY	Berry Pond, Mass.	Whitehead, 1976
C	Crieff Kettle Bog, Ont.	Terasmae, in Karrow, 1963
CH	Charles Lake, Ont.	Bailey, 1969
CR	Crystal Lake, Pa.	Walker and Hartman, 1960
D	Disterhaft Bog, Wis.	West, 1961; Baker, 1970
DO	Dosquet Bog, Que.	Richard, 1973c
DS	Dismal Swamp, Va.	Whitehead, 1965
F	Frains Lake, Mich.	Kerfoot, 1974
GL	Gillis Lake, N.S.	Livingstone, 1958
GR	Green Lake, Mich.	Lawrenz, 1975
H	Harrowsmith Bog, Ont.	Terasmae, 1968
HP	Hack Pond, Va.	Craig, 1969
HL	Hudson Lake, Ind.	Bailey, 1972
J	Jacobson Lake, Minn.	Wright and Watts, 1969
JB	Joncas Lake Bog, Que.	Richard, 1971
K	Kotiranta Lake, Minn.	Wright and Watts, 1969
KE	Kenogami Bog, Que.	Richard, 1973b
KM	Kirschner Marsh, Minn.	Wright *et al.*, 1963
L	Lost Lake, Minn.	Brubaker, 1975
LC	Lake of the Clouds, Minn.	Craig, 1972
LL	Lake Louise, Que.	Vincent, 1973
LM	Lake Mary, Wis.	Webb, 1974b
LR	Lake Rogerine Bog, N.Y.	Nicholas, 1968
M	Montagnais Bog, Que.	Richard, 1973b
MB	Mer Bleue Bog, Que.	Mott and Camfield, 1969
ML	Mirror Lake, N.H.	Likens and Davis, 1975
MO	Moulton Pond, Me.	Davis, R., *et al.*, 1975
MS	Muscotah Marsh, Kansas	Grüger, 1973
MY	Myrtle Lake, Minn.	Janssen, 1968
NU	Nungesser Lake, Ont.	Terasmae, 1967
P	Pickerel Lake, S.D.	Watts and Bright, 1968
PB	Pine Barrens Bog, N.J.	Florer, 1972
PC	Pine Log Camp Bog, N.Y.	Connally and Sirkin, 1971
PL	Pretty Lake, Ind.	Ogden, 1969
PT	Protection Bog, N.Y.	Miller, 1973
Q/P	Quadrangle Lake, Ont. (0–8000 YBP)	Terasmae, 1967
	Prince Lake, Ont. (8000–10,000 YBP)	Saarnisto, 1974
R	Rogers Lake, Ct.	Davis, M.B., 1969
RA	St. Raymond Bog, Que.	Richard, 1973a
RM	Riding Mountain, Man. (F Lake)	Ritchie, 1964 and 1969
RS	Rossburg Bog, Minn.	Wright and Watts, 1969
RZ	Rutz Lake, Minn.	Waddington, 1969
S	Silver Lake, Ohio	Ogden, 1966
SD	Seidel Lake, Wis.	West, 1961
SG	Singletary Lake, N.C.	Frey, 1951

TABLE 1 (*Continued*)

Code	Site	Reference
SL	Silver Lake, N.S.	Livingstone, 1968
T	Terhell Pond, Minn.	McAndrews, 1966
TH	Tiger Hills, Man. (Glenboro Lake)	Ritchie and Lichti-Federovich, 1968
VN	van Nostrand Lake, Ont.	McAndrews, 1970
V	Vestaburg Bog, Mich.	Gilliam *et al.*, 1967
VB	Victoria Rd. Bog, Ont.	Terasmae, 1973
W	Weber Lake, Minn.	Fries, 1962
WB	Woden Bog, Iowa	Durkee, 1971
WG	Wintergreen Lake, Mich.	Bailey, 1973

lished data available (see Acknowledgments).

Tables of the pollen data from each site were stored in computer files. If possible, the actual pollen counts at each depth in a core were entered into the files; but, when counts were not available, the pollen frequencies were extracted from published pollen diagrams. Computer programs were then used to calculate pollen percentages based on a pollen sum of total terrestrial pollen which excluded sedge (Cyperaceae), aquatics, and spores.

Our maps are based on percentage data. Where percentage values are likely to be distorted during times of low pollen influx, e.g., the late-glacial period (Davis, 1969), we have used influx information in order to exclude those percentage values from our maps. The influx data show when postglacial levels of pollen input were first reached, and percentage data from before that time in a given region were not included in mapping. Although this approach minimized a source of distortion on the percentage maps, it also prevented us from mapping the history of the tundra with its characteristically low pollen influx.

The Mapped Pollen Groups

Of the more than 300 pollen types identified within the cores in our data set (Richard, 1970; McAndrews *et al.*, 1973; Birks, 1976), we chose to map the five major pollen groups consisting of 13 commonly identified pollen types. The five groups are: (1) spruce (*Picea*); (2) pine (*Pinus*); (3) oak (*Quercus*); (4) herbs (Gramineae, Chenopodiineae, *Artemisia, Ambrosia,* Compositae, and *Plantago*); and (5) the "BAFT" pollen group which includes *Betula, Acer, Fagus,* and *Tsuga*. These five pollen groups numerically dominate the Holocene pollen records in our study area, and their modern distributions show clear patterns on maps (Davis and Webb, 1975). Studies of modern pollen distributions showed that these groups can be used as indices that roughly define the extent of the boreal forest, the conifer–hardwood forest, the deciduous forest, and the prairie formation (see Figs. 3 and 4; also Webb and McAndrews, 1976). Mapping these pollen types therefore allows monitoring the changes in location of these four vegetational regions during the Holocene period.

Two of the mapped pollen groups require specific definition. The first is the herb group which consists of those nonarboreal pollen types that are characteristic of temperate grasslands, i.e., grasses (Gramineae), composites (Compositae including *Ambrosia* and *Artemisia*), goosefoot and pigweeds (Chenopodiineae), and plantains (*Plantago*). Sedge (Cyperaceae) pollen has been deleted from this pollen group (as well as from the total pollen sum) because outside of the tundra region high postglacial values of this family reflect localized vegetational conditions at the site of deposition (see Wright and Patten, 1963; Webb and McAndrews, 1976; Webb and Yeracaris, in press).

As Fig. 4 shows, the modern distribution of the herb pollen contains a steep gradient along the prairie/forest border; and, east of the prairie region, the 30% contour parallels the area of >60% cultivated land. The herb pollen group is therefore sensitive to three major aspects of the vegetation. These include the distribution of prairie and savanna, the extent of herbaceous openings in forests and woodlands and the impact of cultural disturbance on formerly forested regions.

The second pollen category requiring definition is the BAFT group which represents the combined sum of birch, hemlock, beech, and maple pollen. The BAFT group combines the major pollen types indicative of the northern hardwoods forest type that occurs within the conifer–hardwood formation. In the central Great Lakes region in particular, the BAFT group reflects the distribution of the northern hardwoods forest type (Webb, 1974 a,b). However, to the north, west, and east of this central area (especially beyond the range limits of hemlock and beech) the values of this pollen group mainly reflect the abundance of birch trees. On the maps of this pollen group, sites with rich mixtures of the four genera are shown as solid dots to distinguish them from sites where birch exclusively dominates the BAFT pollen sum.

Although birch has been a major pollen type in the study area throughout the Holocene, the BAFT group as a whole was poorly represented prior to 7000 BP. For this reason we have only produced maps of the BAFT group after this date.

Temporal Control

Radiocarbon dates on the cores were used to determine depth–date relationships for all sites. In cases where authors gave radiocarbon dates corrected for ancient carbonate contamination, these corrected values were used. At many sites, a conspicuous pre- to postsettlement transition was identifiable based on sharp rises in ragweed, pigweed, and grass pollen (Fries, 1962; McAndrews, 1966), and the historical date for local settlement was assigned to that depth. The majority of the cores contained sediments spanning the full Holocene period, but some cores with less complete stratigraphic sections were also used (e.g., Maxwell and Davis, 1972). For cores that were discontinuous, only the sections near or between radiocarbon dates were included. All ages given in this study are expressed as radiocarbon years before present.

For about 20% of the sites, the original authors provided estimated time scales for their cores based on multiple radiocarbon dates (see, for instance, M. Davis, 1969; R. Davis *et al.*, 1975). At all other sites, the age of levels were determined by linearly interpolating between consecutive radiocarbon dates and between the uppermost radiocarbon date and the historically dated presettlement horizon. When possible, sites with single radiocarbon dates were avoided if more extensively dated cores existed in the same area. There was an average of four dates per core with less than one-third of the sites having only a single date.

The Mapped Time Levels

Once time scales have been established for the 62 sites, new tables were computed containing the percentage values of the five pollen groups at each 1000-year level from 11,000 BP to present. Tables containing difference values were then created by subtracting the pollen frequencies between successive time levels. These tables of difference values thus record the magnitude and direction (increases or decreases) of the changes in each pollen group between the different time levels. From these tables, the contoured isopoll and difference maps were then produced.

In order to determine what was the most illustrative yet concise sequence of maps to present, we used the difference values to identify the periods of greatest change.

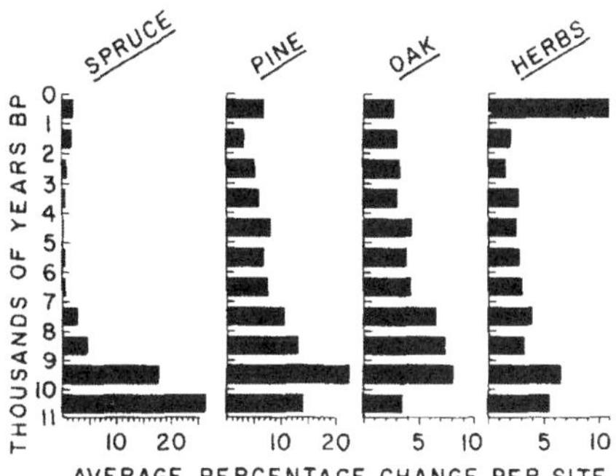

FIG. 2. Graph depicting the average percentage change per site, between each 1000-year level from 11,000 BP to present, for spruce, pine, oak, and herb pollen. The figure shows important shifts in the amount of change these major pollen groups underwent during the Holocene. Values were obtained by summing the total changes (regardless of signs) seen from all mapped sites and then dividing by the number of sites.

Figure 2 shows the average change in percentage values per site for spruce, pine, oak, and herbs between each 1000-year level from 11,000 BP to present. These numbers were calculated by summing the absolute values in the tables of difference values for each pollen group and then dividing by the number of sites.

The largest changes occurred in the early Holocene between 11,000 and 7000 BP (Fig. 2). After 7000 BP, the magnitude of changes decreased except during the last 1000-year interval when European settlement took place. Based on these results, we decided to present maps for every 1000-year level in the early Holocene, whereas the more gradual changes between 7000 and 2000 BP are shown by maps spaced at 2000- and 3000-year intervals. For illustrating the dramatic culturally induced changes in the last 1000 years, maps are presented for 500 BP and for the differences between 500 BP and the present.

CONSTRUCTION OF THE MAPS

The Map Types

Previous cartographic presentations of Holocene pollen data have been of six kinds: (1) contoured isopoll maps showing the continuous distribution of individual pollen types (Szafer, 1935; Firbas, 1949; Birks *et al.*, 1975a; Birks and Saarnisto, 1975); (2) composite maps with pie diagrams of the frequencies for many pollen types at certain time levels (Faegri and Iversen, 1964; van Zeist and Wright, 1967; Ritchie, 1976); (3) dot maps showing the values of pollen percentages for individual pollen types (Neustadt, 1959); (4) difference maps illustrating the patterns of change for pollen types from one time period to the next (Webb, 1973); (5) species-range maps showing the position of the range boundaries for selected taxa and the location of ecotones at specific time intervals (Davis, 1974); and (6) isochrone maps showing the migration of pollen types (Donner, 1963; Markgraf, 1970; Moe, 1970; Moran, 1973; Serebryanny, 1973; Davis, 1976) or the progressive movement of ecotones (Webb and Bernabo, 1974).

Isopoll maps, difference maps, and isochrone maps are utilized in this study. Isopoll maps illustrate the distributional pattern of a given pollen type at a specified time. They are contoured with isopolls that outline regions of equal pollen frequency.

The difference maps are employed in order to compare the changes that occur between different isopoll maps and they are contoured with lines delimiting areas of equal change. Negative changes indicate decreasing pollen frequencies over the time interval and a zero contour is used to separate the areas with increasing values from those with decreasing values. For emphasis regions were shaded that experienced decreases of −10 percentage units or more. A contour interval of 10 percentage units was used on both isopoll and difference maps.

Isochrone maps show the progressive movement of various pollen-defined boundaries over periods spanning many thousands of years. These maps are used to illustrate the movement of major ecotones and the migration of certain genera.

Figure 1 shows that the sample sites are not uniformly distributed and that few sites occur in either the far northern or the far southern portions of the mapped area. For this reason, the placement of contours in these regions must be considered preliminary and will likely be modified as new data become available.

The Paleogeographic Reconstructions

The maps presented in this study include paleogeographic features such as the ice margin, glacial lakes, and marine shoreline. In this way, the changing patterns of pollen distributions could be related to major paleogeographic boundaries, and some analysis of the relationship between the dynamics of the vegetation and other features on the landscape was possible. The position of the Laurentide ice sheet is based on the isochrone maps by Bryson *et al.* (1969), Prest (1969), and Saarnisto (1974). The histories of the glacial lakes and marine transgressions come from works by Elson (1967), Hough (1963), Hughes (1965), Lee (1960), and Terasmae (1959). A radiocarbon chronology compiled in Flint (1971, Table 21-F) was used to establish the sequence of events throughout the mapped region. Comprehensive reconstructions by Prest (1970) aided the synthesis of the various other sources in the central map area. Changes in the Atlantic coastline were reconstructed using bathymetric charts and work by Grant (1970) and Emery and Garrison (1967).

Comparison of Pollen and Tree-Inventory Data

A set of paired contoured maps of modern pollen and tree-inventory data were constructed for spruce, pine, and oak in order to illustrate the relationship between the distributional patterns of these pollen types and the trees producing the pollen (Figs. 3 and

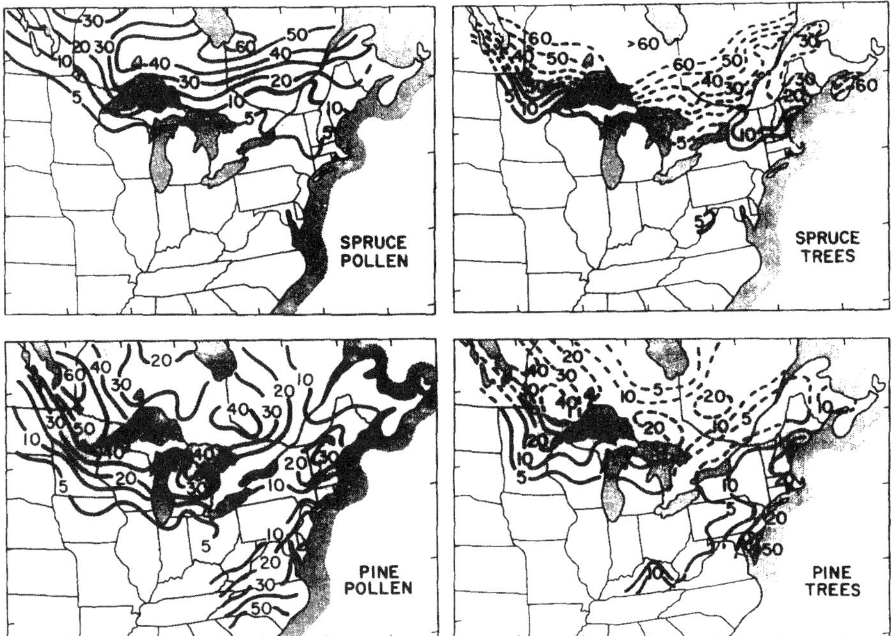

FIG. 3. Maps comparing the modern distributions of spruce pollen with spruce trees and pine pollen with pine trees. The contours are in percent total pollen and percent growing stock of trees. Dashed contours in Canada indicate the less precise nature of the tree-inventory data used there.

4). Tree-inventory data came from 69 forest-survey units published in the U.S.D.A. Forest Service Resource Bulletins available for 20 of the 28 states in our study area. The values plotted are the percentages for each genus within the total growing stock.

Halliday and Brown's maps (1943) provided comparable information on the relative abundances of the trees in Canada. Because the Canadian data are less precise than those in the United States, we have dashed the contours in Canada. In the case of pine trees, Halliday and Brown mapped only the values of jack pine (*Pinus banksiana*). The contours shown on Fig. 3, therefore, underestimate the relative abundance of pine trees in the region of southeastern Canada, which lies within the modern range of white (*Pinus strobus*) and red pines (*Pinus resinosa*).

The maps of modern pollen distributions are based on nearly 400 surface samples presented in Davis and Webb (1975) and Webb and McAndrews (1976). The modern pollen data are contoured as percent of the total terrestrial pollen sum in the same way as the paleopollen data.

Visual comparison of the maps for spruce, pine, and oak demonstrates that, on the scale of this study, the major patterns in the tree-inventory data clearly appear in the pollen data. A numerical comparison between the pairs of maps for spruce, pine, and oak yields Pearson product-moment correlation coefficients of 0.93, 0.78, and 0.93, respectively. These results provide a strong basis for interpreting the pollen data in terms of changes in forest composition. For other demonstrations of how well pollen data reflect vegetational patterns in eastern

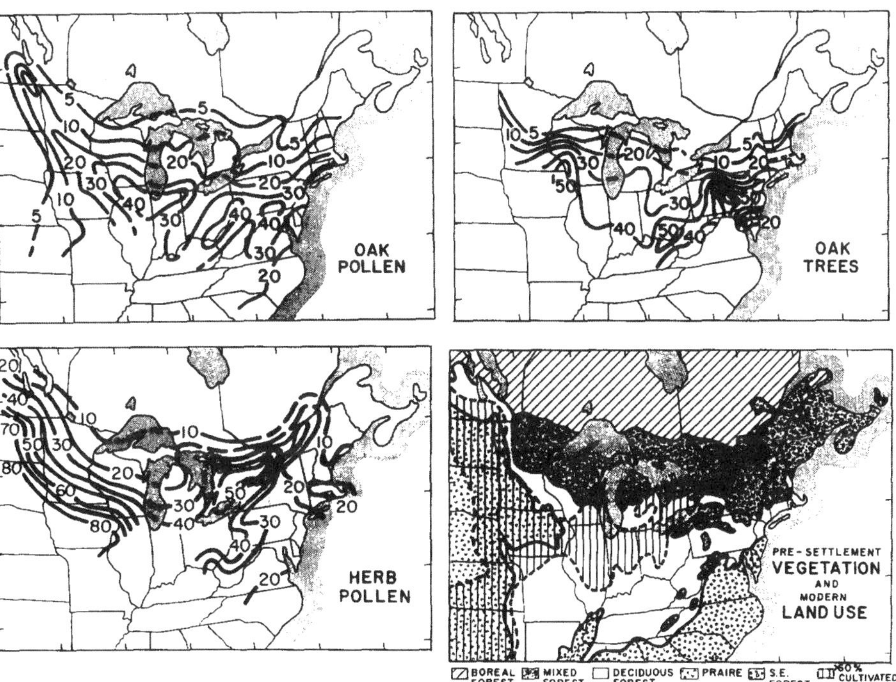

FIG. 4. Maps illustrating the modern distributions of oak pollen and oak trees as well as comparing the pattern of herb pollen to a map of the major vegetational regions with land use data superimposed. Potential vegetation and land use information are from the Oxford World Atlas (1973).

North America, see Davis (1967), Lichti-Federovich and Ritchie (1968), Webb (1974 a,b), Davis and Webb (1975), Birks *et al.* (1975b), and Webb and McAndrews (1976).

DESCRIPTION OF THE HOLOCENE POLLEN MAPS

11,000 to 10,000 BP

Spruce pollen dominated a broad region south of the glacial front 11,000 years ago and extended down the Appalachians in the east (Fig. 5). Gradients of decreasing frequencies of spruce pollen are seen to the north and south of this zone. Macrofossil and pollen evidence suggests that a narrow belt of tundra probably existed where spruce values diminish between the proglacial lakes and late-glacial boreal forest in northern Minnesota (Watts, 1967). Low pollen influx in New England (M. Davis, 1969; R. Davis *et al.*, 1975) and adjacent New Brunswick (Mott, 1975) as well as relatively high herb values in southeastern Quebec (LaSalle, 1966) indicate tundra-like conditions in these areas too.

At this same time, high values of pine pollen were confined to an area from the northern Appalachians southeastward. Oak pollen showed its largest percentages to the far southeast in North Carolina and may also have been a dominant element in the region south of Indiana and Ohio, where few sites are currently available. The greatest frequencies of herb pollen at 11,000 BP occurred west of Minnesota and Iowa. Relatively high percentages of herb pollen also appear south of the glacial Great Lakes and probably indicate relatively open forests in these areas (Davis, 1965).

Between 11,000 and 10,000 BP, spruce pollen underwent a major reduction in abun-

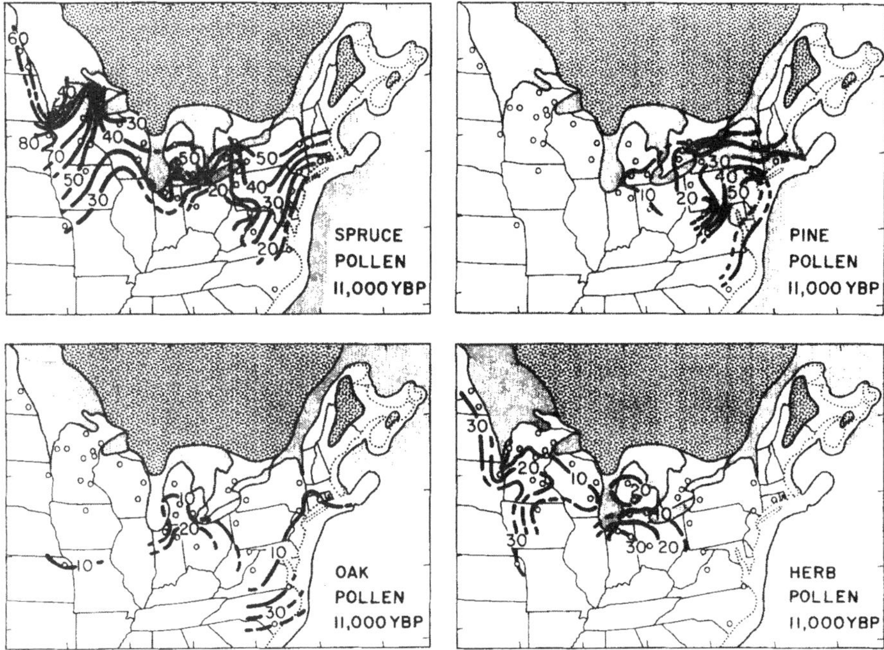

FIG. 5. Isopoll maps showing the distributions of spruce, pine, oak, and herb pollen at 11,000 BP. Contours are in percent total terrestrial pollen. The appropriate paleogeographic features including the ice margin, glacial lakes, and marine shoreline are indicated.

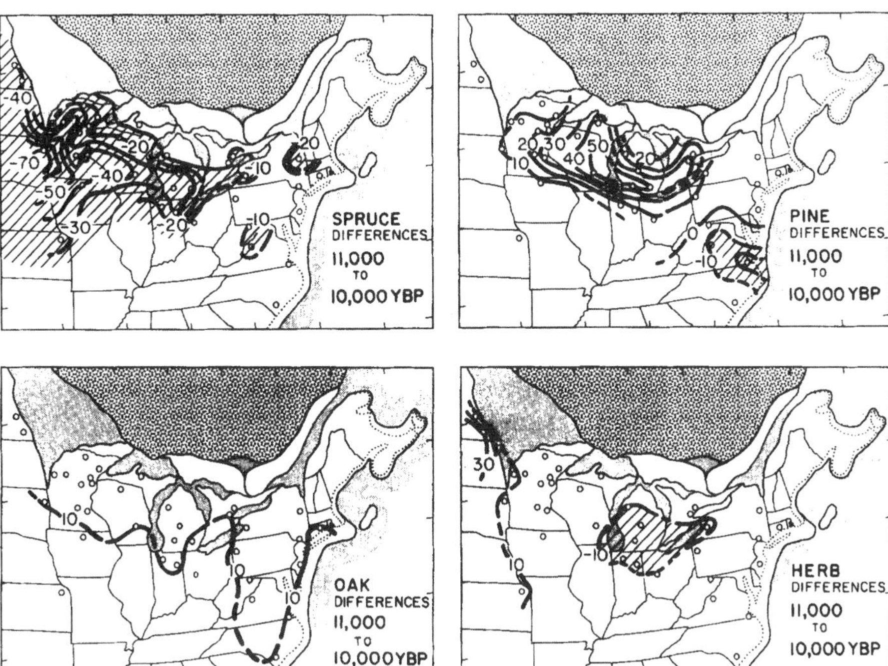

FIG. 6. Difference maps illustrating the patterns of change for spruce, pine, oak, and herb pollen between 11,000 and 10,000 BP. Contours show percentage changes, with negative values indicating decreasing frequencies. Areas with decreases of −10% or more are shaded and the zero contours separate areas of positive and negative change.

dance, while the area dominated by pine pollen shifted westward (Fig. 6). Pine values increased the most in the northeastern portion of the area within which spruce declined, whereas oak and herb pollen replaced spruce pollen in the west. Oak pollen also increased at the expense of pine pollen in the southeastern area of our maps.

Figure 7 shows the results of these changes on the patterns evident at 11,000 BP. By 10,000 BP, spruce pollen no longer shows a continuous band of dominance from east to west and its highest frequencies have shifted northward in the wake of deglaciation (Fig. 7). High values of pine pollen now extend westward across the southern Great Lakes with pine having become the predominant pollen type in a belt that stretches from the Atlantic Coast to Minnesota. The 10 and 20% contours for oak pollen are further north than they were at 11,000 BP, and the 30% contour for herb pollen has moved eastward to a position in eastern South Dakota and western Iowa. Herb values in the east are generally below 10% indicating that the forests were more closed than at 11,000 BP.

10,000 to 9000 BP

Between 10,000 and 9000 BP, spruce pollen decreased over a wide area in the north and was replaced by pine pollen, which in turn decreased throughout the central part of our study area and was itself being displaced by oak pollen (Fig. 8). In the far southeast, however, oak pollen decreased and was being replaced by pine pollen. These changes mark a northward movement of the forest regions in response to the warming

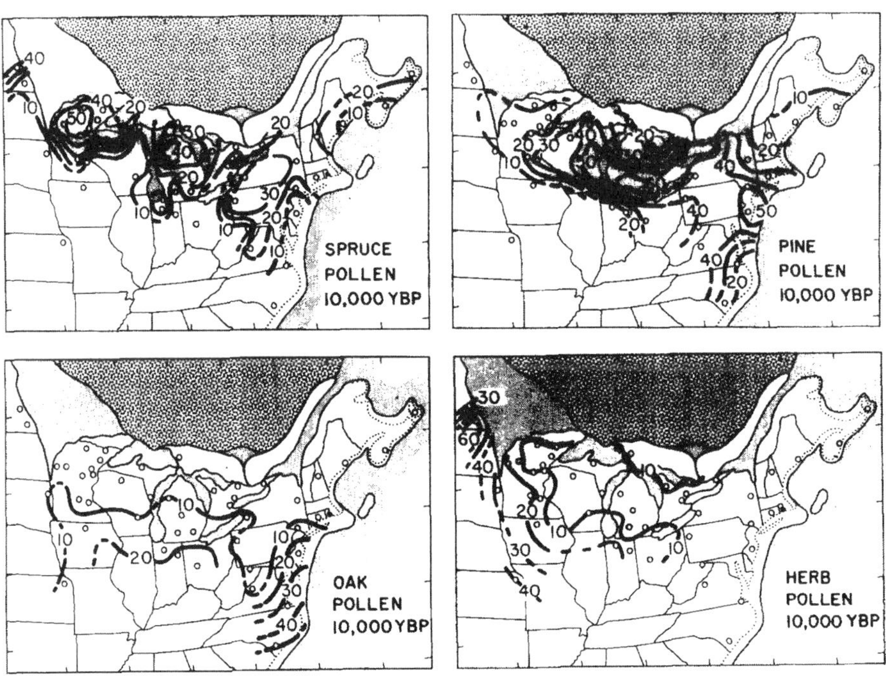

FIG. 7. Isopoll maps showing the distributions of spruce, pine, oak, and herb pollen at 10,000 BP.

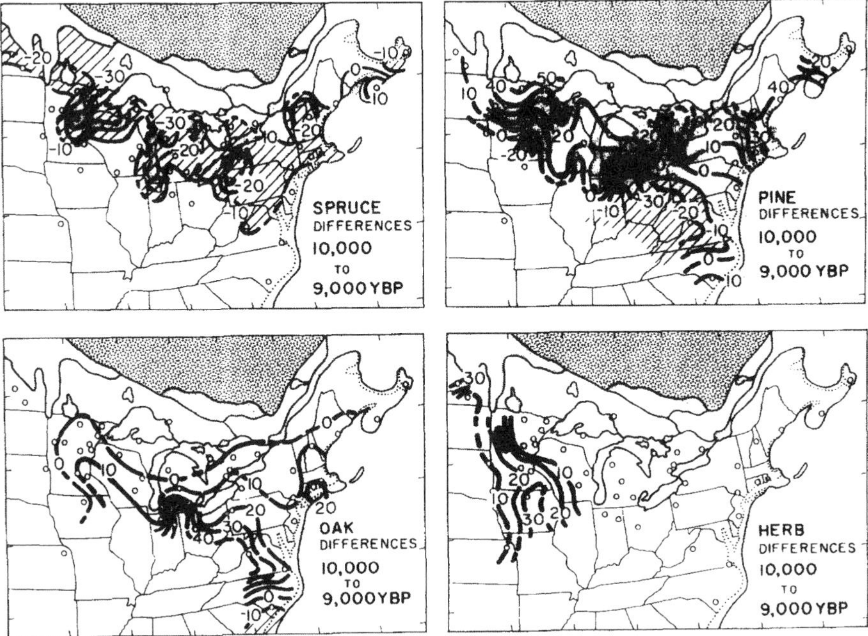

FIG. 8. Difference maps depicting the patterns of change for spruce, pine, oak, and herb pollen between 10,000 and 9000 BP.

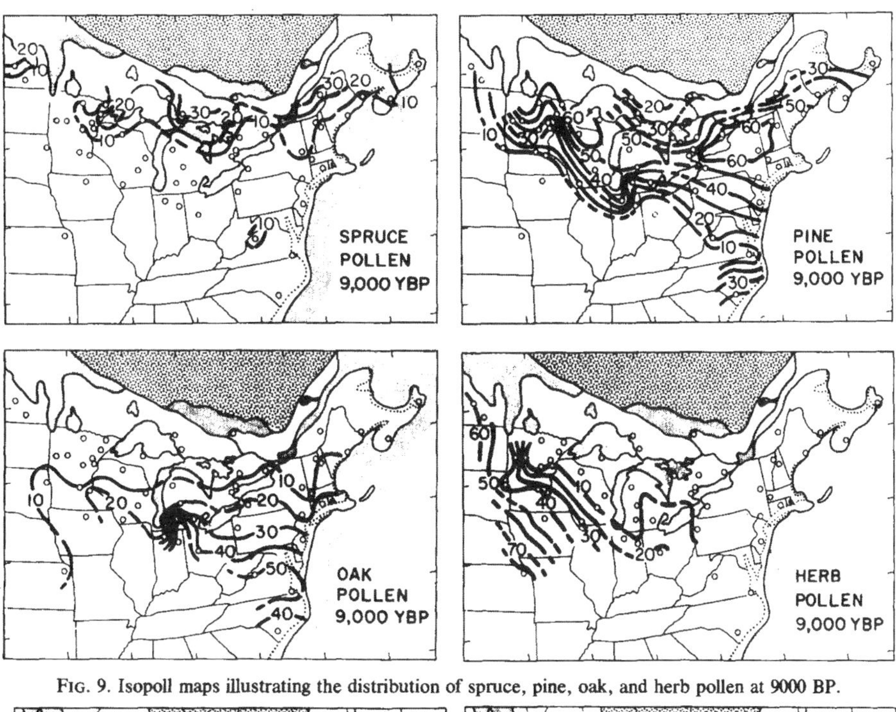

FIG. 9. Isopoll maps illustrating the distribution of spruce, pine, oak, and herb pollen at 9000 BP.

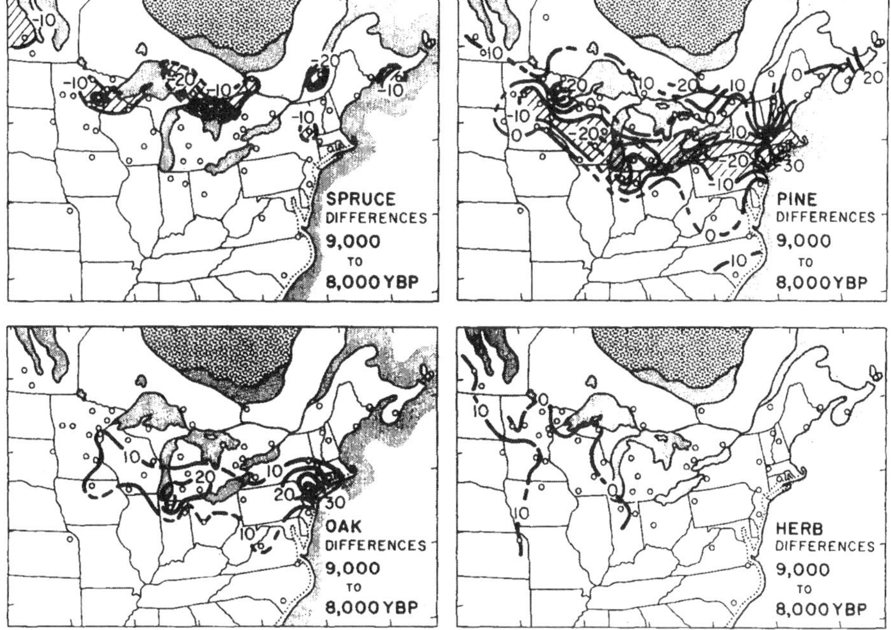

FIG. 10. Difference maps showing the patterns of change for spruce, pine, oak, and herb pollen between 9000 and 8000 BP.

climate and retreating ice. In the west, the frequencies of herb pollen increased dramatically and indicate the most rapid period of eastward prairie migration (Figs. 2 and 8). In southern Manitoba, herb pollen directly replaced spruce pollen, but to the south herb pollen displaced deciduous-tree types such as elm, oak, and ash.

By 9000 BP, spruce was no longer the major pollen type in the Great Lakes region, and its area of dominance was considerably reduced (Fig. 9). Pine was now the major pollen type over much of the area that spruce pollen formally dominated. Values of pine pollen were also high south of the region dominated by oak pollen, which was now a major pollen type reaching 20% or greater as far north as Wisconsin, Michigan, and New York. Frequencies of herb pollen greater than 30% covered most of the western Midwest and extended eastward into southwestern Wisconsin.

9000 to 8000 BP

During the period from 9000 to 8000 BP, the four mapped pollen groups experienced changes of a smaller magnitude than in the preceding 2000 years (see Fig. 2), but continued the trends set during that time interval. The values of spruce pollen decreased in the north where pine frequencies increased (Fig. 10), while just south, the frequencies of pine pollen declined as the values of oak pollen rose. In the western Midwest, the herb frequencies continued to increase.

By 8000 BP, spruce pollen was a minor element over most of the mapped region, with its zone of highest frequencies severely compressed between the area dominated by pine pollen to the south and the proglacial lakes and ice margin in the north (Fig. 11). Unlike spruce, the patterns of pine, oak, and herb pollen by 8000 BP had reached

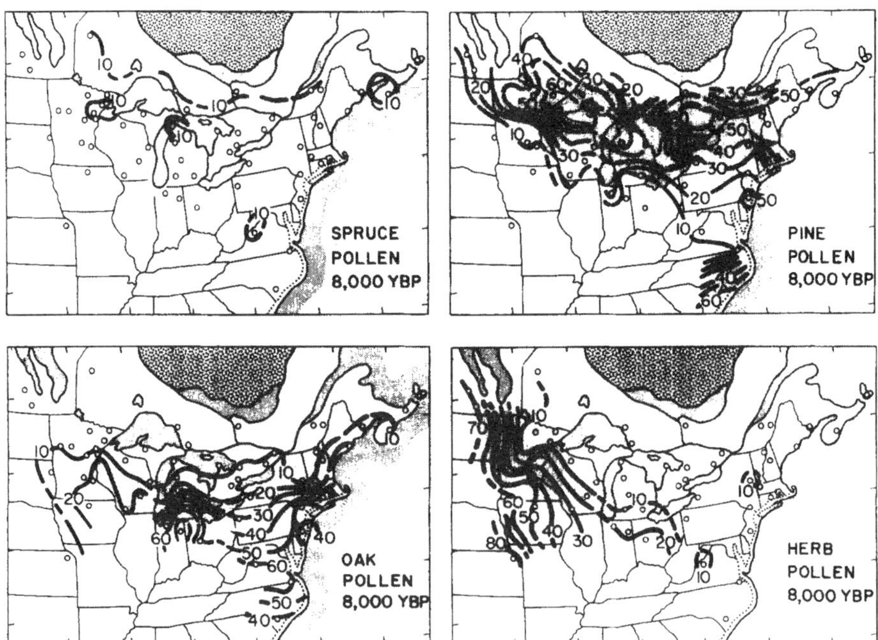

FIG. 11. Isopoll maps depicting the distributions of spruce, pine, oak, and herb pollen at 8000 BP.

configurations grossly similar to their presettlement distributions. Pine was the predominant pollen type throughout a large area from the Great Lakes to New England and also in the region south of Virginia. Oak pollen continued to dominate the area south of the Great Lakes, while high frequencies of herb pollen prevailed west of Wisconsin and Illinois.

8000 to 7000 BP

The Laurentide ice sheet catastrophically collapsed after 8000 BP and had disappeared from the mapped area by 7000 BP (Bryson *et al.*, 1969). The pollen trends evident over the previous 3000 years culminated during this time interval. Spruce pollen essentially disappeared south of about 47°N, and high frequencies of pine and oak pollen shifted further north (Fig. 12). At the same time, herb pollen expanded to its maximum eastward position, and oak savanna became widespread in Minnesota (Wright *et al.*, 1963).

By 7000 years ago, all sites south of the Great Lakes were registering less than 10% spruce pollen (Fig. 13). Because few sites occur north of the Great Lakes, our map may be missing the potentially high values of spruce pollen there. High percentages of pine pollen (>40%) stretched from southwestern Ontario across the northern Great Lakes and into New England, while oak pollen was a major element (>30%) to the south of this area. With its 20% contour as far east as Indiana and southeastern Wisconsin, herb pollen was at its maximum eastward extent prior to European settlement. By this time, BAFT pollen had also become a numerically important pollen group within our region of study (Fig. 23), but only three sites in New York and New England contain major contributions from all four of the pollen types in this group.

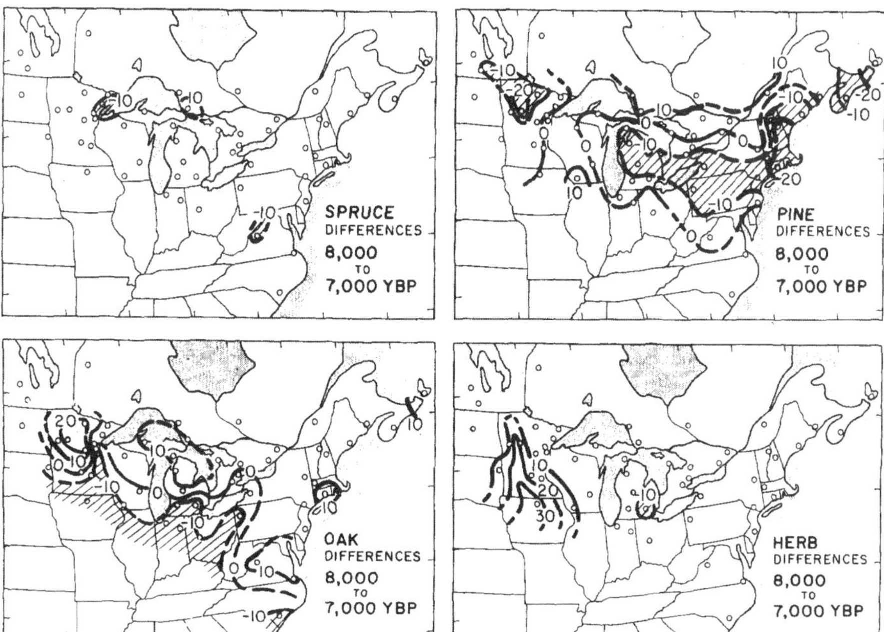

FIG. 12. Difference maps illustrating the patterns of change for spruce, pine, oak, and herb pollen between 8000 and 7000 BP.

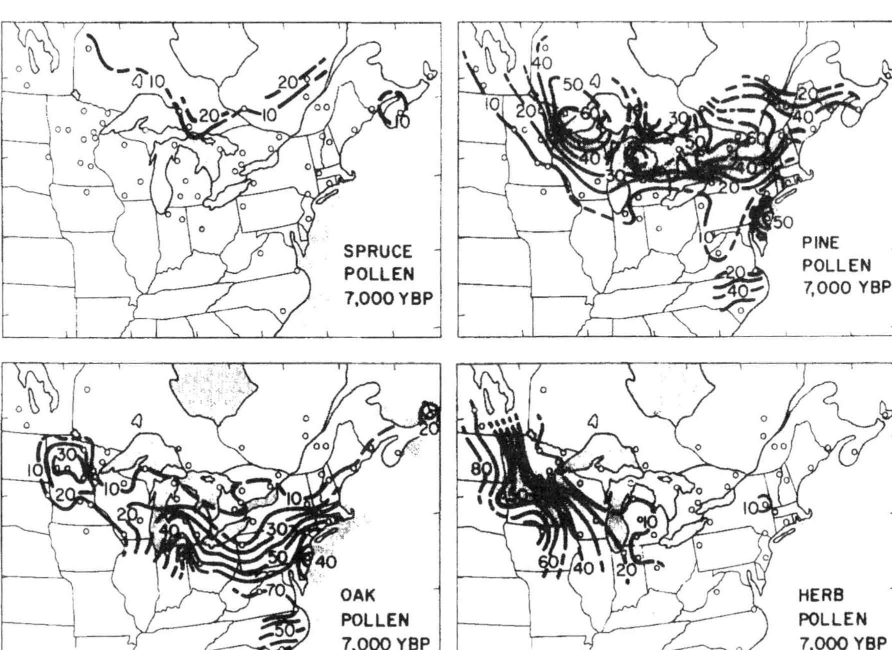

FIG. 13. Isopoll maps showing the distributions of spruce, pine, oak, and herb pollen at 7000 BP.

7000 to 4000 BP

From 7000 to 4000 BP, changes in the distributional patterns of spruce, pine, oak, and herb pollen were more gradual than during the earlier Holocene (Fig. 2). A reversal of many of the previous trends for pine, oak, and herbs took place during this period, and by 4000 BP the cumulative differences are evident on the maps in Fig. 14. Most notable of these reversals are the westward retreat of the high values of herb pollen and the southward recession of high values in pine and oak pollen. In addition, BAFT pollen greatly expanded the region over which it was prominent (Fig. 23).

Just as pine pollen had increased at the expense of spruce pollen from 11,000 to 7000 BP, BAFT pollen increased at the expense of pine pollen from 7000 to 4000 BP. These changes took place in the area between the regions dominated by oak pollen in the south and by spruce pollen in the north (Figs. 14, 15, and 23) and represent the westward expansion of the prominance of birch, maple, beech, and hemlock trees within the conifer–hardwood forest.

As of 4000 years ago, high values of spruce pollen occurred in the far northern portion of the mapped region. This appearance suggests that the boreal forest again covered a wide areal extent (Fig. 15). The zone of the highest values for pine pollen was further south than it was at 7000 BP and occupied a less extensive area, although the values of pine pollen had begun to rise in northwestern Minnesota. Pine percentages in the northern New England region had by 4000 BP dropped to below 40% for the first time since 9000 BP. As just discussed, this change was related to increases in BAFT pollen. The percentages of oak pollen were considerably lower in the far southeast than they had been 3000 years earlier, and the zone with high frequencies

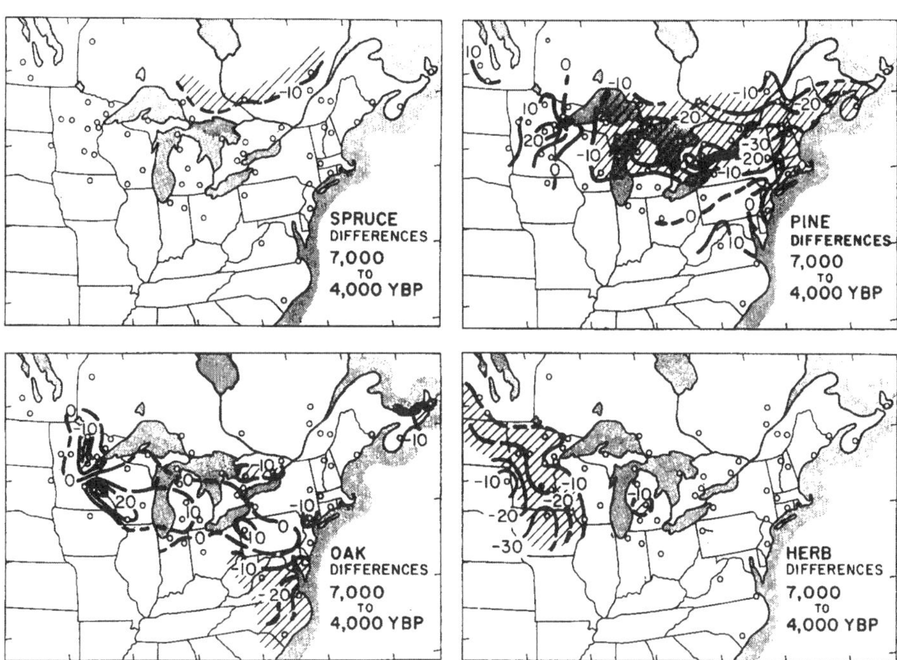

FIG. 14. Difference maps depicting the patterns of change for spruce, pine, oak, and herb pollen between 7000 and 4000 BP.

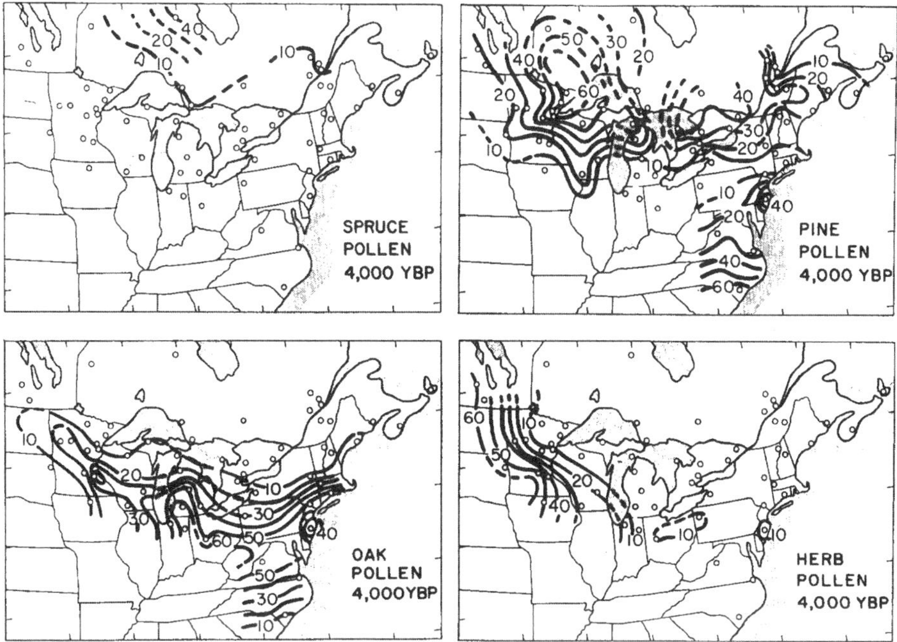

FIG. 15. Isopoll maps illustrating the distributions of spruce, pine, oak, and herb pollen at 4000 BP.

of herb pollen was less extensive, having shifted back westward.

4000 to 2000 BP

From 4000 to 2000 BP, spruce pollen increased its relative abundances just north of the Great Lakes and in southern Manitoba. This event marks the first significant southward movement of the boreal forest during the Holocene. Pine, oak, and herb pollen all continued their reversals of the early Holocene trends (Fig. 16), and BAFT pollen also continued a westward expansion of its area of high values (Fig. 23).

The results of these changes are seen in Figs. 17 and 24. Spruce pollen values over 10% occurred across southern Canada from Manitoba to Quebec. Pine was still the predominant pollen type in southwestern Ontario, northeastern Minnesota, and Quebec as well as to the south of Virginia, but its frequencies were 20% or less throughout the central and eastern Great Lakes region and New England. By 2000 BP, BAFT pollen registered high values (>30%) from lower Michigan eastward into New England. The oak isopolls had shifted further south across the Great Lakes region as the region of high values (>30%) for herb pollen receded slightly further westward. This change resulted in a steepening of the gradient in the Midwest between the areas dominated by nonarboreal and arboreal pollen types.

2000 to 500 BP

In this period just prior to the widespread vegetational disturbances caused by European settlement, the pollen changes were relatively small and often localized. Only spruce pollen shows a regionally consistent pattern with increasing percentages across most of southern Canada and the northernmost United States. Spruce thus continued

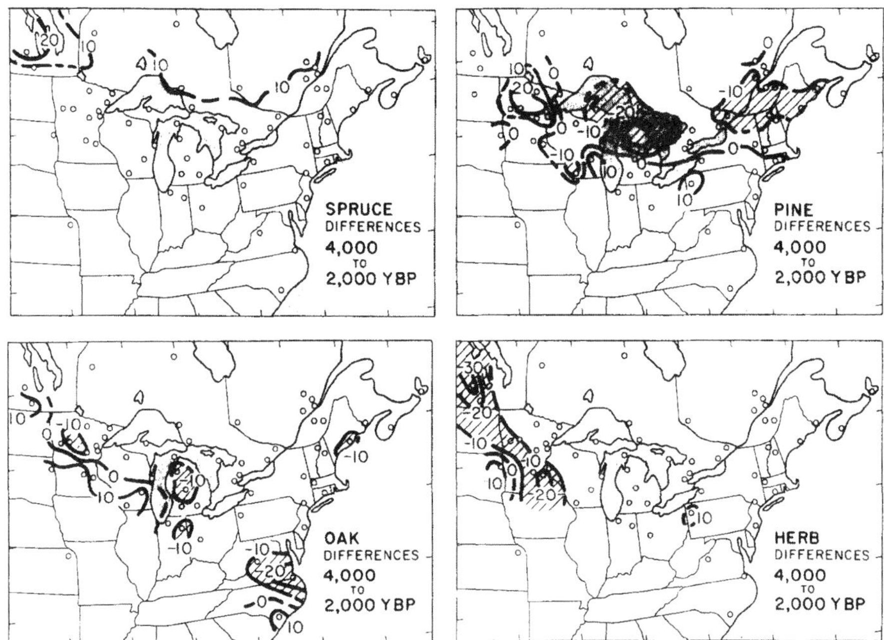

FIG. 16. Difference maps showing the patterns of change for spruce, pine, oak, and herb pollen between 4000 and 2000 BP.

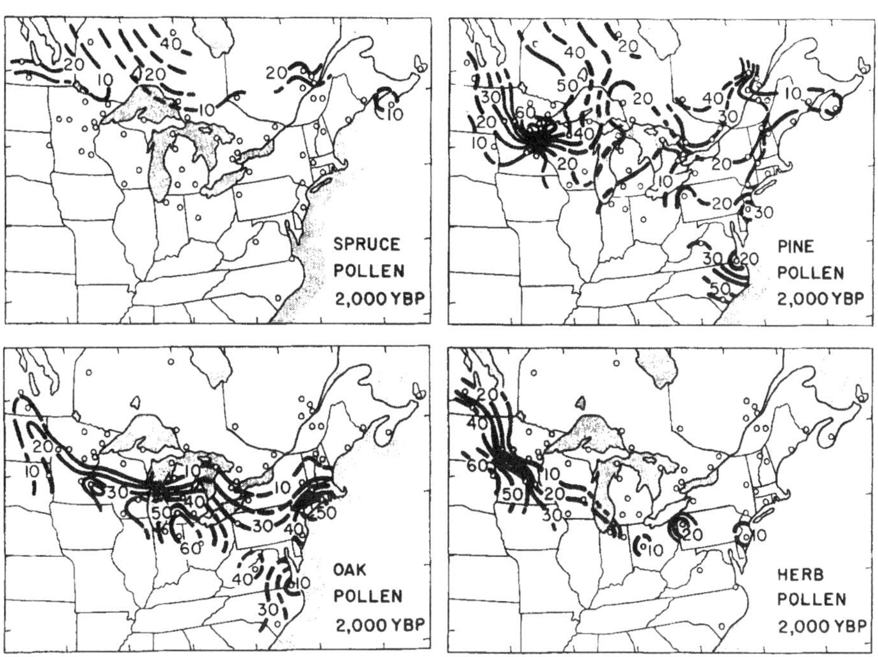

FIG. 17. Isopoll maps depicting the distributions of spruce, pine, oak, and herb pollen at 2000 BP.

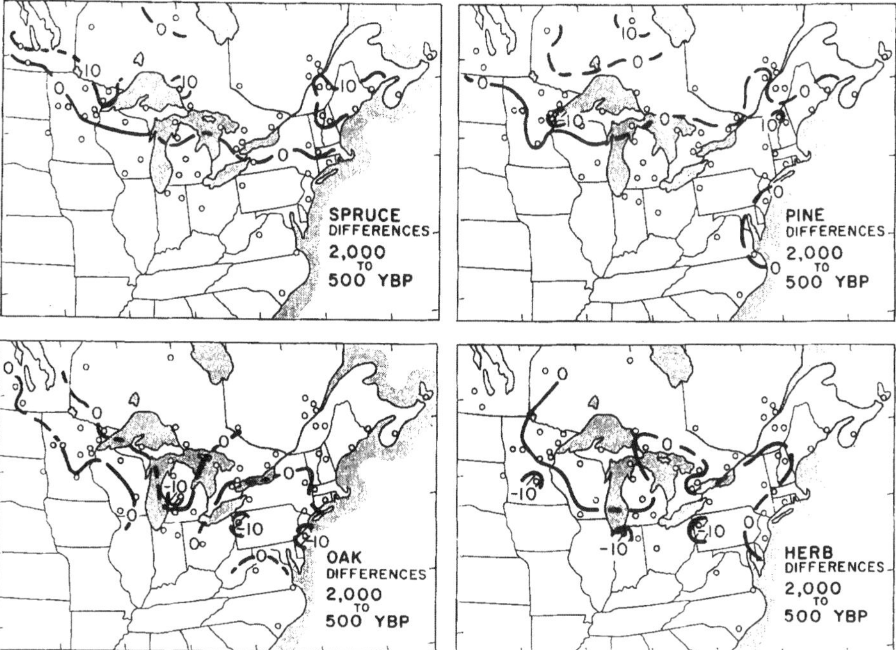

FIG. 18. Difference maps illustrating the patterns of change for spruce, pine, oak, and herb pollen between 2000 and 500 BP.

a trend established 3500 years earlier. The changes in pine and oak pollen are too subdued to show any interpretable trends (Figs. 16 and 18), and the stability of the herb-pollen frequencies indicates a cease to the major westward migration of the prairie border.

The distributional patterns of the major pollen types at 500 BP (Fig. 19) are therefore closely similar to their patterns at 2000 BP. The one exception was the perceptible southward shift of the 10% isopoll for spruce.

500 BP to Present

Unlike any of the previous intervals studied, the changing patterns of pollen distribution from 500 BP to present directly reflect man's impact on the landscape. Because the culturally induced disturbances were complex and metachronous across the mapped area, our difference maps (Fig. 20) do not reveal the full details of the various stages of European settlement. The maps do show, however, the end result of these activities.

From 500 BP to the present, the distribution of spruce pollen remained comparatively stable, but pine and oak pollen decreased as the frequencies of herb pollen increased throughout the central mapped region (Fig. 20). The major increases in herbs occurred where agricultural activities caused nearly complete deforestation (Fig. 4) which resulted in sharp decreases in pine, BAFT, and oak pollen. These changes illustrate the geographical patterns of man's impact on the natural vegetation of northeastern North America.

Summary of the Mapped Patterns

The broadest scale and most rapid changes in the Holocene pollen record occurred during the period of final deglaciation

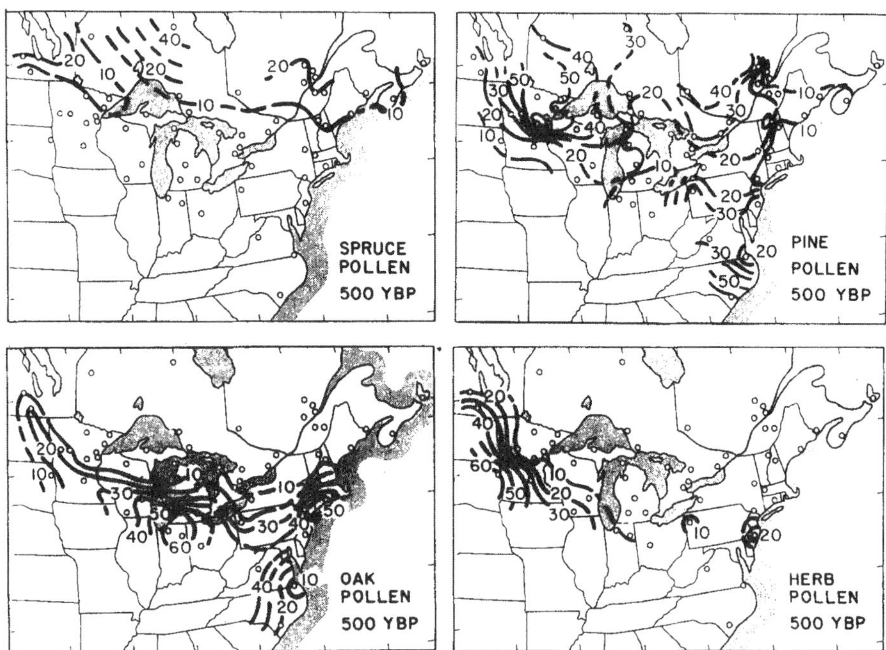

FIG. 19. Isopoll maps showing the distributions of spruce, pine, oak, and herb pollen at 500 BP.

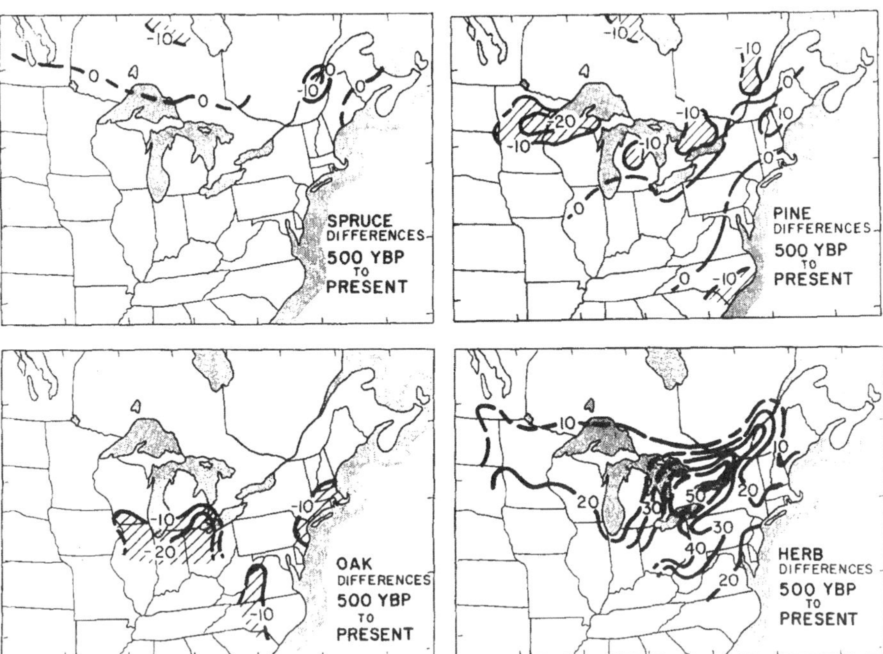

FIG. 20. Difference maps depicting the patterns of change for spruce, pine, oak, and herb pollen between 500 BP and present.

from 11,000 to 7000 BP. Initially, spruce pollen declined over a wide region, and pine moved westward into the Midwest as oak spread northward and prairie herbs expanded eastward. With the continued glacial retreat from 10,000 to 8000 BP, the zone with high values of spruce pollen shifted northward and shrank to a narrow belt, while oak moved north and replaced pine just as pine was moving north and replacing spruce. This northward trend in the region from the Great Lakes eastward continued until 7000 BP and was complemented in the west by an eastward expansion of the area dominated by herb pollen. This expansion was most rapid between 10,000 and 9000 BP, and the prairie attained its maximum eastward extent in the Midwest by 7000 BP.

After 7000 BP, the Laurentide ice sheet quickly disintegrated and the changes in the major pollen types became smaller and opposite to most earlier trends. Between 7000 and 2000 years ago, pine and oak moved southward as the prairie receded westward. The largest expansion occurred in the BAFT group as its pollen expanded in abundance and replaced the high values of pine pollen in the eastern and central Great Lakes region. From about 4000 BP onward, spruce increased in abundance south of a wide front from Manitoba to Quebec. After 2000 BP, only spruce showed a continued southward expansion, and the most dramatic changes appear in the widespread increase of herb pollen as Europeans lumbered, settled, and cultivated the land beginning around 350 BP.

FOUR MAJOR TRENDS IN THE VEGETATION

This section uses isochrone maps for illustrating certain major trends evident on the isopoll and difference maps. These trends include the northward retreat of the late-glacial boreal forest, the movement of the deciduous/conifer–hardwood ecotone, the movement of the prairie/forest ecotone, and the expansion in range and abundance of birch, maple, beech, and hemlock.

The Decline of Spruce

The decline in spruce pollen from 11,000 to 8000 BP represents one of the most widely expressed zonal boundaries in North American pollen profiles (Figs. 6, 8, 10). Moran (1973) illustrated its time-transgressive nature in northeastern North America by mapping isochrones based on 26 dates for the onset of spruce decline. In a similar manner, Fig. 21 shows the times when the values of spruce pollen fell below 15% at the sites used in our study.

Maps of modern pollen (Fig. 3) indicate that spruce pollen generally exceeds 15% in the boreal forest region. The isochrones on Fig. 21, therefore, trace the changing position of the southern edge of the late-glacial spruce-dominated "boreal forest" during its northward movement from 11,000 to 8000 BP. Because this movement occurred somewhat faster than deglaciation was taking place from 11,000 to 9000 BP, the progressive compression of this forest zone left only a thin belt of spruce forest in close proximity to the retreating ice and associated proglacial lakes (Figs. 5, 6, 9). After 9000 BP, the northward progression of the isochrones slowed, and by 8000 BP the southern border of the boreal forest reached its most northern Holocene position. With the rapid disintegration of the ice sheet after 7000 BP, spruce trees quickly spread northward, an event that could also be mapped by isochrones if sufficient data were available.

Like Moran's (1973) map, our map shows an overall time-transgressive trend, but the timing of the isochrones in our map differ from those on Moran's map because two different events were mapped. In contrast to our mapping the time when spruce values had decreased below 15%, Moran (1973) mapped the time of the first significant decrease in spruce pollen.

Advantage may be taken of this difference in the events mapped, because the time lag between the onset of spruce decline and the later decrease of spruce values below 15% provides a rough guide for how rapidly the late-glacial boreal forest was replaced by other types of vegetation. For instance, little difference appears in the location of the isochrones on the two maps at many Midwestern sites, where the spruce decline is

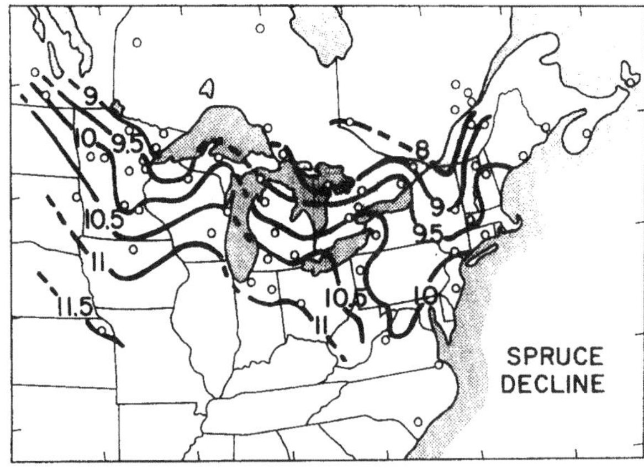

FIG. 21. Isochrones plotting the time, in thousands of years BP, when spruce pollen declined to below 15%.

extremely rapid. Some of these sites show spruce pollen going from 50 to 5% in only a few hundred years (Bryson *et al.*, 1970; Wright, 1971; Watts, 1973). At other sites, however, spruce pollen declines over a period of several thousand years before reaching less than 15%, and the two maps differ markedly in their placement of the isochrones near these sites. Further work is now needed in order to identify the ecological implications of these differences.

Changing Distributions of Pine and Oak

The distribution of pine pollen at 11,000 BP was confined to the east (Fig. 5), and maps constructed by Davis (1976) suggest pine trees were present only as far west as Michigan (see also Cushing, 1965; Wright, 1968b). As the values of spruce pollen declined throughout the Midwest (Fig. 21), pine trees migrated rapidly westward, covering 1000 km in as many years (Figs. 5–7).

By 10,000 BP, the zone of high frequencies of pine pollen occupied the southern Great Lakes region and extended into Michigan, Wisconsin, and Minnesota (Fig 7). The species of pines involved in the westward expansion were *Pinus banksiana* and *Pinus resinosa* (initially probably just *P. banksiana*), while *Pinus strobus* did not arrive in the Midwest until 1000 years later (Wright, 1971; Davis, 1976). At the same time, pine trees were also moving northward in New England, where *Pinus strobus* was replacing spruce (Davis, 1965, 1976). After pine trees completed their major westward expansion, they continued to move northward as the amount of spruce pollen declined. To the south, oak trees were displacing pine trees along the southern edge of the pine-dominated region (Figs. 8, 10, and 12).

Pine trees, especially *Pinus banksiana*, may also have migrated into the Midwest from the west (Ritchie, 1976). The distribution of the cores used in our study, however, only allows us to illustrate the westward expansion of pine.

Maps of modern pollen distributions (Figs. 3 and 4), show that the ecotone between the deciduous forests and the conifer-hardwood forests is marked by the north to south change from the dominance of pine pollen to the dominance of oak pollen. This ecotone stretches from southern New England to Minnesota and is located in a zone where the 20 and 30% isopolls for pine and oak approximately coincide. This same change in pollen dominance appears

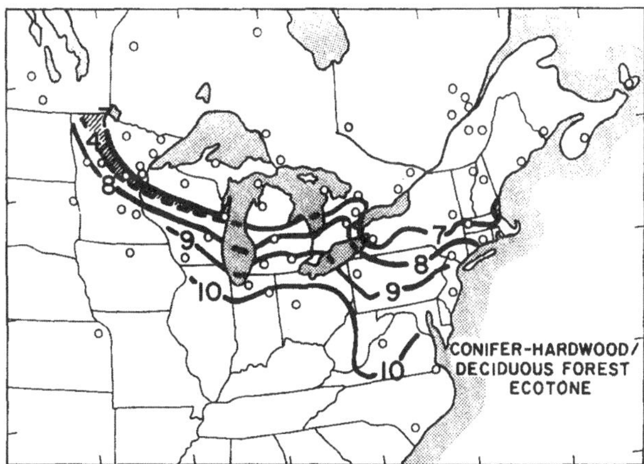

FIG. 22. Isochrones in thousands of years BP depicting the movement of the conifer–hardwood/deciduous-forest ecotone. The position of this boundary was determined on the basis of the distributions of pine and oak pollen on isopoll maps.

on the paleopollen maps, and we have used it to trace the past positions of this ecotone (Fig. 22).

Early in the Holocene, the position of the conifer–hardwood/deciduous forest ecotone shifted rapidly northward, probably in direct response to the fast-changing climatic patterns of that period. After 8000 BP, the movement of this ecotone slowed, and its position moved only slightly further north in New England while edging eastward in Minnesota as the prairie expanded. After 4000 BP, the ecotone shifted back westward in Minnesota as the prairie receded, but little further movement occurred in the east.

In contrast to the relative stability in the position of this ecotone after 7000 BP, the composition of the vegetational regions that it separates continued to be modified as different genera migrated in from the south, east, and west. Beech and hemlock trees did not arrive in the central Great Lakes region until after 8000 BP (Fig. 25), and it was not until several thousand years later that these taxa dramatically increased their abundances in this region (see next section). This latter event distinctly changed the character of the conifer–hardwood formation as these taxa became a dominant pollen group within this region. Likewise, it was not until after 6000 BP that hickory and chestnut trees migrated into the deciduous forests of southern New England and thus modified their composition (Davis, 1976). This type of vegetational change also occurred in other regions. Ritchie (1976) described compositional changes in the boreal forest, which though significant did not alter the identity of that region.

The Expansion of Birch, Maple, Beech, and Hemlock

The development of forests dominated by mixtures of this group of trees can be divided into two phases. The first involves the differential migration of these genera into New England and the Great Lakes region, and the second occurs later when the abundance of these taxa dramatically increased within their range boundaries. The different nature of these two phases aids in interpreting what the environmental factors were that influenced the development of these forests.

In contrast to the widespread early appearances of birch and maple, the northward migration for hemlock and beech was slow. Isochrone maps in Fig. 25 show the migration patterns for hemlock and beech based on the first sustained presence of their pollen at 3% or greater. This criterion for the presence of these trees was established by estimating the average percentages of these pollen types along their present range limits (Davis and Webb, 1975; Webb and McAndrews, 1976). Use of this criterion produced isochrone maps whose patterns are similar to those on the maps that Davis (1976) produced using influx data.

The migration maps show the northward and then westward migration of hemlock and beech from 10,000 to 5000 BP. After the initial immigration of these taxa into the area, the scene was set for the second phase during which the BAFT group increased its prominence in the region where pine had been the dominant pollen type since 9000 BP.

In New England the BAFT group becomes prominent shortly after hemlock and beech arrive, but in the central Great Lakes region this event occurs from 2000 to 5000 years after their initial arrival. This pattern is illustrated by comparing the migration maps to the other isochrone maps on Fig. 25 that plot the location of the 50% isopoll for pine pollen and the 30% isopoll for BAFT pollen.

The delay in the increase of abundances of the BAFT group implies an environmental change occurred after 7000 BP in the Great Lakes region. This change reordered the competitive advantages within the forest communities and allowed the BAFT group to expand its role in regions where these taxa were already established.

The expansion of the area dominated by the BAFT group coincides with the westward retreat of the prairie (Fig. 26).

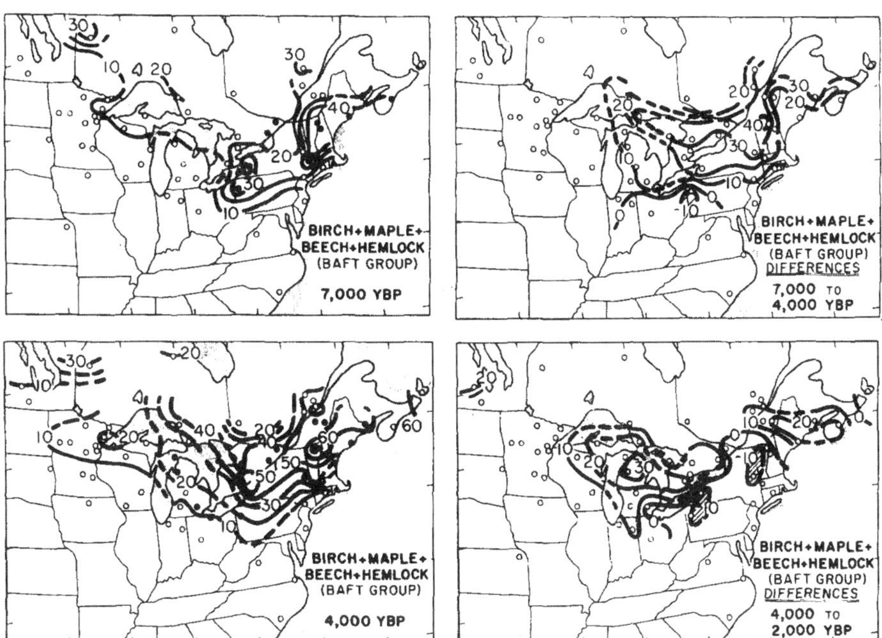

FIG. 23. Isopoll and difference maps showing the changing distribution of the BAFT group (sum of birch, maple, beech, and hemlock pollen) between 7000 and 2000 BP. Sites that have all four BAFT genera well represented are shown as solid dots. Outside these areas, where sites are shown as open dots, BAFT percentages largely reflect the frequencies of birch.

Wright (1968a) and Webb and Bryson (1972) interpret this retreat as indicating a shift to wetter conditions in the northwestern Midwest. Webb and Bryson (1972) also showed an increase in the values of precipitation-minus-potential-evaporation after 4000 BP at Lake Mary, which is located in the modern-day conifer–hardwood forest. These changes

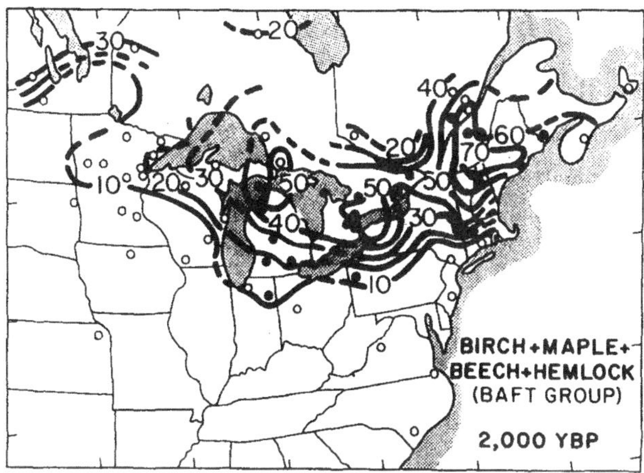

FIG. 24. Isopoll map illustrating the distribution of the BAFT pollen group (birth + maple + beech + hemlock) at 2000 BP.

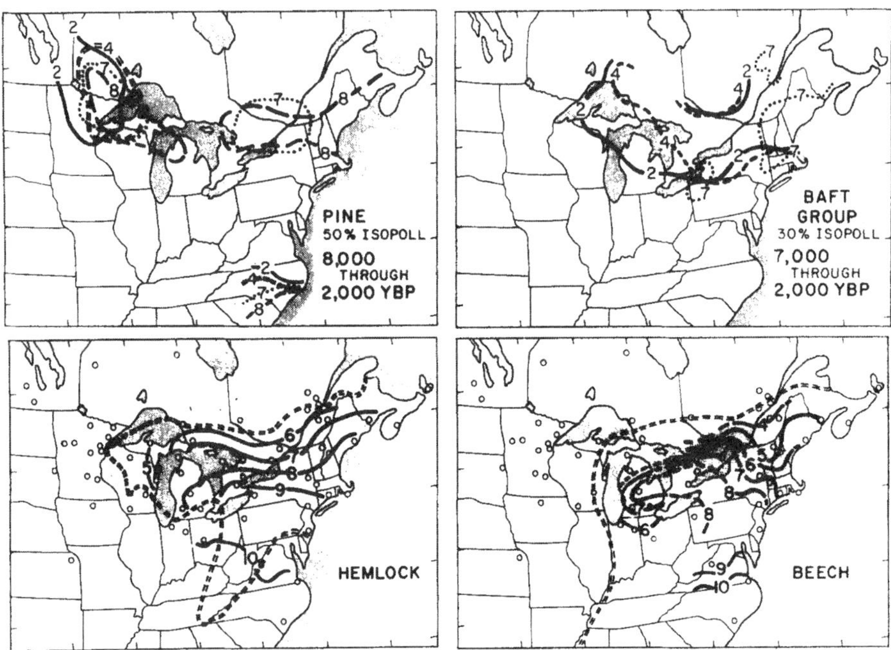

FIG. 25. Isochrone maps showing the location of the pine 50% isopoll, BAFT group 30% isopoll, and the migrations of hemlock and beech. All isochrones are labeled in thousands of years BP. The maps of the pine 50% isopoll and BAFT 30% isopoll show the displacement of the pine-pollen dominance by the expansion of the BAFT group. Double dashed lines on the migration maps indicate the modern range limits of these genera.

would decrease the frequency of fires and thereby create more favorable conditions for increases in the abundance of birch, maple, beech, and hemlock trees.

These results show that both the movements of range limits and the changing patterns of abundances provide biological and environmental information. Each of these aspects of vegetational dynamics should be examined in order to elucidate the factors controlling the development of forest ecosystems.

History of the Prairie/Forest Ecotone

The signs of prairie development in the western Midwest were visible in the pollen record over 11,000 years ago as the vast region formerly occupied by the late-glacial boreal forest began to shrink (Figs. 5 and 21). The values of herb pollen in the eastern portions of South Dakota and Kansas were already beginning to rise as deciduous forest pollen types replaced spruce pollen (Watts and Bright, 1968). At the same time, in southern Manitoba, spruce pollen was being directly replaced by herb pollen (Ritchie and Lichti-Federovich, 1968).

Figure 26 picks up the story from this point onward and summarizes the broad regional history of the prairie/forest ecotone from 11,000 to 2000 BP. This figure illustrates the position of the 30% isopoll for herb pollen at successive 1000-year intervals. Maps of the presettlement distribution of herb pollen indicate that the 30% herb isopoll roughly delineates the boundary between the prairie to the west and the forested region to the east.

The largest eastward shift of the prairie during the Holocene took place between 10,000 and 9000 BP (Fig. 26). Figure 2 also

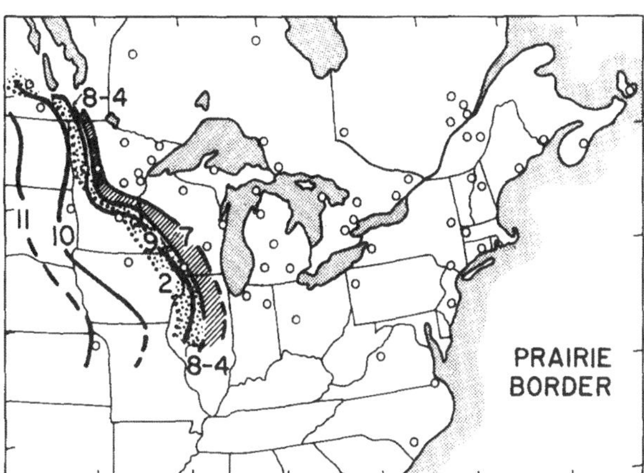

FIG. 26. Isochrones in thousands of years BP illustrating the movements of the prairie/forest ecotone. The position of the prairie border is based on isopoll maps for herb pollen. Shaded areas show the region over which the prairie retreated after reaching its maximum postglacial extent at 7000 BP.

indicates that this time interval experienced the greatest changes in herb pollen prior to those that resulted from European settlement. By 9000 BP, prairie vegetation extended over approximately the same area of the Midwest that it did prior to settlement. The prairie border remained relatively fixed after 9000 BP (Fig. 26), until major increases in herb values from 8000 to 7000 BP resulted in the prairie attaining its maximum eastward extension during the Holocene period (Figs. 12, 13, and 26). Wright (1968a) gives an excellent discussion of this development.

From 7000 to 2000 BP, the prairie/forest border receded westward, but at a slower pace than it had expanded in the early Holocene (Fig. 26). The prairie/forest ecotone has thus been in motion during most of the Holocene, advancing eastward relatively rapidly during the early Holocene and then retreating westward more slowly during the late Holocene.

SUMMARY AND DISCUSSION

General Synopsis

By mapping the distribution of the dominant pollen groups, our paper has traced the changing location and composition of four major vegetational regions during the Holocene epoch. These regions include the boreal forest, conifer–hardwood forest, deciduous forest, and the prairie with their characteristic pollen groups being spruce, pine, oak, and herbs, respectively. Isochrone maps of the ecotones between these vegetational regions indicates that the modern geographic configuration of these formations was established after final deglaciation about 7000 BP. Less dramatic but significant changes continued after this time, particularly the westward retreat of the prairie border. Besides the shifts in the location of these major vegetational regions, their composition was being modified during the entire Holocene epoch as a result of the immigration of new taxa and the changing abundances of already established genera. The history of the BAFT pollen group within the conifer–hardwood region clearly illustrates this aspect of vegetational change.

The various maps presented in our study illustrate the following major vegetational events in the Holocene pollen record of northeastern North America: (1) the progressive shrinkage of the late-glacial boreal forest between 11,000 and 8000 BP to a narrow band just south of the ice sheet with the subsequent redevelopment of an extensive boreal forest after 7000 BP; (2) the expansion of the area dominated by pine from a restricted region in the east at 11,000 BP to a wide belt from Minnesota to New England by 9000 BP, and the subsequent displacement of the pine-dominated forests as the oak-dominated deciduous forests moved up from the south; (3) the continuously changing composition of the deciduous and conifer–hardwood forest regions caused by the successive immigration of taxa from the south throughout most of the Holocene; (4) the progressive expansion of the region dominated by birch, maple, beech, and hemlock trees within the conifer–hardwood region of New England and the Midwest from 7000 to 2000 BP; (5) the eastward movement of the prairie/forest border until 7000 BP and then its subsequent gradual westward retreat; and finally (6) the widespread increase in herb pollen frequencies during the last 300 years related to European settlement.

Scale and Paleoclimatic Implications

Our mapping of the Holocene pollen record has focused on the broad-scale temporal and spatial changes in the vegetation at the formation level. These changes represent vegetational behavior from only one sector within the whole spectrum of vegetational change which stretches from the daily variations in limnic phytoplankton to the evolution of continental floras. The pollen records that we present are too short and possess too general a taxonomy to be sensitive to the evolution of floras, whereas the mixing of pollen in the air and in the sediments removes the imprint of both short-term (daily to seasonal) and fine-scale (pollen production in single trees) changes (Davis, 1968).

Mixing also makes pollen data less precise in time, space, and taxonomy than most standard plant ecological data, but because the mixing results in a smoothing out of certain high-frequency variations, the pollen data gain in their ability to record vegetational patterns at the formation level. We have emphasized the formation-level variation in the data by our choice of map scale, time intervals, and contour intervals. Our maps are thus tuned to the spatial scale of variation that is most likely to reflect macroclimatic changes, because macroclimate plays a dominant role in controlling these broad patterns in the vegetation (Bryson, 1966). The continuity and spatial scale of the patterns which emerge on our difference maps are strikingly similar to anomaly patterns that result from shifts in global atmospheric circulation. This fact suggests that the pollen maps should provide a good basis for constructing paleoclimatic maps when appropriate calibration functions are derived (Webb and Clark, 1976).

Even before such a step is taken, some climatic interpretation is possible from the pollen maps themselves. These maps along with Fig. 2 show that a marked decrease in the magnitude of pollen changes occurred after 7000 BP. This change seems to be tied to the disappearance of the Laurentide ice sheet at this time.

As long as the ice sheet was present south of Hudson's Bay, its position, albedo, and height served as important boundary conditions affecting the general circulation of the atmosphere (Hare, 1973). In summer, the contrast in surface albedo between the ice sheet and the land to the south probably created a steep gradient in surface temperatures near the ice margin. Each movement of the shrinking ice sheet shifted the position of this thermal gradient and thus affected the location of storm tracks during the summer season. Once the ice sheet had receded far to the north, this boundary-condition control within the atmosphere–cryosphere system ceased operation. Shifts in atmospheric circulation were thus less dramatic after the Laurentide ice sheet collapsed, as the general decrease in pollen changes after 7000 BP appears to confirm.

The patterns on the paleopollen maps also

indicate other changes in Holocene climates. The movements of the prairie border, for instance, agree with previous inferences of progressively drier conditions in the Midwest between 10,000 and 7000 BP (Wright, 1968a) and further suggest that the most rapid period of drying was between 10,000 and 9000 BP (Fig. 26). Likewise, the retreat of the prairie border and the simultaneous development of the mesic northern hardwoods (BAFT group) after 7000 BP indicates the development of cooler and/or wetter conditions in the Midwest. Data from the western interior of Canada suggest a similar climatic change is also evident there (Ritchie, 1976).

The Map Perspective

Although other useful summaries of the pollen record already exist (e.g., Wright, 1971), our cartographic approach gives visual expression to this information. The maps make local changes interpretable within the context of regional events and make the dynamic interconnections between the regional events evident. Another consequence of producing the maps is that pollen diagrams with little or no change have as much importance as diagrams with many large changes. Described alone, the "complacent" diagrams have never excited much interest, but on the maps these sites indicate nodal points in the pattern of pollen response to environmental and biogeographical change.

A further consequence of mapping is the useful perspective provided for interpreting the timing and significance of zones in pollen diagrams. The transitions from one major pollen zone to another are in most instances directly associated with the movement of large regional ecotones. The change from the pine-dominated B-zone to the oak-dominated C-zone in southern New England, for instance, reflects the passing of the conifer–hardwood/deciduous-forest ecotone through that region (Fig. 22). The extent to which a given formational ecotone has moved defines the area over which a particular zonal boundary is recognizable. Likewise, the speed of the ecotone's movement controls the degree of synchroneity evident for a zonal boundary throughout the region where it occurs.

The isochrone maps illustrating the Holocene movements of major ecotones (Figs. 21, 22, and 26) show that many different sequences of zones can be found within the mapped area. The zonation seen at a specific site results from its geographical relationship to the movement of these formational ecotones. Second-order subdivisions of the major pollen zones (such as C-1, C-2, etc.) are usually defined by frequency changes in the less prominent pollen types or by the appearance of new taxa. In the latter case, such as the C-3 oak–chestnut zone of southern New England, migration maps can be used to provide information on the nature and timing of these sub-zones (Davis, 1976).

A third consequence of producing the maps grows out of the above discussion and involves their potential for identifying ecologically important features within the vegetation. The paleopollen maps can serve the purpose for paleoecologists that weather maps serve for meteorologists. Just as the weather maps show the storm systems and fronts that form the synoptic conditions within the local weather occurs, the paleopollen maps illustrate the biogeographical setting within which the changes at individual sites occur. The isopoll maps can also help locate the vegetational features such as ecotones or range boundaries, much as isotherm maps aid meteorologists in identifying atmospheric features such as fronts. The movement of these vegetational features can then be traced through time with isochrone maps (Figs. 21–26).

One problem faced in this use of isopoll maps is the lack of established quantitative criteria for defining ecologically significant features such as ecotones. This problem arises because, over broad regions, the modern vegetation has typically been mapped only in qualitative terms. Maps showing different vegetational regions provide little information either about the quantitative nature of the transitions between regions or about the variation of the vegetation within each of the classified regions.

At the moment, at least three different types of criteria exist for locating ecotones on contoured maps. These criteria include: (1) the coincidence of range boundaries of two or more plant species (Davis, 1974); (2) the position of a specified contour (e.g., 30% isopoll) that lies along a recognized ecotone on modern vegetation maps (Figs. 21, 22, and 26); and (3) the position of a steep gradient in the isopolls of one or more pollen types.

Although only the first two criteria have been used for mapping the previous positions of ecotones, the third criterion is particularly attractive. It would be most useful in paleoclimatological studies in those areas where steep gradients in the vegetation are found to be coincident with steep gradients in climatic variables (Bryson, 1966). The use of this criterion, however, requires a more detailed set of fossil data than currently exists. In order to give accurate definition to the zones of sharp change, the sites should be more densely spaced and possess greater time control than the sites in the current data set. Additional studies of modern vegetation are also needed in order to update the maps in Figs. 3 and 4 and to check the coincidence of steep gradients in the tree-inventory and pollen data.

Although we have been able to illustrate many of the major features of the Holocene pollen record using the currently available data, much work needs to be done in order to refine our knowledge of the vegetational history of North America. We hope our work will provide a useful framework in which this refinement can occur.

ACKNOWLEDGMENTS

Our research was supported by the Office for Climate Dynamics, National Science Foundation grants to Brown University (NSF OCD75-14934) and to the Center for Climatic Research, University of Wisconsin, Madison (NSF OCD75-14708) as well as by the International Decade for Ocean Exploration (CLIMAP) NSF Grant to Brown University (NSF IDO76-00398). Thanks are due to R. Laseski, R. Mellor, and Y. Yeracaris for technical assistance; to R. E. Bailey and D. R. Whitehead for use of unpublished data; to R. E. Bailey, R. G. Baker, L. B. Brubaker, M. B. Davis, R. B. Davis, W. C. Kerfoot, J. H. McAndrews, R. J. Mott, P. Richard, J. C. Ritchie, A. M. Swain, J. C. B. Waddington, D. R. Whitehead, and H. E. Wright for supplying pollen counts, and to H. J. B. Birks, M. B. Davis, R. B. Davis, C. J. Heusser, J. Imbrie, J. E. Kutzbach, D. A. Livingstone, J. H. McAndrews, L. J. Maher, A. M. Swain, and W. A. Watts for stimulating discussion during the course of the work.

REFERENCES

Bailey, R. (1969). Charles Lake. (Unpublished data).

Bailey, R. (1972). "Late- and Postglacial Environmental Changes in Northeastern Indiana." Ph.D. Thesis, Indiana University.

Bailey, R. (1973). Wintergreen Lake. (Unpublished data).

Baker, R. G. (1970). A radiocarbon-dated pollen chronology for Wisconsin: Disterhaft Farm Bog revisited. *In* "Geological Society of America, Annual Meeting Abstracts." **2**, 488.

Birks, H. J. B. (1976). Late-Wisconsinan vegetational history at Wolf Creek, central Minnesota. *Ecological Monographs*, **46**, 395–429.

Birks, H. J. B., and Saarnisto, M. (1975). Isopollen maps and principal components analyses of Finnish pollen data for 4,000, 7,000 and 8,000 years ago. *Boreas* **4**, 77–96.

Birks, H. J. B., Deacon, J., and Peglar, S. (1975a). Pollen maps for British Isles 5,000 years ago. *Proceedings Royal Society of London B* **189**, 87–105.

Birks, H. J. B., Webb, T., III, and Berti, A. A. (1975b). Numerical analysis of pollen samples from Central Canada: A comparison of methods. *Review of Palaeobotany and Palynology* **20**, 133–169.

Bryson, R. A. (1966). Air masses, streamlines and the boreal forest. *Geographical Bulletin* **8**, 228–269.

Bryson, R. A., Wendland, W. M., Ives, J. D., and Andrews, J. T. (1969). Radiocarbon isochrones on the disintegration of the Laurentide Ice Sheet. *Arctic and Alpine Research* **1**, 1–14.

Bryson, R. A., Baerreis, D. A., and Wendland, W. M. (1970). The character of late- and postglacial climatic changes. *In* "Pleistocene and Recent Environments of the Central Great Plains" (W. Dort and J. K. Jones, Eds.), pp. 53–74. Univ. of Kansas Press, Lawrence.

Bryson, R. A., and Kutzbach, J. E. (1974). On the analysis of pollen-climate canonical transfer functions. *Quaternary Research* **4**, 162–174.

Brubaker, L. B. (1975). Postglacial forest patterns

associated with till and outwash in northcentral upper Michigan. *Quaternary Research* **5**, 499–527.

CLIMAP (1976). The surface of the Ice-Age earth. *Science* **191**, 1131–1137.

Connally, G. G., and Sirkin, L. A. (1971). Luzerne readvance near Glens Falls, New York. *Geological Society of America Bulletin* **82**, 989–1008.

Craig, A.J. (1969). Vegetational history of the Shenandoah Valley, Virginia. *Geological Society of America Special Papers* **123**, 283–296.

Craig, A. J. (1972). Pollen influx to laminated sediments: A pollen diagram from northeastern Minnesota. *Ecology* **53**, 46–57.

Cushing, E. J. (1965). Problems in the Quaternary phytogeography of the Great Lakes region. *In* "The Quaternary of the United States" (H. E. Wright, Jr. and D. G. Frey, Eds.), pp. 403–416. Princeton Univ. Press, Princeton, N. J.

Davis, A. (1975). "Record of Local and Regional Holocene Environments from the Pollen and Peat Stratigraphies of Some Driftless Area Peat Deposits." Ph.D. Thesis, Univ. of Wisconsin, Madison.

Davis, M. B. (1965). Phytogeography and palynology of northeastern United States. *In* "The Quaternary of the United States" (H. E. Wright, Jr. and D. G. Frey, Eds.), pp. 377–401. Princeton Univ. Press, Princeton, N. J.

Davis, M. B. (1967). Late-glacial climate in northern United States: A comparison of New England and the Great Lakes region. *In* "Quaternary Paleoecology" (E. J. Cushing and H. E. Wright, Jr., Eds.), pp. 11–43. Yale Univ. Press, New Haven.

Davis, M. B. (1968). Pollen grains in lake sediments: redeposition caused by seasonal water circulation. *Science* **162**, 796–799.

Davis, M. B. (1969). Climatic changes in southern Connecticut recorded by pollen deposition at Rogers Lake. *Ecology* **50**, 409–422.

Davis, M. B. (1974). Holocene migrations of ecotones: Continuing northward migration of tree species throughout the Holocene caused changes in community composition and migrations of ecotones. *In* "American Quaternary Association Abstracts of Third Biennial Meeting." pp. 18–22.

Davis, M. B. (1976). Pleistocene biogeography of temperate deciduous forests. *Geoscience and Man* **13**, 13–26.

Davis, R. B., and Webb, T., III (1975). The contemporary distribution of pollen in eastern North America: a comparison with the vegetation. *Quaternary Research* **5**, 395–434.

Davis, R. B., Bradstreet, T. E., Stuckenrath, R., Jr., and Borns, H. W., Jr. (1975). Vegetation and associated environments during the past 14,000 years near Moulton Pond, Maine. *Quaternary Research* **5**, 435–465.

Donner, J. J. (1963). The zoning of the post-glacial pollen diagrams in Finland and the main changes in the forest composition. *Acta Botanica Fennica* **65**, 2–40.

Durkee, L. H. (1971). A pollen profile from Woden Bog, Hancock County, Iowa. *Ecology* **52**, 835–844.

Elson, J. A. (1967). Geology of glacial Lake Agassiz. *In* "Life, Land and Water" (W. J. Mayer-Oakes, Ed.), pp. 37–96. Univ. of Manitoba Press, Winnipeg.

Emery, K. O., and Garrison, L. E. (1967). Sea levels 7,000 to 20,000 years ago. *Science* **157**, 684–687.

Faegri, K., and Iversen, J. (1964). "Textbook of Pollen Analysis." Hafner, New York.

Firbas, F. (1949). "Spät-und nacheiszeitliche Waldegeschichte Mitteleuropas nördlich der Alpen." Verlag von Gustav Fischer, Jena.

Flint, R. F. (1971). "Glacial and Pleistocene Geology." Wiley, New York.

Florer, L. E. (1972). Palynology of a post-glacial bog in the New Jersey Pine Barrens. *Torrey Botanical Bulletin* **99**, 135–138.

Frey, D. G. (1951). Pollen succession in the sediments of Singletary Lake, North Carolina. *Ecology* **32**, 518–533.

Fries, M. (1962). Pollen profiles of Late Pleistocene and recent sediments at Weber Lake, northeastern Minnesota. *Ecology* **43**, 295–308.

Fritts, H. C., Blasing, T. J., Hayden, B. P., and Kutzbach, J. E. (1971). Multivariate techniques for specifying tree-growth and climate relationships and for reconstructing anomalies in paleoclimate. *Journal of Applied Meteorology* **10**, 845–864.

Gilliam, J. A., Kapp, R. O., and Bogue, R. D. (1967). A post-Wisconsin pollen sequence from Vestaburg Bog, Montcalm County, Michigan. *Michigan Academy of Science, Arts and Letters* **52**, 3–17.

Grant, D. R. (1970). Recent coastal submergence of the Maritime provinces, Canada. *Canadian Journal of Earth Sciences* **7**, 676–689.

Grüger, J. (1973). Studies on the late Quaternary vegetation history of northeastern Kansas. *Geological Society of America Bulletin* **84**, 239–250.

Halliday, W. E. D., and Brown, A. W. A. (1943). The distribution of some important forest trees in Canada. *Ecology* **24**, 353–373.

Hare, K. F. (1973). On the climatology of post-Wisconsin events in Canada. *Arctic and Alpine Research* **5**, 169–170.

Hough, J. L. (1963). The prehistoric Great Lakes of North America. *American Scientist* **5**, 84–110.

Hughes, D. L. (1965). Surficial geology of part of the Cochrane District, Ontario, Canada. *Geological Society of America Special Papers* **84**, 535–565.

Imbrie, J., and Kipp, N. G. (1971). A new micropaleontological method for quantitative paleoclimatology: Application to a late Pleistocene Caribbean core. *In* "The Late Cenozoic Glacial Ages" (K. Turekian, Ed.), pp. 71–181. Yale Univ. Press, New Haven.

Janssen, C. R. (1968). Myrtle Lake: A late and postglacial pollen diagram from northern Minnesota. *Canadian Journal of Botany* **46**, 1397–1410.

Kapp, R. O., Bushouse, S., and Foster, B. (1969). A contribution to the geology and forest history of Beaver Island, Michigan. *In* "Proceedings 12th Conference of Great Lakes Research," pp. 225–236.

Karrow, P. F. (1963). "Pleistocene Geology of the Hamilton–Galt Area." Ontario Department of Mines, Toronto, Geological Report 16.

Kerfoot, W. C. (1974). Net accumulation rates and the history of cladoceran communities. *Ecology* **55**, 51–61.

Kipp, N. G. (1976). A new transfer function for estimating past sea-surface conditions from the sea-bed distribution of planktonic foraminiferal assemblages in the North Atlantic. *Geological Society of America Memoir* **145**, 3–42.

LaMarche, V. C., Jr., and Fritts, H. C. (1971). Anomaly patterns of climate over western United States 1700–1930, derived from principal component analysis of tree-ring data. *Monthly Weather Review* **99**, 138–142.

Lasalle, P. (1966). Late Quaternary vegetation and glacial history in the St. Lawrence lowland, Canada. *Leidse Geologische Mededelingen, Leiden* **38**, 91–128.

Lawrenz, R. (1975). "Biostratigraphic Study of Green Lake Michigan." M.S. Thesis, Central Michigan University.

Lee, H. A. (1960). Late-glacial and postglacial Hudson Bay sea episode. *Science* **131**, 1609.

Lichti-Federovich, S., and Ritchie, J. C. (1968). Recent pollen assemblages from the western interior of Canada. *Review of Palaeobotany and Palynology* **7**, 297–344.

Likens, G. E., and Davis, M. B. (1975). Postglacial history of Mirror Lake and its watershed in New Hampshire, U. S. A.: an initial report. *Verhandlunger der Internationalen Vereinigung für theoretische und angewandte Limnologie* **19**, 982–993.

Livingstone, D. A. (1958). Late-glacial and postglacial vegetation from Gillis Lake, Nova Scotia. *American Journal of Science* **256**, 341–359.

Livingstone, D. A. (1968). Some interstadial and postglacial pollen diagrams from eastern Canada. *Ecological Monographs* **38**, 87–125.

Markgaf, V. (1970). Paleohistory of spruce in Switzerland. *Nature (London)* **228**, 249–251.

Maxwell, J. A., and Davis, M. B. (1972). Pollen evidence of Pleistocene and Holocene vegetation on the Allegheny Plateau, Maryland. *Quaternary Research* **2**, 506–530.

McAndrews, J. H. (1966). Postglacial history of prairie, savanna and forest in northeastern Minnesota. *Torrey Botanical Club Memoirs* **22**, 1–72.

McAndrews, J. H. (1970). Fossil pollen and our changing landscape and climate. *Rotunda* **3**, 30–37.

McAndrews, J. H., Berti, A. A., and Norris, G. (1973). "Key to the Quaternary Pollen and Spores of the Great Lakes Region." Life Sciences Miscellaneous Publication, Royal Ontario Museum.

McDowell, L. L., Dole, R. M., Jr., Howard, M., and Farrington, R. A. (1971). Palynology and radiocarbon chronology of Bugbee Wildflower Sanctuary and Natural Area, Caledonia County, Vermont. *Pollen et Spores* **13**, 73–91.

McIntyre, A., and Kipp, N. G. (1976), with Bé, A. W. H., Crowley, T., Kellogg, T., Gardner, J., Prell, W., Ruddiman, W. F. The glacial North Atlantic 18,000 years ago: A CLIMAP reconstruction. *Geological Society of America Memoir* **145**, 43–76.

Miller, N. G. (1973). "Late-Glacial and Postglacial Vegetation Change in Southwestern New York State." New York State Museum and Science Service Bulletin 420.

Moe, D. (1970). The post-glacial immigration of *Picea abies* into Fennoscandia. *Botaniska Notiser* **123**, 61–66.

Moran, J. M. (1973). The late-glacial retreat of 'Arctic' air as suggested by onset of *Picea* decline. *The Professional Geographer* **25**, 373–376.

Mott, R. J. (1975). Palynological studies of lake sediment profiles from southwestern New Brunswick. *Canadian Journal of Earth Sciences* **12**, 273–288.

Mott, R. J., and Camfield, M. (1969). "Palynological Studies in the Ottawa area." Geological Survey of Canada Paper 69.

Neustadt, M. I. (1959). Geschichte der vegetation der USSR im Holozän. *Grana Palynologica* **2**, 67–76.

Nicholas, J. (1968). "Late Pleistocene Palynology of Southeastern New York and Northern New Jersey." Ph.D. Thesis, New York University.

Nichols, H. (1975). "Palynological and Paleoclimatic Study of the Late Quaternary Displacement of the Boreal Forest–Tundra Ecotone in Keewatin and Mackenzie, N.W.T., Canada." Occasional Paper No. 15, Institute of Arctic and Alpine Research, Colorado.

Ogden, J. G., III (1966). Forest history of Ohio, radiocarbon dates and pollen stratigraphy of Silver Lake, Ohio. *Ohio Journal of Science* **66**, 387–400.

Odgen, J. G., III (1969). Correlation of contemporary and late Pleistocene pollen records in the reconstruction of postglacial environments in northeastern North America. *Mitteilungein Internationale Vereinigung Limnologie* **17**, 661–677.

Prest, V. K. (1969). "Retreat of Wisconsin and Recent Ice in North America." Geological Survey of Canada Map 1257A.

Prest, V. K. (1970). Quaternary geology of Canada. *In* "Geology and Economic Minerals of Canada"

(R. J. W. Douglas, Ed.), pp. 676–764, Chap. 12. Geological Survey of Canada, Economic Geology Report 1.

Richard, P. (1970). Atlas pollinique des arbres et de quelque arbustes indigenes du Quebec; I, II, III, IV. *Naturaliste Canadien* **97**, 1–34, 97–161, 241–306.

Richard, P. (1971). Two pollen diagrams from the Quebec City area, Canada. *Pollen et Spores* **13**, 523–559.

Richard, P. (1973a). Historie postglaciaire de la vegetation dans la region de Saint-Raymond de Portneuf, telle que revelee par l'analyse pollinique d'une tourbiere. *Naturaliste Canadien* **100**, 561–575.

Richard, P. (1973b). Historie postglaciaire comparee de la vegetation dans deux localites au nord du Parc des Laurentides, Quebec. *Naturaliste Canadien* **100**, 577–590.

Richard, P. (1973c). Historie postglaciaire comparee de la vegetation dans deux localities au sud de la ville de Quebec. *Naturaliste Canadien* **100**, 591–603.

Ritchie, J. C. (1964). Contributions to the Holocene paleoecology of western Canada: I. The Riding Mountain area. *Canadian Journal of Botany* **42**, 181–196.

Ritchie, J. C. (1969). Absolute pollen frequencies and carbon-14 age of a section of Holocene lake sediment from the Riding Mountain area of Manitoba. *Canadian Journal of Botany* **47**, 1345–1349.

Ritchie, J. C., and Lichti-Federovich, S. (1968). Holocene pollen assemblages from the Tiger Hills, Manitoba. *Canadian Journal of Earch Sciences* **5**, 873–880.

Ritchie, J. C. (1976). The late-Quaternary vegetational history of the western interior of Canada. *Canadian Journal of Botany* **54**, 1793–1818.

Saarnisto, M. (1974). The deglaciation history of the Lake Superior region and its climatic implications. *Quaternary Research* **4**, 316–339.

Saarnisto, M. (1975). Stratigraphic studies on the shoreline displacement of Lake Superior. *Canadian Journal of Earth Sciences* **12**, 300–319.

Serebryanny, L. R. (1973). Post-glacial migration rates of tree species in the northwestern regions of the USSR: palynology and radiocarbon dating. *In* "Palynology: Holocene and Marine Palynology" (N. A. Khotinsky and E. V. Koreneva, Eds.), pp. 14–18. Nauka, Moscow. (in Russian).

Szafer, W. (1935). The significance of isopollen lines for the investigation of the geographical distribution of trees in the postglacial period. *Polska Akademia Umiejetnosci. Wydzial Matematycznoprzyodicy. Bulletin International, S. B.* 235–239.

Terasmae, J. (1959). Notes on the Champlain sea episode in the St. Lawrence lowlands, Quebec. *Science* **130**, 334–335.

Terasmae, J. (1967). Postglacial chronology and forest history in the northern Lake Huron and Lake Superior regions. *In* "Quaternary Paleoecology" (E. J. Cushing and H. E. Wright, Jr., Eds.), pp. 45–48. Yale Univ. Press, New Haven.

Terasmae, J. (1968). A discussion of deglaciation and the boreal forest in the northern Great Lakes region. *Proceedings Entomological Society of Ontario* **99**, 31–43.

Terasmae, J. (1973). Notes on late Wisconsin and early Holocene history of vegetation in Canada. *Arctic and Alpine Research* **5**, 201–222.

Vincent, J.-S. (1973). A palynological study for the Little Clay Belt, northwestern Quebec. *Naturaliste Canadien* **100**, 59–69.

Waddington, J. C. B. (1969). A stratigraphic record of the pollen influx to a Lake in the Big Woods of Minnesota. *Geological Society of America Special Papers* **123**, 263–281.

Walker, P. C., and Hartman, R. I. (1960). The forest sequence of the Hartstown Bog area in western Pennsylvania. *Ecology* **41**, 461–474.

Watts, W. A. (1967). Late-glacial plant macrofossils from Minnesota. *In* "Quaternary Paleoecology" (E. J. Cushing and H. E. Wright, Jr., Eds.), pp. 89–98. Yale Univ. Press, New Haven.

Watts, W. A. (1973). Rates of change and stability in vegetation in the perspective of long periods of time. *In* "Quaternary Plant Ecology" (H. J. B. Birks and R. G. West, Eds.), pp. 195–206. Blackwell, Oxford.

Watts, W. A., and Bright, R. C. (1968). Pollen, seed, and mollusk analysis of a sediment core from Pickerel Lake, northeastern South Dakota. *Geological Society of America Bulletin* **79**, 855–879.

Webb, T., III (1973). A comparison of modern and presettlement pollen from southern Michigan, U. S. A. *Review of Palaeobotany and Palynology* **16**, 137–156.

Webb, T., III (1974a). Corresponding patterns of pollen and vegetation in lower Michigan: a comparison of quantitative data. *Ecology* **55**, 17–28.

Webb, T., III (1974b). A vegetational history from northern Wisconsin: evidence from modern and fossil pollen. *American Midland Naturalist* **92**, 12–34.

Webb, T., III, and Bernabo, J. C. (1974). Palynological evidence of ecotonal migrations in eastern North America. *In* "American Quaternary Association Abstracts of the Third Biennial Meeting," Vol. 26.

Webb, T., III, and Bryson, R. A. (1972). Late- and postglacial climatic change in the northern Midwest, U. S. A.: Quantitative estimates derived from fossil pollen spectra by multivariate statistical analysis. *Quaternary Research* **2**, 70–115.

Webb, T., III, and Clark, D. R. (1976). Calibrating micropaleontological data in climatic terms: A critical review. *Annals New York Academy of Sciences,* **288**, 93–118.

Webb, T., III, and McAndrews, J. H. (1976). Corresponding patterns of contemporary pollen and vegetation in central North America. *Geological Society of America Memoir* **145**, 267–302.

Webb, T., III, and Yeracaris, G. Y. (to appear). Comparison of patterns in pollen data from southeastern Canada and northeastern United States. In "Proceedings of the International Palynological Congress, Lucknow."

West, R. G. (1961). Late and postglacial vegetational history in Wisconsin, particularly changes associated with the Valders readvance. *American Journal of Science* **259**, 766–783.

Whitehead, D. R. (1965). Palynology and Pleistocene phytogeography of unglaciated eastern North America. *In* "The Quaternary of the United States" H. E. Wright, Jr. and D. G. Frey, Eds.), pp. 417–432. Princeton Univ. Press, Princeton, N. J.

Whitehead, D. R. (in preparation). Late-glacial and post-glacial vegetational history of the Berkshires, western Massachusetts.

Wright, H. E., Jr. (1968a). History of the Prairie Peninsula. *In* "The Quaternary of Illinois." (R. E. Bergstrom, Ed.), Special Report 14, pp. 78–88. College of Agriculture, University of Illinois, Urbana.

Wright, H. E., Jr. (1968b). The roles of pine and spruce in the forest history of Minnesota and adjacent areas. *Ecology* **49**, 937–955.

Wright, H. E., Jr. (1971). Late Quaternary vegetation history of North America. *In* "The Late Cenozoic Glacial Ages" (K. Turekian, Ed.), pp. 425–464. Yale Univ. Press, New Haven.

Wright, H. E., Jr. and Patten, H. L. (1963). The pollen sum. *Pollen et Spores* **5**, 526–528.

Wright, H. E., Jr., and Watts, W. A. (1969). Glacial and vegetational history of northeastern Minnesota. *Minnesota Geological Survey* **SP-11**.

Wright, H. E., Jr., Winter, T. C., and Patten, H. L. (1963). Two pollen diagrams from southeastern Minnesota: Problems in the late- and postglacial vegetational history. *Geological Society of America Bulletin* **74**, 1371–1396.

van Zeist, W., and Wright, H. E., Jr. (1967). Über probleme der vegetation und pollenanalyse in Minnesota und angrezenden gebieten (U. S. A.). *In* "Pflanzenzoziologie und Palynologie" (R. Tuxen, Ed.), pp. 121–133. Verlag Dr. W. Junk, Den Haag.

PAPER 28

The Latitudinal Spans of Seaweed Species and Their Patterns of Overlap (1977)

E. C. Pielou

Commentary

CHRISTY M. MCCAIN

In this groundbreaking paper by Evelyn C. Pielou, species ranges and their pattern of overlap among related species is examined statistically along a latitudinal gradient for the first time. Pielou is a British-trained, Canadian researcher with extensive contributions to mathematical ecology and marine ecology, including 10 books and over 60 published articles spanning more than 50 years of scientific productivity. Many of her papers may be more highly cited, but "The Latitudinal Spans of Seaweed Species and Their Patterns of Overlap" remains a basis for much of the macroecological literature on species range limits and size, patterns of overlap among closely and distantly related species, and the latitudinal gradient in species richness (e.g., Colwell and Lees 2000; Gaston 1994; Hengeveld and Haeck 1982; Stevens 1989; Willig and Lyons 1998; Willig et al. 2003).

Pielou tests two conflicting hypotheses of overlap: (1) that closely related species should overlap strongly due to shared ancestry—an idea that is much debated in current literature under the term *phylogenetic niche conservatism* (e.g., Wiens and Graham 2005)—and (2) that closely related species should overlap minimally due to competitive exclusion. Pielou also developed an elegant and easily testable statistical methodology for determining range overlap where all range limits are equally likely, and where all range lengths are randomly located within the latitudinal space. The latter set of analyses set the stage for the future work on mid-domain effects that would emerge more than 20 years later (e.g., Colwell and Hurtt 1994; Colwell and Lees 2000; Willig and Lyons 1998). Using benthic marine algae (684 species), she found that congeneric species overlap in range limits much more often than predicted by competitive exclusion, and that species ranges were independently distributed across the latitudinal gradient. She noted a pervasive trend, now documented in multiple studies, that even though local distributions of algae are highly influenced by competition, geographical distributions are not. She attributed this to allopatric speciation followed by rampant marine dispersal.

Pielou's results also highlighted two now commonly accepted macroecological patterns: that the most common range sizes are the smallest (Anderson 1977; Brown 1984; Gaston 1994) and that latitudinal patterns in diversity are unimodal (Stevens 1989; Willig et al. 2003).

From *Journal of Biogeography* 4:299–311.

Literature Cited

*Anderson, S. 1977. Geographic ranges of North American terrestrial mammals. *American Museum Novitates* 2629:1–15.

*Brown, J. H. 1984. On the relationship between abundance and distribution of species. *American Naturalist* 124:255–79.

Colwell, R. K., and G. C. Hurtt. 1994. Nonbiological gradients in species richness and a spurious Rapoport effect. *American Naturalist* 144: 570–95.

Colwell, R. K., and D. C. Lees. 2000. The mid-domain effect: Geometric constraints on the geography of species richness. *Trends in Ecology and Evolution* 15:70–76.

Gaston, K. J. 1994. Measuring geographic range sizes. *Ecography* 17:198–205.

Hengeveld, R., and J. Haeck. 1982. The distribution of abundance. I. Measurements. *Journal of Biogeography* 9:303–16.

*Stevens, G. C. 1989. The latitudinal gradient in geographical range: How so many species coexist in the tropics. *American Naturalist* 133:240–56.

Wiens, J. J., and C. H. Graham. 2005. Niche conservatism: Integrating evolution, ecology, and conservation biology. *Annual Review of Ecology, Evolution, and Systematics* 36:519–39.

Willig, M. R., D. M. Kaufman, and R. D. Stevens. 2003 Latitudinal gradients of biodiversity: Pattern, process, scale, and synthesis. *Annual Review of Ecology, Evolution, and Systematics* 34:273–309.

Willig, M. R., and S. K. Lyons. 1998 An analytical model of latitudinal gradients of species richness with an empirical test for marsupials and bats in the New World. *Oikos* 81:93–98.

Journal of Biogeography (1977) 4, 299–311

The latitudinal spans of seaweed species and their patterns of overlap

E. C. PIELOU Biology Department, Dalhousie University, Halifax, Nova Scotia, Canada

ABSTRACT. Geographic ranges of groups of related species may overlap one another to a greater or lesser extent. This paper explores the overlap among the ranges ('latitudinal spans') of shoreline species along a north–south trending coast. A method of measuring the overlap of a group of *s* such species is devised and the theoretically expected overlaps, under two contrasted null hypotheses, are derived. According to one hypothesis (H1) the range limits of congeneric species are independently located; according to the other (H2) the ranges themselves, assumed to have their observed lengths, are independently located. The latitudinal spans of 684 species of benthic marine algae, on the western shores of the Atlantic, are then tested for conformity with the hypotheses. Compared with H1, species overlaps exceed expectation very significantly, but the data give no reason to reject H2. It is concluded that the seaweeds' geographic ranges are unaffected by interspecific competition; and (speculatively) that the zonation patterns of seaweeds on the shore are likely to be less developed in regions where, due to recent disappearance of barriers to dispersal, hitherto isolated seaweed communities are mingling.

The variation with latitude of the numbers of seaweed species, and the observed frequency distributions of their latitudinal spans, are also described.

Introduction

Quantitative studies in biogeography have never (so far as I know) attempted to analyse the geometrical patterns formed by the variously-shaped 'patches', some of them overlapping one another, that appear on maps showing the geographic ranges of several species of plants or animals. The task is easier to attempt if the ranges extend through one dimension rather than two, so that the patches representing them on a map degenerate to lines. This is the case with the plants and animals of shores. For a group of such species, a map of their ranges then becomes a sheaf of lines. Each line relates to a single species: the location and length of the line show the position and extent of that species' range. The sheaf as a whole shows the way in which the several species' ranges overlap.

The purpose of this paper is to describe a method of analysing maps of this kind (Part I) and to demonstrate an application of the method to the geographic ranges of the benthic marine algae of the Atlantic coasts of North and South America (Part II). In particular, an answer to the following question is sought. Consider the member species of one genus. One might entertain two conflicting expectations as to their joint geographic 'pattern': on the one hand, because of their common ancestry, one might expect their environmental requirements and hence their geographic ranges to overlap strongly; on the other hand, because of competitive exclusion among them, one might expect their ranges to overlap only slightly, if at all. This enquiry is the topic of Part II. Additional results derived from the same data are given in Part III. The conclusions suggested by the analyses of Parts II and III are brought together in the final section.

I: Theory

In all that follows we shall assume that, for clarity of exposition, the shoreline whose flora and fauna are being studied trends north–south. The geographic range of a species will be called its *span*. The word is used in preference to its commoner synonym *range* so as to avoid ambiguity; the treatment is statistical and hence *range* is best used in its statistical sense only.

The measurement of overlap in a sheaf of spans

We require a method of measuring the overlap, L_s, of a sheaf of s spans. A convenient way of defining, and hence measuring, L_s is as follows. Suppose, first, that only two species are involved. In this case only, write λ for their overlap. Then λ can take the values 0, 1, 1.5 or 2 according to the relative position of the spans, as shown in the upper part of Fig. 1. The lengths of the two spans and the length of their region of overlap are disregarded; all that matters is their relative pattern in a qualitative sense. Values of λ of 1.5 are assigned to the two patterns on the right for the following reason. Strictly speaking, the probability is zero that two species will have the southern (say) limits of their spans at precisely the same latitude. Hence if the observed southern limits of the two species are identical, one can say only that actual, unobserved, patterns yielding $\lambda = 1$ and $\lambda = 2$ are equiprobable; the expectation of their overlap is therefore 1.5.

Now suppose a sheaf contains s spans with $s > 2$. We define the total overlap, L_s, of the spans as

$$L_s = \Sigma\lambda$$

where summation is over all $s(s-1)/2$ possible species pairs. An example is shown in the lower part of Fig. 1, where a sheaf of $s = 6$ spans is shown whose total overlap is $L_s = 16$.

Expected values of L_s given different null hypotheses

It is interesting to derive the theoretical expectations of the variate L_s given a null hypothesis concerning the arrangement of the spans. Two possible hypotheses suggest themselves and I consider them in turn.

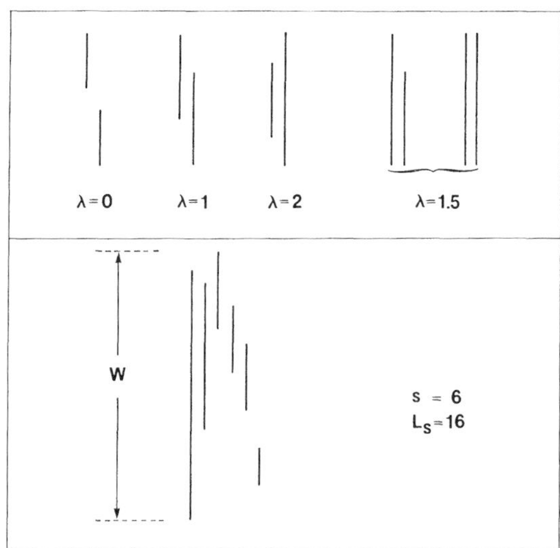

FIG. 1. Above: all the patterns possible for a pair of species-spans, and the overlap, λ, for each pattern. Below: a sheaf of $s = 6$ spans, with a total overlap of $L_s = 16$.

The unconditional hypothesis, H1. According to this hypothesis, the locations of the northern and southern limits of the s spans are entirely at random, subject only to the constraint that for each species its northern limit must, obviously, lie north of its southern limit. Equivalently, suppose one travelled from the northern to the southern extremity of the shoreline concerned, noting the order of the $2s$ 'events' constituted by the northern and southern span limits of each of the s species; then, under H1, all possible orderings of these events are equiprobable. No assumptions whatever are made concerning the magnitudes of the spans.

Given this hypothesis, the expected overlap of a sheaf of s species-spans, say $E(L_s|\text{H1})$, is

$$E(L_s|\text{H1}) = s(s-1)/2. \qquad (1)$$

This result is intuitively obvious. For each species pair, there are three possible and equiprobable values of λ (the pairwise overlap) namely 0, 1 and 2; for any pair, therefore, $E(\lambda|\text{H1}) = 1$. Since L_s is the sum of $s(s-1)/2$ components, each equal to $E(\lambda|\text{H1})$, equation (1) follows immediately.

One may also (Pielou, 1976, 1977) derive the variance of L_s which is:

$$\text{Var}(L_s|\text{H1}) = s(s^2-1)/9;$$

and the general term of the distribution of L_s, in other words the probability that, under H1, the overlap of s species-spans will have any given value. The distribution is symmetrical.

The conditional hypothesis, H2. This hypothesis, unlike H1, takes account of the lengths of the s spans and of the available length of shoreline in which they must be located. Whereas under H1 the lengths of the spans were undefined, under H2 they are assumed to have the lengths actually observed; these spans, of the given lengths, are then assumed to be located independently and at random within the space of length W, say, equal to the observed span of the sheaf as a whole (see Fig. 6).

It may then be shown (see Appendix) that if there are two spans, of lengths x and y with $x \geqslant y$, their expected overlap is

$$E(\lambda|\text{H2}) = \frac{2x(W-x)-y^2}{(W-x)(W-y)} \quad \text{if } W \geqslant x+y \quad (2)$$

or

$$E(\lambda|\text{H2}) = \frac{W+x-2y}{W-y} \quad \text{if } W < x+y. \quad (3)$$

To find the expected overlap of s spans, namely $E(L_s|\text{H2})$, it is necessary to evaluate $E(\lambda|\text{H2})$ for all $s(s-1)/2$ possible pairs of spans, each with its own observed values of x and y, and sum the separate expectations.

II: Within-genus overlap in east coast seaweeds of the Americas

The data

As in nearly all biogeographic investigations, the data came ready-made from the literature; inevitably, much had to be taken on trust. Assuming (as is, no doubt, never justified) that the literature search was exhaustive, the following are the most probable sources of error in recording the spans (geographic ranges) of seaweed species: the extent of a species' span may be underestimated either because specimens were not found, or because no search was made, at places beyond the apparent limit of its span; disjunctions in a span may be overlooked; and the difficulties of algal taxonomy may lead to faulty identifications and to misinterpretations of relationships.

However, if biogeographic studies are to proceed at all, one must assume that the signal-to-noise ratio of the data is high enough to ensure that, by appropriate statistical analysis, the signal may be recovered and correct generalizations derived. This amounts to no more than assuming, as seems reasonable, that the inevitable errors do not dominate the data. Thus what this paper attempts is, in the words of Mattingly (1962), 'the collation of large numbers of facts and their translation into small quantities of knowledge', a necessary task in 'the fact-bestrewn world of today'.

The data summarized and analysed below are based on the recorded northern and southern limits of occurrence, as found in the

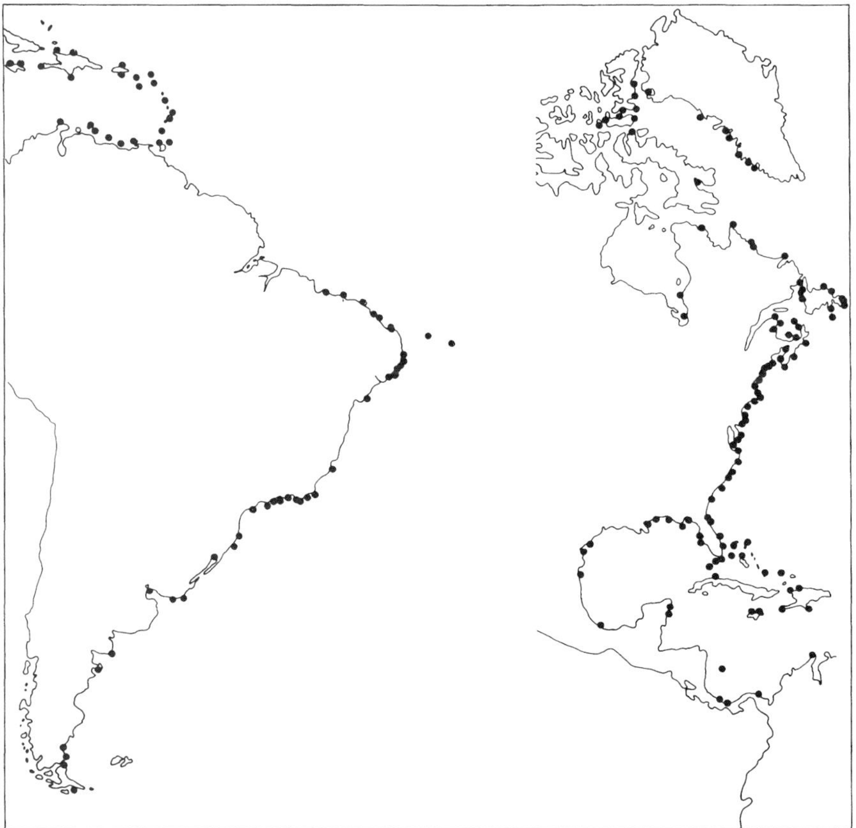

FIG. 2. A sample of the sites from which records were obtained. The maps give a general impression only of the varying density of observing stations; to show all stations in well-studied regions would require large-scale maps.

literature, of 684 species of benthic marine algae (395 species of Rhodophyta, 174 of Chlorophyta, and 115 of Phaeophyta; other classes were excluded). Because the chief purpose of the work is to examine the spatial relations of congeneric species, only those species in genera comprising two or more species are considered.

The span of a species is assumed to be given by its known northern and southern limits, and to contain no disjunctions. A species known only from a single site is defined as having a span of zero length located at that site. The map in Fig. 2 shows the locations of some of the stations from which records were obtained; as may be seen, the coverage is extensive but not equally dense everywhere. The method of analysis is such, however, that the impossibility of discriminating between evidence of absence and absence of evidence cannot lead to seriously faulty conclusions. The worst that can happen is that the evidence may not be strong enough to permit confident rejection of a null hypothesis that ought to be rejected.

To define an algal species' span (or geographic range) in terms of its latitudinal limits is justified on the grounds that the dominant factor controlling the presence of an alga is water temperature, and the trend of the temperature gradient is north–south. Thus the northern shores of the Gulf of Mexico belong, together with the Atlantic shore of Florida north of 28° N (Cape Kennedy), to the temperate phytogeographic region of van den Hoek (1975), whereas the Gulf and

TABLE 1. The observed and expected overlaps of the species within 133 genera of benthic marine algae. (Column headings are defined, and units given, in the text.)

Genus	s	L_S	$E(L_S\|H1)$	$E(L_S\|H2)$	h	W	m
Rhodophyta							
Erythrotrichia	4	9.0	6	7.69	52.0	106.0	40.9
Erythrocladia	4	7.5	6	6.01	37.2	62.3	15.6
Porphyra	5	11.0	10	10.26	76.7	130.7	38.0
Kylinia	3	1.5	3	3.06	34.7	17.0	5.3
Adouinella	10	58.0	45	48.26	78.8	132.8	43.1
Rhodochorton	2	0	1	0.00	16.0	2.9	0
Colaconema	4	10.5	6	10.08	48.4	102.4	52.8
Acrochaetium	27	238.0	351	151.56	48.3	74.2	8.7
Nemalion	3	5.5	3	4.61	49.5	84.1	36.6
Liagora	8	47.5	28	44.38	26.7	50.6	26.7
Galaxaura	12	95.5	66	100.67	33.5	57.4	38.7
Gelidiella	4	10.5	6	10.07	25.7	49.6	27.8
Gelidiopsis	3	3.5	3	5.08	34.6	58.5	45.1
Gelidium	3	5.5	3	5.02	34.6	57.5	33.5
Pterocladia	3	4.0	3	4.53	34.7	69.7	35.0
Rhodophysema	2	1.0	1	1.91	76.7	36.9	21.5
Cruoriopsis	3	0	3	0.27	78.8	37.4	0.9
Peyssonnelia	7	33.0	21	19.96	52.5	56.4	15.1
Clathromorphum	4	9.0	6	7.37	78.8	132.8	49.0
Phymatolithon	2	2.0	1	2.00	52.5	106.5	53.7
Leptophytum	2	1.5*	1	2.00	78.8	37.6	36.4
Lithothamnium	10	19.0	45	17.79	76.2	115.5	12.7
Melobesia	2	2.0	1	2.00	64.2	88.0	72.4
Fosliella	4	7.5	6	8.82	25.4	7.7	3.7
Lithophyllum	12	55.5	66	34.46	45.3	34.5	5.2
Goniolithon	9	63.0	36	37.92	26.7	50.6	16.1
Porolithon	3	4.0	3	4.58	26.7	13.6	5.8
Amphiroa	5	10.5	10	10.92	34.7	74.1	26.1
Corallina	3	5.0	3	5.06	69.8	121.3	76.6
Jania	4	10.0	6	10.32	34.7	69.7	54.4
Kallymenia	3	2.0	3	2.61	76.4	79.3	21.6
Halymenia	11	63.5	55	71.12	34.7	64.8	28.5
Grateloupia	3	5.5	3	5.29	34.7	88.7	49.9
Cryptonemia	2	1.0	1	1.99	34.7	88.7	73.0
Platoma	3	1.0	3	2.32	46.9	69.8	14.8
Gracilaria	15	165.0	105	131.66	47.1	80.1	36.1
Agardhiella	2	1.0	1	1.74	42.6	46.4	27.2
Eucheuma	5	15.0	10	14.58	34.7	57.6	27.3
Catenella	2	2.0	1	2.00	26.7	80.7	61.8
Hypnea	4	10.5	6	9.41	41.6	95.6	62.5
Gymnogongrus	3	3.0	3	4.26	48.0	102.0	40.6
Phyllophora	4	8.0	6	9.98	78.8	40.0	24.6
Gigartina	3	1.0	3	4.33	47.1	82.1	39.9
Fauchea	2	1.5*	1	2.00	22.3	45.2	22.6
Leptofauchea	2	0	1	0.00	12.5	35.9	0
Chrysymenia	3	6.0	3	4.56	34.7	57.6	28.1
Cryptarachne	3	1.0	3	1.99	34.7	47.7	9.5
Botryocladia	4	11.0	6	10.29	34.7	57.8	32.7
Rhodymenia	2	2.0	1	2.00	34.7	57.6	28.8
Lomentaria	3	2.0	3	3.52	46.5	69.7	21.0
Champia	3	4.0	3	5.41	43.8	67.7	50.6
Crouania	2	2.0	1	2.00	26.7	47.6	23.8
Mesothamnion	2	0	1	0.32	25.6	49.5	3.9
Wrangelia	3	3.5	3	5.61	27.8	51.6	36.0
Antithamnion	7	20.0	21	18.99	78.8	134.8	39.0
Callithamnion	7	17.0	21	15.36	49.8	84.7	17.8
Spermothamnion	7	19.0	21	25.44	45.3	68.8	27.3
Griffithsia	6	21.0	15	22.29	47.5	71.4	37.9

TABLE 1 (*contd.*)

Genus	s	L_s	$E(L_s\mid H1)$	$E(L_s\mid H2)$	h	W	m
Ceramium	19	203.5	171	157.10	69.3	125.3	35.3
Seirospora	2	0	1	0.72	42.5	26.5	6.1
Spyridia	3	3.5	3	4.84	43.8	67.4	61.0
Taenioma	2	1.5*	1	2.00	21.9	45.8	44.1
Pantoneura	2	1.0	1	1.92	76.7	34.7	30.6
Dasya	11	78.0	55	40.12	46.5	67.4	14.8
Heterosiphonia	2	1.5*	1	2.00	34.3	58.2	54.4
Rhodomela	2	2.0	1	2.00	66.1	25.6	13.2
Odonthalia	2	2.0	1	2.00	76.2	31.6	19.7
Chondria	12	104.0	66	90.88	46.5	81.5	42.1
Polysiphonia	25	289.5	300	206.65	78.8	132.8	27.8
Bostrychia	4	11.5	6	9.67	38.5	66.8	55.6
Bryocladia	2	2.0	1	2.00	24.7	49.7	49.1
Bryothamnion	2	1.5*	1	2.00	33.6	56.7	52.7
Wrightiella	2	1.0	1	1.96	26.7	39.7	24.6
Dipterosiphonia	2	1.0	1	1.60	25.6	51.7	46.2
Herposiphonia	4	10.5	6	11.27	34.7	58.6	41.8
Lophosiphonia	3	3.0	3	4.66	30.2	34.1	16.4
Acanthophora	2	1.0	1	1.47	28.0	51.8	50.4
Laurencia	9	54.0	36	46.73	34.7	88.7	40.8
Chlorophyta							
Gloeocystis	2	0	1	1.83	68.4	26.9	12.3
Codiolum	3	4.0	3	4.20	68.4	33.9	14.8
Chlorochytrium	4	7.0	6	9.41	78.8	36.6	19.5
Ulothrix	2	2.0	1	2.00	76.3	48.4	39.6
Pilinia	6	15.5	15	12.96	49.5	31.5	7.5
Entocladia	6	16.5	15	12.89	54.4	77.3	18.3
Ochlochaete	2	0	1	1.91	48.5	7.0	3.3
Enteromorpha	12	106.0	66	102.84	78.9	113.8	63.6
Blidingia	2	1.0	1	1.99	75.0	108.4	74.2
Monostroma	3	4.0	3	4.42	76.4	39.5	18.4
Ulva	4	10.5	6	10.33	69.3	104.3	64.3
Ulvaria	2	1.0	1	1.56	78.3	102.2	56.8
Urospora	2	1.0	1	1.74	76.5	80.4	50.9
Chaetomorpha	9	35.5	36	40.47	81.8	105.9	40.9
Rhizoclonium	4	7.5	6	8.69	72.8	96.6	40.9
Cladophora	27	329.0	351	243.47	69.3	104.0	21.5
Spongomorpha	4	10.0	6	10.91	78.3	37.6	27.6
Neomeris	3	4.5	3	4.50	26.7	47.2	20.9
Acetabularia	5	10.0	10	13.96	30.3	54.1	23.0
Valonia	5	15.5	10	15.17	34.6	57.6	39.0
Siphonocladus	2	1.0	1	1.79	26.7	15.8	11.2
Dictyosphaeria	2	2.0	1	2.00	26.7	47.2	42.4
Cladophoropsis	2	2.0	1	2.00	34.3	58.2	45.0
Struvea	4	8.5	6	8.06	34.6	47.5	22.1
Bryopsis	4	10.5	6	9.04	78.8	113.7	56.1
Caulerpa	17	204.0	136	202.86	34.7	58.6	36.6
Arainvillea	7	37.0	21	20.72	34.6	58.4	18.1
Udotea	7	39.5	21	30.41	34.7	55.2	25.1
Penicillus	3	5.5	3	4.83	27.7	48.2	27.9
Halimeda	12	102.5	66	90.79	30.1	53.0	25.1
Codium	7	24.5	21	29.24	42.0	84.8	43.1
Phaeophyta							
Ectocarpus	8	35.5	28	28.75	72.8	125.7	42.0
Giffordia	8	22.5	28	27.14	65.5	89.4	27.4
Streblonema	4	9.0	6	8.49	58.5	30.7	13.5
Sphacelaria	10	34.0	45	38.03	82.5	110.1	29.5
Dictyota	10	74.5	45	65.62	37.6	89.3	55.1

TABLE 1 (*contd.*)

Genus	s	L_s	$E(L_s\|H1)$	$E(L_s\|H2)$	h	W	m
Dictyopteris	4	10.5	6	11.24	25.6	49.6	37.7
Padina	6	20.0	15	21.40	34.7	58.5	26.7
Hecatonema	4	4.0	6	6.12	47.7	65.7	19.8
Myrionema	5	17.0	10	10.95	72.8	54.5	16.9
Ascocyclus	2	0	1	0.39	47.1	29.4	2.8
Ralfsia	6	16.0	15	12.64	66.1	90.0	23.2
Elachistea	4	5.0	6	7.00	76.3	51.7	17.9
Sporochnus	2	1.5*	1	2.00	34.7	57.6	52.5
Desmarestia	2	2.0	1	2.00	81.8	63.8	51.4
Stictyosiphon	4	7.0	6	6.51	78.8	51.0	16.2
Myriotrichia	5	5.5	10	5.95	48.6	30.9	5.7
Punctaria	2	1.0	1	1.72	72.8	38.2	28.8
Rosenvingea	3	5.0	3	5.34	34.7	57.9	32.4
Chorda	2	2.0	1	2.00	72.8	32.1	29.7
Laminaria	5	12.5	10	17.53	78.8	39.9	33.8
Alaria	3	5.5	3	5.87	76.7	34.7	30.6
Fucus	4	11.5	6	10.17	77.8	42.9	23.8
Sargassum	8	40.0	28	41.70	41.6	65.6	43.3
Dictyosiphon	4	9.0	6	6.91	78.8	41.2	14.2

* Cases in which $s = 2$ and $L_s = 1.5$ and (because the larger of the two species' spans is the same as W, the genus-span) $E(L_s|H2) = 2.00$. When this is so it must be assumed that $L_s = E(L_s|H2)$. Therefore these cases have been disregarded in compiling Table 2, which records the number of times $L_s > E(L_s|H2)$.

Atlantic shores of Florida south of that latitude are in the tropical region (Humm, 1969).

One final point to note is that epiphytes have not been treated separately from other species in the analysis. Although the span of a strongly host-specific epiphyte must depend on that of its host, strong host-specificity is rare among the algae considered.

Results*

The seven columns in Table 1 show the following for all 133 genera considered (seventy-eight in Rhodophyta, thirty-one in Chlorophyta, and twenty-four in Phaeophyta): (1) s, the number of species in the genus; (2) L_s, their observed overlap; (3) and (4) $E(L_s|H1)$ and $E(L_s|H2)$, the expected overlaps in the genus under the unconditional and conditional hypotheses respectively; (5) h, the latitude of the most northerly northern limit of the span of any species in the genus; (6) W, the total span of the genus (see Fig. 1); (7) m, the mean span-length of the species in the genus. h, W and m are in degrees and tenths.

* The complete data on which the tables and graphs in this paper are based are too long for publication here. They may be obtained from the author. These data list each species, the latitudes and place names of the northern and southern limits of its span, and the sources of these records with the necessary bibliography.

Table 2 compares the observed overlaps with their expectations under H1 and H2.

Under H1, the distribution of L_s is symmetrical, and therefore the expected median and the expected mean are identical. This makes it possible to test the acceptability of H1 since, given the hypothesis, the observed L_s would be expected to exceed its expected median in 50% of the genera. Table 2 shows the results of testing hypothesis H1 for the three algal classes treated separately; two-tailed tests were done as there was no reason, beforehand, to suppose that deviations from expectation were more likely in one direction than the other. Clearly, we can confidently reject H1 and conclude that congeneric species overlap one another much more strongly than they would if the locations of their boundaries were wholly independent.

Next consider the conditional hypothesis H2. The theoretical distribution of L_s under H2 is not symmetrical and the direction in

TABLE 2. The observed proportions* of genera for which L_S exceeded expectation under the two null hypotheses. In the upper row, the probabilities, P, are for a two-tailed binomial test with $p = 0.5$

		Rhodophyta	Chlorophyta	Phaeophyta
Proportions of genera in which:	$L_S > E(L_S \mid H1)$	51/69 ($P = 0.0002$)	22/26 ($P = 0.0008$)	17/23 ($P = 0.034$)
	$L_S > E(L_S \mid H2)$	29/63	13/27	8/21

* The totals in the denominators are less than the numbers of genera in each class because genera in which $L_S = E(L_S)$ were disregarded.

which it is skewed depends on the relative magnitudes of W and of the s species-spans concerned. Therefore it is impracticable to test the acceptability of H2 in the way that H1 was tested, since one cannot compare the observed L_S values with their expected medians. Indeed, to do an exact test, it would be necessary to compute the separate probabilities of each possible value of L_S for each genus treated individually, a prohibitively lengthy task. However, there is no reason to suppose that the expected distributions of L_S under H2 tend to be skewed positively more often than negatively, or vice versa; that is, it is not unreasonable to postulate that the sign of $E(L_S \mid H2) - \text{Median}(L_S \mid H2)$ will be positive for about half the genera if H2 is correct. This justifies the comparison of observed L_S values with their expectations, $E(L_S \mid H2)$, shown in the second row of Table 2. Although a test based on calculable probabilities cannot be performed, there is obviously no reason to reject H2.

Now recall that both hypotheses specify that certain entities are randomly and independently located. For H1 these entities are the span boundaries, which may be thought of as points on the coast. For H2, the entities are the spans themselves, which may be thought of as segments of coast of given (observed) lengths. Rejection of H1 requires us to conclude that the span of a species is, on average, too large for the locations of its boundaries to be modelled by placing a pair of points at random on a line of length W (the span of the whole genus). Because species-spans tend to be large relative to the spans of their genera, they overlap one another strongly, and small spans are often nested within large ones. At the same time, one might suppose that interspecific competition would cause these large spans to be so arranged as to make their overlap as small as possible consistent with their size. The apparent acceptability of hypothesis H2 suggests that this is not the case. Indeed, acceptance of H2 amounts to concluding that species-spans, treated as entities, are located independently of one another, and are not influenced by between-species competition.

Discussion of the evolutionary implications of these conclusions is deferred to the final section, after other generalizations emerging from the data have been presented.

III: Other observations on species spans

This section brings together summaries of some of the other information contained in Table 1. This information is a byproduct of the data used in studying the overlap of species' spans. Although we are not here concerned with overlap, it should be recalled that since the primary purpose of this investigation is the study of overlap, all the data in Table 1 pertain to genera with two or more species. This restriction seems unlikely to introduce any bias into the results.

In Fig. 3 are histograms showing the distributions of species' spans in the three algal classes. The total numbers of species are: 395 in the Rhodophyta; 174 in the Chlorophyta; 115 in the Phaeophyta. In all three classes the trend is the same: short spans are most numerous, but there is a secondary mode of spans in the 40° to 50° range.

Figure 4 shows plots of σ against latitude where σ is defined as the number of species' spans cut by a given parallel. Thus σ is not necessarily identical with the number of species collected at a station at a given latitude; rather, it is the number of species whose spans are assumed to lie across, and hence to

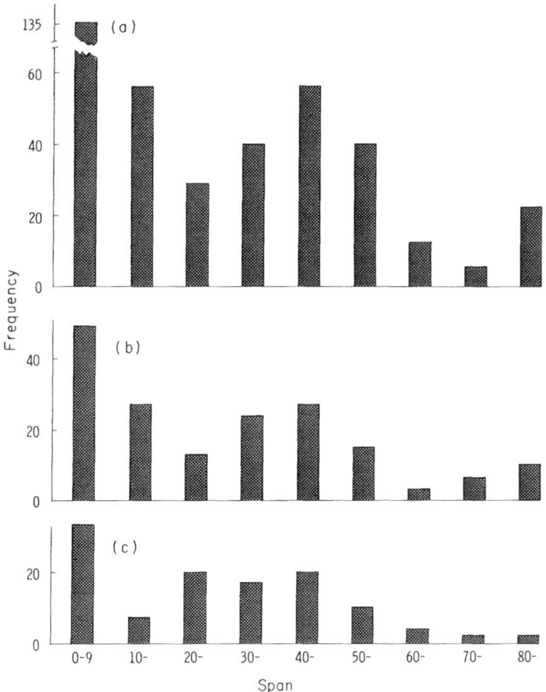

FIG. 3. The frequency distributions of the spans of 684 species. (a) Rhodophyta (395 species); (b) Chlorophyta (174 species); (c) Phaeophyta (115 species).

contain, that latitude. Values of σ were derived for the parallels 55° S, 50° S . . . 70° N, separated by 5° intervals. Therefore the locations of the peaks and troughs in the graphs are the centres of 5°-wide latitude belts that bracket the true peaks and troughs. The similarity of the three graphs is striking. In all three orders, σ attains a local maximum in the latitude belt 15°–25° N, the belt that contains the Greater Antilles. The pronounced trough centred on 35° N is no doubt associated with the Outer Banks of North Carolina where a lack of rocky substrates and the unstable temperature regime of shallow, sandy, shores form a barrier to the north–south dispersal of algae (van den Hoek, 1975; and see Humm, 1969). Cape Hatteras therefore marks a 'break' in the algal flora, the boundary between two very dissimilar phytogeographic regions, and comparatively few species' spans lie across it. The secondary maxima, in the 40°–50° N belt, occur, presumably, in the most species-rich part (the Atlantic Provinces of Canada, and New England) of van den Hoek's western cold-temperate Atlantic–Boreal region. In the southern hemisphere, the pronounced drop in σ south of 30° S is probably due to a shortage of records from far southern latitudes.

Figure 5 combines the data in Table 1 in another way. It relates to all three algal classes taken together. The method of its construction is explained in the legend. An additional point to notice is that if the bivariate histogram mapped in the triangle were projected onto the vertical axis, a histogram of the form of those in Fig. 3 (but relating to all three algal classes combined) would be recovered. If it were projected onto the horizontal axis the result would be a histogram showing the distribution among the species of the latitudes of the midpoints of their spans.

There are obviously two modes, separated by a saddle, in the bivariate histogram. One mode is comprised of spans between 40° and 60° in extent that straddle the equator, though their midpoints are, on average, a few degrees north of it. The other mode is com-

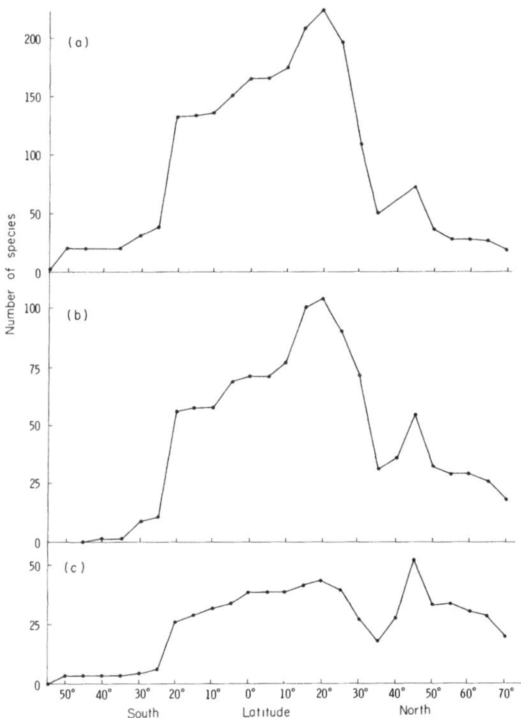

FIG. 4. The number of species-spans, σ, cut by every fifth parallel of latitude from 55° S to 70° N. (a) Rhodophyta; (b) Chlorophyta; (c) Phaeophyta.

prised of spans of small magnitude that are confined to low latitudes in the northern hemisphere. These are, of course, the numerous species found, so far as the western Atlantic is concerned, only in the Caribbean.

Discussion

It was concluded in Part 2 that competition between related algal species has no effect on the locations of their spans, and hence on their geographical zonation pattern. The local, as opposed to geographical, pattern of seaweed distribution, in contrast, appears to be strongly affected by competition. The very precise zonation patterns of intertidal seaweeds, especially on sheltered shores, and the less precise though still conspicuous zonation of subtidal species right down to the limits of light penetration (Dawson, 1966), are presumably the result of intense competition (Chapman, 1973, 1974; Pielou, 1974).

The fact that competition affects the local distributions of algae so strongly, and their geographical distributions apparently not at all, is surprising and thought-provoking. If one assumes that allopatry is a necessary prerequisite for speciation, then the events leading to sympatry of congeneric species must, in outline, be as follows (cf. Croizat, Nelson & Rosen, 1974; Rosen, 1975): a single ancestral population becomes divided into isolated fragments by the appearance of barriers to dispersal; differentiation, and the evolution of reproductive isolation (that is, speciation) take place, allopatrically, in the separated daughter populations; and finally, provided the barriers to dispersal disappear only after speciation is finished, the new species can disperse into each others' geographic ranges and exist sympatrically. However, for sympatry to be complete, two things are necessary. First, the new species, although they originally evolved in, and became adapted to, different geographic

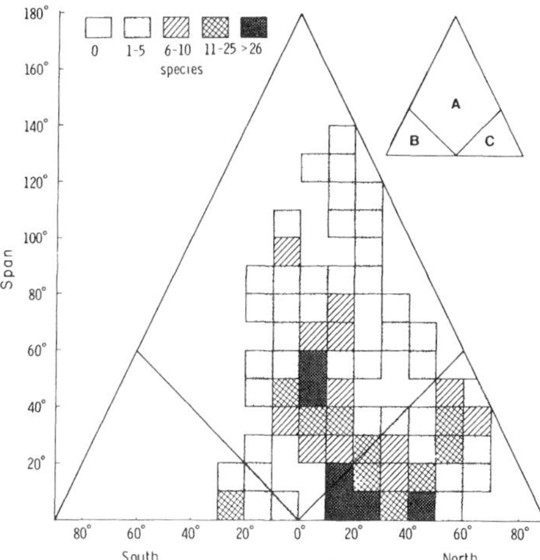

FIG. 5. The joint distribution of the magnitudes of spans, and the latitudes of their midpoints, for all 684 species of algae (data for the Rhodophyta, Chlorophyta and Phaeophyta are combined). The equilateral triangle forming the data space is the set of all points that would be possible assuming a coast stretching from pole to pole and disallowing transpolar spans. Data have been grouped into 10° classes in both variates. Frequencies are grouped into six classes symbolized as shown at the top. *Inset:* To show how the triangular data space is divided into three regions. Region A includes amphitropical species. Regions B and C, respectively, include species whose spans lie wholly in the southern, and wholly in the northern, hemispheres.

regions, must have had time subsequently to adapt to conditions in the regions they are colonizing; that is, each of the species must find in the region of sympatry a habitat in which it can prosper. Secondly, competitive exclusion on a geographic scale must not occur.

These considerations conjure up a situation that is worth trying to visualize. Imagine a once-continuous coastline that later became a row of islands, or was broken up in some other way into isolated blocks separated by barriers to algal dispersal. Consider the fate of one ancestral population, of a single algal species, whose separated daughter populations evolved into new species. Initially there would be one daughter species per island (or block) and this species would not have to compete with its congeners. Because of the absence of competition, it might well have wide ecological amplitude and occupy a wide variety of habitats. Now suppose that the barriers to dispersal disappear; for example, barriers between islands would vanish if a lowering of sea level, during an ice age, led to the disappearance of deepwater channels between neighbouring islands. As the previously-separated species populations mingle, we must suppose that, given time, each will become confined to a narrower range of environmental conditions, that is to a narrower 'realized zone' (Pielou, 1975) in a local, zoned community, than it had formerly occupied.

Thus mingling of congeneric species is presumably accompanied by a 'sorting' process, due to interspecific competition, that causes the species to become restricted to well-defined non-overlapping zones on the shore. How long the sorting takes cannot be estimated. The foregoing arguments suggest that it would be well worth while to compare seaweed zonation patterns in different regions. The argument leads to the following prediction: the more recent the disappearance of barriers to dispersal, the less developed will seaweed zonation be.

An alternative explanation for the results is that speciation in these algae is not (or not usually) allopatric. There appears to be very little evidence for speciation through polyploidy in the marine algae and it would be worth while to investigate whether it, or some other, unsuspected, mechanism can bring about sympatric speciation in this class of plants.

Acknowledgments

My chief thanks are to Anita Regan who single-handedly carried out a herculean literature search, and reconciled conflicting taxonomies. We both thank the following people for responding helpfully to our queries: A.R.O. Chapman, M. Diaz-Piferrer, T. Edelstein, R. Hooper, J. McLachlan, E.C. de Oliveira Filho and G.R. South.

References

Chapman, A.R.O. (1973) A critique of prevailing attitudes towards the control of seaweed zonation on the seashore. *Bot. Mar.* 16, 80–82.

Chapman, A.R.O. (1974) The ecology of macroscopic marine algae. *Ann. Rev. Ecol. Syst.* 5, 65–80.

Croizat, L., Nelson, G.J. & Rosen, D.E. (1974) Centers of origin and related concepts. *Syst. Zool.* 23, 265–287.

Dawson, E.Y. (1966) *Marine Botany. An Introduction.* Holt, Rinehart and Winston, New York.

Hoek, C. van den (1975) Phytogeographic provinces along the coasts of the northern Atlantic Ocean. *Phycologia* 14, 317–330.

Humm, H.J. (1969) Distribution of marine algae along the Atlantic coasts of North America. *Phycologia* 7, 43–53.

Mattingly, P.F. (1962) Towards a zoogeography of the mosquitoes. In: *Taxonomy and Geography* (Ed. by D. Nichols), pp. 17–36. The Systematics Association, London.

Pielou, E.C. (1974) Competition on an environmental gradient. In: *Mathematical Problems in Biology* (Ed. by P. van den Driessche). Springer-Verlag.

Pielou, E.C. (1975) *Ecological Diversity.* Wiley, New York.

Pielou, E.C. (1976) The factual background of ecological models: tapping some unused resources. *Proc. Conf. Environmental Modeling and Simulation*; pp. 668–672. U.S. Environmental Protection Agency.

Pielou, E.C. (1977) *Mathematical Ecology.* Wiley, New York.

Rosen, D.E. (1975) A vicariance model of Caribbean biogeography. *Syst. Zool.* 24, 431–464.

Appendix. *Derivation of* $E(\lambda|H2)$

We wish to find the expected overlap of two spans, of lengths x and y, with $x \geqslant y$, *when they are placed independently and at random in a space of length W.*

Suppose first that $W > x + 2y$. *(Then, for certain locations of x, it would be possible to place y either north or south of it with no overlap.) Denote by a the distance of the northern end of the longer span, x, from the northern end of the space, W. The probabilities that* $\lambda = 0$, *1 or 2 depend on whether both, or only one, of the gaps separating the ends of x from the ends of W exceed y. There are three possibilities: only the gap to the south of x can contain y; the gaps both north and south of x can contain y; only the gap north of x can contain y.*

The first of these possibilities, which obtains when $0 \leqslant a < y$, *is shown in Fig. 6 from which it may be seen that*

$$\left.\begin{aligned} Pr(\lambda = 0|a) &= (W-x-y-a)/(W-y) \\ Pr(\lambda = 1|a) &= (a+y)/(W-y) \\ Pr(\lambda = 2|a) &= (x-y)/(W-y) \end{aligned}\right\} \quad \text{when } 0 \leqslant a < y$$

Analogous arguments show that

$$\left.\begin{aligned} Pr(\lambda = 0|a) &= (W-x-2y)/(W-y) \\ Pr(\lambda = 1|a) &= 2y/(W-y) \\ Pr(\lambda = 2|a) &= (x-y)/(W-y) \end{aligned}\right\} \quad \text{when } y \leqslant a < W-x-y$$

and

$$\left.\begin{aligned} Pr(\lambda = 0|a) &= (a-y)/(W-y) \\ Pr(\lambda = 1|a) &= (W-x+y-a)/(W-y) \\ Pr(\lambda = 2|a) &= (x-y)/(W-y) \end{aligned}\right\} \quad \text{when } W-x-y \leqslant a < W-x$$

The foregoing equations give the probabilities that $\lambda = 0$, *1 or 2 on condition that a has an assigned value. To find the unconditional probabilities, we allow a to be*

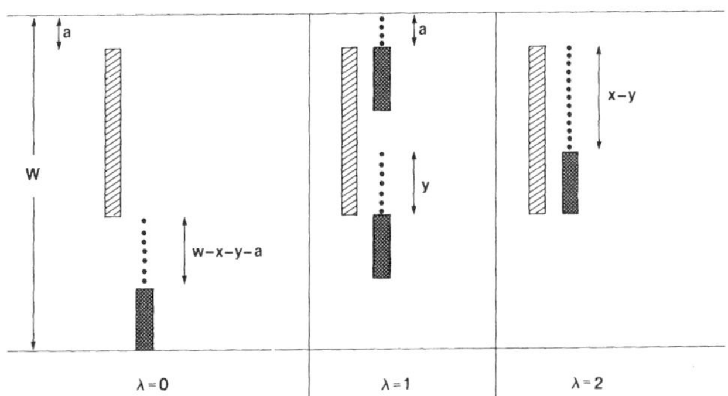

FIG. 6. To illustrate derivation of the probabilities of obtaining $\lambda = 0$, 1 or 2 when $W > x + 2y$ and $0 \leqslant a < y$. The hatched bar is the longer span, x; the black bar is the shorter span, y. The rows of dots show the range of possible locations for the northern end of y that yield overlaps of 0, 1 or 2.

rectangularly distributed in [0, $W-x$] and integrate. Disregarding $Pr(\lambda = 0)$ since it is not required for the determination of $E(\lambda|H2)$, it is seen that

$$Pr(\lambda = 1) = \frac{1}{(W-x)(W-y)}\left\{\int_0^y (a+y)\,da + \int_y^{W-x-y} 2a\,da + \int_{W-x-y}^{W-x} (W-x+y-a)\,da\right\}$$

$$= \frac{y(2W-2x-y)}{(W-x)(W-y)}.$$

Since $Pr(\lambda = 2)$ is independent of a, it is as given in all three sets of equations above, namely

$$Pr(\lambda = 2) = \frac{(x-y)}{(W-y)}.$$

Therefore $E(\lambda|H2) = Pr(\lambda = 1) + 2Pr(\lambda = 2)$

$$= \frac{2x(W-x)-y^2}{(W-x)(W-y)}. \qquad (A1)$$

Suppose next that $x+2y \geqslant W \geqslant x+y$. Then, for all locations of x, it is only possible to place y in (at most) one of the gaps to north or south of it so that there shall be no overlap. Arguing in the same way as before, it is found that $E(\lambda|H2)$ is again as shown in equation (A1). We therefore have, as in equation (2) in the paper,

$$E(\lambda|H2) = \frac{2x(W-x)-y^2}{(W-x)(W-y)} \qquad \text{when } W \geqslant x+y.$$

Lastly, suppose $W < x+y$, so that zero overlap is impossible. It is easy to see that in this case the probabilities are independent of a and that

$$Pr(\lambda = 0) = 0$$
$$Pr(\lambda = 1) = (W-x)/(W-y)$$
$$Pr(\lambda = 2) = (x-y)/(W-y)$$

whence

$$E(\lambda|H2) = \frac{W-x}{W-y} + \frac{2(x-y)}{W-y}$$

$$= \frac{W+x-2y}{W-y} \qquad \text{when } W < x+y$$

which is the same as equation (3) in the paper.

PAPER 29

Seven Forms of Rarity (1981)

D. Rabinowitz

Commentary

KEVIN J. GASTON

Understanding why some species are rare and others are common is a central goal of ecology. Understanding why this is so at large spatial and temporal scales is central to macroecology. Much discussion of this topic had long been founded on very general, and frankly often hazy, notions of just what was meant by rarity and commonness. This paper sought to engender some greater clarity, highlighting foremost the vital importance of recognizing that rare species are not all equal. This has had huge implications for interpretations of rarity.

In this book chapter, Deborah Rabinowitz, then a botanist at the University of Michigan, cogently argued that different forms of rarity can be recognized. Moreover, these different forms likely arise as an outcome of the different pathways by which species become rare, and as a result there are different evolutionary and ecological consequences of being rare. To provide some substance to this argument, she proposed a simple conceptual framework for the different forms of rarity, which has subsequently been reproduced numerous times. This comprised the classification of species into one of eight groups, resulting from the combinations of two states of each of geographic range (large or small), habitat specificity (wide or narrow), and local population size (large, dominant somewhere; or small, nondominant). All but one combination (large range, wide habitat specificity, large local population size) was regarded as a form of rarity, and Rabinowitz suggested some possible examples of most. She highlighted sparse species (those with large ranges, wide habitat specificity, and small local populations) as being of particular interest, memorably describing them as "those, which, when one wants to show the species to a visitor, one can never locate a specimen!" (207). Such species were a topic of much of her own research, attention which arguably has subsequently and lamentably much lapsed.

Deborah Rabinowitz died in 1987 at the age of just 40. Although she somewhat elaborated her ideas on the forms of rarity in a further paper published in 1986, her 1981 book chapter alone has been highly influential in two distinct ways. First, and more trivially, it led to a rash of attempts to categorize assemblages of species according to Rabinowitz's scheme and to compare the outcomes. One suspects that this has often assumed a level of rigor to the process somewhat greater than Rabinowitz herself might have been entirely comfortable with. Nonetheless, it has continued to reinforce the multiplicity of ways in which species can be rare and the need to account for this in a host of contexts (including their conservation). Second, and significantly, this chapter helped stimulate an explosion of empirical investigations into which species were rare, what form this rarity took, and the potential mechanisms that gave rise to it. It has been cited in many

In *The Biological Aspects of Rare Plant Conservation*, edited by H. Synge, pp. 205–17. John Wiley and Sons, New York.

of the other major macroecological works that bear directly or indirectly on the topic of rarity, and presaged the development of key ideas, particularly about the links between species' local abundances, range sizes, niche breadths, and niche positions, and the recognition of general macroecological patterns therein. Debates continue to rage as to which species fall into which parts of abundance-range size-niche space and why.

The Biological Aspects of Rare Plant Conservation
Edited by Hugh Synge

17
Seven forms of rarity

DEBORAH RABINOWITZ *Division of Biological Sciences, University of Michigan, Ann. Arbor, Michigan, USA*

Summary

There are many ways in which a species can become rare and this path has profound evolutionary and ecological consequences. A theoretical framework of an eight-celled table is proposed for the different types of rarity depending on range, habitat specificity and local abundance. Seven forms of rarity are discussed with examples from the North American flora, in particular that of the narrow endemic. Studies of the competitive abilities of sparse and common prairie grasses provide insights into the biological nature of rarity and show that competitive abilities are more critical to persistence than to the regulation of abundance. Natural selection may operate to favor traits which offset the disadvantages of local small population size. We reach conclusions that are both unexpected and relevant to practical conservation philosophies.

Introduction

Perhaps the most common conclusion in this book is that there are many sorts of rare species. This fact is probably because species become rare by several pathways. If rarity has a variety of causes, then the evolutionary and ecological consequences of rarity may be equally diverse.

For instance, a species may be rare because it is especially subject to a density-dependent fungal pathogen, as it is the case with American Chestnut, *Castanea dentata* (Marsh.) Borkh. (Nelson, 1955). Let us contrast this case with one of rarity because of range contraction due to climatic change: *Pelliciera rhizophorae* Tr. & Planch., a mangrove usually placed in a monotypic family, the Pellicieraceae, close to the Theaceae, is now restricted to the Pacific coast from Costa Rica to Colombia, but occurred in Chiapas, southeast Mexico, in the Oligo-miocene (Langenheim *et al.*, 1967). In the case of the chestnut, the fungus has produced a major shift in life history, converting a large tree to a shrub. No such radical morphological or demographic changes accompany the contraction of range in *Pelliciera.* Its local densities remain high, and monospecific stands may still occur, albeit over a smaller area. For *Pelliciera*, we expect that island biogeographic or genetic consequences of drift will predominate; for *Castanea*, the local consequences are ecological and epidemiologic. These two pathways to rarity show remarkably divergent responses. If we can dissect the varieties of rarity, our understanding of rare species may benefit from the provision of a basis for investigating causes and consequences of rarity.

Because authors are often concerned with consistent traits among special sorts of rare species (Griggs, 1940; Stebbins, 1942; Drury, 1974; Smith, 1976), the state of being rare seems rather monolithic from the literature. For instance, Griggs (1940) regards rare species, in his case geographic outliers of the Laurentian shield, as being competitively inferior. Drury (1974) views rare species as those where interbreeding among populations is severely restricted. A great amount of fascinating heterogeneity among rare species is unfortunately obscured by these generalities. In this paper I have two goals, first to construct a general scheme to characterize the varieties of rarity, and secondly, to show how natural selection operates on rare species (Rabinowitz, 1978; Rabinowitz and Rapp, in press).

This classification of rarity differs from the others in this book (see Ayensu Chapter 2), Good and Lavarack (Chapter 5) or Bratton and White (Chapter 39) for three fundamental reasons. First, the aim of drawing up a list of species is not imposed upon me, and so I need not employ the categories to fulfil a legal charge. Secondly, no specific taxa, geographical locality, or administrative units need be kept in mind. Thirdly, the endangered or threatened status of plants is not my central concern. These factors free me from the constraints of pragmatism, and this may contribute some clarity in exploring the biological consequences of rarity. Hopefully, the exercise will permit some new perspectives for people engaged in more practical concerns.

A classification of rare species

To construct flexible categories for rarities, I distinguish three aspects of the situation of a species: geographic range, habitat specificity, and local population size, all of which have been introduced by previous contributors. Most of us would agree that each of these attributes is related to rarity in some way. For instance, illustrating with plants from America, *Andropogon gerardi* Vitman has a huge range – Florida to southern Quebec, westward to northern Mexico and Saskatchewan – whereas *A. niveus* Swallen is restricted to central Florida, and thus is more rare. With respect to habitat specificity, *Solidago canadensis* L. seems quite 'plastic' about where it grows – in thickets, roadsides, forest edges, clearings, prairies, fallow fields, varying soil types, moisture regimes and successional states. *Solidago bartramiana* Fern occurs only on slaty ledges, and *S. sempervirens* L. grows only in brackish conditions of coastal dunes; these two latter species, due to their habitat restriction from whatever causes, are validly regarded as rarer than *S. canadensis*. With respect to the third trait, local population size, *Festuca scabrella* Torr. and *F. idahoensis* Elmer are co-dominants with

Agropyron spicatum (Pursh) Scribn. & Sm. in the Palouse prairies of western North America, and thus their local abundances are large even though this type of grassland is quite limited in geographic extent. In contrast *Festuca paradoxa* Desv. is never dominant or really very common, and because of this 'chronic' local sparsity, we would consider it rarer than the other fescues, despite its more extensive range.

If each of these attributes is dichotomized, a 2 × 2 × 2 or eight-celled block emerges (Figure 1). Although creating the hazard of false reification – that is, converting an idea into an object – such a simple scheme can aid in focusing our thoughts, and this is my intention. The patina – a gloss or incrustation conferred by age – of monolithic rarity may have hindered our understanding of an exceedingly heterogeneous assemblage of organisms. Since the products of rarity are diverse, the causes of rarity and the genetic and population consequences of rarity are undoubtedly equally multiple.

A second caution with such a scheme is that it is a typology of results (by intention) and not a typology of mechanisms or causes (Gould, 1977). Results of similar appearance may mask divergent processes; for instance geographically restricted species may be relictual (Cain, 1940; Ricklefs and Cox, 1978) or incipient (Lewis, 1966). In the absence of the relevant studies, the classification of processes resulting in rarity is a distant goal.

Seven of the eight cells contain rare species in some sense of the word. Only the upper left cell, species with wide ranges, several habitats, and locally high abundances, do not merit the designation. *Chenopodium album* L. is an example: it is circumtropical, nearly circumtemperate, and can occur in dense or sparse stands in weedy and non-weedy situations (Kapoor and Partap, 1979).

Directly beneath is probably the most ignored category of inconspicuous and unspectacular plants, sparse species – those with large ranges, several habitats, but consistently low populations. Such species are familiar (and pedestrian) to most botanists and especially to entomologists. In North America, *Dianthus armeria* L. is a familiar example. One is never really surprised to see Deptford Pink, but one would be quite startled to see it occupy 80 per cent of the biomass in a large field. Sparse plants are those, which, when one wants to show the species to a visitor, one can never locate a specimen! To me, they are the most curious form of rarity because they seem not to have a 'favored' habitat. They almost never appear on lists of 'threatened' or 'endangered' species. Sparse species of prairie grasses in Missouri are the topic of our current studies on the mechanisms of persistence (Rabinowitz, 1978; Rabinowitz and Rapp, in press).

Two of these cells appear to have very few residents, namely species of narrow geographic range but broad habitat specificity. Is this *modus operandi* unfeasible for some evolutionary or ecological reason or do ecologists simply pay little attention to such species? If the former is true, it is of great interest to know why such species either do not arise or have large probabilities of extinction. For instance, demographic stochasticity, which is a process in small populations analagous to

GEOGRAPHIC RANGE	Large		Small	
HABITAT SPECIFICITY	Wide	Narrow	Wide	Narrow
LOCAL POPULATION SIZE				
Large, dominant somewhere	Locally abundant over a large range in several habitats	Locally abundant over a large range in a specific habitat	Locally abundant in several habitats but restricted geographically	Locally abundant in a specific habitat but restricted geographically
Small, non-dominant	Constantly sparse over a large range and in several habitats	Constantly sparse in a specific habitat but over a large range	Constantly sparse and geographically restricted in several habitats	Constantly sparse and geographically restricted in a specific habitat

Figure 1 A typology of rare species based on three characteristics; geographic range, habitat specificity, and local population size

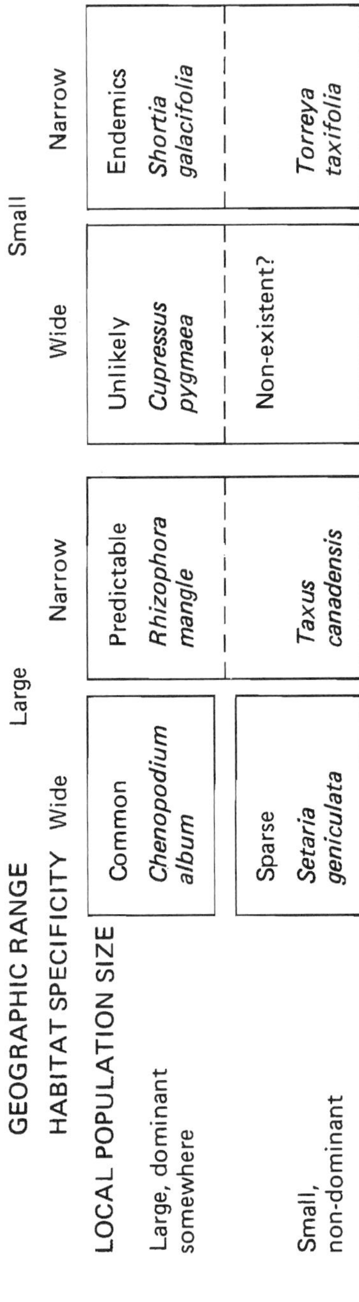

Figure 2 Summary and illustrations of species for the typology of rare species shown in Figure 1

genetic drift and which results in fluctuating population numbers, due to small sample phenomena (May, 1973; Mertz *et al.*, 1976), may cause local extinction. These deletions of populations may reduce the variety of habitats occupied and, in essence, convert a perhaps unstable species into one in the categories on the right in Figure 1, namely an endemic. Examples of such unusual species are *Cupressus pygmaea* (Lemmon) Sarg., a dwarf conifer found on coastal terraces of Mendocino County, California (Westman, 1975; Westman and Whittaker, 1975) and *Fuchsia procumbens* R. Cunn. ex. A. Cunn., a New Zealand plant (Given, Chapter 4), with small range but several habitats.

Species which have wide ranges but are associated with particular habitats are generally quite predictable in their occurrence (especially if you are a good systematist). If one is in a bog, on the strand, or on serpentine soils, one can generally find the plants peculiar to these places with relative confidence. In Caribbean Panama, on calm marine shorelines, for instance, one is very likely to find *Rhizophora mangle* L. and other mangroves, which are characteristically abundant where they occur. These species tend to be precarious as a result of habitat destruction. Mangrove swamps often are endangered because they are a habitat that many people find objectionable for a variety of reasons, usually that the trees are between them and the sea. In contrast to mangroves, which are nearly always locally common, Dr Given (Chapter 4) has given us the example of *Lepidium oleraceum* Forst.f., Cook's scurvygrass, once collected for vitamin C and now found on coastal rocks in several scattered sites around New Zealand but in locally very low densities.

Species with both narrow geographic range and narrow habitat specificity are the classic rarities in the sense of restricted endemics, often endangered or threatened. These rare endemic plants are often showy or newsworthy in some way. *Shortia galacifolia* T. & G., an attractive member of the Diapensiaceae, is endemic to several escarpment gorges of the Appalachian mountains and has endeared itself to the attentions of botanists for over a century (Gray, 1878; Ross, 1936; Davies, 1955, 1960; Rhoades 1966; and Vivian, 1967).

On an autecological level, such species receive a lot of attention. Terrell *et al.* (1978), for instance, have recently provided an excellent comparative study of the endemic aquatic *Zizania texana* Hitchc. and the widespread *Z. aquatica* L. *Zizania texana* lives on only 2.4 km of the upper San Marcos River in Texas in unusual alkaline conditions where water temperatures vary only 5°C annually, in contrast to the more varying conditions of *Z. aquatica* (see also Lucas and Synge, 1978).

The extreme of a restricted rare species is one that is known to have existed, but has been subsequently lost. The intuitive notion of a rare organism is one that is difficult to find, and the most endangered that a species can become is to be declared extinct! Lost species hold a particular fascination, rather like ships lost at sea. A fine example is *Betula uber* (Ashe) Fernald, Ashe's Birch or Virginia Round-leaf Birch, mentioned by Ayensu (Chapter 2), first collected in 1914 from

Smyth County, Virginia. The only other collection, near the first locality, was a single undated specimen rediscovered in 1973 (Mazzeo, 1974). After numerous searches, Johnson (1954) asserted:

> The only conclusion that seems warranted at this time from these several failures to rediscover this birch is that it probably no longer exists as an individual and very likely never did so in the form of a population. Ashe's birch has probably died or been destroyed in the process of urbanization of the community in which he found it 40 years ago. It is probable that this birch variety was founded solely on an aberrant individual and certainly does not appear to deserve further consideration as a species.

Sixty-one years after the original collection, the plant was rediscovered in 1975 by Douglas Ogle, who found the tree by employing an 'if I were a horse' strategy (Preston, 1976). Reasoning that when Ashe collected, the present paved roads did not exist, Ogle searched along traces of logging roads shown him by an elderly resident. The tree is extant in a population of 12 mature trees, some of which were reproductive, one sapling, and 21 seedlings (Ogle and Mazzeo, 1976). *Betula uber* is so rare that it was lost for over 60 years and is an example of the tenacity of botanists, who continued to hunt for living representatives for over half a century, against all reasonable likelihood of its continued existence. Its rediscovery was reported in *The New Yorker* magazine (Kinkead, 1976).

This eight-celled scheme does not include the category of 'pseudo-rare' organisms about which, perhaps, the most sound data exist and which tells us the most about the biological processes occurring in small populations. Species on the margins as opposed to the central portions of their ranges have been an active aspect of evolutionary studies (Stebbins, 1974), especially for *Drosophila* (Lewontin, 1974). In plants, for example, marginal and central populations of *Paeonia californica* Nutt. ex. T. & G. (Stebbins and Ellerton, 1939; Walters, 1942; Grant, 1956, 1975) and more recently of *Hordeum jubatum* L. (Schumaker and Babble, 1980) were compared to assess the relative effects of reproductive isolation, genetic drift, and selection on genetic structure. Ecophysiological and reproductive studies on marginal populations shed light on mechanisms determining or controlling range as shown, for example, by Pigott's studies on *Tilia platyphyllos* Scop. in Britain (Chapter 25). These studies on marginal rarity have the major advantage that they have an automatic control. Monitoring rare species (for instance, Bradshaw's long term assessments of the Teesdale rarities) tells us a lot about the characteristics of these taxa. However, in the absence of comparative data for related common taxa, essentially control species, we cannot judge whether the traits of rare plants are unique to them or are some random sample of plant traits in general and unrelated to the rare state.

Perhaps the least information is available on the fine scale causes of changes in abundance within what seems on casual view to be a homogeneous and appropriate site. Changes of orders of magnitude in population sizes occur on the scale of meters without striking underlying heterogeneity, and this garden-variety variance in density is very puzzling. Greig-Smith and Sagar (Chapter 32) investigated the causes of local rarity in *Carlina vulgaris* L. in a dune site where the plant was locally common very close by. Excluding both the absence of disturbance to produce new sites for establishment and also nutrient deficiencies of the substrate, they found that augmentation by sowing fruits increased local populations and that the likely source of propagule depletion was mammal predation on seeds.

Competitive abilities of sparse species

One aspect of our study of sparse prairie grasses in Missouri is illustrative of the difficulties in dissecting causes versus consequences of rarity. In order to examine the common assumption that rare species are inferior competitors (McNaughton and Wolf, 1970; Schlesinger 1978; Grime, 1979), we established de Wit competition experiments from seed in the glasshouse from May to September (see Harper, 1977, for a general explanation of de Wit plots and Rabinowitz in review for experimental details and a more thorough analysis of the data).

We find the paradoxical result that the sparse species are very nearly uniformly superior competitors to the common grasses. This result is seen in the bottom four graphs (Figure 3) which show the average total yield of a sparse grass on the left of each diagram and the average total yield of a common grass on the right. For the sparse grasses, yield falls above that expected on the basis of the monocultural yield (the dashed line descending to the right). In contrast, for the common grasses, the yields fall below expectation (dashed lines descending to the left). Thus, the convex curves of yield demonstrate the superior competitive abilities of the sparse species.

As a consequence of the superior competitive abilities, however, individuals of the sparse species grow largest when planted in low proportion with a common grass in high proportion (Figure 4). Presumably, this results because the presence of the common grass is more like empty space to a sparse individual than is the presence of other sparse individuals. The two top diagrams (Figure 4) show the dry weight of an individual of a sparse species versus its proportion in a mixture. The identity of the competing species is shown beside each line. For instance, in the upper left diagram, individuals of the sparse species *Festuca paradoxa* are largest when planted as 10 per cent in a mixture with the common grass *Andropogon gerardi* planted as 90 per cent. Individuals of *Sphenopholis obtusata* (upper right diagram) also grow largest when planted in low proportion with either of the common grasses. Contrariwise, the individuals of the common species grow largest when in monoculture or in the presence of other common species (Figure 4, the bottom two diagrams). Thus, an initially paradoxical result is reinterpreted into the Panglossian ('the best of all possible worlds') result (Gould

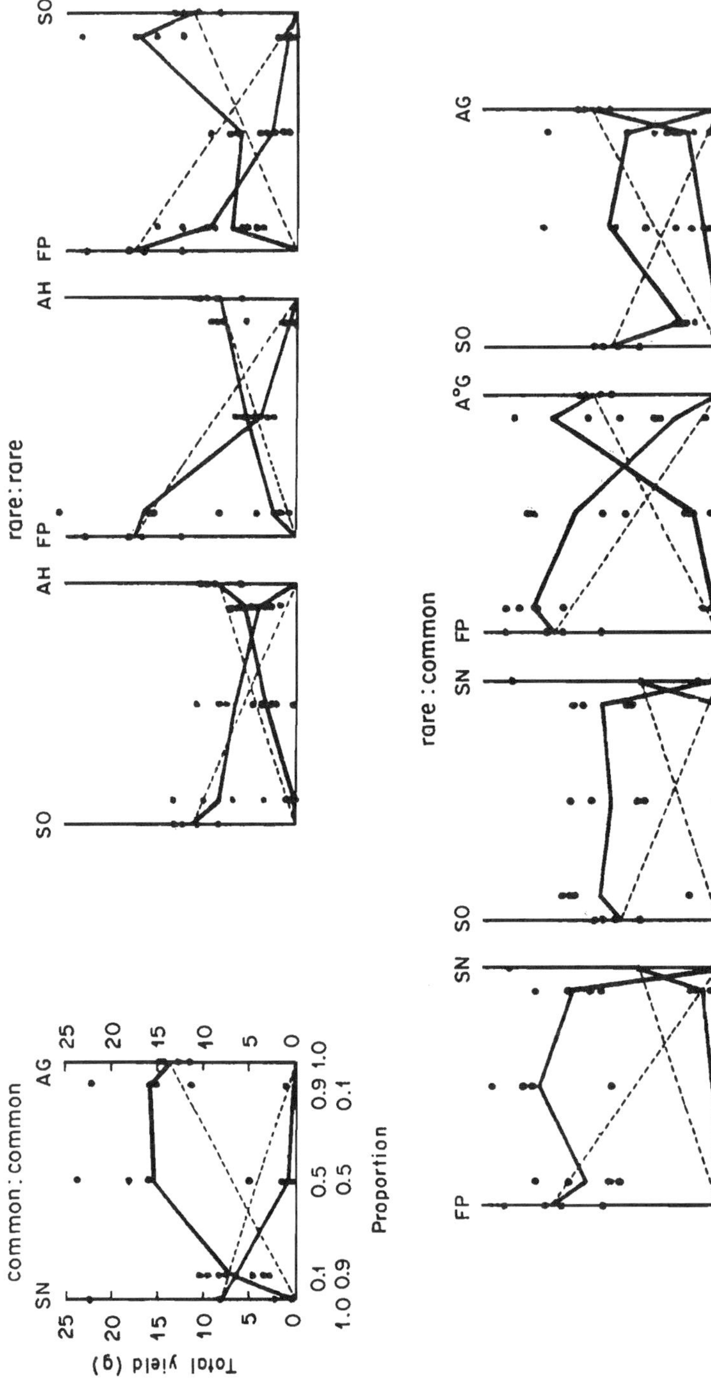

Figure 3 De Wit plots of the total yield from pairwise competition experiments with rare and common species of prairie grasses. Symbols: for common grasses, SN = *Sorghastrum nutans* (L.) Nash, AG = *Andropogon gerardi* Vitman; for sparse grasses, SO = *Sphenopholis obtusata* (Michx.) Scribn., AH = *Agrostis hiemalis* (Walt.) B.S.P., and FP = *Festuca paradoxa* Desv.

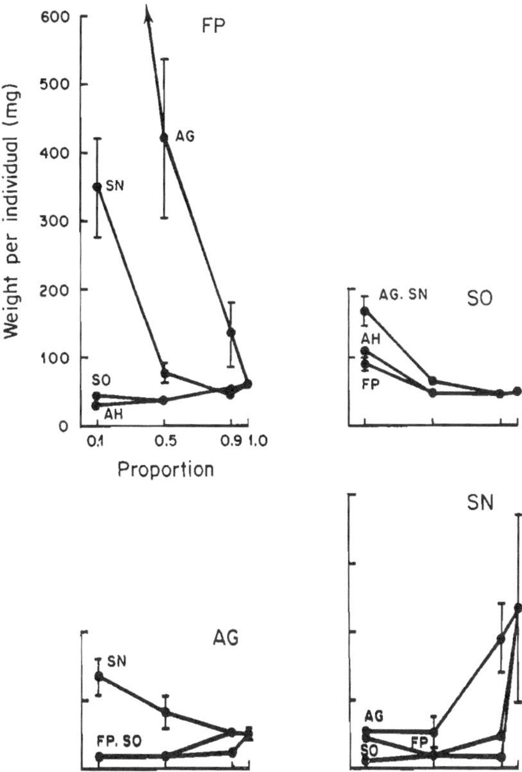

Figure 4 Size of individuals grown in different proportion and with different competitors for rare and common species of prairie grasses. The symbols are the same as for Figure 3. Sparse species are in the top row, common species in the bottom row

and Lewontin, 1979) – that sparse species grow best when sparse, and common species grow best when common.

Natural selection and sparse species

Natural selection cannot, clearly, select for rarity, and it is impossible for rarity to be an adaptive strategy. An individual may be at an advantage because it is rare, for instance, if its herbivores or pathogens cannot find it. Rausher (1980) provides an interesting example for the locally rare *Aristolochia serpentaria* L. and its herbivore *Battus philenor*. As a consequence of the advantage, the individual will reproduce more, become locally more common, and therefore automatically loses the advantage. Thus it is quixotic to say that an organism is adapted to be rare.

But one can assert that an organism may be adapted to the condition or situation of being rare. If an organism is rare for whatever reason (for instance, the fungal pathogen *Endothia parasitica* which infects chestnuts), there are additional disadvantages customarily associated with small population size, for instance being a long way from potential mates. Natural selection can act to favor traits which offset the disadvantages of small local population size, no matter what its cause, and thus render local extinction less likely.

The competitive abilities of the sparse species are best viewed in this light. Since the sparse grasses are competitively superior, the competitive abilities (at least in the short term) cannot be the cause of the sparsity. But given that the species are sparse, the trait that they grow best when surrounded by many individuals of common species is clearly advantageous and will function to render persistence more likely. The competitive abilities are best understood as having nothing to do with the regulation of population size but as a mechanism that offsets a major disadvantage of rarity.

Acknowledgments

I am pleased to acknowledge the assistance of Thora Ellen Thorhallsdottir and Christian Puff. This paper was written while I was a NATO Postdoctoral Fellow at the University College of North Wales, and the research was supported by a grant from the US National Science Foundation (DEB78-11179).

References

Cain, S. A. (1940). 'Some observations on the concept of species senescence', *Ecology*, **21**, 213–15.

Davies, P. A. (1955). 'Distribution and abundance of *Shortia galacifolia*', *Rhodora*, **57**, 189–201.

Davies, P. A. (1960). 'Pollination and seed production in *Shortia galacifolia*', *Castanea*, **25**, 89–96.

Drury, W. H. (1974). 'Rare species', *Biol. Conserv.*, **6**, 162–9.

Gould, S. J. (1977). *Ontogeny and Phylogeny*, Belknap, Cambridge, Mass.

Gould, S. J., and Lewontin, R. C. (1979). 'The spandrels of San Marco and the Panglossian paradigm: a critique of the adaptationist programme', in *The Evolution of Adaptation by Natural Selection* (Eds. J. Maynard Smith and R. Holliday), pp. 581–98, The Royal Society, London.

Grant, V. (1956). 'Chromosome repatterning and adaptation', *Adv. Genet.*, **8**, 89–107.

Grant, V. (1975). *Genetics of Flowering Plants*, Columbia University Press, New York.

Gray, A. (1878). '*Shortia galacifolia* re-discovered', *Amer. J. Sci.*, Ser. III, **16**, 483–5.

Griggs, R. F. (1940). 'The ecology of rare plants', *Bull. Torrey Bot. Club*, 67, 575–94.

Grime, J. P. (1979). *Plant Strategies and Vegetation Processes*, Wiley, Chichester.

Harper, J. L. (1977). *Population Biology of Plants*, Academic Press, London.

Johnson, A. G. (1954). '*Betula lenta* var. *uber* Ashe', *Rhodora*, **56**, 129–31.

Kapoor, P. and Partap, T. (1979). 'New approach to conserve fossil fuels by harnessing efficient energy-capturing systems: under-exploited food plants', *Man-Environment Systems*, **9**, 305–8.

Kinkead, E. (1976). 'Our footloose correspondents: the search for *Betula uber*', *The New Yorker*, **51**, 58–69 (January 12).

Langenheim, J. H., Hackner, B. L., and Bartlett, A. (1967). 'Mangrove pollen at the depositional site of oligo-miocene amber from Chiapas, Mexico', *Bot. Mus. Leaflet, Harvard Univ.*, **21**, 289–324.

Lewis, H. (1966). 'Speciation in flowering plants', *Science*, **152**, 167–72.

Lewontin, R. C. (1974). *The Genetic Basis of Evolutionary Change*, Columbia University Press, New York.

Lucas, G. and Synge, H. (1978). *The IUCN Plant Red Data Book*, IUCN, Morges, Switzerland.

May, R. M. (1973). *Stability and Complexity in Model Ecosystems*, 2nd Ed., Princeton University, Princeton, New Jersey.

Mazzeo, P. M. (1974). '*Betula uber* – what is it and where is it?', *Castanea*, **39**, 273–8.

McNaughton, S. J. and Wolf, L. L. (1970). 'Dominance and the niche in ecological systems', *Science*, **167**, 131–9.

Mertz, D. B., Cawthon, D. A. and Park, T. (1976). 'An experimental analysis of competitive indeterminacy in *Tribolium*', *Proc. Nat. Acad. Sci.*, **73**, 1368–72.

Nelson, T. C. (1955). 'Chestnut replacement in the southern highlands', *Ecology*, **36**, 352–3.

Ogle, D. W. and Mazzeo, P. M. (1976). '*Betula uber*, the Virginia Round-leaf Birch, rediscovered in southwest Virginia', *Castanea*, **41**, 248–56.

Preston, D. J. (1976). 'The rediscovery of *Betula uber*', *Amer. For.*, **82**, 16–20.

Rabinowitz, D. (1978). 'Abundance and diaspore weight in rare and common prairie grasses', *Oecologia (Berlin)*, **37**, 213–19.

Rabinowitz, D. and Rapp, J. K. (in press), 'Dispersal abilities of sparse and common prairie grasses', *Amer. Jl Bot.*

Rausher, M. D. (1980). 'Host abundance, juvenile survival, and oviposition preference in *Battus philenor*', *Evolution*, **34**, 342–55.

Rhoades, M. H. (1966). 'Seed germination of *Shortia galacifolia* T. & G. under controlled conditions', *Rhodora*, **68**, 147–54.

Ricklefs, R. E. and Cox, G. W. (1978). 'Stage of taxon cycle, habitat distribution, and population density in the avifauna of the West Indies', *Amer. Nat.*, **112**, 875–95.

Ross, M. N. (1936). 'Seed reproduction of *Shortia galacifolia*', *J.N.Y. Bot. Garden*, **37**, 208–11.

Schlesinger, W. H. (1978). 'On the relative dominance of shrubs in Okefenokee Swamp', *Amer. Nat.*, **112**, 949–54.

Schumaker, K. M. and Babble, G. R. (1980). 'Patterns of allozymic similarity in ecologically central and marginal populations of *Hordeum jubatum* in Utah', *Evolution*, **34**, 110–16.

Smith, R. L. (1976). 'Ecological genesis of endangered species: the philosophy of preservation', *Ann. Rev. Ecol. Syst.*, **7**, 33–55.

Stebbins, G. L. (1942). 'The genetic approach to problems of rare and endemic species', *Madroño*, **6**, 241–58.

Stebbins, G. L. (1974). *Flowering Plants: Evolution Above the Species Level*, Harvard University Press, Cambridge, Mass. and Edward Arnold, London.

Stebbins, G. L. and Ellerton, S. (1939). 'Structural hybridity in *Paeonia californica* and *P. brownii*', *J. Genet.*, **38**, 1–36.

Terrell, E. E., Emery, W. H. P., and Beaty, H. E. (1978). 'Observations on *Zizania texana* (Texas wildrice), an endangered species', *Bull. Torrey Bot. Club*, **105**, 50–7.

Vivian, V. E. (1967). '*Shortia galacifolia*: its life history and microclimatic requirements', *Bull. Torrey Bot. Club*, **94**, 369–87.

Walters, J. L. (1942). 'Distribution of structural hybrids in *Paeonia californica*', *Amer. J. Bot.*, **29**, 270–5.

Westman, W. E. (1975). 'Edaphic climax pattern of the pygmy forest region of California', *Ecol. Monogr.*, **45**, 109–35.

Westman, W. E. and Whittaker, R. H. (1975). 'The pygmy forest region of northern California: studies on biomass and primary productivity', *J. Ecol.*, **63**, 493–520.

PAPER 30

Dynamics of Regional Distribution: The Core and Satellite Species Hypothesis (1982)

I. Hanski

Commentary

S. KATHLEEN LYONS

The field of ecology is defined as the study of the abundance and distribution of species, and at its core, it seeks to understand why species occur where they do and what limits their population sizes. I. Hanski laid out a model of community structure that posited that stochastic variation in the abundance and distribution of species would have a profound effect on community structure separate from the role of competition. He proposed that the majority of species could be defined as either core species (i.e., species that are regionally common and locally abundant) or satellite species (i.e., species that are locally and regionally rare). This model stood in sharp contrast to the prevailing theory of the day that the frequency distribution of species distributions was unimodal and that community structure was best described by equilibrium dynamics. Indeed, his model provided an explanation for the conflicting results obtained by others when they attempted to determine whether communities were at equilibrium. Hanski argued that core species are expected to interact strongly and to be at equilibrium, whereas satellite species are expected to interact weakly and therefore not to be at equilibrium. Moreover, core and satellite species are expected to show differences in niche spacing and therefore different patterns of character displacement. Core species should be more spread out than satellite species.

This elegant modification of R. Levins's model (1970) set the stage for current understanding of the spatial dynamics of interacting populations and communities. These models provided the basis for theories about source and sink populations and for the roots of the field now known as metapopulation dynamics. In and of itself, that is enough to warrant the inclusion of this paper in a foundational volume. However, this paper used many other macroecological relationships that are commonly thought about today. It provided an example of species occupancy (i.e., the ratio of occupied to unoccupied sites in a sample), first introduced by C. Raunkiaer in 1918 (see Raunkiaer 1934), a measure that has experienced a revival in macroecology. In addition, this paper documented the positive relationship between species abundance and distribution and showed that species that were more regionally widespread and less patchily distributed had larger population sizes. It predicted and documented the negative relationship between geographic range and extinction vulnerability, now a tenet of conservation biology. Indeed, each reading of this classic paper provides new insights into ecological theory, explaining why it has been cited more than 500 times to date in a wide range of papers on spatial and temporal variation in abundance, competition theory, niche theory, neutral theory, metapopulation dynamics, island biogeography, sexual selection, and more (see Google Scholar).

Literature Cited

Levins, R. 1970. Extinction. In *Some mathematical problems in biology: Lectures on mathematics in the life sciences*, edited by M. Gerstenhaber, 2:75–101. American Mathematical Society, Providence, RI.

Raunkiaer, C. 1934. *The life forms of plants and statistical plant geography, being the collected papers of C. Raunkiaer*. Clarendon Press, Oxford.

OIKOS 38: 210–221. Copenhagen 1982

Dynamics of regional distribution: the core and satellite species hypothesis

Ilkka Hanski

Hanski, I. 1982. Dynamics of regional distribution: the core and satellite species hypothesis. – Oikos 38: 210–221.

A new concept is introduced to analyse species' regional distributions and to relate the pattern of distributions to niche relations. Several sets of data indicate that average local abundance is positively correlated with regional distribution, i.e. the fraction of patchily distributed population sites occupied by the species. This observation is not consistent with the assumptions of a model of regional distribution introduced by Levins. A corrected model is now presented, in which the probability of local extinction is a decreasing function of distribution, and a stochastic version of the new model is analysed. If stochastic variation in the rates of local extinction and/or colonization is sufficiently large, species tend to fall into two distinct types, termed the "core" and the "satellite" species. The former are regionally common and locally abundant, and relatively well spaced-out in niche space, while opposite attributes characterize satellite species. This dichotomy, if it exists, provides null hypotheses to test theories about community structure, and it may help to construct better structured theories. Testing the core-satellite hypothesis and its connection to the r-K theory and to Raunkiaer's "law of frequency" are discussed.

I. Hanski, Dept of Zoology, Univ. of Helsinki, P. Rautatiekatu 13, SF-00100 Helsinki 10, Finland.

Предлагается новая концепция для анализа регионального распределения видов и для сравнения характера распределения с соотношением ниш. Несколько серий данных показали, что величина средней локальной численности положительно коррелирует с региональным распространением, то есть с относительным количеством мозаично расположенных видовых стаций, занятых данным видом. Это наблюдение не соотвествует модели регионального распределения, предложенной Левинсом. Здесь предлагается исправленная модель, в которой обсуждается вероятность локального исчезновения вида, как уменьшающаяся функция распространения и стохастическая версия новой модели. Если стохастические колебания скоростей локального исчезновения и/или колонизации достаточного велики, проявляется тенденция разделения видов на два четких типа, называемых "основным" и "сателлитным". Первые обычны в своем регионе, локально многочисленны, а сателлитные виды характеризуются противоположными признаками. Эта дихотомия если она существует, позволяет использовать нуль-гипотезу для проверки теории структуры сообщества и она может помочь в создании более совершенной теории. Обсуждаются результаты проверки гипотезы "основных-сателлитных" видов и ее связи с r-K теорией и биологическими спектрами Раункиера.

Accepted 26 March 1981

1. Introduction

It has been popular to define ecology as the study of abundance and distribution of organisms (e.g. Andrewartha and Birch 1954, Krebs 1972), to the extent that MacArthur and Wilson (1967) state there is no real distinction between ecology and biogeography. For the purposes of this paper, I define *abundance* as the number of individuals at a local population site (for other definitions see Hengeveld 1979). There are good reasons to express abundance as a fraction of the possible maximum numbers sustainable at the site (Andrewartha and Birch 1954), but this may be difficult particularly when dealing with multispecies communities. *Distribution* refers to the number of population sites occupied by the species; this again may be given as the fraction out of the suitable ones within an arbitrary or natural region. Population sites can be discrete units, like true islands; or, like habitat islands, they may have been delimited more arbitrarily from the rest of the environment; or they may be contagious, in which case distribution is simply the proportion of total area occupied. This definition of distribution does not specify the type of spatial patterns, i.e. the locations of the (occupied) sites in space, which is a related but different question.

Theoretical ecology has largely modelled local abundance (e.g. May 1976), while distribution has been left, until recently, to biogeographers, with the exception of the largely descriptive statistical work on animal and plant distributions (Patil et al. 1971, Bartlett 1975, Taylor et al. 1978, Ord et al. 1980). An exception to this rule is Levins's (1970, see also 1969a) model on extinction, which has been followed by a number of studies on interspecific competition (Cohen 1970, Levins and Culver 1971, Horn and MacArthur 1972, Levin 1974, Slatkin 1974, Hanski 1981a; see also Skellam 1951) and predation (Vandermeer 1973, Zeigler 1977) in patchy environments, all of which apply Levins's approach to regional population dynamics and underline the difference between local and regional interactions.

The spatial aspect of population interactions has recently received increasing attention (reviewed by Levin 1976; see also Smith 1974, Levin 1977, 1978, Gurney and Nisbet 1978a, b, Taylor and Taylor 1977, Crowley 1979, Comins and Hassell 1979, Hanski 1981b), and it has become clear that understanding of both spatial processes (distribution of the species in physical space) and resource partitioning (distribution of the species in niche space) are essential components to a satisfactory explanation of the perennial questions: Why are there so many species? Why are there so many rare species? (Wiens 1976, Yodzis 1978, Hanski 1979a). Indeed, some ecologists (e.g. Simberloff 1978) have gone so far as to maintain that, in many or most cases, spatial dynamics in independently developing populations explain most of the "community patterns" (see also Caswell 1976). This contrasts with the approach initiated by MacArthur (summarized in his 1972 book).

Whatever view one holds on the importance of competitive and other biotic interactions in structuring communities, it is an indisputable fact that communities consist of different kinds of species: some are widely distributed while others occur patchily; there exist locally abundant and locally rare species; and in some communities species are, at least apparently, well spaced-out in niche space, while in other communities guilds of similar species coexist. One is tempted to pose the question: Is it possible to find unifying factors to simplify this diversity?

I suggest some narrowing down of this question. It will first be shown that dynamics in local abundance and regional distribution are interdependent. Incorporating this observation into the type of models of regional distribution suggested and first analysed by Levins (1969a, 1970, Levins and Culver 1971) leads to an important structural change in the basic model. The key question in the analysis of the revised model is whether there is an internal equilibrium point on the distribution scale, which most species are approaching, or whether the species are just heading towards either maximal distribution and superabundance, or regional extinction.

2. Local abundance and regional distribution are interdependent

Four examples from different invertebrate taxa are put forward to answer the question, are local abundance and regional distribution independent of each other? The answer is no.

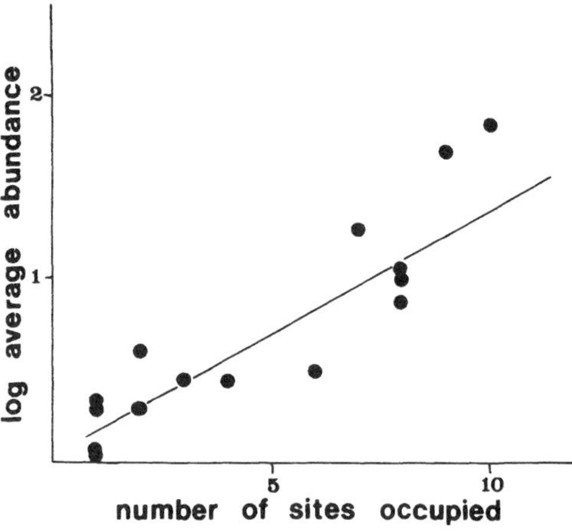

Fig. 1. Relationship between average local abundance and distribution in Anasiewicz's (1971) data on bumblebees from Lublin, Poland. While calculating average abundance only those sites were included from which the species was collected (note logarithmic y-axis). Distribution is the number of sites, maximally 10, occupied by the species. Each dot in this figure represents one species (the line has been drawn by eye).

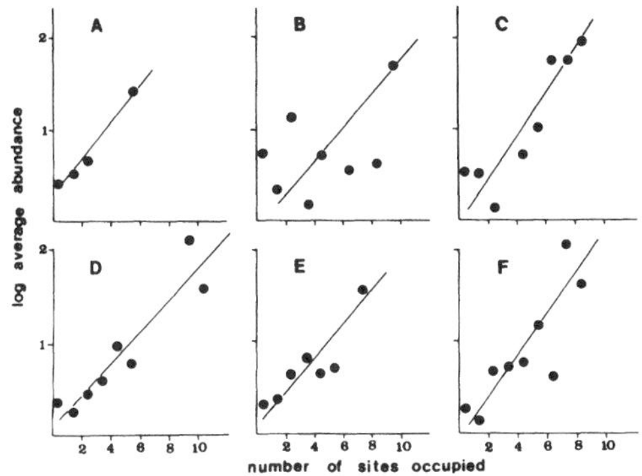

Fig. 2. Relationship between average local abundance and distribution, like in Fig. 1, in Kontkanen's (1950) data on leafhoppers from meadows in East Finland. Each dot in this figure is the average for several species, which had the same distribution. Figures A to F refer to different "communities", representing wet to dry meadows (from left to right) at early (upper row) and late summer (lower row).

Anasiewicz (1971) studied bumblebees in the parks, squares, lawns, etc. of Lublin in Poland – all good examples of discrete habitat islands in the man-made environment. Average local abundance increased with the number of sites from which the species was recorded (Fig. 1; only sites in which the species was present are included in the calculation of average local abundance).

Kontkanen (1950; see also 1937, 1957) sampled leafhoppers from meadows in East Finland. He delimited six "communities" of coexisting species, and, in

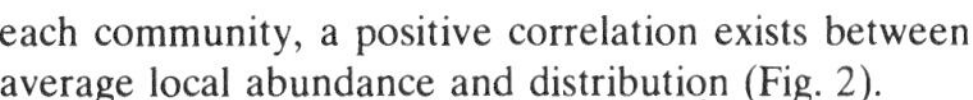
each community, a positive correlation exists between average local abundance and distribution (Fig. 2).

The third example is from Karppinen's (1958) study on soil mites (Oribatei) in two forest types in North Finland. In both habitats, a positive correlation between abundance and distribution is apparent (Fig. 3).

The final example is from my studies on dung and carrion beetles in lowland rain forest in Sarawak (Hanski unpubl.). Trapping was carried out with 10 traps for 4 nights at 12 sites, situated at least 0.5 km from each other in homogeneous virgin forest. I have restricted the analysis to the species-rich genus *Onthophagus* (Scarabaeidae). Once again, a positive correlation exists between the number of trapping sites from which the species was caught and the average catch from one site (Fig. 4). I conclude from these examples that a correlation between abundance and distribution is the rule in nature.

It is beyond the scope of this paper to go into details about the causes of this relationship, but it may be pointed out that the level of between-site movements is clearly of crucial importance. It appears to be common in nature that emigration takes place much before local carrying capacity has been reached, perhaps because of reasons discussed by Lidicker (1962) and Grant (1978).

Datum points in Figs 1 to 4 result from sampling, but because both abundance and distribution are underestimated, an increase in sample size should not change the picture qualitatively. There are, of course, truly rare yet widely distributed species, like the crane *Grus grus* in Finnish marshlands (Järvinen and Sammalisto 1976), but if communities consisting of reasonably similar species are studied, true distribution is expected to be correlated with true average abundance. The contrary

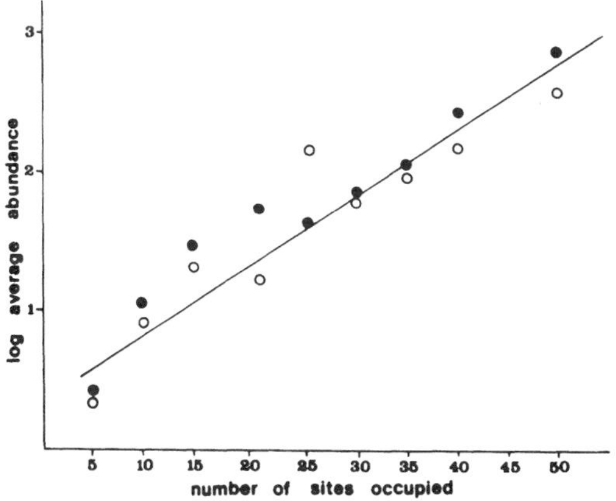

Fig. 3. Relationship between average local abundance and distribution, like in Fig. 1, in Karppinen's (1958) data on soil mites (Oribatei) from two forest types in North Finland. Each dot in this figure is the average for several species, which have been grouped into 10 distribution classes (total number of sites was 50 in both forest types). The two kinds of symbols refer to the two forest types.

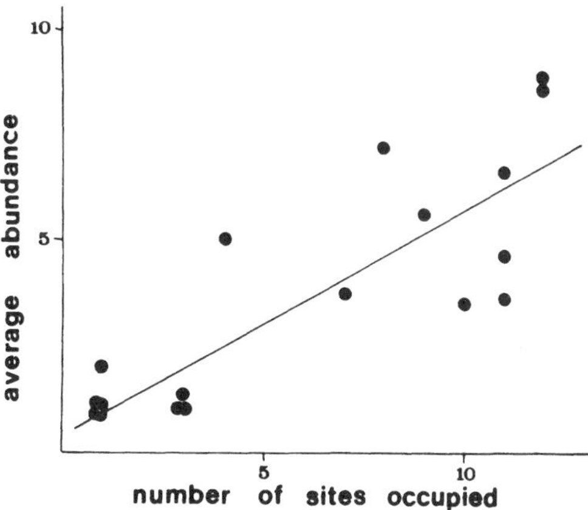

Fig. 4. Relationship between average local abundance and distribution, like in Fig. 1, in *Onthophagus* (Scarabaeidae) in the alluvial forest in Sarawak (Hanski unpubl.). A total of 12 sites was studied in homogeneous virgin forest. Each dot in this figure represents one species (note that y-axis is not logarithmic).

would require spatial variance to decrease by a factor greater than approximately the squared proportional decrease in average abundance, which most certainly is not the case at least in moths and aphids (Taylor et al. 1980, see also 1978). Accepting that spatial variance is proportionally as large in rare as in abundant species, local extinctions are bound to occur (MacArthur and Wilson 1967), and distribution is, at least in rare species, less than the possible maximum.

3. Models of distribution

The paradigm for the dynamics in local abundance is still the logistic equation (Lotka 1925)

$$\frac{dN}{dt} = rN(1-N/K), \quad (1)$$

while perhaps the only similarly general model proposed for distribution is Levins's (1969a)

$$\frac{dp}{dt} = ip(1-p)-ep, \quad (2)$$

where p, a measure of distribution, is the fraction of population sites occupied by the species ($0 \leq p \leq 1$), and i and e are constants for a given species in a given environment. The first term in this equation is the rate of colonization of empty sites, and the second term is the rate of local extinctions. When all suitable sites in the region are occupied, p equals 1. The single internal equilibrium of Eq. (2) is stable, $\hat{p} = 1-e/i$, and regional extinction follows if $e \geq i$.

Levins (1970) subsequently analysed the stochastic version of Eq. (2): the extinction parameter, e, was assumed to be a random variable, with mean $\bar{e}$ and variance σ_e^2. Assuming no autocorrelation ("white noise"), the diffusion equation method (Kimura 1974) may be used to analyse the distribution of p, and gives (Levins 1970),

$$\Phi(p) = Cp^{2(i-\bar{e})/\sigma_e^2-2} \exp(-2ip/\sigma_e^2), \quad (3)$$

as the limiting ($t \to \infty$) distribution of $\Phi(p,t)$. This does not depend on the initial value, $p(0)$. Constant C is necessary to guarantee that $\int_0^1 \Phi(p)dp = 1$. Critical points of Eq. (3) may be found from the equation,

$$2M_{\delta p} - d/dp\ V_{\delta p} = 0, \quad (4)$$

where $M_{\delta p}$ and $V_{\delta p}$ are the mean and variance of the rate of change in the stochastic version of Eq. (2). This gives the condition,

$$i > \bar{e} + \sigma_e^2, \quad (5)$$

for a unimodal distribution $\Phi(p)$ with a peak at $p =$ $1-(\bar{e}+\sigma_e^2)/i$. If (5) does not hold, $\Phi(p)$ is a decreasing function between 0 and 1. The deterministic equilibrium, obtained when $\sigma_e^2 = 0$, is always equal to or greater than the stochastic mode. It should be noted that there are two interpretations for $\Phi(p)$ (Kimura 1964). $\Phi(p)$ gives the distribution of p both for a single species during a long period of time, and for a community of similar species at a given moment.

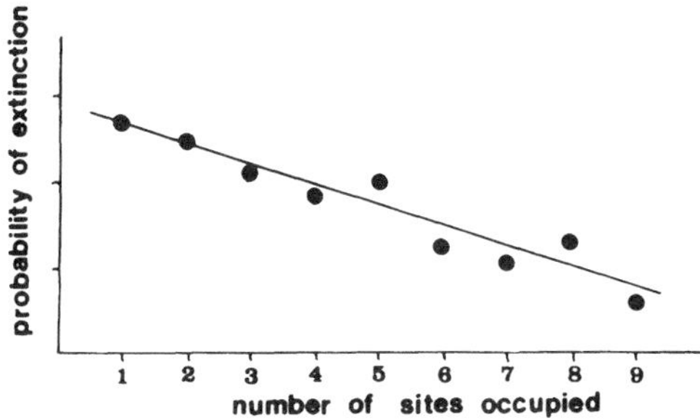

Fig. 5. Relationship between the probability of local extinction (per year) and distribution in Simberloff's (1976) data on mangrove island insects. Maximum number of sites (small mangrove islands) was nine. Each dot represents the average for several species (the line has been drawn by eye).

The assumption that all the local populations are the same, implicit in model (2), is very unrealistic. To model local dynamics at each population site explicitly is out of question (though see DeAngelis et al. 1979), but the relationship found in Sect. 2 provides us with an approximative, yet qualitatively correct, non-constant relationship between p and the "average state" (abundance) in a local population: average local abundance increases with increasing p.

Probability of local extinction increases with decreasing population size (e.g. MacArthur and Wilson 1967, Christiansen and Fenchel 1977). One would expect, therefore (see Sect. 2), that e in Eq. (2) is not constant, but decreases with p. I have found 3 sets of data to test this prediction.

A re-analysis of Simberloff's (1976) results on extinction of local (island) populations of mangrove island insects shows that e in Eq. (2), a parameter related to the probability of local extinction, decreases with increasing p (Fig. 5). The same result was obtained from a similar analysis of Kontkanen's (1950) data on leafhoppers in meadows in East Finland (Fig. 6).

The third example is from Boycott's (1930, see also 1919 and 1936) study on fresh-water molluscs in small ponds in the parish of Aldenham in England. Almost a hundred ponds were surveyed for molluscs and plants in 1915 and 1925. This example is particularly important because the small size of the ponds enabled Boycott (1930: 2–3) to make accurate censuses. The extinctions observed are thus real. (Simberloff (1976) also tried to document all the populations of each island, while

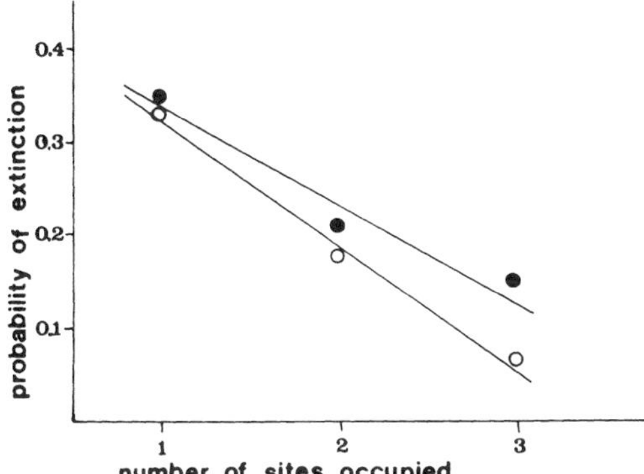

Fig. 6. Relationship between the probability of local extinction and distribution in Kontkanen's (1950) data on leafhoppers in East Finland. Maximum number of sites (meadows) was three. Black dots represent the average for several species and the probability of extinction in five years; open circles give the corresponding annual extinction probabilities (these results indicate that "re-colonization" was frequent, which is partly an artefact of the relatively small sample size, leaving some small populations unnoticed; cf. Kontkanen 1950).

Kontkanen (1950) probably missed many small populations.) My re-analysis (Fig. 7) of Boycott's (1930) data closely agrees with the above results: e is not constant but decreases with p. This result is not quite accurate, because more than one extinction-colonization event may have taken place in 10 years (cf. Diamond and May 1977), but the trend is very clear.

At present we may accept the simplest hypothesis about the rate of extinction: $e'(1-p)p$ (note that $e' \approx e$ when p is small). On this assumption Eq. (2) is replaced by

$$\frac{dp}{dt} = ip(1-p)-e'p(1-p). \qquad (6)$$

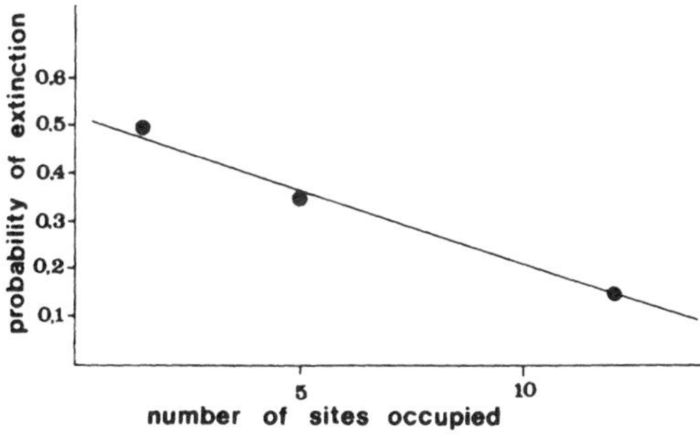

Fig. 7. Relationship between the probability of extinction and distribution in Boycott's (1930) data on fresh-water molluscs in small ponds in England. This figure gives the "net" extinction rate in 10 years. While constructing the figure, I have included the 34 ponds in Boycott's (1930) classes A and B which did not dry up during the study period. Species have been divided into three distribution classes, namely 1 or 2, 3 to 7, and 10 to 14 ponds occupied, and the dots in the figure give the average for each group (altogether there were 16 species).

I shall next explore two simple assumptions about the colonization process.

(1) If i in Eq. (6) does not depend on p, we obtain the logistic equation (1). This has been analysed, in the ecological context, by Levins (1969b; see also May 1973 and Leigh 1975). In the deterministic case, there is one stable equilibrium point, either at $\hat{p} = 1$, if $i > e'$, or at $\hat{p} = 0$, if $i < e'$. For the stochastic version, assume that $s \equiv i - e'$ is a random variable, and that there is no autocorrelation. The diffusion equation method gives,

$$\Phi(p) = Cp^{2(\bar{s}/\sigma_s^2-1)}(1-p)^{-2(\bar{s}/\sigma_s^2+1)}. \qquad (7)$$

$\Phi(p)$ is bimodal if $\sigma_s^2 > \bar{s}$. If, on the other hand, $\bar{s} > \sigma_s^2$, all populations approach maximal distribution, $p = 1$.

(2) Let us assume that the rate of colonization is $(s'p+e')p(1-p)$; the model then becomes,

$$\frac{dp}{dt} = s'p^2(1-p), \quad s' > 0. \qquad (8)$$

Evidently, there is only one stable equilibrium point, $\hat{p} = 1$. If s' is a random variable, and there is no autocorrelation, we can again use the diffusion equation method, which gives,

$$\Phi(p) = C\exp(-2\bar{s}'/\sigma_{s'}^2 p)p^{2(\bar{s}'/\sigma_{s'}^2-2)}(1-p)^{-2(\bar{s}'/\sigma_{s'}^2+1)}. \qquad (9)$$

This distribution is bimodal if $\bar{s}' < \sigma_{s'}^2/3$. If the mean is greater than a third of the variance, all populations become maximally distributed.

A biological justification for assumption (2) about the rate of colonization is the probably increasing number of emigrants with increasing local abundance (e.g. Dempster 1968, Johnson 1969); presumably, more emigrants means more colonizations.

Addendum

During the preparation of this paper it escaped my notice that there may be certain mathematical problems in the use of the diffusion equation technique in the analysis of the models in Sect. 3 (Levins's 1970 analysis is erroneus; see Boorman, S. A. and Levitt, P. R. 1973. Theor. Pop. Biol. 4: 85–128; and see Roughgarden, J. 1979, pp. 384–391. Theory of population genetics and evolutionary ecology: an introduction. MacMillan). A supplementary numerical analysis of a discrete time version of Eq. (6) indicates, nonetheless, that the result presented here is qualitatively correct (Hanski 1982).

4. Ecological appraisal

The present modification of Levins's model led to a radically different conclusion from the one originally drawn by Levins: assuming stochastic variation in the rate of local extinction and/or colonization, populations

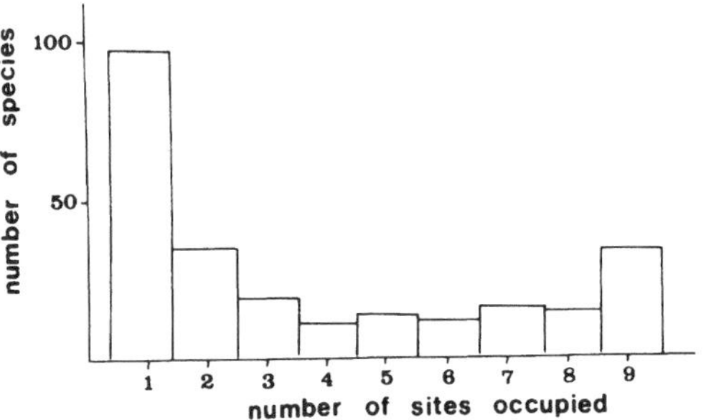

Fig. 8. Frequency distribution of species' distributions in Simberloff's (1976) data on mangrove island insects. Maximum number of sites was nine.

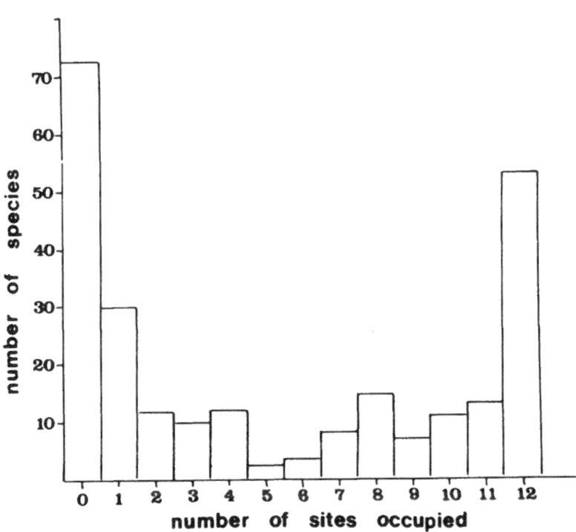

Fig. 10. Frequency distribution of species' distributions in Linkola's (1916) data on anthropochorous plants in Russian Karelia, U.S.S.R. The sites in question are isolated old houses and small villages, numbering 12 (sites 8 to 19 in Linkola 1916).

tend towards either one or the other of the (deterministic) boundary equilibria, $\hat{p} = 0$ and $\hat{p} = 1$, while in Levins's model p would hover around a stable internal equilibrium, $0 \leq \hat{p} \leq 1$. Which result more correctly reflects reality?

We recall that $\Phi(p)$ may be interpreted either as the distribution of p values in one species in a long period of time, or as the distribution of p values in many similar species at one moment of time (cf. Kimura 1964 for analogous interpretations in population genetics). To test the latter qualitatively, we require that all the species may establish local populations at the same sites, and that interspecific influences on model parameters are density- and frequency-independent. My model then predicts bimodality of p's, peaks close to unity and zero, while Levins's model predicts unimodality with the peak not very close to unity or zero.

Simberloff's (1976) data (cf. Fig. 5) support the present model; the distribution of the number of mangrove islands occupied by different species of insects appears bimodal (Fig. 8). To test this formally, we observe that

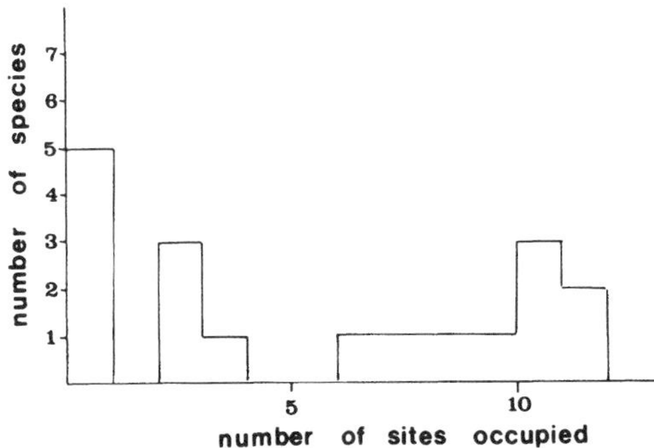

Fig. 9. Frequency distribution of species' distributions in *Onthophagus* in tropical lowland forest in Sarawak (Hanski unpubl.). Maximum number of sites was 12.

12 species were found from 6 islands, and 32 species from all 9 islands. The null hypothesis that the number of species is equal in these two classes is rejected: $\chi^2 = 9.09$, $P < 0.01$.

Onthophagus species in the lowland rain forest in Sarawak (cf. Fig. 4) also show a clear dichotomy into two sets of species (Fig. 9). I shall use the term "core" species for the locally abundant and regionally common species, and the term "satellite" species for locally and regionally rare species. In the case of mangrove island insects (Fig. 8), the same terms can be used, although "intermediate" species are now frequent.

Another data set to test this prediction is due to a study by Linkola (1916) on the occurrence of anthropochorous vascular plants near houses and villages, isolated by natural forest, in Russian Karelia (then Finland, study area ca. 10000 km^2). There was a clear size effect, large villages having more species than small ones (Linkola 1916), and colonization of isolated houses and small villages was perhaps not random, because some species were lacking systematically from them (though some would do so by chance only). For these reasons, I have restricted the analysis to 12 similar sites in Linkola's (1916) material. The frequency distribution of occurrences at the 12 sites is clearly bimodal (Fig. 10), which strongly supports the present model (see Hanski 1982 for a full analysis of Linkola's material).

It suffices to mention here that Kontkanen's (1950) results on leafhoppers and Anasiewicz's (1971) results on bumblebees also support the present model. A full analysis of these two studies will be presented elsewhere (Hanski unpubl.).

5. More about testing the core-satellite hypothesis

The core-satellite species hypothesis is a simple null hypothesis to explain regional rarity. There are two important premises in the model: (1) regional population dynamics *is* important, and (2) stochasticity in regional population dynamics *is* important. Stochastic variation in the parameters of regional dynamics (extinction, colonization) may be due to either demographic or environmental stochasticity.

As the examples given in the previous section showed, testing the main prediction of the model is simple: Is the distribution of species' regional distributions bimodal? If the distribution is clearly bimodal, there are grounds for a dichotomy, and for the use of the concept in a strong sense. Otherwise, one is left with the option of labelling the opposite ends of a continuum, like the r-K species distinction (MacArthur and Wilson 1967) has been replaced by the r-K species continuum (Pianka 1970, 1972, 1974, Southwood 1977).

Other models than the present one may predict a bimodal distribution of regional distributions, and to more rigorously test this model necessitates intensive studies on the rates of local extinction and colonization. If data are available on the rate of change in distribution, validity of Eq. (6) may be tested directly. Alternatively, one may try to document changes in species' status from the core to the satellite class, or vice versa, between regions or in time. Such changes are predicted to occur even if the pattern of environmental stochasticity remains stationary. An alternative model, which we may call an "adaptation" hypothesis, states that core species are better adapted to the environment than are satellite species, and does not predict changes from core to satellite class, or vice versa. Note that also the present model allows for interspecific differences in $\bar{s}$ and σ_s^2 (Sections 3 and 8).

L. R. Taylor's work (1974, 1978, Taylor and Taylor 1977, Taylor et al. 1978, 1980, Taylor and Woiwod 1980) on insect abundance and distribution has demonstrated the ever-changing patterns of regional distributions, anticipated by Andrewartha and Birch (1954), and Taylor's work has shown the interdependence between abundance and distribution. His results imply that local extinctions and colonizations are frequent phenomena (see also Den Boer 1977, Ehrlich 1965, Ehrlich et al. 1980), and that spatial population dynamics is important in all species.

The gravest difficulty in testing the core-satellite species hypothesis in a multispecies context is habitat selection. How does one identify the "population sites" suitable for the species? The reason for insisting on "similar species" in testing the multispecies prediction is just this; if the species are so similar that they have similar habitat selection, there is no problem. Then the number of potentially inhabitable sites is the same and the denominator in calculating the p's is the same for all species. Identical species are, of course, an unattainable abstraction, but in many groups of closely related species habitat selection may be sufficiently similar. An extension of the core-satellite hypothesis into analyses of niche relations (in Sect. 8) requires, in any case, a distinction between habitat and niche (Whittaker et al. 1973).

There is the danger that counter-evidence (unimodal distribution of p's) is dismissed on the basis that the species were not, after all, similar enough. This would be wrong; the question about habitat selection must be resolved before the test is performed. In any case it is safest to include only similar population sites (habitat patches) in the analysis, which, to some extent, removes this problem. These restrictions do not mean that this theory could not work on any species. Careful selection of the species and of the sampling sites is required only because of problems in testing the hypothesis.

To repeat, the core-satellite hypothesis should be tested only with sets of species which may establish populations at the same sites; or, if such data are available, with records for single species in the long course of time. I believe that the former requirement was fulfilled in the above examples. A counter-example is the distributional ecology of water-striders *(Gerris)*. Vepsäläinen has shown, in a series of papers (see especially 1973, 1974a, b, 1978, Järvinen and Vepsäläinen 1976), that *Gerris* species, nine of which occur in Finland, show significant ecological and morphological differences in their adaptations to living in different kinds of lakes, ponds, streams, etc., including wing dimorphism in many species. In this case habitat selection is clearly different in different species, and the core-satellite hypothesis should *not* be used for the whole set of species, though it could be used for each species separately. In the latter context, one could talk about core and satellite populations, and comparisons should be made between regions or times.

Assuming the reality of core and satellite species, one may rephrase Hutchinson's (1959) question and ask: Why, in a given community, are there n core and m satellite species? Constraints on core and satellite species diversity are entirely different, which warrants two questions (n and m) instead of one ($n+m$). Nevertheless, satellite species may become, besides regionally extinct, also core species, and a core species can move to regional extinction only through a stage as a satellite species. Therefore the numbers of species in the two kinds are not independent of each other.

A specific prediction may be derived for the most universal trend in species diversity, namely that the number of species tends to increase with area, whether the region in question is an island or part of the mainland. Because satellite species survive as a set of small populations, their regional existence should hinge on the size of the region (e.g. Hanski 1981a). Therefore, as regions – e.g. islands – become smaller, the proportion of satellite species should decline. If Diamond's (1975, see also 1971, 1973, and others) series of species from

D-tramps to High-S species is matched with the continuum from core to satellite species, the proportion of satellite species indeed decreases with decreasing island size. Diamond (1975: 358), however, considers his D-tramps to be "r-selected", and High-S species to be "K-selected", which is in contrast with the present conceptualization – core species are certainly not "r-selected".

6. The core-satellite hypothesis and r-K theory

There exists a common basis for the core-satellite hypothesis and the by now well established r-K species theory (MacArthur and Wilson 1967, Gadgil and Bossert 1970, Pianka 1970, 1972; but see Wilbur et al. 1974, Southwood 1977, Christiansen and Fenchel 1977, Schaffer 1979). Both hypotheses stem from the same model – the logistic equation – which has been applied at the level of local abundance in the r-K model, and at the level of regional distribution in the core-satellite model.

Nonetheless, the r-K species concept is used in a deterministic fashion to predict properties of species from the properties of their environment (e.g. Pianka 1970, 1974, Southwood 1977, Vepsäläinen 1978), while the core-satellite distinction is caused in the model by stochastic variation in spatial population dynamics. Unlike the satellite species, r-species are thought to be frequently locally abundant in comparison with K-species, but this does not follow from the mathematical model (logistic equation).

Although the two concepts are fundamentally different, core species are related to K-species, and, less obviously, satellite species are related to r-species.

7. A historical perspective

It is common in ecology that authors – or their readers – find "new" ideas preceded by earlier workers (Hutchinson 1978, McIntosh 1980), and nowadays preferable by Darwin. This may be viewed as a mark of soundness in the argument – or is McIntosh (1962) correct in claiming that "certain ideas seem to be invulnerable to attack and persist although subjected to multiple executions"? The core-satellite hypothesis is not an exception to the rule. The irony here is that the idea McIntosh was executing in 1962 was nothing else but bimodality of the distribution of spatial occurrences – the very prediction from the models in Sect. 3.

G. F. Gause (1936a: 323, see also 1936b) wrote: "The most important structural property of biocoenosis is the existence of definite quantitative relations between the abundant species and the rarer ones." One such relation, which Gause (1936a) discussed at length, is Raunkiaer's "law of frequency" (Raunkiaer 1913, 1918, 1934; a pioneering work by Jaccard in 1902), which has been much used especially in plant ecology until the 1960's (e.g. Oosting 1956, Hanson and Churchill 1961, Mueller-Dombois and Ellenberg 1974), and which is of special relevance here. To see this, divide p into 5 segments of equal length (–0.2, 0.21–0.4, etc.), and denote by A to E the numbers of species falling into the 5 classes. Raunkiaer's "law of frequency" states that $A > B > C \gtreqless F < E$. Quite unexpectedly, the simple theory suggested in Sect. 3 predicted Raunkiaer's "law of frequency".

Nonetheless, with papers by Gleason (1920, 1929) and Romell (1930), criticism of the "law of frequency" started to accumulate (Gause 1936a, Preston 1948, Williams 1950), culminating in the above-mentioned "execution" by McIntosh (1962). It had been shown that the frequency distribution of species' frequencies depends, in Williams's (1950) words, on "the number of quadrats, the size of the quadrats, and on the Index of Diversity of the population." This criticism is justified. In view of the connection to the core-satellite hypothesis, one significant difference between the "laws" should be pointed out (see also Hanski 1982).

Frequency is the fraction of (usually small) samples, typically quadrats, in which the species occurs, all samples having been taken from the same homogeneous community. *Distribution,* as it was defined in the introduction and used in the models (Sect. 3), is a measure of occurrence on the between-site scale. Although the "true" population level may be difficult to specify (for an extreme example see Brussard and Ehrlich 1970a, b), the distinction between distribution and frequency is an important one whenever regional population dynamics are important, i.e. whenever many local populations are studied. It has been pointed out that the highest (E) of Raunkiaer's frequency classes is more inclusive than the lower ones, because the frequency classes include unequal density classes (Gleason 1929, Ashby 1935, McIntosh 1962). But unlike between density (abundance) and frequency classes in homogeneous communities, there is no simple statistical relationship between distribution and local abundance, the correlations in Figs 1 to 4 (Sect. 2) being due to ecological processes (notwithstanding problems of sampling; Sect. 2). In fact, the purpose of using the "law of frequency" was to determine the homogeneity of a stand of vegetation (or a community of animals; see e.g. Kontkanen 1950); bimodality ($D < E$) was namely expected only in homogeneous stands, which is an interesting convergence to my independently thought requirement of similar habitat selection in the species to be analysed (Sect. 5).

8. Concluding remarks: visiting Hutchinson's niche space

After these observations and theorizing, the reader may ask: What is gained by calling regionally common and locally abundant species core species, and rare species satellite species?

My answer is twofold. If the frequency distribution of species' regional distributions is indeed bimodal, this is interesting for its own sake, because it appears not to be the null hypothesis for many ecologists, who rather expect the kind of unimodality predicted by Levins's model. Secondly, and more importantly, if such a dichotomy exists in many natural communities, this should help us to provide a functional explanation for patterns of abundance and distribution. To take an example, if the core-satellite hypothesis is upheld, one may proceed by restricting the application of the equilibrium theory (MacArthur 1972, May 1973, 1976) to the core species, and employing appropriate non-equilibrium models for the satellite species. Caswell (1978) presumably had a similar idea in mind when he, after discussing the virtues of equilibrium and non-equilibrium models in ecology, conjectured: "Perhaps a community consists of a core of dominant species, which interact strongly enough among themselves to arrive at equilibrium, surrounded by a larger set of non-equilibrium species playing their roles against the background of the equilibrium species."

In the introduction I referred to a third structural property of communities besides abundance relations (Engen 1978) and spatial distributions (Simberloff 1978): distribution of species, or strictly speaking their "niches", in Hutchinson's (1957) niche space. The perennial question is how well spaced-out niches are in niche space. Intuition says and theory (e.g. MacArthur 1972, Lawlor and Maynard Smith 1976) predicts that interspecific competition causes better spacing-out, and ultimately and ideally leads to a uniform distribution of niches in niche space. In view of the controversy about the importance of competition in structuring communities (Paine 1966, Harper 1969, Janzen 1970, Dayton 1971, Connell 1975, 1978, Caswell 1976, Glasser 1979), this is an important question. The problem is that, in practice, other factors besides competition come into the play, making any "test" difficult. It is not surprising, therefore, that this kind of argument has led to widely varying conclusions (MacArthur 1972, Schoener 1974, Sale 1974, Inger and Colwell 1977, Southwood 1978, Strong et al. 1979, Pianka et al. 1979, Hanski 1979a, Lawlor 1980). The difficulty is in the formulation of a proper null hypothesis (see especially Lawlor 1980; the null hypothesis is *not* necessarily a random distribution of niches in niche space, Hanski 1979a, Grant and Abbott 1980), and in the multitude of factors potentially – and in practice – causing changes in niche position.

This is where the core-satellite hypothesis may prove useful. The following heuristic argument shows that there exists, after all, at least one unequivocal null hypothesis: if interspecific competition is important in structuring communities, core species should be better spaced-out in niche space than satellite species.

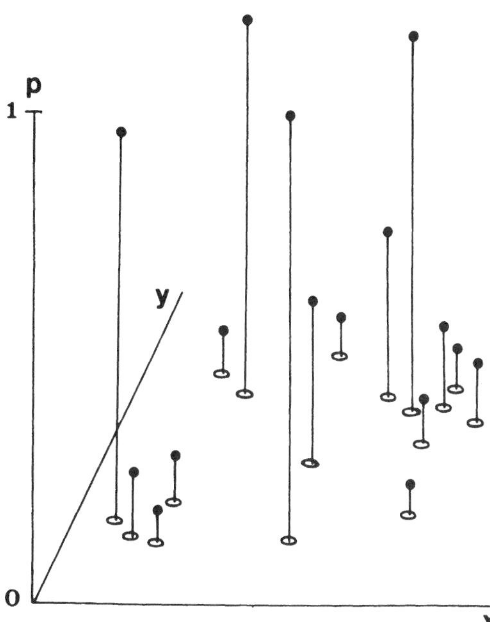

Fig. 11. A schematic representation of the hypothesis, explained in Sect. 8, that core species are better spaced-out in niche space than a random sub-set of the same size of all species, and better spaced-out than satellite species. x and y are two niche dimensions, and p denotes distribution, as in the rest of the paper, which varies from 0 to 1. Open circles represent niche positions, and the black dots give the position of species in the 3-dimensional space. It should be recalled that the argument is stochastic (see text), and the situation depicted in this figure is a static picture of a dynamic process.

Recall that the model is:

$$\frac{dp}{dt} = s_i p_i (1 - p^i), \quad i = 1, \ldots, n \qquad \text{(Eq. 6)}$$

where n is the number of species, and s_i is a random variable with mean $\bar{s}_i$ and variance $\sigma^2_{s_i}$. The probability that species i is a core species at a given time is an increasing function of $\bar{i}_s/\sigma^2_{s_i}$. How does interspecific competition influence this ratio? Competition should increase the rate of extinction, it should decrease the rate of colonization – hence competition will decrease $\bar{s}_i$ – and it will probably increase $\sigma^2_{s_i}$. Consequently, $\bar{s}_i/\sigma^2_{s_i}$ will decrease, and the probability of species i staying/becoming a satellite species increases. The closer the competitor(s), the stronger the effect. Naturally, if there are two close competitors, both of which are core species, this model only predicts that one of them is likely to become a satellite species. The stochastic nature of the single species model is preserved in the multispecies context.

We may visualize species in a space constructed of Hutchinson's (1957) niche space and of one extra axis, giving the extent of spatial distribution, p, which correlates, as we saw in Sect. 2, with abundance (Fig. 11). The greater the density of other species in the neighbourhood of any species in niche space, the smaller the probability that this species is, at a given time, a core species. Hence, core species are not expected to be a random sub-set of all the species with respect to niche position; we expect core species to comprise such a subset within which species are better spaced-out from each other than species are within a truly random sub-set. It follows that core species are also better spaced-out than satellite species.

A within-community analysis, such as suggested here, provides a concrete point of reference to test null hypotheses about community structure. Still, the present model cannot be but one step towards better understanding of community structure and its evolution. For instance, one could argue that even if core species are better spaced-out than satellite species in a community, this is perhaps not due to competition but to predation. Other studies are necessary to establish which explanation is correct. Theoretical work is also needed to clarify the expectations in multispecies communities. The value of the present approach can only be judged after several ecologists have tried it independently on their own data. Tests on dung beetles (Hanski 1980a) and bumblebees (Hanski unpubl.) have given encouraging results (see also Hanski 1979b, 1980b).

Acknowledgements – I wish to thank several colleagues for their useful comments on the different versions of the manuscript: R. Alatalo, J. Cartes, E. Connor, B. Don, Y. Haila, O. Järvinen (twice), Y. Mäkinen, E. Ranta, U. Safriel, D. Simberloff, and K. Vepsäläinen. O. Prŷs-Jones drew my attention to papers by Gause.

References

Anasiewicz, A. 1971. Observations on the bumble-bees in Lublin. – Ekol. Polska 19: 401–417.

Andrewartha, H. G. and Birch, L. C. 1954. The distribution and abundance of animals. – Chicago Univ. Press, Chicago.

Ashby, E. 1935. The quantitative analysis of vegetation. – Ann. Botany 49: 779–802.

Bartlett, M. S. 1975. The statistical analysis of spatial patterns. – Monogr. Appl. Prob. Statistics, Chapman and Hall, London.

Boycott, A. E. 1919. The freshwater mollusca of the parish of Aldenham: an introduction to the study of their oecological relationships. – Trans. Hertf. Nat. Hist. Soc. 17: 153–200.

– 1930. A re-survey of the fresh-water mollusca of the parish of Aldenham after ten years with special reference to the effect of drought. – Trans. Hertf. Nat. Hist. Soc. 19: 1–25.

– 1936. The habitats of fresh-water Mollusca in Britain. – J. Anim. Ecol. 5: 116–180.

Brussard, P. F. and Ehrlich, P. R. 1970a. Adult behaviour and population structure in *Erebia epipsodea* (Lepidoptera: Satyridae). – Ecology 51: 880–885.

– and Ehrlich, P. R. 1970b. The population structure of *Erebia epipsodea* (Lepidoptera: Satyridae). – Ecology 51: 119–129.

Caswell, H. 1976. Community structure: a neutral model analysis. – Ecol. Monogr. 46: 327–354.

– 1978. Predator-mediated coexistence: a non-equilibrium model. – Am. Nat. 112: 127–154.

Christiansen, F. B. and Fenchel, T. M. 1977. Theories of populations in biological communities. – Springer, Berlin.

Cohen, J. E. 1970. A Markov contingency-table model for replicated Lotka-Volterra systems near equilibrium. – Am. Nat. 104: 547–560.

Comins, H. N. and Hassell, M. P. 1979. The dynamics of optimally foraging predators and parasitoids. – J. Anim. Ecol. 48: 335–353.

Connell, J. H. 1975. Some mechanisms producing structure in natural communities. – In: Cody, M. L. and Diamond, J. M. (eds.), Ecology and evolution of communities, The Belknap Press of Harvard Univ. Press, Cambridge.

– 1978. Diversity in tropical rain forests and coral reefs. – Science 199: 1302–1310.

Crowley, P. H. 1979. Predator-mediated coexistence: an equilibrium interpretation. – J. Theor. Biol. 80: 129–144.

Dayton, P. K. 1971. Competition, disturbance, and community organization: the provision and subsequent utilization of space in a rocky intertidal community. – Ecol. Monogr. 41: 351–389.

DeAngelis, D. L., Travis, C. C. and Post, W. M. 1979. Persistance and stability of seed-dispersed species in a patchy environment. – Theor. Pop. Biol. 16: 107–125.

Dempster, J. P. 1968. Intra-specific competition and dispersal: as exemplified by a psyllid and its anthocorid predator. – In: Southwood, T. R. E. (ed.), Insect abundance. Symp. R. Ent. Soc. London, No. 4, Blackwell Scient. Publ., Oxford.

Den Boer, P. J. 1977. Dispersal power and survival. Carabids in a cultivated countryside. – Misc. papers 14, Land. Wageningen, The Netherlands.

Diamond, J. M. 1971. Comparison of faunal equilibrium turnover rates on a tropical and a temperate islands. – Proc. Natl Acad. Sci. USA 68: 2742–2745.

– 1973. Distributional ecology of New Guinea birds. – Science 179: 759–769.

– 1975. Assembly of species communities. – In: Cody, M. L. and Diamond, J. M. (eds.), Ecology and evolution of communities, The Belknap Press of Harvard Univ. Press, Cambridge.

– and May, R. M. 1977. Species turnover rates on islands: dependence on census interval. – Science 197: 266–276.

Engen, S. 1978. Stochastic abundance models. – Chapman and Hall, London.

Ehrlich, P. R. 1965. The population biology of the butterfly *Euphydryas editha* II. The structure of the Jasper Ridge colony. – Evolution 19: 327–336.

– Murphy, D. D., Singer, M. C., Sherwood, C. B., White, R. R. and Brown, I. L. 1980. Extinction, reduction, stability and increase: the responses of checkerspot butterfly *(Euphydryas)* populations to the California drought. – Oecologia (Berl.) 46: 101–105.

Gadgil, M. and Bossert, W. 1970. Life history consequences of natural selection. – Am. Nat. 104: 1–24.

Gause, G. F. 1936a. The principles of biocoenology. – Quart. Rev. Biol. 11: 320–336.

– 1936b. Some basic problems in biocoenology. (In Russian, summary in English) – Zool. Zh. 15: 363–381.

Glasser, J. W. 1979. The role of predation in shaping and maintaining the structure of communities. – Am. Nat. 113: 631–641.

Gleason, H. A. 1920. Some applications of the quadrat method. – Bull. Torrey Bot. Club 47: 21–33.

– 1929. The significance of Raunkiaer's law of frequency. – Ecology 10: 406–408.

Grant, P. R. 1978. Dispersal in relation to carrying capacity. – Proc. Natl Acad. Sci. USA 75: 2854–2858.

– and Abbott, I. 1980. Interspecific competition, island biogeography and null hypotheses. – Evolution 34: 332–341.

Gurney, W. S. C. and Nisbet, R. M. 1978a. Predator-prey fluctuations in patchy environments. – J. Anim. Ecol. 47: 85–102.
– and Nisbet, R. M. 1978b. Single-species population fluctuations in patchy environments. – Am. Nat. 112: 1075–1090.
Hanski, I. 1979a. The community of coprophagous beetles. – D. Phil. thesis, Univ. of Oxford, Oxford (unpubl.).
– 1979b. Structure of communities. (In Finnish, summary in English.) – Luonnon Tutkija 83: 132–137.
– 1980a. The community of coprophagous beetles (Coleoptera, Scarabaeidae and Hydrophilidae) in northern Europe. – Ann. Ent. Fenn. 46: 57–74.
– 1980b. Structure of insect communities: the core and the satellite species. – In: Abstracts of the XVI Int. Congress of Entomology, Kyoto.
– 1981a. Exploitative competition in transient habitat patches. – In: Chapman, D. G., Gallucci, V. and Williams, F. M. (eds.), Quantitative population dynamics. Statistical ecology, Vol. 13, Intern. Co-op. Publ. House, Fairland, Maryland.
– 1981b. Coexistence of competitors in patchy environment with and without predation. – Oikos 37: 306–312.
– 1982. Distributional ecology of anthropochorous plants in villages surrounded by forest. – Ann. Bot. Fenn. 19 (in press).
Hanson, H. C. and Churchill, E. D. 1961. The plant community. – Reinhold, New York.
Harper, J. L. 1969. The role of predation in vegetation diversity. – Brookhaven Symp. Biol. 22: 48–62.
Hengeveld, R. 1979. On the use of abundance and species-abundance curves. – In: Ord, J. K., Patil, G. P. and Taillie, C. (eds.), Statistical distributions in ecological work. Statistical ecology, Vol. 4, Intern. Co-op. Publ. House, Fairland, Maryland.
Horn, H. S. and MacArthur, R. H. 1972. Competition among fugitive species in a harlequin environment. – Ecology 53: 749–752.
Hutchinson, G. E. 1957. Concluding remarks. – Cold Spring Harbour Symp. Quant. Biol. 22: 415–427.
– 1959. Homage to Santa Rosalia or Why are there so many kinds of animals? – Am. Nat. 93: 145–159.
– 1978. An introduction to population ecology. – Yale Univ. Press., New Haven, Conn.
Inger, R. F. and Colwell, R. K. 1977. Organization of contiguous communities of amphibians and reptiles in Thailand. – Ecol. Monogr. 47: 229–253.
Jaccard, P. 1902. Lois de distribution florale dans la zone alpine. – Bull. Soc. Vand. Sci. Nat. (Lausanne), 38.
Janzen, D. H. 1970. Herbivores and the number of tree species in tropical forests. – Am. Nat. 104: 501–528.
Järvinen, O. and Sammalisto, L. 1976. Regional trends in the avifauna of Finnish peatland bogs. – Ann. Zool. Fenn. 13: 31–43.
– and Vepsäläinen, K. 1976. Wing dimorphism as an adaptive strategy in water-striders *(Gerris)*. – Hereditas 84: 61–68.
Johnson, C. G. 1969. Migration and dispersal of insects by flight. – Methuen, London.
Karppinen, E. 1958. Untersuchungen über die Oribatiden (Acari) der Waldböden von Hylocomium-Myrtillus-Typ in Nord-Finnland. – Ann. Ent. Fenn. 24: 149–168.
Kimura, M. 1964. Diffusion models in population genetics. – J. Appl. Prob. 1: 177–232.
Kontkanen, P. 1937. Quantitative Untersuchungen über die Insektenfauna der Feldschicht auf einigen Wiesen in Nord-Karelien. – Ann. Zool. Soc. 3, No. 4.
– 1950. Quantitative and seasonal studies on the leafhopper fauna of the field stratum on open areas in North Karelia. – Ann. Zool. Soc. 13, No. 8.
– 1957. On the delimitation of communities in research on animal biocoenotics. – Cold Spring Harbour Symp. Quan. Biol. 22: 373–378.
Krebs, C. J. 1972. Ecology: the experimental analysis of distribution and abundance. – Harper & Row, New York.
Lawlor, L. R. 1980. Structure and stability in natural and randomly constructed competitive communities. – Am. Nat. 116: 394–408.
– and Maynard Smith, J. 1976. The coevolution and stability of competing species. – Am. Nat. 111: 79–99.
Leigh, E. G. Jr. 1975. Population fluctuations, community stability and environmental variability. – In: Cody, M. L. and Diamond, J. M. (eds.), Ecology and evolution of communities, The Belknap Press of Harvard Univ. Press, Cambridge.
Levin, S. A. 1974. Dispersion and population interactions. – Am. Nat. 108: 207–228.
– 1976. Population dynamic models in heterogeneous environment. – Ann. Rev. Ecol. Syst. 7: 287–310.
– 1977. Spatial patterning and the structure of ecological communities. – Some mathematical questions in biology, VII, Vol. 8. Am. Math. Soc., Providence, R. I.
– 1978. Pattern formation in ecological communities. – In: Steele, J. H. (ed.), Spatial pattern in plankton communities. Plenum, New York.
Levins, R. 1969a. Some demographic and genetic consequences of environmental heterogeneity for biological control. – Bull. Ent. Soc. Am. 15: 237–240.
– 1969b. The effect of random variation of different types of population growth. – Proc. Natl Acad. Sci. U.S. 62: 1061–1065.
– 1970. Extinction. – In: Gerstenhaber, M. (ed.), Some mathematical problems in biology. Lectures on mathematics in the life sciences 2. Am. Math. Soc., Providence, R. I.
– and Culver, D. 1971. Regional coexistence of species and competition between rare species. – Proc. Natl Acad. Sci. U.S. 68: 1246–1248.
Lidicker, W. Z. Jr. 1962. Emigration as a possible mechanism permitting the regulation of population density below carrying capacity. – Am. Nat. 96: 29–33.
Linkola, K. 1916. Studien über den Einfluss der Kultur auf die Flora in den Gegenden nördlich vom Ladogasee. I. – Acta Soc. Fauna et Flora Fenn. 45.
Lotka, A. J. 1925. Elements of physical biology. – Williams and Wilkins, Baltimore.
MacArthur, R. H. 1972. Geographical ecology. – Harper & Row, New York.
– and Wilson, E. O. 1967. The theory of island biogeography. – Princeton Univ. Press, Princeton.
May, R. M. 1973. Stability and complexity in model ecosystems. – Princeton Univ. Press, Princeton.
– (ed.) 1976. Theoretical ecology. Principles and applications. – Blackwell Sci. Publ., Oxford.
McIntosh, R. P. 1962. Raunkiaer's "law of frequency". – Ecology 43: 533–535.
– 1980. The background and some current problems in theoretical ecology. – Synthese 43: 195–255.
Mueller-Dombois, D. and Ellenberg, H. 1974. Aims and methods of vegetation ecology. – John Wiley, New York.
Oosting, H. J. 1956. The study of plant communities. – Freeman, San Francisco.
Ord, J. K., Patil, G. P. and Taillie, C. (eds.) 1980. Statistical distributions in ecological work. – Statistical ecology, Vol. 4. Intern. Co-op. Publ. House, Fairland, Maryland.
Paine, R. T. 1966. Food web complexity and species diversity. – Am. Nat. 100: 65–75.
Patil, G. P., Pielou, E. C. and Waters, W. E. (eds.) 1971. Spatial patterns and statistical distributions. – Statistical ecology, Vol. 1. Penn. State Univ. Press, Pennsylvania.
Pianka, E. R. 1970. On r- and K-selection. – Am. Nat. 104: 592–597.

– 1972. r and K selection or b and d selection? – Am. Nat. 106: 581–588.
– 1974. Evolutionary ecology. – Harper & Row, New York.
– Huey, R. B. and Lawlor, L. R. 1979. Niche segregation in desert lizards. – In: Horn, D. J., Mitchell, R. D. and Stairs, G. R. (eds), Analysis of ecological systems. Ohio State Univ. Press, Columbus.
Preston, F. W. 1948. The commonness and rarity of species. – Ecology 29: 254–283.
Raunkiaer, C. 1913. Formationsstätistiske Undersøgelser paa Skagens Odde. – Bot. Tidskr. Kobenhavn 33: 197–228.
– 1918. Recherches statistiques sur les formations vegetales. – Biol. Medd. 1: 1–80.
– 1934. The life forms of plants and statistical plant geography. – Clarendon Press, Oxford.
Romell, L. G. 1930. Comments on Raunkiaer's and similar methods of vegetation analysis and the "law of frequency". – Ecology 11: 589–596.
Sale, P. F. 1974. Overlap in resource use, and interspecific competition. – Oecologia (Berl.) 17: 245–256.
Schaeffer, W. M. 1979. The theory of life-history evolution and its application to Atlantic salmon. – Symp. Zool. Soc. London 44: 307–326.
Schoener, T. W. 1974. Resource partitioning in ecological communities. – Science 185: 27–39.
Simberloff, D. 1976. Experimental zoogeography of islands: effects of island size. – Ecology 57: 629–648.
– 1978. Colonization of islands by insects: immigration, extinction, and diversity. – In: Mound, L. A. and Waloff, N. (eds.), Diversity of insect faunas. Symp. R. Ent. Soc. London, No. 9, Blackwell Sci. Publ., Oxford.
Skellam, J. G. 1951. Random dispersal in theoretical populations. – Biometrika 38: 196–218.
Slatkin, M. 1974. Competition and regional coexistence. – Ecology 55: 128–134.
Smith, A. T. 1974. The distribution and dispersal of pika: consequences of insular population structure. – Ecology 55: 1112–1119.
Southwood, T. R. E. 1977. Habitat, the templet for ecological strategies. – J. Anim. Ecol. 46: 337–365.
– 1978. The components of diversity. – In: Mound, L. A. and Waloff, N. (eds.), Diversity of insect faunas. Symp. R. Ent. Soc. London, No 9, Blackwell Sci. Publ., Oxford.
Strong, D. R., Szyska, L. A. and Simberloff, D. S. 1979. Tests of community-wide character displacement against null hypotheses. – Evolution 33: 897–913.
Taylor, L. R. 1974. Monitoring change in the distribution and abundance of insects. – Report Rothamsted Exp. Station for 1973, Part 2.
– 1978. Bates, Williams, Hutchinson – a variety of diversities. In: Mound, L. A. and Waloff, N. (eds.), Diversity of insect faunas. Symp. R. Ent. Soc. London, No. 9, Blackwell Sci. Publ., Oxford.
– and Taylor, R. A. J. 1977. Aggregation, migration and population mechanics. – Nature, Lond. 265: 415–421.
– and Woiwod, I. P. 1980. Temporal stability as a density-dependent species characteristics. – J. Anim. Ecol. 49: 209–224.
– Woiwod, I. P. and Perry, J. N. 1978. The density-dependence of spatial behaviour and the rarity of randomness. – J. Anim. Ecol. 47: 383–406.
– Woiwod, I. P. and Perry, J. N. 1980. Variance and the large scale spatial stability of aphids, moths and birds. – J. Anim. Ecol. 49: 831–854.
Vandermeer, J. H. 1973. On the regional stabilization of locally unstable predator-prey relationship. – J. Theor. Biol. 41: 161–170.
Vepsäläinen, K. 1973. The distribution and habitats of *Gerris* Fabr. species (Heteroptera, Gerridae) in Finland. – Ann. Zool. Fenn. 10: 419–444.
– 1974a. The wing lengths, reproductive strategies and habitats of Hungarian *Gerris* Fabr. species (Heteroptera, Gerridae). – Ann. Acad. Sci. Fenn. A 202.
– 1974b. The life cycles and wing lengths of Finnish *Gerris* Fabr. species (Heteroptera, Gerridae). – Acta Zool. Fenn. 141.
– 1978. Wing dimorphism and diapause in *Gerris* – determination and adaptive significance. – In: Dingle, H. (ed.), The evolution of migration and diapause in insects. Springer, New York.
Whittaker, R. H., Levin, S. A. and Root, R. B. 1973. Niche, habitat, and ecotope. – Am. Nat. 107: 321–338.
Wiens, J. A. 1976. Population responses to patchy environments. – Ann. Rev. Ecol. Syst. 7: 81–120.
Wilbur, H. M., Tinkle, D. W. and Collins, J. P. 1974. Environmental certainty, trophic level, and resource availability in life history evolution. – Am. Nat. 108: 805–817.
Williams, C. B. 1950. The application of the logarithmic series to the frequency of occurrence of plant species in quadrats. – J. Ecol. 38: 107–138.
Yodzis, P. 1978. Competition for space and the structure of ecological communities. – Lecture notes in Biomath., Springer, Berlin.
Zeigler, B. P. 1977. Persistence and patchiness of predator-prey systems induced by discrete event population exchange mechanisms. – J. theor. Biol. 67: 677–686.

PAPER 31

On the Relationship between Abundance and Distribution of Species (1984)

J. H. Brown

Commentary

CHRISTY M. MCCAIN

J. H. Brown's paper, one of the most-cited foundational papers in macroecology, has had an enormous impact on the fields of ecology, evolutionary biology, biodiversity conservation, biogeography, and spatial modeling of populations. In fact, the citations continue to rise, by about 2.5 more citations a year, culminating with 64 citations in 2008 alone, and many of the papers that cite it are themselves pivotal works (50 papers have more than 100 citations each). What makes this paper such a cornerstone to modern theory and empirical research?

In a nutshell, this paper established many of the fundamental processes and underlying theories of species distributions and abundance patterns that are tested, assumed, and discussed in many biological fields. As Brown stated in the first paragraph, "Although it has long been recognized that abundance and distribution are intimately interrelated, the nature of this relationship has not been investigated systematically over the range of spatial scales from local populations to entire geographic ranges of species." This goal and the process of its exploration has now become a large part of the field of macroecology (Brown and Maurer 1989), and many of these themes have been revisited in subsequent books: Brown's *Macroecology* (1995) and K. J. Gaston and T. M. Blackburn's *Pattern and Process in Macroecology* (2000).

The first topic Brown tackled was the patterns and theory underlying the spatial variation in abundance within species. Building on earlier work (Grinnell 1922; Hengeveld and Haeck 1982; Whittaker 1960), he used various sources of data to demonstrate that population density across a species' range is greatest toward the range center and declines toward the range boundaries. He insightfully noted that the generality of the "abundant center" pattern varies based on sampling, gradient truncation, species abundance, environmental discontinuities, and competitor overlap. Brown stated two linked assumptions in his explanation: (1) a species' range corresponds to its environmental niche, and (2) those environmental niche variables are spatially autocorrelated. He argued that if these assumptions are true, then the range center encompasses the optimal environmental conditions and supports the highest abundances. At increasing distance from the range center the environmental conditions stray from the optimum and abundance declines. This "abundant center" pattern, he argued, can be defined by a normal probability density function. This became the underlying assumption of many spatially explicit population models and biodiversity conservation models, and was widely tested empirically, with recent work suggesting less generality than previously thought (Sagarin and Gaines 2002).

The second topic Brown tackled was the patterns and theory underlying the positive correlation between abundance and distribution among related species (Hanski 1982; Bock and Ricklefs 1983). He used various studies to document the generality that species with high population densities are widespread, whereas

From *American Naturalist* 124:255–79. *The American Naturalist* © 1984 The University of Chicago. Reprinted with permission from The University of Chicago.

rare species have restricted geographical distributions. After dismissing an explanation due to a statistical artifact, Brown stated an alternative: if a species is able to garner enough resources to support high numbers in a particular site, then such a species should be able to sustain smaller numbers in a large number of sites. Whereas a related species with a similar niche that cannot garner resources to support high numbers, then, should be restricted to fewer sites. These ideas were based on the two assumptions above and a third: similar and related species share substantial portions of their niches (i.e., niche conservatism). He noted that such an abundance-distribution relationship is mediated by relatedness, trophic level, and body size. These ideas and predictions have been widely tested empirically and theoretically, and are also the foundation of many population, biogeographic, and conservation models. His final discussion then proposed multiple tests and extensions of these ideas for population ecology, community ecology, and evolutionary biology that have captivated many researchers. Brown advocated how this large-scale approach deviates from previous ecological theory, an approach that has been embraced across scientific disciplines where employing a macroscale lens to their patterns and processes has led to powerful and interesting insights.

Literature Cited

Bock, C. E., and R. E. Ricklefs. 1983. Range size and local abundance of some North American songbirds: A positive correlation. *American Naturalist* 122:295–99.

Brown, J. H. 1995. *Macroecology*. University of Chicago Press, Chicago.

*Brown, J. H., and B. A. Maurer. 1989. Macroecology: The division of food and space among species on continents. *Science* 243:1145–50.

Gaston, K. J., and T. M. Blackburn. 2000. *Pattern and process in macroecology*. Blackwell Science, Oxford.

Grinnell, J. 1922. The role of the "accidental." *Auk* 39:373–80.

*Hanski, I. 1982. Dynamics of regional distribution: The core and satellite species hypothesis. *Oikos* 38:210–21.

Hengeveld, R., and J. Haeck. 1982 The distribution of abundance. I. Measurements. *Journal of Biogeography* 9:303–16.

Sagarin, R. D., and S. D. Gaines. 2002. The 'abundant centre' distribution: To what extent is it a biogeographic rule? *Ecology Letters* 5:137–47.

Whittaker, R. H. 1960. Vegetation of the Siskiyou Mountains, Oregon and California. *Ecological Monographs* 30:279–338.

Vol. 124, No. 2 The American Naturalist August 1984

ON THE RELATIONSHIP BETWEEN ABUNDANCE AND DISTRIBUTION OF SPECIES

James H. Brown

Department of Ecology and Evolutionary Biology, University of Arizona, Tucson, Arizona 85721

Submitted July 8, 1983; Accepted February 27, 1984

How environmental conditions and population processes determine the abundance and distribution of species is a central problem of ecology and biogeography. Although it has long been recognized that abundance and distribution are intimately interrelated, the nature of this relationship has not been investigated systematically over the range of spatial scales from local populations to entire geographic ranges of species. On a local scale, i.e., the small habitat patches that constitute most ecologists' study areas, the relationship between population density and spatial distribution of individuals has been studied by many population and community ecologists (e.g., Andrewartha and Birch 1954; Krebs 1978, and numerous references therein). Distribution on a large geographic scale has usually been regarded as the special province of biogeography, whose practitioners often have little experience or interest in population ecology (but see, e.g., Grinnell 1922; MacArthur 1972; Walter 1979; Rapoport 1982; Brown and Gibson 1983). Thus few investigators have systematically studied variation in population density over the geographic range of species. Recently, however, several authors have presented data that suggest a general relationship between local population density and spatial distribution on a geographic scale (e.g., Rabinowitz 1981; Hanski 1982*a*, 1982*b*, 1982*c*; Bock and Ricklefs 1983; J. T. Emlen et al., MS).

Here I reanalyze and synthesize some of the diverse information available on the relationship between abundance and distribution. These data suggest extremely general patterns within and among species that appear to hold for organisms as diverse as vascular plants, intertidal invertebrates, terrestrial arthropods, planktonic crustaceans, and terrestrial vertebrates. I develop a general theory to explain these relationships. This conceptual construct and the empirical observations that motivated it focus attention on problems that span the boundaries between the traditional disciplines of population ecology, community ecology, biogeography, population genetics, and evolution.

Clarification of terminology should facilitate understanding of what follows. The paper is concerned with the relationship between two attributes of populations and species: the density of individuals within a local area and the extent of the distribution of individuals in space. I shall often use the term *abundance* to

Am. Nat. 1984. Vol. 124, pp. 255–279.

refer to local population density, and the terms *rare* and *common* (or *abundant)* to describe extremes of density. Similarly, I shall often use *distribution* or *range* to refer to spatial distribution, and the terms *restricted* (or *local)* and *widespread* to describe the extremes.

THE PATTERNS

Spatial Variation in Abundance Within Species

We are all aware that within their geographic ranges all species are relatively numerous in some habitats and regions, whereas they are scarce or absent in others; the limit of the geographic range occurs where population density over large areas declines to zero. Is there any general pattern of spatial variation in abundance within the area in which a species normally occurs? The answer appears to be yes, density is greatest near the center of the range and declines, usually gradually, toward the boundaries. This pattern holds both within steep, geographically restricted gradients of environmental change, such as on mountainsides and within the intertidal zone, and over the entire geographic ranges of widespread species.

Examples of variation in population density over ecological gradients within local regions are shown in figure 1. These confirm the general pattern noted by Whittaker (1956, 1960, 1965) in his classical papers on patterns of vegetation: although individual species attain different maximum densities in different parts of the gradient, abundances of most species decline relatively gradually and symmetrically with increasing distance in either direction from their peaks. Of course some species attain their highest densities near one end of the measured gradient, so one tail (and sometimes probably the peak as well) of the distribution is missing. The generality of the pattern is demonstrated by the fact that it holds not only for plants in gradients of both moisture and elevation in different geographic regions (e.g., figs. 1*a*, 1*b*, 1*c*, 3, and other data in Whittaker [1956, 1960] and Beals [1969]), but also for invertebrates within the gradient of intertidal exposure, such as on a rocky shore in the northern Gulf of California (fig. 1*d;* see other data in Field and Robb [1970]).

Almost all of these plots of local density as a function of distribution in a gradient resemble normal curves, as indicated by the fact that Whittaker routinely fitted his data on plant distributions with Gaussian curves. Of course, sometimes at least one tail is missing when the distribution of the species extends beyond the measured gradient. Although not all these curves may represent exact normal distributions, they exhibit a strong central tendency and are neither highly skewed nor strongly leptokurtic or platykurtic. These empirical distributions obviously contain sampling error, so it is comforting that the curves appear smoother and more normal as the sample size is increased. The curves plotted in figure 1 are for abundant species, and as expected, distributions of the rarer species exhibit much more sampling error. Many of the distributions for common species do not differ significantly from normal distributions, whereas they are highly significantly different from random distributions (i.e., uniform, but with sampling error) of individuals within the gradients.

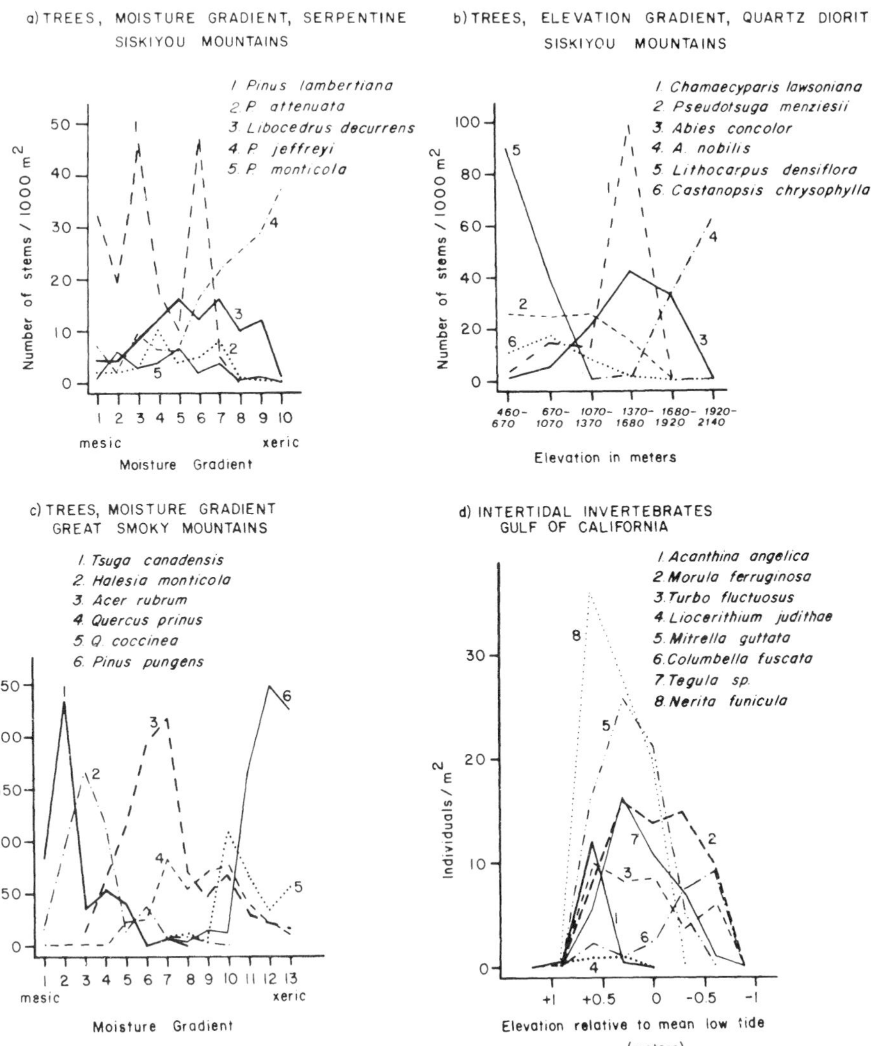

FIG. 1.—Distributions of abundances of terrestrial plant and intertidal mollusk species in local gradients of rapid ecological change: *a* and *b* data replotted from Whittaker (1960); *c* data replotted from Whittaker (1956); and *d* from E. H. Boyer, unpublished data. Note that most of the distributions seem to be surprisingly unimodal and symmetrical, except for those of species that obviously extend well beyond the measured gradient. The most abundant species in each gradient are depicted here; species were not selected on the basis of the pattern of the distribution of density.

The data on local population densities also suggest a gradual decline in abundance from the center to the boundaries of the geographic ranges of widespread species. The best data set available is for birds (but see Delcourt et al. 1981). D. Bystrak of the U. S. Fish and Wildlife Service has prepared maps, based on censuses from the Breeding Bird Survey, showing variation in abundance over the geographic ranges of many North American species. Two of these are shown in figure 2 (see also Bystrak 1979, fig. 5). Note that the greatest abundance of each species occurs near the center of its range, and population density declines gradually toward most boundaries. Even when an abrupt barrier of unsuitable habitat, such as the ocean, limits the distribution, abundance usually declines gradually as this boundary is approached. This general pattern of variation in abundance can be shown graphically by plotting density as a function of distance along four transects through the widest part of the range in four major compass directions and then averaging the values (fig. 2). A similar pattern was reported by J. T. Emlen et al. (MS) who sampled bird populations along a 850-km north-south transect of riparian forest habitat in the Mississippi Valley. They noted that, "The census data revealed convex density profiles for each species, curves that fluctuated considerably from station to station but tended to be level across range centers and slope peripherally to north and south boundaries at rates of up to 30% per degree of latitude." For 7 of the 19 species with northern or southern range boundaries within or close to their survey area, more than 50% of the variation in local density within the transect was related to latitude, with abundance always decreasing toward the northern or southern boundary. In contrast, for only 2 of 22 more widely distributed species that did not reach their northern or southern limits near the survey area was there an equally close correlation between density and latitude.

Although it is difficult to obtain quantitative data for organisms other than birds, the avian distributions exhibit a general pattern long recognized by naturalists (e.g., see Grinnell 1922): along a transect from the center to a boundary of its range a species tends to inhabit a progressively smaller proportion of local sample areas and habitats, and even without local regions where it does occur its average population density declines. Thus Rapoport (1982, fig. 6.3) used aerial photographs to document the decline in both density and frequency of occurrence of the palm tree *(Copernicia alba)* in 1-km^2 sample areas along a 113-km east-west transect through the boundary of its range in Argentina. It is hardly surprising that density within sites and frequency of occurrence among sites are closely interrelated, since both depend on the spatial scale of sampling. This is also true on more local scales, such as within steep ecological gradients. Figure 3*a* plots some of Whittaker's (1960) data for frequency of occurrence of the commonest herb species in replicated 1-m^2 samples along a moisture gradient in the Siskiyou Mountains of Oregon and California. Note that each species occurs in many sample sites in the center of its range, but becomes much more patchily distributed toward the periphery. Rare species (fig. 3*b*) show qualitatively similar patterns, but they inhabit only a small proportion of the sites even in the center of their distributions, and of course they exhibit much more sampling error.

That the pattern of abundance is greatest in the center of the distribution and

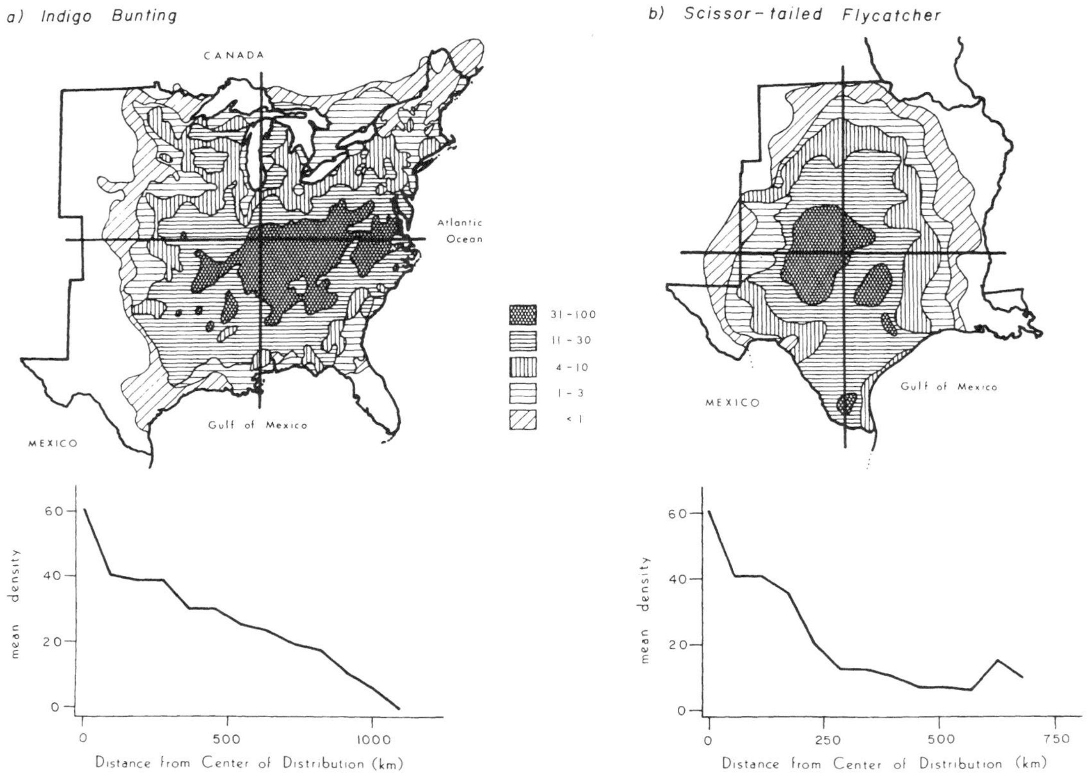

FIG. 2.—Distribution of abundance of two bird species, indigo bunting (*Passerina cyanea*) and scissortailed flycatcher (*Muscivora forficata*) over their geographic ranges. Above, mean population density per standardized census. Below, mean population density along four arbitrary transects running from the center of the range to the periphery as shown above. Compiled from maps drawn by D. Bystrak (see also Bystrak 1979, 1981) and based on the U.S. Fish and Wildlife Service's Breeding Bird Survey.

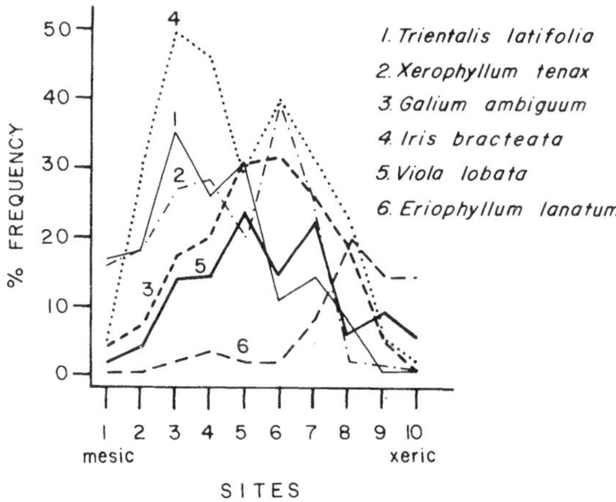

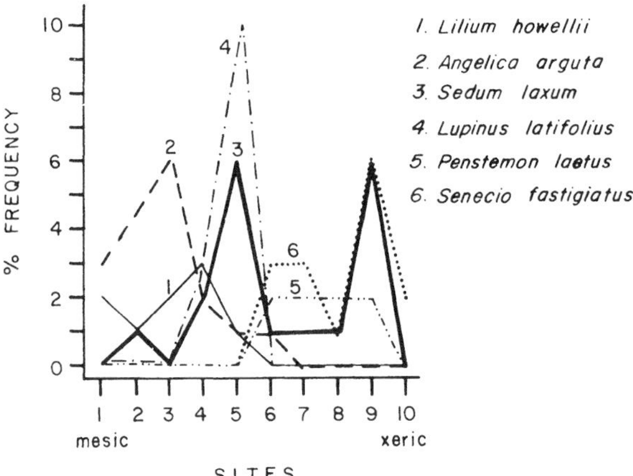

FIG. 3.—Distributions (% frequency in many small replicated samples) of (*a*) common and (*b*) rare herbs in the same moisture gradient in the Siskiyou Mountains, plotted from data in Whittaker (1960). Note that the species that occur in a large proportion of samples and presumably also attain high population densities exhibit approximately normal-shaped curves (allowing for missing tails that presumably extend beyond the measured gradient), whereas the rare species show considerable sampling error. The data were not selected except to choose the 6 most abundant species and an arbitrary 6 rare species.

declines gradually toward the boundaries has been emphasized by other investigators (e.g., Hengeveld and Haeck 1982), and appears to be very general: it is observed in a wide variety of organisms and over a range of spatial scales. This is not to say that there are no exceptions. If one looks for them, it is easy to find examples of discontinuous variation in abundance and precipitous declines in population density at a range boundary. Often when such examples have been described, they appear to be related to an abrupt change in a single environmental variable: usually either a physical factor or the population density of an intensively interacting species of competitor, predator, or prey. Thus a striking exception to the gradual decline of bird populations toward the boundaries of their ranges in the Mississippi Valley is the abrupt replacement of two chickadee species (southern *Parus carolinensis* by northern *P. atricapillus*) that J. T. Emlen et al. (MS) attribute to competitive exclusion. For other examples of abrupt distributional boundaries apparently caused by competitive or predator-prey interactions, see Krebs (1978) and Brown and Gibson (1983). Equally rapid changes in density are sometimes associated with abrupt discontinuities in the physical environment. For example, terrestrial plant populations often decline precipitously at the edges of bodies of water or in the region of rapid transition between different soil types. When suitable habitat occurs in isolated patches, such as the montane forests inhabited by boreal plants and animals in desert regions, then there are multiple modes in the distribution of abundance over space. On a sufficiently small scale, the environment of most organisms is patchy, so population density should exhibit a multimodal distribution over space. I do not want to minimize the importance of such discontinuous or multimodal spatial variation in abundance, but I note that often these patterns disappear when data from many replicated sites are averaged or when some other technique is used to analyze the distribution over a wider area. At the extreme of fine spatial and temporal scales all distributions are discontinuous, because a single individual is either present or absent, but this does not tell us much of general interest about how species vary in abundance over their ranges.

Correlation Between Abundance and Distribution Among Species

Several recent studies (e.g., Hanski 1982*a*, 1982*b*, 1982*c*; Bock and Ricklefs 1983) demonstrate a positive relationship between local abundance and geographic distribution among closely related, ecologically similar species. When closely related plants of the same life form or animals of the same guild are compared, those species that have the highest local population densities tend to inhabit a greater proportion of sample sites within a region and to have wider geographic ranges; conversely, species that are always rare, also have restricted spatial distributions. Thus Hanski (1982*c*) showed that among species in several guilds of terrestrial arthropods, from soil mites in Finland to scarab beetles in Sarawak, there was a highly significant positive correlation between average density within a site (counting only sample sites where the species occurred) and the number of different local sites (within a few kilometers) where the species was found. Bock and Ricklefs (1983) demonstrated the same pattern in birds on a much

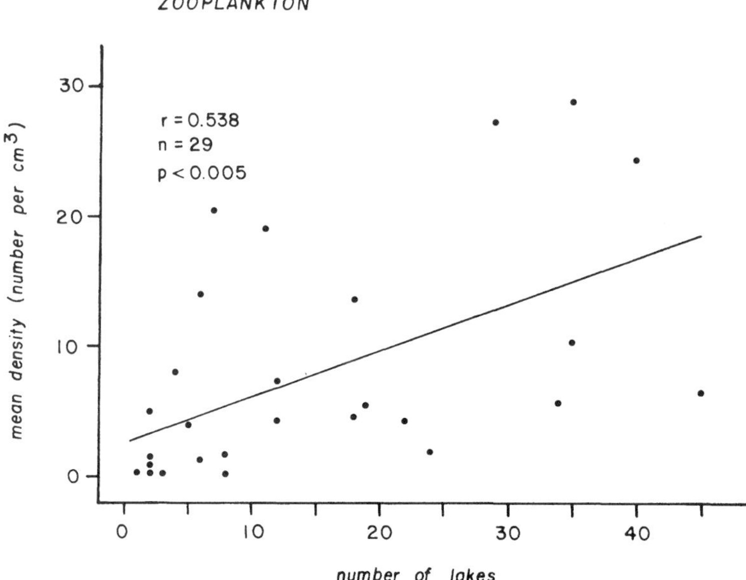

FIG. 4.—Relationship between abundance and distribution of zooplankton species in 45 nearby lakes in northwestern Ontario, plotted from data in Patalas (1971). When mean population density per lake (counting only lakes where the species occurs) is plotted against the number of lakes inhabited, a highly significant positive correlation is obtained.

larger geographic scale. Using Christmas Bird Counts of finch (family Emberizidae) species, they found a highly significant positive correlation between average abundance within local censuses (counting only censuses in which the species occurred) and the area of the geographic range.

The generality of this pattern can be demonstrated by analysis of other data sets. Plotting Patalas' (1971) data on the abundance and distribution of planktonic crustaceans in 45 lakes in Ontario, Canada, reveals the same relationship described by Hanski: the number of lakes inhabited by a species is positively correlated with its average density in those lakes where it occurs (fig. 4; see also fig. 9*a*). Reanalysis of Whittaker's (1960) data on plant species distribution in gradients of moisture and elevation in the Siskiyou and Great Smoky Mountains shows highly significant positive correlations between average local density within sites and the range of the species in the gradient (fig. 5). Regardless of plant life form (herb or tree), soil type (diorite, serpentine, or olivine gabbro), or kind of gradient (moisture or elevation), locally abundant species consistently have wide ranges whereas rare species are restricted to a narrow region of the gradient. These correlations are probably considerably more precise than they appear, because Whittaker's data include many species (especially abundant, widely distributed ones) whose distributions apparently extend well beyond the measured gradient. I restricted my analysis to those species that reached a peak

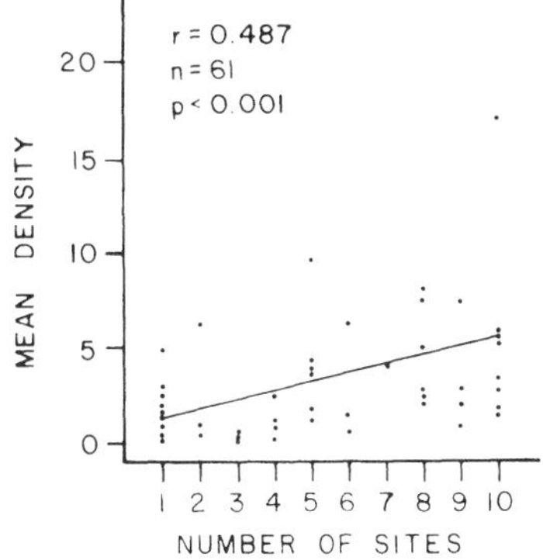

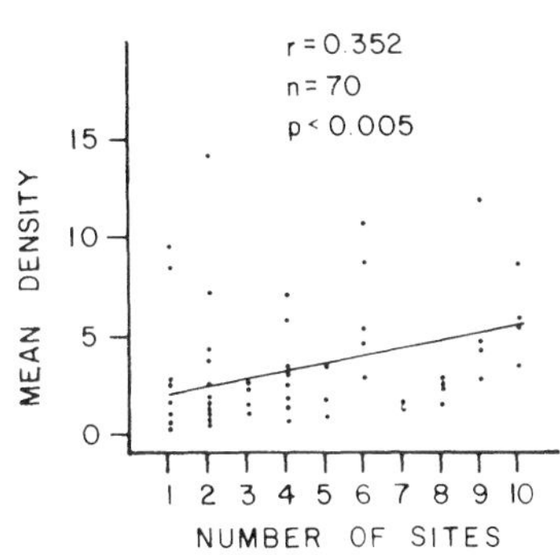

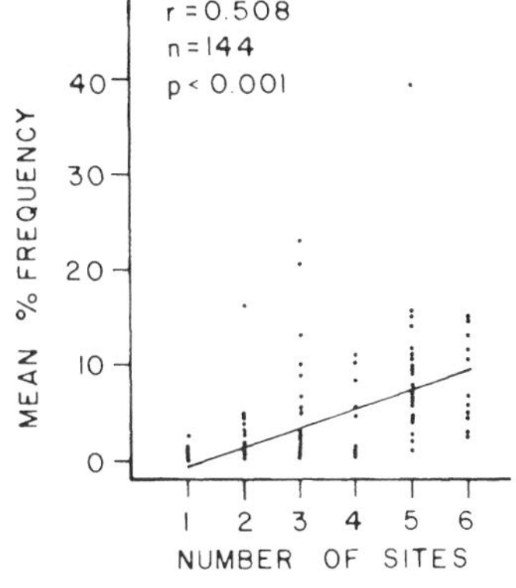

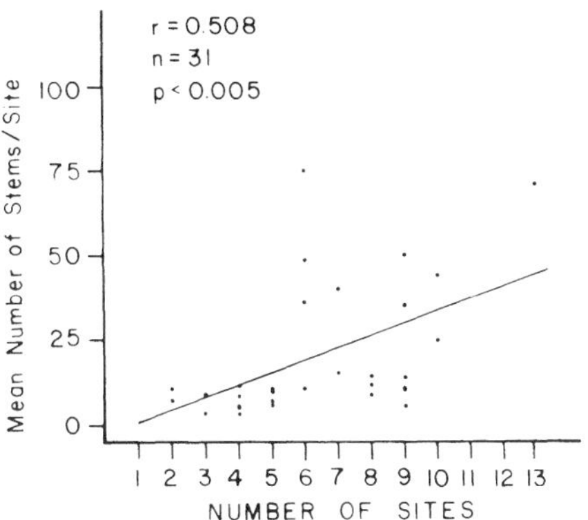

Fig. 5.—Relationship between average local abundance (counting only sites where the species occurred) and number of sites inhabited along a local gradient of rapid ecological change. Data are for terrestrial plants: (*a*), (*b*), and (*c*) replotted from data in Whittaker (1960), and (*d*) in Whittaker (1956). All species were plotted except those which attained peak abundance at either end of the gradient, because these presumably had distributions that extended well beyond the measured gradient. Note the highly significant positive correlations in all cases.

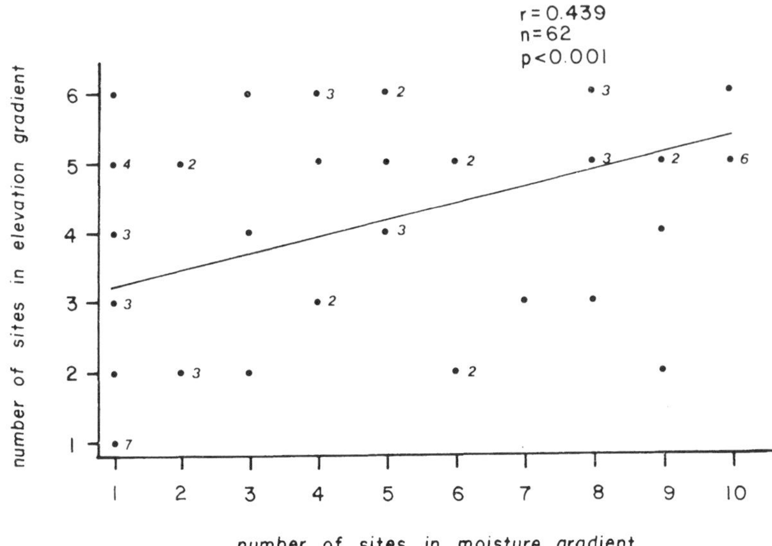

FIG. 6.—Relationship between number of sites inhabited along an elevation gradient and number of sites inhabited along a moisture gradient by herb species in the Siskiyou Mountains. Data are replotted from Whittaker (1960). Numerals beside the points indicate numbers of repeated observations. Note the highly significant positive correlation.

density within the gradient, but nevertheless almost half of the distributions of some species may not have been measured. The effect of this on goodness-of-fit can be seen by comparing the correlation coefficients for the largest data set, herbs in a moisture gradient on serpentine soil, using first the 102 species that occurred most frequently in one of the central 8 of the 10 sites ($r = 0.26$) and then just the 30 species that occurred most frequently in the central 4 of the 10 sites ($r = 0.49$). Because of the overall correlation between abundance and distribution, it is not surprising that species that are widely distributed in a gradient of one environmental variable (moisture) also tend to range more widely in an independent gradient of a second variable (elevation; fig. 6). Finally, compilation of data on another group of North American birds, hawks (families Accipitridae and Falconidae), demonstrates a positive correlation between average local population density and area of the geographic range (fig. 7).

One possible explanation of the correlations presented above is that they are simply the result of statistical sampling processes: the greater the average abundance of a species, the more likely it would be to appear in samples and the greater would be its apparent range. This explanation is inadequate to account for the pattern, however, for at least two reasons. First, although there is a well-documented tendency in statistics for the range of values observed to increase

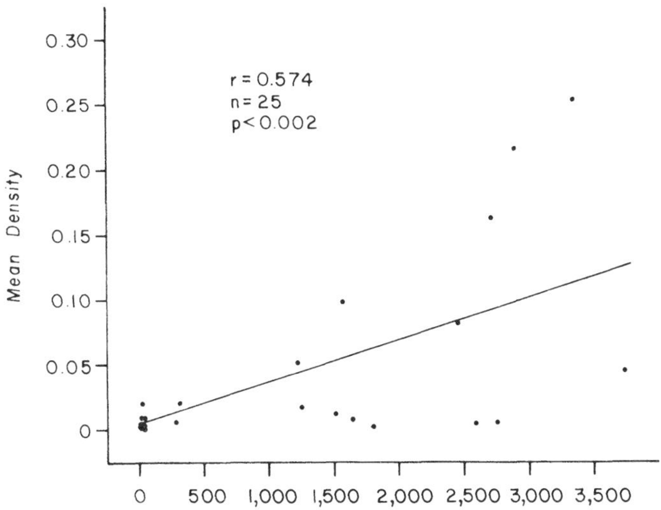

FIG. 7.—Relationship between average local population density (including only sites where the species occurred) and area of the geographic winter range for North American diurnal birds of prey. Densities from Christmas Bird Counts (courtesy of C. E. Bock) and areas from range maps (courtesy of C. S. Robbins). Only species with the majority of their winter ranges in North America were included. Note the highly significant positive correlation.

with the sample size (e.g., for the normal distribution see Tippett [1925]), this effect is much too small to account for the magnitude of change in the geographic range with average population density. Sampling from the same theoretical population, the relative change in the range decreases rapidly with increasing sample size. Thus increasing sample size from 100 to 1,000 increases the expected range by a factor of only 1.30. In contrast, for the data on hawks (all species of which have total population sizes of at least several hundred) shown in figure 7, increasing local densities by a factor of 10 increases the linear dimensions of their geographic ranges (square root of areas) by a factor of approximately 45. The second reason that the pattern cannot be simply a statistical artifact is that in some groups, such as North American land birds, the limits of geographic ranges are known with sufficient accuracy to insure that they are real boundaries. Beyond these range limits local density is not just low; populations simply do not occur.

THE THEORY

I now propose a single general explanation for both of these patterns: the relatively symmetrical, monotonic decrease in abundance from the center of the distribution toward all boundaries, and the positive correlation between local population density and extent of spatial distribution among similar species. This theory is based on three major assumptions.

1. This assumption concerns the ecological requirements of species. The abundance and distribution of each species is determined by combinations of many physical and biotic variables that are required for survival and reproduction of its individuals. These requirements define the dimensions of Hutchinson's (1957) multidimensional niche for each species. Variations in population density of a species over space is assumed to reflect the probability density distribution of the required combinations of environmental variables.

2. This assumption concerns the pattern of spatial variation in the environment, which has both stochastic and deterministic components. Some sets of variables (factors) are distributed independently of each other and there is a significant degree of apparently random local variation. Environmental variation also is autocorrelated, so that the probability of sites having similar combinations of environmental variables is an inverse function of the distance between them.

3. This assumption concerns the extent to which species vary in their requirements. I assume that closely related, ecologically similar species differ substantially in only one or a very small number of niche dimensions. This limited differentiation reflects evolutionary constraints on morphology, physiology, and behavior as a result of relatively recent descent from a common ancestor.

Spatial Variation in Abundance Within Species

From the first two assumptions it follows that population density should be highest near the center of a species range and should decline toward the boundaries. For each species there should be one most favorable site where population density should be greatest because the combination of environmental variables most closely corresponds to the requirements of the species. If spatial variation in the environment is autocorrelated, then with increasing distance from this site the environment will become progressively more different, niche requirements of the species will be met less frequently, and abundance will decline. There will be a decreasing number of local sites where individuals can occur at all, and even within these patches population densities will tend to be lower because resources are scarce and/or conditions approach the limits that can be tolerated.

The exact form of spatial variation in abundance will depend on the number and kind of environmental factors that comprise the niche and on the spatial pattern of variation of these variables. That so many of the empirical distributions resemble normal curves can be explained as follows. The normal probability density function is the limit distribution of a sum of random variables. Therefore it follows that if there are many different niche dimensions which interact additively to determine population density and if these variables are distributed independently of each other in space, then density should approximate a normal distribution along any spatial dimension. This is analogous to the way that many different genes, acting more or less additively and independently tend to produce a normal distribution of a trait in quantitative genetics (Falconer 1960). As in the case of genetics, it is not necessary that the effect of the environmental variables be exactly additive or random; if there is a sufficiently large number of more or less independent factors that have small effects these variables will tend to result in a normal probability density distribution as a consequence of the central limit theorem.

The additional assumptions required to develop a model that predicts normal distributions of abundance over space are generally consistent with Hutchinson's (1957) formulation of the multidimensional niche. Many different environmental variables, including both physical factors such as temperature, sunlight, water, salinity, pH, and nutrient concentrations, and biotic factors such as competition, availability of prey and mutualists, and ability to avoid predators and pathogens, act in combination to determine local population density and some of these vary independently of each other in space. One feature of the multidimensional niche that at first does not appear consistent with the model is the fact that Hutchinson defined the niche in terms of set theory so that (theoretically) variables interact multiplicatively to determine presence or absence in an all or nothing fashion. "It is supposed that all points in each fundamental niche imply equal probability of persistence of the species, all points outside each niche, zero probability of survival . . . there will however be an optimal part of the niche with markedly suboptimal conditions near the boundaries" (Hutchinson 1957, p. 417). Thus a realistic model of the multidimensional niche and of environmental variation in the niche variables predicts a normal distribution of population density along any transect through the species range. This is consistent with empirical data sets and with the fact that Whittaker (1956, 1960, 1965, and elsewhere) and others have represented such distributions as normal-shaped curves.

Although I suggest the normal probability density distribution as a useful general model of spatial variation in abundance, I admit that there are many exceptions. One of the chief values of the model, however, is its ability to account for even these exceptions. There are two main classes of exceptions (fig. 8), and each corresponds to a case in which a different one of the assumptions of the model is conspicuously violated. Multimodal patterns of abundance occur when the assumption that environmental similarity is a continuously decreasing function of the distance between sites is violated. This happens whenever environmental conditions are patchily distributed, which is almost always true when abundance is analyzed on a sufficiently small spatial scale. The second kind of exceptions, abrupt changes in abundance over a short distance, occurs when, instead of density being limited by the combined effects of many variables, a rapid environmental change causes one factor (or several covarying factors) to assume overwhelming importance. Examples include changes in both the physical environment, such as abrupt interfaces between terrestrial and aquatic habitats, and the biotic environment, such as may be caused by the presence or absence of severe competitors or predators.

Many empirical studies of the influence of specific environmental factors on the abundance and distribution of species have naturally focused on situations in which one particular variable changes abruptly and discontinuously while all others vary so gradually so as to remain essentially constant (e.g., see Krebs 1978;

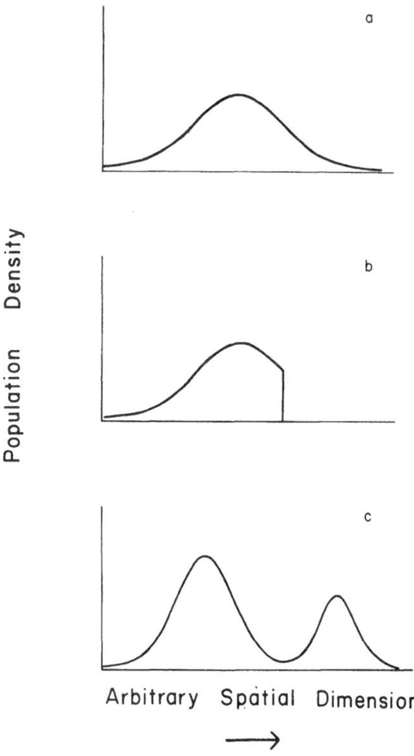

FIG. 8.—Hypothetical distributions of population density along transects through the species range when: *a*, all assumptions of the simple model are met and the distribution approaches a normal curve; *b*, a single limiting variable changes abruptly and the distribution is truncated; and *c*, spatially separated sites have similar combinations of variables and the distribution is multimodal, combined from more than one approximately normal curve.

Brown and Gibson 1983). Although such studies are useful for identifying important dimensions of the niche, in retrospect, I believe they have diverted attention from the multidimensional nature of the niche and importance of this concept for understanding the fundamental relationship between abundance and distribution. Although a single variable can exert a threshold effect to determine whether or not a species occurs on a site, if a population is present its abundance will depend on the contributions of several relatively independent factors. As a consequence of their multiple niche requirements, almost all species have highly restricted distributions; they are confined to only a part of a continent or ocean and they are found in only one or a small number of habitat types. Furthermore, they tend to be limited by different factors at different boundaries of their geographic ranges. For example, different environmental variables and biological processes must cause the death and reproductive failure at opposite extremes of the highly symmetrical distributions of species in environmental gradients on mountainsides or in the intertidal (fig. 1).

Correlation Between Abundance and Distribution Among Species

If niches are multidimensional and if spatial variation in the environment tends to be autocorrelated, then there should be a positive correlation between abundance and distribution for those species that differ in only a very few niche dimensions. Changes in requirements for one or a small number of variables that increase local abundance within the range should also enable the species to colonize new areas at the periphery of its distribution where those factors were previously limiting. Stated more generally, those species that can tolerate conditions and acquire sufficient resources so as to attain high densities in some places, should also be able to occur (albeit often at lower densities) in many other sites over a relatively large area. On the other hand, species that are otherwise similar, but have such narrow requirements that they cannot attain high abundances anywhere, will necessarily be restricted to the few sites within the limited geographic region where they can satisfy their needs. Thus there are positive correlations between maximum density, average density over the area where the species occurs, the number of sites inhabited within a local region, and the area of the geographic range, but only for closely related, ecologically similar species.

These correlations break down when distantly related, ecologically dissimilar species are compared, because such organisms have evolved niches of entirely different configuration. Even among such species, however, there appear to be some predictable relationships between abundance and distribution that can be explained in terms of their niche requirements. For example, the laws of thermodynamics dictate that populations of predators be less dense than prey of comparable body size and reproductive rate. If predators and prey are constrained by the dynamics of speciation and extinction processes (see MacArthur and Wilson 1967; Rosenzweig 1975; Brown 1981) to have similar distributions of population sizes (or at least similar minimum population sizes) among species, then it follows that, on average, individual consumers should use the environment on a larger spatial scale and their populations should exhibit lower densities and wider distributions than producers. Similarly, organisms of large body size interact with the environment on a larger scale than smaller species that can better exploit the spatial heterogeneity that Hutchinson (1959, p. 155) termed the "mosaic nature of the environment." Thus larger organisms tend to have less dense, more widely distributed populations, whereas smaller ones attain higher densities and exhibit more local and patchy distributions (see Brown 1981).

The empirical relationship between abundance and distribution was called to my attention by the papers of Hanski (1982*a*, 1982*c*) and Bock and Ricklefs (1983). The latter presents no detailed explanation for the pattern, and I disagree with Hanski's interpretation, which he calls the core and satellite species hypothesis. From data primarily on the distribution of terrestrial arthropods among local regions of superficially similar habitat, Hanski noted not only that frequency of occurrence among sites is positively correlated with average abundance (fig. 9*a*), but also that the distribution among species of number of sites inhabited is bimodal (fig. 9*b*). He assumed that all sites are equally suitable for all species and that species distributions among those sites vary randomly over time. He then

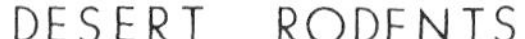

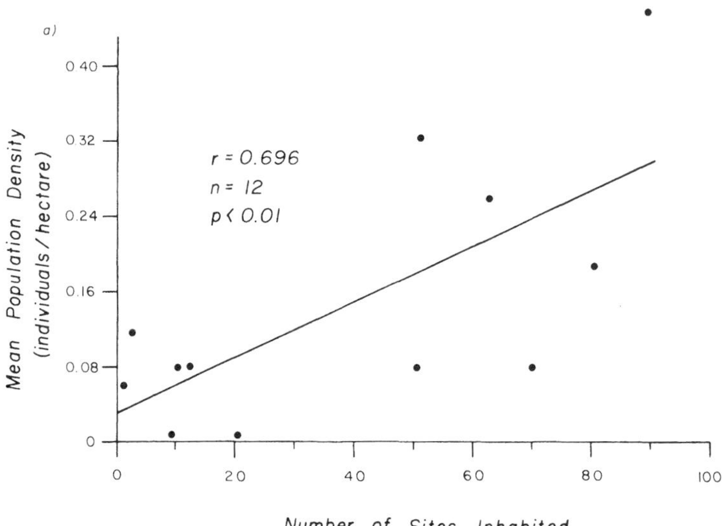

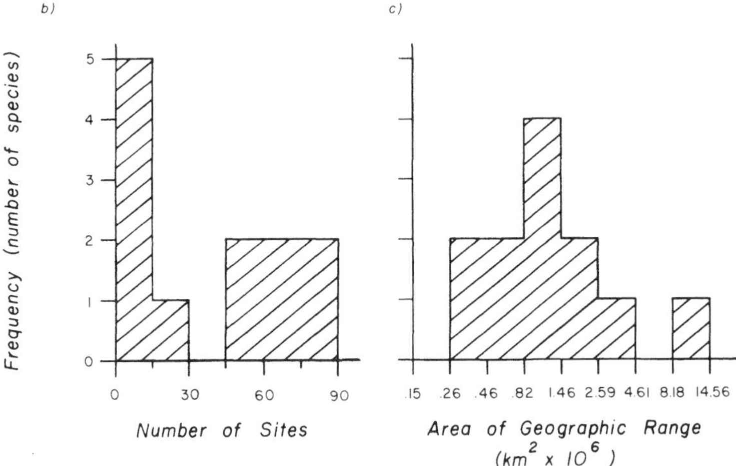

FIG. 9.—Distributions and abundance of 12 desert rodent species that occur together in southern Nevada (plotted from data of Jorgensen and Hayward [1965]). *a*, Relationship between mean population density per site (counting only sites where the species occurred) and the number of sites (out of a total of 115) inhabited. Note the significant positive correlation. *b*, Frequency distribution of the 12 species with respect to the number of sites inhabited. Such a bimodal distribution is the basis of Hanski's (1982*a*, 1982*b*, 1982*c*) division of a community into core and satellite species (right and left modes, respectively). *c*, Frequency distribution of the 12 species with respect to the areas of their geographic ranges (plotted on a logarithmic scale from measurements of range maps in Hall [1981]). Note the unimodal, approximately lognormal distribution.

developed a simple model which suggests that communities are comprised of two kinds of species: core species, which are locally abundant and inhabit virtually all suitable sites, and satellite species, which are rare and distributed essentially at random among a few sites. I have three criticisms of Hanski's model. First, it is highly unlikely that all sites are equally favorable for all species. It is much more realistic to assume that the differences in local distribution and abundance of closely related, ecologically similar species, although they may have a stochastic component, are primarily the result of different requirements and tolerances. Otherwise, what determines the population densities and spatial distributions? Second, Hanski's model is based on a stochastic process analogous to genetic drift affecting allele frequency, which assumes that the distributions of species vary randomly in time between two stable boundary conditions: ubiquitous (inhabiting all local sites) and extinct throughout the region. Therefore, Hanski must invoke colonization from outside the local area to account for the high frequency of species that are found in only a very few sites instead of none at all as the random walk model would predict. This seems contrived and unrealistic, because a large scale perspective would show that many of these rare species are unlikely to produce such colonists because they are just as uncommon in surrounding regions; they are species of limited geographic distribution and low average abundance. A final and related point is that the bimodal distribution of frequency of occurrence is almost certainly an artifact of sampling a small number of sites with a local region. Other evidence suggests that both local population densities and areas of entire geographic ranges exhibit unimodal, approximately lognormal distributions among species (e.g., Preston 1962; Williams 1964; Rapoport 1982). Since these two variables are positively correlated with each other, this means that there are many rare, restricted species and few abundant, widespread ones. Now if one samples a limited number of sites within a local region, one finds a frequency distribution with two modes, one, a result of the rare species that are restricted to a few sites, and the other, a result of the common species that inhabit many or all of the sites (fig. 9*b;* see also Hanski 1982*c,* figs. 8, 9, 10). If the spatial frame of reference were expanded to include the entire geographic ranges of all the species, however, the second mode would disappear because the frequency distributions of areas of geographic ranges are unimodal and approximately lognormal (e.g., fig. 9*c).*

GENERAL DISCUSSION

The ideas developed in this paper constitute a very general, qualitative kind of theory. The empirical patterns of variation in abundance over space within species and correlation between local density and geographic distribution among species can be explained with simple arguments that follow logically from assumptions that seem to be robust empirical generalizations. I have shown that the theory can be derived more formally by redefining the concept of the multidimensional niche so that the distribution of population density over space should approach a normal probability density distribution. If these ideas have merit, eventually it should be possible to develop more precise mathematical models that will generate specific testable predictions about niche relationships within and among species. Such models would differ from those currently in vogue (but see MacArthur and Wilson 1967), because they would focus on variation in abundance over space rather than over time and because they would be based explicitly on the concept of multidimensional niches.

My synthesis builds on themes developed by Hutchinson (1957, 1959; see also Whittaker 1965; McNaughton and Wolf 1970) in his classical papers on the multidimensional niche and diversity of species. I find it interesting that these papers of Hutchinson have so often been cited, yet the ideas I have resurrected here seem to have had little direct impact on the development of ecological and biogeographic theory. Clearly we have much to learn before we can answer Hutchinson's question, Why are there so many kinds of organisms? I believe the statistical approach I have begun to develop here focuses attention on some of the important questions, but it does not yet provide many satisfying answers. The species in different taxonomic groups, life forms, and guilds exhibit different relationships between abundances and distribution. Aside from a few general comments on the consequences of trophic status and body size (see Hutchinson 1959; Brown 1981), I have not explored these relationships, although the central questions of population and community ecology can be phrased in these terms: How many different factors constitute major dimensions of the niche of a species and how do these vary in space and time so as to limit abundance and distribution? How do the multidimensional niches of coexisting species relate to each other? How does this organization vary as the identity and diversity of species changes over the surface of the earth? Elsewhere (Brown 1981) I have advocated what I now call a statistical approach to ecology that focuses on the distributions of variables among many discrete units, either individuals within species or species within local communities and larger geographic regions. The present paper provides an example of this approach, which is focused here on the spatial distribution of individuals within and between closely related populations.

Some critics may object that the perspective that I have adopted ignores the really interesting and important problem: what environmental variables and population attributes determine the abundance and limit the distribution of particular local populations? Indeed, we need a better understanding of these mechanisms in order to pursue the approach that I advocate. On the other hand, there are over two million species in the world, each has its own unique geographic range, and its local abundance is determined by different combinations of environmental variables in different parts of its range. Obviously, to describe and explain all this diversity in detail is an impossible task. Sometimes in such cases scientific progress can be made by finding a frame of reference in which the apparently idiosyncratic features of small systems are revealed as general statistical patterns of larger systems that then suggest common explanations. There appears to be such a general relationship between abundance and distribution for most species and closely related groups of species that has important implications for many areas of ecology and evolutionary biology.

First consider the number of and degree of independence among the environmental variables that importantly affect the abundance and distribution of each species. Preliminary investigations suggest that the number of independent niche dimensions is modest, perhaps between five and 10 for most species. This would be large enough to produce the fairly regular spatial patterns described above, and small enough to produce the kinds of exceptions often observed (fig. 8). On the one hand, ecological constraints appear to limit the number of dimensions. Computer simulations (performed in collaboration with M. Sanderson and A. Harvey) show that as the number of independent limiting factors increases, it becomes increasingly difficult for the species to exist anywhere, because favorable combinations of all variables become increasingly unlikely to occur for purely statistical reasons. On the other hand, when only one or two factors limit distribution and abundance, populations should be subject to strong directional selection to adapt to these conditions, thereby diminishing their negative effects until other factors become limiting and the intensity of selection resulting from any one factor is reduced. Such evolution of the niche should reinforce the tendency of independent limiting factors to result in symmetrical, normal-shaped curves of abundance over the range of species. The ecological and evolutionary consequences of the dimensionality of niches seem to offer opportunities for both empirical and theoretical research.

A related problem also concerns the dimensionality of the niche and its consequences for the abundance and distribution of species. Many of the ideas of current evolutionary ecology are based on widespread acceptance of the trade-off principle. The data and ideas presented in this paper (see also Hanski 1982*c*) appear to challenge currently held notions about the ecological consequences of the purported trade-off between specialization and generalization implied by the saying "jack-of-all-trades-master-of-none." This dictum would predict that specialists with narrow tolerances should be more efficient in exploiting a more limited range of resources, and hence should have more restricted distributions but higher local abundances than generalists. I can find no evidence of such a trade-off, at least among closely related, ecologically similar species. There is a very general tendency for species that have restricted distributions (presumably because they are specialized and able to tolerate only a narrow range of conditions) to be rare, whereas more widespread species attain higher local population densities. At least some data (e.g., fig. 6) also suggest a lack of trade-offs among different niche dimensions; species that can tolerate wide variation in one factor also tend to be tolerant of other factors, and hence to be both locally abundant and spatially widespread. There are apparent exceptions—species that attain high densities within particular habitat types but are restricted to small areas (e.g., see Rabinowitz 1981)—but I believe that these species usually violate an assumption of my model. They differ from their nearest wide-ranging relatives along many niche dimensions, and this is indicated by the fact that they are often members of well-differentiated endemic taxa. For example, a suite of special adaptations to several niche variables must be required to exploit highly localized, physically harsh environments such as salt marshes, hypersaline lakes, and hot springs, and the few species that can inhabit these are often both locally abundant and well differentiated from their relatives in more benign habitats (Terborgh 1973; Brown 1981; Brown and Gibson 1983). Thus my model makes predictions about the ecological and evolutionary consequences of niche dimensionality that can be tested by evaluating further the relationships among environmental limiting factors, patterns of abundance and distribution, and degree of evolutionary differentiation in taxa for which adequate data are available.

In order to account completely for the general relationships between abundance and distribution, it is necessary to understand not only the multidimensional nature of the niche but also the spatial variation in the dynamics of population growth and regulation. Other investigators have noted and proposed explanations for the gradual decline in population density toward the edge of the range. For example, Grinnell (1922) and Wiens and Rotenberry (1981) suggested that the geographic distributions of bird populations may represent a dynamic equilibrium between the exportation of emigrants from source areas (usually regions near the center of the geographic range where birth rates exceed death rates) and the importation of these individuals into sink areas (usually regions at the periphery of the range where this continual immigration sustains local populations whose death rates exceed birth rates). Although such centrifugal dispersal might account, at least in part, for spatial variation in population density in organisms as vagile as birds (but see J. T. Emlen et al. [MS] for alternative hypotheses), it probably cannot explain similar patterns in less mobile organisms, such as plants with limited seed dispersal. Nevertheless, it should be possible to develop and test mathematical models that account for spatial variation in population dynamics by considering the effects of both population density and extrinsic environmental variables on the rates of birth, death, immigration, and emigration (Hanski 1982*c*, Holt 1983; B. A. Maurer and J. H. Brown, MS).

The fact that local abundance may depend to some extent on the pattern of distribution on a larger spatial scale has important implications for community ecology. It has long been recognized that the frequency distribution of population densities among species within a local region exhibits a regular pattern in which there are many relatively rare species and only a few common ones (e.g., MacArthur 1957; Preston 1962; Williams 1964; Whittaker 1965), but this has long defied a satisfying biological explanation (but see May 1975; Sugihara 1980). The present paper focuses attention on the fact that many of the locally rare species represent the tails of the spatial distributions of species that are more common in other regions. This appears to support Gleason's (1926) classical "individualistic" concept of species distribution and community organization: assuming dispersal is not limiting, species tend to occur relatively independently of most other species wherever environmental conditions are suitable, and local population densities are determined by the extent to which the local environment meets the requirements of individuals. Over their geographic ranges most species coexist with many different combinations of other species, and even those that are sufficiently abundant to dominate their guilds and communities in some regions are rare and relatively unimportant in others. The limited study areas of most community ecologists are too small to be representative of the variety of conditions that have shaped the evolution of the niches of most of the species. Note that this Gleasonian view does not necessarily imply, as some authors have suggested, that interspecific interactions are unimportant in determining the abundance and distribution of species and the organization of communities. It simply implies that communities are not highly integrated units comprised of many species that have coevolved to interact specifically with each other. The general approach that I apply here to analyze the statistical distribution of individuals within species and among closely related species, might also be used in an analogous way to investigate the density and distribution of species within and among communities.

So far in this paper I have treated species and their niches as if they were constant over space, even though I am aware of substantial geographic variation in some species. This simplifying assumption seems justified so long as the ecological variation within species is small relative to the differences between species. This must be true generally, because it is the basis of classifying closely related populations into species.

Consideration of this problem, however, raises important questions about the evolutionary dynamics of speciation and extinction that underlie the general relationship between abundance and distribution of species. The apparent absence of a trade-off between generalists and specialists suggests a wide distribution of evolutionary success among species. If I define success of species as the analogue of fitness of individuals, the probability of leaving descendants over evolutionary time, then in general the abundant, widespread species must be more successful than the rare, restricted ones. If the latter have substantially higher extinction rates, however, whence come the new species that replace them? Do the abundant, widely distributed species continually bud off small, isolated differentiating populations around the margins of their geographic ranges? This mechanism of allopatric speciation seems consistent not only with much evolutionary theory (e.g., see Mayr 1963; Futuyma and Mayer 1980; Wiley 1981), but also with the ecological patterns and processes developed here.

In particular the ecological relationships appear to support some but not all of Brown's (1957, 1959) ideas about general adaptation and centrifugal speciation. Central populations of widespread, abundant species would seem relatively resistant to rapid, directional evolutionary change. Not only do these populations so dominate their guilds and communities that relatively little improvement in ecological performance is likely, they are also distributed relatively continuously over a variety of local environments so that there is little opportunity for spatial isolation to facilitate genetic differentiation of locally adapted populations. Thus most of the selection will be stabilizing selection that tends to maintain the generalized adaptations. In contrast, peripheral populations of the same widespread species will tend to be not only rare, but also restricted to isolated patches of suitable habitat. If this spatial isolation reduces gene flow sufficiently, these peripheral populations can respond to directional selection, adapt to local conditions, and eventually differentiate into new species. Over evolutionary time such newly formed species could increase substantially in abundance and distribution in two situations: if the environment changes so as to favor forms with their special adaptations, or if they are able to evolve to increase their share of limited resources, perhaps in part by increasing their ability to compete with ancestral and other closely related species.

This evolutionary scenario, based on ecological and genetic processes operating at the population level, would seem to go a long way toward explaining many macroevolutionary phenomena.

1. The apparent stasis observed in many fossil species over long periods of evolutionary time (Eldredge and Gould 1972; Stanley 1979) can be attributed to stabilizing selection acting on the central populations of widespread, abundant species. Representatives of these populations should dominate the fossil record because of the statistics of sampling, so that rate of evolution within most species should appear to be very slow.

2. The association between substantial evolutionary change and speciation events (Eldredge and Gould 1972; Stanley 1979) can be explained in terms of strong directional selection resulting in the local adaptation and genetic differentiation of isolated peripheral populations. In order to become sufficiently abundant and widespread to appear in the fossil record, newly formed species would usually have to benefit from environmental change or undergo enough ecological (and usually also morphological) differentiation to compete successfully with ancestral and other closely related species. This would also account for the commonly observed pattern (Hennig 1979) that after a speciation event one of the resulting sister species is usually extremely similar to the ancestral species whereas the other is highly differentiated.

3. Many macroevolutionary trends that appear to be the result of different speciation or extinction rates within different phyletic lines might be explained in terms of the positive correlation between abundance and distribution among closely related species. Compared to their rare, restricted relatives, abundant widespread species should have lower extinction rates. Whether they should also exhibit higher speciation rates is not so clear, especially since most speciation should occur in small, peripheral, isolated populations, so it may be difficult to detect in the fossil record. Despite these kinds of difficulties it should be possible to develop and test macroevolutionary hypotheses based on the ecological relationships between abundance and distribution. For example, Hansen (1980) and Jablonski and Lutz (1983) describe relationships between speciation and extinction rates, area of geographic range, and mode of larval dispersal in fossil mollusks: species with planktonic larvae have lower rates of both speciation and extinction, and larger geographic ranges than species that brood their young. If this partly results from the fact that these species are distributed along a spectrum from rare and restricted to abundant and widespread, then the planktonic forms should on average have maintained higher local population densities than their nonplanktonic relatives. This could be tested if the fossil record can be used to estimate relative population densities.

4. Finally, I would expect major evolutionary innovations to evolve initially as special adaptations of rare, restricted populations (usually at the periphery of the geographic or ecological ranges of their more abundant and widespread relatives) in response to directional selection caused by unusual local environmental conditions. This seems to fit well with recent empirical findings of Jablonski et al. (1983) that new phyletic lines appear to have evolved primarily in marginal habitats.

Because of all these implications, I encourage further investigation of macroscopic, statistical patterns in the spatial and temporal distributions of individuals within species and among closely related species. In trying to characterize these patterns and understand the mechanistic processes that are responsible for them, it should be possible to make progress in integrating the disciplines of ecology, biogeography, population genetics, and evolution.

SUMMARY

There appears to be a general relationship between abundance and distribution that has two parts. First, within species, population density tends to be greatest in the center of the range and to decline gradually toward the boundaries. This pattern holds over a range of spatial scales from steep environmental gradients within local regions to the entire geographic range. Exceptions include: (1) abrupt changes in abundance that usually correspond to sharp, discontinuous changes in single environmental variables; and (2) multimodal patterns of abundance that are caused by environmental patchiness. The second general relationship is that among closely related, ecologically similar species spatial distribution is positively correlated with average abundance. Again this pattern holds over a variety of spatial scales from local regions to entire geographic ranges. These empirical patterns have already been reported in the literature, but their generality is demonstrated by analysis of additional data for diverse kinds of organisms.

A single general theory accounts for these observations and follows logically from three assumptions. First, the abundance and distribution of each species are limited by the combination of physical and biotic environmental variables that determines the multidimensional niche. Second, spatial variation in these environmental variables is somewhat stochastic but autocorrelated, so that nearby sites tend to have more similar environmental conditions than more distant ones. Third, closely related, ecologically similar species differ in no more than a very few niche dimensions. A more formal model can be developed that predicts that under these assumptions the distribution of population density over space should approximate a normal probability density distribution. Most exceptions to this predicted pattern can be explained as cases in which assumptions of the model are clearly violated.

This paper represents an example of a statistical approach that should be useful for investigating complex ecological systems comprised of many components, such as species of many individuals or communities of many species. The general relationships between abundance and distribution developed here eventually should contribute to our understanding of the biogeography, population genetics, and evolution of species as well as the ecological attributes of populations and communities.

ACKNOWLEDGMENTS

I thank numerous students and colleagues who have stimulated and clarified my thinking about these problems: C. Bock, J. Emlen, J. Endler, A. Harvey, R. Holt, A. Kodric-Brown, E. Rapoport, P. Price, K. Rusterholz, M. Sanderson, W. Schaffer, W. Porter, and especially B. Maurer. M. Kurzius did most of the data analysis. C. Bock, E. Boyer, D. Bystrak, J. Emlen, and C. Robbins kindly made available unpublished data. The research was supported in part by NSF Grant BSR-8021535.

LITERATURE CITED

Andrewartha, H. G., and L. C. Birch. 1954. The distribution and abundance of animals. University of Chicago Press, Chicago.

Beals, E. W. 1969. Vegetational change along altitudinal gradients. Science 165:981–985.

Bock, C. E., and R. E. Ricklefs. 1983. Range size and local abundance of some North American songbirds: a positive correlation. Am. Nat. 122:295–299.

Brown, J. H. 1981. Two decades of homage to Santa Rosalia: toward a general theory of diversity. Am. Zool. 21:877–888.

Brown, J. H., and A. C. Gibson. 1983. Biogeography. Mosby, St. Louis, Mo.

Brown, W. L. 1957. Centrifugal speciation. Q. Rev. Biol. 32:247–277.

———. 1959. General adaptation and evolution. Syst. Zool. 7:157–168.

Bystrak, D. 1979. The breeding bird survey. Sialia 1:74–79.

———. 1981. The North American Breeding Bird Survey. Stud. Avian Biol. 6:34–41.

Delcourt, H. R., D. C. West, and P. A. Delcourt. 1981. Forests of the southeastern United States: quantitative maps for aboveground woody biomass, carbon, and dominance of major tree taxa. Ecology 62:879–887.

Eldredge, N., and S. J. Gould. 1972. Punctuated equilibria: an alternative to phylogenetic gradualism. Pages 82–115 *in* T. J. M. Schopf, ed. Models in paleobiology. Freeman Cooper, San Francisco.

Falconer, D. S. 1960. Introduction to quantitative genetics. Oliver & Boyd, Edinburg.

Field, J. G., and F. T. Robb. 1970. Numerical methods in marine ecology: gradient analysis of rocky shore samples from False Bay. Zool. Afr. 5:191–210.

Futuyma, D. J., and G. C. Mayer. 1980. Non-allopatric speciation in animals. Syst. Zool. 29:254–271.

Gleason, H. A. 1926. The individualistic concept of plant associations. Bull. Torrey Bot. Club 53:7–26.

Grinnell, J. 1922. The role of the "accidental." Auk 39:373–380.

Hall, E. R. 1981. The mammals of North America. Wiley, New York.

Hansen, T. A. 1980. Influence of larval dispersal and geographic distribution on species longevity in neogastropods. Paleobiology 6:193–207.

Hanski, I. 1982*a*. Communities of bumblebees: testing the core-satellite species hypothesis. Ann. Zool. Fenn. 19:65–73.

———. 1982*b*. Distributional ecology of anthropochorus plants in villages surrounded by forest. Ann. Bot. Fenn. 19:1–15.

———. 1982*c*. Dynamics of regional distribution: the core and satellite species hypothesis. Oikos 38:210–221.

Hengeveld, R., and J. Haeck. 1982. The distribution of abundance. I. Measurements. J. Biogeogr. 9:303–316.

Hennig, W. 1979. Phylogenetic systematics. 3d ed. Transl. by D. D. Davis and R. Zanderl. University of Illinois Press, Urbana.

Holt, R. 1983. Immigration and the dynamics of peripheral populations. Pages 680–694 *in* A. Rhodin and K. Miyata, eds. Advances in herpetology and evolutionary biology. Museum of Comparative Zoology, Harvard University, Cambridge, Mass.

Hutchinson, G. E. 1957. Concluding remarks. Cold Spring Harbor Symp. Q. Biol. 22:415–427.

———. 1959. Homage to Santa Rosalia, or why are there so many kinds of animals? Am. Nat. 93:145–159.

Jablonski, D., and R. A. Lutz. 1983. Larval ecology of marine benthic invertebrates: paleobiological implications. Biol. Rev. 58:21–89.

Jablonski, D., J. J. Sepkoski, D. J. Bottjer, and P. M. Sheehan. 1983. Onshore-offshore patterns in the evolution of Phanerozoic shelf communities. Science 222:1123–1125.

Jorgensen, C. D., and C. L. Hayward. 1965. Mammals of the Nevada test site. Brigham Young Univ. Sci. Bull. Biol. Ser. 6(3):1–81.

Krebs, C. J. 1978. Ecology: the experimental analysis of distribution and abundance. Harper & Row, New York.

MacArthur, R. H. 1957. On the relative abundance of bird species. Proc. Natl. Acad. Sci. USA 43:293–295.

———. 1972. Geographical ecology. Harper & Row, New York.

MacArthur, R. H., and E. O. Wilson. 1967. The theory of island biogeography. Princeton University Press, Princeton, N.J.

McNaughton, S. J., and L. L. Wolf. 1970. Dominance and the niche in ecological systems. Science 167:131–139.

May, R. M. 1975. Patterns of species abundance and diversity. Pages 81–120 *in* M. L. Cody and J. M. Diamond, eds. Ecology and evolution of communities. Harvard University Press, Cambridge, Mass.

Mayr, E. 1963. Animal species and evolution. Harvard University Press, Cambridge, Mass.

Patalas, K. 1971. Crustacean plankton in forty-five lakes in the Experimental Lakes Area, northwestern Ontario. J. Fish. Res. Board Can. 28:231–244.

Preston, F. W. 1962. The canonical distribution of commonness and rarity. Ecology 43:185–215, 410–432.

Rabinowitz, D. 1981. Seven forms of rarity. Pages 205–215 *in* H. Synge, ed. The biological aspects of rare plant conservation. Wiley, New York.

Rapoport, E. H. 1982. Areography: geographical strategies of species. Pergamon, Oxford.

Rosenzweig, M. L. 1975. On continental steady states of species diversity. Pages 121–141 *in* M. L. Cody and J. M. Diamond, eds. Ecology and evolution of communities. Harvard University Press, Cambridge, Mass.

Stanley, S. M. 1979. Macroevolution. Freeman, San Francisco.

Sugihara, G. 1980. Minimal community structure: an explanation of species abundance patterns. Am. Nat. 116:770–787.

Terborgh, J. 1973. On the notion of favorableness in plant ecology. Am. Nat. 107:481–501.

Tippett, L. H. C. 1925. On the extreme individuals and the range of samples taken from a normal population. Biometrika 17:364–387.

Walter, H. 1979. Vegetation of the earth and ecological systems of the geo-biosphere. Springer-Verlag, Berlin.

Whittaker, R. H. 1956. Vegetation of the Great Smoky Mountains. Ecol. Monogr. 22:1–44.

———. 1960. Vegetation of the Siskiyou Mountains, Oregon and California. Ecol. Monogr. 30:279–338.

———. 1965. Dominance and diversity in land plant communities. Science 147:250–259.

Wiens, J. A., and J. T. Rotenberry. 1981. Censusing and evaluation of avian habitat occupancy. Stud. Avian Biol. 6:522–532.

Wiley, E. O. 1981. Phylogenetics, the theory and practice of phylogenetic systematics. Wiley, New York.

Williams, C. E. 1964. Patterns in the balance of nature and related problems in quantitative ecology. Academic Press, New York.

SPECIES DIVERSITY

Edited by Jessica Theodor, Alison G. Boyer, and David J. Currie

Species diversity patterns have long been of interest to both ecologists and paleontologists. Historically, however, these disciplines operated somewhat independently of each other, with ecologists focusing on spatial gradients of diversity and paleontologists on temporal ones. However, diversity, like many other large-scale patterns, is the result of processes operating in both ecological and evolutionary time. The integration of these divergent perspectives broadens the understanding of the current distributions of organisms: diversity is generated by evolutionary processes, and maintained and regulated by ecological ones. Macroecology, especially when examined at a large spatial scale, often facilitates comparisons of contemporary patterns and processes with data from the fossil record, whose limitations necessitate a wider spatial focus, and whose temporal span offers critical information on the historical development of ecosystems. Papers in this section reflect the close interweaving and concurrent development of macroecological and macroevolutionary perspectives on diversity over the past few decades.

Two major themes emerge from these papers: (1) How is diversity generated and regulated? And (2) how are species distributed in the world? Because the temporal scales of these questions are quite different, it is no surprise that the question of how biodiversity is generated and regulated has been primarily the domain of paleontology. Conversely, the question of how species subdivide the world requires the more detailed knowledge of contemporary organisms available to ecologists.

The late 1970s and 1980s were a tremendously vibrant time in paleobiology. David Raup, Steve Gould, Jack Sepkoski, Jim Valentine, David Jablonski, and others expanded the outlook pioneered by G. G. Simpson and Norman Newell, bringing a strong quantitative focus and ideas from ecological and evolutionary theory to a discipline that had often been castigated as "stamp collecting." Simpson's *Tempo and Mode in Evolution* (1944) brought paleontology into the modern synthesis of evolutionary theory. His influence was also felt in ecology. Simpson and Newell, both at the American Museum of Natural History and Columbia University, brought to paleontology a focus on ecology, an understanding of statistics, and a strong awareness that the fossil record was not simply a record of lineal ancestors and descendants but a record of populations sampled over time. The papers chosen here reflect the emergence of what might be identified as macroecological themes: global mass extinction (D. M. Raup and J. J. Sepkoski), taxonomic differentiation over geographic space (J. J. Sepkoski), evolutionary innovation (K. J. Niklas, B. H. Tiffney, and A. H. Knoll), and taxonomic diversification (J. Valentine).

A central focus in macroecology has been to identify factors that govern the distribution of species over the globe, allowing coexistence of more species in the tropics than at the poles, a pattern that has remained provocative since latitudinal gradients were first quantified (Fischer 1960; Simpson 1964). The papers included here, though by no means exhaustive, provide foundational macroecological perspectives on several essential components of diversity patterns: body size (R. M. May; K. P. Dial and J. M. Marzluff), spatial partitioning of geographic ranges (G. C. Stevens), and energy availability and allocation (J. H. Brown; D. H. Wright). E. R. Pianka (1966) reviewed a list of six hypotheses proposed to explain latitudinal diversity gradients: time, spatial heterogeneity, competition, predation, climatic stability, and productivity. Pianka suggested that the time hypothesis was "not readily amenable to conclusive tests" (1966, 35), and many authors have concentrated on abiotic determinants of diversity gradients (Willig et al. 2003; Currie et al. 2004; Hawkins et al. 2004; Field et al. 2009). Subsequent work has flourished in testing multiple—more than Pianka's six—explanations for latitudinal gradients. Some (competition, predation) have been abandoned, while others (environmental stability and spatial heterogeneity) have not been well supported by the evidence, or are dependent on the spatial scale of the study (see reviews by Rohde 1992; Field et al. 2009). In spite of the difficulties of testing the time hypothesis, many newer hypotheses for the pattern take into account the fundamental evolutionary processes that influence distributions of all species—speciation, dispersal, and extinction (see Rosenzweig 1995; Hawkins and Porter 2001; Wiens and Donohue 2004; Jablonski et al. 2006; Algar et al. 2007; Valentine et al. 2008; Krug et al. 2009).

By way of example, the out-of-the-tropics model is based on an examination of the origination rates in marine fossil invertebrates, which show clearly that the tropics generate significantly more diversity than high-latitude environments (Jablonksi et al. 2006). Subsequent refinements of this model, highlighting the important role of niche incumbency and temperature (Valentine et al. 2008), represent the ongoing dialogue between paleobiology and ecology in trying to explain the species diversity patterns seen today. Testing hypotheses with the spatial extent and detail available for extant organisms and the temporal range available in the fossil record will undoubtedly contribute to a better understanding of species diversity patterns in both space and time.

Literature Cited

Algar, A. C., J. T. Kerr, and D. J. Currie. 2007. A test of metabolic theory as the mechanism underlying broad-scale species-richness gradients. *Global Ecology and Biogeography* 16:170–78.

Currie, D. J., G. G. Mittelbach, H. V. Cornell, R. Field, J. F. Guégan, B. A. Hawkins, D. M. Kaufman, et al. 2004. Predictions and tests of climate-based hypotheses of broad-scale variation in taxonomic richness. *Ecology Letters* 7 (12): 1121–34.

Field, R., B. A. Hawkins, H. V. Cornell, D. J. Currie, J. A. F. Diniz-Filho, J. F. Guégan, D. M. Kaufman, J. T. Kerr, G. G. Mittelbach, T. Oberdorff. 2009. Spatial species-richness gradients across scales: A meta-analysis. *Journal of Biogeography* 36 (1): 132–47.

Fischer, A. G. 1960. Latitudinal variation in organic diversity. *Evolution* 14:64–81.

Hawkins, B. A., and J. A. F. Diniz-Filho. 2004. 'Latitude' and geographic patterns in species richness. *Ecography* 27 (2): 269–72.

Hawkins, B. A., and E. E. Porter. 2001. Area and the latitudinal diversity gradient for terrestrial birds. *Ecology Letters* 4:595–601.

Jablonski, D., K. Roy, and J. W. Valentine. 2006. Out of the tropics: Evolutionary dynamics of the latitudinal diversity gradient. *Science* 314:102–6.

Krug, A. Z., D. Jablonski, J. W. Valentine, and K. Roy. 2009. Generation of Earth's first-order biodiversity pattern. *Astrobiology* 9:113–24.

Pianka, E. R. 1966. Latitudinal gradients in species diversity: A review of concepts. *American Naturalist* 100:33–46.

Rohde, K. 1992. Latitudinal gradients in species diversity: The search for the primary cause. *Oikos* 65:514–27.

Rosenzweig, M. L. 1995. *Species diversity in space and time*. Cambridge University Press, New York.

Simpson, G. G. 1944. *Tempo and mode in evolution*. Columbia University Press, New York.

———. 1964. Species density of North American recent mammals. *Systematic Zoology* 13:57–73.
Valentine, J. W., D. Jablonski, A. Z. Krug, and K. Roy. 2008. Incumbency, diversity, and latitudinal gradients. *Paleobiology* 34 (2): 169–78.
Wiens, J. J., and M. J. Donoghue. 2004. Historical biogeography, ecology and species richness. *Trends in Ecology and Evolution* 19:639–44.
Willig, M. R., D. M. Kaufman, and R. D. Stevens. 2003. Latitudinal gradients of biodiversity: Pattern, process, scale, and synthesis. *Annual Review of Ecology, Evolution, and Systematics* 34 (1): 273–309.

PAPER 32

The Dynamics and Diversity of Insect Faunas (1978)

R. M. May

Commentary

ALLEN H. HURLBERT

In 1978, Robert May's concluding remarks to the Ninth Symposium of the Royal Entomological Society of London were published in a symposium volume on the diversity of insect faunas. Rather than pondering the question that many had posed before of why there are so many species, May instead asked, why are so many of all species insects? This subtle shift in the line of questioning led him, after considering the main hypotheses of the day, to conclude that most of the explanation for the relative diversity of insects "lies simply in the[ir] small size . . . which allows them to divide their environment much more finely than larger animals can" (194).

In exploring the relationship between body size and species richness, this paper was foundational in several respects. Most importantly, after presenting data for a number of vertebrate and invertebrate groups, May displayed in figure 12.7 the first attempt to characterize the size-richness relationship for all terrestrial animals. This figure illustrates what is now considered to be a fairly universal pattern: the diversity of small-bodied species, with the possible exception of the smallest size classes, is far greater than that of large-bodied ones. This work helped stimulate a large number of papers, including at least three in this volume (K. P. Dial and J. M. Marzluff 1988 [paper no. 38]; D. R. Morse, N. E. Stork, and J. T. Lawton 1988 [paper no. 16]; and J. H. Brown and P. F. Nicoletto 1991 [paper no. 18]), which sought to more thoroughly explore the nature of and reasons for this relationship.

The importance May ascribed to the relationship between body size and the grain at which organisms perceive their environment later became a central thesis for those applying fractal geometry to basic ecological patterns (e.g., Morse et al. 1985; Ritchie and Olff 1999). May's focus in this paper on the power law–like form of the size-richness relationship, and his brief consideration of other macroecological scaling relationships (richness to abundance, size to abundance), laid the groundwork for much current research that seeks to identify theoretical linkages among not only these patterns but also with species-area relationships, the scaling of home range to body size, and other patterns (e.g., Southwood et al. 2006).

Literature Cited

*Brown, J. H., and P. F. Nicoletto. 1991. Spatial scaling of species composition: Body masses of North American land mammals. *American Naturalist* 138:1478–1512.

*Dial, K. P., and J. M. Marzluff. 1988. Are the smallest organisms the most diverse? *Ecology* 69:1620–24.

Morse, D. R., J. H. Lawton, M. M. Dodson, and M. H. Wil-

In *Diversity of Insect Faunas*, edited by L. A. Mound and N. Waloff, pp. 188–204. Royal Entomological Society of London Symposium 9. Blackwell Scientific, Oxford.

liamson. 1985. Fractal dimension of vegetation and the distribution of arthropod body lengths. *Nature* 314:731–33.

*Morse, D. R., N. E. Stork, and J. H. Lawton. 1988. Species number, species abundance and body length relationships of arboreal beetles in Bornean lowland rain forest trees. *Ecological Entomology* 13:25–37.

Ritchie, M. E., and H. Olff. 1999. Spatial scaling laws yield a synthetic theory of biodiversity. *Nature* 400:557–60.

Southwood, T. R. E., R. M. May, and G. Sugihara. 2006. Observations on related ecological exponents. *Proceedings of the National Academy of Sciences* 103:6931–33.

12 • The dynamics and diversity of insect faunas

ROBERT M. MAY

Biology Department, Princeton University, Princeton, N.J., 08540

and

Imperial College Field Station, Silwood Park, Ascot, Berks.

> 'Entomologists can speak with fervor of the intricacies of taxonomic investigation. Among all biologists, they seem to have made the worst mess of it, characterizing so many families, genera, and species that they have far outstripped the whole field of their taxonomist brethren. This is not really their fault; it is merely a feeble attempt to sort out the avalanche of insects that Nature has lavished on the Earth.'
>
> Brues (1946, p. viii)

Several contributors to the present symposium have already remarked, with pardonable entomological chauvinism, that most species of living things are insects: insects account for something like 50–60% of all currently denumerated species (plants and animals), and around 90% of terrestrial animal species.

Any attempt to understand this diversity of insect faunas must ultimately deal both with absolute diversity (why are there around 10^6 species of organisms, rather than 10^8 or 10^4?), and with relative diversity (why are so many of these species insects?). My paper is addressed almost exclusively to the question of relative diversity.

The first part of the paper aims to review various of the dynamical and evolutionary factors that influence species diversity: these include single-species dynamics, competition, predation, food web structure, numerical abundance of the species, short- and long-term aspects of evolutionary rates, and evolutionary genetics. Some of these have been treated in greater depth earlier in the symposium. For each such factor, my emphasis is on determining in what way, if at all, it favours insects over other animals. The gist of the discussion is that, although all these factors help mould absolute diversity, most of them have little bearing on the diversity of insects relative to other organisms. It is argued that the main reason for the relative diversity of insects is their small size; this, coupled with their dispersal ability, enables them to carve the world into niches that are smaller (in both space and time), and thus more numerous. The point has, of course, been made by previous authors, and it is discussed more fully in Southwood's paper.

This conclusion leads to the second part of my paper, which is a preliminary attempt to determine the empirical relation between numbers of species of terrestrial animals and their size (length or weight). I think this is a fundamental relationship, and that it must be documented and understood before we can hope for any quantitative explanation of the relative diversity of insect faunas.

1: Factors influencing diversity

In his classic *Homage to Santa Rosalia*, Hutchinson (1959) asked 'why are there so many kinds of animals?' In what follows, we focus rather on the question 'why are there so many more kinds of insects than of other animals?'

1.1: SINGLE-SPECIES DYNAMICS

The past few years have seen significant advances in our understanding of the extraordinary richness of dynamical behaviour that is latent within the simplest of non-linear deterministic models (Li & Yorke, 1975; May, 1974*a*, 1976*a*); earlier work had tended to be confined to the analysis of small disturbances about equilibrium values. In particular, essentially all the first-order difference equations that have been propounded in the biological literature (as deterministic models for the behaviour of single-species populations with non-overlapping generations) have been shown to exhibit stable points, or sustained stable cyclic oscillations, or apparently chaotic fluctuations, depending on the values of intrinsic growth rates and other biological parameters (May & Oster, 1976). This work has innate mathematical interest. More importantly for biologists, the fact that simple deterministic models can lead to dynamical trajectories indistinguishable from the sample function of a random process holds disturbing implications both for the analysis and interpretation of data, and for hopes of long-term population forecasting.

Various people (Hassell *et al*, 1976; Stubbs, 1977) have attempted to estimate the parameter values in such models, by fitting them to population data. The parameter values thus estimated for field populations tend to lie in the domain corresponding to a stable equilibrium. Although this is a comforting conclusion, and one that is perhaps plausible on evolutionary grounds, it is not final. Quite apart from technical difficulties in estimating the relevant parameters, there are no truly single-species situations in the real world; once one has a multi-species situation, chaotic dynamical behaviour is likely to be more common (Guckenheimer *et al*, 1977), with all its attendant problems for data analysis and long-term predictions.

Despite these complications, it is often possible to estimate the population parameters for particular organisms, and to show how they correlate with life history strategies (for good reviews, see Southwood, 1976, 1977*a*). This permits the synthesis of a great deal of natural history information, and also paves the way for codifying the types of control strategies that are appropriate to particular kinds of insect pests (Southwood, 1977*b*; Conway, 1976; May, 1976*b*).

Reviewing this whole body of work, we see that the range of dynamical behaviour manifested by insect populations is not strikingly different from that of other organisms. Instances of, e.g., stable cycles can be drawn evenhandedly from insects or mammals: witness the almost physics-like fit between Nicholson's blowfly data and a simple model, and the use of the same model to give a somewhat more general explanation of the 3-to-4 year cycles of mice, voles and lemmings (May, 1976*c*, ch. 2). Examples of systems with multiple stable states are similarly ecumenical (May, 1977).

Insofar as insect dynamics differ from the dynamics of larger animals, it is that their comparatively small size and short generation time endow them with the capacity for relatively high intrinsic rates of population growth, r. This makes it easier for them to escape from superior competitors by accepting the boom-and-bust economy of the 'fugitive species' (Hutchinson, 1951; Southwood, 1977*a*). We will return to this below.

1.2: COMPETITION

In the 1920s and 1930s, the theoretical work of Lotka and Volterra, and the experiments of Gause, Park and others led to the enunciation of the competitive exclusion principle: species that make their livings in identical ways cannot coexist. More recently, Hutchinson, MacArthur and others have posed the more contentful question of how similar can species be, yet coexist? What are the limits to similarity, the limits to niche overlap and species packing?

Although most competitive situations are too complicated to unravel, insight can be gained from those special situations where a set of competitors sort themselves out mainly along one resource axis (such as food size, or foraging height). Hutchinson (1959) observed that there are many examples, including both vertebrates and invertebrates, of sequences of competing species in which each is roughly twice as massive as the next. This leads to ratios of around 1.3 in the linear dimensions of successive species. Many other examples that conform to the 1.3 ratio have since been given: e.g., for birds by MacArthur (1972), Diamond (1975), Cody (1974), for spiders by Uetz (1977); other examples are mentioned by Southwood and by Halkka (this symposium). Indeed Dyar (1890) long ago noted the closely related fact that successive larval instars of many insects have weight ratios of 2 and linear ratios of 1.3. This provoked much discussion (which seems to have been forgotten), and even speculation that the underlying mechanism was a doubling of the number of cells between instars (for a review, see Bodenheimer, 1933).

This empirical relation is still not understood. There is a growing body of theory (for a review, see May, 1976*c*, ch. 8) which suggests that, with regard to the size of food items, the average difference between two species should not be significantly less than the characteristic range of food sizes utilised by either species. This, however, does not explain why the average intraspecific range of food items typically spans a weight ratio of around 2.

It is worth noting that the Dyar–Hutchinson rule also holds for instars of children's bicycles, for sets of kitchen skillets, for ensembles of recorders, and for consorts of viols (Horn & May, 1977). The rule may well have more to do with assembling sets of tools, than with anything directly biological.

As far as the Dyar–Hutchinson rule goes, insects show no significant differences from vertebrates or other invertebrates. Although the limits to similarity and niche overlap are clearly central in determining absolute diversity, they appear to have little to do with the relative diversity of insect faunas.

1.3: PREDATION

As many writers have stressed (e.g., Connell, 1975), predation can profoundly modify the outcome of competition among the prey, thus increasing diversity.

A generalist predator, acting impartially on prey species that all have roughly equal intrinsic growth rates, makes no difference to competitive coexistence among the prey. Most vertebrate predators, however, concentrate their attacks disproportionately upon the prey species that happens to be most abundant at any one time. This 'switching' behaviour, which commonly derives from the way vertebrate predators form a 'search image', can enable the coexistence of prey species, of which some would otherwise be competitively excluded (Holling, 1959; Murdoch, 1969; Roughgarden & Feldman, 1975). Empirical and theoretical work has recently shown that many invertebrate predators also exhibit attack patterns that result in the predation rate on any one prey species being of the sigmoidal 'Type III' or 'vertebrate' form (Murdoch, 1977, and references therein; Hassell *et al*, 1977; Hassell, 1978). One pervasive mechanism whereby this can come about is if the spatial distribution of prey is patchy (either for a single prey species, or for a mixture of several prey species), and if the predators have searching behaviour that leads to differential aggregation in patches of high prey density; in a multi-prey situation, the upshot is tantamount to predator switching. (Incidentally, Beddington *et al* (1978) have made a convincing case that this combination of spatial heterogeneity and predator aggregation is the key mechanism in maintaining those arthropod prey-predator or host-parasitoid systems in which the prey population is well below the level set simply by the environmental carrying capacity.)

There is a second important way in which predation can enhance diversity, even in the absence of 'switching' effects. Generalised predation can promote coexistence if, among the prey, the inferior competitors have higher intrinsic growth rates.

We noted earlier that insects and other small animals can have relatively large values of r, which facilitates their playing the role of fugitive species. In addition, we now see that large r-values enable insects to exploit the second (non-switching) mechanism whereby predation can enhance diversity. These two factors, either separately or in conjunction, thus contribute to the relative diversity of insect faunas. It is my view, however, that this contribution is a comparatively minor one.

1.4: FOOD WEB STRUCTURE

As observed by Hutchinson (1959), trophic structure as such contributes little to species diversity. There are but few levels in even the longest food chains.

In this context, it should be noted that overall dynamical stability may be the main factor limiting trophic complexity (see, e.g., Pimm & Lawton, 1977, and references therein); this contrasts with the conventional explanation of the number of trophic levels, which sees them as set by considerations of energy flow. A corollary of this view is that complex ecosystems are likely often to have evolved as loosely coupled assemblies of simpler subsystems. This notion, which was originally advanced on abstractly theoretical grounds (May, 1974*b*), has gained some support from the empirical studies of Gilbert (1975 and this symposium), Lawton (this symposium) and others, and from D. S. Wilson's (1978) studies of models for the evolution of ecosystems. The notion is appealing, because it carries the implication that studies of interacting populations may have direct relevance at the ecosystem level (Lawton, 1976).

None of these broad aspects of food web topology, however, make appreciable contributions to the diversity of insects relative to other animals. Among the exceptions to this sweeping statement are the insect parasitoids. The biochemistry and life history of parasitoids are closely matched to their hosts, so that, from the standpoint of many of the host's predators, parasitised and unparasitised hosts are indistinguishable; the parasitoids have, in a sense, succeeded in slipping in an extra trophic level. But I think this is a tactical detail, contributing relatively little to overall insect diversity.

1.5: SPECIES VERSUS NUMBER OF INDIVIDUALS

The number of individual insects is, by virtue of their small size, vastly greater than the number of individuals of larger animals. May this numerical abundance, of itself, account for the larger number of insect species?

A crude answer may be given by borrowing the methods used in deriving the Preston–MacArthur–Wilson species-area relation. This theoretical relation is obtained by: (i) assuming a particular distribution of species relative abundance, (see, e.g., Taylor, this symposium), which then gives a relation between the total number of individuals, N, and the number of species, S; and (ii) assuming that N is linearly proportional to area A. Here we need only the first, and less dubious, of these assumptions. Preston (1962) and MacArthur and Wilson (1967) assumed a special ('canonical') member of the 1-parameter class of lognormal distributions of species relative abundance. This gives a complicated relation between N and S, which for large values of $S (S > 30$ or so) reduces to

$$S \sim N^z \tag{1}$$

with $z = 0.25$. More generally, relinquishing the 'canonical' hypothesis, reasonable lognormal distributions give eq (1), with z having values in the range $z \simeq 0.20$ to 0.35 (for a much more full discussion, see May, 1975). In short, the number of species does depend on the total number of individuals in a given taxonomic class, but only weakly (typically as the 1/4 power).

By making some further very crude generalisations, we can relate the total number of individuals, N, to their typical linear dimension, L. First, we recall Odum's (1968; see also Van Valen, 1973) observation that, within a given trophic level, the net productivity, P, is very roughly the same for organisms spanning a wide range of sizes (bacteria to deer): $P \sim$ constant. Second, biomass B may be roughly approximated as $B \sim PT_g$, where T_g is of the order of the generation time of the organism. Third, numerical abundance, N, scales as biomass divided by the mass of an individual, and hence, again roughly, $N \sim B/L^3$. Finally, noting that T_g scales approximately as L (Bonner, 1965), we arrive at $N \sim (\text{constant})/L^2$. Eq. (1) now takes the form

$$S \sim L^{-x} \tag{2}$$

with x somewhere in the neighbourhood of $x \simeq 0.5$.

I strongly emphasise that the species-size relation (2) pays regard *only* to statistical generalities about the distribution of relative abundance of individuals among species. It *ignores* ecological aspects of species' size (which is dealt with separately, below). Eq (2) suggests that a 100-fold decrease in the characteristic length of a group of organisms (from, say, 30 cm to 3 mm) will, by virtue of these statistical generalities alone, produce something of the order of a 10-fold increase in the number of species. As is shown in detail below (Figs. 12.1 to 12.7), this is nowhere near enough to explain the observed increase in species number with decreasing size.

1.6: EVOLUTIONARY RATES: A GRAND ARGUMENT

Compared with larger creatures, insects tend to have short generation times and high mortality rates. These two factors can accelerate evolutionary processes. As pointed out at different times by Mayr (1976), evolutionary rates may also be faster in very large populations or, alternatively, in species with many scattered subpopulations; both of these phenomena are preeminently exhibited by insects.

In short, the evolutionary clock ticks faster for insects and other small organisms. This is clearly seen in disease and pest systems, where significant evolutionary changes have taken place in decades, or even years, in response to antibiotics and insecticides.

If the living world showed a pattern of yet increasing and unsaturated diversification of organic forms, the faster 'evolutionary clock' of insects could account for their relatively great diversity. But the evidence assembled by Simpson (1953, 1969), and more recently reviewed by Raup (1977 and references therein; see also Coope, this symposium), shows a pattern of saturation which, for terrestrial animals, reaches back (with major and minor fluctuations, and with much relay and replacement) to the Permian, and possibly beyond. Indeed for insects Carpenter (1977, p. 69) has gone so far as to write 'In terms of diversity of form and the association of generalized and specialized species, the fauna of the Permian was probably the most diverse in the history of the Insecta'.

If it is accepted that the diversity of terrestrial faunas is roughly in an equilibrium or saturated state on a geological time scale, then insects' potential for faster evolutionary rates cannot be invoked as a direct explanation of their relative diversity.

1.7: EVOLUTIONARY RATES: A MORPHOLOGICAL ARGUMENT

A more detailed variation on the above theme is that insects' potential for fast evolutionary changes permits them to indulge in a greater amount of morphological 'fine-tuning' than is possible for taxa of bigger animals. The great diversification of insect mouth-parts, which contrasts with the evolutionary conservatism of the vertebrate mouth, could for example be attributed to this mechanism.

However, I think this puts the cart before the horse. The diversity of insect mouth-parts is more likely to be a secondary effect, deriving from many more niches being available to creatures of small size (see section 1.9 below). It seems unlikely that the tourist experience on the Serengeti plains would have been enriched by the presence of many more species, had the evolution of the vertebrate jaw manifested less conservatism; when the occasion has arisen, vertebrate mouth-parts have shown considerable adaptability (witness the baleen whales). It can be contended that vertebrate feet show as much functional diversity as insect feet.

This whole question is a complicated and unresolved one. It has, for example, been argued that the remarkable adaptive radiation and diversity of cichlid fishes stems from an 'evolutionary breakthrough' in the morphology of mouth-parts (see, e.g., Fryer & Iles, 1972, and references therein; for further discussion, see Sage & Selander, 1975). These issues are related to those treated in the next section.

1.8: EVOLUTION AND POLYMORPHISM

In a most interesting paper, Selander and Kaufman (1973) have compiled data from electrophoretic studies of protein polymorphisms in a variety of animals. They show that the average invertebrate individual is heterozygous at 15% of its loci, whereas the mean level of heterozygosity for vertebrates is 5%. Selander and Kaufman review possible causes and implications of their findings (see also Southwood, this symposium).

For our present purposes, it is sufficient to note that such intraspecific variability is the raw stuff with which evolution works. If this difference in polymorphism levels between vertebrates and invertebrates is taken at face value, it reinforces the arguments outlined above, to the effect that insects have the capacity for relatively rapid evolutionary responses. But there is an important caveat. Electrophoretic techniques test for qualitative differences between proteins coded by structural genes; they tell nothing about quantitative differences at the molecular level due to variation in regulatory genes (see, e.g., Feldman, 1978). This point is underlined by King and Wilson's (1974) demonstration that chimpanzees and humans differ to about the same degree as sibling species of *Drosophila*, if the difference is measured by structural genes (as revealed by gel electrophoresis). On the other hand, chimpanzees and humans do indeed differ morphologically at the taxonomic level conventionally attributed to them. This example supports the contention that in advanced groups, such as vertebrates in general and primates in particular, evolution primarily involves regulatory rather than structural genes.

Our current inability to assess variability at the genetic loci that code for timing and development means that we cannot say whether the overall genetic variability of insects is greater, less, or much the same as that of vertebrates. This unresolved problem lies at the heart of evolutionary genetics.

1.9: THE SIZE OF ANIMALS AND THE NUMBER OF NICHES

Each of the above factors has been argued to play little or no role in explaining the *relative* diversity of insects. I think most of the explanation lies simply in the small size of insects, which allows them to divide their environment much more finely than larger animals can; a given species of tree may be only part of an elephant's resource base, whereas it can be subdivided into a myriad of niches for tiny animals. This notion has of course been propounded by many people, including MacArthur and Levins (1964; see also MacArthur, 1971, and Levins, 1968), who introduced the concept of 'grain size'. Southwood (this symposium; see also Lawton, this symposium) has developed some of these ideas in a more quantitative way.

There have, however, been very few attempts to pursue these notions to get some estimate of the relative numbers of species of animals in given size classes. The main thing I know of is a difficult paper by Hutchinson and MacArthur (1959): within a given biotype, their model implies 'rapid increase in number of species up to a modal size and a slow decline in number, ideally asymptotic to unity, as the size increases'. At the large end of the spectrum, the number of species, S, falls off with the characteristic linear dimension of an animal, L, roughly as

$$S \sim L^{-2} \tag{3}$$

Such a relation implies that a 10-fold reduction in length (e.g., 3 cm to 3 mm) will see a 100-fold increase in the number of species. Hutchinson and MacArthur give some data, mainly from North American mammals, which tend to be in support. Other discussions of species-size relations have been largely confined to particular groups, and are reviewed below.

In essence, eq (3) comes from the assumption that animals see their environment as a 2-dimensional mosaic (e.g., the total surface of a tree) to be partitioned up, on a scale set by their perceived grain size (which goes as L^2). More generally, one could argue that it is volumes rather than surfaces that are divided, and that time is also a resource axis to be divided on a scale set by generation times (which tend to scale as L); thus a generalised version of eq (3) is

$$S \sim L^{-y} \tag{4}$$

with y somewhere in the neighbourhood of 2 to 3.

These fanciful speculations should not be pursued in advance of the facts. What we need is more empirical information than is currently available concerning the number of species as a function of size. Some such information is presented below. In each of the figures I have indicated (by a light, dashed line) the form of the theoretical relationship (3), $S \sim L^{-2}$; I did this more for entertainment than from any sense of conviction.

2: Number of species versus size

This part begins with some fairly exact results for the number of species versus size in particular groups of animals, and ends with a very crude estimate of the overall species-size relation obtained by pooling all terrestrial animals. For any one group (e.g., beetles), ecological aspects of the species-size relation tend to be masked by the group blending into ecologically similar, but taxonomically different, groups at both low and high ends of its size range.

2.1: PREVIOUS WORK

Van Valen (1973) has pulled together a vast amount of data, to discuss the relation between number of species (and genera) of mammals, birds and flowering plants and their sizes. Van Valen breaks up the groups in various ways, and discusses the implications of his results: this paper should be read. Figs. 12.1 and 12.2 are based on his results.

Schoener and Janzen (1968; see also Janzen and Schoener, 1968) have gathered data, based on some 10^4 sweep net samples, on the number of species, number of individuals and sizes of insects in several tropical (Costa Rica) and temperate (Massachusetts) locations. Their main aim was to understand how the faunal composition, and size, of insects varied among these different regions. Basing their analysis on Hemmingson's (1934) conclusion that, for a given taxon, the size distribution of species in a given region often approximates a lognormal, Schoener and Janzen give the mean and variance of the length of insects in various orders (weighted according to the number of individuals, or to the number of species). The lognormal, however, provides a Procrustean fit to the data, and the authors end up using a '3-parameter lognormal' to accommodate to the facts that the species-size distributions suggest a 'minimum size limit or threshold', and have 'relatively long tails to the right'. Janzen (1973*a,b*; 1977) has subsequently presented further results, based on a similarly heroic set of 800 sweeps at each of 25 sites in Costa Rica and on Caribbean islands. For beetles, bugs, and a lumped class of all other arthropods, he catalogues the number of individuals (but not the number of species) according to size classes. It would be nice to see all this, or similar, data reanalysed to give species-size distribution patterns of the kind shown below (rather than only means and diversity indices). One problem here is that sweep net samples may not give an unbiassed estimate of the species-size distribution in a given biotype; if relatively large-sized species range more widely (Southwood, this symposium), they may tend to be over-represented in such samples.

2.2: MAMMALS

Fig. 12.1 shows a log-log plot of the total number of terrestrial mammals (excluding bats) as a function of the animal's weight. The figure is obtained by aggregating Van Valen's (1973) results. Fig. 12.1 also gives the analogous histogram for all British mammals (again excluding bats), based on Southern (1964).

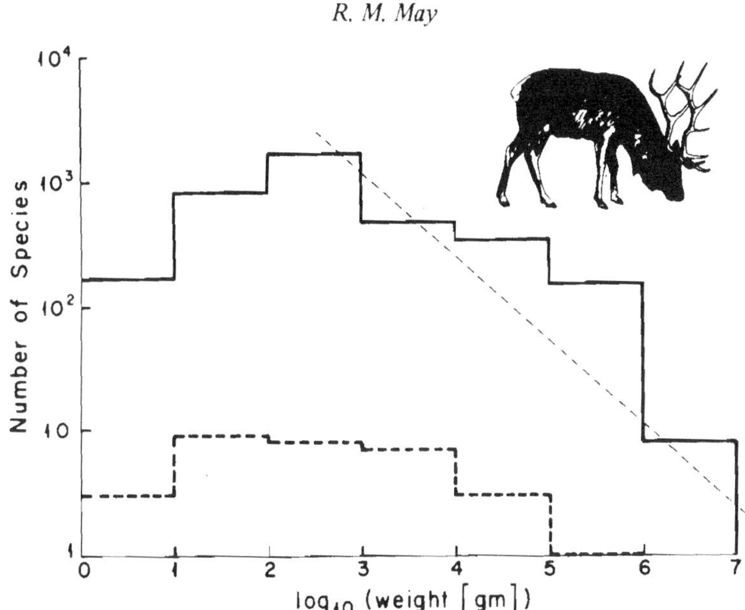

FIG. 12.1. Number of species of all terrestrial mammals (solid histogram) and of British mammals (dashed histogram), excluding bats, according to weight classes. Note the doubly logarithmic scale. The thin dashed line illustrates the shape of the relation (3), $S \sim L^{-2}$ (assuming weight scales as L^3 : see McMahon, 1973, and references therein).

Notice that the species-size distribution for British mammals is not very different from that for all terrestrial mammals, uniformly scaled down to allow for the much smaller number of species.

[Marine mammals have been omitted. So have bats, partly because their species-size distribution is more like that for passerine birds (Van Valen, 1973). The distribution for British bats is again close to a scaled-down version of the global distribution (not shown here). But Britain contains a larger fraction of the world's species of bats than of other mammals, and so a figure that combined all terrestrial mammals (including bats) would show the average British mammal to be noticeably smaller than the global average.]

2.3: BIRDS

Fig. 12.2 is also based on Van Valen's (1973) work, and shows the number of species of terrestrial birds according to weight classes. Birds whose way of life is mainly aquatic have been omitted; such birds are systematically heavier than most, and although their inclusion in Fig. 12.2 would run counter to the spirit of the exercise, it would give a much nicer fit to the dashed ($S \sim L^{-2}$) line. I have resisted this temptation.

2.4: BRITISH COLEOPTERA

As we move from large and aesthetically appealing animals down to small ones, we encounter the problem that collectors and systematists have tended to give more attention to larger insects. Thus a current count of Coleopteran species stands around 350 000, but it has been suggested the true count may be over one million.

One way of minimising this difficulty is to confine attention to relatively well-studied

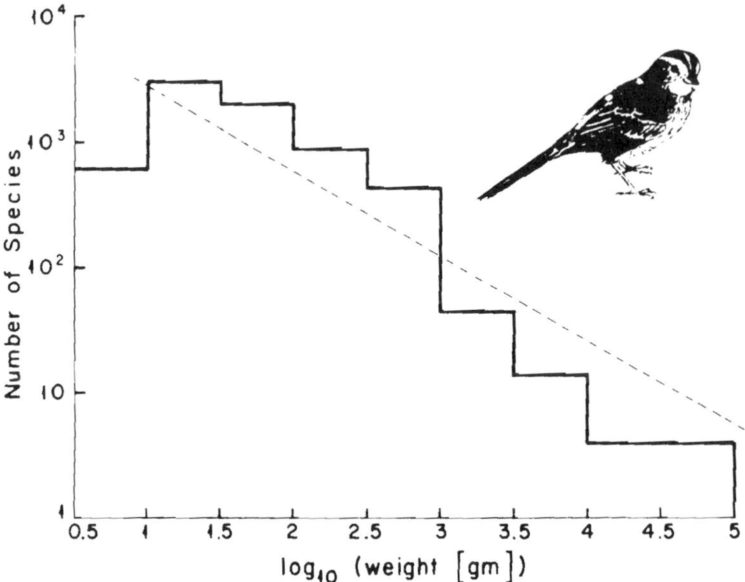

FIG. 12.2. The number of species of non-aquatic birds, classified according to weight. The dashed line is as in Fig. 12.1. (Data compiled by Van Valen, 1973).

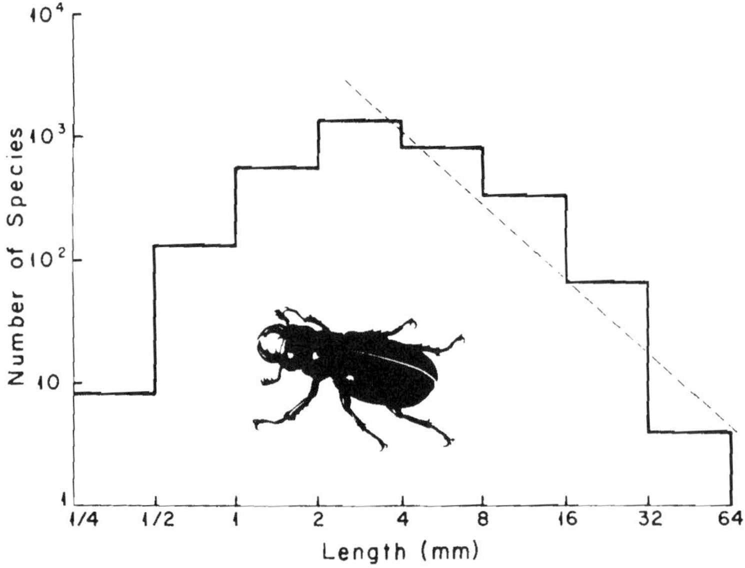

FIG. 12.3. The number of species of British Coleoptera, classified according to length. (Information taken mainly from Fowler, 1887.) The dashed line illustrates the shape of the relation (3), $S \sim L^{-2}$.

groups of insects. British Coleoptera are one such group, comprising some 3500 species. Fig. 12.3 is a log-log plot of British beetle species according to length classes, which I have compiled from Fowler's (1887) monograph. This figure invites comparison with Fig. 12.1 (all terrestrial mammals): the numbers of species on which the two figures are based are roughly equal; they span similar ranges (over two orders of magnitude in length classes); and they manifest similar shapes.

The purpose of Fig. 12.4 is to show that patterns are harder to discern if one gives a linear plot of species against size.

2.5: BRITISH LEPIDOPTERA

I have derived Fig. 12.5 mainly from Meyrick's (1927) monograph on the British Lepidoptera, another well-studied group. The figure displays the 2200-odd species according to length classes, using the linear dimension conventionally employed to characterise these creatures (the wing-spread or 'expanse of wings', defined by Meyrick as twice 'the distance from the tip of the forewing to the centre of the thorax'). This group shows a smaller size range (less than a factor of 100) than the British beetles.

2.6: BUTTERFLIES

Moving from Britain into the more biologically exciting tropical world, I have fastened on butterflies as likely to be the best-studied group of insects. Fig. 12.6 shows the species-size relation for butterflies (superfamily Papilionoidea, not including the Hesperioidea) in the Australian geographical realm, compiled with ruler and patience from D'Abrera (1971: even here the systematics of the smaller species is still in a state of flux, particularly among the Lycaenidae, which comprise more than one third of the butterfly species in the region).

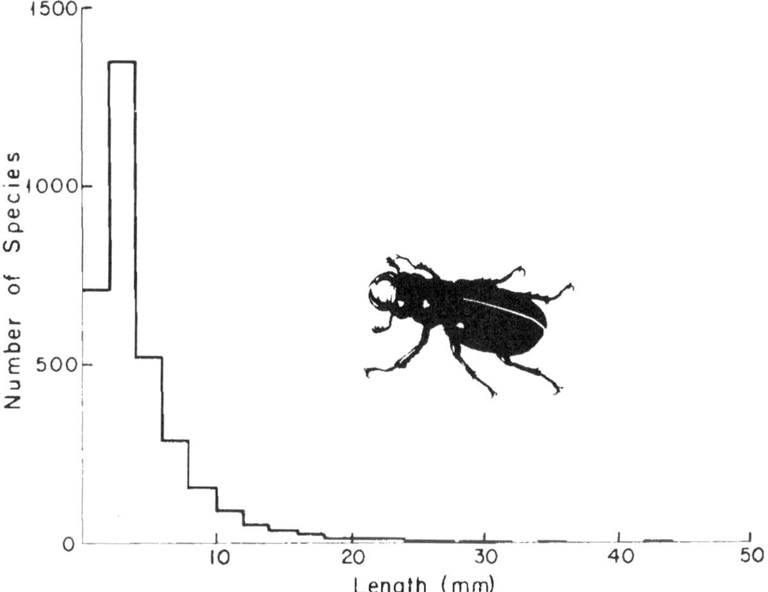

FIG. 12.4. Species-length histogram for British Coleoptera, as in Fig. 12.3, except here linear axes are used instead of logarithmic ones.

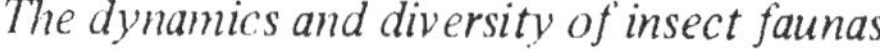

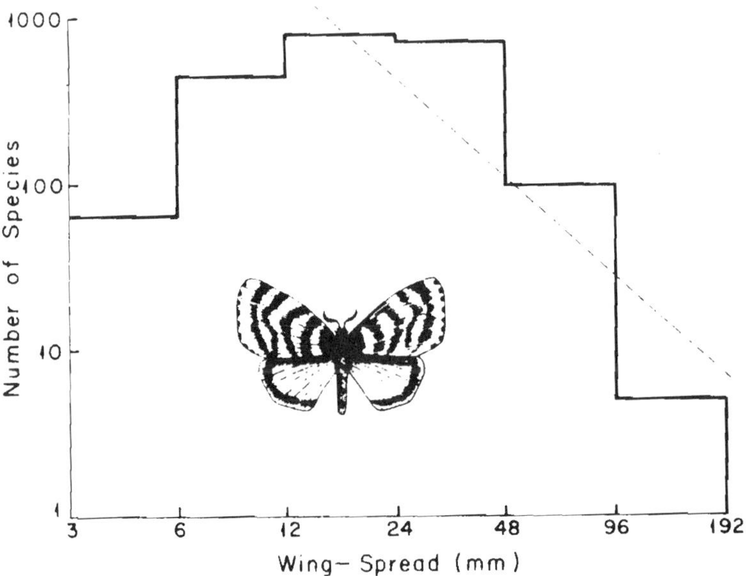

FIG. 12.5. The number of species of British Lepidoptera, classified according to wing-spread, as defined in the text. (Information taken mainly from Meyrick, 1927.)

The size distribution of the 60 species of British butterflies is shown for comparison. As in Fig. 12.1, the British distribution is not very different from a uniformly scaled-down version of that for the Australian realm. From such a naive scaling, the numbers of British species in the ascending size classes of Fig. 12.6 would be 2.6, 30.3, 23.5, 3.1, 0.5, which is to be compared with the actual numbers of 1, 31, 27, 1, 0.

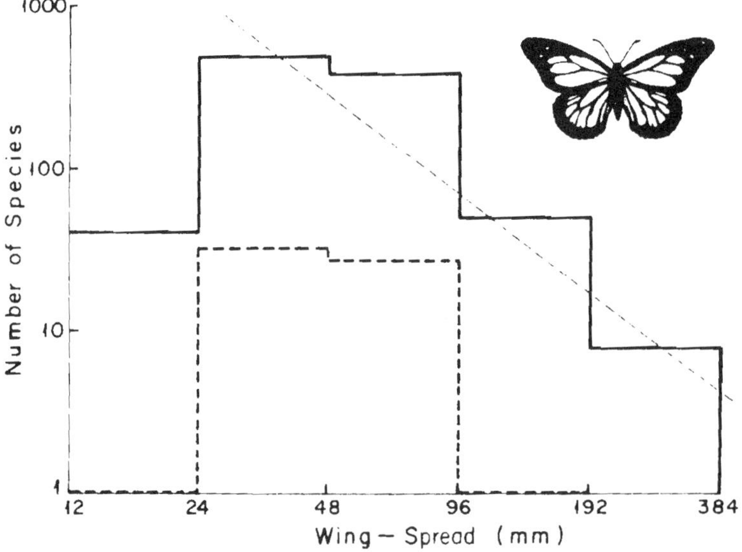

FIG. 12.6. The number of species of butterflies in the Australian geographical realm (solid histogram) and in Britain (dashed histogram), classified according to wing-spread. For discussion, see the text.

2.7: SIZE DIFFERENCES BETWEEN TROPICAL AND TEMPERATE INSECTS

Figs. 12.1 and 12.6 suggest the species-size distribution of British mammals and butterflies may be uniformly scaled-down versions of the global or tropical distributions, so that the mean sizes are the same. This is contrary to many people's intuition, which is that within most taxonomic groups British species are systematically smaller than tropical ones (e.g., Humphries, 1972). Such an intuition may derive in part from collections and monographs that give more thorough attention to large species of tropical insects than to small ones; some collections and writings treat races of larger species as carefully as they treat the smaller species themselves.

There is, however (to the contrary of Figs. 12.1 and 12.6), data to support the view that the typical species of tropical insect is larger than its temperate counterpart. Waloff (1954) has noted that, among the Acridoidea (a superfamily of Orthopterans), the average length among 31 temperate species is 25 mm, while for 60 tropical species it is 34 mm; this is only a fraction of the tropical species, and the sample could be biassed toward larger ones. The above-mentioned work of Janzen and Schoener gives a detailed discussion of the average sizes of several groups of insects in a variety of tropical and temperate environments. These results are very crudely summarised in Table 12.1. They exhibit a tendency for species to be bigger in the tropics. Janzen (1973*a,b*; 1975; 1977; see also Elton, 1973 and Enders, 1975) has advanced arguments to explain why this is so. But, as was noted above, sweep net samples could well be biassed toward a more complete representation of larger insect species, especially in the tropics. Such samples are not exhaustive lists in the way Figs. 12.1–6 are.

In short, it is not clear whether there are systematic average differences in sizes between tropical and temperate species of insects, nor whether such differences are present in some taxa and some regions and not in others. These questions deserve further exploration, in the field and in the museum.

2.8: SPECIES VERSUS SIZE FOR TERRESTRIAL ANIMALS

As was mentioned above, if we want a species-size relation in which the ecological aspects are not hidden by details of taxonomy and classification, we need to combine data for all

Table 12.1. The average species length (weighted according to number of species, not number of individuals) for various taxa of insects in tropical versus temperate habitats†

Taxon	Average tropical length (mm)	Average temperate length (mm)
Coleoptera	3.0	3.7
Formicidae	5.1	3.8
Other Hymenoptera	3.9	2.7
Homoptera	4.9	3.5
Hemiptera	9.3	6.3
Diptera	3.0	3.1
All insects	5.1	3.5

†I have obtained the average tropical length by the dubious procedure of taking an unweighted average over Schoener and Janzen's (1968) tropical Areas I, II and III, and the average temperate length by an unweighted average over their numbers for the months June–October.

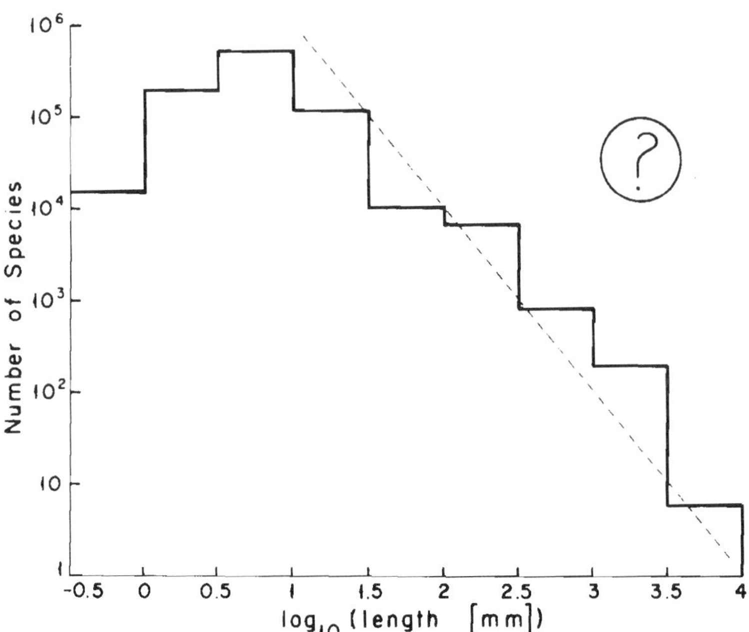

FIG. 12.7. The number of species of all terrestrial animals, classified according to length. The question mark serves to emphasise that this is a tentative figure, based on very crude approximations. The dashed line illustrates the shape of the relation (3), $S \sim L^{-2}$.

terrestrial animal species. Such information is hard to come by. Fig. 12.7 offers a first attempt at such a synoptic species-size distribution. This figure is based on crude approximations and outright guesses; it may be regarded as a cartoon, both in the older sense of a preliminary sketch, and in the contemporary sense of something risible.

Some of the main problems underlying Fig. 12.7 are as follows.

Bacteria, viruses, etc.,

As one drops below 1 mm and beyond, the number of species (protozoans, bacteria, viruses) decreases markedly. This could be a real phenomenon, and Janzen (1977) has given an interesting explanation of how it could come about. An alternative explanation is that conventional taxonomy begins to break down. The way bacteria exchange genetic information across species and generic lines, and the way R-factors and the like are incorporated in the bacterial chromosome, suggest that the concept of species is less well defined for bacteria than for larger animals. One could argue that at the bacteria-virus level one has a continuum of organic forms rather than distinct species, or alternatively one could argue that rates of evolution are so fast that the species concept is a loose one. In either event, it could be that the decrease in numbers of species at very small sizes is more apparent than real.

Small animals

It is widely acknowledged that the systematics of small arthropods and other invertebrates is, in most instances, in a rudimentary state. As Mayr (1969, p. 13) has written 'We must take it for granted that a large part of the mite fauna of the world will remain unsampled, unnamed, and unclassified [not to mention unwept, unhonoured, and unsung] for decades to come.' This could mean the size classes below 1 cm in Fig. 12.7 are underestimated by a factor 2 or more.

Other worries

The shape of Fig. 12.7 is determined almost entirely by the mammals, birds, reptiles and arthropods. For all except the five largest arthropod orders, I simply assigned a characteristic size to the order. For the five orders with the most species, I made a crude partitioning into size classes along lines guessed with the help of Figs. 12.3–6. (Detailed notes on the construction of Fig. 12.7 are available.) The saving grace in all this is that Fig. 12.7 is painted with a broad brush; a change of a factor of 2 in the number of species in any one size class would hardly alter the overall shape.

The relation sought in Fig. 12.7 is of fundamental biological importance. The pattern appears indeed to be very roughly described by eq (4), $S \sim L^{-y}$, with y a little less than 2. Although it would be silly to try to specify a precise y-value, it can be said that the figure is inconsistent with y-values as small as 1 or as large as 3.

The next step, I think, is to try for a more accurate version of Fig. 12.7. This should be possible, at least for the terrestrial animals in well-studied regions, such as Britain or (maybe) North America. Then the stage would be set for an explanation of the underlying ecological mechanism, and thence for an explanation of the relative diversity of insect faunas.

Although some people will find it dissatisfying, I find it pleasing that this symposium ends with several large questions, rather than with answers.

Acknowledgements

I have been helped by more people than I can list. Preeminent were M. P. Hassell, C. Istock, J. H. Lawton, S. McNeill, O. W. Richards, R. K. Selander, T. R. E. Southwood, D. W. Tonkyn, G. K. Waage, and N. Waloff. This work was supported in part by the US National Science Foundation, under grant DEB 77-01565.

References

Beddington J. R., Free C. A. & Lawton J. H. (1978) Modelling biological control: on the characteristics of successful natural enemies. *Nature, Lond.* **273**, 513–519.

Bodenheimer F. S. (1933) The progression factor in insect growth. *Q. Rev. Biol.* **8**, 92–95.

Bonner J. T. (1965) *Size and Cycle: an Essay on the Structure of Biology*. Princeton University Press, Princeton, N.J.

Brues C. T. (1946) *Insect Dietary* xxvi + 466 pp. Harvard University Press, Cambridge, Mass.

Carpenter F. M. (1977) Geological history and evolution of the insects. In, *Proc. XVth Int. Congr. Ent. Washington D.C.* 1976: pp. 63–70.

Cody M. L. (1974) *Competition and the structure of bird communities*. Princeton University Press, Princeton, N.J.

Connell J. H. (1975) Some mechanisms producing structure in natural communities. In *Ecology and Evolution of Communities*, Ed. M. L. Cody & J. M. Diamond, pp. 460–490. Harvard University Press. Cambridge, Mass.

Conway G. R. (1976) Man versus pests. In *Theoretical Ecology: Principles and Applications*, Ed. R. M. May, pp. 257–281. Blackwell Scientific Publications, Oxford.

D'Abrera B. (1971) *Butterflies of the Australian Region* 415 pp. Lansdowne Press, Melbourne.

Diamond, J. M. (1975) Assembly of species communities. In *Ecology and Evolution of Communities*, Ed. M. L. Cody & J. M. Diamond, pp. 342–444. Harvard University Press, Cambridge, Mass.

Dyar H. G. (1890) The number of moults of Lepidopterous larvae. *Psyche, Camb.* **5**, 420–422.

Elton C. S. (1973) The structure of invertebrate populations inside neotropical rain forests. *J. anim. Ecol.* **42**, 55–104.

Enders F. (1975) The influence of hunting manner on prey size, particularly in spiders with long attack distances. *Am. Nat.* **109**: 737–763.

Feldman M. W. (1978) *Genetic Variation in Natural Populations.* (in preparation).

Fowler W. W. (1887) *The Coleoptera of the British Isles* (6 Vols.). Reeve and Co., London.

Fryer G. & Iles T. D. (1972) *The Cichlid Fishes of the Great Lakes of Africa.* T. F. H. Publications, Neptune City, N.J.

Gilbert L. E. (1975) Ecological consequences of a coevolved mutualism between butterflies and plants. In *Coevolution of Animals and Plants*, Ed. L. E. Gilbert & P. H. Raven. University of Texas Press, Austin, Texas.

Guckenheimer J., Oster G. F. & Ipaktchi A. (1977) The dynamics of density dependent population models. *J. Math. Biol.* **4**, 101–147.

Hassell M. P. (1978) *Arthropod Predator-Prey System*. Princeton University Press, Princeton, N.J.

Hassell M. P., Lawton J. H. & May R. M. (1976) Patterns of dynamical behaviour in single-species populations. *J. anim. Ecol.* **45**, 471–486.

Hassell M. P., Lawton J. H. & Beddington J. R. (1977) Sigmoid functional responses by invertebrate predators and parasitoids. *J. anim. Ecol.* **46**, 249–262.

Hemmingsen A. M. (1934) A statistical analysis of the differences in body size of related species. *Vidensk. Meddr. dansk naturh. Foren.* **98**, 125–160.

Holling C. S. (1959) The components of predation as revealed by a study of small-mammal predation of the European pine sawfly. *Can. Ent.* **91**, 293–320.

Horn H. S. & May R. M. (1977) Limits to similarity among coexisting competitors. *Nature, Lond.* **270**, 660–661.

Humphries B. (1972) *The Wonderful World of Barry McKenzie*. Deutsch, London.

Hutchinson G. E. (1951) Copepodology for the ornithologist. *Ecology* **32**, 571–577.

Hutchinson G. E. (1959) Homage to Santa Rosalia, or why are there so many kinds of animals? *Am. Nat.* **93**, 145–159.

Hutchinson G. E. & MacArthur R. H. (1959) A theoretical ecological model of size distributions among species of animals. *Am. Nat.* **93**, 117–125.

Janzen D. H. (1973*a*) Sweep samples of tropical foliage insects: description of study sites, with data on species abundance and size distributions. *Ecology* **54**, 659–686.

Janzen D. H. (1973*b*) Sweep samples of tropical foliage insects: effects of seasons, vegetation types, elevation, time of day and insularity. *Ecology* **54**, 687–708.

Janzen D. H. (1975) *Ecology of Plants in the Tropics*. E. Arnold, London.

Janzen D. H. (1977) Why are there so many species of insects? *In, Proc. XV Int. Congr. Ent. Washington D.C.* pp. 84–94.

Janzen D. H. & Schoener T. W. (1968) Differences in insect abundance and diversity between wetter and drier sites during a tropical dry season. *Ecology* **49**, 96–110.

King M.-C. & Wilson A. C. (1974) Evolution at two levels in humans and chimpanzees. *Science* **188**, 107–116.

Lawton J. H. (1976) Mathematical models in ecology (book review). *Nature, Lond.* **264**, 138–139.

Levins R. (1968) *Evolution in Changing Environments*. Princeton University Press, Princeton, N.J.

Li T-Y. & Yorke J. A. (1975) Period three implies chaos. *Am. Math. Monthly* **82**, 985–992.

MacArthur R. H. (1971) Patterns of Terrestrial Bird Communities. In *Avian Biology, Vol. I.* pp. 189–221. Academic Press, New York.

MacArthur R. H. (1972) *Geographical Ecology*. Harper and Row, New York, N.Y.

MacArthur R. H. & Levins R. (1964) Competition, habitat selection, and character displacement in a patchy environment. *Proc. natn. Acad. Sci. U.S.A.* **51**, 1207–1210.

MacArthur R. H. & Wilson E. O. (1967) *The Theory of Island Biogeography*. Princeton University Press, Princeton, N.J.

McMahon T. (1973) Size and shape in biology. *Science* 179, 1201–1204.

May R. M. (1974*a*) Biological populations with nonoverlapping generations: stable points, stable cycles, and chaos. *Science* 186, 645–647.

May R. M. (1974*b*) *Stability and Complexity in Model Ecosystems*. Princeton University Press, Princeton, N.J.

May R. M. (1975) Patterns of species abundance and diversity. In *Ecology and Evolution of Communities*, Ed. M. L. Cody & J. M. Diamond, pp. 81–120. Harvard University Press, Cambridge, Mass.

May R. R. (1976*a*) Simple mathematical models with very complicated dynamics. *Nature, Lond.* **261**, 459–467.

May R. M. (1976*b*) Coexistence with insect pests. *Nature, Lond.* **264**, 211–212.

May R. M. (ed.) (1976*c*) *Theoretical Ecology: Principles and Applications*. Blackwell Scientific Publications, Oxford.

May R. M. (1977) Thresholds and breakpoints: ecosystems with a multiplicity of stable states. *Nature, Lond.* **269**, 471–477.

May R. M. & Oster G. F. (1976) Bifurcations and dynamic complexity in simple ecological models. *Am. Nat.* **110**, 573–599.

Mayr E. (1969) *Principles of Systematic Zoology*. McGraw-Hill, New York, N.Y.

Mayr E. (1976) *Evolution and the Diversity of Life: Selected Essays*. Harvard University Press, Cambridge, Mass.

Meyrick E. (1927) *A Revised Handbook of British Lepidoptera*. Watkins and Doncaster, London.

Murdoch W. W. (1969) Switching in general predators: experiments on predator specificity and stability of prey populations. *Ecol. Monogr.* **39**, 335–354.

Murdoch W. W. (1977) Stabilizing effects of spatial heterogeneity in predator-prey systems. *Theor. Pop. Biol.* **11**, 252–273.

Odum E. P. (1968) Energy flow in ecosystems: a historical review. *Am. Zool.* **8**, 11–18.

Pimm S. L. & Lawton J. H. (1977) Number of trophic levels in ecological communities. *Nature, Lond.* **268**, 329–331.

Preston F. W. (1962) The canonical distribution of commonness and rarity. *Ecology* **43**, 185–215 and 410–432.

Raup D. M. (1977) Probabilistic models in evolutionary biology. *Am. Sci. ent.* **65**, 50–57.

Roughgarden J. & Feldman M. (1975) Species packing and predation pressure. *Ecology* **56**, 489–492.

Sage R. D. & Selander R. K. (1975) Trophic radiation through polymorphism in cichlid fishes. *Proc. natn. Acad. Sci. U.S.A.* **72**, 4669–4673.

Schoener T. W. & Janzen D. H. (1968) Notes on environmental determinants of tropical versus temperate insect size patterns. *Am. Nat.* **101**, 207–224.

Selander R. K. & Kaufman D. W. (1973) Genic variability and strategies of adaptation in animals. *Proc. natn. Acad. Sci. U.S.A.* **70**, 1875–1877.

Simpson G. G. (1953) *Evolution and Geography: an Essay on Historical Biogeography with Special Reference to Mammals*. Oregon Univ. Press, Eugene, Oregon.

Simpson G. G. (1969) The first three billion years of community evolution. In *Diversity and Stability in Ecological Systems*, pp. 162–177. U.S. Department of Commerce, Springfield, Va.

Southern H. N. (ed.) (1964) *The Handbook of British Mammals*. Blackwell Scientific Publications, Oxford.

Southwood T. R. E. (1976) Bionomic strategies and population parameters. In *Theoretical Ecology: Principles and Applications*, Ed. R. M. May, pp. 26–48. Blackwell Scientific Publications, Oxford.

Southwood T. R. E. (1977*a*) Habitat, the templet for ecological strategies? *J. anim. Ecol.* **46**, 337–366.

Southwood T. R. E. (1977*b*) Entomology and mankind (opening address). In *Proc. XV Int. Congr. Ent. Washington D.C.* pp. 36–51.

Stubbs M. (1977) Density dependence in the life-cycles of animals and its importance in K- and r-strategies. *J. anim. Ecol.* **46**, 677–688.

Uetz G. W. (1977) Coexistence in a guild of wandering spiders. *J. anim. Ecol.* **46**, 531–542.

Van Valen L. (1973) Body size and numbers of plants and animals. *Evolution* **27**, 27–35.

Waloff N. (1954) The number and development of ovarioles of some Acridoidea (Orthoptera) in relation to climate. *Physiol. Comp. Oecol.* **3**, 370–390.

Wilson D. S. (1978) *Evolution on the Level of Populations and Communities*. (in preparation).

PAPER 33

Determinants of Diversity in Higher Taxonomic Categories (1980)

J. W. Valentine

Commentary

DAVID JABLONSKI

This paper outlines a conceptual model for one of the most striking large-scale evolutionary patterns in the fossil record: the early appearance of higher taxa (and, by inference, body plans or major adaptations) early in diversifications followed by a decline of that evolutionary creativity over time. Such patterns, entailing a strong temporal shift in the production of new higher taxa relative to the origin of species or genera, occur not only in the great Cambrian diversification of metazoan phyla, but also in the diversifications of marine invertebrate classes and orders and in major terrestrial radiations. J. W. Valentine models diversification as the spread of a clade through a landscape of ecological niches or adaptive zones. Originations are drawn from a strongly right-skewed size-frequency distribution of step sizes: most evolutionary steps are small, but a few are relatively large; small jumps establish new species, with increasingly large jumps establishing taxa of increasingly higher rank, all the way up to phyla. (Valentine is not claiming massive phenotypic leaps in the founding of higher taxa, but relatively rapid and extensive changes promoted by changes in gene regulation.) Occupied cells, or tesserae, cannot be invaded by new taxa, and the founding of higher taxa requires correspondingly larger open spaces. This process does not set an equilibrium in the strict sense but creates negative feedback that will tend to damp the production of higher taxa over time. Major, ecologically clumped extinctions will boost origination probabilities above the species or genus level, but even those will tend to create gaps filled in from the side via small jumps (i.e., via speciation from ecologically similar forms). Because the landscape is dynamic, with cells expanding and contracting, merging and splitting, the model is termed an adaptive kaleidoscope.

Valentine's highly influential paper provided both a conceptual framework for the study of diversifications—a hierarchical approach that broke free of simple population-genetic strictures—and a null model that could allow tests for the role of additional factors. Tests and extensions of the taxic approach into multivariate morphometric treatments (morphospaces), and into spaces defined by discrete character states, confirmed the widespread discordance between phenotypic evolution and species accumulation early in diversification (Foote 1997; Ruta et al. 2006; Erwin 2007; Jablonski 2007). The model is difficult to parameterize, but quantitative versions of the adaptive kaleidoscope model or related concepts have appeared (e.g., Walker and Valentine 1984; Valentine and Walker 1986; Gavrilets 1999). Following Valentine's lead, some authors have argued that the size-frequency distribution of step sizes may not have been constant through time, or that positive feedbacks on diversity via ecosystem engineering might figure more strongly in the dynamics, and these views should prompt new analyses. Thus the classic tension between genomic and ecological explanations for the cross-level discordance persists.

From *Paleobiology* 6:444–50. Reprinted with permission from the Paleontological Society.

Macroecologists have explored hierarchical models of diversity to a lesser extent. Higher taxa have been used as proxies for species-level richness, but functional, morphological, and taxic diversity can be partially decoupled in interesting ways (Wainwright 2007; Jablonski 2008). For example, species-to-genus ratios vary with latitude, bathymetry, and altitude, and the discordance may be stronger at higher taxonomic levels or with respect to morphospace occupation, with major implications for the assembly of biotas and the maintenance of their evolutionary structures.

Literature Cited

Erwin, D. H. 2007. Disparity: Morphologic pattern and developmental context. *Palaeontology* 50:57–73.

Foote, M. 1997. The evolution of morphological diversity. *Annual Review of Ecology and Systematics* 28:129–52.

Gavrilets, S. 1999. Dynamics of clade diversification on the morphological hypercube. *Proceedings of the Royal Society of London B* 266:817–24.

Jablonski, D. 2007. Scale and hierarchy in macroevolution. *Palaeontology* 50:87–109.

———. 2008. Biotic interaction and macroevolution: Extensions and mismatches across scales and levels. *Evolution* 62:715–39.

Ruta, M., P. J. Wagner, and M. I. Coates. 2006. Evolutionary patterns in early tetrapods. I: Rapid initial diversification followed by decrease in rates of character change. *Proceedings of the Royal Society of London B* 273:2107–11.

Valentine, J. W., and T. D. Walker. 1986. Diversity trends within a model taxonomic hierarchy. *Physica D* 22:31–42.

Wainwright, P. C. 2007. Functional versus morphological diversity in macroevolution. *Annual Review of Ecology, Evolution, and Systematics* 38:381–401.

Walker, T. D., and J. W. Valentine. 1984. Equilibrium models of evolutionary species diversity and the number of empty niches. *American Naturalist* 124:887–99.

Paleobiology, 6(4), 1980, pp. 444–450

Determinants of diversity in higher taxonomic categories

James W. Valentine

Abstract.—It is often assumed that, if a few species are introduced into a relatively empty environment, the subsequent diversification will take the form of a logistic growth curve, rising to an equilibrium level of species richness. The diversifications of taxa in higher categories commonly resemble logistic curves, although there are no well-defined theoretical bases for such a resemblance.

A model of diversification of taxa in higher categories is based on the notion that many taxa originate rapidly. Relatively small changes leading to new species occur at a high frequency, while larger changes leading to progressively higher taxa occur with progressive rarity. During diversification in an empty environment, few large changes will occur before the environment is filled. The rate of filling, relative to the rate of production of higher taxa, determines the richness of taxa in higher categories and gives the diversification curves a logistic appearance although the maximum level achieved is not an equilibrium. Subsequently, opportunities for diversification will generally lead only to the appearance of taxa in progressively lower categories.

James W. Valentine. Department of Geological Sciences, University of California, Santa Barbara, California 93106

Accepted: May 27, 1980

Introduction

This paper is an attempt to examine some of the factors that regulate the levels and patterns of diversity or richness of animal taxa within the highest categories. The factors that regulate the patterns of species diversity or richness, both locally and globally, have been widely discussed (a few general references include Connell and Orias 1964; MacArthur 1965; Sanders 1969; Valentine 1973; Pielou 1975; and Huston 1979). Many factors that affect local species richness have been identified, but there is little agreement as to their effects on global patterns. Authors that treat the geographic richness patterns of higher taxa usually consider them to be derivable from the richness patterns of species. Indeed, because the fossil record of species is particularly fragmentary, paleontologists are usually forced to use paleobiogeographic patterns of generic or familial richness as clues to the patterns of species (for example, most papers in Hallam 1973). The biogeographic patterns of the living biota provide empirical justification for some of these practices (Stehli 1970; Campbell and Valentine 1977).

Changes in levels of species richness through time have received much attention also. Again, because the fossil record is more fragmentary at the species level, the generic and familial levels are commonly employed as an estimate of species trends (Newell 1967; Valentine 1969; many papers in Hallam 1977). However, there is no theoretical requirement that the ratio of species to higher taxa remain constant, and indeed there is evidence that it has not (Valentine 1969). Nevertheless, it is usually considered reasonable to use genus or family richness as an indication of species richness (Valentine 1970; Sepkoski 1978; Raup 1979).

Richness trends at the very highest taxonomic levels—phylum and class—seem poorly correlated with species richness trends. There are so few taxa at these higher levels, and the range of their species contents varies so greatly, that a mean phylum/species or class/species ratio is not useful. Even if we restrict ourselves to the sea, and thus avoid the inflation of arthropod species numbers by the Insecta, phyla today vary in number of species from around 10 (Priapulida) to a hundred or so (Pogonophora) to thousands (Bryozoa) to tens of thousands (Mollusca)—a range of between three and four or-

0094-8373/80/0604–0008/$1.00

ders of magnitude. Extinction of a selected thousand marine species, less than 0.3%, could lower the number of phyla by 8 or 9, about 30%. The numbers of species within some phyla (such as the Brachiopoda) and classes (such as the Crinoidea), and orders (such as the Nautiloidea) and the proportions of species numbers within higher taxa, have changed greatly during the Phanerozoic.

The inclusiveness and rank of taxa within higher categories is partly a matter of tradition and taste. Different taxa within a category are not strictly comparable units, and unusual levels of monographic attention and/or atypical taxonomic practices can warp the taxonomic hierarchy of one group relative to others. Some of these difficulties are avoided here since we are concerned with patterns of branching and of morphological distance and not with any particular phylogenetic hypothesis.

Equilibrium Numbers of Higher Taxa?

Much theory concerning the richness of species assumes that at least at a given time there is some species capacity that cannot be overreached, or not for long. As this capacity is approached during evolutionary diversification, a feedback operates to slow diversification and/or to increase extinction rates. A diversity curve should rise from a founding lineage in an empty environment to the level of that environment's capacity according to a logistic pattern. Whether there is an equilibrium number of taxa in higher categories is a little-discussed question. Sepkoski (1978, 1979) has applied an equilibrium explanation to the patterns of early Phanerozoic diversity in families and orders. His results are quite interesting and suggest that the concept is useful. Can it be justified biologically?

One approach to this last question is to model the processes of diversification. To this end we employ a scenario based on present understanding of evolutionary processes and phenomena cast in the form of a game (Valentine 1980). The game is played with lineages on a mosaic board that represents environmental conditions. Each mosaic tile or *tessera* represents a relatively uniform region; conditions change more abruptly at boundaries than within tesserae. Some boundaries are quite sharp (as between tesserae representing land and those representing sea). In effect the board is an adaptive mosiac. Only one lineage at any time may occupy a given tessera, making the game an equilibrium model for lineages.

Diversification begins from a single founding lineage. Studies of diversification rates in the fossil record have provided evidence that there is more than one mode by which evolution has produced novelties. In many lineages, significant morphological change is compressed into short intervals of time associated with speciation (morphospeciation actually), with the remainder of the species' history spent in morphological stasis (as modeled by Eldredge and Gould 1972). Some genera have a similar pattern, in that the morphological change that signals the advent of a genus may occur in a shorter period of time than would be required to attain the morphological distance involved if it were due simply to morphospeciation events similar to those which created species within the genus (see Stanley 1978). That is, the rate of origination of genera commonly seems accelerated relative to the rate of origination of species. The same pattern appears to hold for some higher taxa up to and including phyla (Cloud 1949; Valentine 1972; Valentine and Campbell 1975; Gould 1977); the rate of morphological change involved in the appearance of many higher taxa exceeds the rates of morphological change involved in the appearance of their principal sub-taxa.

Classical microevolution, which must include changes in structural genes and in regulatory genes that have relatively small phenotypic effects, occurs in small steps. This is because large changes in the properties of gene products are less likely to have high fitness than are small steps. When microevolution can be traced in the fossil record it is a slow process, capable of transforming one species to another only over long time spreads and seemingly incapable of producing the rapid, large morphological changes required to found some new species, genera, and taxa in higher categories (see Stanley 1979). These more rapid changes are lumped as *macroevolution* and can be ascribed chiefly to changes in the genetic regulatory apparatus which elicit new patterns of expression among the structural genes (Britten and Davidson

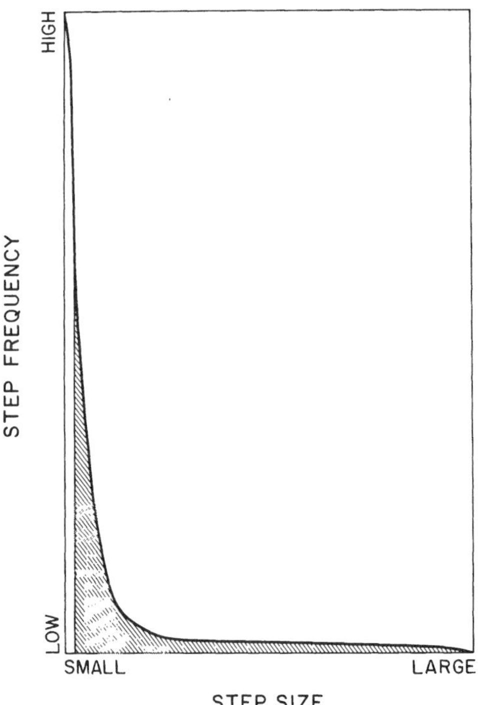

FIGURE 1. A frequency distribution of morphological steps based on the assumption that it is much easier to achieve viable small steps than viable large ones. The smallest changes (at the left) do not create new taxa but act to fine tune species to their environments. The diagonally-ruled area represents those evolutionary steps large enough to produce new taxa, from species (on the left) to phyla (on the right).

1971; Wilson 1975; Valentine and Campbell 1975). These changes must often involve heterochronies, such as neoteny. Microevolutionary and macroevolutionary processes are various, complex, and intergrading.

To return to the adaptive mosaic game, we permit diversification to occur in a range of modes and rates, from very small steps representing speciation events up to giant steps representing the creation of a new body plan at the phylum level. It is not implied that new phyla (or even species) necessarily arise during a single evolutionary event but that they can arise rapidly along a macroevolutionary pathway. It is assumed that it is much easier to achieve a small step than a large one. Therefore a distribution of steps is chosen as depicted in Fig. 1, with many small steps, few intermediates, and rare large steps. In real life, phylum-level steps must require a longer time to evolve than do, say, genus-level steps, but in the model all steps are taken in equal time. Distortions of diversification patterns do not necessarily result, however, for differential times of origin are accommodated by the differential rates of formation, in this case of phyla and genera, as controlled by the frequency distribution of step size in the pool of steps.

As the first move in the game, we draw a step at random (with replacement) from the pool of steps represented in Fig. 1. It is of course most likely to be a small step. A correspondingly small distance is marked off on the board from the parent lineage, along a direction chosen at random with each play. The tessera that we reach (most likely a tessera adjoining the one occupied by the founding lineage) is now occupied by the first daughter lineage. In the second move of the game, we draw another step and direction for the founding lineage and also a step and direction for the first daughter lineage, which also becomes a locus for diversification. Whether or not a daughter lineage alights successfully on a tessera is determined by simple rules. All daughters require an area free of other lineages in order to be viable. Daughter lineages resulting from larger steps require larger areas; daughter lineages resulting from the smallest (species) steps require only an empty tessera. If in the same move two or more lineages arrive at the same tessera, then the one representing the smaller step endures with the other(s) extinguished; ties are broken arbitrarily. If the products of the second move alight on tesserae where they are all viable, there are then four loci to generate steps in the third move. The game continues in this fashion until the board is filled.

A common pattern is such a game will be for the area of occupied tesserae to expand irregularly and gradually around the founding lineage by small steps. However, an occasional step will be drawn that is intermediate or even large and thus will found an outlying center of diversification from which spreading by small steps will proceed. In keeping with the pattern in the fossil record, the largest (phylum) steps must be very large relative to species steps so that even at their low frequency of appearance the average rate of spreading across the board is greater

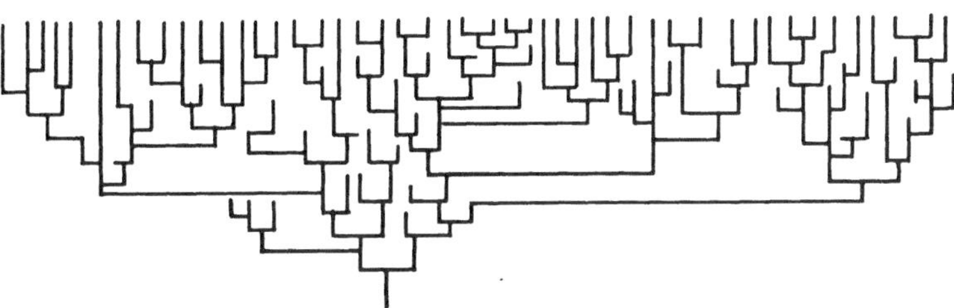

FIGURE 2. Trends of diversification through time. The vertical bars are lineages. The horizontal bars are steps; the longer the horizontal bar, the larger the step. Large steps occur more commonly during the early history of diversification. As time goes by and the environment fills, larger steps become increasingly rare. For clarity, the number of small steps envisioned relative to the number of large steps, is grossly underrepresented in the diagram.

at the phylum than at the species level. Progress across the board must come chiefly at the higher though rarer step sizes (taxonomic levels). Thus when the board margins are reached there is still a varied patchwork of many unoccupied tesserae scattered across the board. A cartoon of the succession of steps in a completed game displays the pattern in Fig. 2. As the tesserae become filled, the chances that a large step will be successful fall drastically. Since they are rare in any case, large steps will stop appearing, and intermediate steps will become rarer. Eventually, diversification will proceed only via the very common small steps. The resulting phylogenetic pattern resembles the pattern that we infer from the fossil record, and the various branches may be classified into a hierarchy of categories like the taxonomic hierarchy.

When this game is played on a given board, the number of higher taxa—larger steps—is regulated by the number of tesserae, the size of the larger steps, and the frequency with which large steps occur relative to the very much higher frequency of smaller steps. There is not an equilibrium of richness to which the number of higher taxa in any category will rise, or (as we see later) to which it will return if perturbed. Nevertheless, there is a certain range of richness within which any given higher category will usually fall, if the game is repeated from the beginning. The actual richness level is stochastic, varying about the same mean. The pattern of branching would nearly always resemble the one in Fig. 2. If the game board is quite large with respect to the size of the large steps, then there would of course be more large steps taken and the levels of richness of large steps in repeated games would tend to cluster relatively more closely about a mean than when the game board is small.

In earth history we have had only one game board and one game, although conditions on the board have varied much through time. If we assume that the filling of a nearly empty environment is a reasonable description of what went on during the metazoan radiation of late Precambrian and Cambrian times, then the logistic property of the curves for higher taxa originating during this time is explained. Although there is not an equilibrium level in the normal sense, there is a most likely target level (based on the frequency distribution of step sizes) which the actual level tends to approximate. As this level was approached, feedback from the environment (the shutting down of opportunities owing to the filling of tesserae by lineages) would cause a lowering of the rate of appearance of new higher taxa.

Changes in the Numbers of Higher Taxa

As mentioned above, real earth environments change in time, and these changes are at times so severe as to cause large-scale extinctions. In order to judge how this would affect the diversity of higher taxa, we must transform the game from a static to a dynamic basis. Let us modify the game board so that it changes through time, with tesserae expanding or contracting; some boundaries grow weak and disappear so that some tesserae merge, while other boundaries pinch in and split tesserae into two or more divisions. The adaptive mosaic now becomes an adaptive kaleidoscope. Lineages occupying the game board may become extinct, as when their tesserae disappear or change abruptly.

Some changes will be random and independent, so that tesserae vacated by extinct lineages are scattered across the board. Such scattered vacancies would nearly all be filled by spreading of descendants from occupied neighboring tesserae. Occasionally extinctions will be more contagious, sweeping whole domains of tesserae clear of occupants. Again the opportunity for diversification occurs, but again the open areas will be rapidly filled in from the margin. It would take a large area indeed before there is much chance that a successful step of the largest kind would find the open target before it fills in.

The extraordinary richness of pattern of taxonomic diversity that can arise from random events has been beautifully demonstrated by Raup et al. (1973). They generated clades by tracing founding lineages through a series of stages; at each stage lineages could either split, terminate or continue. The percentages of each of these events were imposed, but the outcome for any lineage was chosen at random for each stage. The resulting patterns of diversification, extinction and levels of diversity were of great variety and resembled patterns recorded by fossil clades.

The present model involves the whole hierarchy of taxa, but even phylum-level clades would occasionally be small ones, as for example when their growth was constrained by a high level of occupation of tesserae near the founding lineage. Small phyla and classes might occasionally be lost (Anderson 1974; Raup 1979). So long as the rate of appearance of the largest steps remains sufficiently lower than the rates of appearance of small steps, the space emptied by lost phylum-level clades would hardly ever be refilled by large steps but by small or perhaps intermediate ones. Only when extinctions returned the board to a state resembling the earliest stages of the game would new phyla be likely to appear. Similarly, new classes would tend to appear early though modally later than phyla and would be more likely than phyla to appear following extinction waves. Orders would peak later still and would be still more likely to arise in tesserae cleared by extinction.

If the adaptive kaleidoscope is allowed to operate until a large number of extinctions have occurred, the history of the pattern of higher taxa is easily described in outline. The phyla will be most diverse following initial filling of the environment and will then decline without replacement as extinctions happen to carry them off. Classes will peak later and will then suffer declines from extinctions. As we descend to lower taxonomic categories we will reach a point when the frequency of new steps of appropriate magnitude will be sufficiently high that some rediversification will usually occur after extinctions. For some still smaller class of step, the rediversification may restore diversity levels to previous highs. The point on the hierarchy that these distinct points will be reached depends, for a given extinction history, on the frequency of appearance of step size (Fig. 1). The actual levels achieved are not precisely predictable in any trial because the game involves stochastic processes, but the common trends are as described.

In this model the taxa that escape extinction tend to be the more diverse (or the luckier). Except for failed diversification attempts that alight upon or too near to occupied tesserae, extinction patterns are imposed independently of diversification. Therefore there is a strong tendency for clades to become extinct near the times of their origin when they are not very diverse, mitigated somewhat when they happen to appear long before a significant extinction event. Thus in the model, clades founded by phylum- or class-level changes will go extinct as frequently as clades founded by lesser changes, clade size for clade size. The times of peak diversification rates of taxa within a given category, which would be after earliest diversification but before the environment approaches saturation, should tend to be the times of greatest extinction rates for those taxa. Therefore peak class extinctions should tend to occur after peak phylum extinctions and so on down the taxonomic hierarchy. The relatively numerous higher taxa that appear and become extinct in the lower Paleozoic, sometimes regarded as "adaptive experiments," may simply represent this effect. Of course waves of mass extinction can cause extinction peaks at any time, and these may mask this effect, especially for taxa in lower categories.

Additional Considerations

The adaptive kaleidoscope game can generate diversities within a hierarchy of clades that behave much as the fossil record indicates that the actual taxonomic hierarchy behaved historically. To get this result requires a fairly complex game, but it is still far simpler than the processes it attempts to model. The game could be modified so as to incorporate further complications and to become more realistic. For example, the adaptive breadth of lineages can be varied so that some would occupy more than one tessera, thus gaining added protection from extinction. Since animals themselves provide adaptive opportunities, the game board might be expanded as lineages become numerous. Other improvements can easily be suggested. For example, the game does not consider those qualitative aspects of the lineages that might cause extinctions or diversifications to be nonrandom. However, if the game as it stands were quantified it would provide an indication of how much of the history of higher taxon richness can be explained as if it *were* random, an approach discussed by Raup et al. (1973). Many other plausible modifications will occur to the interested reader.

Our last qualitative point should be mentioned. It has been claimed that novelties often arise from rather "primitive" or "generalized" stocks rather than from "advanced" or "specialized" ones (especially by Cope 1896). A common explanation is that the primitive forms are more adaptively flexible and can be modified most easily. The advanced forms are more likely to be bundles of finely honed coadaptations suited to life in a narrow range of conditions and thus to form unpromising material for major modification. The point has been heavily disputed by some (see Mayr 1963). At any rate the adaptive kaleidoscope model does not consider this possibility. To the extent that it is true, the source lineages (for the larger steps at least) would be greatly restricted in the model, to founding lineages and their most immediate descendants—the more primitive members of their clade. This would certainly have a damping effect on diversification rates in higher categories, at least after the initial diversification had been well underway. It would increase the differential likelihood of the appearance of large steps between the early diversification and the later rediversifications that follow extinctions. Not only would rediversification at high levels be restricted to "primitive" sources, but these sources would tend to become fewer in time as extinctions took an accumulating toll of them. Thus, even if it is assumed that there has been no change in the frequency distribution of genetic changes that underlie micro- and macroevolution, the workings of history will tend to bias the distribution of potentially adaptive products of these changes, progresssively in favor of the smaller steps.

Conclusion

The adaptive kaleidoscope is not meant to provide an accurate simulation of the diversity of the taxonomic hierarchy during the Phanerozoic but rather to generate plausible explanations for the trends and patterns there, which can then be tested from details of the fossil record. If the model has some general validity, it indicates that three factors are chiefly responsible for diversity levels among the higher taxonomic categories: (1) the structure of the physical environment (including organisms viewed as habitats); (2) the degree of filling of the environment, or to put it another way, the pattern and breadth of opportunity presented by the unoccupied portion of the environment; and (3) the size-frequency distribution of evolutionary steps. This latter factor has not been considered as particulary important under the paradigm of the new synthetic theory of evolution but may be the most significant single factor regulating diversity at higher taxonomic levels.

Acknowledgments

Dr. J. K. S. Walker, University of Michigan, kindly reviewed game details and was instrumental in achieving biologically plausible game rules. For critical review and additional improvement of this manuscript, I am grateful to Stanley Awramik, University of California, Santa Barbara; Cathryn A. Campbell; David M. Raup, Field Museum of Natural History, Chicago; Charlotte Schreiber, Lamont-Doherty Geological Observatory, Palisades, New York; and J. J. Sepkoski, University of Chicago. The image of a kaleidoscope as an environmental model was suggested long ago by Daniel Axelrod. These ideas have grown from a talk prepared for the Conference on Intelligent Life in the Universe, sponsored by NASA at Ames Research Center, Moffett Field, California.

Literature Cited

ANDERSON, S. 1974. Patterns of faunal evolution. Q. Rev. Biol. *49*:311–332.

BRITTEN, R. J. AND E. H. DAVIDSON. 1971. Repetitive and non-repetitive DNA sequences and a speculation on the origins of evolutionary novelty. Q. Rev. Biol. *46*:111–133.

CAMPBELL, C. A. AND J. W. VALENTINE. 1977. Comparability of modern and ancient marine faunal provinces. Paleobiology. *3*:49–57.

CLOUD, P. E. 1949. Some problems and patterns of evolution exemplified by fossil invertebrates. Evolution. *2*:322–350.

CONNELL, J. H. AND E. ORIAS. 1964. The ecological regulation of species diversity. Am. Nat. *98*:399–414.

COPE, E. D. 1896. The Primary Factors of Evolution. Open Court; Chicago, Illinois.

ELDREDGE, N. AND S. J. GOULD. 1972. Punctuated equilibria: an alternative to phyletic gradualism. Pp. 82–115. In: Schopf, T. J. M., ed. Models in Paleobiology. Freeman; San Francisco, California.

GOULD, S. J. 1977. Ontogeny and Phylogeny. 501 pp. Harvard Univ. Press; Cambridge, Massachusetts.

HALLAM, A., ED. 1973. Atlas of Palaeobiogeography. 531 pp. Elsevier Sci. Publ. Co.; Amsterdam, London and New York.

HALLAM, A., ED. 1977. Patterns of Evolution as Illustrated by the Fossil Record. 591 pp. Elsevier Sci. Publ. Co.; Amsterdam, London and New York.

HUSTON, M. 1979. A general hypothesis of species diversity. Am. Nat. *113*:81–101.

MACARTHUR, R. H. 1965. Patterns of species diversity. Biol. Rev. *40*:510–533.

MAYR, E. 1963. Animal Species and Evolution. 797 pp. Harvard Univ. Press; Cambridge, Massachusetts.

NEWELL, N. D. 1967. Revolutions in the history of life. Geol. Soc. Am. Spec. Pap. *89*:63–91.

PIELOU, E. C. 1975. Ecological Diversity. Wiley; New York.

RAUP, D. M. 1979. Size of the Permo-Triassic bottleneck and its evolutionary implications. Science. *206*:217–218.

RAUP, D. M., S. J. GOULD, T. J. M. SCHOPF, AND D. S. SIMBERLOFF. 1973. Stochastic models of phylogeny and the evolution of diversity. J. Geol. *81*:525–542.

SANDERS, H. L. 1969. Benthic marine diversity and the stability-time hypothesis. Brookhaven Symp. Biol. *22*:71–81.

SEPKOSKI, J. J., JR. 1978. A kinetic model of Phanerozoic taxonomic diversity. I. Analysis of marine orders. Paleobiology. *4*:223–251.

SEPKOSKI, J. J., JR. 1979. A kinetic model of Phanerozoic taxonomic diversity. II. Early Phanerozoic families and multiple equilibria. Paleobiology. *5*:222–251.

STANLEY, S. M. 1978. Chronospecies' longevities, the origin of genera, and the punctuational model of evolution. Paleobiology. *4*:26–40.

STANLEY, S. M. 1979. Macroevolution, Pattern and Process. 332 pp. Freeman; San Francisco, California.

STEHLI, F. G. 1970. A test of the earth's magnetic field during Permian time. J. Geophys. Res. *75*:3325–3342.

VALENTINE, J. W. 1969. Patterns of taxonomic and ecological structure of the shelf benthos during Phanerozoic time. Paleontology. *12*:684–709.

VALENTINE, J. W. 1970. How many marine invertebrate species? A new approximation. J. Paleontol. *44*:410–415.

VALENTINE, J. W. 1973. Evolutionary Paleoecology of the Marine Biosphere. 472 pp. Prentice-Hall; Englewood Cliffs, New Jersey.

VALENTINE, J. W. 1980. The emergence and radiation of multicellular organisms. In: Billingham, J., ed. Life in the Universe. M.I.T. Press; Cambridge, Massachusetts. In Press.

VALENTINE, J. W. AND C. A. CAMPBELL. 1975. Genetic regulation and the fossil record. Am. Sci. *63*:673–680.

WILSON, A. C. 1975. Evolutionary importance of gene regulation. Stadler Symp. *7*:117–133.

PAPER 34

Two Decades of Homage to Santa Rosalia: Toward a General Theory of Diversity (1981)

J. H. Brown

Commentary

WALTER JETZ

Five decades after G. E. Hutchinson's influential "Homage to Santa Rosalia" and three decades since J. H. Brown's passionate reflections, both papers and the quest for understanding biodiversity remain as topical as ever. In a symposium with several mathematically inclined theoreticians as fellow speakers, Brown delivered a bold call for an essentially macroecological understanding of biodiversity patterns that in some ways was as much ahead of its time as Hutchinson's paper was 20 years earlier.

Few if any papers in the 20th century had as much influence on community ecology as Hutchinson's concluding remarks to the 1958 metting of the American Society of Naturalists. But not all of the work that followed (much of it by R. H. MacArthur and his students) and that then became the ecological theory of the 1960s and 1970s offered equally rewarding avenues in helping answer Hutchinson's original question of why there are so many species. Brown offered a clear verdict: in this regard, studies into the dynamics of coexistence and population interactions, ever popular at the time, have disappointed. But another approach that is more heuristic and instead takes an equilibrium view has highlighted a road of success: island biogeography theory—to this day one of the most successful attempts to integrate evolutionary and ecological processes at the broad scale. Brown elegantly proceeded to deliver a wake-up call to the many ecologists at the time who may have become disillusioned, and encouraged them to look beyond single communities and to refocus on Hutchinson's theme of energetics that until then had been relatively neglected. He offered two clear pathways: the way in which energy availability determines the capacity for a place to support diversity, and the way in which this capacity is apportioned ("allocated") across constituent members.

Brown's ideas build on previous observations about the geographic variation in richness by Hutchinson, MacArthur (1972), J. Terborgh (1973), J. J. Schall and E. R. Pianka (1978), and others. But they are radical in their synthesis, vision, and heuristic drive. The productivity-diversity relationship plots presented in figure 1, until today a dominant theme in ecology, were the first of their kind. These patterns, as much as the intuitive appeal of their presentation, injected fresh energy in the search for the causes of biodiversity gradients, and stimulated extensive work in the three decades since that confirmed the strength and generality of productivity-richness relationships worldwide.

Data sets have come a long way since the days of the "computer-generated map" used for Brown's figure 1, but this does not take away from the original vision. The conceptual and empirical links between individuals, their abundance and energetic needs, species and available energy, were later developed further by Brown's student D. H. Wright (1983) and others. These form the classic ingredients of a macroecological approach. Brown's paper sets the agenda for a macroecological perspective

From *American Zoologist* 21:877–88. Reprinted with permission from Oxford University Press.

on the why and where of biodiversity that proves fruitful to this day. In contrast to other presenters at this symposium, this paper advocated, over a more mathematical algorithmic approach, the empirically driven heuristic perspective that then came to characterize macroecology. This sort of debate about the right ingredients of a theory characterizes ecology to this day. As does the pursuit to answer Hutchinson's 50-year-old question.

Literature Cited

Hutchinson, G. E. 1959. Homage to Santa Rosalia; or, Why are there so many kinds of animals? *American Naturalist* 93:145–59.

MacArthur, R. H. 1972. *Geographical ecology*. Harper and Row, New York.

Schall, J. J., and E. R. Pianka. 1978. Geographical trends in numbers of species. *Science* 201: 679–86.

Terborgh, J. 1973. On the notion of favorableness in plant ecology. *American Naturalist* 107:481–501.

*Wright, D. H. 1983. Species-energy theory: An extension of species-area theory. *Oikos* 41:496–506.

Amer. Zool., 21:877–888 (1981)

Two Decades of Homage to Santa Rosalia: Toward a General Theory of Diversity[1]

James H. Brown

Department of Ecology and Evolutionary Biology, University of Arizona, Tucson, Arizona 85721

Synopsis. In 1959, in his seminal paper "Homage to Santa Rosalia," G. E. Hutchinson asked, Why are there so many kinds of organisms? This paper focused attention on problems of species diversity and community organization that have occupied many theoretical and empirical ecologists for the last two decades. In the present paper I evaluate the attempt to answer Hutchinson's question by considering three topics. First, I reexamine the main themes which Hutchinson developed in "The Homage" and call attention to the central importance of energetic relationships in his view of ecological communities. Second, I examine the development of theoretical community ecology over the last two decades in an attempt to determine why some avenues of investigation, such as competition theory, have proven disappointing, whereas others, such as the theory of island biogeography, have enjoyed at least modest success. Finally, I suggest that future attempts to understand patterns of species diversity might focus on developing two kinds of theoretical constructs: capacity rules, which describe how characteristics of the physical environment determine its capacity to support life, and allocation rules, which describe how limited energetic resources are subdivided among species.

Introduction

One of the greatest remaining challenges in biology is to explain the diversity of living things. What determines the number and kinds of animals, plants and microbes that live together in one place? What causes variation in diversity from one place to another? What accounts for changes in the abundance and identity of species over time? How do individual species contribute to the diversity and stability of the natural world? These are largely problems of community and ecosystem ecology. Historical evolution and biogeography provide descriptive information about the development of diversity, and the process of natural selection accounts for the adaptations of individual species to their particular environments. But the capacity of the environment to support species is an ecological property. This capacity is determined ultimately by the physical environment: by availability of energy and other physical and chemical requisites for life and by physical heterogeneity in time and space. More proximally, the capacity of the environment to support species depends on how essential resources are apportioned among species: on relationships between organisms and their physical environment, and on interspecific interactions among the organisms themselves. Each individual species plays a unique role in the ecosystem as a whole; it potentially influences every other species through its interspecific interactions and its effects on the physical environment.

The attempt to provide a general, theoretical explanation for the diversity of living things has a long, distinguished history. Darwin and Wallace made the first great advances, and Lotka and Volterra, Grinnell and Gause, Clements and Shelford, Elton, Lack and Lindemann all made important contributions. In 1959 G. E. Hutchinson in his "Homage to Santa Rosalia" asked, Why are there so many kinds of living things? This seminal paper provided a general synthetic overview of the problems posed by organic diversity and suggested where we might look for the answers. The last two decades have witnessed a tremendous burst of activity as numerous scientists, from field naturalists to mathematicians, have grappled with Hutchinson's question.

How successful has been this effort to develop a general theory of organic diversity? Agreeing to participate in the present symposium has forced me to address this

[1] From the Symposium on *Theoretical Ecology* presented at the Annual Meeting of the American Society of Zoologists, 27–30 December 1980, at Seattle, Washington.

question. It has been a difficult, but valuable experience. I have concluded that the record of the last two decades has been very mixed. There have been both successes and failures; some approaches have proven disappointing, but others continue to be promising. Perhaps the greatest disappointment has been emotional. Many of the brightest and most ambitious scientists of the last generation committed their careers to theoretical ecology. Their feverish activity led to great expectations, which have not yet been fulfilled. Despite two decades of intensive investigation, Hutchinson's question remains largely unanswered. Perhaps it is a good time to reflect on the record of the past and to assess the prospects for the future.

THE HOMAGE: AN EMPHASIS ON ENERGETICS

"Homage to Santa Rosalia" is well worth reading, or rereading, more than two decades after its publication. If one considers the state of community ecology in 1959, it was an exceptionally original and insightful contribution. If one considers the present state of the discipline, it still has much to say. Although Hutchinson scarcely mentions the word energy, his basic message is that to understand the diversity of life we should investigate how usable energy is acquired by and apportioned among species. In his final introductory paragraph (p. 147) he states, "In any study of evolutionary ecology, food relations appear as one of the most important aspects of the system of animate nature. There is quite obviously much more to living communities than the raw dictum 'eat or be eaten,' but in order to understand the higher intricacies of any ecological system, it is most easy to start from this crudely simple point of view." In the remainder of the paper Hutchinson applies this dictum to speculate on the causes and limits of organic diversity.

Hutchinson develops five major themes. These are worth reexamining briefly in view of recent work. First, Hutchinson considers the length and number of food chains. He suggests that the length of food chains is limited by the attenuation of energy as it is passed with low efficiency from one trophic level to the next. He points out that body size and life history characteristics limit the trophic roles that certain species can play. He notes that the enormous diversity of terrestrial animals, which is much greater than that of aquatic ones, can probably be attributed not only to the large number of terrestrial plant species, but also to the architectural and functional diversity of individual plants. Both of these characteristics of terrestrial plants encourage the evolution of herbivorous animals, especially insects, specialized to feed on particular plant species and on certain parts of individuals of those species. Hutchinson then proceeds to discuss the weblike interrelations among food chains and to consider the relationship between structural complexity and dynamic stability. He incorrectly states (p. 149) that MacArthur (1955) has provided ". . . a formal proof of the increase in stability of a community as the number of links in its food web increases." But his main point is that many simple communities should be subject to invasion by new species as a result of colonization or speciation processes, so these communities would increase in diversity and complexity until some upper limit to diversity is reached.

Hutchinson's third theme concerns the effects of productivity and habitable area in limiting diversity. In a very perceptive passage (p. 150) he states "If we can have one or two species of a large family adapted to the rigors of Arctic existence, why can we not have more? It is reasonable to suppose that the total biomass may be involved. If the fundamental productivity of an area is limited by a short growing season to such a degree that the total biomass is less than under more favorable conditions, then the rarer species in a community may be so rare that they do not exist." He goes on to note that area itself may exert a similar influence, so that small islands typically contain fewer species with larger niches than nearby larger islands or mainlands. He attributes these patterns in part to the effects of interspecific competition affecting the allocation of limited energy sources among species. This theme is developed further in the next section, where Hutchinson considers how similar species can be in their requirements and still coexist. He notes that there appears to be a definite limit, which among closely related species is often reflected in particular ratios (which are now called Hutchinson's ratios) of body size or dimensions of trophic appendages.

Hutchinson's final theme concerns what he calls the "mosaic nature of the environment." In a few well chosen sentences he anticipates much of what has later been said about the concept of environmental grain. He points out that within the constraints inherent in any taxonomic group or trophic level, small organisms almost always outnumber large ones, because (p. 155) ". . . small size, by permitting animals to become specialized to the conditions offered by small diversified elements in the environmental mosaic, clearly makes possible a degree of diversity quite unknown among groups of larger organisms."

Hutchinson summarized his arguments by concluding (p. 155) that ". . . the reason why there are so many species of animals is at least partly because a complex trophic organization of a community is more stable than a simple one, but that limits are set by the tendency of food chains to shorten or become blurred, by unfavorable physical factors, by space, by the fineness of possible subdivision of niches, and by those characteristics of the environmental mosaic which permit a greater diversity of small than of large allied species."

Subsequent Community Theory: Energetics Ignored

Most of the themes which Hutchinson developed in "The Homage" were not totally new, but the synthesis of these ideas to address the problem of diversity represented a major innovative achievement. Hutchinson used energetics as the basic currency of ecological interaction to provide a unified conceptual foundation for his arguments.

During the last two decades most of the ideas in "The Homage" have been explored by theoretical ecologists, but the central importance of energetics has largely been ignored. The reason for this appears to have much to do with personalities. By the beginning of the 1960s a growing number of bright, highly motivated scientists were trying to understand the structure and function of complex ecological systems containing large numbers of species. These investigators rapidly sorted themselves into two schools, each influenced by a single dominant scientist; these schools divided the enormous task confronting them into two largely nonoverlapping spheres of influence. One of these schools, led by Eugene Odum, followed the intellectual traditions of Clements, Shelford, Lindemann, and Elton, and viewed the ecosystem, if not as a superorganism in a strict Clementsian sense, at least as a complete holistic system with interesting and important emergent properties. This school, the ecosystem ecologists, concentrated on the flow of energy and matter through the community and emphasized interactions between organisms and the physical environment. In focusing on the ecosystem as a whole, they largely ignored the diversity of the component species and the unique role that each plays. The other school followed Hutchinson's student Robert MacArthur and the traditions of Lotka, Volterra, Gause, Grinnell and Lack. This school, the evolutionary ecologists, concentrated on the ecological and evolutionary interactions between species. They were concerned directly with understanding community organization and species diversity, and they tried to expand the competition models of Lotka and Volterra and the newly developed models of predation (Rosenzweig and MacArthur, 1963) to account for the coexistence of species in increasingly complex communities. Their models were based on changes in population size of one species as a function of population density of other species, however, and they virtually ignored energetics as an explicit currency of ecological interaction. For the past 20 years, evolutionary ecologists have tried to answer Hutchinson's question, "Why are there so many species?" without adopting the synthetic viewpoint based on energetics, that Hutchinson had used to frame the question and to suggest possible answers.

My general assessment is that evolutionary ecologists have learned a great deal about the adaptive strategies of particular species and the basic kinds of interspecific interactions: competition, predation and mutualism. They have been much less successful in determining how these species and their interactions contribute to the structure and function of communities. I believe there are two reasons for this. One is the failure to appreciate the fundamental underlying role of energetics. The other is the inherent difficulty in trying to understand the organization of complex systems by working from the bottom up, attempting to recreate the whole system by assembling the component parts in proper relationship to each other. A brief examination of the history of science suggests that success in understanding complex systems usually comes from dealing with them on their own terms, taking them apart from the top down, inducing the processes underlying their organization from patterns in the relationships of the components to each other. The alternative approach of trying to recreate the entire system by assembling the components rarely works because, if the system is really complex, there is an overwhelming number of possibilities.

The recent history of theoretical ecology provides numerous examples of how these two problems have plagued efforts to develop a general theory of community organization and organic diversity. For example, May (1973) investigated the relationship between species diversity and community stability. He pointed out that there is nothing about complexity or diversity *per se* which tends to promote stability. He set up model communities in which species interacted at random and tested their stability by simulation. He concluded that complex communities containing diverse species tended to be highly unstable and that the obvious stability of natural communities was probably owing to highly structured (nonrandom) patterns of interspecific interaction. Lawlor (1978) subsequently reexamined May's model communities and pointed out that the assumption that species interact at random violates certain biological constraints which operate on natural communities. The most obvious of these constraints come from trophic structure: The flow of energy through the community is virtually unidirectional; plants (except insectivorous ones) do not eat herbivores, and herbivores rarely feed on carnivores. Lawlor pointed out that because of these constraints, in thousands of simulations of species interacting at random, May would have been unlikely to have obtained a single community with realistic trophic structure. The relationship between species diversity, functional organization, and stability of communities remains a challenging problem.

Before going on to assess two kinds of community theory in some detail, it is worth digressing to point out one area where the theory of evolutionary ecology has been quite successful: the development of optimal foraging theory. I suggest there are two reasons for this success. First, from the initial models of MacArthur and Pianka (1966) and Emlen (1966) through later, more sophisticated and realistic treatments (see Pyke *et al.*, 1977 for a review), energy has been used as an implicit, and usually as an explicit currency. The dictum "eat or be eaten" does so dominate the lives of many animals that much of their behavior can be understood in terms of selection to maximize rates of energy intake while minimizing risk of predation (Sih, 1980). Second, there has been little attempt to develop theories of community organization from the foraging behavior of animal species. Instead, ecologists interested in optimal foraging theory have been content to dissect the foraging behavior of animals to learn the selective and mechanistic processes by which animals make "decisions" about where to look for food, how long to stay in an area, how to search, and what foods to pursue and eat. In this they have been very successful.

COMPETITION THEORY: A DISAPPOINTMENT

In "The Homage" Hutchinson developed what he called "the Volterra-Gause principle" to ask how similar two species can be in their utilization of limiting resources and still avoid interspecific competition sufficiently to coexist in the same community. Of all the ideas which Hutchinson discussed, this one has received by far the most attention from theoretical ecologists. Virtually all of their endeavors are based on the Lotka-Volterra models of interspecific competition. These equations express competition in terms of α_{ij}, the effect on the population growth rate of species i of an individual of species j relative to the effect of a conspecific individual. Attempts to provide a direct answer to Hutchinson's question led to theories of limiting similarity (MacArthur and Levins, 1967; May and MacArthur, 1972; Abrams, 1975). Levins (1968) extended the Lotka-Volterra model to express the pairwise interactions among all species in a community as the entries α_{ij} in the so-called community matrix. MacArthur (1972) created models of species packing in which he investigated the relationship between coexistence and the availability and utilization of limited resources. Theoreticians were not alone in their enthusiasm for these models. Many more empirical ecologists spent much time studying interspecific competition and trying to measure the α_{ij}'s required to test the theories. Among both theoretical and field ecologists there was widespread belief that interspecific competition was the primary factor which limits diversity, and that working out the mechanisms of competitive interaction was the key to understanding the organization of communities.

In the last few years enthusiasm has given way to disappointment as this approach has proven unproductive. It is worth inquiring into the reasons for this failure so that we may avoid making the same mistakes in the future. There are several problems. First, the models and their predictions are highly sensitive to the underlying assumptions. The results of the theories are crucially dependent not only on the assumptions of the community models themselves, but also on the assumptions of the underlying Lotka-Volterra equations and the logistic equation of population growth. Relaxation of these assumptions in biologically realistic ways may lead to totally different results. Abrams (1975) perhaps has shown this most clearly for the theory of limiting similarity, but similar problems plague most of competition theory. Second, the models are not empirically operational. Not only do they not yield unambiguous, testable (which in ecology usually means qualitative) predictions, they also fail to give the field naturalist a clear idea of what variables he should measure. For example, MacArthur and Levins' (1967) model of limiting similarity suggests that two species cannot coexist if $\alpha_{ij} = \alpha_{ji} > 0.54$. How does one falsify this prediction? How do we interpret an empirically estimated α_{ij} of 0.60 or 0.70? What result is close enough to be construed as supporting the theory and what is sufficient to reject it? More importantly, how does the field ecologist measure α_{ij}? It is generally conceded that it is impractical to measure α_{ij} directly by experimentally manipulating the density of species j and recording the effects on the population growth rate of species i. Besides, if this is done in the field in a community of many species, this method has its own conceptual difficulties (see Schaffer, 1981). Many ecologists have suggested ways of estimating α_{ij} indirectly by measuring overlap in the use of limiting resources (*e.g.*, MacArthur and Levins, 1967; Cody, 1974) or spatial or temporal variation in the relative abundance of species (*e.g.*, Levins, 1968; Crowell and Pimm, 1976; Hallett and Pimm, 1979). All of these methods invoke additional assumptions, which are often difficult to verify in the field. These problems undermine the entire value of the theory. There have been numerous attempts to measure α_{ij} for species interacting in nature, but I doubt that there would be general agreement on a single instance in which it has been measured correctly.

Perhaps the most serious problem with using this kind of competition theory to investigate community organization is the inherent difficulty in trying to understand complex systems by putting together the basic components in proper relationship to each other. As stated earlier, this approach is usually unproductive because there are just too many possibilities. Consider the implications of Holt's (1977; see also Levine, 1976; Lawlor, 1979) conclusion that even a simple community matrix potentially conceals a large number of indirect interactions which may be as important as the direct, pairwise interactions in determining the structure and dynamics of the community. Current theory suggests that it is very difficult to put even a few species together in such a way as to obtain a biologically realistic, complex system that is stable, yet we are surrounded by natural communities that contain far more species and which persist for long periods of time. The implication is that there must be a set of rules above and beyond those that govern the pairwise interactions among species which dictate the assembly of species into communities.

I suspect that at least some of these rules embody the principles of energetics, which Hutchinson stressed in "The Homage," but which have been largely ignored in subsequent models of interspecific competition. Both individual organisms and ecological communities are thermodynamically unlikely. They maintain their organization only through the continual intake and expenditure of energy. Interspecific competition is important because it enables each species in the community to obtain a share of the limited energy and thus to persist. But I suggest that we would learn more about the structure and diversity of communities by focusing on the patterns of energy allocation among the species than by concentrating, as we have for the last two decades, on the effects of interspecific competition on population dynamics.

Island Biogeography Theory: A Heuristic Success

One body of recent theory has, in my opinion, contributed substantially to explaining organic diversity. This is MacArthur and Wilson's (1963, 1967) equilibrium theory of island biogeography and the subsequent work which it stimulated. The value of this theory has been largely heuristic. It has been tested repeatedly, often rejected, and not yet to my knowledge proven to be both necessary and sufficient to account for the diversity of a single insular biota (see Brown, 1978; Gilbert, 1980). Nevertheless, the theory provides an exceptionally useful conceptual framework for investigating the patterns of species diversity and the underlying mechanisms which produce these patterns.

The MacArthur-Wilson model has three characteristics which I believe any successful general theory of diversity must possess. First, it is an equilibrium model. It deliberately ignores the effects of unique historical events, and seeks an ecological explanation for diversity. While it recognizes that successional processes may occur (as in the recolonization of Krakatau), it explains the ultimate limit on diversity in terms of an equilibrium between opposing rates of colonization and extinction. This is not to say that history is unimportant. Geological and climatic changes have had major, long lasting effects on the composition and diversity of biotas. But the patterns of diversity at equilibrium which are predicted by the model, but often not observed in nature, have proven extremely valuable for interpreting the influence of historical events on insular communities (*e.g.*, Brown, 1971).

The second important feature of the model is that it confronts the problem of diversity directly. Number of species is the primary currency of the model. Furthermore, the theory attempts to account for diversity in terms of both relationships between biological processes (in this case, colonization and extinction) and characteristics of the physical environment (island size and isolation). The theory not only recognizes that at equilibrium, diversity must be limited by constraints of the inanimate environment, it also attempts to understand the biological mechanisms through which these constraints affect the number and kinds of species which comprise insular communities.

The third desirable characteristic of the model is that it is empirically operational. It makes robust, qualitative predictions which can be tested rigorously with the kinds of data field ecologists and biogeographers can be expected to obtain. The MacArthur-Wilson model is one of the few models of community ecology that have been repeatedly and unequivocally falsified (*e.g.*, Brown, 1971). Although some might argue that there must be little merit in an idea that has so often been proven wrong (or at least not quite right), to do so would be to underestimate the heuristic value of the theory. The model has changed the way we think about diversity, and in testing and rejecting it we have learned a great deal about the effects of historical events, physical factors, and ecological process on species diversity.

The MacArthur-Wilson model represents an encouraging beginning, but it is clearly not the ultimate answer. In recent years its deficiencies, even as a heuristic tool, have become increasingly apparent. Perhaps the most serious is the failure of the model to consider the biological mechanisms underlying the processes of colonization and extinction. The model implies that the determination of diversity is a very stochastic process. Species continually colonize and go extinct, the biota at equilibrium is constantly changing species composition, and all of these processes occur essentially at random. MacArthur and Wilson knew that this really was not so, but given the state of biogeography and community ecology in the mid-1960s it was a useful simplifying assumption. If we hope to continue to make progress, however, we must delve deeper and seek more deterministic theories of species diversity and community organization.

Some of the themes of "The Homage" suggest promising places to start. Consider the effect of island area on species diversity. It seems likely, as MacArthur and Wilson suggested, that the influence of area is mediated largely through increasing rates of extinction as island size decreases. What is the biological cause of these extinctions? Actually, there are probably two interrelated causes. As island size decreases so do both habitat diversity and total productivity (on a per island, not a per unit area, basis). A consideration of the relationship between energetics and the mosaic nature of the environment would suggest that the pattern of extinction would be far from random. We would predict that carnivores should go extinct before herbivores of comparable size; further, within the same trophic level, species of large body size should disappear before their smaller relatives, and habitat specialists should go extinct before generalists. In at least one case, these deterministic patterns are exactly what are observed (Brown, 1971). Such results lend encouragement to the effort to develop a general theory of diversity.

Prospects for the Future: Ingredients of a General Theory

It would be unrealistic to suggest that we are now in a position to advance such a theory. We still have much to learn about the limits of diversity and the organization of communities. Much of this knowledge can only come from additional empirical studies. Nevertheless, I suspect that the lack of information is less critical than the insight required to select and assemble existing data and ideas in new and productive ways. What is necessary to answer Hutchinson's question: Why are there so many species?

I will assume that the goal is to construct an equilibrium theory for reasons both heuristic and practical. Historical geological and climatic events have had profound influence on the composition and diversity of some biotas, but, as argued earlier, it is easier to understand the effects of historical perturbations from a conceptual framework that assumes eventual equilibration of rates of origination (either colonization or speciation) and extinction. The assumption of equilibrium is not even terribly unrealistic when applied to much of the earth's present biota. Given a long period of geological and climatic stability (say 10 million or even 100 million years), most ecologists would probably expect to find pretty much the same general patterns of diversity as at present. The Arctic tundra, the Andean altiplano, salt marshes, hot springs, desert oases and small oceanic islands would still support many fewer species than the Amazon rain forest or the Great Barrier Reef.

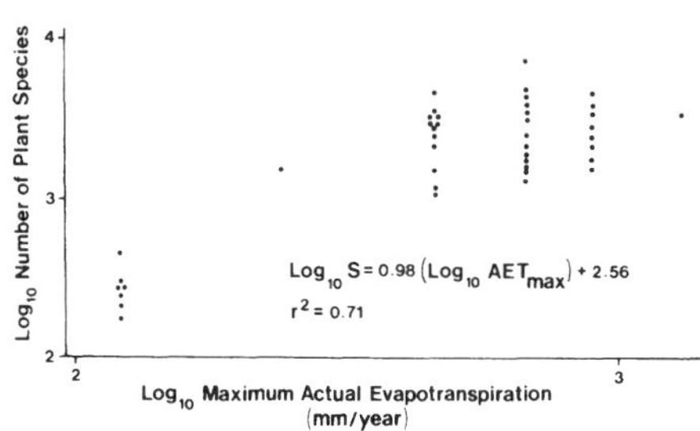

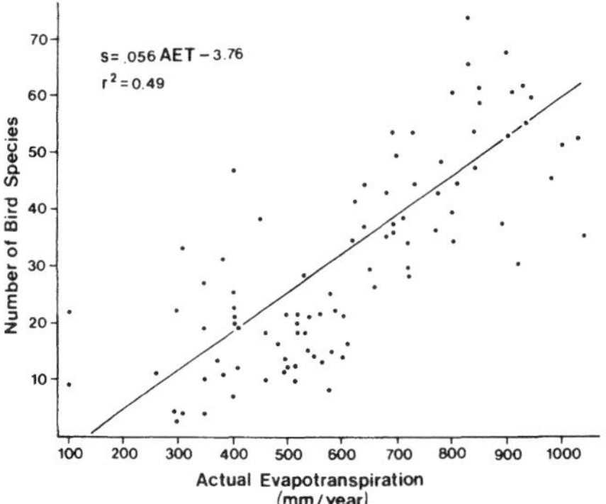

FIG. 1. The relationship between species diversity and productivity for two groups of organisms: above, for terrestrial plants, data compiled from North American local and regional floras by D. H. Wright; below, for birds, data compiled from Christmas counts from non-coastal sites in North America by J. B. Dunning, Jr. and J. Taylor. Productivity was estimated as actual evapotranspiration, which was obtained from a computer-generated map supplied by M. L. Rosenzweig.

Capacity rules

I suggest that a general equilibrium theory of diversity must contain two kinds of constructs, which I shall call capacity rules and allocation rules. The capacity rules define the physical characteristics of environments which determine their capacity to support life. The most important ingredient of the capacity rules is the availability of usable energy. Usable energy can be defined as any essential substance which organisms can potentially (given their constraints) extract from their environment and use to do the useful work of surviving and reproducing. However, for entire ecological communities, availability of usable energy can be measured as the rate of primary production, because the photosynthetic process produces essentially all of the energy used, not only by green plants, but by all other organisms as well. Except in a few environments where physical conditions prevent the oxidation of organic carbon, all of the energy fixed in photosynthesis is utilized by organisms (Hairston *et al.*, 1960). Darwin's "struggle for existence" is largely the struggle of all organisms to obtain usable energy that can be used to produce offspring. The more energy is available in usable form, the more organisms and hence, the more species the environment can support.

That there is a positive, causal relationship between productivity and diversity is not a new idea. Hutchinson, in the passage quoted earlier, suggested it in "The Homage," and it has subsequently been advanced by Connell and Orias (1964) and MacArthur (1972). Our work on communities of seed-eating desert animals has provided empirical support for this concept (Brown, 1973; Brown *et al.*, 1979). Despite these studies, the idea has not been widely accepted.

To show that the correlation between productivity and diversity has some generality, in Figure 1 I present some data analyzed by three of my students. For both terrestrial vascular plants and wintering land birds in North America actual evapotranspiration, a readily obtainable, but crude estimate of primary productivity (see Rosenzweig, 1968; Leith and Whittaker, 1975), accounts for half or more of the variation in number of species. Actual evapotranspiration is a reasonably good predictor of primary productivity because it incorporates the effects of temperature and water availability on photosynthesis, but it neglects other factors, such as soil chemistry, which can have major effects on plant productivity and community organization. Of course for birds, which are primarily carnivores, primary production provides only an indirect, and perhaps a poor estimate of availability of usable energy.

The second component of the capacity rules must be a measure of variation of the physical environment in time and space.

Some highly productive environments, such as salt marshes and hot springs, contain few species. These habitats typically are characterized by extreme physical conditions; ecologists, especially plant ecologists, often refer to them as harsh. But this characterization does not answer Hutchinson's question: if some species can adapt and live there, why cannot others invade? Terborgh (1973) in a very insightful paper proposed an answer. Harshness is indeed relative, but common features of harsh, productive habitats normally include not only unusual physical conditions, but also small size, spatial isolation, and (sometimes) ephemeral existence. Harsh environments contain few species because they have low rates of colonization, a consequence of their spatial isolation and the fact that most species available to colonize from surrounding habitats are unable to tolerate the extreme physical conditions, and high extinction rates, a consequence of small population sizes resulting from inhabiting restricted areas and sometimes from environmental fluctuations. Consequently, it is necessary to distinguish between harsh environments, such as desert oases, hot springs and salt marshes, which are physically distinctive, small and isolated but also productive, and other habitats, such as tundras and deserts, which are abundant and widespread but also contain few species, in this case because they are unproductive.

Although I feel confident that the most important ingredients of the capacity rules are availability of usable energy and the pattern of spatial and temporal variation that determines effective harshness, I am not at all sure how these elements interact with each other and with the organisms to affect diversity. Perhaps the most difficult problem is to develop an accurate measure of harshness, which requires that we assess the effects of spatial and temporal heterogeneity on the capacity of environments to maintain populations in the face of extinction. It will be easier to measure available energy, although for particular guilds of species this must be assessed in terms of usable energy resources. For fruit-eating animals, available energy must be measured as the availability of suitable fruits. In this case, the fact that frugivorous mammals are abundant and diverse only in tropical habitats where fruits are available throughout the year (Fleming, 1973) is certainly consistent with the arguments developed above.

To illustrate a few of the difficulties involved in formulating capacity rules, consider a specific example which Krebs (1978) has used to argue against the generality of the relationship between productivity and diversity. Several studies (*e.g.*, Whiteside and Harmsworth, 1967) show an inverse relationship between the number of zooplankton species and the productivity of temperate lakes. This is true when primary productivity is measured on a per unit area basis, because the highest diversity of zooplankton is found in large, oligotrophic lakes. But what maintains persistent populations of zooplankton, the productivity of a unit of area or the productivity of the entire lake? If, as we might suppose, it is the latter, then the productivity-diversity relationship is supported. My student, D. H. Wright (unpublished reanalysis of the original data), has shown that number of zooplankton species is positively correlated with both area of lake and total productivity (productivity/unit area × area of lake).

Allocation rules

In addition to knowing how the constraints of the physical environment determine the availability of usable energy to organisms, in order to understand diversity we must also learn how available energy is apportioned among species. I call the general patterns and processes of energy subdivision allocation rules, because it is only by obtaining a sufficient share of the total usable energy that a particular species is able to maintain its population and persist as a member of the community. Thus the mechanisms of allocation interact with the capacity rules to determine the number and kinds of species which can be supported at equilibrium. Clearly the ultimate processes underlying the allocation rules are the basic interactions among species populations: competition, predation and

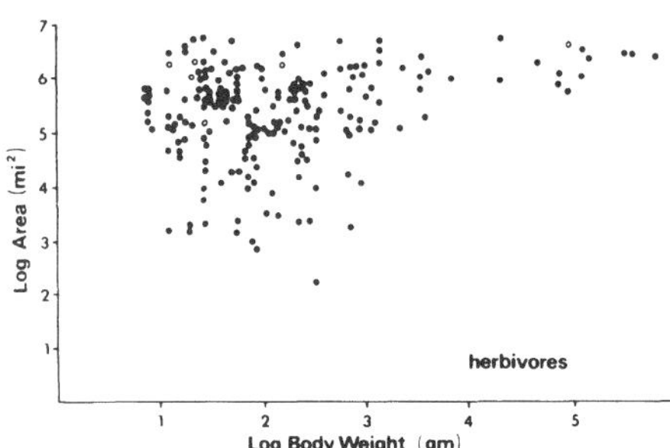

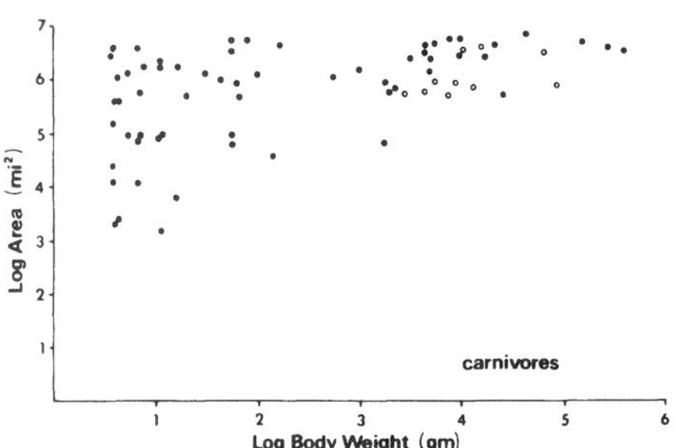

FIG. 2. The relationship between area of the species geographic range and body size for North American land mammals. Areas were measured from the range maps in Hall and Kelson (1959) by planimetry. The taxonomy follows Burt and Grossenheider (1964). Unshaded circles represent species which have larger ranges than the size indicated because their ranges extend beyond the maps into South America. Bats were excluded from the analysis; marsupials, insectivores, and armadillos and carnivores were assumed to be carnivores, and all others were classified as herbivores.

mutualism. But earlier I suggested that it is impractical to attempt to understand community-wide patterns of resource allocation and coexistence in terms of the population dynamics of interacting species. Is there an alternative? I think there is, but the empirical and theoretical bases for the allocation rules are at present even less well developed than those for the capacity rules.

I suggest we start by searching for empirical patterns in characteristics of species which affect their utilization of energy: local abundance, body size, geographic range, and trophic status, and then by developing testable mechanistic hypotheses to account for these patterns. Sufficient work has been done to demonstrate that interesting general patterns are present, but so far these lack a synthetic, mechanistic explanation.

Perhaps the most long recognized patterns are in the distribution of abundance and body size among species within local communities. The common observation that only a few species are common, whereas most are relatively rare has been quantified. MacArthur (1957) fitted the relative abundance of bird species to a broken stick distribution, whereas Preston (1962) noted that the distribution of abundances of many kinds of organisms often is lognormal. Similarly, the fact that within a taxonomic group or trophic level, species of small body size are more numerous than those of large size, was considered by Hutchinson and MacArthur (1959) and later by May (1978), who showed that size distributions also tended to be lognormal.

The community level consequences of these patterns remain to be explored completely. In retrospect, it is perhaps unfortunate that MacArthur ceased to investigate size and abundance distributions, and perhaps discouraged others with his public repudiation of his broken stick models (1966). One recurrent problem, that will face any who pursue this approach, is that of spatial scale. Hairston (1969) showed that the pattern of species abundance distributions varies with the size or spatial area of the community sampled. This is not surprising because species have different spatial distributions depending on their body size, trophic status and other characteristics. Figure 2 depicts the areas of the geographic ranges of North American mammal species. Two patterns are immediately apparent. First, there is a minimum area for the ranges of species which increases with increasing body size. Second, this minimum area is about an order of magnitude larger for carnivores than for herbivores. These patterns suggest that the minimum size of a species geographic range is determined by the probability of extinction. We do not see species with smaller ranges because these species have gone extinct. The probability of extinction appears to be related to population size which, in turn is affected by body size and trophic position, because these factors determine to a large extent the ability of energy resources to support populations. These patterns of geographic ranges have two other important consequences for community ecologists which should be noted in passing. Most species have much broader ranges than the study areas of community ecologists, and these wide ranges appear to be important in enabling the species to persist over evolutionary time. Also, since the sizes of geographic ranges vary, particular species must coexist with different combinations of other species to form different communities in different parts of their geographic range.

It is not at all clear to me how these patterns of abundance, body size, geographic distribution, trophic position and other characteristics which affect energy utilization should be synthesized and conceptualized to develop useful allocation rules. But the fact that these patterns exist and that they appear to have a common basis in ecological energetics suggests that there may be general mechanisms of energy allocation which limit diversity and determine community organization. The existence of these patterns raises the hope that we can derive general allocation rules to account for species diversity which are not dependent on the details of the population dynamics of particular interacting species.

Concluding Remarks

Among my colleagues who call themselves community ecologists I detect widespread pessimism and disappointment. Many of them seem to feel that ecological theory has promised far more than it has delivered and that it has not contributed very much that has proven useful for understanding the natural world. Hutchinson's question, Why are there so many species? remains largely unanswered despite a great deal of theoretical and empirical work.

Now, more than two decades after publication of "Homage to Santa Rosalia," I suggest we would still be well advised to pursue the ideas that Hutchinson advanced there. In particular I recommend adopting Hutchinson's emphasis on the fundamental role of energetics in evolutionary and community ecology. The acquisition and utilization of energy in accordance with the laws of thermodynamics remains the best place to start "to understand the higher intricacies of any ecological system . . ." (p. 147).

There are those who will disagree. Some will continue to try to build a theory of community ecology based on population interactions. Others will argue that communities are so complex and unstructured that it is unrealistic to hope to develop a general theory of diversity. I am reluctant to argue strongly against these points of view. Diversity among ecologists, as among other organisms, makes life interesting and leads ecologists to new ideas. The intense activity in both theoretical and empirical ecology has made the last two decades an exciting time, even if the ultimate answers have eluded a generation of bright, dedicated ecologists. Perhaps the lack of immediate success should not be surprising. Ecological communities are perhaps the most complex of biological structures. Who ever thought it would be easy to find out why there are so many species?

Acknowledgments

Although I assume sole responsibility for the contents of this paper, my viewpoint and ideas have been influenced by numerous students and colleagues. In particular, I thank the graduate students in my Evolutionary Ecology Seminar in the Fall of 1979, and W. M. Schaffer, and M. L. Rosenzweig for valuable discussions. D. H. Wright, J. B. Dunning, Jr., and J. Taylor kindly allowed me to use their data and analyses in Figure 1. The National Science Foundation has generously supported my research, most recently with Grant DEB 76-83858.

References

Abrams, P. 1975. Limiting similarity and the form of the competition coefficient. Theor. Pop. Biol. 8:356–375.

Brown, J. H. 1971. Mammals on mountaintops: Nonequilibrium insular biogeography. Amer. Natur. 105:467–478.

Brown, J. H. 1973. Species diversity of seed-eating desert rodents in sand dune habitats. Ecology 54:775–787.

Brown, J. H. 1978. The theory of insular biogeography and the distribution of boreal birds and mammals. *In* K. T. Harper and J. L. Reveal (eds.), *Intermountain biogeography: A symposium.* Great Basin Naturalist Mem. 2:209–227.

Brown, J. H., O. J. Reichman, and D. W. Davidson. 1979. Granivory in desert ecosystems. Ann. Rev. Ecol. Syst. 10:201–227.

Burt, W. H. and R. P. Grossenheider. 1964. *A field guide to the mammals.* Houghton-Mifflin, Boston.

Cody, M. L. 1974. *Competition and the structure of bird communities.* Princeton University Press, Princeton, N.J.

Connell, J. H. and E. Orias. 1964. The ecological regulation of species diversity. Amer. Natur. 98:399–414.

Crowell, K. L. and S. L. Pimm. 1976. Competition and niche shifts of mice introduced onto islands. Oikos 27:251–258.

Emlen, J. M. 1966. The role of time and energy in food preference. Amer. Natur. 100:611–617.

Fleming, T. H. 1973. Numbers of mammal species in North and Central American forest communities. Ecology 54:555–563.

Gilbert, F. S. 1980. The equilibrium theory of island biogeography: Fact or fiction? J. Biogeog. 7:209–235.

Hairston, N. G. 1969. On the relative abundance of species. Ecology 50:1091–1094.

Hairston, N. G., F. E. Smith, and L. B. Slobodkin. 1960. Community structure, population control, and competition. Amer. Natur. 94:421–425.

Hall, E. R. and K. R. Kelson. 1959. *The mammals of North America.* Ronald Press, New York.

Hallett, J. G. and S. L. Pimm. 1979. Direct estimation of competition. Amer. Natur. 113:593–600.

Holt, R. D. 1977. Predation, apparent competition, and the structure of prey communities. Theor. Pop. Biol. 12:197–229.

Hutchinson, G. E. 1959. Homage to Santa Rosalia, or why are there so many kinds of animals? Amer. Natur. 93:145–159.

Hutchinson, G. E. and R. H. MacArthur. 1959. A theoretical model of size distributions among species of animals. Amer. Natur. 93:117–125.

Krebs, C. J. 1978. *Ecology: The experimental analysis of distribution and abundance.* Harper and Row, New York.

Lawlor, L. R. 1978. A comment on randomly constructed model ecosystems. Amer. Natur. 112: 445–447.

Lawlor, L. R. 1979. Direct and indirect effects of n-species competition. Oecologia 43:355–364.

Leith, H. and R. H. Whittaker. 1975. *Primary productivity of the biosphere.* Springer-Verlag, New York.

Levine, S. 1976. Competitive interactions in ecosystems. Amer. Natur. 110:903–910.

Levins, R. 1968. *Evolution in changing environments.* Princeton University Press, Princeton, N.J.

MacArthur, R. H. 1955. Fluctuations of animal populations, and a measure of community stability. Ecology 36:533–536.

MacArthur, R. H. 1957. On the relative abundance of bird species. Proc. Nat. Acad. Sci. U.S.A. 43:293–295.

MacArthur, R. H. 1966. Note on Mrs. Pielou's comments. Ecology 47:1074.

MacArthur, R. H. 1972. *Geographical ecology.* Harper and Row, New York.

MacArthur, R. H. and R. Levins. 1967. The limiting similarity, convergence, and divergence of coexisting species. Amer. Natur. 101:377–385.

MacArthur, R. H. and E. R. Pianka. 1966. On optimal use of a patchy environment. Amer. Natur. 100:603–609.

MacArthur, R. H. and E. O. Wilson. 1963. An equilibrium theory of insular biogeography. Evolution 17:373–387.

MacArthur, R. H. and E. O. Wilson. 1967. *The theory of island biogeography.* Princeton University Press, Princeton, N.J.

May, R. M. 1973. *Stability and complexity in model ecosystems.* Princeton University Press, Princeton, N.J.

May, R. M. 1978. The dynamics and diversity of insect faunas. *In* L. A. Mound and N. Waloff (eds.), *Diversity of insect faunas,* pp. 188–204. Blackwell, New York.

May, R. M. and R. H. MacArthur. 1972. Niche overlap as a function of environmental variability. Proc. Nat. Acad. Sci. U.S.A. 69:1109–1113.

Preston, F. W. 1962. The canonical distribution of commonness and rarity. Ecology 43:185–215, 410–432.

Pyke, G. H., H. R. Pulliam, and E. L. Charnov. 1977. Optimal foraging: A selective review of theory and tests. Quart. Rev. Biol. 52:137–154.

Rosenzweig, M. L. 1968. Net primary productivity of terrestrial communities: Prediction from climatological data. Amer. Natur. 102:67–74.

Rosenzweig, M. L. and R. H. MacArthur. 1963. Graphical representation and stability conditions of predator-prey interactions. Amer. Natur. 97:209–223.

Schaffer, W. M. 1981. Ecological abstraction: The consequences of reduced dimensionality in ecological models. Ecology. (In press)

Sih, A. 1980. Optimal behavior: Can foragers balance two conflicting demands? Science 210: 1041–1043.

Terborgh, J. 1973. On the notion of favorableness in plant ecology. Amer. Natur. 107:481–501.

Whiteside, M. C. and R. V. Harmsworth. 1967. Species diversity in Chydorid (Cladocera) communities. Ecology 48:664–667

PAPER 35

Mass Extinctions in the Marine Fossil Record (1982)

D. M. Raup and J. J. Sepkoski Jr.

Commentary

JESSICA THEODOR

D. M. Raup and J. J. Sepkoski Jr.'s paper contains one of the most widely reproduced graphs in paleontology—it shows marine extinctions over time, highlighting the "Big Five" mass extinctions—and the paper has been cited, on average, 30 times per year since publication (according to statistics from Google Scholar). This paper represented a colossal effort of data collection, establishing a global data set for marine taxa with statistical analysis to illustrate global patterns of marine extinction through the Phanerozoic. The impact of this paper in paleontology is obvious—it stimulated a revolution in analytical methods (Marshall 1990; Sepkoski and Kendrick 1993; Alroy 1996, 1998, 2000; Foote 1999, 2001; Alroy et al. 2001), and helped to begin a tidal wave of studies that used the fossil record to ask major questions about organismal diversity and distribution (Raup and Sepkoski 1984; Flessa and Jablonski 1985; Miller 1988; Miller and Sepkoski 1988; Raup 1991, 1992; Bambach 1993; Brett and Baird 1995; Wagner 1995; Ivany and Schopf 1996; Benton 1997; Westrop and Adrain 1998). However, this paper also had an impact on macroecology, by bringing the tempo and mode of extinction to the fore. Certainly ecologists had recognized extinction, but this analysis established beyond doubt the statistical reality of mass extinctions, showing that they had happened repeatedly through geologic time, and established the magnitude of some of these events (in the worst case, the end-Permian, as much as 90% of all marine diversity was lost). This analysis also showed that global standing diversity had increased over time, primarily caused by decreasing background extinction rather than by any increase in origination rates, an important point in understanding that diversity dynamics are strongly affected by history.

This paper also emphasized an approach that has become a hallmark of macroecology: the use of statistical analysis of very large data sets, not only temporally but spatially, to answer basic ecological questions (Walker and Valentine 1984; Shaffer 1985; Herbold and Moyle 1986; Rhode 1988, 1992; Maurer 1989; Cornell and Laughton 1992; Tokeshi 1993; Sheil 1996; Cowling 2001; Jablonski 2002). This paper caused a revolution in paleontological research, but it also had a clear effect on macroecology in both methodology and in concepts.

Literature Cited

Alroy, J. 1996. Constant extinction, constrained diversification, and uncoordinated stasis in North American mammals. *Palaeogeography, Palaeoclimatology, Palaeoecology* 127:285–311.

———. 1998. Cope's Rule and the dynamics of body mass evolution in North American fossil mammals. *Science* 280:731–34.

———. 2000. New methods for quantifying macro-

From *Science* 215:1501–3. Reprinted with permission from AAAS.

evolutionary patterns and processes. *Paleobiology* 26:707–33.

Alroy, J., C. R. Marshall, R. K. Bambach, K. Bezusko, M. Foote, F. T. Fürsich, T. A. Hansen, et al. 2001. Effects of sampling standardization on estimates of Phanerozoic marine diversification. *Proceedings of the National Academy of Sciences* 98:6261–66.

Bambach, R. K. 1993. Seafood through time: Changes in biomass, energetics, and productivity in the marine ecosystem. *Paleobiology* 19:372–97.

Benton, M. J. 1997. Models for the diversification of life. *Trends in Ecology and Evolution* 12:490–95.

Brett, C. E., and G. Baird. 1995. Coordinated stasis and evolutionary ecology of Silurian to Middle Devonian faunas in the Appalachian Basin. In *Speciation in the fossil record*, edited by R. Anstey and D. H. Erwin, 285–315. Columbia University Press, New York.

Cornell, H. V., and J. H. Laughton. 1992. Species interactions, local and regional processes, and limits to the richness of ecological communities: A theoretical perspective. *Journal of Animal Ecology* 61:1–12.

Cowling, S. A. 2001. Plant carbon balance, evolutionary innovation and extinction in land plants. *Global Change Biology* 7:231–39.

Flessa, K., and D. Jablonski. 1985. Declining Phanerozoic background extinction rates: Effect of taxonomic structure? *Nature* 313:216–18.

Foote, M. 1999. Morphological diversity in the evolutionary radiation of Paleozoic and post-Paleozoic crinoids. *Paleobiology* 25:1–116.

———. 2001. Inferring temporal patterns of preservation, origination, and extinction from taxonomic survivorship analysis. *Paleobiology* 27:602–30.

Herbold, B., and P. B. Moyle. 1986. Introduced species and vacant niches. *American Naturalist* 128:751–60.

Ivany, L. C., and Schopf, K. M., eds. 1996. New perspectives on faunal stability in the fossil record. *Palaeogeography, Palaeoclimatology, Palaeoecology* 127.

Jablonski, D. 2002. Survival without recovery after mass extinctions. *Proceedings of the National Academy of Sciences* 99:8139–44.

Marshall, C. R. 1990. Confidence intervals on stratigraphic ranges. *Paleobiology* 16:1–10.

Maurer, B. A. 1989. Diversity dependent species dynamics: Incorporating the effects of population level processes on species dynamics. *Paleobiology* 15:133–46.

Miller, A. I. 1988. Spatio-temporal transitions in Paleozoic Bivalvia: An analysis of North American fossil assemblages. *Historical Biology* 1:251–73.

Miller, A. I., and J. J. Sepkoski Jr. 1988. Modeling bivalve diversification: The effect of interaction on a macroevolutionary system. *Paleobiology* 14:364–69.

Raup, D. M. 1991. A kill curve for Phanerozoic marine species. *Paleobiology* 17:37–48.

———. 1992. Large-body impact and extinction in the Phanerozoic. *Paleobiology* 18:80–88.

Raup, D. M., and J. J. Sepkoski Jr. 1984. Periodicity of extinctions in the geologic past. *Proceedings of the National Academy of Sciences* 81:801–5.

Rohde, K. 1998. Latitudinal gradients in species diversity: Area matters, but how much? *Oikos* 82:184–90.

———. 1992. Latitudinal gradients in species diversity: The search for the primary cause. *Oikos* 65:514–27.

Sepkoski, J. J., Jr., and D. C. Kendrick. 1993. Numerical experiments with model monophyletic and paraphyletic taxa. *Paleobiology* 19:168–84.

Shaffer, M. L., and F. B. Samson. 1985. Population size and extinction: A note on determining critical population sizes. *American Naturalist* 125:144–52.

Sheil, D. 1996. Species richness, tropical forest dynamics and sampling: Questioning cause and effect. *Oikos* 76:587–90.

Tokeshi, M. 1993. Species abundance patterns and community structure. *Advances in Ecological Research* 24:111–86.

Wagner, P. J. 1995. Diversity patterns among early gastropods: Contrasting taxonomic and phylogenetic descriptions. *Paleobiology* 21:410–39.

Westrop, S. R., and J. M. Adrain. 1998. Trilobite alpha diversity and the reorganization of Ordovician benthic marine communities. *Paleobiology* 24:1–16.

Walker, T. D., and J. W. Valentine. 1984. Equilibrium models of evolutionary species diversity and the number of empty niches. *American Naturalist* 124:887–99.

Atmosphere (Academic Press, New York, 1977).
6. WMO/ICSU, *Report of the Joint Organizing Committee Study Conference on Climate Models: Performance, Intercomparison and Sensitivity Studies* [GARP (Global Atmospheric Research Program) Publication Series 22, World Meteorological Organization, Geneva, 1979], vol. 1.
7. J. Shukla, D. Randall, D. Straus, Y. Sud, L. Marx, *Winter and Summer Simulations with the GLAS Climate Model* (National Aeronautics and Space Administration/Goddard Space Flight Center, Greenbelt, Md., in press).
8. J. Charney, W. J. Quirk, S. H. Chow, J. Kornfield, *J. Atmos. Sci.* 34, 1366 (1977).

2 December 1981

Mass Extinctions in the Marine Fossil Record

Abstract. *A new compilation of fossil data on invertebrate and vertebrate families indicates that four mass extinctions in the marine realm are statistically distinct from background extinction levels. These four occurred late in the Ordovician, Permian, Triassic, and Cretaceous periods. A fifth extinction event in the Devonian stands out from the background but is not statistically significant in these data. Background extinction rates appear to have declined since Cambrian time, which is consistent with the prediction that optimization of fitness should increase through evolutionary time.*

A number of mass extinctions have "reset" major parts of the evolutionary system during the Phanerozoic. However, the precise timing and magnitude of these events has been difficult to measure because data from the fossil record are fragmentary. Comprehensive and accurate data on extinct species have always been unobtainable, and therefore most workers have been forced to investigate extinctions at the level of genera, families, and orders, with family-level data generally preferred as the best compromise between sampling limitations and taxonomic uncertainty (*1*). Historically, the three best summaries of familial data from the fossil record have been those of Newell (*2*), Cutbill and Funnell (*3*), and Valentine (*4*). But even with these data sets, identification of specific mass extinctions has been difficult and often subjective because of taxonomic problems and especially stratigraphic imprecision. Many macroevolutionary phenomena including mass extinctions have characteristic time scales that are geologically rather short (less than several tens of millions years) and can become lost or grossly distorted when analyzed without adequate stratigraphic control.

We now present a new analysis of extinctions based on a more comprehensive and accurate data set for marine animal families. Marine vertebrates as well as invertebrates and protozoans are included, and the data benefit from compilation of taxonomic and stratigraphic investigations far beyond traditional sources (*5*). The compilation encompasses approximately 3300 fossil marine families, of which about 2400 are extinct. Times of extinction for 87 percent of the families have been resolved to the level of the stratigraphic stage (mean duration, 7.4×10^6 years), and most of the remaining data has been resolved to stratigraphic series (mean duration, 20×10^6 years).

The rates of extinction calculated from the familial data plotted against geologic time are illustrated in Fig. 1. Each point was calculated as follows: the number of families that became extinct in each of the 76 post-Tommotian (early Lower Cambrian) stages (*6*) was divided by the estimated duration of the stage (*7*); these initial rates were then modified by adding extinction rates calculated from the lower resolution series-level data to the appropriate stages. Calculations were made separately for "shelly" taxa and for rarely preserved taxa (*8*). The effect of this segregation was negligible in most cases so that the data for rarely preserved animals are not included with most points in Fig. 1. For four stages, however, addition of rarely preserved families increased calculated extinction rates by more than 0.5 family per million years. These are the stages that contain the four major Lagerstätten of the Phanerozoic marine record: Burgess Shale (Cambrian, Templetonian), Hunsrück Shale (Devonian, Siegenian), Mazon Creek concretions (Carboniferous, Moscovian), and Solnhofen Limestone (Jurassic, "Tithonian"). The combined rates for shelly and rarely preserved families for these four stages are indicated in Fig. 1 by X's with the rates for shelly families shown below. Only the Burgess Shale (Templetonian) stands out on the plot.

The distribution of the 76 points for shelly animals in Fig. 1 suggests that two rates of extinction have been operative through the Phanerozoic. (i) Normal, or background, extinction: the majority of points fall in a rather tight cluster at extinction rates less than 8.0 extinctions per million years. (ii) Mass extinction: several points stand out as being considerably higher than the background and show a maximum of 19.3 familial extinctions per million years.

The problem of determining rigorously which points in Fig. 1 should be considered mass extinctions can be approached as a simple data analysis problem of identifying trends and outliers. As an initial step, we computed a linear regression (not shown) for all 76 extinction points as a function of geologic time and then searched for significant departures from this line. Four points (or 5 percent of the data) fell above the one-sided 99 percent confidence interval. These points, which are circled in Fig. 1, are (per million years) the Ashgillian (19.3 fm), Guadalupian (14.0 fm), Dzhulfian (15.7 fm), and Maestrichtian (16.3 fm). A fifth point, the Norian (10.8 fm), fell

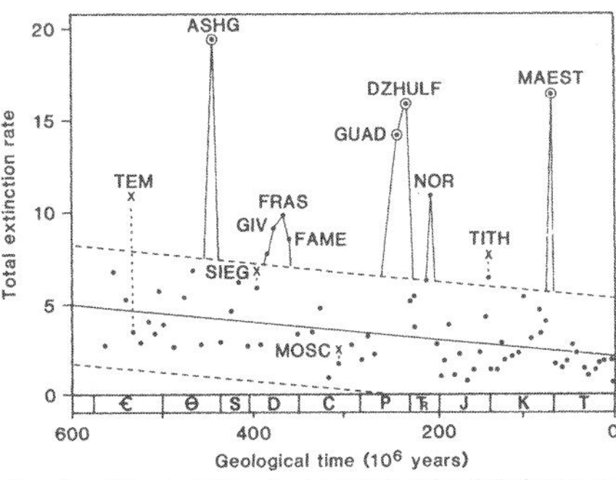

Fig. 1. Total extinction rate (extinctions per million years) through time for families of marine invertebrates and vertebrates. The plot shows statistically significant mass extinctions late in the Ordovician (ASHG), Permian (GUAD-DZHULF), Triassic (NOR), and Cretaceous (MAEST). An extinction event in the late Devonian (GIV-FRAS-FAME) is noticeable but not statistically significant. Circled points are those where the departure from the main cluster is highly significant ($P < .01$); X's indicate those cases where inclusion of rarely preserved animal groups substantially increases the calculated extinction rate (the point directly below the X is the rate calculated without the rarely preserved groups). The figure also shows a general decline in background extinction rate through time. The regression line is fit to the 67 points having extinction rates less than eight families per 10^6 years, and the dashed lines define the 95 percent confidence band for the regression. Abbreviations: TEM, Templetonian; ASHG, Ashgillian; SIEG, Siegenian; GIV, Givetian; FRAS, Frasnian; FAME, Famennian; MOSC, Moscovian; GUAD, Guadalupian; DZHULF, Dzhulfian; NOR, Norian; TITH, Tithonian; MAEST, Maestrichtian.

0036-8075/82/0319-1501$01.00/0

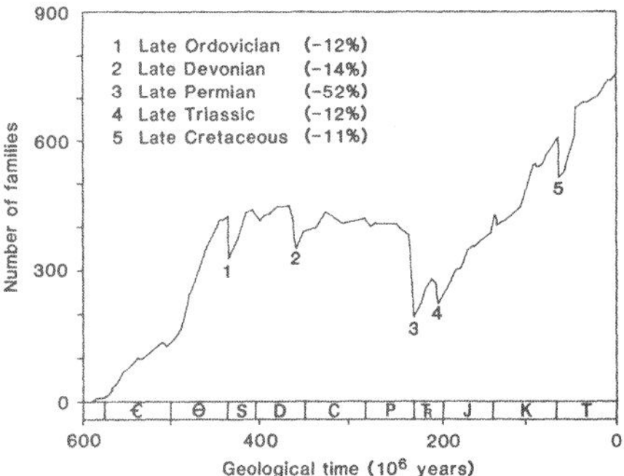

Fig. 2. Standing diversity through time for families of marine vertebrates and invertebrates. Rarely preserved groups are not included. Five mass extinctions, indicated by numerals, are recognizable by abrupt drops in the diversity curve. The relative magnitudes of these drops (measured from the stage before to the stage after the extinction event) are given in parentheses in the upper left. All mass extinctions but No. 2 (Devonian) are statistically significant in Fig. 1 and three (Nos. 1, 3, and 5) are highly significant ($P < .01$).

above the one-sided 95 percent confidence interval. Some or all of these mass extinctions have been recognized previously but without consistency and without statistical testing.

The outlying points identified above also can be recognized as major perturbations in marine diversity. The Phanerozoic diversity curve compiled from the familial data is shown in Fig. 2. Five extinction events are seen as sharp drops in standing diversity. Four of these (counting the Guadalupian and Dzhulfian as a single event) match the statistically significant outliers in Fig. 1. The fifth, labeled "2" in Fig. 2, is a late Devonian extinction that has been recognized by previous workers. This extinction does not appear as a statistically significant event in Fig. 1 because the family extinctions are distributed over two stages, the Frasnian and the preceding Givetian, which have a combined duration of about 15 million years (*9*). This smearing of extinctions may represent sampling error in that failure to identify the actual time of extinction will almost always push apparent extinctions backward in time. Alternatively, the smearing may reflect a real phenomenon—an extinction "event" that took place over millions of years. The continuation of high extinction rates into the Famennian is consistent with this hypothesis. However, it should be noted that, on the basis of other information, McLaren (*10*) suggested a meteorite impact as one possible explanation for the Frasnian extinctions.

In summary, five mass extinctions are clearly defined in the familial data. These extinctions occurred in the Late Ordovician (Ashgillian), Late Devonian (Givetian-Frasnian), Late Permian (Guadalupian-Dzhulfian), Late Triassic (Norian), and Late Cretaceous (Maestrichtian). The occurrence of these major extinctions near the ends of geologic periods simply reflects the fact that the stratigraphers who established the geologic time scale in the first half of the 19th century chose major faunal breaks as boundaries for the principal subdivisions.

With the major Phanerozoic events isolated in Fig. 1, a more accurate assessment of the nature of background extinction can be made. Although some smaller but well-known extinction events may remain hidden in Fig. 1 (*11*), the residual cluster of points suggests that background rates have been declining since the early Paleozoic. The solid line in Fig. 1 is a linear regression fitted to the 67 extinction rates for shelly animals after removal of the major extinction events; the dashed lines, which envelop nearly all these points, represent the 95 percent confidence band for the regression. The correlation coefficient for the regression is .47, which can be considered statistically significant if problems of time series and data selection are ignored. The slope of the regression line is nontrivial and indicates that the total rate of background extinction has decreased from about 4.6 to 2.0 fm per million years since the Early Cambrian. This is surprising in view of the fact that the rates are not normalized for standing diversity, which has increased substantially since the Cambrian (Fig. 2). The decline in extinction rates could be just an artifact of the "pull of the Recent" (*12*). In contrast, a decrease in extinction rate is predictable from first principles if one argues that general optimization of fitness through evolutionary time should lead to prolonged survival. This is speculative but it is worthy of further consideration because broad predictions of progressive change in evolutionary dynamics are so rarely realized when tested with data.

The decline in background extinction rate from the Early Cambrian to the Recent means that approximately 710 family extinctions did not occur that would have if the Cambrian rate had been sustained. This number is essentially identical to the amount by which familial diversity increased over that interval (680 families) (Fig. 2). This suggests that the net increase in standing diversity through the Phanerozoic may have been more an effect of decrease in extinction than increase in origination.

In conclusion, our analysis shows that major mass extinctions are far more distinct from background extinction than has been indicated by previous analyses of other data sets. Four mass extinctions are statistically significant events and are likely to represent phenomena qualitatively different from the background. The data do not tell us, of course, what stresses caused the mass extinctions. The extinctions were short-lived events in geological time, but the data do not have the resolving power to show whether the events were also short-lived in human or ecological time.

DAVID M. RAUP
Field Museum of Natural History, Chicago, Illinois 60605
J. JOHN SEPKOSKI, JR.
Department of Geophysical Sciences, University of Chicago, Chicago 60637

References and Notes

1. J. W. Valentine, *J. Paleontol.* **48**, 549 (1974).
2. N. D. Newell, *Sci. Am.* **208**, 77 (1963); *Geol. Soc. Am. Spec. Pap.* **89**, 63 (1967).
3. J. L. Cutbill and B. M. Funnel, in *The Fossil Record*, W. B. Harland *et al.*, Eds. (Geological Society of London, London, 1967), p. 791.
4. J. W. Valentine, *Palaeontology* **12**, 684 (1969).
5. The data were compiled from R. C. Moore *et al.*, Eds. *Treatise on Invertebrate Paleontology* (Geological Society of America and Univ. of Kansas Press, Lawrence, 1953–1979); W. B. Harland *et al.*, Eds., *The Fossil Record* (Geological Society of London, London, 1967); A. S. Romer, *Vertebrate Paleontology* (Univ. of Chicago Press, Chicago, 1966); also, 380 additional papers and monographs. A complete listing of these data is scheduled to appear in J. J. Sepkoski, Jr., *Milwaukee Pub. Mus. Contrib. Biol. Geol.*, in press.
6. Vendian and Tommotian points were excluded from the analysis because these intervals have exceptionally low diversities and therefore exceptionally low extinction rates.
7. The geologic time scale used is a composite based on a number of recently published stage-level time scales and differs only slightly from that used by J. J. Sepkoski, Jr., *Paleobiology* **5**, 222 (1979).
8. Four general kinds of animals were considered to have low fossilization potential: (i) soft-bodied animals without mineralized skeletons (for example: Nemertina, Priapulida, and Sipunculida as well as many Hydrozoa, Scyphozoa, and Polychaeta); (ii) animals with lightly sclerotized skeletons (such as many Crustacea); (iii) rarely reported fossil animals with multielement skeletons that dissociate rapidly after death (such as Octocorallia and Holothuroidea as well as some Asterozoa and Osteichthyes); and (iv) deep-sea animals with extremely poor fossil records (such as some Crinoidea, Chondrichthyes, and Osteichthyes).
9. The Frasnian might still be considered statistically distinct in Fig. 1. If each point in that figure is considered an independent event, then the probability that three of the nine highest points would be clustered about the Frasnian point is quite low ($P = .002$).
10. D. J. McLaren, *J. Paleontol.* **44**, 801 (1970).
11. "Minor" mass extinctions, which do not appear as noticeable perturbations in Figs. 1 and 2,

include at least three to five events in the Cambrian [A. R. Palmer, *J. Paleontol.* **39**, 149 (1965); M. E. Taylor, *ibid.* **42**, 1319 (1968); J. H. Stitt, *ibid.* **45**, 178 (1971)], an Early Jurassic (Toarcian) event [A. Hallam, *Paleobiology* **3**, 58 (1977)], a terminal Eocene event [H. Tappan and A. R. Loeblich, *Geol. Soc. Am. Spec. Pap.* **127**, 247 (1971); A. G. Fischer and M. A. Arthur, *Soc. Econ. Paleontol. Mineral* **25**, 19 (1977)], and possible events in the latest Jurassic (Tithonian) and early Late Cretaceous (Cenomanian) (see A. G. Fisher and M. A. Arthur, above).

12. D. M. Raup, *Carnegie Mus. Nat. Hist. Bull.* **13**, 85 (1979).
13. We thank R. K. Bambach for critical comments on this work.

28 September 1981; revised 31 November 1981

Lumber Spill in Central California Waters: Implications for Oil Spills and Sea Otters

Abstract. *A large quantity of lumber was spilled in the ocean off central California during the winter of 1978, and it spread through most of the range of the threatened California sea otter population within 4 weeks. The movement rates of lumber were similar to those of oil slicks observed elsewhere. These observations indicate that a major oil spill could expose significant numbers of California sea otters to oil contamination.*

The California population of the sea otter [*Enhydra lutris nereis* (Merriam)] was listed as "threatened" in 1977 (*1*) pursuant to the Endangered Species Act of 1973 (*2*). The listing was based on the possibility that a major oil spill could occur within the sea otter range and could kill a significant portion of the population, placing it in danger of extinction. This concern arises from the known sensitivity of the species to oil contamination (*3*). A spill of gasoline and diesel oil nearshore in the Kurile Islands, U.S.S.R., spread through 40 km of coastline and killed over 100 sea otters (*4*). Concern for the status of the California sea otter is heightened by the lack of evidence of significant population growth since 1973 (*5*).

It is difficult to project the critical day-to-day movements of floating oil near the sea otter range on the basis of existing oceanographic data. Surface current patterns off central California (San Francisco to Point Conception) have been examined with several techniques (*6–9*). The principal result is the description of mean flow patterns on a seasonal scale. However, studies of drogues and remote imagery have shown that short-term departures from mean seasonal drift may be frequent in the California current system (*6, 7, 10, 11*). Such departures involve tidal oscillations and mesoscale meanders and eddies (*6, 7, 10, 11*). The prediction of the direction of drift of floating oil is further complicated by the dominant role of wind stress at the air-sea interface (*12*). As far as we know, there are no records of major oil spills off central California on which to base predictions of oil drift. We know of no published studies of day-to-day movements of other floating materials off central California over an appropriately small time scale.

In this report we describe the movements and beaching of a large volume of lumber spilled off central California in the winter of 1978. Floating materials such as drift cards and plastic sheets have been used successfully by others in modeling the movements of oil on the sea surface (*13*). Our data provide a first approximation of the disposition of the floating component of a large oil spill occurring under similar conditions of weather and sea. Information of this kind is needed if we are to understand the potential impacts of oil spills on the California sea otter population and to develop management plans for improving the status of the population, now numbering about 1800 animals (*5*).

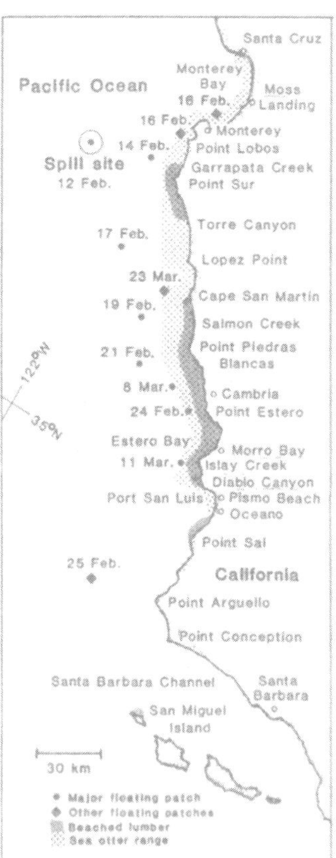

Fig. 1. Sightings of the major patch and smaller patches of floating lumber and areas of significant lumber beaching in and near the range of the sea otter population in California. Observations were made after the spillage of 2×10^6 board feet of lumber off Point Sur on 12 February 1978. All sightings of beached lumber were recorded between 12 February and 31 March 1978. The times and the positions corresponding to each sighting of the major patch of floating lumber are as follows: 12 February, 0650 GMT 36°15′N, 122°25′W (spill site); 14 February, 1930 GMT, 36°19′N, 122°07′W; 17 February, 0130 GMT, 35°55′N, 121°51′W; 19 February, 2030 GMT, 35°40′N, 121°37′W; 21 February, 0136 GMT, 35°29′N, 121°29′W; 24 February, 2100 GMT, 35°27′N, 121°05′W; 8 March, 1816 GMT, 35°30′N, 121°16′W; 11 March, 2006 GMT, 35°18′N, 120°58′W. Mean wind speed (meters per second) and the direction (relative to true north) between sightings of the major floating patch, during the period when the patch was closest to shore, were as follows: 21 to 24 February, 3.57, 317°; 24 February to 8 March, 0.90, 154°; 8 to 11 March, 4.30, 308° [wind data are from the shore station at Point Piedras Blancas (*18, 19*)].

A cargo of 2×10^6 board feet of finished lumber (volume equivalent to 2.9×10^4 barrels of oil) spilled from a barge under tow in heavy weather 40 km west of Point Sur on 12 February 1978 (*14*). The spilled lumber was hazardous to navigation (*15*) and was therefore monitored by aircraft and merchant vessels in subsequent weeks (*16*). We compiled a record of observations of the floating lumber at sea along with sightings of lumber washed ashore after the spill.

Much of the spilled lumber remained in a single large patch that moved first toward the coast and then southeastward, parallel to the shoreline, during the first 10 days after the spill (Fig. 1). By 24 February the major patch was within 7 km of shore near Point Estero and remained relatively close to shore for the balance of the observation period (through March). Other patches of floating lumber were seen off Monterey, Point Lobos, Cape San Martin, and Point Arguello during the survey period. Beached lumber was found throughout two sections of coastline within the sea otter range, a northern section of about

0036-8075/82/0319-1503$01.00/0

PAPER 36

Patterns in Vascular Land Plant Diversification (1983)

K. J. Niklas, B. H. Tiffney, and A. H. Knoll

Commentary

PETER WILF

Are there large-scale patterns of biodiversity through deep time? This question is fundamental to understanding present-day diversity and ecology. The fossil record is the only direct source of information about taxon richness through geologic time and how it is affected by evolutionary innovations, environmental disturbances, and other factors. Because of their high preservation potential, marine, hard-shelled, benthic invertebrates are the source of most of the existing data on diversity through time, and hundreds of papers on the topic have been produced since the era of seminal observations by J. W. Valentine, D. M. Raup, J. J. Sepkoski Jr., and others (e.g., Valentine 1969; Raup 1972; Sepkoski 1976; Sepkoski et al. 1981; Alroy et al. 2008). Perhaps the most remarkable contribution from this time, and the setting for the K. J. Niklas, B. H. Tiffney, and A. H. Knoll paper featured here, was Sepkoski's (1981) statistical recognition through factor analysis of three major evolutionary faunas of marine invertebrates through the Phanerozoic: the Cambrian, Paleozoic, and Modern (Mesozoic-Cenozoic), the radiation of each coupled to the decline in diversity of the preceding. Although still debated (Alroy 2004; Stanley 2007), Sepkoski's evolutionary faunas remain one of the most influential ideas in all of paleontology.

Niklas, Tiffney, and Knoll, via factor analysis of a global compilation of land-plant occurrences, asked whether land plants, though fundamentally different from benthic invertebrates, analogously demonstrate distinct evolutionary associations. If so, that would suggest generalities in evolution that transcend animal versus plant and sea versus land. Their contribution is part of a classic series that still constitutes the majority of papers on plant diversity through time, and which shows a trio of paleobotanists engaged in quantifying the plant-fossil record during the heyday of the marine invertebrate studies (Niklas 1977, 1988; Knoll et al. 1979; Niklas et al. 1980, 1985; Tiffney 1981; Knoll et al. 1984; Knoll 1986; Niklas and Tiffney 1994).

Niklas, Tiffney, and Knoll discriminated four successive land-plant radiations, each broadly associated with a morphological or reproductive innovation and an initial burst in evolutionary rates and turnover (lettering as in the paper): (*a*) a Silurian-Devonian initial radiation of early vascular plants; (*b*) a Devonian-Carboniferous radiation of free-sporing plants; (*c*) a radiation of seed plants from the late Devonian to the late Paleozoic, leading to a gymnosperm-dominated Mesozoic flora; and (*d*) the massive angiosperm radiation from the Early Cretaceous. Radiations *b* and *d* were each associated with large increases in species diversity. Further, the transitions *a* to *b* and *c* to *d* were each associated with the decline of the older assemblage. Niklas, Tiffney, and Knoll concluded that the overall pattern of statistically recognizable associations of organisms with common evolutionary innovations, their appearances tied to the decline of the older associations, was very similar in nature to that

Reprinted by permission from Macmillan Publishers Ltd: *Nature* 303:614–16, copyright 1983.

of Sepkoski's evolutionary marine faunas and thus suggested "generalized patterns in the evolution of higher taxa" (614). Is this conclusion still plausible, nearly three decades later?

First, it must be noted that biases on observed fossil richness are numerous for all fossil groups and much discussed (e.g., Sepkoski 1981; Peters 2005), and these issues were well known to Niklas, Tiffney, and Knoll (e.g., Knoll et al. 1979; Niklas and Tiffney 1994). Examples include inconsistent sampling and undersampling, variation in preservational environment and the outcrop area and sediment volume of fossiliferous units, obsolete or inconsistent taxonomy, and change through time in climate, sea level, and continental positions.

These biases affect plant macrofossils considerably more than marine invertebrates because they are mostly deposited on the erosive continents (Raymond and Metz 1995; Wing and DiMichele 1995; Rees 2002). Fossil plants may be abundant locally, but at a broader scale, they are extremely patchy in time and space. Many time periods are so undersampled in the Niklas, Tiffney, and Knoll compilations that the addition of well-sampled, rich floras subsequently published (e.g., Johnson 2002; Anderson and Anderson 2003) would probably alter the overall pattern significantly. The various plant organs (wood, leaves, flowers, fruits and seeds, pollen, etc.) preserve in very different ways and are hard to correlate as whole plants. The most diverse and abundant organ group, angiosperm leaves, is heavily burdened with obsolete taxonomy. Thus confidence in global observations of diversity change is low in comparison to data from stratigraphically and paleoenvironmentally controlled, regional macrofloral studies, which also have far greater potential to reveal the rapid response of biodiversity to climate change and extinction (Johnson et al. 1989; Wing and Harrington 2001; McElwain et al. 2007).

Second, decades of new data have altered the understanding of the specific evolutionary units (*a–d*) and transitions proposed by Niklas, Tiffney, and Knoll. Regarding the Silurian-Devonian transition (*a* to *b*), phylogenetic evidence suggests that many of the clades in *b* had, quite plausibly, evolved during the Silurian (Kenrick and Crane 1997). However, most of the records from this time come from Euramerica, which was predominantly underwater during the Silurian and emerging into the Devonian; sampling of the Silurian is thus very uneven (Raymond and Metz 1995; Kenrick and Crane 1997). In sum, the *a*-to-*b* transition remains poorly understood. The transition from spore plants to seed plants (*b* to *c*) is now better characterized in terms of landscape heterogeneity and plant lineages tracking their biomes separately in a climate-driven manner (seed plants can occupy much drier habitats than spore plants), rather than competitive displacement within biomes (DiMichele and Aronson 1992; DiMichele et al. 2008; DiMichele et al. 2009; Falcon-Lang et al. 2009). The transition to the angiosperm radiation (*c* to *d*) remains remarkable in terms of species numbers (Lidgard and Crane 1990), and the concurrent extinction of many gymnosperm clades is clear. Troubling issues are the lack of evidence for ecological dominance of angiosperms prior to the latest Cretaceous, despite their high diversity (Wing et al. 1993), and the unsettling fact that the angiosperm ancestors, and thus their contribution to diversity, remain completely unknown. Molecular phylogenetic data indicate an angiosperm stem reaching far into the Paleozoic (e.g., Palmer et al. 2004).

The issues listed above are sufficient, in my view, to render the specific suggestion of "generalized patterns in the evolution of higher taxa" as far from proven, though still plausible. Nevertheless, these and other observations of Niklas, Tiffney, and Knoll are foundational, because they still form the only baseline of observation for future work, and they suggest large-scale patterns of plant diversification that are still widely discussed. No other workers have before or since put forth a global compilation and analysis of the Phanerozoic land-plant record. Their direct comparisons to the marine invertebrate record, considered here, were forward-thinking, highly novel, and daring considering the much smaller quantity of data available.

Literature Cited

Alroy, J. 2004. Are Sepkoski's evolutionary faunas dynamically coherent? *Evolutionary Ecology Research* 6:1–32.

Alroy, J., M. Aberhan, D. J. Bottjer, M. Foote, F. T. Fürsich, P. J. Harries, A. J. W. Hendy, et al. 2008. Phanerozoic trends in the global diversity of marine invertebrates. *Science* 321:97–100.

Anderson, J. M., and H. M. Anderson. 2003. *Heyday of the gymnosperms: Systematics and biodiversity of the Late Triassic Molteno fructifications. Strelitzia*, vol. 15. National Botanical Institute, Pretoria, South Africa.

DiMichele, W. A., and R. B. Aronson. 1992. The Pennsylvanian-Permian vegetational transition: A terrestrial analogue to the onshore-offshore hypothesis. *Evolution* 46:807–24.

DiMichele, W. A., H. Kerp, N. J. Tabor, and C. V. Looy. 2008. The so-called "Paleophytic-Mesophytic" transition in equatorial Pangea: Multiple biomes and vegetational tracking of climate change through geological time. *Palaeogeography, Palaeoclimatology, Palaeoecology* 268:152–63.

DiMichele, W. A., I. P. Montañez, C. J. Poulsen, and N. J. Tabor. 2009. Climate and vegetational regime shifts in the late Paleozoic ice age earth. *Geobiology* 7:200–226.

Falcon-Lang, H. J., W. J. Nelson, S. Elrick, C. V. Looy, P. R. Ames, and W. A. DiMichele. 2009. Incised channel fills containing conifers indicate that seasonally dry vegetation dominated Pennsylvanian tropical lowlands. *Geology* 37:923–26.

Johnson, K. R. 2002. The megaflora of the Hell Creek and lower Fort Union formations in the western Dakotas: Vegetational response to climate change, the Cretaceous-Tertiary boundary event, and rapid marine transgression. *Geological Society of America Special Paper* 361:329–91.

Johnson, K. R., D. J. Nichols, M. Attrep Jr., and C. J. Orth. 1989. High-resolution leaf-fossil record spanning the Cretaceous-Tertiary boundary. *Nature* 340:708–11.

Kenrick, P., and P. R. Crane. 1997. The origin and early diversification of land plants: A cladistic study. Smithsonian Institution, Washington, DC.

Knoll, A. H. 1986. Patterns of change in plant communities through time. In *Community ecology*, edited by J. Diamond and T. J. Case, 126–44. Harper and Row, New York.

Knoll, A. H., K. J. Niklas, P. G. Gensel, and B. H. Tiffney. 1984. Character diversification and patterns of evolution in early vascular plants. *Paleobiology* 10:34–47.

Knoll, A. H., K. J. Niklas, and B. H. Tiffney. 1979. Phanerozoic land-plant diversity in North America. *Science* 206:1400–1402.

Lidgard, S., and P. R. Crane. 1990. Angiosperm diversification and Cretaceous floristic trends: A comparison of palynofloras and leaf macrofloras. *Paleobiology* 16:77–93.

McElwain, J. C., M. E. Popa, S. P. Hesselbo, M. Haworth, and F. Surlyk. 2007. Macroecological responses of terrestrial vegetation to climatic and atmospheric change across the Triassic/Jurassic boundary in East Greenland. *Paleobiology* 33:547–73.

Niklas, K. J. 1977. Theoretical evolutionary rates in plant groups and the fossil record. *Brittonia* 29:241–54.

———. 1988. Patterns of vascular plant diversification in the fossil record: Proof and conjecture. *Annals of the Missouri Botanical Garden* 75:35–54.

Niklas, K. J., and B. H. Tiffney. 1994. The quantification of plant biodiversity through time. *Philosophical Transactions of the Royal Society of London B* 345:35–44.

Niklas, K. J., B. H. Tiffney, and A. H. Knoll. 1980. Apparent changes in the diversity of fossil plants. In *Evolutionary biology*, edited by W. C. Steere, M. K. Hecht, and B. Wallace, 12:1–89. Plenum, New York.

———. 1985. Patterns in vascular land plant diversification: An analysis at the species level. In *Phanerozoic diversity patterns: Profiles in macroevolution*, edited by J. W. Valentine, 97–108. Princeton University Press, Princeton, NJ.

Palmer, J. D., D. E. Soltis, and M. W. Chase. 2004. The plant tree of life: An overview and some points of view. *American Journal of Botany* 91:1437–45.

Peters, S. E. 2005. Geologic constraints on the macroevolutionary history of marine animals. *Proceedings of the National Academy of Sciences* 102:12326–31.

Raup, D. M. 1972. Taxonomic diversity during the Phanerozoic. *Science* 177:1065–71.

Raymond, A., and C. Metz. 1995. Laurussian land-plant diversity during the Silurian and Devonian: Mass extinction, sampling bias, or both? *Paleobiology* 21:74–91.

Rees, P. M. 2002. Land-plant diversity and the end-Permian mass extinction. *Geology* 30:827–30.

Sepkoski, J. J., Jr. 1976. Species diversity in the Phanerozoic: Species-area effects. *Paleobiology* 2:298–303.

———. 1981. A factor analytic description of the Phanerozoic marine fossil record. *Paleobiology* 7:36–53.

Sepkoski, J. J., Jr., R. K. Bambach, D. M. Raup, and J. W. Valentine. 1981. Phanerozoic marine diversity and the fossil record. *Nature* 293:435–37.

Stanley, S. M. 2007. An analysis of the history of marine animal diversity. *Paleobiology* 33:1–55.

Tiffney, B. H. 1981. Diversity and major events in the evolution of land plants. In *Paleobotany, paleoecology, and evolution*, edited by K. J. Niklas, 2:193–230. Praeger Press, New York.

Valentine, J. W. 1969. Patterns of taxonomic and ecological structure of the shelf benthos during Phanerozoic time. *Palaeontology* 12:684–709.

Wing, S. L., and W. A. DiMichele. 1995. Conflict between local and global changes in plant diversity through geological time. *Palaios* 10:551–64.

Wing, S. L., and G. J. Harrington. 2001. Floral response to rapid warming in the earliest Eocene and implications for concurrent faunal change. *Paleobiology* 27:539–63.

Wing, S. L., L. J. Hickey, and C. C. Swisher. 1993. Implications of an exceptional fossil flora for Late Cretaceous vegetation. *Nature* 363:342–44.

614 LETTERS TO NATURE NATURE VOL. 303 16 JUNE 1983

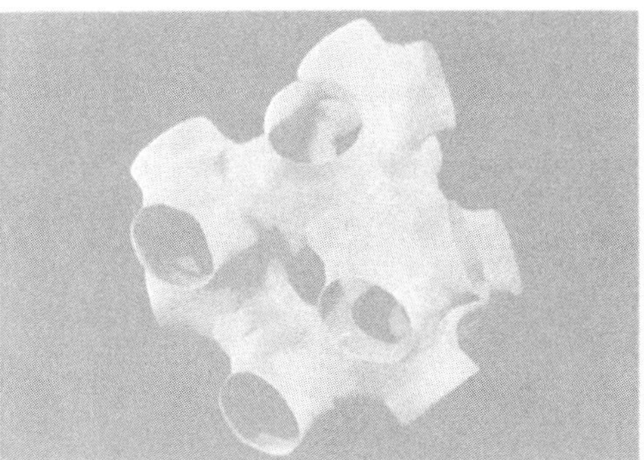

Fig. 4 Model of part of Schwarz's tetrahedral periodic surface, made of strips of paper glued together and painted. This surface would lie within the lipid bilayer which separates the water channels into two equal sub-volumes.

value as an upper bound for the diameter of the cylinders, we find that the calculated maximum volume of the lipid in the model would be 42% of the cell, which is much too low. The conclusion is that the channels are filled with water.

The space between the water labyrinths must be occupied by lipid. This space forms a single, continuous region which contains Schwarz's tetrahedral minimal surface. If the hydrophilic head groups contact the water in the channels with the hydrocarbon tails pointing away from the water, then the lipid forms a bilayer which can be accommodated by the dimensions and symmetry of this space. A model of part of this surface is illustrated in Fig. 4. The surface can be theoretically built up of units, each composed of that surface of minimal area which is bounded by four straight lines connecting the vertices of a regular tetrahedron. The mathematical expression for this surface and illustrations of a model are given by Schwarz[4], who also explains how the unit can be infinitely repeated to form the complete periodic surface. Each edge of the crystallographic cell passes normally through the centre of a two-fold saddle-surface, while normal to the body diagonals there are three-fold 'monkey saddles'[14].

The viscous isotropic phase of monoglycerides has been found among the digestion products of fat[15]. We know of no other naturally occurring example of Schwarz's periodic tetrahedral minimal surface.

Supported by NIH grant GM27278. We thank Dr M. K. Reedy for the use of X-ray equipment and Mr Leonidas Cordova for technical assistance.

Received 6 December 1982; accepted 12 April 1983.

1. Tardieu, A. & Luzzati, V. *Biochim. biophys. Acta* **219**, 11–17 (1970).
2. Luzzati, V. & Spegt, P. A. *Nature* **220**, 485–488 (1968).
3. Scriven, L. E. *Nature* **263**, 123–125 (1976); in *Micellization, Solubilization and Microemulsions* Vol. 2 (ed. Mittal, K. L.) (Plenum, New York, 1977).
4. Schwarz, H. A. *Gesammelte Mathematische Abhandlung* Vol. 1 (Springer, Berlin, 1890).
5. Schoen, A. H. *NASA Technical Note* D-5541, Washington DC (1970).
6. Larsson, K., Fontell, K. & Krog, N. *Chem. Phys. Lipids* **27**, 321–328 (1980).
7. Lindblom, G., Larsson, K., Johansson, L., Fontell, K. & Forsen, S. J. *Am. chem. Soc.* **101**, 5465–5470 (1969).
8. Lutton, E. S. *J. Am. Oil chem. Soc.* **42**, 1068–1070 (1965).
9. Luzzati, V. in *Biological Membranes* (ed. Chapman, D.) (Academic, New York, 1967).
10. *International Tables for X-ray Crystallography* Vol. II (Kynoch, Birmingham, 1967).
11. Tardieu, A. thesis, Univ. Paris Sud (1972).
12. Bragg, Sir L., Claringbull, G. F. & Taylor, W. H. *Crystal Structures of Minerals* (Cornell University Press, New York, 1965).
13. Klug, A., Finch, J. T. & Franklin, R. E. *Biochim. biophys. Acta* **25**, 242–252 (1957).
14. Nitsche, J. C. C. *Vorlesungen uber Minimalflachen* (Springer, Berlin, 1975).
15. Patton, J. S. & Carey, M. C. *Science* **204**, 145–148 (1979).
16. Lipson, H. and Steeple, H. *Interpretation of Powder Diffraction Patterns* (Macmillan, London, 1970).

0028–0836/83/240614—03$01.00

Patterns in vascular land plant diversification

Karl J. Niklas*, Bruce H. Tiffney† & Andrew H. Knoll‡

* Division of Biological Sciences, Section of Plant Biology, Cornell University, Ithaca, New York 14850, USA
† Peabody Museum of Natural History and Department of Biology, Yale University, New Haven, Connecticut 06511, USA
‡ Harvard University Herbaria, Harvard University, Cambridge, Massachusetts 02139, USA

Statistical analyses indicate that the Phanerozoic history of vascular land plants (tracheophytes) may be interpreted in terms of the successive radiations of four major plant groups, each characterized by a common morphological and/or reproductive grade. Following initial invasion of the land, the diversification of each group coincides with a decline in species numbers of the previously dominant group. Analyses indicate that each group is initially characterized by a high rate of appearance of new species but short species durations. However, within a group through time, new species appear less frequently but species durations increase. Parallels between the diversification patterns observed for tracheophytes and those previously noted for marine invertebrates[1] suggest there are generalized patterns in the evolution of higher taxa.

Studies of the taxonomic diversity of Phanerozoic invertebrates have shown a pattern of changes in diversity through time, and that this pattern reflects an evolutionary phenomenon strong enough to overcome the presumed biases of the fossil record[1]. *Q*-mode factor analyses of marine metazoa further indicate that the overall pattern can be explained in terms of the expansion and subsequent contraction of three sequential, discrete 'evolutionary faunas'[2]. From these studies, quantitative estimates of the rates and magnitudes of major biological radiations and extinctions have been obtained for marine organisms. Palaeobotanists have also been aware of similar patterns in the Phanerozoic history of tracheophytes. Eras of geological time have been identified palaeobotanically as the Palaeophytic, Mesophytic and Cenophytic, corresponding to those periods dominated by the pteridophytes, gymnosperms and angiosperms, respectively[3]. Our quantitative studies of the patterns of diversification of vascular plants indicate that the fossil record of tracheophytes is as amenable to analysis as that of marine invertebrates[4–6] and offer the opportunity to characterize major patterns in tracheophyte evolution. Because of the fundamental differences in the biology of invertebrates and vascular plants, as well as differences in the habitats in which they evolved, comparisons between the patterns of diversification for these two groups could provide an opportunity to evaluate the possible existence of unifying patterns in the evolution of life.

Our analysis at the species level (Fig. 1) indicates that the overall pattern of terrestrial plant diversification can be separated into four distinct evolutionary phases: (1) a Silurian–mid-Devonian proliferation of early vascular plants that are characterized by a simple and presumably primitive morphology (Fig. 1*a*); (2) a subsequent late Devonian-Carboniferous radiation of derived pteridophytic linages (ferns, lycopods, sphenopsids, progymnosperms) (Fig. 1*b*); (3) the appearance of seed plants in the late Devonian and their adaptive radiation in the late Palaeozoic, culminating in a gymnosperm-dominated Mesozoic flora (Fig. 1*c*); and (4) the appearance and rise of flowering plants in the Cretaceous and Tertiary (Fig. 1*d*). Two of these four phases of tracheophyte evolution are associated with significant increases in the total species diversity of land plants. Following the establishment of a land flora, the adaptive radiation of the advanced pteridophytes (derived from the earliest tracheophytes), together with the initial expansion of

NATURE VOL. 303 16 JUNE 1983 LETTERS TO NATURE 615

early seed plants, resulted in a fourfold increase in total species numbers from the late Devonian to the late Carboniferous. The second major increase is associated with the radiation of the angiosperms, resulting in a threefold increase in observed species diversity by the Neogene. The Permo-Triassic boundary marks a period in which known species numbers actually declined by about 20%, and in which the sedimentary record of plant-bearing rocks is poor, at least in Europe and North America. Although the physiognomic and taxonomic changes in terrestrial vegetation accompanying this transition were marked, overall species diversity did not subsequently increase significantly (Fig. 1*b*, *c*). While a number of time-dependent and time-independent biases influence our estimates of diversity[4], our summation of the proportional representation of different groups at any one geological time should not be distorted by such biases.

Based on our analysis of tracheophyte history, it is evident that in two cases (the transition of group *a* to *b* and group *c* to *d*) the appearance of a new plant grade was associated with a levelling off and subsequent decline in the number of species in the preceding dominant group. This pattern is similar to that seen for the evolutionary 'succession' of major marine invertebrate groups ('evolutionary faunas'[1]) during the Phanerozoic[2]. In parallel, we believe that this pattern suggests the 'competitive displacement' of older, less specialized taxa by newer and presumably more specialized taxa. In a third case (the replacement of group *b* by group *c*), the transition coincides with environmental changes of the late Palaeozoic and earliest Mesozoic. Here, circumstances suggest that the transition is, to some degree, a function of the partial extinction of group *b* followed by the radiation of members of group *c* into the vacated ecological space. It should be noted, however, that species numbers do not necessarily reflect the extent of ecological dominance.

Although stable numbers of species were attained at times in the past, new rounds of diversification accompanied major evolutionary innovations, particularly those involving reproductive features. The data are consistent with ecological theory which suggests that plant species increasingly subdivide their environment through the evolution of more diverse resource use until constrained by the inherent limits of their morphological, physiological and reproductive capabilities[7]. Adaptive radiations of new major groups may involve the removal of one or more of these constraints, permitting new strategies of environmental fragmentation and thus, further diversification. This general pattern is supported by within-flora studies[4] which similarly suggest that it is innovations in reproductive biology, permitting escape from species saturation in land plant communities, that have characterized the evolution of new plant groups.

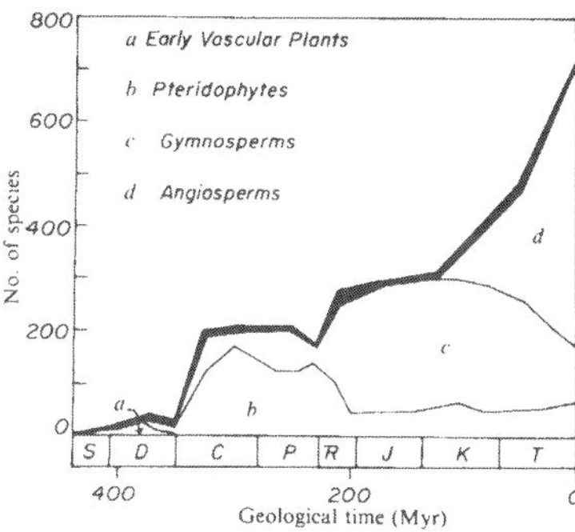

Fig. 1 An analysis of vascular land plant diversity reveals the presence of four groups which have successively dominated the terrestrial flora. Each group is composed of plants sharing a common structural and/or reproductive grade. Following the invasion of land by primitive tracheophytes (group *a*), the remaining three groups successively replace each other, the radiation of one group terminating the dominance of its predecessor. Each radiation involves a major compositional change in the terrestrial flora, and in the cases of groups *b* and *d*, a major rise in overall species number. The transition from group *b* to group *c* is unique in its coincidence with a major period of change in the physical environment. The solid black line at the top of the graph represents non-vascular land plants and *incertae sedis*. Approximately 18,000 fossil plant species citations were compiled and computerized by taxonomic affinity (species, genus, family), age and location. Although the citations were collected worldwide, the majority of the data come from European and North American sources. The data were subjected to a *Q*-mode factor analysis (ref. 2) using the CABFAC Q-mode vector analysis program[9-12]: S, Silurian; D, Devonian; C, Carboniferous; Tr, Triassic; J, Jurassic; K, Cretaceous; T, Tertiary.

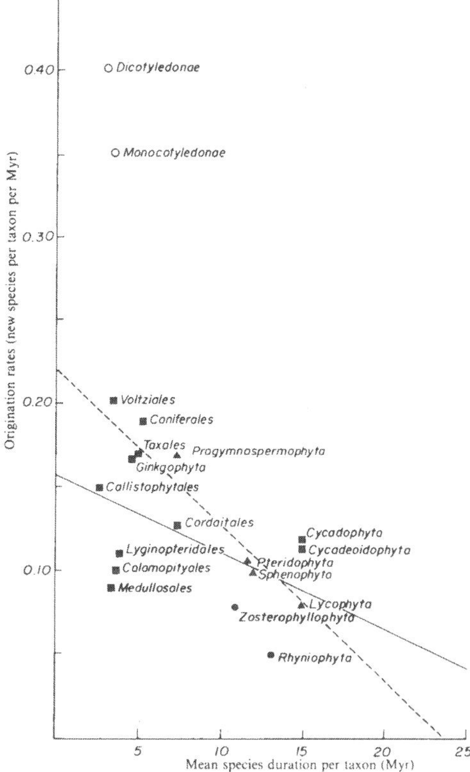

Fig. 2 Summed species origination rates for suprageneric plant taxa versus mean species duration of the same suprageneric taxa. The oldest taxa (rhyniophytes, zosterophyllophytes, lycopods) have the lowest origination rates and longest durations, while the most recent taxa (Monocotyledonae, Dicotyledonae) have the highest origination rates and some of the shortest durations. The dotted line indicates a linear regression of origination rates versus durations for all groups plotted ($r = 0.47$); the solid line is a regression of the same data, but excluding the Monocotyledonae and Dicotyledonae ($r = 0.49$). Suprageneric groups represented by dots belong to group *a* of Fig. 1, those represented by triangles belong to group *b*, those represented by squares belong to group *c*, and those represented by open circles belong to group *d* of Fig. 1. Origination rates (R) for the various suprageneric groups are based on the following computation: $R = r_s - r_e$, where $r_s = S/Dt$, $r_e = E/Dt$; S is the number of new species, E is the number of extinctions and D is the total diversity (number of taxa in each group) per time interval t. The relationships given are crude estimates of R and D, but are probably an accurate first-order approximation.

In addition to the broad patterns of overall diversity and group replacement, we observe patterns of change in rates of evolution both between and within major groups. Among each successive major group shown in Fig. 1, there is a general trend for an increase in the mean rate of appearance of new species and a decrease in the mean species duration (see Fig. 2). The slowest rates of species origination and longest durations are seen in groups *a* and *b*, while the highest rates and short durations are seen in the angiosperms. Even if the short species durations calculated for the flowering plants falsely result from their relatively recent appearance, their high species origination rates are not influenced by the same bias, and are an order of magnitude above those of the lowest valued group. The rates of species origination and the lengths of species duration also change in a consistent manner within each of the four major groups.

The overall pattern of increasing rates of species origination and decreasing species durations between the major groups are statistically significant. A linear regression of the average species origination rate and the mean species duration for each of the 19 suprageneric groups of Fig. 2 that go to make up the four major groups of Fig. 1 gives $r = 0.47$; elimination of the high values of the angiosperms yields $r = 0.49$ ($n = 17$). Both r-values fall between the 95–99% confidence interval, indicating a high degree of correlation between species origination rates and species durations. Studies indicate that differences in apparent speciation rates have a high correlation with plant breeding strategies and karyotypic variation[8]. Possible plant breeding strategies favouring high speciation rates involve small effective population sizes and specialized mechanisms of pollen and/or propagule dispersal, features commonly found in the most rapidly speciating angiosperm families. By contrast, large effective population size and generalist pollen and propagule dispersal features (commonly observed in gymnosperms) may favour slower speciation rates[8].

Clearly, the many factors that have influenced the diversification of tracheophytes on the one hand, and marine invertebrates on the other, were both quantitatively and qualitatively different. Similarly, the extent of environmental stability and heterogeneity experienced in the marine and terrestrial habitats are quite distinct. Yet the patterns of diversity observed for these two groups are similar. Both indicate that there have been relatively few changes through time in the structure of the Earth's biota, and that such changes that have occurred have accompanied the radiations of major taxa and have frequently resulted in significant increases in total diversity. Further, following a period of rapid diversification, each evolutionary fauna and flora has tended to approach a levelling-off in the increase in the number of species and ultimately a decline in number as a new group radiates. However, throughout this history, the overall trend has been for a continued increase in total diversity. This suggests that similar types of interactions between developmental potential and ecological constraints may have controlled the evolution of diversity on land and in shallow marine environments.

We acknowledge the support of NSF grants DEB-81-18416 and EG-81-18749 (K.J.N.) and DEB-79-05081 (B.H.T.).

Received 17 December 1982; accepted 26 April 1983.

1. Sepkoski, J. J. Jr., Bambach, R. K., Kaup, D. M. & Valentine, J. W. *Nature* **293**, 435–437 (1981).
2. Sepkoski, J. J. Jr., Bambach, R. K., Kaup, D. M. & Valentine, J. W. *Paleobiology* **7**, 36 (1981).
3. Gothan, W. & Weyland, H. *Lehrbuch der Paleobotanik* (Akademic, Berlin, 1954).
4. Knoll, A. H., Niklas, K. J. & Tiffney, B. H. *Science* **206**, 1400–1402 (1979).
5. Niklas, K. J., Tiffney, B. H. & Knoll, A. H. in *Evolutionary Biology* Vol. 12 (eds Hecht, M. K., Steere, W. C. & Wallace, B.) 1–89 (Plenum, New York, 1980).
6. Tiffney, B. H. in *Paleobotany, Paleoecology and Evolution* Vol. 2 (ed. Nitlas, K. J.) 193–230 (Praeger, New York, 1981).
7. Whittaker, R. H. in *Evolutionary Biology* Vol. 10 (eds Hecht, M. K., Steere, W. C. & Wallace, B.) 1–68 (Plenum, New York, 1977).
8. Levin, D. A. & Wilson, A. C. *Proc. natn. Acad. Sci. U.S.A.* **73**, 2086–2090 (1976).
9. Klovan, J. E. & Imbrie, J. *Math. Geol.* **3**, 61–77 (1971).
10. Flessa, K. & Levinton, J. S. *J. Geol.* **83**, 239–248 (1975).
11. Joreskog, K. G., Klovan, J. E. & Regment, R. A. *Geological Factor Analysis* (Elsevier, Amsterdam, 1976).
12. Smith, C. A. F. III *Paleobiology* **3**, 41–48 (1977).

0028–0836/83/240616—03$01.00

Colour-generating interactions across the corpus callosum

Edwin H. Land, David H. Hubel, Margaret S. Livingstone, S. Hollis Perry & Michael M. Burns

Rowland Institute for Science, Cambridge, Massachusetts 02142, USA and Department of Neurobiology, Harvard Medical School, Boston, Massachusetts 02115, USA

Human vision has the remarkable property that, over a wide range, changes in the wavelength composition of the source light illuminating a scene result in very little change in the colour of any of the objects[1]. This colour constancy can be explained by the retinex theory, which predicts the colour of a point on any object from a computed relationship between the radiation from that point and the radiation from all the other points in the field of view (Fig. 1). Thus the computations for colour perception occur across large distances in the visual field. It has not been clear, however, whether these long-range interactions take place in the retina or the cortex. Reports that long-range colour interactions can be reproduced binocularly when one band of wavelengths enters one eye and a different band enters the other[2] might seem to establish the cortex as the site of the computation. Many observers, however, see very unsatisfactory colour or no colour at all in this binocular situation, suggesting that the cortex may not be the only site at which the computation is carried out, or even the most important site. We have now tested the role of the cortex in a human subject in whom the nerve fibres connecting cortical areas subserving two separate parts of the visual field had been severed, and find that the cortex is necessary for long-range colour computations.

We obtained the subject (J.W.) with the help of Drs Norman Geschwind and Alexander Reeves, and as a preparatory step, we tested his colour vision, which proved to be normal. The subject was a 20-yr-old male who had had intractable epileptic seizures (complex limbic). In 1979 he underwent, in two stages, a complete resection of his corpus callosum, which markedly decreased the number and severity of his seizures. The subject had no obvious intellectual or physical impairment. He had been extensively tested in connection with the callosal cut. In this subject, if the site of the computation is indeed cortical, events in one visual half-field should have no influence on the appearance of objects in the other half-field. If an influence is found, the computation responsible for that influence must be retinal.

Our objective was to construct a visual scene in which there was one region, of constant luminous flux and wavelength composition, whose colour was to be reported on. This test region would span the midline of the visual field; a large second region whose illumination could be varied would be confined to one visual half-field or to the other.

We finally adopted the set-up shown in Fig. 2, where F, the point of fixation, lay within a test spot, T (Color-Aid paper RV Tint 1). A Mondrian[3] of variously coloured Munsell papers lay to the left of the test spot and hence was entirely in the subject's left visual field (Fig. 2*a*). The parts of the field to the right of the borders of the Mondrian and test spot were black velvet. To keep the Mondrian in the subject's left visual field, we set the fixation point at 3.7° to the right of the right-hand border of the Mondrian. Beyond the experimental display, in the subject's extreme visual periphery all the surfaces in the room were low-reflection blacks.

PAPER 37

Species-Energy Theory: An Extension of Species-Area Theory (1983)

D. H. Wright

Commentary

DAVID J. CURRIE

David H. Wright's 1983 *Oikos* paper is, in my opinion, a model of how macroecology should work. This paper has been cited over 750 times (see Google Scholar), and it provided a methodological model for a generation of subsequent work that explores global-scale variation in species richness.

Wright's study began with a classic biological question: what determines the number of species in an assemblage? The hypothesized answer—energy supply—was also a classic that had been suggested, but not actually tested, by G. Evelyn Hutchinson in 1959. The literature at the time of Wright's study contained many competing hypotheses, all of which were apparently supported by at least some observations (or, in other words, there were at least some observations consistent with each of those hypotheses).

Wright's study was critically important because it derived some very strong predictions from Hutchinson's hypothesis and then tested them using macroecological methods. To wit, if richness depends upon the amount of energy available, then richness should vary in proportion to area, since bigger areas collect more energy. This prediction is coincident with one of the main predictions of R. H. MacArthur and E. O. Wilson's classic island biogeography theory. Further, richness should also vary in proportion to productivity per unit area. To test these predictions, Wright gathered observations from most of the islands larger than 1,000,000 square kilometers worldwide. He showed that the product of area and areal actual evapotranspiration (a surrogate for primary productivity) could statistically account for 80% of the variation in species richness among these islands.

Thus Wright's/Hutchinson's hypothesis had survived a very strong test. While many others have looked for observations (sometimes anecdotal) that *support* a particular hypothesis, Wright systematically surveyed a body of evidence to *test* whether the predictions of his hypothesis were consistent with nature. From a falsificationist point of view, surviving a strong test provides strong support for a hypothesis, whereas finding particular observations that are consistent with a hypothesis provides no support at all.

Wright's study is also important because of the way it used correlations. Since Wright's hypothesis predicted a correlation, he could use that correlation to test the hypothesis. In contrast, there is a great temptation in macroecology to look for correlations and to ask, after the fact, what processes generated them. There is nothing wrong with using correlations to help formulate hypotheses. However, it is clearly logically unjustified to infer processes from correlations. Wright's study is a brilliant example of how a correlation that is predicted by a hypothesis can be used to test that hypothesis.

Literature Cited

Hutchinson, G. E. 1959. Homage to Santa Rosalia; or, Why there are so many kinds of animals. *American Naturalist* 93:145–59.

OIKOS 41: 496–506. Copenhagen 1983

Species-energy theory: an extension of species-area theory

David Hamilton Wright

Wright, D. H. 1983. Species-energy theory: an extension of species-area theory. – Oikos 41: 496–506.

A more general biogeographic theory of island species number is produced by replacing area with a more direct measure of available energy in the models of MacArthur and Wilson and Preston. This theory, species-energy theory, extends beyond species-area theory in that it applies to islands that differ in their per-unit-area productivity due to differences in physical environment, such as climate. Examination of data on species number of angiosperms and of land and freshwater birds on islands worldwide, ranging from Greenland and Spitsbergen to New Guinea and Jamaica, demonstrates that species-energy theory can explain 70 to 80% of the variation in species number on such widely varying islands, and further suggests the existence of regular geographic trends in resource utilization or species-abundance patterns. The concepts embodied in species-energy theory can in principle be used to develop predictions of species' abundances and probabilities of occurrence on an island. Species-energy theory may also provide a unified basis for understanding a broad set of observations of patterns in species diversity.

D. H. Wright, Dept of Ecology and Evolutionary Biology, Univ. of Arizona, Tucson, AZ 85721, USA

Более общая биогеографическая теория числа островных видов создана путем репликации территории, где проводились прямые измерения использования энергии к моделям Мак Артура и Уилсона и Пристона. Эта теория, теория энергетики видов, выходит за рамки теории ареала вида, в которой она использовалась для островов, различающихся по продуктивности на единицу площади в результате различий таких физических факторов, как климат. Анализ данных по числу видов покрытосемянных растений и наземных и пресноводных видов птиц на островах всего мира от Гренландии и Шпицбергена до Новой Гвинеи и Ямайки показал, что теория энергетики видов может объяснить 70-80% вариаций в числе видов на таких сильно различающихся отсровах и подтверждает наличие закономерных географических тенденций в характере утилизации ресурсов или видовом разнообразии. Концепции, включенные в теорию энергетики видов, могут в принципе использоваться для предсказания видового обилия и вероятности встречаемости на острове. Теория энергетики видов может также представить общую основу для понимания широкого круга исследований типов видового разнообразия.

Accepted 2 August 1983

Introduction

Ecologists have long pondered how to predict how many species will occur in a community. Recent work (DeAngelis 1980, Brown 1981, Yodzis 1981) suggests, in agreement with earlier authors (Hutchinson 1959, Connell and Orias 1964, MacArthur 1965, 1972: 183) that the energy supply supporting a community limits the capacity of that community to contain species. This paper considers the relevance of a broad set of observations relating energy supply and species diversity and abundance, and attempts to provide a unified basis for understanding these patterns. To do this, I modify two important ecological models of species number and show how energy supply can limit species number. The modified theory has additional explanatory power and also raises new questions.

One of the most productive ecological models which addresses patterns of species number has been the equilibrium theory of island biogeography, proposed by MacArthur and Wilson (1963, 1967), which portrays the regulation of species diversity as a dynamic process where immigration opposes extinction. The primary factors affecting insular immigration and extinction rates identified by MacArthur and Wilson were area and isolation. Reasoning that larger islands would have larger populations and thus lower extinction rates, and that more isolated islands would have lower immigration rates, they generated a number of predictions in accord with empirical observations. In particular, islands of larger area were expected to support more species than smaller islands.

This conclusion, that species number should increase with increasing island area, was also reached by Preston (1962), who proposed an equation for the species-area curve. By regarding larger islands as having larger numbers of individuals, and by assuming a lognormal species-abundance distribution on all islands, Preston derived the approximate form of the species-area relationship as

$$S = CA^z \qquad (1),$$

where S is the number of species, A is area, C is a constant related to population density, and z is a constant. This equation is an approximation of the actual theoretical relationship between S and A, which is not a simple function (May 1975). Preston arrived at the power function approximation by fitting theoretical points to a line by least squares regression. The power function form was chosen for convenience and good fit over a reasonable range of S and A.

Species-area theory, both in the formulation of MacArthur and Wilson (1963, 1967) and that of Preston (1962), rests on the fundamental assumption that increasing island areas allow increased population sizes. However, it is clear that area itself usually has no direct effect on organisms, and that area is in fact a convenient secondary correlate which measures more proximate factors. Two such factors are frequently mentioned: first, increased area implies a greater total amount of habitat, and so a greater total amount of resources, capable of supporting larger populations. Second, larger islands or insular areas may contain a greater variety of habitats or resource types, thus supporting populations of a greater variety of species (MacArthur and Wilson 1967, MacArthur 1972: 102, Connor and McCoy 1979, Brown 1981). In both cases energy, in the form of resources, is a parameter of interest.

The accuracy of area as a measure of either the total amount of resources or the variety of resource types available depends on the set of islands examined. In general, area is accurate if the islands are fairly uniform in their climate, topography, and geology (Preston 1962, MacArthur and Wilson 1967: 8, 13), but is less satisfactory when the islands are varied in these respects (Johnson et al. 1968, Power 1972, Johnson 1975). In order to extend the species-area models to apply more generally, then, it is necessary to adopt a more direct measure of the important proximate factors than area provides. I will present a modification of the models of Preston and MacArthur and Wilson that replaces the area parameter with available energy, explore the implications of the theory resulting from this replacement, which I have called species-energy theory, and present relevant data for angiosperms and land and freshwater birds on islands. The potential for using these concepts to predict species occurrence and abundance as well as the number of species on an island, and the value of the theory in providing a unified basis for interpreting diverse observations, are also discussed.

The meaning of available energy in species-energy theory

Species-energy theory is obtained essentially by replacing "area" with "available energy" in the models of MacArthur and Wilson (1963, 1967) and Preston (1962). This substitution is illustrated in the following section. This section explains the meaning of available energy and why its placement in these models is desirable. The estimation of available energy in practice is discussed in the section "Testing species-energy theory".

Available energy on an island is the rate at which resources available to the species of interest are produced on the island as a whole. In other words, available energy measures the total amount of available resource production on an island, which is one of the likely proximate factors affecting species population sizes. Area can only measure total available resource production accurately when the per-unit-area productivity of different islands is similar.

Available energy, like area, does not directly measure the variety of resource types present on an island, but is correlated with it. That is, the larger the total resource production on an island, the greater the variety of resource types is likely to be.

Thus available energy serves as a more general measure of total resource production on an island than does area, and also provides some information on the variety of resource types present. Species-energy theory should therefore prove more general than species-area theory, particularly by being applicable to islands which vary in their per-unit-area resource productivity, due to factors such as climate, topography, or soil chemistry.

The effect of available energy in the equilibrium model, and species-energy curves

The equilibrium theory of available energy and species number is completely analogous to the MacArthur-Wilson model (1963, 1967). The number of species on an island is represented as the result of a dynamic immigration-extinction process, and, following MacArthur and Wilson, I assume that smaller populations are more likely to suffer extinction (see also Jones and Diamond 1976, Terborgh and Winter 1980, Leigh 1981). If the island as a whole produces little energy that is available to the species in question, then species population sizes will be small, and the extinction rate on the island will be high. On the other hand, islands with large amounts of available energy will support larger populations of all species, and so will have lower extinction rates. Thus available energy has the same effect on extinction rate that MacArthur and Wilson proposed for area. Also, like area, available energy should often have negligible effect on immigration rates. For islands of similar isolation, then, the islands with greater available energy will have higher equilibrium species numbers. Fig. 1 presents the familiar graphical representation of this result.

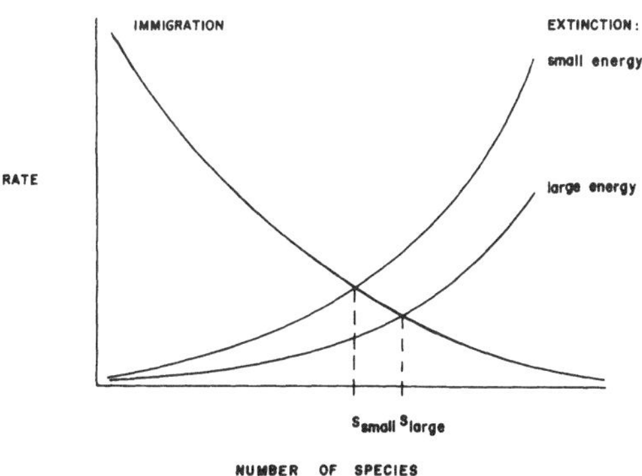

Fig. 1. Immigration and extinction rates versus number of species present on an island. Islands with larger total amounts of available energy have lower extinction rates, resulting in a higher equilibrium number of species ($S_{large} > S_{small}$).

By assuming that a specific form of species-abundance distribution holds on all islands, it is possible to derive a more precise, though possibly less general, form of the relationship between available energy and species number, analogous to the species-area curve derived by Preston (1962; see May 1975 for a more exhaustive treatment). Some interesting ties exist between species-energy and species-area curves.

A basic assumption in Preston's formulation of the species-area curve was that the total number of individuals of all species on an island should be proportional to its area, i.e., that $N = \delta A$, where δ is the total density of individuals per unit area. Similarly, I assume here that the total number of individuals on an island, N, is proportional to the total production of available energy on the island, E, i.e.,

$$N = \varrho E$$

where ϱ is the number of individuals supported per unit of available energy. That ϱ is geographically constant for a given taxonomic group must be subject to empirical test. For the present I will assume that it is constant. The predicted form of the species-energy curve is then approximately (Preston 1962, May 1975)

$$S = kE^z \qquad (2)$$

where k is a constant related to ϱ, and S and z are as in Eq. (1).

In the species-area relation (1), the coefficient C is shorthand for

$$C = b \left(\frac{\delta}{m}\right)^z$$

where b is a fitted constant from the power function approximation to the theoretical species-area relationship, and m is the population size of the rarest species (Preston 1962). The constant k, on the other hand, is

$$k = b \left(\frac{\varrho}{m}\right)^z.$$

C and k are closely related. If we let r be the per-unit-area productivity of available energy on an island, then, since $\varrho r = \delta$,

$$C = kr^z \qquad (3).$$

This implies that C should vary with per-unit-area productivity, as has been previously suggested (MacArthur and Wilson 1967: 17), and that if we examined C values from species-area curves for a given taxonomic group and compared these with the per-unit-area productivity values for the corresponding sets of islands, we would expect the linear relationship

$$\log C = \log k + z \log r$$

This makes good sense, since C is larger under conditions where δ, the total density of individuals, is higher; and the total density of individuals is generally higher where the per-unit-area productivity of the habitat is higher. Connor and McCoy (1979) provided an indirect test of this prediction by correlating C values, as well as mean number of species for sets of islands (which differs from C for a given sets of islands only due to choice of units for area, see Gould 1979), with latitude. Unfortunately, latitude is not a very accurate indicator of per-unit-area productivity of available energy. Most of the correlation coefficients calculated by Connor and McCoy were non-significant but negative; a significant negative correlation was found between latitude and mean number of species for birds. In any collection of data, a possible relationship between C and r could be obscured by differences in the isolation, ranges of area covered, and differing z values of different sets of islands.

The relationship suggested by Eq. (3) can be restated in a more qualitative, intuitive way: among islands of similar isolation, islands with low per-unit-area productivity are expected to fall below islands with higher productivity in a logarithmic plot of species number against area. This is what Lassen (1975) found for snails in oligotrophic and eutrophic lakes.

The relationship between C and r also suggests an alternative method of conceptualizing the species-energy curve. Think of a graph of log S versus log A on which sets of islands differing in their per-unit-area productivity of available energy (r) are plotted. For the sake of argument, assume that the slopes of the various species-area curves are approximately the same. Then these relations plot as more or less parallel lines with slopes of about z, with sets of islands with higher r above those of lower productivity (Fig. 2). Their higher C, the log of which is the intercept on the log S axis, reflects this arrangement. If log S were instead plotted against log E = log A + log r, each line would be shifted rightward by an amount log r. Since the higher lines are shifted correspondingly more to the right, this results in the separate species-area curves coalescing into a single species-energy curve.

In both the equilibrium model and species-energy curves, species-area theory appears as a special case of species-energy theory. When islands of similar per-unit-area productivity are examined, area serves as an excellent relative measure of available energy, and the species-energy models collapse to the species-area models of MacArthur and Wilson and Preston. Thus species-energy theory does not require the abandonment of species-area investigations, nor does it necessarily invalidate the use of area as a parameter. For the purposes that it has served to date, namely the comparison of islands with similar per-unit-area productivity, species-area theory is completely valid.

To summarize at this point, species-energy theory recognizes available energy as a proximate factor affecting species population sizes. Many of the predictions made by MacArthur and Wilson and by Preston apply to species-energy theory, since available energy can easily be substituted for area in both models. Doing so results in the following qualitative and quantitative predictions: it should be possible to compare islands differing radically in their per-unit-area productivity of available energy and area, and among islands of similar isolation, islands with greater amounts of available energy should support more species than islands with less available energy. More precisely, if species-abundance distributions on islands are lognormal and of similar form, we expect the approximate relationship $S = kE^z$. If many species-area relationships of island groups of similar isolation are examined, the coefficient C of the species-area curve should vary with per-unit-area productivity according to the relationship $C = kr^z$. Species-area theory continues to be interesting and useful, and is contained as a special case within species-energy theory.

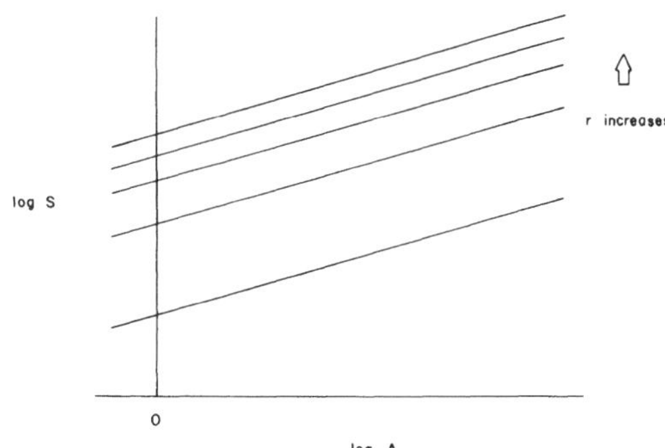

Fig. 2. Hypothetical species-area curves for islands of similar isolation but different per-unit-area productivity of available energy (r).

Testing species-energy theory

Testing the predictions of species-energy theory usually involves the problem of measuring available energy. The predictions of Eqs (2) and (3), in particular, rely on this quantity.

Available energy should be estimated by considering what amount of energy production in general on an island is available to a given group of species, due to the unique requirements and constraints of that group. For example, a measurement of energy available to plants might recognize solar energy as the ultimate source of plant energy, and then also take into account constraints on the usability of raw solar energy due to factors such as lack of water or nutrients. Energy available to animals consists of the production of food items that can be included in the diets of the group in question.

500

Tab. 1. Number of species of angiosperms and breeding land and freshwater bird on islands worldwide.

Island	Location	Area (10^3 km^2)	Average AET[a] (cm yr^{-1})	Average NPP[b] (kg m$^{-2}\cdot$ yr^{-1})	Total AET[c] (km^3 yr^{-1})	Total NPP[d] (10^9 t yr^{-1})	Number of[e] angiosperms	Number of[f] birds
1. Australia	25°S, 135°E	7705	55	0.6	4240	4.62	12,000	529
2. New Guinea	5°S, 142°E	809	125	2.3	1010	1.86	9000	540
3. Borneo	0°, 115°E	751	–	2.0	–	1.50	–	420
4. Madagascar	20°S, 47°E	591	88	1.5	517	0.89	6000	184
5. Philippines	11°N, 123°E	300	125	–	375	–	7620	–
excluding Palawan		288	–	2.1	–	0.60	–	325
6. Japan	37°N, 139°E	359	63	1.1	226	0.39	3417	216
7. Sulawesi	2°S, 122°E	179	–	2.1	–	0.38	–	220
8. New Zealand	42°S, 173 E	269	63	–	169	–	1725	–
9. Java	7°S, 110°E	124	–	2.1	–	0.26	–	337
10. Cuba	22°N, 79°W	114	125	1.7	143	0.19	8000	124
11. Britain	54°N, 2°W	230	50	1.1	115	0.25	1730	172
12. Hispaniola	19°N, 71°W	76	–	1.7	–	0.13	–	106
13. Sri Lanka	7°N, 81°E	66	88	1.8	58.1	0.12	3000	235
14. Taiwan	24°N, 121°E	36	125	–	45.0	–	3265	–
15. Tasmania	42°S, 147°E	67	63	–	42.2	–	1200	–
16. Solomon Is.	7°S, 157°E	30	125	–	37.5	–	1650	–
17. Iceland	65°N, 18°W	103 (85[g])	38	–	32.3	–	375	–
18. Ireland	53°N, 8°W	84	38	1.1	32.0	0.092	1200	108
19. Timor	9°S, 125°E	34	–	2.0	–	0.068	–	137
20. Newfoundland	48°N, 56°W	112	–	0.6	–	0.067	–	103
21. Baffin	68°N, 73°W	476 (453)	7	0.1	31.7	0.045	261	40
22. Sakhalin	50°N, 142°E	78	38	–	29.6	–	1166	–
23. Greenland	73°N, 40°W	2176 (384)	7	–	26.9	–	400	–
24. Vancouver	49°N, 126°W	32	–	1.2	–	0.038	–	124
25. Flores	9°S, 121°E	17	–	2.1	–	0.036	–	143
26. New Caledonia	20°S, 165°E	19	100	1.8	19.0	0.034	2600	68
27. Fiji	17°S, 178°E	18	100	1.8	18.0	0.032	1250	61
28. Palawan	10°N, 118°E	12	–	2.1	–	0.025	–	131
29. Sumba	10°S, 120°E	11	–	2.0	–	0.022	–	103
30. Victoria	71°N, 110°W	212	7	0.1	14.8	0.021	222	43
31. Jamaica	18°N, 77°W	12	100	1.7	12.0	0.020	2888	99
32. Ellesmere	80°N, 80°W	196 (108)	–	0.1	–	0.011	–	22
33. Banks	73°N, 120°W	64	–	0.1	–	0.0064	–	34
34. Falkland Is.	52°S, 59°W	16	32	0.4	5.1	0.0064	170	33
35. Novaya Zemlya	74°N, 56°E	83 (62)	7	–	4.3	–	200	–
36. Spitsbergen	78°N, 20°E	61 (25)	7	0.1	1.7	0.0025	137	9

a. Average actual evapotranspiration, in cm of water per year. Estimated from Lieth (1975) for ice-free areas.
b. Average net primary productivity, in kg of dry matter per m^2 per year. Estimated from Lieth (1975) for ice-free areas.
c. Obtained by multiplying average Aet by the area of the island free from permanent ice cover.
d. Obtained by multiplying average NPP by the area of the island free from permanent ice cover.
e. From Good (1974), except 5 (Merrill 1926), 6 (Numata 1974), 21 and 30 (Porsild 1957), and 31 (Adams 1972).
f. From Pizzey (1980), Rand and Gilliard (1967), Delacour and Mayr (1946), Moreau (1966), Anon. (1975), Preston (1962), Parslow (1973), Godfrey (1966), Mayr (1945), Woods (1975), and Løvenskiold (1963).
g. In parentheses is area free from permanent ice cover. From Encyclopædia Brittanica or estimated from maps.

OIKOS 41:3 (1983)

Ideally, available energy would be measured in units of energy per unit time, e.g. joules per year. However, in practice, any relative measure of available energy can serve, as long as it bears a consistent proportionality to available energy for the set of islands examined. The units of a relative measure will not necessarily be those of energy/time. A disadvantage of using a variety of relative methods to estimate available energy is that the use of different units precludes certain comparisons between data sets, such as comparison of intercept values.

To test the generality of the equation $S = kE^z$, I have compiled data on the number of species of angiosperms and of land and freshwater birds on islands worldwide (Tab. 1). All world islands with areas greater than 10^5 km^2, for which data on species numbers of angiosperms or land and freshwater birds were available to me, were tabulated; with the exception that several islands in the Canadian arctic archipelago were excluded, in order to avoid over-representation of these high latitude islands in the sample, and because many are not well-explored.

For both angiosperms and birds, relative measures of available energy were used. For angiosperms, actual evapotranspiration (AET) was used to produce a measure of energy available to plants. Despite its name, AET does not involve actual measurements of evaporation or transpiration rates, but is calculated from climatic data for any particular site. AET estimates total incident solar energy by mean monthly temperatures above 0°C, and corrects for the amount of this energy which is unavailable to plants due to lack of water by taking into account monthly mean precipitation and standardized estimates of evaporation, transpiration, and soil water storage (Major 1963, Rosenzweig 1968). Its utility is demonstrated by the observation that AET is the best known single predictor of global patterns of terrestrial primary productivity (Lieth 1975).

To estimate available energy for angiosperms, total AET (TAET) for an island was calculated by multiplying the annual rate of AET, averaged over the whole island, by the area of the island. Annual AET rates, in cm H_2O yr^{-1}, were obtained from a global map of AET developed by E. Box (Lieth 1975). TAET, in km^3 H_2O yr^{-1}, is plotted with angiosperm species number on logarithmic coordinates in Fig. 3. The resulting relationship is fit by the curve $S = 123\ (TAET)^{0.62}$, explaining 70% of the variation in the logarithm of angiosperm species number. Island area does not explain a significant proportion of the variance ($S = 475\ A^{0.26}$, A in 10^3 km^2, $r = 0.26$, $P > 0.05$), and the fit obtained using log TAET is significantly better (test for equality of correlation coefficients, $t = 2.88$, $P < 0.005$; Sokal and Rohlf 1969: 521).

For land and freshwater birds, total net primary production (TNPP) was used as a relative measure of available energy. TNPP was estimated by multiplying the average per-unit-area net primary productivity on the island, in kg dry matter $\cdot$ m^{-2} $\cdot$ yr^{-1} (Lieth 1975), by the area of the island. Regressing log S on log TNPP yields the relation $S = 358\ (TNPP)^{0.47}$, TNPP in 10^9 t dry matter yr^{-1}, explaining 80% of the variation in log S (Fig. 4). The species-area regression is significant in this case ($S = 32.8\ A^{0.27}$, $r = 0.44$, $P < 0.05$), but area again explains significantly less of the variance in species number ($t = 3.35$, $P < 0.001$).

In both plots, islands at high latitudes, which are

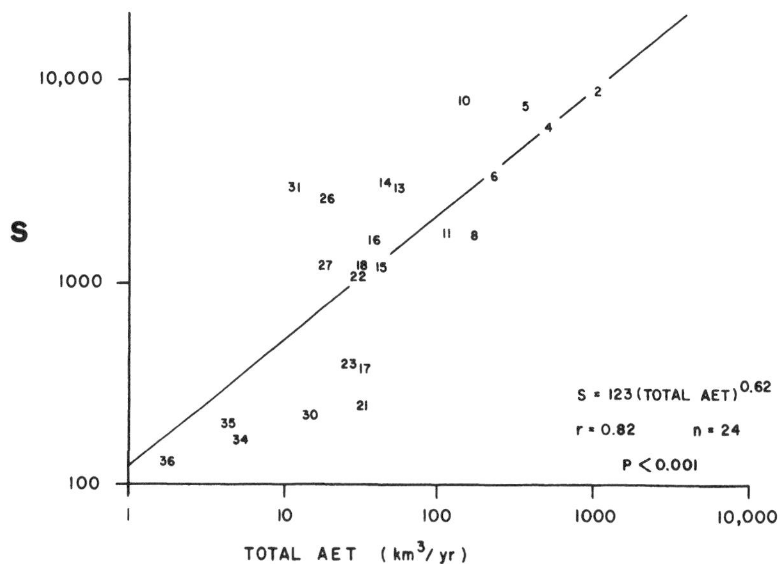

Fig. 3. Number of angiosperm species (S) on 24 islands worldwide, plotted against the total actual evapotranspiration (Total AET) from the island annually. Numbers in the plot refer to islands listed in Tab. 1.

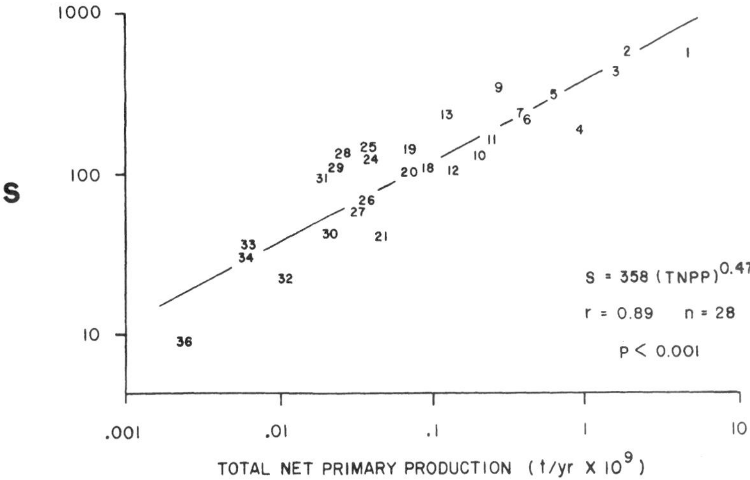

Fig. 4. Number of breeding land and freshwater bird species (S) on 28 islands worldwide, plotted against the total net primary production on the island annually, in billions of metric tons of dry matter. Numbers in the plot refer to islands listed in Tab. 1.

typified by low per-unit-area rates of AET and net primary productivity, still tend to have lower species numbers than islands with the same TAET or TNPP but with higher per-unit-area rates. There are many possible interpretations of this result, for example:

1) It may be that TAET and TNPP are not consistently proportional to available energy, but overestimate it for low per-unit-area productivity habitats.
2) It may be that ϱ, the number of individuals supported per unit of available energy, is not constant, but decreases with decreasing r.
3) Species-abundance distributions may differ in form in areas with different r. For example, if the standard deviation of lognormal species-abundance distributions increases in low per-unit-area productivity environments (Preston 1980), then z values for these environments will tend to be lower (May 1975: eq. A16, Schoener 1976, Martin 1981). This will result in the failure of species-area curves to coalesce in a species-energy plot, with the species-energy curves of low-r islands falling below those of high-r islands.
4) It might be that the low species diversity of the source pools for high latitude islands depresses their equilibrium species number. If this is the case, we would expect the depression to be more marked on larger islands.

All of these possibilities are quite amenable to empirical test, and suggest the existence of interesting geographical patterns in resource utilization and species-abundance distributions. In any case, it should be noted that the combined tendency of low per-unit-area productivity islands to fall below and to the left of high productivity islands in Figs 3 and 4 probably has artificially increased the estimates of z.

In both examples in this section, a relative rather than a direct measure of available energy was used, and this will often necessarily be the case in tests of species-energy theory. However, it is also possible to identify situations in which direct measurements of available energy could be made. Freshwater ecosystems seem most promising; for instance, species number of zooplankton could be compared with available energy in lakes and ponds, by converting measurements of total net primary production to units of energy/time. Such a comparison would be especially relevant to species-energy theory if variation in per-unit-area productivity among the lakes and ponds, due to oligotrophy, eutrophy, or factors such as elevation, were large. In terrestrial ecosystems, total net primary production, again converted to energy/time, may provide a first order estimate of available energy for herbivorous insects. Available energy for web-building spiders or bats could perhaps be estimated using insect light-trap data.

The attempts to apply species-energy theory that I have presented in this section have shown that available energy can be used to explain variation in species number when islands with widely different physical environments are compared. In addition, the specific quantitative prediction that the species-energy curves of different sets of islands should fall on a single line has apparently been falsified. This falsification has suggested some interesting questions, investigation of which could potentially lead to a refinement of the theory, and to a greater understanding of general patterns of diversity.

Discussion

The concepts embodied in species-energy theory and equilibrium biogeography can potentially be used not only to predict patterns of species number but also to address in detail the abundances of individual species on islands and their probabilities of being present or absent. To do so naturally requires more detailed information about individual species and islands than is necessary merely to predict a rough species number. Predicting species abundances and probabilities of occurrence will require, at least, knowledge about individual species' resource requirements and about the production of resources on islands.

Detailed information about the total production on an island of resources available to a particular species; which would ideally include some consideration of how much was unavailable due to physical environmental conditions, competitors, and predators; and about the resource requirements of individuals of the species, would allow an estimation of the species' abundance if present. To oversimplify, a species should be abundant if it has abundant available resources on an island.

Because of the assumed relationship between population size and extinction rate, the abundance that a species can maintain should affect the probability that it will occur on an island. Simberloff (1969) suggested that the probability that a species A occurs on an island is

$$\Pr(A) = \frac{1}{1 + \frac{E_A(n)}{I_A(d)}} \quad (4)$$

where $I_A(d)$ is A's immigration rate and $E_A(n)$ is A's extinction rate when present on the island. I have indicated that I_A is a function of the distance, d, of the island from sources of immigrants, while E_A is a function of the population size, n, that A can maintain on the island (Jones and Diamond 1976, Terborgh and Winter 1980). For any particular species the forms of I(d) and E(n) would have to be measured to obtain explicit probabilities of occurrence; however, qualitative predictions about relative probabilities of occurrence can easily be made by comparing the magnitudes of the ratio I/E for different species or taxa (Wright 1981).

In essence, species-energy considerations lead to the expectation that a species with abundant resources on an island should be both more abundant on the island if it does occur and, consequently, more likely to occur than an equally dispersive species with few resources.

These notions about the occurrence and abundance of species, in conjunction with the equilibrium and species-energy curve models of species-energy theory, provide a framework that unifies a broad set of observations of patterns in species diversity. None of these observations are new, and some of the patterns are so intuitive that there seems little need for a theoretical explanation. The virtue of the theory in such instances lies in its ability to interrelate previously unconnected observations, and to point out new areas of interest and new avenues of approach. I will briefly list some of the observations of diversity patterns for which species-energy theory may provide an explanation.

The diversity of trophic levels. Due to the progressive diminution of available energy at higher trophic levels, species-energy theory predicts that lower trophic levels should be more diverse; for example, on any given island there should be more herbivores than carnivores. An examination of Tab. 1 shows that birds are always less diverse than angiosperms on any island. This is consistent with species-energy theory, since all birds are at least 1 trophic level above angiosperms. Such ideas date as far back as Elton's early discussion of the pyramid of numbers (1927). This prediction is not exact, but interacts with the factors discussed below.

Metabolic requirements and diversity. Other things being equal, species with greater metabolic requirements should in general be less diverse, less abundant, and occur less frequently than species with lower requirements. Because total metabolic rate scales approximately as body mass to the power of 0.75 for organisms ranging from homeothermic and poikilothermic vertebrates to trees and bacteria (Schmidt-Nielsen 1979: 186), species-energy theory may serve as a partial explanation for the observed size-diversity relationship (Van Valen 1973, May 1978); that is, why small organisms are roughly 100 times as diverse as similar organisms 10 times their length. (For example, assuming no bias in the amount of energy available to organisms of different sizes in a habitat, the environment would supply $10^{0.75}$ times as many individual-equivalents of energy to organisms weighing 1 g as to those weighing 10 g, implying a species-supporting capacity $(10^{0.75})^z$ times as great for the smaller organisms.) Species of large body size or high metabolic requirement are often among the first to be extirpated from newly formed islands or habitat areas reduced in size (Brown 1971, Willis 1974, Terborgh and Winter 1980, Wright 1981).

It is tempting to use this argument to compare homeothermic and poikilothermic vertebrates; however, these groups differ greatly and with systematic biases in other important respects, such as the resources available to them, their susceptibility to predation, their response to climatic conditions, and their potential for immigration, so that such comparisons will be difficult to interpret (but see Wilcox 1980, Wright 1981).

It is interesting to note that the effect of body size is likely to interact with the effect of trophic level. Lindemann (1942) suggested that the relative sizes of producers and consumers would be of profound importance in understanding ecosystems. In marine ecosystems, the producers are small and primary consumers typically somewhat larger, whereas in terrestrial systems producers are much larger, and primary consumers may be considerably smaller than the organisms they consume. This may in part account for the diversity of herbivorous

Tab. 2. Frequency of occurrence and average population density, when present, of crustacean zooplankters in 34 lakes in northwestern Ontario (from Patalas 1971, 1968 sample data). Spearman rank correlation $r_s = 0.53$, $P < 0.05$.

Species	Average population density (cm^{-2})	Frequency of occurrence (No. lakes present/34)
1. *Cyclops bicuspidatus*	17.17	0.59
2. *Diaptomus minutus*	9.61	0.68
3. *Diaptomus sicilis*	5.79	0.12
4. *Orthocyclops modestus*	4.56	0.12
5. *Diaptomus oregonensis*	2.86	0.26
6. *Mesocyclops edax*	2.43	0.71
7. *Tropocyclops prasinus*	2.38	0.76
8. *Diaptomus leptopus*	2.02	0.15
9. *Daphnia retrocurva*	1.84	0.26
10. *Cyclops vernalis*	1.83	0.24
11. *Bosmina longirostris*	1.73	0.88
12. *Diaphanosoma leuchtenbergianum*	1.54	0.47
13. *Daphnia catawba*	1.43	0.15
14. *Daphnia galeata*	1.37	0.35
15. *Chydorus sphaericus*	1.21	0.18
16. *Diaphanosoma brachyurum*	1.16	0.38
17. *Holopedium gibberum*	0.86	0.68
18. *Daphnia longiremis*	0.60	0.06
19. *Epischura lacustris*	0.36	0.47
20. *Ceriodaphnia lacustris*	0.24	0.15
21. *Senecella calanoides*	0.23	0.06
22. *Daphnia schoedleri*	0.19	0.03
23. *Limnocalanus macrurus*	0.09	0.09
24. *Ceriodaphnia pulchella*	0.07	0.03
25. *Leptodora kindtii*	0.02	0.21
26. *Streblocerus serricaudatus*	0.004	0.03

insects being greater than the diversity of their plant hosts. Insect parasitoids also constitute an interesting case.

Rarity and patchy distribution. Eq. (4) implies that rare species (species which have low abundance when they occur), because they are more prone to extinction events, should occur less frequently than more abundant species on a large set of islands. Tab. 2 demonstrates this relationship for crustaceans in lakes in northwestern Ontario; Diamond (1980) suggests that this is also qualitatively the case for New Guinea birds. Species may be rare for a great variety of ecological reasons, such as susceptibility to predators, competitors, or environmental conditions, or rarity of requisite resources on the island. Several authors have previously suggested a relationship between extreme resource specialization and species extinction rates (Connell and Orias 1964, Brown 1971, 1978, Willis 1974, Terborgh and Winter 1980).

Seasonal variation in diversity. As would be predicted by species-energy theory, seasonal trends in diversity mirror seasonal trends in energy availability (Connell and Orias 1964, Shapiro 1975). This pattern is especially marked in the temperate zones, where many birds emigrate, and many mammals, reptiles, amphibians, insects, and plants enter periods of dormancy.

Latitudinal diversity gradients. Species-energy theory would predict latitudinal gradients in diversity if lower latitudes were also characterized by greater total available energy. As mentioned above, a rough correlation exists between latitude and measures of per-unit-area productivity, such as AET and net primary productivity (Lieth 1975). Terborgh (1973) has argued that the area of low latitude habitats is also much greater than that of high latitude habitats, but this is less than clear for regions such as central America and northern tundra and taiga. Quantitative estimates of habitat areas and productivity are needed to address this question.

Finally, a comment on the applicability of species-energy theory is in order. As presented here, species-energy theory assumes that the population size of a group of species in an insular habitat is affected by the total amount of available energy production (i.e. $\partial N/\partial E = \varrho$, $\varrho > 0$). In some cases this may not be true, for example, fishes on coral reefs or seabirds on high latitude islands might be so severely limited by the availability of hiding holes or nesting sites that available energy has no effect on the number of individuals in the community. A simple experimental test of this assumption would be to see whether the number of individuals in the community changed in response to an increased or decreased food supply. Overall, however, it is clear that species-energy theory lacks universal applicability in that energy need not necessarily affect population sizes. While it might be possible in any particular instance to identify the most important non-energetic factors affecting a community's population size and so its species number, these factors vary from place to place and with the taxa being considered; thus such an approach lacks the generality of species-energy theory in explaining geographic patterns in species diversity. Ideally, we seek a "species-limiting factors" theory which can both account for a variety of factors, potentially including energy, that affect population sizes; and apply across taxonomic and geographic distances. Species-energy theory is only a step in this direction.

Acknowledgements – I am indebted to Jim Brown, without whose stimulus these ideas might never have materialized, for numerous fruitful discussions of the topic. Jim Brown, Tom Gibson, Kate Iverson, and Eric Larsen read and made helpful comments on a draft of the manuscript. Michael Rosenzweig and a reviewer also made substantive suggestions.

References

Adams, C. D. 1972. Flowering plants of Jamaica. – Univ. of the West Indies, Mona, Jamaica.

Anonymous 1975. Check-list of Japanese birds. – Orn. Soc. Japan, Gakken, Tokyo.

Brown, J. H. 1971. Mammals on mountaintops: nonequilibrium insular biogeography. – Am. Nat. 105: 467–478.

– 1978. The theory of insular biogeography and the distribution of boreal birds and mammals. – In: Harper, K. T. and Reveal, J. L. (eds.), Intermountain biogeography: a symposium. Great Basin Naturalist Memoirs 2: 209–227.

– 1981. Two decades of homage to Santa Rosalia: toward a general theory of diversity. – Am. Zool. 21: 877–888.

Connell, J. H. and Orias, E. 1964. The ecological regulation of species diversity. – Am. Nat. 98: 399–414.

Connor, E. F. and McCoy, E. D. 1979. The statistics and biology of the species-area relationship. – Am. Nat. 113: 791–833.

DeAngelis, D. L. 1980. Energy flow, nutrient cycling, and ecosystem resilience. – Ecology 61: 764–771.

Delacour, J. and Mayr, E. 1946. Birds of the Philippines. – MacMillan, New York.

Diamond, J. M. 1980. Patchy distributions of tropical birds. – In: Soule, M. E. and Wilcox, B. A. (eds.), Conservation biology: an evolutionary-ecological perspective. Sinauer, Sunderland, MA, pp. 57–74.

Elton, C. 1927. Animal ecology. – Sidgwick and Jackson, London.

Godfrey, W. E. 1966. The birds of Canada. – National Museums of Canada Bull. 203, Biological Ser. No. 73, Ottawa, Ontario.

Good, R. 1974. The geography of the flowering plants. – Longman, London.

Gould, S. J. 1979. An allometric interpretation of species-area curves: the meaning of the coefficient. – Am. Nat. 114: 335–343.

Hutchinson, G. E. 1959. Homage to Santa Rosalia, or why are there so many kinds of animals. – Am. Nat. 93: 145–159.

Johnson, M. P., Mason, L. G., and Raven, P. H. 1968. Ecological parameters and plant species diversity. – Am. Nat. 102: 297–306.

Johnson, N. K. 1975. Controls of number of bird species on montane islands in the Great Basin. – Evolution 29: 545–567.

Jones, H. L. and Diamond, J. M. 1976. Short-time-base studies of turnover in breeding bird populations on the California Channel Islands. – Condor 78: 526–549.

Lassen, H. H. 1975. The diversity of freshwater snails in view of the equilibrium theory of island biogeography. – Oecologia (Berl.) 19: 1–8.

Leigh, E. G. 1981. The average lifetime of a population in a varying environment. – J. Theor. Biol. 90: 213–239.

Lieth, H. 1975. Modeling the primary productivity of the world. – In: Lieth, H. and Whittaker, R. H. (eds.), Primary productivity of the biosphere. Springer, New York, pp. 237–263.

Lindemann, R. L. 1942. The trophic-dynamic aspect of ecology. – Ecology 23: 399–418.

Løvenskiold, H. L. 1963. Avifauna Svalbardensis. – Norsk Polarinst. No. 129, Oslo.

MacArthur, R. H. 1965. Patterns of species diversity. – Biol. Rev. 40: 510–533.

– 1972. Geographical ecology. – Harper and Row, New York.

– and Wilson, E. O. 1963. An equilibrium theory of insular zoogeography. – Evolution 17: 373–387.

– and Wilson, E. O. 1967. The theory of island biogeography. – Princeton Univ. Press, Princeton, NJ.

Major, J. 1963. A climatic index to vascular plant activity. – Ecology 44: 485–498.

Martin, T. E. 1981. Species-area slopes and coefficients: a caution on their interpretation. – Am. Nat. 118: 823–837.

May, R. M. 1975. Patterns of species abundance and diversity. – In: Cody, M. L. and Diamond, J. M. (eds.), Ecology and evolution of communities. Harvard Univ. Press, Cambridge, MA, pp. 81–120.

– 1978. The dynamics and diversity of insect faunas. – In: Mound, L. A. and Waloff, N. (eds.), Diversity of insect faunas. Symp. R. Ent. Soc. London, No. 9. Blackwell, New York, pp. 188–204.

Mayr, E. 1945. Birds of the southwest Pacific. – MacMillan, New York.

Merrill, E. D. 1926, reprinted 1967. An enumeration of Philippine flowering plants, Vol. IV. – A. Asher, Amsterdam.

Moreau, R. E. 1966. The bird faunas of Africa and its islands. – Academic Press, New York.

Numata, M. (ed.) 1974. The flora and vegetation of Japan. – Elsevier, New York.

Parslow, J. 1973. Breeding birds of Britain and Ireland. – Poyser, Berkhamsted, England.

Patalas, K. 1971. Crustacean plankton communities in 45 lakes in the Experimental Lakes Area, northwestern Ontario. – J. Fish. Res. Bd Canada 28: 231–244.

Pizzey, G. 1980. A field guide to the birds of Australia. – Collins, Sydney.

Porsild, A. E. 1957. Illustrated flora of the Canadian Arctic archipelago. – National Museums of Canada Bull. 146, Biological Ser. No. 50. Ottawa, Ontario.

Power, D. M. 1972. Numbers of bird species on the California islands. – Evolution 26: 451–463.

Preston, F. W. 1962. The canonical distribution of commonness and rarity: I and II. – Ecology 43: 185–215 and 410–432.

– 1980. Noncanonical distributions of commonness and rarity. – Ecology 61: 88–97.

Rand, A. L. and Gilliard, E. T. 1967. Handbook of the New Guinea birds. – Weidenfeld and Nicholson, London.

Rosenzweig, M. L. 1968. Net primary production of terrestrial communities: prediction from climatological data. – Am. Nat. 102: 67–74.

Schmidt-Nielsen, K. 1979. Animal physiology: adaptation and environment. 2nd ed. – Cambridge Univ. Press, London.

Schoener, T. W. 1976. The species-area relationship within archipelagos: models and evidence from island land birds. – In: Firth, H. J. and Calaby, J. H. (eds.), Proc. 16th Int. Orn. Congr. Australian Academy of Science, Canberra, pp. 629–642.
Shapiro, A. M. 1975. The temporal component of butterfly species diversity. – In: Cody, M. L. and Diamond, J. M. (eds.), Ecology and evolution of communities. Harvard Univ. Press, Cambridge, MA, pp. 181–195.
Simberloff, D. S. 1969. Experimental zoogeography of islands: a model of insular colonization. – Ecology 50: 296–314.
Sokal, R. R. and Rohlf, F. J. 1969. Biometry. – Freeman, San Francisco.
Terborgh, J. 1973. On the notion of favorableness in plant ecology. – Am. Nat. 107: 481–501.
– and Winter, B. 1980. Some causes of extinction. – In: Soule, M. E. and Wilcox, B. A. (eds.), Conservation biology: an evolutionary-ecological perspective. Sinauer, Sunderland, MA, pp. 119–134.
Van Valen, L. 1973. Body size and numbers of plants and animals. – Evolution 27: 27–35.
Wilcox, B. A. 1980. Insular ecology and conservation. – In: Soule, M. E. and Wilcox, B. A. (eds.), Conservation biology: an evolutionary-ecological perspective. Sinauer, Sunderland, MA, pp. 95–118.
Willis, E. O. 1974. Populations and local extinctions of birds on Barro Colorado Island, Panama. – Ecol. Monogr. 44: 153–169.
Woods, R. W. 1975. The birds of the Falkland Islands. – Compton, Salisbury, England.
Wright, S. J. 1981. Intra-archipelago vertebrate distributions: the slope of the species-area relation. – Am. Nat. 118: 726–748.
Yodzis, P. 1981. The structure of assembled communities. – J. Theor. Biol. 92: 103–117.

PAPER 38

Are the Smallest Organisms the Most Diverse? (1988)

K. P. Dial and J. M. Marzluff

Commentary

ALISON G. BOYER

Are the smallest organisms the most diverse? Before this paper, species diversity was thought to be higher in smaller-bodied than in larger-bodied organisms. G. E. Hutchinson and R. H. Macarthur (1959) and R. M. May (1978) argued that this must be the case since small organisms necessarily subdivide the environment more finely. This would allow them to pack more species into a given area. However, when K. P. Dial and J. M. Marzluff examined a set of species body size–frequency distributions, the most diverse size classes tended to be smallish to medium sized, and not the very smallest size, within the group. This right-skewed, yet modal, shape is now recognized as a prevalent pattern, and explaining the shape of the body-size distribution has become a classic question in macroecology (see Allen et al. 2006). Perhaps the most influential aspect of their paper, though, comes in the last few paragraphs. They proposed a conceptual model of the evolution of the species body-size distribution based on size-biased speciation and extinction rates. If speciation is highest in small-bodied groups and extinction is high in both the largest and the very smallest groups, then a right-skewed, modal body-size distribution will result over time.

By compiling virtually every published body-size data set available, this study moved the argument concerning body size and diversity from the realm of speculation, based on a few body-size distributions, into the "macroecological" realm, where large data sets reign. This paper is now more influential than ever as there has been a resurgence of interest in the relationship between body size and diversity with the description of body-size distributions for a diversity of insects, marine invertebrates, plants, fish, reptiles, and amphibians (Enquist and Niklas 2001; Boback and Guyer 2003; McClain 2004; Ulrich 2006; Olden et al. 2007).

Over the past two decades, the macroecological approach has seen a shift from description of large-scale patterns to a search for mechanistic explanations (Brown 1999). By outlining specific mechanisms for evolution of a skewed body-size distribution, Dial and Marzluff set up a research agenda for subsequent authors to examine and test these mechanisms. The relation of body size to speciation and extinction rates has been an area of growing interest, receiving attention both with regard to molecular clocks (Allen et al. 2006) and macroevolutionary trends (Liow et al. 2008). Observations from this paper contributed to concepts of "optimal body size" (Brown et al. 1993; Blackburn and Gaston 1996) and have also played a role in the fast-growing area of microbial diversity (Green and Bohannon 2006).

From *Ecology* 69:1620–24. Reprinted with permission from the Ecological Society of America.

Literature Cited

Allen, C. R., A. S. Garmestani, T. D. Havlicek, P. A. Marquet, G. D. Peterson, C. Restrepo, C. A. Stow, and B. E. Weeks. 2006. Patterns in body mass distributions: Sifting among alternative hypotheses. *Ecology Letters* 9:630–43.

Blackburn, T. M., and K. J. Gaston. 1996. On being the right size: Different definitions of "right." *Oikos* 75:551–57.

Boback, S. M., and C. Guyer. 2003. Empirical evidence for an optimal body size in snakes. *Evolution* 57:345–51.

Brown, J. H. 1999. Macroecology: Progress and prospects. *Oikos* 87:3–14.

Brown, J. H., P. A. Marquet, and M. L. Taper. 1993. Evolution of body size: Consequences of an energetic definition of fitness. *American Naturalist* 142:573–84.

Enquist, B. J., and K. J. Niklas. 2001. Invariant scaling relations across tree-dominated communities. *Nature* 410:655–60.

Green, J. L., and B. J. M. Bohannan. 2006. Spatial scaling of microbial biodiversity. *Trends in Ecology and Evolution* 21:501–7.

*Hutchinson, G. E., and R. H. MacArthur. 1959. A theoretical ecological model of size distributions among species of animals. *American Naturalist* 93:117–25.

Liow, L. H., M. Fortelius, E. Bingham, K. Lintulaakso, H. Mannila, L. Flynn, and N. C. Stenseth. 2008. Higher origination and extinction rates in larger mammals. *Proceedings of the National Academy of Sciences* 105:6097–6102.

*May, R. M. 1978. The dynamics and diversity of insect faunas. In *Diversity of insect faunas*, edited by L. A. Mound and N. Waloff, pp. 188–204. Royal Entomological Society of London Symposium 9. Blackwell Scientific, Oxford.

McClain, C. R. 2004. Connecting species richness, abundance and body size in deep-sea gastropods. *Global Ecology and Biogeography* 13:327–34.

Olden, J. D., Z. S. Hogan, and M. J. Vander Zanden. 2007. Small fish, big fish, red fish, blue fish: Size-biased extinction risk of the world's freshwater and marine fishes. *Global Ecology and Biogeography* 16:694–701.

Ulrich, W. 2006. Body weight distributions of European Hymenoptera. *Oikos* 114:518–28.

Literature Cited

Chapin, F. S., III. 1980. The mineral nutrition of wild plants. Annual Review of Ecology and Systematics **11**:233–260.

Grime, J. P. 1974. Vegetation classification by reference to strategies. Nature **250**:26–31.

———. 1979. Plant strategies and vegetation processes. John Wiley and Sons, Chichester, England.

———. 1985. Towards a functional description of vegetation. Pages 503–514 *in* J. White, editor. The population structure of vegetation. Dr. W. Junk, Dordrecht, The Netherlands.

———. 1988. The C-S-R model of primary plant strategies—origins, implications and tests. Pages 371–393 *in* L. D. Gottlieb and K. S. Jain, editors. Plant evolutionary biology. Chapman and Hall, London, England.

Grime, J. P., and R. Hunt. 1975. Relative growth-rate: its range and adaptive significance in a local flora. Journal of Ecology **63**:393–422.

Grime, J. P., R. Hunt, and W. J. Krzanowski. 1987. Evolutionary physiological ecology of plants. Pages 105–125 *in* P. Calow, editor. Evolutionary physiological ecology. Cambridge University Press, Cambridge, England.

Grubb, P. J. 1985. Plant populations and vegetation in relation to habitat, disturbance and competition: problems of generalization. Pages 595–621 *in* J. White, editor. The population structure of vegetation. Dr. W. Junk, Dordrecht, The Netherlands.

Loehle, C. 1988. Problems with the triangular model for representing plant strategies. Ecology **69**:284–286.

Shepherd, S. A. 1981. Ecological strategies in a deep water red algal community. Botanica Marina **24**:457–463.

[1] *Manuscript received 3 September 1987; revised and accepted 13 November 1987.*

[2] *Unit of Comparative Plant Ecology (NERC), Department of Botany, The University of Sheffield, Sheffield S1O 2TN, England*

Ecology, 69(5), 1988, pp. 1620–1624

ARE THE SMALLEST ORGANISMS THE MOST DIVERSE?[1]

Kenneth P. Dial[2] and John M. Marzluff[3]

Body size has repeatedly been implicated as an important correlate of taxonomic diversity (e.g., Stanley 1973, Van Valen 1973*a*, May 1978, Bock and Farrand 1980). Small organisms are usually found to be more diverse than large ones, and this is purported to stem from an ability of small organisms to subdivide their environment more finely. Greater subdivision may permit small organisms to speciate and/or avoid extinction more than large ones (Hutchinson 1959, Hutchinson and MacArthur 1959, May 1978, Southwood 1978). Recently, in his MacArthur Award Lecture, May (1986) reiterated this thesis. In an earlier paper (May 1978) he stated explicitly the existence of a strong, inverse, monotonic relationship between diversity and relative body size within a taxonomic assemblage. Curiously, in every distribution that May (1978, 1986) presented, the smallest body-size category is never the most abundant. May (1978) reasoned that the smallest body-size categories are underrepresented because of insufficient sampling. Although this may be partially true for certain assemblages (e.g., insect groups), it is an inadequate explanation for the more thoroughly surveyed mammalian and avian taxa.

May (1986) stated the need for "both more empirical data about actual species–size relations in species assemblies in a range of geographical areas, and new ideas about possible mechanisms." In this paper we provide data from an extensive survey of taxonomic assemblages for patterns of relative body size in selected invertebrate, vertebrate, and plant taxa.

Methods

We use three terms to discuss taxonomic diversity. A *subunit* is the taxonomic category counted within a *unit* (e.g., species within a genus, where species are the subunits and genus is the unit). An *assemblage* is an entire collection of units and their subunits (e.g., all species/genera within taxonomically related families).

We collated body-size data for 46 assemblages, including mammals, birds, fish, reptiles, amphibians, plants, insects, and marine invertebrates. Only those assemblages for which body sizes are known for all or nearly all (>90%) members were used. We assumed that the range of body sizes measured within an assemblage represents the true range for that assemblage. We attempted to obtain the most representative size of subunits by using measurement scales traditionally employed by researchers in particular disciplines. Therefore, we report mean body masses of mammals and birds, median of a length distribution of invertebrates and plants, and maximum lengths of fish. Our analyses only concern relative sizes; thus the fact that body mass is a cubed function of length (Peters 1983) does not affect relative size within an assemblage.

The relative body size of the taxonomic unit showing the greatest diversity within an assemblage was calculated by dividing the number of units with a smaller average body size by the total number of units. Multiplication by 100 produces a scale from 0 to 100%. For example, within the family Sciuridae (containing nine genera), the genus *Spermophilus* is the most diverse (15 species, $\bar{X}$ mass = 368 g) and ranks fifth in terms of smallest average body mass of a member species. Therefore, the calculated relative body size for this unit is: (4/9) × 100 = 44%. A value of 0% indicates that the most diverse group is also the smallest in terms of body size, whereas a value of 100% indicates it is the largest. Relative-body-size measures enable us to compare the body size–diversity relationships of many assemblages, including those that are distantly related. Each frequency distribution for an assemblage (e.g., a single histogram from Figs. 12.1–12.3, 12.5–12.7 in May 1978; Figs. 3–5 in May 1986) yields a single measure of relative body size for the most diverse unit within that assemblage.

Results

A monotonic relationship between body size and taxonomic diversity does not exist. The most diverse unit was, on average, larger than 38% of the units in its assemblage (Fig. 1). There was considerable variation in relative size of the dominant unit (SD = 30.03), but two important points were evident: (1) only 13% (6/46) of the assemblages were dominated by the smallest unit, and (2) only 11% (5/46) were dominated by the largest unit. Small to medium sized units were most likely to dominate assemblages. In particular, half (50%) of the assemblages were dominated by a unit 16–40% larger than the smallest unit in the assemblage. The distribution of body sizes of dominant units differed from a random (uniform) distribution ($\chi^2_{(4)}$ = 15.23, $P < .005$). Nearly 90% of the time the smallest unit did not dominate an assemblage.

This relationship was independent of assemblage size, taxonomic affinity, and sampling area. Assemblage size was not correlated with relative size of the dominant group ($r = -0.10$, $N = 45$, $P \gg .05$). Relative size of dominant units averaged 35.9% for mammals (N = 15), 42.6% for birds (N = 11), 25% for fish (N = 4), 57.1% for invertebrates (N = 9), and 22% for amphibians and reptiles (N = 4). The difference in the two most disparate groups (invertebrates vs. amphibians and reptiles) approached significance (W = 75.7, P = .06, Mann-Whitney U test); however, all other pairwise comparisons between groups were nonsignificant. In those assemblages for which data were adequate to make the comparison, the size of the sampling area (e.g., considered globally or within a continent) did not greatly influence the position of the dominant units with respect to their relative body size within an assemblage. For example, within mammals, world-wide assemblages ($\bar{X}$ = 27.5) were dominated by units of relative size equal to those that dominated North American assemblages ($\bar{X}$ = 29.0) (W = 66.0, P = .86, Mann-Whitney U test).

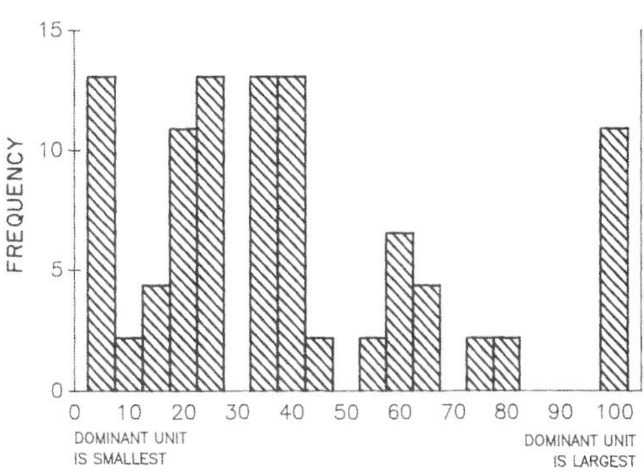

FIG. 1. Frequency distribution of percent relative body size of the most diverse unit for 46 taxa. Note: if, in fact, the smallest organisms were the most diverse within a taxonomic assemblage, then the histogram would result in packing on the left side of the scale.

Discussion

Contrary to general belief, the smallest organisms are usually not the most speciose members within a taxonomic category. Our results extend those of Stanley (1973), Van Valen (1973*b*), Bock and Farrand (1980), and Glazier (1987) by relating body size to diversity in a wide range of assemblages. Our results are consistent with May's data (1978, 1986), but not with his conclusion that the smallest organisms are most speciose. The relatively fewer number of small species is not necessarily an artifact of sampling. Birds and mammals have been thoroughly sampled and well documented, yet only within the carnivorous mammals are the smallest members the most diverse (Table 1). We think May (1978, 1986) presents an accurate version of the relationship of body size and diversity, but fails to recognize that it is not the smallest organisms but on average those which are 38% larger (Fig. 1) that are the most diverse within an assemblage. We do not believe body sizes and taxonomic diversity are randomly related within an assemblage, because 50% of the most diverse groups fell within a narrow size range (relative size of 16–40%). This is twice the number expected within this body size range by chance alone.

Why are not all assemblages dominated by small-to-medium-sized taxa? Spatial and temporal changes in environmental variability and the manner in which taxa of different sizes respond to this variability (i.e., environmental graininess) may influence the pattern of body size and taxonomic diversity that we have observed. Rapidly speciating small-to-medium-sized taxa may give rise to most new taxa (Stanley 1973), and they may always be relatively abundant, but depending

TABLE 1. Summary statistics of assemblages used in body size analyses.

Assemblage	Subunit/ unit*	Dominant unit	Percent relative body size	References
Mammals				
(Worldwide)				
Artiodactyla	(s/f)	Bovidae	80	Eisenberg 1981/Walker 1975
Carnivora	(s/f)	Viverridae	0	Eisenberg 1981/Walker 1975
Cetacea	(s/f)	Delphinidae	64	Eisenberg 1981/Walker 1975
Chiroptera	(s/f)	Vespertilionidae	16	Eisenberg 1981/Walker 1975
Marsupialia	(s/f)	Didelphidae	31	Eisenberg 1981/Walker 1975
Primata	(s/f)	Cercopithecidae	73	Eisenberg 1981/Walker 1975
Rodentia	(s/f)	Cricetidae	24	Eisenberg 1981/Walker 1975
All orders	(s/o)	Rodentia	22	Eisenberg 1981/Walker 1975
(North American)				
Carnivora	(s/f)	Mustelidae	0	J. H. Brown (*personal communication*)
Carnivora	(s/g)	*Felis*	56	J. H. Brown (*personal communication*)
Cricetidae	(s/g)	*Peromyscus*	37	J. H. Brown (*personal communication*)
Chiroptera	(s/f)	Vespertilionidae	22	J. H. Brown (*personal communication*)
Heteromyidae	(s/g)	*Perognathus*	25	J. H. Brown (*personal communication*)
Insectivora	(s/o)	*Sorex*	40	J. H. Brown (*personal communication*)
Lagomorpha	(s/g)	*Sylvilagus*	25	J. H. Brown (*personal communication*)
Sciuridae	(s/g)	*Spermophilus*	44	J. H. Brown (*personal communication*)
Birds				
(Worldwide)				
Corvidae	(s/g)	*Corvus*	100	Goodwin 1976
All orders	(s/o)	Passeriformes	21	Dunning 1984
(North American)				
Accipitridae	(s/g)	*Buteo*	55	Dunning 1984
Anseriformes	(s/g)	*Anas*	20	Bellrose 1978
Ardeidae	(s/g)	*Egretta*	38	Dunning 1984
Emberizidae	(s/g)	*Dendroica*	13	Dunning 1984
Galliformes	(s/g)	*Callipepla*	33	Johnsgard 1973
Laridae	(s/g)	*Larus*	100	Dunning 1984
Picidae	(s/g)	*Picoides*	17	Dunning 1984
Rallidae	(s/g)	*Rallus*	63	Dunning 1984
Scolopacidae	(s/g)	*Calidris*	33	Dunning 1984
Tyrannidae	(s/g)	*Empidonax*	10	Dunning 1984
Overall	(s/f)	Emberizidae	21	Dunning 1984
Reptiles				
(North American)				
Chelonia	(s/g)	*Malaclemys*	31	Cochran and Goin 1970
Amphibians				
(North American)				
Anura	(g/f)	Hylidae	17	Cochran and Goin 1970
Hylidae	(s/g)	*Hyla*	40	Cochran and Goin 1970
Ranidae	(s/g)	*Rana*	0	Cochran and Goin 1970
Fishes				
(Worldwide)				
Cypriniformes	(g/f)	Cyprinidae	100	Nelson 1984
Etheostomatini	(s/g)	*Etheostoma*	0	Kuehne and Barbour 1983
Lamniformes	(g/f)	Carcharhinidae	0	Nelson 1984
Squaliformes	(g/f)	Squalidae	0	Nelson 1984
Invertebrates				
(North American)				
Echinodermata	(s/c)	Asteroidea	100	Ranson 1981
Lepidoptera	(s/f)	Lycaenidae	20	Pyle 1981
Mytiloida	(s/g)	*Modiolus*	100	Ranson 1981
Myoida	(g/f)	Pholadidae	60	Ranson 1981
Papilionidae	(s/g)	*Papilio*	15	Pyle 1981
Phaladomyoida	(g/f)	Lyonsiidae	60	Collins 1959
Phaladomyoida	(s/f)	*Pandora*	40	Collins 1959
Polyplacophora	(s/g)	*Mopalia*	32	Collins 1959
Plants				
(Worldwide)				
Gymnosperms	(s/g)	*Pinus*	33	Fowells 1965

* s = species, g = genus or genera, f = family, o = order, c = class.

upon prevailing conditions, either very small or very large taxa may be at a selective advantage. The fossil record and recent extinctions suggest that although body size increases through time, very large size is more of a liability than an asset (Stanley 1973, Van Valen 1973*b*), perhaps because environmental conditions fluctuate rapidly and favor smaller organisms with short generation times. During stable conditions, however, large-sized taxa may have an advantage over smaller taxa in their dominance of resources. As conditions in one locale change through time, different body sizes will be favored, but shifts between stable and variable conditions will tend to cull extreme body sizes. Likewise, at one point in time, different locales, or different taxa in the same locale, experience variable degrees of environmental fluctuation, and thus some groups become rich in large forms whereas others proliferate into small species. We have summarized in a model how environmental fluctuation may lead to extinction of very large and very small taxa (Fig. 2). High extinction rates of extreme-sized taxa coupled with high speciation, radiation, and colonization rates of small taxa lead to the prediction that small to medium-sized taxa will have the highest rates of diversification. In a forthcoming paper we discuss life history traits of taxa that may affect extinction, speciation, and diversification.

Acknowledgments: We are grateful to Drs. S. Anderson, J. B. Dunning, Jr., D. S. Glazier, G. E. Goslow, Jr., F. A. Jenkins, Jr., B. D. Patterson, R. Raikow, two anonymous reviewers, and especially J. H. Brown and P. W. Price for their comments and discussion on this project. We thank H. O. Hooper, Organized Research Committee, Northern Arizona University, for travel money. We thank C. S. Marzluff and K. J. Dial for their understanding and support.

Literature Cited

Bellrose, F. C. 1978. Ducks, geese, and swans of North America. Second edition. Stackpole, Harrisburg, Pennsylvania, USA.

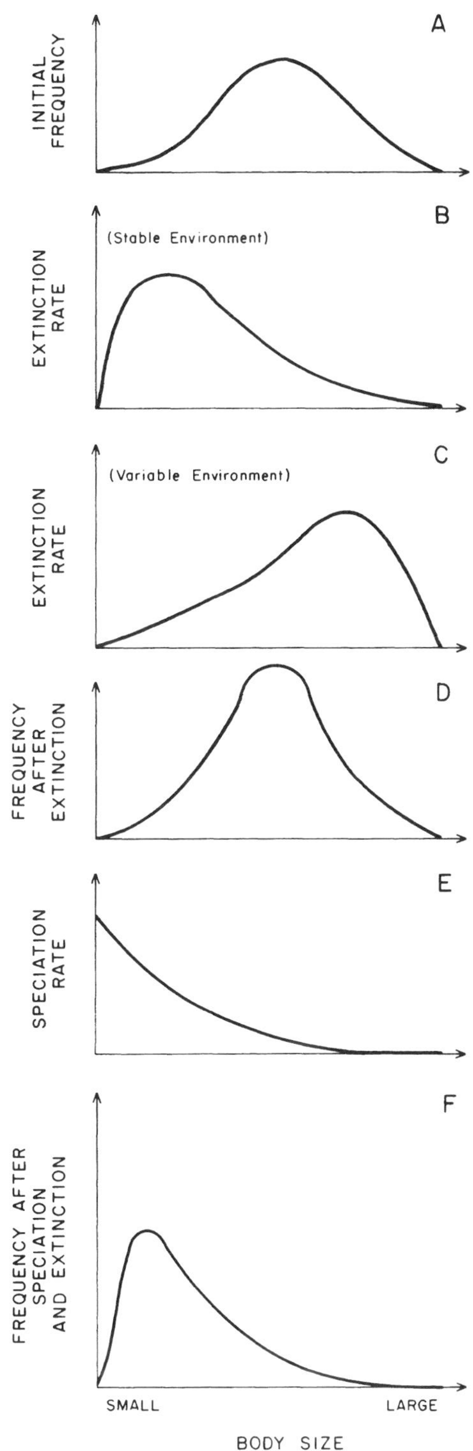

FIG. 2. Model showing how size-specific extinction and size-specific speciation may produce a preponderance of small to medium-sized taxa. Beginning with an arbitrary distribution of body sizes (e.g., a normal distribution [A]), stable environments lead to high rates of extinction for small taxa (B) and variable environments lead to high rates of extinction for large taxa (C). Through time, shifts between stable and variable environments cull small and large taxa, resulting in a preponderance of intermediate-sized taxa (D). Speciation, radiation, and colonization rates are highest for small organisms (E) and this results in a shift from intermediate-sized taxa (D) toward small-sized taxa (F). Through evolutionary time, extinction of taxa with extreme sizes and radiation of small taxa leads to a preponderance of small, but not the smallest, sized taxa (F).

Bock, W. J., and J. Farrand, Jr. 1980. The number of species and genera of recent birds: a contribution to comparative systematics. American Museum Novitates **2703**:1–29.
Cochran, D. M., and C. J. Goin. 1970. The new field guide of reptiles and amphibians. G. P. Putnam Sons, New York, New York, USA.
Collins, H. H., Jr. 1959. Harper and Row's complete field guide to North American wildlife. Eastern edition. Harper and Row, New York, New York, USA.
Dunning, J. B., Jr. 1984. Body weights of 686 species of North American birds. Western Bird Banding Association Monograph Number 1. Eldon, Cave Creek, Arizona, USA.
Eisenberg, J. F. 1981. The mammalian radiations. University of Chicago Press, Chicago, Illinois, USA.
Fowells, H. A. 1965. Silvics of forest trees of the United States. United States Department of Agriculture, Agriculture Handbook Number **271**.
Glazier, D. S. 1987. Energetics and taxonomic patterns of species diversity. Systematic Zoology **36**:62–71.
Goodwin, D. 1976. Crows of the world. British Museum Natural History Publications, London, England.
Hutchinson, G. E. 1959. Homage to Santa Rosalia, or why are there so many kinds of animals? American Naturalist **93**:145–159.
Hutchinson, G. E., and R. H. MacArthur. 1959. A theoretical ecological model of size distributions among species of animals. American Naturalist **93**:117–125.
Johnsgard, P. A. 1973. Grouse and quail of North America. University of Nebraska Press, Lincoln, Nebraska, USA.
Kuehne, R. A., and R. W. Barbour. 1983. The American darters. University Press of Kentucky, Lexington, Kentucky, USA.
May, R. M. 1978. The dynamics and diversity of insect faunas. Pages 188–204 *in* L. A. Mound and N. Waloff, editors. Diversity of insect faunas. Blackwell Scientific, New York, New York, USA.
———. 1986. The search for patterns in the balance of nature: advances and retreats. Ecology **67**:1115–1126.
Nelson, J. S. 1984. Fishes of the world. Second edition. J. Wiley and Sons, New York, New York, USA.
Peters, R. H. 1983. The ecological implications of body size. Cambridge University Press, Cambridge, England.
Pyle, R. M. 1981. The Audubon Society field guide to North American butterflies. Alfred A. Knopf, New York, New York, USA.
Ranson, J. E. 1981. Harper and Row's complete field guide to North American wildlife. Western edition. Harper and Row, New York, New York, USA.
Southwood, T. R. E. 1978. The components of diversity. Pages 19–40 *in* L. A. Mound and N. Waloff, editors. Diversity of insect faunas. Blackwell Scientific, New York, New York, USA.
Stanley, S. M. 1973. An explanation for Cope's rule. Evolution **27**:1–26.
Van Valen, L. 1973*a*. A new evolutionary law. Evolutionary Theory **1**:1–30.
———. 1973*b*. Body size and numbers of plants and animals. Evolution **27**:27–35.
Walker, E. P. 1975. Mammals of the world. Third edition. Johns Hopkins University Press, Baltimore, Maryland, USA.

[1] *Manuscript received 3 November 1987; revised 26 February 1988; accepted 29 February 1988.*
[2] *Museum of Comparative Zoology, Harvard University, Cambridge, Massachusetts 02138 USA.*
[3] *Department of Biological Sciences, Box 5640, Northern Arizona University, Flagstaff, Arizona 86011 USA.*

Ecology, 69(5), 1988, pp. 1624–1627

AGONISTIC INTERACTIONS IN A KEYSTONE PREDATORY STARFISH[1]

Stephen R. Palumbi[2] and Leonard A. Freed[2]

A major goal of community ecology is to elucidate mechanisms that determine species richness and diversity in natural communities. Because competitive exclusion can eliminate species, much attention has been focused on mechanisms by which competitors can coexist. One class of mechanisms deals with the ways in which competitors partition resources (Schoener 1974). A second class of mechanisms deals with predator-mediated coexistence of competitors. Studies in marine (Connell 1961, Paine 1966, 1969, Paine and Vadas 1969, Dayton 1971, Lubchenco 1983, 1985, Paine et al. 1985), aquatic (Hall et al. 1970, Morin 1981, 1983), and terrestrial (Harper 1969) habitats have shown that predators (or herbivores) can indirectly increase species richness and diversity by consuming competitively superior prey. Such predators and herbivores have been termed keystone species because of their fundamental effect on community structure (Paine 1969).

Theoretical models (Caswell 1978, Noy-Meir 1981) and empirical studies (Paine and Vadas 1969, Lubchenco 1983, 1985) of predator-mediated coexistence have shown that the effect of a keystone species is generally greatest at intermediate predation intensities. This means that mechanisms that regulate predation intensity can have a dramatic effect on prey diversity. Powerful factors that alter predation rate in many systems are the social interactions among predators. Territoriality and aggression may prevent local populations of predators from reaching carrying capacity (Brown 1969, Waser and Wiley 1979, Anderson et al. 1982). Even when density is not reduced, social interactions can decrease the effective predation rate (Ens and Goss-Custard 1984, Lipcius and Hines 1986). Consequently, social interactions may alter the impact of a keystone species on prey community structure.

PAPER 39

Alpha, Beta, or Gamma: Where Does All the Diversity Go? (1988)

J. J. Sepkoski Jr.

Commentary

PETER WAGNER

Sifting through the works of an outstanding scientist is not unlike going through the recordings of a great rock band. There are, of course, the big "hits" that everyone associates with the name. However, there always are a number of works that make one stop and say, "I'd forgotten that he/she/they had done this one, too." Jack Sepkoski's "Alpha, Beta, or Gamma: Where Does All the Diversity Go?" represents just such a paper. Sepkoski's illustrious career is best known for his comprehensive summaries of diversity patterns over the Phanerozoic (i.e., the last 520 million years). Particular important and well-known studies delimited the "Three Faunas" (showing a shift from trilobite-dominated seas to brachiopod-dominated seas to mollusk-dominated seas). Others quantified major extinction patterns, such as the "Big 5" mass extinctions (five intervals of significantly elevated extinction, including the event that terminated such iconic groups as dinosaurs, ammonites, and trilobites) and the demonstration that pulses of extinction occur nonrandomly at an astronomical timescale.

Sepkoski's works, including the ones mentioned above, were strongly influenced by ecological models and theory. However, this particular study directly examined how basic macroecological parameters (environmental and biogeographic differentiation of faunas) affect basic macroevolutionary parameters (i.e., global diversity and diversification over time). Richard Bambach already had shown that alpha diversity (i.e., richness within particular habitats) increased over time. However, that might have been only part of the story: beta diversity (i.e., differentiation among adjacent environments and regions) and gamma diversity (i.e., differentiation among different faunal provinces) might also have contributed. Using just the Paleozoic (520–245 million years ago), Sepkoski demonstrated that this was very much the case. Increased alpha diversity did contribute to the marked increase in diversity over the course of the Ordovician (490–445 million years ago); however, alpha diversity did not increase anywhere near as much as did global diversity. However, environmental differentiation (beta diversity) greatly increased in that time. In other words, yes, there were a few more species per habitat, but what was really important was the increase in the number of different habitats. Alpha, beta, and gamma also played major roles in the loss of diversity: the end-Ordovician mass extinction (possibly the second greatest mass extinction) did not just reduce alpha diversity, but it also left an early Silurian world with low beta and gamma diversity.

"Alpha, Beta, or Gamma" stands out in one other way. Whereas much of Sepkoski's work was based on reports of the first and last occurrences of genera and families, this work was based on faunal lists from particular localities. Much as some of the songs on *Revolver* anticipate the musical styles that the Beatles never

From *Paleobiology* 14:221–34. Reprinted with permission from the Paleontological Society.

fully explored, Sepkoski's methods here anticipate the large database studies that began in the next decade. The synthesis of macroevolutionary and macroecological theory is still ripe for development and will require tools such as the Paleobiology Database. Sepkoski's analysis might well be Elvis or the Beatles next to the latest modern databases, but the fact that most people know who the Beatles and Elvis are testifies to its importance.

Paleobiology, 14(3), 1988, pp. 221–234

Alpha, beta, or gamma: where does all the diversity go?

J. John Sepkoski, Jr.

Abstract.—Global taxonomic richness is affected by variation in three components: within-community, or alpha, diversity; between-community, or beta, diversity; and between-region, or gamma, diversity. A data set consisting of 505 faunal lists distributed among 40 stratigraphic intervals and six environmental zones was used to investigate how variation in alpha and beta diversity influenced global diversity through the Paleozoic, and especially during the Ordovician radiations. As first shown by Bambach (1977), alpha diversity increased by 50 to 70 percent in offshore marine environments during the Ordovician and then remained essentially constant for the remainder of the Paleozoic. The increase is insufficient, however, to account for the 300 percent rise observed in global generic diversity. It is shown that beta diversity among level, soft-bottom communities also increased significantly during the early Paleozoic. This change is related to enhanced habitat selection, and presumably increased overall specialization, among diversifying taxa during the Ordovician radiations. Combined with alpha diversity, the measured change in beta diversity still accounts for only about half of the increase in global diversity. Other sources of increase are probably not related to variation in gamma diversity but rather to appearance and/or expansion of organic reefs, hardground communities, bryozoan thickets, and crinoid gardens during the Ordovician.

J. John Sepkoski, Jr. Department of the Geophysical Sciences, University of Chicago, 5734 South Ellis Avenue, Chicago, Illinois 60637

Accepted: May 13, 1988

Introduction

Global taxonomic diversity can be divided into three components: *alpha diversity*, the richness of taxa at a single locality or in a particular community; *beta diversity*, the taxonomic differentiation of fauna or flora between sites or communities; and *gamma diversity*, the taxonomic differentiation between geographic regions (Whittaker 1960, 1972, 1975; Tramer 1974; Cody 1975; Brown and Gibson 1983). Alpha diversity measures packing within a community and thus reflects how finely species are dividing ecological resources. Beta diversity, on the other hand, can be used to measure the amount of turnover in species composition along environmental gradients; thus, it can reflect the extent of habitat selection or specialization. Finally, gamma diversity is similar to beta diversity, but is measured at much larger spatial scales; it reflects the degree of provinciality or endemicity in the biota. (Whittaker [1977] actually reserved the term "gamma diversity" for the combination of alpha and beta diversity, which he also called "landscape diversity" [see also Whittaker 1960, 1972]; he used "delta diversity" for between-region, or provincial, differentiation.)

Both alpha and gamma diversity have been discussed in relation to the evolution of global marine diversity. Bambach (1977), in his pivotal study of Phanerozoic species richness, analyzed patterns of alpha diversity in a large sample of fossil communities of Cambrian through Cenozoic age. He demonstrated that within open marine environments, alpha diversity neither remained constant nor increased continuously through the Phanerozoic; rather, it increased episodically, in steps, followed by long intervals of near constancy. Sepkoski et al. (1981) later showed that this pattern is shared by various estimates of global taxonomic diversity. All exhibit low diversity in the Cambrian, higher but not persistently increasing diversity in the later Paleozoic; low diversity again in the early Mesozoic; and increasing diversity thereafter. Seilacher's (1974, 1977) data on within-facies trace fossil taxa, which provide another estimate of alpha diversity, also share this basic pattern.

0094-8373/88/1403-0001/$1.00

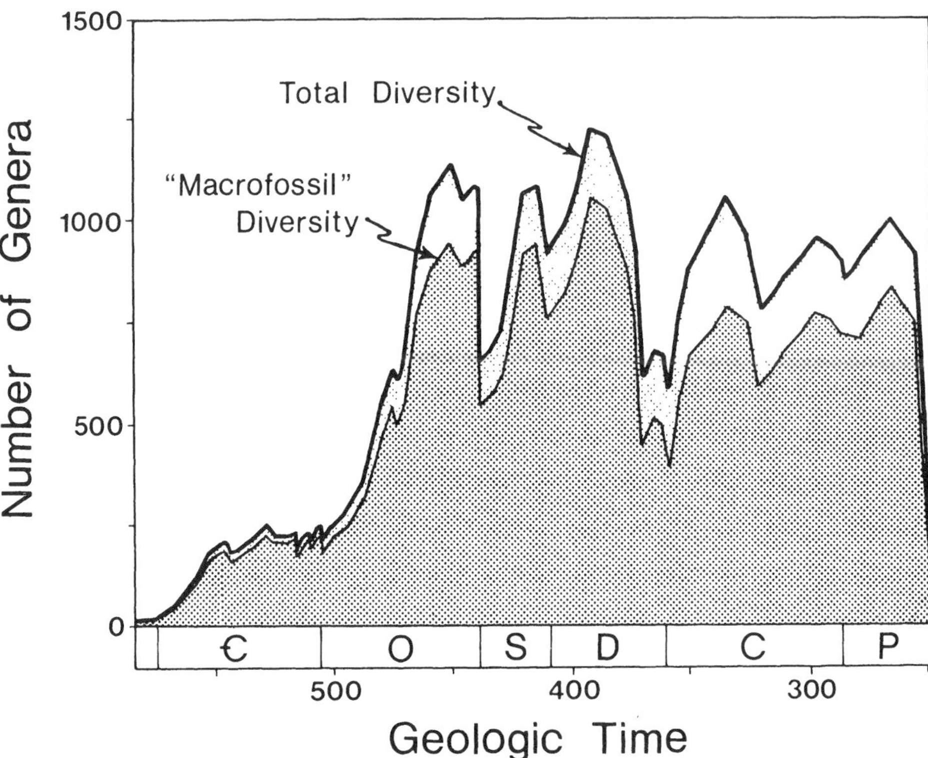

FIGURE 1. Global generic diversity for marine animals through the Paleozoic Era. The upper curve is for all marine taxa, whereas the lower is for macrofossil taxa only (i.e., excluding all radiolarians and conodonts and most foraminifers and ostracodes; see text). Diversity was computed for intervals (mostly substages) averaging 5.5 m.y. in duration, using the compilation of genera described in Sepkoski (1986). The pattern of diversity exhibited by the genera is similar to that seen for families (Sepkoski 1979, 1981), but with the various extinction events (drops in diversity) accentuated.

The correspondence between measures of alpha diversity and global diversity is only qualitative, however, and some major quantitative discrepencies exist. In particular, the measured increase in alpha diversity across the Ordovician radiations, between the Cambrian and later Paleozoic, is only about 50 percent (Bambach 1977), whereas the concurrent increase in global diversity is on the order of 300 percent at both the familial (Sepkoski and Sheehan 1983) and the generic (Fig. 1) taxonomic levels. There is, thus, a problem of "missing diversity"—a hidden source of taxa that produces the great increase in global diversity. Some authors, particularly Valentine (1970, 1971, 1973), have argued that provinciality plays a key role in diversification (see also Valentine and Moores 1970, 1972; Flessa and Imbrie 1973; Valentine et al. 1978; Schopf 1979; Cracraft 1982); thus, gamma diversity could be an important source of the excess diversity at the global level. But beta diversity, or increasing differentiation of local communities, could also play an important role.

The purpose of this paper is to investigate changing levels of beta diversity during the Paleozoic and to assess its contribution to global diversification during the Ordovician radiations. I begin by describing the data base for this investigation and use it to reproduce Bambach's (1977) results on Paleozoic alpha diversity. I then generate estimates of beta diversity for marine shelf faunas through the

Paleozoic Era. Finally, I discuss the implications of trends in beta diversity for global diversification and for the evolution of ecological specialization.

Data

The data for this study consist of 505 fossil marine faunal assemblages of Cambrian through Permian age, taken primarily from the published literature. These data are the same as used in an investigation of environmental patterns of extinction (Sepkoski 1987) and represent an extension of the data sets of Sepkoski and Sheehan (1983) and Sepkoski and Miller (1985); many of the details of data collection and analytic design appear in those papers.

The present data set differs from the earlier ones in two important aspects: (1) it is considerably larger and more evenly sampled with respect to time and environment; (2) it incorporates a more rigidly defined environmental framework. This framework, used for arraying the assemblages along an onshore-offshore gradient, is schematically illustrated in Fig. 2 for both carbonate (or low-sedimentation) and siliciclastic (high-sedimentation) settings. The two gradients in the figure are vertically aligned so as to show the approximate analogy between constituent environments. These gradients are, of course, extreme simplifications of the habitat "hyperspace" of marine organisms, ignoring the many possible variations in water quality, circulation, nutrients, substrate consistency, etc., that can occur along an onshore-offshore transect. The compression of this complexity into a one-dimensional scheme simply permits ready comparison of trends and patterns over long intervals of time.

Also illustrated in Fig. 2 and characterized more completely in Table 1 are six environmental zones used to group the fossil assemblages for the purpose of analysis. These zones are broadly comparable to Boucot's (1975) "Marine Benthic Assemblages." However, in order to avoid circularity, fossil assemblages were assigned to zones on the basis of their lithologic context rather than faunal content. Criteria for assignment included the sedimentary features listed in Table 1 as well as

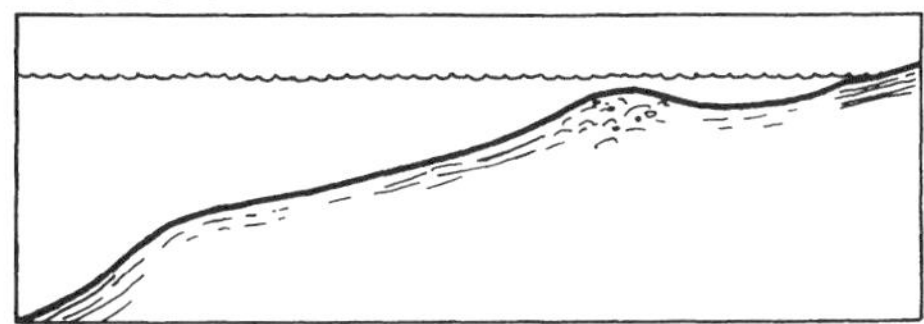

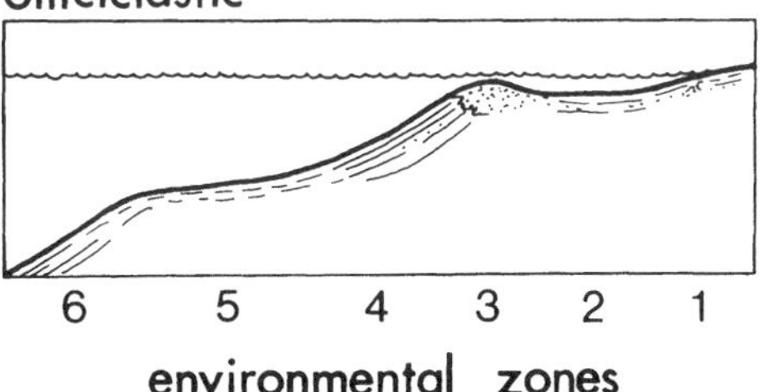

FIGURE 2. Environmental framework used to array fossil assemblages along an onshore-offshore gradient. The upper diagram is for low-sedimentation, dominantly carbonate settings, and the lower diagram is for high-sedimentation (deltaic), siliciclastic settings. The six environmental zones indicated at the bottom of the figure were used to group the assemblages for analysis (see Table 1).

stratigraphic position in transgressive or regressive sequences, geographic location with respect to shelf edge, and critically assessed environmental interpretations of other authors. But even with these multiple criteria, assignments were not always unambiguous, and numerous complications were encountered. In some stratigraphic situations, one or more of the environmental zones are missing, as, for example, zone 3 on many shelves with moderate slope or on deltas with moderate sedimentation rates. In other cases, lithologic and stratigraphic environmental signatures can be interpreted in varying ways, as best exemplified by Carboniferous cyclothem sequences (see Heckel [1977, 1980, 1986] vs. Merrill and Martin [1976]; Heckel's interpretations were employed in this study).

The fossil assemblages themselves are represented by faunal lists taken from paleoecologic and biostratigraphic studies. Lists were selected that included all dominant and subdominant macrofossil taxa sampled from well-circumscribed lithofacies. Lists that omitted

TABLE 1. Description of generalized environmental zones used to array fossil assemblages into an onshore-offshore gradient (see Fig. 2).

Zone 1:	peritidal environments with lithofacies characterized by desiccation features, stromatolites, flaser bedding, large-scale cross-bedding, and/or beach and tidal channel deposits; in high-sedimentation deltaic environments, includes shallow estuarine facies.
Zone 2:	nearshore, protected subtidal environments including shelf lagoons and delta-platform settings characterized by frequently heterolithic fine-grained lithologies with storm layers.
Zone 3:	offshore wave-agitated environments including bars, oolite shoals, bioherm-rich areas, and delta-front sands.
Zone 4:	shallow open-shelf (low-sedimentation) and prodelta (high-sedimentation) environments with evidence of frequent storm influence and, where reported, occurrence of benthic algae.
Zone 5:	"deep" open-shelf and fore-delta environments with fine-grained sediments, low frequencies of storm reworking, and little or no benthic algae.
Zone 6:	deepwater environments including slope and basinal settings with turbidite sequences and/or black shales; several cratonic basins with black shales deposited under low-oxygen conditions also included here.

rare taxa were not necessarily excluded, however, in contrast to Bambach (1977). Microfossils, operationally defined as taxa with maximum adult dimensions under 5 mm, were deleted from the faunal lists because of inconsistent documentation among the published sources (cf. Sepkoski and Miller 1985); this eliminated all radiolarians and conodonts and most foraminifers (except fusulinids and several others) and ostracodes (except leperditicopids). No other taxonomic or ecologic groups were removed, and both benthic and pelagic taxa were analyzed together (although the proportion of demonstrably pelagic taxa was small in most assemblages). However, reefs, mounds, and hardgrounds were avoided in the collection of data, and only level, soft-bottom assemblages were used.

Every attempt was made during data compilation to obtain as even a distribution of assemblages with respect to time and environment as possible. This involved adding more than 200 assemblages to the data set of Sepkoski and Miller (1985) while eliminating some redundant or poorly documented assemblages. Most assemblages were derived from the Laurentian portion of North America, but approximately 7 percent were from Europe for situations in which there were insufficient North American data. But even with this concession, it was impossible to obtain absolutely even sampling; some environmental zones (particularly zones 2 to 4) have a slightly denser concentration of assemblages than others, although this does not vary appreciably among the geologic systems.

Patterns of Alpha Diversity

The new data set on marine fossil assemblages can be used both to corroborate and to elaborate Bambach's (1977) analysis of within-community diversity. The new data include more points for the Paleozoic (505 in contrast to 211), contain points for the Permian (lacking in Bambach's study), and provide greater environmental resolution (with six as opposed to three environmental zones). However, there are several limitations: the new data do not necessarily include all rare taxa, as noted above, and the faunal lists are resolved only to the taxonomic level of genus, because of inconsistent definition and identification of species among the published sources. Both shortcomings will tend to diminish alpha diversities relative to Bambach's data, but neither should alter relative trends or patterns. Furthermore, since the vast majority of genera are represented by only a single species in the faunal lists with high taxonomic resolution, the genus and species levels are nearly congruent in this study. Still, it would be desirable to obtain results similar to those of Bambach's species-level analysis, since this would not only corroborate Bambach's results but would also test the veracity of the new data.

The matrix in Fig. 3 presents average alpha diversities computed for each environmental zone and each geologic period. Arithmetic means, rather than the medians employed by Bambach (1977), are used simply so that error estimates, shown in the lower right of each cell, can be included in the figure. This analytic difference is of minor consequence, since median values for the new data show essentially identical patterns.

The environmental trend in average alpha

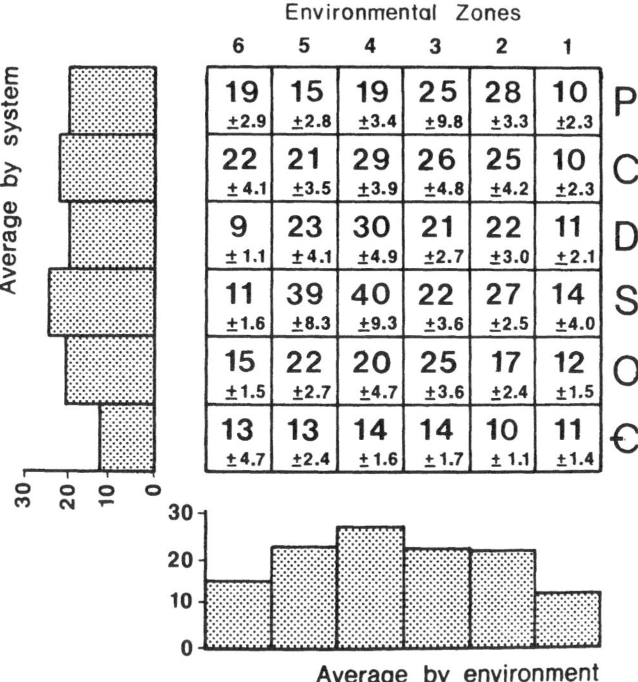

FIGURE 3. Average generic alpha diversities for each environmental zone and each geologic period of the Paleozoic. The integers in the matrix are mean numbers of genera per assemblage; the numbers in the lower right corners of the cells are standard errors of estimate about the means. The histogram below the matrix displays the mean alpha diversity for each environmental zone across the Paleozoic, and the histogram to the left displays the mean for each period across all environmental zones.

diversities is summarized by the lower histogram in Fig. 3. The basic trend is similar to that documented by other workers in more detailed paleoecologic studies (e.g., Ludvigsen 1978; Hurst 1979; Watkins 1979; Hurst and Watkins 1981; Yancey and Stevens 1981; Lockley 1983). Diversity is lowest in nearshore zone 1, where the mean is 11 genera (median = 9). This zone largely overlaps Bambach's (1977) "high stress" environmental category but also includes some of his "variable nearshore" environments. As found by Bambach, average alpha diversities within this zone remain virtually unchanged through the Paleozoic.

Offshore from zone 1, alpha diversity tends to increase to a maximum in zone 4, which is included in Bambach's "open marine" category. This increase is subdued in the Cambrian but quite pronounced in later periods. The increase is not always completely regular, however, and differences in diversity between zone 2, which overlaps Bambach's "variable nearshore" category, and zone 3 are generally minor. It is not clear whether this similarity between the two zones is real, reflecting true standing alpha diversities, or a taphonomic artifact, reflecting post-mortem mixing of heterogeneous, patchy low-diversity communities which seem to be characteristic of at least some portions of zone 2 (cf. Johnson 1972; MacDonald 1976).

Beyond zone 4, alpha diversities tend to decline across the outer shelf (zone 5) and into deeper, off-shelf waters (zone 6). This pattern is actually consistent with Bambach's results, since he included some offshore low-oxygen biofacies in his "high stress" category (R. K. Bambach, pers. comm.). It should be noted that alpha diversities can be considerably higher in some Paleozoic deeper-water environments when these occasionally become well oxygenated, as during the glacial episode of the Late Ordovician (P. M. Sheehan, pers. comm.; see, for example, Nilsson 1977).

Through time, all environmental zones offshore from zone 1 exhibit increased alpha diversity since the Cambrian, as indicated by the lefthand histogram in Fig. 3. However, none shows persistently increasing diversity throughout the Paleozoic. This, again, is consistent with Bambach's (1977) result and is qualitatively similar to the pattern for global generic diversity illustrated in Fig. 1. The highest average alpha diversities in Fig. 3 occur in the Silurian; this results from inclusion of about a dozen very large composite faunal lists from New York State (based on unpublished work of Ziegler et al. 1971), which also greatly increase the variance in alpha diversity in zones 4 and 5 of the Silurian. Average alpha diversities for the Permian in Fig. 3 tend to be slightly lower than for earlier periods, especially in zones 4 and 5; this reflects the effect of assemblages sampled from the Phosphoria Formation (Yochelson 1968), which had low diversities presumably because of excessive nutrient enrichment resulting from upwelling (cf. Rosenzweig 1971; Valentine 1973; Tilman 1982). Alpha diversity in off-shelf zone 6 is high in the Permian, as well as the Carboniferous, relative to the earlier Paleozoic. This mean increase is probably real, reflecting the establishment of benthic molluscan communities in a variety of deeper-water environments during the late Paleozoic (Yancey and Stevens 1981; Malinky and Mapes 1982; Boardman et al. 1984; Kammer et al. 1986).

Overall, the patterns of alpha diversity evident in Fig. 3 are very similar to those found by Bambach (1977). This pertains also to the problem of "missing diversity." The increase in generic alpha diversity summarized in the left-hand histogram of Fig. 3 is about 70 percent from the Cambrian to the later Paleozoic. This is comparable to Bambach's 50 percent increase in species richness within open marine environments, but is far below the 300 percent increase in global generic diversity illustrated in Fig. 1. Even if the low diversities of zones 1 and 6 are ignored, the observed increase in generic alpha diversity during the Ordovician radiations is still less than 100 percent. However, there was more diversity on the Paleozoic shelves than the within-community numbers reveal; it is hidden, in part, in between-community beta diversity, as argued below.

Patterns of Beta Diversity

Measurement.—Beta diversity is essentially the amount of diversity gained when two or more samples or communities are combined. There are several ways of measuring this quantity, as outlined by Whittaker (1977). Perhaps the simplest measure is the ratio of the total number of taxa, T_t, in a pooled set of samples to the mean alpha diversity, or average number of taxa, T, in each:

$$BD = T_t / \bar{T}. \qquad (1)$$

This index, however, is sensitive to the number of samples in the pooled set which, in this study, could be rather variable. It was found in most instances that in order to gain sample numbers large enough to produce stable values of BD, samples had to be pooled over large intervals of time, leading to conflation of evolutionary turnover with ecological differentiation.

Another measure of beta diversity is the "Coefficient of Community," or Jaccard Coefficient:

$$S_J = T_c / (T_1 + T_2 - T_c), \qquad (2)$$

where T_1 is the number of taxa in one sample or community, T_2 is the number in a second, and T_c is the number of taxa in common between the two. Some authors have used the Sørenson Coefficient, also known as the Czekanowski or Dice Coefficient (Sepkoski 1974), as the "Coefficient of Community" (Whittaker 1972; Pielou 1975). The Sørenson Coefficient, which is defined as $S_D = 2T_c/(T_1 + T_2)$, is actually redundant with the Jaccard Coefficient, since it can be shown that $S_D = 2S_J/(1 + S_J)$.

The Coefficient of Community in essence measures the proportion of redundant taxa in the pooled biota of two samples. It is thus an inverse measure of beta diversity, decreasing as beta diversity increases. I found that for the data analyzed here, the Coefficient of Community was somewhat less sensitive to variation in sample size than was *BD*, and therefore I used it in most analyses.

In the ideal situation, beta diversity is measured in a single time plane. However, there were insufficient numbers of contemporaneous fossil assemblages in the data set for this to be feasible. Therefore, beta diversity was measured over discrete time intervals. The data set was divided into 40 stratigraphic intervals, equivalent to stages and subseries as listed in Table 2. (These intervals are similar to those used in Sepkoski [1987] to investigate environmental patterns of extinction.) With an average duration of 7.5 m.y., these intervals are shorter than the mean duration of genera in all Paleozoic periods, so that most problems of accumulated evolutionary turnover are avoided. (Based on the data in Sepkoski [1987], the Cambrian contains the shortest mean generic durations, ranging from 8.4 to 18.9 m.y. among the environmental zones, as computed from the reciprocal of average per-genus rates of extinction [Raup 1985].)

With 40 stratigraphic intervals and six environmental zones, the data could be converted into a 40-by-6 matrix for analysis. Assemblages within each cell of the matrix were pooled, and the Coefficient of Community was used to estimate beta diversity between pairs of zones within the same stratigraphic interval. In order to minimize problems associated with small sample sizes, only cells that contained two or more fossil assemblages were used in computing coefficients. This pooling masked some beta diversity, since no two assemblages within the same cell were identical. In order to test whether this within-zone beta diversity might be changing through time, the index *BD* was computed for all cells containing two or more assemblages; one-way analysis of variance detected no significant differences among mean indices for the six periods of the Paleozoic ($F = 1.92$; 5,30 *df*; $P > 0.05$).

Results.—Trends found for between-zone beta diversity are very similar to those seen at the alpha level: beta diversity is lowest in the Cambrian and higher, but not continuously increasing, in the later Paleozoic. Table 3 lists mean values of the Coefficient of Community measured between adjacent environmental zones for each period of the Paleozoic. The Cambrian has the highest value at 0.31, indicating the largest proportion of genera shared between zones and therefore the lowest beta diversity. This result obtains despite the fact that the Cambrian has the shortest generic durations, as noted above; high evolutionary turnover within the measurement intervals should decrease the Coefficient of Community given that all faunal lists pooled

TABLE 2. Stratigraphic intervals used in the measurement of beta diversity. A mixture of North American and European intervals are used. Numbers in parentheses are the number of intervals within each system.

Permian (5):	lower Wolfcampian; upper Wolfcampian; Leonardian; Wordian; Capitanian and Ochoan.
Carboniferous (8):	Kinderhookian; Osagean; Meramecian; Chesterian; Morrowan and Atokan; Desmoinesian; Missourian; Virgilian.
Devonian (7):	Lochkovian; Pragian; Emsian; Eifelian; Givetian; Frasnian; Famennian.
Silurian (4):	lower and middle Llandoverian; upper Llandoverian; Wenlockian; Ludlovian and Pridolian.
Ordovician (8):	Gasconadan; Beekmantownian; Whiterockian; Chazyan; Blackriverian; Trentonian; Edenian; Maysvillian and Richmondian.
Cambrian (8):	*Fallotaspis* and *Nevadella* Zones; *Bonnia-Olenellus* Zone; *Plagiura-Poliella* and *Albertella* Zones; *Glossopleura* and *Bathyuriscus-Elrathina* Zones; *Bolaspidella* Zone; *Cedaria* and *Crepicephalus* Zones; Steptoean (upper Dresbachian and lower Franconian); Sunwaptan (upper Franconian and Trempealeauan).

TABLE 3. Mean values of the Coefficient of Community (Jaccard Coefficient) between adjacent environmental zones for periods of the Paleozoic.

Period	Mean coefficient	Standard error	Number of cell pairs
Permian	0.20	±0.028	14
Carboniferous	0.22	±0.029	16
Devonian	0.22	±0.036	16
Silurian	0.22	±0.016	14
Ordovician	0.19	±0.026	19
Cambrian	0.31	±0.037	8

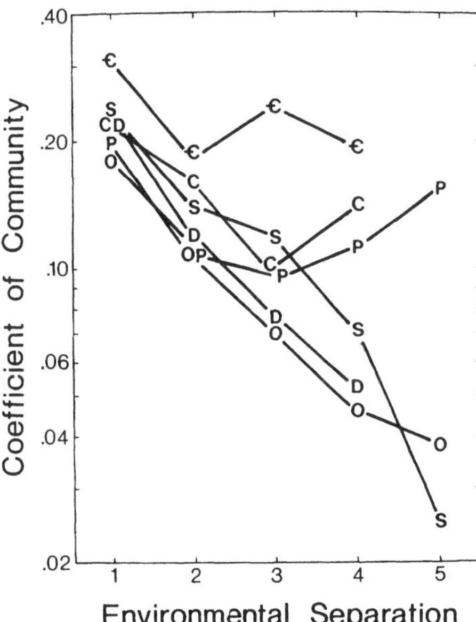

FIGURE 4. Beta diversity (Coefficient of Community) as a function of environmental separation. Mean values of the Coefficient for each geologic period (indicated by standard symbols) are plotted on a logarithmic ordinate against separation of the environmental zones on the abscissa (1 = adjacent zones, 2 = one zone in between, etc.). Only averages based on two or more points are plotted (hence the missing values for the Cambrian, Devonian, and Carboniferous at separations of five zones). The figure shows that the Cambrian Period has the highest average values of the Coefficient of Community over all separations and displays virtually no decay beyond adjacent zones, indicating comparatively low beta diversity.

within adjacent environmental zones are not of precisely the same age. The mean value of the Coefficient of Community for the Cambrian is significantly higher than the mean of 0.21 for the five later Paleozoic periods (t-test: $t = 2.35$; 85 df; $P < 0.02$). No significant differences are detectable among these later periods (one-way ANOVA: $F = 0.231$; 4,72 df; $P \gg 0.05$). Thus, it would appear that the Ordovician radiations led to a substantial increase in marine beta diversity.

This change in beta diversity was not confined simply to adjacent environmental zones but extended across the shelf. Fig. 4 illustrates how mean values of the Coefficient of Community vary with increasing environmental difference. The Cambrian exhibits significantly higher mean values than all other geologic periods, regardless of the separation between environmental zones (Friedmann test: $\chi^2 = 17.00$; 5 df; $P < 0.01$). Furthermore, beyond adjacent zones, there is essentially no decay in mean Coefficients in the Cambrian; rather, the means hover around 0.18 for separations of two to five zones. This indicates that there is a substantial component of the Cambrian fauna, approaching 20%, that was without habitat specialization and living in all environmental zones across the shelf. Inspection of the raw data indicates that this widespread component includes many of the trilobite genera that lend their names to platform-facies biostratigraphic zones (e.g., *Bonnia, Olenellus, Bathyuriscus, Elrathina, Cedaria,* and others) as well as some inarticulate brachiopods (e.g., *Acrothele, Lingulella, Micromitra,* and *Paterina*) and other taxa (e.g., *Hyolithes*).

The situation for the Cambrian contrasts with that for the later Paleozoic, especially the Ordovician through Devonian. Mean values of the Coefficient of Community for these three periods exhibit continuous decay with increasing environmental separation, as shown in Fig. 4. This pattern is similar to the decline in biotic similarity with habitat difference observed in some more carefully sampled modern communities (e.g., Whittaker 1960). It indicates that there is no substantial set of genera in the mid-Paleozoic that could live everywhere along the shelf gradient.

Coefficients of Community for the Carboniferous and Permian exhibit a pattern of change intermediate between those of the Cambrian and the mid-Paleozoic. The mean values decline over environmental separations of one to three zones, similar to the mid-Paleozoic, but then increase again over separations of four and five zones. This pattern reflects the occurrence of molluscan communities in both shallow and deep waters in the late Paleozoic. Representatives of several genera of bivalves (e.g., *Dunbarella, Edmondia,* and *Nuculopsis*) and gastropods (e.g., *Ianthinopsis, Straparollus,* and *Worthenia*) are found in both nearshore zones 1 and 2 and offshore zone 6 in the Carboniferous and Permian (see, for example, Boardman et al. 1984; Kammer et al. 1986). This increases the faunal similarity of the extremes of the gradient, as evident in Fig. 4. It does not, however, reflect a cadre of genera living in all environments, devoid of habitat specialization, as observed for the Cambrian.

The sources of beta diversity with respect to environment are illustrated in Fig. 5. This graph displays mean values of the Coefficient of Community for adjacent environmental zones across the shelf for the Ordovician through Permian. Faunal similarity tends to be highest, and thus beta diversity lowest, in the mid-shelf region, particularly between zones 3 and 4. This is the region dominated by the Paleozoic evolutionary fauna following the Ordovician radiations (Sepkoski and Miller 1985). Mean values of the Coefficient decline toward shallow water, dominated by molluscs in the post-Cambrian Paleozoic (Bretsky 1968, 1969), and even more steeply toward deep water, dominated by mixed taxa in the mid-Paleozoic and by molluscs in the late Paleozoic. The rise of these differentiated faunas at the margins of the environmental gradient is not the sole source of increased beta diversity after the Cambrian, however; the Coefficient of Community tends to be higher at all points along the shelf in the Cambrian, indicating a universally lower beta diversity.

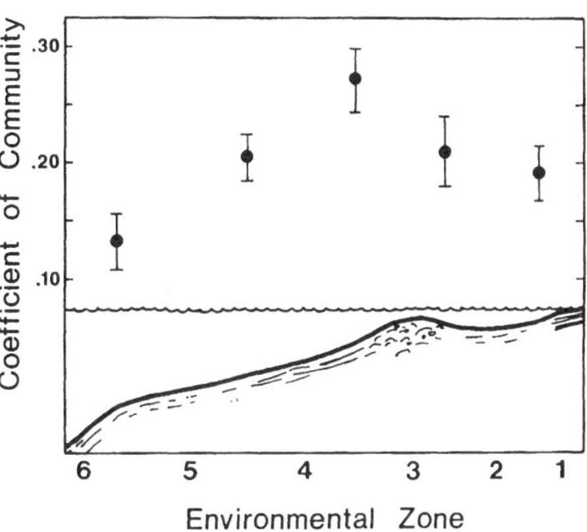

FIGURE 5. Average beta diversities (Coefficients of Community) for adjacent pairs of environmental zones across the post-Cambrian Paleozoic shelf. Dots indicate the arithmetic means for the Ordovician through Permian, and error bars indicate one standard error of estimate on either side. The pattern shows that average faunal similarity is highest, and therefore beta diversity lowest, between zones 3 and 4, and that similarity is lowest, and beta diversity highest, between outer shelf zone 5 and deep-water zone 6.

Discussion

Alpha and beta diversity have been described as reflecting species' differentiation of niche and habitat, respectively (Whittaker 1975, 1977). The patterns seen in the data for Paleozoic marine assemblages suggest that both kinds of ecological differentiation were evolving simultaneously during the Ordovician radiations. The increase in generic alpha diversity and, by inference, local species richness (cf. Bambach 1977) indicate increased niche packing and/or exploitation of marginal resources and new "ecospace." The latter has been argued to be important by Bambach (1983, 1985) in his guild analysis of marine communities; it can also be seen in the expanded infaunalization (Bambach and Sepkoski 1979; Sepkoski 1982; Larson and Rhoads 1983; Droser 1987; Droser and Bottjer 1987) and epifaunal tiering (Ausich and Bottjer 1982, 1985; Bottjer and Ausich 1986) associated with the Ordovician radiations.

The increase in habitat differentiation indicated by the measured values of beta diversity implies a general increase in ecological specialization between the Cambrian and the later Paleozoic. This supports earlier claims based on studies of global diversification and faunal change (Sepkoski 1979, 1981). In those, I suggested that members of the Cambrian evolutionary fauna, the group of taxa dominant during the Cambrian Period, tended to be more generalized in terms of morphology and ecology than were members of the succeeding Paleozoic evolutionary fauna. This suggestion echoed Valentine's (1973) gestalt impression that Cambrian animals were "grubby," having plain, unspecialized morphologies and larger modal niche sizes than later species (see also Valentine 1969; Cisne 1974). This impression, however, was based upon a qualitative and essentially subjective assessment of what constitutes generalized or specialized morphology. This is true of most claims of evolutionary change in specialization (cf. Flessa et al. 1975): the choice of what characters are important and how the degree of specialization is to be measured necessarily involves subjective elements that often are biased by knowledge of how characters evolve through a clade's history. Beta diversity avoids such pitfalls in providing a measure of one aspect of specialization that is operationally independent of morphology (other than what is needed to identify taxa).

In addition to specialization, the changes in beta diversity during the Ordovician radiations imply increased differentiation of what are recognized as ecological communities. This also corroborates qualitative impressions of the nature of early Paleozoic communities (Sepkoski 1979, 1981), which have been criticized by some workers (e.g., Ludvigsen and Westrop 1983). Most Cambrian fossil communities indeed do seem broadly similar across environmental gradients, usually sharing substantial numbers of taxa (especially among trilobites); often they can be distinguished only at low taxonomic levels. Ordovician and later Paleozoic communities are usually (although not invariably) more distinct, especially at higher taxonomic levels, and are therefore considerably easier to define and identify.

Change in community differentiation, or packing, may be a general aspect of diversity change, at least in the Paleozoic. This is suggested by various observations concerning community evolution during the Early Silurian, following the terminal Ordovician (Ashgillian) mass extinction (Sheehan 1975, 1980, 1982; Boucot 1978, 1983; Cocks and McKerrow 1978; Cocks et al. 1984). Global diversity experienced a sharp decline during the Ashgillian event, as illustrated in Fig. 1, and was comparatively low during the Early Silurian (although still higher than in the Cambrian). At the opening of the Silurian, most benthic communities had low alpha diversities and were not clearly differentiated across the shelf (for exceptions see Baarli 1987). Through the Early Silurian, as global diversity rebounded, communities became more diverse and differentiated, attaining the familiar five community status of Ziegler (1965)—*Lingula, Eocoelia, Pentamerus, Stricklandia*, and *Clorinda*—by the middle Llandoverian. Apparently, at least one more easily recognizable unit, the Marginal *Clorinda* Community, became differentiated by the latest Llandoverian (Cocks and Rickards 1969; Cocks and McKerrow 1978).

Concurrent increase in alpha and beta diversity over evolutionary time scales was the expectation of Whittaker (1975, 1977). He predicated this on the premise that there is at least some, perhaps low-level, competition among species living together. Therefore, new species cannot be packed into communities at random but must enter either by (1) exploiting resources not utilized by existing members of the community (cf. Bambach 1983) or (2) wedging into the habitat (or other resource) gradient at points between overlap of two or more species (cf. MacArthur 1965, 1969; Valentine 1969). The first alternative would increase principally alpha diversity, whereas the second would increase beta diversity as well, especially if accompanied by some habitat contraction among existing species (Fig. 6). Whittaker expected that evolutionary increase in alpha and beta diversity would be indeterminate, without a saturation level, and possibly self-augmenting, with new species providing resources (e.g., food, shelter) for yet additional species. The data presented here, however, indicate that this is not true, at least for marine animals, and that there can be geologically long intervals of time without significant changes in either alpha or beta diversity.

Alpha and beta diversity are not the only possible sources of global taxonomic richness. Biogeographic differentiation, or gamma diversity, can also be important. The question is, are the measured changes in alpha and beta diversity sufficient to account for the global diversification during the Ordovician radiations? The simple answer is no. The decrease in average faunal similarity between adjacent environmental zones from 0.3 in the Cambrian to 0.2 in the later Paleozoic translates into somewhat less than 50 percent more genera packed across the shelf. Coupled with the 50 percent increase in alpha diversity, this accounts for only a twofold increase in total diversity on the shelf, far less than the fourfold increase observed for global generic richness. There is still missing diversity.

The obvious source for the remaining missing diversity would be in biogeographic differentiation. But there is no evidence for increasing provinciality or endemism from the Cambrian into the Ordovician. Data summarized by Valentine et al. (1978) show no change in the number of provinces (or realms) between the two periods. In fact, Jaanusson (1979) indicates that there was declining provinciality in the Late Ordovician. Provinciality does not seem to increase substantially until the Early Devonian (Boucot 1975; Ziegler et al. 1981), when global generic diversity reaches its Paleozoic maximum (Fig. 1). But the rise in diversity to this maximum is small compared to the increase associated with the Ordovician radiations. Following the Devonian, there is an increase from roughly six marine provinces late in the Early Carboniferous to perhaps 12 in the Permian (Ziegler et al. 1981) without concomitant increase in alpha, beta, or global diversity; this again indicates a subordinate role of gamma diversity, at least as measured at the generic level.

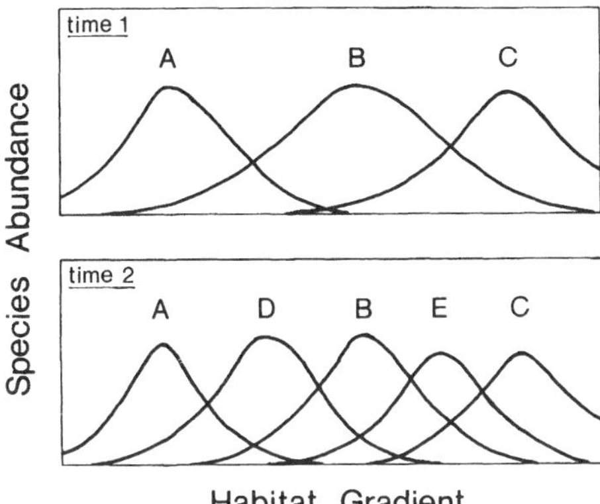

FIGURE 6. A model for the evolution of alpha and beta diversity along a habitat (or environmental) gradient, after Whittaker (1975, 1977). The bell-shaped curves in the panels represent the abundance of each lettered species along the gradient. At *time 1*, three species, *A*, *B* and *C*, coexist along the gradient. At *time 2*, two new species, *D* and *E*, have interjected themselves along the gradient between previous species, causing some habitat contraction among them. The result is more faunal turnover, or beta diversity, along the gradient but also higher alpha diversity at any point as a result of more overlapping species.

Alternative sources of global diversity involve elements of beta diversity not represented in the analyzed data set. The data are for level, soft-bottom communities, which make up most of the areal expanse of the shelf and compose a major volume of the rock record. But there are other kinds of communities that are small in expanse but rich in species. Most obvious among these are reef communities, which generally harbor far more species than surrounding level-bottom areas (cf. Bambach 1977). Metazoan-constructed reefs are unknown from the Middle and Upper Cambrian. Algal-sponge mounds appear in the Lower Ordovician (Toomey 1981), and reefs appear and diversify in the Middle and Upper Ordovician (Newell 1971; Pitcher 1971; Wilson 1975; Sheehan 1985; Fagerstrom 1987). This new community type (or class of types), situated frequently in environmental zone 3, may have contributed substantially to global diversity. Other benthic community types that either appear *de novo* or diversify greatly during the Ordovician radiations include hardground communities (Palmer 1982), bryozoan thickets, and crinoid "gardens," the last of which Bambach (1977) noted constituted the richest species assemblages in his Paleozoic data set (see also Ausich et al. 1979). Thus, there is a variety of other ecological sources for the missing diversity of the Ordovician radiations. And these sources, like the level-bottom communities, involved evolution of taxa with specialized habitat requirements that increased beta diversity across the Paleozoic shelves and therefore global diversity summed over the world ocean.

Summary

The principal arguments of this paper can be summarized as follows:

1. Global taxonomic diversity measured at the genus level increased by a factor of four during the Ordovician radiations, between the Cambrian and the later Paleozoic.

2. A small portion of this increase can be accounted for in alpha diversity, or within-community richness, which exhibits a 50 to 70 percent increase at the species (Bambach 1977) and genus levels during the Ordovician.

3. Another part of global diversity can be found in beta diversity, or between-community differentiation, which also increases by about 50 percent at the generic level for soft-bottom communities during the Ordovician.

4. The increase in beta diversity reflects increasing habitat specialization of taxa as the Paleozoic evolutionary fauna replaced the Cambrian fauna.

5. Remaining global diversity that cannot be attributed to measured changes in alpha and beta diversity probably was not contributed by increasing provinciality, or gamma diversity, during the Ordovician but rather by "hidden" sources of beta diversity, namely the appearance or expansion of new community types, including reefs, hardground communities, bryozoan thickets, and crinoid gardens.

Acknowledgments

I thank A. I. Miller for invaluable assistance in gathering and collating data for this study, P. H. Heckel for advice on environmental interpretations, and M. E. Patzkowsky for help with initial data analysis. Ideas expressed benefitted from discussions with R. K. Bambach and A. M. Ziegler. R. K. Bambach, D. Jablonski, A. I. Miller, D. W. McShea, and P. M. Sheehan made suggestions for improving the manuscript. Research received partial support from NSF grant DEB 81-08890 and NASA grant NAG 2-282.

Literature Cited

AUSICH, W. I. AND D. J. BOTTJER. 1982. Tiering in suspension-feeding communities on soft substrata throughout the Phanerozoic. Science *216*:173–174.

AUSICH, W. I. AND D. J. BOTTJER. 1985. Phanerozoic tiering in suspension-feeding communities on soft substrata: implications for diversity. Pp. 255–274. *In* Valentine, J. W. (ed.), Phanerozoic Diversity Patterns: Profiles in Macroevolution. Princeton University Press and Pacific Division, American Association for the Advancement of Science; Princeton, New Jersey.

AUSICH, W. I., T. W. KRAMMER, AND N. G. LANE. 1979. Fossil communities of the Borden (Mississippian) delta in Indiana and northern Kentucky. Journal of Paleontology *53*:1182–1196.

BAARLI, B. G. 1987. Benthic faunal associations in the Lower Silurian Solvik Formation of the Oslo-Asker Districts, Norway. Lethaia *20*:75–90.

BAMBACH, R. K. 1977. Species richness in marine benthic habitats through the Phanerozoic. Paleobiology *3*:152–167.

BAMBACH, R. K. 1983. Ecospace utilization and guilds in marine communities through the Phanerozoic. Pp. 719–746. *In* Tevesz, M. J. S. and P. L. McCall (eds.), Biotic Interactions in Recent and Fossil Benthic Communities. Plenum; New York.

BAMBACH, R. K. 1985. Classes and adaptive variety: the ecology of diversification in marine faunas through the Phanerozoic. Pp. 191–253. *In* Valentine, J. W. (ed.), Phanerozoic Diversity Patterns: Profiles in Macroevolution. Princeton University Press and Pacific Division, American Association for the Advancement of Science; Princeton, New Jersey.

BAMBACH, R. K. AND J. J. SEPKOSKI, JR. 1979. The increasing influence of biologic activity on sedimentary stratification through the Phanerozoic. Geological Society of America Abstracts with Program *11*:383.

BOARDMAN, D. R., II, R. H. MAPES, T. E. YANCEY, AND J. M. MALINKY. 1984. A new model for allogenic community succession within North American Pennsylvanian cyclothems and implications on the black shale problem. Pp. 141–182. *In* Hyne, N. J. (ed.), Limestones of the Mid-Continent. Tulsa Geological Society Special Publication No. 2.

BOTTJER, D. J. AND W. I. AUSICH. 1986. Phanerozoic development of tiering in soft substrata suspension-feeding communities. Paleobiology *12*:400–420.

BOUCOT, A. J. 1975. Evolution and Extinction Rate Controls. Elsevier; Amsterdam. 427 pp.

BOUCOT, A. J. 1978. Community evolution and rates of cladogenesis. Evolutionary Biology *11*:545–655.

BOUCOT, A. J. 1983. Does evolution take place in an ecological vacuum? II. Journal of Paleontology *56*:1–30.

BRETSKY, P. W. 1968. Evolution of Paleozoic marine invertebrate communities. Science *159*:1231–1233.

BRETSKY, P. W. 1969. Evolution of Paleozoic benthic marine invertebrate communities. Palaeogeography, Palaeoclimatology, Palaeoecology *6*:45–59.

BROWN, J. H. AND A. C. GIBSON. 1983. Biogeography. C. V. Mosby, Company; St. Louis. 643 pp.

CISNE, J. L. 1974. Evolution of the world fauna of aquatic free-living arthropods. Evolution *28*:337–366.

COCKS, L. R. M. AND W. S. MCKERROW. 1978. Silurian. Pp. 93–124. *In* McKerrow, W. S. (ed.), The Ecology of Fossils. MIT Press; Cambridge, Massachusetts.

COCKS, L. R. M. AND R. B. RICKARDS. 1969. Five boreholes in Shropshire and the relationships of shelly and graptolitic facies in the Lower Silurian. Quarterly Journal of the Geological Society of London *124*:213–238.

COCKS, L. R. M., H. WOODOCOCKS, R. B. RICKARDS, J. T. TEMPLE, AND P. D. LANE. 1984. The Llandovery Series of the type area. British Museum (Natural History) Bulletin, Geology *38*:131–182.

CODY, M. L. 1975. Towards a theory of continental species diversities. Pp. 214–257. *In* Cody, M. L. and J. M. Diamond (eds.), Ecology and Evolution of Communities. Belknap Press; Cambridge, Massachusetts.

CRACRAFT, J. 1982. A nonequilibrium theory for the rate-control of speciation and extinction and the origin of macroevolutionary patterns. Systematic Zoology *31*:348–365.

DROSER, M. L. 1987. Trends in extent and depth of bioturbation in Great Basin Precambrian–Ordovician strata, California, Nevada and Utah. Unpublished Ph.D. dissertation, University of Southern California; Los Angeles. 365 pp.

DROSER, M. L. AND D. J. BOTTJER. 1987. Early Phanerozoic stepwise increase in bioturbation. Geological Society of America Abstracts with Program *19*:647–648.

FAGERSTROM, J. A. 1987. The Evolution of Reef Communities. John Wiley and Sons; New York. 600 pp.

FLESSA, K. AND J. IMBRIE. 1973. Evolutionary pulsations: evidence from Phanerozoic diversity patterns. Pp. 247–285. *In* Tarling, D. H. and S. K. Runcorn (eds.), Implications of Continental Drift to the Earth Sciences, vol. 1. Academic Press; London.

FLESSA, K. W., K. V. POWERS, AND J. L. CISNE. 1975. Specialization and evolutionary longevity in the Arthropoda. Paleobiology *1*:71-81.

HECKEL, P. H. 1977. Origin of phosphatic black shale facies in Pennsylvanian cyclothems of mid-continent North America. American Association of Petroleum Geologists Bulletin *61*:1045-1068.

HECKEL, P. H. 1980. Paleogeography of eustatic model for deposition of midcontinent upper Pennsylvanian cyclothems. *In* Fouch, T. D. and E. R. Magathan (eds.), Paleozoic Paleogeography of West-Central United States. Rocky Mountain Section, Society of Economic Paleontologists and Mineralogists.

HECKEL, P. H. 1986. Sea-level curve for Pennsylvanian eustatic marine transgressive-regressive cycles along midcontinent outcrop belt, North America. Geology *14*:330-334.

HURST, J. M. 1979. Evolution, succession and replacement in the type Upper Caradoc (Ordovician) benthic faunas of England. Palaeogeography, Palaeoclimatology, Palaeoecology *27*: 189-246.

HURST, J. M. AND R. WATKINS. 1981. Lower Paleozoic clastic, level-bottom community organization and evolution based on Caradoc and Ludlow comparisons. Pp. 69-100. *In* Gray, J., A. J. Boucot, and W. B. N. Berry (eds.), Communities of the Past. Hutchinson Ross Publishing Company; Stroudsburg, Pennsylvania.

JAANUSSON, V. 1979. Ordovician. Pp. A136-A166. *In* Robison, R. A. and C. Teichert (eds.), Treatise on Invertebrate Paleontology, Part A, Introduction. Geological Society of America; Boulder, Colorado.

JOHNSON, R. G. 1972. Conceptual models of benthic marine communities. Pp. 148-159. *In* Schopf, T. J. M. (ed.), Models in Paleobiology. Freeman, Cooper and Company; San Francisco.

KAMMER, T. W., C. E. BRETT, D. R. BOARDMAN, II, AND R. H. MAPES. 1986. Ecologic stability of the dysaerobic biofacies during the late Paleozoic. Lethaia *19*:109-121.

LARSON, D. W. AND D. C. RHOADS. 1983. The evolution of infaunal communities and sedimentary fabrics. Pp. 627-648. *In* Tevesz, M. J. S. and P. L. McCall (eds.), Biotic Interactions in Recent and Fossil Benthic Communities. Plenum; New York.

LOCKLEY, M. G. 1983. A review of brachiopod-dominated palaeocommunities from the type Ordovician. Palaeontology *26*: 111-145.

LUDVIGSEN, R. 1978. Middle Ordovician trilobite biofacies, southern MacKenzie Mountains. Pp. 1-33. *In* Stelck, C. R. and B. D. E. Chatterton (eds.), Western and Arctic Canadian Biostratigraphy. Geological Association of Canada Special Paper *18*.

LUDVIGSEN, R. AND S. R. WESTROP. 1983. Trilobite biofacies of the Cambrian-Ordovician boundary interval in northern North America. Alcheringa *7*:301-319.

MACARTHUR, R. H. 1965. Patterns of species diversity. Biological Reviews *40*:510-533.

MACARTHUR, R. H. 1969. Patterns of communities in the tropics. Biological Journal of the Linnean Society *1*:19-30.

MACDONALD, K. B. 1976. Paleocommunities: toward some confidence limits. Pp. 87-106. *In* Scott, R. W. and R. R. West (eds.), Structure and Classification of Paleocommunities. Dowden, Hutchinson and Ross Publishing Company; Stroudsburg, Pennsylvania.

MALINKY, J. M. AND R. H. MAPES. 1982. A test of the lagoonal versus offshore depositional models for midcontinent Pennsylvanian black and dark gray shales. Third North American Paleontological Convention, Proceedings *2*:347-352.

MERRILL, G. K. AND M. D. MARTIN. 1976. Environmental control of conodont distribution in the Bond and Mattoon Formations (Pennsylvanian, Missourian) northern Illinois. Pp. 243-271. *In* Barnes, C. R. (ed.), Conodont Paleoecology. Geological Association of Canada Special Publication *15*.

NEWELL, N. D. 1971. An outline of tropical organic reefs. American Museum of Natural History Novitates *2465*:1-37.

NILSSON, R. 1977. A boring through Middle and Upper Ordovician strata at Koängen in western Scania, southern Sweden. Sveriges Geologiska Undersökning, Series C, *71*(8).

PALMER, T. J. 1982. Cambrian to Cretaceous changes in hard ground communities. Lethaia *15*:309-323.

PIELOU, E. C. 1975. Ecological Diversity. Wiley; New York. 165 pp.

PITCHER, M. 1971. Middle Ordovician reef assemblages. North American Paleontological Convention, Chicago, 1969, Proceedings, Part J:1341-1357.

RAUP, D. M. 1985. Mathematical models of cladogenesis. Paleobiology *11*:42-52.

ROSENZWEIG, M. 1971. Paradox of enrichment: destabilization of exploitation ecosystems in ecological time. Science *171*:385-387.

SCHOPF, T. J. M. 1979. The role of biogeographic provinces in regulating marine faunal diversity through geologic time. Pp. 449-457. *In* Gray, J. and A. J. Boucot (eds.), Historical Biogeography, Plate Tectonics, and the Changing Environment. Oregon State University Press; Corvallis, Oregon.

SEILACHER, A. 1974. Flysch trace fossils: evolution of behavioral diversity in the deep-sea. Neues Jahrbuch für Geologie und Paläontologie Monatshefte *4*:233-245.

SEILACHER, A. 1977. Evolution of trace fossil communities. Pp. 359-376. *In* Hallam, A. (ed.), Patterns of Evolution. Elsevier; Amsterdam.

SEPKOSKI, J. J., JR. 1974. Quantified coefficients of association and measurement of similarity. Mathematical Geology *6*:135-152.

SEPKOSKI, J. J., JR. 1979. A kinetic model of Phanerozoic taxonomic diversity, II. Early Phanerozoic families and multiple equilibria. Paleobiology *5*:222-252.

SEPKOSKI, J. J., JR. 1981. The uniqueness of the Cambrian fauna. Pp. 203-207. *In* Taylor, M. E. (ed.), Short Papers for the Second International Symposium on the Cambrian System. United States Geological Survey Open-File Report 81-743.

SEPKOSKI, J. J., JR. 1982. Flat-pebble conglomerates, storm deposits, and the Cambrian bottom fauna. Pp. 371-385. *In* Einsele, G. and A. Seilacher (eds.), Cyclic and Event Stratification. Springer-Verlag; Berlin.

SEPKOSKI, J. J., JR. 1986. Phanerozoic overview of mass extinction. Pp. 277-295. *In* Raup, D. M. and D. Jablonski (eds.), Patterns and Processes in the History of Life. Springer-Verlag; Berlin.

SEPKOSKI, J. J., JR. 1987. Environmental trends in extinction during the Phanerozoic. Science *235*:64-66.

SEPKOSKI, J. J., JR. AND A. I. MILLER. 1985. Evolutionary faunas and the distribution of Paleozoic benthic communities in space and time. Pp. 153-190. *In* Valentine, J. W. (ed.), Phanerozoic Diversity Patterns: Profiles in Macroevolution. Princeton University Press and Pacific Division, American Association for the Advancement of Science; Princeton, New Jersey.

SEPKOSKI, J. J., JR. AND P. M. SHEEHAN. 1983. Diversification, faunal change, and community replacement during the Ordovician radiations. Pp. 673-717. *In* Tevesz, M. J. S. and P. M. McCall (eds.), Biotic Interactions in Recent and Fossil Benthic Communities. Plenum; New York.

SEPOSKI, J. J., JR., R. K. BAMBACH, D. M. RAUP, AND W. VALENTINE. 1981. Phanerozoic marine diversity and the fossil record. Nature *293*:435-437.

SHEEHAN, P. M. 1975. Brachiopod synecology in a time of crisis (Late Ordovician-Early Silurian). Paleobiology *1*:205-212.

SHEEHAN, P. M. 1980. Paleogeography and marine communities of the Silurian carbonate shelf in Utah and Nevada. Pp. 19-37. *In* Fouch, T. D. and E. R. Magathan (eds.), Paleozoic Paleogeography of West-Central United States. Rocky Mountain Section, Society of Economic Paleontologists and Mineralogists.

SHEEHAN, P. M. 1982. Brachiopod macroevolution at the Ordovician-Silurian boundary. Third North American Paleontological Convention, Proceedings *2*:477–481.

SHEEHAN, P. M. 1985. Reefs are not so different—they follow the evolutionary pattern of level-bottom communities. Geology *13*:46–49.

TILMAN, D. 1982. Resource Competition and Community Structure. Monographs in Population Biology *17*. Princeton University Press; Princeton, New Jersey. 296 pp.

TOOMEY, D. F. 1981. Organic-buildup constructional capability in Lower Ordovician and late Paleozoic mounds. Pp. 35–68. *In* Gray, J., A. J. Boucot, and W. B. N. Berry (eds.), Communities of the Past. Hutchinson Ross Publishing Company; Stroudsburg, Pennsylvania.

TRAMER, E. J. 1974. On latitudinal gradients in avian diversity. Condor *76*:123–130.

VALENTINE, J. W. 1969. Patterns of taxonomic and ecological structure of the shelf benthos during Phanerozoic time. Palaeontology *12*:684–709.

VALENTINE, J. W. 1970. How many marine invertebrate species? A new approximation. Journal of Paleontology *44*:410–415.

VALENTINE, J. W. 1971. Plate tectonics and shallow marine diversity and endemism, an actualistic model. Systematic Zoology *20*:253–264.

VALENTINE, J. W. 1973. Evolutionary Paleoecology of the Marine Biosphere. Prentice-Hall; Englewood Cliffs, New Jersey. 511 pp.

VALENTINE, J. W. AND E. M. MOORES. 1970. Plate tectonic regulation of biotic diversity and sea level: a model. Nature *228*: 657–659.

VALENTINE, J. W. AND E. M. MOORES. 1972. Global tectonics and the fossil record. Journal of Geology *80*:167–184.

VALENTINE, J. W., T. C. FOIN, AND D. PEART. 1978. A provincial model of Phanerozoic marine diversity. Paleobiology *4*:55–66.

WATKINS, R. 1979. Benthic community organization in the Ludlow Series of the Welsh Borderland. British Museum (Natural History) Bulletin, Geology *31*:175–280.

WHITTAKER, R. H. 1960. Vegetation of the Siskiyou Mountains, Oregon and California. Ecological Monographs *30*:279–338.

WHITTAKER, R. H. 1972. Evolution and measurement of species diversity. Taxon *21*:213–251.

WHITTAKER, R. H. 1975. Communities and Ecosystems, 2nd ed. Macmillan; New York. 385 pp.

WHITTAKER, R. H. 1977. Evolution of species diversity in land communities. Evolutionary Biology *10*:1–67.

WILSON, J. L. 1975. Carbonate Facies in Geologic History. Springer-Verlag; New York. 471 pp.

YANCEY, T. E. AND C. H. STEVENS. 1981. Early Permian fossil communities in northeastern Nevada and northwestern Utah. Pp. 243–269. *In* Gray, J., A. J. Boucot, and W. B. N. Berry (eds.), Communities of the Past. Hutchinson Ross Publishing Company; Stroudsburg, Pennsylvania.

YOCHELSON, E. L. 1968. Biostratigraphy of the Phosphoria, Park City, and Shedhorn Formations. United States Geological Survey Professional Paper *313D*:D571–D660.

ZIEGLER, A. M. 1965. Silurian marine communities and their environmental significance. Nature *207*:270–272.

ZIEGLER, A. M., R. K. BAMBACH, J. T. PARRISH, S. F. BARRETT, E. H. GIERLOWSKI, W. C. PARKER, A. RAYMOND, AND J. J. SEPKOSKI, JR. 1981. Paleozoic biogeography and climatology. Pp. 231–266. *In* Niklas, K. J. (ed.), Paleobotany, Paleoecology, and Evolution, vol. 2. Praeger; New York.

ZIEGLER, A. M., G. NEWALL, M. S. HALLECK, AND R. K. BAMBACH. 1971. Repeated community-sediment patterns in the Silurian of the northern Appalachian Basin. Geological Society of America Abstracts with Program *3*:760–761.

PAPER 40

The Latitudinal Gradient in Geographical Range: How So Many Species Coexist in the Tropics (1989)

G. C. Stevens

Commentary

DAVID J. CURRIE

G. C. Stevens's article is a classic in macroecology, cited by over 900 subsequent publications (see Google Scholar). It introduced English-speaking ecologists to Eduardo Rapoport's observation that the latitudinal ranges of species tend to be broader at high latitudes than in the tropics. Stevens coined the moniker "Rapoport's rule" for this correlation, and he presented data showing that the "rule" applied—more or less—in several taxonomic groups in North America.

Stevens went on to propose that Rapoport's rule occurs because the annual range of climatic conditions is much smaller in the tropics than at high latitudes. In principle, stable climatic conditions could permit evolutionary specialization on very specific conditions since, "in a tropical setting, within a few kilometers, or just a few tens of meters of elevation, there exists a set of climatic conditions that the grandparents of a given organism never experienced" (245). In contrast, wide annual swings of climatic variables at high latitudes preclude evolutionary specialization. The climatic stability hypothesis had been discussed in earlier papers by P. H. Klopfer (1959) and E. R. Pianka (1966). But Stevens's was the first (as far as I know) to suggest that specialization might lead both to small ranges and to high species richness in the tropics. Scores of subsequent studies went on to examine whether Rapoport effects occurred in other taxa and other places.

Why did this paper make such a splash? The paper suggested that a hitherto little-noted biogeographic pattern might be quite general, which invited further testing. And Stevens gave the correlation a grand and citable name.

Stevens's paper is interesting from an epistemological point of view. Was Stevens interested in *testing* the climatic stability hypothesis? The first step of a test is to derive predictions from a hypothesis, and he derived an interesting one: that migratory species avoid wide climatic shifts. Their ranges should not, therefore, show the same latitudinal trend shown by nonmigratory species. Stevens noted that bird ranges in the former Soviet Union do not show the Rapoport pattern but that subsets of species that are nonmigratory do. But the Siberian pattern only superficially resembles the patterns that Stevens reported from North America. He does not comment on this. Nor does he comment on North American migratory versus nonmigratory birds. Neither did Stevens's paper address the most obvious prediction of his hypothesis: that the range size of species in an area should be correlated with the range of climatic conditions. Arguably, Stevens was more interested in "interpreting" the correlations between diversity, range size, and latitude than in hypothesis tests or predictive models.

The greatest fault of Stevens's study, in my view, is that it did not attempt to predict the patterns of interest—diversity and range size—from the hypothesized driver—climatic variability. Subsequent work has shown that Rapoport patterns are not as ubiquitous as first thought, and that spatial variation in richness is more strongly correlated with mean climate than with annual climatic variation. A great strength of Stevens's study, in contrast, is that it generated a great deal of interest in the sizes of species' geographic ranges and the processes that control them.

Literature Cited

Klopfer, P. H. 1959. Environmental determinants of faunal diversity. *American Naturalist* 93: 337–42.

Pianka, E. R. 1966. Latitudinal gradients in species diversity: A review of concepts. *American Naturalist* 100:33–46.

Vol. 133, No. 2 The American Naturalist February 1989

THE LATITUDINAL GRADIENT IN GEOGRAPHICAL RANGE: HOW SO MANY SPECIES COEXIST IN THE TROPICS

George C. Stevens

Department of Biology, Gustavus Adolphus College, Saint Peter, Minnesota 56082

Submitted October 17, 1986; Revised March 20 and December 4, 1987; Accepted May 6, 1988

The tendency for species richness to increase with decreasing latitude is well known (Wallace 1878; Dobzhansky 1950; Fischer 1960; Pianka 1966; see also tables 1–4) but poorly understood. Its impact on the thinking of biologists is reflected in the large literature associated with the gradient and in the current debate over the 12 possible explanations for the phenomenon (the 10 listed in Pianka 1978, one in Huston 1979, and one in Terborgh 1985). Since most ecology textbooks (MacArthur 1972; Colinvaux 1973; Emlen 1973; Krebs 1978; Ricklefs 1979; Brown and Gibson 1983) review this debate, there is no need to rework that here. My intent is to introduce a simple observation into the discussion to suggest a new approach to the problem.

After presenting evidence for a second important latitudinal correlate (called "Rapoport's rule"), I give an overview of the data that form our perception of the latitudinal gradient in species richness. This overview emphasizes the exceptions to the gradient and demonstrates that Rapoport's rule and the latitudinal gradient in species richness have coincident exceptional taxa. Given this coincidence, I hypothesize that both are an outcome of the same process. Focusing attention on Rapoport's rule, instead of the more complicated question of species richness, sheds light on the origin of both latitudinal gradients.

THE LATITUDINAL GRADIENT IN SIZE OF GEOGRAPHICAL RANGE, OR RAPOPORT'S RULE

When the latitudinal extent of the geographical range of organisms occurring at a given latitude is plotted against latitude, a simple positive correlation is found (figs. 1–5). This pattern can be found by rounding to the nearest 5° the northernmost and southernmost extremes of the geographical ranges of individual species and then calculating the average north-to-south extent of species found at each 5° band of latitude. I suggest that this correlation between geographical range and latitude be called "Rapoport's rule" after Eduardo H. Rapoport, who made passing reference to the correlation while describing the degree of geographical overlap between the distributions of subspecies (Rapoport 1975, 1982). Remark-

Am. Nat. 1989. Vol. 133, pp. 240–256.

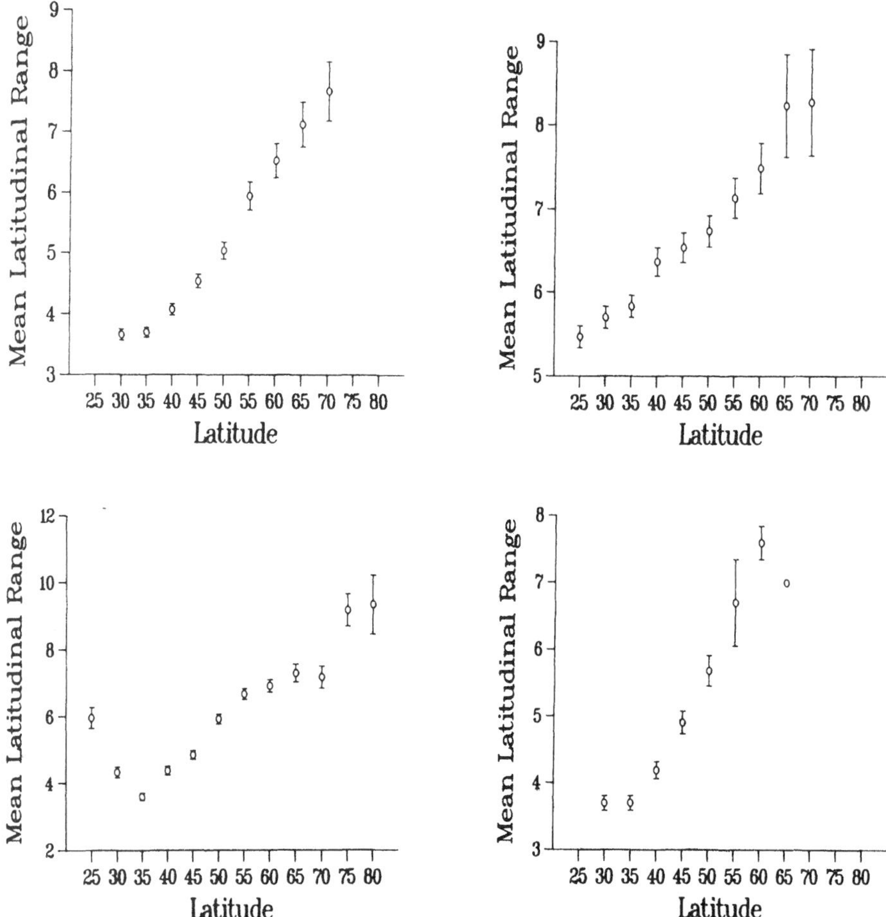

FIG. 1 (*top, left*).—Mean latitudinal extent of North American trees native to various latitudes. Sample sizes (left to right): 267, 324, 273, 182, 118, 47, 29, 17, 6. Latitudinal extent in this and all subsequent figures is the number of degrees of latitude, rounded to the nearest 5°, through which the native geographical range of the species passes. The error bars are one standard error of the mean. Data from Brockman 1968.

FIG. 2 (*top, right*).—Mean latitudinal extent of North American marine mollusks with hard body parts. Sample sizes (left to right): 510, 474, 467, 266, 203, 181, 133, 97, 41, 40. Annotations as in figure 1. Data from Rehder 1981.

FIG. 3 (*bottom, left*).—Mean latitudinal extent of North American freshwater and coastal fishes. Sample sizes (left to right): 147, 362, 588, 435, 302, 162, 107, 79, 51, 41, 6, 3. Annotations as in figure 1. Data from Lee et al. 1980 et seq. The apparent increase in the mean latitudinal extent of fishes in the southern United States is due to an edge effect. At the southernmost sites of the United States (at about 30° to 25° latitude), there is little land per latitudinal band and few freshwater species. As a result, most of the species listed are coastal (often marine) and have wide ranges.

FIG. 4 (*bottom, right*).—Mean latitudinal extent of North American reptiles and amphibians. Sample sizes (left to right): 205, 210, 145, 80, 38, 10, 5, 1. Annotations as in figure 1. Data from Conant 1958.

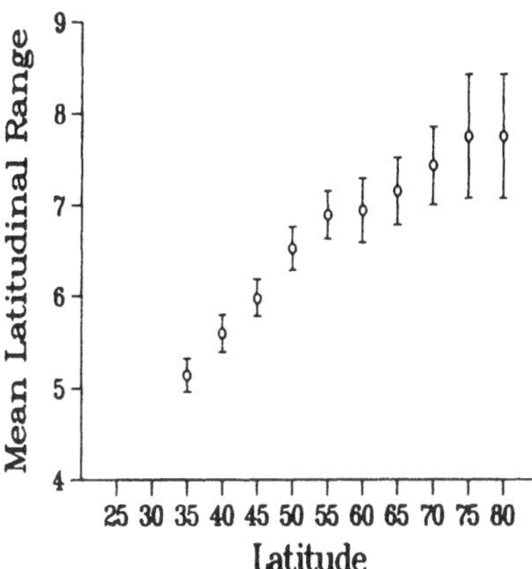

FIG. 5.—Mean latitudinal extent of North American mammals. Sample sizes (left to right): 214, 176, 148, 117, 95, 71, 53, 40, 8, 8. Annotations as in figure 1. Data from Burt 1964.

ably, this correlation (i.e., Rapoport's rule) is found in all higher taxa whose geographical ranges are well known (except migratory birds; a fact discussed later). This is not to say that all low-latitude organisms have small geographical ranges (e.g., *Bufo marinus*, the marine toad, is a good counterexample to that claim), but on the average, the ranges of organisms decline with declining latitude.

The aspect of size of the geographical range that is being compared is the length of the north-south axis. Total areal extent does not yield as clear a correlation as does latitudinal extent, possibly because the area of the geographical range does not reflect the climatic challenges a species must face as well as does latitudinal extent. Two points separated only by longitudinal differences do not as consistently show climatic differences as two points separated by degrees of latitude. To see this, consider how climatic variables change with latitude. Temperature extremes show a simple relation with latitude (fig. 6); the range in temperature readings over the course of many years is a positive function of latitude. Rainfall as a function of latitude is more complicated (figs. 7, 8), but there is still an easily explained pattern. The range in annual precipitation values for a single station over many years shows the greatest absolute range of values in the tropics (fig. 7), but this is a consequence of the magnitude of accumulated rainfall. A better measure of the environmental stresses faced by a resident at the different latitudes might be the variation relative to the mean experienced at that site (fig. 8). In any case, the point is that climatic variables show a rather simple relationship with latitude. Similar plots of longitude versus climatic conditions do not yield such regular patterns. For this reason, the width of the geographical range contains little information about climatic variability within the range of a species. Therefore, the north-south axis of the geographical range was used as the major information-containing variable in this analysis.

Rapoport (1975) compared the geographical distributions of mammalian subspecies of the same species and found that subspecies at lower latitudes tend to have smaller distributions than their higher-latitude counterparts. His compari-

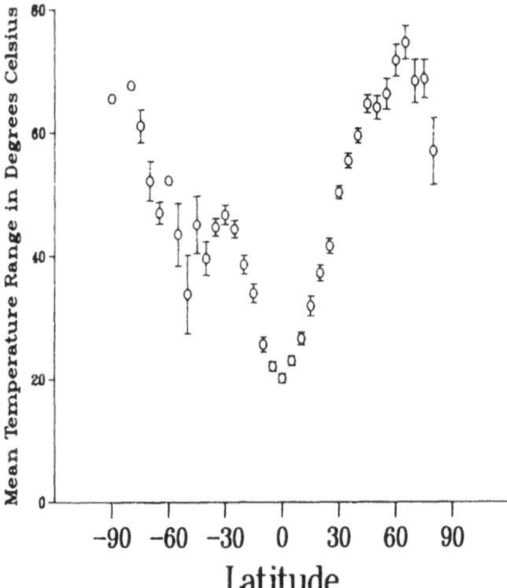

FIG. 6.—Mean of absolute range in temperature as a function of latitude. Negative latitudes are in the Southern Hemisphere. The data represent record highs and lows (not maximum daily extremes) over 30 yr (1931 to 1960) or less (then up to 1970) depending on the particular station. Annotations as in figure 1. Data are from 1056 stations (slightly biased toward European stations) spread over the globe and compiled in Müller 1982.

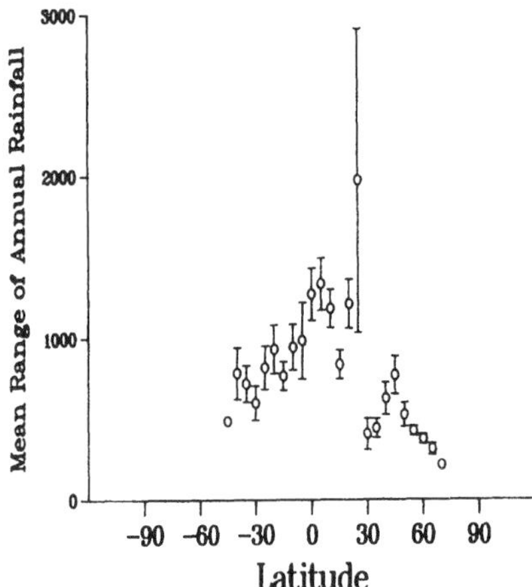

FIG. 7.—Mean of absolute range in accumulated annual rainfall as a function of latitude. Periods of data collection and annotations as in figure 6.

sons included 136 species from North America, 30 from Central America, and 31 species whose ranges entered both North and Central America. Rapoport restricted his analysis to mammals that had been well studied in order to avoid major biases resulting from differences in the completeness of mammalian censuses in different latitudes. His work confirms the pattern presented in figures 1–5 and also allows the correlation shown by those figures to be extended to tropical latitudes. Since the range limits of tropical organisms are less well known than those of

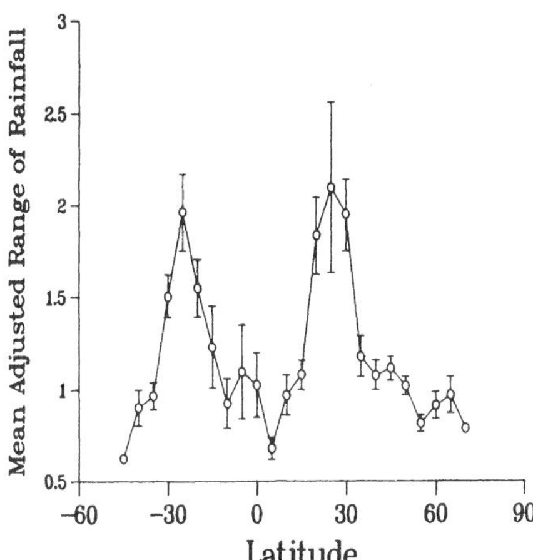

FIG. 8.—Mean of relative range in accumulated annual rainfall expressed as the difference between the annual maximum and annual minimum divided by the mean annual rainfall. Data and annotations as in figure 6.

North American organisms, estimates derived from plots like those found in figures 1–5 would tend to be misleading if broadly applied to tropical organisms. For example, using the distributional data from Holdridge and Poveda (1975) for trees of Santa Rosa National Park in northwestern Guanacaste Province, Costa Rica (ca. 10° N latitude), an average of 11.45° ($N = 73$) of mean latitudinal extent is found. This value is certainly overinflated, because narrowly distributed species are those most likely to have missing data. Yet 11.45° is less than that found for any of the other latitudes presented in figure 1. Rapoport's method is the only way of extending my survey into tropical latitudes without introducing unacceptable error into the correlation. His method succeeds because it is applied only to subspecies whose geographical ranges are well known.

Rapoport's work (1975) also provides evidence that the smaller geographical ranges of tropical organisms are not an outcome of the smaller landmass of tropical lands (i.e., Central America). By plotting the average meridional expansion of mammals throughout the North American continent, he found that the percentage of the width of the landmass used by the average mammal decreases from 93.1% in the far north to 64.2% in Costa Rica (noted in another context in McCoy and Connor 1980). Thus, the reduction in the size of the geographical range of mammals is not exclusively due to the reduction in the landmass of North America toward the tropics. Marine mollusks show the same pattern as terrestrial organisms (fig. 2), also suggesting that Rapoport's rule is not a simple matter of the size of the landmass.

The emphasis on the latitudinal gradient in geographical-range size within the North Temperate Zone serves another purpose. As argued below, Rapoport's rule and the latitudinal gradient in species richness may arise from similar ecological processes. Generally, the high species richness of tropical latitudes is taken as a tropical phenomenon, something to be studied by tropical biologists. What must be emphasized is that the processes that produce these two latitudinal correlates are at work within North America. They are not due to some abnormality of tropical climates, glacial disturbances of temperate areas, or some antibiotic agent in polar habitats. These latitudinal gradients can be studied at all latitudes.

WHY RAPOPORT'S RULE EXISTS

Consider a spruce tree living in interior Alaska. This individual tree must have the ability to survive temperatures as cold as −55°C in the winter and as warm as +34°C during the summer. The seasonal variation in temperature or moisture content of any given piece of forest precludes ecological specialization on some smaller set of climatic conditions. The tolerance of any individual organism must span the range of conditions to which it is exposed throughout its life. The large latitudinal extent of high-latitude organisms is a simple consequence of the selective advantage to those individuals with wide climatic tolerances, tolerances that are needed for the successful exploitation of any given high-latitude location. For tropical organisms, individuals that have wide climatic tolerances derive no great advantage. The full breadth of their potential tolerances are never tested by natural selection. This does not preclude the evolution of different climate-tolerant races within the total geographical range of the tropical species, but selection for wide tolerance of single individuals would not be expected. Such broad tolerances would not be selectively advantageous to the individual possessing them and might even be detrimental if they reduce the efficiency of exploitation of particular microclimatic conditions. These ideas are not new, having been discussed in reference to barriers to gene flow by Janzen (1967) and having found empirical support in the work of Huey (1978) and others (MacArthur 1965, 1969). Pielou used similar reasoning in interpreting the correlation between latitudinal and altitudinal range in *Pinus* (1979, p. 218).

It does not follow, and the biogeographers cited above did not claim, that phenotypic plasticity alone explains the boundaries of the geographical range of a species. Examples of geographical variants can be found in all the taxa listed here for both high and low latitudes. The point is that in a tropical setting, within a few kilometers, or just a few tens of meters in elevation, there exists a set of climatic conditions that the grandparents of a given organism never experienced. This situation is much more difficult to find in a temperate or an arctic area. The spatial scale of distinctively different microhabitats is smaller in the tropics. This makes the evolution of tropical geographical variants less likely. There simply is not enough physical space to support a large (and therefore persistent) population of genetically distinctive members of a microclimate-tolerant race.

In summary, two points differing in latitude are likely to experience different ranges of climatic conditions, with tropical areas showing a narrower range of temperature and rainfall and extratropical areas showing a greater annual range (at least in temperature if not in rainfall). Temperate and polar areas may experience conditions on a given day that are similar to those found in tropical areas, but their annual range of climatic conditions far exceeds that ever experienced by organisms restricted to the tropics. Natural selection favors the wide climatic tolerance

TABLE 1

COMPILATIONS OF REGIONAL SURVEYS SHOWING A LATITUDINAL GRADIENT IN SPECIES RICHNESS

Organism or Guild	Region	Source
Vertebrates		
Non-oceanic birds	New World	Dobzhansky 1950
	New World	MacArthur 1969
	New World	Cook 1969
	Nearctic	Tramer 1974
	Palearctic	Järvinen 1979
Mammals	New World	Simpson 1964
	New World	Wilson 1974
Fish	Nearctic	Horn & Allen 1978
Reptiles	Nearctic	Kiester 1971
Anurans	global	Arnold 1972
Lizards	global	Arnold 1972
	Nearctic	Schall & Pianka 1978
Snakes	New World	Dobzhansky 1950
	global	Arnold 1972
Invertebrates		
Papilionid butterflies	global	Scriber 1973, 1984
Sphingid moths	New World	Schreiber 1978
Dragonflies	global	Tillyard 1917, cited in Williams 1964
Wood-boring Scolytidae and Platypodidae	global	Beaver 1979
Planktonic foraminiferans	Nearctic	Stehli et al. 1969
Permian brachiopods	Nearctic	Stehli et al. 1969
Corals	Australian	Wells 1955, cited in Fischer 1960
	global	Stehli & Wells 1971
Tunicates	global	Hartmeyer 1911, cited in Fischer 1960
Calanid crustaceans	global	Brodskij 1959, cited in Fischer 1960
Mollusks	Nearctic	Fischer 1960
Plants		
Trees	Palearctic	Silvertown 1985
Orchids	New World	Dressler 1981

of individuals in high-latitude areas and shows no preferential treatment of the same in low latitudes. As a consequence, individual organisms of high latitudes are less restricted in their habitat use; their distribution thus shows greater latitudinal extent than that of species in low latitudes.

THE LATITUDINAL GRADIENT IN SPECIES RICHNESS

There is an ecological connection between the correlation of geographical-range size and latitude with the correlation between species richness and latitude. Illustrating this connection requires a review of the data giving rise to the idea that the tropics support more species of organisms than do higher latitudes. The exceptions to this pattern deserve special attention, since I show that exceptions to one latitudinal correlation are also exceptions to the other.

From tables 1 and 2, it is clear that the latitudinal gradient in species richness applies to the same broad array of taxa in which Rapoport's rule exists. These tables divide the data into those generated by range maps (table 1) and those

TABLE 2

Compilations of Point Samples Showing a Latitudinal Gradient in Species Richness

Organism or Guild	Region	Source
Vertebrates		
Non-oceanic birds	New World	Karr 1971
	New World	Karr & Roth 1971
	Nearctic	Tramer 1974
Mammals	New World	Fleming 1973
Lizards	Nearctic	Pianka 1967
Freshwater fish	global	Barbour & Brown 1974
Invertebrates		
Arthropod communities	Nearctic	Teraguchi et al. 1981
Litter mites	New World	Stanton 1979
Stream invertebrates	Nearctic	Stout & Vandermeer 1975
Marine invertebrates	New World	Heck 1979
Lepidoptera	New World	Ricklefs & O'Rourke 1975
Ants	global	Kusnezov 1957
	New World	MacArthur 1972
Marine copepods	Nearctic	Turner 1981
Polychaetes	global	Sanders 1968
	Old World	Ben-Eliahu & Safriel 1982
Gastropods	New World	Spight 1977
	Nearctic	MacDonald 1969
Marine bivalves	global	Sanders 1968
Epizooplankton	Nearctic	Grice & Hart 1962
Plants		
Trees	New World	Dobzhansky 1950
	Nearctic	Monk 1967
	Nearctic	Glenn-Lewin 1977

produced by counting species that occur at particular sample points (table 2). Several times (MacArthur 1965, 1969; Whittaker 1969) it has been pointed out that habitats (however defined by the investigator) in tropical latitudes generally support more species than similarly defined habitats in the Temperate Zone. Without doubt, the high species richness of tropical latitudes is due to more than just a greater variety of distinctively different habitats in the tropics. In a given tropical habitat, more species coexist than in analogous extratropical sites.

The two most frequently cited exceptions to the latitudinal gradient in species richness are both cases in which recent reviewers have failed to update the interpretations of earlier workers. According to Stout and Vandermeer (1975), the findings of Patrick (1966) probably represent incomplete sampling of rare species and should not be taken as a counterexample to the latitudinal gradient in species richness of aquatic communities. The work of Thorson (1951) is also often used as counterevidence, but he is reported to have changed his mind after more data collection (Sanders 1968). Other often-cited studies are clearly too narrow in geographical extent to be considered a test of the latitudinal pattern (e.g., the state of Texas, Rogers 1976; the deserts of the United States, Brown 1973, Brown and Davidson 1977), and the authors did not intend for their data to be used in this way.

TABLE 3

PSEUDO-EXCEPTIONS TO THE LATITUDINAL GRADIENT IN SPECIES RICHNESS

Organism or Guild	Region	Source
Vertebrates		
Rodents	Nearctic	Brown 1973
	Nearctic	Brown & Davidson 1977
Fish	New World	Patrick 1966, cited in MacArthur 1969
	Nearctic	Miller 1958
Reptiles	Nearctic	Rogers 1976
Invertebrates		
Freshwater invertebrates	New World	Patrick 1966, cited in MacArthur 1969
Rocky-intertidal invertebrates	New World	Paine 1966
Basommatophoran mollusks	global	Hubendick 1962
Decapod crustaceans	Neotropical	Abele 1974
Estuarine polychaetes and bivalves	global	Sanders 1968
Deep-sea polychaetes and bivalves	global	Sanders 1968
Marine infauna	global	Thorson 1951
Apoidea	global	Michener 1979

TABLE 4

EXTRATROPICAL PEAKS OR CONTRADICTORY PATTERNS OF SPECIES RICHNESS

Organism or Guild	Region	Source
Vertebrates		
Non-oceanic birds	Nearctic	Cody 1966
	Australia	Schall & Pianka 1978
Lizards	Australia	Schall & Pianka 1978
Invertebrates		
Ichneumonid parasitoids	Old World	Owen & Owen 1974
	Nearctic	Janzen 1981
Collembola	global	Rapoport 1975

Other pseudo-exceptions (table 3) to the latitudinal gradient in species richness include cases in which an unclear statement of the phenomenon has led to false expectations. It should not come as any surprise that some organisms (penguins, conifers, willows, brown algae, seals, etc.) have greater species richness at high latitudes than at low latitudes. These cases are not relevant to the latitudinal gradient in species richness because in each situation the particular ecological role filled by the organism is filled by many more species at lower latitudes (e.g., willows are replaced by several genera of shrubby plants at lower latitudes). The latitudinal gradient in species richness is an observation about the number of species in an assemblage of species, not the number of species in a genus. Similar but less obvious kinds of ecological replacements may account for the exceptional findings of the remaining entries in table 3. These authors have restricted their studies to narrowly defined sets of taxa without adequate study of potential ecological analogues.

The remaining exceptions (table 4) do not pose a serious threat to the generality

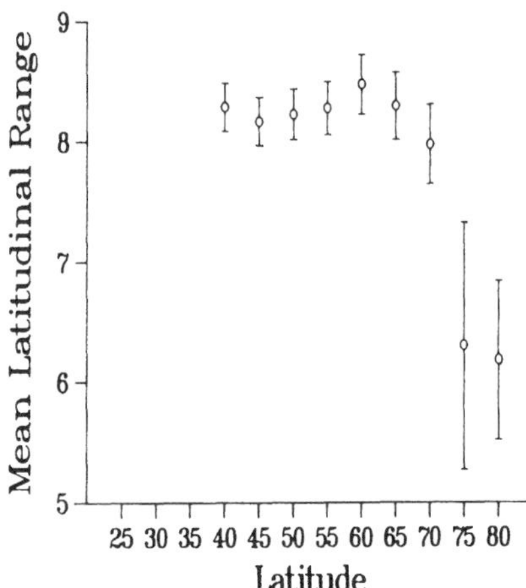

FIG. 9.—Mean latitudinal extent of the breeding ranges of Soviet non-oceanic birds as a function of latitude. Sample sizes (left to right): 440, 441, 395, 315, 254, 190, 124, 20, 11. Annotations as in figure 1. Data from Dement'ev et al. 1951–1954.

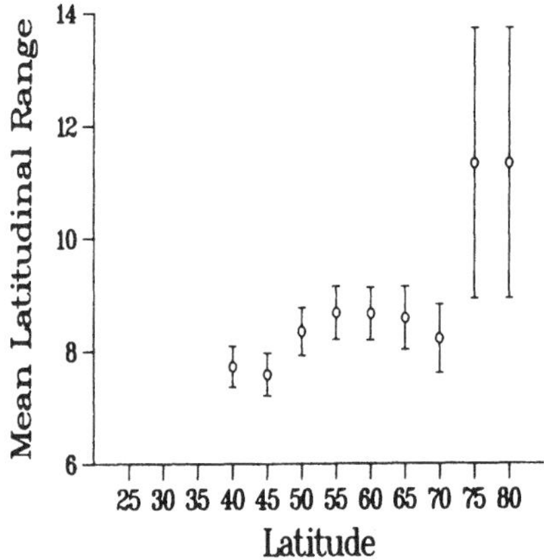

FIG. 10.—Data from figure 9 but only for nonmigratory birds. Sample sizes (left to right): 112, 97, 76, 54, 41, 31, 18, 3, 3. Annotations as in figure 1.

of the latitudinal gradient in species richness, but they are of interest because they provide a key to what types of ecological processes produce the pattern. Compare these exceptions to organisms that do not show the Rapoport phenomenon.

THE EXCEPTIONS TO RAPOPORT'S RULE

The pattern of latitudinal extent of Soviet bird breeding distributions is exceptional (fig. 9). No positive correlation between latitudinal extent and latitude can be seen when the breeding ranges of migratory and nonmigratory birds are lumped together in a single graph. When only the nonmigratory birds are displayed (fig. 10), the now-familiar Rapoport phenomenon returns. (This also shows that

Rapoport's rule applies to Old World species as well as to those of the New World.) Since migratory birds do not experience the full range of climatic conditions that a breeding ground offers over the seasons, they are not selected for the same breadth of climatic tolerance as are nonmigratory birds breeding in the same area. Migration allows birds to live in a narrow range of environmental conditions, and hence their breeding distributions do not match Rapoport's rule. The observation that organisms migrating long distances do not show the Rapoport phenomenon strengthens its climate-based explanation discussed earlier.

Although the data are not yet available, I would hypothesize that Rapoport's rule would not be found in organisms that act like migrants in their exploitation of the environment. Insects that are dormant for much of the year "migrate" in that they avoid some of the environmental extremes the seasons offer. Annual plants that reside in the seed stage during much of the year, or organisms that actually migrate or hibernate, should not show variation in their geographical ranges consistent with Rapoport's rule. Internal parasites that are buffered from the external environment may also not show the Rapoport phenomenon, provided that their host use is so generalized that their distribution is not limited to the distribution of a single host species. The difficulties in testing these hypotheses are that the taxa being studied must be sufficiently species-rich to allow statistical comparison and that their geographical ranges must be equally well known in all of the latitudes being compared.

HOW RAPOPORT'S RULE MAY PRODUCE THE LATITUDINAL GRADIENT IN SPECIES RICHNESS

Looking back over the counterexamples to the latitudinal gradient in species richness in table 4, it should now be noted that strongly seasonal or migratory organisms found in that table should not be expected to show the Rapoport phenomenon. Take, for example, Ichneumonidae, which do not show a simple latitudinal gradient in species richness but instead show a peak at mid-latitudes. These parasitoids make their living during the warm months of the year and, in a sense, live in the tropics, no matter what latitude they call home. Because these species are freed from selection to broaden their environmental tolerances, other ecological factors (outlined in Janzen 1981) become the major determinants of their distribution. Similar reasoning leads one to conclude that all the taxa of table 4 are groups that should not be expected to show the Rapoport phenomenon for the same reasons that migratory Soviet birds (fig. 9) are exceptional. Both the migratory birds and the exceptional taxa in table 4 experience only a small portion of the climatic variation their breeding latitudes have to offer.

The coincidence of the exceptional taxa for these two latitudinal correlates does not demonstrate that they both have the same cause, but the coincidence begs for some simple explanation. The remainder of this paper is devoted to clarifying the connection between these two latitudinal gradients.

A prediction of Rapoport's rule is that organisms from low latitudes have narrower tolerances for climatic conditions than do high-latitude species (provided that the above explanation of Rapoport's rule is correct). The consequence of this prediction is that tropical latitudes appear as a finer mosaic of distinctive microclimates to a tropical organism than to a temperate or polar organism (or to climate generalists like humans). This difference arises because a climate change that is minor to an organism from a high latitude is a major (possibly life-threatening) change to an organism from a lower latitude, even though the magnitude of the change in climate is the same. As mentioned earlier, several reviews have made the point that greater habitat heterogeneity of tropical areas does not account for all of the gradient in species richness because even comparable habitat types support more species in tropical than in extratropical latitudes (MacArthur 1965, 1969; Whittaker 1969). The increased environmental sensitivity of low-latitude organisms does not result in an increase in the number of obvious ecotones in the tropics but produces greater heterogeneity in the success of organisms exploiting a given location. This heterogeneity may allow for species coexistence that might otherwise be impossible.

If the microhabitat requirements of tropical organisms are narrowly defined, then the dispersal powers of individuals near the edge of their preferred microhabitat may extend to unfavorable areas. Often this would result in the arrival of individuals in areas where they are able to survive but unable to maintain their population. Shmida and Wilson considered the various determinants of species richness in plant communities and presented a new category called "mass effect" (1985, p. 2). This enrichment of species number in a given habitat is the result of "the establishment of species in sites where they cannot be self-maintaining" (p. 2). From their surveys of desert washes, they found that valleys contain many species of hillside plants whose populations cannot persist in the valley bottoms without the constant input of seeds from the hillsides. This is precisely the phenomenon that I suggest inflates the species richness of tropical forests (some anecdotal evidence suggests that this inflation process occurs in low latitudes; Heck 1979).

The existence of Rapoport's rule suggests that tropical organisms have narrower environmental tolerances than temperate or polar organisms. Their narrower tolerances would lead to greater spatial heterogeneity and discontinuity of the areas where they do well. The sites where a species is successful become source areas for colonists that sometimes arrive in habitats to which they are poorly suited. Populations of these poorly suited arrivals cannot be excluded through competition with better-adapted locals because their population dynamics depend on the proximity of areas where they do well, not on local conditions (for a similar argument, see the "rescue effect" in Brown and Kodric-Brown 1977). The greater species richness of tropical habitats would then be a result of prolonging the coexistence of species whose traits would otherwise lead to competitive exclusion.

Several researchers (Connell 1978; Hubbell 1979, 1980; Huston 1979) have proposed a nonequilibrium hypothesis to account for the high species richness of tropical forests. In its usual form, this approach involves some kind of disturbance to the community. The problem for proponents of these explanations is that even with nonequilibrium conditions, competitively inferior species are eventually lost in disturbance models (see especially Hubbell 1980). The loss occurs because these models calculate the probability that a particular species will colonize a newly open resource as a function of the local abundance of the potential colonist. When species presence is decoupled from the success or failure of propagules, even competitively inferior species can stay in the system. From the explanation of the latitudinal gradient in species richness given above, it should be noted that a mechanism for the continued existence of competitively inferior species already exists. The "rescue effect" of Brown and Kodric-Brown (1977) or the "mass effect" of Shmida and Wilson (1985), coupled with the proximity of narrowly defined source areas (as expected from the Rapoport effect), will allow more species to coexist than will considering the forest patch a closed system.

If tropical communities are more "open" than originally imagined, then a reevaluation of the boundaries of tropical communities must be considered. This reevaluation leads to several testable predictions. The sensitivity of tropical communities to isolating disturbance (i.e., the creation of habitat islands) should be higher than that of similar temperate or polar systems. A substantially larger fraction of the species in tropical communities should be very rare, and, of those species, a smaller proportion should be "globally rare" (i.e., rare in all parts of their distribution) than in other latitudes. Each of these predictions is based on the idea that many of the species in tropical communities are not locally self-sustaining but are maintained as rare populations through continued immigration from areas where they are successful.

CLARIFICATION

There are several pitfalls I have encountered in trying to present the ideas in this paper. Generally, these problems arise because an attempt is made to use the latitudinal gradient in species richness to explain Rapoport's rule instead of the reverse. There is no way to determine cause and effect here except to try it both ways and look for logical inconsistencies. At the outset, both competition and differences in stability seem to give promise of explaining Rapoport's rule, but ultimately neither explanation is satisfactory.

The existence of Rapoport's rule should not be taken as an indication that competition forces species to specialize in a way that reduces the size of their geographical range. Anderson and Koopman (1981) expanded upon Rosenzweig's (1975) suggestion that competition influences the size of an organism's geographical range, but they were unable to show that a correlation between species richness of the local community and the size of the geographical range of the interacting species was a general phenomenon. Even though it is difficult to separate the species richness of the community from latitude, Rapoport's rule should not automatically be given a competitive interpretation. Given the large number of species encountered by an organism with a large geographical range, it is unlikely that conditions in just one of the communities of which it is a part will strongly influence the evolution of the suite of traits that determine its geographical range (Janzen 1985).

Rapoport's rule also does not depend on differences in the stability of the habitats at different latitudes. It may seem that the variability in climatic conditions is just another way of measuring environmental stability, but this is not my intent. Rapoport's rule tells us little about the degree of specialization expected among organisms interacting through a diverse resource base. The climate axis of the resource hypervolume itself is narrower in low latitudes, but the relative widths of exploitation of that axis by particular organisms are not. I do not predict how organisms will divide up the available resources.

SUMMARY

The latitudinal gradient in species richness is paralleled by a latitudinal gradient in geographical-range size called Rapoport's rule. It is suggested that the greater annual range of climatic conditions to which individuals in high-latitude environments are exposed relative to what low-latitude organisms face has favored the evolution of broad climatic tolerances in high-latitude species. This broad tolerance of individuals from high latitudes has led to wider latitudinal extent in the geographical range of high-latitude species than of lower-latitude species.

The existence of Rapoport's rule suggests yet another way of looking at the latitudinal gradient in species richness. If low-latitude species typically have narrower environmental tolerances than high-latitude species, then equal dispersal abilities in the two groups would place more tropical organisms out of their preferred habitat than higher-latitude species out of their preferred habitat. It is hypothesized that a larger number of "accidentals" (i.e., species that are poorly suited for the habitat) occur in tropical assemblages. The constant input of these accidentals artificially inflates species numbers and inhibits competitive exclusion.

ACKNOWLEDGMENTS

W. S. Armbruster, R. Bellig, E. E. Berg, F. S. Chapin, J. F. Fox, D. H. Janzen, K. K. Matthew, M. Parkinson, E. H. Rapoport, and R. E. Ricklefs critically reviewed and certainly improved the manuscript. The Institute of Arctic Biology, University of Alaska, and National Science Foundation grants (DEB 76-21409 and DEB 78-09149 to S. P. Hubbell; DEB 77-04889, DEB 80-11558, and BSR 83-08388 to D. H. Janzen; and DEB 81-13409 to G.C.S.) provided logistic and financial support for this work. Without the assistance and foresight of the Servicio Parques Nacionales of Costa Rica, and especially the help of Eliezer Arce, F. Cortés, and J. A. Salazar, this work would not have been possible.

LITERATURE CITED

Abele, L. G. 1974. Species diversity of decapod crustaceans in marine habitats. Ecology 55:156–161.

Anderson, S., and K. F. Koopman. 1981. Does interspecific competition limit the sizes of ranges of species? Am. Mus. Novit. 2716:1–10.

Arnold, S. J. 1972. Species densities of predators and their prey. Am. Nat. 106:220–236.

Barbour, C. D., and J. H. Brown. 1974. Fish species diversity in lakes. Am. Nat. 108:473–489.

Beaver, R. A. 1979. Host specificity of temperate and tropical animals. Nature (Lond.) 281:139–141.

Ben-Eliahu, M. N., and U. N. Safriel. 1982. A comparison between species diversities of polychaetes from tropical and temperate structurally similar rocky intertidal habitats. J. Biogeogr. 9: 371–390.

Brockman, C. F. 1968. Trees of North America; a field guide to the major native and introduced species north of Mexico. H. S. Zim, ed. Golden Press, New York.

Brodskij, A. K. 1959. Leden in der Tiefe des Polarbeckens. Naturwiss. Rundsch. 12:52–56.

Brown, J. H. 1973. Species diversity of seed-eating desert rodents in sand dune habitats. Ecology 54:775–787.

Brown, J. H., and D. W. Davidson. 1977. Competition between seed-eating rodents and ants in desert ecosystems. Science (Wash., D.C.) 196:880–882.

Brown, J. H., and A. C. Gibson. 1983. Biogeography. Mosby, St. Louis.

Brown, J. H., and A. Kodric-Brown. 1977. Turnover rates in insular biogeography: effect of immigration on extinction. Ecology 58:445–449.

Burt, W. H. 1964. A field guide to the mammals; field marks of all species found north of the Mexican boundary. Houghton Mifflin, Boston.

Cody, M. L. 1966. The consistency of intra- and inter-continental grassland bird species counts. Am. Nat. 100:371–376.

Colinvaux, P. A. 1973. Introduction to ecology. Wiley, New York.

Conant, R. 1958. A field guide to reptiles and amphibians of the United States and Canada east of the 100th meridian. Houghton Mifflin, Boston.

Connell, J. H. 1978. Diversity in tropical forests and coral reefs. Science (Wash., D.C.) 199:1302–1310.

Cook, R. E. 1969. Variation in species density of North American birds. Syst. Zool. 18:63–84.

Dement'ev, G. P., N. A. Gladkov, E. S. Ptushenkov, E. P. Spangenberg, and A. M. Sudilovskaya. 1951–1954. Birds of the Soviet Union (Ptitsy Sovetskogo Soyuza). Vols. I–VI. G. P. Dement'ev and N. A. Gladkov, eds. A. Birron and Z. S. Cole, transl. Z. S. Cole, ed. Israel Program for Scientific Translations, Jerusalem 1966. U.S. Department of Commerce, Clearinghouse for Federal Scientific and Technical Information, Springfield, Va.

Dobzhansky, T. 1950. Evolution in the tropics. Am. Sci. 38:209–221.

Dressler, R. L. 1981. The orchids: natural history and classification. Harvard University Press, Cambridge, Mass.

Emlen, J. M. 1973. Ecology: an evolutionary approach. Addison-Wesley, Reading, Mass.

Fischer, A. G. 1960. Latitudinal variations in organic diversity. Evolution 14:64–81.

Fleming, T. H. 1973. Numbers of mammal species in North and Central American forest communities. Ecology 54:555–563.

Glenn-Lewin, D. C. 1977. Species diversity in North American temperate forests. Vegetatio 33:153–162.

Grice, G. D., and A. D. Hart. 1962. The abundance, seasonal occurrence and distribution of the epizooplankton between New York and Bermuda. Ecol. Monogr. 32:287–309.

Hartmeyer, R. 1911. Tunicata. Chap. XVII. Die geographische Verbreitung. Pages 1498–1726 *in* H. G. Bronn, ed. Dr. H. G. Bronn's Klassen und Ordnungen des Tier-Reiches: wissenschaftlich dargestellt in Wort und Bild. III, Suppl. Winter, Leipzig.

Heck, K. L. 1979. Some determinants of the composition and abundance of motile macroinvertebrate species in tropical and temperate turtlegrass (*Thalassia testudinum*) meadows. J. Biogeogr. 6:183–200.

Holdridge, L. R., and L. J. Poveda A. 1975. Arboles de Costa Rica. Vol. I. Palmas, otras monocotiledóneas arboreas y arboles con hojas compuestas o lobuladas. Centro Científico Tropical, San José, Costa Rica.

Horn, M. H., and L. G. Allen. 1978. A distributional analysis of California coastal marine fishes. J. Biogeogr. 5:23–42.

Hubbell, S. P. 1979. Tree dispersion, abundance, and diversity in a tropical forest. Science (Wash., D.C.) 203:1299–1309.

———. 1980. Seed predation and the coexistence of tree species in tropical forests. Oikos 35:214–229.

Hubendick, B. 1962. Aspects on the diversity of the fresh-water fauna. Oikos 13:249–261.

Huey, R. B. 1978. Latitudinal pattern of between-altitude faunal similarity: mountains might be "higher" in the tropics. Am. Nat. 112:225–229.

Huston, M. 1979. A general hypothesis of species diversity. Am. Nat. 113:81–101.
Janzen, D. H. 1967. Why mountain passes are higher in the tropics. Am. Nat. 101:233–249.
———. 1981. The peak in North American ichneumonid species richness lies between 38° and 42° N. Ecology 62:532–537.
———. 1985. On ecological fitting. Oikos 45:308–310.
Järvinen, O. 1979. Geographical gradients of stability in European land bird communities. Oecologia (Berl.) 38:51–69.
Karr, J. R. 1971. Structure of avian communities in selected Panama and Illinois habitats. Ecol. Monogr. 41:207–233.
Karr, J. R., and R. R. Roth. 1971. Vegetation structure and avian diversity in several New World areas. Am. Nat. 105:423–435.
Kiester, A. R. 1971. Species density of North American amphibians and reptiles. Syst. Zool. 20: 127–137.
Krebs, C. J. 1978. Ecology, the experimental analysis of distribution and abundance. Harper & Row, New York.
Kusnezov, N. 1957. Numbers of species of ants in faunae of different latitudes. Evolution 11:298–299.
Lee, D. S., C. R. Gilbert, C. H. Hocutt, R. E. Jenkins, D. E. McAllister, and J. R. Stauffer, Jr. 1980 et seq. Atlas of North American freshwater fishes. Publ. North Carolina Biological Survey 1980-12. North Carolina State Museum of Natural History, Raleigh, N.C.
MacArthur, R. H. 1965. Patterns of species diversity. Biol. Rev. Camb. Philos. Soc. 40:510–533.
———. 1969. Patterns of communities in the tropics. Biol. J. Linn. Soc. 1:19–30.
———. 1972. Geographical ecology; patterns in the distribution of species. Harper & Row, New York.
MacDonald, K. B. 1969. Quantitative studies of salt marsh mollusc faunas from the North American Pacific Coast. Ecol. Monogr. 39:33–60.
McCoy, E. D., and E. F. Connor. 1980. Latitudinal gradients in the species diversity of North American mammals. Evolution 34:193–203.
Michener, C. D. 1979. Biogeography of the bees. Ann. Mo. Bot. Gard. 66:277–347.
Miller, R. R. 1958. Origin and affinities of the freshwater fish fauna of western North America. Pages 187–222 *in* C. L. Hubbs, ed. Zoogeography. Publ. 51. American Association for the Advancement of Science, Washington, D.C.
Monk, C. D. 1967. Tree species diversity in the eastern deciduous forest with particular reference to north central Florida. Am. Nat. 101:173–187.
Müller, M. J. 1982. Selected climatic data for a global set of standard stations for vegetation science. Junk, The Hague.
Owen, D. F., and J. Owen. 1974. Species diversity in temperate and tropical Ichneumonidae. Nature (Lond.) 249:583–584.
Paine, R. T. 1966. Food web complexity and species diversity. Am. Nat. 100:65–75.
Patrick, R. 1966. The Catherwood Foundation Peruvian Amazon expedition: limnological and systematic studies. Monogr. Acad. Nat. Sci. Phila. 14:1–495.
Pianka, E. R. 1966. Latitudinal gradients in species diversity: a review of concepts. Am. Nat. 100: 33–46.
———. 1967. On lizard species diversity: North American flatland deserts. Ecology 48:333–351.
———. 1978. Evolutionary ecology. 2d ed. Harper & Row, New York.
Pielou, E. C. 1979. Biogeography. Wiley, New York.
Rapoport, E. H. 1975. Areografía: estrategias geográficas de las especies. Fondo de Cultura Económica, Mexico City.
———. 1982. Areography: geographical strategies of species. 1st English ed. B. Drausal, transl. Publ. Fundación Bariloche. Vol. 1. Pergamon, New York.
Rehder, H. A. 1981. The Audubon Society field guide to North American seashells. Knopf, New York.
Ricklefs, R. E. 1979. Ecology. 2d ed. Chiron, New York.
Ricklefs, R. E., and K. O'Rourke. 1975. Aspect diversity in moths: a temperate-tropical comparison. Evolution 29:313–324.
Rogers, J. S. 1976. Species density and taxonomic diversity of Texas amphibians and reptiles. Syst. Zool. 25:26–40.
Rosenzweig, M. L. 1975. On continental steady states of species diversity. Pages 121–140 *in* M. L.

Cody and J. M. Diamond, eds. Ecology and evolution of communities. Belknap Press of Harvard University Press, Cambridge, Mass.

Sanders, H. L. 1968. Marine benthic diversity: a comparative study. Am. Nat. 102:243–282.

Schall, J. J., and E. R. Pianka. 1978. Geographical trends in numbers of species. Science (Wash., D.C.) 201:679–686.

Schreiber, H. 1978. Dispersal centres of Sphingidae (Lepidoptera) in the Neotropical region. Biogeographica 10. Junk, The Hague.

Scriber, J. M. 1973. Latitudinal gradients in larval feeding specialization of the world Papilionidae (Lepidoptera). Psyche (Camb., Mass.) 80:355–373.

———. 1984. Larval foodplant utilization by the world Papilionidae (Lep.): latitudinal gradients reappraised. Tokurana (Acta Rhopalocerologica) Mishima-shi, Japan, 6/7:1–50.

Shmida, A., and M. V. Wilson. 1985. Biological determinants of species diversity. J. Biogeogr. 12: 1–20.

Silvertown, J. 1985. History of a latitudinal diversity gradient: woody plants in Europe 13,000–1000 years B.P. J. Biogeogr. 12:519–525.

Simpson, G. G. 1964. Species density of North American Recent mammals. Syst. Zool. 13:57–73.

Spight, T. M. 1977. Diversity of shallow-water gastropod communities on temperate and tropical beaches. Am. Nat. 111:1077–1097.

Stanton, N. L. 1979. Patterns of species diversity in temperate and tropical litter mites. Ecology 60:295–304.

Stehli, F. G., and J. W. Wells. 1971. Diversity and age patterns in hermatypic corals. Syst. Zool. 20:115–126.

Stehli, F. G., R. G. Douglas, and N. D. Newell. 1969. Generation and maintenance of gradients in taxonomic diversity. Science (Wash., D.C.) 164:947–949.

Stout, J., and J. Vandermeer. 1975. Comparison of species richness for stream-inhabiting insects in tropical and mid-latitude streams. Am. Nat. 109:263–280.

Teraguchi, S., J. Stenzel, J. Sedlacek, and R. Deininger. 1981. Arthropod-grass communities: comparison of communities in Ohio and Alaska. J. Biogeogr. 8:53–65.

Terborgh, J. 1985. The vertical component of plant species diversity in temperate and tropical forests. Am. Nat. 126:760–776.

Thorson, G. 1951. Zur jetzigen Lage der marinen Bodetier-Ökologie. Verh. Dtsch. Zool. Ges. 1951. Zool. Anz. Suppl. 16:276–327.

Tillyard, R. J. 1917. The biology of dragonflies. Cambridge University Press, Cambridge.

Tramer, E. J. 1974. On latitudinal gradients in avian diversity. Condor 76:123–130.

Turner, J. T. 1981. Latitudinal patterns of calanoid and cyclopoid copepod diversity in estuarine waters of eastern North America. J. Biogeogr. 8:369–382.

Wallace, A. R. 1878. Tropical nature and other essays. Macmillan, London.

Wells, J. W. 1955. A survey of the distribution of reef coral genera in the Great Barrier Reef region. Pages 21–29 *in* Reports of the Great Barrier Reef Committee. IV, Pt. 2. Brisbane, Queensland, Australia.

Whittaker, R. H. 1969. Evolution of diversity in plant communities. Pages 178–196 *in* G. M. Woodwell and H. H. Smith, eds. Diversity and stability in ecological systems. Brookhaven National Laboratory, Upton, N.Y.

Williams, C. B. 1964. Patterns in the balance of nature and related problems in quantitative ecology. Academic Press, New York.

Wilson, J. W., III. 1974. Analytical zoogeography of North American mammals. Evolution 28: 124–140.

METHODOLOGICAL ADVANCES

Edited by John L. Gittleman

Any science is only as solid as its methodological foundation. So true is this that, in considering pioneering papers on macroecological methodology, one finds that most of the themes that were of influence historically have remained so in the current literature. Recent reviews and critiques focus on conceptual, or null-model, approaches, data availability, the geographical and taxonomic scale of sampling, statistical analysis, and phylogenetic correlation (e.g., Brown 1995; Blackburn and Gaston 1998; Gaston and Blackburn 1999). These are recurrent matters in the papers gathered in this section that form the classical foundation for macroecology.

Perhaps the most vexing methodological problem is how to conceptually tackle a macroecological problem. Consensus holds that a null model of expectation should be formulated to test against the hypothesis being examined (R. K. Colwell and D. W. Winkler). This not only encourages a more rigorous statistical test (E. F. Connor and E. D. McCoy), but the formulated hypothesis will then be objective, quantitative, and repeatable (J. A. Wolfe). For example, when a conceptual approach for species-abundance patterns involves a null model with a functional relationship of given slope and intercept, this will discourage any correlation verifying the hypothesis in a post hoc manner; the use of competing structural statistics for slope and intercept calculation, such as major axis versus regression (see below), will also tighten the null expectation and resultant test (Connor and McCoy; P. H. Harvey and G. M. Mace). Development of a conceptual, or null, model may also involve taxonomic or phylogenetic structure in the data: if it is assumed that related species are more similar to one another ecologically, then a null test may be developed to see whether a given macroecological hypothesis can be tested against this null phylogenetic relationship (Harvey 1996).

One other reason suggests the importance of a null model. The logic of hypothesis testing often goes both ways in terms of a pattern verifying a prediction and a hypothesis agreeing with the data (Connor and McCoy). A null model at the outset, such as an explicit power function in a regression model, will develop a logical flow from hypothesis generation to test, helping to break this circularity. However, a word of caution is necessary: when a rigorous, large-scale analysis generates a new hypothesis, either of a general form where the causal mechanism is unclear or of a specific form where variables are causally understood, then macroecological study is making valuable, steady scientific progress that will produce a priori tests (Harvey and Mace).

Because macroecological approaches necessarily incorporate big diverse databases, data quality and quantity are at the heart of the science (Harvey and Pagel 1989). First and foremost, it is critical that variables are well defined, easily replicated across studies, and quantified as best as possible, so that information can be used in repeated studies and added to by future data collection (Clutton-Brock and Harvey 1977). Statistical description of variables can reveal biases that result from intrinsic properties of measurement and/or taxonomic spread.

For example, life-history studies of mammals reveal that gestation lengths of mammals surprisingly cluster around three, six, or nine months. This obviously has little to do with natural variance in gestation lengths of mammals but does have a lot to do with rounding error from days to months (Gittleman 1988). Clearly, much of the power and progress of macroecology is due to the brute force of having lots of data. Beginning in the 1970s many ecological and organismal journals began insisting on the publication of data sets within papers or the placment of data on file in some repository (J. A. WOLFE); the *American Naturalist* often filed databases at the University of Michigan. This trend has now culminated in the fast-paced, comprehensive informatics area. Today, it seems surprising, almost quaint, that there was an insistence on variables being measured, quantified, and published (Clutton-Brock and Harvey 1977; Wolfe). It was indeed novel that a macroecological, comparative study could be carried out on quantitative measures of behavioral, ecological, physiological, and evolutionary variables, and there was resistance that this would mask true patterns or ignore intrinsic variance within species (e.g., Jolly 1972; Clutton-Brock 1974). Such skepticism changed with the accumulation of data and the success of cross-taxonomic studies, giving a greater identity to the macroecological field as a whole.

All of the papers chosen here describe problems of sampling and taxonomic representation. Sampling is such a critical issue that a subfield of study (rarefaction) emerged out of a concern for ensuring that local populations are well represented to reflect global measures (Wolfe), that sampling of geographic scale (and in paleobiological analysis, timescale) is widespread (D. M. RAUP), and that taxonomic coverage is reasonably complete to show a general trend for the lineage being studied (Clutton-Brock and Harvey; Harvey and Mace; Connor and McCoy; Raup). The general message from these early papers is that modeling or empirical analysis must show prior to hypothesis testing to what extent, if any, sampling bias influences the observed macroecological patterns. This could be carried out either by using measures of sampling effort as an additional hypothesis (Connor and McCoy) or by modeling the effect of sampling by taking the residuals of sampling on the target variable (e.g., species number, range size) and testing with the relative variable (Harvey and Mace).

Once a hypothesis is formed, ideally with a sound null model underlying it, and a robust comparative database is gathered, the next step is to select appropriate statistical methods for testing. Macroecological studies typically involve bivariate or multivariate relations, with the former being more commonplace both because of the data available and macroecologists' inherent ability to develop clear predictive multivariate tests. Thus bivariate relations especially involving scaling patterns have focused on statistical problems of underlying structural relations and establishing lines of best fit. Although it was clear in the biostatistics literature, the distinction between the strength (i.e., correlation) and form (i.e., estimates of best-fit lines) of a relationship was a critical point in these early methodological guides (Harvey and Mace; Connor and McCoy). The correlation coefficient often may indicate the statistical power of a trend or relationship—possibly even tipping the decision of having a high profile paper or not—but this should not be confused in calculating the line of best fit. Even though classic papers (e.g., Jolicoeur 1968) showed that regression was not the most suitable statistic for producing a line of best fit when variables are measured with error, regression has continued as the model of choice, largely because it is easier to calculate than major axis or reduced major axis; in addition, scales or measurement such as with log transformation produces more consistent results with a regression model. The key methodological point from these papers is that with a bivariate relationship the final conclusion often rests with how the line of best fit was calculated, yet the statistic is quite vulnerable to a host of problems intrinsic to the variables of study. In addition, even though a correlation coefficient and slope are independent statistical measures, a very high correlation will often designate that selection of a method for establishing a line of best fit is moot. These are easily understood guidelines but difficult to always put into practice.

The final but by no means least important methodological theme is that a bias (or possible insight) may emerge when considering the taxonomic or phylogenetic structure of the data in a macroecological study. Assuming that a taxonomy is stable and a phylogeny reliable (Wolfe, J. Felsenstein), one can use the similarity of most ecological characteristics among related species as a predicted null model for analyzing trends (Connor and McCoy), even though the underlying causation for such phylogenetic "constraints" or "inertia" is uncertain (Felsenstein, Harvey and Mace). Considering taxonomy or phylogeny in a macroecological study has gone through many phases (see Harvey and Pagel 1991), beginning with doing nothing and simply treating species points as independent, to considering the variance across taxa by using a nested analysis of variance, to calculating statistically independent points (from a given evolutionary model) at each node in a phylogeny. Certainly, the watershed paper was by J. Felsenstein, who most clearly described the statistical problem of independent data points, why this revealed the importance of a phylogeny, and how to develop a methodological solution based on the hierarchical structure of phylogenetic inheritance. However, at the time, implementation was almost impossible because of lack of available phylogenies. Indeed, Felsenstein's paper was viewed by some as "rather nihilistic" (Felsenstein, p. 14) because accurate and/or complete phylogenies were so rare. Nevertheless, many positive activities in systematics, informatic databasing, and comparative statistical methods were jump-started by Felsenstein's theoretical approach. Phylogenetic information is now accumulating at a rapid pace. Further, an insistence on looking at the phylogenetic structure of macrodata encouraged many to diagnose the variance of taxonomic differences prior to testing any macrohypothesis (Gittleman and Kot 1981). As with other approaches that integrate new methodologies, data, and scientific fields (e.g., macroecology and evolution), the application of phylogenetic comparative methods was increasingly an accepted protocol that almost any analysis above the species level considered. The request to transform data, even when there might be little if any phylogenetic structure in ecological traits, became extreme and certainly did not agree with the original problem described by Felsenstein (see Gittleman and Luh 1992; Losos 1999). Today, there is a more balanced perspective for both diagnosing the original problem and using multiple evolutionary models and phylogenies for statistically satisfying tests.

The papers included in this section reveal an overarching concern from macroecology's inception about data quality and availability, null models, and statistical rigor. The future will undoubtedly progress for finer resolution with these issues while also developing greater speed, analytical capacity, global databasing, and predictive macromodels. Macroecological methodologies will be of increasing sophistication as global problems and solutions become more critical in a global scientific world.

Literature Cited

Blackburn, T. M., and K. J. Gaston. 1998. Some methodological issues in macroecology. *American Naturalist* 151:68–83.

Brown, J. H. 1995. *Macroecology*. University of Chicago Press, Chicago.

Clutton-Brock, T. H. 1974. Primate social organization and ecology. *Nature* 250:539–42.

Clutton-Brock, T. H., and P. H. Harvey. 1977. Primate ecology and social organization. *Journal of Zoology* 183:1–39.

Gaston, K. J., and T. M. Blackburn. 1999. A critique for macroecology. *Oikos* 84:353–68.
Gittleman, J. L. 1988. The comparative approach in ethology: Aims and limitations. In *Perspectives ethology*, edited by P. P. G. Bateson and P. H. Klopfer, 8:55–83. Plenum Press, New York.
Gittleman, J. L., and M. Kot. 1981. Adaptation: Statistics and a null model for testing phylogenetic effects. *Systematic Zoology* 39:227–41.
Gittleman, J. L., and H. Kwang-Luh. 1992. On comparing comparative methods. *Annual Review of Ecology and Systematics* 23:383–404.
Harvey, P. H. 1996. Phylogenies for ecologists. *Journal of Animal Ecology* 65:255–62.
Harvey, P. H., and M. D. Pagel. 1989. Comparative studies in evolutionary ecology: Using the data base. In *Toward a more exact ecology*, edited by P. J. Grubb and J. B. Whittaker, 209–27. Blackwell, Oxford.
———. 1991. *The comparative method in evolutionary biology*. Oxford University Press, Oxford.
Jolicoeur, P. 1968. Interval estimation of the slope of the major axis of a bivariate normal distribution in the case of a small sample. *Biometrics* 24: 679–82.
Jolly, A. 1972. *The evolution of primate behavior*. Macmillan, New York.
Losos, J. B. 1999. Uncertainty in the reconstruction of ancestral character states and limitations on the use of phylogenetic comparative methods. *Animal Behaviour* 58:1319–24.

PAPER 41

Tertiary Climatic Fluctuations and Methods of Analysis of Tertiary Floras (1971)

J. A. Wolfe

Commentary

SCOTT L. WING

This paper was part of a pitched battle over the paleoclimatic interpretation of Cenozoic fossil floras. In the 1960s Jack Wolfe and coauthors had begun reconstructing paleotemperatures using leaf-margin analysis, a technique relying on the strong positive correlation in living vegetation between mean annual temperature and the proportion of dicot species with smooth, or entire-margined, leaves. Although the correlation of leaf-margin state and mean annual temperature had been observed by E. W. Sinnott and I. W. Bailey in the early 20th century, Wolfe's use of the method broke from a long tradition of inferring paleoclimate from the climatic preferences of living relatives of fossil taxa, a method practiced extensively by D. I. Axelrod and others in the mid-20th century and before. In 1967 Wolfe and D. M. Hopkins concluded from leaf-margin analyses that middle and high-middle latitudes had experienced tropical climates during the warmest part of the Cenozoic, and that there was a major cooling event at the end of the Eocene (early Oligocene by then-current timescales). In 1969 Axelrod and H. P. Bailey published a long criticism of the Wolfe and Hopkins (1967) paper. The paper reprinted here was Wolfe's counterattack.

Wolfe's critique was incisive and highly influential. He showed that floristic interpretations of paleoclimate were, up to then, largely qualitative and strongly influenced by incorrect identifications of fossil plants. In contrast leaf-margin analysis was easily applied, quantitative, and insensitive to changing botanical identifications. Floristic inferences about paleoclimate began to wane from this date. This paper was also the first of Wolfe's compilations of the modern climatic distribution of vegetation and leaf features, which culminated in 1979 with a US Geological Survey Professional Paper documenting a correlation of leaf-margin percent with mean annual temperature that is still used in estimating paleotemperatures. Wolfe's defense of a mid-Cenozoic climatic cooling was also borne out by later studies showing rapid global cooling at the end of the Eocene in other terrestrial paleoclimate proxies and oxygen isotope analysis of planktic marine organisms. The cooling at the Eocene-Oligocene boundary is now equated with the onset of major glaciation on Antarctica, and recognized to have had global biogeographic effects through severing the Holarctic ranges of cold-intolerant organisms.

Literature Cited

Axelrod, D. I., and H. P. Bailey. 1969. Paleotemperature analysis of Tertiary floras. *Palaeogeography, Palaeoclimatology, and Palaeoecology* 6:163–95.

Wolfe, J. A., and D. M. Hopkins. 1967. Climatic changes recorded by Tertiary land floras in northwestern North America. *Symposium 11th Pacific Science Congress* 25:67–76.

Palaeogeography, Palaeoclimatology, Palaeoecology
Elsevier Publishing Company, Amsterdam – Printed in The Netherlands

TERTIARY CLIMATIC FLUCTUATIONS AND METHODS OF ANALYSIS OF TERTIARY FLORAS

JACK A. WOLFE

U.S. Geological Survey, Menlo Park, Calif. (U.S.A.)
Museum of Paleontology, University of California, Berkeley, Calif. (U.S.A.)

(Received June 9, 1970)

ABSTRACT

On theoretical grounds, an analysis of the physiognomy of a Tertiary leaf assemblage is more direct and reliable than a circuitous floristic analysis in assigning thermal regimes to fossil assemblages. Using primarily foliar physiognomy and secondarily floristic composition, it can be shown that: (*1*) some middle latitude Tertiary assemblages probably lived under meteorologically tropical climates; (*2*) a major and rapid climatic deterioration occurred in the Oligocene; and (*3*) a major climatic fluctuation probably occurred in the Late Eocene.

These analyses thus substantiate the conclusions of several other paleobotanists regarding climatic fluctuations. Recent criticisms of these analyses are shown to be invalid and to be based largely on misinterpretations.

INTRODUCTION

Previous studies of the climatic implications of Tertiary plant assemblages from North America have typically agreed that major climatic fluctuations occurred during the Tertiary (e.g., DORF, 1959, 1963; WOLFE and HOPKINS, 1967). These studies have recently been criticized for a purported failure to consider floras that are: (*1*) documented by published systematic work; (*2*) in stratigraphic superposition; and (*3*) confined to a limited area; an additional criticism is "... the use of imprecise thermal criteria to interpret plants as temperature indicators" (AXELROD and BAILEY, 1969, p.165).

AXELROD (1966a, b) has attempted to demonstrate that many Paleogene assemblages from western North America are definitely not indicative of a tropical climate (mean of the cold month at least 18°C). He has thus suggested (1966b, p.45) that the differing aspects of such assemblages as the Clarno and the overlying Bridge Creek in Oregon represent only a moderate decline in temperature concommitant with moderate uplift. The same sequence, however, was interpreted by WOLFE and HOPKINS (1967) as a major climatic deterioration from a marginally tropical to a warm temperate climate.

The current report is concerned primarily with discussing how to arrive at the most accurate (i.e., the most valid) and precise (i.e., ability to reproduce) estimates of paleotemperatures for Tertiary assemblages, whether some middle latitude assemblages indicate tropical climates, and whether available evidence indicates major temperature fluctuations during the Tertiary. Secondarily the criticisms of WOLFE and HOPKINS' (1967) analysis will be examined for validity.

The thesis developed by WOLFE and HOPKINS (1967, p.67) is that the foliar physiognomy (particularly the leaf margin, i.e., entire vs. non-entire) of fossil assemblages is a sensitive index of the environment in which the assemblage lived and that this index "... is largely independent of the supraspecific taxonomy of fossil leaf species involved...". This concept is, of course, not new, and has, for example, been discussed by BAILEY and SINNOTT (1915, 1916), MACGINITIE (1941), and WOLFE and BARGHOORN (1960). WOLFE and HOPKINS (1967), however, applied this method of analysis to 52 fossil plant assemblages in the western United States. Many of the assemblages in local stratigraphic sections show significant changes in foliar physiognomy.

A DEFENSE OF WOLFE AND HOPKINS (1967)

A previous discussion of Tertiary climatic change (WOLFE and HOPKINS, 1967) has been particularly criticized by AXELROD and BAILEY (1969) on various grounds (Fig.1). These may be considered in turn:

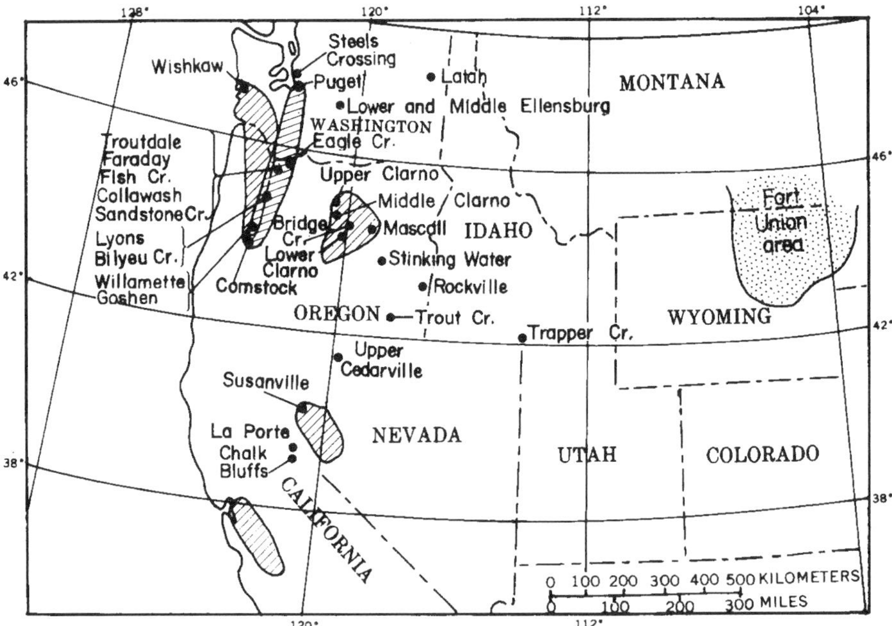

Fig.1. Location of some Tertiary plant assemblages in the western conterminous United States. Diagonally lined areas show local sequences of floras analyzed by AXELROD and BAILEY (1969). Horizontally lined area includes 21 assemblages analyzed by WOLFE and HOPKINS (1967).

(*1*) *Use of unpublished floras.* Twenty-eight (54%) of the floras analyzed by Wolfe and Hopkins were referenced to publications listed in the bibliography. Several of the others had also been published or have been published since. This documentation compares favorably with Axelrod and Bailey's analysis, in which eleven (46%) of the floras have been documented by publication. What is more significant, physiognomic analyses of modern vegetation (e.g., WEBB, 1959) are typically documented by the publication of statistical analyses (the procedure used by WOLFE and HOPKINS, 1967) and not by publication of the taxonomic treatment of the floristic components of the vegetation.

(*2*) *Lack of superposition.* Sixty percent of the assemblages analyzed by Wolfe and Hopkins were stated to be in stratigraphic successions, some of which (e.g., items 21, 26, 33, 42, 51 of WOLFE and HOPKINS, 1967, pp.70–71) are very localized.

(*3*) *Geographic factors.* The possibility of the differences between certain assemblages resulting largely from local environmental factors such as altitude or a coastal versus an interior location were fully appreciated by WOLFE and HOPKINS (1967). Where possible, coastal assemblages were compared with one another (e.g., items 42 and 51) and interior assemblages were compared with one another (e.g., items 44 and 50). Differences in altitude were similarly considered by Wolfe and Hopkins; indeed, the altitudinal gradation of Miocene floras suggested by AXELROD (1964) and AXELROD and BAILEY (1969) is a reflection not primarily of altitude but primarily of age, as indicated by age assignments independent of paleobotanical grounds (Fig.2; cf. WOLFE, 1969b, p.88).

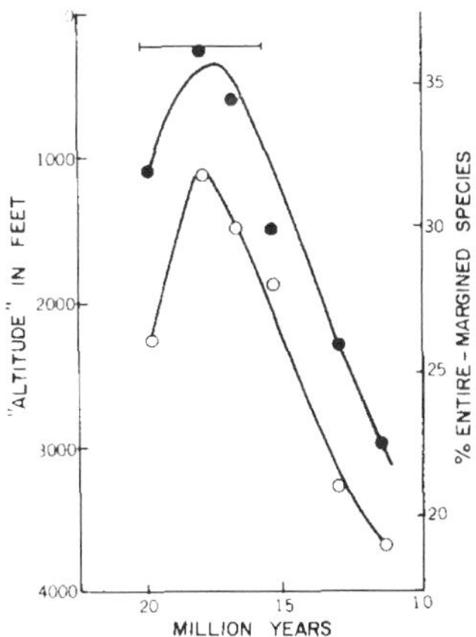

Fig.2. Comparison of "altitudinal" gradations with time. Altitudes (solid circles) are those suggested by AXELROD (1964) for the Upper Cedarville, Latah, "Succor Creek" (= Rockville), Mascall, Trout Creek, and Trapper Creek floras. Leaf-margin percentages (open circles) are those given by WOLFE and HOPKINS (1967) for the same floras. Ages and age ranges assigned are in accordance with available radiometric data, except for the Trapper Creek, the age of which is based on detailed geologic mapping, stratigraphy, mammals, diatoms, and fresh-water molluscs (MAPEL and HAIL, 1959).

(*4*) *Use of imprecise thermal criteria.* I do not agree with Axelrod and Bailey's conclusion that the distribution of *Picea* is not an important thermal criterion. A change to a spruce-dominated forest from a broad-leaved deciduous forest is largely controlled by summer heat, as expressed by the 21 °C July isotherm. Certainly the thermal requirements of the different species of a genus such as *Picea* are varied, but in North America *Picea* is not native to areas of high summer heat; this is shown by the "approximate coincidences" (WOLFE and LEOPOLD, 1967, p.200) of the southern boundary of various species of *Picea* with the 21 °C July isotherm. Exceptions to these coincidences are, as noted by WOLFE and LEOPOLD (1967, p.201), primarily in areas of low precipitation. Such areas obviously include coastal central California; there, *P. sitchensis* extends southward only to the Fort Ross area (ca. 1,180 mm annual precipitation) and does not occur, for example, near San Francisco (ca. 520 mm annual precipitation).

DETERMINATION OF PALEOCLIMATES

The basic disagreement between the conclusions of AXELROD and BAILEY (1969) and those of WOLFE and HOPKINS (1967) devolves on the respective methods utilized, both in theory and application. No disagreement exists on the widely accepted concept that climate is the major controlling factor in the distribution of vegetational types. The disagreement is almost entirely on how, in the analysis of an assemblage of fossil plants, a worker can determine the vegetational type represented by that assemblage and thence the climate.

Floristic analyses of climate

Floristic analyses attempt to define extant vegetation in terms of its floristic composition. Such a procedure has the inherent tendency of confusing the fundamentally different concepts of vegetation and flora. Although some vegetational terms incorporate taxonomic categories (e.g., oak–laurel forest), it is clear that the vegetational type is distinguished because of its physiognomic peculiarities and not because of its taxonomic composition. Oaks and laurels are, for example, associated in broad-leaved deciduous forests such as the "Mixed Mesophytic forest", but the term "oak–laurel forest" indicates a broad-leaved evergreen forest that has certain other unique physiognomic features. On the other hand, the forests of the Amazon lowlands and Malaya, despite their many taxonomic differences, both represent "Tropical Rain forest".

The approach used by AXELROD (1966b) is fundamentally qualitative. Various subjective weighting is given to the different genera composing an assemblage, and the assumption is made that the thermal requirements of most genera have remained unchanged during the Tertiary. Once the present distributions in extant vegetational types of the extant relatives of the species or genera represented are determined, the worker must determine what vegetational type is represented. Relying further on the distribution of taxa, the worker then must determine what particular part of the vegetational formation is represented, e.g., whether within the Broad-leaved Evergreen forest formation, a Tropical Rain forest, or an Evergreen Sclerophyllous Broad-leaved (oak-laurel) forest. Going even further, and still relying on the present distributions of the taxa, the worker must judge what particular part of the climatic range of a given vegetational type is represented. Once all this has been done, the worker can finally estimate an effective temperature *(ET)*, mean annual temperature *(T)*, and mean annual range of temperature *(A)* for the assemblage.

The assignment of purportedly precise fractional *ET*'s to fossil assemblages (AXELROD, 1966a, b) is of dubious value. MACGINITIE (1962, p.91) has aptly noted that "Reconstructions of Tertiary climates on the basis of comparisons with living plant associations and existing climates can hardly be more than approximations".

This method must clearly be based on valid determinations, and hence

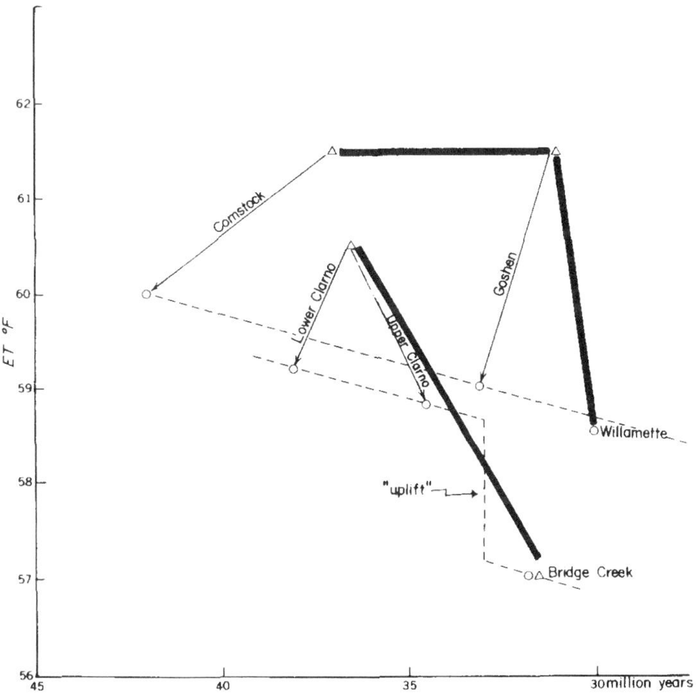

Fig.3. The migration of "*ET*'s" in value and time. Triangles represent values assigned by AXELROD (1966b, pp.41–45), and circles represent values assigned by AXELROD and BAILEY (1969, fig.13) for the same assemblages (Willamette not discussed previously by Axelrod). Solid thick lines represent climatic change if new estimates were not made (compare with WOLFE and HOPKINS, 1967, p.72), and broken lines represent climatic change postulated by AXELROD and BAILEY (1969). The arrows show direction of change of values.

the documentation of these determinations is significant. AXELROD (1964, p.122), for example, considers that the occurrence of *Persea* in the conifer-dominated, Miocene Trapper Creek flora of Idaho can be correlated with a high degree of temperateness (equability); the illustrated specimen (AXELROD, 1964, plate XIII, fig.12), however, has craspedodrome secondary veins and small teeth as in *Alnus* (*Persea* has camptodrome secondary veins and an entire margin). Thus, one of the fundamental bases for assigning a high "temperateness index" and a low effective temperature to the Trapper Creek assemblage is invalid.

The very basis for AXELROD and BAILEY's (1969) analysis of the climatic change during the Oligocene—their estimation of *ET*'s for various assemblages—is highly suspect. Aside from the data presented below that indicate that their estimates are probably inaccurate, the previous estimates for the same assemblages by AXELROD (1966a, b, 1968) are at strong variance with those presented by AXELROD and BAILEY (1969; see Fig.3). No reason has been given for these significant changes of opinion; note that if the estimates of AXELROD (1966a, b) were not revised, the data would indicate a considerable climatic deterioration during the Oligocene, which is what WOLFE and HOPKINS (1967) stated.

If *ET* estimates for the same assemblages can be so drastically revised, it is pertinent to ask not only what is the basis for the revision but also what was the original basis for the estimate. In only two instances has Axelrod stated what the basis was for assigning an *ET* to a lowland Paleogene assemblage: "The Goshen has a tropical aspect, yet the presence of *Phoebe*, *Alangium*, *Cercidiphyllum*, and *Platanus* suggests that it was not tropical (*ET* + 64.4°F) but warm temperate, that is *ET* 61.5°" (AXELROD, 1966b, p.42). Both *Phoebe* and *Alangium* are important constituents of the lowland tropical rain forest in the Philippine Islands (BROWN, 1919). *Platanus* is in fact known from meteorologically tropical forests near Vientiane, Laos (GAGNEPAIN, 1939), and is known from elevations as low as 200 m in eastern Mexico (MIRANDA and SHARP, 1950), which must also be meteorologically tropical. Of these four genera, only *Cercidiphyllum*—an obviously relictual monotypic genus—does not today enter the meteorological tropics. Similarly, regarding the Comstock AXELROD (1966b, p.42) stated: "Recognizing several taxonomic changes made by MacGinitie (1941), the following genera are now considered to be in the flora... The occurrence of *Cercidiphyllum*, *Chaetoptelea*, *Platanophyllum*, and *Koelreuteria* suggests that the area was warm temperate, probably with an *ET* near 61.5°F". In fact, no one (including MACGINITIE, 1941, p.84) has previously recognized *Chaetoptelea* in the Comstock assemblage, and the "*Koelreuteria*" was considered by MACGINITIE (1941, p.142) to represent *Rhus*. Even so, *Chaetoptelea* is a member of tropical lowland vegetation in Central America (KNAPP, 1965, p.275), and *Koelreuteria* occurs in "broad-leaved forest at low altitudes" on Taiwan (LI, 1963, p.497); this forest is in part meteorologically tropical. AXELROD (1966b, p.42) goes on to state: "This is consistent with the presence of other genera that occur chiefly in warm temperate rain forests (*Mag-*

nolia, Anona, Persea, Viburnum, Cryptocarya), not in the tropical lowlands". In fact, *Annona (= Anona)* and *Persea* are primarily lowland tropical in the new world (KNAPP, 1965, p.289). In the old world, *Persea* (*Machilus* of authors) is dominantly a low altitude, tropical genus (LI, 1963, pp.224–227), and *Cryptocarya* is also typically lowland tropical (see various lists in BROWN, 1919). *Magnolia* is typically not a lowland tropical genus, but some members of the genus are "primary components" of the lowland rain forest in meteorologically tropical Hainan (WANG, 1961, p.161). *Viburnum* is also dominantly not tropical, but species such as *V. odoratissimum* do occur in meteorologically tropical forests from the southern part of Taiwan through southeast Asia to India (LI, 1963, p.891). Thus all the genera (except for the relictual *Cercidiphyllum*) that AXELROD (1966b, p.42) considered to indicate less than tropical conditions for the Goshen and Comstock are known in meteorologically tropical forests today and several of the genera are dominantly tropical. If other *ET*'s are based on erroneous data and invalid interpretations such as these, the changing estimates of *ET*'s perhaps become understandable but their validity and precision remain suspect.

Estimates of mean annual ranges of temperature for Neogene and Paleogene assemblages are subject to considerable variation. Previously AXELROD (1966a, p.164) has suggested a mean annual range of temperature of about 20°F (or 15°F: AXELROD, 1964, p.59) for the lowland region of western Oregon during the Middle Miocene, but the analysis of AXELROD and BAILEY (1969, fig.12) leads them to estimate the Eocene mean annual range at 20°F and the Middle Miocene mean annual range at 25°F. The estimate for the Eocene has also been drastically revised, because previously AXELROD (1966a, p.161; 1966b, p.49) has suggested an annual range for the Comstock and other lowland assemblages of 10°F. The new estimate thus represents a change of 100%. Certainly the changes of *ET* and *A* values for the same assemblages far surpass the limits suggested by AXELROD (1966a, p.168): "Slightly different assumptions as to *ET* (i.e., 61° vs. 62°) or *A* values (10° vs. 12°) by different investigators are to be expected...". The fact that the same worker has not been able to reproduce precisely (or even approximately) various temperature estimates for the same assemblages indicates that the techniques and perhaps also the theories of analysis on which the estimates are based are invalid.

Not only can the same worker apparently not consistently reproduce the same estimates of *ET* and *A* for the same assemblages, the estimates of other workers do not always agree for the same assemblages. The Troutdale assemblage, for example, was thought by CHANEY (1944, p.334) to have lived under a climate approximating that of coastal Oregon today. Taking the data from a coastal station such as Astoria, Oregon, mean annual temperature is about 51°F and mean annual range of temperature is about 20°F. AXELROD and BAILEY (1969, fig.12), however, assign for those respective temperatures values of approximately 56°F and 23°F. Both sets of estimates are based on the same fundamental principles of vegetational analysis through floristic composition. Similarly, CHANEY (1959, p.58) suggests a mean annual temperature of about 62°F for the Mascall climate, but AXELROD and BAILEY (1969, fig.12) suggest a value of about 55°F; again, both estimates are based on modern distributions of taxa related to those in the fossil assemblage. AXELROD and BAILEY (1969, p.173) state that: "The precision of estimating T and A of a Tertiary flora will naturally be governed by adequate sampling, proper identification of taxa, and a thorough familiarity with related modern forests and the conditions under which they live". Inasmuch as the first two governing factors are not involved in the differences between the opinions of Chaney vs. the opinions of Axelrod and Bailey, the third and highly subjective factor—"thorough familiarity"—must be the basis for disagreement.

Physiognomic analyses of climate

TABLE I

PERCENTAGES OF SPECIES THAT HAVE ENTIRE-MARGINED LEAVES IN SOME MODERN FLORAS AND COMPARISONS WITH TEMPERATURE DATA

Flora	*Percent entire*	*M.A.T.*	*M.A.R.*	*Vegetation*
Malaya	86	28	1	Tropical Rain forest
Philippine Islands (200 m)	82	26	4	Tropical Rain forest
Ceylon (lowland)	81	27	2	Tropical Rain forest
Manilla	81	27	3	Tropical Rain forest
East Indies	77	26	3	Tropical Rain forest
Philippine Islands (450 m)	76	26	5	Tropical Rain forest
Hawaii (lowland)	75	24	4	Paratropical Rain forest
Ceylon (upland)	73			Submontane Rain forest
Philippine Islands (700 m)	72	24	1	Submontane Rain forest
Hong Kong	72	22	13	Paratropical Rain forest
Hainan (lowland)	70	24	11	Paratropical Rain forest
Philippine Islands (1,100 m)	69	23	2	Montane Rain forest
Taiwan (0–500 m)	61	21	11	Paratropical Rain forest
Hawaii (upland)	57	16	4	Montane Rain forest
Hainan (upland)	55			subtropical forest
Fukien (upland)	50	19	19	subtropical forest
North Kwangsi	49	19	12	subtropical forest
Taiwan (500–1,000 m)	47			subtropical forest
Taiwan (1,000–2,000 m)	41			subtropical forest
North Kiangsi	38	11	22	Mixed Mesophytic forest
South Anhwei	36			Mixed Mesophytic forest
North Chekiang	34	11	26	Mixed Mesophytic forest
East Szechuan-West Hupeh	30			Mixed Mesophytic forest
South Kiangsu	24			Mixed Mesophytic forest
Northern China Plain	22	11	30	deciduous oak forest
Shensi	22			deciduous oak forest
Manchuria	10	4	40	Mixed Northern Hardwood forest

M.A.T. = Mean annual temperature; M.A.R. = mean annual range of temperature. Leaf margin data after BAILEY and SINNOTT (1916) and BROWN (1919), or original compilations based on WANG (1961) and LI (1963).

The leaf-margin analysis used by WOLFE and HOPKINS (1967) is, on the other hand, fundamentally quantitative. Forty-six regional floras containing over 100,000 dicotyledonous species were analyzed by BAILEY and SINNOTT (1916) with respect to leaf margins, and, of course, many other data are available. The correlation between leaf-margin percentages (i.e., the percentages of species that have entire-margined leaves) and vegetational types is striking (Table I). There is clearly some adaptive significance between the type of leaf margin and climate, although the exact physiological relationships are not known.

Neobotanists concerned with the relationship between environment and physiognomy have also recognized the importance of analyzing Tertiary floras *independent of taxonomy*: "... conclusions as to the climates of Tertiary floras are perhaps more firmly based when they are drawn from a statistical study of leaf sizes and similar features... than when ... they ... rest on the *taxonomic affinities* of the fossil flora rather than its physiognomy, which appears, at least as far as modern vegetation is concerned, to be a very sensitive index of environmental conditions" (RICHARDS, 1952, p.154).

The independence of taxonomy and foliar physiognomy is important. For example, species described by AXELROD (1956) as *Salix* and *Comptonia* were referred by WOLFE (1964) to, respectively, the rosaceous *Peraphyllum* and *Lyonothamnus* (these changes were accepted by AXELROD, 1967, p.280). The ecologic significance of *Salix* vs. *Peraphyllum* and *Comptonia* vs. *Lyonothamnus* is, of course, great. No matter, however, to which families the species are validly referred, the climatic analysis based on foliar physiognomy remains unchanged. I agree that valid determinations are necessary in a vegetational analysis based on floristic composition such as those of AXELROD (1956, 1966b), but such supraspecific determinations are irrelevant to physiognomic analyses.

The reproducibility of leaf-margin percentages by various workers is significant. For example, CHANEY (1959) listed 50 dicotyledonous leaf species from the Mascall, of which 26% are entire-margined. In revising various Miocene assemblages from Alaska and the Oregon Cascades, I have also revised some species (synonymizing some, splitting others) in the Mascall and the resulting leaf-margin percentage is 28% (WOLFE and HOPKINS, 1967). Leaf-margin percentages for fossil assemblages are thus subject to some variation. This arises from two main factors: (*1*) small numbers of species in some of the assemblages; and (*2*) variation in the systematic treatment of species, i.e., whether certain species are combined or maintained as separate entities. The second factor has a significant impact on small assemblages but almost no effect on large assemblages (30 or more species). The Late Miocene Faraday assemblage, for example, was originally stated by WOLFE and BARGHOORN (1960) to have 27 leaf species, of which 22% were entire-margined. Further collecting and analysis of species already known in this assemblage resulted in the recognition of 30 species, of which only 13% were entire-margined (WOLFE and HOPKINS, 1967). Further work now indicates that

32 species are present, of which 16% are entire-margined. Note, however, that all these percentages fall within the range of the northern broad-leaved deciduous forests (Table I), so the concept of the type of vegetation represented by the Faraday assemblage has not changed. In some instances, small assemblages may yield results that are substantiated by further work.

An assemblage of 30 or more species appears through experience to be a highly reliable statistical base, i.e., the leaf-margin percentage can be reproduced to within a few percent despite the collection of representatives of new species or the revision of the original species. For the purposes of obtaining a reliable leaf-margin percentage, the number of *specimens* is of little significance; it is the number of *species* represented that is significant. BAILEY and SINNOTT (1916) noted that the preponderance of entire-margined leaves in tropical forests is even greater than the preponderance of entire-margined species, and consequently it might be assumed that a count of leaves that had entire vs. non-entire margins in a fossil assemblage might yield a better datum. The over-representation of probable stream-side types (MACGINITIE, 1953, p.46) in fossil assemblages, coupled with the over-representation of non-entire-margined species in stream-side vegetation (WOLFE, 1969a, pp.37–39), indicate that specimen counts could yield highly unreliable results in vegetational interpretations. The abundance of leaves of a particular species in deposits has, in fact, not been shown to have a significant correlation with the representation of organisms (or the more significant basal area) of that species in the vegetation (WOLFE, 1969b, pp.86–87). The term "adequate sample" must be defined not in relation to the number of leaves or other megafossils recovered but in terms of the number of species represented. I suggest from experience that percentages based on 30 or more species are probably reproducible with further collecting or revision to ± 5%, percentages based on 20–29 species are probably reproducible to ± 10%, and percentages based on less than 20 species should be regarded as highly tentative.

I emphasize that the total woody dicotyledonous flora (excluding water plants) must be included in the computations; the list for the Latah flora (CHANEY, 1959) was based largely on the published work of other authors, but examination of the U.S. Geological Survey collections indicated that these authors had not described many other species, most of which had entire margins. In the original description of the Latah flora, KNOWLTON (1926) described 57 leaf species, of which 35% have entire margins; this percentage is remarkably close to the 32% published by WOLFE and HOPKINS (1967) and based on 70 species. Workers who followed Knowlton apparently concentrated on publishing the more characteristic non-entire-margined species, and this led to an invalid interpretation of the Latah by WOLFE and BARGHOORN (1960).

Individual assemblages conceivably could have leaf-margin percentages that do not accurately reflect that of the regional vegetation. This situation arise from the concentration of non-entire-margined species in stream- and lake-side vegetation relative to the slope or interfluve vegetation. It is possible—particularly in regard to small assemblages—that stream-side types could be over-represented in the flora, and hence the leaf-margin percentage should be considered as a minimal figure. In the instances of large floras, however, the leaf-margin percentage is assumed to reflect both the stream-side and slope vegetation. That this may be so is indicated by the ten Late Miocene floras from the Pacific Northwest used by WOLFE and HOPKINS (1967). The percentages ranged from 13–24%; if these floras are combined and a more reliable statistical base of slightly over 100 species is used, the leaf-margin percentage is 21%. This figure can probably be taken as adequately reflecting the leaf-margin percentage for the regional flora during the Late Miocene. Seven of the ten floras have lower percentages, two are also 21%, and only one is slightly higher. Interestingly, the two lowland floras (Faraday and Weyerhauser) have percentages below 21%, as do some upland floras (Hidden Lake, Trapper Creek, and Thorn Creek).

The claim is made (AXELROD and BAILEY, 1969, p.181) that the entire margin is the result of a lack of frost and is not directly related to the level of heat. It is, therefore, suggested that areas of high equability have a high percentage of entire-margined species.

The vegetation of middle elevations (500–2,000 m) on Taiwan is an example of a broad-leaved evergreen (oak–laurel) forest that lives under a highly equable climate. LI (1963) gives altitudinal ranges for most species, and based on his work I (WOLFE, 1969a) compiled leaf-margin percentages for each 200 m increment. The percentages varied from 47 to 48% entire at lower elevations in the oak–laurel forest to a minimum of about 40% entire at higher elevations. Similar percentages, based on the work of WANG (1961), can be obtained for the upland broad-leaved evergreen forest of Yunnan. The highest percentages obtained for the oak–laurel forest are in areas of higher temperature or lower equability such as Fukien (about 50%) and upland Hainan (55%).

In truly tropical vegetation, a similar decrease in entire-margined leaves has been noted as temperature decreases but equability increases. BROWN's (1919) study of the vegetation of the Philippines indicates that at 200 m (a mean annual range of 4°C) the percentage is 82 and at 1,100 m (a mean annual range of 2°C) the percentage is 69. BAILEY and SINNOTT (1916, pp.27–28) have previously demonstrated that the more equable tropical and subtropical uplands have lower percentages of entire-margined species than do the less equable and hotter lowlands. Thus, it is indeed temperature and not solely equability that is related to leaf characteristics such as the entire margin.

Temperate forests in the Southern Hemisphere (e.g., New Zealand, Australia) have a high percentage of entire-margined species, most of which are microphyllous. BAILEY and SINNOTT (1916) suggested that this abnormally high proportion of entire-margined species was due to historical factors, i.e., the lack of large land areas of a cold temperate mesophytic climate that would have produced a deciduous forest; indeed, deciduous species are rare in the temperate rain forests of Australia. A parallel adaptation of evergreen species in Northern Hemisphere temperate climates can be demonstrated. In lowland western Oregon and Washington, for example, 70% of the evergreen woody species are entire-margined as opposed to 22% of the deciduous woody species. The few notophyllous or mesophyllous species (e.g., *Acer*) are primarily deciduous, as opposed to the several nanophyllous or microphyllous species of evergreen Ericaceae. It is apparent that leaf-margin data from Tertiary floras in the Northern Hemisphere—floras that contain deciduous plants—should not be compared to the leaf-margin percentages obtained for Southern Hemisphere temperate vegetation unless probable deciduous species are omitted from the comparative data.

An analysis of leaf-margin data in terms of deciduous vs. evergreen can be useful in the interpretation of a fossil assemblage. The Late Miocene (Homerian) flora of the Cook Inlet region of Alaska (latitude 59°–63°N) contains almost 60 dicotyledonous leaf species (WOLFE, 1966). The leaf-margin analysis indicates that 35% are entire-margined; such a percentage could indicate either a warm temperate Mixed Mesophytic forest (24–38%) entire or a forest similar to that of the Cascade Range in the Pacific Northwest (35%; WOLFE, 1969a, p.38). The probable deciduous species (largely members of Salicaceae, Betulaceae, and Rosaceae), however, have a leaf-margin percentage of about 22, i.e., similar to the extant deciduous vegetation of the Pacific Northwest. Floristic comparisons (WOLFE, 1969b) also indicate such a relationship.

Leaf-margin analyses cannot, of course, be the sole criterion for determination of past climates. Other features, e.g., leaf size and the presence or absence of drip tips (WOLFE and HOPKINS, 1967, p.67), are useful but are difficult to quantify in regard to fossil assemblages. The difficulty stems primarily from selectivity of the depositional environment; large leaves obviously will, in many environments, tend to be fragmented by turbulent currents and hence be under-represented. Another problem is that fossil assemblages may contain an over-representation of streamside plants (MACGINITIE, 1953, p.46), and thus stenophyllous plants that typically fall into the low size-classes (RICHARDS, 1952) may be over-represented. Such over-representation would yield an analysis indicating a cooler climate in the instance of a leaf-size analysis, as it may also in the instance of the leaf-margin analysis (WOLFE, 1969a, pp.37–39). The distribution of leaf-size classes in a fossil assemblage is a valuable adjunct to the leaf-margin analysis but can hardly be considered as free from selectivity of the depositional environment as can the leaf-margin analysis.

One point that must be stressed is that although leaf-margin percentages offer a significant datum on magnitude of climatic change, they cannot be directly translated into temperature regimes. The percentages, along with other physiognomic features of leaves and wood, allow a correlation to be made with extant vegetational types. The floristic analysis can, in a general way, serve as a check on such correlations. Then an approximate temperature regime can be assigned based on the regimes of similar extant vegetation. Whether, however, two workers can agree on the precise temperature regime that a fossil assemblage lived under (based on physiognomic criteria) is for the purposes of climatic fluctuations irrelevant; the leaf-margin data in themselves give an idea of the magnitude of climatic change.

Basically foliar physiognomy is a direct reflection of the vegetation and hence the climate. Inferences based on the thermal requirements of extant taxa are avoided. It is theoretically impossible for all lineages to have maintained the same thermal requirements through time (MACGINITIE, 1962, p.91; WOLFE, 1969b, p.85). Temperate lineages, which are relied on by AXELROD (1966b) to indicate cool climates, are the lineages most likely to have changed their thermal requirements if, as comparative morphology and anatomy indicate, the temperate species and genera are more advanced than and hence probably derived from tropical species and genera. I concur with VAN STEENIS (1962, p.290) that there is no reason to assume that the temperature requirements of the majority of tropical genera have changed significantly during the Tertiary, but theoretically the temperature requirements of *now temperate* lineages must have changed. With these concepts in mind, I suggest that physiognomic analyses of Tertiary leaf assemblages—particularly those of Paleogene age—are more reliable than the floristic analyses used by AXELROD (1966b).

ANALYSES OF SOME PALEOGENE FLORAS

Leaf-margin percentages in the range of 60–75 are characteristic of a number of Paleogene floras in western North America (Steels Crossing, Kushtaka, Upper Clarno, Comstock). Preliminary size measurements (Table II) indicate that the assemblages (except for the John Day Gulch—see below) were dominantly in the notophyll to mesophyll size-classes. What were the climatic conditions that produced vegetation of this type?

In regard to the leaf-margin percentages, several extant vegetational types have leaf-margin percentages in the 60–75% range: Paratropical Rain forest (the vegetation of lowland Taiwan and Hainan; see WOLFE, 1969a, p.40), Submontane Rain forest, and Montane Rain forest ("mossy forest"). The two climatic features that characterize the climates for these vegetational types are: (*1*) a typical lack of frost; and (*2*) ample precipitation during all parts of the year. The Submontane Rain forest is typically truly tropical, i.e., the mean of the coldest month is at least 18 °C, and both the Montane and Paratropical Rain forests are, in places, also truly tropical.

In leaf size, however, these Paleogene assemblages cannot represent small-leaved Montane Rain forest, but are instead allied to the large-leaved Submontane and Paratropical Rain forests. Both these forests live under a mean annual temperature of about 20°–25 °C, whereas the dissimilar Montane Rain forest has a

TABLE II

LEAF-MARGIN PERCENTAGES AND DISTRIBUTION OF LEAF SIZE-CLASSES IN SOME PALEOGENE FLORAS IN THE WESTERN UNITED STATES

Flora	*No. of spp.*	*Percent of species that are:*			
		entire-margined	*microphyllous*	*notophyllous*	*mesophyllous*
Middle Oligocene					
Bilyeu Creek, Oregon	37	57	14	51	35
Goshen, Oregon	49	55	7	64	29
Early Oligocene					
La Porte, California	41	67	36	45	19
Upper Clarno, Oregon	48	68	25	45	30
Comstock, Oregon	27	67	14	59	27
Later Eocene					
Middle Clarno, Oregon	33	40	60	28	12
Cashman (Puget Group, loc. 9731), Washington	41	62	10	56	34
Kushtaka, Alaska	62	65	16	50	34
Steels Crossing, Washington	36	75	11	39	50
Lower Clarno, Oregon	55	69	not analyzed		
Susanville, California	22	68	14	36	50
Earlier Eocene					
Chalk Bluffs, California	71	45	20	58	22

lower mean annual temperature. The foliar physiognomy of these Paleogene assemblages thus indicates: (*1*) abundant precipitation throughout the year; (*2*) a typical lack of frost; and (*3*) a mean annual temperature of 20°–25°C.

The floristic composition of these Paleogene assemblages corroborates in a general way the physiognomic analysis. Several families and genera that are now dominantly tropical occur in these assemblages; these same genera and families, although extending today into the lowland paratropical regions (a high mean annual temperature), do not extend into the Montane Rain forest of tropical mountains (a low mean annual temperature but a frostless climate). The Upper Clarno assemblage, for example, contains *Olax*, *Erythropalum*, *Tinomiscium*, *Tinospora*, *Dracontomelon*, and *Iodes* (R. A. Scott in CHANDLER, 1964, p.58). If, as AXELROD (1964, p.58) suggests, such genera were able to exist under low heat levels in a frostless climate, why do they not do so today and enter the Montane Rain forest? Many of the "temperate" genera that occur in these same Paleogene assemblages are today either clearly relictual (*Platycarya*, *Cercidiphyllum*, and *Euptelea*, for example), or actually do participate in Paratropical or Submontane Rain forests (*Alnus*, *Acer*, and *Carya*, for example).

The vegetation of upland Mexico (the *Tierra Templada*) has been compared to some Paleogene assemblages because of the admixture in this extant vegetation of many broad-leaved deciduous and evergreen plants. From the lists given by KNAPP (1965) of the dominants, the percentage of species that have entire-margined leaves is approximately 50, and is hence considerably different from assemblages such as the Steels Crossing, Lower and Upper Clarno, and La Porte. In specific areas of the *Tierra Templada* that receive regular frosts, such as in Tamaulipas (HERNANDEZ et al., 1951) and near Huauchinango, Puebla (MIRANDA and SHARP, 1950), the percentages range from 36–38.

An additional physiognomic characteristic that differentiates the lowland tropical and paratropical vegetation from montane forests such as those of the *Tierra Templada* is that in the lowland forests lianes are diverse and conspicuous (WANG, 1961, p.159; KNAPP, 1965, p.259) in contrast to the large epiphytic element in the upland vegetation (KNAPP, 1965, p.291). Leaves of lianes tend to conform to a peculiar morphologic pattern (RICHARDS, 1952, p.107), and applying this criterion to fossil leaves, it is apparent that the liane element in assemblages such as the Steels Crossing and isochronous assemblages in Alaska was notably diverse (WOLFE, 1969a, p.50). Menispermaceae, Icacinaceae, and Vitaceae are well represented by many genera in the Upper Clarno assemblage (R. A. Scott in CHANDLER, 1964, p.58), and these families contain mostly lianes today. Woods that show the obvious anatomical adaptations to the liane habit have also been found at the Upper Clarno locality (R. A. Scott, personal communication, October, 1969). The diversity of probable lianes in these floras is a strong indication that the vegetation represented was similar to Tropical and Paratropical Rain forests and not to upland "temperate" rain forests.

The climate under which some assemblages such as the late Middle Eocene Steels Crossing assemblage lived was almost certainly truly tropical in the meteorological definition. The leaf-margin percentage for the Steels Crossing flora is 75, which is borderline between Tropical and Paratropical Rain forests and is close to that (76) for the meteorologically tropical Submontane Rain forest at 700 m elevation in the Philippine Islands (BROWN, 1919). In leaf size (Table II) the Steels Crossing flora compares well with the meteorologically tropical Mesophyll Vine and Simple Mesophyll Vine forests of eastern Australia (WEBB, 1959).

The Alaskan assemblages isochronous with the Steels Crossing represent, both physiognomically and floristically, Paratropical Rain forest (WOLFE, 1969a). The Alaskan assemblages include, for example, members of Menispermaceae (7 spp.), Annonaceae, Icacinaceae (7 spp.), and Dipterocarpaceae, all of which are dominantly tropical families, as well as three species of palms. The foliar physiognomy indicates the presence of a diverse liane element, a characteristic of Tropical, Submontane, and Paratropical Rain forests. The leaves were dominantly broad-leaved evergreen, the percentage of entire-margined leaf species is 65, and numerous species had drip-tips. It would be difficult to construct a climatic model that would allow Paratropical Rain forest to grow at latitude 60 N and not allow tropical climates to expand significantly poleward beyond their present limits. Whether the Steels Crossing assemblage represents warm Paratropical or Tropical Rain forest is climatically not significant; warm Paratropical Rain forest (lowland Hainan, southern lowland Taiwan) is meteorologically tropical.

The Susanville assemblage has generally been considered to be of Late Eocene age (AXELROD, 1966b, pp.41–42; 1968, p.733), an age in accordance with the occurrence in the flora of *Platanophyllum angustiloba* (LESQ.) MACG., which is thought to be descended from a characteristic Early to Middle Eocene species (MACGINITIE, 1941, p.127). The Susanville assemblage cannot be of Paleocene age, as thought by some workers (AXELROD and BAILEY, 1969; fig.13). The one—and, insofar as I am aware, only—collection from the main Susanville locality contains less than 100 specimens (L. Lesquereux in DILLER, 1889, p.420) distributed among 23 species. Other localities in this area have yielded mostly fragmentary specimens, but the stratigraphic relationships of these localities to the main Susanville locality are unknown. Clearly the Susanville assemblage has not been adequately sampled, whether or not one regards leaf counts as significant. The few genera thus far determined include *Hemitelia*, *Platanophyllum*, *Saurauia*, and *Laurophyllum*; these genera are insufficient for any attempt to determine vegetation from floristic composition. These genera could be expected in vegetation ranging from Tropical Rain forest to various subtropical forests. A tentative determination of the vegetational type represented can, however, be based on foliar physiognomy. The leaf-margin percentage is 68%; if allowance is made for the sample size, the assemblage could represent Tropical, Paratropical, or Montane Rain forest. The leaves, however, are typically in the mesophyll or notophyll size-class, and three of the "mesophyll" species are actually in the macrophyll size-class. Based on the leaf-size data, the Susanville assemblage can best be regarded as representing vegetation similar to the complex mesophyll vine or simple mesophyll vine forests of Australia (WEBB, 1959, p.555), both of which are meteorologically tropical. The coriaceous texture of the leaves also indicates an evergreen habit for all but one of the species represented. The available physiognomic data indicate that the Susanville assemblage probably lived under a meteorologically tropical climate.

Conifers may be represented in some (others, e.g., Susanville and Upper Clarno, lack conifers) of these meteorologically tropical assemblages. This is not surprising in view of the fact that the Paratropical Rain forest of southern China, which straddles the meteorological tropics, contains conifers such as *Amentotaxus* (known in the La Porte flora; AXELROD, 1966b, p.59), the endemic *Glyptostrobus* (commonly in many Paleogene assemblages), and, of course, *Pinus*.

CLIMATIC CHANGES

The Oligocene deterioration

At least three sequences of assemblages (Upper Clarno–Bridge Creek;

Comstock–Goshen–Willamette; Bilyeu Creek–Lyons) in the Pacific Northwest, as well as assemblages in Alaska, show pronounced changes in both foliar physiognomy and floristic composition in the Oligocene. WOLFE and HOPKINS (1967) consider that such changes reflect a pronounced climatic deterioration; radiometric ages cited by WOLFE and HOPKINS (1967) indicate that the deterioration occurred within two (possibly within one) million years. AXELROD and BAILEY (1969), on the other hand, interpret the changes as only reflecting a slight vegetational and hence climatic change. This change, they suggest, was largely related to the beginning of the regular incidence of frosts; in one instance, however, uplift is called upon as well to explain the climatic shift.

The three main sequences of assemblages that indicate an Oligocene deterioration are: (*1*) the Upper Clarno (34 million years, Early Oligocene) and Bridge Creek (31–32 million years, Middle? Oligocene) in central Oregon; (*2*) the Comstock (latest Eocene or earliest Oligocene), Goshen (Middle Oligocene), and Willamette (31 million years, Middle? Oligocene); and (*3*) the Bilyeu Creek (Middle Oligocene), and Lyons (Middle? Oligocene). As noted above (p.32), data published previously by AXELROD (1966a, b) could be interpreted as indicating a major climatic shift involving some of these floras, but, without explanation, temperature estimates have been revised for some of these assemblages so that no major climatic shift is needed.

The difference between the Upper Clarno and Bridge Creek assemblages is not, as stated by AXELROD (1966b, p.45), a change from a broad-leaved evergreen forest similar to the subtropical oak–laurel forest of China to a broad-leaved deciduous forest similar to that of the uplands bordering the Yangtze Valley. The oak–laurel forest has a leaf-margin percentage of about 40–50% in areas of high equability, whereas the percentage for the Mixed Mesophytic forest of the Yangtze is about 25–38. A change from one vegetation type to the other would thus be reflected in a change of about 2% in leaf-margin percentages. The Upper Clarno (68% entire) differs by 43% from the Bridge Creek (25% entire). Floristically the oak–laurel and Mixed Mesophytic forests are transitional and many genera are common to the two forest types. WANG (1961) lists over 150 genera as occurring in the oak–laurel forest; over 100 of the same genera he lists as represented in the Mixed Mesophytic forest. Over 40 genera of plants are known from the Upper Clarno (SCOTT, 1954; and R. A. Scott in CHANDLER, 1964, p.58) and at least an equal number are known from the Bridge Creek (CHANEY, 1952). Only five genera are now known to be common to these two assemblages. The physiognomic and floristic differences between the Comstock and Willamette assemblages are equally as great; the Goshen–Willamette and the Bilyeu Creek–Lyons differences are not as great because the Goshen and Bilyeu Creek represent a stage near the beginning of the deterioration. Thus the undocumented statement that "In living vegetation, temperature differences much smaller than those called for by Wolfe and Hopkins bring about similar changes in leaf form and floral composition" (AXELROD and BAILEY, 1969, p.190) is not in accord with available evidence.

The Bilyeu Creek–Lyons sequence, located at the western margin of the Cascade Range and within a few kilometers of Middle Oligocene marine beds, shows the strong floristic and physiognomic change during the Oligocene deterioration. Of the 37 dicotyledonous species (KLUCKING,, 1962), 57% are entire-margined; in size, the leaves are dominantly mesophyll (35%) and notophyll (51%). The leaf-margin percentages indicate vegetation falling near the boundary between the oak–laurel and Paratropical Rain forests (i.e., probably frostless but not meteorologically tropical), and the leaf-size percentages indicate a similarity to the Complex Notophyll Vine forest (WEBB, 1959), which lives under a climate that is frostless but may or may not be meteorologically tropical. The evidence thus indicates that the Bilyeu Creek assemblage lived under a frostless climate that was not meteorologically tropical. The Lyons assemblage was obtained from beds about 180 m stratigraphically higher than the Bilyeu Creek beds. Only 12% of the species fall into the notophyll size-class and none in the mesophyll. The leaf-margin percentage is 28%. Whereas over three-fourths of the Bilyeu Creek species represent genera that are now exclusively evergreen broad-leaved, this category is represented in the Lyons by only 12% (all microphylls) of the species. Gymnosperms, which are represented in the Bilyeu Creek flora only by the ubiquitous *Metasequoia*, are represented by five species in the Lyons (J. A. Wolfe and R. W. Brown in PECK et al., 1964, p.22). Only three genera are common to the two floras. Both the physiognomic and floristic change between the Bilyeu Creek and Lyons assemblages in a short stratigraphic interval indicate a major climatic deterioration during the Oligocene.

The climatic data utilized by AXELROD and BAILEY (1969, fig.15) to delineate the boundary between the evergreen broad-leaved forest of southern China and the broad-leaved deciduous (Mixed Mesophytic) forest are, in part, inaccurate. Stations such as En-shih (= Engshih; western Hupeh, altitude 469 m), Han-chou (= Hangchow; Chekiang, altitude 10 m), and Nan-ching (= Nanking; Kiangsu, altitude 68 m), actually apply to the broad-leaved evergreen forest and not to the deciduous forest. In western Hupeh and eastern Szechuan, the Mixed Mesophytic forest is not known below 1,000 m altitude and WANG (1961, pp.110–112) also states that in Hupeh "... the original vegetation on the lower elevations was of the evergreen broad-leaved type". He (WANG, 1961, p.103) also notes that the deciduous forest in Chekiang and Kiangsu begins "... at about 500 m or higher..." and that the original lowland vegetation was probably broad-leaved evergreen. Similarly, data from I-ch'ang (a city at low elevation in the Yangtze Valley of Hupeh) and Shang-hai (near the mouth of the Yangtze) cannot be considered to apply to the broad-leaved deciduous forest. The only station cited that does apply to the deciduous forest is Lu-shan (= Kuling; Kiangsi, 1,070 m), and to this can be added data from T'ien-mu-shan (Chekiang, 1,060 m). Both these stations are about 250–300 m above the lower boundary of the forest, but the T'ien-mu-shan station is near the upper limit of the forest and Lu-shan is about 430 m below the upper limit. In Japan and Korea, stations in the Mixed Mesophytic forest have a range of mean annual temperature similar to that of the Chinese stations.

It is emphasized that the southern boundary of the Mixed Mesophytic forest is not entirely clear. At high elevations on Kyushu and Taiwan and in the uplands of Kweichow and western Szechuan, the Evergreen Sclerophyllous Broad-leaved forest contains many broad-leaved deciduous trees that are also members of the Mixed Mesophytic forest. At higher altitudes broad-leaved deciduous trees become increasingly common, and, in western Szechuan, this transitional zone (2,200–2,500 m) includes numerous conifers. At even higher altitudes, the forest becomes dominantly coniferous. This transitional zone could be considered an extension of the Mixed Mesophytic forest, but WANG (1961, pp.113–114) notes that the vegetation is typified by a mixture of trees that are broad-leaved evergreen, broad-leaved deciduous, and needle-leaved. This transitional zone, therefore, is not physiognomically a part of the Mixed Mesophytic forest, Evergreen Sclerophyllous Broad-leaved forest, or a coniferous forest. The transitional zone is increasingly restricted southwards and is typically lacking or sparsely represented in Yunnan and the Himalayas. Climatological data for the bulk of the zone are not known, but data for Simla, India, and Unzendake are representative. In Szechuan, Ya-an, Mo-mien, and Mien-ssu-chen are in the Evergreen Sclerophyllous forest, whereas K'ang-ting is near the upper boundary of the transitional zone; the climatological data for these stations are, however, based on short-term observations.

The boundary between the broad-leaved deciduous and broad-leaved evergreen forests is almost certainly temperature-related, but the relationship suggested by AXELROD and BAILEY (1969, p.191) is probably invalid. "In China, a mean of 50°F (10°C) for the coldest month of the year also approximates the boundary between the broad-leaved evergreen and mixed mesophytic forests (see WANG, 1961, fig.5, 11)". The data in Fig.4 illustrate strikingly that the 10°C cold month mean has no relation to the deciduous-evergreen boundary.

The northern and upper altitudinal limits of mesophytic, broad-leaved evergreen forest appear to be related to three main temperature factors (Fig.5): (*1*) mean annual temperature of about 13°C or higher; (*2*) annual range of temperature less than about 25° to 26°C; and (*3*) frosts not frequent or of long duration.

The third factor is the most difficult to pinpoint. In the southeastern United States, areas such as the Piedmont of Georgia and the Carolinas satisfy the first two temperature factors but the forest is primarily deciduous oak and pine. Extreme minimal temperatures on the Piedmont are, in many instances, not as great as those known in Memphis, Tennessee, which is approximately the northernmost extent of broad-leaved evergreen forest up the Mississippi Valley. The frost penetration, however, is greater on the Piedmont than at Memphis, indicating that at

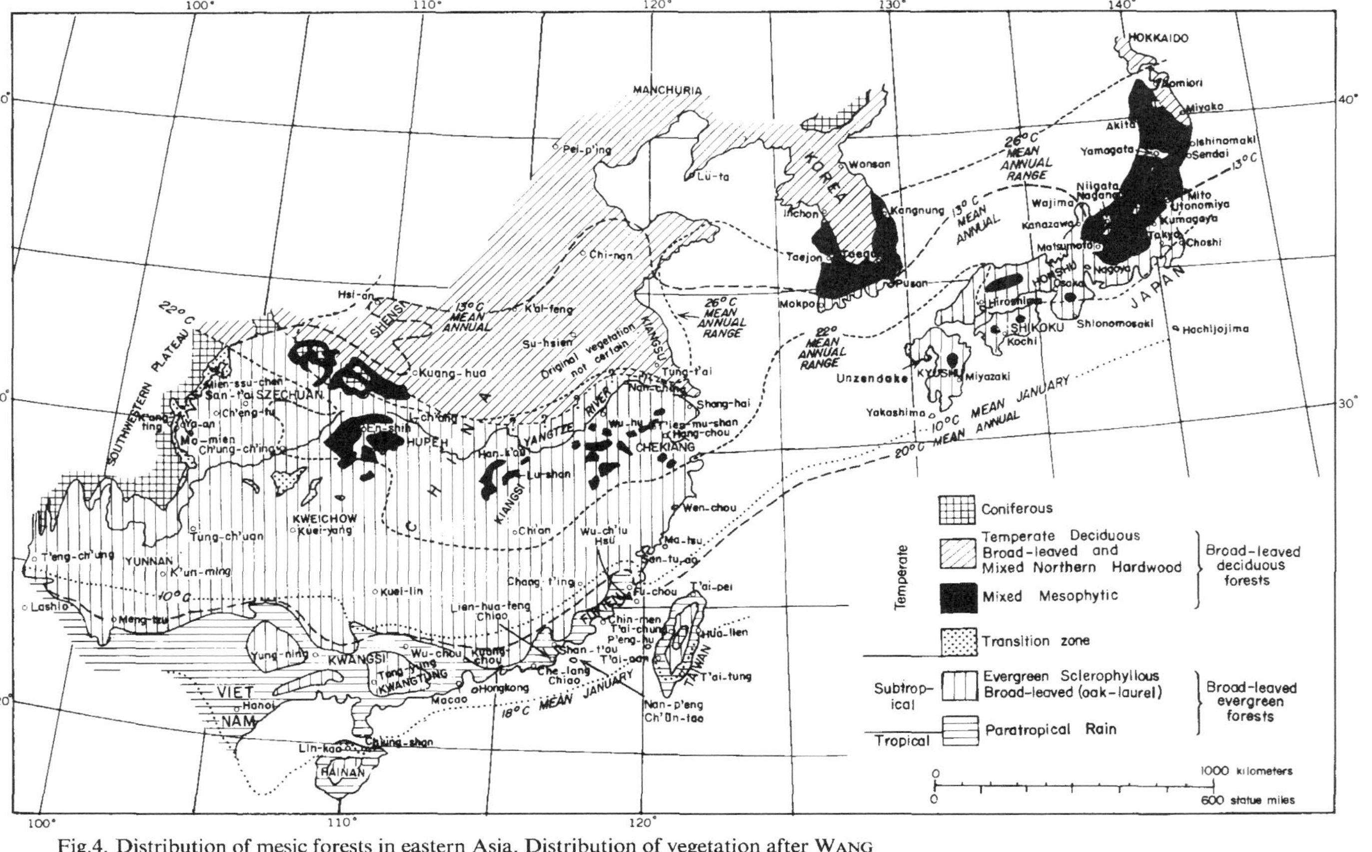

Fig.4. Distribution of mesic forests in eastern Asia. Distribution of vegetation after WANG (1961, fig.5), except that, in accordance with WANG'S (1961) discussion: (*1*) the upland broad-leaved forest of western Szechuan and northern Kweichow is shown as a transitional forest of broad-leaved evergreen and deciduous plants and conifers; (*2*) the vegetation of lowland Szechuan and Hupeh is shown as broad-leaved evergreen; and (*3*) the vegetation of the lower reaches of the Yangtze Valley is also shown as broad-leaved evergreen. The nature of the original vegetation of the northern Chinese plains is problematic because of extensive agriculture.

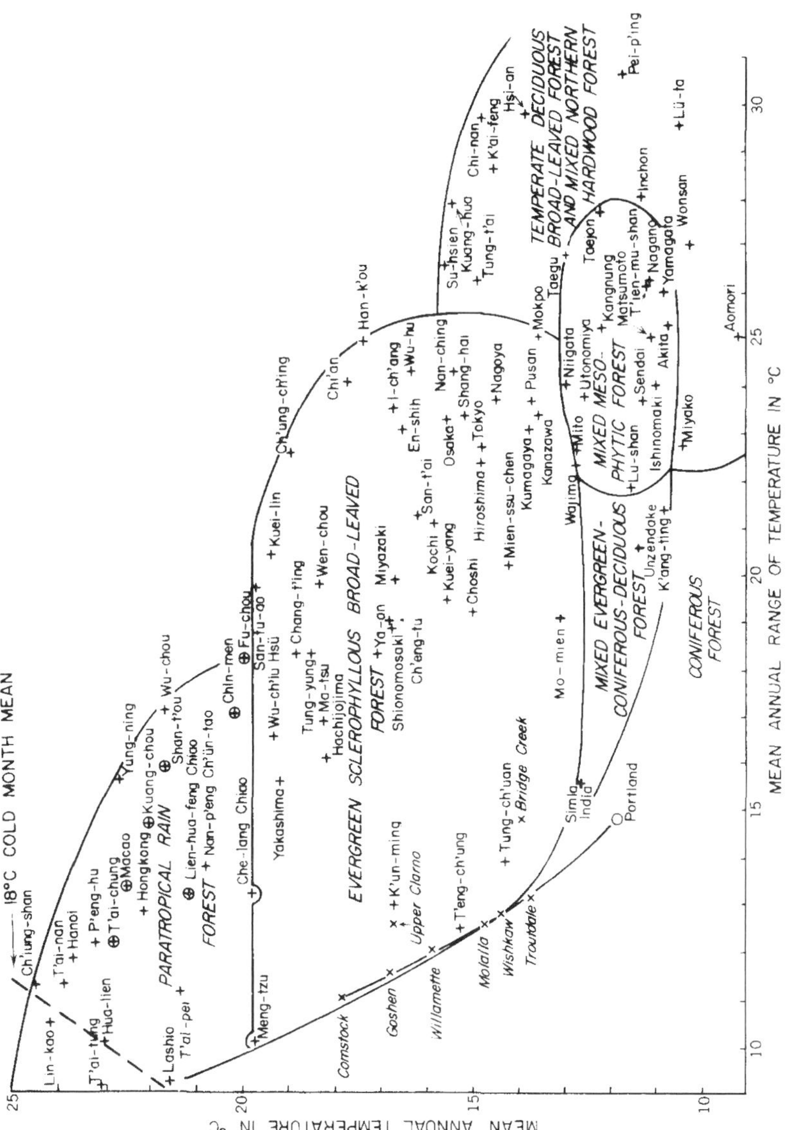

Fig.5. Climatic data for some stations in the mesic forests of eastern Asia. Vegetational data from Fig.4: stations from that figure denoted by "+". Climatic regimes assigned by AXELROD and BAILEY (1969, fig.12) to the Upper Clarno, Bridge Creek, and floras in the "western Oregon" sequence, denoted by "x". Note that none of the thermal parameters assigned come close to those for extant broad-leaved deciduous forests. A "+" for stations in the paratropical rain forest indicates that temperatures below 0°C have been recorded; both Lin-kao and Ch'iung-shan probably had sub-zero temperatures during the winter of 1955, but no data have been seen that substantiate this.

Memphis the frosts are probably not as prolonged or frequent as on the Piedmont. This factor of frost does not appear to be operative in limiting the broad-leaved evergreen forest in eastern Asia, because climates in the broad-leaved deciduous forest there have less severe frosts than at Memphis. Deciduous or coniferous forest is largely the result of the lack of any one of these three temperature requirements in a mesophytic climatic regime, i.e., the coniferous or broad-leaved deciduous forests are present because the mesophyllous evergreen habit is not favorable. Deciduous broad-leaved and coniferous plants are diversely represented in broad-leaved evergreen forests, but broad-leaved evergreens (except for nanophyllous or microphyllous types simulating conifers) are typically poorly represented in deciduous or coniferous forests.

The climatic regime (mean annual temperature of 14 °C, annual range of 15 °C) suggested by some workers for the Bridge Creek flora is, therefore, not known to be present in any area of the Mixed Mesophytic forest today and in fact is not known to be present in any mesophytic broad-leaved deciduous forest. In order to replace a forest such as that at K'un-ming (about the same climatic regime as inferred by AXELROD and BAILEY, 1969, for the Upper Clarno) with a forest such as the Mixed Mesophytic (which the Bridge Creek resembles physiognomically), it is necessary to lower the mean annual temperature by only about 4 °C, but, most importantly to raise the mean annual range by over 10 °C. If the Upper Clarno climate was truly frostless, as AXELROD (1966b, p.50) has previously suggested, then the change from the Upper Clarno to Bridge Creek involves a drop of at least 4 °–8 °C in the mean annual temperature. If, as the physiognomic and floristic evidence indicates (p.40), the Upper Clarno vegetation corresponds to Paratropical Rain forest, a minimal drop of 7 °C is needed in the mean annual temperature.

If the Upper Clarno assemblage lived under a climatic regime similar to that at K'un-ming, Yunnan, a moderate rise in elevation (1,000 ft.) coupled with only a slight decrease in mean annual temperature should logically produce vegetation similar to that at altitudes of perhaps 1,500–2,000 ft. above K'un-ming. Certainly the temperature regime assigned to the Bridge Creek assemblage would be similar to that at 2,650 m, which is about 2,200 ft. (ca. 700 m) above K'un-ming (1,922 m), assuming the prevailing lapse rate in that area (ca. 0.4 °C/100 m). According to WANG (1961, pp.148–151), this vegetation up to an altitude of 3,000 m is broad-leaved evergreen, and is dominated by many species and several genera of evergreen Fagaceae; other elements include evergreen Magnoliaceae, Lauraceae, and Theaceae. Computations from the short lists given by WANG (1961, p.150) indicate that the percentage of entire-margined species is about 45.

Does the Bridge Creek assemblage resemble this upland Yunnan vegetation, either floristically or physiognomically? All paleobotanists (e.g., CHANEY, 1952; AXELROD, 1966b, p.45) who have analyzed the Bridge Creek assemblage are in agreement that it represents a broad-leaved deciduous forest. The most diversely represented families in the Bridge Creek assemblage are Betulaceae, Juglandaceae, and Aceraceae. Some Fagaceae are represented, but at some localities (e.g., Twickenham) they are few or lacking; one species of Lauraceae and no species of Magnoliaceae or Theaceae are present. Only 25% of the leaf species have entire margins. Clearly, the Bridge Creek assemblage is not similar floristically or physiognomically to the vegetation of upland Yunnan.

Uplift of eastern Oregon cannot be used to explain the differences between the Upper Clarno and Bridge Creek assemblages; as noted above, the differences are as strong as those today between the Paratropical Rain forest and a forest that has many deciduous species. These two vegetational types are separated altitudinally on Taiwan by about 1,500 m, and no geologic data indicate that an uplift of that magnitude occurred in eastern Oregon in the two-million-year interval between Upper Clarno and Bridge Creek time. Eastern Oregon was tectonically active during the Tertiary, but the geologic data do not indicate whether the region was uplifted or downwarped during the Upper Clarno-Bridge Creek interval; either could have happened (R. L. Hay, personal communication, October, 1969), but certainly uplift of the magnitude needed to obtain the vegetational change demonstrated would have left an unmistakable record.

The same major temperature shift can be demonstrated in Alaska. The Alaskan Early Oligocene floras (WOLFE, 1969a) represent a subtropical laurel forest, whereas, as pointed out by WOLFE and HOPKINS (1967) the succeeding floras represent a broad-leaved deciduous forest; the leaf-margin percentage declined from 56 (new data on larger collections lower the percentage to 55; WOLFE, 1969a, p.49) to 6. The fact that major changes occurred in Alaska as well as in Oregon indicates that slight changes in local climates cannot explain the climatic change. The evidence thus best suits a major climatic deterioration during the Oligocene.

A Late Eocene fluctuation

The physiognomic analysis (Table II) of the foliage of the Middle Clarno (John Day Gulch) assemblage in eastern Oregon is in strong contrast to the Lower and Upper Clarno assemblages. Whereas on both physiognomic and floristic grounds the sub- and suprajacent assemblages represent Paratropical Rain forest, the Middle Clarno assemblage appears to represent a broad-leaved evergreen forest that had a large broad-leaved deciduous element. Determination of all the genera in the Middle Clarno is not complete, but many characteristic genera have been determined: cf. *Zamia*, *Pinus*, *Sequoia*, *Chamaecyparis*, *Betula*, *Lithocarpus/Pasania*, *Mahonia*, *Cinnamomophyllum*, *Laurophyllum* (3 spp.), *Platanophyllum*, *Platanus*, *Prunus*, *Rubus*, *Ailanthus*, *Rhus*, *Toxicodendron*, *Dipteronia*, *Vitis*, *Colubrina*, and *Alangium*. Some of the genera are also represented in the Upper and Lower Clarno assemblages, but most are not. The strong admixture of evergreen oaks, laurels, cycads, conifers, and deciduous plants in the Middle Clarno is characteristic of the transitional region between the broad-leaved evergreen and coniferous forests in upland Yunnan and Taiwan; the latter forest has a leaf-margin percentage almost identical to that of the Middle Clarno. The climatic shift must involve a considerable reduction (perhaps as much as 8 °C) in the mean annual temperature.

A second area in which this Late Eocene fluctuation has been recorded is at the head of the Gulf of Alaska (WOLFE, 1969a). There a broad-leaved evergreen forest (65% entire-margined leaves) was closely succeeded by an oak–laurel forest (54%), then a probable broad-leaved deciduous forest (20%), and, in the Early Oligocene by a broad-leaved evergreen forest (55%). The leaf-margin percentage for the cool assemblage should be taken as a minimum, because it is based on an assemblage of only twenty species, which include six members of the typically stream-side Salicaceae and Betulaceae. These floras all occur in an interdigitating marine-non-marine sequence of extensive coal-bearing rocks in a limited geographic area.

In the Puget Group of Washington, a similar sequence can be demonstrated based on new research. Floras such as the Steels Crossing (and its equivalents in the type section of the Lower Ravenian) are succeeded by a forest dominantly broad-leaved evergreen but that also included several broad-leaved deciduous plants; this cool assemblage in the Upper Ravenian is in turn overlain by rocks that have yielded a broad-leaved evergreen forest similar to that of the upper part of the Clarno Formation. In terms of leaf-margin percentages, the sequence in the Puget is 75–62 to 38–65. Originally the cool flora was included in the "Upper Puget flora" by WOLFE and HOPKINS (1967); the cool flora is depauperate compared to the Upper Puget, and thus the significance of the then few species from the cool interval was submerged. Note that the vegetational changes in either the Gulf of Alaska or main Puget sequence cannot be attributed to: (*1*) uplift; or (*2*) changes of successional nature due to vulcanism; the Gulf of Alaska sequence lacks volcanic rocks and the main sequence in the type section of the Ravenian of the Puget Group (WOLFE, 1968, fig.2) contains only a few intrusive volcanic rocks.

The duration and exact timing of the cool period is unknown, except that it falls within the Late Eocene, and correlates with the Auversian or Bartonian of Europe. Most, if not all, the upland conifer assemblages discussed by AXELROD (1966a, b) may belong to this cool interval; several of these assemblages have reliable radiometric ages ranging from about 38 to 42 million years (some of the radiometric ages that indicate older ages are suspect; see WOLFE, 1969a, pp.67–68).

DISCUSSION

The fact that a significant climatic shift during the Oligocene took place in such widely separated areas as Europe (NEMEJČ, 1959), western Siberia (ZHILIN, 1966), and New Zealand (DEVEREUX, 1967; JENKINS, 1968) indicates that the shifts recorded in Alaska and Oregon by WOLFE and HOPKINS (1967) are not due to local factors. The rapidity of the shift in New Zealand (DEVEREUX, 1967, p.1004; JENKINS, 1968, p.34) as well as in western North America indicates that the shift is not due to a gradually deteriorating climate. Notable also is the fact that the warming near the Oligocene–Miocene boundary has also been detected in eastern Siberia (VLASOV, 1964), central Europe (NEMEJČ, 1959), southwestern Europe (SITTLER, 1967), and New Zealand (BEU and MAXWELL, 1968), and such a warming has been detected in Alaska (WOLFE and HOPKINS, 1967) and the in western conterminous United States on both plant and marine molluscan data (ADDICOTT, 1969). The Middle Miocene warming is very well documented in Japan (TANAI and HUZIOKA, 1967), New Zealand (DEVEREUX, 1967), and Europe (MAI, 1964), as well as in western North America (WOLFE and HOPKINS, 1967; ADDICOTT, 1969). The evidence presented by these authors clearly indicates that unequivocal evidence for major, global-wide temperature fluctuation has been demonstrated and documented.

Considerable similarity exists between the climatic curves of DORF (1963) and WOLFE and HOPKINS (1967) that were constructed for the western United States. These similarities include the cooling during the Paleocene, maximum warmth in the later Eocene, a strong decline in the Oligocene, and warming in the Middle Miocene. The amount of warmth, however, that DORF (1959, 1963) postulates for western North America during the Early Tertiary is considerably less than I would suggest. The differences stem in large part from DORF's (1959) adherence to the Arcto-Tertiary concept, thus considering temperate high latitude assemblages to be of Eocene age. More recent work (WOLFE, 1966; 1969a) has shown that the only independently dated Eocene assemblages (except for the cool Late Eocene assemblage) from high latitudes are paratropical to subtropical and that the assemblages Dorf considered to be of Eocene age are largely either Paleocene or Neogene in age. Despite such differences in regard to high latitude assemblages, the general agreement between DORF (1963) and WOLFE and HOPKINS (1967) on Tertiary climatic fluctuations is significant.

The fluctuations do represent distinct changes in mean annual temperature. The amount of change in mean annual range of temperature is not known, but the greatest mean annual range today in Paratropical Rain forest is about 16–17°C, and the meteorologically tropical part of this forest has a maximum range of 12°C. The present mean annual range in western Oregon, for example, is about 16°C, and thus an increase of at least 4°C is indicated. The change, however, appears to be far more complex than that. Mesophytic broad-leaved deciduous forest in eastern Asia (eastern North America is not considered because of the high frost frequencies and penetration) has a minimal mean annual range of 22°C, and it is clear that the Early–Middle Miocene vegetation of much of the northwest Pacific represented a Mixed Mesophytic forest (CHANEY, 1959; WOLFE, 1969b). Thus the evidence indicates that the mean annual range of temperature increased during at least part of the Paleogene and has decreased since at least the

Middle Miocene by at least 6 °C. A similar pattern must have prevailed in west-central Europe and southern Alaska; in these areas, the Early–Middle Miocene vegetation was mesophytic broad-leaved deciduous, but the present mean annual ranges are lower than those found in the Asian mesophytic broad-leaved deciduous vegetation.

AXELROD (1966a, b) has elaborated on MACGINITIE's (1941) concept that some genera during the Tertiary lived under a lower mean annual temperature and mean annual range of temperature than do the same genera today. AXELROD (1966b) has, however, extended this concept to vegetation. It is improbable that this extension is valid. The floristic composition of various vegetational types probably was different, i.e., vegetational types are continually changing in regard to floristic composition. The vegetational (physiognomic) types, however, are the result of the physiological and phenotypic responses of plants to various environmental parameters. During the Tertiary, no vegetational type could have lived under conditions different from those that limit the same vegetational type today.

The framework of vegetational zonation during the Paleogene theorized by AXELROD (1966a, b, 1968) is based on Axelrod's analysis of the relative ages of various plant assemblages and their temperature requirements. Thus he postulates that during the Paleogene upland forests in northern Nevada descended altitudinally northwards so that the same vegetation type was present at sea level in southern Alaska (latitude 60°). Unless climatic fluctuations are incorporated into this framework, however, Axelrod's concept of vegetational zonation cannot be valid. The Alaskan vegetation, as noted previously (p.000), was at times during the Paleogene paratropical and clearly warmer than the upland Nevada assemblages such as the Copper Basin and even warmer than some lowland Eocene assemblages in the Pacific Northwest such as the type Upper Ravenian or the John Day Gulch (Middle Clarno).

Part of the problem in AXELROD's (1966a) analysis of Paleogene vegetational zonation is that the purported zonation is based on some assemblages which are actually of Neogene age (Kenai; WOLFE, 1966, 1969a, b), on other assemblages that are of Paleocene age (Kupreanof Island; WOLFE et al., 1966), of Eocene age (Copper Basin; AXELROD, 1966b), and Oligocene age (Goshen; MACGINITIE, 1953). As GOOD (1953, pp.266–267) has noted, any attempts to establish vegetational zonation must be based on isochronous assemblages, and AXELROD's (1966a) analysis was based on assemblages that represent a 40-million-year time-span.

In order to substantiate climatic trends, it is indeed necessary to base such trends on assemblages (plants or animal) that are closely spaced stratigraphically. If an analysis of Tertiary climatic trends in a given area is based on as few as six assemblages—some 30 million years apart— is it unlikely that the resulting conclusions can be considered as indicating the lack of fluctuations. Some assemblages that are widely spaced in time, however, may show the general direction of climatic change. For example, the only described Paleogene floras of east-central California are the Early Eocene Chalk Bluffs (MACGINITIE, 1941) and the Early Oligocene La Porte (POTBURY, 1935); as pointed out previously, the Chalk Bluffs and La Porte assemblages are about 30 km apart. The La Porte flora has "... a more tropical aspect than that of the Chalk Bluffs flora" (MACGINITIE, 1941,p.85), which clearly indicates that the climate in central California did not generally cool in the Eocene. And, in fact, this warming has been previously pointed out by AXELROD (1966b, p.43); the Montgomery Creek (including the Susanville) "... shows relationship to the older Chalk Bluffs flora that is less tropical in aspect...".

Any model that seeks to explain Tertiary climatic change must take into account at least the following:

(*1*) The existence of broad-leaved evergreen Paratropical Rain forest at latitude 60°–61°N and of marginally Tropical Rain forest at 52°N during part of the Paleogene.

(*2*) The pronounced fluctuations during the Paleogene, and less pronounced fluctuations during the Neogene.

(*3*) The increase in mean annual range of temperature during at least part of the Paleogene and subsequent decrease.

(*4*) The summer-wet regimes on west coasts during part of the Tertiary.

(*5*) The development of deserts during the later Neogene.

(*6*) The apparently contradictory patterns of climatic change depending on latitude.

The last point needs amplification. The fossil plant assemblages of eastern Asia equatorward from about latitude 40°N consistently show an over-all warming since at least sometime in the Neogene. This is true for assemblages from southern Honshu and Kyushu (TANAI, 1961, p.145), Taiwan (CHANEY and CHUANG, 1968), and Viet Nam (MCHEDLISHVILI, 1960). The assemblages poleward from about latitude 45°N show a definite cooling trend for the same interval, including assemblages from Hokkaido (TANAI, 1961, p.148) and Kamchatka (CHELEBAEVA, 1968). These divergent patterns stretching over many degrees of latitude clearly cannot be explained by tectonic features. It is interesting that, given a decrease in the inclination of the earth's axis of rotation, areas poleward of latitude 43° have progressively less insolation, whereas areas equatorward have progressively greater insolation (MILANKOVICH, 1939; VAN WOERKOM, 1953). The idea of an over-all cooling climate during the Neogene was developed primarily from studies of fossil assemblages north of latitude 43°N and the idea may be applicable only to this region. It is also significant that DEVEREUX (1967) and GILL (1968) both show some Early Miocene of Oligocene temperatures lower than present temperatures at latitudes equatorward from latitude 43°S, despite the fact that the glaciation of Antarctica has probably depressed present temperatures.

I stress the fact that "The general climatic history of the earlier half of

Paleogene time remains poorly known..." (WOLFE and HOPKINS, 1967, p.69). Almost all paleobotanists who have been concerned with Tertiary climates have agreed that the Eocene was the warmest epoch of the Tertiary (e.g., REID and CHANDLER, 1933; VAKHRAMEYEV, 1966; CHANEY, 1967) and definitely warmer than the Paleocene. The problem in deciphering Paleocene and Eocene climates is in large part a problem of sampling in stratigraphic successions. More recent work (WOLFE, 1969a) in well-controlled sections in Alaska, Washington, and Oregon indicates that there probably were major climatic fluctuations in the Eocene. These fluctuations have been obscured previously by the inclusion of depauperate, cool assemblages with diverse, warm assemblages (as was done in the instances of the Puget floras by WOLFE and HOPKINS, 1967) as one assemblage. Whether Tertiary climatic fluctuations were periodic is uncertain, but some radiometric ages indicate such a possibility (WOLFE, 1969a, pp.62–69).

CONCLUSIONS

Physiognomic and floristic analysis of Tertiary leaf assemblages from the western United States indicates that there were climatic fluctuations during the Tertiary and that some of the climatic shifts were rapid. These fluctuations appear to be correlative with fluctuations recorded from other areas of the world. Although local factors may have influenced the relative intensity of these fluctuations, the climatic changes appear to be due to some factor(s) of world-wide significance.

Tertiary climatic changes are indeed complex. Many more studies are needed in stratigraphically and latitudinally critical areas. During the Tertiary and particularly the Neogene continental drift and other major tectonic movements of large land masses are of less significance than in progressively older periods in interpreting paleoclimatic data. Paleoclimatologists should eventually be able to construct models of Tertiary climates and climatic changes; if such changes are periodic, the periodicity should be applicable to the reconstruction of climates of more ancient periods.

ACKNOWLEDGEMENTS

Persons who have contributed significantly to this paper in regard to both extensive discussions of the ideas presented and critically reading the manuscript are: J. W. Durham, H. D. MacGinitie, and H. E. Schorn (University of California, Berkeley), and W. O. Addicott (U.S. Geological Survey). Publication of this paper has been authorized by the Director, U.S. Geological Survey.

REFERENCES

ADDICOTT, W. O., 1969. Tertiary climatic change in the marginal northeast Pacific Ocean. *Science*, 165: 583–586.

AXELROD, D. I., 1956. Mio-Pliocene floras from west-central Nevada. *Univ. Calif. (Berkeley) Publ., Bull. Dept. Geol. Sci.*, 33: 1–322.

AXELROD, D. I., 1964. The Miocene Trapper Creek flora of southern Idaho. *Univ. Calif. (Berkeley) Publ., Bull. Dept. Geol. Sci.*, 51: 1–181.

AXELROD, D. I., 1966a. A method for determining the altitudes of Tertiary floras. *Palaeobotanist*, 14: 144–171.

AXELROD, D. I., 1966b. The Eocene Copper Basin flora of northeastern Nevada. *Univ. Calif. (Berkeley) Publ., Bull. Dept. Geol. Sci.*, 59: 1–125.

AXELROD, D. I., 1967. Geologic history of the Californian insular flora. *Proc. Symp. Biology Californian Islands, Santa Barbara Botany Garden*, pp.267–315.

AXELROD, D. I., 1968. Tertiary floras and topographic history of the Snake River basin, Idaho. *Geol. Soc. Am., Bull.*, 79: 713–734.

AXELROD, D. I. and BAILEY, H. P., 1969. Paleotemperature analysis of Tertiary floras. *Palaeogeography, Palaeoclimatol., Palaeoecol.*, 6: 163–195.

BAILEY, I. W. and SINNOTT, E. W., 1915. A botanical index of Cretaceous and Tertiary climates. *Science*, 41: 831–834.

BAILEY, I. W. and SINNOTT, E. W., 1916. The climatic distribution of certain types of angiosperm leaves. *Am. J. Botany*, 3: 24–39.

BEU, A. G. and MAXWELL, P. A., 1968. Molluscan evidence for Tertiary sea temperatures in New Zealand: a reconsideration. *Tuatara*, 16(1): 68–74.

BROWN, W. H., 1919. Vegetation of Philippine mountains. *Manila, Publ. Dept. Agricult. Nat. Resources*, 1: 1–434.

CHANDLER, M. E. J., 1964. *The Lower Tertiary Floras of Southern England*. British Museum (Nat. Hist.), London, 4: 151 pp.

CHANEY, R. W., 1944. The Troutdale flora. *Carnegie Inst. Wash. Publ.*, 553: 323–351.

CHANEY, R. W., 1952. Conifer dominants in the Middle Tertiary of the John Day Basin, Oregon. *Palaeobotanist*, 1: 105–113.

CHANEY, R. W., 1959. Miocene floras of the Columbia Plateau, I. Composition and interpretation. *Carnegie Inst. Wash. Publ.*, 617: 1–134.

CHANEY, R. W., 1967. Miocene forests of the Pacific Basin: their ancestors and their descendants. *Jubilee Publ. Commemorating Prof. Sasa, 60th Birthday*, pp.209–239.

CHANEY, R. W. and CHUANG, C. C., 1968. An oak-laurel forest in the Miocene of Taiwan, I. *Geol. Soc. China*, 11: 3–18.

CHELEBAEVA, A. I., 1968. Neogenovaya flora reki Levoi Pirozhnikovoi na Kamchatke (The Neogene flora of the River Pirozhnikovoi in Kamchatka). *Akad. Nauk S.S.S.R., Botan. Zh.*, 53: 737–748.

DEVEREUX, I., 1967. Oxygen isotope paleotemperature measurements on New Zealand Tertiary fossils. *New Zealand J. Sci.*, 10: 988–1011.

DILLER, J. S., 1889. Geology of the Lassen Peak district. *U.S., Geol. Surv., 8th Ann. Rept.*, 1: 395–432.

DORF, E., 1959. Climatic changes of the past and present. *Univ. Michigan Mus., Palaeontol. Contrib.*, 13(8): 181–210.

DORF, E., 1963. The use of fossil plants in paleoclimatic interpretations. In: A. E. M. NAIRN (Editor), *Problems in Palaeoclimatology*. Interscience, London, pp.13–31.

FISHER, R. V., 1968. Pyrogenic mineral stability, lower member of John Day Formation, eastern Oregon. *Univ. Calif. (Berkeley) Publ., Bull. Dept. Geol. Sci.*, 75: 1–36.

GAGNEPAIN, F., 1939. Un genre nouveau de Butomacées et quelques espèces nouvelles d'Indo-Chine. *Bull. Soc. Botan. France*, 86: 300–303.

GILL, E. D., 1968. Oxygen isotope paleotemperature determinations from Victoria, Australia. *Tuatara*, 16(1): 56–61.

GOOD, R., 1953. *The Geography of the Flowering Plants*. Longmans and Green, London, 452 pp.

HERNÁNDEZ, X. E., CRUM, H., FOX, W. B. and SHARP, A. J.,1951. A unique vegetational area in Tamalaulipas. *Torrey Botan. Club Bull.*, 78(6): 458–463.

JENKINS, D. G., 1968. Planktonic Foraminifera as indicators of New Zealand Tertiary paleotemperatures. *Tuatara*, 16(1): 32–37.

KLUCKING, E. P., 1962. *An Oligocene Flora from the Western Cascades*. Thesis, University of California, Berkeley, Calif., 241 pp.

KNAPP, R., 1965. *Die Vegetation von Nord- und Mittelamerika und der Hawaii-Inseln*. Fischer, Stuttgart, 373 pp.

KNOWLTON, F. H., 1926. Flora of the Latah Formation of Spokane, Washington, and Coeur d'Alene, Idaho. *U.S., Geol. Surv., Profess. Papers*, 140-A: 17–81.

LI, H. L., 1963. *Woody Flora of Taiwan*. Livingston and Morris Arboretum, Narbeth, Pa., 974 pp.

MACGINITIE, H. D., 1941. A Middle Eocene flora from the central Sierra Nevada. *Carnegie Inst. Wash. Publ.*, 543: 1–178.

MACGINITIE, H. D., 1953. Fossil plants of the Florissant beds, Colorado. *Carnegie Inst. Wash. Publ.*, 599: 1–198.

MACGINITIE, H. D., 1962. The Kilgore flora. *Univ. Calif. (Berkeley), Publ., Bull. Dept. Geol. Sci.*, 35(2): 67–158.

MAI, D. H., 1964. Die Mastixioideen-Floren im Tertiär der Oberlausitz. *Palaontol. Abhandl., Abt. B*, 2: 1–92.

MAPEL, W. J. and HAIL, W. J., 1959. Tertiary geology of the Goose Creek district, Cassia County, Idaho, Box Elder County, Utah, and Elko County, Nevada. *U.S., Geol. Surv., Bull.*, 1055-H: 217–254.

MCHEDLISHVILI, P. A., 1960. New data on the Tertiary flora of North Viet-Nam. *Dokl. Akad. Nauk S.S.S.R.*, 135(3): 694–697. (English translation published by *Am. Geol. Inst.*, 1961: 1159–1161.)

MILANKOVITCH, M., 1938. Astronomische Mittel zur Erforschung der erdgeschichlichen Klimate. *Handbuch Geophys.*, 9(3): 593–698.

MIRANDA, F. and SHARP, A. J., 1950. Characteristics of the vegetation in certain temperate regions of eastern Mexico. *Ecology*, 31(3): 313–333.

NEMEJČ, F., 1964. Biostratigraphic sequence of floras in the Tertiary of Czechoslovakia. *Časopis Mineral. Geol.*, 9: 107–109.

PECK, D. L., GRIGGS, A. B., SCHLICKER, H. G., WELLS, F. G. and DOLE, H. M., 1964. Geology of the central and northern parts of the Western Cascade Range in Oregon. *U.S., Geol. Surv., Profess. Papers*, 449: 1–56.

POTBURY, S. S., 1935. The La Porte flora of Plumas County, California. *Carnegie Inst. Wash. Publ.*, 465: 29–82.

REID, M. E. and CHANDLER, M. E. J., 1933. *The London Clay Flora*. British Mus. (Nat. Hist.), London, 561 pp.

RICHARDS, P. W., 1952. *The Tropical Rain Forest*. Univ. Cambridge Press, Cambridge, Mass., 450 pp.

SCOTT, R. A., 1954. Fossil fruits and seeds from the Eocene Clarno formation of Oregon. *Palaeontographica, B*, 96: 66–97.

SITTLER, C., 1967. Mise en évidence d'un rechauffement climatique a la limite de l'Oligocene et du Miocene. *Rev. Palaeobotan. Palynol.*, 2: 163–172.

TANAI, T., 1961. Neogene floral change in Japan. *Hokkaido Univ. Fac. Sci. J., Ser. 4*, 11(2): 199–298.

TANAI, T. and HUZIOKA, K., 1967. Climatic implications of Tertiary floras in Japan. In: K. HATAI (Editor), *Tertiary Correlations and Climatic Changes in the Pacific*. Sasaki, Sendai, pp. 89–94.

VAKHRAMEYEV, V. A., 1966. Botanic–geographic zonality in the geologic past and the evolution of the plant kingdom. *Paleontol. Zh.*, 1: 6–81 (English translation available from Telberg Book Company, New York, N.Y.)

VAN STEENIS, C. G. G. J., 1962. The land-bridge theory in botany. *Blumea*, 11: 235–372.

VAN WOERKOM, A. J. J., 1953. The astronomical theory of climatic changes. In: H. SHAPLEY (Editor), *Climatic Change*. Harvard Univ. Press, Cambridge, Mass., pp.147–157.

VLASOV, G. M., 1964. Paleogene and Neogene climatic fluctuations in the Far East. *Dokl. Akad.*

Nauk S.S.S.R., 157: 589–592. (English translation published by *Am. Geol. Inst.*, 1965, pp.17–20.)

WANG, C. W., 1961. The forests of China. *Harvard Univ., Publ. Maria Moors Cabot Found.*, 5: 1–313.

WEBB, L. J., 1959. Physiognomic classification of Australian rain forests. *J. Ecol.*, 47: 551–570.

WOLFE, J. A., 1964. Miocene floras from Fingerrock Wash, southwestern Nevada. *U.S., Geol. Surv., Profess. Papers*, 454-N: N1–N36.

WOLFE, J. A., 1966. Tertiary plants from the Cook Inlet region, Alaska. *U.S., Geol. Surv., Profess. Paperes*, 398-B: B1–B32.

WOLFE, J. A., 1968. Paleogene biostratigraphy of non-marine rocks in King County, Washington. *U.S., Geol. Surv., Profess. Papers*, 571: 1–33.

WOLFE, J. A., 1969a. Paleogene floras from the Gulf of Alaska region. *U.S., Geol. Surv., Open-file Rept.*, 114 pp.

WOLFE, J. A., 1969b. Neogene floristic and vegetational history of the Pacific Northwest. *Madroño*, 20: 83–110.

WOLFE, J. A. and BARGHOORN, E. S., 1960. Generic change in Tertiary floras in relation to age. *Am. J. Sci.*, 258-A: 388–399.

WOLFE, J. A. and HOPKINS, D. M., 1967. Climatic changes recorded by Tertiary land floras in northwestern North America. In: K. HATAI (Editor), *Tertiary Correlations and Climatic Changes in the Pacific—Symp. Pacific Sci. Congr., 11th, Tokyo, Aug.–Sept. 1966*, 25: 67–76.

WOLFE, J. A. and LEOPOLD, E. B., 1967. Neogene and Early Quaternary vegetation of northwestern North America and northeastern Asia. In: D. M. HOPKINS (Editor), *The Bering Land Bridge*. Stanford Univ. Press, Stanford, Calif., pp.193–206.

WOLFE, J. A., HOPKINS, D. M. and LEOPOLD, E. B., 1966. Tertiary stratigraphy and paleobotany of the Cook Inlet region, Alaska. *U.S., Geol. Surv., Profess. Papers*, 398-A: A1–A29.

ZHILIN, S. G., 1966. A new species of *Carya* from the Late Oligocene. *Paleontol. Zh.*, 4: 104–108. (English translation available from Telberg Book Company, New York, N.Y.)

PAPER 42

Taxonomic Diversity Estimation Using Rarefaction (1975)

D. M. Raup

Commentary

ANDREW M. BUSH

Charles Darwin (1859) infamously disparaged the completeness of the fossil record, and his attitude influenced generations of paleontologists. However, by the 1970s, paleontologists increasingly aspired to contribute to ecological and evolutionary theory (the paleobiology movement), so this lack of confidence in the record posed a fundamental problem. Some paleobiologists optimistically charged ahead with new theories and approaches, while others suspected that they were wasting their time. Approaches were needed that reconciled these viewpoints.

David M. Raup was firmly in the paleobiology camp, but he shared the skeptics' concerns that the fossil record was a flawed manuscript—incomplete, biased, and impossible to read literally. However, his skepticism did not lead to fatalism, and he saw no reason that a flawed fossil record could not be read and understood if the biases were also understood. Moreover, he was analytically clever enough to try to distill biological signals from the sprawling, messy record. His philosophy—deeply skeptical of the record, but optimistic that its flaws could be understood and overcome—was possibly his greatest contribution to paleontology.

In the fossil record, marine biodiversity appears to increase from the late Mesozoic to the present, but Raup (in one of his bolder claims) argued that the increase was merely an artifact of increased sample sizes resulting from increased rock exposure (Raup 1972). If true, diversity was historically much more stable than was believed, and a first-order pattern of the fossil record was entirely illusory. If the primary signal in the fossil record was unreliable, how then to measure historical trends in diversity? In the following paper, Raup argued that analytical standardization could overcome natural variations in the data. Specifically, he reasoned that sampling intensity in different time periods could be standardized using a technique such as rarefaction, recently proposed by H. L. Sanders (1968) and refined by others. He calculated a standardized diversity curve for echinoids spanning the past 250 million years, and concluded that the number of echinoid families did increase, although not as much as indicated by the raw data. True to form, he did not present his diversity curve as the "true" curve; rather, he pointed out that other biases still might lurk in the data, although they would be easier to uncover now that simple sampling bias had been removed.

Although Raup's work on the biases that afflict diversity curves was extremely influential, the use of rarefaction and similar methods to standardize these curves did not become widespread until several of his students reignited interest in these biases and began compiling large databases for diversity analysis (Alroy 1996; Miller and Foote 1996). Sampling-standardized diversity analysis of the marine fossil record is currently a central focus of the Paleobiology Database project (Alroy et al. 2008) and of research into paleontological diversity dynamics. Raup's attitude is evident in these efforts—the fossil record may be imperfect, but it is not incorrigible.

From *Paleobiology* 1:333–42. Reprinted with permission from the Paleontological Society.

Literature Cited

Alroy, J. 1996. Constant extinction, constrained diversification, and uncoordinated stasis in North American mammals. *Palaeogeography, Palaeoclimatology, Palaeoecology* 127:285–311.

Alroy, J., M. Aberhan, D. J. Bottjer, M. Foote, F. T. Fürsich, P. J. Harries, A. J. W. Hendy, et al. 2008. Phanerozoic trends in the global diversity of marine invertebrates. *Science* 321:97–100.

Darwin, C. 1859. *On the origin of species by means of natural selection, or The preservation of favoured races in the struggle for life.* John Murray, London.

Miller, A. I., and M. Foote. 1996. Calibrating the Ordovician Radiation of marine life: Implications for Phanerozoic diversity trends. *Paleobiology* 22:304–9.

Raup, D. M. 1972. Taxonomic diversity during the Phanerozoic. *Science* 177:1065–71.

*Sanders, H. L. 1968. Marine benthic diversity: A comparative study. *American Naturalist* 102:243–82.

Paleobiology. 1975. vol. 1, pp. 333–342.

Taxonomic diversity estimation using rarefaction

David M. Raup

Abstract.—Benthic ecologists have successfully applied rarefaction techniques to the problem of compensating for the effect of sample size on apparent species diversity (= species richness). The same method can be used in studies of diversity at higher taxonomic levels (families and orders) in the fossil record where samples represent world-wide distributions of species or genera over long periods of geologic time.

Application of rarefaction to several large samples of post-Paleozoic echinoids (totaling 7,911 species) confirms the utility of the method. Rarefaction shows that the observed increase in the number of echinoid families since the Paleozoic is real in the sense that it cannot be explained solely by the increase in numbers of preserved species. There has been no statistically significant increase in the number of families since mid-Cretaceous, however. At the order level, echinoid diversity may have been nearly constant since late Triassic or early Jurassic.

David M. Raup, Department of Geological Sciences, University of Rochester, Rochester, N. Y. 14627

Accepted: July 9, 1975

Introduction

Estimation of the number of fossil taxa present in a stratigraphic interval is plagued by sampling problems. A large sample yields, on the average, more higher taxa than a small sample. As more specimens or species are discovered, more higher taxa are added but not in direct proportion. Thus, a low value for the number of higher taxa may be just an artifact of small sample size. The problem can be avoided only when samples are consistently large enough to contain all or virtually all higher taxa: a rare event in paleontology except at the highest taxonomic level. (It requires many fewer species to find all phyla than to find all genera.) (See Raup 1972 for mathematical treatment).

Benthic ecologists have faced the same sort of problem. If a bottom sample or dredge haul contains several thousand specimens, the number of species in the resulting faunal list will almost certainly be larger than if the sample contained only a few tens or hundreds of specimens. Sanders (1968) developed a method he called "rarefaction" to cope with this problem, and he was spectacularly successful in using the method to show that species diversity in the deep-sea is as high as or higher than in many shallow water habitats. Rarefaction is basically an interpolation technique making it possible to estimate how many species would have been found had the sample been smaller than it actually was. The technique is derived from an analysis of the relative frequencies of specimens within species and makes it possible to compare estimated diversities at a constant sample size.

Rarefaction has been used occasionally in fossil situations (Stanton and Evans 1972 for example). The context was the same as that of Sanders; that is, specimens within species (or genera) were counted, and each sample was "rarefied" so that assemblages of different sizes from different biofacies could be compared.

The present paper will extend the application of rarefaction to higher taxa and to larger blocks of geologic time. Data from post-Paleozoic echinoids will be used to test the applicability of the method.

Methods

In the present context, we are concerned with counting the total number of taxa present in an area or in a habitat or during a period of time. As such, we are concerned with taxonomic "richness" as this term has been used in the literature of diversity (see Hurlbert 1971 for a discussion of semantic problems). Figure 1 shows rarefaction curves for four

US ISSN 0094–8373

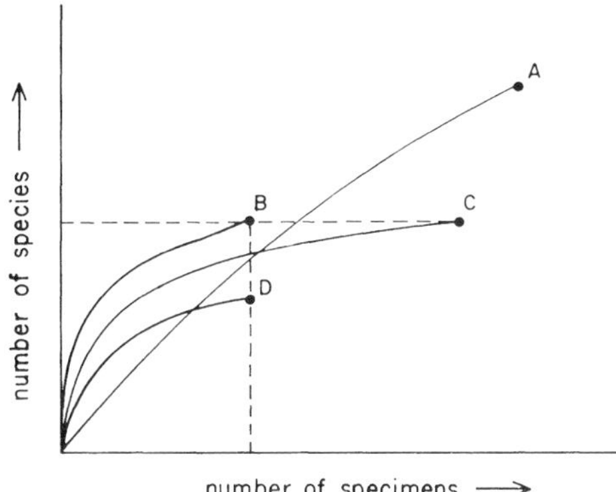

FIGURE 1. Hypothetical rarefaction curves. The labeled dots represent the actual numbers of specimens and species in each sample. The curves leading back to the origin show the probable decrease in species with reduction in sample size.

hypothetical samples. The right hand end of each curves represents the observed number of specimens and the observed number of species. The curves extending back to the left are based on the numbers of specimens in each species (the species abundance distribution). The shape of a given rarefaction curve is a complex function of the "evenness" of species frequency and the average number of specimens per species. If several samples are being treated, as in Figure 1, and the shapes of the rarefaction curves are approximately the same, diversities can be compared readily. The curves labeled *B*, *C* and *D* satisfy this. Samples B and C have the same number of species, but C has many more specimens, and its rarefaction curve is consistently below that of B. Thus, for any arbitrary sample size, the diversity of C will be less than that of B, a fact not determinable from the simple count of the number of species present. Sample D has fewer species than B, but the same number of specimens. The rarefaction curves show that the actual diversity of D must in fact be less than that of B (and also less than that of C). Where rarefaction curves cross, as in the case of A with C and A with D, diversity assessments are difficult or impossible to make. For large samples, A has more species than C, but for small samples the reverse is true. Since all specimen-species curves must ultimately level out at the number of species actually present, it is tempting to try to extrapolate curves A and C to such a sample size. But, as has been pointed out by Sanders (1968) and subsequent authors, the rarefaction technique does not allow extrapolation of this sort, but only interpolation between samples or toward smaller samples. Thus, the relative species diversities of A and C in Figure 1 must remain in doubt. This indicates a fundamental constraint in the application of rarefaction to actual cases: the several samples must be sufficiently similar that rarefaction curves display minimal crossing.

In the time since Sanders' 1968 paper, several authors have pointed out minor problems with his technique. Simberloff (1972) and Hurlbert (1971) independently noted that Sanders' technique is not completely independent of sample size (see also Fager 1972). Simberloff and Hurlbert offered identical solutions to the problem and their refinements will be used here.

Given species abundance data, the following equation may be used to calculate the estimated number of species $E(S_n)$ for a given number of specimens (n):

$$E(S_n) = \sum_{i=1}^{S}\left[1 - \frac{\binom{N-N_i}{n}}{\binom{N}{n}}\right] \qquad (1)$$

where N is the number of individuals in the original sample, S is the number of species in the original sample, and N_i is the number of individuals in the ith species. The equation is taken from Hurlbert (1971, Eq. 13) but is algebraically identical to one presented by Simberloff (1974). In addition, Heck, Van Belle, and Simberloff (1975) have developed an analytical expression for the variance of $E(S_n)$, as follows:

$$\mathrm{Var}(S_n) = \binom{N}{n}^{-1}\left[\sum_{i=1}^{S}\binom{N-N_i}{n}\left(1 - \frac{\binom{N-N_i}{n}}{\binom{N}{n}}\right) + 2\sum_{\substack{j=2\\ i<j}}^{S}\left(\binom{N-N_i-N_j}{n} - \frac{\binom{N-N_i}{n}\binom{N-N_j}{n}}{\binom{N}{n}}\right)\right] \qquad (2)$$

After solving these equations for a series of arbitrary values of n (less than the actual sample size), it is possible to plot the rarefaction curve and its confidence limits. Alternatively, values of $E(S_n)$ can be calculated for several samples at a single n value equal to or less than the N of the smallest sample being compared. This has the disadvantage that the crossing of rarefaction curves is not evident, and thus comparative results may not be as confidently obtained as if the analysis were based on the full rarefaction curves. It should be empha-

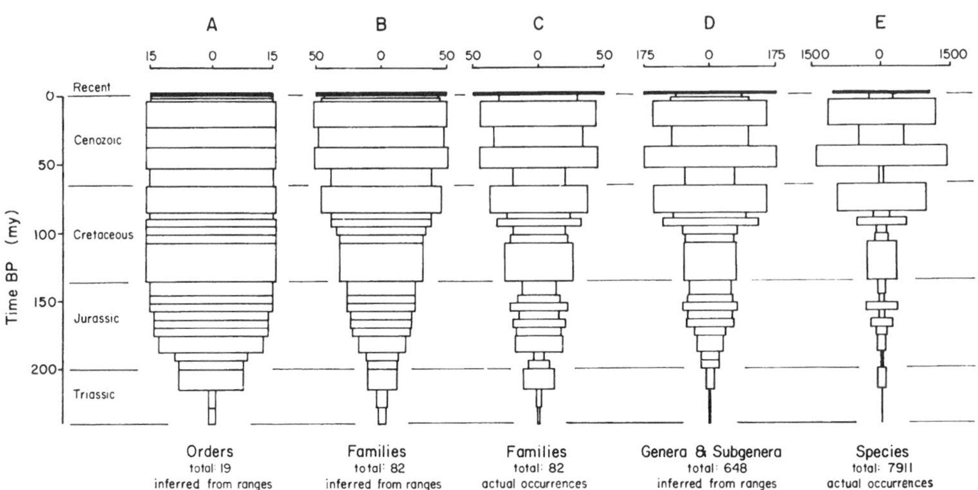

FIGURE 2. Post-Paleozoic echinoid diversity at several taxonomic levels. The raw data are given in Table 1.

sized that any comparisons made using rarefaction are done on a relative rather than an absolute basis. The true (complete) diversities of the faunas are generally unknown because sampling is inevitably incomplete.

Sanders and subsequent authors have pointed out several limitations of the rarefaction method. Ideally, the technique should be applied only where habitats and biologic groups are the same from sample to sample, and sampling procedures are uniform. The degree to which this is satisfied must remain a matter of judgment. Sanders limited his analysis to certain groups of benthic invertebrates and to soft, fine-grained sediments. In paleontology, rarefaction is not appropriate for comparing a graptolitic fauna of a black shale with a shelly fauna of a limestone, but may be appropriate for comparing several graptolitic assemblages in shales.

In the present paleontological application, I will use data on the occurrence of species or genera within higher taxa rather than specimens within species. This shift of taxonomic rank does not change the method mathematically or conceptually. The important ingredient is that the frequency distribution of a small unit in a classificatory hierarchy is known within some larger unit.

When working in an evolutionary rather than an ecological context, an additional constraint is important. This has to do with evolutionary turnover. Suppose, for example, that during an interval of geologic time, the number of co-existing families is constant, but there is turnover such that some families become extinct and others originate. The whole time interval will thus have more families than any part of it. Rarefaction curves for parts of the time interval will be lower—indicating lower diversity—than the curve for the interval as a whole. Thus, the results of rarefaction are partially time dependent. This difficulty can be avoided by rarefying samples which represent approximately equal durations of geologic time.

There are undoubtedly other evolutionary phenomena that can mimic sampling effects and thus influence rarefaction results. For example, as a higher taxon undergoes adaptive radiation, it may evolve from a rare faunal element (few species) to a common one (many species). The implications of this should be explored further. For the time being, however, the simple correction for turnover described above treats the most significant evolutionary effect.

Material

Post-Paleozoic echinoids provide an ideal testing ground for rarefaction because the higher category taxonomy is relatively stable and because excellent data on species distribution are available. Also, Kier (1974) has reviewed many important aspects of the history

Table 1. Data for Figure 2.

	Age at base (my)	Orders (inferred)	Families (inferred)	Families (actual)	Genera & Subgenera (inferred)	Species (actual)
RECENT		15	50	50	175	1058
CENOZ						
Pleistocene	1.8	15	44		88	
Pliocene	5	15	45	29	106	260
Miocene	22.5	16	51	44	153	1187
Oligocene	37.5	16	48	34	103	469
Eocene	53.5	16	51	45	176	1438
Paleocene	65	15	39	19	69	44
CRET						
Senonian	85	16	46	37	150	978
Turonian	90	15	38	24	97	169
Cenomanian	95	16	38	32	126	543
Albian	101	16	34	20	72	116
Aptian	107	16	32	22	67	138
Neocomian	135	16	32	25	69	320
JURASSIC						
Portlandian	146	15	26	13	56	73
Kimmeridgian	152	15	26	15	59	74
Oxfordian	158	15	26	22	73	347
Callovian	164	14	23	15	59	76
Bathonian	170	14	23	20	64	225
Bajocian	176	14	22	17	43	136
Pliensb.-Toarc.	188	13	18	18	36	104
Sinemurian	194	9	12	4	25	16
Hettangian	200	8	11	8	24	32
TRI						
U. Triassic	215	8	11	11	11	104
M. Triassic	227.5	1	4	2	2	2
L. Triassic	240	1	3	2	2	2

of echinoid diversity in relation to the nature of the fossil record.

In the post-Paleozoic, there are approximately 7,500 fossil echinoid species currently recognized and about 1,050 living species. These totals are based on a compilation made from Lambert & Thiery (1909–1925) by Susan Wunder of the University of Rochester combined with a compilation of species described since 1925 made by Mary Lawson of the U.S. National Museum (Kier, personal communication 1975). Of the 7,500 fossil species, all but about 650 are well enough placed stratigraphically and taxonomically for detailed studies of diversity. Thus, the sample used for this study (hereafter called the Wunder-Lawson data) consists of about 6,850 fossil species and about 1,050 Recent species. All have been assigned to families and orders using the classification of the *Treatise on Invertebrate Paleontology* (Moore 1966). In addition, data on all recognized genera and subgenera (about 650) are available directly from the *Treatise*.

Figure 2 shows several ways of displaying the diversity history of post-Paleozoic echinoids. In each case, the width of the bar is the number of taxa and the height is absolute time. The stratigraphic intervals are the standard series and stages used in the *Treatise*, and the time scale is from Berggren (1972) for the

TABLE 2. Intervals of geologic time used for rarefaction.

Duration (my)	Biostratigraphic stages or series
37.5	Oligocene-Pleistocene ("late" Cenozoic)
27.5	Paleocene-Eocene ("early" Cenozoic)
36	Albian-Senonian ("late" Cretaceous)
34	Neocomian-Aptian ("early" Cretaceous)
29	Callovian-Portlandian ("late" Jurassic)
36	Hettangian-Bathonian ("early" Jurassic)
40	Scythian-Rhaetian (Triassic)

(mean duration: 34.3 million years)

Cenozoic and from Harland et al. (1964) for the Mesozoic (as modified by Lambert [1971] for the major boundaries). The data for Figure 2 are given in Table 1.

Figures 2A and 2B show the diversity histories of orders and families, respectively. Both use the "inferred range" method: that is, a stratigraphic interval is credited with a taxon if the interval is within the range of the taxon. Thus, the number of taxa given for an interval is usually more than the number actually preserved and found in that interval. For comparison, Figure 2C shows family diversity based only on actual species occurrences. Diversity is much more variable with this method of plotting because of the vagaries of preservation. The data have not been normalized for duration of stratigraphic stages—as might be appropriate at certain taxonomic levels (Rohr and Boucot 1974). For the present purposes, a simple plot of raw data is sufficient.

Figure 2D shows diversity of genera and subgenera computed by the inferred range method (using *Treatise* data), and Figure 2E shows species diversity by number of actual occurrences (using the Wunder-Lawson data). Virtually all fossil species in the sample are known from only a single stratigraphic stage—a fact probably reflecting taxonomic practice more than evolutionary turnover.

Figure 2 as a whole shows increase in diversity since the beginning of the Triassic at all taxonomic levels—though the patterns differ. At the ordinal level, the increase is concentrated in the Triassic and Jurassic (as also

TABLE 3. Example of species/family frequency data and its rarefaction. The total number of families is 27, and of species is 458.

Family	# of Species in Neocomian-Aptian	Species (n)	Estimated Families E(F_n)	Variance
Acrosaleniidae	4			
Arbaciidae	22	450	26.9	.07
Cassidulidae	2	400	26.5	.47
Cidaridae	77	350	25.9	.84
Clypeidae	10	300	25.3	1.18
Collyritidae	2	250	24.6	1.51
Conulidae	12	200	23.6	1.86
Diplocidaridae	10	150	22.3	2.30
Disasteridae	17	100	20.1	2.97
Discoididae	5	50	15.7	3.98
Echinidae	1	40	14.2	4.12
Echinometridae	8	30	12.4	4.11
Glyphocyphidae	3	20	10.1	3.84
Hemiasteridae	4	10	6.8	3.16
Hemicidaridae	15			
Holasteridae	9			
Holectypidae	9			
Nucleolitidae	70			
Orthopsidae	6			
Pedinidae	1			
Phymosomatidae	13			
Pseudodiadematidae	64			
Pygasteridae	1			
Saleniidae	17			
Stomechinidae	7			
Temnopleuridae	1			
Toxasteridae	68			

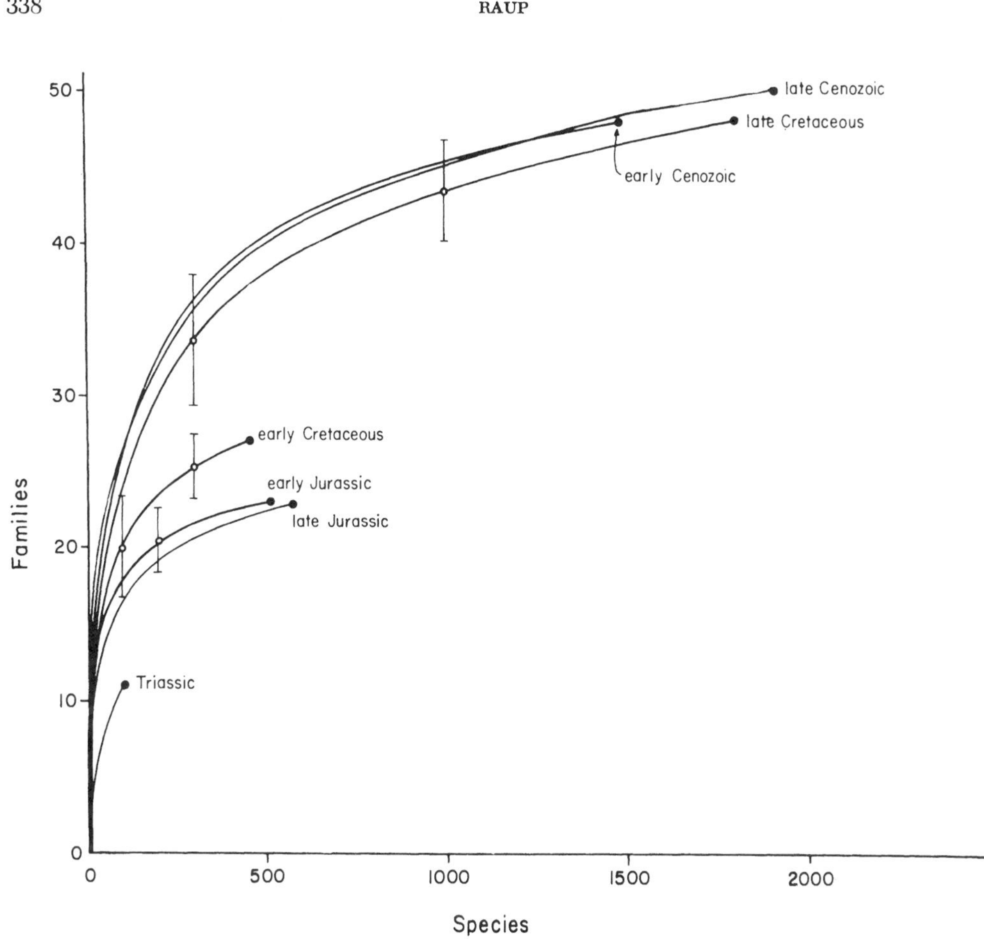

FIGURE 3. Species/family rarefaction curves for echinoids in seven post-Paleozoic intervals (see Table 2 for definition of the intervals). Vertical error bars include the estimated number of families ± 1.96 times the standard deviation of the estimated number (95% confidence limits).

shown by data published by Kier, 1974). In the latest Cenozoic, the number of orders actually declined slightly. At the family level, the rise is more or less continuous from Triassic through Eocene, after which the number of families seems to level off.

For genera and subgenera and for species, the data contain substantially more noise, but the percentage increase after the Triassic is clearly much greater than at higher taxonomic levels. Figures 2D and 2E both show a peak in the Eocene followed by a decline.

The Central Question for Rarefaction

As has been discussed by Valentine (1973) and other authors, the evolution of diversity should be expected to be different at different taxonomic levels. One could easily construct biologically plausible *post hoc* explanations for the patterns seen in Figure 2. Two fundamental questions must be answered first, however: (1) is the general diversity increase due to non-biologic, time-dependent biases which make a post-Paleozoic increase in diversity almost inevitable (see Raup 1972 for discussion), and (2) regardless of preservational and other such biases, are the observed changes in diversity just a consequence of changes in sample size (specimens and/or species)? The purpose of the rarefaction method is to answer the second of these questions. Because data on numbers of specimens are not available, we must use species or genera as the basic mea-

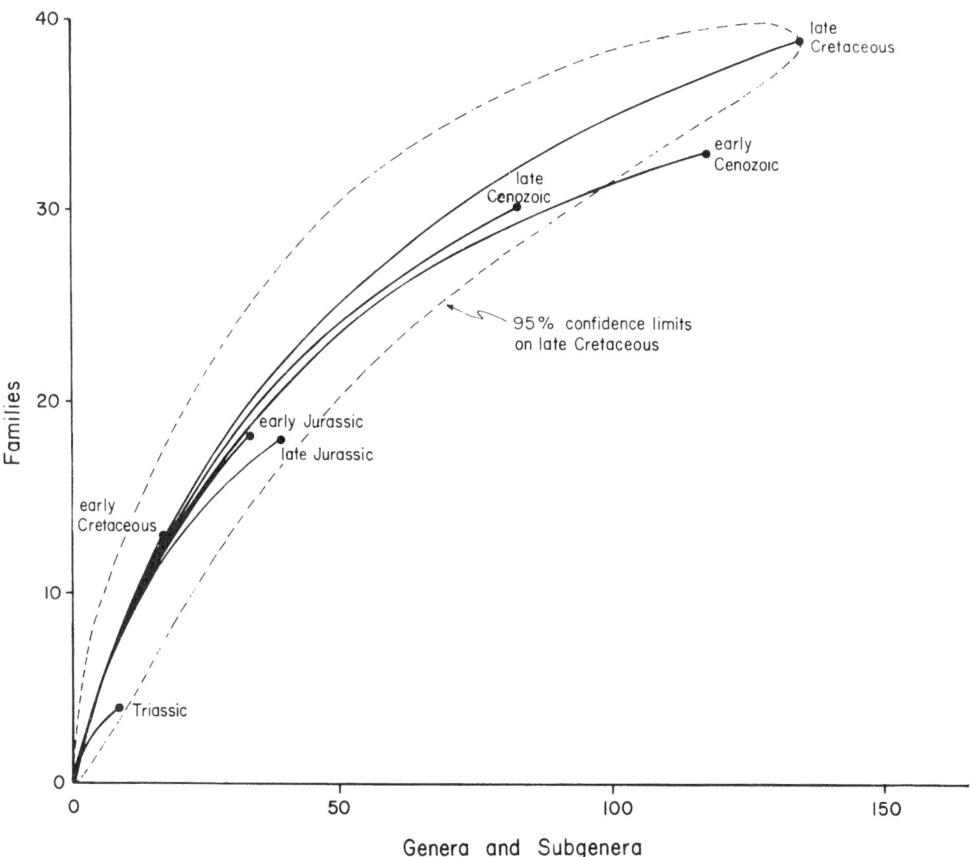

FIGURE 4. Genus-subgenus/family rarefaction curves for post-Paleozoic echinoids.

sure of sample size and explore diversity at higher taxonomic levels. The kind of question rarefaction is expected to answer is: "if the Eocene sample were as small as the Upper Triassic sample, would the Eocene still have more families (or orders) than the Upper Triassic?"

Rarefaction of Echinoid Data

Rarefaction was carried out at three taxonomic levels: (1) species within families, (2) genera within families, and (3) species within orders. A fourth logical combination—species within genera—was not possible because the Wunder-Lawson data set does not yet provide the proper generic affiliations of all the species.

Because it is important that the samples to be rarefied come from approximately equal lengths of time (see above), the post-Paleozoic was divided into seven reasonably equal intervals. These are listed with their durations in Table 2. Because of the necessity of using established stage boundaries, the seven intervals could not be made exactly equal. Furthermore, the uncertainties of absolute dating, particularly at the stage level, are such that the durations given in Table 2 are subject to considerable error. But the seven-fold breakdown is probably the best that can be made in the present state of radiogenic dating. Furthermore, minor alterations in the breakdown—as would be demanded if another time scale were used (such as that of Braziunas 1975)—do not affect the rarefaction results significantly. Although none of the seven intervals cross system boundaries, it should be noted that they do cross established series boundaries. For example, the Cretaceous is divided

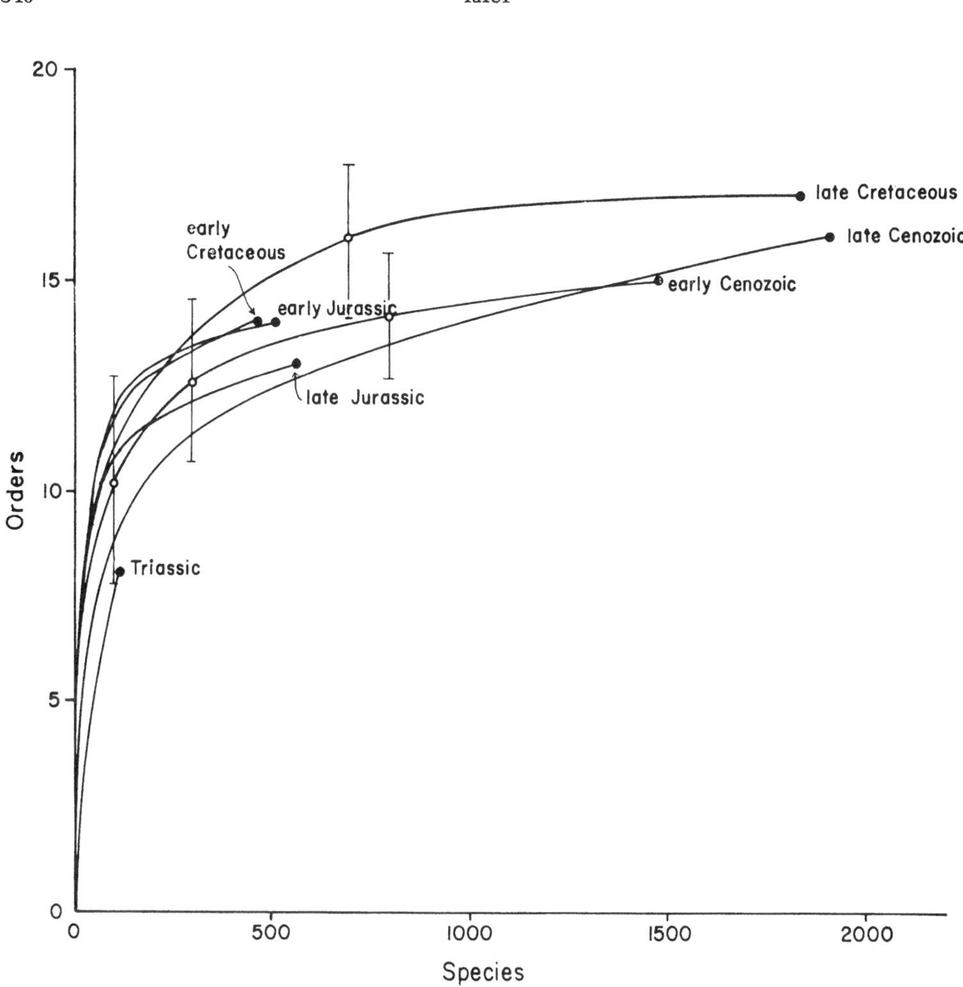

FIGURE 5. Species/order rarefaction curves for post-Paleozoic echinoids. Error bars as in Figure 3.

at the Aptian-Albian boundary so that "late" Cretaceous includes the Albian through the Senonian.

Table 3 gives the raw data for one of the echinoid samples and the results of the rarefaction computations. In the Neocomian-Aptian interval, there were 458 species occurrences among 27 families. The remaining 55 post-Paleozoic families either did not exist during this part of the Cretaceous *or* have not been found in Neocomian-Aptian rocks. The input for rarefaction is the list of frequencies: the presence or absence of specific families is immaterial. The right-hand portion of Table 3 gives the estimated number of families, $E(F_n)$, that *would* have been found for various arbitrary sample sizes (n) less than 458. For each estimate, a variance for the number of families is also given.

In Figure 3, the rarefaction data for the Neocomian-Aptian interval are plotted along with curves for the other six intervals. Vertical bars showing the 95% confidence limits are included periodically for several of the rarefaction curves. The error bars extend a distance equal to 1.96 times the standard deviation of the estimated number of families in each direction. As can be anticipated from the variance data in Table 3, the confidence band is narrowest at each end of the rarefaction curve (approaching zero at the extremes).

Without resorting to elaborate statistical testing, it is clear that the curves for the seven geologic intervals fall into distinct groups: (1)

low family diversity (Triassic), (2) intermediate diversity (Jurassic and "early" Cretaceous), and (3) high diversity ("late" Cretaceous and Cenozoic). Within the latter two groups, numbers of families are not statistically distinguishable although the separation between the Jurassic curves and that for the "early" Cretaceous may be real.

The important conclusion to be drawn from Figure 3 is that the post-Paleozoic rise in number of families is real in the sense that it cannot be discounted on the basis of sampling alone.

When *genera* within families are rarefied, the results are somewhat different (Figure 4). The rarefaction curves are much more tightly clustered: nearly all lie within the 95% confidence band of the "late" Cretaceous curve, indicating that the observed separations could easily have occurred by chance. On the other hand, the curves are arrayed in a plausible order (Triassic curve lowest, Jurassic curves next highest, etc.), and the plot may be taken to *suggest* that family diversity reached a maximum sometime in the Cretaceous. This conclusion is not in conflict with Figure 3.

Of the two approaches to estimating family diversity, that based on species (Figure 3) is probably much more dependable than that employing genera (Figure 4): there are about ten times more species than genera and greater resolution can thus be achieved.

Figure 5 shows rarefaction of species within orders. Here, there is considerable crossing of curves—not observed in the earlier plots—and the error bars indicate that most of the separations could easily be due to chance. The Triassic curve is low but is within the confidence limits of the "early" Cenozoic curve (see error bar) and well within the confidence limits of the "late" Cenozoic curve (error bar not shown). The data are compatible with the proposition that there has been no significant change in the number of orders since the Triassic and in particular, that the Jurassic rise in ordinal diversity seen in Figure 2 may be an artifact of sampling. It follows from this that as more species are found in the Jurassic (and even in the Triassic), early records of orders not now known from these rocks will be found.

In summary, the rarefaction analysis shows (1) the observed post-Paleozoic increase in number of families is not simply a matter of sample size (number of species found), (2) the post-Paleozoic increase is not as great as suggested by the raw data (Figure 2), (3) there has been no significant change in the number of families since mid-Cretaceous, and (4) there has been no significant change in the number of orders since the beginning of the Jurassic (and possibly not since well back into the Triassic).

It should be emphasized that conclusion (1), above, does not make the claim that the post-Paleozoic increase in family diversity is biologically real! This can only be determined when other, non-sampling biases are fully evaluated.

Rarefaction of Data From the Recent

For each of the three pairs of taxonomic levels, the Recent data were also treated by rarefaction. The results have not been included here because the samples are not comparable to fossil data: the Recent sample includes some species and higher taxa which are virtually unpreservable as fossils by virtue of their habitat (deep-sea) or fragile skeletal structures. This causes the species/family curve, for example, to plot above the highest fossil curve—even though the number of families actually found is not higher. To compare rarefaction curves for Recent and fossil assemblages is as improper as to use the method for comparing graptolitic shales and shelly limestones!

Discussion

The rarefaction technique, as applied to higher taxa of fossil echinoids, has successfully compensated for differences in sample size. The resulting picture of diversity is one in which we can have more confidence, but it does not preclude the possibility that biases, other than those related to sample size, are operating to distort the evolutionary record. A clearer look at these biases can be accomplished, however, now that the most obvious effects of sampling have been eliminated.

Acknowledgments

I wish to thank J. John Sepkoski, Jr. and D. S. Simberloff for valuable advice at various stages of this work. Special thanks go to Porter M. Kier, Mary Lawson, and Susan

Wunder for help in data acquisition, without which the echinoid analysis would have been impossible. The research was supported by the Earth Sciences Section, National Science Foundation, NSF Gant DES75-03870.

Literature Cited

BERGGREN, W. A. 1972. A Cenozoic time-scale—some implications for regional geology and paleobiogeography. Lethaia. *5*:195–215.

BRAZIUNAS, T. F. 1975. A geological duration chart. Geology. *3*:342–343.

FAGER, E. W. 1972. Diversity: a sampling study. Am. Nat. *106*:293–310.

HARLAND, W. B., A. G. SMITH, AND B. WILCOCK, EDS. 1964. The Phanerozoic Time-scale. 458 pp. Geol. Soc. Lond.; London.

HECK, K. L., JR., G. VAN BELLE, AND D. S. SIMBERLOFF. 1975. Explicit calculation of the rarefaction diversity measurement and the determination of sufficient sample size. Ecology. Submitted.

HURLBERT, S. H. 1971. The nonconcept of species diversity: a critique and alternative parameters. Ecology. *52*:577–586.

KIER, P. M. 1974. Evolutionary trends and their functional significance in the post-Paleozoic echinoids. Paleontol. Soc. Mem. 5. J. Paleontol. *48* (suppl. to no. 3):1–95.

LAMBERT, J. AND P. THIERY. 1909–1925. Essai de nomenclature raisonnee des Echinides. 607 pp. Chaumont, France.

LAMBERT, R. ST. J. 1971. The pre-Pleistocene Phanerozoic time-scale—a review. pp. 9–31. In: Harland, W. B. and E. H. Francis, eds. The Phanerozoic Time-scale, A Supplement (Part 1). Geol. Soc. Lond. Spec. Publ. 5.

MOORE, R. C., ED. 1966. Treatise on Invertebrate Paleontology, Part U, Echinodermata 3. 695 pp. Geol. Soc. Am. and Univ. Kans. Press.

RAUP, D. M. 1972. Taxonomic diversity during the Phanerozoic. Science. *177*:1065–1071.

ROHR, D. M. AND A. J. BOUCOT. 1974. Evolutionary patterns in the Paleozoic Bivalvia: documentation and some theoretical considerations: discussion. Geol. Soc. Am. Bull. *85*:665–666.

SANDERS, H. L. 1968. Marinebenthic diversity: a comparative study. Am. Nat. *102*:243–282.

SIMBERLOFF, D. S. 1972. Properties of the rarefaction diversity measurement. Am. Nat. *106*:414–418.

SIMBERLOFF, D. S. 1974. Permo-Triassic extinctions: effects of area on biotic equilibrium. J. Geol. *82*:267–274.

STANTON, R. J., JR. AND I. EVANS. 1972. Community structure and sampling requirement in paleoecology. J. Paleontol. *46*:845–858.

VALENTINE, J. W. 1973. Evolutionary Paleoecology of the Marine Biosphere. 511 pp. Prentice-Hall, Inc.; Englewood Cliffs, New Jersey.

PAPER 43

The Statistics and Biology of the Species-Area Relationship (1979)

E. F. Connor and E. D. McCoy

Commentary

BRIAN A. MAURER

In 1948, Frank Preston developed the idea that abundance of species within a local community followed a characteristic pattern that could be described by a lognormal probability distribution. In 1962, he extended his approach to derive an estimate of the exponent of a power relationship between species richness and area. His model suggested that this exponent should be around 0.25. R. H. MacArthur and E. O. Wilson used Preston's approach to develop the "equilibrium theory of island biogeography," maintaining that species number on an island was a steady state mediated by immigration and extinction.

E. F. Connor and E. D. McCoy, in their seminal paper, examined these ideas by collecting 100 data sets from a variety of sources that related species richness to island area. They demonstrated that there was a large variance in the values of species-area exponents using the power-function model, and also showed that the power-function model was not the exclusively best-fit model to these data sets. The fundamental structure of their study was prescient of the techniques used by macroecologists a decade later. They started with a well-formed expectation, gathered information across a large number of systems that this expectation should apply to, and then studied the resultant patterns among systems. It is conceivable that Connor and McCoy could have collected data on many more systems, but it is doubtful that their fundamental conclusions would be materially different. Their meta-analytic approach placed the discussion regarding species diversity more firmly on empirical grounds, ultimately ushering in more modern approaches incorporating explicit assumptions regarding the dynamical processes underlying spatial variation in diversity.

Literature Cited

Preston, F. W. 1948. The commonness, and rarity, of species. *Ecology* 29:254–83.

———. 1962. The canonical distribution of commonness and rarity: Part I. *Ecology* 43:185–215.

Vol. 113, No. 6 The American Naturalist June 1979

THE STATISTICS AND BIOLOGY OF THE SPECIES-AREA RELATIONSHIP

EDWARD F. CONNOR AND EARL D. MCCOY*†

Department of Biological Science, Florida State University, Tallahassee, Florida 32306

Regional differences in species number have puzzled naturalists since the early 1800's, and explanations account for a large part of modern ecological research. Two venerable observations form the cornerstone of our knowledge on the subject: The number of species within a taxonomic group tends to increase with decreasing latitude (see Fischer 1960; Pianka 1966); and the number of species within a taxonomic group tends to increase with increasing area (see Preston 1960, 1962; Williams 1964; MacArthur and Wilson 1967; Simberloff 1972). Despite early research on the latter trend (the species-area relationship), ecologists have studied it intensely only in the last 50 yr. The relationship was originally envisioned as an empirical tool and used in three principle ways: (1) to determine optimal sample size and sample number, (2) to determine the minimum area of a "community," and (3) to predict the number of species in areas larger than those sampled. All three uses are discussed by Kilburn (1966).

More recently interest in the species-area relationship has focused on mechanistic explanations, its precise mathematical descriptions, and interpretations of parameters derived from these mathematical descriptions. Williams (1964) and Preston (1960, 1962) have proposed that the exponential and power function models ("exponential model" throughout this paper also refers to the species/log area transformation, and "power function" also refers to the log species/log area transformation) of the species-area relationship result from the way in which individuals are distributed among species. Williams' (1964) exponential model, which emphasizes habitat heterogeneity, was considered important by many plant ecologists but is now largely ignored. Preston's (1960, 1962) power function model was based on the assumption of a dynamic equilibrium of species exchanges between islands in an archipelago. This assumption led to the equation of the power function model with the idea of a dynamic equilibrium as expounded by MacArthur and Wilson (1963, 1967), such that an adequate fit of this model to observed species numbers has been viewed as support of the equilibrium hypothesis (Grant 1970; Diamond 1973; Simpson 1974). The interplay of the equilibrium hypothesis and the power function model of the species-area relationship has led to interpretation of the slope and intercept of the power function model exclusively in the context of the equilibrium hypothesis. In particular, specific values of the slope of the power function are often construed to

* Order of authorship determined by the toss of a coin.

† Present address: Department of Biology, University of South Florida, Tampa, Florida 33620.

Am. Nat. 1979. Vol. 113, pp. 791–833.
 0003-0147/79/1306-0002$03.26

indicate the presence or absence of equilibrium (e.g., Preston 1962; Brown 1971; Diamond 1973).

Our concern with the use and interpretation of species-area curves derives from the post facto and ad hoc nature of the inferences and interpretations drawn from them. Not only is the power function model of the species-area relationship construed as evidence of equilibrium, but equilibrium is also considered to imply the power function: It is admittedly easier to collect species numbers than to examine the processes that determine them. Although the power-function model of the species-area relationship may be consistent with the equilibrium hypothesis view of the determination of species numbers, it by no means constitutes disproof of alternative mechanisms (Simberloff 1972, 1976*b*). In an effort to clarify the relationship between the equilibrium hypothesis and the power function model of the species-area relationship, we pose three questions regarding the basis, use, and interpretation of species-area curves. (1) Does the equilibrium model provide a unique theoretical basis for the species-area relationship? (2) Is the power function model (log/log), derived from equilibrium theory, the best model of the species-area relationship? (3) Can the parameters of the power function or other species-area models be interpreted biologically?

IS THERE A UNIQUE THEORETICAL BASIS FOR THE SPECIES-AREA RELATIONSHIP?

Two principal hypotheses have been advanced to account for the significant positive correlation often observed between numbers of species and area. The first, termed the "habitat-diversity hypothesis," was developed by Williams (1964) who proposed that as the amount of area sampled is increased new habitats with their associated species are encountered, and thus species number increases with area. The second hypothesis, termed the "area–per se hypothesis," was developed by Preston (1960, 1962) and MacArthur and Wilson (1963, 1967), and is derived as a prediction of the equilibrium theory of island biogeography. This hypothesis deemphasizes the importance of habitat diversity and instead explains species number as a function of immigration and extinction rates (see Simberloff 1972). Immigration rates are assumed to be dependent upon the distance of the area in question from the species source pool, but independent of island size; extinction rates are assumed inversely proportional to population sizes, which in turn are assumed directly proportional to area. Thus, if distance is held constant population sizes in small areas should be relatively small (other things being equal), implying high probabilities of species extinction; while population sizes in large areas should be relatively large, implying low probabilities of species extinction. It follows, then, that at any particular time one should observe more individuals and species in large areas, and therefore a positive correlation between species number and area. Sets of mathematical arguments have been developed, again mainly by Preston (1960, 1962) and Williams (1964), which predict the exact form of the species-area relationship. These mathematical arguments are independent of the hypotheses described above, but have become entwined with them; they are discussed in the following section.

A simple alternative to these two hypotheses is that species number is controlled by passive sampling from the species pool, larger areas receiving effectively larger samples than smaller ones, and ultimately containing more species. This sampling hypothesis could also generate the observed positive correlation between species number and area, but denies the importance of habitat differences and population processes in generating species numbers. The important distinction between the sampling hypothesis and either the habitat-diversity or area–per se hypotheses is that under this hypothesis the correlation between species number and area is viewed solely as a sampling phenomenon, rather than the result of biological processes such as diversification through specialized habitat utilization or the balancing of species immigrations and extinctions. The idea that the species-area relationship is purely a sampling phenomena should be considered a null hypothesis, and all hypotheses invoking biological processes to explain the species-area relationship should be considered alternatives.

Abele (1974), Harman (1972), and Dexter (1972) have all demonstrated a positive correlation between species number and number of habitats; Abele and Patton (1976) and Simberloff (1976*a*) have demonstrated the feasibility of the area–per se hypothesis; and Osman (1977) has shown that passive sampling is probably very important in determining the number of species found on different-sized boulders in the subtidal. Thus, each mechanism is probably important in determining the correlation between species number and area in one or another species assemblage, but practically it is difficult to assess their proportional contribution in any particular study. (For an illustration of the problems involved see McCoy and Connor [1976].) Most studies have failed to eliminate alternative hypotheses, although the experiments of Simberloff (1976*a*) are a step toward this end. Each hypothesis can be tested only by direct experimentation, and not by comparing post facto the consistency of empirical observations (species numbers) with hypothesized predictions. To conclude that habitat diversity alone is the cause of the species-area relationship one must not only demonstrate the effects of such diversity on numbers of species, but also the lack of any relationship between extinction probabilities and area. On the other hand, to conclude that area alone can influence the number of species, one must identify a species-area effect in a truly homogeneous habitat. Additional experimental designs are needed to eliminate the remaining alternatives.

Clearly, all three explanations (and perhaps more) should be kept in mind. At the same location some species may occur only on large areas because their particular habitat requirements are only found there (Whitehead and Jones 1969), for some species a critical population size above which extinction becomes unlikely may obtain only on large areas (Mertz 1971), and more random immigrants may be found on large than on small areas. The reasons underlying local diversity patterns can be elucidated only by sound biological examination and experimentation, not by the invocation of currently-accepted dogma.

IS THERE A BEST MODEL OF THE SPECIES-AREA RELATIONSHIP?

It is clear, even from the earliest observations, that species-area curves become asymptotic for large areas. Plant ecologists first attempted to elucidate the exact form of this curvilinear relationship early in the present century (Jaccard 1908, 1912;

Arrhenius 1921, 1923*a*, 1923*b*; Gleason 1922, 1925), although Watson implied in 1835 that species-area curves are inherently logarithmic. Arrhenius (1921) postulated that the relationship is a power function: ,

$$S = kA^z, \tag{1}$$

which is often approximated by a double logarithmic transformation:

$$\log S = \log k + z \log A. \tag{2}$$

Gleason (1922) noted that Arrhenius' equation gave impossibly high estimates of species number when extrapolated to large areas. He proposed instead that the relationship is exponential:

$$S = \log k + z \log A. \tag{3}$$

In early work, the exponential model received the most attention, especially from plant ecologists (e.g., Pidgeon and Ashby 1940; Evans et al. 1955; Hopkins 1955), and seemed to fit data reasonably well. Dony (1963), however, was an early champion among plant ecologists of the power function model. The exponential model derived theoretical underpinnings from Fisher et al. (1943) and Williams (1943, 1944, 1947), who demonstrated that, if one assumes population sizes to be proportional to area, a log-series relative abundance distribution leads directly to the exponential form of the species-area relationship. Contemporary work by Preston (1948, 1960, 1962), however, derived the log-normal relative abundance distribution, which with similar assumptions leads to the power function form of the species-area relationship. Preston (1962) and Bliss (1965) also showed that the log-series distribution apparently present in many studies was more likely a sampling distribution derived from a truncated underlying log-normal distribution. Preston's work has subsequently led to the near-uniform acceptance of the power function as the best model of the species-area relationship.

It is logical to ascribe the status "best model" to the one fitting the data best. Goodall (1952, p. 217), for instance, states, "A decision between the two proposed forms of the species-area curve cannot be made on a priori grounds, but must rest on observational data." This sound warning has frequently been ignored. Based on theoretical considerations, the power function has been treated as if it were a paradigm (*sensu* Kuhn 1962), usually escaping comparison with other models, and often has been fitted to species-area data ignoring important underlying assumptions (see Preston 1960, 1962). Thus, we feel it necessary to examine whether or not there is justification for the assumed universality of the power function.

To do so, we obtained from an extensive and growing literature 100 data sets detailing the numbers of species of various taxa from circumscribed areas (see Appendix; the literature survey was completed in early 1976). For a majority of these studies, the original author(s) fitted some species-area model (usually the power function) to their data. In the remaining instances the analyses are entirely ours. The logspecies/logarea (power function), species/logarea (exponential), logspecies/area, and species/area (untransformed) models were fitted to each data set as the data were reported in the literature. In some cases the data sets were modified by excluding outliers (see McCoy and Connor 1976). The SPSS package, version 5.18, run on a

CDC 6400 computer at the Florida State University Computing Center was used for all statistical computations (Appendix).

The rationale for fitting the power function to all species-area data without testing the fit of other models appears to be a profound and perhaps unwarranted confidence that the species in question demonstrate a log-normal relative abundance distribution. This confidence hardly seems justified, however, since the conditions which led Preston (1960, 1962) to propose a log-normal relative abundance distribution are often not met: i.e., the areas are "true isolates" (independent and never contiguous), the log-normal distribution is totally "unveiled," and the number of species is large (at least 50–100) to avoid "contagion." Even though these criteria are not satisfied, the power function may show a significant correlation between species number and area because it can closely approximate both the untransformed model and the exponential model, especially when there is a great deal of variance around the regression line. Unfortunately, approximating these models with the power function may mask valuable biological information (May 1975). The inference that a significant fit necessarily implies an underlying log-normal distribution is therefore ill-founded. Clearly, a more reasonable course is to search out the model giving the best statistical fit.

The reasons for transforming the independent and/or dependent variable(s) in regression analysis (see Sokal and Rohlf 1969, pp. 476) are to transform a curvilinear relationship into a linear one and to normalize the residuals and make them homoscedastic. The procedure usually allows an increase in the proportion of variance explained. Keeping these criteria in mind, the best model was determined by visual inspection of graphical plots of each data set for the untransformed and all transformed models, as suggested by Sokal and Rohlf (1969). The model that adequately linearized the relationship and reduced the deviation of points around the regression line was categorized as the best model (Appendix). If neither the untransformed model nor any of the log-transformations linearized the relationship, no best model was designated. If two or more models linearized the relationship and reduced the scatter of points about the line, the model with the highest r was considered the best model. Often two models fit a data set equally well (r's differing by less than 5%), and in these instances both models were considered best models.

Of the 100 data sets, 35 are best fit by the untransformed model (table 1), so that log-transformation of either of the variables is statistically inappropriate. Only 36 of the remaining 65 data sets are best fit by the log/log approximation of the power function. Most importantly, the reason the log/log model fits such a large number of data sets is that it turns virtually any monotonic function into a straight line (Preston 1962). Thus, 75 of the 100 data sets show no substantial lack-of-fit when log/log transformed (that is, no systematic pattern in the residuals can be detected by visual inspection). Recall though, that only 36 of these 75 log/log transformations are considered best models.

Dony (unpublished manuscript kindly supplied to us by F. H. Perring) has compared the fit of the power function and of the exponential model to a number of species-area relationships derived from plant quadrat studies, and concluded that the power function is usually superior in linearizing species-area relationships. This result is consistent with our findings, but our analyses indicate that although the

TABLE 1

SUMMARY OF THE "BEST MODEL" ANALYSES; (A) FOR ALL STUDIES AND (B) WITH THOSE BEST FIT BY THE UNTRANSFORMED MODEL (35 studies) REMOVED

	MODELS			
	S/A	S/LA	LS/A	LS/LA
A				
Highest r	50	52	24	53
No "lack-of-fit"	47	38	22	75
Both (Best fit)	35	27	14	43
B				
Highest r	...	32	11	45
No "lack-of-fit"	...	24	7	44
Both (Best fit)	...	19	5	36

NOTE.—Entries indicate the number of times a particular model possessed the highest r, no "lack-of-fit," or both these characteristics. There were studies for which two or more models fit equally well, since we did not discriminate between correlation coefficients that differed by less than 5%. As a result, the rows do not sum to 100. S/A = untransformed model; S/LA = species/logarea model; LS/A = logspecies/area model; LS/LA = logspecies/logarea model.

power function may often be superior to the exponential model it does not provide a better fit substantially more frequently than does the untransformed model.

We can discern no apparent pattern that seems to predict when the log/log model will be the best fit. As noted previously, studies meeting Preston's two assumptions (i.e., true isolates and large total species number) should be best fit by the log/log model. However, when only such studies are considered, less than half (14 of a total 32) are best fit by the power function exclusively (see fig. 1). From the work of Preston (1960), Williams (1964), and May (1975) we might expect the log/log model to fit studies with relatively large area ranges better, as a consequence of higher total species numbers. However, this pattern is not apparent when relationships among the area ranges of these 32 data sets and their best fit models are examined (fig. 1). Neither number of orders of magnitude of area that a data set covers, nor the particular orders of magnitude that are covered, indicate which model should be the best fit.

The apparent linearity of the relationship between species number and area may be the result of sampling a narrow range of areas. A few researchers (e.g., Archibald 1949; Vestal 1949; Niering 1963; Whitehead and Jones 1969; Abbott 1973; Lassen 1975) have noted that the species-area curves for their data sets possess multiple inflection points when a wide range of area is sampled. This observation is a restatement of the concept of breaks in the species-area relationship noted by Cain (1938). In these instances species-area plots are sigmoidal and are not linearized by the transformations we considered. Thus, in order to depict accurately the distribution of species number with area and select a best model, one must sample a wide range of area (Diamond and Mayr 1976).

If log-normal relative abundance distributions predominate in nature then the power function may have theoretical justification. However, since both the log-normal distribution and the power function are so robust their ability to approxi-

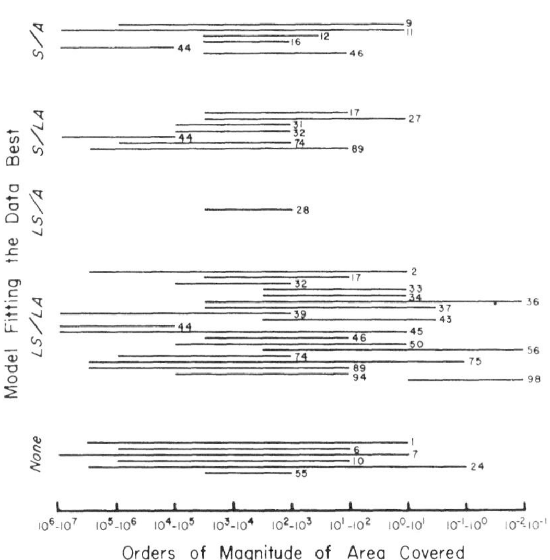

FIG. 1.—Area ranges covered by the 32 studies meeting Preston's criteria (i.e., true isolates and large numbers of species). The studies are grouped by their best-fit models in order to show the lack of relationship between area range and best-fit model. Each line represents the area range of a single species-area curve. The numbers placed at the ends of the lines refer to the studies as numbered in the Appendix.

mate the distribution of abundances and species numbers may reflect nothing more than the central limit theorem (May 1975). These properties are a strong practical justification for the use of the power function, yet cloud its biological interpretation.

CAN THE PARAMETERS OF THE POWER FUNCTION BE INTERPRETED STATISTICALLY AND BIOLOGICALLY?

Prior to 1960, discussion centered on the best-fit model of the species-area relationship and accurate prediction of species number. Many recent analyses, however, have attempted to interpret the slope and intercept parameters of the power function. Gleason (1922, 1925) and Arrhenius (1921, 1923) originally considered the parameters of the species-area relationship to be arbitrary fitted constants. Concomitant with the hegemony achieved by Preston's power function model was the development of the idea that the parameters of the power function possessed biological significance. This concept was first manifested by Preston's (1962) prediction of a "canonical" 0.262 value of the slope parameter of the power function caused by the hypothesized log-normal distribution of individuals into species. Subsequently, most publications of estimated values of the parameters of the power function have suggested biological interpretations and attempted to compare these parameter values. In disciplines other than ecology the power function has frequently been applied to the description of biological phenomena. It has been used widely in morphological (Huxley 1932; Gould 1966), fisheries (Ricker 1973), physiological (Gunther and Guerra 1955; von Bertalanffy 1957) and other analytical contexts (see

Gould 1966; Zar 1968), in many of which biological interpretations are suggested for its parameters. The parameters of the exponential species-area model, although receiving some attention from plant ecologists, have generally been ignored along with the parameters of the untransformed and logspecies/area models. Before discussing the substance of these interpretations and comparisons, we describe the techniques used to obtain these parameter estimates and detail statistically correct procedures for comparing and drawing inferences from them.

In practice data are seldom fitted to the power function per se, but are usually fitted to its log/log transformation. In both the exponent z is the slope of the line. The power function has an assumed y-intercept ($A = 0$) of 0, while its log/log transformation has a y-intercept ($A = 1$) of log k (see eq. [1] and [2]). As pointed out by Zar (1968), fitting data to the log/log transformation yields only approximate estimates of the parameters of the power function, and may in fact produce significantly different estimates of z, especially when $r \ll 1$ (which often occurs in species-area analyses). Nevertheless, the log/log transformation is assumed equivalent to, and has been used to estimate k and z values from, the power function in species-area relationships with, we believe, only one exception (Sepkoski and Rex 1974).

The exponential or species/logarea model possesses a slope of z and a y-intercept of log k. The untransformed and the logspecies/area models, which are of the forms

$$S = zA + k \tag{4}$$

and

$$\log S = zA + k, \tag{5}$$

respectively, have slopes z and y-intercepts k. Neither the untransformed nor the logspecies/area models are in use in simple species-area analyses (see however, Moore and Hooper 1975; Strong et al. 1977), but have been included in multiple regression analyses of species number (Johnson and Simberloff 1974; Strong et al. 1977). The exponential model, as stated previously, was originally proposed by Gleason (1922) and has commonly been employed in botanical studies.

Estimates of the parameter values (z, k, log k) have always been obtained from model I least-squares regression. In model I regression, only the dependent variable is assumed to be subject to measurement error. However, it is quite common in species-area relationships to encounter a sizable error in the measurement of the independent variable, area. When the assumption of no measurement error in the independent variable is violated, least-squares regression will systematically underestimate the slope (Ricker 1973). To alleviate this problem two alternatives are available, the "Berksen case" and model II regression. In the Berksen case (Ricker 1973), measurement error is permitted but controlled by the experimenter (e.g., island areas selected a priori; 10 km^2, 100 km^2, 1,000 km^2, etc.). In species-area studies, the measurement error in area is uncontrolled, and therefore model II regression should be used. Ricker (1973), who provides an excellent review of the problem, recommends the reduced-major-axis (geometric mean) regression method, although others (Jolicoeur 1968; Pilbeam and Gould 1974) prefer major-axis regression (the first principal component). We have computed both least-squares and reduced-major-axis parameter estimates for our analyses, although we will discuss only least-squares estimates because similar trends in slopes and intercepts were obtained using both techniques. Reduced-major-axis parameter values are, however, consistently higher than least-squares values (tables 2 and 3).

Regardless of the particular model, the interpretation and comparison of parameter estimates is constrained by the prerequisites and assumptions of the formal statistical procedures used in their estimation. The slope parameter (z) and intercept parameter (k or $\log k$) may be compared to some hypothesized value (e.g., Preston's canonical 0.262 for z or 1 for $\log k$) through the application of the appropriate t test (Sokal and Rohlf 1969). Comparisons among z values, although slightly more difficult, may also be accomplished by the application of the appropriate t test or by analysis of covariance (Sokal and Rohlf 1969), but additionally require that the range of values of the independent variable (area in this case) overlap considerably between studies (i.e., if islands in archipelago A range in area between, say, 1 and 10^5 km^2, then the island areas of archipelago B must either be completely included within, or comprise a majority of, this range). The comparison of intercept parameters between regressions is similarly constrained and the appropriate t test is identical to that for the comparison of slopes between regressions, with the values of the intercepts and their standard errors appropriately substituted. However, the slope and intercept of the power function are interdependent parameters, and as a result only intercepts from regressions of equal slopes can be compared (White and Gould 1965). Tests for differences in intercepts are only available for parallel lines, since no sure technique to separate the effects of the correlation between slope and intercept on the intercept from real differences in the intercept is available.

In some models either the slope or the intercept parameters depend upon the measurement units of the independent variable, area. The estimate of the slope is unaffected by the measurement units of area in the power function and exponential models; they need not be in the same units in two regressions for comparison purposes. However, the intercepts, k in the power function and $\log k$ in the exponential model, depend upon the units of area measurement. In the untransformed and logspecies/area models, the intercept is independent of and the slope dependent upon the units in which area is measured.

An additional problem in estimating and comparing intercepts arises when small areas have not been included in the regression (Diamond and Mayr 1976). For the untransformed and logspecies/area models this means islands approaching 0 area; and for the power function and exponential models islands at least as small as 1 unit of area. When such points are not included in the regression, estimating and interpreting the intercept values amounts to extrapolating beyond the ends of the regression line, where the confidence intervals flare dramatically (Haas 1975). As pointed out by Sokal and Rohlf (1969, p. 426–427), "... one should be very cautious about extrapolating from a regression equation if one has any doubts about the linearity of the relationship." The inherently asymptotic behavior and possible sigmoidal form of the species-area relationship raise such doubts.

Interpretation of the Slope Parameter

A particularly interesting characteristic common to all models of the species-area relationship is the rate at which species accumulate with increments in area. In linear

TABLE 2

MEANS, STANDARD DEVIATIONS, MINIMUMS, AND MAXIMUMS OF LEAST-SQUARES AND REDUCED-MAJOR-AXIS SLOPE VALUES FOR EACH OF THE FOUR MODELS

	Slope Values				
	Mean	SD	Minimum	Maximum	No.
Untransformed model					
Least-squares	40.130	281.497	−.000	2,645.093	90
RMA	62.248	467.443	.000	4,415.848	90
Log/log model					
Least-squares	.310	0.227	−.276	1.132	90
RMA	.468	0.285	.114	1.700	90
Species/logarea model					
Least-squares	38.831	98.587	−442.640	486.430	90
RMA	81.014	181.005	2.088	1,361.969	90
Logspecies/area model					
Least-squares	1.083	4.493	−.000	31.411	90
RMA	1.715	7.967	0	65.033	90

NOTE.—Ten of the 100 studies are not included in this analysis since the area measurements were in linear, cubic, or other measurements not readily converted to km^2. The studies deleted are listed in the Appendix as numbers 91–100. Values of "−.000" indicate small negative numbers.

TABLE 3

MEANS, STANDARD DEVIATIONS, MINIMUMS, AND MAXIMUMS OF LEAST-SQUARES AND REDUCED-MAJOR-AXIS INTERCEPT VALUES FOR EACH OF THE FOUR MODELS

	Intercept Values				
	Mean	SD	Minimum	Maximum	No.
Untransformed model					
Least-squares	69.852	214.990	−23.672	1,626.268	90
RMA	50.651	157.737	−84.548	1,060.492	90
Log/log model					
Least-squares	.704	1.153	−4.402	3.695	90
RMA	.274	1.518	−8.728	3.652	90
Species/logarea model					
Least-squares	8.405	446.869	−733.762	3,887.370	90
RMA	−172.285	655.154	−5,734.608	375,062	90
Logspecies/area model					
Least-squares	1.163	.668	−.440	3.142	90
RMA	1.055	.681	−1.070	3.121	90

NOTE.—Ten of the 100 studies are not included in this analysis since the area measurements were in linear, cubic, or other measurements not readily converted to km^2. The studies deleted are listed in the Appendix as numbers 91–100.

models this rate of accumulation is represented by a single parameter, the slope of the line, and as a consequence of the assumed linearity of the model is a constant value. Curvilinear models treat the rate of accumulation of species as a constantly changing value (hence the inherent curvilinearity of the model) described by one to a few parameters. Because of the relative ease of manipulation and interpretation, linear models and linear approximations to curvilinear models have naturally been preferred. Of the four linear models we have examined, only the parameters of the log/log approximation to the power function have been the subject of considerable interpretive effort. The following discussion of interpretations of the slope parameter will predominantly concern the log/log model with only passing references to the other models.

The averages and ranges of least-squares and reduced-major-axis estimates of slope values encountered in our set of 100 species-area curves from the four linear models are presented in table 2. In all four models, large positive values indicate high rates of species accumulation with increments in area, whereas small values indicate low species accumulation rates and negative values an absolute impoverishment of large areas relative to small ones. In the log/log model, a slope value of 1.0 indicates that species number and area are "isometric" (*sensu* Gould 1966). Slope values above 1.0 indicate a relatively greater number of species per unit area in large than in small areas, and slope values between 0.0 and 1.0 indicate a diminishing return in species number per unit area (Abele and Connor 1978).

Preston's canonical 0.262 *slope and the regularity of observed z-values.*—The first statement concerned with the pattern in the value of the slope parameter was Preston's prediction of a canonical 0.262 slope value in the log/log model; many empirically obtained values were consistent with this figure. Although Preston (1962) noted that the logspecies/logarea curve derived from his canonical log-normal relative abundance distribution has a slope of 0.262, errors in sampling and other factors cause variation about this canonical value. Thus, Preston (1962) considered values of about 0.17 to 0.33 to be within the canonical range, while MacArthur and Wilson (1967) accepted values of about 0.20 to 0.35. Preston's "canonical hypothesis" was that the parameter γ of the underlying log-normal distribution is 1, which yields his predicted slope value. May (1975), using a set of realistic but noncanonical log-normal relative abundance distributions ($\gamma = 0.60$–1.70), derived slopes in the range of 0.15 to 0.39. Finally, Schoener's (1976) modification of the equilibrium model leads to slopes between 0 and 0.50. It has become axiomatic that a slope within the circumscribed range noted above (about 0.20 to 0.40) is a singular consequence of deriving a logspecies/logarea relationship from an underlying log-normal relative abundance distribution. However, a few researchers (May 1975, Schoener 1976) have suggested that the result may more likely be a mathematical coincidence. We agree that coincidence is involved, and illustrate here why the slopes of the log/log curves fall regularly between 0.20 and 0.40.

Consider the equation relating the regression coefficient, or slope of the regression line z, to the correlation coefficient r:

$$z = r(s_y/s_x), \tag{6}$$

TABLE 4

CONSTRUCTION OF THE EXPECTED VALUES OF THE REGRESSION COEFFICIENT (z) WITH THE CONSTRAINTS $0 \leq r \leq 1$ AND $0 \leq s_y/s_x \leq 1$ (see eq. [6]).

	s_y/s_x									
r	.1	.2	.3	.4	.5	.6	.7	.8	.9	1.0
.1	.01	.02	.03	.04	.05	.06	.07	.08	.09	.10
.2	.02	.04	.06	.08	.10	.12	.14	.16	.18	.20
.3	.03	.06	.09	.12	.15	.18	.21	.24	.27	.30
.4	.04	.08	.12	.16	.20	.24	.28	.32	.36	.40
.5	.05	.10	.15	.20	.25	.30	.35	.40	.45	.50
.6	.06	.12	.18	.24	.30	.36	.42	.48	.54	.60
.7	.07	.14	.21	.28	.35	.42	.49	.56	.63	.70
.8	.08	.16	.24	.32	.40	.48	.56	.64	.72	.80
.9	.09	.18	.27	.36	.45	.54	.63	.72	.81	.90
1.0	.10	.20	.30	.40	.50	.60	.70	.80	.90	1.00

where s_y and s_x are the standard deviations of the dependent and independent variables, respectively (Draper and Smith 1966, p. 35). Allowing that the value of r falls between 0 and 1 (as it must for a positive correlation) and that $s_y < s_x$ (because of the asymptotic behavior of species number), we construct the relationship shown in table 4 simply by multiplying the marginal values of r and s_y/s_x to yield slope values (eq. [6]).

Even with these conservative assumptions, 30% of the slopes are expected to fall between 0.20 and 0.40. However, of the 100 species-area curves we examined, 45% had log/log slope values between 0.20 and 0.40, (see fig. 2). Since the ranges of r and s_y/s_x of our 100 species-area relationships, and we assume of most analyses, tend to be much smaller, then slope values between 0.20 and 0.40 should be, and are, more frequently observed. The question most germane to this problem is why do r and s_y/s_x have such narrow ranges?

Values of the correlation coefficient r are usually above 0.50 for logspecies/logarea regressions, most likely because insignificant correlation coefficients are not published, and because both variables are log-transformed. The observed narrow range of s_y/s_x (usually between 0.20 and 0.60) is a consequence of the asymptotic behavior of species number; once species number becomes asymptotic, area can be increased virtually indefinitely, and concurrently s_y/s_x and the slope will decline. In other words, since species-area curves are characterized by inherently larger ranges of areas than species numbers, the numerator of the term s_y/s_x will always be smaller (usually much smaller) than the denominator. Hence, the small fractional values of s_y/s_x multiplied by r (see eq. [6]) produce lower slopes the larger the area range.

In essence, our contention is that the narrow range of observed slope values (0.20–0.40) is more parsimoniously explained to result from the characteristics of the regression system, and not from underlying log-normal relative-abundance distributions. One might argue that the observation of 45/100 slope values between 0.20 and 0.40 merely confirms May's (1975) observation on the robust nature of the noncanonical log-normal relative abundance distribution and does not really demonstrate

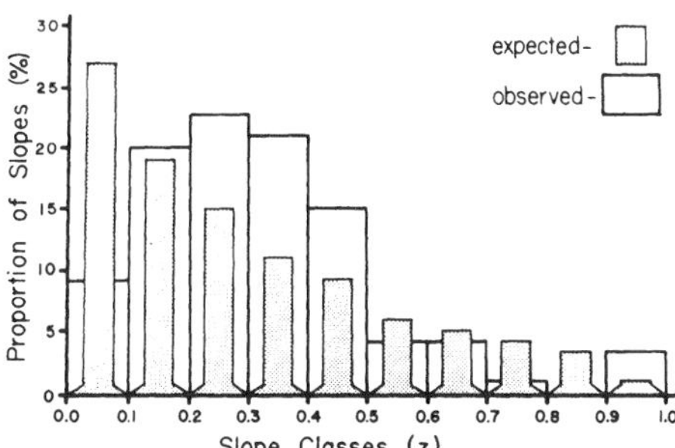

FIG. 2.—Comparison of expected and observed log/log slope values. Expected proportions of slope values for particular classes were generated by summing the entries in table 4 and dividing by 100 for each class. Observed proportions were similarly derived from the data in the Appendix. Slope values exceeding 1 or less than 0 (2 values each) were tabulated within the highest and lowest slope-value classes, respectively.

any mathematical coincidence. We counter this by noting that of the 36 data sets best fit by the log/log model (see table 1), only 15 have slopes between 0.20 and 0.40. This observation means that a slope between 0.20 and 0.40 is often obtained even when fitting the log/log model to data probably lacking an underlying log-normal relative abundance distribution.

Furthermore, in a completely unrelated discipline, slopes between 0.20 and 0.40 also show up consistently. In brain weight-to-body weight allometric regressions, intraspecific plots uniformly show a slope of 0.20 and 0.40 (Pilbeam and Gould 1974 and included references). This functional relationship is maintained by organisms displaying similar body plans over a wide size range. Interspecific plots of animals having an allometric relationship of brain weight to body weight display a higher slope (nearly always 0.66), and those with increased cephalization, an even higher one (greater than 1). Here again, in the intraspecific plots the range of the dependent variable is always much less than that of the independent variable (s_y/s_x exhibits small fractional values), r's are very high (usually greater than 0.90), and the slope almost always falls in the interval 0.20 to 0.40. In interspecific plots the range of the dependent variable is automatically increased (because of greater variability in size between adults of different species than among adults of the same species), therefore s_y/s_x and the slope increase also.

The regular occurrence of slope values between 0.20 and 0.40 thus seems to be an expected characteristic of any regression system with a high r value and a small range in the dependent variable relative to that in the independent variable. Although species-area curves derived from an underlying log-normal relative abundance distribution also display a similar narrow range of values, slope values in this range can be expected regardless of the underlying relative-abundance distribution. When interpreting slope values we suggest, to borrow a phrase from Gould (1971), that slopes in the 0.20 to 0.40 range (approximately) be considered as a "criterion of subtraction," or as the null hypothesized range of slope values, perhaps indicating correlation between species number and area without a functional relationship. It may be that only slope values deviating from this range possess biological significance.

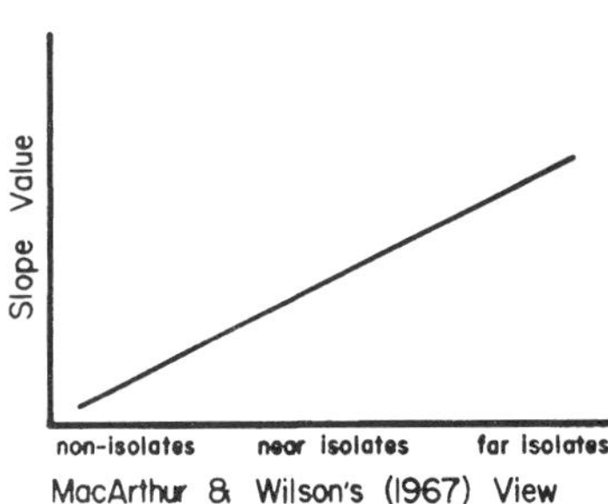

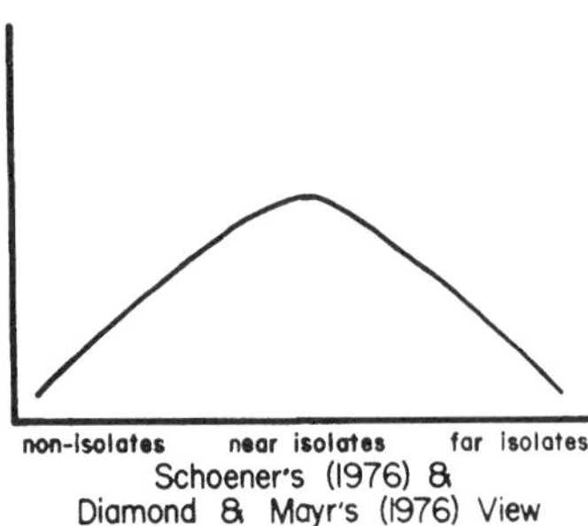

FIG. 3.—Diagrammatic representation of MacArthur and Wilson's (1967), Schoener's (1976), and Diamond and Mayr's (1976) hypotheses concerning the relationship of the slope value of the power function to isolation.

Island versus continental differences in the slope parameter.—We have seen that, based on the assumption of an underlying log-normal relative-abundance distribution, Preston (1960, 1962) predicted that the slope value of the log/log model for true isolates should be in the range 0.20–0.40. Deviations in observed slope values from the theoretical value were attributed to increases in habitat diversity (higher values) or to sampling nonisolated areas (lower values). Preston (1960) envisioned sampling from nonisolated areas as sampling from a truncated log-normal relative abundance distribution in which the ratio of species to individuals is much higher than in the complete log-normal distribution characteristic of an isolate. As a result, small areas would be overrich in species and the slope of the species-area curve would be depressed below the canonical value. Preston (1960) made his original observation of these low slope values in species-area curves for the Nearctic ($z = 0.12$) and Neotropical ($z = 0.16$) avifaunas. MacArthur and Wilson (1967) restated Preston's (1960) idea, proposing that slope values derived from nonisolated areas, either within islands or within continents, should fall in the range 0.12–0.19. They argue that since many transients will be encountered in the nonisolated areas, independent of area, species numbers in small areas will be inflated, depressing the slope of the logspecies/logarea curve (see fig. 3). Although not suggested by MacArthur and Wilson (1967), it is also wise to confine predictions to comparisons within taxa or other groupings of species with similar dispersal abilities.

Preston (1960) and MacArthur and Wilson's (1967) prediction of lower continental than island slope values can be interpreted literally or liberally. Their hypothesis could be considered falsified if the predicted pattern of slope values, in the specified ranges (0.12–0.19 for continents and 0.20–0.40 for islands) does not obtain. Alternatively, we could consider their hypothesis at least qualitatively supported if the predicted differences in slopes occur even though they do not segregate into the specified ranges.

Simberloff's (1976*a*) experimental work has shown that for nonisolated areas within islands species numbers are in fact inflated for small areas, suggesting that the transient hypothesis is sound. However, he made no attempt to relate his results to the slope value of the logspecies/logarea curve, since his sample size was small and the use of serially self-contained sample areas violates the assumption in regression that each measurement of the independent variable be derived independently. In addition, Goodall (1952), Greig-Smith (1964), and Kobayashi (1974, 1976) believe that slopes derived by combining random samples will be higher than those derived from the continuous expansion of a single sample, an effect independent of the transient hypothesis.

Adequate data to examine the effect of the transient hypothesis on the slope of the species-area curve are unavailable, but Johnson et al.'s (1968) analysis of the floras of the California Channel Islands and mainland southern California bears on this problem. Johnson et al. (1968) report a slope value for the Channel Islands of 0.472 and a slope value of 0.158 for mainland areas. This result appears to fit Preston's and MacArthur and Wilson's prediction at least qualitatively; however, the area ranges of the island (0.02–134 mile^2) and mainland (5.9–24,000 mile^2) regressions barely overlap, so the slopes cannot be compared properly. When we compare slope values generated from Johnson et al.'s (1968) data, but with similar area ranges (i.e., deleting islands with areas less than 1 mile^2 and mainland sites with areas greater than 529 mile^2) the island (0.06) and mainland (0.27) slope values differ as per MacArthur and Wilson's and Preston's prediction. However, these values still do not segregate into the predicted ranges. Preston's (1960) original observations of low slope values in the nonisolated Nearctic and Neotropical avifaunas are subject to the same criticism. The area ranges covered by these continental studies are tremendously greater than those of any island archipelago. The behavior of these slope values could result from depression of species numbers in large areas, because of the asymptotic nature of species numbers, rather than the inflation of species numbers caused by more transients in small areas. Brown's (1971) study of the montane mammals of the great basin also appears to support the transient hypothesis and its effect on the slope of the logspecies/logarea curve. However, Brown's mainland (nonisolated) slope value was based on four sample areas, none of which were within the range of area covered by the comparable small isolates, exactly the range critical to a test of the transient hypothesis.

Low slope values have also been obtained for truly insular situations (isolates). Case (1975) reported a slope of 0.166 for the lizards of the California Channel Islands, Baroni-Urbani (1971) a slope of 0.188 for the ants of the Tuscan archipelago, and Harris (1973) a slope of 0.157 for the birds of the Galapagos. This evidence falsifies MacArthur and Wilson's (1967) prediction of isolate slopes falling exclusively in the 0.20–0.40 range (or at least not below 0.20), but remains open to the interpretation that were the slopes for those taxa known for adjacent nonisolated mainland areas, they would be comensurately lower.

The evidence indicates that the postulated effect of transients on slope values from nonisolated areas remains testable when interpreted broadly. Although slope values from some isolated areas fall within the predicted range for nonisolated areas, actual slopes from nonisolated areas could be lower yet. The relatively low correlations ($r < .9$) observed between species numbers and area in most instances, and their considerable range, could possibly mask this pattern if it exists.

Isolation and the slope parameter.—It has long been known that geographically isolated archipelagos possess depauperate biotas. Hamilton et al. (1963) and later others (Simpson 1974; Power 1972; Johnson and Simberloff 1974; Johnson et al.

1968, etc.) have demonstrated that isolation explains a significant amount of the variation in species number. Utilizing stepwise multiple regression analyses, each of these workers concluded that isolation accounts for the reduced numbers of species after the effect of area has been factored out.

In view of this pattern, MacArthur and Wilson (1967) proposed a parallel phenomenon for the slope of the species-area relationship. Their prediction, based on equilibrium theory, was that the slope of the species-area curve would be higher for distant or isolated archipelagos (fig. 3). This explanation is an extension of the transient hypothesis offered for island versus continent differences in the slope parameter (Preston 1960; MacArthur and Wilson 1967). The idea that isolated archipelagos have fewer transients caused by lower immigration rates has been challenged by Abbott and Grant (1976).

MacArthur and Wilson (1963) were able to muster little evidence to support their prediction; and subsequently Hamilton and Armstrong (1965) observed a decreased slope with isolation, exactly opposite MacArthur and Wilson's prediction (fig. 3). Schoener (1976) provides the best and most complete analysis of this question to date. He plotted the slope values obtained for land and freshwater birds from 23 archipelagos versus isolation and confirmed the result of Hamilton and Armstrong, that the slope decreases with isolation. We performed analyses similar to Schoener's and show an identical trend. For the total birds subset (17 studies, see section on the Latitudinal dependence of the species-area relationship for a detailed explanation concerning how this subset was constructed) Spearman correlation coefficients were computed between the slope parameter and isolation distance. The results show that the log/log slope is significantly negatively correlated with isolation ($r = -.6872$, $P = .004$).

Schoener's explanation for this relationship is that the slope of the species-area curve is dependent upon the size of the source pool of species, which in the case of distant archipelagos will be small, therefore lowering the slope. However, Schoener's explanation may not apply to all taxa since distant archipelagos may have smaller source pools without having lower slopes if the intercept also changes with isolation (fig. 4). We can see from this problem that although trends in the slope or intercept with isolation may be observed, we have no means of predicting their form. Even if the pattern observed by Schoener (1976) and Hamilton and Armstrong (1965) was determined to be ubiquitous, it reveals little more than has long been established: Distant archipelagos have depauperate biotas.

Equilibrium theory explanations of variation in slope.—Numerous authors have attempted to explain variation in the log/log slope value in terms of the "equilibrium theory" proposed by Preston (1960) and MacArthur and Wilson (1963, 1967). Equilibrium theory considers species number to be the result of a dynamic balance between immigration and extinction of species. Species number may be affected by either process individually (varying immigration or extinction rates) or both simultaneously. An interrelationship between immigration and extinction rates and the parameters of the species-area curve, although never fully explored, has been assumed to exist (Ricklefs and Cox 1972). As previously stated, MacArthur and Wilson (1967) first predicted that high immigration rates would decrease the slope of the species-area relationship. Subsequently Brown (1971), Terborgh (1973), and Strong and Levin (1975) have interpreted empirically derived estimates of the slope

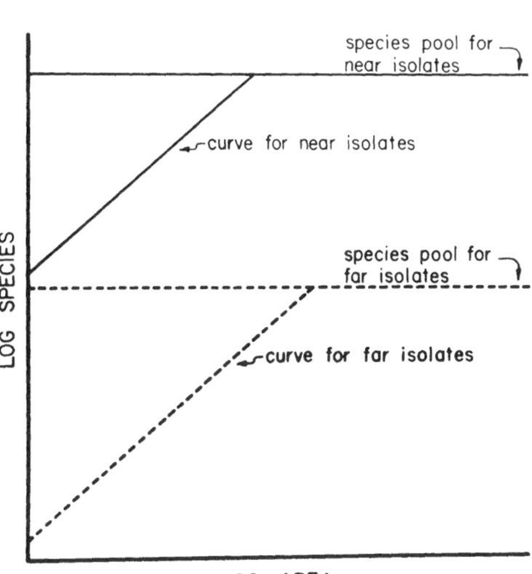

FIG. 4.—Illustration of how the slope of the species-area curve is potentially independent of the size of the source pool of species. In this example the slopes of the hypothetical species-area curves are equal even though the source pool of the distant archipelago is smaller than that of the near archipelago, because the y-intercept value of the curve for the distant archipelago is changed.

(z) in such a manner. However, Johnson and Simberloff (1974) point out that even within the equilibrium theory context low z values are not uniquely explained by high immigration rates, but likewise by low extinction rates or by a combination of high immigration and low extinction rates. Thus, three alternative hypotheses can be generated from a single theoretical framework (equilibrium theory), whose uncritical acceptance has been criticized by Lynch and Johnson (1974) and Simberloff (1976*b*).

An additional problem is that of establishing ultimate causality. Strong and Levin (1975), for example, postulate that the relatively low z value for the parasitic fungi of British trees compared to that of the phytophagous insects of British trees is due to high immigration rates for fungi. Their logic derives from the anemochorous dispersal of fungal spores. Even given this dispersal characteristic, the ultimate cause of the low z value for fungi may be due to a depauperate species pool, inasmuch as the high dispersibility of fungal spores would inhibit diversification through allopatric speciation. This latter alternative, that of an evolutionary difference in insect and fungal diversification caused by dispersibility, and Strong and Levin's equilibrium theory model must be viewed as competing hypotheses.

As discussed by Simberloff (1976*b*), equilibrium theory, like the log/log model of the species-area relationship, has been elevated to the status of a paradigm. Moreover, the ascendency of equilibrium theory as the major underlying theoretical framework in biogeography and population ecology has motivated many workers to interpret their results within the framework and to consider successful interpretation prima facie evidence of the veracity of the interpretation. Equilibrium theory and ideas interpreted within its framework must be restated as testable hypotheses, not accepted as proven.

Interpretation of the Intercept Parameter

The intercept parameter (*y*-intercept value) has been virtually ignored as a quantity deserving biological or statistical explanation, or as a basis for biological inferences. MacArthur and Wilson (1967) consider it solely as a fitted constant relating to local environmental conditions. Unlike the slope parameter, no regularly recurring values have been reported and no "canonical" value hypothesized. MacArthur's (1965, 1969) treatment of the latitudinal relationship of the species-area effect and Johnson and Raven's (1970) view that the intercept will decrease with increasing latitude are the only attempts to explain geographic patterns (in this case purely hypothetical) in the intercept parameter.

The averages and ranges of least-squares and reduced-major-axis estimates of intercept values encountered in our set of 100 species-area curves from the four linear models are presented in table 3. As mentioned previously, the untransformed and logspecies-area intercept parameters are not dependent on the measurement units of area, whereas the log/log and species/logarea parameters are. Biologically realistic values of the intercept parameter in the untransformed and logspecies/area models are values of 0.0 and below; positive values of the intercept parameter in these models would indicate the unlikely situation that in a sample of no area there exists some number of species. In practice, parameter values greater than zero are commonly found (see Appendix), and as a result are uninterpretable in these instances. Biologically realistic values of the intercept parameter in the log/log and species/logarea models contain a large range of real numbers. Positive values indicate that some number of species (if 1.0 or greater) will be found or that a probability of finding species (if between 0.0 and 1.0) exists when a sample of one unit of area is examined. Negative or zero values of the intercept parameter in these models indicate that no species will be found in a sample of one unit of area.

Heatwole (1975) suggests that, because of the uninterpretable values often obtained for the *y*-axis or species-intercept, we abandon attempts to attach biological significance to it and use instead the *x*-axis or area-intercept. Heatwole considers the *x*-intercept to be an indication of the "minimal area" necessary to support a breeding population of the particular taxon being studied. Hopkins (1957) previously discussed the term "minimal area" in plant community analyses; however, his usage is completely different from Heatwole's. Currently, Heatwole's suggestion remains an unexplored possibility.

The intercept parameter may, in fact, be affected by local environmental conditions or other factors (MacArthur and Wilson 1967), but concrete demonstration of these relationships and an assessment of their proportional contribution to its variation would be enormously difficult, since the proper analysis must follow the procedures described above. Assembling a large enough subset of intercept values from species-area curves with homogenous slopes that simultaneously vary with respect to the environmental conditions under study would probably be impossible. The same factors that may potentially cause variation in the intercept are likely to have similar effects on the slope parameter, thereby precluding the examination of their relationship to the intercept parameter. Because of these analytical problems, and also the lack of any a priori theoretical framework for its biological significance, the intercept parameter must be considered simply a fitted constant.

The Partitioning of Alpha and Beta Diversities into the Slope and Intercept Parameters

Whittaker (1960) first introduced the concept of alpha, or within habitat, and beta, or between habitat, diversity in 1960 as an attempt to partition diversity into independent components. MacArthur (1965) attempted, in part, to unify conceptual treatments of diversity using the species-area curve as an analytical tool by suggesting that the intercept parameter was a measure of alpha diversity and the slope parameter a measure of beta diversity.

Inasmuch as the concepts of alpha and beta diversity treated these components as independent, the attempt to establish their proportionality to the parameters of the log/log species-area model was doomed from the start. Since the slope and intercept of the power-function are algebraically interdependent parameters (White and Gould 1965, Gould 1966, 1971), when slope changes occur (caused, according to MacArthur, by adding or deleting habitats) it is impossible to compare the newly generated intercept to the pre–slope-change intercept since no statistical procedure exists to separate differences between intercepts caused either by the slope or by real changes in the intercept. Therefore, in MacArthur's system a change in slope precludes identifying a change in intercept.

Beyond the critique on statistical grounds, some empirical observations on the slope parameter are also pertinent. Several workers have prepared species-area curves for "single-species habitat islands"; Strong (1974*b*) and Strong et al. (1977) for phytophagous insects on host plant islands and Abele (1976) and Abele and Patton (1976) for decapod crustaceans on "coral head islands." Southwood (1960), Janzen (1968), and Strong (1974*a*) all contend that many phytophagous insects view single plant species as a habitat. Abele and Patton (1976) give convincing evidence that single-species coral heads are a single habitat by demonstrating that all decapod associates are found on a complete size range of coral heads. If the slope from the log/log species-area model is a measure of between-habitat diversity, as suggested by MacArthur, we would expect slope values of zero for these within-habitat studies. Instead, we observe values of z ranging from 0.327 to 0.370 (all significantly different from zero, $P < .05$). In essence, as we add area of the same type of habitat we add species and therefore generate a "within-habitat slope" (which is consistent with the area–per se hypothesis). Although it is possible that slope values would be higher if habitats were added, it is evident that between-habitat diversity does not account completely for observed slope values.

As shown above, even for simple systems some component of the slope is probably due to within-habitat diversity. For more interesting cases, such as archipelagos of true islands, we have no way of enumerating the numbers of habitats or their respective areas in order to attribute differences in slopes or intercepts to changes in alpha or beta diversities. We therefore consider it logically and practically impossible to apportion alpha and beta diversities to the intercept and slope parameters.

The Latitudinal Dependence of the Species-Area Relationship

MacArthur (1965, 1969) predicted that concomitant with latitudinal gradients in species number (either total or mean species number for equal sized areas) one should observe latitudinal gradients in either or both of the parameters (slope and intercept) of the power function. This prediction, in tandem with his attempt to apportion within-habitat and between-habitat diversity to the intercept and slope parameters, led him to conclude that an investigation of the latitudinal dependence of the slope and intercept of the species-area relationship would enable one to discriminate between three alternative explanations for the existence of latitudinal diversity gradients. MacArthur reasoned that (1) if only the intercept was inversely correlated with latitude then latitudinal diversity gradients could be attributed to increased within-habitat diversity in the tropics, (2) if only the slope was inversely correlated with latitude then latitudinal diversity gradients were due to increased between-habitat diversity in the tropics, and (3) if both the intercept and slope were inversely correlated with latitude then latitudinal diversity gradients were due to increases in both within-and between-habitat in the tropics. However, as suggested above, within- and between-habitat diversity cannot be apportioned to the intercept and slope parameters for both statistical and biological reasons. A further problem stems from the lack of any technique for comparing intercepts between studies with unequal slopes. Thus, if a relationship exists between the slope and latitude, it precludes detecting any relationship between the intercept and latitude. As a result, MacArthur's third alternative, given contemporary analytical methods in parametric regression, could not be demonstrated even if it were the correct alternative.

Although the theoretical framework suggested by MacArthur for the interpretation of trends in the relationship between the slope or intercept of the log/log species-area model and latitude seems incorrect, the original prediction that a trend will exist is still worthy of examination. The basic question is: Given that we observe latitudinal gradients in total species number and mean number of species per unit area, should we expect to observe similar trends in the parameters (slope and intercept) of an empirically fitted model of the entire distribution of species number with area? To answer this question we again examine our set of 100 species-area curves, contrasting MacArthur's predictions as a set of alternative hypotheses against the null hypothesis that no trends exist. We will consider the relationship between the slope parameter and latitude, the intercept parameter and latitude, and, although not a part of MacArthur's prediction, the linear correlation coefficient and latitude.

Slope and latitude.—In order to examine the relationship between the slope parameter and latitude, we obtained subsets of studies within which valid comparisons of slopes could be made. To compare slopes from two species-area curves, each study must span similar area ranges or at least overlap considerably. To this constraint we added the requirement that comparisons be made only within taxonomic levels (orders, families, etc.). Since lower taxonomic levels are inherently less diverse than higher ones, for the same area range their slopes will automatically be lowered and could therefore generate spurious correlations or mask real correlations between the slope parameter and latitude. For example, slopes of species-area curves for vascular plants should not be compared to slopes of species-area curves for grasses only. The same problems could occur if studies of mixed taxonomic groupings (e.g., mammals, vascular plants, insects, and fish) were compared, since each taxa does not represent a constant proportion of the biota.

Given these two constraints, we determined that out of 100 species-area relation-

TABLE 5

RELATIONSHIP BETWEEN THE SLOPE PARAMETER AND LATITUDE

	VALUES OF r				
MODEL	Fish	Insects	Total Birds	Land Birds	Land and FW Birds†
Untransformed	−1.000*	−.4000	−.2108	.5000	−.6000
log/log	−.4000	0	−.0833	−.4000	.4000
Species/logarea	−.4000	0	−.4926*	−1.000*	0
logspecies/area	−.8000	−.4000	−.1386	−.5000	0

* Spearman's correlation coefficients between slope values and latitude means for each study in a subgroup (significant correlations, $P < .05$, are indicated by an asterisk; for a listing of studies comprising each subgroup see Appendix.

† FW = freshwater.

TABLE 6

CORRELATION (Spearman's) OF MEAN AND MAXIMUM SPECIES NUMBER WITH LATITUDE FOR TAXONOMIC SUBGROUPS OF SIMILAR AREA RANGE

Subgroup	Mean No.	logmean No. of	Max No. of	logmax No. of
Total birds (17)	−.6005*	−.5956*	−.5294*	−.5294*
Land birds (5)	−1.000*	−.9000*	−.9000*	−.9000*
Land & freshwater birds (4)	0	0	0	0
Insects	.8000	.8000	.2000	.2000
Fish (4)	−.6377	−.5218	−.4478	−.4478

NOTE.—The procedures used in constructing the subgroups are described in the text. For a listing of the studies included within each subgroup see Appendix.

* $P < .05$.

ships including numerous taxa, only five subsets fulfilling these requirements could be constructed; total birds (17 studies), land birds (5 studies), land and freshwater birds (4 studies), fish (4 studies), and insects (5 studies). This paucity of comparable studies illustrates the need for the continued examination and enumeration of species-area relationships.

Nonparametric correlation coefficients (Spearman's) were computed between the slope parameter and latitude for each of the four models of the species-area relationship being considered. The results of these analyses are presented in table 5. Both the mean and maximum number of species in each species-area relationship are significantly negatively correlated with latitude in only two of these subgroups, total birds and land birds (table 6). For land and freshwater birds, insects, and fish neither mean nor maximum number of species is correlated with latitude; in other words, no latitudinal gradient in species diversity is demonstrated by these three groups. This is not to say that in actuality land and freshwater birds, insects, and fish exhibit no latitudinal diversity gradient, only that for these particular species-area curves they do not. Since these three subgroups display no latitudinal diversity gradient, it is unlikely, although possible, that pattern in their slope values could be due to latitude. Thus, we attribute little significance to the correlation between the least-squares estimate of the slope parameter in the linear model and latitude for the fish subgroup (table 6).

Interestingly enough, for the two groups that display latitudinal gradients in mean and maximum species number, significant correlations between the slope parameter and latitude were not demonstrated for the log/log model but were evident for the exponential (species/logarea) model (table 6). When the species-area curves comprising these two groups are examined, either the species/logarea or the log/log are the best-fit models, indicating that the lack of relationship between the slope of the log/log model and latitude cannot be attributed to these subsets' being anomalous groupings, which are relatively poorly fit by the log/log model. For those subsets demonstrating latitudinal gradients in mean and maximum species number, only the slope in the exponential (species/logarea) model was significantly correlated with latitude.

Intercept and latitude.—Since intercepts can only be compared among groups of species-area curves with homogeneous slopes, we first constructed subsets by comparing slopes for all possible pairs of species-area curves for each of the four models in both the total birds and land birds subgroups. For the total birds subgroup this amounted to 136 t values per model and for the land birds subgroup 10 t values for each model.

In each subset of values, no slope differed significantly ($P < .05$) from any other member of the subset, and no other studies meeting these criteria could be added to the subset. For the 17 total bird studies, one subset of six studies in the untransformed model, three subsets of six in the log/log model, two subsets of five in the species/logarea model, and two subsets of six in the logspecies/area model could be constructed. For those models with multiple subsets, the subsets differed in composition from between one and four studies, but never were completely different. No subset of homogeneous slope values common to each of the four models could be constructed. For the five studies in the land birds subgroup, all slopes were significantly different in the untransformed model, one subset of three studies could be constructed in the log/log model, and one subset each of two studies could be constructed in the species/logarea and logspecies/area models. These subsets of the land birds grouping were considered too small for further analysis.

The relationship between intercept and latitude for the total-birds grouping of homogeneous slopes was investigated using Spearman's correlation coefficient. The results of these analyses are presented in table 7. No relationship between intercept and latitude was identified for either the untransformed, log/log, or species/logarea models. For the species/logarea model, where a relationship between slope and latitude had previously been identified, this analysis was actually superfluous since the existence of a slope trend precludes identifying an intercept trend. The results obtained for the logspecies/area model are equivocal. A significant relationship was identified in only one of the two subsets. Again, more and larger subgroups are needed for a complete analysis.

The linear correlation coefficient, r, and latitude.—Several workers (Preston 1962; Schoener 1976; Dony unpublished manuscript) have indicated that there may be an effect of geographic location on the fit of different models of the species-area relationship. To test this proposition we plotted the correlation coefficient derived

TABLE 7

RELATIONSHIP BETWEEN THE INTERCEPT PARAMETER AND LATITUDE FOR HOMOGENOUS SUBSETS OF SLOPE VALUES IN THE TOTAL BIRDS SUBGROUP

Subgroup	*r*	*P*	No.	Source Studies
Untransformed	−.2571	.312	6	(3, 14, 21, 39, 64, 74)
log/log	.0286	.479	6	(14, 15, 59, 74, 81, 89)
	−.1429	.394	6	(14, 27, 59, 75, 78, 79)
	−.1429	.394	6	(14, 27, 59, 75, 78, 81)
Species/logarea	−.6000	.143	5	(15, 21, 59, 60, 89)
	−.5000	.196	5	(15, 21, 59, 60, 79)
logspecies/area	−.6000	.105	6	(3, 14, 15, 21, 24, 64)
	−.7714	.037	6	(14, 21, 39, 60, 81, 89)

NOTE.—r = Spearman's correlation coefficient and P = level of significance. Subset composition indicated in parentheses refers to studies numbered in Appendix.

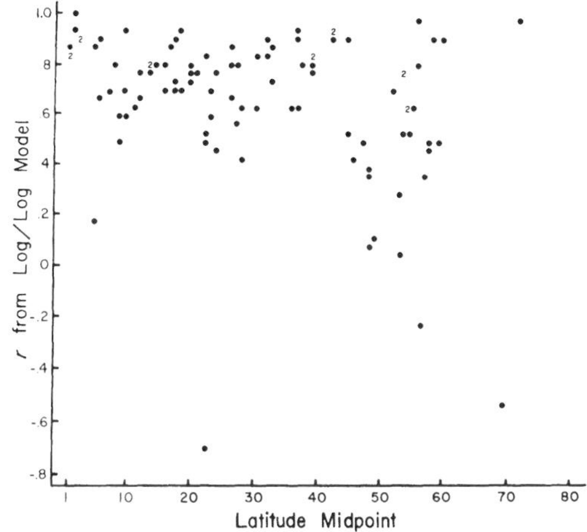

FIG. 5.—Relationship between the linear correlation coefficient of the log/log species-area model and the latitudinal midpoint of each study ($r = -.3183$, $P < .001$, $N = 100$). The relationship remains significant even when negative correlation values are removed.

from the log/log model of our 100 data sets versus latitude. Figure 5 shows that these correlation coefficients are negatively correlated with latitude ($r = -.3183$, $P < .001$); that is, log-area explains more of the variance in log-species at low latitudes that it does at high latitudes. The linear correlation coefficient is also significantly negatively correlated with latitude in each of the other three models. It might be suspected that this correlation is spurious, derived from a possible correlation between latitude and the number of data points contained in each study. However, the number of data points in a study is not correlated with latitude ($r = .0459$, $P = .325$).

Although the correlation between r and latitude is highly significant, 92% of the variance in r remains unexplained. This is partially due to the heterogeneity of the set of species-area relationships utilized. For example, habitat islands (eight studies) show no relationship between r and latitude, whereas for distant archipelagos (35 studies) latitude explains 49% of the variance in *r*. Longitudinal variance in species number also contributes to the large residual variance; for instance, studies performed in the British Isles and the Mediterranean region tend to have *r*'s that are higher than those from other regions at the same latitude.

Biologically, the lower correlation between species number and area at high latitudes may be the result of the relatively small source pool of species (as evidenced by latitudinal gradients in species number) and to each species' having on the average a relatively wider distribution than low latitude species (McCoy and Connor, in prep.). Hence, given the few species available to colonize a particular area and their wide distribution, species number rapidly becomes asymptotic for small areas and fails to increase when large areas are examined. Further, stochastic fluctuations in climate serve to maintain disequilibria in species numbers (Abbot and Grant 1976), resulting in a poor relationship between species number and area. Our analyses have revealed that there is no latitudinal dependence of the parameters of the species-area relationship, contrary to MacArthur's prediction. We do, however, confirm his intuition that there is a latitudinal dependence of the species-area relationship, but that it is manifested by the degree of correlation between species number and area, not the slope and intercept parameters.

SUMMARY AND CONCLUSIONS

We have discussed three basic questions concerning the species-area relationship. We now briefly summarize our conclusions, and discuss their ramifications for the future use of the species-area relationship, both its methods and interpretation.

Is there a unique theoretical basis for the species-area relationship?—Our discussion of the theoretical basis of the species-area relationship was basically inconclusive. The two most frequently proposed hypotheses, habitat diversity and area per se are both possibly correct, yet the result of either mechanism is neither qualitatively nor quantitatively different. One virtually always observes a positive correlation between species number and area, regardless of the mechanism. On the other hand, this result can also be explained as a consequence of isolates passively obtaining samples from some species pool, large isolates receiving effectively larger samples and ultimately containing more species than small isolates. It seems plausible that the habitat-diversity hypothesis could be tested by looking at equal sized areas with various numbers of habitats, assuming that habitats could be defined objectively. The area-per se hypothesis requires that one actually demonstrate decreased extinction rates for larger islands (heretofore taken to be a logical assumption), and the sampling hypothesis requires that we demonstrate a direct proportionality between immigration rates and area. There may be at least a grain of truth in each of these mechanisms. Each of these three, and possibly others, may play a role in producing the observed positive correlation between species number and area.

Is there a best-fit model of the species-area relationship?—Our analyses of 100 species-area curves indicates that there is no single best-fit model. The best-fit model for a particular species-area curve can only be determined empirically. Of the four linear models we examined, the power function and the untransformed models provide good fits most frequently. Curvilinear models were not examined, even though when a wide range of areas is sampled the species-area relationship can become sigmoidal. Comparing species-area curves from curvilinear models is inherently more complicated, and it is uncertain that any additional benefit would be derived. We suggest continued use of the power function and other linear models because of the relative ease which they can be compared, and their past and present wide usage.

Can the parameters of a particular model, specifically the power function, be interpreted?—In general, we have found that published predictions and interpretations concerning both the slope and intercept parameters are not supported by the available evidence. Many other predictions and interpretations are either logically untestable or require additional data for an adequate test. Because of these results, we are skeptical that any biological significance can be attached to these parameters and recommend that they be viewed simply as fitted constants devoid of specific biological meanings.

Species-area relationships: methods.—A discussion of the methods used in obtaining parameter estimates and comparing parameter values was presented. The use of either model I or model II regression in biology as a whole has usually been a matter of taste left to the discretion of each worker. However, in species-area analyses the degree of error in the independent variable, area, is great enough to warrant considering application of model II regression methods uniformly. The results of comparisons involving least-squares parameter estimates remain unchanged when using model II estimates. In this respect one has some leeway in choosing model I or model II regression, since model II yields more accurate estimates of the parameters, whereas the results of comparing parameter values are the same regardless of whether model I or model II estimates are employed. Obviously one should use model II when attempting to obtain accurate estimates of the parameters, and either model when comparing parameter values.

Perhaps a more fundamental question is whether regression or correlation should be used in species-area analyses. The particular problem under investigation dictates which method is appropriate. Correlation only allows the assessment of the degree of relationship between species number and area, and regression yields parameter values permitting comparisons of the bivariate distribution of species number with area. If one is interested only in the degree of relatedness between species number and area, correlation is the appropriate method. If one wishes to compare two or more bivariate distributions, then regression is the proper technique.

We recommend that each of these methods be used exclusively for the purposes described above. This is actually no more than recommending that biologists use statistics correctly. We especially encourage the publication of nonsignificant correlation coefficients between species number and area, values that now are probably either eliminated by the review process, by an author's disbelief in his own results, or the thought that they are uninteresting. Such examples are as informative about species-area relationships as are significant positive correlations, if not more so.

Species-area relationships: interpretations.—The interpretation of species-area relationships can be based on three criteria, (1) which model is the best fit, (2) the strength of correlation between species-number and area, and (3) how the parameter values compare to other published values.

Although Williams (1964), Preston (1960), and May (1975) have demonstrated that the log-series and log-normal relative abundance distributions are best fit by the exponential and power functions, respectively, the converse is not true. A species-area relationship best fit by an exponential or power function does not indicate an underlying log-series or log-normal relative abundance distribution. Relative abundance distributions can only be determined empirically; although one can predict, as May (1975) has done, that as a consequence of the statistics of large numbers relative abundance distributions are most likely to be log-normal for large species numbers. It may be premature to conclude that demonstrating that a particular model is the best-fit model is uninteresting, but as yet no significance can be attached to any particular model.

The degree of correlation or relatedness between species number and area could potentially be affected by numerous factors. The observation of an inverse relationship between the linear correlation coefficient and latitude may be due to high latitude species' possessing greater geographical ranges; hence few new species are encountered when one examines large versus small areas. No other pattern in the degree of correlation between species number and area has yet been identified. However, only a limited set of coefficients has been published (only significant correlations). The possibility exists that some pattern has been obscured by this practice.

The species-area relationship has unfortunately been used as a justification for conservation practices in which large areas are preserved in preference to small areas, since large areas are considered to contain more species (Terborgh 1974, 1975; Diamond 1975; Wilson and Willis 1975). Although we agree in principle with the preservation of large areas, the species-area relationship does not provide an unambiguous justification. As discussed by Preston (1962), Simberloff (1972), Simberloff and Abele (1976), and Abele and Connor (1978), it is both conceptually and actually possible that a group of small preserves contain more species than a single large preserve of equal total area. The dependence of the linear correlation coefficient of species-area curves on latitude also clouds the broad application of recommendations based on the species-area relationship without reference to geographic location. We agree with Simberloff and Abele (1976) that conservation areas should be designed with specific goals in mind, providing the particular habitat requirements for the species to be preserved.

Ultimately, species-area curves will be most useful in comparing diversities between geographical regions, habitats, or taxa over a range of sample sizes, or between different sized samples. Classical diversity measures compare diversities based on a single sample size, whereas species-area curves permit the comparison of the entire distribution of species number with area. Species-area curves can also be used to "factor out" the effect of area on diversity, so that the effects of other variables on species numbers can be determined. Strong (1974*a*, 1974*b*) has done this with phytophagous insect diversities in order to examine the effects of time, and Abele (1976) has done this to examine the effects of environmental stability on coral-inhabiting decapod crustaceans. Simberloff (1974), Raup (1976), and Sepkoski (1976) have employed species-area curves to explain the Permo-Triassic extinctions and Phanerozoic diversity trends in shallow-water marine invertebrates, although Raup views the effect of area (volume of sedimentary rocks) purely as a sampling phenomenon. It is through these comparisons of species-area curves, and only indirectly so, that the parameters of the power function or any other model have biological significance. In the absence of a priori theoretical bases for predictions concerning parameter values, such values must be considered simple fitted constants devoid of biological meaning.

ACKNOWLEDGMENTS

This manuscript benefitted greatly from the careful reading and critical comments of K. L. Heck, D. R. Strong, D. Simberloff, L. G. Abele, and G. Morrison, as well as the consultation of D. Meeter. We thank R. Ricklefs, F. H. Perring, J. Rey, and L. G. Abele for providing unpublished data for our analysis, P. Liepshutz, T. Berman, J. White, and L. Maddox for typing another tedious manuscript, Arlee Montalvo for preparing the figures, and Mike Auerbach for photographic assistance. This research was supported in part by NSF grant DEB 76-07330 to Daniel Simberloff and NSF grant DEB 76-24415 to Donald R. Strong.

APPENDIX

The following table is a list of each of the 100 species-area curves used in our analyses. We have included the source of each data set, the number of localities used to compute each curve, correlation and regression coefficients from all four models discussed in text, and other data pertinent to particular analyses or to the construction of subgroups for analysis. Data are not reported for studies not utilized in a particular group of analyses. For example, study number 1, Abbott's (1974) analysis of the land plants of sub-Antartctic islands, is not given a taxonomic subgroup classification since a large enough subgroup of studies with similar area ranges could not be constructed for plants.

In studies marked with an asterisk, the author(s) only provided species lists and did not perform species-area analyses. For these studies, we obtained areas and latitudes from various gazetteers and atlases and performed all species-area analyses. Those studies not marked with an asterisk are those where the author(s) performed some type of species-area analysis. We subsequently reanalyzed each of these studies using all four models discussed in text. Two studies were modified by the exclusion of outliers; Diamond's (1972) study of birds of the New Guinea islands and Johnson and Simberloff's (1974) study of plants in the British Isles. In Diamond's study New Guinea data were deleted and in Johnson and Simberloff's data from Britain were deleted.

Only least-squares estimates of regression coefficients are reported, although reduced-major-axis (RMA) estimates may be simply computed as RMA slope = least-squares slope/correlation coefficient, and RMA intercept = mean number of species − (RMA slope × mean area).

Further explanatory notes and keys to abbreviations are provided below.

Taxon—Taxonomic grouping as listed by the original author(s).

Location—General region or name of archipelago.

Habitat classification—Each study was classified as either a near archipelago (NA), distant archipelago (DA), aquatic study (AQ), habitat island (HI), or quadrat study (QUAD).

Best-fit model—The best-fit models given are based on the criteria described in text. Blanks indicate conflicting results on the criteria used and these studies were deleted from analyses of best models.

Area range—Letter designations indicate area range subgroups in which each study was included (A, 10^{-2}–10^{1} km²; B, 10^{-1}–10^{2} km²; C, 0–10^{4} km²; D, 10–10^{4} km²; E, 0–10^{5} km²; F,

TABLE 1A

Source	Taxon†	Location	Habitat Classification	Best-Fit Model	Area Range	Orders of Magnitude of Area	Taxonomic Subgroup Classification
1. Abbott (1974)	Land plants	Sub-Antarct. islands	DA	None	C, E, F, G	5	...
2. Abbott (1974)	Insects	Sub-Antarct. islands	DA	LS/LA	C, E, F, G	5	Insects
3. Abbott (1974)	Breeding passerine birds	Sub-Antarct. islands	DA	...	C, E, F, G	5	Total birds
4. Abele (unpublished)*	FW decapods	W. Indies	AQ	S/A, S/LA	F, G	4	...
5. Abele (unpublished)*	Marine shrimps	W. Indies	AQ	S/LA	C	4	...
6. Barbour & Brown (1974)	FW fish (lakes)	Afr.	AQ	None	F, G	4	Fish
7. Barbour & Brown (1974)	FW fish (lakes)	USSR	AQ	None	C, E, F, G	6	Fish
8. Barbour & Brown (1974)	FW fish (lakes)	Am. (low lat.)	AQ	S/LA	C	4	Fish
9. Barbour & Brown (1974)	FW fish (lakes)	Am. (high lat.)	AQ	S/A	C, E, F, G	5	Fish
10. Barbour & Brown (1974)	FW fish (lakes)	World-wide (low lat.)	AQ	None	F, G	4	...
11. Barbour & Brown (1974)	FW fish (lakes)	World-wide (high lat.)	AQ	S/A	C, E, F, G	6	...
12. Beard (1949)*	Plants	Leeward & windward islands	DA	S/A	D, G	2	...
13. Brown (1971)	Mammals (montane)	Great Basin, N. Am.	HI	LS/LA	D, G	2	...
14. Carrick & Ingham (1970)*	Breeding seabirds	Sub-Antarct. islands	DA	All	D, G	2	Total birds

15. Simberloff (1970)*	Birds	Canary Is.	DA	All	D, G	1	Total birds
16. Lems (1960)*	Vascular plants	Canary Is.	DA	S/A	D, G	1	...
17. Exell (1944)*	Angiosperms	Gulf of Guinea Is.	DA	S/LA, LS/LA	D, G	2	...
18. Case (1975)	Lizards	Gulf of Calif. islands	NA	S/LA	B	3	...
19. Case (1975)	Perennial plants	Gulf of Calif. islands	NA	S/LA, LS/LA	B	3	...
20. Glassman (1965)*	Palms	W. Indies	DA	S/A	C	4	...
21. Cook (1974)	Birds (tepuis)	Venezuela	HI	LS/A, LS/LA	...	3	Total birds
22. Culver et al. (1973) ...	Aquatic cave fauna	W. Va.	AQ	S/A, LS/A, S/LA	D, G	1	...
23. Culver et al. (1973) ...	Terrestrial cave fauna	W. Va.	DA	LS/LA	D, G	1	...
24. Diamond (1972)	Land & FW birds	New Guinea islands	NA	None	C, E, F, G	6	Total birds, land & FW birds
25. Amerson (1971)	Vascular plants	Fr. Frigate Shoals, Hi.	DA	S/A	B	2	...
26. Amerson (1971)	Resident seabirds	Fr. Frigate Shoals, Hi.	DA	S/A, LS/A	B	2	...
27. Greenslade (1968)	Land and FW birds	Solomon Is.	DA	S/LA	C	3	Total birds, land & FW birds
28. Gressitt (1970)*	Insects	Sub-Antarct. islands	DA	LS/A	D, G	1	Insects
29. Gressitt (1965)*	Spiders & harvestmen	Sub-Antarct. islands	DA	S/A, S/LA	D, G	2	...
30. Harrison & Hendrickson (1963)* ..	Microchiroptera	Straits of Malacca Is.	NA	S/A, LS/LA	...	3	...
31. Carlquist (1974)*	Insects	Hi.	DA	S/LA	D, G	2	...
32. Carlquist (1974)*	Angiosperms	Hi.	DA	S/LA, LS/LA	D, G	2	...
33. Johnson et al. (1968) ..	Plants	Calif. Channel Is. (islands)	NA	LS/LA	...	2	...
34. Johnson et al. (1968) ..	Plants	Calif. Channel Is. (islands & island groups)	NA	LS/LA	...	2	...
35. Johnson et al. (1968) ..	Plants	S. Calif. mainland	NA	LS/LA	F, G	3	...

(*Continued*)

TABLE 1A (*Continued*)

Source	Taxon†	Location	Habitat Classification	Best-Fit Model	Area Range	Orders of Magnitude of Area	Taxonomic Subgroup Classification
36. Johnson & Raven (1973)	Vascular plants	Galapagos Is.	DA	LS/LA	C	5	...
37. Johnson & Simberloff (1974)	Vascular plants	British Isles	NA	LS/LA	C	4	...
38. Koopman (1958)*	Bats	Lesser Antilles	NA	S/A	D, G	1	...
39. MacArthur & Wilson (1967)	Land and FW birds	Lesser Sunda Is.	NA	LS/LA	...	4	Total birds, land & FW birds
40. Niering (1956, 1963)	Vascular plants	Kapingamarengi Atoll	DA	LS/LA	A	1	...
41. Opler (1974)	Microlepidoptera (oaks)	Calif.	HI	S/A, LS/LA	...	2	...
42. Power (1972)	Breeding & summer land birds	Calif. Channel Is.	NA	S/LA	...	3	...
43. Power (1972)	Plants	Calif. Channel Is.	NA	LS/LA	...	3	...
44. Preston (1962)	Breeding birds	E. Indies	DA	S/A, LS/LA, S/LA	H	2	...
45. Preston (1962)	Land vertebrates	Islands in Lake Mich.	DA	LS/LA	C	5	...
46. Hamilton & Armstrong (1965)	Birds	Gulf of Guinea islands	DA	S/A, LS/LA	D, G	2	Total birds
47. Rey (unpublished, references in)*	Carabid beetles	Low lat.	QUAD	S/A	F, G	5	...
48. Rey (unpublished references in)*	Carabid beetles	High lat.	QUAD	S/A	...	3	...
49. Seidenfaden and Sorensen (1937)*	Plants	Greenl.	QUAD	LS/A	...	1	...
50. Hulten (1960)*	Vascular plants	Aleutian Is.	NA	LS/LA	C	4	...
51. Sepkoski & Rex (1974)	FW mussels	Eastern USA	AQ	LS/LA	D, G	1	...

52. Strong (1974*a*)	Insects (trees)	G.B.	HI	LS/LA	D, G	2	Insects
53. Strong & Levin (1975)	Fungi (trees)	G.B.	HI	S/A, LS/LA	D, G	2	. . .
54. Strong (1974*b*)	Insects (cacao)	World-wide	HI	S/A	C	3	Insects
55. Thornton (1967)	Psocids	Hi.	DA	None	D, G	1	. . .
56. Baroni-Urbani (1971)	Phanerogamic plants	Tuscan archipelago	NA	LS/LA	. . .	4	. . .
57. Baroni-Urbani (1971)	Orthoptera	Tuscan archipelago	NA	LS/LA	. . .	4	. . .
58. Baroni-Urbani (1971)	Ants	Tuscan archipelago	NA	LS/LA	. . .	4	. . .
59. Vuilleumier (1970)	Land & FW birds (paramos)	N. Andes	HI	S/A	D, G	2	Total birds, land & FW birds
60. Watson (1964)	Breeding land birds	Aegean Is.	NA	S/LA	F, G	4	Total birds
61. Wilson (1961)	Ants	Melanesia	DA	S/A	C, F, G	4	. . .
62. Wilson & Taylor (1967)	Ants	Polynesia	DA	LS/LA	. . .	4	. . .
63. Heatwole (1975)	Reptiles	New Guinea Cays	NA	None	. . .	4	. . .
64. Harris (1973)	Breeding passerines	Galapagos Is.	DA	LS/LA, S/LA	C	3	Total birds
65. Abbott (1973)*	Breeding passerines	Bass Strait Is.	NA	None	C	5	. . .
66. Hope (1973)*	Mammals	Bass Strait Is.	NA	. . .	C	4	. . .
67. Levins & Heatwole (1963)	Reptiles & amphibians	W. Indies	DA	S/A	C, E, F, G	6	. . .
68. Hall and Kelson (1959)*	Mammals	N. Am. (high lat.)	QUAD	None	H	2	. . .
69. Hall and Kelson (1959)*	Mammals	N. Am. (low lat.)	QUAD	S/A	H	2	. . .
70. Hall and Kelson (1959)*	Quadrupeds	N. Am. (high lat.)	QUAD	None	H	2	. . .

(*Continued*)

TABLE 1A (*Continued*)

Source	Taxon†	Location	Habitat Classification	Best-Fit Model	Area Range	Orders of Magnitude of Area	Taxonomic Subgroup Classification
71. Hall and Kelson (1959)*	Quadrupeds	N. Am. (low lat.)	QUAD	S/A, LS/A	H	2	...
72. Hall and Kelson (1959)*	Bats	N. Am. (high lat.)	QUAD	None	H	2	...
73. Hall and Kelson (1959)*	Bats	N. Am. (low lat.)	QUAD	S/A	H	2	...
74. Terborgh (1973)	Land birds	W. Indies	DA	S/LA, LS/LA	...	3	Total birds, land birds
75. Schoener (1976)	Land birds	Malaysian region	NA	LS/LA	C, E, F, G	6	Total birds, land birds
76. Lassen (1975)	FW snails (oligotrophic) lakes	Den.	AQ	LS/LA	B	4	...
77. Lassen (1975)	FW snails (eutrophic) lakes	Den.	AQ	None	B	4	...
78. Schoener (1976)	Birds	Shetland Is.	NA	S/LA	B	4	Total birds, land birds
79. Simberloff (1970)*	Birds	Orkney Is.	NA	S/LA, LS/LA	...	6	Total birds, land birds
80. Strong et al. (1977) ...	Insects (sugar cane)	World-wide	HI	LS/LA	C	6	Insects
81. Simberloff (1970)*	Birds	G.B.	NA	...	C, F, G	2	Total birds, land birds
82. Levins & Heatwole (1963)	Orchids	W. Indies	DA	S/A, LS/A	C	4	...
83. Levins & Heatwole (1963)	Sedges	W. Indies	DA	S/A, LS/A	C	4	...

84. Levins & Heatwole (1963)	Grasses	W. Indies	DA	S/A, LS/A	D, G	2	...
85. Amerson (1975)	Breeding seabirds	Pearl & Hermes Reef, Hi.	DA	S/LA, LS/A	A	1	...
86. Amerson (1975)	Vascular plants	Pearl & Hermes Reef, Hi.	DA	S/A, S/LA	A	1	...
87. Luther (1961)	Vascular plants	Gulf of Finl. islands	NA	...	A	1	...
88. Weissman & Rentz (1976)	Orthoptera	Cal. Channel Is.	NA	S/A, LS/A	...	2	...
89. Ricklefs & Cox (1972)	Land & raptorial birds	W. Indies	DA	S/LA, LS/LA	F, G	4	Total birds
90. Malyshev (1969)*	Plants	USSR	QUAD	S/LA	...	...	...
91. Abele (unpublished, from Patton 1974)*	Decapods (corals)	Heron Is., Aust.	AQ	LS/LA	...	...	...
92. Abele (1976)	Decapods (corals)	Uva Is., Panama	AQ	S/A	...	...	...
93. Abele (1976)	Decapods (corals)	Perlis Is., Panama	AQ	S/A, S/LA	...	...	...
94. Croasdale (1973)*	Algae (ponds)	Ellesmere Island, Can.	AQ	LS/LA	...	3	...
95. Ellis (1960)*	Infauna	Baffin Is.	AQ	LS/LA	...	2	...
96. Patrick (1967)	Diatoms	Pa.	AQ	S/LA	...	...	...
97. Vuilleumier (1973)	Aquatic cave fauna	Switzerland	AQ	S/A	...	...	...
98. Simberloff (1976*a*)*	Arboreal arthropods (mangroves)	Fla.	NA	LS/LA	...	...	...
99. Cairns & Ruthven (1970)	Protozoans	Douglas Lake, Mich.	AQ	S/LA, LS/LA	...	...	...
100. Connor (unpublished)*	Vascular plants (spoil islands)	SW Fla.	DA	LS/A, LS/LA	...	...	...

* Source performed no species-area analyses.
† FW = freshwater.

TABLE 2A

	No. of Cases	*R* lin	*B* lin	*Z* lin	*R* log	*B* log	*Z* log	*R* SE1	*B* SE1	*Z* SE1	*R* SE2	*B* SE2	*Z* SE2	SD *y*	SD *x*
1.	19	.462	37.935	.008	.364	.918	.223	.396	8.693	19.119	.308	1.289	.000	.654	1.065
2.	19	.142	51.986	.004	.395	.993	.217	.287	14.300	20.771	.290	1.364	.000	.586	1.065
3.	19	.685	.837	.001	.091	.127	.025	.247	.376	.542	.523	.117	.000	.295	1.065
4.	18	.756	5.516	.000	.721	−.014	.248	.750	−8.134	5.080	.659	.662	.000	.344	1.001
5.	18	.586	22.593	.005	.797	.811	.258	.771	1.930	12.245	.420	1.268	.000	.326	1.008
6.	14	.769	39.780	.003	.700	.430	.354	.688	−150.457	63.458	.719	1.509	.000	.423	.837
7.	16	.853	17.763	.000	.494	.919	.120	.549	−16.369	13.762	.744	1.220	.000	.363	1.499
8.	12	.487	17.799	.003	.500	1.040	.117	.596	10.622	6.426	.442	1.167	.000	.299	1.283
9.	21	.624	28.835	.001	.487	1.161	.116	.573	.462	14.501	.587	1.375	.000	.340	1.432
10.	29	.799	26.424	.003	.698	.922	.206	.597	−18.387	25.296	.684	1.337	.000	.424	1.441
11.	41	.590	28.600	.000	.459	1.070	.110	.536	−4.033	13.061	.495	1.345	.000	.347	1.456
12.	8	.908	121.021	.050	.815	1.754	.159	.829	1.752	56.535	.882	2.090	.000	.078	.397
13.	17	.703	3.552	.003	.822	−.463	.429	.808	−7.199	4.870	.685	.494	.000	.290	.556
14.	10	.377	15.107	.001	.352	.817	.127	.359	5.923	3.966	.371	1.111	.000	.253	.702
15.	8	.654	27.439	.005	.628	1.150	.122	.633	5.813	9.146	.641	1.440	.000	.069	.354
16.	7	.590	336.841	.225	.439	2.030	.238	.511	−605.451	403.663	.500	2.587	.000	.182	.335
17.	4	.976	190.221	.302	.998	1.578	.398	.967	−334.924	321.205	.887	2.265	.000	.376	.944
18.	24	.373	4.751	.003	.659	.447	.166	.657	2.907	1.909	.376	.607	.000	.256	1.016
19.	24	.524	24.425	.023	.793	1.029	.265	.834	10.675	14.661	.411	1.291	.000	.339	1.016
20.	23	.848	3.826	.001	.767	−.478	.379	.626	−21.147	10.907	.676	.462	.000	.541	1.096
21.	13	.152	42.069	.009	.202	1.508	.052	.134	38.377	3.152	.220	1.570	.000	.204	.791
22.	6	.701	4.269	.007	.639	.356	.178	.655	1.430	1.952	.652	.620	.001	.128	.458
23.	7	.780	1.619	.028	.911	−.667	.659	.783	−7.023	6.603	.818	.232	.002	.411	.568
24.	50	.439	50.646	.007	.885	1.932	.232	.764	5.451	25.726	.423	1.611	.000	.299	1.138
25.	10	.907	.223	187.428	.709	1.250	.389	.738	12.898	4.195	.788	.105	15.696	.441	.805
26.	10	.790	.476	121.479	.606	.971	.302	.653	8.790	2.763	.741	.059	13.450	.402	.805
27.	26	.679	35.203	.007	.803	.906	.271	.869	−4.936	20.404	.522	1.460	.000	.322	.953
28.	6	.201	34.198	.002	.121	1.430	.036	.069	32.284	1.461	.266	1.489	.000	.191	.639
29.	9	−.123	7.042	−.000	.043	.512	.029	−.137	10.344	−1.300	.080	.572	.000	.479	.721
30.	5	.977	3.264	.111	.919	.360	.387	.912	1.807	6.616	.886	.511	.006	.384	.911
31.	7	.585	83.651	.012	.773	−1.694	1.132	.877	−241.465	116.216	.406	1.533	.000	.828	.565
32.	7	.434	63.626	.005	.793	−.018	.584	.822	−107.748	59.943	.398	1.651	.000	.416	.565
33.	14	.710	93.137	1.080	.898	1.484	.454	.810	36.650	133.149	.703	1.715	.003	.578	1.131
34.	18	.742	87.977	1.101	.856	1.571	.416	.819	33.794	132.763	.693	1.770	.003	.518	1.061

35.	10	.680	1,194.512	.019	.873	2.605	.158	.810	−119.837	486.430	.595	3.050	.000	.213	1.174
36.	29	.616	65.289	.082	.884	1.231	.404	.786	42.858	60.407	.428	1.437	.000	.690	1.510
37.	41	.516	295.965	.238	.629	2.100	.209	.654	79.002	157.841	.486	2.389	.000	.293	.884
38.	7	.863	2.683	.008	.639	−.535	.519	.680	−46.481	21.304	.737	.687	.000	.432	.532
39.	21	.885	91.528	.001	.928	.553	.371	.893	−266.282	104.300	.637	1.877	.000	.398	.996
40.	28	.863	9.188	303.130	.854	1.969	.435	.882	47.821	16.604	.778	.968	7.400	.244	.480
41.	15	.924	2.935	.000	.949	−1.155	.465	.856	−22.356	7.213	.894	.523	.000	.311	.634
42.	16	.571	10.113	.041	.643	.786	.209	.629	6.974	5.464	.534	.919	.001	.319	.980
43.	16	.746	81.541	.826	.847	1.435	.445	.847	13.510	114.291	.662	1.734	.003	.515	.980
44.	10	.912	155.394	.000	.954	.610	.351	.929	−733.762	200.309	.847	2.185	.000	.253	.688
45.	12	.836	33.612	.066	.926	1.269	.261	.915	26.672	20.823	.567	1.411	.001	.430	1.524
46.	5	.916	19.675	.049	.945	.371	.500	.829	−61.729	47.977	.803	1.309	.000	.436	.823
47.	12	.989	71.400	.000	.931	.081	.451	.766	−642.185	230.990	.651	1.593	0	.722	1.490
48.	8	.892	14.700	.002	.285	.589	.197	.588	−151.522	72.832	.631	.950	.000	.908	1.315
49.	7	−.552	218.296	−.000	−.519	3.685	−.276	−.454	663.698	−93.102	−.616	2.364	−.000	.068	.127
50.	40	.724	39.216	.060	.698	.000	.653	.737	−78.767	73.744	.527	1.111	.000	.819	.876
51.	44	.552	7.817	.000	.646	−.405	.348	.597	−19.793	7.757	.533	.851	.000	.251	.466
52.	26	.727	14.901	.022	.789	−2.108	1.120	.656	−291.531	112.536	.589	1.118	.000	.668	.471
53.	23	.570	10.182	.002	.533	.117	.304	.547	−22.631	11.463	.515	.996	.000	.233	.409
54.	21	.829	15.269	.026	.610	.498	.370	.716	−24.505	32.335	.642	.970	.000	.591	.975
55.	6	.462	37.068	.005	.725	.468	.402	.771	−32.471	27.289	.408	1.499	.000	.274	.494
56.	17	.782	208.612	4.608	.899	2.319	.309	.903	368.037	183.877	.545	2.098	.005	.542	1.577
57.	12	.770	11.934	.165	.931	.993	.269	.840	14.815	7.653	.558	.950	.004	.432	1.496
58.	20	.640	9.247	.120	.952	1.056	.188	.900	14.855	5.002	.557	.859	.004	.337	1.707
59.	15	.798	13.775	.012	.676	.540	.296	.718	−17.706	16.387	.686	1.124	.000	.296	.676
60.	39	.841	12.109	.000	.841	−.171	.351	.900	−25.798	12.013	.561	.978	.000	.424	1.017
61.	25	.974	10.160	.000	.777	.197	.242	.616	−33.096	13.847	.718	1.002	.000	.320	1.027
62.	36	.331	10.183	.001	.671	.683	.162	.662	3.458	4.227	.340	.941	.000	.249	1.032
63.	16	.885	2.583	.041	.707	.480	.140	.716	4.027	1.495	.656	.365	.003	.314	1.584
64.	15	.550	14.382	.003	.853	.870	.157	.874	6.064	5.177	.478	1.128	.000	.182	.989
65.	33	.617	5.506	.021	.803	.315	.390	.811	2.762	6.941	.533	.485	.001	.544	1.120
66.	15	.964	1.968	.001	.767	−.059	.246	.816	1.117	2.774	.814	.233	.000	.421	1.316
67.	46	.913	9.132	.001	.747	.333	.280	.720	−9.649	11.782	.547	.825	.000	.461	1.228
68.	35	.644	−23.672	.000	.650	−1.438	.593	.482	−393.892	84.846	.778	1.163	.000	.276	.302
69.	30	.588	148.003	.000	.528	1.899	.060	.528	46.701	22.667	.586	2.170	0	.051	.442
70.	35	.646	−15.220	.000	.652	−1.225	.548	.488	−321.961	72.043	.777	1.183	.000	.254	.302
71.	30	.913	79.298	.000	.843	1.013	.200	.833	−145.697	50.229	.888	1.912	.000	.105	.442
72.	35	.627	−8.319	.000	.544	−4.402	.925	.449	−62.436	12.924	.707	−.440	.000	.514	.302

(*Continued*)

TABLE 2A (*Continued*)

	No. of Cases	*R* lin	*B* lin	*Z* lin	*R* log	*B* log	*Z* log	*R* SE1	*B* SE1	*Z* SE1	*R* SE2	*B* SE2	*Z* SE2	SD *y*	SD *x*
73.	30	−.765	67.784	−.000	−.722	2.896	−.236	−.720	185.649	−26.195	−.777	1.838	−.000	.144	.442
74.	19	.777	31.699	.001	.864	.979	.187	.913	−15.501	18.170	.661	1.470	.000	.214	.990
75.	34	.615	29.646	.010	.903	.878	.320	.826	−1.247	25.504	.507	1.291	.000	.445	1.258
76.	19	.893	4.102	.110	.964	.666	.227	.908	5.723	3.016	.736	.573	.006	.289	1.227
77.	68	.612	6.679	1.082	.820	1.074	.146	.826	12.425	2.296	.473	.728	.054	.339	1.903
78.	47	.522	6.460	.020	.906	.721	.322	.955	7.370	4.630	.330	.672	.001	.392	1.104
79.	18	.788	11.276	.141	.899	.778	.311	.863	2.726	10.561	.735	1.043	.004	.185	.535
80.	75	.422	24.959	.007	.708	.409	.369	.559	−11.456	21.863	.388	1.051	.000	.592	1.135
81.	26	.736	24.766	.000	.777	.900	.243	.865	5.421	15.579	.418	1.230	.000	.464	1.487
82.	12	.868	.973	.004	.788	−.578	.419	.659	−9.176	5.018	.892	.307	.000	.249	.468
83.	12	.962	1.080	.003	.802	−.664	.423	.784	−7.068	3.943	.938	.221	.000	.247	.468
84.	13	.953	4.252	.005	.799	.202	.255	.739	−5.161	5.081	.924	.693	.000	.183	.573
85.	8	.794	.316	173.544	.811	1.914	.782	.794	20.423	8.781	.800	.130	15.224	.444	.461
86.	8	.705	−.208	170.090	.581	1.603	.661	.658	18.522	8.036	.632	.056	14.204	.524	.461
87.	22	.599	21.358	2,645.093	.499	2.551	.482	.598	131.735	39.172	.483	1.202	31.411	.482	.499
88.	8	.896	5.129	.109	.739	.681	.280	.687	.288	10.283	.898	.835	.003	.309	.815
89.	30	.666	26.561	.000	.912	.965	.184	.924	−5.636	14.268	.566	1.386	.000	.209	1.034
90.	51	−.198	1,626.268	−.000	−.238	3.695	−.108	−.325	3,887.370	−442.640	−.124	3.142	0	.255	.566
91.	39	.323	6.189	.000	.478	−.061	.237	.436	−4.146	2.999	.344	.759	.000	.149	.301
92.	109	.460	6.985	.001	.485	.078	.252	.498	−7.499	4.873	.414	.835	.000	.145	.279
93.	35	.677	7.003	.001	.599	−.316	.356	.644	−19.386	8.212	.566	.844	.000	.204	.343
94.	15	−.121	41.962	−.001	.341	1.258	.106	.255	22.793	6.066	−.012	1.567	−.000	.210	.678
95.	20	.941	27.763	31.139	.989	1.780	.403	.986	59.663	38.256	.892	1.457	.309	.144	.353
96.	6	.856	11.172	.049	.832	−.107	.667	.960	−14.048	20.474	.654	.756	.001	.673	.840
97.	48	.561	5.206	.071	.426	.119	.434	.485	−3.746	8.672	.394	.595	.003	.438	.430
98.	32	.652	55.878	.031	.759	1.279	.221	.774	−9.123	31.520	.623	1.737	.000	.111	.380
99.	10	.599	13.893	.000	.549	.946	.075	.589	7.044	2.770	.554	1.131	.000	.125	.919
100.	6	.803	.137	.002	.891	−2.486	.957	.792	−156.141	45.242	.894	.824	.000	.214	.199

TABLE 3A

	Lat. Midpoint	Lat. Range	Mean No. Species	Max. No. Species	Isolation Distance	Mean Area
1.	49.00	24.00	47.368	...	...	1,132.158
2.	49.00	24.00	56.316	300	...	1,132.158
3.	49.00	24.00	1.474	9	1,000	1,132.158
4.	16.50	10.00	7.944	...	...	17,033.032
5.	16.50	10.00	26.056	...	...	763.468
6.	7.40	14.80	79.643	245	...	12,958.000
7.	58.25	32.50	32.500	156	...	65,834.500
8.	23.10	23.20	20.250	48	...	857.833
9.	48.10	11.80	42.762	114	...	14,962.330
10.	17.60	17.10	49.724	245	...	6,958.759
11.	58.25	32.50	36.098	156	...	33,370.878
12.	15.00	6.00	152.000	...	...	622.892
13.	38.50	5.00	5.706	...	...	855.915
14.	57.24	22.47	17.500	27	1,000	2,105.401
15.	28.75	2.50	32.125	40	100	982.625
16.	28.75	2.50	580.143	...	...	1,082.143
17.	1.88	3.75	443.250	...	...	837.991
18.	27.00	6.00	5.250	...	...	183.433
19.	27.00	6.00	28.667	...	...	183.433
20.	21.00	22.00	9.783	...	...	10,255.970
21.	5.25	2.90	44.308	78	...	240.231
22.	37.50	.10	5.333	...	...	151.983
23.	37.50	.10	5.286	...	...	132.714
24.	5.00	10.00	57.180	158	50	7,667.668
25.	23.50	1.00	3.500	...	...	.018
26.	23.50	1.00	2.600	...	...	.018
27.	8.25	6.50	44.500	80	805	1,411.414
28.	49.30	8.60	36.333	53	...	1,400.833
29.	53.60	2.60	6.444	...	...	2,896.758
30.	3.00	2.00	7.200	...	...	35.483
31.	20.55	3.50	112.286	...	...	2,364.289
32.	20.55	3.50	74.714	...	...	2,364.289
33.	32.70	8.80	209.857	...	...	108.040
34.	32.70	8.80	181.333	...	...	84.765
35.	33.05	9.70	1,480.100	...	...	15,177.304
36.	1.00	2.00	87.345	...	...	270.748
37.	56.20	10.10	361.171	...	...	274.027
38.	11.25	2.50	13.714	...	...	1,311.274
39.	6.00	2.00	139.143	447	40	67,638.696
40.	1.00	.01	15.286	...	...	.020
41.	37.00	4.00	8.667	...	...	38,107.358
42.	30.95	6.10	13.938	...	...	94.049
43.	30.95	6.10	159.188	...	...	94.049
44.	9.50	19.00	261.600	...	...	239,566.130
45.	45.50	1.00	47.000	...	...	203.373
46.	2.13	4.25	56.600	133	30	757.934
47.	19.00	30.00	203.750	...	...	659,132.330
48.	53.15	23.70	86.250	...	...	47,431.747
49.	69.80	19.50	174.429	...	...	187,175.611
50.	53.00	4.00	71.400	...	...	540.108

(*Continued*)

TABLE 3A (*Continued*)

	Lat. Midpoint	Lat. Range	Mean No. Species	Max. No. Species	Isolation Distance	Mean Area
51.	36.50	19.00	11.477	...	...	17,066.903
52.	54.25	8.50	69.269	284	...	2,522.808
53.	54.25	8.50	14.870	...	...	2,697.435
54.	10.00	20.00	31.381	153	...	624.276
55.	20.55	3.50	42.833	...	...	1,053.000
56.	42.80	1.25	303.588	...	...	20.612
57.	42.80	1.25	16.750	...	...	29.192
58.	42.80	1.25	11.350	...	...	17.520
59.	5.75	11.50	23.667	65	...	842.667
60.	39.45	7.10	17.180	53	1	38,823.218
61.	12.00	24.00	16.720	...	...	42,238.284
62.	12.00	24.00	10.917	...	...	631.990
63.	10.00	3.00	3.438	...	...	21.084
64.	1.00	2.00	15.800	24	933	521.318
65.	40.00	2.00	7.697	...	...	102.327
66.	40.00	2.00	4.000	...	...	2,022.127
67.	17.75	10.50	13.544	...	...	5,086.002
68.	55.00	35.00	89.571	...	...	431,279.196
69.	22.50	30.00	163.100	...	...	200,487.544
70.	55.00	35.00	80.057	...	...	431,279.196
71.	22.50	30.00	112.233	...	...	200,487.544
72.	55.00	35.00	9.686	...	...	431,279.196
73.	22.50	30.00	51.133	...	...	200,487.544
74.	17.50	11.00	37.842	79	112	12,218.610
75.	3.00	6.00	37.941	141	...	794.917
76.	56.50	3.00	5.842	...	...	15.822
77.	56.50	3.00	8.074	...	...	1.289
78.	60.33	.85	7.064	22	160	30.246
79.	59.13	.75	15.611	29	10	30.818
80.	19.00	38.00	30.867	247	...	893.107
81.	54.25	8.50	29.846	115	35	12,201.889
82.	14.00	8.00	4.167	...	...	726.487
83.	14.00	8.00	3.417	...	...	726.487
84.	14.00	8.00	7.846	...	...	627.421
85.	27.50	1.00	5.375	...	...	.029
86.	27.50	1.00	4.750	...	...	.029
87.	59.75	.50	42.591	...	...	.008
88.	33.45	1.10	17.500	...	...	113.188
89.	18.25	16.50	29.767	69	112	7,447.033
90.	57.00	40.00	1,527.412	...	...	436,158.824
91.	24.00	1.00	7.154	...	...	...
92.	9.00	.01	8.798	...	...	...
93.	9.00	.01	10.571	...	...	...
94.	81.70	.10	40.400	...	...	...
95.	73.00	.10	50.650	...	...	...
96.	39.51	.01	22.167	...	...	...
97.	46.00	1.00	8.042	...	...	...
98.	24.75	.33	68.281	...	...	...
99.	45.39	.01	15.500	...	...	...
100.	27.00	1.00	28.333	...	...	...

10–10^5 km^2; and G, 10–10^4 km^2, and 10–10^5 km^2). Dots indicate studies that could not be grouped into these categories because they covered narrow or peculiar ranges of area.

Orders of magnitude of area—The numbers of orders of magnitude of area covered by each data set. Dots indicate studies comprising less than one order of magnitude of area or studies in which area was measured in units other than square kilometers (i.e., studies 91, 92, and 93 in cm^3, study 96 in mm^2, etc.).

Taxonomic subgroup classification—Taxonomic subgroups as utilized in our analyses of the latitudinal dependence of the species-area relationship. Categorizations are provided only for those studies used in the analyses.

Number of Cases—Numbers of areas (i.e., islands, quadrats, etc.) used in each study.

R lin, *B* lin, *Z* lin—Respectively, the correlation coefficient, the intercept, and the slope from the untransformed model.

R log, *B* log, *Z* log—Respectively, the correlation coefficient, intercept, and slope from the log/log model.

R SE2, B SE1,. Z SE1—Respectively, the correlation coefficient, intercept, and slope from the species/log-area model.

R SE2, B SE2, Z SE2—Respectively, the correlation coefficient, intercept, and slope of the log-species/area model.

SD y and SD x—The standard deviation of species number (SD *y*) and area (SD *x*) for each species-area curve.

Latitudinal midpoint—The sum of the maximum and minimum latitudes of localities included in a species-area curve divided by 2 (values in ° lat.).

Latitudinal range—Total range of latitude (in degrees) covered by each study.

Mean number of species—Average number of species included in each species-area regression.

Maximum number of species—Largest number of species on a single locality in each study. Data are included only for those studies used in analyses of the latitudinal dependence of slope values.

Isolation distance—Distance in kilometers from the nearest hypothesized source area. Data are included only for the "total birds" taxonomic subgroup.

Mean area—Average size of areas included in each species-area regression (km^2).

LITERATURE CITED

Abbott, I. 1973. Birds of Bass Strait. Proc. R. Soc. Victoria 85:197–223.

———. 1974. Numbers of plant, insect, and land bird species on nineteen remote islands in the Southern Hemisphere. Biol. J. Linn. Soc. 6:143–152.

Abbott, I., and P. R. Grant. 1976. Nonequilibrial bird faunas on islands. Am. Nat. 110:507–528.

Abele, L. G. 1974. Species diversity of decapod crustaceans in marine habitats. Ecology 55:156–161.

———. 1976. Comparative species richness in fluctuating and constant environments: coral-associated decapod crustaceans. Science 192:461–463.

Abele, L. G., and E. F. Connor. 1978. Application of island biogeography theory to refuge design: making the right decision for the wrong reasons. Proc. 1st Conf. Sci. Res. Natl. Parks, U.S. Department of the Interior, National Parks Service (in press).

Abele, L. G., and W. K. Patton. 1976. The size of coral heads and the community biology of associated decapod crustaceans. J. Biogeogr. 3:35–47.

Amerson, A. B. 1971. The natural history of French Frigate Schoals, northwestern Hawaiian Islands. Atoll Res. Bull. 150.

———. 1975. Species richness on the nondisturbed northwestern Hawaiian Islands. Ecology 56:435–444.

Archibald, E. E. A. 1949. The specific character of plant communities II. A quantitative approach. J. Ecol. 37:274–288.

Arrhenius, O. 1921. Species and area. J. Ecol. 9:95–99.

———. 1923*a*. On the relation between species and area—a reply. Ecology 4:90–91.

———. 1923*b*. Statistical investigations in the constitution of plant associations. Ecology 4:68–73.

Barbour, C. D., and J. H. Brown. 1974. Fish species diversity in lakes. Am. Nat. 108:473–489.

Baroni-Urbani, C. 1971. Studien zur Ameisenfauna Italiens. XI. Die Ameisen des Taskanischen Archipels. Betrachtungen zer Herkunft dur Inselfaunen. Rev. Suisse Zool. 78:1037–1067.
Beard, J. S. 1949. The natural vegetation of the windward and leeward islands. Clarendon, Oxford.
Bertalanffy, L. von. 1957. Quantitative laws in metabolism and growth. Q. Rev. Biol. 32:217–231.
Bliss, C. I. 1965. An analysis of some insect trap records. Pages 385–397 *in* G. P. Patil, ed. Classical and contagious discrete distributions. Statistical Publishing Society, Calcutta.
Brown, J. H. 1971. Mammals on mountaintops: nonequilibrium insular biogeography. Am. Nat. 105:467–478.
Cain, S. A. 1938. The species-area curve. Am. Midl. Nat. 19:573–581.
Cairns, J., and J. A. Ruthven. 1970. Artificial microhabitat size and the number of colonizing protozoan species. Trans. Am. Microsc. Soc. 89:100–109.
Carlquist, S. 1974. Island biology. Columbia University Press, New York.
Carrick, R., and S. E. Ingham. 1970. Ecology and population dynamics of Antarctic seabirds. Pages 505–525. *in* M. W. Holdgate, ed. Antarctic ecology. Academic Press, London.
Case, T. J. 1975. Species numbers, density compensation, and colonizing ability of lizards on islands in the Gulf of California. Ecology 56:3–18.
Cook, R. E. 1974. Origin of the highland avifauna of southern Venezuela. Syst. Zool. 23:257–264.
Croasdale, H. 1973. Freshwater algae of Ellesmere Island, N.W.T. Natl. Mus. Can. Publ. Bot., no. 3.
Culver, D., J. R. Holsinger, and R. Bargody. 1973. Toward a predictive cave biogeography: the Greenbriar Valley as a case study. Evolution 27:689–695.
Dexter, D. 1972. Comparison of the community structure in a Pacific and Atlantic Panamanian sandy beach. Bull. Mar. Sci. 22:449–462.
Diamond, J. M. 1972. Biogeographic kinetics: estimation of relaxation times for avifaunas of southwest Pacific islands. Proc. Natl. Acad. Sci. USA 69:3199–3203.
———. 1973. Distributional ecology of New Guinea birds. Science 179:759–769.
———. 1975. The island dilemma: lessons of modern biogeographic studies for the design of natural reserves. Biol. Conserv. 7:129–146.
Diamond, J. M., and E. Mayr. 1976. Species-area relation for birds of the Solomon Archipelago. Proc. Natl. Acad. Sci. USA 73:262–266.
Dony, J. G. 1963. The expectation of plant records from prescribed areas. Watsonia 5:377–385.
Draper, N., and H. Smith. 1966. Applied regression analysis. Wiley, New York.
Ellis, D. V. 1960. Marine infaunal benthos in Arctic North America. Arctic Inst. North Am. Tech. Pap. 5.
Evans, F. C., P. J. Clark, and R. H. Brand. 1955. Estimation of the number of species present on a given area. Ecology 36:342–343.
Exell, A. W. 1944. Catalogue of the vascular plants of S. Tome (with Principe and Annobon). British Museum (Natural History), London.
Fischer, A. G. 1960. Latitudinal variations in organic diversity. Evolution 14:64–81.
Fisher, R. A., A. S. Corbet, and C. B. Williams. 1943. The relation between the number of species and the number of individuals in a random sample of an animal population. J. Anim. Ecol. 12:42–58.
Glassman, S. F. 1965. Geographic distribution of the New World palms. Principes 8:47–49.
Gleason, H. A. 1922. On the relation between species and area. Ecology 3:158–162.
———. 1925. Species and area. Ecology 6:66–74.
Goodall, D. W. 1952. Quantitative aspect of plant distribution. Biol. Rev. 27:194–245.
Gould, S. J. 1966. Allometry and size in ontogeny and phylogeny. Biol. Rev. 41:587–640.
———. 1971. Geometric similarity in allometric growth: a contribution to the problem of scaling in the evolution of size. Am. Nat. 105:113–136.
Grant, P. R. 1970. Colonization of islands by ecologically dissimilar species of mammals. Can. J. Zool. 48:545–553.
Greenslade, P. J. M. 1968. Island patterns in the Solomon Islands bird fauna. Evolution 22:751–761.
Greig-Smith, P. 1964. Quantitative plant ecology. Butterworth's, London.
Gressitt, J. L. 1965. Biogeography and ecology of land arthropods of Antarctica. Pages 431–490. *in* J. van Mieghem and P. Vanoye, eds. Biogeography and ecology in Antarctica. Junk, The Hague.
———. 1970. Subantarctic entomology and biogeography. Pac. Inst. Monogr. 23:295–374.
Gunther, B., and E. Guerra. 1955. Biological similarities. Acta. Physiol. Lat. Am. 5:169–186.
Haas, P. H. 1975. Some comments on use of the species-area curve. Am. Nat. 109:371–373.

Hall, E. R., and K. R. Kelson. 1959. The mammals of North America. Vols. 1–2. Ronald, New York.
Hamilton, T. H., and N. E. Armstrong. 1965. Environmental determination of insular variation in bird species abundance in the Gulf of Guinea. Nature 207:148–151.
Hamilton, T. H., I. Rubinoff, C. H. Barth, and G. L. Bush. 1963. Species abundance: natural regulation of insular variation. Science 142:1575–1577.
Harman, W. N. 1972. Benthic substrates: their effect on fresh-water mollusca. Ecology 53:271–277.
Harris, M. P. 1973. The Galapagos avifauna. Condor 75:265–278.
Harrison, J. L., and J. R. Hendrickson. 1963. The fauna of the islands of the Straits of Malacca. Pages 543–555 *in* J. L. Gressitt, ed. Pacific Basin biogeography. Bishop Museum, Honolulu.
Heatwole, H. 1975. Biogeography of reptiles on some of the islands and cays of eastern Papua-New Guinea. Atoll Res. Bull. 180.
Hope, J. H. 1973. Mammals of the Bass Strait Islands. Proc. R. Soc. Victoria 85:163–195.
Hopkins, B. 1955. The species-area relations of plant communities. J. Ecol. 43:409–426.
———. 1957. The concept of minimal area. J. Ecol. 45:441–449.
Hulten, E. 1960. Flora of the Aleutian Islands. Cramer, Weinheim.
Huxley, J. S. 1932. Problems of relative growth. MacVeagh, London.
Jaccard, P. 1908. Nouvelles recherches sur la distribution florale. Bull. Soc. Vaudoise Sci. Nat. 44:223.
———. 1912. The distribution of the flora in the alpine zone. New Phytol. 11:37–50.
Janzen, D. H. 1968. Host plants as islands in evolutionary and contemporary time. Am. Nat. 102:592–595.
Johnson, M. P., and P. H. Raven. 1970. Natural regulation of plant species diversity. Evol. Biol. 4:127–162.
———. 1973. Species number and endemism: the Galapagos Archipelago revisited. Science 179:893–895.
Johnson, M. P., L. G. Mason, and P. H. Raven. 1968. Ecological parameters and plant species diversity. Am. Nat. 102:297–306.
Johnson, M. P., and D. S. Simberloff. 1974. Environmental determinants of island species numbers in the British Isles. J. Biogeogr. 1:149–154.
Jolicoeur, P. 1968. Interval estimation of the slope of the major axis of a bivariate normal distribution in the case of a small sample. Biometrics 24:679–682.
Kilburn, P. D. 1966. Analysis of the species-area relation. Ecology 47:831–843.
Kobayashi, S. 1974. The species-area relation. I. A model for discrete sampling. Res. Popul. Ecol. 15:223–237.
———. 1976. The species-area relation. III. A third model for delimited community. Res. Popul. Ecol. 17:243–254.
Koopman, K. F. 1958. Land bridges and ecology in bat distribution on islands off the northern coast of South America. Evolution 12:429–439.
Kuhn, D. S. 1962. The structure of scientific revolutions. Foundations of the Unity of Science. Vol. II, no. 2. University of Chicago Press, Chicago.
Lassen, H. H. 1975. The diversity of freshwater snails in view of the equilibrium theory of island biogeography. Oecologia 19:1–8.
Lems, C. 1960. Floristic botany of the Canary Islands. Sarracenia 5:1–94.
Levins, R., and H. Heatwole. 1963. On the distribution of organisms on islands. Caribb. J. Sci. 3:173–177.
Luther, H. 1961. Veranderungen in der gefasspflanzen flora der Meeresfelsen von Tvarminne. Acta Bot. Fenn. 62.
Lynch, J. F., and N. K. Johnson. 1974. Turnover and equilibria in insular avifaunas, with special reference to the California Channel Islands. Condor 76:370–384.
MacArthur, R. 1965. Patterns of species diversity. Biol. Rev. 40:510–533.
———. 1969. Patterns of communities in the tropics. Biol. J. Linn. Soc. 1:19–30.
MacArthur, R., and E. O. Wilson. 1963. An equilibrium theory of insular zoogeography. Evolution 17:373–387.
———. 1967. The theory of island biogeography. Princeton University Press, Princeton, N.J.
McCoy, E. D., and E. F. Connor. 1976. Environmental determinants of island species number in the British Isles: a reconsideration. J. Biogeogr. 3:381–382.
Malyshev, L. I. 1969. The dependence of the species abundance of a flora on the environmental and historical factors. Acad. Sci. (USSR) Bot. J. 54:1137–1147.
May, R. M. 1975. Patterns of species abundance and diversity. Pages 81–120 *in* M. L. Cody and J. M. Diamond, eds. Ecology and evolution of communities, Belknap, Cambridge, Mass.

Mertz, D. B. 1971. The mathematical demography of the California condor population. Am. Nat. 105:437–453.
Moore, N. W., and M. D. Hooper. 1975. On the number of bird species in British woods. Biol. Conserv. 8:239–250.
Niering, W. A. 1956. Bioecology of Kapingamarangi Atoll, Caroline Islands: terrestrial aspects. Atoll Res. Bull. 49.
———. 1963. Terrestrial ecology of Kapingamarangi Atoll, Caroline Islands. Ecol. Monogr. 33:131–160.
Opler, P. A. 1974. Oaks as evolutionary islands for leaf-mining insects. Am. Sci. 62:67–73.
Osman, R. W. 1977. The establishment and development of a marine epifaunal community. Ecol. Monogr. 47:37–63.
Patrick, R. 1967. The effect of invasion rates, species pool, and size of area on the structure of the diatom community. Proc. Natl. Acad. Sci. USA 58:1335–1342.
Patton, W. K. 1974. Community structure among the animals inhabiting the coral *Pocillopora damicornis* at Heron Island, Australia. Pages 219–243 *in* W. B. Vernberg, ed. Symbiosis in the sea. University of South Carolina Press, Columbia.
Pianka, E. 1966. Latitudinal gradients in species diversity: a review of concepts. Am. Nat. 100:33–46.
Pidgeon, I. M., and E. Ashby. 1940. Studies in applied ecology. I. A statistical analysis of regeneration following protection from grazing. Proc. Linn. Soc. N.S.W. 65:123–143.
Pilbeam, D., and S. J. Gould. 1974. Size and scaling in human evolution. Science 186:892–901.
Power, D. M. 1972. Numbers of bird species on the California Islands. Evolution 26:451–463.
Preston, F. W. 1948. The commonness, and rarity, of species. Ecology 29:254–283.
———. 1960. Time and space and the variation of species. Ecology 41:611–627.
———. 1962. The canonical distribution of commonness and rarity. Ecology 43:185–215, 410–432.
Raup, D. M. 1976. Species diversity in the Phanerozoic: an interpretation. Paleobiology 2:289–297.
Ricker, W. E. 1973. Linear regression in fishery research. J. Fish. Res. Board Can. 30:409–434.
Ricklefs, R. E., and G. W. Cox. 1972. Taxon cycles in the West Indian avifauna. Am. Nat. 106:195–219.
Schoener, T. W. 1976. The species-area relation within archipelagos: models and evidence from island land birds. Pages 629–642 *in* H. J. Firth and J. H. Calaby, eds. Proceedings of the 16th International Ornithological Conference. Australian Academy of Science, Canberra.
Siedenfaden, G., and T. Sorensen. 1937. A summary of the vascular plants found in eastern Greenland. Medd. Gronl. 101:141–215.
Sepkoski, J. J. 1976. Species diversity in the Phanerozoic: species-area effects. Paleobiology 2:298–303.
Sepkoski, J. J., and M. A. Rex. 1974. Distribution of freshwater mussels: coastal rivers as biogeographic islands. Syst. Zool. 23:165–188.
Simberloff, D. S. 1970. Taxonomic diversity of island biotas. Evolution 24:23–47.
———. 1972. Models in biogeography. Pages 160–191 *in* T. J. M. Schopf, ed. Models in paleobiology. Freeman, San Francisco.
———. 1974. Permo-Triassic extinctions: effects of area on biotic equilibrium. J. Geol. 82:267–274.
———. 1976*a*. Experimental zoogeography of islands: effects of island size. Ecology 57:629–648.
———. 1976*b*. Species turnover and equilibrium island biogeography. Science 194:572–578.
Simberloff, D. S., and L. G. Abele. 1976. Island biogeography theory and conservation practice. Science 191:285–286.
Simpson, B. B. 1974. Glacial migration of plants: island biogeographical evidence. Science 185:698–700.
Sokal, R. R., and F. J. Rohlf. 1969. Biometry. Freeman, San Francisco.
Southwood, T. R. E. 1960. The number and species of insects associated with various trees. J. Anim. Ecol. 30:1–8.
Strong, D. R. 1974*a*. Nonasymptotic species richness models and the insects of British trees. Proc. Natl. Acad. Sci. USA 71:2766–2769.
———. 1974*b*. Rapid asymptotic species accumulation in phytophagous insect communities: the pests of cacao. Science 185:1064–1066.
Strong, D. R., and D. A. Levin. 1975. Species richness of the parasitic fungi of British trees. Proc. Natl. Acad. Sci. USA 72:2116–2119.
Strong, D. R., E. D. McCoy, and J. R. Rey. 1977. Time and the number of herbivore species: the pests of sugar cane. Ecology 58:167–175.

Terborgh, J. 1973. Chance, habitat and dispersal in the distribution of birds in the West Indies. Evolution 27:338–349.

———. 1974. Preservation of natural diversity: the problem of extinction prone species. Bioscience 24:715–722.

———. 1975. Faunal equilibria and the design of wildlife preserves. Pages 369–380 *in* F. B. Golley and E. Medina, eds. Tropical ecological systems, trends in terrestrial and aquatic research. Springer-Verlag, New York.

Thornton, I. W. B. 1967. The measurement of isolation on archipelagos, and its relation to insular faunal size and endemism. Evolution 21:842–849.

Vestal, A. G. 1949. Minimum areas for different vegetations. Their determination from species-area curves. Ill. Biol. Monogr. 20(3).

Vuilleumier, F. 1970. Insular biogeography in continental regions. I. The northern Andes of South America. Am. Nat. 104:373–388.

———. 1973. Insular biogeography in continental regions. II. Cave faunas from Tessin, southern Switzerland. Syst. Zool. 22:64–76.

Watson, G. 1964. Ecology and evolution of passerine birds on the islands of the Aegean Sea. Vols. 1–2 Ph.D. diss., Yale University.

Watson, H. C. 1835. Remarks on the geographical distribution of British plants. n.p., London.

Weissman, D. B., and D. E. Rentz. 1976. Zoogeography of the grasshoppers and their relatives (Orthoptera) on the California Channel Islands. J. Biogeogr. 3:105–114.

White, J. F., and S. J. Gould. 1965. Interpretation of the coefficient in the allometric equation. Am. Nat. 99:5–18.

Whitehead, D. R., and C. E. Jones. 1969. Small islands and the equilibrium theory of insular biogeography. Evolution 23:171–179.

Whittaker, R. H. 1960. Vegetation of the Siskiyou Mountains, Oregon and California. Ecol. Monogr. 30:279–338.

Williams, C. B. 1943. Area and number of species. Nature 152:264–267.

———. 1944. Some applications of the logarithmic series and the index of diversity to ecological problems. J. Ecol. 32:1–44.

———. 1947. The logarithmic series and its application to biological problems. J. Ecol. 34:253–272.

———. 1964. Patterns in the balance of nature. Academic Press, London.

Wilson, E., and E. O. Willis. 1975. Applied biogeography. Pages 522–534 *in* M. L. Cody and J. M. Diamond, eds. Ecology and evolution of communities. Belknap, Cambridge, Mass.

Wilson, E. O. 1961. The nature of the taxon cycle in the Melanesian ant fauna. Am. Nat. 95:169–193.

Wilson, E. O., and R. W. Taylor. 1967. An estimate of the potential evolutionary increase in species density in the Polynesian ant fauna. Evolution 21:1–10.

Zar, J. H. 1968. Calculation and miscalculation of the allometric equation as a model in biological data. Bioscience 18:1118–1120.

PAPER 44

Comparisons between Taxa and Adaptive Trends: Problems of Methodology (1982)

P. H. Harvey and G. M. Mace

Commentary

TIM M. BLACKBURN

One of the attractions of macroecology is the grand view it provides of the natural world. By bringing together data on multiple populations or species from censuses and surveys across national, continental, and even global scales, it has revealed many novel and surprising patterns in, and relationships between, a range of key biological variables. Plots of abundance versus distributional extent, distributional extent versus latitude, body mass versus abundance, and many others, led to a surge of ecological hypotheses and associated tests. Arguably, macroecology grew so rapidly in the 1980s and 1990s because of the power of these pictures.

Yet the devil is in the details, and relationships are not necessarily what they seem. Even before the flowering of modern macroecology, P. H. Harvey and G. M. Mace had made this very point in a classic book chapter from 1982. Using simple examples and simulations, they showed how violating the assumptions of statistical tests, notably of the independence of data and distribution of error variance, leads to incorrect estimates of statistical parameters. For example, ordinary least-squares regression assumes that there is no error variance in the independent variable and will underestimate the regression slope (and overestimate the intercept) if there is. Data points that are autocorrelated because they derive from related species will also affect regression parameter estimates. This results in potentially erroneous conclusions about form and function. Harvey and Mace concentrated on linear comparisons between pairs of variables, but their message applies to all macroecological analyses: methodology matters. Yet this message was not universally heard, and many of the early studies in macroecology are flawed as a result. Given ongoing debates about the form of structural relationships (e.g., in relation to energy use), Harvey and Mace's chapter should be compulsory reading for the current generation of macroecologists.

In *Current Problems in Sociobiology*, edited by King's College Sociobiology Group, pp. 343–61. Cambridge University Press, Cambridge.

16

Comparisons between taxa and adaptive trends: problems of methodology

PAUL H. HARVEY AND GEORGINA M. MACE

Introduction

Many morphological and behavioural characters vary between taxa in similar ways. Some associations are not at all surprising, such as that between warning coloration and toxicity in the insects. Others are more perplexing, such as the repeated correlation shown between sexual dimorphism in body size and body size itself across many animal groups from insects to mammals (Rensch, 1959). Comparisons between traits which vary among taxa has been a useful tool for both generating and testing hypotheses about the functional or adaptive significance of morphological and behavioural variation (see Clutton-Brock & Harvey, 1979). However, such comparisons face methodological problems that are often ignored, or barely mentioned in passing. In this chapter, we shall be concerned with those problems, together with some possible solutions to them. We restrict our discussion to bivariate analyses of continuously distributed variables; as will be apparent, multivariate analyses face all the difficulties discussed here, plus some additional ones to which we allude at the end of the chapter.

The quantitative methods used will depend upon the purpose of the analysis, and in the first section we outline various uses of comparative studies in biological investigations. We then present some criteria to help determine a suitable taxonomic level at which to analyse the available data.

Once the purpose of the analysis is decided, and the appropriate taxonomic level selected, we must turn to the analysis itself. Three underlying statistical models are generally used for dealing with linear comparisons between normalised variables. However, the frequency with which they are used is not closely related to their suitabilites. We present a simple data set and use it to demonstrate differences in outcome between the three techniques. Finally, we discuss additional methodological problems (to some of which we can offer no solution) together with examples of recent papers that have encountered and sometimes fallen foul of such problems.

Hypothesis testing and hypothesis generation

Comparisons across taxa can be used in at least four ways. We deal briefly with each. First, they can be used to test general hypotheses; second, to test specific hypotheses; third, to describe new variables; and fourth, to identify the *form* of the relationship between variables. The first three may be used to test hypotheses and the latter two, as we shall see, are often useful for hypothesis generation.

Testing general hypotheses

As a science moves towards quantification, hypotheses to be tested often have only a general form. Biologists may forsee the outcome of general hypotheses before more precise engineering or energetic predictions can be formulated. An example can be drawn from the theory of sexual selection. We know that, across a variety of taxa, larger animals are more likely to win fights than smaller conspecifics. We also know that there are metabolic costs associated with increasing body size. The trade off between the two is likely to result in some optimal body size, given genetic and physiological constraints. Sexual selection theory predicts that the sex with the higher variance in reproductive success will compete among itself for access to mates. We therefore expect the divergence in body size between the sexes to increase with the difference in variance of reproductive success between the sexes. As a consequence, we predict that (within some vertebrate groups) sexual dimorphism in body size will increase with the degree of polygyny, males becoming larger relative to females. But, we can make no precise predictions about the nature of the relationship because we can measure neither the costs associated with increasing body size nor the benefits of being larger, and because the degree of polygyny does not necessarily reflect the difference in variance of reproductive success between the sexes. At best, we can expect a positive correlation between the degree of polygyny and the extent of sexual dimorphism (Alexander *et al.*, 1979). Other examples of this kind are legion and illustrate the generality of the approach – the relationship between island size and numbers of species (MacArthur & Wilson, 1967), the positive correlation between prey size and some aspect of predator size (Hespenheide, 1973) and between territory quality and degree of polygyny in some birds (Verner, 1964).

Testing specific hypotheses

Hypotheses vary in their predictive power. If we are examining a linear relationship between two variables we may have some idea of the expected slope. Hypothesis testing then involves producing a line of best fit and examining whether the estimated confidence limits of the slope embrace the predicted value.

For example, Gould (1975*a*) predicted that in a number of vertebrate groups the logarithm of tooth area would increase with the logarithm of body size producing a slope of 0.75 (scaling at the same rate as metabolic costs) rather than 0.66 (scaling with surface area). Similarly, many of the scaling costs of locomotion can be formulated in precise terms from mechanical and energetic considerations and the actual performances may be compared with predicted values (Alexander, 1977).

Alternatively, hypotheses may be insufficiently formulated to predict the form of the relationship (in terms of slope or slope and elevation) but differences in slope or elevation between groups with different characteristics might be expected. For example, home range area relates to body size with the same exponent in mammalian 'browsers' and 'grazers' and we can predict a higher elevation for 'browsers' which depend on a less-abundant food supply. Mechanical arguments predict that the costs of running at different speeds scale differently with body size in quadrupeds and bipeds (Taylor, 1977).

Describing new variables

One of the all pervading problems facing cross-taxonomic comparisons is that of confounding variables (Clutton-Brock & Harvey, 1979). We are often interested in the relationship between two variables after the effects of a third have been removed. When that third variable is (body) size, as it often is, then we are dealing with deviations from an allometric relationship (Gould, 1966). The deviations either from an empirical line of best fit, or from some hypothetical relationship can be treated as measures of a new variable for further quantitative investigation. The relationship of these deviations to other scalar or ordinal measures can be used for both hypothesis generation and hypothesis testing.

A commonly used variable of this kind is the *encephalisation quotient*. Deviations from a best-fit line of the logarithmic plot of brain weight against body weight measure the brain size of an animal when the effects of the underlying relationship with body size have been removed. Encephalisation quotients have been used to test predictions about the differences in relative brain size between predators and prey (Jerison, 1973), and about ecological and behavioural correlates of relative brain size (Bauchot & Stephan, 1969; Pirlot & Stephan, 1970; Clutton-Brock & Harvey, 1980; Mace, Harvey & Clutton-Brock, 1981). Another example is *relative male canine size* (Harvey, Kavanagh & Clutton-Brock, 1978). It was predicted that once the effects of body size had been removed, the canine size typical of males from different primate species should be related to both breeding system (a sexual selection hypothesis) and mate defence strategies (an anti-predator hypothesis). Treating the measure of relative male canine size as a newly defined normal variate, tests of both hypotheses could be made using cross-species comparisons.

Identifying the form of a relationship

The scaling of one variable on another may provide important clues about the underlying cause of the relationship. For example, it is now widely acknowledged that in warm-blooded animals metabolic costs scale against body weight to the power of 0.75 (Kleiber, 1961). Thus the classical 'surface law' whereby the total energetic costs incurred by an animal should be proportional to it's surface area (i.e. to the power of 0.66) has had to be discarded (Gould, 1966) and, based on necessary distortions from geometric similarity between different sized animals, other testable causes for the relationship have been suggested (McMahon, 1973; Wilkie, 1977).

Choosing the taxonomic level for analysis

If we employ quantitative methods for comparative studies, we are bound by assumptions associated with statistical analysis and we have to make certain decisions about the validity of taxonomic groupings. Transcending taxonomic levels is unusual in biometry and as a consequence statistical assumptions may be unwittingly violated. In this section we mention some pitfalls and discuss ways to avoid them. An initial assumption is that the same taxonomic level is equivalent in different groups. While this may not be reasonable across very different phyla, the assumption is generally justified in more closely knit groups. For example, in a study of inter-family differences among the mammals, we would assume that the Hyaenidae are equivalent to the Muridae. We are then faced with two problems. First, the choice of taxon used to produce statistically independent points (e.g. species, genus, family); and second, the range of forms over which to collect the data. We refer to these as the choice of lower and upper taxonomic level respectively.

Choice of lower taxonomic level

The main problem here is the independence of data points. If our upper limit of analysis was the order, then we might use species, generic or family estimates for comparing relationships between variables within the order. When we are dealing with real data, species are not evenly distributed across genera, nor genera across families. If species within a genus tend to have similar characteristics due to phylogenetic constraints, then the analysis of species data will be statistically biased by those genera containing large numbers of species. An example may illustrate this point more clearly. Harestad & Bunnell (1979) undertook a study of home-range size and body size in mammals. They used species data, and compared relationships between dietetic groups. The four largest species in their herbivore sample were all from the genus *Odocoileus* and had similar body weights and home ranges sizes to each other, but smaller home ranges than other similar-sized species. Because of this single genus, the slope of the herbivore line was considerably reduced. Now, if these four congeneric species have similar characteristics due to phylogenetic constraints, they are not statistically independent and the analysis is biased by their presence – they really represent only one point. It is arguable, however, that they have similar body sizes and home ranges due to selection and convergence because they live in similar habitats; we cannot conclude *a priori* that similarity results from phylogenetic inertia. But, since we rarely have any external evidence to justify the latter interpretation, we should use the lowest taxonomic level that can be justified on statistical grounds, i.e. the lowest taxonomic level at which maximum variance is exhibited in our measured variable after the data have been normalized. Clutton-Brock & Harvey (1977*a*) suggest the use of nested analysis of variance to identify that level: in an analysis of primates, they were able to use genera as their lower taxonomic level for analysing seven of nine variables since no additional variance was revealed in comparisons among families over that found among genera within families.

It is not easy to produce hard and fast rules over this matter. Taking the practice of statistical conservatism to its extremes may only lead to other problems: sample sizes may become prohibitively small or analyses may become blurred through inclusion of very diverse groups. In general, the pitfalls encountered by taking too low a taxonomic level are serious enough to justify statistical safety. And subspecific points should not be combined with data from higher levels. The two are qualitatively different since subspecies or populations generally share the same gene pool.

Choice of upper taxonomic level

Relationships found within one taxon may be of a different form in other taxa. For instance, in species plots of brain weight against body weight, the elevations of best fit lines differ among genera, and the slopes of best fit lines vary among taxonomic levels (Gould, 1975*b*). Before amalgamating successive taxonomic levels for analysis, statistical tests should always be employed to ensure that no such differences exist.

Methods of comparison

Normality and linearity

For some models of best fit we assume that the data are normally distributed on both axes and that relationships are linear; standard tests of both assumptions are available from statistical texts. If correlation coefficients are low and normalising the data (by transformation) results in non-linearity, then serious problems can arise in attempting to describe the form of the relationship. We shall return to these and other problems after we have considered the methods available for establishing lines of best fit.

The line of best fit

Three models are generally used to estimate straight lines of best fit. They are regression analysis, major axis (= principal axis = principal component) analysis and reduced major axis analysis. It was claimed by Kermack & Haldane (1950) that reduced major axis is the technique most suitable for estimating lines of 'organic correlation', but for the historical reason that regression analysis is applicable to various models of experimental design, where one of the variables is measured without error, regression analysis has often and erroneously been used for comparative studies.

Before we distinguish among the three models, we should emphasise the distinction between the *strength* (i.e. correlation) and the *form* (i.e. parameters of the best fit line) of a relationship. As an increasing amount of statistical error variance is incorporated into measures on the two axes, so the correlation coefficient will decrease. The correlation coefficient (or its squared value, the coefficient of determination) is a useful adjunct to bivariate analysis and should usually be quoted. However, there is no *a priori* reason why the correlation coefficient should be a component variable in the calculation of the best fit line, despite the protestations of Jolicoeur (1968) and Jolicoeur & Mosimann (1968).

Each method of producing a line of best fit minimises some measure of the deviation of points around the line, Fig. 16.1 illustrates the quantity

Fig. 16.1. Lines of best fit produced by a set of data (points are the black dots). Lines drawn from the points A1 and A2 show the distances minimised for regression analysis and major axis analysis, and the areas minimised for reduced major axis analysis. Imbrie (1956) gives a similar figure. For details see text.

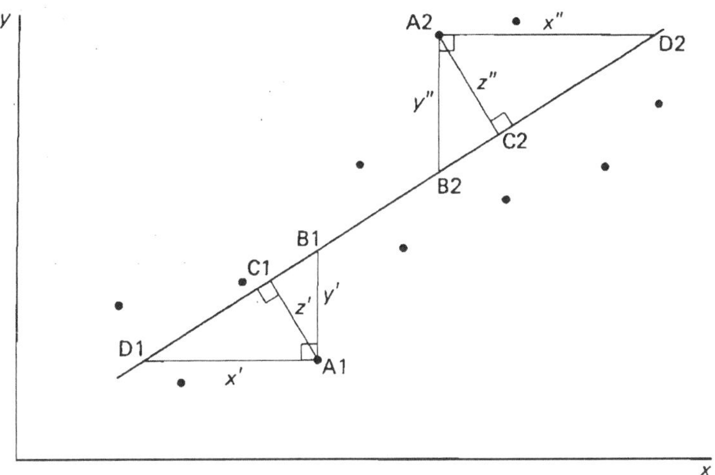

Regression of y on x produces that line which minimises the sum of squared distances of points from the line, when those distances are measured perpendicular to the x axis (Fig. 16.1: y' and y''), whereas the regression of x on y minimises the sum of squared distances measured perpendicular to the y axis (x' and x'').

Major axis analysis minimises the sum of squared distances of points from the line when the distances are measured perpendicular to the line of best fit (z' and z''). This line of best fit, therefore, is that which has the minimum variance of points about itself and consequently accounts for the maximum amount of variance in the data.

Reduced major axis analysis minimises the sum of the areas of triangles bounded by the line of best fit and lines drawn from it to the data points and parallel with the two original axes ($A_1 B_1 D_1$ and $A_2 B_2 D_2$ on Fig. 16.1; see also Imbrie, 1956).

Regression analysis only produces an unbiased estimate of the line of best fit when variables are without error* on the abscissa. As error is

* The error referred to throughout this chapter, unless qualified, has two possible sources. First, error due to inaccurate measurement. Second, error resulting from lack of dependence. If two variables are each, in part, determined by a third variable but not by each other, then they cannot be said to be independent and dependent variables. And, if one variable determines a second, but only in part, then the second will have error resulting from a lack of total dependence on the first. The hypothesized sources of error may determine the best-fit analysis performed.

introduced into the values on the x axis, so the slope becomes shallower when y is regressed on x. The regression of y on x is not affected when error is introduced onto the y axis. So long as the error variances on the two axes are equal, major axis and reduced major axis analyses do not suffer from this problem.

However, major axis analysis suffers from a different disadvantage: the relationship between the variables changes in a non-intuitive way with changes in the scale of measurement. For instance, if we perform a major axis analysis on a set of data where one variable is measured in centimetres and then repeat the analysis on the same data set with the centimetres changed to millimetres simply by multiplying the original values by ten, the slopes in the two analyses will not differ by a factor of ten, unless the correlation coefficient is unity. In contrast, they would differ by the scaling factor of ten if the analyses were by either reduced major axis or regression techniques. Under one transformation this disadvantage does not occur: when changes in scale are performed prior to logarithmically transforming the data for analysis, the slope is not affected (Jolicoeur & Mosimann, 1968). This caveat is important because logarithmically transformed axes are often used in comparative studies and tests for heterogeneity among more than two slopes are available for major axis but not for reduced major axis (see below).

In order to demonstrate these points as clearly as possible, we give the results of a simple simulation study that was designed to demonstrate the different outcomes from the three types of analysis.

An example

Three arrays of 100 numbers were each taken randomly from a normal distribution with a mean of zero and a variance of one. Data set A consisted of the first array, data set B of the first plus the second array, and data set C of the first plus the third array. Sets B and C are therefore correlated both with each other and with A, but have some added error variance which is equal in both data sets. The 'underlying' relationship between all three data sets is a slope of one when each is plotted against any other since there is a one : one correspondence through data set A.

Table 16.1 gives the results of comparing set A (without error) with set B (with error), and of comparing set B with set C (also with error). Clearly, the correlation between B and C is less than that between A and B because of the additional error variance associated with the first comparison.

Regression analysis of B on A reveals the 'correct' slope of one since A is without error. The other regression slopes are influenced by the error terms in B and C and underestimate the slope of one.

Reduced major axis and major axis analysis give similar answers to each other and give the 'correct' slope of about one when the error variance is equal on both axes (comparing sets B and C). However, the use of set A, because it is without error, renders both techniques inappropriate.

We now introduce data set D which is the same as set B, but with each value multiplied by ten. We use this set to demonstrate that major axis analysis results in a slope that is 'not invariant under changes of scale' (Kermack & Haldane, 1950). Depending on which data set is used to create the y axis, we would expect the slopes of D plotted against C to be either one-tenth or ten times greater than the slopes of C plotted against B. This is so for both regression analysis and for reduced major axis analysis, but not for major axis analysis where the scaling factor is not ten but about sixteen (Table 16.2). But if B were a data set that had been logarithmically

Table 16.1. *Slopes of best-fit lines relating pairs of data sets*

Ordinate	Abscissa	Regression	Reduced major axis	Major axis	Correlation coefficient
A	B	0.57	0.72	0.67	0.79
B	A	1.09	1.38	1.50	
B	C	0.67	1.04	1.07	0.64
C	B	0.62	0.96	0.94	

Data set A has no error variance while B and C have equal error variance which is the same as the total variance in A (see text). Comparisons giving the 'expected' slope of 1 are underlined.

Table 16.2. *Slopes of best-fit lines relating pairs of data sets*

Ordinate	Abscissa	Regression	Reduced major axis	Major axis	Correlation coefficient
D	C	6.71	10.44	16.14	0.64
C	D	0.06	0.10	0.61	
E	C	0.67	1.04	1.04	0.64
C	E	0.62	0.96	0.96	

Data sets B and C are as in Table 16.1, D is the same as B with all the values multiplied by 10 and E is B after the transformation $\log_e(e^B \times 10)$. Comparisons producing the 'expected' slopes are underlined.

transformed, then changing the scale before the transformation would not have influenced the slope. We can see this by examining the relationship between E and C, where the values of E are the natural exponents (i.e. antilogarithms) of the values of B multiplied by ten and then logged (Table 16.2).

Finally, we demonstrate the effects of unequal error variance on the two axes. We use data set F which consists of the first array plus the values of the second array adjusted so that their variance is halved while their mean remains the same. F is therefore data set B with less (one half) error variance. In the comparisons between C and F, even reduced major axis produces unsuitable results (Table 16.3). The variance of B is 2 while that of F is 1.5. Therefore, reduced major axis gives us a slope approximately equal to $(2)^{\frac{1}{2}}/(1.5)^{\frac{1}{2}}$, that is the ratio of the standard deviations. In this case we know the error variances of B and F to be 1 and 0.5 respectively, therefore we could have subtracted these prior to calculating the reduced major axis slope and establishing the underlying slope of one. In practice, we do occasionally have estimates of the error variance in our data and these can be similarly incorporated into our analysis to give a more reliable result (see Kermack, 1954, for a biological example).

In summary, when there is a perfect underlying relationship between two variables but error is present on both axes and of equal magnitude, then reduced major axis analysis produces the most satisfactory line of best fit. Regression is only appropriate when values on the abscissa are without error, and major axis analysis is only suitable for use on logarithmically transformed data because otherwise the line of best fit so produced changes unpredictably under changes of scale.

Additional problems

Before amalgamating taxonomic levels for analysis, it is always advisable to test whether the relationships have the same form in different taxa, or in different groups under test. Assuming normality and linearity of

Table 16.3. *Slope of best-fit lines relating data sets B and F*

Ordinate	Abscissa	Regression	Reduced major axis	Major axis	Correlation coefficient
B	F	0.83	1.17	1.25	0.71
F	B	0.60	0.85	0.80	

F is the same as C in Table 16.1 but with only one half the error variance.

relationships *within* taxa, two parameters – the slope and the elevation – may vary *between* taxa. If slopes differ between taxa, there is little value in comparing elevations, although techniques are available for doing so for a small overlap range between two empirical distributions (Imbrie, 1956). We shall concentrate on methods for comparing slopes, and then elevations for cases where slopes are assumed to be equal.

Comparing slopes. The slope of the reduced major axis is calculated as the ratio of the standard deviations of points measured on the y axis and on the x axis. If the standard deviation on the ordinate is greater than that on the abscissa then the slope is larger than one, and if it is less then the slope is below one. The value given (a ratio of square roots) is, however, always positive and the sign of the slope can only be found by examining the sign of the sum of cross-products (or the correlation coefficient). Comparing pairs of slopes (where the total sample size of the combined data is above 35) is straightforward: since standard errors of reduced major axes are the same as those of the corresponding regression coefficients (Tessier, 1948), simple z-tests can be employed (Imbrie, 1956).

Testing for heterogeneity between more than two slopes is not easy. We can suggest two approaches. First, the analogue of the analysis of covariance used to compare slopes in regression analysis (Snedecor, 1956). Analysis of covariance measures squared deviations from three types of regression line: within sample lines, the overall line ('total') going through all sample points, and the 'common' regression whose slope is a weighted average of within sample slopes (geometrically, it transforms each sample to have the same mean on both axes, leaves their variances unaltered, and calculates another 'total' line). The analogue of squared deviations in reduced major axis analysis is the sum of the areas of the triangles described above (see Fig. 16.1). Calculation of slopes presents no problem since the same variances can be used as in covariance analysis (see Snedecor, 1956), nor does the calculation of the summed triangle area, but statistical testing is difficult. Covariance analysis utilises variance ratio (F) tests, calculated from the F distribution, whereas using our suggested method produces a statistic whose underlying distribution is unknown (at least to us). Nevertheless, simple simulations which incorporate the relevant means, variances and covariances of our observed data set can provide a reliable statistic in any particular case. Clearly, a set of tables would be useful.

Second, since the disadvantage of the major axis technique which results from changes of scale disappears when we use logarithmically transformed data (Jolicoeur & Mosimann, 1968; see above), it is often possible to use major axis analysis to calculate lines of best fit for biological data. Commonly, logarithmic transformations are necessary to normalise the data, and the resulting transformations are linear. Statistical determination of heterogeneity among two or more samples is possible and we suggest using a maximum likelihood method (see Appendix).

Comparing elevations. If we are persuaded that slopes do not differ between taxa, then we may wish to test for differences in elevation. This entails the calculation of a 'common' reduced major axis slope (see above). Within samples, sums of squares are calculated and summed, and the ratio of the square roots of these values for the ordinate and the abscissa provides us with a common slope. Alternatively, when using major axis analysis on logarithmically transformed data, the common slope is determined during the procedure for testing slope heterogeneity (see Appendix). For convenience, a line can then be imposed which goes through the total sample mean on each axis, and deviations from the line can be compared using standard analysis of variance. The form of deviations used depends on the type of analysis involved in producing the line of best fit (see Fig. 16.1), and possibly on the reason for the analysis in the first place. If we were attempting to remove the effects of a particular variable, we might be justified in using deviations perpendicular to that axis *after* we had produced a line of best fit.

Discussion

In a methodological paper of this sort it is all too easy to highlight trivial errors in the work of others. That is not our intention. Rather, we wish to emphasise that the errors discussed above are widespread. Where possible, we cite our own work, or alternatively major papers which have been or can be expected to be widely quoted in the field of behavioural ecology or sociobiology. For convenience, we focus on certain topics (social behaviour, ranging behaviour and allometric relationships) to which we return as we develop our discussion in terms of the methodological framework outlined above. Our choice of topic is idiosyncratic, but the interested reader will identify parallel papers in other areas of comparative study.

Choice of taxonomic level

We have already mentioned how a limitation on the quantity of available data can tempt workers to inflate sample sizes by incorporating data from taxonomic levels below that at which the comparison contains statistically independent points. We discussed one example from the home range and body-size literature, but there are many others. We have chosen another example from a different kind of study to discuss here because it clearly illustrates the problem, and therefore it can be used to suggest a solution. Sherman (1979) demonstrates an association between chromosome number and eusociality in the Hymenoptera. He presents data from 382 species across 20 families. Three of the families are primarily eusocial and, treating species within families as independent points for analysis, he 'statistically compares' chromosome numbers between families. As Sherman himself notes, if there are phylogenetic constraints, (see above) on chromosome number then this procedure is clearly invalid. And there may be such constraints, so we should justify *statistical* independence of species data in the absence of any biological information. Nested analysis of variance (Sokal & Rohlf, 1969) reveals heterogeneity as high as the superfamily level within suborders, with no additional variation among suborders (both the within and among superfamily variances are greater than the additional variance between suborders).

If we were to be reasonably sure of the effect being examined, data such as Sherman's should be used only for comparisons within the taxonomic level immediately above that being employed for analysis. That is, interfamily comparisons must be restricted within a superfamily, and species should not be used as independent points to compare families within an order.

Another example comes from the allometric relationship between brain size and body size among mammals. It is well established that brain size increases to about the 0.67 to 0.75 power of body weight among adult mammals from diverse taxa. However, in lower-level taxa the exponent is smaller (see Gould, 1971, 1975*b*) so that within genera it lies between 0.2 and 0.4 (Lande, 1979). If we were comparing deviations from some line of best fit (as when producing the encephalisation quotients discussed above), quite different estimates would result from measuring species deviations from the generic or from the family line. Again, we suggest using the taxonomic level immediately above the one in question; thus, species deviations should be measured from the generic lines, and generic deviations from the family lines. If lines of best fit for brain-weight against body-weight relationships were being compared across different families, then slopes set by generic points should be used since the uneven distribution of species among genera can produce quite irrelevant differences in slope or elevation.

The line of best fit

As the variable on the abscissa is measured with increasing error, regression analysis produces a progressively lower slope. Usually, variables in comparative studies are measured with considerable error and correlation coefficients are far from unity. Regression then provides an unsuitable model and any estimate of slope calculated by regression analysis will be too low. Nevertheless, regression analysis has been used repeatedly in such studies and the error variances on the abscissas have been conveniently ignored. For example, since McNab (1963) first used regression analysis on logarithmically transformed data to investigate the relationship between home range size and body size in small mammals, the technique has been routinely used by later workers studying other groups (e.g. birds: Schoener (1968); lizards: Turner, Jeinrich & Weintraub (1969); primates: Milton & May (1976), Clutton-Brock & Harvey (1977*a*); these and other vertebrate and invertebrate groups: G.E. Belovsky & J.B. Slade (in preparation)). The slopes involved have been the subject of both discussion and controversy, but although it has been widely acknowledged that body weights have considerable error variance (e.g. see Turner *et al.*, 1969) it has apparently never been appreciated that the slopes underestimate the true values since correlations tend to be of the order 0.5 to 0.8. For example, Turner *et al.* (1969) quote a regression slope of 0.88 for the logarithm of home range regressed on the logarithm of body weight across female lizard species, but the correlation is 0.71 and a reduced major axis estimate of the slope would be 1.23. (We are not arguing that reduced major axis is the correct model for analysis here. Clearly, since home range size probably depends on body size (via metabolic needs), a regression model *that incorporates measurement error on the abscissa* would be more appropriate (Sokal and Rohlf, 1969).) The authors were interested in comparing the slope to that relating metabolic costs to body weight, but in the absence of any information on the extent of the error variances in the two analyses, the comparison becomes meaningless. Discussion of the biological interpretation of regression slopes (e.g. Clutton-Brock & Harvey, 1977*a*) now seems to us to be rather a vacuous exercise because they may have little biological relevance. The message is that when a perfect underlying relationship is postulated (as in the simulations), reduced major axis should be used wherever possible after removing estimates of the error variance from both variables (see above and Kermack, 1954).

Another disadvantage of regression is its ability, through extrapolation, to produce artificial differences in elevation of relationships measured across different taxa. Fig. 16.2 provides a hypothetical example; reduced major axis analysis would reveal no differences in elevation between the two data sets while regression analysis would indicate a difference. Imbrie (1956) provides a more detailed discussion of this point, together with an example taken from allometric relationships in brachiopod morphology. Those readers familiar with the brain : body size literature will recognise Fig. 16.2 as very similar to that found when brain size is plotted against body size on logarithmically transformed axes. In fact, in that case the differences in elevation still exist when major axis analysis is used (Clutton-Brock & Harvey, 1980; Mace *et al.*, 1981). As more data become available, and groups of animals are compared by plots of life history variables on body weight analysed by regression techniques (e.g. Western, 1979; Millar, 1977), we caution against uncritical acceptance of apparent slope or elevation differences when correlation coefficients are low.

The effect of changes of scale on slopes when major axis analysis is used on data that have not been logarithmically transformed is a source of serious potential error, and we need only turn to one of the major textbooks in the biological sciences (Sokal & Rohlf, 1969) to find an example of the technique being used incorrectly.

In a comment on the study of allometry, Gould (1975*b*) argued that the

Fig. 16.2. Lines of best fit produced by two data sets A and B. Regression lines are broken, and reduced major axis lines unbroken. The range of values in each data set lies between the ends of the lines on the *x*-axis. Intercepts for the two regression lines from data sets A and B are shown by points *ra* and *rb* respectively. The intercept for both lines produced by reduced major axis analysis is shown by the point *rma*.

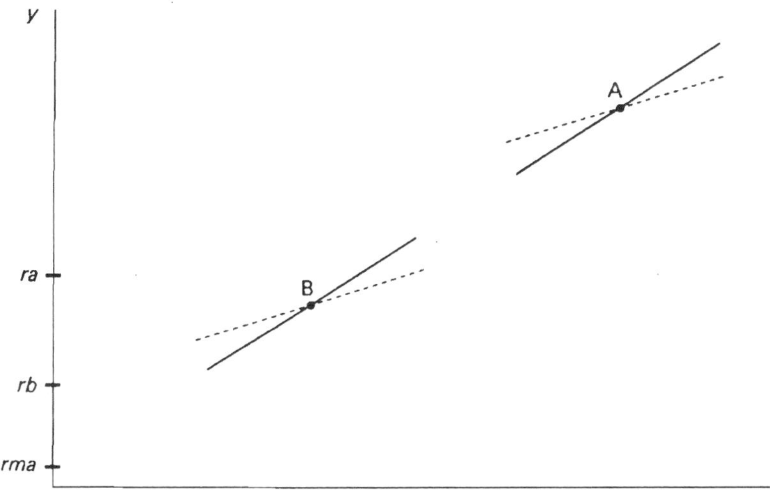

analysis of simple bivariate or double logarithmic plots would be relegated to 'minor significance if not historical oblivion' as 'multivariate techniques supercede bivariate studies'. That, hopefully, will be so. But as more independent variables are incorporated into regression equations so that we deal with a multiple regression model, then the problems concerning error discussed above are compounded. In addition, further bias as well as difficulties of interpretation are introduced when there is correlation between the independent variables (see Post, 1981). Nevertheless, papers in behavioural ecology employ multiple regression without considering the problems of error and correlation (e.g. Jorde & Spuhler, 1974; Baker & Parker, 1979). In some cases, the procedures used are so questionable that any conclusions drawn should either be discarded or treated with extreme caution. The problem does not end with regression models: principal component (major axis) analysis is commonplace nowadays, yet even the simple bivariate case has, as we have shown, problems of interpretation associated with it. How much more so for a multivariate analysis?

Summary

1. Behavioural and morphological variation across taxa provides a rich source of comparative data for both generating and testing adaptationist hypotheses.
2. Such comparative studies face a variety of problems that are often ignored. Choosing a suitable taxonomic level for study and the correct statistical technique for analysis are cases in point.
3. This chapter discusses a variety of methodological problems and pitfalls that are frequently encountered in the sociobiological literature. Solutions to several of these problems are illustrated with a simple simulation study.

This chapter is affectionately dedicated to King's College Sociobiology Group. We wish to take this opportunity to thank everyone associated with the Group for the hospitality, encouragement and academic stimulation shown during our numerous visits to Cambridge over the past few years. We should also like to thank John Maynard Smith for his continued encouragement at Sussex. Finally, we are grateful to those people from King's College, Cambridge, The University of Sussex, Harvard University, and The University of Washington at Seattle who helped at various stages during the preparation of this chapter.

Appendix. To test for heterogeneity of slopes using major axis analysis

The test depends on major axis analysis rotating the original axes through angle θ so that $(x, y) \rightarrow (x^*, y^*)$ and the correlation of points in the new coordinates (r_i for group i of size n_i) is zero. If we have g sets of data (groups), we investigate whether the covariance matrices of each group are diagonal (whether by roating we have removed all correlation). Morrison (1967, pp. 111 onwards) gives such a test for a single group. It computes the likelihood ratio

$$\lambda = |\hat{R}|^{\frac{1}{2}n}$$

where $\hat{R}$ is the observed correlation matrix. We perform g such tests so that

$$\lambda = g_{i=1} \prod |\hat{R}_i|^{\frac{1}{2}n_i}$$

(where for group i the observed correlation matrix is $\hat{R}_i$, so $|\hat{R}_i| = 1 - r_i^2$).

The degrees of freedom will be $gp(p - 1)/2$ which, since the number of variables p is 2, reduces to g. However, if we iterate on a common estimate of θ for all groups so as to minimise the likelihood ratio, this removes one further degree of freedom.

We can look up $-2\ln\lambda$ on a χ^2 table with $g-1$ degrees of freedom. Morrison points out that Bartlett suggests the use of

$$\chi^2 = -\sum_{i=1}^{g} (n_i - 15/6) \ln |\hat{R}_i|$$

rather than

$$\chi^2 = -\sum_{i=1}^{g} (n_i) \ln |\hat{R}_i|$$

as an improved value which converges to a true χ^2 distribution more quickly as all the n_i approach ∞.

If χ^2 is significant, we reject the null hypothesis that slopes are equal. If χ^2 is not significant, we use sine θ as our estimate of a common slope.

We thank Professor J. Felsenstein for developing this test.

References

Alexander, R.D., Hoogland, J.L., Howard, R., Noonan, K.M. & Sherman, P.W. (1979). Sexual dimorphism and breeding systems in pinnipeds, ungulates, primates and humans. In *Evolutionary Biology and Human Social Behavior*, ed. N.A. Chagnon & W.D. Irons, pp. 402–35. Duxbury Press: North Scituate, Mass.

Alexander, R.M. (1977). Mechanics and scaling of terrestrial locomotion. In *Scale Effects in Animal Locomotion*, ed. T.J. Pedley, pp. 93–110. Academic Press: London.

Baker, R.R. & Parker, G.A. (1979). The evolution of bird colouration. *Philosophical Transactions of the Royal Society of London*, B, **287**, 63–130.

Bauchot, R. & Stephan, H. (1969). Encephalisation et niveau évolutif chez les Simiens. *Mammalia*, **33**, 225–75.

Belovsky, G.E. & Slade, J.B. (in preparation). Body size – home range area patterns and an energy maximising explanation.

Clutton-Brock, T.H. & Harvey, P.H. (1977*a*). Primate ecology and social organisation. *Journal of Zoology, London*, **183**, 1–39.

Clutton-Brock, T.H. & Harvey, P.H. (1977*b*). Species differences in feeding and ranging behaviour in primates. In *Primate Ecology: Studies of Feeding and Ranging Behaviour in Lemurs, Monkeys and Apes*, ed. T.H. Clutton-Brock, pp. 557–84. Academic Press: London.

Clutton-Brock, T.H. & Harvey, P.H. (1979). Comparison and adaptation. *Proceedings of the Royal Society*, B, **205**, 547–65.

Clutton-Brock, T.H. & Harvey, P.H. (1980). Primates, brains and ecology. *Journal of Zoology, London*, **190**, 309–23.

Gould, S.J. (1966). Allometry and size in ontogeny and phylogeny. *Biological Reviews*, **41**, 587–640.

Gould, S.J. (1971). Geometric similarity in allometric growth: a contribution to the problem of scaling in the evolution of size. *American Naturalist*, **105**, 113–36.

Gould, S.J. (1975*a*). On scaling of tooth size in mammals. *American Zoologist*, **15**, 351–62.

Gould, S.J. (1975*b*). Allometry in primates with an emphasis on the scaling and evolution of the brain. In *Approaches to Primate Paleobiology*, ed. F. Szalay, pp. 244–92. Karger: Basel.

Harestad, A.S. & Bunnell, F.L. (1979). Home range and body weight – a re-evaluation. *Ecology*, **60**, 389–402.

Harvey, P.H., Kavanagh, M. & Clutton-Brock, T.H. (1978). Sexual dimorphism in primate teeth. *Journal of Zoology, London*, **186**, 475–86.

Hespenheide, H.A. (1973). Ecological inferences from morphological data. *Annual Review of Ecology and Systematics*, **4**, 213–29.

Imbrie, J. (1956). Biometrical methods in the study of invertebrate fossils. *Bulletin of the American Museum of Natural History*, **108**, 217–52.

Jerison, H.J. (1973). *Evolution of the Brain and Intelligence*. Academic Press: New York.

Jolicoeur, P. (1968). Interval estimation of the slope of the major axis of a bivariate normal distribution in the case of a small sample. *Biometrics*, **24**, 679–82.

Jolicoeur, P. & Mosimann, J.E. (1968). Intervalles de confiance pour la pente de l'axe majeur d'une distribution bidimensionelle. *Biometrie-Praximetrie*, **9**, 121–40.

Jorde, L.B. & Spuhler, J.N. (1974). A statistical analysis of selected aspects of primate demography, ecology and social behaviour. *Journal of Anthropological Research*, **30**, 199–224.

Kermack, K.A. (1954). Biometrical study of *Micraster coranguinum* and *M. (Isomicraster) senonensis*. *Proceedings of the Royal Society of London*, **B**, **237**, 375–428.

Kermack, K.A. & Haldane, J.B.S. (1950). Organic correlation and allometry. *Biometrika*, **37**, 30–41.

Kleiber, M. (1961). *The Fire of Life: an Introduction to Animal Energetics*. Wiley: New York.

Lande, R. (1979). Quantitative genetic analysis of multivariate evolution applied to brain : body size allometry. *Evolution*, **33**, 402–16.

MacArthur, R.H. & Wilson, E.O. (1967). *The Theory of Island Biogeography*. Princeton University Press: New Jersey.

Mace, G.M., Harvey, P.H. & Clutton-Brock, T.H. (1981). Brain size and ecology in small mammals. *Journal of Zoology, London*, **193**, 333–54.

McMahon, T. (1973). Size and shape in biology. *Science*, **179**, 1201–4.

McNab, B.W. (1963). Bioenergetics and the determination of home range size. *American Naturalist*, **97**, 133–40.

Millar J.S. (1977). Body size and reproduction in terrestrial eutherian mammals. *Evolution*, **31**, 370–86.

Milton, K. & May, M.L. (1976). Body weight, diet and home range area in primates. *Nature*, **259**, 459–62.

Morrison, D.F. (1967). *Multivariate Statistical Methods*. McGraw Hill: New York.

Pirlot, P. & Stephan, H. (1970). Encephalisation in Chiroptera. *Canadian Journal of Zoology*, **48**, 433–44.

Post, D.G. (1981). Sexual dimorphism in the anthropoid primates: some thoughts on causes, correlates and the relationship to body size. In *Sexual Dimorphism in Primates*, ed. C. Eastman & P. Heisler. Garland Press: New York. (In press.)

Rensch, B. (1959). *Evolution Above the Species Level*. Methuen: London.

Schoener, T.W. (1968). Sizes of feeding territories among birds. *Ecology*, **49**, 123–41.

Sherman, P.W. (1979). Insect chromosome number and eusociality. *American Naturalist*, **113**, 925–35.

Snedecor, G.W. (1956). *Statistical Methods*. State University Press: Iowa.

Sokal, R.R. & Rohlf, F.J. (1969). *Biometry*. Freeman: San Francisco.

Taylor, C.R. (1977). The energetics of terrestrial locomotion and body size of the vertebrates. In *Scale Effects in Animal Locomotion*, ed. T.J. Pedley, pp. 127–41. Academic Press: London.

Tessier, G. (1948). La relation d'allometrie: sa signification statistique et biologique. *Biometrics*, **4**, 14–53.

Turner, F.B., Jeinrich, R.I. & Weintraub, J.D. (1969). Home ranges and body size of lizards. *Ecology*, **50**, 1076–81.

Verner, J. (1964). Evolution of polygyny in the long billed marsh wren. *Evolution*, **18**, 252–61.

Western, D. (1979). Life history and ecological implications of size in mammals. *African Journal of Ecology*, **17**, 185–204.

Wilkie, D.R. (1977). Metabolism and body size. In *Scale Effects in Animal Locomotion*, ed. T.J. Pedley pp. 23–36. Academic Press: London.

PAPER 45

A Null Model for Null Models in Biogeography (1984)

R. K. Colwell and D. W. Winkler

Commentary

NICHOLAS J. GOTELLI

R. K. Colwell and D. W. Winkler's contribution is a deserved classic in macroecology that was written at a time when community ecologists and biogeographers were engaged in an intense debate over the role of interspecific competition and chance processes in producing biogeographic patterns. Proponents of competition theory (e.g., Diamond 1975) argued that competition for limited resources at the local scale could be strong enough to lead to biogeographic patterns such as checkerboard distributions (pairs of species that never occur together), forbidden species combinations (particular combinations of species that were locally unstable), and character displacement (divergence in sympatry versus allopatry of body size and morphology of competing species). Opponents of competition theory (e.g., Connor and Simberloff 1979) argued that the evidence for these competitive processes was flimsy, and that the patterns in many assemblages were no different than would be expected by chance. These patterns were analyzed with "null models," statistical tests in which observed data matrices of species occurrences or body sizes were randomized or reshuffled to try and establish biogeographic patterns that might be expected in the absence of competition (Gotelli and Graves 1996).

Colwell and Winkler's approach was to frame the controversy this way: "If we could seed a series of virgin, replicate earths with primordial life and set the level of interspecific competition differently in each, could we tell them apart three billion years later by looking at biogeographical patterns?" (344). To answer this question, Colwell and Winkler built a comprehensive computer simulation, GOD, to do exactly this. GOD used a hierarchical, branching model of stochastic speciation, extinction, and character evolution that had first been developed by paleontologists (Raup and Gould 1974). But GOD introduced cleverly titled algorithms and subroutines to assemble sympatric communities (WALLACE), permit interspecific competition to operate based on the similarity of evolved morphological characters (GAUSE), allow species to colonize a set of islands ("The Lack Archipelago"), and then reshuffle the resulting distributions as in a null-model analysis ("The Tallahassee Archipelago"). Other model experiments included creating local assemblages unstructured by competition ("The Kropotkin Archipelago") and assemblages in which dispersal ability was correlated with morphological characters ("The Icarus Archipelago"). The results of these simulation experiments suggested that the extent of interspecific competition might be underestimated by null-model tests based on randomization of current empirical patterns ("The Narcissus Effect"), and that analyzing an entire assemblage this way might obscure strong patterns of negative association between competing, closely related species ("The J. P. Morgan Effect," also known as "the dilution effect"; Diamond and Gilpin 1982).

Excerpt pp. 344–59 from *Ecological Communities: Conceptual Issues and the Evidence* edited by D. R. Strong Jr., D. Simberloff, L. G. Abele, and A. B. Thistle.

Colwell and Winkler's contribution was important not only in the context of the then-raging null-model debate (the collection of papers in Strong et al.'s [1984] book were derived from a 1981 symposium that brought together all the major players in the null-models controversy). It clearly demonstrated the necessity of benchmarking the performance and statistical behavior of randomization tests, something that would not begin to happen in the co-occurrence literature for another 16 years (Gotelli 2000). Second, Colwell and Winkler's GOD model foreshadowed an exciting research front in contemporary macroecology, in which spatially explicit, mechanistic simulation models are being used to study large-scale patterns of species richness (Storch et al. 2006; Rahbek et al. 2007). In this area of macroecology, Colwell pioneered a new kind of null model that generates a "mid-domain effect"—a peak of species richness in the center of a biogeographic domain that arises simply from geometric constraints on the stochastic placement of species ranges (Colwell and Hurtt 1994; Colwell and Lees 2000). In addition to geometric constraints, these simulation models are beginning to incorporate evolutionary branching processes as in GOD (Rangel et al. 2007), but so far they have not included species interactions. There is still much we can learn from Colwell and Winkler's key contribution.

Literature Cited

Colwell, R. K., and G. C. Hurtt. 1994. Nonbiological gradients in species richness and a spurious Rapoport effect. *American Naturalist* 144:570–95.

Colwell, R. K., and D. C. Lees. 2000. The mid-domain effect: Geometric constraints on the geography of species richness. *Trends in Ecology and Evolution* 15:70–76.

Connor, E. F., and D. Simberloff. 1979. The assembly of species communities: Chance or competition? *Ecology* 60:1132–40.

Diamond, J. M. 1975. Assembly of species communities. In *Ecology and evolution of communities*, edited by M. L. Cody and J. M. Diamond, 342–444. Harvard University Press, Cambridge, MA.

Diamond, J. M., and M. E. Gilpin. 1982. Examination of the "null" model of Connor and Simberloff for species co-occurrences on islands. *Oecologia* 52:64–74.

Gotelli, N. J. 2000. Null model analysis of species co-occurrence patterns. *Ecology* 81:2606–21.

Gotelli, N. J., and G. R. Graves. 1996. *Null models in ecology*. Smithsonian Institution Press, Washington, DC.

Rahbek, C., N. J. Gotelli, R. K. Colwell, G. L. Entsminger, T. F. L. V. B. Rangel, and G. R. Graves. 2007. Predicting continental patterns of bird species richness with spatially explicit models. *Proceedings of the Royal Society B* 274:165–74.

Rangel, T. F. L. V. B., J. A. Diniz-Filho, and R. K. Colwell. 2007. Species richness and evolutionary niche dynamics: A spatial pattern-oriented simulation experiment. *American Naturalist* 170:602–16.

Raup, D. M., and S. J. Gould. 1974. Stochastic simulation and evolution of morphology: Towards a nomothetic paleontology. *Systematic Zoology* 23:305–32.

Storch D., R. G. Davies, S. Zajicek, C. D. L. Orme, V. Olson, G. H. Thomas, T. S. Ding, et al. 2006. Energy, range dynamics and global species richness patterns: Reconciling mid-domain effects and environmental determinants of avian diversity. *Ecology Letters* 9:1308–20.

Strong, D. R., Jr., D. Simberloff, L. G. Abele, and A. B. Thistle. 1984. *Ecological communities: Conceptual issues and the evidence*. Princeton University Press, Princeton, NJ.

20.

A Null Model for Null Models in Biogeography

ROBERT K. COLWELL

Department of Zoology, University of California, Berkeley, California 94720

DAVID W. WINKLER

Museum of Vertebrate Zoology and Department of Zoology, University of California, Berkeley, California 94720

If we could seed a series of virgin, replicate earths with primordial life and set the level of interspecific competition differently in each, could we tell them apart three billion years later by looking at biogeographical patterns? In this paper we present the results of an effort to approximate this experiment by computer simulation, with the purpose of examining the potentials and the pitfalls of several methods of biogeographical analysis. Since we control the intensity of competitive exclusion as a variable in the simulation, biases and limitations in the construction of null models in biogeographical studies can be studied directly. We will show that the effects of competitive exclusion on island biotas are likely to be underestimated or even obscured by several directional biases that are difficult to estimate or avoid in the real world. However, far from counseling despair, we hope that the model we have developed may lead to a better understanding of the complex interactions between evolution, ecology, and chance events in biogeography.

The principal message we hope to convey is that the setting of constraints and assumptions in the development of null models is not an arbitrary matter. The null hypothesis tested in any analysis of biogeographical data (*e.g.* Connor and Simberloff, 1978; Strong *et al.*, 1979) is not that empirical patterns do not differ from random ones, but that they do not differ from patterns generated by a particular model of the world. The design of such a "null model" is critical, but is necessarily based on biological judgments. We will proceed by describing a general model of evolution, biogeography, and ecological interactions within which many null models for biogeographical patterns can be defined.

The Model: Phylogenesis, Morphological Evolution, and Taxonomy

The model consists of two parts. The first, described in this section, builds upon the models of Raup *et al.* (1973) and Raup and Gould (1974) for the stochastic generation of phylogenies. with "phenotypic" characters evolving stochastically at each speciation event. Implemented in a program called "GOD," this part of our model, like that of Raup and Gould (1974), produces "biotas" of any desired size that mimic many of the commonly observed properties of real-world phylogenies (Figure 20.1). The input variables for the program are simply the probabilities of speciation, extinction, and character change and a seed for the random number generator. (The number of characters and the size of the tree must also

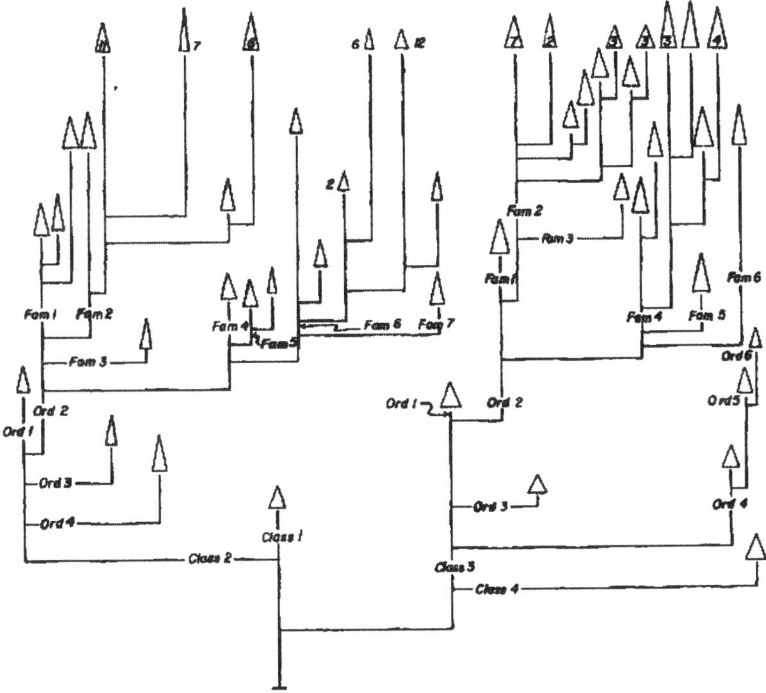

Fig. 20.1. Example of a stochastically generated phylogenetic tree ("Beta"), based entirely on three probabilities: speciation (.20), extinction (.10), and character change (.80). Evolution was stopped at 100 species, 51 of which had become extinct, leaving 49 living species (at the top of the tree) as a "mainland pool" for populating imaginary archipelagoes. The triangles represent the "bills" of some computer creature that partitions a food resource (as every programmer knows, larger bugs mean larger bills). Bill length and bill width are linear transformations of independently evolving Characters 3 and 4, respectively. Each triangle represents a genus, monotypic if no number appears; otherwise, the numbers in or near triangles give the number of species in that genus. The computer-generated taxonomy is based exclusively on the topology of the tree (not on characters) according to a cladistic algorithm.

be set, but they do not affect patterns of evolution.) The tree is initiated with a single species whose "characters" all have initial values of zero. At each iteration of a counter for evolutionary time, this "root" species and each of its living descendant species gives rise to a daughter species with probability "PSPN," goes extinct with probability "PEXT," or continues to exist unchanged with probability (1 − PSPN − PEXT). The process continues until a predefined number of speciation events have occurred, until a predefined number of time units have passed, or until all species are extinct. (When PEXT approximates or exceeds PSPN, many trees die young. Even when PSPN is twice PEXT, some trees become extinct in early stages, but are generally safe from extinction once some "epidemic threshold" has been passed.)

Morphological evolution is by founder effect: each character in a daughter species is independently subject to random change at the time of speciation, but is fixed ever after. The parent species retains its previous phenotype. (It is not the point of this paper to defend this view of evolution; we have adopted it in building on the results of Raup and Gould, 1974). At speciation, each character in the daughter species gets "larger" with probability PEVOL/2, "smaller" with probability PEVOL/2, or stays the same with probability (1 − PEVOL). Change in each character is entirely independent of change in the others. The amount of change is a (uniform) random number between zero and one, in contrast to the model of Raup and Gould (1974), in which all changes are of unit magnitude.

We measure the phenotypic dissimilarity of a pair of species as the euclidean distance between them in character space. Allowing the characters to evolve on a continuous scale effectively eliminates ties in distance arising by evolutionary convergence and parallelism. GOD computes the euclidean distance in character space between all pairs of species living at the time the tree is terminated. A specifiable subset of the characters can be excluded from the distance computation. (We use this option to separate a character or characters for "vagility" from "morphological" characters upon which coexistence among successful colonists is to depend.)

Since one of the techniques we evaluate is sampling within taxonomically constrained biotas, it is necessary to produce a hierarchical classification of the species in each tree. There are, of course, a great number of algorithms available for classifying sets of species on the basis of their phenotypic characters, but these are not appropriate (or necessary) in this case, since the actual phylogeny is completely known. The tree-generating program GOD outputs a taxonomic "name" for each species (as a string of 1's and 2's) that completely specifies a path from that species to the root of the tree, in dichotomous fashion. (Thus, species "2122" is the second species on the tip of the second twig on the first branch of the second trunk of the phylogenetic tree, if all possible branches evolve.) Sister groups at the time of branching end in "1" and "2," respectively, and share all digits to the left. When the tree is complete, some species (including most extinct ones) will have fewer digits in their names than others, depending on the number of nodes between the species in question and the root of the tree. At this point all names shorter than the longest name in the tree are right-filled with 1's, producing monobasic taxa at various hierarchical levels. This step is necessary to preserve the cladistic hierarchy, and to ensure that each species has a unique "name."

The Model: Biogeography, Ecology, Classification, and Sampling

For each evolutionary tree generated, program GOD outputs a list of all species, with information on when each originated; when it went extinct, in the case of "fossil" species; the identity of its immediate ancestor; its score for each character; its dichotomous cladistic "name"; and, for each living species, the euclidean distance from its location in character space to the location of each other living species. This information is used by a separate program, "WALLACE," to carry out a variety of neontological operations. In effect, the living species produced by GOD are assumed to have evolved in a world free of selective forces, including species interactions. Alternatively, one may imagine that selection has operated, but in a way indistinguishable from stochastic character change and stochastic speciation (*i.e.* that the species in a given tree have all lived their lives allopatrically in some highly dissected continent.)

In this study we designate the set of living species in a large phylogeny produced by GOD to be a "mainland" pool of potential dispersers. From this pool WALLACE assembles sympatric communities. The probability of each species' being chosen from the pool may be a constant, or may be weighted by one or more of its characters representing "vagility." Archipelagoes of island communities can be assembled with a specified number of species per island (sample size) and a specified number of islands (samples)—a great advantage over natural archipelagoes. Interspecific competition can be imposed with adjustable force, either before colonization (in the "mainland" pool) or after assignment of species to islands, by eliminating a specifiable proportion of the species present as victims of competitive elimination, their vulnerabilities based on morphological similarity to other species on the island. Finally, various sampling schemes can be used on archipelagoes with known histories of colonization and competition to investigate the power and biases of each scheme in revealing the forces that produced the biogeographical and morphological patterns in an archipelago.

Before describing the "treatments" applied to communities and archipelagoes, we must explain the algorithms used to eliminate species by "competition" and to classify species hierarchically for taxonomically constrained sampling.

"GAUSE," a subroutine of WALLACE, ranks all the species in any sample sent to it according to their vulnerability to direct and diffuse competition, based on a specifiable subset of the characters of each species. The algorithm used to produce this ranking is as follows. First, the matrix of euclidean distances in morphological space between all pairs of species in the sample is searched for the smallest distance. Then, one or the other of the two species involved is ranked for competitive elimination and struck from the matrix; the smallest distance value in the matrix (and, of course, all other distances involving that particular species as well) is thereby removed from the matrix. Then the next smallest distance is found, one of the two species involved is ranked and eliminated from consideration, and so on, until all species are ranked. (Both members of the last remaining pair are automatically given the last rank.)

A simple criterion determines which of the two species is to be ranked and eliminated from the matrix at each cycle. The species whose removal would produce a greater increase in the mean of all remaining distances in the matrix is chosen—put another way, this is the species whose "diffuse similarity" to the remaining species is greater. This choice is easily accomplished; the program consults (and constantly updates) a vector of marginal totals for the distance matrix, and simply asks which of the two species has the smaller sum of distances to all others. In case of a tie, one of the two species is chosen at random to be ranked for elimination (ties are very scarce when two or more characters are involved and characters evolve on a continuous scale). Figure 20.2 shows examples of the operation of GAUSE on island samples.

"HENNIG," another subroutine of WALLACE, takes the dichotomous code names for all the species in a tree (including the extinct species) and produces a hierarchical classification, according to a specifiable variable "KUT" that tells HENNIG the maximum number of lineal nodes per taxonomic level. For example, if KUT is set at 3, up to 3 levels of branching can be embraced in a single taxonomic level. If all possible speciations have occurred in all lineages, each taxon at this level would include 2-to-the-KUT-power (8) subtaxa. (This is, of course, only rarely the case in all lineages.) Thus KUT sets the maximum number of species per genus, genera per family, and so on. (Subroutine HENNIG appears in WALLACE, rather than in GOD, to permit a series of reclassifications to be performed with different values of KUT, without the need to rerun GOD each time.) Finally, HENNIG produces nomenclature, assigning consecutive numerals to subtaxa within each taxon, as names.

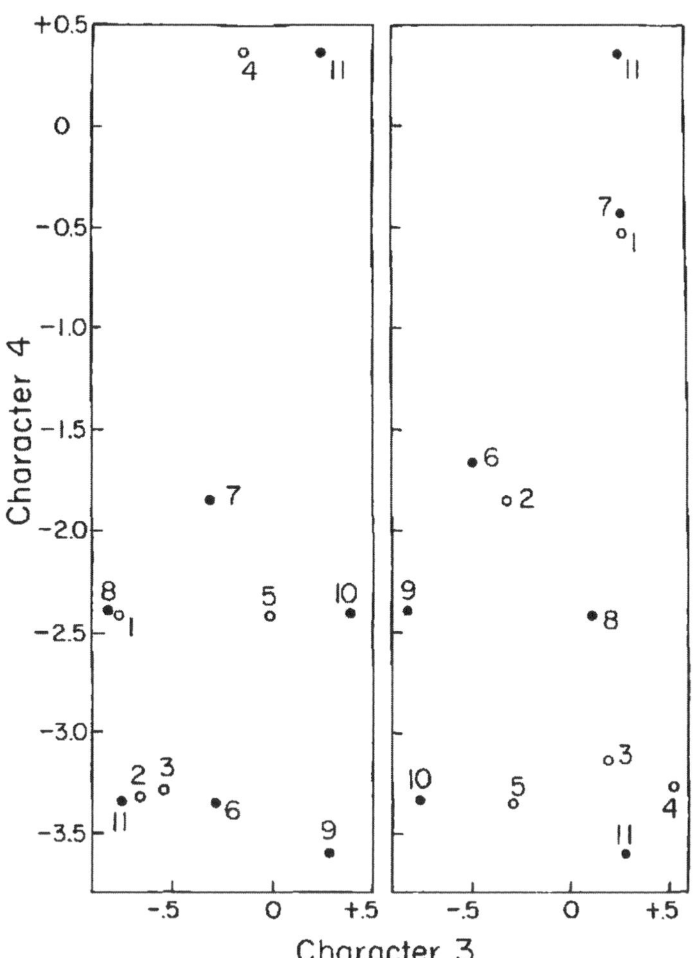

Fig. 20.2. Ranking of island species according to their vulnerability to extinction. The two boxes represent two of the islands of an Icarus-Lack Archipelago, each colonized by twelve living species from tree eta. The numbers are the ranks, from most vulnerable to competitive elimination (1) to least vulnerable (11), within islands. (The ranking algorithm is explained in the text.) If five of the species on each island are eliminated by competition in rank order (PKILL = .42, Figure 20.4, line C), the open circles will be removed, leaving the solid dots. Characters 3 and 4 correspond to the "bill" characters of Figure 20.1, untransformed.

Treatment 1: The Lack Archipelago

From the "mainland pool" (all the living species in a single tree) "*K*" samples of "*N*" different species each are drawn at random. (In other words, sampling is without replacement within samples, with replacement between samples.) Each sample is the biota of an island in a *K*-island archipelago. Next, GAUSE is called to rank the species within each island biota for competitive vulnerability. (We use Characters 3 and 4—"bill length" and "bill width," let us say—to compute distances in two-dimensional character space.) A parameter "PKILL" then determines the proportion of each biota to be eliminated; in effect, PKILL sets the limiting level of similarity scaled to the overall similarity on the island. Next, the portion of each island biota to be eliminated by competition, set by PKILL, is marked for local extinction (Figure 20.2). The surviving species on each island represent the biota of the Lack Archipelago, a place where the ravages of competition can be tuned experimentally. (No migration among the islands of the archipelago is permitted.)

Treatment 2: The Tallahassee Archipelago

This treatment involves a reshuffling of the species in the Lack Archipelago, in which, after competition, each island has $N(1 - \text{PKILL}) =$ "NLEFT" species remaining. Any species that survived competition on at least one island is placed in a pool for resampling. Now, K samples of NLEFT different species each are drawn at random from this pool as the biotas of the K islands in the Tallahassee Archipelago.

Treatment 3: The Kropotkin Archipelago

The K islands of this archipelago are each populated with NLEFT species chosen at random from the mainland pool. All species are mutually congenial; no competition occurs here.

Treatment 4: The Icarus Archipelago

In this treatment, we use Characters 1 and 2 of each species in the mainland pool to compute an index of vagility, or dispersal potential. The index is simply the product of the two characters after a constant is added to each character to bring the minimum character state up to zero (since characters may evolve into negative values). Two characters were combined to reduce the frequency of evolutionary convergence and parallelism. The probability of any particular species' being chosen to populate the Icarus Archipelago is proportional to its value for the vagility index, in contrast to Treatments 1, 2, and 3, in which uniform probabilities are used. The K islands of the Icarus Archipelago are each populated with NLEFT different species chosen from the mainland pool by this method.

Treatment 5: The Icarus-Lack Archipelago

This archipelago is populated in the same way as the Lack Archipelago, except that the colonists are chosen from the mainland pool as for the Icarus Archipelago—by vagility-weighted probabilities. Competitive elimination after colonization is the same as in the Lack Archipelago.

Treatment 6: The Icarus-Tallahassee Archipelago

This archipelago is populated by the same means as the Tallahassee Archipelago, but species are drawn from the repooled survivors of competition in the Icarus-Lack Archipelago. (The resampling is done uniformly, not with vagility-weighted probabilities.)

Treatment 7: The Linnaean Quadrille

This treatment is a resampling of the post-competitive biota of the Lack Archipelago with taxonomic constraints, contrasted with a taxonomically matched sample from the mainland pool. The sampling algorithm begins with the (uniform) random selection of one of the K islands in the Lack Archipelago. Next, a species on that island is chosen at random (the "propositus"), and its taxonomic identity is determined at a designated taxonomic level ("genus," "family," "order," etc.). Now a list is made of all other species on that island that are "contaxonic" (congeneric, confamilial, etc., depending on the designated level) with the propositus species, and one of them is chosen at random. (If there is no other species on that island that is contaxonic with the propositus, a new island and a new propositus are chosen at random.) In this way, a pair of contaxonic, coexisting species is found in the Lack Archipelago; call it the Island Pair. Now a list is made of all species in the mainland pool that are contaxonic with the Island Pair, and a Mainland Pair is chosen from the list. (There must, of course, be at least one such pair, since the archipelago fauna is a subset of the mainland pool.) This process is repeated until 20 matched sets of pairs have been selected. Any redundant sets are then removed before statistical analysis.

RESULTS AND CONCLUSIONS

Several trees were generated by program GOD, some using the same set of input probabilities but with different random number seeds to get an idea of the variability to be expected, some with different values of the input probabilities to explore their effects on the phylogenies and biogeographical analyses.

Tree beta, shown in Figure 20.1, had an extinction probability only half that of the speciation probability, producing rapid radiation and only a few species per lineage in extinct groups. Trees generated with extinction probability equal to speciation probability are less "radiating," with major groups becoming extinct. Tree eta (which produced the raw data for Figures 20.2, 20.3, and 20.4) is such a tree. Raup and Gould (1974) show that there are often strong correlations among characters for the species in a stochastic tree. The branching process itself reproduces certain character state combinations in the descendant species of unusually prolific ancestral species, while other lineages become extinct. Character correlations tend to be higher in "equilibrium" trees (PEXT = PSPN) than in rapidly radiating trees (PSPN > PEXT) because of less homogeneous filling of morphological space.

Even though the amount and direction of character change in our model is a uniform random variate at each speciation event, the frequency distribution of any given character among species in our imaginary biotas typically has a strong mode, usually not located at zero, as also shown by Raup and Gould (1974) for unit character changes. Like many other effects that we will discuss, the existence of such modes is a consequence of the differential propagation, by chance, of certain lineages that represent variations on a theme. Substituting a biologically more realistic rule for character change, such as a normal random distribution, will only emphasize the modality. The tests used by Simberloff and Boecklen (1981) to assess the effects of competitive displacement on characters of coexisting species assume a uniform frequency of character states as the null distribution. Since the average (and minimum) distance between random draws from a uniform distribution is always larger than the distance between random draws from a modal distribution with the same range, their tests are biased to an unknown degree against finding any effects of character displacement, which acts in the same direction on similarity. The uniform random distribution, however appealing as a null hypothesis for character states, cannot be produced by any stochastic evolutionary process known to us, except as a pathological limiting case. Thus any distribution of character states from nature that is uniform (or log-uniform) is, in itself, strong evidence for character displacement.

By comparing the effects of treatments 1–7 on patterns of morphological similarity among "coexisting" species on our imaginary islands and in mainland pools, we can demonstrate three different inherent biases in some techniques of biogeographical analysis designed to test for the effects of interspecific competition. All three biases tend to obscure or even reverse any competitive effects, seriously weakening any inferences drawn using the techniques that produce them.

1. *The Narcissus effect: Sampling from a post-competition pool underestimates the role of competition, since its effect is already reflected in the pool.* When the species of an archipelago are pooled and then sampled at random to test for the role that competition may have played in eliminating closely similar sympatric species, there is a consistent underestimation of the effect of competition, since the most vulnerable species have been eliminated from the entire archipelago (or never evolved in the first place, in the case of a local radiation). In our model, this situation is represented by the comparison of a Lack Archipelago with a Tallahassee

Archipelago. The correct comparison (assuming it to be possible) is between the empirical island pattern (the Lack Archipelago) and a sample from a truly pre-competitive pool—in the model, the Kropotkin Archipelago, sampled at random from the mainland pool, with no competition.

Figure 20.3 shows the Narcissus effect for tree eta. The values plotted are the mean distances in two-character morphology space (Figure 20.2) for first through sixth nearest neighbors, averaged over islands (see Inger and Colwell, 1977). Pooling all pairwise distances obscures the relationship between closely similar species, which is the focus of competition. (Gilpin and Diamond (this volume), in a similar vein, decry the pooling of diverse guilds in sampling for null models.) In Figure 20.3, the dissimilarity of coexisting species in the Lack Archipelago (dashed line) is consistently greater than in the Kropotkin Archipelago (dotted line)—significantly greater for first nearest neighbors. The Tallahassee Archipelago (solid line) differs in the right direction for first nearest neighbors (though not significantly), but quickly unites with the dashed line for more distant neighbors. Pooling and resampling consistently underestimates any role that competition may have played (Grant and Abbott, 1980; Case and Sidell, 1983). It is important to recognize this directional bias, even if it is

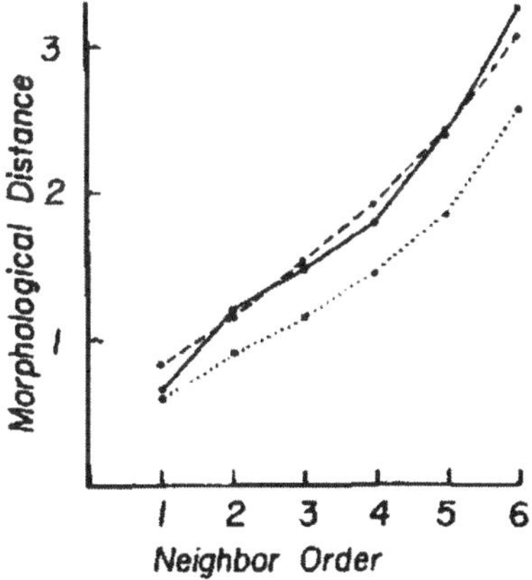

Fig. 20.3. The Narcissus effect for tree eta. Mean euclidean distances to first through sixth nearest neighbor in character space (Characters 3 and 4) for the seven species on each of five islands (each point is the mean of five means of seven scores), contrasted for three treatments. Coexisting species of the Lack Archipelago (42% eliminated by competition—dashed line) are consistently less similar to one another than are the species in the Kropotkin Archipelago (no competition—dotted line), but the effect of competition is only weakly detectable by comparison of the Lack islands with those of the Tallahassee Archipelago (solid line), formed by pooling and resampling the survivors of competition in the Lack Archipelago. Sampling from the mainland pool was uniform, not vagility-weighted, for this analysis.

usually impossible to avoid it. Narcissus could not see the bottom of the pool for his own image, and could not guess its depth.

2. *The Icarus effect: Correlations between vagility and morphology can obscure the effects of competition in morphological comparisons of mainland and island biotas.* When comparisons are made between the morphological characteristics of mainland and island biotas, the assumption is often made that the species in the mainland pool do not differ significantly among themselves in vagility (Simberloff, 1978a). This assumption ignores the fact that phylogenies, both real and random (Raup and Gould, 1974), result in character correlations throughout the phenotype. For example, the morphological spectrum of the subset of bird species colonizing any group of islands will include relatively many swallow bills and relatively few tinamou bills, not because of any intrinsic property of their bills, but because swallows have much higher vagility than tinamous, and in the course of their evolution, either because of selection or not, the bills of tinamous have become and have remained invariably different from those of swallows. To the extent that similarities among the high-vagility species colonizing an archipelago *increase* the mean similarity in island faunas over that of mainland pools, the effects of any subsequent or concomitant interspecific interactions that *decrease* the similarity of coexisting species on the island will be confounded. Simberloff (1978a), Grant and Abbott (1980), and Gilpin and Diamond (this volume) discuss this and related problems.

The results for tree eta show this effect (Figure 20.4). Species in the pool selected preferentially according to vagility (line B in Figure 20.4, the Icarus Archipelago) are significantly more similar to one another morphologically than those drawn at random from the mainland pool ignoring their vagilities (line A in Figure 20.4, the Kropotkin Archipelago). When the vagility-weighted pool of colonists is subjected to strong competition (line C in Figure 20.4, an Icarus-Lack Archipelago), the effect of competitive elimination of the most similar species, which increases the mean morphological distance between species, swamps the weaker, opposing effect of vagility-biased colonization, which decreases the mean morphological distance as a result of correlation between morphology and vagility. With competition at this very high level (42% of colonists eliminated), a comparison of samples from the archipelago pool (line D in Figure 20.4, an Icarus-Tallahassee Archipelago) reveals a significant effect of competition. However, when the proportion of species eliminated by competition is reduced to 22% (line E in Figure 20.4)—still a rather high level of competition—the opposing effects are closely balanced, and a comparison with either the mainland pool (line A in Figure 20.4) or with random samples from the archipelago pool (line E in Figure 20.4) reveals no significant difference. The character correlations in our data

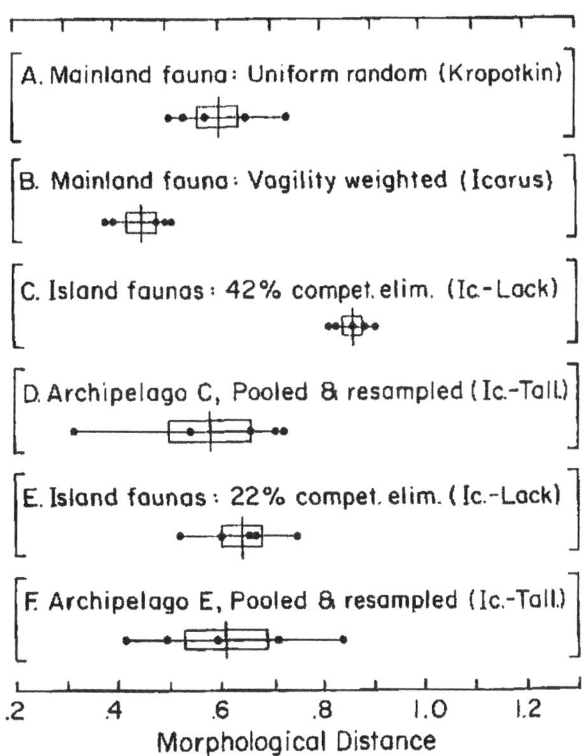

Fig. 20.4. The Icarus effect for tree eta, and its interaction with the Narcissus effect. In each treatment, 7 (NLEFT) species coexist on each island, and there are 5 (K) islands per archipelago. Each point is the mean euclidean distance for one of the five islands from each of the species on that island to its nearest neighbor in character space. Boxes represent one standard error above and below the treatment mean. *A priori* comparisons between treatments were made with Mann-Whitney U-tests (one-tailed) on the plotted points. When morphology (Characters 3 and 4) is correlated with vagility (Characters 1 and 2), the Icarus effect (A vs. B, $p = .006$) and the effect of competition (B vs. C, $p = .004$; or B vs. E, $p = .004$) are opposed in direction, and may completely cancel each other out (A vs. E, $p = .210$, ns). Under these conditions, competition must be extremely strong to be detected by pooling and resampling (for C vs. D, $p = .004$; but for E vs. F, $p = .345$, ns). For the generation of tree eta, PSPN = PEXT = .30, PEVOL = .70.

are the consequence of the statistical properties of conservative branching processes (the "spreading effect" of Raup and Gould, 1974), and so, presumably, are many of the character correlations in real phylogenies. But to this inevitable background level of intercorrelation are added the potentially powerful effects of selection on functionally integrated phenotypes. Had Icarus had hollow bones and air sacs, the wings he fashioned might have carried him further.

3. *The J. P. Morgan effect: The weaker the taxonomic constraints on sampling, the harder it becomes to detect competition.* Results from the Linnaean Quadrille analysis for three trees with identical input probabilities are given in Figure 20.5. The pattern is quite consistent: as sampling is constrained to successively higher taxonomic levels, the difference in morphological distance between contaxonic species in the Lack Archipelago and in the mainland pool decreases monotonically (lower

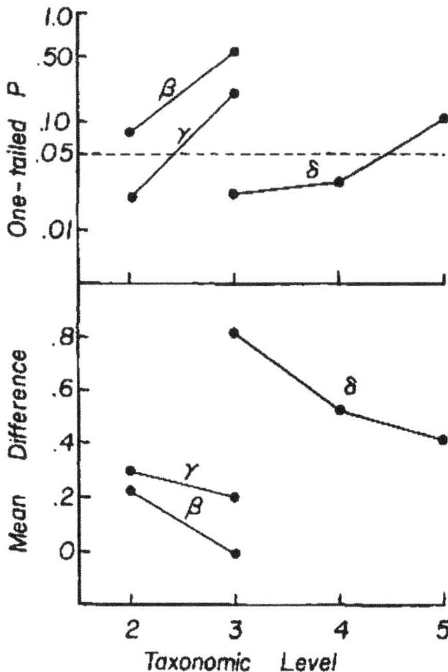

Fig 20.5 The J. P. Morgan effect. If post-competitive archipelago biotas are sampled at random within specified taxonomic limits, then matched with similarly constrained samples from the mainland (source) pool, the effects of competition become less and less detectable as the constraints are weakened (higher taxonomic levels sampled). In the lower graph, we plot the mean difference between the interspecific morphological distance of island species and the interspecific distance of their mainland counterparts (based on Characters 3 and 4). Sample sizes (number of paired pairs) range from 17 to 20. The upper graph shows the significance of one-tailed, paired *t*-tests corresponding to the mean differences in the lower graph. Data are plotted for three trees (beta, delta, and gamma), generated from different random number seeds, but using the same control probabilities (PSPN = .20, PEXT = .10, PEVOL = .80). The island samples are from Lack Archipelagoes (no vagility weighting), with each island losing five of its original twelve species by competition (PKILL = .42), as in Figure 20.2. The computer-generated classification is based on 3 nodes per taxonomic level (KUT = 3). See the text for details.

part of Figure 20.5). The statistical consequences are shown in the upper part of the figure. The null hypothesis of no difference cannot be rejected for higher levels. The cause of this effect is simply that evolution is a conservative branching process: close relatives are more similar than distant ones, and are thus more likely to be incompatible when competition is intense. Including distantly related species in a null sampling pool simply drowns out the signal with noise, progressively weakening the power of the design to detect competition. Others have pointed out this problem (Grant and Abbott, 1980), or related ones (Gilpin and Diamond, this volume), on the basis of biological arguments. We wish to emphasize that it is a general methodological difficulty, demonstrable with imaginary biotas generated under the simplest biological assumptions.

DISCUSSION

The development of neutral models in community ecology has progressed rapidly since the pioneering effort of Caswell (1976). As a result mainly of the efforts of the "Tallahassee School," stochastic models have been proposed as null hypotheses to test for the effect of deterministic features in structuring the composition of isolated biotas (*e.g.* Simberloff, 1970, 1978a, this volume; Connor and Simberloff, 1978, 1979; Strong, this volume; Wilbur and Travis, this volume) and the morphology of their component species (*e.g.* Strong *et al.*, 1979; Ricklefs and Travis, 1980; Schoener, this volume). The development of these models and their application to various aspects of community ecology are forcing a healthy reexamination of the assumptions and conclusions of much of community ecological theory, and the data believed to corroborate it, of the past 20 years. But to insure the most permanent advances, a careful examination of the null hypotheses and their use is as important as a test of their alternatives.

The approach of these studies is that of classical statistical inference. A null hypothesis is chosen, its predictions are compared to the observed data, and the magnitude of the dissimilarity between observed and predicted patterns is evaluated statistically to estimate the probability that the dissimilarity occurred purely by chance. Hypothesis testing at the level of communities differs from more familiar applications only in the difficulty of construction of the null hypothesis.

One of the most important aspects of ecological hypothesis testing and the one that has received the least attention from ecologists is the relevance of considering both Type I and Type II errors in the establishment of statistical criteria for rejection of the null hypothesis. When th investigator has no *a priori* reason to "hedge bets" toward either one o two hypotheses, every effort should be made to equalize alpha and bet and to make them both as small as circumstances allow. Unfortunately, in almost all ecological situations the alternative hypothesis is of the composite type. That is, there is an amorphous class of alternative hypotheses instead of a single, well-defined alternative. When the alternative hypotheses cannot be defined precisely, it is impossible to estimate the probability of a Type II error, beta. In the face of unknown values of beta, establishment of criteria for null hypothesis rejection on the basis of alpha alone becomes totally arbitrary. Furthermore, the custom of reducing the critical (rejection) value of alpha to very low levels unavoidably makes the probability of a Type II error correspondingly higher. In the case of ecological null hypotheses, tests of low power are often used with small sample sizes and low alpha levels (*e.g.* Strong *et al.*, 1979; Connor and McCoy, 1979; Connor and Simberloff, 1978), or probabilities of independent tests of the same (null) hypothesis are not combined (Simberloff and Connor, 1981), producing an unmeasured but potentially overwhelming bias against the rejection of false null hypotheses. Thus, the tests of deterministic community-level effects that have been conducted to date have often been biased against the rejection of the null hypothesis as a result of the way in which the actual statistical comparisons of the null hypotheses with the empirical data have been conducted. These problems are largely a shortcoming of the framework of statistical hypothesis testing, and many of them (*e.g.* the calculation of beta) may not be resolvable. The best that can be done in many situations may be the choice of a test of the highest power available and a setting of liberally high alpha values. (See also Neyman, 1977; Hodges and Lehman, 1970; Grant and Abbott, 1980; Green, 1979; and Hendrickson, 1981, for a discussion of some of these problems.)

We have designed our "null model for null models" to explore the effects of biases of a different kind, namely, the biases inherent in the design of the null hypothesis. Our results demonstrate that construction of an appropriate, unbiased, null hypothesis is very difficult. The counteracting effects of competitive elimination and correlations between vagility and morphology (the Icarus effect) can produce communities that are, with current techniques, indistinguishable from random assemblages of the component species. This problem is exacerbated by the fact that it is impossible to know the evolutionary histories of real communities—even simulated lineages are inconsistent in the degree to which they display this effect. No matter how carefully data from living species are analyzed, ecologists conducting community-level studies will always be plagued by the difficulty of confronting "the ghost of competition past" (Connell, 1980). Our model explicates the danger (the Narcissus effect) pointed out by Grant and Abbott (1980) in using an archipelago biota as the source pool for a null hypothesis implemented in the same archipelago (Strong *et al.*, 1979). The very same danger is associated with the use of real-world mainland biotas as well. In general, the choice of *any* null hypothesis for morphological studies of competition is plagued by this inability to gauge the contribution of current *and* historical competition to morphological differences among community members. Lacking a complete fossil record of the community in question, the ecologist can neither measure the magnitude of this bias nor correct for it in the analysis.

Apart from the problems associated with the construction of null hypotheses *per se*, the general strategy of community-level analysis designed to investigate interspecific processes has problems of its own. If interspecific competition occurs in a community it presumably will be strongest and hence most detectable in a small subset of ecologically similar species. Including other less similar species in an analysis runs the risk of obscuring the effects of competition. Community-level analyses are of great heuristic value in detecting community patterns (Diamond, 1975; Inger and Colwell, 1977). The patterns thus detected, however, must not be overinterpreted as proof of process. The assessment of any interspecific process such as competition will be most profitably pursued and the results most firmly established by an in-depth analysis of small groups of ecologically similar species identified in the community-level analysis (Colwell, 1979).

Acknowledgments

We are grateful to David Raup for sharing the program that does the phylogenetic bookkeeping and tree-plotting in our program GOD. We also wish to thank Ted Case, Hal Caswell, Lloyd Goldwasser, Peter Grant, David Jablonsky, Mary-Claire King, Paul Licht, the students of Zoology 244 (fall, 1979), and the participants in this symposium for discussion and assistance. This work was supported by NSF grant DEB78-12038 (to R.K.C.), and by the U. C. Berkeley Committee on Research.

Note

The programs described in this paper are available upon request from the authors. They are written in FORTRAN, heavily annotated to make modification easy.

PAPER 46

Phylogenies and the Comparative Method (1985)

J. Felsenstein

Commentary

T. JONATHAN DAVIES

The tendency for closely related species to resemble each other in their morphology and ecology has long been recognized and is a fundamental tenet of modern taxonomic classifications. Nonetheless, this inconvenient truth was largely ignored in the comparative literature until the 1980s, which witnessed the publication of a series of classic papers by, among others, P. Harvey, G. Mace, J. Gittleman, and J. Felsenstein, and the birth of a new statistical methodology using information from phylogenetic trees. Prior to this, traditional comparative approaches usually involved correlating phenotypes, or phenotype and an environment, across a range of species—for example, metabolic rate and body mass, or environmental temperature and metabolic rate—and species were regarded as statistically independent. But because of the hierarchically clustered structure of phylogenetic trees, species are statistically nonindependent. Felsenstein provided a simple illustration, suppose we envisage a case in which the phylogeny for a large number of species is represented by two species clusters (clades), which differ in their mean phenotype for two traits, X and Y, but for which there is no relationship between phenotypes within each cluster. Regressing X against Y, (i.e., treating species as independent) will result in an artifactual correlation between the two traits, a product of their shared evolutionary history. The lack of any widely applicable methodology to correct for evolutionary covariance was a major stumbling block. In the paper that follows, Felsenstein described a solution using differences between sister taxa (independent contrasts). Rather than treating species as independent units, Felsenstein demonstrated how contrasts across nodes on a phylogeny represent evolutionarily independent events. All that was required was a phylogenetic tree linking the species of interest and an assumed evolutionary model of trait change along its branches. Despite recent advances, these requirements still place limits on the method. Automated gene sequencing has resulted in the rapidly increasing availability of phylogenetic data, but a comprehensive estimate of the tree of life is far from complete. A Brownian motion model of trait evolution, in which traits evolve independently in a manner analogous to a random walk, is frequently assumed when employing independent contrast, yet alternative evolutionary models are rarely tested. Nevertheless, Felsenstein noted that even poorly formed models that attempt to correct for nonindependence would be "immeasurably superior to simply treating the species as if independently evolved" (7–8).

From *American Naturalist* 125:1–15.

Vol. 125, No. 1 The American Naturalist January 1985

PHYLOGENIES AND THE COMPARATIVE METHOD

JOSEPH FELSENSTEIN

Department of Genetics SK-50, University of Washington, Seattle, Washington 98195

Submitted November 30, 1983; Accepted May 23, 1984

Recent years have seen a growth in numerical studies using the comparative method. The method usually involves a comparison of two phenotypes across a range of species or higher taxa, or a comparison of one phenotype with an environmental variable. Objectives of such studies vary, and include assessing whether one variable is correlated with another and assessing whether the regression of one variable on another differs significantly from some expected value. Notable recent studies using statistical methods of this type include Pilbeam and Gould's (1974) regressions of tooth area on several size measurements in mammals; Sherman's (1979) test of the relation between insect chromosome numbers and social behavior; Damuth's (1981) investigation of population density and body size in mammals; Martin's (1981) regression of brain weight in mammals on body weight; Givnish's (1982) examination of traits associated with dioecy across the families of angiosperms; and Armstrong's (1983) regressions of brain weight on body weight and basal metabolism rate in mammals.

My intention is to point out a serious statistical problem with this approach, a problem that affects all of these studies. It arises from the fact that species are part of a hierarchically structured phylogeny, and thus cannot be regarded for statistical purposes as if drawn independently from the same distribution. This problem has been noticed before, and previous suggestions of ways of coping with it are briefly discussed. The nonindependence can be circumvented in principle if adequate information on the phylogeny is available. The information needed to do so and the limitations on its use will be discussed. The problem will be discussed and illustrated with reference to continuous variables, but the same statistical issues arise when one or both of the variables are discrete, in which case the statistical methods involve contingency tables rather than regressions and correlations.

THE PROBLEM

Suppose that we have examined eight species and wish to know whether their brain size (Y) is proportional to their body size (X). We may wish to test whether the slope of the regression of Y on X (or preferably of $log\ Y$ on $log\ X$) is unity. Figure 1 shows a scatter diagram of hypothetical data. It is tempting to simply do

Am. Nat. 1985. Vol. 125, pp. 1–15.

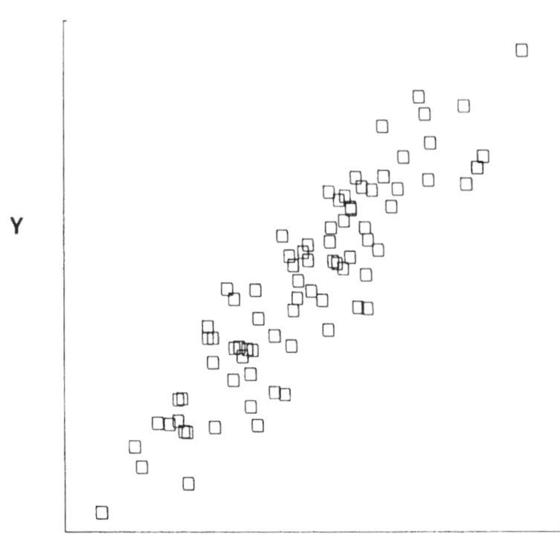

Fig. 1.—Scatter diagram of hypothetical data from 8 species, showing the relationship between Y and X.

an ordinary regression and see whether the confidence limits on the slope include unity. (Since there is random error in both X and Y this is a questionable procedure, but we leave that issue aside here.)

If we were to do such a regression, what would be the implicit statistical model on which it was based? The simplest assumption would seem to be that the points in figure 1 were drawn independently from a bivariate normal (Gaussian) distribution. What evolutionary model could result in such a distribution? The simplest is shown in figure 2: the eight species resulted from a single explosive adaptive radiation. Along each lineage there were changes in both characters. If evolution in each lineage were independent, and the changes in the two characters were drawn from a bivariate normal distribution, then our distributional assumptions would be justified. The individual species could be regarded as independent samples from a single bivariate normal distribution.

Let us accept for the moment that the changes of a set of characters in different branches of a phylogeny can be reasonably well approximated as being drawn from a multivariate normal distribution, and that changes in distinct branches are independent. Even given those assumptions, a problem arises based on the unlikelihood that the phylogeny has the form shown in figure 2.

Consider instead the phylogeny shown in figure 3. In it, the eight species consist of four pairs of close relatives. Suppose that the changes in the two characters in each branch of the tree can be regarded as drawn from a bivariate normal distribution with some degree of correlation between the characters. We might not expect the same amount of change in short branches of the tree, such as the eight terminal branches, as in longer branches such as the four that arise from the original radiation. Let us assume that the variance of the distribution of change in a branch is proportional to the length in time of the branch, much as it would be if the characters were undergoing bivariate, and correlated, Brownian motion.

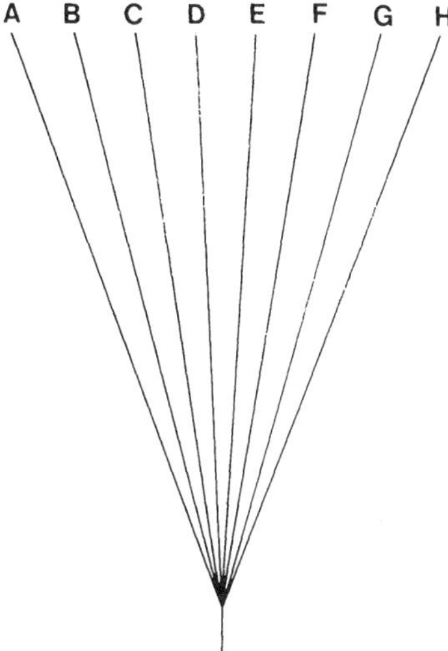

Fig. 2.—One phylogeny for the 8 species, showing a burst of adaptive radiation with each lineage evolving independently from a common starting point.

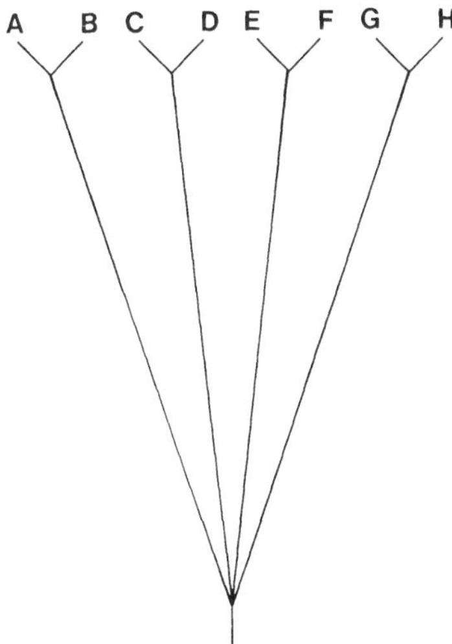

Fig. 3.—Another phylogeny for the 8 species, showing a radiation that gives rise to 4 pairs of closely related species.

This would produce a distribution like that shown in figure 4. It is apparent that instead of eight independent points we have four pairs of close relatives. If we were to carry out a statistical test based on the assumption of independence, say a test of the hypothesis that the slope of the regression of Y on X was zero, we would imagine ourselves to have 6 degrees of freedom $(8 - 2)$. In fact, we very nearly have only four independent points, so that the effective number of degrees of freedom is closer to 2 $(4 - 2)$. A test of the significance of the slope, or of the

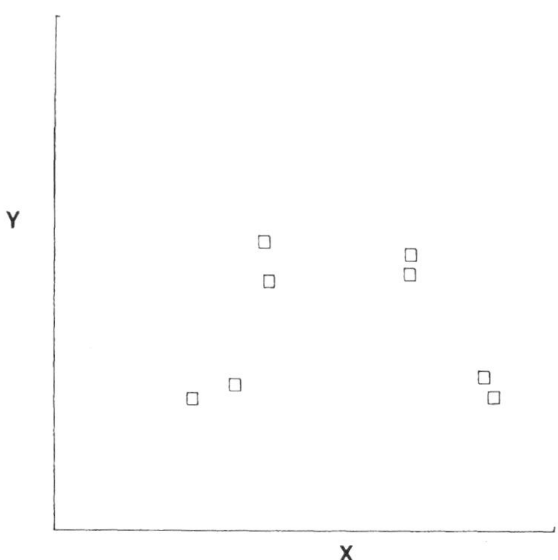

FIG. 4.—A data set simulated using the phylogeny of fig. 3, under a model of random, normally distributed, independent change in each character, where the change in each branch is drawn independently from a normal distribution with mean zero and variance proportional to the length of the branch.

extent of its difference from any preassigned value, will be excessively likely to show significance if the number of degrees of freedom is taken to be 6 rather than 2.

A worst case of sorts for the naive analysis is shown in figure 5, where the phylogeny shows that a large number of species actually consist of two groups of moderately close relatives. Suppose that the data turned out to look like that in figure 6. There appears to be a significant regression of Y on X. If the points are distinguished according to which monophyletic group they came from (fig. 7), we can see that there are two clusters. Within each of these groups there is no significant regression of one character on the other. The means of the two groups differ, but since there are only two group means they must perforce lie on a straight line, so that the between-group regression has no degrees of freedom and cannot be significant. Yet a regression assuming independence of the species finds a significant slope ($P < .05$). It can be shown that there are more nearly 3 than 40 independent points in the diagram.

One might imagine that the problem could be avoided by use of robust nonparametric statistics. In fact, nonparametric methods, unless specifically designed to cope with the problem of nonindependence, are just as vulnerable to the problem as are parametric methods. For the data of figure 6, a Spearman rank correlation finds a nearly significant correlation ($P < .065$) between the two variables, showing that it has little better ability to cope with nonindependence than do parametric methods.

One might also imagine that we could escape from the problem simply by ensuring that we sample the species at random from the species that form the tips of the phylogeny, and thus somehow escape from the nonrandomness of the pool

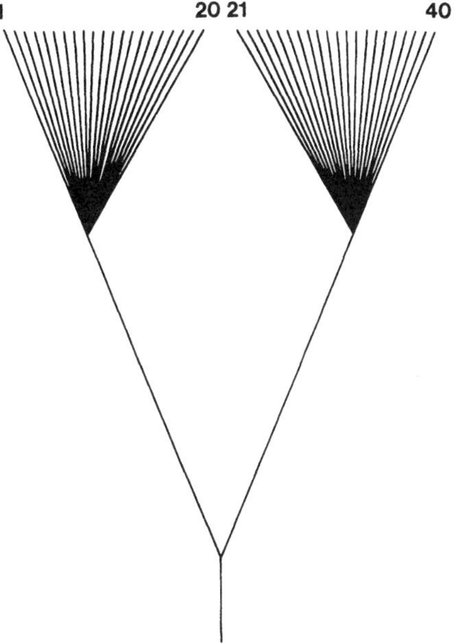

FIG. 5.—A "worst case" phylogeny for 40 species, in which there prove to be 2 groups each of 20 close relatives.

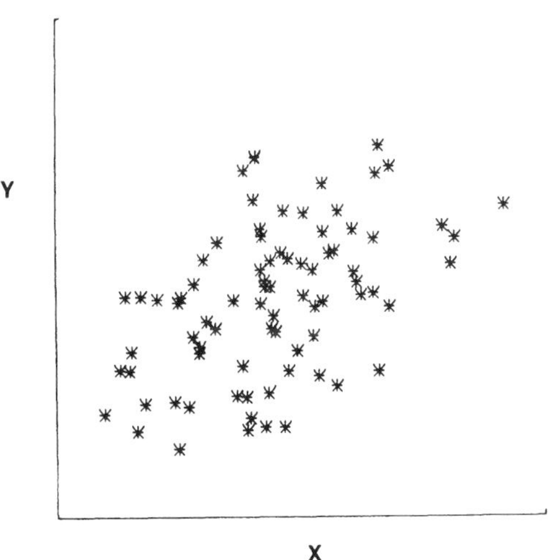

FIG. 6.—A typical data set that might be generated for the phylogeny in fig. 5 using the model of independent Brownian motion (normal increments) in each character.

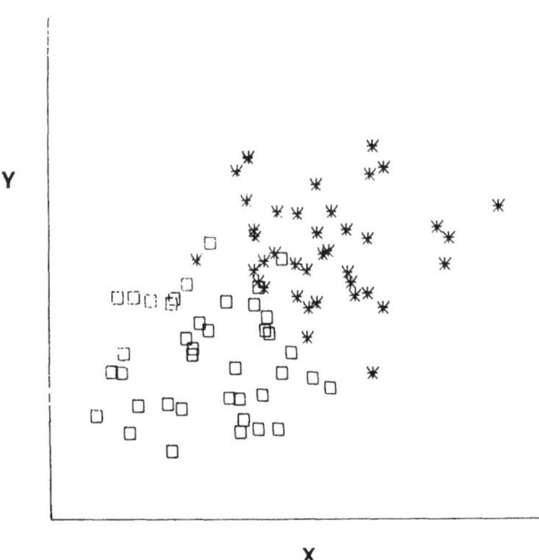

FIG. 7.—The same data set, with the points distinguished to show the members of the 2 monophyletic taxa. It can immediately be seen that the apparently significant relationship of fig. 6 is illusory.

of species from which we are sampling. This does not work. Imagine two species that have diverged some time ago, and thus have diverged in both brain and body weight. Clearly the correlation between those characters cannot be significant, since there are only two points. Now if each species gives rise to a group of 100 daughter species, essentially identical to it, we now have two clusters of 100 species each. Sampling species from this pool of 200 species, we are actually sampling from two species, but do not know it. Correctly analyzed, no data from this group could possibly achieve significance, but if we draw (say) 50 species at random from the 200 and analyze that data as if the points were independent we will probably conclude that there is a significant correlation between brain and body weight.

There is one case in which the problem does not arise. That is when the characters respond essentially instantaneously to natural selection in the current environment, so that phylogenetic inertia is essentially absent. In that case we could correlate a phenotype with the environment. We could also correlate two characters with each other, provided that we realized that their correlation might simply reflect response to a common environmental factor. It may be doubted how often phylogenetic inertia is effectively absent. In any case the presumption of the absence of phylogenetic inertia should be acknowledged whenever it is proposed to do comparative studies without taking account of the phylogeny.

PREVIOUS APPROACHES

The problem of correcting for nonindependent evolutionary origins has not gone unnoticed by previous workers in comparative biology. Clutton-Brock and Harvey (1977, pp. 6–8) pointed out that "if phylogenetic inertia is strong, the potential adaptations that related species may evolve will be similarly constrained, with the effect that species cannot be regarded as independent of each other." They used a nested analysis of variance to find that taxonomic level (in their case it was genera) which accounted for as much of the variation as possible, and then tried to correct for nonindependence by using genera rather than species as the units of their statistical analysis. A more complete exposition of their approach is given by Harvey and Mace (1982). Baker and Parker (1979), who were analyzing bird coloration, also pointed out the problem, and tried to correct for it by seeing whether the same relationships held within different families. Sherman (1979) and Givnish (1982) discussed the problem, though without attempting to correct for it.

Gittleman (1981) used a parsimony method to reconstruct and count the number of times parental care had evolved independently on a phylogeny derived from the classification of bony fishes. Ridley (1983) has discussed the problem of nonindependence in considerable detail, also proposing that parsimony be used to reconstruct the placements of changes on the phylogeny, to enable tests of whether the occurrence of changes in two characters are independent.

There are two problems with using a parsimony approach as suggested by Gittleman and by Ridley. The most serious is that one is usually forced to rely on a presumed monophyly of taxa in the Linnean classification system, in the absence of external evidence as to the phylogenetic relationships between the groups. The traditional classification system is, of course, only partly a monophyletic one: only about half of the classes of the Chordata are thought to be monophyletic, for example, and even within the Mammalia orders such as Insectivora and Carnivora are believed not be monophyletic. The classification system is not sufficiently detailed to show all of the structure in the phylogeny, even if all its groups were monophyletic: the relationships between some orders of mammals are undoubtedly closer than between others. The phylogenetic meaning of a given category varies from group to group: the insect genus *Drosophila* is thought to be as old as the mammalian order Primates.

A second problem with parsimony assignments is that they have only partial statistical justification: I have argued elsewhere (Felsenstein 1978) that when parsimony is used to reconstruct the phylogeny, it can have undesirable statistical properties when evolutionary rates are not small and differ sufficiently in different lineages (for a review, see Felsenstein [1983]). Even when the phylogeny is known, and parsimony is used only to reconstruct the placement of changes of character state, it is well known that this can lead to biases; for example, if two changes occur in parallel in sister lineages, the reconstruction will instead show one change occurring in their immediate ancestor. If changes in one character occurred in parallel in the sister lineages, but the change in another character occurred in the immediate common ancestor, then, although the reconstructed changes appear coincident, the actual changes in those characters were not in fact coincident.

The seriousness of the additional statistical error and statistical biases that this may cause in comparative studies has never, to my knowledge, been investigated. Nevertheless, assigning changes of characters by parsimony on a known phylogeny would be immeasurably superior to simply treating the species as if independently evolved.

A POSSIBLE SOLUTION

If we know the phylogeny and have a model of evolutionary change, it should be possible in principle to correct for the nonindependence of taxa. To see how, first let us consider the highly symmetrical phylogeny in figure 8, supposing that we know that it is the true phylogeny. Recalling that each character is being assumed to be evolving by a Brownian motion that is independent in each lineage, then taking X_i to be the phenotype X in species i it is easy to see that differences between pairs of adjacent tips, such as $X_1 - X_2$ and $X_3 - X_4$, must be independent. This is so because the difference $X_1 - X_2$ depends only on events in branches 1 and 2, while $X_3 - X_4$ depends only on events in branches 3 and 4, and these two sets of events are independent.

Brownian motion is a random process modeling the wanderings of a molecule affected by thermal noise. If we measure the position of the molecule along one axis, its successive displacements are independent. This has the effect that the displacement after time v has elapsed is the sum of a large number of small displacements, each of which is equally likely to be either positive or negative. The result is that the total displacement is drawn from a normal distribution with mean zero and a variance proportional to v. In the present model the different characters undergo Brownian motion at different rates, so that after one unit of time the change in X has variance s_X^2 and the (possibly correlated) change in Y has variance s_Y^2. After v units of time their variances are, respectively, s_X^2v and s_Y^2v.

Given this model, it is straightforward to show (Felsenstein 1973, 1981*b*) that the contrast $X_1 - X_2$ has expectation zero and variance $2s_X^2v_1$. Since we assume that we know the v_i, we can scale the contrast by dividing by its standard deviation, obtaining a variate that should have expectation zero and unit variance. We can similarly scale the other three contrasts $X_3 - X_4$, $X_5 - X_6$, and $X_7 - X_8$ by dividing each by the square root of $2\ s_X^2\ v_1$. Even more contrasts are available. It will be less obvious, but nevertheless true, that $(X_1 + X_2)/2 - (X_3 + X_4)/2$ is a contrast independent of the others. It will have expectation zero and variance $s_X^2(v_1 + 2v_9)$. We can continue down the tree in similar fashion, obtaining two more contrasts, $(X_5 + X_6)/2 - (X_7 + X_8)/2$ and $(X_1 + X_2 + X_3 + X_4)/4 - (X_5 + X_6 + X_7 + X_8)/4$. Their expectations are also zero, and their variances are, respectively, $s_X^2(v_1 + 2v_9)$ and $s_X^2(2v_{13} + v_9 + v_1/2)$. They too can be scaled to have unit variance.

We have now extracted from this tree seven independent contrasts on the X scale, each of which can be regarded as drawn from a normal distribution with mean zero and variance one. We can carry out the same process in the variable Y, and obtain seven independent contrasts in the same way. The X contrasts will be independent of each other but not of the Y contrasts. It can be shown that the coresponding contrasts $X_1 - X_2$ and $Y_1 - Y_2$ have covariance

$$\mathrm{Cov}[X_1 - X_2, Y_1 - Y_2] = 2v_1\ s_X\ s_Y\ r_{XY}, \quad (1)$$

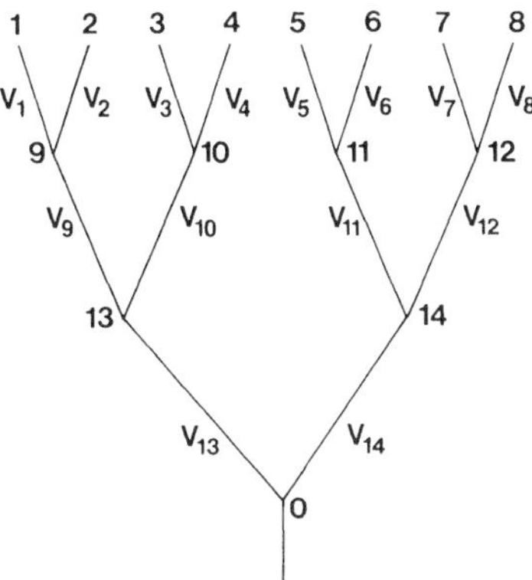

FIG. 8.—An example of a phylogeny, assumed known, from which we can define independent contrasts between taxa. This tree is highly symmetric, so that $v_1 = v_2 = v_3 = v_4 = v_5 = v_6 = v_7 = v_8$, $v_9 = v_{10} = v_{11} = v_{12}$, and $v_{13} = v_{14}$.

so that these two contrasts have the same correlation as the original variates. Since contrasts such as $X_1 - X_2$ and $X_3 - X_4$ are independent, a fortiori $X_1 - X_2$ will be independent of $Y_3 - Y_4$.

The quantities s_X and s_Y are important. It would be unreasonable to assume that characters X and Y had the same rates of evolution: s_X and s_Y are the scaling constants that convert from Brownian motion to the scales on which X and Y actually evolve. Thus X is undergoing a Brownian motion, with variance s_X^2 accumulating per unit time, and Y is undergoing a (possibly correlated) Brownian motion with variance s_Y^2 accumulating per unit time. We leave aside for the moment the problem of estimating s_X and s_Y, and assume that they are known.

By dividing each contrast by its standard deviation, we have obtained from the original eight species seven pairs of contrasts that can be regarded as drawn independently from a bivariate normal distribution with means zero, variances unity, and an unknown correlation r_{XY} between the members of a pair. Testing independence of the evolution of X and Y reduces to simply testing whether this correlation is zero. If instead we wanted to know the regression of changes in one variable on changes in another, we could use s_X and s_Y to compute

$$b_{Y.X} = s_Y\, r_{XY}/s_X \tag{2a}$$

and

$$b_{X.Y} = s_X\, r_{XY}/s_Y. \tag{2b}$$

These are not the usual equations for interconverting correlations and regressions, since s_X and s_Y are not observed standard deviations of X and Y, but scaling constants that are merely proportional to the standard deviations of the variables X and Y. Even though they are not standard deviations, they do allow us to correctly convert correlations into regressions. Other methods of analysis, such as principal components, can be carried out in similar fashion.

The preceding computation of contrasts depended on the phylogeny having a particular, and very unlikely, symmetric structure. Fortunately a more general procedure exists, of which the above was a special case. I have discussed its elements elsewhere (Felsenstein 1973) as part of a computational method for obtaining the likelihood of a given phylogeny. The general prescription for computing these contrasts is repeated applications of the following steps: (1) Find two tips on the phylogeny that are adjacent (say nodes i and j) and have a common ancestor, say node k. (2) Compute the contrast $X_i - X_j$. This has expectation zero and variance proportional to $v_i + v_j$. (3) Remove the two tips from the tree, leaving behind only the ancestor k, which now becomes a tip. Assign it the character value

$$X_k = \frac{(1/v_i)\, X_i + (1/v_j)\, X_j}{1/v_i + 1/v_j}. \tag{3}$$

the weighted average of X_i and X_j, the weights being proportional to the inverses of the variances v_i and v_j. (4) Lengthen the branch below node k by increasing its length from v_k to $v_k + v_i v_j/(v_i + v_j)$. This lengthening occurs because the weighted average that computes X_k in equation (3) does not compute the phenotype of the ancestor but only estimates it, and does so with an error that is statistically indistinguishable from an extra burst of evolution after node k.

After one pass through steps 1–4, we have found one contrast and reduced the number of tips on the tree by one. We continue to repeat steps 1–4 until there is only one tip left on the tree. This will extract $n - 1$ contrasts if there were originally n species. Each contrast can be divided by the square root of its variance to bring them to a common variance. Since the v_i are arbitrary, this procedure can be used on a phylogeny of any shape whatsoever, even on ones that contain multifurcations, since those can always be represented as a series of bifurcations having some branch lengths zero.

Figure 9 shows a nonsymmetric phylogeny, and table 1 the contrasts extracted from it by the above algorithm. The reader may want to try steps 1–4 on the symmetric phylogeny of figure 8, to verify the correctness of the contrasts and variances given above.

DIFFICULTIES

One might imagine that, with the ability to compute independent contrasts from any phylogeny, we have an acceptable method of correcting for the presence of the phylogeny. Unfortunately, this is not the case. A number of difficulties intervene that leave us with much work remaining to be done.

How Do We Reconstruct the Phylogeny?

In practice, we will hardly ever know the phylogeny in advance in sufficient detail to use it to obtain the contrasts. There are three sources of information likely to be used to reconstruct the phylogeny.

1. *Gene frequencies.*—A number of electrophoretic loci, blood group loci, or DNA restriction polymorphisms may be chosen and the frequencies of the alleles

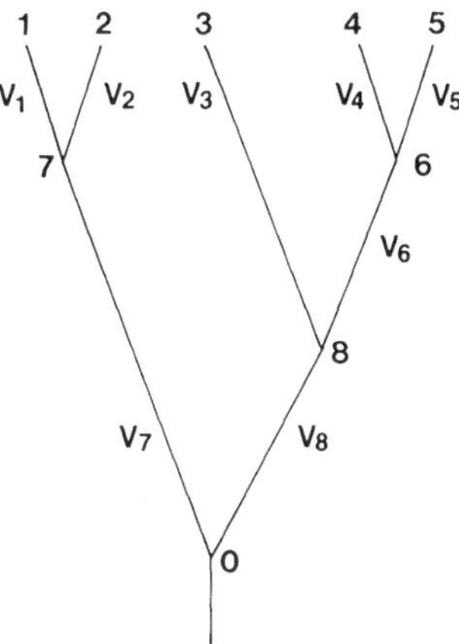

FIG. 9.—A less symmetrical phylogeny. The independent contrasts for this phylogeny are given in table 1.

TABLE 1

THE FOUR CONTRASTS EXTRACTED FROM THE PHYLOGENY SHOWN IN FIGURE 9, EACH WITH ITS VARIANCE, ALL COMPUTED USING STEPS 1–4 IN THE TEXT

CONTRAST	VARIANCE
$X_1 - X_2$	$v_1 + v_2$
$X_4 - X_5$	$v_4 + v_5$
$X_3 - X_6$	$v_3 + v_6'$
$X_7 - X_8$	$v_7' + v_8'$

where

$$X_6 = \frac{v_4 X_5 + v_5 X_4}{v_4 + v_5}$$

$$v_6' = v_6 + v_4 v_5/(v_4 + v_5)$$

$$X_7 = \frac{v_2 X_1 + v_1 X_2}{v_1 + v_2}$$

$$v_7' = v_7 + v_1 v_2/(v_1 + v_2)$$

$$X_8 = \frac{v_6' X_3 + v_3 X_6}{v_3 + v_6}$$

$$v_8' = v_7' + v_3 v_6'/(v_3 + v_6')$$

at these loci determined in a sample from each species. This permits us to make an estimate of the phylogeny. The estimate has an error that must be taken into account when using it (see discussion below). It depends on the assumption that the evolutionary changes at the loci are predominantly due to genetic drift, and can therefore be modeled by Brownian motion of suitably transformed gene frequencies. This is more plausible the lower the taxonomic level at which we are working. A review of methods for inferring phylogenies from gene frequencies will be found in my recent paper on that subject (Felsenstein 1981*b*).

2. *Molecular sequences.*—As nucleic acid sequences become available for a wider range of organisms, it will become practical to infer the phylogeny from these, on the assumption that a "molecular clock" is valid for those sequences. I have described elsewhere (Felsenstein 1981*a*) a maximum likelihood method for inferring phylogenies from nucleic acid sequences.

3. *Quantitative characters.*—Most morphological taxonomic studies do not collect gene frequencies, but collect many morphological measurements besides those that we want to correlate. Could we use those quantitative characters to infer the phylogeny? We could do so in principle if we had a probability model for the evolution of the characters. The difficulty is that quantitative characters will evolve at different rates, and in a correlated fashion (as in the important case of size correlations). If we could find a transformation of the characters to a new set of coordinates that could be modeled as evolving independently by Brownian motion, with equal rates of accumulation of variance, we could apply the maximum likelihood methods developed for gene-frequency data. There would be no circularity involved: the new coordinates would be independent of each other and of the characters we were investigating. Unfortunately, there is no obvious way to get the information needed to untangle the skein of character correlations. Within-population samples, even when available, do not necessarily give us the required information. They allow us to estimate phenotypic covariances rather than additive genetic covariances. The latter control the covariances of evolutionary changes in characters if the characters change by natural selection or genetic drift. Even if the additive genetic covariances could be obtained by means of breeding experiments, they could not tell us whether the selection pressures for different characters were correlated. Cold weather, for example, might impose selection for large size and dark coloration while warm weather might select for small size and light coloration, leading to changes in these characters being correlated even if there were no genetic correlation between them. For the time being, transforming to remove within-population phenotypic covariance is the best that can be done, but this is necessarily an approximation, whose adequacy is unknown.

4. *The characters we are investigating.*—Sometimes the only characters available to us are the two whose relationship we are investigating. This is not only a severely limited amount of information with which to infer a phylogeny, but could result in some circularity since the phylogeny is inferred from some of the same information that is being used to reconstruct the changes of the characters along it. The matter needs a careful statistical investigation, but preliminary indications are that when we are trying to infer both the phylogeny and character correlations there is confounding between these, so that we can infer one or the other but not both.

How Do We Put Confidence Intervals on Our Inferences?

If we were given a phylogeny known to be the true one, and were willing to trust the Brownian motion model of character change, we could obtain the appropriate contrasts from the phylogeny as outlined above, and use standard formulas to place confidence limits on the inferred correlations or slopes. But the phylogeny is never known without error. How are we to take the uncertainty of our knowledge of the phylogeny into account when constructing confidence intervals?

In principle this can be done by considering the phylogeny T and the set of correlations (or slopes) C to be a single multivariate quantity (T, C) being estimated by maximum likelihood. The estimate is that pair (T^*, C^*) resulting in the highest likelihood, and an approximate confidence interval is the set of all points (T, C) whose likelihood is an acceptable fraction of the maximum likelihood, as judged by the likelihood ratio test. If a single correlation or slope is being estimated, the 95% confidence interval will be all values of C for which there is a phylogeny T such that their likelihood $L(T, C) \geq 0.1465\ L(T^*, C^*)$, since this is the ratio that would just reach significance in the likelihood ratio test with 1 degree of freedom.

It may be possible in some cases to discard the information about the phylogeny that we are least certain of, and use only those features in which we have reasonably high confidence. A student (whose name is unfortunately not known to me; see the ACKNOWLEDGMENTS section below) has pointed out to me that we could use contrasts between pairs of species that we were fairly sure had a common ancestor not shared with any member of another pair, and that these contrasts could then be safely assumed to be independent. For example, in a study of mammals we could use pairs consisting of two seals, two whales, two bats, two deer, etc. These contrasts would be independent, but would not necessarily have equal variances. We therefore could not simply compute correlations or regressions from the pairs. We could use certain nonparametric methods such as a sign test to test whether the changes in the two characters were correlated, but other nonparametric methods such as Spearman's rank correlation would not be valid because we could not assume that the pairs were drawn from a common distribution, even though they are independent. This sign test is essentially the same as that used by Baker and Parker (1979) to test whether regressions were similar in different bird families. It should be obvious that there is much statistical work remaining to be done on robust methods of using partial knowledge of phylogenies to make inferences about regressions and correlations of characters.

What If We Lack an Acceptable Statistical Model of Character Change?

All of the above has been predicated on the acceptance of the Brownian motion model as a realistic statistical model of character change. There are certainly many reasons for being skeptical of its validity. Persistence of selection pressures over time may lead to correlation of changes in successive branches of the phylogeny, and common selective regimes experienced by different populations owing to common environmental factors such as weather or predators may lead to changes in different lineages being correlated. Since the lengths of the branches of the phylogeny are given not in time units, but rather in units of expected variance of change (the v_i) the model already allows rates of change to differ in different lineages, and would allow, for example, change to be faster after speciation events than during later periods, as assumed to many punctuationists.

There is no reason to believe that the normal distribution is particularly plausible as the distribution from which changes in individual branches of the phylogeny are drawn, except insofar as the net change in a branch is the resultant of a series of bursts of change and thus might be approximately normal.

The matter of the model is an obvious point for future development and (to the extent that this is possible) empirical study. One rather serious problem that confronts comparative studies is that the relationship under study may change through time. Harvey and Mace (1982) have discussed the problem of change of the slope of the relationship between two variables with taxonomic level, which appears to be quite common. It should be possible to use the current model to study statistically whether there is any connection between the variance of a contrast and the slope of the regression of one variable on another.

What if We Do Not Take the Phylogeny into Consideration?

Some reviewers of this paper felt that the message was "rather nihilistic," and suggested that it would be much improved if I could present a simple and robust method that obviated the need to have an accurate knowledge of the phylogeny. I entirely sympathize, but do not have a method that solves the problem. The best we can do is perhaps to use pairs of close relatives as suggested above, although this discards at least half of the data. Comparative biologists may understandably feel frustrated upon being told that they need to know the phylogenies of their groups in great detail, when this is not something they had much interest in knowing. Nevertheless, efforts to cope with the effects of the phylogeny will have to be made. Phylogenies are fundamental to comparative biology; there is no doing it without taking them into account.

SUMMARY

Comparative studies of the relationship between two phenotypes, or between a phenotype and an environment, are frequently carried out by invalid statistical methods. Most regression, correlation, and contingency table methods, including nonparametric methods, assume that the points are drawn independently from a common distribution. When species are taken from a branching phylogeny, they are manifestly nonindependent. Use of a statistical method that assumes independence will cause overstatement of the significance in hypothesis tests. Some illustrative examples of these phenomena have been given, and limitations of previous proposals of ways to correct for the nonindependence have been discussed.

A method of correcting for the phylogeny has been proposed. It requires that we know both the tree topology and the branch lengths, and that we be willing to allow the characters to be modeled by Brownian motion on a linear scale. Given these conditions, the phylogeny specifies a set of contrasts among species, contrasts that are statistically independent and can be used in regression or correlation studies. The considerable barriers to making practical use of this technique have been discussed.

ACKNOWLEDGMENTS

The suggestion that pairs of closely related organisms could be used in a way that avoided the need to know the full phylogeny was made by a student during discussion following my seminar at the Department of Genetics, University College, London. Unfortunately, I have been unable to discover her name. I wish to thank Ray Huey, John Gittleman, Robert Martin, and Mart Ridley for helpful discussions and/or access to their unpublished work. I am particularly grateful to Paul Harvey for bringing this particular piece of biological real estate to my attention and for many helpful conversations about it. I also wish to thank the reviewers of this paper for many helpful suggestions and for saving me from myself on at least one point. This work was supported by Task Agreement no. DE-AT06-76EV71005 of contract number DE-AM06-76RL02225 between the U.S. Department of Energy and the University of Washington.

LITERATURE CITED

Armstrong, E. 1983. Relative brain size and metabolism in mammals. Science 220:1302–1304.

Baker, R. R., and Parker, G. A. 1979. The evolution of bird colouration. Philos. Trans. R. Soc. Ser. *B* 287:63–130.

Clutton-Brock, T. H., and P. H. Harvey. 1977. Primate ecology and social organization. J. Zool. 183:1–39.

Damuth, J. 1981. Population density and body size in mammals. Nature 290:699–700.

Felsenstein, J. 1973. Maximum-likelihood estimation of evolutionary trees from continuous characters. Am. J. Hum. Genet. 25:471–492.

———. 1978. Cases in which parsimony or compatibility methods will be positively misleading. Syst. Zool. 27:401–410.

———. 1981*a*. Evolutionary trees from DNA sequences: a maximum likelihood approach. J. Mol. Evol. 17:368–376.

———. 1981*b*. Evolutionary trees from gene frequencies and quantitative characters: finding maximum likelihood estimates. Evolution 35:1229–1242.

———. 1983. Parsimony in systematics: biological and statistical issues. Annu. Rev. Ecol. Syst. 14:313–333.

Gittleman, J. 1981. The phylogeny of parental care in fishes. Anim. Behav. 29:936–941.

Givnish, T. J. 1982. Outcrossing versus ecological constraints in the evolution of dioecy. Am. Nat. 119:849–851.

Harvey, P. H., and G. Mace. 1982. Comparisons between taxa and adaptive trends: problems of methodology. Pages 343–361 *in* Current problems in sociobiology. King's College Sociobiology Group, ed. Cambridge University Press, Cambridge.

Martin, R. D. 1981. Relative brain size and basal metabolic rate in terrestrial vertebrates. Nature 293:57–60.

Pilbeam, D., and S. J. Gould. 1974. Size and scaling in human evolution. Science 186:892–901.

Ridley, M. 1983. The explanation of organic diversity. The comparative method and adaptations of mating. Clarendon, Oxford.

Sherman, P. W. 1979. Insect chromosome numbers and eusociality. Am. Nat. 113:925–935.

Contributors

Ford Ballantyne IV
Odum School of Ecology
140 E. Green Street
The University of Georgia
Athens, GA 30602

Tim M. Blackburn
Institute of Zoology
Zoological Society of London
Regent's Park, London, United Kingdom
NW1 4RY

Alison G. Boyer
Department of Ecology and Evolutionary Biology
University of Tennessee
447 Dabney Hall
Knoxville, TN 37996

James H. Brown
Department of Biology
University of New Mexico
Albuquerque, NM 87131

Andrew M. Bush
Department of Ecology and Evolutionary Biology / Center for Integrative Geosciences
University of Connecticut
Storrs, CT 06269

Andrew Clarke
Department of Biological Sciences
British Antarctic Survey
Cambridge, United Kingdom
CB3 OET

Daniel P. Costa
Department of Ecology and Evolutionary Biology / Institute of Marine Sciences
University of California, Santa Cruz
Santa Cruz, CA 95064

David J. Currie
Biology Department
University of Ottawa
Ottawa, ON, Canada
K1N 6N5

T. Jonathan Davies
Department of Biology
McGill University
Montreal, QC, Canada
H3A 1BA

S. K. Morgan Ernest
Department of Biology / The Ecology Center
Utah State University
Logan, UT 84322

Alistair Evans
School of Biological Sciences
Monash University
VIC 3800 Australia

Michael Foote
Department of Geophysical Sciences
University of Chicago
5734 South Ellis Avenue
Chicago, IL 60637

Kevin J. Gaston
Biodiversity and Macroecology Group
Department of Animal and Plant Sciences
University of Sheffield
Sheffield, United Kingdom
S10 2TN

John L. Gittleman
Odum School of Ecology
140 E. Green Street
University of Georgia
Athens, GA 30602

Nicholas J. Gotelli
Department of Biology
University of Vermont
Burlington, VT 05405

Allen H. Hurlbert
Department of Biology
University of North Carolina at Chapel Hill
CB# 3280, Coker Hall
Chapel Hill, NC 27599

David Jablonski
Department of Geophysical Sciences
University of Chicago
5734 South Ellis Avenue
Chicago, IL 60637

Walter Jetz
Department of Ecology and Evolutionary Biology
Yale University
165 Prospect Street
New Haven, CT 06520

Douglas A. Kelt
Department of Wildlife, Fish, and Conservation Biology
1083 & 1077 Academic Surge
University of California, Davis
One Shields Avenue
Davis, CA 95616

Matthew A. Kosnik
Department of Biological Sciences
Macquarie University
NSW 2109 Australia

S. Kathleen Lyons
Department of Paleobiology
National Museum of Natural History
Smithsonian Institution [NHB, MRC 121]
PO Box 37012
Washington, DC 20013

Brian A. Maurer
Department of Fisheries and Wildlife
Michigan State University
East Lansing, MI 48824

Christy M. McCain
Department of Ecology and Evolutionary Biology
Colorado University Natural History Museum
Campus Box 265
University of Colorado at Boulder
Boulder, CO 80309

Brian J. McGill
School of Natural Resources
University of Arizona
325 Biosciences East
Tucson, AZ 85721

Karl J. Niklas
Department of Plant Biology
Cornell University
Ithaca, NY 14853

Richard M. Sibly
School of Biological Sciences
University of Reading
Reading, Berkshire, United Kingdom
RG6 6AS

Felisa A. Smith
Department of Biology
University of New Mexico
Albuquerque, NM 87131

David Storch
Center for Theoretical Study, Charles University / Academy of Sciences of the Czech Republic
Viničná 7, 128 44 Praha 2, Czech Republic

Jessica Theodor
Department of Biological Sciences
University of Calgary
Calgary, AB, Canada
T2N 1N4

Mark D. Uhen
Department of Atmospheric, Oceanic and Earth Sciences
George Mason University
Fairfax, VA 22030

Peter Wagner
Department of Paleobiology
National Museum of Natural History
Smithsonian Institution [NHB, MRC 121]
PO Box 37012
Washington, DC 20013

Ethan P. White
Department of Biology / The Ecology Center
Utah State University
Logan, UT 84322

Peter Wilf
Department of Geosciences
509 Deike Building
Pennsylvania State University
University Park, PA 16802

John W. (Jack) Williams
Department of Geography
University of Wisconsin–Madison
550 North Park Street
Madison, WI 53706

Scott L. Wing
Department of Paleobiology
National Museum of Natural History
Smithsonian Institution [NHB, MRC 121]
PO Box 37012
Washington, DC 20013

Index